Triangles and Angles

Right Triangle

Triangle has one 90°
(right) angle.

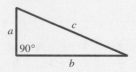

Pythagorean Formula
(*for right triangles*)

$c^2 = a^2 + b^2$

Right Angle

Measure is 90°.

Isosceles Triangle

Two sides are equal.

$AB = BC$

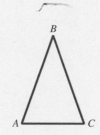

Straight Angle

Measure is 180°.

Equilateral Triangle

All sides are equal.

$AB = BC = CA$

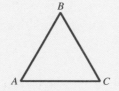

Complementary Angles

The sum of the measures of
two complementary
angles is 90°.

Angles ① and ②
are complementary.

Sum of the Angles of Any Triangle

$A + B + C = 180°$

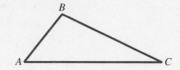

Supplementary Angles

The sum of the
measures of two
supplementary
angles is 180°.

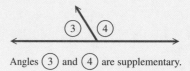

Angles ③ and ④ are supplementary.

Similar Triangles

Corresponding angles are
equal; corresponding sides
are proportional.

$A = D, B = E, C = F$

$\dfrac{AB}{DE} = \dfrac{AC}{DF} = \dfrac{BC}{EF}$

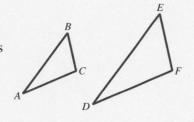

Vertical Angles

Vertical angles have
equal measures.

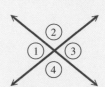

Angle ① = Angle ③

Angle ② = Angle ④

Beginning and Intermediate Algebra

THIRD EDITION

Beginning and Intermediate Algebra

THIRD EDITION

Margaret L. Lial

American River College

John Hornsby

University of New Orleans

Terry McGinnis

Boston   San Francisco   New York
London   Toronto   Sydney   Tokyo   Singapore   Madrid
Mexico City   Munich   Paris   Cape Town   Hong Kong   Montreal

Publisher: Greg Tobin

Editor in Chief: Maureen O'Connor

Project Editor: Suzanne Alley

Assistant Editor: Jolene Lehr

Managing Editor: Ron Hampton

Production Supervisor: Kathleen A. Manley

Production Services: Elm Street Publishing Services, Inc.

Compositor: Beacon Publishing Services

Marketing Manager: Dona Kenly

Marketing Coordinator: Lindsay Skay

Prepress Supervisor: Caroline Fell

Manufacturing Buyer: Evelyn Beaton

Text Designer: Susan Carsten Raymond

Cover Designer: Dennis Schaefer

Cover Photograph: Daryl Benson/Masterfile

Media Producer: Sara Anderson

Software Development: Alicia Anderson

For permission to use copyrighted material, grateful acknowledgment is made to the copyright holders on page I-15, which is hereby made part of this copyright page.

Library of Congress Cataloging-in-Publication Data
Lial, Margaret L.
    Beginning and intermediate algebra.—3rd ed. / Margaret L. Lial, John Hornsby, Terry McGinnis.
       p.   cm.
    Includes index.
    ISBN 0-321-12715-3 (alk. paper)
    1. Algebra.   I. Hornsby, John.   II. McGinnis, Terry.   III. Title.

QA152.3.L52   2004
512.9—dc21

                       2002042902

6 7 8 9 10—QWD—07 06

Contents

Chapter 4   Exponents and Polynomials   **243**

Chapter 5   Factoring and Applications   **315**

Chapter 6   Rational Expressions and Applications   **377**

List of Applications

Preface

The third edition of *Beginning and Intermediate Algebra* continues our ongoing commitment to provide the best possible text and supplements package to help instructors teach and students succeed. To that end, we have tried to address the diverse needs of today's students through a more open design, updated figures and graphs, helpful features, careful explanations of topics, and a comprehensive package of supplements and study aids. We have also taken special care to respond to the suggestions of users and reviewers and have added many new examples and exercises based on their feedback. Students who have never studied algebra—as well as those who require further review of basic algebraic concepts before taking additional courses in mathematics, business, science, nursing, or other fields—will benefit from the text's student-oriented approach.

After many years of benefiting from her behind-the-scenes assistance, we are pleased to welcome Terry McGinnis as coauthor of this series, which includes this text as well as the following books:

- *Beginning Algebra,* Ninth Edition, by Lial, Hornsby, and McGinnis

- *Intermediate Algebra,* Ninth Edition, by Lial, Hornsby, and McGinnis

- *Algebra for College Students,* Fifth Edition, by Lial, Hornsby, and McGinnis

WHAT'S NEW IN THIS EDITION?

We believe students and instructors will welcome the following new features.

New Real-Life Applications   We are always on the lookout for interesting data to use in real-life applications. As a result, we have included many new or updated examples and exercises throughout the text that focus on real-life applications of mathematics. These applied problems provide a modern flavor that will appeal to and motivate students. (See pp. 189, 640, and 852.) A comprehensive List of Applications appears at the beginning of the text. (See pp. x i–xiv.)

New Figures and Photos   Today's students are more visually oriented than ever. Thus, we have made a concerted effort to add mathematical figures, diagrams, tables, and graphs whenever possible. (See pp. 135, 153, and 475.) Many of the graphs use a style similar to that seen by students in today's print and electronic media. Photos have been incorporated to enhance applications in examples and exercises. (See pp. 155, 409, and 797.)

Increased Emphasis on Problem Solving   Introduced in Chapter 2, our six-step problem-solving method has been refined and integrated throughout the text. The six steps, *Read, Assign a Variable, Write an Equation, Solve, State the Answer,* and *Check,* are emphasized in boldface type and repeated in examples and exercises to reinforce the problem-solving process for students. (See pp. 115, 356, and 557.) Special boxes that include additional problem-solving information and tips are interspersed throughout the text. (See pp. 44, 121, and 145.)

Chapter Openers   New chapter openers feature real-world applications of mathematics that are relevant to students and tied to specific material within the chapters. Examples of topics include the stock market, the Olympics, and credit card debt. (See pp. 1, 93, and 187—Chapters 1, 2, and 3.)

Now Try Exercises   To actively engage students in the learning process, each example now concludes with a reference to one or more parallel exercises from the corresponding exercise set. In this way, students are able to immediately apply and reinforce the concepts and skills presented in the examples. (See pp. 129, 489, and 652.)

Summary Exercises   Based on user feedback, we have more than doubled the number of in-chapter summary exercises. These special exercise sets provide students with the all-important *mixed* review problems they need to master topics. Summaries of solution methods or additional examples are often included. (See pp. 114, 346, and 427.)

Glossary   A comprehensive glossary of key terms from throughout the text is included at the back of the book. (See pp. G-1 to G-8.)

WHAT FAMILIAR FEATURES HAVE BEEN RETAINED?

We have retained the popular features of previous editions of the text, some of which follow.

Learning Objectives   Each section begins with clearly stated, numbered objectives, and the included material is directly keyed to these objectives so that students know exactly what is covered in each section. (See pp. 188, 337, and 930.)

Cautions and Notes   One of the most popular features of previous editions, **CAUTION** and **NOTE** boxes warn students about common errors and emphasize important ideas throughout the exposition. (See pp. 108, 117, and 418.) There are more of these in the third edition than in the second, and the new text design makes them easier to spot.

Connections   Connections boxes have been streamlined. They continue to provide connections to the real world or to other mathematical concepts, historical background, and thought-provoking questions for writing or class discussion. (See pages 263, 476, and 616.)

Ample and Varied Exercise Sets   The text contains a wealth of exercises to provide students with opportunities to practice, apply, connect, and extend the algebraic skills they are learning. Numerous illustrations, tables, graphs, and photos have been added to the exercise sets to help students visualize the problems they are solving. Problem types include writing ✍, estimation, graphing calculator ▦, and challenging exercises that go beyond the examples as well as applications and multiple-choice, matching, true/false, and fill-in-the-blank problems. (See pp. 27, 638, and 893.)

Relating Concepts Exercises   These sets of exercises help students tie together topics and develop problem-solving skills as they compare and contrast ideas, identify and describe patterns, and extend concepts to new situations. (See pp. 222, 554, and 601.) These exercises make great collaborative activities for pairs or small groups of students.

Technology Insights Exercises   We assume that all students of this text have access to scientific calculators. *While graphing calculators are not required for this text,* some students may go on to courses that use them. For this reason, we have included Technology Insights exercises in selected exercise sets. These exercises provide an opportunity for students to interpret typical results seen on graphing calculator screens. Actual calculator screens from the Texas Instruments TI-83 Plus graphing calculator are featured. (See pp. 223, 496, and 893.)

Group Activities   Appearing at the end of each chapter, these real-data activities allow students to apply the mathematical content of the chapter in a collaborative setting. (See pp. 173, 437, and 578.)

Ample Opportunity for Review   Each chapter concludes with an extensive Chapter Summary that features Key Terms, New Symbols, Test Your Word Power, and a Quick Review of each section's content with additional examples. A comprehensive set of Chapter Review Exercises, keyed to individual sections, is included, as are Mixed Review Exercises and a Chapter Test. Beginning with Chapter 2, each chapter concludes with a set of Cumulative Review Exercises that cover material going back to Chapter 1. (See pp. 368–376 and 695–706.)

WHAT CONTENT CHANGES HAVE BEEN MADE?

We have worked hard to fine-tune and polish presentations of topics throughout the text based on user and reviewer feedback. Some of the content changes include the following:

- Former Section 2.1 on the addition and multiplication properties of equality has been split into two sections to allow each property to be treated individually before combining their use in Section 2.3. This chapter also includes linear inequalities in one variable in Section 2.8.

- Chapter 3 on linear equations in two variables now covers forms of the equations of a line in Section 3.4.

- Chapter 4 on exponents and polynomials has been reorganized so that the sections on exponents are covered early in the chapter.

- For increased flexibility, Chapter 7 begins with two review sections on graphing, slopes, and equations of lines, in addition to new material on linear models. The presentation on functions in Section 7.3 has been rewritten. A new section on variation is also included.

- Topics on inequalities and absolute value including compound inequalities, absolute value equations and inequalities, and linear inequalities in two variables are now covered in Chapter 9.

- Rational exponents are covered earlier in Chapter 10. The presentation on complex numbers has been moved to the end of this chapter so that complex solutions can be found when solving quadratic equations in Sections 11.1–11.4.

- Additional graphs of functions, operations on functions, and composition are covered in new Section 12.1.

- Four new appendices have been included. Appendix A provides an introduction to calculators, Appendix D includes all new material on mean, median, and mode, Appendix E provides a review of the metric system, and Appendix H covers Determinants and Cramer's Rule (former Section 9.7).

WHAT SUPPLEMENTS ARE AVAILABLE?

Our extensive supplements package includes an Annotated Instructor's Edition, testing materials, solutions manuals, tutorial software, videotapes, and a state-of-the-art Web site. For more information about any of the following supplements, please contact your Addison-Wesley sales consultant.

For the Student

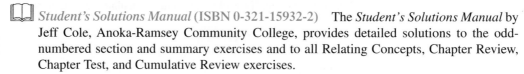 *Student's Solutions Manual* (ISBN 0-321-15932-2)   The *Student's Solutions Manual* by Jeff Cole, Anoka-Ramsey Community College, provides detailed solutions to the odd-numbered section and summary exercises and to all Relating Concepts, Chapter Review, Chapter Test, and Cumulative Review exercises.

Addison-Wesley Math Tutor Center   The Addison-Wesley Math Tutor Center is staffed by qualified college mathematics instructors who tutor students on examples and exercises from the textbook. Tutoring is provided via toll-free telephone, toll-free fax, e-mail, and the Internet. Interactive Web-based technology allows students and tutors to view and listen to live instruction—in real-time over the Internet. The Math Tutor Center is accessed through a registration number that can be packaged with a new textbook or purchased separately. (*Note:* MyMathLab students obtain access to the Math Tutor Center through their MyMathLab access code.)

InterAct Math® Tutorial Software (ISBN 0-321-15924-1)   Available on CD-ROM, this interactive tutorial software provides algorithmically generated practice exercises that are correlated at the objective level to the odd-numbered exercises in the text. Every exercise in the program is accompanied by an example and a guided solution designed to involve students in the solution process. Selected problems also include a video clip to help students visualize concepts. The software tracks student activity and scores and can generate printed summaries of students' progress.

MathXL   MathXL is an on-line testing, homework, and tutorial system that uses algorithmically generated exercises correlated to the textbook. Instructors can assign tests and homework provided by Addison-Wesley or create and customize their own tests and homework assignments. Instructors can also track their students' results and tutorial work in an on-line gradebook. Students can take chapter tests and receive personalized study plans that diagnose weaknesses and link students to areas they need to study and retest. Students can also work unlimited practice problems and receive tutorial instruction for areas in which they need improvement. MathXL can be packaged with new copies of *Beginning and Intermediate Algebra,* Third Edition. Please contact your Addison-Wesley sales representative for details.

Videotape Series (ISBN 0-321-15928-4)   This series of videotapes, created specifically for *Beginning and Intermediate Algebra,* Third Edition, features an engaging team of lecturers who provide comprehensive lessons on every objective in the text. The videos include a stop-the-tape feature that encourages students to pause the video, work through the example presented on their own, and then resume play to watch the video instructor go over the solution.

Digital Video Tutor (ISBN 0-321-16824-0)   This supplement provides the entire set of videotapes for the text in digital format on CD-ROM, making it easy and convenient for students to watch video segments from a computer, either at home or on campus. Available for purchase with the text at minimal cost, the Digital Video Tutor is ideal for distance learning and supplemental instruction.

MyMathLab   MyMathLab is a complete on-line course for Addison-Wesley mathematics textbooks that provides interactive, multimedia instruction correlated to the textbook content. MyMathLab is easily customizable to suit the needs of students and instructors and provides a comprehensive and efficient on-line course-management system that allows for diagnosis, assessment, and tracking of students' progress.

MyMathLab features:

- Chapter and section folders in the on-line course mirror the textbook Table of Contents and contain a wide range of multimedia instruction, including video lectures, tutorial software, and electronic supplements.

- Actual pages of the textbook are loaded into MyMathLab, and as students work through a section of on-line text, they can link to multimedia resources—such as video and audio clips, tutorial exercises, and interactive animations—that are correlated directly to examples and exercises in the text.

- Hyperlinks take the user directly to on-line testing, diagnosis, tutorials, and tracking in MathXL—Addison-Wesley's tutorial and testing system for mathematics and statistics.

- Instructors can create, copy, edit, assign, and track all tests and homework for their course as well as track students' results and practice work.

- With push-button ease, instructors can remove, hide, or annotate Addison-Wesley preloaded content, add their own course documents, or change the order in which material is presented.

- Using the communication tools found in MyMathLab, instructors can hold on-line office hours, host a discussion board, create communication groups within their class, send e-mail, and maintain a course calendar.

- Print supplements are available on-line, side-by-side with the textbook.

For more information, visit our Web site at www.mymathlab.com or contact your Addison-Wesley sales representative for a live demonstration.

For the Instructor

CLASSROOM EXAMPLE

Simplify $\dfrac{2(7 + 8) + 2}{3 \cdot 5 + 1}$.

Answer: 2

TEACHING TIP Warn students that $\dfrac{4}{15} + \dfrac{5}{9}$ is NOT EQUAL TO $\dfrac{9}{24}$.

Annotated Instructor's Edition (ISBN 0-321-12762-5) For immediate access, the Annotated Instructor's Edition provides answers to all text exercises in the margin or next to the corresponding exercise, as well as Classroom Examples (formerly Chalkboard Examples) and Teaching Tips, printed in blue for easy visibility. Based on user feedback, we have increased the number of Classroom Examples and Teaching Tips.

Exercises designed for writing and graphing calculator use are indicated in both the Student Edition and the Annotated Instructor's Edition.

Instructor's Solutions Manual (ISBN 0-321-15923-3) The *Instructor's Solutions Manual,* by Jeff Cole, Anoka-Ramsey Community College, provides complete solutions to all text exercises.

Answer Book (ISBN 0-321-15927-6) The *Answer Book* provides answers to all the exercises in the text.

Printed Test Bank (ISBN 0-321-15926-8) Written by Jon Becker, Indiana University Northwest, the *Printed Test Bank* contains two diagnostic pretests, four free-response and two multiple-choice test forms per chapter, and two final exams. Additional practice exercises for almost every objective of every section of the text are also included. A conversion guide from the second to the third edition is also included.

Adjunct Support Manual (ISBN 0-321-19744-5)   This manual includes resources designed to help both new and adjunct faculty with course preparation and classroom management as well as offering helpful teaching tips.

Adjunct Support Center   The Adjunct Support Center offers consultation on suggested syllabi, helpful tips on using the textbook support package, assistance with textbook content, and advice on classroom strategies from qualified mathematics instructors with over fifty years of combined teaching experience. The Adjunct Support Center is available Sunday through Thursday evenings from 5 P.M. to midnight. Phone: 1-800-435-4084; E-mail: adjunctsupport@awl.com; Fax: 1-877-262-9774.

TestGen with QuizMaster (ISBN 0-321-15929-2)   TestGen enables instructors to build, edit, print, and administer tests using a computerized bank of questions developed to cover all the objectives of the text. Instructors can modify test bank questions or add new questions by using the built-in question editor, which allows users to create graphs, import graphics, insert math notation, and insert variable numbers or text. Tests can be printed or administered on-line via the Web or other network. TestGen comes packaged with QuizMaster, which allows students to take tests on a local area network. The software is available on a dual-platform Windows/Macintosh CD-ROM.

MathXL   MathXL is an on-line testing, homework, and tutorial system that uses algorithmically generated exercises correlated to the textbook. Instructors can assign tests and homework provided by Addison-Wesley or create and customize their own tests and homework assignments. Instructors can also track their students' results and tutorial work in an on-line gradebook. Students can take chapter tests and receive personalized study plans that diagnose weaknesses and link students to areas they need to study and retest. Students can also work unlimited practice problems and receive tutorial instruction for areas in which they need improvement. MathXL can be packaged with new copies of *Beginning and Intermediate Algebra,* Third Edition. Please contact your Addison-Wesley sales representative for details.

MyMathLab   MyMathLab is a complete on-line course for Addison-Wesley mathematics textbooks that provides interactive, multimedia instruction correlated to the textbook content. MyMathLab is easily customizable to suit the needs of students and instructors and provides a comprehensive and efficient on-line course-management system that allows for diagnosis, assessment, and tracking of students' progress.

MyMathLab features:

- Chapter and section folders in the on-line course mirror the textbook Table of Contents and contain a wide range of multimedia instruction, including video lectures, tutorial software, and electronic supplements.

- Actual pages of the textbook are loaded into MyMathLab, and as students work through a section of on-line text, they can link to multimedia resources—such as video and audio clips, tutorial exercises, and interactive animations—that are correlated directly to examples and exercises in the text.

- Hyperlinks take the user directly to on-line testing, diagnosis, tutorials, and tracking in MathXL—Addison-Wesley's tutorial and testing system for mathematics and statistics.

- Instructors can create, copy, edit, assign, and track all tests and homework for their course as well as track students' results and practice work.

- With push-button ease, instructors can remove, hide, or annotate Addison-Wesley preloaded content, add their own course documents, or change the order in which material is presented.

- Using the communication tools found in MyMathLab, instructors can hold on-line office hours, host a discussion board, create communication groups within their class, send e-mail, and maintain a course calendar.

- Print supplements are available on-line, side-by-side with the textbook.

For more information, visit our Web site at www.mymathlab.com or contact your Addison-Wesley sales representative for a live demonstration.

ACKNOWLEDGMENTS

The comments, criticisms, and suggestions of users, nonusers, instructors, and students have positively shaped this textbook over the years, and we are most grateful for the many responses we have received. The feedback gathered for this revision of the text was particularly helpful, and we especially wish to thank the following individuals who provided invaluable suggestions:

Mary Kay Abbey, *Montgomery College*
Marwan Abu-Sawwa, *Florida Community College, Jacksonville*
Jose Alvarado, *University of Texas, Pan American*
Sonya Armstrong, *West Virginia State College*
Mohammad Aslam, *Georgia Perimeter College*
Rajappa Asthagiri, *Miami University, Middletown*
Mary Lou Baker, *Columbia State Community College*
James J. Ball, *Indiana State University*
Dixilee Blackinton, *Weber State University*
Bob Bohac, *North Idaho College*
Tammy Borren, *Columbus State Community College*
Frances Brewer, *Vance-Granville Community College*
Marc Campbell, *Daytona Beach Community College*
Lisa Cuneo, *Pennsylvania State University, DuBois*
Charles Curtis, *Missouri Southern State College*
Jo Dobbin, *Fayetteville Technical Community College*
Sharon Edgmon, *Bakersfield College*
Lucy Edwards, *Las Positas College*
Carol Flakus, *Lower Columbia College*
Reginald W. Fulwood, *Palm Beach Community College*
Joe Howe, *St. Charles Community College*
Matthew Hudock, *St. Philip's College*
Dale Hughes, *Johnson County Community College*
Nancy R. Johnson, *Manatee Community College*
Judy Kasabian, *El Camino College*
Mike Keller, *St. Johns River Community College*
Nancy Ketchum, *Moberly Area Community College*
Mike Kirby, *Tidewater Community College, Virginia Beach*
Linda Kodama, *Kapi'olani Community College*
Laura Lowrey, *Georgia Perimeter College*
Tony Masci, *Notre Dame College*

Timothy McLendon, *East Central College*
Jean P. Millen, *Georgia Perimeter College, Central Campus*
Kausha Miller, *Lexington Community College*
Molly Misko, *Gadsden State Community College*
Kathy Nickell, *College of DuPage*
Linda M. Partlow, *Itawamba Community College*
Marilyn Platt, *Gaston College*
Larry Pontaski, *Pueblo Community College*
Tammy Potter, *Gadsden State Community College*
Joan Prymas, *Herkimer County Community College*
Nelissa Rutishauser, *Mohawk Valley Community College*
Gwen Terwilliger, *University of Toledo*
Timothy Thompson, *Oregon Institute of Technology*
Mark Tom, *College of the Sequoias*
Bettie Truitt, *Black Hawk College*
Virginia Urban, *Fashion Institute of Technology*
Tony Vavra, *West Virginia Northern Community College*
Mansoor Vejdani, *University of Cincinnati*
Laura M. Villarreal, *University of Texas at Brownsville* and *Texas Southmost College*
Gail Wiltse, *St. Johns River Community College*
Jackie Wing, *Angelina College*
Mary Wolyniak, *Broome Community College*

Over the years, we have come to rely on an extensive team of experienced professionals. Our sincere thanks go to these dedicated individuals at Addison-Wesley, who worked long and hard to make this revision a success: Maureen O'Connor, Suzanne Alley, Dona Kenly, Kathy Manley, Dennis Schaefer, Susan Raymond, Jolene Lehr, and Lindsay Skay.

Abby Tanenbaum did an outstanding job checking the answers to exercises and also provided invaluable assistance during the production process. Steven Pusztai provided his customary excellent production work. Thanks are due Jeff Cole, who supplied accurate, helpful solutions manuals, and Jon Becker, who provided the comprehensive *Printed Test Bank*. We are most grateful to Paul Van Erden for yet another accurate, useful index; and Becky Troutman for preparing the comprehensive List of Applications.

As an author team, we are committed to the goal stated earlier in this Preface—to provide the best possible text and supplements package to help instructors teach and students succeed. We are most grateful to all those over the years who have aspired to this goal with us. As we continue to work toward it, we would welcome any comments or suggestions you might have. Please feel free to send your comments via e-mail to math@aw.com.

Margaret L. Lial
John Hornsby
Terry McGinnis

Feature Walkthrough

Linear Equations in Two Variables — 3

3.1 Reading Graphs; Linear Equations in Two Variables

3.2 Graphing Linear Equations in Two Variables

3.3 The Slope of a Line

3.4 Equations of a Line

U.S. debt from credit cards continues to increase. In recent years, college campuses have become fertile territory as credit card companies pitch their plastic to students at bookstores, student unions, and sporting events. As a result, three out of four undergrads now have at least one credit card and carry an average balance of $2748. (*Source:* Nellie Mae.) In Example 6 of Section 3.2, we use the concepts of this chapter to investigate credit card debt.

Chapter Opener

Each chapter opens with an application and section outline. The application in the opener is tied to specific material within the chapter.

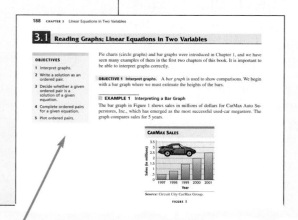

Learning Objectives

Each section opens with a highlighted list of clearly stated, numbered learning objectives. These learning objectives are reinforced throughout the section by restating the learning objective where appropriate so that students always know exactly what is being covered.

Notes

Important ideas are emphasized in *Note* boxes that appear throughout the text.

Cautions

Students are warned of common errors through the use of *Caution* boxes that are found throughout the text.

3.4 Equations of a Line

OBJECTIVES

1 Write an equation of a line given its slope and y-intercept.
2 Graph a line given its slope and a point on the line.
3 Write an equation of a line given its slope and any point on the line.
4 Write an equation of a line given two points on the line.

In the previous section we found the slope (steepness) of a line from the equation of the line by solving the equation for y. In that form, the slope is the coefficient of x. For example, the slope of the line with equation $y = 2x + 3$ is 2, the coefficient of x. What does the number 3 represent? If $x = 0$, the equation becomes

$$y = 2(0) + 3 = 0 + 3 = 3.$$

Since $y = 3$ corresponds to $x = 0$, $(0, 3)$ is the y-intercept of the graph of $y = 2x + 3$. An equation like $y = 2x + 3$ that is solved for y is said to be in *slope-intercept form* because both the slope and the y-intercept of the line can be read directly from the equation.

Slope-Intercept Form

The **slope-intercept form** of the equation of a line with slope m and y-intercept $(0, b)$ is

$$y = mx + b.$$

Remember that the intercept in the slope-intercept form is the *y-intercept*.

NOTE The slope-intercept form is the most useful form for a linear equation because of the information we can determine from it. It is also the form used by graphing calculators and the one that describes a *linear function*, an important concept in mathematics.

CAUTION A common error in factoring a difference of cubes, such as $x^3 - y^3 = (x - y)(x^2 + xy + y^2)$, is to try to factor $x^2 + xy + y^2$. It is easy to confuse this factor with a perfect square trinomial, $x^2 + 2xy + y^2$. Because there is no 2 in $x^2 + xy + y^2$, it is very unusual to be able to further factor an expression of the form $x^2 + xy + y^2$.

Classroom Examples and Teaching Tips

The Annotated Instructor's Edition provides answers to all text exercises and Group Activities in color in the margin or next to the corresponding exercise. In addition, *Classroom Examples* and *Teaching Tips* are included to assist instructors in creating examples to use in class that are different from what students have in their textbooks. *Teaching Tips* offer guidance on presenting the material at hand.

New! Now Try Exercises

Now Try exercises are found after each example to encourage active learning. This feature asks students to work exercises in the exercise sets that parallel the example just studied.

TEACHING TIP On some occasions, students may try to "reduce" $\frac{5^8}{5^5}$ as $\frac{1^8}{1^5}$. Stress that the expression does not say "5 divided by 5."

CAUTION A common **error** is to write $\frac{5^8}{5^4} = 1^{8-4} = 1^4$. Notice that by the quotient rule, the quotient should have the *same base*, 5. That is,

$$\frac{5^8}{5^4} = 5^{8-4} = 5^4.$$

If you are not sure, use the definition of an exponent to write out the factors:

$$5^8 = 5 \cdot 5 \cdot 5 \cdot 5 \cdot 5 \cdot 5 \cdot 5 \cdot 5 \quad \text{and} \quad 5^4 = 5 \cdot 5 \cdot 5 \cdot 5.$$

Then it is clear that the quotient is 5^4.

CLASSROOM EXAMPLE

Simplify. Write answers with positive exponents.

(a) $\frac{4^7}{4^3}$

(b) $\frac{4^8}{4^3}$

(c) $\frac{x^{-6}}{x^{-12}}$

(d) $\frac{8^4 m^5 n^{-2}}{8^2 m^{-n} n^2}$ ($m, n \neq 0$)

Answer: (a) 4^4 or 16

(b) $\frac{1}{4^2}$ or $\frac{1}{16}$ (c) x^6

(d) $\frac{1}{8mn^3}$

EXAMPLE 4 Using the Quotient Rule

Simplify, using the quotient rule for exponents. Write answers with positive exponents.

(a) $\frac{5^6}{5^5} = 5^{6-5} = 5^2$ or 25

(b) $\frac{4^2}{4^9} = 4^{2-9} = 4^{-7} = \frac{1}{4^7}$

(c) $\frac{5^{-3}}{5^{-7}} = 5^{-3-(-7)} = 5^4$ or 625

(d) $\frac{q^5}{q^{-3}} = q^{5-(-3)} = q^8$ ($q \neq 0$)

(e) $\frac{3^2 x^5}{3^4 x^3} = \frac{3^2}{3^4} \cdot \frac{x^5}{x^3} = 3^{2-4} \cdot x^{5-3} = 3^{-2} x^2 = \frac{x^2}{3^2}$ ($x \neq 0$)

(f) $\frac{(m+n)^{-2}}{(m+n)^{-4}} = (m+n)^{-2-(-4)} = (m+n)^{-2+4} = (m+n)^2$ ($m \neq -n$)

(g) $\frac{7x^{-3}y^2}{2^{-1}x^2y^{-5}} = \frac{7 \cdot 2^1 y^2 y^5}{x^2 x^3} = \frac{14y^7}{x^5}$ ($x, y \neq 0$)

Now Try Exercises 29, 41, 45, and 49.

The definitions and rules for exponents given in this section and Section 4.1 are summarized here.

TEACHING TIP Stress that the product, quotient, and power rules are the same for positive and negative exponents.

Definitions and Rules for Exponents

For any integers m and n:

		Examples
Product rule	$a^m \cdot a^n = a^{m+n}$	$7^4 \cdot 7^5 = 7^9$
Zero exponent	$a^0 = 1$ ($a \neq 0$)	$(-3)^0 = 1$
Negative exponent	$a^{-n} = \frac{1}{a^n}$ ($a \neq 0$)	$5^{-3} = \frac{1}{5^3}$
Quotient rule	$\frac{a^m}{a^n} = a^{m-n}$ ($a \neq 0$)	$\frac{2^2}{2^5} = 2^{2-5} = 2^{-3} = \frac{1}{2^3}$
Power rules (a)	$(a^m)^n = a^{mn}$	$(4^2)^3 = 4^6$
(b)	$(ab)^m = a^m b^m$	$(3k)^4 = 3^4 k^4$
(c)	$\left(\frac{a}{b}\right)^m = \frac{a^m}{b^m}$ ($b \neq 0$)	$\left(\frac{2}{3}\right)^2 = \frac{2^2}{3^2}$

(continued)

Connections

Connections boxes provide connections to the real world or to other mathematical concepts, historical background, and thought-provoking questions for writing or class discussion.

Writing Exercises

Writing exercises abound in the Lial series through the *Connections* boxes and also in the exercise sets. Some writing exercises require only short written answers, and others require lengthier journal-type responses where students are asked to fully explain terminology, procedures and methods, document their understanding using examples, or make connections between topics.

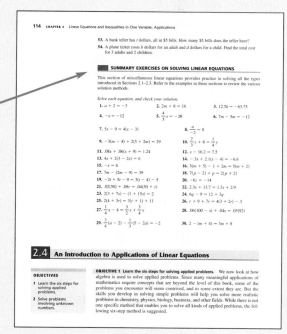

Problem Solving

The Lial *six-step problem-solving* method is clearly explained in Chapter 2 and is then continually reinforced in examples and exercises throughout the text to aid students in solving application problems.

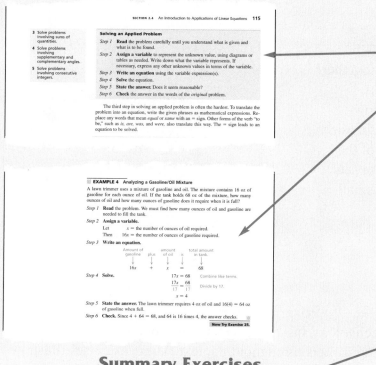

Summary Exercises

Summary Exercises appear in many chapters to provide students with *mixed* practice problems needed to master topics.

Ample and Varied Exercise Sets

Algebra students require a large number of varied practice exercises to master the material they have just learned. This text contains thousands of exercises including summary and review exercises, numerous conceptual and writing exercises, and challenging exercises that go beyond the examples. Multiple-choice, matching, true/false, and completion exercises help to provide variety. Exercises suitable for graphing calculator use are marked with an icon.

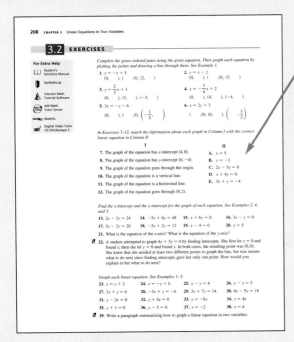

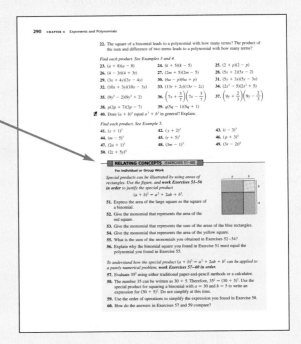

Relating Concepts

Found in selected exercise sets, these exercises tie together topics and highlight the relationships among various concepts and skills. For example, they may show how algebra and geometry are related or how a graph of a linear equation in two variables is related to the solution of the corresponding linear equation in one variable. These sets of exercises make great collaborative activities for small groups of students.

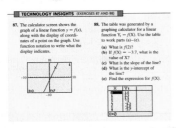

Technology Insights

Technology Insights exercises are found in selected exercise sets throughout the text. These exercises illustrate the power of graphing calculators and provide an opportunity for students to interpret typical results seen on graphing calculator screens. (A graphing calculator is *not* required to complete these exercises.)

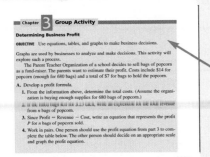

Group Activities

Appearing at the end of each chapter, these activities allow students to work collaboratively to solve a problem related to the chapter material.

Ample Opportunity for Review

One of the most popular features of the Lial textbooks is the extensive and well thought-out end-of-chapter material. At the end of each chapter, students will find:

Key Terms and *New Symbols* that are keyed back to the appropriate section for easy reference and study.

Test Your Word Power helps students understand and master mathematical vocabulary; key terms from the chapter are presented with four possible definitions in multiple-choice format.

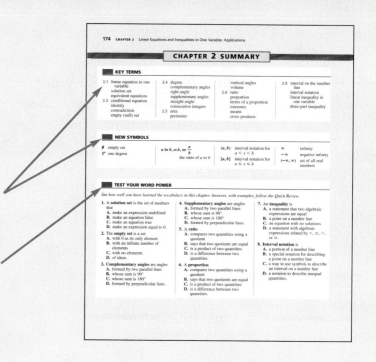

Quick Review sections give students not only the main concepts from the chapter (referenced back to the appropriate section) but also an adjacent example of each concept.

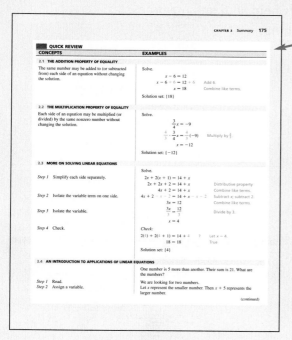

Review Exercises are keyed to appropriate sections so that students can refer to examples of that type of problem if they need help.

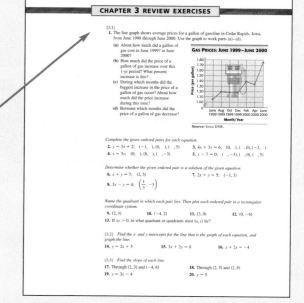

Mixed Review Exercises require students to solve problems without the help of section references.

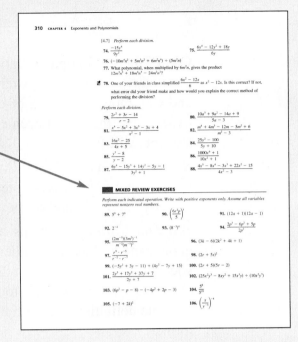

Chapter Tests help students practice for the real thing.

Cumulative Review Exercises gather various types of exercises from preceding chapters to help students remember and retain what they are learning throughout the course.

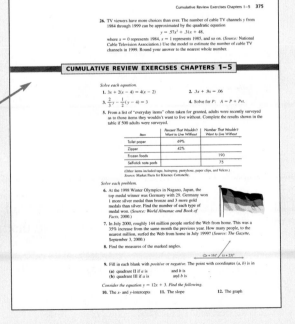

The Real Number System

1

The table indicates the figures for several stocks listed on the NASDAQ stock exchange for Wednesday, May 29, 2002. The final column represents the change in the price of the stock from the previous day. These changes are indicated by signed numbers. A positive number, such as +.75, means that the price of one share of the stock rose 75 cents from the previous day's closing; a negative number, such as −2.24, means that the price fell $2.24. In the exercises for Sections 1.1 and 1.4, we apply the concepts of this chapter to investigate situations involving stock prices.

Stock	Vol.	Close	Chg.
AAONs	14	31.85	+.75
ABC Bcp	7	14.55	
AC Moore	58	40.35	−2.24
ACT Tele	400	3.30	+.14
ADC Tel	4752	3.49	−.12
AFC Ent	225	30.40	−.96
AHL Sev	79	2.03	−.21

Source: Times Picayune.

1.1   Fractions

As preparation for the study of algebra, we begin this section with a brief review of arithmetic. In everyday life, the numbers seen most often are the **natural numbers,**

$$1, 2, 3, 4, \ldots,$$

the **whole numbers,**

$$0, 1, 2, 3, 4, \ldots,$$

and **fractions,** such as

$$\frac{1}{2}, \quad \frac{2}{3}, \quad \text{and} \quad \frac{15}{7}.$$

The parts of a fraction are named as follows.

$$\text{Fraction bar} \rightarrow \frac{4 \quad \leftarrow \text{Numerator}}{7 \quad \leftarrow \text{Denominator}}$$

As we will see later, the fraction bar represents division $\left(\frac{a}{b} = a \div b\right)$ and also serves as a grouping symbol.

If the numerator of a fraction is less than the denominator, we call it a **proper fraction.** A proper fraction has a value less than 1. If the numerator is greater than or equal to the denominator, the fraction is an **improper fraction.** An improper fraction that has a value greater than 1 is often written as a **mixed number.** For example, $\frac{12}{5}$ may be written as $2\frac{2}{5}$. In algebra, we prefer to use the improper form because it is easier to work with. In applications, we usually convert answers to mixed form, which is more meaningful.

OBJECTIVE 1 Learn the definition of *factor.*   In the statement $2 \times 9 = 18$, the numbers 2 and 9 are called **factors** of 18. Other factors of 18 include 1, 3, 6, and 18. The result of the multiplication, 18, is called the **product.**

There are several ways of representing the product of two numbers. For example,

$$6 \times 3, \quad 6 \cdot 3, \quad (6)(3), \quad 6(3), \quad \text{and} \quad (6)3$$

all represent the product of 6 and 3.

The number 18 is **factored** by writing it as the product of two or more numbers. For example, 18 can be factored as $6 \cdot 3$, $18 \cdot 1$, $9 \cdot 2$, or $3 \cdot 3 \cdot 2$. (In algebra, a raised dot $\cdot$ is often used instead of the $\times$ symbol to indicate multiplication.)

A natural number (except 1) is **prime** if it has only itself and 1 as factors. "Factors" are understood here to mean natural number factors. (By agreement, the number 1 is not a prime number.) The first dozen primes are

$$2, 3, 5, 7, 11, 13, 17, 19, 23, 29, 31, 37.$$

A natural number (except 1) that is not prime, such as 4, 6, 8, 9, and 12, is a **composite** number.

It is often useful to find all **prime factors** of a number—those factors that are prime numbers. For example, the only prime factors of 18 are 2 and 3.

EXAMPLE 1 Factoring Numbers

Write each number as the product of prime factors.

(a) 35

Write 35 as the product of the prime fractors 5 and 7, or as

$$35 = 5 \cdot 7.$$

(b) 24

One way to begin is to divide by the smallest prime, 2, to get

$$24 = 2 \cdot 12.$$

Now divide 12 by 2 to find factors of 12.

$$24 = 2 \cdot 2 \cdot 6$$

Since 6 can be written as $2 \cdot 3$,

$$24 = 2 \cdot 2 \cdot 2 \cdot 3,$$

where all factors are prime.

Now Try Exercises 9 and 19.

NOTE It is not necessary to start with the smallest prime factor, as shown in Example 1(b). In fact, no matter which prime factor we start with, we will *always* obtain the same prime factorization.

OBJECTIVE 2 Write fractions in lowest terms. We use prime numbers to write fractions in *lowest terms*. A fraction is in **lowest terms** when the numerator and denominator have no factors in common (other than 1). We use the **basic principle of fractions** to write a fraction in lowest terms.

Basic Principle of Fractions

If the numerator and denominator of a fraction are multiplied or divided by the same nonzero number, the value of the fraction is not changed.

For example, $\frac{12}{16}$ can be written in lowest terms as follows:

$$\frac{12}{16} = \frac{3 \cdot 4}{4 \cdot 4} = \frac{3}{4} \cdot \frac{4}{4} = \frac{3}{4} \cdot 1 = \frac{3}{4}.$$

This procedure uses the rule for multiplying fractions (covered in the next objective) and the multiplication property of 1 (covered in Section 1.7).

To write a fraction in lowest terms, use these steps.

Writing a Fraction in Lowest Terms

Step 1 Write the numerator and the denominator as the product of prime factors.

Step 2 Divide the numerator and the denominator by the **greatest common factor,** the product of all factors common to both.

EXAMPLE 2 Writing Fractions in Lowest Terms

Write each fraction in lowest terms.

(a) $\dfrac{10}{15} = \dfrac{2 \cdot 5}{3 \cdot 5} = \dfrac{2 \cdot 1}{3 \cdot 1} = \dfrac{2}{3}$

Since 5 is the greatest common factor of 10 and 15, dividing both numerator and denominator by 5 gives the fraction in lowest terms.

(b) $\dfrac{15}{45} = \dfrac{3 \cdot 5}{3 \cdot 3 \cdot 5} = \dfrac{1 \cdot 1}{3 \cdot 1 \cdot 1} = \dfrac{1}{3}$

The factored form shows that 3 and 5 are the common factors of both 15 and 45. Dividing both 15 and 45 by $3 \cdot 5 = 15$ gives $\frac{15}{45}$ in lowest terms as $\frac{1}{3}$.

Now Try Exercises 25 and 29.

We can simplify this process by finding the greatest common factor in the numerator and denominator by inspection. For instance, in Example 2(b), we can use 15 rather than $3 \cdot 5$.

$$\frac{15}{45} = \frac{15}{3 \cdot 15} = \frac{1}{3 \cdot 1} = \frac{1}{3}$$

CAUTION Errors may occur when writing fractions in lowest terms if the factor 1 is not included. To see this, refer to Example 2(b). In the equation

$$\frac{3 \cdot 5}{3 \cdot 3 \cdot 5} = \frac{?}{3},$$

if 1 is not written in the numerator when dividing out common factors, you may make an error. The ? should be replaced by 1.

OBJECTIVE 3 Multiply and divide fractions. We multiply two fractions by first multiplying their numerators and then multiplying their denominators. This rule is written in symbols as follows.

Multiplying Fractions

If $\dfrac{a}{b}$ and $\dfrac{c}{d}$ are fractions, then $\qquad \dfrac{a}{b} \cdot \dfrac{c}{d} = \dfrac{a \cdot c}{b \cdot d}.$

EXAMPLE 3 Multiplying Fractions

Find each product, and write it in lowest terms.

(a) $\dfrac{3}{8} \cdot \dfrac{4}{9} = \dfrac{3 \cdot 4}{8 \cdot 9}$ 　　Multiply numerators.
　　　　　　　　　　　　Multiply denominators.

$\phantom{\text{(a)}} = \dfrac{3 \cdot 4}{2 \cdot 4 \cdot 3 \cdot 3}$ 　　Factor.

$\phantom{\text{(a)}} = \dfrac{1}{2 \cdot 3} = \dfrac{1}{6}$ 　　Write in lowest terms.

(b) $2\dfrac{1}{3} \cdot 5\dfrac{1}{2} = \dfrac{7}{3} \cdot \dfrac{11}{2}$ Write as improper fractions.

$\qquad\qquad = \dfrac{77}{6}$ or $12\dfrac{5}{6}$ Multiply numerators and denominators.

Now Try Exercises 35 and 39.

Two fractions are **reciprocals** of each other if their product is 1. For example, $\frac{3}{4}$ and $\frac{4}{3}$ are reciprocals since

$$\frac{3}{4} \cdot \frac{4}{3} = \frac{12}{12} = 1.$$

Also, $\frac{7}{11}$ and $\frac{11}{7}$ are reciprocals of each other, as are $\frac{1}{6}$ and 6. Because division is the opposite (or inverse) of multiplication, we use reciprocals to divide fractions. To divide two fractions, multiply the first fraction by the reciprocal of the second fraction.

Dividing Fractions

For the fractions $\dfrac{a}{b}$ and $\dfrac{c}{d}$, $\dfrac{a}{b} \div \dfrac{c}{d} = \dfrac{a}{b} \cdot \dfrac{d}{c}.$

That is, to divide by a fraction, multiply by its reciprocal.

We will explain why this method works in Chapter 6. However, as an example, we know that $20 \div 10 = 2$ and $20 \cdot \frac{1}{10} = 2$. The answer to a division problem is called a **quotient.** For example, the quotient of 20 and 10 is 2.

■ **EXAMPLE 4**   **Dividing Fractions**

Find each quotient, and write it in lowest terms.

(a) $\dfrac{3}{4} \div \dfrac{8}{5} = \dfrac{3}{4} \cdot \dfrac{5}{8} = \dfrac{3 \cdot 5}{4 \cdot 8} = \dfrac{15}{32}$

Multiply by the reciprocal of the second fraction.

(b) $\dfrac{3}{4} \div \dfrac{5}{8} = \dfrac{3}{4} \cdot \dfrac{8}{5} = \dfrac{3 \cdot 8}{4 \cdot 5} = \dfrac{3 \cdot 4 \cdot 2}{4 \cdot 5} = \dfrac{6}{5}$ or $1\dfrac{1}{5}$

(c) $\dfrac{5}{8} \div 10 = \dfrac{5}{8} \div \dfrac{10}{1} = \dfrac{5}{8} \cdot \dfrac{1}{10} = \dfrac{1}{16}$

Write 10 as $\frac{10}{1}$.

(d) $1\dfrac{2}{3} \div 4\dfrac{1}{2} = \dfrac{5}{3} \div \dfrac{9}{2}$ Write as improper fractions.

$\qquad\qquad = \dfrac{5}{3} \cdot \dfrac{2}{9}$ Multiply by the reciprocal of the second fraction.

$\qquad\qquad = \dfrac{10}{27}$

Now Try Exercises 43, 45, and 47.

OBJECTIVE 4 Add and subtract fractions. The result of adding two numbers is called the **sum** of the numbers. For example, $2 + 3 = 5$, so 5 is the sum of 2 and 3. To find the sum of two fractions having the same denominator, add the numerators and keep the same denominator.

Adding Fractions

If $\dfrac{a}{b}$ and $\dfrac{c}{b}$ are fractions, then $\qquad \dfrac{a}{b} + \dfrac{c}{b} = \dfrac{a + c}{b}.$

EXAMPLE 5 Adding Fractions with the Same Denominator

Add.

(a) $\dfrac{3}{7} + \dfrac{2}{7} = \dfrac{3 + 2}{7} = \dfrac{5}{7}$ Add numerators; keep the same denominator.

(b) $\dfrac{2}{10} + \dfrac{3}{10} = \dfrac{2 + 3}{10} = \dfrac{5}{10} = \dfrac{1}{2}$ Write in lowest terms.

Now Try Exercise 51.

If the fractions to be added do not have the same denominators, we can still use the procedure above, but only *after* we rewrite the fractions with a common denominator. For example, to rewrite $\frac{3}{4}$ as a fraction with denominator 32,

$$\frac{3}{4} = \frac{?}{32},$$

we find the number that can be multiplied by 4 to give 32. Since $4 \cdot 8 = 32$, we use the number 8. By the basic principle of fractions, we multiply the numerator and the denominator by 8.

$$\frac{3}{4} = \frac{3 \cdot 8}{4 \cdot 8} = \frac{24}{32}$$

Finding the Least Common Denominator

To add or subtract fractions with different denominators, find the **least common denominator (LCD)** as follows.

Step 1 Factor both denominators.

Step 2 For the LCD, use every factor that appears in any factored form. If a factor is repeated, use the largest number of repeats in the LCD.

The next example shows this procedure.

EXAMPLE 6 Adding Fractions with Different Denominators

Add. Write the sums in lowest terms.

(a) $\dfrac{4}{15} + \dfrac{5}{9}$

To find the least common denominator, first factor both denominators.

$$15 = 5 \cdot 3 \quad \text{and} \quad 9 = 3 \cdot 3$$

Since 5 and 3 appear as factors, and 3 is a factor of 9 twice, the LCD is

$$15 \quad 9$$

$$5 \cdot 3 \cdot 3 \quad \text{or} \quad 45.$$

Write each fraction with 45 as denominator.

$$\frac{4}{15} = \frac{4 \cdot 3}{15 \cdot 3} = \frac{12}{45} \quad \text{and} \quad \frac{5}{9} = \frac{5 \cdot 5}{9 \cdot 5} = \frac{25}{45}$$

Now add the two equivalent fractions.

$$\frac{4}{15} + \frac{5}{9} = \frac{12}{45} + \frac{25}{45} = \frac{37}{45}$$

(b) $3\frac{1}{2} + 2\frac{3}{4}$

We add mixed numbers using either of two methods.

Method 1

Rewrite both numbers as improper fractions.

$$3\frac{1}{2} = 3 + \frac{1}{2} = \frac{3}{1} + \frac{1}{2} = \frac{6}{2} + \frac{1}{2} = \frac{6+1}{2} = \frac{7}{2}$$

$$2\frac{3}{4} = 2 + \frac{3}{4} = \frac{8}{4} + \frac{3}{4} = \frac{8+3}{4} = \frac{11}{4}$$

Now add. The common denominator is 4.

$$3\frac{1}{2} + 2\frac{3}{4} = \frac{7}{2} + \frac{11}{4} = \frac{14}{4} + \frac{11}{4} = \frac{25}{4} \quad \text{or} \quad 6\frac{1}{4}$$

Method 2

Write $3\frac{1}{2}$ as $3\frac{2}{4}$. Then add vertically.

$$
\begin{array}{cc}
3\dfrac{1}{2} & 3\dfrac{2}{4} \\
\rightarrow & \\
+\,2\dfrac{3}{4} & +\,2\dfrac{3}{4} \\
\hline
 & 5\dfrac{5}{4}
\end{array}
$$

Since $\frac{5}{4} = 1\frac{1}{4}$,

$$5\frac{5}{4} = 5 + 1\frac{1}{4} = 6\frac{1}{4} \quad \text{or} \quad \frac{25}{4}.$$

Now Try Exercises 53 and 55.

The **difference** between two numbers is found by subtracting the numbers. For example, $9 - 5 = 4$, so the difference between 9 and 5 is 4. Subtraction of fractions is similar to addition. Just subtract the numerators instead of adding them; again, keep the same denominator.

> **Subtracting Fractions**
>
> If $\dfrac{a}{b}$ and $\dfrac{c}{b}$ are fractions, then $\dfrac{a}{b} - \dfrac{c}{b} = \dfrac{a-c}{b}$.

EXAMPLE 7 Subtracting Fractions

Subtract. Write the differences in lowest terms.

(a) $\dfrac{15}{8} - \dfrac{3}{8} = \dfrac{15-3}{8}$ Subtract numerators; keep the same denominator.

$\qquad\qquad = \dfrac{12}{8} = \dfrac{3}{2}$ or $1\dfrac{1}{2}$ Lowest terms

(b) $\dfrac{7}{18} - \dfrac{4}{15}$

Here, $18 = 2 \cdot 3 \cdot 3$ and $15 = 3 \cdot 5$, so the LCD is $2 \cdot 3 \cdot 3 \cdot 5 = 90$.

$$\dfrac{7}{18} - \dfrac{4}{15} = \dfrac{7 \cdot 5}{2 \cdot 3 \cdot 3 \cdot 5} - \dfrac{4 \cdot 2 \cdot 3}{2 \cdot 3 \cdot 3 \cdot 5} = \dfrac{35}{90} - \dfrac{24}{90} = \dfrac{11}{90}$$

(c) $\dfrac{15}{32} - \dfrac{11}{45}$

Since $32 = 2 \cdot 2 \cdot 2 \cdot 2 \cdot 2$ and $45 = 3 \cdot 3 \cdot 5$, there are no common factors, and the LCD is $32 \cdot 45 = 1440$.

$$\dfrac{15}{32} - \dfrac{11}{45} = \dfrac{15 \cdot 45}{32 \cdot 45} - \dfrac{11 \cdot 32}{45 \cdot 32}$$ Get a common denominator.

$$= \dfrac{675}{1440} - \dfrac{352}{1440}$$

$$= \dfrac{323}{1440}$$ Subtract.

Now Try Exercises 57 and 59.

OBJECTIVE 5 Solve applied problems that involve fractions. Applied problems often require work with fractions. For example, when a carpenter reads diagrams and plans, he or she often must work with fractions whose denominators are 2, 4, 8, 16, or 32, as shown in the next example.

EXAMPLE 8 Adding Fractions to Solve an Applied Problem

The diagram in Figure 1 appears in the book *Woodworker's 39 Sure-Fire Projects*. It is the front view of a corner bookcase/desk. Add the fractions shown in the diagram to find the height of the bookcase/desk to the top of the writing surface.

We must add the following measures (" means inches):

$$\dfrac{3}{4}, \quad 4\dfrac{1}{2}, \quad 9\dfrac{1}{2}, \quad \dfrac{3}{4}, \quad 9\dfrac{1}{2}, \quad \dfrac{3}{4}, \quad 4\dfrac{1}{2}.$$

We begin by changing $4\dfrac{1}{2}$ to $4\dfrac{2}{4}$ and $9\dfrac{1}{2}$ to $9\dfrac{2}{4}$, since the common denominator is 4. Then we use Method 2 from Example 6(b).

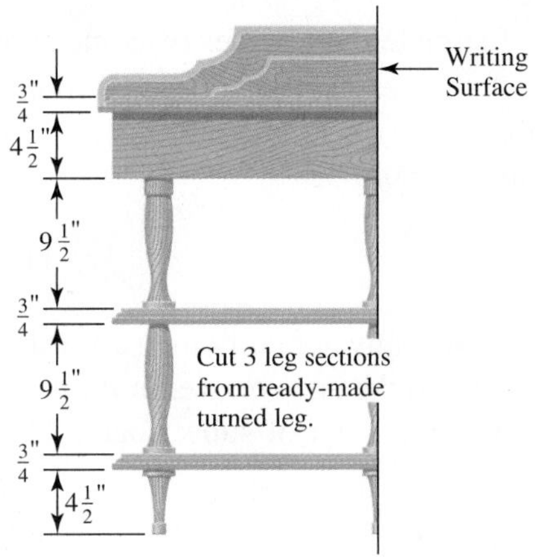

$$\begin{array}{cc} \dfrac{3}{4} & \dfrac{3}{4} \\[6pt] 4\dfrac{1}{2} & 4\dfrac{2}{4} \\[6pt] 9\dfrac{1}{2} & 9\dfrac{2}{4} \\[6pt] \dfrac{3}{4} \;\rightarrow\; & \dfrac{3}{4} \\[6pt] 9\dfrac{1}{2} & 9\dfrac{2}{4} \\[6pt] \dfrac{3}{4} & \dfrac{3}{4} \\[6pt] +\,4\dfrac{1}{2} & 4\dfrac{2}{4} \\[6pt] \hline & 26\dfrac{17}{4} \end{array}$$

FIGURE 1

Since $\frac{17}{4} = 4\frac{1}{4}$, $26\frac{17}{4} = 26 + 4\frac{1}{4} = 30\frac{1}{4}$. The height is $30\frac{1}{4}$ in. It is best to give answers as mixed numbers in applications like this.

Now Try Exercise 69.

OBJECTIVE 6 Interpret data in a circle graph. A **circle graph** or **pie chart** is often used to give a pictorial representation of data. A circle is used to indicate the total of all the categories represented. The circle is divided into sectors, or wedges (like pieces of pie), whose sizes show the relative magnitudes of the categories. The sum of all the fractional parts must be 1 (for 1 whole circle).

EXAMPLE 9 Using a Circle Graph (Pie Chart) to Interpret Information

The 1999 market share for satellite-TV home subscribers is shown in the circle graph in Figure 2. The number of subscribers reached 12 million in August 1999.

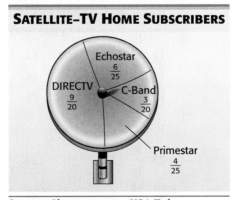

SATELLITE–TV HOME SUBSCRIBERS

Source: Skyreport.com; *USA Today.*

FIGURE 2

(a) Which provider had the largest share of the home subscriber market in August 1999? What was that share?

In the circle graph, the sector for DIRECTV is the largest, so DIRECTV had the largest market share, $\frac{9}{20}$.

(b) Estimate the number of home subscribers to DIRECTV in August 1999.

A market share of $\frac{9}{20}$ can be rounded to $\frac{10}{20}$, or $\frac{1}{2}$. We multiply $\frac{1}{2}$ by the total number of subscribers, 12 million. A good estimate for the number of DIRECTV subscribers would be

$$\frac{1}{2}(12) = 6 \text{ million.}$$

(c) How many actual home subscribers to DIRECTV were there?

To find the answer, we multiply the actual fraction from the graph for DIRECTV, $\frac{9}{20}$, by the number of subscribers, 12 million:

$$\frac{9}{20}(12) = \frac{9}{20} \cdot \frac{12}{1} = \frac{27}{5} = 5\frac{2}{5}.$$

Thus, $5\frac{2}{5}$ million, or 5,400,000, homes subscribed to DIRECTV. This is reasonable given our estimate in part (b).

Now Try Exercises 75 and 77.

1.1 EXERCISES

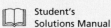

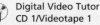
Decide whether each statement is true *or* false. *If it is false, say why.*

1. In the fraction $\frac{3}{7}$, 3 is the numerator and 7 is the denominator.

2. The mixed number equivalent of $\frac{41}{5}$ is $8\frac{1}{5}$.

3. The fraction $\frac{17}{51}$ is in lowest terms.

4. The reciprocal of $\frac{8}{2}$ is $\frac{4}{1}$.

5. The product of 8 and 2 is 10.

6. The difference between 12 and 2 is 6.

Identify each number as prime, composite, *or* neither. *If the number is composite, write it as the product of prime factors. See Example 1.*

7. 19	**8.** 31	**9.** 64	**10.** 99
11. 3458	**12.** 1025	**13.** 1	**14.** 0
15. 30	**16.** 40	**17.** 500	**18.** 700
19. 124	**20.** 120	**21.** 29	**22.** 83

Write each fraction in lowest terms. See Example 2.

23. $\frac{8}{16}$ **24.** $\frac{4}{12}$ **25.** $\frac{15}{18}$ **26.** $\frac{16}{20}$

27. $\frac{18}{60}$ **28.** $\frac{16}{64}$ **29.** $\frac{144}{120}$ **30.** $\frac{132}{77}$

31. One of the following is the correct way to write $\frac{16}{24}$ in lowest terms. Which one is it?

A. $\dfrac{16}{24} = \dfrac{8+8}{8+16} = \dfrac{8}{16} = \dfrac{1}{2}$ **B.** $\dfrac{16}{24} = \dfrac{4 \cdot 4}{4 \cdot 6} = \dfrac{4}{6}$

C. $\dfrac{16}{24} = \dfrac{8 \cdot 2}{8 \cdot 3} = \dfrac{2}{3}$ **D.** $\dfrac{16}{24} = \dfrac{14+2}{21+3} = \dfrac{2}{3}$

32. For the fractions $\frac{p}{q}$ and $\frac{r}{s}$, which one of the following can serve as a common denominator?

   A. $q \cdot s$ **B.** $q + s$ **C.** $p \cdot r$ **D.** $p + r$

Find each product or quotient, and write it in lowest terms. See Examples 3 and 4.

33. $\dfrac{4}{5} \cdot \dfrac{6}{7}$ **34.** $\dfrac{5}{9} \cdot \dfrac{10}{7}$ **35.** $\dfrac{1}{10} \cdot \dfrac{12}{5}$ **36.** $\dfrac{6}{11} \cdot \dfrac{2}{3}$

37. $\dfrac{15}{4} \cdot \dfrac{8}{25}$ **38.** $\dfrac{4}{7} \cdot \dfrac{21}{8}$ **39.** $2\dfrac{2}{3} \cdot 5\dfrac{4}{5}$ **40.** $3\dfrac{3}{5} \cdot 7\dfrac{1}{6}$

41. $\dfrac{5}{4} \div \dfrac{3}{8}$ **42.** $\dfrac{7}{6} \div \dfrac{9}{10}$ **43.** $\dfrac{32}{5} \div \dfrac{8}{15}$ **44.** $\dfrac{24}{7} \div \dfrac{6}{21}$

45. $\dfrac{3}{4} \div 12$ **46.** $\dfrac{2}{5} \div 30$ **47.** $2\dfrac{5}{8} \div 1\dfrac{15}{32}$ **48.** $2\dfrac{3}{10} \div 7\dfrac{4}{5}$

✎ **49.** Write a summary explaining how to multiply and divide two fractions. Give examples.

✎ **50.** Write a summary explaining how to add and subtract two fractions. Give examples.

Find each sum or difference, and write it in lowest terms. See Examples 5–7.

51. $\dfrac{7}{12} + \dfrac{1}{12}$ **52.** $\dfrac{3}{16} + \dfrac{5}{16}$ **53.** $\dfrac{5}{9} + \dfrac{1}{3}$ **54.** $\dfrac{4}{15} + \dfrac{1}{5}$

55. $3\dfrac{1}{8} + 2\dfrac{1}{4}$ **56.** $5\dfrac{3}{4} + 3\dfrac{1}{3}$ **57.** $\dfrac{13}{15} - \dfrac{3}{15}$ **58.** $\dfrac{11}{12} - \dfrac{3}{12}$

59. $\dfrac{7}{12} - \dfrac{1}{9}$ **60.** $\dfrac{11}{16} - \dfrac{1}{12}$ **61.** $6\dfrac{1}{4} - 5\dfrac{1}{3}$ **62.** $8\dfrac{4}{5} - 7\dfrac{4}{9}$

Solve each applied problem. See Example 8.

Use the table, which appears on a package of Quaker Quick Grits, to answer Exercises 63 and 64.

	Microwave		Stove Top	
Servings	1	1	4	6
Water	$\frac{3}{4}$ cup	1 cup	3 cups	4 cups
Grits	3 Tbsp	3 Tbsp	$\frac{3}{4}$ cup	1 cup
Salt (optional)	Dash	Dash	$\frac{1}{4}$ tsp	$\frac{1}{2}$ tsp

63. How many cups of water would be needed for eight microwave servings?

64. How many tsp of salt would be needed for five stove top servings? (*Hint:* 5 is halfway between 4 and 6.)

Stock prices are now reported using decimal numerals. However, prior to April 9, 2001, fractions were used for stock prices. With this in mind, answer Exercises 65 and 66.

65. On Tuesday, February 10, 1998, Earthlink stock on the NASDAQ exchange closed the day at $4\frac{5}{8}$ (dollars) ahead of where it had opened. It closed at $38\frac{5}{8}$ (dollars). What was its opening price?

66. A report in *USA Today* on February 10, 1998, stated that Teva Pharmaceutical skidded $9\frac{9}{16}$ (dollars) to $37\frac{1}{2}$ (dollars) after the Israeli drugmaker said fourth-quarter net income was likely to be below expectations. What was its price before the skid?

67. A hardware store sells a 40-piece socket wrench set. The measure of the largest socket is $\frac{3}{4}$ in., while the measure of the smallest socket is $\frac{3}{16}$ in. What is the difference between these measures?

68. Two sockets in a socket wrench set have measures of $\frac{9}{16}$ in. and $\frac{3}{8}$ in. What is the difference between these two measures?

69. A motel owner has decided to expand his business by buying a piece of property next to the motel. The property has an irregular shape, with five sides as shown in the figure. Find the total distance around the piece of property. (This is called the *perimeter* of the figure.)

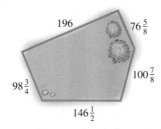

196   $76\frac{5}{8}$

$98\frac{3}{4}$ $100\frac{7}{8}$

$146\frac{1}{2}$

Measurements in feet

70. Find the perimeter of the triangle in the figure.

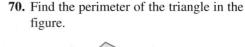

$5\frac{1}{4}$ ft $7\frac{1}{2}$ ft

$10\frac{1}{8}$ ft

71. A piece of board is $15\frac{5}{8}$ in. long. If it must be divided into 3 pieces of equal length, how long must each piece be?

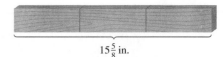

$15\frac{5}{8}$ in.

72. Under existing standards, most of the holes in Swiss cheese must have diameters between $\frac{11}{16}$ and $\frac{13}{16}$ in. To accommodate new high-speed slicing machines, the USDA wants to reduce the minimum size to $\frac{3}{8}$ in. How much smaller is $\frac{3}{8}$ in. than $\frac{11}{16}$ in.? (*Source:* U.S. Department of Agriculture.)

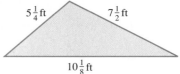

73. Tex's favorite recipe for barbecue sauce calls for $2\frac{1}{3}$ cups of tomato sauce. The recipe makes enough barbecue sauce to serve 7 people. How much tomato sauce is needed for 1 serving?

74. A cake recipe calls for $1\frac{3}{4}$ cups of sugar. A caterer has $15\frac{1}{2}$ cups of sugar on hand. How many cakes can he make?

More than 8 million immigrants were admitted to the United States between 1990 and 1997. The circle graph gives the fractional number from each region of birth for these immigrants. Use the graph to answer the following questions. See Example 9.

75. What fractional part of the immigrants were from Other regions?

76. What fractional part of the immigrants were from Latin America or Asia?

77. How many (in millions) were from Europe?

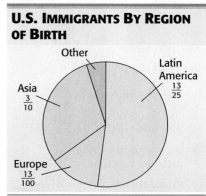

U.S. IMMIGRANTS BY REGION OF BIRTH

Other

Latin America $\frac{13}{25}$

Asia $\frac{3}{10}$

Europe $\frac{13}{100}$

Source: U.S. Bureau of the Census.

78. At the conclusion of the Addison Wesley softball league season, batting statistics for five players were as follows.

Player	At-Bats	Hits	Home Runs
Alison Romike	40	9	2
Jennifer Crum	36	12	3
Jason Jordan	11	5	1
Greg Tobin	16	8	0
Greg Erb	20	10	2

Use the table to answer each question, using estimation skills as necessary.

(a) Which player got a hit in exactly $\frac{1}{3}$ of his or her at-bats?

(b) Which player got a hit in just less than $\frac{1}{2}$ of his or her at-bats?

(c) Which player got a home run in just less than $\frac{1}{10}$ of his or her at-bats?

(d) Which player got a hit in just less than $\frac{1}{4}$ of his or her at-bats?

(e) Which two players got hits in exactly the same fractional parts of their at-bats? What was the fractional part, expressed in lowest terms?

79. For each description, write a fraction in lowest terms that represents the region described.

(a) The dots in the rectangle as a part of the dots in the entire figure

(b) The dots in the triangle as a part of the dots in the entire figure

(c) The dots in the overlapping region of the triangle and the rectangle as a part of the dots in the triangle alone

(d) The dots in the overlapping region of the triangle and the rectangle as a part of the dots in the rectangle alone

80. Estimate the best approximation for the following sum:

$$\frac{14}{26} + \frac{98}{99} + \frac{100}{51} + \frac{90}{31} + \frac{13}{27}.$$

A. 6   **B.** 7   **C.** 5   **D.** 8

1.2 Exponents, Order of Operations, and Inequality

OBJECTIVES

1. Use exponents.

2. Use the order of operations rules.

3. Use more than one grouping symbol.

4. Know the meanings of $\neq, <, >, \leq$, and $\geq$.

5. Translate word statements to symbols.

6. Write statements that change the direction of inequality symbols.

7. Interpret data in a bar graph.

OBJECTIVE 1 Use exponents.   In the prime factored form of 81, written

$$81 = 3 \cdot 3 \cdot 3 \cdot 3,$$

the factor 3 appears four times. In algebra, repeated factors are written with an *exponent*. For example, in $3 \cdot 3 \cdot 3 \cdot 3$, the number 3 appears as a factor four times, so the product is written as 3^4, and is read "3 to the fourth power."

$$\underbrace{3 \cdot 3 \cdot 3 \cdot 3}_{\text{4 factors of 3}} = 3^{\overset{\text{Exponent}}{4}}$$
Base

The number 4 is the **exponent** or **power** and 3 is the **base** in the **exponential expression** 3^4. A natural number exponent, then, tells how many times the base is used as a factor. A number raised to the first power is simply that number. For example, $5^1 = 5$ and $\left(\frac{1}{2}\right)^1 = \frac{1}{2}$.

EXAMPLE 1 Evaluating Exponential Expressions

Find the value of each exponential expression.

(a) $5^2 = \underline{5 \cdot 5} = 25$
      5 is used as a factor 2 times.

Read 5^2 as "5 squared" or "the square of 5."

(b) $6^3 = \underline{6 \cdot 6 \cdot 6} = 216$
      6 is used as a factor 3 times.

Read 6^3 as "6 cubed" or "the cube of 6."

(c) $2^5 = 2 \cdot 2 \cdot 2 \cdot 2 \cdot 2 = 32$ 2 is used as a factor 5 times.
Read 2^5 as "2 to the fifth power."

(d) $\left(\frac{2}{3}\right)^3 = \frac{2}{3} \cdot \frac{2}{3} \cdot \frac{2}{3} = \frac{8}{27}$ $\frac{2}{3}$ is used as a factor 3 times.

Now Try Exercises 5 and 17.

CAUTION   Squaring, or raising a number to the second power, is *not* the same as doubling the number. For example,

$$3^2 \quad \text{means} \quad 3 \cdot 3, \quad not \quad 2 \cdot 3.$$

Thus $3^2 = 9$, not 6. Similarly, cubing, or raising a number to the third power, does *not* mean tripling the number.

OBJECTIVE 2 Use the order of operations rules. Many problems involve more than one operation. To indicate the order in which the operations should be performed, we often use **grouping symbols.** If no grouping symbols are used, we apply the order of operations rules discussed below.

Consider the expression $5 + 2 \cdot 3$. To show that the multiplication should be performed before the addition, we can use parentheses to write

$$5 + (2 \cdot 3) = 5 + 6 = 11.$$

If addition is to be performed first, the parentheses should group $5 + 2$ as follows.

$$(5 + 2) \cdot 3 = 7 \cdot 3 = 21$$

Other grouping symbols used in more complicated expressions are brackets $[\ ]$, braces $\{\ \}$, and fraction bars. $\left(\text{For example, in } \frac{8 - 2}{3}, \text{ the expression } 8 - 2 \text{ is considered to be grouped in the numerator.}\right)$

To work problems with more than one operation, we use the following **order of operations.** This order is used by most calculators and computers.

Order of Operations

If grouping symbols are present, simplify within them, innermost first (and above and below fraction bars separately), in the following order.

Step 1 Apply all **exponents.**

Step 2 Do any **multiplications** or **divisions** in the order in which they occur, working from left to right.

Step 3 Do any **additions** or **subtractions** in the order in which they occur, working from left to right.

If no grouping symbols are present, start with Step 1.

When no operation is indicated, as in $3(7)$ or $(-5)(-4)$, multiplication is understood.

■ EXAMPLE 2 Using the Order of Operations

Find the value of each expression.

(a) $9(6 + 11)$

Using the order of operations given above, work first inside the parentheses.

$$9(6 + 11) = 9(17) \qquad \text{Work inside parentheses.}$$
$$= 153 \qquad \text{Multiply.}$$

(b) $6 \cdot 8 + 5 \cdot 2$

Do the multiplications, working from left to right, and then add.

$$6 \cdot 8 + 5 \cdot 2 = 48 + 10 \qquad \text{Multiply.}$$
$$= 58 \qquad \text{Add.}$$

(c) $2(5 + 6) + 7 \cdot 3 = 2(11) + 7 \cdot 3 \qquad \text{Work inside parentheses.}$
$$= 22 + 21 \qquad \text{Multiply.}$$
$$= 43 \qquad \text{Add.}$$

(d) $9 - 2^3 + 5$

Find 2^3 first.

$$
\begin{aligned}
9 - 2^3 + 5 &= 9 - 2 \cdot 2 \cdot 2 + 5 \qquad &\text{Use the exponent.}\\
&= 9 - 8 + 5 \qquad &\text{Multiply.}\\
&= 1 + 5 \qquad &\text{Subtract.}\\
&= 6 \qquad &\text{Add.}
\end{aligned}
$$

(e)
$$
\begin{aligned}
72 \div 2 \cdot 3 + 4 \cdot 2^3 - 3^3 &= 72 \div 2 \cdot 3 + 4 \cdot 8 - 27 \qquad &\text{Use the exponents.}\\
&= 36 \cdot 3 + 4 \cdot 8 - 27 \qquad &\text{Divide.}\\
&= 108 + 32 - 27 \qquad &\text{Multiply.}\\
&= 140 - 27 \qquad &\text{Add.}\\
&= 113 \qquad &\text{Subtract.}
\end{aligned}
$$

Notice that the multiplications and divisions are performed from left to right *as they appear;* then the additions and subtractions should be done from left to right, *as they appear.*

Now Try Exercises 23, 27, and 29.

OBJECTIVE 3   Use more than one grouping symbol.   An expression with double (or *nested*) parentheses, such as $2(8 + 3(6 + 5))$, can be confusing. For clarity, square brackets, [   ], often are used in place of one pair of parentheses. Fraction bars also act as grouping symbols.

EXAMPLE 3   Using Brackets and Fraction Bars as Grouping Symbols

Simplify each expression.

(a) $2[8 + 3(6 + 5)]$

Work first within the parentheses, and then simplify inside the brackets until a single number remains.

$$
\begin{aligned}
2[8 + 3(6 + 5)] &= 2[8 + 3(11)]\\
&= 2[8 + 33]\\
&= 2[41]\\
&= 82
\end{aligned}
$$

(b) $\dfrac{4(5 + 3) + 3}{2(3) - 1}$

The expression can be written as the quotient

$$[4(5 + 3) + 3] \div [2(3) - 1],$$

which shows that the fraction bar groups the numerator and denominator separately. Simplify both numerator and denominator, then divide, if possible.

$$
\begin{aligned}
\frac{4(5 + 3) + 3}{2(3) - 1} &= \frac{4(8) + 3}{2(3) - 1} \qquad &\text{Work inside parentheses.}\\[2mm]
&= \frac{32 + 3}{6 - 1} \qquad &\text{Multiply.}
\end{aligned}
$$

$$= \frac{35}{5} \qquad \text{Add and subtract.}$$

$$= 7 \qquad \text{Divide.}$$

Now Try Exercises 31 and 35.

NOTE Parentheses and fraction bars are used as grouping symbols to indicate an expression that represents a single number. That is why we must first simplify within parentheses and above and below fraction bars.

OBJECTIVE 4 Know the meanings of ≠, <, >, ≤, and ≥. So far, we have used the symbols for the operations of arithmetic and the symbol for equality (=). The symbols ≠, <, >, ≤, and ≥ are used to express an **inequality,** a statement that two expressions are not equal. The equality symbol with a slash through it, ≠, means "is not equal to." For example,

$$7 \neq 8$$

indicates that 7 is not equal to 8.

If two numbers are not equal, then one of the numbers must be less than the other. The symbol < represents "is less than," so that "7 is less than 8" is written

$$7 < 8.$$

The word *is* in the phrase "is less than" is a verb, which actually makes $7 < 8$ a sentence. This is not the case for expressions, like $7 + 8$, $7 - 8$, and so on.

The symbol > means "is greater than." Write "8 is greater than 2" as

$$8 > 2.$$

To keep the meanings of the symbols < and > clear, remember that the symbol always *points to the lesser number.* For example, write "8 is less than 15" by pointing the symbol toward the 8:

$$\text{Lesser number} \longrightarrow 8 < 15.$$

Two other symbols, ≤ and ≥, also represent the idea of inequality. The symbol ≤ means "is less than or equal to," so

$$5 \leq 9$$

means "5 is less than or equal to 9." This statement is true, since $5 < 9$ is true. *If either the < part or the = part is true, then the inequality ≤ is true.*

The symbol ≥ means "is greater than or equal to." Again,

$$9 \geq 5$$

is true because $9 > 5$ is true. Also, $8 \leq 8$ is true since $8 = 8$ is true. But it is not true that $13 \leq 9$ because neither $13 < 9$ nor $13 = 9$ is true.

EXAMPLE 4 Using Inequality Symbols

Determine whether each statement is true or false.

(a) $6 \neq 5 + 1$

The statement is false because 6 *is equal to* $5 + 1$.

(b) $5 + 3 < 19$

Since $5 + 3$ represents a number (8) that is less than 19, this statement is true.

(c) $15 \leq 20 \cdot 2$

The statement $15 \leq 20 \cdot 2$ is true, since $15 < 40$.

(d) $25 \geq 30$

Both $25 > 30$ and $25 = 30$ are false. Because of this, $25 \geq 30$ is false.

(e) $12 \geq 12$

Since $12 = 12$, this statement is true.

(f) $9 < 9$

Since $9 = 9$, this statement is false.

Now Try Exercises 41 and 43.

OBJECTIVE 5 **Translate word statements to symbols.**   An important part of algebra deals with translating words into algebraic notation.

PROBLEM SOLVING

As we will see throughout this book, the ability to solve problems using mathematics is based on translating the words of the problem into symbols. The next example is the first of many that illustrate such translations.

EXAMPLE 5   **Translating from Words to Symbols**

Write each word statement in symbols.

(a) Twelve equals ten plus two.
$$12 = 10 + 2$$

(b) Nine is less than ten.
$$9 < 10$$

(c) Fifteen is not equal to eighteen.
$$15 \neq 18$$

(d) Seven is greater than four.
$$7 > 4$$

(e) Thirteen is less than or equal to forty.
$$13 \leq 40$$

(f) Eleven is greater than or equal to eleven.
$$11 \geq 11$$

Now Try Exercises 55 and 57.

OBJECTIVE 6 **Write statements that change the direction of inequality symbols.**   Any statement with $<$ can be converted to one with $>$, and any statement with $>$ can be converted to one with $<$. We do this by reversing the order of the numbers and the direction of the symbol. For example, the statement $6 < 10$ can be written with $>$ as $10 > 6$. Similarly, the statement $4 \leq 10$ can be changed to $10 \geq 4$.

EXAMPLE 6   **Converting between Inequality Symbols**

The following examples show the same statement written in two equally correct ways.

(a) $9 < 16$ $16 > 9$

(b) $5 > 2$ $2 < 5$

(c) $3 \leq 8$ $8 \geq 3$

(d) $12 \geq 5$ $5 \leq 12$

Note that in each pair of inequalities, the inequality symbol points toward the smaller number.

Now Try Exercise 71.

Here is a summary of the symbols discussed in this section.

Symbols of Equality and Inequality

$=$	is equal to	$\neq$	is not equal to
$<$	is less than	$>$	is greater than
$\leq$	is less than or equal to	$\geq$	is greater than or equal to

---

CAUTION   The equality and inequality symbols are used to write mathematical *sentences* that describe the relationship between two numbers. On the other hand, the symbols for operators $(+, -, \times, \div)$ are used to write mathematical *expressions* that represent a single number. For example, compare the sentence $4 < 10$ with the expression $4 + 10$, which represents the number 14.

OBJECTIVE 7   Interpret data in a bar graph.   **Bar graphs** are often used to summarize data in a concise manner.

EXAMPLE 7   Interpreting Inequality Concepts Using a Bar Graph

The bar graph in Figure 3 shows the federal budget outlays for national defense during the 1990s.

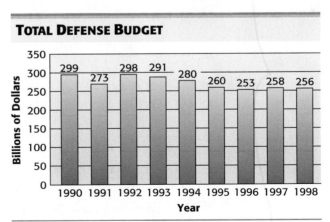

TOTAL DEFENSE BUDGET

Source: U.S. Office of Management and Budget.

FIGURE 3

(a) Which years had outlays greater than 280 billion dollars?

Look for numbers greater than 280 at the tops of the columns. In 1990, 1992, and 1993, the outlays were 299 billion, 298 billion, and 291 billion dollars, respectively.

(b) Which years had outlays greater than 300 billion dollars?

Since the tops of all the bars are below the line representing 300 billion, there were no such years. This is reinforced by the fact that all of the numbers at the tops of the bars are less than 300 (billion).

Now Try Exercise 79.

1.2 EXERCISES

 Decide whether each statement is true *or* false. *If it is false, explain why.*

1. When evaluated, $4 + 3(8 - 2)$ is equal to 42.

2. $3^3 = 9$

3. The statement "4 is 12 less than 16" is interpreted $4 = 12 - 16$.

4. The statement "6 is 4 less than 10" is interpreted $6 < 10 - 4$.

Find the value of each exponential expression. See Example 1.

5. 7^2　　　　**6.** 4^2　　　　**7.** 12^2　　　　**8.** 14^2

9. 4^3　　　　**10.** 5^3　　　　**11.** 10^3　　　　**12.** 11^3

13. 3^4　　　　**14.** 6^4　　　　**15.** 4^5　　　　**16.** 3^5

17. $\left(\dfrac{2}{3}\right)^4$　　**18.** $\left(\dfrac{3}{4}\right)^3$　　**19.** $(.04)^3$　　**20.** $(.05)^4$

 21. Explain in your own words how to evaluate a power of a number, such as 6^3.

22. Explain why any power of 1 must be equal to 1.

Find the value of each expression. See Examples 2 and 3.

23. $64 \div 4 \times 2$

24. $250 \div 5 \times 2$

25. $\dfrac{1}{4} \cdot \dfrac{2}{3} + \dfrac{2}{5} \cdot \dfrac{11}{3}$

26. $\dfrac{9}{4} \cdot \dfrac{2}{3} + \dfrac{4}{5} \cdot \dfrac{5}{3}$

27. $9 \cdot 4 - 8 \cdot 3$

28. $11 \cdot 4 + 10 \cdot 3$

29. $3(4 + 2) + 8 \cdot 3$

30. $9(1 + 7) + 2 \cdot 5$

31. $5[3 + 4(2^2)]$

32. $6[2 + 8(3^3)]$

33. $3^2[(11 + 3) - 4]$

34. $4^2[(13 + 4) - 8]$

35. $\dfrac{6(3^2 - 1) + 8}{8 - 2^2}$

36. $\dfrac{2(8^2 - 4) + 8}{29 - 3^3}$

37. $\dfrac{4(6 + 2) + 8(8 - 3)}{6(4 - 2) - 2^2}$

38. $\dfrac{6(5 + 1) - 9(1 + 1)}{5(8 - 6) - 2^3}$

39. Write an explanation of how you would use the order of operations to simplify $4 + 3(2^2 - 1)^3$.

40. When evaluating $(4^2 + 3^3)^4$, what is the *last* exponent that would be applied?

First simplify both sides of each inequality. Then tell whether the given statement is true *or* false. *See Examples 2–4.*

41. $9 \cdot 3 - 11 \leq 16$

42. $6 \cdot 5 - 12 \leq 18$

43. $5 \cdot 11 + 2 \cdot 3 \leq 60$

44. $9 \cdot 3 + 4 \cdot 5 \geq 48$

45. $0 \geq 12 \cdot 3 - 6 \cdot 6$

46. $10 \leq 13 \cdot 2 - 15 \cdot 1$

47. $45 \geq 2[2 + 3(2 + 5)]$

48. $55 \geq 3[4 + 3(4 + 1)]$

49. $[3 \cdot 4 + 5(2)] \cdot 3 > 72$

50. $2 \cdot [7 \cdot 5 - 3(2)] \leq 58$

51. $\dfrac{3 + 5(4 - 1)}{2 \cdot 4 + 1} \geq 3$

52. $\dfrac{7(3 + 1) - 2}{3 + 5 \cdot 2} \leq 2$

53. $3 \geq \dfrac{2(5 + 1) - 3(1 + 1)}{5(8 - 6) - 4 \cdot 2}$

54. $7 \leq \dfrac{3(8 - 3) + 2(4 - 1)}{9(6 - 2) - 11(5 - 2)}$

Write each word statement in symbols. See Example 5.

55. Fifteen is equal to five plus ten.

56. Twelve is equal to twenty minus eight.

57. Nine is greater than five minus four.

58. Ten is greater than six plus one.

59. Sixteen is not equal to nineteen.

60. Three is not equal to four.

61. Two is less than or equal to three.

62. Five is less than or equal to nine.

Write each statement in words and decide whether it is true *or* false.

63. $7 < 19$

64. $9 < 10$

65. $3 \neq 6$

66. $9 \neq 13$

67. $8 \geq 11$

68. $4 \leq 2$

69. Construct a true statement that involves an addition on the left side, the symbol $\geq$, and a multiplication on the right side.

70. Construct a false statement that involves subtraction on the left side, the symbol $\leq$, and a division on the right side. Then tell why the statement is false and how it could be changed to become true.

Write each statement with the inequality symbol reversed while keeping the same meaning. See Example 6.

71. $5 < 30$

72. $8 > 4$

73. $12 \geq 3$

74. $25 \leq 41$

75. What English-language phrase is used to express the fact that one person's age *is less than* another person's age?

76. What English-language phrase is used to express the fact that one person's height *is greater than* another person's height?

77. $12 \geq 12$ is a true statement. Suppose that someone tells you the following: "$12 \geq 12$ is false, because even though 12 is equal to 12, 12 is not greater than 12." How would you respond to this?

78. The table shows results of a science literacy survey by Jon Miller of the International Center for the Advancement of Science Literacy in Chicago.

Country	Science Literacy Index
United States	56
Netherlands	52
France	50
Canada	45
Greece	38
Japan	36

(a) Which countries scored more than 50?

(b) Which countries scored at most 40?

(c) For which countries were scores not less than 50?

The bar graph shows world coal production by year for the years 1993 through 1999. Use the graph to answer Exercises 79 and 80. See Example 7.

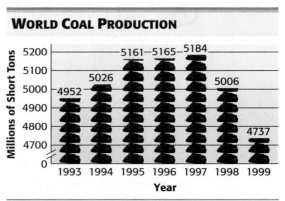

WORLD COAL PRODUCTION

Source: U.S. Energy Information Administration, *International Energy Review,* 1999.

79. In two of the years represented, coal production was less than production in the previous year. What were these two years?

80. What was the first year in which production was greater than 5100 millions of short tons?

81. According to an article in the April 1, 2002 issue of *USA Today,* the stock for Clear Channel, the largest radio station owner in the country, closed at $36.80 per share on September 20, 2001. A year prior to that date, the same stock had closed at $53.76.

  (a) By how much had the stock price decreased?

  (b) To find the percent of decrease, divide the amount from part (a) by 53.76 and convert to a percent. What was the percent of decrease? (Round to the nearest tenth of a percent.)

82. In March 2001, farmers were paid $15.00 per hundred pounds of lettuce. Due to a cold snap in Arizona and California, one year later the price had soared to $86.50. (*Source:* "Prices Soar as Cold Snap Shreds Iceberg Lettuce Supply," *USA Today,* April 1, 2001.)

  (a) By how much had the price increased?

  (b) What was the percent of increase? (Round to the nearest percent.) (*Hint:* See Exercise 81(b).)

Insert one pair of parentheses to make the left side of each equation equal to the right side.

83. $3 \cdot 6 + 4 \cdot 2 = 60$

84. $2 \cdot 8 - 1 \cdot 3 = 42$

85. $10 - 7 - 3 = 6$

86. $15 - 10 - 2 = 7$

87. $8 + 2^2 = 100$

88. $4 + 2^2 = 36$

1.3 Variables, Expressions, and Equations

OBJECTIVE 1 Define *variable,* and find the value of an algebraic expression, given the values of the variables. A **variable** is a symbol, usually a letter such as x, y, or z, used to represent any unknown number. An **algebraic expression** is a collection of numbers, variables, operation symbols, and grouping symbols (such as parentheses). For example,

$$6(x + 5), \qquad 2m - 9, \qquad \text{and} \qquad 8p^2 + 6p + 2$$

are all algebraic expressions. In the algebraic expression $2m - 9$, the expression $2m$ indicates the product of 2 and m, just as $8p^2$ shows the product of 8 and p^2. Also, $6(x + 5)$ means the product of 6 and $x + 5$. An algebraic expression has different numerical values for different values of the variable.

EXAMPLE 1 Evaluating Expressions

Find the numerical value of each algebraic expression when $m = 5$.

(a) $8m$

Replace m with 5, to get

$$8m = 8 \cdot 5 = 40.$$

(b) $3m^2$

For $m = 5$,

$$3m^2 = 3 \cdot 5^2 = 3 \cdot 25 = 75.$$

Now Try Exercises 15 and 17.

CAUTION In Example 1(b), notice that $3m^2$ means $3 \cdot m^2$; it *does not* mean $3m \cdot 3m$. The product $3m \cdot 3m$ is indicated by $(3m)^2$.

EXAMPLE 2 Evaluating Expressions

Find the value of each expression when $x = 5$ and $y = 3$.

(a) $2x + 7y$

Replace x with 5 and y with 3. Follow the order of operations: Multiply first, then add.

$$2x + 7y = 2 \cdot 5 + 7 \cdot 3 \qquad \text{Let } x = 5 \text{ and } y = 3.$$
$$= 10 + 21 \qquad \text{Multiply.}$$
$$= 31 \qquad \text{Add.}$$

(b) $\dfrac{9x - 8y}{2x - y} = \dfrac{9 \cdot 5 - 8 \cdot 3}{2 \cdot 5 - 3} \qquad \text{Let } x = 5 \text{ and } y = 3.$

$$= \frac{45 - 24}{10 - 3} \qquad \text{Multiply.}$$

$$= \frac{21}{7} \qquad \text{Subtract.}$$

$$= 3 \qquad \text{Divide.}$$

(c) $x^2 - 2y^2 = 5^2 - 2 \cdot 3^2$ Let $x = 5$ and $y = 3$.

$= 25 - 2 \cdot 9$ Use the exponents.

$= 25 - 18$ Multiply.

$= 7$ Subtract.

Now Try Exercises 27, 35, and 37.

OBJECTIVE 2 **Convert phrases from words to algebraic expressions.**

PROBLEM SOLVING

Sometimes variables must be used to change word phrases into algebraic expressions. Such translations are used in problem solving.

EXAMPLE 3 **Using Variables to Change Word Phrases to Algebraic Expressions**

Change each word phrase to an algebraic expression. Use x as the variable.

(a) The sum of a number and 9

"Sum" is the answer to an addition problem. This phrase translates as

$$x + 9 \qquad \text{or} \qquad 9 + x.$$

(b) 7 minus a number

"Minus" indicates subtraction, so the answer is $7 - x$.

(c) 7 less than a number

Write 7 less than a number as $x - 7$. In this case $7 - x$ would not be correct, because "less than" means "subtracted from."

(d) The product of 11 and a number

$$11 \cdot x \qquad \text{or} \qquad 11x$$

As mentioned earlier, $11x$ means 11 times x. No symbol is needed to indicate the product of a number and a variable.

(e) 5 divided by a number

This translates as $\frac{5}{x}$. The expression $\frac{x}{5}$ would *not* be correct here.

(f) The product of 2, and the sum of a number and 8

We are multiplying 2 times another number. This number is the sum of x and 8, written $x + 8$. Using parentheses for this sum, the final expression is

$$2(x + 8).$$

Now Try Exercises 43, 49, and 53.

CAUTION In Example 3(b), the response $x - 7$ would *not* be correct; this statement translates as "a number minus 7," not "7 minus a number." The expressions $7 - x$ and $x - 7$ are rarely equal. For example, if $x = 10$, $10 - 7 \neq 7 - 10$. ($7 - 10$ is a *negative number*, discussed in Section 1.4.)

OBJECTIVE 3 Identify solutions of equations. An **equation** is a statement that two algebraic expressions are equal. Therefore, an equation *always* includes the equality symbol, $=$. Examples of equations are

$$x + 4 = 11, \qquad 2y = 16, \qquad \text{and} \qquad 4p + 1 = 25 - p.$$

To **solve** an equation means to find the values of the variable that make the equation true. Such values of the variable are called the **solutions** of the equation.

■ **EXAMPLE 4 Deciding Whether a Number Is a Solution of an Equation**

Decide whether the given number is a solution of the equation.

(a) $5p + 1 = 36;$ 7

Replace p with 7.

$$
\begin{aligned}
5p + 1 &= 36 \\
5 \cdot 7 + 1 &= 36 \qquad ? \qquad \text{Let } p = 7. \\
35 + 1 &= 36 \qquad ? \\
36 &= 36 \qquad\qquad \text{True}
\end{aligned}
$$

The number 7 is a solution of the equation.

(b) $9m - 6 = 32;$ 4

$$
\begin{aligned}
9m - 6 &= 32 \\
9 \cdot 4 - 6 &= 32 \qquad ? \qquad \text{Let } m = 4. \\
36 - 6 &= 32 \qquad ? \\
30 &= 32 \qquad\qquad \text{False}
\end{aligned}
$$

The number 4 is not a solution of the equation.

Now Try Exercise 61.

OBJECTIVE 4 Identify solutions of equations from a set of numbers. A **set** is a collection of objects. In mathematics, these objects are most often numbers. The objects that belong to the set, called **elements** of the set, are written between **set braces.** For example, the set containing the numbers 1, 2, 3, 4, and 5 is written as

$$\{1, 2, 3, 4, 5\}.$$

For more information about sets, see Appendix C at the back of this book.

In some cases, the set of numbers from which the solutions of an equation must be chosen is specifically stated. One way of determining solutions is direct substitution of all possible replacements. The ones that lead to a true statement are solutions.

■ **EXAMPLE 5 Finding a Solution from a Given Set**

Change each word statement to an equation. Use x as the variable. Then find all solutions of the equation from the set

$$\{0, 2, 4, 6, 8, 10\}.$$

(a) The sum of a number and four is six.

The word *is* suggests "equals." If x represents the unknown number, then translate as follows.

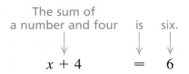

Try each number from the given set {0, 2, 4, 6, 8, 10}, in turn, to see that 2 is the only solution of $x + 4 = 6$.

(b) 9 more than five times a number is 49.

Use x to represent the unknown number. Start with $5x$ and then add 9 to it. The word *is* translates as $=$.

$$5x + 9 = 49$$

Try each number from {0, 2, 4, 6, 8, 10}. The solution is 8, since $5 \cdot 8 + 9 = 49$.

(c) The sum of a number and 12 is equal to four times the number.

If x represents the number, "the sum of a number and 12," is represented by $x + 12$. The translation is

$$x + 12 = 4x.$$

Trying each replacement leads to a true statement when $x = 4$, since $4 + 12 = 4(4) = 16$.

Now Try Exercise 69.

OBJECTIVE 5 Distinguish between an *expression* and an *equation.* Students often have trouble distinguishing between equations and expressions. Remember that an equation is a sentence; an expression is a phrase.

EXAMPLE 6 Distinguishing between Equations and Expressions

Decide whether each of the following is an equation or an expression.

(a) $2x - 5y$

There is no equals sign, so this is an expression.

(b) $2x = 5y$

Because an equals sign is present, this is an equation.

Now Try Exercises 77 and 81.

1.3 EXERCISES

Fill in each blank with the correct response.

1. If $x = 3$, then the value of $x + 7$ is _____.

2. If $x = 1$ and $y = 2$, then the value of $4xy$ is _____.

3. The sum of 12 and x is represented by the expression _____. If $x = 9$, the value of that expression is _____.

4. If x can be chosen from the set $\{0, 1, 2, 3, 4, 5\}$, the only solution of $x + 5 = 9$ is _____.

5. Will the equation $x = x + 4$ ever have a solution? _____.

6. $2x + 3$ is an _____, while $2x + 3 = 8$ is an _____.
    (equation/expression)   (equation/expression)

Exercises 7–12 cover some of the concepts introduced in this section. Give a short explanation for each.

7. Explain why $2x^3$ is not the same as $2x \cdot 2x \cdot 2x$.

8. Why are "5 less than a number" and "5 is less than a number" translated differently?

9. Explain why, when evaluating the expression $4x^2$ for $x = 3$, 3 must be squared *before* multiplying by 4.

10. What value of x would cause the expression $2x + 3$ to equal 9? Explain your reasoning.

11. There are many pairs of values of x and y for which $2x + y$ will equal 6. Name two such pairs and describe how you determined them.

12. Suppose that for the equation $3x - y = 9$, the value of x is given as 4. What would be the corresponding value of y? How do you know this?

*Find the numerical value if (**a**) $x = 4$ and (**b**) $x = 6$. See Example 1.*

13. $x + 9$ **14.** $x - 1$ **15.** $5x$ **16.** $7x$ **17.** $4x^2$

18. $5x^2$ **19.** $\dfrac{x + 1}{3}$ **20.** $\dfrac{x - 2}{5}$ **21.** $\dfrac{3x - 5}{2x}$ **22.** $\dfrac{4x - 1}{3x}$

23. $3x^2 + x$ **24.** $2x + x^2$ **25.** $6.459x$ **26.** $.74x^2$

*Find the numerical value if (**a**) $x = 2$ and $y = 1$ and (**b**) $x = 1$ and $y = 5$. See Example 2.*

27. $8x + 3y + 5$ **28.** $4x + 2y + 7$ **29.** $3(x + 2y)$ **30.** $2(2x + y)$

31. $x + \dfrac{4}{y}$ **32.** $y + \dfrac{8}{x}$ **33.** $\dfrac{x}{2} + \dfrac{y}{3}$ **34.** $\dfrac{x}{5} + \dfrac{y}{4}$

35. $\dfrac{2x + 4y - 6}{5y + 2}$ **36.** $\dfrac{4x + 3y - 1}{x}$ **37.** $2y^2 + 5x$ **38.** $6x^2 + 4y$

39. $\dfrac{3x + y^2}{2x + 3y}$ **40.** $\dfrac{x^2 + 1}{4x + 5y}$ **41.** $.841x^2 + .32y^2$ **42.** $.941x^2 + .2y^2$

Change each word phrase to an algebraic expression. Use x as the variable to represent the number. See Example 3.

43. Twelve times a number **44.** Nine times a number

45. Seven added to a number **46.** Thirteen added to a number

47. Two subtracted from a number

48. Eight subtracted from a number

49. A number subtracted from seven

50. A number subtracted from fourteen

51. The difference between a number and 6

52. The difference between 6 and a number

53. 12 divided by a number

54. A number divided by 12

55. The product of 6, and four less than a number.

56. The product of 9, and five more than a number.

57. In the phrase "Four more than the product of a number and 6," does the word *and* signify the operation of addition? Explain.

58. Suppose that the directions on a test read "Solve the following expressions." How would you politely correct the person who wrote these directions? What alternative directions might you suggest?

Decide whether the given number is a solution of the equation. See Example 4.

59. $5m + 2 = 7$; 1

60. $3r + 5 = 8$; 1

61. $2y + 3(y - 2) = 14$; 3

62. $6a + 2(a + 3) = 14$; 2

63. $6p + 4p + 9 = 11$; $\dfrac{1}{5}$

64. $2x + 3x + 8 = 20$; $\dfrac{12}{5}$

65. $3r^2 - 2 = 46$; 4

66. $2x^2 + 1 = 19$; 3

67. $\dfrac{z + 4}{2 - z} = \dfrac{13}{5}$; $\dfrac{1}{3}$

68. $\dfrac{x + 6}{x - 2} = \dfrac{37}{5}$; $\dfrac{13}{4}$

Change each word statement to an equation. Use x as the variable. Find all solutions from the set {2, 4, 6, 8, 10}. See Example 5.

69. The sum of a number and 8 is 18.

70. A number minus three equals 1.

71. Sixteen minus three-fourths of a number is 13.

72. The sum of six-fifths of a number and 2 is 14.

73. One more than twice a number is 5.

74. The product of a number and 3 is 6.

75. Three times a number is equal to 8 more than twice the number.

76. Twelve divided by a number equals $\frac{1}{3}$ times that number.

Identify each as an expression *or an* equation. *See Example 6.*

77. $3x + 2(x - 4)$

78. $5y - (3y + 6)$

79. $7t + 2(t + 1) = 4$

80. $9r + 3(r - 4) = 2$

81. $x + y = 3$

82. $x + y - 3$

A **mathematical model** *is an equation that describes the relationship between two quantities. For example, based on data from the United States Olympic Committee, the winning distances for the men's discus throw can be approximated by the equation*

$$y = 1.2304x - 2224.5,$$

where x is the year (between 1896 and 1996) and y is in feet. Use this model to approximate the winning distances for the following years.

83. 1912

84. 1936

85. 1960

86. 1992

1.4 Real Numbers and the Number Line

OBJECTIVES

1 Classify numbers and graph them on number lines.

2 Tell which of two different real numbers is smaller.

3 Find additive inverses and absolute values of real numbers.

4 Interpret the meanings of real numbers from a table of data.

OBJECTIVE 1 Classify numbers and graph them on number lines. In Section 1.1 we introduced two important sets of numbers, the *natural numbers* and the *whole numbers.*

Natural Numbers*

$\{1, 2, 3, 4, \ldots\}$ is the set of **natural numbers.**

Whole Numbers

$\{0, 1, 2, 3, 4, \ldots\}$ is the set of **whole numbers.**

NOTE The three dots show that the list of numbers continues in the same way indefinitely.

These numbers, along with many others, can be represented on **number lines** like the one in Figure 4. The **graph** of a number is the point on the number line associated with that number. The number is called the **coordinate** of the point. We draw a number line by choosing any point on the line and labeling it 0. Then we choose any point to the right of 0 and label it 1. The distance between 0 and 1 gives a unit of measure used to locate other points, as shown in Figure 4. The points labeled in Figure 4 correspond to the first few whole numbers.

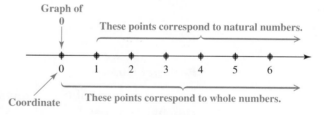

FIGURE 4

The natural numbers are located to the right of 0 on the number line. But numbers may also be placed to the left of 0. For each natural number we can place a corresponding number to the left of 0. These numbers, written $-1, -2, -3, -4$, and so on, are shown in Figure 5 on the next page. Each is the **opposite** or **negative** of a natural number. The natural numbers, their opposites, and 0 form a new set of numbers called the *integers.*

Integers

$\{\ldots, -3, -2, -1, 0, 1, 2, 3, \ldots\}$ is the set of **integers.**

---

*The symbols { and } are braces used in conjunction with sets. See Appendix C.

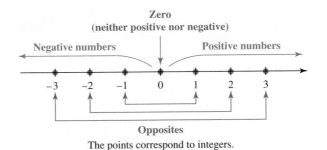

Opposites

The points correspond to integers.

FIGURE 5

Positive numbers and negative numbers are called **signed numbers.** There are many practical applications of negative numbers. For example, a Fahrenheit temperature on a cold January day might be $-10°$, and a business that spends more than it takes in has a negative "profit."

EXAMPLE 1 Using Negative Numbers in Applications

Use an integer to express the number in italics in each application.

(a) The lowest Fahrenheit temperature ever recorded in meteorological records was *129°* below zero at Vostok, Antarctica, on July 21, 1983. (*Source: World Almanac and Book of Facts,* 2000).

Use $-129°$ because "below zero" indicates a negative number.

(b) The shore surrounding the Dead Sea is *1340* ft below sea level. (*Source: Microsoft Encarta Encyclopedia 2000*).

Again, "below sea level" indicates a negative number, -1340.

Now Try Exercises 3 and 5.

Fractions, introduced in Section 1.1, are examples of *rational numbers.*

Rational Numbers

$\{x \mid x$ is a quotient of two integers, with denominator not 0$\}$ is the set of **rational numbers.**

   (Read the part in the braces as "the set of all numbers x such that x is a quotient of two integers, with denominator not 0.")

NOTE The set symbolism used in the definition of rational numbers,

$$\{x \mid x \text{ has a certain property}\},$$

is called **set-builder notation.** This notation is convenient to use when it is not possible to list all the elements of a set.

Since any integer can be written as the quotient of itself and 1, all integers also are rational numbers. Figure 6 shows a number line with the graphs of several rational numbers.

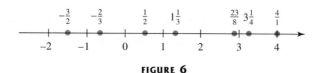

FIGURE 6

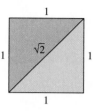

FIGURE 7

Although many numbers are rational, not all are. For example, a square that measures one unit on a side has a diagonal whose length is the square root of 2, written $\sqrt{2}$. See Figure 7. It can be shown that $\sqrt{2}$ cannot be written as a quotient of integers. Because of this, $\sqrt{2}$ is not rational; it is *irrational*. Other examples of **irrational numbers** are $\sqrt{3}$, $\sqrt{7}$, $-\sqrt{10}$, and π (the ratio of the *circumference* of a circle to its diameter).

Both rational and irrational numbers can be represented by points on the number line and together form the set of *real numbers*.

Real Numbers

$\{x \mid x$ is a rational or an irrational number$\}$ is the set of **real numbers.**

Real numbers can be written as decimals. Any rational number will have a decimal that either comes to an end (terminates) or repeats in a fixed "block" of digits. For example, $\frac{2}{5} = .4$ and $\frac{27}{100} = .27$ are rational numbers with terminating decimals; $\frac{1}{3} = .333\ldots$ and $\frac{3}{11} = .27272727\ldots$ are repeating decimals, often written as $.\overline{3}$ and $.\overline{27}$ with a bar over the repeating digit(s). The decimal representation of an irrational number will neither terminate nor repeat.

An example of a number that is not a real number is the square root of a negative number like $\sqrt{-5}$. These numbers are discussed in Chapter 10.

Two ways to represent the relationships among the various types of numbers are shown in Figure 8. Part (a) also gives some examples. Notice that every real number is either a rational number or an irrational number.

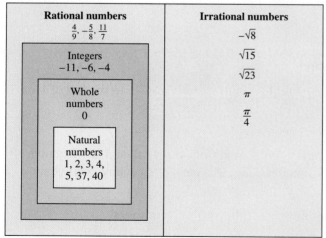

All numbers shown are real numbers.

(a)

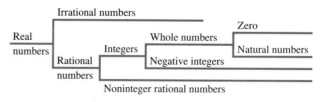

(b)

FIGURE 8

EXAMPLE 2   Determining Whether a Number Belongs to a Set

List the numbers in the set

$$\left\{-5, -\frac{2}{3}, 0, \sqrt{2}, 3\frac{1}{4}, 5, 5.8\right\}$$

that belong to each set of numbers.

(a) Natural numbers

The only natural number in the set is 5.

(b) Whole numbers

The whole numbers consist of the natural numbers and 0. So the elements of the set that are whole numbers are 0 and 5.

(c) Integers

The integers in the set are $-5, 0,$ and 5.

(d) Rational numbers

The rational numbers are $-5, -\frac{2}{3}, 0, 3\frac{1}{4}, 5, 5.8,$ since each of these numbers *can* be written as the quotient of two integers. For example, $3\frac{1}{4} = \frac{13}{4}$ and $5.8 = \frac{58}{10}$.

(e) Irrational numbers

The only irrational number in the set is $\sqrt{2}$.

(f) Real numbers

All the numbers in the set are real numbers.

> **Now Try Exercise 19.**

OBJECTIVE 2 Tell which of two different real numbers is smaller. Given any two whole numbers, you probably can tell which number is smaller. But what happens with negative numbers, as in the set of integers? Positive numbers decrease as the corresponding points on the number line go to the left. For example, $8 < 12$ because 8 is to the left of 12 on the number line. This ordering is extended to all real numbers by definition.

Ordering of Real Numbers

For any two real numbers a and b, ***a* is less than *b*** if a is to the left of b on the number line.

This means that any negative number is smaller than 0, and any negative number is smaller than any positive number. Also, 0 is smaller than any positive number.

EXAMPLE 3   Determining the Order of Real Numbers

Is it true that $-3 < -1$?

To decide whether the statement is true, locate both numbers, -3 and -1, on a number line, as shown in Figure 9. Since -3 is to the left of -1 on the number line, -3 is smaller than -1. The statement $-3 < -1$ is true.

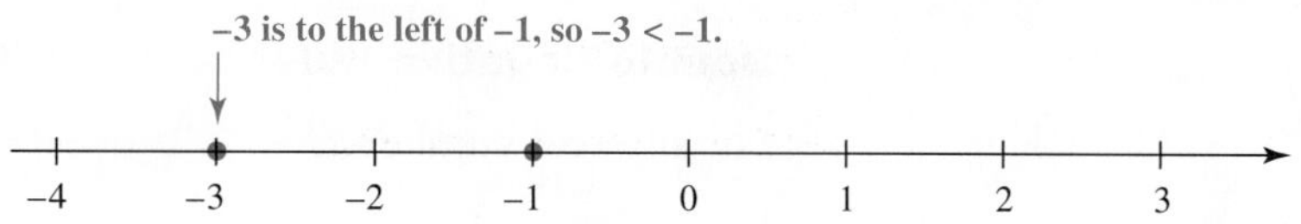

−3 is to the left of −1, so −3 < −1.

FIGURE 9

Now Try Exercise 53.

NOTE   In Section 1.2 we saw how it is possible to rewrite a statement involving < as an equivalent statement involving >. The question in Example 2 can also be worded as follows: Is it true that $-1 > -3$? This is, or course, also a true statement.

We can also say that for any two real numbers a and b, **a is greater than b** if a is to the right of b on the number line.

OBJECTIVE 3   Find additive inverses and absolute values of real numbers.   By a property of the real numbers, for any real number x (except 0), there is exactly one number on the number line the same distance from 0 as x but on the opposite side of 0. For example, Figure 10 shows that the numbers 1 and -1 are each the same distance from 0 but are on opposite sides of 0. The numbers 1 and -1 are called *additive inverses,* or *opposites,* of each other.

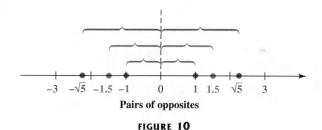

Pairs of opposites

FIGURE 10

Additive Inverse

The **additive inverse** of a number x is the number that is the same distance from 0 on the number line as x, but on the opposite side of 0.

The additive inverse of a number can be indicated by writing the symbol $-$ in front of the number. With this symbol, the additive inverse of 7 is written -7. The additive inverse of -3 is written $-(-3)$, and can be read "the opposite of -3" or "the negative of -3." Figure 10 suggests that 3 is the additive inverse of -3. A number can have only one additive inverse, and so the symbols 3 and $-(-3)$ must represent the same number, which means that

$$-(-3) = 3.$$

This idea can be generalized as follows.

Double Negative Rule*

For any real number x, $\qquad\qquad$ $-(-x) = x$.

Number	Additive Inverse
-4	$-(-4)$ or 4
0	0
19	-19
$-\dfrac{2}{3}$	$\dfrac{2}{3}$
$.52$	$-.52$

The table in the margin shows several numbers and their additive inverses. It suggests that the additive inverse of a number is found by changing the sign of the number. An important property of additive inverses will be studied later in this chapter: $a + (-a) = (-a) + a = 0$ for all real numbers a.

As previously mentioned, additive inverses are numbers that are the same distance from 0 on the number line. See Figure 10. We can also express this idea by saying that a number and its additive inverse have the same absolute value. The **absolute value** of a real number can be defined as the distance between 0 and the number on the number line. The symbol for the absolute value of the number x is $|x|$, read "the absolute value of x." For example, the distance between 2 and 0 on the number line is 2 units, so

$$|2| = 2.$$

Because the distance between -2 and 0 on the number is also 2 units,

$$|-2| = 2.$$

Since distance is a physical measurement, which is never negative, ***the absolute value of a number is never negative.*** For example, $|12| = 12$ and $|-12| = 12$, since both 12 and -12 lie at a distance of 12 units from 0 on the number line. Also, since 0 is a distance of 0 units from 0, $|0| = 0$.

In symbols, the absolute value of x is defined as follows.

Absolute Value

$$|x| = \begin{cases} x & \text{if } x \geq 0 \\ -x & \text{if } x < 0 \end{cases}$$

By this definition, if x is a positive number or 0, then its absolute value is x itself. For example, since 8 is a positive number, $|8| = 8$. However, if x is a negative number, then its absolute value is the additive inverse of x. This means that if $x = -9$, then $|-9| = -(-9) = 9$, since the additive inverse of -9 is 9.

CAUTION The definition of absolute value can be confusing if it is not read carefully. The "$-x$" in the second part of the definition *does not* represent a negative number. Since x is negative in the second part, $-x$ represents the opposite of a negative number, that is, a positive number. *The absolute value of a number is never negative.*

---

*This rule is justified by interpreting $-(-x)$ as $-1 \cdot (-x) = [(-1)(-1)]x = 1 \cdot x = x$. This requires concepts covered in later sections of this chapter.

EXAMPLE 4 Finding Absolute Value

Simplify by finding the absolute value.

(a) $|5| = 5$

(b) $|-5| = -(-5) = 5$

(c) $-|5| = -(5) = -5$

(d) $-|-14| = -(14) = -14$

(e) $|8 - 2| = |6| = 6$

(f) $-|8 - 2| = -|6| = -6$

Now Try Exercises 35, 37, and 39.

Parts (e) and (f) of Example 4 show that absolute value bars are also grouping symbols. You must perform any operations that appear inside absolute value symbols before finding the absolute value.

OBJECTIVE 4 Interpret the meanings of real numbers from a table of data. The Producer Price Index is the oldest continuous statistical series published by the Bureau of Labor Statistics. It measures the average changes in prices received by producers of all commodities produced in the United States. The next example shows how signed numbers can be used to interpret such data.

EXAMPLE 5 Interpreting Data

The table shows the percent change in the Producer Price Index for selected commodities from 1998 to 1999 and from 1999 to 2000. Use the table to answer each question.

Commodity	Change from 1998 to 1999	Change from 1999 to 2000
Fresh fruits and melons	13.1	−12.8
Gasoline	11.3	30.1
X-ray and electromedical equipment	−2.4	−2.5
Prepared animal feeds	−9.6	4.6
Office and store machines and equipment	0	.6

Source: U.S. Bureau of Labor Statistics.

(a) What commodity in which year represents the greatest percent decrease?

We must find the negative number with the greatest absolute value. The number that satisfies this condition is −12.8; the greatest percent decrease was shown by fresh fruits and melons from 1999 to 2000.

(b) Which commodity in which year represents the least change?

In this case, we must find the number (either positive, negative, or zero) with the least absolute value. From 1998 to 1999, office and store machines and equipment showed no change, as represented by 0.

Now Try Exercises 65 and 67.

1.4 EXERCISES

In the applications in Exercises 1–6, use an integer to express each number in italics representing a change. In Exercises 7 and 8, use a rational number. See Example 1.

1. Between 1995 and 2000, the number of U.S. citizens 65 years of age or older increased by *1,198,000.* (*Source:* U.S. Bureau of the Census.)

2. Between 1990 and 2000, the mean SAT verbal score for Florida residents increased by *3,* while the mathematics score increased by *7.* (*Source:* The College Board.)

3. From 1990 to 2000, the number of cable TV systems in the United States went from 9575 to 10,500, representing an increase of *925.* (*Source: Television and Cable Factbook.*)

4. From 1990 to 1998, the population of Glen Ellyn, Illinois, went from 25,956 to 24,919, representing a decrease of *1037* people. (*Source:* U.S. Bureau of the Census.)

5. In 1935, there were 15,295 banks in the United States. By 1999, this number was 10,221, representing a decrease of *5074* banks. (*Source:* Federal Deposit Insurance Corporation.)

6. Death Valley lies *282* feet below sea level. (Sea level is considered 0 feet.) (*Source:* U.S. Geological Survey, Department of the Interior.)

7. On Thursday, May 30, 2002, the Dow Jones Industrial Average closed at 9911.69. On the previous day it had closed at 9923.04. Thus on Thursday it closed down *11.35.* (*Source: Times Picayune.*)

8. On Thursday, May 30, 2002, the NASDAQ closed at 1631.92. On the previous day it had closed at 1624.39. Thus on Thursday it closed up *7.53.* (*Source: Times Picayune.*)

In Exercises 9–14, give a number that satisfies the given condition.

9. An integer between 3.5 and 4.5

10. A rational number between 3.8 and 3.9

11. A whole number that is not positive and is less than 1

12. A whole number greater than 4.5

13. An irrational number that is between $\sqrt{11}$ and $\sqrt{13}$

14. A real number that is neither negative nor positive

In Exercises 15–18, decide whether each statement is true *or* false.

15. Every natural number is positive.   16. Every whole number is positive.

17. Every integer is a rational number.   18. Every rational number is a real number.

For Exercises 19 and 20, see Example 2.

19. List all numbers from the set

$$\left\{ -9, -\sqrt{7}, -1\frac{1}{4}, -\frac{3}{5}, 0, \sqrt{5}, 3, 5.9, 7 \right\}$$

that are

(a) natural numbers; (b) whole numbers; (c) integers;
(d) rational numbers; (e) irrational numbers; (f) real numbers.

20. List all numbers from the set

$$\left\{ -5.3, -5, -\sqrt{3}, -1, -\frac{1}{9}, 0, 1.2, 1.8, 3, \sqrt{11} \right\}$$

that are

(a) natural numbers; (b) whole numbers; (c) integers;
(d) rational numbers; (e) irrational numbers; (f) real numbers.

21. Explain in your own words the different sets of numbers introduced in this section, and give an example of each kind.

22. What two possible situations exist for the decimal representation of a rational number?

Graph each group of numbers on a number line. See Figures 5 and 6.

23. $0, 3, -5, -6$ **24.** $2, 6, -2, -1$

25. $-2, -6, -4, 3, 4$ **26.** $-5, -3, -2, 0, 4$

27. $\frac{1}{4}, 2\frac{1}{2}, -3\frac{4}{5}, -4, -1\frac{5}{8}$ **28.** $5\frac{1}{4}, 4\frac{5}{9}, -2\frac{1}{3}, 0, -3\frac{2}{5}$

29. Match each expression in Column I with its value in Column II. Choices in Column II may be used once, more than once, or not at all.

I	II
(a) $\lvert -7 \rvert$	**A.** 7
(b) $-(-7)$	**B.** -7
(c) $-\lvert -7 \rvert$	**C.** Neither A nor B
(d) $-\lvert -(-7) \rvert$	**D.** Both A and B

30. Fill in the blanks with the correct values: The opposite of -2 is _____, while the absolute value of -2 is _____. The additive inverse of -2 is _____, while the additive inverse of the absolute value of -2 is _____.

Find (a) the opposite (or additive inverse) of each number and (b) the absolute value of each number.

31. -2 **32.** -8 **33.** 6 **34.** 11

Simplify by finding the absolute value. See Example 4.

35. $\lvert -6 \rvert$ **36.** $\lvert -12 \rvert$ **37.** $-\lvert -12 \rvert$

38. $-\lvert -6 \rvert$ **39.** $\lvert 6 - 3 \rvert$ **40.** $-\lvert 6 - 3 \rvert$

Select the smaller of the two given numbers. See Examples 3 and 4.

41. $-12, -4$ **42.** $-9, -14$ **43.** $-8, -1$

44. $-15, -16$ **45.** $3, \lvert -4 \rvert$ **46.** $5, \lvert -2 \rvert$

47. $\lvert -3 \rvert, \lvert -4 \rvert$ **48.** $\lvert -8 \rvert, \lvert -9 \rvert$ **49.** $-\lvert -6 \rvert, -\lvert -4 \rvert$

50. $-\lvert -2 \rvert, -\lvert -3 \rvert$ **51.** $\lvert 5 - 3 \rvert, \lvert 6 - 2 \rvert$ **52.** $\lvert 7 - 2 \rvert, \lvert 8 - 1 \rvert$

Decide whether each statement is true *or* false. *See Examples 3 and 4.*

53. $-5 < -2$ **54.** $-8 > -2$ **55.** $-4 \le -(-5)$

56. $-6 \le -(-3)$ **57.** $|-6| < |-9|$ **58.** $|-12| < |-20|$

59. $-|8| > |-9|$ **60.** $-|12| > |-15|$ **61.** $-|-5| \ge -|-9|$

62. $-|-12| \le -|-15|$ **63.** $|6 - 5| \ge |6 - 2|$ **64.** $|13 - 8| \le |7 - 4|$

The table shows the percent change in the Producer Price Index for selected construction commodities from 1998 to 1999 and from 1999 to 2000.* Use the table to answer Exercises 65–68. See Example 5.

Commodity	Change from 1998 to 1999	Change from 1999 to 2000
Softwood plywood	32.1	−33.2
Outdoor lighting equipment	−.5	2.4
Steel pipe and tubes	−6.9	4.3
Gypsum products	30.4	−6.6
Paving mixtures and blocks	.4	17.4

*2000 data are preliminary.
Source: U.S. Bureau of Labor Statistics.

65. Which commodity in which year represents the greatest percentage increase?

66. Which commodity in which year represents the greatest percentage decrease?

67. Which commodity in which year represents the least change?

68. Which commodity represents an increase for both years?

Give three numbers between −6 *and* 6 *that satisfy each given condition.*

69. Positive real numbers but not integers   **70.** Real numbers but not positive numbers

71. Real numbers but not whole numbers   **72.** Rational numbers but not integers

73. Real numbers but not rational numbers   **74.** Rational numbers but not negative numbers

75. Students often say "Absolute value is always positive." Is this true? Explain.   **76.** True or false: If a is negative, then $|a| = -a$.

1.5   Adding and Subtracting Real Numbers

OBJECTIVES

1 Add two numbers with the same sign.

2 Add positive and negative numbers.

In this and the next section, we extend the rules for operations with positive numbers to the negative numbers.

OBJECTIVE 1 Add two numbers with the same sign. A number line can be used to illustrate adding real numbers.

3 Use the definition of subtraction.

4 Use the order of operations with real numbers.

5 Interpret words and phrases involving addition and subtraction.

6 Use signed numbers to interpret data.

■ **EXAMPLE 1** Adding Numbers on a Number Line

(a) Use a number line to find the sum $2 + 3$.

Add the positive numbers 2 and 3 on the number line by starting at 0 and drawing an arrow 2 units to the *right,* as shown in Figure 11. This arrow represents the number 2 in the sum $2 + 3$. Then, from the right end of this arrow draw another arrow 3 units to the right. The number below the end of this second arrow is 5, so $2 + 3 = 5$.

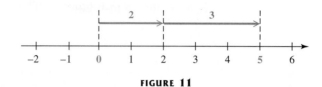

FIGURE 11

(b) Use a number line to find the sum $-2 + (-4)$. (Parentheses are placed around the -4 to avoid the confusing use of $+$ and $-$ next to each other.)

Add the negative numbers -2 and -4 on the number line by starting at 0 and drawing an arrow 2 units to the *left,* as shown in Figure 12. The arrow is drawn to the left to represent the addition of a *negative* number. From the left end of the first arrow, draw a second arrow 4 units to the left. The number below the end of this second arrow is -6, so $-2 + (-4) = -6$.

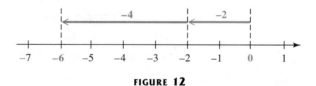

FIGURE 12

Now Try Exercise 1.

In Example 1, the sum of the two negative numbers -2 and -4 is a negative number whose distance from 0 is the sum of the distance of -2 from 0 and the distance of -4 from 0. That is, *the sum of two negative numbers is the negative of the sum of their absolute values.*

$$-2 + (-4) = -(|-2| + |-4|) = -(2 + 4) = -6$$

Adding Numbers with the Same Sign

To add two numbers with the *same* sign, add the absolute values of the numbers. The sum has the same sign as the numbers being added.

■ **EXAMPLE 2** Using the Rule to Add Two Negative Numbers

Find each sum.

(a) $-2 + (-9) = -(|-2| + |-9|) = -(2 + 9) = -11$

(b) $-8 + (-12) = -20$ **(c)** $-15 + (-3) = -18$

Now Try Exercise 7.

OBJECTIVE 2 Add positive and negative numbers. We can use a number line to illustrate the sum of a positive number and a negative number.

EXAMPLE 3 Adding Numbers with Different Signs

Use a number line to find the sum $-2 + 5$.

Find the sum $-2 + 5$ on the number line by starting at 0 and drawing an arrow 2 units to the left. From the left end of this arrow, draw a second arrow 5 units to the right, as shown in Figure 13. The number below the end of the second arrow is 3, so $-2 + 5 = 3$.

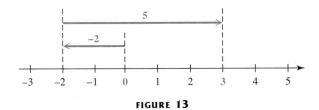

FIGURE 13

Now Try Exercise 3.

Adding Numbers with Different Signs

To add two numbers with *different* signs, subtract the smaller absolute value from the larger absolute value. The answer has the sign of the number with the larger absolute value.

For example, to add -12 and 5, find their absolute values: $|-12| = 12$ and $|5| = 5$. Then find the difference between these absolute values: $12 - 5 = 7$. Since $|-12| > |5|$, the sum will be negative, so the final answer is $-12 + 5 = -7$.

While a number line is useful in showing the rules for addition, it is important to be able to find sums mentally.

EXAMPLE 4 Adding Mentally

Check each answer, trying to work the addition mentally. If you have trouble, use a number line.

(a) $7 + (-4) = 3$

(b) $-8 + 12 = 4$

(c) $-\dfrac{1}{2} + \dfrac{1}{8} = -\dfrac{4}{8} + \dfrac{1}{8} = -\dfrac{3}{8}$ Remember to find a common denominator first.

(d) $\dfrac{5}{6} + \left(-\dfrac{4}{3}\right) = -\dfrac{1}{2}$

(e) $-4.6 + 8.1 = 3.5$

(f) $-16 + 16 = 0$

(g) $42 + (-42) = 0$

Now Try Exercises 15 and 27.

Parts (f) and (g) in Example 4 suggest that the sum of a number and its additive inverse is 0. This is always true, and this property is discussed further in Section 1.7.

The rules for adding signed numbers are summarized as follows.

Adding Signed Numbers

Same sign Add the absolute values of the numbers. The sum has the same sign as the given numbers.

Different signs Find the difference between the larger absolute value and the smaller. The sum has the sign of the number with the larger absolute value.

OBJECTIVE 3 Use the definition of subtraction. We can illustrate subtraction of 4 from 7, written $7 - 4$, using a number line. As seen in Figure 14, we begin at 0 and draw an arrow 7 units to the right. From the right end of this arrow, we draw an arrow 4 units to the left. The number at the end of the second arrow shows that $7 - 4 = 3$.

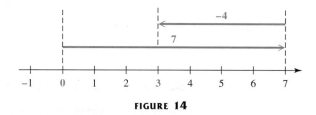

FIGURE 14

The procedure used to find $7 - 4$ is exactly the same procedure that would be used to find $7 + (-4)$, so

$$7 - 4 = 7 + (-4).$$

This suggests that *subtracting* a positive number from a larger positive number is the same as *adding* the additive inverse of the smaller number to the larger. This result leads to the definition of subtraction for all real numbers.

Subtraction

For any real numbers x and y,

$$x - y = x + (-y).$$

That is, to *subtract* y from x, *add the additive inverse* (or opposite) of y to x.

The definition gives the following procedure for subtracting signed numbers.

Subtracting Signed Numbers

Step 1 Change the subtraction symbol to the addition symbol.
Step 2 Change the sign of the number being subtracted.
Step 3 Add.

EXAMPLE 5   Using the Definition of Subtraction

Subtract.

Change − to +.

No change

Additive inverse of 3

(a) $12 - 3 = 12 + (-3) = 9$

(b) $5 - 7 = 5 + (-7) = -2$

Change − to +.

No change

Additive inverse of −5

(c) $-3 - (-5) = -3 + (5) = 2$

(d) $-6 - 9 = -6 + (-9) = -15$

(e) $\dfrac{4}{3} - \left(-\dfrac{1}{2}\right) = \dfrac{4}{3} + \dfrac{1}{2} = \dfrac{8}{6} + \dfrac{3}{6} = \dfrac{11}{6}$ or $1\dfrac{5}{6}$

Now Try Exercises 43, 51, and 55.

We have now used the symbol − for three purposes:

1. to represent subtraction, as in $9 - 5 = 4$;

2. to represent negative numbers, such as -10, -2, and -3;

3. to represent the opposite (or negative) of a number, as in "the opposite (or negative) of 8 is -8."

We may see more than one use of − in the same problem, such as $-6 - (-9)$, where -9 is subtracted from -6. The meaning of the − symbol depends on its position in the algebraic expression.

OBJECTIVE 4 Use the order of operations with real numbers.   As before, with problems that have grouping symbols, first do any operations inside the parentheses and brackets. Work within the innermost set of grouping symbols first, and then work outward.

EXAMPLE 6   Adding and Subtracting with Grouping Symbols

Perform each indicated operation.

(a)
$$-6 - [2 - (8 + 3)] = -6 - [2 - 11] \qquad \text{Add.}$$
$$= -6 - [2 + (-11)] \qquad \text{Definition of subtraction}$$
$$= -6 - (-9) \qquad \text{Add.}$$
$$= -6 + (9) \qquad \text{Definition of subtraction}$$
$$= 3 \qquad \text{Add.}$$

(b) $5 + [(-3 - 2) - (4 - 1)] = 5 + [(-3 + (-2)) - 3]$

$$= 5 + [(-5) - 3]$$
$$= 5 + [(-5) + (-3)]$$
$$= 5 + (-8)$$
$$= -3$$

(c) $\dfrac{2}{3} - \left[\dfrac{1}{12} - \left(-\dfrac{1}{4} \right) \right] = \dfrac{8}{12} - \left[\dfrac{1}{12} - \left(-\dfrac{3}{12} \right) \right]$ Find a common denominator.

$$= \dfrac{8}{12} - \left[\dfrac{1}{12} + \dfrac{3}{12} \right]$$ Definition of subtraction

$$= \dfrac{8}{12} - \dfrac{4}{12}$$ Add.

$$= \dfrac{4}{12}$$ Subtract.

$$= \dfrac{1}{3}$$ Lowest terms

(d) $|4 - 7| + 2|6 - 3| = |-3| + 2|3|$ Work within absolute value bars.

$$= 3 + 2 \cdot 3$$ Evaluate absolute values.

$$= 3 + 6$$ Multiply.

$$= 9$$ Add.

Now Try Exercises 65, 75, and 79.

OBJECTIVE 5 Interpret words and phrases involving addition and subtraction.

PROBLEM SOLVING

As we mentioned earlier, problem solving often requires translating words and phrases into symbols. The word *sum* indicates addition. The table lists some of the words and phrases that also signify addition.

Word or Phrase	Example	Numerical Expression and Simplification
Sum of	The *sum of* −3 and 4	−3 + 4 = 1
Added to	5 *added to* −8	−8 + 5 = −3
More than	12 *more than* −5	−5 + 12 = 7
Increased by	−6 *increased by* 13	−6 + 13 = 7
Plus	3 *plus* 14	3 + 14 = 17

▓ EXAMPLE 7 Interpreting Words and Phrases Involving Addition

Write a numerical expression for each phrase and simplify the expression.

(a) The sum of -8 and 4 and 6

$$-8 + 4 + 6 = -4 + 6 = 2 \qquad \text{Add in order from left to right.}$$

(b) 3 more than -5, increased by 12

$$-5 + 3 + 12 = -2 + 12 = 10$$

Now Try Exercise 87.

▓ PROBLEM SOLVING ▓

To solve problems that involve subtraction, we must be able to interpret key words and phrases that indicate subtraction. *Difference* is one of them. Some of these are given in the table.

Word, Phrase, or Sentence	Example	Numerical Expression and Simplification
Difference between	The *difference between* -3 and -8	$-3 - (-8) = -3 + 8 = 5$
Subtracted from	12 *subtracted from* 18	$18 - 12 = 6$
From..., subtract....	From 12, subtract 8.	$12 - 8 = 12 + (-8) = 4$
Less	6 *less* 5	$6 - 5 = 1$
Less than	6 *less than* 5	$5 - 6 = 5 + (-6) = -1$
Decreased by	9 *decreased by* -4	$9 - (-4) = 9 + 4 = 13$
Minus	8 *minus* 5	$8 - 5 = 3$

CAUTION When you are subtracting two numbers, it is important that you write them in the correct order, because, in general, $a - b \neq b - a$. For example, $5 - 3 \neq 3 - 5$. For this reason, *think carefully before interpreting an expression involving subtraction.* (This difficulty did not arise for addition.)

▓ EXAMPLE 8 Interpreting Words and Phrases Involving Subtraction

Write a numerical expression for each phrase and simplify the expression.

(a) The difference between -8 and 5

In this book, we will write the numbers in the order they are given when "difference between" is used.*

$$-8 - 5 = -8 + (-5) = -13$$

(b) 4 subtracted from the sum of 8 and -3

Here addition is also used, as indicated by the word *sum*. First, add 8 and -3. Next, subtract 4 *from* this sum.

$$[8 + (-3)] - 4 = 5 - 4 = 1$$

---

*In some cases, people interpret "the difference between" (at least for two positive numbers) to represent the larger minus the smaller. However, we will not do so in this book.

(c) 4 less than −6

Be careful with order. Here, 4 must be taken *from* −6.

$$-6 - 4 = -6 + (-4) = -10$$

Notice that "4 less than −6" differs from "4 *is less than* −6." The second of these is symbolized $4 < -6$ (which is a false statement).

(d) 8, decreased by 5 less than 12

First, write "5 less than 12" as $12 - 5$. Next, subtract $12 - 5$ from 8.

$$8 - (12 - 5) = 8 - 7 = 1$$

Now Try Exercises 91 and 97.

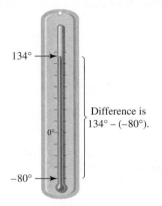

FIGURE 15

EXAMPLE 9 Solving a Problem Involving Subtraction

The record high temperature in the United States was 134° Fahrenheit, recorded at Death Valley, California in 1913. The record low was −80°F, at Prospect Creek, Alaska, in 1971. See Figure 15. What is the difference between these highest and lowest temperatures? (*Source: World Almanac and Book of Facts, 2000.*)

We must subtract the lowest temperature from the highest temperature.

$$134 - (-80) = 134 + 80 \quad \text{Definition of subtraction}$$
$$= 214 \quad \text{Add.}$$

The difference between the two temperatures is 214°F.

Now Try Exercise 113.

OBJECTIVE 6 Use signed numbers to interpret data.

EXAMPLE 10 Using a Signed Number in Data Interpretation

The bar graph in Figure 16 gives the Producer Price Index (PPI) for crude materials between 1994 and 1999.

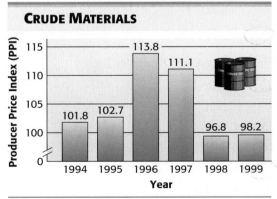

Source: U.S. Bureau of Labor Statistics, Producer Price Indexes, monthly and annual.

FIGURE 16

(a) Use a signed number to represent the change in the PPI from 1995 to 1996.

To find this change, we start with the index number from 1996 and subtract from it the index number from 1995.

$$\underbrace{113.8}_{\text{1996 index}} - \underbrace{102.7}_{\text{1995 index}} = \underbrace{+11.1}_{\substack{\text{A positive number}\\\text{indicates an increase.}}}$$

(b) Use a signed number to represent the change in the PPI from 1996 to 1997. We use the same procedure as in part (a).

$$\underbrace{111.1}_{\text{1997 index}} - \underbrace{113.8}_{\text{1996 index}} = \underbrace{111.1 + (-113.8) = -2.7}_{\substack{\text{A negative number}\\\text{indicates a decrease.}}}$$

Now Try Exercises 99 and 101.

1.5 EXERCISES

Fill in each blank with the correct response.

1. The sum of two negative numbers will always be a _____ number. (positive/negative)
Give a number-line illustration using the sum $-2 + (-3)$.

2. The sum of a number and its opposite will always be ____.

3. If I am adding a positive number and a negative number, and the negative number has the larger absolute value, the sum will be a _____ number. Give (positive/negative)
a number-line illustration using the sum $-4 + 2$.

4. To simplify the expression $8 + [-2 + (-3 + 5)]$, I should begin by adding ____ and ____, according to the rule for order of operations.

5. Explain in words how to add signed numbers. Consider the various cases and give examples.

6. Explain in words how to subtract signed numbers.

Find each sum. See Examples 1–6.

7. $-6 + (-2)$ **8.** $-8 + (-3)$ **9.** $-3 + (-9)$

10. $-11 + (-5)$ **11.** $5 + (-3)$ **12.** $11 + (-8)$

13. $6 + (-8)$ **14.** $3 + (-7)$ **15.** $12 + (-8)$

16. $10 + (-2)$ **17.** $4 + [13 + (-5)]$ **18.** $6 + [2 + (-13)]$

19. $8 + [-2 + (-1)]$ **20.** $12 + [-3 + (-4)]$ **21.** $-2 + [5 + (-1)]$

22. $-8 + [9 + (-2)]$ **23.** $-6 + [6 + (-9)]$ **24.** $-3 + [11 + (-8)]$

25. $[(-9) + (-3)] + 12$ **26.** $[(-8) + (-6)] + 10$ **27.** $\dfrac{1}{6} + \dfrac{2}{3}$

28. $\dfrac{9}{10} + \left(-\dfrac{3}{5}\right)$ **29.** $\dfrac{5}{8} + \left(-\dfrac{17}{12}\right)$ **30.** $-\dfrac{6}{25} + \dfrac{19}{20}$

31. $2\frac{1}{2} + \left(-3\frac{1}{4}\right)$

32. $-4\frac{3}{8} + 6\frac{1}{2}$

33. $-6.1 + [3.2 + (-4.8)]$

34. $-9.4 + [-5.8 + (-1.4)]$

35. $[-3 + (-4)] + [5 + (-6)]$

36. $[-8 + (-3)] + [-7 + (-6)]$

37. $[-4 + (-3)] + [8 + (-1)]$

38. $[-5 + (-9)] + [16 + (-21)]$

39. $[-4 + (-6)] + [(-3) + (-8)] + [12 + (-11)]$

40. $[-2 + (-11)] + [12 + (-2)] + [18 + (-6)]$

Find each difference. See Examples 1–6.

41. $3 - 6$

42. $7 - 12$

43. $5 - 9$

44. $8 - 13$

45. $-6 - 2$

46. $-11 - 4$

47. $-9 - 5$

48. $-12 - 15$

49. $6 - (-3)$

50. $12 - (-2)$

51. $-6 - (-2)$

52. $-7 - (-5)$

53. $2 - (3 - 5)$

54. $-3 - (4 - 11)$

55. $\frac{1}{2} - \left(-\frac{1}{4}\right)$

56. $\frac{1}{3} - \left(-\frac{4}{3}\right)$

57. $-\frac{3}{4} - \frac{5}{8}$

58. $-\frac{5}{6} - \frac{1}{2}$

59. $\frac{5}{8} - \left(-\frac{1}{2} - \frac{3}{4}\right)$

60. $\frac{9}{10} - \left(\frac{1}{8} - \frac{3}{10}\right)$

61. $3.4 - (-8.2)$

62. $5.7 - (-11.6)$

63. $-6.4 - 3.5$

64. $-4.4 - 8.6$

Perform each indicated operation. See Examples 1–6.

65. $(4 - 6) + 12$

66. $(3 - 7) + 4$

67. $(8 - 1) - 12$

68. $(9 - 3) - 15$

69. $6 - (-8 + 3)$

70. $8 - (-9 + 5)$

71. $2 + (-4 - 8)$

72. $6 + (-9 - 2)$

73. $|-5 - 6| + |9 + 2|$

74. $|-4 + 8| + |6 - 1|$

75. $|-8 - 2| - |-9 - 3|$

76. $|-4 - 2| - |-8 - 1|$

77. $-9 + [(3 - 2) - (-4 + 2)]$

78. $-8 - [(-4 - 1) + (9 - 2)]$

79. $-3 + [(-5 - 8) - (-6 + 2)]$

80. $-4 + [(-12 + 1) - (-1 - 9)]$

81. $-9.1237 + [(-4.8099 - 3.2516) + 11.27903]$

82. $-7.6247 - [(-3.9928 + 1.42773) - (-2.80981)]$

Write a numerical expression for each phrase and simplify. See Example 7.

83. The sum of -5 and 12 and 6

84. The sum of -3 and 5 and -12

85. 14 added to the sum of -19 and -4

86. -2 added to the sum of -18 and 11

87. The sum of -4 and -10, increased by 12

88. The sum of -7 and -13, increased by 14

89. 4 more than the sum of 8 and -18

90. 10 more than the sum of -4 and -6

Write a numerical expression for each phrase and simplify. See Example 8.

91. The difference between 4 and -8

92. The difference between 7 and -14

93. 8 less than -2

94. 9 less than -13

95. The sum of 9 and −4, decreased by 7

96. The sum of 12 and −7, decreased by 14

97. 12 less than the difference between 8 and −5

98. 19 less than the difference between 9 and −2

The bar graph shows federal budget outlays for the U.S. Treasury Department for the years 1998 through 2001. Use a signed number to represent the change in outlay for each time period. See Example 10.

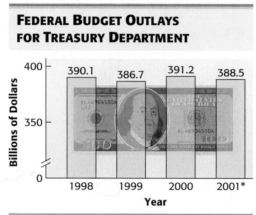

FEDERAL BUDGET OUTLAYS FOR TREASURY DEPARTMENT

Billions of Dollars (vertical axis)

Year	1998	1999	2000	2001*
	390.1	386.7	391.2	388.5

*Estimated

Source: U.S. Office of Management and Budget.

99. 1998 to 1999

100. 1999 to 2000

101. 2000 to 2001

102. 1998 to 2001

The two tables show the heights of some selected mountains and the depths of some selected trenches. Use the information given to answer Exercises 103–108.

Mountain	Height (in feet)	Trench	Depth (in feet, as a negative number)
Foraker	17,400	Philippine	−32,995
Wilson	14,246	Cayman	−24,721
Pikes Peak	14,110	Java	−23,376

Source: World Almanac and Book of Facts, 2000.

103. What is the difference between the height of Mt. Foraker and the depth of the Philippine Trench?

104. What is the difference between the height of Pikes Peak and the depth of the Java Trench?

105. How much deeper is the Cayman Trench than the Java Trench?

106. How much deeper is the Philippine Trench than the Cayman Trench?

107. How much higher is Mt. Wilson than Pikes Peak?

108. If Mt. Wilson and Pikes Peak were stacked one on top of the other, how much higher would they be than Mt. Foraker?

Pikes Peak

Solve each problem. See Example 9.

109. Based on census population projections for 2020, New York will lose 5 seats in the U.S. House of Representatives, Pennsylvania will lose 4 seats, and Ohio will lose 3. Write a signed number that represents the total number of seats these three states are projected to lose. (*Source:* Population Reference Bureau.)

110. Michigan is projected to lose 3 seats in the U.S. House of Representatives and Illinois 2 in 2020. The states projected to gain the most seats are California with 9, Texas with 5, Florida with 3, Georgia with 2, and Arizona with 2. Write a signed number that represents the algebraic sum of these changes. (*Source:* Population Reference Bureau.)

111. On January 23, 1943, the temperature rose 49°F in two minutes in Spearfish, South Dakota. The starting temperature was −4°F. What was the temperature two minutes later? (*Source: Guinness Book of World Records,* 2002.)

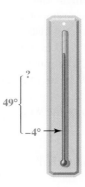

112. The lowest temperature ever recorded in Arkansas was −29°F. The highest temperature ever recorded there was 149°F more than the lowest. What was this highest temperature? (*Source: World Almanac and Book of Facts,* 2000.)

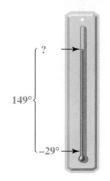

113. The coldest temperature recorded in Chicago, Illinois, was −35°F in 1996. The record low in South Dakota was set in 1936 and was 23°F lower than Chicago's record low. What was the record low in South Dakota? (*Source: World Almanac and Book of Facts,* 2000.)

114. The top of Mt. Whitney, visible from Death Valley, has an altitude of 14,494 ft above sea level. The bottom of Death Valley is 282 ft below sea level. Using 0 as sea level, find the difference between these two elevations. (*Source: World Almanac and Book of Facts,* 2000.)

115. No one knows just why humpback whales heave their 45-ton bodies out of the water, but leap they do. (This activity is called *breaching.*) Mark and Debbie, two researchers based on the island of Maui, noticed that one of their favorite whales, "Pineapple," leaped 15 ft above the surface of the ocean while her mate cruised 12 ft below the surface. What is the difference between these two distances?

116. The surface, or rim, of a canyon is at altitude 0. On a hike down into the canyon, a party of hikers stops for a rest at 130 m below the surface. They then descend another 54 m. Write the new altitude as a signed number.

117. A pilot announces to the passengers that the current altitude of their plane is 34,000 ft. Because of turbulence, the pilot is forced to descend 2100 ft. Write the new altitude as a signed number.

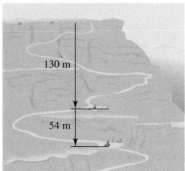

118. A welder working with stainless steel must use precise measurements. Suppose a welder attaches two pieces of steel that are each 3.60 in. long, and then attaches an additional three pieces that are each 9.10 in. long. She finally cuts off a piece that is 7.60 in. long. Find the length of the welded piece of steel.

119. Jennifer owes $153 to a credit card company. She makes a $14 purchase with the card and then pays $60 on the account. What is her current balance as a signed number?

120. A female polar bear weighed 660 lb when she entered her winter den. She lost 45 lb during each of the first two months of hibernation, and another 205 lb before leaving the den with her two cubs in March. How much did she weigh when she left the den?

1.6 Multiplying and Dividing Real Numbers

OBJECTIVES

1 Find the product of a positive number and a negative number.

2 Find the product of two negative numbers.

3 Identify factors of integers.

In this section we learn how to multiply with positive and negative numbers. We already know how to multiply positive numbers and that the product of two positive numbers is positive. We also know that the product of 0 and any positive number is 0, so we extend that property to all real numbers.

Multiplication by Zero

For any real number x,　　$x \cdot 0 = 0.$

4 Use the reciprocal of a number to apply the definition of division.

5 Use the order of operations when multiplying and dividing signed numbers.

6 Evaluate expressions involving variables.

7 Interpret words and phrases involving multiplication and division.

8 Translate simple sentences into equations.

OBJECTIVE 1 Find the product of a positive number and a negative number. To define the product of a positive and a negative number so that the result is consistent with the multiplication of two positive numbers, look at the following pattern.

$$3 \cdot 5 = 15$$
$$3 \cdot 4 = 12$$
$$3 \cdot 3 = 9$$
$$3 \cdot 2 = 6$$
$$3 \cdot 1 = 3$$
$$3 \cdot 0 = 0$$
$$3 \cdot (-1) = ?$$

The products decrease by 3.

What should $3(-1)$ equal? The product $3(-1)$ represents the sum

$$-1 + (-1) + (-1) = -3,$$

so the product should be -3. Also,

$$3(-2) = -2 + (-2) + (-2) = -6$$

and

$$3(-3) = -3 + (-3) + (-3) = -9.$$

These results maintain the pattern in the list, which suggests the following rule.

Multiplying Numbers with Different Signs

For any positive real numbers x and y,

$$x(-y) = -(xy) \qquad \text{and} \qquad (-x)y = -(xy).$$

That is, the product of two numbers with opposite signs is negative.

EXAMPLE 1 Multiplying a Positive Number and a Negative Number

Find each product using the multiplication rule given in the box.

(a) $8(-5) = -(8 \cdot 5) = -40$ **(b)** $5(-4) = -(5 \cdot 4) = -20$

(c) $-7(2) = -(7 \cdot 2) = -14$ **(d)** $-9(3) = -(9 \cdot 3) = -27$

Now Try Exercise 13.

OBJECTIVE 2 Find the product of two negative numbers. The product of two positive numbers is positive, and the product of a positive and a negative number is negative. What about the product of two negative numbers? Look at another pattern.

$$-5(4) = -20$$
$$-5(3) = -15$$
$$-5(2) = -10$$
$$-5(1) = -5$$
$$-5(0) = 0$$
$$-5(-1) = ?$$

The products increase by 5.

The numbers on the left of the equals sign (in color) decrease by 1 for each step down the list. The products on the right increase by 5 for each step down the list. To maintain this pattern, $-5(-1)$ should be 5 more than $-5(0)$, or 5 more than 0, so

$$-5(-1) = 5.$$

The pattern continues with

$$-5(-2) = 10$$
$$-5(-3) = 15$$
$$-5(-4) = 20$$
$$-5(-5) = 25,$$

and so on, which suggests the next rule.

Multiplying Two Negative Numbers

For any positive real numbers x and y,

$$-x(-y) = xy.$$

That is, the product of two negative numbers is positive.

EXAMPLE 2 Multiplying Two Negative Numbers

Find each product using the multiplication rule given in the box.

(a) $-9(-2) = 9 \cdot 2 = 18$ **(b)** $-6(-12) = 6 \cdot 12 = 72$

(c) $-8(-1) = 8 \cdot 1 = 8$ **(d)** $-15(-2) = 15 \cdot 2 = 30$

Now Try Exercise 11.

A summary of multiplying signed numbers is given here.

Multiplying Signed Numbers

The product of two numbers having the *same* sign is *positive,* and the product of two numbers having *different* signs is *negative.*

OBJECTIVE 3 Identify factors of integers. In Section 1.1 the definition of a *factor* was given for whole numbers. (For example, since $9 \cdot 5 = 45$, both 9 and 5 are factors of 45.) The definition can now be extended to integers.

If the product of two integers is a third integer, then each of the two integers is a *factor* of the third. For example, $(-3)(-4) = 12$, so -3 and -4 are both factors of 12. The integer factors of 12 are $-12, -6, -4, -3, -2, -1, 1, 2, 3, 4, 6$, and 12.

The table on the next page shows several integers and the factors of those integers.

Integer	18	20	15	7	1
Pairs of factors	1, 18	1, 20	1, 15	1, 7	1, 1
	2, 9	2, 10	3, 5	−1, −7	−1, −1
	3, 6	4, 5	−1, −15		
	−1, −18	−1, −20	−3, −5		
	−2, −9	−2, −10			
	−3, −6	−4, −5			

Now Try Exercise 29.

OBJECTIVE 4 Use the reciprocal of a number to apply the definition of division.   In Section 1.5 we saw that the difference between two numbers is found by adding the additive inverse of the second number to the first. Similarly, the *quotient* of two numbers is found by *multiplying* by the *reciprocal,* or *multiplicative inverse.* By definition, since

$$8 \cdot \frac{1}{8} = \frac{8}{8} = 1 \qquad \text{and} \qquad \frac{5}{4} \cdot \frac{4}{5} = \frac{20}{20} = 1,$$

the reciprocal or multiplicative inverse of 8 is $\frac{1}{8}$, and of $\frac{5}{4}$ is $\frac{4}{5}$.

Reciprocal or Multiplicative Inverse

Pairs of numbers whose product is 1 are **reciprocals,** or **multiplicative inverses,** of each other.

Number	Multiplicative Inverse (Reciprocal)
4	$\frac{1}{4}$
$.3 = \frac{3}{10}$	$\frac{10}{3}$
−5	$\frac{1}{-5}$ or $-\frac{1}{5}$
$-\frac{5}{8}$	$-\frac{8}{5}$
0	None
1	1
−1	−1

The table in the margin shows several numbers and their multiplicative inverses. Why is there no multiplicative inverse for the number 0? Suppose that k is to be the multiplicative inverse of 0. Then $k \cdot 0$ should equal 1. But $k \cdot 0 = 0$ for any number k. Since there is no value of k that is a solution of the equation $k \cdot 0 = 1$, the following statement can be made.

0 has no multiplicative inverse.

By definition, the quotient of x and y is the product of x and the multiplicative inverse of y.

Division

For any real numbers x and y, with $y \neq 0$, $\dfrac{x}{y} = x \cdot \dfrac{1}{y}.$

The definition of division indicates that y, the number to divide by, cannot be 0. The reason is that 0 has no multiplicative inverse, so that $\frac{1}{0}$ is not a number. **Because 0 has no multiplicative inverse,** *division by 0 is undefined.* **If a division problem turns out to involve division by 0, write "undefined."**

NOTE While division *by* 0 $\left(\frac{a}{0}\right)$ is undefined, we may divide 0 by any nonzero number. In fact, if $a \neq 0$,

$$\frac{0}{a} = 0.$$

Since division is defined in terms of multiplication, all the rules for multiplying signed numbers also apply to dividing them.

EXAMPLE 3 Using the Definition of Division

Find each quotient using the definition of division.

(a) $\dfrac{12}{3} = 12 \cdot \dfrac{1}{3} = 4$

(b) $\dfrac{-10}{2} = -10 \cdot \dfrac{1}{2} = -5$

(c) $\dfrac{-14}{-7} = -14\left(\dfrac{1}{-7}\right) = 2$

(d) $-\dfrac{2}{3} \div \left(-\dfrac{4}{5}\right) = -\dfrac{2}{3} \cdot \left(-\dfrac{5}{4}\right) = \dfrac{5}{6}$

(e) $\dfrac{0}{13} = 0\left(\dfrac{1}{13}\right) = 0 \qquad \dfrac{0}{a} = 0 \quad (a \neq 0)$

(f) $\dfrac{-10}{0}$ Undefined

> **Now Try Exercises 33, 39, and 45.**

When dividing fractions, multiplying by the reciprocal works well. However, using the definition of division directly with integers is awkward. It is easier to divide in the usual way, then determine the sign of the answer.

Dividing Signed Numbers

The quotient of two numbers having the *same* sign is *positive;* the quotient of two numbers having *different* signs is *negative.*

Note that these are the same as the rules for multiplication.

EXAMPLE 4 Dividing Signed Numbers

Find each quotient.

(a) $\dfrac{8}{-2} = -4$

(b) $\dfrac{-4.5}{-.09} = 50$

(c) $-\dfrac{1}{8} \div \left(-\dfrac{3}{4}\right) = -\dfrac{1}{8} \cdot \left(-\dfrac{4}{3}\right) = \dfrac{1}{6}$

> **Now Try Exercises 35, 41, and 43.**

From the definitions of multiplication and division of real numbers,

$$\frac{-40}{8} = -40 \cdot \frac{1}{8} = -5, \quad \text{and} \quad \frac{40}{-8} = 40\left(\frac{1}{-8}\right) = -5, \quad \text{so}$$

$$\frac{-40}{8} = \frac{40}{-8}.$$

Based on this example, the quotient of a positive and a negative number can be expressed in any of the following three forms.

For any positive real numbers x and y, $\quad \dfrac{-x}{y} = \dfrac{x}{-y} = -\dfrac{x}{y}.$

Similarly, the quotient of two negative numbers can be expressed as a quotient of two positive numbers.

For any positive real numbers x and y, $\quad \dfrac{-x}{-y} = \dfrac{x}{y}.$

OBJECTIVE 5 Use the order of operations when multiplying and dividing signed numbers.

EXAMPLE 5   Using the Order of Operations

Perform each indicated operation.

(a) $-9(2) - (-3)(2)$

First find all products, working from left to right.

$$-9(2) - (-3)(2) = -18 - (-6)$$
$$= -18 + 6$$
$$= -12$$

(b) $-6(-2) - 3(-4) = 12 - (-12)$
$$= 12 + 12$$
$$= 24$$

(c) $-5(-2 - 3) = -5(-5) = 25$

(d) $\dfrac{5(-2) - 3(4)}{2(1 - 6)}$

Simplify the numerator and denominator separately. Then write in lowest terms.

$$\frac{5(-2) - 3(4)}{2(1 - 6)} = \frac{-10 - 12}{2(-5)} = \frac{-22}{-10} = \frac{11}{5}$$

Now Try Exercises 53 and 67.

We summarize the rules for operations with signed numbers on the next page.

Operations with Signed Numbers

ADDITION

Same sign   Add the absolute values of the numbers. The sum has the same sign as the numbers.

$$-4 + (-6) = -10$$

Different signs   Subtract the number with the smaller absolute value from the one with the larger. Give the sum the sign of the number having the larger absolute value.

$$4 + (-6) = -(6 - 4) = -2$$

SUBTRACTION

Add the opposite of the second number to the first number.

$$8 - (-3) = 8 + 3 = 11$$

MULTIPLICATION AND DIVISION

Same sign   The product or quotient of two numbers with the same sign is positive.

$$-5(-6) = 30 \qquad \text{and} \qquad \frac{-36}{-12} = 3$$

Different signs   The product or quotient of two numbers with different signs is negative.

$$-5(6) = -30 \qquad \text{and} \qquad \frac{18}{-6} = -3$$

Division by 0 is undefined.

OBJECTIVE 6 Evaluate expressions involving variables.   The next examples show numbers substituted for variables where the rules for multiplying and dividing signed numbers must be used.

EXAMPLE 6   Evaluating Expressions for Numerical Values

Evaluate each expression, given that $x = -1$, $y = -2$, and $m = -3$.

(a) $(3x + 4y)(-2m)$

   First substitute the given values for the variables. Then use the order of operations to find the value of the expression.

$$(3x + 4y)(-2m) = [3(-1) + 4(-2)][-2(-3)] \qquad \text{Put parentheses around the value for each variable.}$$

$$= [-3 + (-8)][6] \qquad \text{Multiply.}$$

$$= (-11)(6) \qquad \text{Add inside the parentheses.}$$

$$= -66 \qquad \text{Multiply.}$$

(b) $2x^2 - 3y^2$

Use parentheses as shown.

$$
\begin{aligned}
2(-1)^2 - 3(-2)^2 &= 2(1) - 3(4) && \text{Apply the exponents.}\\
&= 2 - 12 && \text{Multiply.}\\
&= -10 && \text{Subtract.}
\end{aligned}
$$

(c) $\dfrac{4y^2 + x}{m}$

$$
\begin{aligned}
\frac{4(-2)^2 + (-1)}{-3} &= \frac{4(4) + (-1)}{-3} && \text{Apply the exponent.}\\[2mm]
&= \frac{16 + (-1)}{-3} && \text{Multiply.}\\[2mm]
&= \frac{15}{-3} && \text{Add.}\\[2mm]
&= -5 && \text{Divide.}
\end{aligned}
$$

Notice how the fraction bar was used as a grouping symbol.

Now Try Exercises 77 and 85.

OBJECTIVE 7 Interpret words and phrases involving multiplication and division. Just as there are words and phrases that indicate addition and subtraction, certain words also indicate multiplication and division.

■ PROBLEM SOLVING ■

The word *product* refers to multiplication. The table gives other key words and phrases that indicate multiplication.

Word or Phrase	Example	Numerical Expression and Simplification
Product of	The *product of* -5 and -2	$-5(-2) = 10$
Times	13 *times* -4	$13(-4) = -52$
Twice (meaning "2 times")	*Twice* 6	$2(6) = 12$
Of (used with fractions)	$\dfrac{1}{2}$ *of* 10	$\dfrac{1}{2}(10) = 5$
Percent of	12% *of* -16	$.12(-16) = -1.92$
As much as	$\dfrac{2}{3}$ *as much as* 30	$\dfrac{2}{3}(30) = 20$

■ EXAMPLE 7 Interpreting Words and Phrases Involving Multiplication

Write a numerical expression for each phrase and simplify. Use the order of operations.

(a) The product of 12 and the sum of 3 and -6

Here, 12 is multiplied by "the sum of 3 and -6."

$$12[3 + (-6)] = 12(-3) = -36$$

(b) Twice the difference between 8 and -4

$$2[8 - (-4)] = 2[8 + 4] = 2(12) = 24$$

(c) Two-thirds of the sum of -5 and -3

$$\frac{2}{3}[-5 + (-3)] = \frac{2}{3}[-8] = -\frac{16}{3}$$

(d) 15% of the difference between 14 and -2
Remember that 15% = .15.

$$.15[14 - (-2)] = .15[14 + 2] = .15(16) = 2.4$$

(e) Double the product of 3 and 4

$$2 \cdot (3 \cdot 4) = 2 \cdot 12 = 24$$

Now Try Exercises 89, 93, and 101.

PROBLEM SOLVING

The word *quotient* refers to the answer in a division problem. In algebra, quotients are usually represented with a fraction bar; the symbol $\div$ is seldom used. When translating applied problems involving division, use a fraction bar. The table gives some key phrases associated with division.

Phrase	Example	Numerical Expression and Simplification
Quotient of	The *quotient of* -24 and 3	$\frac{-24}{3} = -8$
Divided by	-16 *divided by* -4	$\frac{-16}{-4} = 4$
Ratio of	The *ratio of* 2 to 3	$\frac{2}{3}$

It is customary to write the first number named as the numerator and the second as the denominator when interpreting a phrase involving division, as shown in the next example.

EXAMPLE 8   Interpreting Words and Phrases Involving Division

Write a numerical expression for each phrase and simplify the expression.

(a) The quotient of 14 and the sum of -9 and 2
   "Quotient" indicates division. The number 14 is the numerator and "the sum of -9 and 2" is the denominator.

$$\frac{14}{-9 + 2} = \frac{14}{-7} = -2$$

(b) The product of 5 and -6, divided by the difference between -7 and 8
   The numerator of the fraction representing the division is obtained by multiplying 5 and -6. The denominator is found by subtracting -7 and 8.

$$\frac{5(-6)}{-7 - 8} = \frac{-30}{-15} = 2$$

Now Try Exercise 95.

OBJECTIVE 8 Translate simple sentences into equations. In this section and the previous one, important words and phrases involving the four operations of arithmetic have been introduced. We can use these words and phrases to interpret sentences that translate into equations.

EXAMPLE 9 Translating Words into Equations

Write each sentence in symbols, using x as the variable, and guess or use trial and error to find the solution. All solutions come from the list of integers between -12 and 12, inclusive.

(a) Three *times* a number *is* -18.

The word *times* indicates multiplication, and the word *is* translates as the equals sign (=).

$$3x = -18$$

Since the integer between -12 and 12 inclusive that makes this statement true is -6, the solution of the equation is -6.

(b) The *sum* of a number and 9 *is* 12.

$$x + 9 = 12$$

Since $3 + 9 = 12$, the solution of this equation is 3.

(c) The *difference between* a number and 5 *is* 0.

$$x - 5 = 0$$

Since $5 - 5 = 0$, the solution of this equation is 5.

(d) The *quotient* of 24 and a number *is* -2.

$$\frac{24}{x} = -2$$

Here, x must be a negative number, since the numerator is positive and the quotient is negative. Since $\frac{24}{-12} = -2$, the solution is -12.

Now Try Exercises 103 and 107.

CAUTION It is important to recognize the distinction between the types of problems found in Examples 7 and 8 and Example 9. In Examples 7 and 8, the *phrases* translate as *expressions,* while in Example 9, the *sentences* translate as *equations.* Remember that an equation is a sentence with an = sign, while an expression is a phrase.

$$\frac{5(-6)}{-7 - 8} \qquad 3x = -18$$

$$\text{Expression} \qquad \text{Equation}$$

1.6 EXERCISES

Fill in each blank with one of the following: greater than 0, less than 0, equal to 0.

1. A positive number is _____.

2. A negative number is _____.

3. The product or the quotient of two numbers with the same sign is _____.

4. The product or the quotient of two numbers with different signs is _____.

5. If three negative numbers are multiplied together, the product is _____.

6. If two negative numbers are multiplied and then their product is divided by a negative number, the result is _____.

7. If a negative number is squared and the result is added to a positive number, the final answer is _____.

8. The reciprocal of a negative number is _____.

9. If three positive numbers, five negative numbers, and zero are multiplied, the product is _____.

10. The fifth power of a negative number is _____.

Find each product. See Examples 1 and 2.

11. $-3(-4)$ **12.** $-3(4)$ **13.** $3(-4)$ **14.** $-2(-8)$

15. $-10(-12)$ **16.** $9(-5)$ **17.** $3(-11)$ **18.** $3(-15)$

19. $15(-11)$ **20.** $-9(-4)$ **21.** $-\dfrac{3}{8} \cdot \left(-\dfrac{10}{9}\right)$ **22.** $-\dfrac{5}{4} \cdot \left(-\dfrac{5}{8}\right)$

23. $\left(-1\dfrac{1}{4}\right)\left(\dfrac{2}{15}\right)$ **24.** $\left(\dfrac{3}{7}\right)\left(-1\dfrac{5}{9}\right)$ **25.** $(-8)\left(-\dfrac{3}{4}\right)$ **26.** $(-6)\left(-\dfrac{5}{3}\right)$

Find all integer factors of each number.

27. 32   **28.** 36   **29.** 40   **30.** 50   **31.** 31   **32.** 17

Find each quotient. See Examples 3 and 4.

33. $\dfrac{15}{5}$ **34.** $\dfrac{25}{5}$ **35.** $\dfrac{-30}{6}$ **36.** $\dfrac{-28}{14}$

37. $\dfrac{-28}{-4}$ **38.** $\dfrac{-35}{-7}$ **39.** $\dfrac{96}{-16}$ **40.** $\dfrac{38}{-19}$

41. $-\dfrac{4}{3} \div \left(-\dfrac{1}{8}\right)$ **42.** $-\dfrac{5}{6} \div \left(-\dfrac{15}{7}\right)$ **43.** $\dfrac{-8.8}{2.2}$ **44.** $\dfrac{-4.6}{-.23}$

45. $\dfrac{0}{-2}$ **46.** $\dfrac{0}{-8}$ **47.** $\dfrac{12}{0}$ **48.** $\dfrac{6}{0}$

Perform each indicated operation. See Examples 5(a), (b), and (c).

49. $7 - 3 \cdot 6$ **50.** $8 - 2 \cdot 5$ **51.** $-10 - (-4)(2)$

52. $-11 - (-3)(6)$ **53.** $-7(3 - 8)$ **54.** $-5(4 - 7)$

55. $(12 - 14)(1 - 4)$ **56.** $(8 - 9)(4 - 12)$ **57.** $(7 - 10)(10 - 4)$

58. $(5 - 12)(19 - 4)$ **59.** $(-2 - 8)(-6) + 7$ **60.** $(-9 - 4)(-2) + 10$

61. $3(-5) + |3 - 10|$ **62.** $4(-8) + |4 - 15|$

Perform each indicated operation. See Example 5(d).

63. $\dfrac{-5(-6)}{9 - (-1)}$ **64.** $\dfrac{-12(-5)}{7 - (-5)}$ **65.** $\dfrac{-21(3)}{-3 - 6}$

66. $\dfrac{-40(3)}{-2 - 3}$ **67.** $\dfrac{-10(2) + 6(2)}{-3 - (-1)}$ **68.** $\dfrac{8(-1) - |(-4)(-3)|}{-6 - (-1)}$

69. $\dfrac{-27(-2) - |6 \cdot 4|}{-2(3) - 2(2)}$ **70.** $\dfrac{-13(-4) - (-8)(-2)}{(-10)(2) - 4(-2)}$ **71.** $\dfrac{-5(2) + [3(-2) - 4]}{-3 - (-1)}$

72. Explain the method you would use to evaluate $3x + 2y$ when $x = -3$ and $y = 4$.

73. If x and y are both replaced by negative numbers, is the value of $4x + 8y$ positive or negative?

74. Repeat Exercise 73, but replace the first word *negative* with *positive*.

Evaluate each expression if $x = 6$, $y = -4$, and $a = 3$. See Example 6.

75. $5x - 2y + 3a$ **76.** $6x - 5y + 4a$

77. $(2x + y)(3a)$ **78.** $(5x - 2y)(-2a)$

79. $\left(\dfrac{1}{3}x - \dfrac{4}{5}y\right)\left(-\dfrac{1}{5}a\right)$ **80.** $\left(\dfrac{5}{6}x + \dfrac{3}{2}y\right)\left(-\dfrac{1}{3}a\right)$

81. $(-5 + x)(-3 + y)(3 - a)$ **82.** $(6 - x)(5 + y)(3 + a)$

83. $-2y^2 + 3a$ **84.** $5x - 4a^2$

85. $\dfrac{2y^2 - x}{a + 10}$ **86.** $\dfrac{xy + 8a}{x - y}$

Write a numerical expression for each phrase and simplify. See Examples 7 and 8.

87. The product of -9 and 2, added to 9

88. The product of 4 and -7, added to -12

89. Twice the product of -1 and 6, subtracted from -4

90. Twice the product of -8 and 2, subtracted from -1

91. Nine subtracted from the product of 1.5 and -3.2

92. Three subtracted from the product of 4.2 and -8.5

93. The product of 12 and the difference between 9 and -8

94. The product of -3 and the difference between 3 and -7

95. The quotient of -12 and the sum of -5 and -1

96. The quotient of -20 and the sum of -8 and -2

97. The sum of 15 and -3, divided by the product of 4 and -3

98. The sum of -18 and -6, divided by the product of 2 and -4

99. Twice the sum of 8 and 9

100. Three-fourths of the sum of -8 and 12

101. 20% of the product of -5 and 6

102. $\frac{2}{3}$ as much as the difference between 8 and -1

Write each statement in symbols, using x as the variable, and find the solution by guessing or by using trial and error. All solutions come from the set of integers between −12 and 12, inclusive. See Example 9.

103. The quotient of a number and 3 is −3.

104. The quotient of a number and 4 is −1.

105. 6 less than a number is 4.

106. 7 less than a number is 2.

107. When 5 is added to a number, the result is −5.

108. When 6 is added to a number, the result is −3.

*To find the **average** of a group of numbers, we add the numbers and then divide the sum by the number of terms added. For example, to find the average of 14, 8, 3, 9, and 1, we add them and then divide by 5:*

$$\frac{14 + 8 + 3 + 9 + 1}{5} = \frac{35}{5} = 7.$$

The average of these numbers is 7.

Exercises 109–114 involve finding the average of a group of numbers.

109. Find the average of 23, 18, 13, −4, and −8.

110. Find the average of 18, 12, 0, −4, and −10.

111. What is the average of all integers between −10 and 14, inclusive of both?

112. What is the average of all even integers between −18 and 4, inclusive of both?

113. If the average of a group of numbers is 0, what is the sum of all the numbers?

114. Suppose there is a group of numbers with some positive and some negative. Under what conditions will the average be a positive number? Under what conditions will the average be negative?

The operation of division is used in divisibility tests. A divisibility test allows us to determine whether a given number is divisible (without remainder) by another number. For example, a number is divisible by 2 if its last digit is divisible by 2, and not otherwise.

115. Tell why **(a)** 3,473,986 is divisible by 2 and **(b)** 4,336,879 is not divisible by 2.

116. An integer is divisible by 3 if the sum of its digits is divisible by 3, and not otherwise. Show that **(a)** 4,799,232 is divisible by 3 and **(b)** 2,443,871 is not divisible by 3.

117. An integer is divisible by 4 if its last two digits form a number divisible by 4, and not otherwise. Show that **(a)** 6,221,464 is divisible by 4 and **(b)** 2,876,335 is not divisible by 4.

118. An integer is divisible by 5 if its last digit is divisible by 5, and not otherwise. Show that **(a)** 3,774,595 is divisible by 5 and **(b)** 9,332,123 is not divisible by 5.

119. An integer is divisible by 6 if it is divisible by both 2 and 3, and not otherwise. Show that **(a)** 1,524,822 is divisible by 6 and **(b)** 2,873,590 is not divisible by 6.

120. An integer is divisible by 8 if its last three digits form a number divisible by 8, and not otherwise. Show that **(a)** 2,923,296 is divisible by 8 and **(b)** 7,291,623 is not divisible by 8.

121. An integer is divisible by 9 if the sum of its digits is divisible by 9, and not otherwise. Show that **(a)** 4,114,107 is divisible by 9 and **(b)** 2,287,321 is not divisible by 9.

122. An integer is divisible by 12 if it is divisible by both 3 and 4, and not otherwise. Show that **(a)** 4,253,520 is divisible by 12 and **(b)** 4,249,474 is not divisible by 12.

Addition

Same sign   Add the absolute values of the numbers. The sum has the same sign as the numbers.

Different signs   Subtract the number with the smaller absolute value from the one with the larger. Give the sum the sign of the number having the larger absolute value.

Subtraction

Add the opposite of the second number to the first number.

Multiplication and Division

Same sign   The product or quotient of two numbers with the same sign is positive.

Different signs   The product or quotient of two numbers with different signs is negative.

Division by 0 is undefined.

Perform each indicated operation.

1. $14 - 3 \cdot 10$

2. $-3(8) - 4(-7)$

3. $(3 - 8)(-2) - 10$

4. $-6(7 - 3)$

5. $7 - (-3)(2 - 10)$

6. $-4[(-2)(6) - 7]$

7. $(-4)(7) - (-5)(2)$

8. $-5[-4 - (-2)(-7)]$

9. $40 - (-2)[8 - 9]$

10. $\dfrac{5(-4)}{-7 - (-2)}$

11. $\dfrac{-3 - (-9 + 1)}{-7 - (-6)}$

12. $\dfrac{5(-8 + 3)}{13(-2) + (-7)(-3)}$

13. $\dfrac{6^2 - 8}{-2(2) + 4(-1)}$

14. $\dfrac{16(-8 + 5)}{15(-3) + (-7 - 4)(-3)}$

15. $\dfrac{9(-6) - 3(8)}{4(-7) + (-2)(-11)}$

16. $\dfrac{2^2 + 4^2}{5^2 - 3^2}$

17. $\dfrac{(2 + 4)^2}{(5 - 3)^2}$

18. $\dfrac{4^3 - 3^3}{-5(-4 + 2)}$

19. $\dfrac{-9(-6) + (-2)(27)}{3(8 - 9)}$

20. $|-4(9)| - |-11|$

21. $\dfrac{6(-10 + 3)}{15(-2) - 3(-9)}$

22. $\dfrac{(-9)^2 - 9^2}{3^2 - 5^2}$

23. $\dfrac{(-10)^2 + 10^2}{-10(5)}$

24. $-\dfrac{3}{4} \div \left(-\dfrac{5}{8}\right)$

25. $\dfrac{1}{2} \div \left(-\dfrac{1}{2}\right)$

26. $\dfrac{8^2 - 12}{(-5)^2 + 2(6)}$

27. $\left[\dfrac{5}{8} - \left(-\dfrac{1}{16}\right)\right] + \dfrac{3}{8}$

28. $\left(\dfrac{1}{2} - \dfrac{1}{3}\right) - \dfrac{5}{6}$

29. $-.9(-3.7)$

30. $-5.1(-.2)$

31. $-3^2 - 2^2$

32. $|-2(3) + 4| - |-2|$

33. $40 - (-2)[-5 - 3]$

Evaluate each expression if $x = -2$, $y = 3$, and $a = 4$.

34. $-x + y - 3a$

35. $(x + 6)^3 - y^3$

36. $(x - y) - (a - 2y)$

37. $\left(\dfrac{1}{2}x + \dfrac{2}{3}y\right)\left(-\dfrac{1}{4}a\right)$

38. $\dfrac{2x + 3y}{a - xy}$

39. $\dfrac{x^2 - y^2}{x^2 + y^2}$

40. $-x^2 + 3y$

1.7 Properties of Real Numbers

OBJECTIVES

1 Use the commutative properties.

2 Use the associative properties.

3 Use the identity properties.

4 Use the inverse properties.

5 Use the distributive property.

If you were asked to find the sum $3 + 89 + 97$, you might mentally add $3 + 97$ to get 100, and then add $100 + 89$ to get 189. While the rule for order of operations says to add from left to right, we may change the order of the terms and group them in any way we choose without affecting the sum. These are examples of shortcuts that we use in everyday mathematics. These shortcuts are justified by the basic properties of addition and multiplication, discussed in this section. In these properties, a, b, and c represent real numbers.

OBJECTIVE 1 Use the commutative properties.   The word *commute* means to go back and forth. Many people commute to work or to school. If you travel from home to work and follow the same route from work to home, you travel the same distance each time. The **commutative properties** say that if two numbers are added or multiplied in any order, the result is the same.

Commutative Properties

$$a + b = b + a$$
$$ab = ba$$

EXAMPLE 1   Using the Commutative Properties

Use a commutative property to complete each statement.

(a) $-8 + 5 = 5 +$ _____

By the commutative property for addition, the missing number is -8, since $-8 + 5 = 5 + (-8)$.

(b) $(-2)7 =$ _____(-2)

By the commutative property for multiplication, the missing number is 7, since $(-2)7 = 7(-2)$.

Now Try Exercises 1 and 3.

OBJECTIVE 2 Use the associative properties.   When we *associate* one object with another, we tend to think of those objects as being grouped together. The **associative properties** say that when we add or multiply three numbers, we can group the first two together or the last two together and get the same answer.

Associative Properties

$$(a + b) + c = a + (b + c)$$
$$(ab)c = a(bc)$$

■ EXAMPLE 2   Using the Associative Properties

Use an associative property to complete each statement.

(a) $8 + (-1 + 4) = (8 + \underline{\hspace{1cm}}) + 4$

The missing number is -1.

(b) $[2 \cdot (-7)] \cdot 6 = 2 \cdot \underline{\hspace{1cm}}$

The completed expression on the right should be $2 \cdot [(-7) \cdot 6]$.

Now Try Exercises 5 and 7.

By the associative property of addition, the sum of three numbers will be the same no matter how the numbers are "associated" in groups. For this reason, parentheses can be left out in many addition problems. For example, both

$$(-1 + 2) + 3 \qquad \text{and} \qquad -1 + (2 + 3)$$

can be written as

$$-1 + 2 + 3.$$

In the same way, parentheses also can be left out of many multiplication problems.

■ EXAMPLE 3   Distinguishing between the Associative and Commutative Properties

(a) Is $(2 + 4) + 5 = 2 + (4 + 5)$ an example of the associative property or the commutative property?

The order of the three numbers is the same on both sides of the equals sign. The only change is in the *grouping,* or association, of the numbers. Therefore, this is an example of the associative property.

(b) Is $6(3 \cdot 10) = 6(10 \cdot 3)$ an example of the associative property or the commutative property?

The same numbers, 3 and 10, are grouped on each side. On the left, the 3 appears first, but on the right, the 10 appears first. Since the only change involves the *order* of the numbers, this statement is an example of the commutative property.

(c) Is $(8 + 1) + 7 = 8 + (7 + 1)$ an example of the associative property or the commutative property?

In the statement, both the order and the grouping are changed. On the left the order of the three numbers is 8, 1, and 7. On the right it is 8, 7, and 1. On the left the 8 and 1 are grouped, and on the right the 7 and 1 are grouped. Therefore, both the associative and the commutative properties are used.

Now Try Exercises 15 and 23.

■ EXAMPLE 4   Using the Commutative and Associative Properties

Find the sum $23 + 41 + 2 + 9 + 25$.

The commutative and associative properties make it possible to choose pairs of numbers whose sums are easy to add.

$$23 + 41 + 2 + 9 + 25 = (41 + 9) + (23 + 2) + 25$$
$$= 50 + 25 + 25$$
$$= 100$$

Now Try Exercise 37.

OBJECTIVE 3 Use the identity properties. If a child wears a costume on Halloween, the child's appearance is changed, but his or her *identity* is unchanged. The identity of a real number is left unchanged when identity properties are applied. The **identity properties** say that the sum of 0 and any number equals that number, and the product of 1 and any number equals that number.

Identity Properties

$$a + 0 = a \quad \text{and} \quad 0 + a = a$$
$$a \cdot 1 = a \quad \text{and} \quad 1 \cdot a = a$$

The number 0 leaves the identity, or value, of any real number unchanged by addition. For this reason, 0 is called the **identity element for addition,** or the **additive identity.** Since multiplication by 1 leaves any real number unchanged, 1 is the **identity element for multiplication,** or the **multiplicative identity.**

EXAMPLE 5 Using the Identity Properties

These statements are examples of the identity properties.

(a) $-3 + 0 = -3$

(b) $1 \cdot \dfrac{1}{2} = \dfrac{1}{2}$

> **Now Try Exercise 31.**

We use the identity property for multiplication to write fractions in lowest terms and to find common denominators.

EXAMPLE 6 Using the Identity Property to Simplify Expressions

Simplify each expression.

(a) $\dfrac{49}{35} = \dfrac{7 \cdot 7}{5 \cdot 7}$ Factor.

$= \dfrac{7}{5} \cdot \dfrac{7}{7}$ Write as a product.

$= \dfrac{7}{5} \cdot 1$ Divide.

$= \dfrac{7}{5}$ Identity property

(b) $\dfrac{3}{4} + \dfrac{5}{24} = \dfrac{3}{4} \cdot 1 + \dfrac{5}{24}$ Identity property

$= \dfrac{3}{4} \cdot \dfrac{6}{6} + \dfrac{5}{24}$ Find a common denominator.

$= \dfrac{18}{24} + \dfrac{5}{24}$ Multiply.

$= \dfrac{23}{24}$ Add.

> **Now Try Exercise 27.**

OBJECTIVE 4 Use the inverse properties.   Each day before you go to work or school, you probably put on your shoes before you leave. Before you go to sleep at night, you probably take them off, and this leads to the same situation that existed before you put them on. These operations from everyday life are examples of *inverse* operations. The **inverse properties** of addition and multiplication lead to the additive and multiplicative identities, respectively. Recall that $-a$ is the **additive inverse, or opposite,** of a and $\frac{1}{a}$ is the **multiplicative inverse, or reciprocal,** of the nonzero number a. The sum of the numbers a and $-a$ is 0, and the product of the nonzero numbers a and $\frac{1}{a}$ is 1.

Inverse Properties

$$a + (-a) = 0 \quad \text{and} \quad -a + a = 0$$

$$a \cdot \frac{1}{a} = 1 \quad \text{and} \quad \frac{1}{a} \cdot a = 1 \quad (a \neq 0)$$

EXAMPLE 7   Using the Inverse Properties

The following statements are examples of the inverse properties.

(a) $\dfrac{2}{3} \cdot \dfrac{3}{2} = 1$

(b) $-5\left(-\dfrac{1}{5}\right) = 1$

(c) $-\dfrac{1}{2} + \dfrac{1}{2} = 0$

(d) $4 + (-4) = 0$

> **Now Try Exercise 19.**

In the next example, we show how the various properties are used to simplify an expression.

EXAMPLE 8   Using Properties to Simplify an Expression

Simplify $-2x + 10 + 2x$, using the properties discussed in this section.

$$
\begin{aligned}
-2x + 10 + 2x &= (-2x + 10) + 2x && \text{Order of operations} \\
&= [10 + (-2x)] + 2x && \text{Commutative property} \\
&= 10 + [(-2x) + 2x] && \text{Associative property} \\
&= 10 + 0 && \text{Inverse property} \\
&= 10 && \text{Identity property}
\end{aligned}
$$

Note that for *any* value of x, $-2x$ and $2x$ are additive inverses; that is why we can use the inverse property in this simplification.

> **Now Try Exercise 43.**

NOTE   The detailed procedure shown in Example 8 is seldom, if ever, used in practice. We include the example to show how the properties of this section apply, even though steps may be skipped when actually doing the simplification.

OBJECTIVE 5 Use the distributive property.   The everyday meaning of the word *distribute* is "to give out from one to several." An important property of real number operations involves this idea.

Look at the value of the following expressions.

$$2(5 + 8) = 2(13) = 26$$
$$2(5) + 2(8) = 10 + 16 = 26$$

Since both expressions equal 26,

$$2(5 + 8) = 2(5) + 2(8).$$

This result is an example of the *distributive property,* the only property involving *both* addition and multiplication. With this property, a product can be changed to a sum or difference. This idea is illustrated by the divided rectangle in Figure 17.

The area of the left part is 2(5) = 10.
The area of the right part is 2(8) = 16.
The total area is 2(5 + 8) = 2(13) = 26 or the total area is
2(5) + 2(8) = 10 + 16 = 26.
Thus, 2(5 + 8) = 2(5) + 2(8).

FIGURE 17

The **distributive property** says that multiplying a number a by a sum of numbers $b + c$ gives the same result as multiplying a by b and a by c and then adding the two products.

Distributive Property

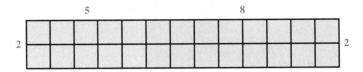

$$a(b + c) = ab + ac \qquad \text{and} \qquad (b + c)a = ba + ca$$

As the arrows show, the a outside the parentheses is "distributed" over the b and c inside. Another form of the distributive property is valid for subtraction.

$$a(b - c) = ab - ac \qquad \text{and} \qquad (b - c)a = ba - ca$$

The distributive property also can be extended to more than two numbers.

$$a(b + c + d) = ab + ac + ad$$

The distributive property can be used "in reverse." For example, we can write

$$ac + bc = (a + b)c.$$

EXAMPLE 9 Using the Distributive Property

Use the distributive property to rewrite each expression.

(a) $5(9 + 6) = 5 \cdot 9 + 5 \cdot 6$ Distributive property

$ = 45 + 30$ Multiply.

$ = 75$ Add.

(b) $4(x + 5 + y) = 4x + 4 \cdot 5 + 4y$ Distributive property

$\qquad\qquad\qquad\quad = 4x + 20 + 4y$ Multiply.

(c) $-2(x + 3) = -2x + (-2)(3)$ Distributive property

$\qquad\qquad\qquad = -2x - 6$ Multiply.

(d) $3(k - 9) = 3k - 3 \cdot 9$ Distributive property

$\qquad\qquad\quad = 3k - 27$ Multiply.

(e) $8(3r + 11t + 5z) = 8(3r) + 8(11t) + 8(5z)$ Distributive property

$\qquad\qquad\qquad\qquad = (8 \cdot 3)r + (8 \cdot 11)t + (8 \cdot 5)z$ Associative property

$\qquad\qquad\qquad\qquad = 24r + 88t + 40z$ Multiply.

(f) $6 \cdot 8 + 6 \cdot 2 = 6(8 + 2)$ Distributive property

$\qquad\qquad\qquad\; = 6(10) = 60$ Add, then multiply.

(g) $4x - 4m = 4(x - m)$ Distributive property

(h) $6x - 12 = 6 \cdot x - 6 \cdot 2 = 6(x - 2)$ Distributive property

Now Try Exercises 57, 63, 65, and 69.

The symbol $-a$ may be interpreted as $-1 \cdot a$. Similarly, when a negative sign precedes an expression within parentheses, it may also be interpreted as a factor of -1. Thus, we can use the distributive property to remove parentheses from expressions such as $-(2y + 3)$. We do this by first writing $-(2y + 3)$ as $-1 \cdot (2y + 3)$.

$-(2y + 3) = -1 \cdot (2y + 3)$

$\qquad\qquad = -1 \cdot 2y + (-1) \cdot 3$ Distributive property

$\qquad\qquad = -2y - 3$ Multiply.

■ **EXAMPLE 10   Using the Distributive Property to Remove Parentheses**

Write without parentheses.

(a) $-(7r - 8) = -1(7r - 8)$

$\qquad\qquad\quad = -1(7r) + (-1)(-8)$ Distributive property

$\qquad\qquad\quad = -7r + 8$ Multiply.

(b) $-(-9w + 2) = -1(-9w + 2)$

$\qquad\qquad\qquad = 9w - 2$

Now Try Exercise 75.

The properties discussed here are the basic properties that justify how we do algebra. Here is a summary of these properties.

Properties of Addition and Multiplication

For any real numbers a, b, and c, the following properties hold.

Commutative Properties $a + b = b + a$ $ab = ba$

Associative Properties $(a + b) + c = a + (b + c)$

$\qquad\qquad\qquad\qquad\quad (ab)c = a(bc)$ (continued)

Identity Properties	There is a real number 0 such that
	$$a + 0 = a \qquad \text{and} \qquad 0 + a = a.$$
	There is a real number 1 such that
	$$a \cdot 1 = a \qquad \text{and} \qquad 1 \cdot a = a.$$
Inverse Properties	For each real number a, there is a single real number $-a$ such that
	$$a + (-a) = 0 \qquad \text{and} \qquad (-a) + a = 0.$$
	For each nonzero real number a, there is a single real number $\frac{1}{a}$ such that
	$$a \cdot \frac{1}{a} = 1 \qquad \text{and} \qquad \frac{1}{a} \cdot a = 1.$$
Distributive Properties	$$a(b + c) = ab + ac \qquad (b + c)a = ba + ca$$

1.7 EXERCISES

Use the commutative or the associative property to complete each statement. State which property is used. See Examples 1 and 2.

1. $-12 + 6 = 6 + $ _____

2. $8 + (-4) = -4 + $ _____

3. $-6 \cdot 3 = $ _____ $\cdot (-6)$

4. $-12 \cdot 6 = 6 \cdot $ _____

5. $(4 + 7) + 8 = 4 + ($ _____ $+ 8)$

6. $(-2 + 3) + 6 = -2 + ($ _____ $+ 6)$

7. $8 \cdot (3 \cdot 6) = ($ _____ $\cdot 3) \cdot 6$

8. $6 \cdot (4 \cdot 2) = (6 \cdot $ _____ $) \cdot 2$

9. Match each item in Column I with the correct choice(s) from Column II. Choices may be used once, more than once, or not at all.

I	**II**
(a) Identity element for addition	**A.** $(5 \cdot 4) \cdot 3 = 5 \cdot (4 \cdot 3)$
(b) Identity element for multiplication	**B.** 0
(c) Additive inverse of a	**C.** $-a$
(d) Multiplicative inverse, or reciprocal, of the nonzero number a	**D.** -1
(e) The number that is its own additive inverse	**E.** $5 \cdot 4 \cdot 3 = 60$
(f) The two numbers that are their own multiplicative inverses	**F.** 1
	G. $(5 \cdot 4) \cdot 3 = 3 \cdot (5 \cdot 4)$
(g) The only number that has no multiplicative inverse	**H.** $5(4 + 3) = 5 \cdot 4 + 5 \cdot 3$
(h) An example of the associative property	**I.** $\frac{1}{a}$
(i) An example of the commutative property	
(j) An example of the distributive property	

10. Explain the difference between the commutative and associative properties.

Decide whether each statement is an example of the commutative, associative, identity, inverse, *or distributive property. See Examples 1, 2, 3, 5, 6, 7, and 9.*

11. $7 + 18 = 18 + 7$

12. $13 + 12 = 12 + 13$

13. $5 \cdot (13 \cdot 7) = (5 \cdot 13) \cdot 7$

14. $-4 \cdot (2 \cdot 6) = (-4 \cdot 2) \cdot 6$

15. $-6 + (12 + 7) = (-6 + 12) + 7$

16. $(-8 + 13) + 2 = -8 + (13 + 2)$

17. $-6 + 6 = 0$

18. $12 + (-12) = 0$

19. $\frac{2}{3}\left(\frac{3}{2}\right) = 1$

20. $\frac{5}{8}\left(\frac{8}{5}\right) = 1$

21. $2.34 + 0 = 2.34$

22. $-8.456 + 0 = -8.456$

23. $(4 + 17) + 3 = 3 + (4 + 17)$

24. $(-8 + 4) + 12 = 12 + (-8 + 4)$

25. $6(x + y) = 6x + 6y$

26. $14(t + s) = 14t + 14s$

27. $-\frac{5}{9} = -\frac{5}{9} \cdot \frac{3}{3} = -\frac{15}{27}$

28. $\frac{13}{12} = \frac{13}{12} \cdot \frac{7}{7} = \frac{91}{84}$

29. $5(2x) + 5(3y) = 5(2x + 3y)$

30. $3(5t) - 3(7r) = 3(5t - 7r)$

31. The following conversation actually took place between one of the authors of this book and his son, Jack, when Jack was 4 years old:

> DADDY: "Jack, what is 3 + 0?"
>
> JACK: "3."
>
> DADDY: "Jack, what is 4 + 0?"
>
> JACK: "4. And Daddy, *string* plus zero equals *string*!"

What property of addition did Jack recognize?

32. The distributive property holds for multiplication with respect to addition. Is there a distributive property for addition with respect to multiplication? If not, give an example to show why.

33. Write a paragraph explaining in your own words the identity and inverse properties of addition and multiplication.

34. Write a paragraph explaining in your own words the distributive property of multiplication with respect to addition. Give examples.

Find each sum. Use the commutative and associative properties to make your work easier. See Example 4.

35. $97 + 13 + 3 + 37$

36. $49 + 199 + 1 + 1$

37. $1999 + 2 + 1 + 8$

38. $2998 + 3 + 2 + 17$

39. $159 + 12 + 141 + 88$

40. $106 + 8 + (-6) + (-8)$

41. $843 + 627 + (-43) + (-27)$

42. $1846 + 1293 + (-46) + (-93)$

Use the properties of this section to simplify each expression. See Examples 7 and 8.

43. $6t + 8 - 6t + 3$

44. $9r + 12 - 9r + 1$

45. $\frac{2}{3}x - 11 + 11 - \frac{2}{3}x$

46. $\frac{1}{5}y + 4 - 4 - \frac{1}{5}y$

47. $\left(\frac{9}{7}\right)(-.38)\left(\frac{7}{9}\right)$

48. $\left(\frac{4}{5}\right)(-.73)\left(\frac{5}{4}\right)$

49. $t + (-t) + \frac{1}{2}(2)$

50. $w + (-w) + \frac{1}{4}(4)$

51. Evaluate $25 - (6 - 2)$ and evaluate $(25 - 6) - 2$. Do you think subtraction is associative?

52. Evaluate $180 \div (15 \div 3)$ and evaluate $(180 \div 15) \div 3$. Do you think division is associative?

53. Suppose that a student shows you the following work.

$$-3(4 - 6) = -3(4) - 3(6) = -12 - 18 = -30$$

The student has made a very common error. Explain the student's mistake, and work the problem correctly.

54. Explain how the procedure of changing $\frac{3}{4}$ to $\frac{9}{12}$ requires the use of the multiplicative identity element, 1.

Use the distributive property to rewrite each expression. Simplify if possible. See Example 9.

55. $5(9 + 8)$ **56.** $6(11 + 8)$ **57.** $4(t + 3)$

58. $5(w + 4)$ **59.** $-8(r + 3)$ **60.** $-11(x + 4)$

61. $-5(y - 4)$ **62.** $-9(g - 4)$ **63.** $-\frac{4}{3}(12y + 15z)$

64. $-\frac{2}{5}(10b + 20a)$ **65.** $8z + 8w$ **66.** $4s + 4r$

67. $7(2v) + 7(5r)$ **68.** $13(5w) + 13(4p)$ **69.** $8(3r + 4s - 5y)$

70. $2(5u - 3v + 7w)$ **71.** $-3(8x + 3y + 4z)$ **72.** $-5(2x - 5y + 6z)$

73. $5x + 15$ **74.** $9p + 18$

Use the distributive property to write each expression without parentheses. See Example 10.

75. $-(4t + 3m)$ **76.** $-(9x + 12y)$ **77.** $-(-5c - 4d)$

78. $-(-13x - 15y)$ **79.** $-(-3q + 5r - 8s)$ **80.** $-(-4z + 5w - 9y)$

81. The operations of "getting out of bed" and "taking a shower" are not commutative. Give an example of another pair of everyday operations that are not commutative.

82. The phrase "dog biting man" has two different meanings, depending on how the words are associated:

(dog biting) man   dog (biting man)

Give another example of a three-word phrase that has different meanings depending on how the words are associated.

RELATING CONCEPTS (EXERCISES 83–86)

For Individual or Group Work

In Section 1.6 we used a pattern to see that the product of two negative numbers is a positive number. In the group of exercises that follows, we show another justification for determining the sign of the product of two negative numbers. **Work Exercises 83–86 in order.**

83. Evaluate the expression $-3[5 + (-5)]$ by using the order of operations.

84. Write the expression in Exercise 83 using the distributive property. Do not simplify the products.

85. The product $-3(5)$ should be part of the answer you wrote for Exercise 84. Based on the results in Section 1.6, what is this product?

86. In Exercise 83, you should have obtained 0 as an answer. Now, consider the following, using the results of Exercises 83 and 85.

$$-3[5 + (-5)] = -3(5) + (-3)(-5)$$
$$0 = -15 + ?$$

The question mark represents the product $(-3)(-5)$. When added to -15, it must give a sum of 0. Therefore, how must we interpret $(-3)(-5)$?

87. Consider the statement

$$-2(5 \cdot 7) = (-2 \cdot 5) \cdot (-2 \cdot 7).$$

(a) Is this statement true?

(b) If it is not true, tell which property you think was erroneously applied.

88. Consider the statement

$$4 + (3 \cdot 5) = (4 + 3) \cdot (4 + 5).$$

(a) Is this statement true?

(b) If it is not true, tell which property you think was erroneously applied.

1.8   Simplifying Expressions

OBJECTIVES

1 Simplify expressions.

2 Identify terms and numerical coefficients.

3 Identify like terms.

4 Combine like terms.

5 Simplify expressions from word phrases.

OBJECTIVE 1 Simplify expressions.   In this section we show how to simplify expressions using the properties of addition and multiplication introduced in the previous section.

EXAMPLE 1   Simplifying Expressions

Simplify each expression.

(a) $4x + 8 + 9$

Since $8 + 9 = 17$,

$$4x + 8 + 9 = 4x + 17.$$

(b) $4(3m - 2n)$

Use the distributive property first.

$$4(3m - 2n) = 4(3m) - 4(2n) \qquad \text{Arrows denote distributive property.}$$
$$= (4 \cdot 3)m - (4 \cdot 2)n \qquad \text{Associative property}$$
$$= 12m - 8n$$

(c) $6 + 3(4k + 5) = 6 + 3(4k) + 3(5)$ Distributive property
$$= 6 + (3 \cdot 4)k + 3(5) \qquad \text{Associative property}$$
$$= 6 + 12k + 15$$
$$= 6 + 15 + 12k \qquad \text{Commutative property}$$
$$= 21 + 12k$$

(d) $5 - (2y - 8) = 5 - 1 \cdot (2y - 8)$ $-a = -1 \cdot a$
$$= 5 - 1(2y) - 1(-8) \qquad \text{Distributive property}$$
$$= 5 - 2y + 8$$
$$= 5 + 8 - 2y \qquad \text{Commutative property}$$
$$= 13 - 2y$$

Now Try Exercises 3 and 5.

NOTE   In Examples 1(c) and 1(d), a different use of the commutative property would have resulted in answers of $12k + 21$ and $-2y + 13$. These answers also would be acceptable.

The steps using the commutative and associative properties will not be shown in the rest of the examples, but you should be aware that they are usually involved.

OBJECTIVE 2 Identify terms and numerical coefficients.   A **term** is a number, a variable, or a product or quotient of numbers and variables raised to powers.* Examples of terms include

$$-9x^2, \quad 15y, \quad -3, \quad 8m^2n, \quad \frac{2}{p}, \quad \text{and} \quad k.$$

The **numerical coefficient** of the term $9m$ is 9, the numerical coefficient of $-15x^3y^2$ is -15, the numerical coefficient of x is 1, and the numerical coefficient of 8 is 8. In the expression $\frac{x}{3}$, the numerical coefficient of x is $\frac{1}{3}$ since $\frac{x}{3} = \frac{1x}{3} = \frac{1}{3}x$.

Now Try Exercises 9 and 15.

Term	Numerical Coefficient
$-7y$	-7
$34r^3$	34
$-26x^5yz^4$	-26
$-k$	-1
r	1
$\frac{3x}{8} = \frac{3}{8}x$	$\frac{3}{8}$

NOTE It is important to be able to distinguish between *terms* and *factors*. For example, in the expression $8x^3 + 12x^2$, there are two *terms*, $8x^3$ and $12x^2$. On the other hand, in the one-term expression $(8x^3)(12x^2)$, $8x^3$ and $12x^2$ are *factors*.

Examples of terms and their numerical coefficients are shown in the table in the margin.

OBJECTIVE 3 Identify like terms.   Terms with exactly the same variables that have the same exponents are **like terms.** For example, $9m$ and $4m$ have the same variable and are like terms. Also, $6x^3$ and $-5x^3$ are like terms. The terms $-4y^3$ and $4y^2$ have different exponents and are **unlike terms.**

Here are some additional examples.

$5x$ and $-12x$	$3x^2y$ and $5x^2y$	Like terms
$4xy^2$ and $5xy$	$-7w^3z^3$ and $2xz^3$	Unlike terms

Now Try Exercises 19 and 25.

OBJECTIVE 4 Combine like terms.   Recall the distributive property.

$$x(y + z) = xy + xz.$$

As seen in the previous section, this statement can also be written "backward" as

$$xy + xz = x(y + z).$$

This form of the distributive property may be used to find the sum or difference of like terms. For example,

$$3x + 5x = (3 + 5)x = 8x.$$

This process is called **combining like terms.**

NOTE Remember that *only like terms may be combined.*

*Another name for certain terms, **monomial**, is introduced in Chapter 4.

EXAMPLE 2 Combining Like Terms

Combine like terms in each expression.

(a) $9m + 5m$

Use the distributive property as given above.

$$9m + 5m = (9 + 5)m = 14m$$

(b) $6r + 3r + 2r = (6 + 3 + 2)r = 11r$ Distributive property

(c) $4x + x = 4x + 1x = (4 + 1)x = 5x$ (Note: $x = 1x$.)

(d) $16y^2 - 9y^2 = (16 - 9)y^2 = 7y^2$

(e) $32y + 10y^2$ cannot be combined because $32y$ and $10y^2$ are unlike terms. We cannot use the distributive property here to combine coefficients.

Now Try Exercises 31, 35, and 45.

When an expression involves parentheses, the distributive property is used both "forward" and "backward" to combine like terms, as shown in the following example.

EXAMPLE 3 Simplifying Expressions Involving Like Terms

Combine like terms in each expression.

(a) $14y + 2(6 + 3y) = 14y + 2(6) + 2(3y)$ Distributive property

$\qquad\qquad\qquad\quad = 14y + 12 + 6y$ Multiply.

$\qquad\qquad\qquad\quad = 20y + 12$ Combine like terms.

(b) $9k - 6 - 3(2 - 5k) = 9k - 6 - 3(2) - 3(-5k)$ Distributive property

$\qquad\qquad\qquad\qquad\quad = 9k - 6 - 6 + 15k$ Multiply.

$\qquad\qquad\qquad\qquad\quad = 24k - 12$ Combine like terms.

(c) $-(2 - r) + 10r = -1(2 - r) + 10r$ $-(2 - r) = -1(2 - r)$

$\qquad\qquad\qquad\quad = -1(2) - 1(-r) + 10r$ Distributive property

$\qquad\qquad\qquad\quad = -2 + 1r + 10r$ Multiply.

$\qquad\qquad\qquad\quad = -2 + 11r$ Combine like terms.

(d) $100[.03(x + 4)] = [(100)(.03)](x + 4)$ Associative property

$\qquad\qquad\qquad\quad = 3(x + 4)$ Multiply.

$\qquad\qquad\qquad\quad = 3x + 12$ Distributive property

(e) $5(2a - 6) - 3(4a - 9) = 10a - 30 - 12a + 27$ Distributive property

$\qquad\qquad\qquad\qquad\quad = -2a - 3$ Combine like terms.

Now Try Exercises 49, 51, 53, and 55.

Example 3(e) suggests that the commutative property can be used with subtraction by treating the subtracted terms as the addition of their additive inverses.

NOTE Examples 2 and 3 suggest that like terms may be combined by adding or subtracting the coefficients of the terms and keeping the same variable factors.

OBJECTIVE 5 Simplify expressions from word phrases. Earlier we saw how to translate words, phrases, and statements into expressions and equations. Now we can simplify translated expressions by combining like terms.

▮ **EXAMPLE 4 Translating Words to a Mathematical Expression**

Translate to a mathematical expression, and simplify: The sum of 9, five times a number, four times the number, and six times the number.

The word "sum" indicates that the terms should be added. Use x to represent the number. Then the phrase translates as

$$9 + 5x + 4x + 6x, \qquad \text{Write with symbols.}$$

which simplifies to

$$9 + 15x. \qquad \text{Combine like terms.} \qquad ▮$$

> **Now Try Exercise 67.**

CAUTION In Example 4, we are dealing with an expression to be simplified, *not* an equation to be solved.

1.8 EXERCISES

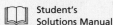

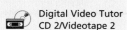
Simplify each expression. See Example 1.

1. $4r + 19 - 8$ **2.** $7t + 18 - 4$ **3.** $5 + 2(x - 3y)$

4. $8 + 3(s - 6t)$ **5.** $-2 - (5 - 3p)$ **6.** $-10 - (7 - 14r)$

7. $6 + (4 - 3x) - 8$ **8.** $-12 + (7 - 8x) + 6$

Give the numerical coefficient of each term.

9. $-12k$ **10.** $-23y$ **11.** $5m^2$ **12.** $-3n^6$ **13.** xw

14. pq **15.** $-x$ **16.** $-t$ **17.** 74 **18.** 98

Identify each group of terms as like *or* unlike.

19. $8r, -13r$ **20.** $-7a, 12a$ **21.** $5z^4, 9z^3$ **22.** $8x^5, -10x^3$

23. $4, 9, -24$ **24.** $7, 17, -83$ **25.** x, y **26.** t, s

Simplify each expression by combining like terms. See Examples 1–3.

27. $9y + 8y$ **28.** $15m + 12m$

29. $-4a - 2a$ **30.** $-3z - 9z$

31. $12b + b$ **32.** $30x + x$

33. $2k + 9 + 5k + 6$ **34.** $2 + 17z + 1 + 2z$

35. $-5y + 3 - 1 + 5 + y - 7$ **36.** $2k - 7 - 5k + 7k - 3 - k$

37. $-2x + 3 + 4x - 17 + 20$ **38.** $r - 6 - 12r - 4 + 6r$

39. $16 - 5m - 4m - 2 + 2m$ **40.** $6 - 3z - 2z - 5 + z - 3z$

41. $-10 + x + 4x - 7 - 4x$ **42.** $-p + 10p - 3p - 4 - 5p$

43. $1 + 7x + 11x - 1 + 5x$ **44.** $-r + 2 - 5r + 3 + 4r$

45. $6y^2 + 11y^2 - 8y^2$ **46.** $-9m^3 + 3m^3 - 7m^3$

47. $2p^2 + 3p^2 - 8p^3 - 6p^3$ **48.** $5y^3 + 6y^3 - 3y^2 - 4y^2$

49. $2(4x + 6) + 3$ **50.** $4(6y - 9) + 7$

51. $100[.05(x + 3)]$ **52.** $100[.06(x + 5)]$

53. $-4(y - 7) - 6$ **54.** $-5(t - 13) - 4$

55. $-5(5y - 9) + 3(3y + 6)$ **56.** $-3(2t + 4) + 8(2t - 4)$

57. $-3(2r - 3) + 2(5r + 3)$ **58.** $-4(5y - 7) + 3(2y - 5)$

59. $8(2k - 1) - (4k - 3)$ **60.** $6(3p - 2) - (5p + 1)$

61. $-2(-3k + 2) - (5k - 6) - 3k - 5$ **62.** $-2(3r - 4) - (6 - r) + 2r - 5$

63. $-4(-3k + 3) - (6k - 4) - 2k + 1$ **64.** $-5(8j + 2) - (5j - 3) - 3j + 17$

65. $-7.5(2y + 4) - 2.9(3y - 6)$ **66.** $8.4(6t - 6) + 2.4(9 - 3t)$

Translate each phrase into a mathematical expression. Use x as the variable. Combine like terms when possible. See Example 4.

67. Five times a number, added to the sum of the number and three

68. Six times a number, added to the sum of the number and six

69. A number multiplied by -7, subtracted from the sum of 13 and six times the number

70. A number multiplied by 5, subtracted from the sum of 14 and eight times the number

71. Six times a number added to -4, subtracted from twice the sum of three times the number and 4 (*Hint: Twice* means two times.)

72. Nine times a number added to 6, subtracted from triple the sum of 12 and 8 times the number (*Hint: Triple* means three times.)

▍ **RELATING CONCEPTS** (EXERCISES 73–80) ▍

For Individual or Group Work

Work Exercises 73–80 in order. *They will help prepare you for graphing later in the text.*

73. Evaluate the expression $x + 2$ for the values of x shown in the table.

x	$x + 2$
0	
1	
2	
3	

74. Based on your results from Exercise 73, complete the following statement: For every increase of 1 unit for x, the value of $x + 2$ increases by _____ unit(s).

75. Repeat Exercise 73 for these expressions:

    (a) $x + 1$ **(b)** $x + 3$ **(c)** $x + 4$ (continued)

76. Based on your results from Exercises 73 and 75, make a conjecture (an educated guess) about what happens to the value of an expression of the form $x + b$ for any value of b, as x increases by 1 unit.

77. Repeat Exercise 73 for these expressions:

    (a) $2x + 2$ **(b)** $3x + 2$ **(c)** $4x + 2$

78. Based on your results from Exercise 77, complete the following statement: For every increase of 1 unit for x, the value of $mx + 2$ increases by _____ units.

79. Repeat Exercise 73 and compare your results to those in Exercise 77 for these expressions:

    (a) $2x + 7$ **(b)** $3x + 5$ **(c)** $4x + 1$

80. Based on your results from Exercises 73–79, complete the following statement: For every increase of 1 unit for x, the value of $mx + b$ increases by _____ units.

81. There is an old saying, "You can't add apples and oranges." Explain how this saying can be applied to the goal of Objective 4 in this section.

82. Explain how the distributive property is used in combining $6t + 5t$ to get $11t$.

83. Write the expression $9x - (x + 2)$ using words, as in Exercises 67–72.

84. Write the expression $2(3x + 5) - 2(x + 4)$ using words, as in Exercises 67–72.

Chapter **1** **Group Activity**

Comparing Floor Plans of Houses

OBJECTIVE   Use arithmetic skills to make comparisons.

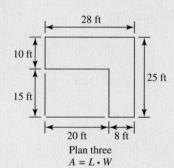

This activity explores perimeters and areas of different shaped homes. Floor plans for three different homes are given below.

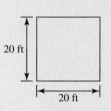

Plan one
$A = \pi r^2$
$C = 2\pi r$

Plan two
$A = s^2$

Plan three
$A = L \cdot W$

A. As a group, look at the dimensions of the given floor plans. Considering only the dimensions, which plan do you think has the greatest area?

B. Now have each student in your group pick one floor plan. For each plan, find the following. (Round all answers to the nearest whole number. In Plan one, let $\pi = 3.14$.)

1. The area of the plan

2. The perimeter or circumference (the distance around the outside) of the plan

C. Share your findings with the group and answer the following questions.

1. What did you determine about the areas of the three floor plans?

2. Which plan has the smallest perimeter? Which has the largest perimeter?

3. Why do you think houses with round floor plans might be more energy efficient?

4. What advantages do you think houses with square or rectangular floor plans have? Why do you think floor plans with these shapes are most common for homes today?

CHAPTER **1** SUMMARY

KEY TERMS

1.1 natural numbers
whole numbers
fractions
numerator
denominator
proper fraction
improper fraction
mixed number
factor
product
factored
prime number
composite number
greatest common
    factor
lowest terms
basic principle of
    fractions

reciprocal
quotient
sum
least common
    denominator (LCD)
difference
1.2 exponent (power)
base
exponential expression
grouping symbols
order of operations
inequality
1.3 variable
algebraic expression
equation
solution
set
element

1.4 number line
graph
coordinate
negative numbers
positive numbers
signed numbers
integers
rational numbers
set-builder notation
irrational numbers
real numbers
additive inverse
    (opposite)
absolute value
1.6 multiplicative inverse
    (reciprocal)
1.7 commutative property
associative property

identity property
identity element for
    addition (additive
    identity)
identity element for
    multiplication
    (multiplicative
    identity)
inverse property
distributive property
1.8 term
numerical coefficient
like terms
unlike terms
combining like terms

NEW SYMBOLS

a^n n factors of a

[ ] square brackets

$=$ is equal to

$\neq$ is not equal to

$<$ is less than

$>$ is greater than

$\leq$ is less than or equal
   to

$\geq$ is greater than or
   equal to

{ } set braces

$\{x \mid x$ **has a certain**
   property$\}$
      set-builder notation

$-x$ the additive inverse, or
   opposite, of x

$|x|$ absolute value of x

$\dfrac{1}{x}$ the multiplicative
   inverse, or reciprocal,
   of the nonzero
   number x

$a(b), (a)b, (a)(b), a \cdot b,$
   or ab a times b

$\dfrac{a}{b}$ **or** a/b a divided by b

TEST YOUR WORD POWER

See how well you have learned the vocabulary in this chapter. Answers, with examples, follow the Quick Review.

1. A **factor** is
   A. the answer in an addition problem
   B. the answer in a multiplication
      problem
   C. one of two or more numbers that
      are added to get another number
   D. one of two or more numbers that
      are multiplied to get another
      number.

2. A number is **prime** if
   A. it cannot be factored
   B. it has just one factor

   C. it has only itself and 1 as
      factors
   D. it has at least two different
      factors.

3. An **exponent** is
   A. a symbol that tells how many
      numbers are being multiplied
   B. a number raised to a power
   C. a number that tells how many
      times a factor is repeated
   D. one of two or more numbers that
      are multiplied.

4. A **variable** is
   A. a symbol used to represent an
      unknown number
   B. a value that makes an equation
      true
   C. a solution of an equation
   D. the answer in a division
      problem.

5. An **integer** is
   A. a positive or negative number
   B. a natural number, its opposite, or
      zero

C. any number that can be graphed on a number line
D. the quotient of two numbers.

6. A **coordinate** is
   A. the number that corresponds to a point on a number line
   B. the graph of a number
   C. any point on a number line
   D. the distance from 0 on a number line.

7. The **absolute value** of a number is
   A. the graph of the number
   B. the reciprocal of the number
   C. the opposite of the number
   D. the distance between 0 and the number on a number line.

8. A **term** is
   A. a numerical factor
   B. a number or a product or quotient of numbers and variables raised to powers

C. one of several variables with the same exponents
D. a sum of numbers and variables raised to powers.

9. A **numerical coefficient** is
   A. the numerical factor in a term
   B. the number of terms in an expression
   C. a variable raised to a power
   D. the variable factor in a term.

QUICK REVIEW

CONCEPTS	EXAMPLES

1.1 FRACTIONS

Operations with Fractions

Addition/Subtraction:
1. To add/subtract fractions with the same denominator, add/subtract the numerators and keep the same denominator.
2. To add/subtract fractions with different denominators, find the LCD and write each fraction with this LCD. Then follow the procedure above.

Multiplication: Multiply numerators and multiply denominators.

Division: Multiply the first fraction by the reciprocal of the second fraction.

Perform each operation.

$$\frac{2}{5} + \frac{7}{5} = \frac{2+7}{5} = \frac{9}{5}$$

$$\frac{2}{3} - \frac{1}{2} = \frac{4}{6} - \frac{3}{6} \qquad \text{6 is the LCD.}$$

$$= \frac{4-3}{6} = \frac{1}{6}$$

$$\frac{4}{3} \cdot \frac{5}{6} = \frac{20}{18} = \frac{10}{9}$$

$$\frac{6}{5} \div \frac{1}{4} = \frac{6}{5} \cdot \frac{4}{1} = \frac{24}{5}$$

1.2 EXPONENTS, ORDER OF OPERATIONS, AND INEQUALITY

Order of Operations

Simplify within parentheses or above and below fraction bars first, in the following order.

Step 1 Apply all exponents.

Step 2 Do any multiplications or divisions from left to right.

Step 3 Do any additions or subtractions from left to right.

If no grouping symbols are present, start with Step 1.

Simplify $36 - 4(2^2 + 3)$.

$$36 - 4(2^2 + 3) = 36 - 4(4 + 3)$$
$$= 36 - 4(7)$$
$$= 36 - 28$$
$$= 8$$

(continued)

CONCEPTS	EXAMPLES

1.3 VARIABLES, EXPRESSIONS, AND EQUATIONS

Evaluate an expression with a variable by substituting a given number for the variable.

Evaluate $2x + y^2$ if $x = 3$ and $y = -4$.

$$2x + y^2 = 2(3) + (-4)^2$$
$$= 6 + 16$$
$$= 22$$

Values of a variable that make an equation true are solutions of the equation.

Is 2 a solution of $5x + 3 = 18$?

$$5(2) + 3 = 18 \qquad ?$$
$$13 = 18 \qquad \text{False}$$

2 is not a solution.

1.4 REAL NUMBERS AND THE NUMBER LINE

The Ordering of Real Numbers
a is less than b if a is to the left of b on the number line.

Graph -2, 0, and 3.

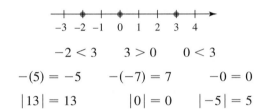

$$-2 < 3 \qquad 3 > 0 \qquad 0 < 3$$

The additive inverse of x is $-x$.

The absolute value of x, $|x|$, is the distance between x and 0 on the number line.

$$-(5) = -5 \qquad -(-7) = 7 \qquad -0 = 0$$
$$|13| = 13 \qquad |0| = 0 \qquad |-5| = 5$$

1.5 ADDING AND SUBTRACTING REAL NUMBERS

Rules for Adding Signed Numbers
To add two numbers with the same sign, add their absolute values. The sum has that same sign.

To add two numbers with different signs, subtract their absolute values. The sum has the sign of the number with larger absolute value.

Add.

$$9 + 4 = 13$$
$$-8 + (-5) = -13$$
$$7 + (-12) = -5$$
$$-5 + 13 = 8$$

Definition of Subtraction

$$x - y = x + (-y)$$

Subtract.

$$5 - (-2) = 5 + 2 = 7$$

Rules for Subtracting Signed Numbers
1. Change the subtraction symbol to the addition symbol.
2. Change the sign of the number being subtracted.
3. Add, using the rules for addition.

$$-3 - 4 = -3 + (-4) = -7$$

$$-2 - (-6) = -2 + 6 = 4$$

$$13 - (-8) = 13 + 8 = 21$$

CONCEPTS	EXAMPLES

1.6   MULTIPLYING AND DIVIDING REAL NUMBERS

Rules for Multiplying and Dividing Signed Numbers

The product (or quotient) of two numbers having the *same sign* is *positive*; the product (or quotient) of two numbers having *different signs* is *negative*.

Multiply or divide.

$$6 \cdot 5 = 30 \qquad -7(-8) = 56 \qquad \frac{20}{4} = 5$$

$$\frac{-24}{-6} = 4 \qquad -6(5) = -30 \qquad 6(-5) = -30$$

$$\frac{-18}{9} = -2 \qquad \frac{49}{-7} = -7$$

Definition of Division

$$\frac{x}{y} = x \cdot \frac{1}{y}, \quad y \neq 0$$

Division by 0 is undefined.

0 divided by a nonzero number equals 0.

$$\frac{10}{2} = 10 \cdot \frac{1}{2} = 5$$

$$\frac{5}{0} \text{ is undefined.}$$

$$\frac{0}{5} = 0$$

1.7   PROPERTIES OF REAL NUMBERS

Commutative Properties

$$a + b = b + a$$
$$ab = ba$$

$$7 + (-1) = -1 + 7$$
$$5(-3) = (-3)5$$

Associative Properties

$$(a + b) + c = a + (b + c)$$
$$(ab)c = a(bc)$$

$$(3 + 4) + 8 = 3 + (4 + 8)$$
$$[-2(6)]4 = -2[(6)4]$$

Identity Properties

$$a + 0 = a \qquad 0 + a = a$$
$$a \cdot 1 = a \qquad 1 \cdot a = a$$

$$-7 + 0 = -7 \qquad 0 + (-7) = -7$$
$$9 \cdot 1 = 9 \qquad 1 \cdot 9 = 9$$

Inverse Properties

$$a + (-a) = 0 \qquad -a + a = 0$$
$$a \cdot \frac{1}{a} = 1 \qquad \frac{1}{a} \cdot a = 1 \quad (a \neq 0)$$

$$7 + (-7) = 0 \qquad -7 + 7 = 0$$
$$-2\left(-\frac{1}{2}\right) = 1 \qquad -\frac{1}{2}(-2) = 1$$

Distributive Properties

$$a(b + c) = ab + ac$$
$$(b + c)a = ba + ca$$
$$a(b - c) = ab - ac$$

$$5(4 + 2) = 5(4) + 5(2)$$
$$(4 + 2)5 = 4(5) + 2(5)$$
$$9(5 - 4) = 9(5) - 9(4)$$

(continued)

CONCEPTS	EXAMPLES

1.8　SIMPLIFYING EXPRESSIONS

Only like terms may be combined.

$$-3y^2 + 6y^2 + 14y^2 = 17y^2$$

$$4(3 + 2x) - 6(5 - x)$$
$$= 12 + 8x - 30 + 6x \qquad \text{Distributive property}$$
$$= 14x - 18$$

Answers to Test Your Word Power

1. D; *Example:* Since $2 \times 5 = 10$, the numbers 2 and 5 are factors of 10; other factors of 10 are $-10, -5, -2, -1, 1,$ and 10.
2. C; *Examples:* 2, 3, 11, 41, 53　**3.** C; *Example:* In 2^3, the number 3 is the exponent (or power), so 2 is a factor three times;
$2^3 = 2 \cdot 2 \cdot 2 = 8$.　**4.** A; *Examples:* a, b, c　**5.** B; *Examples:* $-9, 0, 6$　**6.** A; *Example:* The point graphed 3 units to the right of 0
on a number line has coordinate 3.　**7.** D; *Examples:* $|2| = 2$ and $|-2| = 2$　**8.** B; *Examples:* $6, \frac{x}{2}, -4ab^2$　**9.** A; *Examples:* The
term 3 has numerical coefficient 3, $8z$ has numerical coefficient 8, and $-10x^4y$ has numerical coefficient -10.

CHAPTER 1 REVIEW EXERCISES

[1.1]　*Perform each indicated operation.* *

1. $\dfrac{8}{5} \div \dfrac{32}{15}$

2. $\dfrac{3}{8} + 3\dfrac{1}{2} - \dfrac{3}{16}$

3. The circle graph illustrates how 400 people responded to a survey that asked "Do you think that the CEOs of corporations guilty of accounting fraud should go to jail?" What fractional part of the group did not have an opinion?

4. Based on the graph in Exercise 3, how many people responded "yes"?

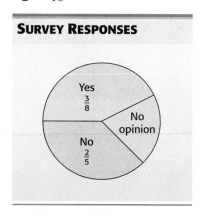

SURVEY RESPONSES

[1.2]　*Find the value of each exponential expression.*

5. 5^4　　**6.** $\left(\dfrac{3}{5}\right)^3$　　**7.** $(.02)^5$　　**8.** $(.001)^3$

---

*For help with the Review Exercises in this text, refer to the appropriate section given in brackets.

Find the value of each expression.

9. $8 \cdot 5 - 13$

10. $7[3 + 6(3^2)]$

11. $\dfrac{9(4^2 - 3)}{4 \cdot 5 - 17}$

12. $\dfrac{6(5 - 4) + 2(4 - 2)}{3^2 - (4 + 3)}$

Tell whether each statement is true *or* false.

13. $12 \cdot 3 - 6 \cdot 6 \le 0$

14. $3[5(2) - 3] > 20$

15. $9 \le 4^2 - 8$

Write each word statement in symbols.

16. Thirteen is less than seventeen.

17. Five plus two is not equal to ten.

18. The bar graph shows the number of worldwide airline fatalities for the years 1990 to 1999.

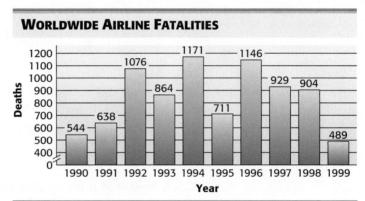

WORLDWIDE AIRLINE FATALITIES

Source: International Civil Aviation Organization.

(a) In which years were *fewer than* 700 people killed?
(b) In which years were *at least* 905 people killed?
(c) How many people were killed in the 5 yr having the largest numbers of deaths?

[1.3]   *Find the numerical value of each expression if $x = 6$ and $y = 3$.*

19. $2x + 6y$

20. $4(3x - y)$

21. $\dfrac{x}{3} + 4y$

22. $\dfrac{x^2 + 3}{3y - x}$

Change each word phrase to an algebraic expression. Use x as the variable to represent the number.

23. Six added to a number

24. A number subtracted from eight

25. Nine subtracted from six times a number

26. Three-fifths of a number added to 12

Decide whether the given number is a solution of the given equation.

27. $5x + 3(x + 2) = 22;\quad 2$

28. $\dfrac{t + 5}{3t} = 1;\quad 6$

Change each word statement to an equation. Use x as the variable. Then find the solution from the set $\{0, 2, 4, 6, 8, 10\}$.

29. Six less than twice a number is 10.

30. The product of a number and 4 is 8.

[1.4] *Graph each group of numbers on a number line.*

31. $-4, -\dfrac{1}{2}, 0, 2.5, 5$

32. $-2, |-3|, -3, |-1|$

Classify each number, using the sets natural numbers, whole numbers, integers, rational numbers, irrational numbers, real numbers.

33. $\dfrac{4}{3}$

34. $\sqrt{6}$

Select the smaller number in each pair.

35. $-10, 5$ **36.** $-8, -9$ **37.** $-\dfrac{2}{3}, -\dfrac{3}{4}$ **38.** $0, -|23|$

Decide whether each statement is true *or* false.

39. $12 > -13$ **40.** $0 > -5$ **41.** $-9 < -7$ **42.** $-13 \geq -13$

For each number, **(a)** *find the opposite of the number and* **(b)** *find the absolute value of the number.*

43. -9 **44.** 0 **45.** 6 **46.** $-\dfrac{5}{7}$

Simplify each number by removing absolute value symbols.

47. $|-12|$ **48.** $-|3|$ **49.** $-|-19|$ **50.** $-|9 - 2|$

[1.5] *Perform each indicated operation.*

51. $-10 + 4$ **52.** $14 + (-18)$ **53.** $-8 + (-9)$

54. $\dfrac{4}{9} + \left(-\dfrac{5}{4}\right)$ **55.** $-13.5 + (-8.3)$ **56.** $(-10 + 7) + (-11)$

57. $[-6 + (-8) + 8] + [9 + (-13)]$ **58.** $(-4 + 7) + (-11 + 3) + (-15 + 1)$

59. $-7 - 4$ **60.** $-12 - (-11)$

61. $5 - (-2)$ **62.** $-\dfrac{3}{7} - \dfrac{4}{5}$

63. $2.56 - (-7.75)$ **64.** $(-10 - 4) - (-2)$

65. $(-3 + 4) - (-1)$ **66.** $-(-5 + 6) - 2$

Write a numerical expression for each phrase and simplify the expression.

67. 19 added to the sum of -31 and 12   **68.** 13 more than the sum of -4 and -8

69. The difference between -4 and -6 **70.** Five less than the sum of 4 and -8

Find the solution of each equation from the set $\{-3, -2, -1, 0, 1, 2, 3\}$ *by guessing or by trial and error.*

71. $x + (-2) = -4$ **72.** $12 + x = 11$

Solve each problem.

73. Like many people, Otis Taylor neglects to keep up his checkbook balance. When he finally balanced his account, he found the balance was −$23.75, so he deposited $50.00. What is his new balance?

74. The low temperature in Yellowknife, in the Canadian Northwest Territories, one January day was −26°F. It rose 16° that day. What was the high temperature?

75. Eric owed his brother $28. He repaid $13 but then borrowed another $14. What positive or negative amount represents his present financial status?

76. If the temperature drops 7° below its previous level of −3°, what is the new temperature?

77. A football team gained 3 yd on the first play from scrimmage, lost 12 yd on the second play, and then gained 13 yd on the third play. How many yards did the team gain or lose altogether?

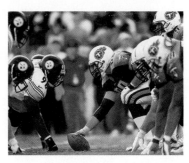

78. On Wednesday, May 29, 2002, the Dow Jones Industrial Average closed at 9923.04, down 58.54 from the previous day. What was the closing value the previous day? (*Source: Times Picayune.*)

[1.6] *Perform each indicated operation.*

79. $(-12)(-3)$　　**80.** $15(-7)$　　**81.** $-\dfrac{4}{3}\left(-\dfrac{3}{8}\right)$　　**82.** $(-4.8)(-2.1)$

83. $5(8 - 12)$　　**84.** $(5 - 7)(8 - 3)$　　**85.** $2(-6) - (-4)(-3)$

86. $3(-10) - 5$　　**87.** $\dfrac{-36}{-9}$　　**88.** $\dfrac{220}{-11}$

89. $-\dfrac{1}{2} \div \dfrac{2}{3}$　　**90.** $-33.9 \div (-3)$　　**91.** $\dfrac{-5(3) - 1}{8 - 4(-2)}$

92. $\dfrac{5(-2) - 3(4)}{-2[3 - (-2)] - 1}$　　**93.** $\dfrac{10^2 - 5^2}{8^2 + 3^2 - (-2)}$　　**94.** $\dfrac{(.6)^2 + (.8)^2}{(-1.2)^2 - (-.56)}$

Evaluate each expression if $x = -5$, $y = 4$, and $z = -3$.

95. $6x - 4z$　　**96.** $5x + y - z$　　**97.** $5x^2$　　**98.** $z^2(3x - 8y)$

Write a numerical expression for each phrase and simplify the expression.

99. Nine less than the product of −4 and 5

100. Five-sixths of the sum of 12 and −6

101. The quotient of 12 and the sum of 8 and −4

102. The product of −20 and 12, divided by the difference between 15 and −15

Write each sentence in symbols, using x as the variable, and find the solution by guessing or by trial and error. All solutions come from the list of integers between −12 and 12.

103. 8 times a number is −24.

104. The quotient of a number and 3 is −2.

Find the average of each group of numbers.

105. 26, 38, 40, 20, 4, 14, 96, 18

106. $-12, 28, -36, 0, 12, -10$

[1.7] *Decide whether each statement is an example of the* commutative, associative, identity, inverse, *or* distributive *property.*

107. $6 + 0 = 6$

108. $5 \cdot 1 = 5$

109. $-\frac{2}{3}\left(-\frac{3}{2}\right) = 1$

110. $17 + (-17) = 0$

111. $5 + (-9 + 2) = [5 + (-9)] + 2$

112. $w(xy) = (wx)y$

113. $3x + 3y = 3(x + y)$

114. $(1 + 2) + 3 = 3 + (1 + 2)$

Use the distributive property to rewrite each expression. Simplify if possible.

115. $7y + 14$

116. $-12(4 - t)$

117. $3(2s) + 3(5y)$

118. $-(-4r + 5s)$

119. Evaluate $25 - (5 - 2)$ and $(25 - 5) - 2$. Use this example to explain why subtraction is not associative.

120. Evaluate $180 \div (15 \div 5)$ and $(180 \div 15) \div 5$. Use this example to explain why division is not associative.

[1.8] *Combine like terms whenever possible.*

121. $2m + 9m$

122. $15p^2 - 7p^2 + 8p^2$

123. $5p^2 - 4p + 6p + 11p^2$

124. $-2(3k - 5) + 2(k + 1)$

125. $7(2m + 3) - 2(8m - 4)$

126. $-(2k + 8) - (3k - 7)$

In Exercises 127–130, choose the letter of the correct response.

127. Which one of the following is true for all real numbers x?

A. $6 + 2x = 8x$ **B.** $6 - 2x = 4x$
C. $6x - 2x = 4x$ **D.** $3 + 8(4x - 6) = 11(4x - 6)$

128. Which one of the following is an example of a pair of like terms?

A. $6t, 6w$ **B.** $-8x^2y, 9xy^2$ **C.** $5ry, 6yr$ **D.** $-5x^2, 2x^3$

129. Which one of the following is an example of a term with numerical coefficient 5?

A. $5x^3y^7$ **B.** x^5 **C.** $\frac{x}{5}$ **D.** 5^2xy^3

130. Which one of the following is a correct translation for "six times a number, subtracted from the product of eleven and the number" (if x represents the number)?

A. $6x - 11x$ **B.** $11x - 6x$ **C.** $(11 + x) - 6x$ **D.** $6x - (11 + x)$

MIXED REVIEW EXERCISES*

Perform each indicated operation.

131. $[(-2) + 7 - (-5)] + [-4 - (-10)]$

132. $\left(-\frac{5}{6}\right)^2$

*The order of exercises in this final group does not correspond to the order in which topics occur in the chapter. This random ordering should help you prepare for the chapter test in yet another way.

133. $\dfrac{6(-4) + 2(-12)}{5(-3) + (-3)}$

134. $\dfrac{3}{8} - \dfrac{5}{12}$

135. $\dfrac{8^2 + 6^2}{7^2 + 1^2}$

136. $-16(-3.5) - 7.2(-3)$

137. $2\dfrac{5}{6} - 4\dfrac{1}{3}$

138. $-8 + [(-4 + 17) - (-3 - 3)]$

139. $-\dfrac{12}{5} \div \dfrac{9}{7}$

140. $(-8 - 3) - 5(2 - 9)$

141. $5x^2 - 12y^2 + 3x^2 - 9y^2$

142. $-4(2t + 1) - 8(-3t + 4)$

143. Write a sentence or two explaining the special considerations involving 0 when dividing.

144. "Two negatives give a positive" is often heard from students. Is this correct? Use more precise language in explaining what this means.

145. Use x as the variable and write an expression for "the product of 5, and the sum of a number and 7." Then use the distributive property to rewrite the expression.

146. The highest temperature ever recorded in Albany, New York, was 99°F, while the lowest was 112° less than the highest. What was the lowest temperature ever recorded in Albany? (*Source: World Almanac and Book of Facts*, 2000.)

CHAPTER 1 TEST

1. Write $\dfrac{63}{99}$ in lowest terms.

2. Add: $\dfrac{5}{8} + \dfrac{11}{12} + \dfrac{7}{15}$.

3. Divide: $\dfrac{19}{15} \div \dfrac{6}{5}$.

4. The circle graph indicates the market share of different means of intercity transportation in a recent year, based on 1230 million passengers carried.

   (a) How many of these passengers used air travel?

   (b) How many of these passengers did not use the bus?

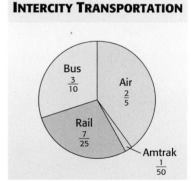

INTERCITY TRANSPORTATION

Bus $\frac{3}{10}$ Air $\frac{2}{5}$ Rail $\frac{7}{25}$ Amtrak $\frac{1}{50}$

Source: Eno Transportation Foundation, Inc.

5. Decide whether $4[-20 + 7(-2)] \le 135$ is true or false.

6. Graph the group of numbers $-1, -3, |-4|, |-1|$ on a number line.

7. To which of the following sets does $-\dfrac{2}{3}$ belong: natural numbers, whole numbers, integers, rational numbers, irrational numbers, real numbers?

8. Explain how a number line can be used to show that -8 is less than -1.

9. Write in symbols: The quotient of -6 and the sum of 2 and -8. Simplify the expression.

Perform each indicated operation.

10. $-2 - (5 - 17) + (-6)$

11. $-5\dfrac{1}{2} + 2\dfrac{2}{3}$

12. $-6 - [-7 + (2 - 3)]$

13. $4^2 + (-8) - (2^3 - 6)$

14. $(-5)(-12) + 4(-4) + (-8)^2$

15. $\dfrac{-7 - (-6 + 2)}{-5 - (-4)}$

16. $\dfrac{30(-1 - 2)}{-9[3 - (-2)] - 12(-2)}$

Find the solution for each equation from the set $\{-6, -4, -2, 0, 2, 4, 6\}$ by guessing or by trial and error.

17. $-x + 3 = -3$

18. $-3x = -12$

Evaluate each expression, given $x = -2$ and $y = 4$.

19. $3x - 4y^2$

20. $\dfrac{5x + 7y}{3(x + y)}$

Solve each problem.

21. The highest elevation in Argentina is Mt. Aconcagua, which is 6960 m above sea level. The lowest point in Argentina is the Valdes Peninsula, 40 m below sea level. Find the difference between the highest and lowest elevations.

22. For a certain system of rating major league baseball relief pitchers, 3 points are awarded for a save, 3 points are awarded for a win, 2 points are subtracted for a loss, and 2 points are subtracted for a blown save. If Mariano Rivera of the New York Yankees has 4 saves, 3 wins, 2 losses, and 1 blown save, how many points does he have?

23. The bar graph shows the federal budget outlays for national defense for the years 1990 through 1998. Use a signed number to represent the change in outlays for each time period. For example, the change from 1995 to 1996 was $253.3 − $259.6 = −$6.3 billion.

(a) 1991–1992
(b) 1993–1994
(c) 1996–1997
(d) 1997–1998

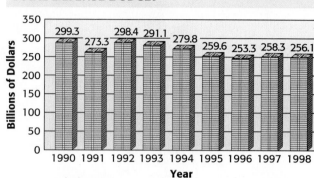

TOTAL DEFENSE BUDGET

Source: U.S. Office of Management and Budget.

Match each property in Column I with the example of it in Column II.

I	II
24. Commutative property	**A.** $3x + 0 = 3x$
25. Associative property	**B.** $(5 + 2) + 8 = 8 + (5 + 2)$
26. Inverse property	**C.** $-3(x + y) = -3x + (-3y)$
27. Identity property	**D.** $-5 + (3 + 2) = (-5 + 3) + 2$
28. Distributive property	**E.** $-\dfrac{5}{3}\left(-\dfrac{3}{5}\right) = 1$

29. What property is used to show that $3(x + 1) = 3x + 3$?

30. Consider the expression $-6[5 + (-2)]$.
   (a) Evaluate it by first working within the brackets.
   (b) Evaluate it by using the distributive property.
   (c) Why must the answers in items (a) and (b) be the same?

Simplify by combining like terms.

31. $8x + 4x - 6x + x + 14x$ **32.** $5(2x - 1) - (x - 12) + 2(3x - 5)$

Linear Equations and Inequalities in One Variable; Applications

From a gathering of 13 nations in Athens in 1896, the Olympic Games have truly become a worldwide event. The XIX Olympic Winter Games in Salt Lake City in 2002 attracted athletes from 82 countries and entertained 3.5 billion viewers around the world.

One ceremonial aspect of the games is the flying of the Olympic flag with its five interlocking rings of different colors on a white background. First introduced at the 1920 Games in Antwerp, Belgium, the five rings on the flag symbolize unity among the nations of Africa, the Americas, Asia, Australia, and Europe. (*Source: Microsoft Encarta Encyclopedia 2002;* www.olympic-usa.org) Throughout this chapter we use linear equations to solve applications about the Olympics.

2.1 The Addition Property of Equality

Recall from Section 1.3 that an *equation* is a statement that two algebraic expressions are equal. The simplest type of equation is a *linear equation*.

OBJECTIVE 1 Identify linear equations.

Linear Equation in One Variable

A **linear equation in one variable** can be written in the form

$$Ax + B = C$$

for real numbers A, B, and C, with $A \neq 0$.

For example,

$$4x + 9 = 0, \qquad 2x - 3 = 5, \qquad \text{and} \qquad x = 7$$

are linear equations in one variable (x). The final two can be written in the specified form using properties developed in this chapter. However,

$$x^2 + 2x = 5, \qquad \frac{1}{x} = 6, \qquad \text{and} \qquad |2x + 6| = 0$$

are *not* linear equations.

As we saw in Section 1.3, a *solution* of an equation is a number that makes the equation true when it replaces the variable. An equation is solved by finding its **solution set,** the set of all solutions. Equations that have exactly the same solution sets are **equivalent equations.** Linear equations are solved by using a series of steps to produce a simpler equivalent equation of the form

$$x = \textbf{a number} \qquad \text{or} \qquad \textbf{a number} = x.$$

OBJECTIVE 2 Use the addition property of equality. In the equation $x - 5 = 2$, both $x - 5$ and 2 represent the same number because this is the meaning of the equals sign. To solve the equation, we change the left side from $x - 5$ to just x. We do this by adding 5 to $x - 5$. We use 5 because 5 is the opposite (additive inverse) of -5, and $-5 + 5 = 0$. To keep the two sides equal, we must also add 5 to the right side.

$$
\begin{aligned}
x - 5 &= 2 && \text{Given equation} \\
x - 5 + 5 &= 2 + 5 && \text{Add 5 to each side.} \\
x + 0 &= 7 && \text{Additive inverse property} \\
x &= 7 && \text{Additive identity property}
\end{aligned}
$$

The solution of the given equation is 7. We check by replacing x with 7 in the original equation.

Check:
$$
\begin{aligned}
x - 5 &= 2 && \text{Original equation} \\
7 - 5 &= 2 \quad ? && \text{Let } x = 7. \\
2 &= 2 && \text{True}
\end{aligned}
$$

Since the final equation is true, 7 checks as the solution and {7} is the solution set.

To solve the equation, we added the same number to each side. The **addition property of equality** justifies this step.

> ### Addition Property of Equality
>
> If A, B, and C are real numbers, then the equations
> $$A = B \quad \text{and} \quad A + C = B + C$$
> are equivalent equations.
>
>   That is, we can add the same number to each side of an equation without changing the solution.

  In the addition property, C represents a real number. This means that any quantity that represents a real number can be added to each side of an equation to change it to an equivalent equation.

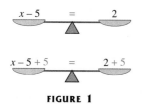

$$x - 5 = 2$$

$$x - 5 + 5 = 2 + 5$$

FIGURE 1

NOTE   Equations can be thought of in terms of a balance. Thus, adding the same quantity to each side does not affect the balance. See Figure 1.

EXAMPLE 1   Using the Addition Property of Equality

Solve $x - 16 = 7$.

  Our goal is to get an equivalent equation of the form $x = $ a number. To do this, we use the addition property of equality and add 16 to each side.

$$x - 16 = 7$$
$$x - 16 + 16 = 7 + 16 \qquad \text{Add 16 to each side.}$$
$$x = 23 \qquad \text{Combine like terms.}$$

Note that we combined the steps that change $x - 16 + 16$ to $x + 0$ and $x + 0$ to x. We will combine these steps from now on. We check by substituting 23 for x in the *original* equation.

Check:
$$x - 16 = 7 \qquad \text{Original equation}$$
$$23 - 16 = 7 \quad ? \qquad \text{Let } x = 23.$$
$$7 = 7 \qquad \text{True}$$

Since a true statement results, $\{23\}$ is the solution set.

Now Try Exercise 7.

EXAMPLE 2   Using the Addition Property of Equality

Solve $x - 2.9 = -6.4$.

  To get x alone on the left side, we must eliminate the -2.9. We use the addition property of equality and add 2.9 to each side.

$$x - 2.9 = -6.4$$
$$x - 2.9 + 2.9 = -6.4 + 2.9 \qquad \text{Add 2.9 to each side.}$$
$$x = -3.5$$

Check: $x - 2.9 = -6.4$ Original equation

$-3.5 - 2.9 = -6.4$?   Let $x = -3.5$.

$-6.4 = -6.4$ True

Since a true statement results, the solution set is $\{-3.5\}$.

Now Try Exercise 21.

The addition property of equality says that the same number may be *added* to each side of an equation. In Chapter 1, subtraction was defined as addition of the opposite. Thus, we can also use the following rule when solving an equation.

> The same number may be subtracted from each side of an equation without changing the solution.

▋ EXAMPLE 3   Using the Addition Property of Equality

Solve $-7 = x + 22$.

Here the variable x is on the right side of the equation. To get x alone on the right, we must eliminate the 22 by subtracting 22 from each side.

$$-7 = x + 22$$
$$-7 - 22 = x + 22 - 22 \qquad \text{Subtract 22 from each side.}$$
$$-29 = x \quad \text{or} \quad x = -29$$

Check: $-7 = x + 22$ Original equation

$-7 = -29 + 22$?   Let $x = -29$.

$-7 = -7$ True

Thus, the solution set is $\{-29\}$.

Now Try Exercise 17.

▋ EXAMPLE 4   Subtracting a Variable Expression

Solve $\frac{3}{5}k + 17 = \frac{8}{5}k$.

To get all terms with variables on the same side of the equation, subtract $\frac{3}{5}k$ from each side.

$$\frac{3}{5}k + 17 = \frac{8}{5}k$$
$$\frac{3}{5}k + 17 - \frac{3}{5}k = \frac{8}{5}k - \frac{3}{5}k \qquad \text{Subtract } \tfrac{3}{5}k \text{ from each side.}$$
$$17 = 1k \qquad \text{Combine like terms; } \tfrac{5}{5}k = 1k.$$
$$17 = k \qquad \text{Multiplicative identity property}$$

(From now on we will skip the step that changes $1k$ to k.) Check the solution by replacing k with 17 in the original equation. The solution set is $\{17\}$.

Now Try Exercise 25.

What happens if we solve the equation in Example 4 by first subtracting $\frac{8}{5}k$ from each side?

$$\frac{3}{5}k + 17 = \frac{8}{5}k$$

$$\frac{3}{5}k + 17 - \frac{8}{5}k = \frac{8}{5}k - \frac{8}{5}k \qquad \text{Subtract } \tfrac{8}{5}k \text{ from each side.}$$

$$17 - k = 0 \qquad \text{Combine like terms; } -\tfrac{5}{5}k = -1k = -k.$$

$$17 - k - 17 = 0 - 17 \qquad \text{Subtract 17 from each side.}$$

$$-k = -17 \qquad \text{Combine like terms; additive inverse}$$

This result gives the value of $-k$, but not of k itself. However, it does say that the additive inverse of k is -17, which means that k must be 17, the same result we obtained in Example 4.

$$-k = -17$$
$$k = 17$$

(This result can also be justified using the multiplication property of equality, covered in Section 2.2.)

OBJECTIVE 3 Simplify equations, and then use the addition property of equality. Sometimes an equation must be simplified as a first step in its solution.

___

◼ EXAMPLE 5 Simplifying an Equation before Solving

Solve $3k - 12 + k + 2 = 5 + 3k + 2$.

Begin by combining like terms on each side of the equation to get

$$4k - 10 = 7 + 3k.$$

Next, get all terms that contain variables on the same side of the equation and all terms without variables on the other side. One way to start is to subtract $3k$ from each side.

$$4k - 10 - 3k = 7 + 3k - 3k \qquad \text{Subtract } 3k \text{ from each side.}$$

$$k - 10 = 7 \qquad \text{Combine like terms.}$$

$$k - 10 + 10 = 7 + 10 \qquad \text{Add 10 to each side.}$$

$$k = 17 \qquad \text{Combine like terms.}$$

Check: Substitute 17 for k in the original equation.

$$3k - 12 + k + 2 = 5 + 3k + 2 \qquad \quad \text{Original equation}$$

$$3(17) - 12 + 17 + 2 = 5 + 3(17) + 2 \quad ? \quad \text{Let } k = 17.$$

$$51 - 12 + 17 + 2 = 5 + 51 + 2 \quad ? \quad \text{Multiply.}$$

$$58 = 58 \qquad \qquad \text{True}$$

The check results in a true statement, so the solution set is $\{17\}$.

Now Try Exercise 41.

EXAMPLE 6   **Using the Distributive Property to Simplify an Equation**

Solve $3(2 + 5x) - (1 + 14x) = 6$.

$$3(2 + 5x) - (1 + 14x) = 6$$

$$3(2 + 5x) - 1(1 + 14x) = 6 \qquad -(1 + 14x) = -1(1 + 14x)$$

$$3(2) + 3(5x) - 1(1) - 1(14x) = 6 \qquad \text{Distributive property}$$

$$6 + 15x - 1 - 14x = 6 \qquad \text{Multiply.}$$

$$x + 5 = 6 \qquad \text{Combine like terms.}$$

$$x + 5 - 5 = 6 - 5 \qquad \text{Subtract 5 from each side.}$$

$$x = 1 \qquad \text{Combine like terms.}$$

Check by substituting 1 for x in the original equation. The solution set is $\{1\}$.

Now Try Exercise 49.

CAUTION   Be careful to apply the distributive property correctly in a problem like that in Example 6, or a sign error may result.

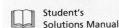   **EXERCISES**

1. Which pairs of equations are equivalent equations?
   A. $x + 2 = 6$ and $x = 4$ **B.** $10 - x = 5$ and $x = -5$
   C. $x + 3 = 9$ and $x = 6$ **D.** $4 + x = 8$ and $x = -4$

2. Decide whether each of the following is an expression or an equation. If it is an expression, simplify it. If it is an equation, solve it.
   (a) $5x + 8 - 4x + 7$ **(b)** $-6y + 12 + 7y - 5$
   (c) $5x + 8 - 4x = 7$ **(d)** $-6y + 12 + 7y = -5$

3. In your own words, state the addition property of equality. Give an example.

4. Explain how to check a solution of an equation.

Solve each equation, and check your solution. See Examples 1–4.

5. $x - 4 = 8$	**6.** $x - 8 = 9$	**7.** $y - 12 = 19$
8. $t - 15 = 25$	**9.** $x - 5 = -8$	**10.** $x - 7 = -9$
11. $r + 9 = 13$	**12.** $y + 6 = 10$	**13.** $x + 26 = 17$
14. $x + 45 = 24$	**15.** $7 + r = -3$	**16.** $8 + k = -4$
17. $2 = p + 15$	**18.** $3 = z + 17$	**19.** $-2 = x - 12$
20. $-6 = x - 21$	**21.** $x - 8.4 = -2.1$	**22.** $y - 15.5 = -5.1$
23. $t + 12.3 = -4.6$	**24.** $x + 21.5 = -13.4$	**25.** $\dfrac{2}{5}w - 6 = \dfrac{7}{5}w$
26. $-\dfrac{2}{7}z + 2 = \dfrac{5}{7}z$	**27.** $5.6x + 2 = 4.6x$	**28.** $9.1x - 5 = 8.1x$

29. $3p + 6 = 10 + 2p$ **30.** $8b - 4 = -6 + 7b$ **31.** $1.2y - 4 = .2y - 4$

32. $7.7r + 6 = 6.7r + 6$ **33.** $\dfrac{1}{2}x + 2 = -\dfrac{1}{2}x$ **34.** $\dfrac{1}{5}x - 7 = -\dfrac{4}{5}x$

35. $3x + 7 - 2x = 0$ **36.** $5x + 4 - 4x = 0$

37. Which of the following are *not* linear equations in one variable?

    A. $x^2 - 5x + 6 = 0$ **B.** $x^3 = x$

    C. $3x - 4 = 0$ **D.** $7x - 6x = 3 + 9x$

Define a linear equation in one variable in words.

38. Refer to the definition of linear equation in one variable given in this section. Why is the restriction $A \neq 0$ necessary?

Solve each equation, and check your solution. See Examples 5 and 6.

39. $5t + 3 + 2t - 6t = 4 + 12$ **40.** $4x + 3x - 6 - 6x = 10 + 3$

41. $6x + 5 + 7x + 3 = 12x + 4$ **42.** $4x - 3 - 8x + 1 = -5x + 9$

43. $5.2q - 4.6 - 7.1q = -.9q - 4.6$ **44.** $-4.0x + 2.7 - 1.6x = -4.6x + 2.7$

45. $\dfrac{5}{7}x + \dfrac{1}{3} = \dfrac{2}{5} - \dfrac{2}{7}x + \dfrac{2}{5}$ **46.** $\dfrac{6}{7}s - \dfrac{3}{4} = \dfrac{4}{5} - \dfrac{1}{7}s + \dfrac{1}{6}$

47. $(5y + 6) - (3 + 4y) = 10$ **48.** $(8r - 3) - (7r + 1) = -6$

49. $2(p + 5) - (9 + p) = -3$ **50.** $4(k - 6) - (3k + 2) = -5$

51. $-6(2b + 1) + (13b - 7) = 0$ **52.** $-5(3w - 3) + (1 + 16w) = 0$

53. $10(-2x + 1) = -19(x + 1)$ **54.** $2(2 - 3r) = -5(r - 3)$

55. $-2(8p + 2) - 3(2 - 7p) - 2(4 + 2p) = 0$

56. $-5(1 - 2z) + 4(3 - z) - 7(3 + z) = 0$

57. $4(7x - 1) + 3(2 - 5x) - 4(3x + 5) = -6$

58. $9(2m - 3) - 4(5 + 3m) - 5(4 + m) = -3$

59. Write an equation that requires the use of the addition property of equality, where 6 must be added to each side to solve the equation, and the solution is a negative number.

60. Write an equation that requires the use of the addition property of equality, where $\frac{1}{2}$ must be subtracted from each side, and the solution is a positive number.

Write an equation using the information given in the problem. Use x as the variable. Then solve the equation.

61. Three times a number is 17 more than twice the number. Find the number.

62. One, added to three times a number, is three less than four times the number. Find the number.

63. If six times a number is subtracted from seven times the number, the result is -9. Find the number.

64. If five times a number is added to three times the number, the result is the sum of seven times the number and 9. Find the number.

2.2 The Multiplication Property of Equality

OBJECTIVE 1 Use the multiplication property of equality. The addition property of equality alone is not enough to solve some equations, such as $3x + 2 = 17$.

$$3x + 2 = 17$$
$$3x + 2 - 2 = 17 - 2 \qquad \text{Subtract 2 from each side.}$$
$$3x = 15 \qquad \text{Combine like terms.}$$

Notice that the coefficient of x on the left side is 3, not 1 as desired. Another property is needed to change $3x = 15$ to an equation of the form

$$x = \text{a number.}$$

If $3x = 15$, then $3x$ and 15 both represent the same number. Multiplying both $3x$ and 15 by the same number will also result in an equality. The **multiplication property of equality** states that we can multiply each side of an equation by the same nonzero number without changing the solution.

Multiplication Property of Equality

If A, B, and C ($C \neq 0$) represent real numbers, then the equations

$$A = B \qquad \text{and} \qquad AC = BC$$

are equivalent equations.

That is, we can multiply each side of an equation by the same nonzero number without changing the solution.

This property can be used to solve $3x = 15$. The $3x$ on the left must be changed to $1x$, or x, instead of $3x$. To isolate x, we multiply each side of the equation by $\frac{1}{3}$. We use $\frac{1}{3}$ because $\frac{1}{3}$ is the reciprocal of 3, and $\frac{1}{3} \cdot 3 = \frac{3}{3} = 1$.

$$3x = 15$$
$$\frac{1}{3}(3x) = \frac{1}{3} \cdot 15 \qquad \text{Multiply each side by } \tfrac{1}{3}.$$
$$\left(\frac{1}{3} \cdot 3\right)x = \frac{1}{3} \cdot 15 \qquad \text{Associative property}$$
$$1x = 5 \qquad \text{Multiplicative inverse property}$$
$$x = 5 \qquad \text{Multiplicative identity property}$$

We check by substituting 5 for x in the original equation. The solution set of the equation is $\{5\}$. From now on, we will combine the last two steps shown in the preceding example.

Just as the addition property of equality permits *subtracting* the same number from each side of an equation, the multiplication property of equality permits *dividing* each side of an equation by the same nonzero number. For example, the equation $3x = 15$, which we just solved by multiplying each side by $\frac{1}{3}$, could also be solved by dividing each side by 3.

$$3x = 15$$

$$\frac{3x}{3} = \frac{15}{3} \qquad \text{Divide each side by 3.}$$

$$x = 5$$

We can divide each side of an equation by the same nonzero number without changing the solution. Do not, however, divide each side by a variable, as that may result in losing a valid solution.

NOTE In practice, it is usually easier to multiply on each side if the coefficient of the variable is a fraction, and divide on each side if the coefficient is an integer. For example, to solve

$$-\frac{3}{4}x = 12,$$

it is easier to multiply by $-\frac{4}{3}$, the reciprocal of $-\frac{3}{4}$, than to divide by $-\frac{3}{4}$. On the other hand, to solve

$$-5x = -20,$$

it is easier to divide by -5 than to multiply by $-\frac{1}{5}$.

EXAMPLE 1 Dividing Each Side of an Equation by a Nonzero Number

Solve $25p = 30$.

Transform the equation so that p (instead of $25p$) is on the left by using the multiplication property of equality. Divide each side of the equation by 25, the coefficient of p.

$$25p = 30$$

$$\frac{25p}{25} = \frac{30}{25} \qquad \text{Divide each side by 25.}$$

$$p = \frac{30}{25} = \frac{6}{5} \qquad \text{Write in lowest terms.}$$

To check, substitute $\frac{6}{5}$ for p in the original equation.

Check:
$$25p = 30$$

$$\frac{25}{1}\left(\frac{6}{5}\right) = 30 \qquad ? \qquad \text{Let } p = \frac{6}{5}.$$

$$30 = 30 \qquad \text{True}$$

The solution set is $\left\{\frac{6}{5}\right\}$.

Now Try Exercise 25.

▨ **EXAMPLE 2** Solving an Equation with Decimals

Solve $-2.1x = 6.09$.

$$-2.1x = 6.09$$

$$\frac{-2.1x}{-2.1} = \frac{6.09}{-2.1} \qquad \text{Divide each side by } -2.1.$$

A calculator will simplify the work at this point.

$$x = -2.9 \qquad \text{Divide.}$$

Check:
$$-2.1x = 6.09 \qquad \text{Original equation}$$
$$-2.1(-2.9) = 6.09 \qquad ? \quad \text{Let } x = -2.9.$$
$$6.09 = 6.09 \qquad \text{True}$$

The solution set is $\{-2.9\}$.

Now Try Exercise 39.

In the next two examples, multiplication produces the solution more quickly than division would.

▨ **EXAMPLE 3** Using the Multiplication Property of Equality

Solve $\dfrac{a}{4} = 3$.

Replace $\frac{a}{4}$ by $\frac{1}{4}a$, since dividing by 4 is the same as multiplying by $\frac{1}{4}$. To get a alone on the left, multiply each side by 4, the reciprocal of the coefficient of a.

$$\frac{a}{4} = 3$$

$$\frac{1}{4}a = 3 \qquad \text{Change } \tfrac{a}{4} \text{ to } \tfrac{1}{4}a.$$

$$4 \cdot \frac{1}{4}a = 4 \cdot 3 \qquad \text{Multiply by 4.}$$

$$a = 12 \qquad \begin{array}{l}\text{Multiplicative inverse property;}\\ \text{multiplicative identity property}\end{array}$$

Check that 12 is the solution.

Check:
$$\frac{a}{4} = 3 \qquad \text{Original equation}$$

$$\frac{12}{4} = 3 \qquad ? \quad \text{Let } a = 12.$$

$$3 = 3 \qquad \text{True}$$

The solution set is $\{12\}$.

Now Try Exercise 43.

▨ **EXAMPLE 4**   **Using the Multiplication Property of Equality**

Solve $\frac{3}{4}h = 6$.

To transform the equation so that h is alone on the left, multiply each side of the equation by $\frac{4}{3}$. Use $\frac{4}{3}$ because $\frac{4}{3} \cdot \frac{3}{4}h = 1 \cdot h = h$.

$$\frac{3}{4}h = 6$$

$$\frac{4}{3}\left(\frac{3}{4}h\right) = \frac{4}{3} \cdot 6 \qquad \text{Multiply by } \tfrac{4}{3}.$$

$$1 \cdot h = \frac{4}{3} \cdot \frac{6}{1} \qquad \text{Multiplicative inverse property}$$

$$h = 8 \qquad \text{Multiplicative identity property; multiply fractions.}$$

Check by substituting 8 for h in the original equation. The solution set is $\{8\}$.   ▨

Now Try Exercise 47.

In Section 2.1, we obtained the equation $-k = -17$ in our alternate solution to Example 4. We reasoned that since this equation says that the additive inverse (or opposite) of k is -17, then k must equal 17. We can also use the multiplication property of equality to obtain the same result, as shown in the next example.

▨ **EXAMPLE 5**   **Using the Multiplication Property of Equality When the Coefficient of the Variable Is −1**

Solve $-k = -17$.

On the left side, change $-k$ to k by first writing $-k$ as $-1 \cdot k$.

$$-k = -17$$

$$-1 \cdot k = -17 \qquad -k = -1 \cdot k$$

$$-1(-1 \cdot k) = -1(-17) \qquad \text{Multiply by } -1, \text{ since } -1(-1) = 1.$$

$$[-1(-1)] \cdot k = 17 \qquad \text{Associative property; multiply.}$$

$$1 \cdot k = 17 \qquad \text{Multiplicative inverse property}$$

$$k = 17 \qquad \text{Multiplicative identity property}$$

Check: $\qquad\quad -k = -17 \qquad \text{Original equation}$

$$-(17) = -17 \qquad ? \quad \text{Let } k = 17.$$

$$-17 = -17 \qquad \text{True}$$

The solution, 17, checks, so $\{17\}$ is the solution set.   ▨

Now Try Exercise 53.

From Example 5, we see that the following is true.

$$\text{If} \quad -x = a, \quad \text{then} \quad x = -a.$$

OBJECTIVE 2 Simplify equations, and then use the multiplication property of equality. In the next example, it is necessary to simplify the equation before using the multiplication property of equality.

EXAMPLE 6 Simplifying an Equation

Solve $5m + 6m = 33$.

$$5m + 6m = 33$$
$$11m = 33 \qquad \text{Combine like terms.}$$
$$\frac{11m}{11} = \frac{33}{11} \qquad \text{Divide by 11.}$$
$$1m = 3 \qquad \text{Divide.}$$
$$m = 3 \qquad \text{Multiplicative identity property}$$

Check this proposed solution. The solution set is $\{3\}$.

Now Try Exercise 57.

CONNECTIONS

The use of algebra to solve equations and applied problems is very old. The 3600-year-old Rhind Papyrus includes the following "word problem." "Aha, its whole, its seventh, it makes 19." This brief sentence describes the equation

$$x + \frac{x}{7} = 19.$$

The solution of this equation is $16\frac{5}{8}$. The word *algebra* is from the work *Hisab al-jabr m'al muquabalah,* written in the ninth century by Muhammed ibn Musa Al-Khowarizmi. The title means "the science of transposition and cancellation." From Latin versions of Khowarizmi's text, "al-jabr" became the broad term covering the art of equation solving.

2.2 EXERCISES

For Extra Help

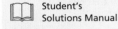 Student's Solutions Manual

 MyMathLab

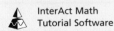 InterAct Math Tutorial Software

 AW Math Tutor Center

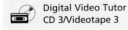 MathXL

Digital Video Tutor CD 3/Videotape 3

1. In your own words, state the multiplication property of equality. Give an example.

2. In the statement of the multiplication property of equality in this section, there is a restriction that $C \neq 0$. What would happen if you multiplied each side of an equation by 0?

3. Which equation does *not* require the use of the multiplication property of equality?

A. $3x - 5x = 6$ **B.** $-\frac{1}{4}x = 12$ **C.** $5x - 4x = 7$ **D.** $\frac{x}{3} = -2$

4. Tell whether you would use the addition or multiplication property of equality to solve each equation. Explain your answer.

(a) $3x = 12$ **(b)** $3 + x = 12$

5. A student tried to solve the equation $4x = 8$ by dividing each side by 8. Why is this wrong?

6. State how you would find the solution of a linear equation if your next-to-last step reads "$-x = 5$."

By what number is it necessary to multiply both sides of each equation to get just x on the left side? Do not actually solve these equations.

7. $\frac{2}{3}x = 8$ **8.** $\frac{4}{5}x = 6$ **9.** $\frac{x}{10} = 3$ **10.** $\frac{x}{100} = 8$

11. $\frac{9}{2}x = -4$ **12.** $-\frac{8}{3}x = -11$ **13.** $-x = .36$ **14.** $-x = .29$

By what number is it necessary to divide both sides of each equation to get just x on the left side? Do not actually solve these equations.

15. $6x = 5$ **16.** $7x = 10$ **17.** $-4x = 13$ **18.** $-13x = 6$

19. $.12x = 48$ **20.** $.21x = 63$ **21.** $-x = 23$ **22.** $-x = 49$

Solve each equation, and check your solution. See Examples 1–5.

23. $5x = 30$ **24.** $7x = 56$ **25.** $2m = 15$ **26.** $3m = 10$

27. $3a = -15$ **28.** $5k = -70$ **29.** $-3x = 12$ **30.** $-4x = 36$

31. $10t = -36$ **32.** $4s = -34$ **33.** $-6x = -72$ **34.** $-8x = -64$

35. $2r = 0$ **36.** $5x = 0$ **37.** $.2t = 8$ **38.** $.9x = 18$

39. $-2.1m = 25.62$ **40.** $-3.9a = 31.2$ **41.** $\frac{1}{4}y = -12$ **42.** $\frac{1}{5}p = -3$

43. $\frac{z}{6} = 12$ **44.** $\frac{x}{5} = 15$ **45.** $\frac{x}{7} = -5$ **46.** $\frac{k}{8} = -3$

47. $\frac{2}{7}p = 4$ **48.** $\frac{3}{8}y = 9$ **49.** $-\frac{5}{6}t = -15$ **50.** $-\frac{3}{4}k = -21$

51. $-\frac{7}{9}c = \frac{3}{5}$ **52.** $-\frac{5}{6}d = \frac{4}{9}$ **53.** $-y = 12$ **54.** $-t = 14$

55. $-x = -\frac{3}{4}$ **56.** $-y = -\frac{1}{2}$

Solve each equation, and check your solution. See Example 6.

57. $4x + 3x = 21$ **58.** $9x + 2x = 121$ **59.** $3r - 5r = 10$

60. $9p - 13p = 24$ **61.** $5m + 6m - 2m = 63$ **62.** $11r - 5r + 6r = 168$

63. $-6x + 4x - 7x = 0$ **64.** $-5x + 4x - 8x = 0$ **65.** $9w - 5w + w = -3$

66. $10y - 6y + 3y = -4$

67. Write an equation that requires the use of the multiplication property of equality, where each side must be multiplied by $\frac{2}{3}$, and the solution is a negative number.

68. Write an equation that requires the use of the multiplication property of equality, where each side must be divided by 100, and the solution is not an integer.

Write an equation using the information given in the problem. Use x as the variable. Then solve the equation.

69. When a number is multiplied by 4, the result is 6. Find the number.

70. When a number is multiplied by −4, the result is 10. Find the number.

71. When a number is divided by −5, the result is 2. Find the number.

72. If twice a number is divided by 5, the result is 4. Find the number.

2.3 More on Solving Linear Equations

OBJECTIVE 1 Learn and use the four steps for solving a linear equation. In this section, we use the addition and multiplication properties together to solve more complicated equations. We use the following four-step method.

Solving a Linear Equation

Step 1 **Simplify each side separately.** Clear parentheses using the distributive property, if needed, and combine like terms.

Step 2 **Isolate the variable term on one side.** Use the addition property if necessary so that the variable term is on one side of the equation and a number is on the other.

Step 3 **Isolate the variable.** Use the multiplication property if necessary to get the equation in the form $x = $ a number.

Step 4 **Check.** Substitute the proposed solution into the *original* equation to see if a true statement results.

▌ **EXAMPLE 1 Using the Four Steps to Solve an Equation**

Solve $3r + 4 - 2r - 7 = 4r + 3$.

Step 1	$3r + 4 - 2r - 7 = 4r + 3$	
	$r - 3 = 4r + 3$	Combine like terms.
Step 2	$r - 3 + 3 = 4r + 3 + 3$	Add 3.
	$r = 4r + 6$	Combine like terms.
	$r - 4r = 4r + 6 - 4r$	Subtract 4r.
	$-3r = 6$	Combine like terms.
Step 3	$\dfrac{-3r}{-3} = \dfrac{6}{-3}$	Divide by −3.
	$r = -2$	$\frac{-3}{-3} = 1; 1r = r$

Step 4   Substitute -2 for r in the original equation to check.

$$3r + 4 - 2r - 7 = 4r + 3 \qquad \text{Original equation}$$
$$3(-2) + 4 - 2(-2) - 7 = 4(-2) + 3 \quad ? \quad \text{Let } r = -2.$$
$$-6 + 4 + 4 - 7 = -8 + 3 \quad ? \quad \text{Multiply.}$$
$$-5 = -5 \qquad \text{True}$$

The solution set of the equation is $\{-2\}$.

Now Try Exercise 7.

NOTE   In Step 2 of Example 1, we added and subtracted the terms in such a way that the variable term ended up on the left side of the equation. Choosing differently would have put the variable term on the right side of the equation. Either way, the same solution results.

EXAMPLE 2   Using the Four Steps to Solve an Equation

Solve $4(k - 3) - k = k - 6$.

Step 1   Before combining like terms, use the distributive property to simplify $4(k - 3)$.

$$4(k - 3) - k = k - 6$$
$$4(k) + 4(-3) - k = k - 6 \qquad \text{Distributive property}$$
$$4k - 12 - k = k - 6$$
$$3k - 12 = k - 6 \qquad \text{Combine like terms.}$$

Step 2
$$3k - 12 - k = k - 6 - k \qquad \text{Subtract } k.$$
$$2k - 12 = -6 \qquad \text{Combine like terms.}$$
$$2k - 12 + 12 = -6 + 12 \qquad \text{Add 12.}$$
$$2k = 6 \qquad \text{Combine like terms.}$$

Step 3
$$\frac{2k}{2} = \frac{6}{2} \qquad \text{Divide by 2.}$$
$$k = 3$$

Step 4   Check by substituting 3 for k in the original equation. Work inside the parentheses first.

$$4(k - 3) - k = k - 6 \qquad \text{Original equation}$$
$$4(3 - 3) - 3 = 3 - 6 \quad ? \quad \text{Let } k = 3.$$
$$4(0) - 3 = 3 - 6 \quad ?$$
$$0 - 3 = 3 - 6 \quad ?$$
$$-3 = -3 \qquad \text{True}$$

The solution set of the equation is $\{3\}$.

Now Try Exercise 11.

■ **EXAMPLE 3** Using the Four Steps to Solve an Equation

Solve $8a - (3 + 2a) = 3a + 1$.

Step 1 Clear parentheses using the distributive property.

$$8a - (3 + 2a) = 3a + 1$$
$$8a - 1(3 + 2a) = 3a + 1 \qquad \text{Multiplicative identity property}$$
$$8a - 3 - 2a = 3a + 1 \qquad \text{Distributive property}$$
$$6a - 3 = 3a + 1$$

Step 2 $\quad 6a - 3 - 3a = 3a + 1 - 3a \qquad \text{Subtract } 3a.$

$$3a - 3 = 1$$
$$3a - 3 + 3 = 1 + 3 \qquad \text{Add 3.}$$
$$3a = 4$$

Step 3 $\qquad \dfrac{3a}{3} = \dfrac{4}{3} \qquad \text{Divide by 3.}$

$$a = \dfrac{4}{3}$$

Step 4 Check this solution in the original equation.

$$8a - (3 + 2a) = 3a + 1 \qquad \text{Original equation}$$

$$8\left(\dfrac{4}{3}\right) - \left[3 + 2\left(\dfrac{4}{3}\right)\right] = 3\left(\dfrac{4}{3}\right) + 1 \qquad ? \qquad \text{Let } a = \tfrac{4}{3}.$$

$$\dfrac{32}{3} - \left[3 + \dfrac{8}{3}\right] = 4 + 1 \qquad ?$$

$$\dfrac{32}{3} - \left[\dfrac{9}{3} + \dfrac{8}{3}\right] = 5 \qquad ?$$

$$\dfrac{32}{3} - \dfrac{17}{3} = 5 \qquad ?$$

$$5 = 5 \qquad \text{True}$$

The check shows that $\left\{\dfrac{4}{3}\right\}$ is the solution set. ■

Now Try Exercise 9.

CAUTION Be very careful with signs when solving an equation like the one in Example 3. When clearing parentheses in the expression

$$8 - (3 + 2a),$$

remember that the $-$ sign acts like a factor of -1 and affects the sign of *every* term within the parentheses. Thus,

$$8 - (3 + 2a) = 8 + (-1)(3 + 2a)$$
$$= 8 - 3 - 2a.$$

Change to $-$ in *both* terms.

EXAMPLE 4   Using the Four Steps to Solve an Equation

Solve $4(8 - 3t) = 32 - 8(t + 2)$.

Step 1
$$4(8 - 3t) = 32 - 8(t + 2)$$
$$32 - 12t = 32 - 8t - 16 \qquad \text{Distributive property}$$
$$32 - 12t = 16 - 8t$$

Step 2
$$32 - 12t + 12t = 16 - 8t + 12t \qquad \text{Add 12t.}$$
$$32 = 16 + 4t$$
$$32 - 16 = 16 + 4t - 16 \qquad \text{Subtract 16.}$$
$$16 = 4t$$

Step 3
$$\frac{16}{4} = \frac{4t}{4} \qquad \text{Divide by 4.}$$
$$4 = t \quad \text{or} \quad t = 4$$

Step 4   Check:
$$4(8 - 3t) = 32 - 8(t + 2) \qquad \text{Original equation}$$
$$4(8 - 3 \cdot 4) = 32 - 8(4 + 2) \qquad ? \qquad \text{Let } t = 4.$$
$$4(8 - 12) = 32 - 8(6) \qquad ?$$
$$4(-4) = 32 - 48 \qquad ?$$
$$-16 = -16 \qquad \text{True}$$

Since a true statement results, the solution set is $\{4\}$.

Now Try Exercise 13.

OBJECTIVE 2 **Solve equations with fractions or decimals as coefficients.**   We can clear an equation of fractions by multiplying each side by the least common denominator (LCD) of all the fractions in the equation. It is a good idea to do this *before* starting the four-step method to avoid working with fractions.

EXAMPLE 5   Solving an Equation with Fractions as Coefficients

Solve $\dfrac{2}{3}x - \dfrac{1}{2}x = -\dfrac{1}{6}x - 2$.

The LCD of all the fractions in the equation is 6, so multiply each side by 6.

$$\frac{2}{3}x - \frac{1}{2}x = -\frac{1}{6}x - 2$$

$$6\left(\frac{2}{3}x - \frac{1}{2}x\right) = 6\left(-\frac{1}{6}x - 2\right) \qquad \text{Multiply by 6.}$$

$$6\left(\frac{2}{3}x\right) + 6\left(-\frac{1}{2}x\right) = 6\left(-\frac{1}{6}x\right) + 6(-2) \qquad \text{Distributive property}$$

$$4x - 3x = -x - 12$$

Now use the four steps to solve this equivalent equation.

Step 1
$$x = -x - 12 \qquad \text{Combine like terms.}$$

Step 2 $\qquad$ $x + x = -x - 12 + x$ $\qquad$ Add x.

$\qquad\qquad\qquad$ $2x = -12$ $\qquad$ Combine like terms.

Step 3 $\qquad$ $\dfrac{2x}{2} = \dfrac{-12}{2}$ $\qquad$ Divide by 2.

$\qquad\qquad\qquad$ $x = -6$

Step 4 *Check:*

$$\frac{2}{3}x - \frac{1}{2}x = -\frac{1}{6}x - 2 \qquad \text{Original equation}$$

$$\frac{2}{3}(-6) - \frac{1}{2}(-6) = -\frac{1}{6}(-6) - 2 \quad ? \quad \text{Let } x = -6.$$

$$-4 + 3 = 1 - 2 \qquad ?$$

$$-1 = -1 \qquad \text{True}$$

The solution set of the equation is $\{-6\}$.

Now Try Exercise 23.

CAUTION When clearing an equation of fractions, be sure to multiply *every* term on each side of the equation by the LCD.

The multiplication property is also used to clear an equation of decimals.

EXAMPLE 6 Solving an Equation with Decimals as Coefficients

Solve $.1t + .05(20 - t) = .09(20)$.

The decimals are expressed as tenths and hundredths. Choose the smallest exponent on 10 needed to eliminate the decimals; in this case, use $10^2 = 100$. A number can be multiplied by 100 by moving the decimal point two places to the right.

$$.10t + .05(20 - t) = .09(20) \qquad .1 = .10$$

$$10t + 5(20 - t) = 9(20) \qquad \text{Multiply by 100.}$$

Now use the four steps.

Step 1 $\qquad$ $10t + 5(20) + 5(-t) = 180$ $\qquad$ Distributive property

$\qquad\qquad\qquad$ $10t + 100 - 5t = 180$

$\qquad\qquad\qquad$ $5t + 100 = 180$ $\qquad$ Combine like terms.

Step 2 $\qquad$ $5t + 100 - 100 = 180 - 100$ $\qquad$ Subtract 100.

$\qquad\qquad\qquad$ $5t = 80$ $\qquad$ Combine like terms.

Step 3 $\qquad$ $\dfrac{5t}{5} = \dfrac{80}{5}$ $\qquad$ Divide by 5.

$\qquad\qquad\qquad$ $t = 16$

Step 4 Check that $\{16\}$ is the solution set by substituting 16 for t in the original equation.

Now Try Exercise 29.

OBJECTIVE 3 Solve equations with no solution or infinitely many solutions. Each equation that we have solved so far has had exactly one solution. An equation with exactly one solution is a **conditional equation** because it is only true under certain conditions. As the next examples show, linear equations may also have no solution or infinitely many solutions. (The four steps are not identified in these examples. See if you can identify them.)

EXAMPLE 7 Solving an Equation That Has Infinitely Many Solutions

Solve $5x - 15 = 5(x - 3)$.

$$5x - 15 = 5(x - 3)$$
$$5x - 15 = 5x - 15 \qquad \text{Distributive property}$$
$$5x - 15 - 5x = 5x - 15 - 5x \qquad \text{Subtract 5x.}$$
$$-15 = -15$$
$$-15 + 15 = -15 + 15 \qquad \text{Add 15.}$$
$$0 = 0 \qquad \text{True}$$

The variable has "disappeared." Since the last statement ($0 = 0$) is true, *any* real number is a solution. (We could have predicted this from the line in the solution that says $5x - 15 = 5x - 15$, which is certainly true for *any* value of x.) An equation with both sides exactly the same, like $0 = 0$, is called an **identity.** An identity is true for all replacements of the variables. We indicate this by writing the solution set as {all real numbers}.

Now Try Exercise 17.

CAUTION When solving an equation like the one in Example 7, do not write {0} as the solution set. Although 0 is a solution, there are infinitely many other solutions.

EXAMPLE 8 Solving an Equation That Has No Solution

Solve $2x + 3(x + 1) = 5x + 4$.

$$2x + 3(x + 1) = 5x + 4$$
$$2x + 3x + 3 = 5x + 4 \qquad \text{Distributive property}$$
$$5x + 3 = 5x + 4$$
$$5x + 3 - 5x = 5x + 4 - 5x \qquad \text{Subtract 5x.}$$
$$3 = 4 \qquad \text{False}$$

Again, the variable has disappeared, but this time a false statement ($3 = 4$) results. Whenever this happens in solving an equation, it is a signal that the equation has no solution. An equation with no solution is called a **contradiction.** Its solution set is the **empty set,** or **null set,** symbolized $\emptyset$.

Now Try Exercise 21.

The table on the next page summarizes the solution sets of the three types of linear equations.

Type of Linear Equation	Final Equation in Solution	Number of Solutions	Solution Set
Conditional	$x =$ a number	One	{a number}
Identity	A true statement with no variable, such as $0 = 0$	Infinite	{all real numbers}
Contradiction	A false statement with no variable, such as $3 = 4$	None	$\emptyset$

OBJECTIVE 4 Write expressions for two related unknown quantities. Next, we continue our work with translating from words to symbols.

─── ■ **PROBLEM SOLVING** ■ ───────────────────

Often we are given a problem in which the sum of two quantities is a particular number, and we are asked to find the values of the two quantities. Example 9 shows how to express the unknown quantities in terms of a single variable.

■ **EXAMPLE 9 Translating a Phrase into an Algebraic Expression**

Two numbers have a sum of 23. If one of the numbers is represented by k, find an expression for the other number.

First, suppose that the sum of two numbers is 23, and one of the numbers is 10. How would you find the other number? You would subtract 10 from 23 to get 13; $23 - 10 = 13$. So instead of using 10 as one of the numbers, use k as stated in the problem. The other number would be obtained in the same way. You must subtract k from 23. Therefore, an expression for the other number is $23 - k$.

Now Try Exercise 47.

NOTE The approach used in Example 9, first writing an expression with a trial number, is also useful when translating applied problems to equations.

2.3 EXERCISES

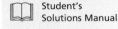

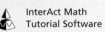

1. In your own words, give the four steps used to solve a linear equation. Give an example to demonstrate the steps.

2. After working correctly through several steps of the solution of a linear equation, a student obtains the equation $7x = 3x$. Then the student divides each side by x to get $7 = 3$, and gives $\emptyset$ as the answer. Is this correct? If not, explain why.

3. Which one of the following linear equations does *not* have {all real numbers} as its solution set?

   A. $5x = 4x + x$ **B.** $2(x + 6) = 2x + 12$ **C.** $\dfrac{1}{2}x = .5x$ **D.** $3x = 2x$

4. Based on the discussion in this section, if an equation has decimals or fractions as coefficients, what additional step will make the work easier?

Solve each equation, and check your solution. See Examples 1–4, 7, and 8.

5. $3x + 8 = 5x + 10$

6. $10p + 6 = 12p - 4$

7. $12h - 5 = 11h + 5 - h$

8. $-4x - 1 = -5x + 1 + 3x$

9. $-2p + 7 = 3 - (5p + 1)$

10. $4x + 9 = 3 - (x - 2)$

11. $3(4x + 2) + 5x = 30 - x$

12. $5(2m + 3) - 4m = 8m + 27$

13. $6(3w + 5) = 2(10w + 10)$

14. $4(2x - 1) = -6(x + 3)$

15. $6(4x - 1) = 12(2x + 3)$

16. $6(2x + 8) = 4(3x - 6)$

17. $3(2x - 4) = 6(x - 2)$

18. $3(6 - 4x) = 2(-6x + 9)$

19. $7r - 5r + 2 = 5r - r$

20. $9p - 4p + 6 = 7p - 3p$

21. $11x - 5(x + 2) = 6x + 5$

22. $6x - 4(x + 1) = 2x + 4$

Solve each equation by clearing fractions or decimals. Check your solution. See Examples 5 and 6.

23. $\dfrac{3}{5}t - \dfrac{1}{10}t = t - \dfrac{5}{2}$

24. $-\dfrac{2}{7}r + 2r = \dfrac{1}{2}r + \dfrac{17}{2}$

25. $-\dfrac{1}{4}(x - 12) + \dfrac{1}{2}(x + 2) = x + 4$

26. $\dfrac{1}{9}(y + 18) + \dfrac{1}{3}(2y + 3) = y + 3$

27. $\dfrac{2}{3}k - \left(k + \dfrac{1}{4}\right) = \dfrac{1}{12}(k + 4)$

28. $-\dfrac{5}{6}q - \left(q - \dfrac{1}{2}\right) = \dfrac{1}{4}(q + 1)$

29. $.20(60) + .05x = .10(60 + x)$

30. $.30(30) + .15x = .20(30 + x)$

31. $1.00x + .05(12 - x) = .10(63)$

32. $.92x + .98(12 - x) = .96(12)$

33. $.06(10,000) + .08x = .072(10,000 + x)$

34. $.02(5000) + .03x = .025(5000 + x)$

Solve each equation, and check your solution. See Examples 1–8.

35. $10(2x - 1) = 8(2x + 1) + 14$

36. $9(3k - 5) = 12(3k - 1) - 51$

37. $-(4y + 2) - (-3y - 5) = 3$

38. $-(6k - 5) - (-5k + 8) = -3$

39. $\dfrac{1}{2}(x + 2) + \dfrac{3}{4}(x + 4) = x + 5$

40. $\dfrac{1}{3}(x + 3) + \dfrac{1}{6}(x - 6) = x + 3$

41. $.10(x + 80) + .20x = 14$

42. $.30(x + 15) + .40(x + 25) = 25$

43. $4(x + 8) = 2(2x + 6) + 20$

44. $4(x + 3) = 2(2x + 8) - 4$

45. $9(v + 1) - 3v = 2(3v + 1) - 8$

46. $8(t - 3) + 4t = 6(2t + 1) - 10$

Write the answer to each problem as an algebraic expression. See Example 9.

47. Two numbers have a sum of 11. One of the numbers is q. Find the other number.

48. The product of two numbers is 9. One of the numbers is k. What is the other number?

49. A football player gained x yd rushing. On the next down he gained 7 yd. How many yards did he gain altogether?

50. A baseball player got 65 hits one season. He got h of the hits in one game. How many hits did he get in the rest of the games?

51. Mary is a yr old. How old will she be in 12 yr? How old was she 5 yr ago?

52. Tom has r quarters. Find the value of the quarters in cents.

53. A bank teller has t dollars, all in $5 bills. How many $5 bills does the teller have?

54. A plane ticket costs b dollars for an adult and d dollars for a child. Find the total cost for 3 adults and 2 children.

▨ SUMMARY EXERCISES ON SOLVING LINEAR EQUATIONS

This section of miscellaneous linear equations provides practice in solving all the types introduced in Sections 2.1–2.3. Refer to the examples in these sections to review the various solution methods.

Solve each equation, and check your solution.

1. $a + 2 = -3$

2. $2m + 8 = 16$

3. $12.5k = -63.75$

4. $-x = -12$

5. $\dfrac{4}{5}x = -20$

6. $7m - 5m = -12$

7. $5x - 9 = 4(x - 3)$

8. $\dfrac{a}{-2} = 8$

9. $-3(m - 4) + 2(5 + 2m) = 29$

10. $\dfrac{2}{3}y + 8 = \dfrac{1}{4}y$

11. $.08x + .06(x + 9) = 1.24$

12. $x - 16.2 = 7.5$

13. $4x + 2(3 - 2x) = 6$

14. $-.3x + 2.1(x - 4) = -6.6$

15. $-x = 6$

16. $3(m + 5) - 1 + 2m = 5(m + 2)$

17. $7m - (2m - 9) = 39$

18. $7(p - 2) + p = 2(p + 2)$

19. $-2t + 5t - 9 = 3(t - 4) - 5$

20. $-6z = -14$

21. $.02(50) + .08r = .04(50 + r)$

22. $2.3x + 13.7 = 1.3x + 2.9$

23. $2(3 + 7x) - (1 + 15x) = 2$

24. $6q - 9 = 12 + 3q$

25. $2(4 + 3r) = 3(r + 1) + 11$

26. $r + 9 + 7r = 4(3 + 2r) - 3$

27. $\dfrac{1}{4}x - 4 = \dfrac{3}{2}x + \dfrac{3}{4}x$

28. $.06(100 - x) + .04x = .05(92)$

29. $\dfrac{3}{4}(a - 2) - \dfrac{1}{3}(5 - 2a) = -2$

30. $2 - (m + 4) = 3m + 8$

2.4   An Introduction to Applications of Linear Equations

OBJECTIVES

1 Learn the six steps for solving applied problems.

2 Solve problems involving unknown numbers.

OBJECTIVE 1   Learn the six steps for solving applied problems.   We now look at how algebra is used to solve applied problems. Since many meaningful applications of mathematics require concepts that are beyond the level of this book, some of the problems you encounter will seem contrived, and to some extent they are. But the skills you develop in solving simple problems will help you solve more realistic problems in chemistry, physics, biology, business, and other fields. While there is not one specific method that enables you to solve all kinds of applied problems, the following six-step method is suggested.

3 Solve problems involving sums of quantities.

4 Solve problems involving supplementary and complementary angles.

5 Solve problems involving consecutive integers.

Solving an Applied Problem

Step 1 **Read** the problem carefully until you understand what is given and what is to be found.

Step 2 **Assign a variable** to represent the unknown value, using diagrams or tables as needed. Write down what the variable represents. If necessary, express any other unknown values in terms of the variable.

Step 3 **Write an equation** using the variable expression(s).

Step 4 **Solve** the equation.

Step 5 **State the answer.** Does it seem reasonable?

Step 6 **Check** the answer in the words of the *original* problem.

The third step in solving an applied problem is often the hardest. To translate the problem into an equation, write the given phrases as mathematical expressions. Replace any words that mean *equal* or *same* with an $=$ sign. Other forms of the verb "to be," such as *is, are, was,* and *were,* also translate this way. The $=$ sign leads to an equation to be solved.

OBJECTIVE 2 Solve problems involving unknown numbers. Some of the simplest applied problems involve unknown numbers.

EXAMPLE 1 Finding the Value of an Unknown Number

The product of 4, and a number decreased by 7, is 100. What is the number?

Step 1 **Read** the problem carefully. Decide what you are being asked to find.

Step 2 **Assign a variable** to represent the unknown quantity. In this problem, we are asked to find a number, so we write

Let $x =$ the number.

There are no other unknown quantities to find.

Step 3 **Write an equation.**

$$
\begin{array}{ccccccc}
\text{The product} & & \text{a} & \text{decreased} & & & \\
\text{of 4,} & \text{and} & \text{number} & \text{by} & 7, & \text{is} & 100. \\
\downarrow & & \downarrow & \downarrow & \downarrow & \downarrow & \downarrow \\
4\;\cdot & & (x & - & 7) & = & 100
\end{array}
$$

Because of the commas in the given problem, writing the equation as $4x - 7 = 100$ is incorrect. The equation $4x - 7 = 100$ corresponds to the statement "The product of 4 and a number, decreased by 7, is 100."

Step 4 **Solve** the equation.

$$4(x - 7) = 100$$

$$4x - 28 = 100 \qquad \text{Distributive property}$$

$$4x - 28 + 28 = 100 + 28 \qquad \text{Add 28.}$$

$$4x = 128 \qquad \text{Combine like terms.}$$

$$\frac{4x}{4} = \frac{128}{4} \qquad \text{Divide by 4.}$$

$$x = 32$$

Step 5 **State the answer.** The number is 32.

Step 6 **Check.** When 32 is decreased by 7, we get $32 - 7 = 25$. If 4 is multiplied by 25, we get 100, as required. The answer, 32, is correct.

Now Try Exercise 5.

OBJECTIVE 3 Solve problems involving sums of quantities. A common type of problem in elementary algebra involves finding two quantities when the sum of the quantities is known. In Example 9 of the previous section, we prepared for this type of problem by writing mathematical expressions for two related unknown quantities.

▌ PROBLEM SOLVING ▐

In general, to solve problems involving sums of quantities, choose a variable to represent one of the unknowns. Then represent the other quantity in terms of the *same* variable, using information from the problem. Write an equation based on the words of the problem.

▌ EXAMPLE 2 Finding Numbers of Olympic Medals

In the 2002 Winter Olympics in Salt Lake City, the United States won 10 more medals than Norway. The two countries won a total of 58 medals. How many medals did each country win? (*Source:* U.S. Olympic Committee.)

Step 1 **Read** the problem. We are given information about the total number of medals and asked to find the number each country won.

Step 2 **Assign a variable.**

Let $\quad\quad\quad\quad x =$ the number of medals Norway won.

Then $\quad\quad x + 10 =$ the number of medals the U.S. won.

Step 3 **Write an equation.**

The total	is	the number of medals Norway won	plus	the number of medals the U.S. won.
↓	↓	↓	↓	↓
58	=	x	+	$(x + 10)$

Step 4 **Solve** the equation.

$$58 = 2x + 10 \qquad \text{Combine like terms.}$$
$$58 - 10 = 2x + 10 - 10 \qquad \text{Subtract 10.}$$
$$48 = 2x \qquad \text{Combine like terms.}$$
$$\frac{48}{2} = \frac{2x}{2} \qquad \text{Divide by 2.}$$
$$24 = x \quad \text{or} \quad x = 24$$

Step 5 **State the answer.** The variable x represents the number of medals Norway won, so Norway won 24 medals. Then the number of medals the United States won is $x + 10 = 24 + 10 = 34.$

Step 6 **Check.** Since the United States won 34 medals and Norway won 24, the total number of medals was $34 + 24 = 58$. Because $34 - 24 = 10$, the United States won 10 more medals than Norway. This information agrees with what is given in the problem, so the answer checks.

Now Try Exercise 11.

NOTE The problem in Example 2 could also be solved by letting x represent the number of medals the United States won. Then $x - 10$ would represent the number of medals Norway won. The equation would be

$$58 = x + (x - 10).$$

The solution of this equation is 34, which is the number of U.S. medals. The number of Norwegian medals would be $34 - 10 = 24$. The answers are the same, whichever approach is used.

EXAMPLE 3   Finding the Number of Orders for Tea

The owner of P. J.'s Coffeehouse found that on one day the number of orders for tea was $\frac{1}{3}$ the number of orders for coffee. If the total number of orders for the two drinks was 76, how many orders were placed for tea?

Step 1 **Read** the problem. It asks for the number of orders for tea.

Step 2 **Assign a variable.** Because of the way the problem is stated, let the variable represent the number of orders for coffee.

Let $x =$ the number of orders for coffee.

Then $\frac{1}{3}x =$ the number of orders for tea.

Step 3 **Write an equation.** Use the fact that the total number of orders was 76.

$$
\begin{array}{ccccc}
\text{The total} & \text{is} & \text{orders for coffee} & \text{plus} & \text{orders for tea.} \\
\downarrow & \downarrow & \downarrow & \downarrow & \downarrow \\
76 & = & x & + & \dfrac{1}{3}x
\end{array}
$$

Step 4 **Solve** the equation.

$$76 = \frac{4}{3}x \qquad x = 1x = \tfrac{3}{3}x; \text{ Combine like terms.}$$

$$\frac{3}{4}(76) = \frac{3}{4}\left(\frac{4}{3}x\right) \qquad \text{Multiply by } \tfrac{3}{4}.$$

$$57 = x$$

Step 5 **State the answer.** In this problem, *x does not represent the quantity that we are asked to find.* The number of orders for tea was $\frac{1}{3}x$. So $\frac{1}{3}(57) = 19$ is the number of orders for tea.

Step 6 **Check.** The number of coffee orders (x) was 57 and the number of tea orders was 19; 19 is one-third of 57, and $19 + 57 = 76$. Since this agrees with the information given in the problem, the answer is correct.

Now Try Exercise 23.

> ### PROBLEM SOLVING
>
> In Example 3, it was easier to let the variable represent the quantity that was *not* asked for. This required extra work in Step 5 to find the number of orders for tea. In some cases, this approach is easier than letting the variable represent the quantity that we are asked to find. Experience in solving problems will indicate when this approach is useful, and experience comes only from solving many problems.

EXAMPLE 4 Analyzing a Gasoline/Oil Mixture

A lawn trimmer uses a mixture of gasoline and oil. The mixture contains 16 oz of gasoline for each ounce of oil. If the tank holds 68 oz of the mixture, how many ounces of oil and how many ounces of gasoline does it require when it is full?

Step 1 **Read** the problem. We must find how many ounces of oil and gasoline are needed to fill the tank.

Step 2 **Assign a variable.**

Let x = the number of ounces of oil required.

Then $16x$ = the number of ounces of gasoline required.

Step 3 **Write an equation.**

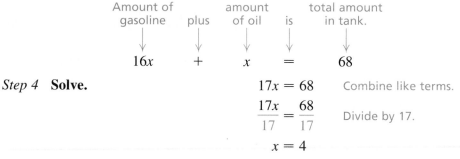

Step 4 **Solve.**
$$17x = 68 \qquad \text{Combine like terms.}$$
$$\frac{17x}{17} = \frac{68}{17} \qquad \text{Divide by 17.}$$
$$x = 4$$

Step 5 **State the answer.** The lawn trimmer requires 4 oz of oil and $16(4) = 64$ oz of gasoline when full.

Step 6 **Check.** Since $4 + 64 = 68$, and 64 is 16 times 4, the answer checks.

Now Try Exercise 25.

> ### PROBLEM SOLVING
>
> Sometimes it is necessary to find three unknown quantities in an applied problem. Frequently the three unknowns are compared in *pairs*. When this happens, it is usually easiest to let the variable represent the unknown found in both pairs.

EXAMPLE 5 Dividing a Board into Pieces

The instructions for a woodworking project call for three pieces of wood. The longest piece must be twice the length of the middle-sized piece, and the shortest piece must be 10 in. shorter than the middle-sized piece. Maria Gonzales has a board 70 in. long that she wishes to use. How long can each piece be?

Step 1 **Read** the problem. There will be three answers.

Step 2 **Assign a variable.** Since the middle-sized piece appears in both pairs of comparisons, let x represent the length, in inches, of the middle-sized piece. We have

$$x = \text{the length of the middle-sized piece,}$$
$$2x = \text{the length of the longest piece, and}$$
$$x - 10 = \text{the length of the shortest piece.}$$

A sketch is helpful here. See Figure 2.

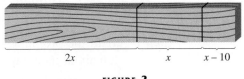

FIGURE 2

Step 3 **Write an equation.**

Longest	plus	middle-sized	plus	shortest	is	total length.
$2x$	$+$	x	$+$	$(x - 10)$	$=$	70

Step 4 **Solve.**

$$4x - 10 = 70 \qquad \text{Combine like terms.}$$
$$4x - 10 + 10 = 70 + 10 \qquad \text{Add 10.}$$
$$4x = 80 \qquad \text{Combine like terms.}$$
$$\frac{4x}{4} = \frac{80}{4} \qquad \text{Divide by 4.}$$
$$x = 20$$

Step 5 **State the answer.** The middle-sized piece is 20 in. long, the longest piece is $2(20) = 40$ in. long, and the shortest piece is $20 - 10 = 10$ in. long.

Step 6 **Check.** The sum of the lengths is 70 in. All conditions of the problem are satisfied.

Now Try Exercise 29.

OBJECTIVE 4 Solve problems involving supplementary and complementary angles. An angle can be measured by a unit called the **degree** (°), which is $\frac{1}{360}$ of a complete rotation. Two angles whose sum is 90° are said to be **complementary,** or complements of each other. An angle that measures 90° is a **right angle.** Two angles whose sum is 180° are said to be **supplementary,** or supplements of each other. One angle *supplements* the other to form a **straight angle** of 180°. See Figure 3. If x represents the degree measure of an angle, then

$90 - x$ represents the degree measure of its complement,

and $180 - x$ represents the degree measure of its supplement.

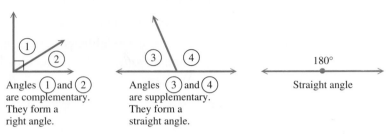

FIGURE 3

$10 + 2 \cdot (90 - x)$

■ **EXAMPLE 6   Finding the Measure of an Angle**

Find the measure of an angle whose supplement is 10° more than twice its complement.

Step 1   **Read** the problem. We are to find the measure of an angle, given information about its complement and its supplement.

Step 2   **Assign a variable.**

Let $x =$ the degree measure of the angle.

Then $90 - x =$ the degree measure of its complement,

and $180 - x =$ the degree measure of its supplement.

Step 3   **Write an equation.**

Supplement	is	10	more than	twice	its complement.
↓	↓	↓	↓	↓	↓
$180 - x$	$=$	10	$+$	$2 \cdot$	$(90 - x)$

Step 4   **Solve.**

$$180 - x = 10 + 180 - 2x \qquad \text{Distributive property}$$
$$180 - x = 190 - 2x \qquad \text{Combine like terms.}$$
$$180 - x + 2x = 190 - 2x + 2x \qquad \text{Add } 2x.$$
$$180 + x = 190 \qquad \text{Combine like terms.}$$
$$180 + x - 180 = 190 - 180 \qquad \text{Subtract 180.}$$
$$x = 10$$

Step 5   **State the answer.** The measure of the angle is 10°.

Step 6   **Check.** The complement of 10° is 80° and the supplement of 10° is 170°. 170° is equal to 10° more than twice 80° ($170 = 10 + 2(80)$ is true); therefore, the answer is correct.

Now Try Exercise 43.

OBJECTIVE 5 Solve problems involving consecutive integers.   Two integers that differ by 1 are called **consecutive integers.** For example, 3 and 4, 6 and 7, and -2 and -1

are pairs of consecutive integers. In general, if x represents an integer, $x + 1$ represents the next larger consecutive integer.

Consecutive *even* integers, such as 8 and 10, differ by 2. Similarly, consecutive *odd* integers, such as 9 and 11, also differ by 2. In general, if x represents an even integer, $x + 2$ represents the next larger consecutive even integer. The same holds true for odd integers; that is, if x is an odd integer, $x + 2$ is the next larger odd integer.

PROBLEM SOLVING

When solving consecutive integer problems, if $x =$ the first integer, then for any

two consecutive integers, use	$x, \quad x + 1;$
two consecutive *even* integers, use	$x, \quad x + 2;$
two consecutive *odd* integers, use	$x, \quad x + 2.$

EXAMPLE 7   Finding Consecutive Integers

Two pages that face each other in this book have 569 as the sum of their page numbers. What are the page numbers?

Step 1 **Read** the problem. Because the two pages face each other, they must have page numbers that are consecutive integers.

Step 2 **Assign a variable.**

Let $\qquad x =$ the smaller page number.

Then $\quad x + 1 =$ the larger page number.

Step 3 **Write an equation.** Because the sum of the page numbers is 569, the equation is

$$x + (x + 1) = 569.$$

Step 4 **Solve.** $\qquad$

$2x + 1 = 569$	Combine like terms.
$2x = 568$	Subtract 1.
$x = 284$	Divide by 2.

Step 5 **State the answer.** The smaller page number is 284, and the larger page number is $284 + 1 = 285$.

Step 6 **Check.** The sum of 284 and 285 is 569. The answer is correct.

Now Try Exercise 47.

In the final example, we do not number the steps. See if you can identify them.

EXAMPLE 8   Finding Consecutive Odd Integers

If the smaller of two consecutive odd integers is doubled, the result is 7 more than the larger of the two integers. Find the two integers.

Let x be the smaller integer. Since the two numbers are consecutive *odd* integers, then $x + 2$ is the larger. Now we write an equation.

If the smaller is doubled, the result is   7   more than   the larger.

$$2x = 7 + (x + 2)$$

$2x = 9 + x$ Combine like terms.

$x = 9$ Subtract x.

The first integer is 9 and the second is $9 + 2 = 11$. To check, we see that when 9 is doubled, we get 18, which is 7 more than the larger odd integer, 11. The answers are correct.

Now Try Exercise 49.

CONNECTIONS

George Polya (1888–1985), a native of Budapest, Hungary, wrote the modern classic *How to Solve It.* In this book he proposed a four-step process for problem solving:

1. Understand the problem.
2. Devise a plan.
3. Carry out the plan.
4. Look back and check.

For Discussion or Writing

Compare Polya's four-step process with the six steps given in this section. Identify which of the steps in our list match Polya's four steps. Trial and error is also a useful problem-solving tool. Where does this method fit into Polya's steps?

2.4 EXERCISES

For Extra Help

Student's Solutions Manual

MyMathLab

InterAct Math Tutorial Software

Tutor Center
AW Math Tutor Center

MathXL MathXL

Digital Video Tutor CD 3/Videotape 3

1. In your own words, write the general procedure for solving applications as outlined in this section.

2. List some of the words that translate as "=" when writing an equation to solve an applied problem.

3. Suppose that a problem requires you to find the number of cars on a dealer's lot. Which one of the following would *not* be a reasonable answer? Justify your answer.

   A. 0   **B.** 45   **C.** 1   **D.** $6\frac{1}{2}$

4. Suppose that a problem requires you to find the number of hours a light bulb is on during a day. Which one of the following would *not* be a reasonable answer? Justify your answer.

   A. 0   **B.** 4.5   **C.** 13   **D.** 25

Solve each problem. See Example 1.

5. If 2 is added to five times a number, the result is equal to 5 more than four times the number. Find the number.

6. If four times a number is added to 8, the result is three times the number added to 5. Find the number.

7. If 2 is subtracted from a number and this difference is tripled, the result is 6 more than the number. Find the number.

8. If 3 is added to a number and this sum is doubled, the result is 2 more than the number. Find the number.

9. The sum of three times a number and 7 more than the number is the same as the difference between -11 and twice the number. What is the number?

10. If 4 is added to twice a number and this sum is multiplied by 2, the result is the same as if the number is multiplied by 3 and 4 is added to the product. What is the number?

Solve each problem. See Example 2.

11. The number of drive-in movie screens has declined steadily in the United States since the 1960s. California and New York were two of the states with the most remaining drive-in movie screens in 2001. California had 11 more screens than New York, and there were 107 screens total in the two states. How many drive-in movie screens remained in each state? (*Source:* National Association of Theatre Owners.)

12. Thursday is the most-watched night for the major broadcast TV networks (ABC, CBS, NBC, and Fox), with 20 million more viewers than Saturday, the least-watched night. The total for the two nights is 102 million viewers. How many viewers of the major networks are there on each of these nights? (*Source:* Nielsen Media Research.)

13. The U.S. Senate has 100 members. During the 106th session (1999–2001), there were 10 more Republicans than Democrats. How many Democrats and Republicans were there in the Senate? (*Source: World Almanac and Book of Facts,* 2000.)

14. The total number of Democrats and Republicans in the U.S. House of Representatives during the 106th session was 434. There were 12 more Republicans than Democrats. How many members of each party were there? (*Source: World Almanac and Book of Facts,* 2000.)

15. The rock band U2 generated top revenue on the concert circuit in 2001. U2 and second-place 'N Sync together took in $196.5 million from ticket sales. If 'N Sync took in $22.9 million less than U2, how much revenue did each generate? (*Source:* Pollstar.)

16. The Toyota Camry was the top-selling passenger car in the United States in 2000, followed by the Honda Accord. Honda Accord sales were 18 thousand less than Toyota Camry sales, and 828 thousand of these two cars were sold. How many of each make of car were sold? (*Source:* Ward's Communications.)

17. In the 2001–2002 NBA regular season, the Sacramento Kings won two less than three times as many games as they lost. The Kings played 82 games. How many wins and losses did the team have? (*Source:* nba.com)

18. In the 2000–2001 regular baseball season, the Oakland Athletics won 18 less than twice as many games as they lost. They played 162 regular season games. How many wins and losses did the team have? (*Source: World Almanac and Book of Facts,* 2002.)

Solve each problem. See Examples 3 and 4.

19. A September 2000 issue of *Coin World* listed the value of a "Mint State-65" (uncirculated) 1950 Jefferson nickel minted at Denver at $\frac{7}{6}$ the value of a similar condition 1945 nickel minted at Philadelphia. Together the total value of the two coins is $26.00. What is the value of each coin?

20. The largest sheep ranch in the world is located in Australia. The number of sheep on the ranch is $\frac{8}{3}$ the number of uninvited kangaroos grazing on the pastureland. Together, herds of these two animals number 88,000. How many sheep and how many kangaroos roam the ranch? (*Source: Guinness Book of Records.*)

21. In 1988, a dairy in Alberta, Canada, created a sundae with approximately 1 lb of topping for every 83.2 lb of ice cream. The total of the two ingredients weighed approximately 45,225 lb. To the nearest tenth of a pound, how many pounds of ice cream and how many pounds of topping were there? (*Source: Guinness Book of Records.*)

22. A husky running the Iditarod (a thousand-mile race between Anchorage and Nome, Alaska) burns $5\frac{3}{8}$ calories in exertion for every 1 calorie burned in thermoregulation in extreme cold. According to one scientific study, a husky in top condition burns an amazing total of 11,200 calories per day. How many calories are burned for exertion, and how many are burned for regulation of body temperature? Round answers to the nearest whole number.

23. A pharmacist found that at the end of the day she had $\frac{4}{3}$ as many prescriptions for antibiotics as she did for tranquilizers. She had 42 prescriptions altogether for these two types of drugs. How many did she have for tranquilizers?

24. In a mixture of concrete, there are 3 lb of cement mix for every 1 lb of gravel. If the mixture contains a total of 140 lb of these two ingredients, how many pounds of gravel are there?

 25. A mixture of nuts contains only peanuts and cashews. For every 1 oz of cashews there are 5 oz of peanuts. If the mixture contains a total of 27 oz, how many ounces of each type of nut does the mixture contain?

26. An insecticide contains 95 cg of inert ingredient for every 1 cg of active ingredient. If a quantity of the insecticide weighs 336 cg, how much of each type of ingredient does it contain?

Solve each problem. See Example 5.

27. In one day, Akilah Cadet received 13 packages. Federal Express delivered three times as many as Airborne Express, while United Parcel Service delivered 2 fewer than Airborne Express. How many packages did each service deliver to Akilah?

28. In his job at the post office, Eddie Thibodeaux works a 6.5-hr day. He sorts mail, sells stamps, and does supervisory work. One day he sold stamps twice as long as he sorted mail, and he supervised .5 hr longer than he sorted mail. How many hours did he spend at each task?

29. The United States earned an all-time high of 34 medals at the 2002 Winter Olympics. The number of gold medals earned was 1 less than the number of bronze medals. The number of silver medals earned was 2 more than the number of bronze medals. How many of each kind of medal did the United States earn? (*Source:* U.S. Olympic Committee.)

30. Nagaraj Nanjappa has a party-length submarine sandwich 59 in. long. He wants to cut it into three pieces so that the middle piece is 5 in. longer than the shortest piece and the shortest piece is 9 in. shorter than the longest piece. How long should the three pieces be?

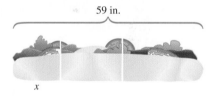

31. Venus is 31.2 million mi farther from the sun than Mercury, while Earth is 57 million mi farther from the sun than Mercury. If the total of the distances from these three planets to the sun is 196.2 million mi, how far away from the sun is Mercury? (All distances given here are *mean* (*average*) distances.) (*Source: Universal Almanac,* 1997.)

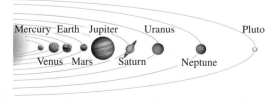

32. Together, Saturn, Jupiter, and Mars have a total of 36 known satellites (moons). Jupiter has 2 fewer satellites than Saturn, and Mars has 16 fewer satellites than Saturn. How many known satellites does Mars have? (*Source: World Almanac and Book of Facts,* 2000.)

33. The sum of the measures of the angles of any triangle is 180°. In triangle *ABC,* angles *A* and *B* have the same measure, while the measure of angle *C* is 60° larger than each of *A* and *B.* What are the measures of the three angles?

34. In triangle *ABC,* the measure of angle *A* is 141° more than the measure of angle *B.* The measure of angle *B* is the same as the measure of angle *C.* Find the measure of each angle. (*Hint:* See Exercise 33.)

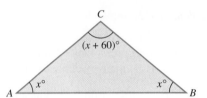

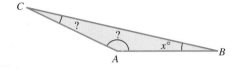

Use the concepts of this section to answer each question.

35. If the sum of two numbers is *k,* and one of the numbers is *m,* how can you express the other number?

36. If the product of two numbers is *r,* and one of the numbers is *s* (*s* ≠ 0), how can you express the other number?

37. Is there an angle whose supplement is equal to its complement? If so, what is the measure of the angle?

38. Is there an angle that is equal to its supplement? Is there an angle that is equal to its complement? If the answer is yes to either question, give the measure of the angle.

39. If *x* represents an integer, how can you express the next smaller consecutive integer in terms of *x*?

40. If *x* represents an integer, how can you express the next smaller even integer in terms of *x*?

Solve each problem. See Example 6.

41. Find the measure an an angle whose complement is four times its measure.

42. Find the measure of an angle whose supplement is three times its measure.

43. Find the measure of an angle whose supplement measures 39° more than twice its complement.

44. Find the measure of an angle whose supplement measures 38° less than three times its complement.

45. Find the measure of an angle such that the difference between the measures of its supplement and three times its complement is 10°.

46. Find the measure of an angle such that the sum of the measures of its complement and its supplement is 160°.

Solve each problem. See Examples 7 and 8.

47. The numbers on two consecutively numbered gym lockers have a sum of 137. What are the locker numbers?

48. The sum of two consecutive checkbook check numbers is 357. Find the numbers.

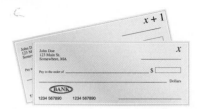

49. Find two consecutive even integers such that the smaller added to three times the larger gives a sum of 46.

50. Find two consecutive odd integers such that twice the larger is 17 more than the smaller.

51. Two pages that are back-to-back in this book have 203 as the sum of their page numbers. What are the page numbers?

52. Two houses on the same side of the street have house numbers that are consecutive even integers. The sum of the integers is 58. What are the two house numbers?

53. When the smaller of two consecutive integers is added to three times the larger, the result is 43. Find the integers.

54. If five times the smaller of two consecutive integers is added to three times the larger, the result is 59. Find the integers.

55. If the sum of three consecutive even integers is 60, what is the smallest even integer? (*Hint:* If x and $x + 2$ represent the first two consecutive even integers, how would you represent the third consecutive even integer?)

56. If the sum of three consecutive odd integers is 69, what is the largest odd integer?

57. If 6 is subtracted from the largest of three consecutive odd integers, with this result multiplied by 2, the answer is 23 less than the sum of the first and twice the second of the integers. Find the integers.

58. If the first and third of three consecutive even integers are added, the result is 22 less than three times the second integer. Find the integers.

Apply the ideas of this section to solve Exercises 59 and 60, which are based on the graphs.

59. In a recent year, the funding for Head Start programs increased by .55 billion dollars from the funding in the previous year. The following year, the increase was .20 billion dollars more. For this 3-year period, the total funding was 9.64 billion dollars. How much was funded in each of these years? (*Source:* U.S. Department of Health and Human Services.)

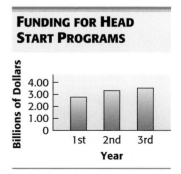

60. According to data provided by the National Safety Council for a recent year, the number of serious injuries per 100,000 participants in football, bicycling, and golf is illustrated in the graph. There were 800 more in bicycling than in golf, and there were 1267 more in football than in bicycling. Altogether there were 3179 serious injuries per 100,000 participants. How many such serious injuries were there in each sport?

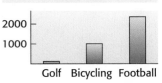

SERIOUS INJURIES
(per 100,000 participants)

Golf Bicycling Football

2.5 Formulas and Applications from Geometry

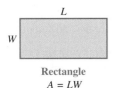

Rectangle
$A = LW$

FIGURE 4

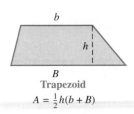

Trapezoid
$A = \frac{1}{2}h(b + B)$

FIGURE 5

Many applied problems can be solved with formulas. For example, formulas exist for geometric figures such as squares and circles, for distance, for money earned on bank savings, and for converting English measurements to metric measurements. The formulas used in this book are given on the inside covers.

OBJECTIVE 1 Solve a formula for one variable, given the values of the other variables. Given the values of all but one of the variables in a formula, we can find the value of the remaining variable by using the methods introduced in this chapter.

In Example 1, we use the idea of *area*. The **area** of a plane (two-dimensional) geometric figure is a measure of the surface covered by the figure.

■ **EXAMPLE 1** Using Formulas to Evaluate Variables

Find the value of the remaining variable in each formula.

(a) $A = LW$; $A = 64$, $L = 10$

As shown in Figure 4, this formula gives the area of a rectangle with length L and width W. Substitute the given values into the formula and then solve for W.

$$A = LW$$
$$64 = 10W \qquad \text{Let } A = 64 \text{ and } L = 10.$$
$$\frac{64}{10} = \frac{10W}{10} \qquad \text{Divide by 10.}$$
$$6.4 = W$$

The width of the rectangle is 6.4. Since $10(6.4) = 64$, the given area, the answer checks.

(b) $A = \frac{1}{2}h(b + B)$; $A = 210$, $B = 27$, $h = 10$

This formula gives the area of a trapezoid with parallel sides of lengths b and B and distance h between the parallel sides. See Figure 5. Again, begin by substituting the given values into the formula.

$$A = \frac{1}{2}h(b + B)$$
$$210 = \frac{1}{2}(10)(b + 27) \qquad A = 210, h = 10, B = 27$$

Now solve for b.

$$210 = 5(b + 27) \qquad \text{Multiply.}$$
$$210 = 5b + 135 \qquad \text{Distributive property}$$
$$210 - 135 = 5b + 135 - 135 \qquad \text{Subtract 135.}$$
$$75 = 5b \qquad \text{Combine like terms.}$$
$$\frac{75}{5} = \frac{5b}{5} \qquad \text{Divide by 5.}$$
$$15 = b$$

Check that the length of the shorter parallel side, b, is 15.

Now Try Exercises 19 and 23.

OBJECTIVE 2 Use a formula to solve an applied problem. As the next examples show, formulas are often used to solve applied problems. *It is a good idea to draw a sketch when a geometric figure is involved.* Example 2 uses the idea of perimeter. The **perimeter** of a plane (two-dimensional) geometric figure is the distance around the figure, that is, the sum of the lengths of its sides. We use the six steps introduced in the previous section.

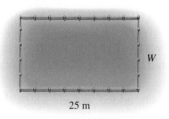

25 m

FIGURE 6

EXAMPLE 2 Finding the Width of a Rectangular Lot

A rectangular lot has perimeter 80 m and length 25 m. Find the width of the lot.

Step 1 **Read.** We are told to find the width of the lot.

Step 2 **Assign a variable.** Let W = the width of the lot in meters. See Figure 6.

Step 3 **Write an equation.** The formula for the perimeter of a rectangle is

$$P = 2L + 2W.$$

We find the width by substituting 80 for P and 25 for L in the formula.

$$80 = 2(25) + 2W \qquad P = 80, L = 25$$

Step 4 **Solve** the equation.

$$80 = 50 + 2W \qquad \text{Multiply.}$$
$$80 - 50 = 50 + 2W - 50 \qquad \text{Subtract 50.}$$
$$30 = 2W \qquad \text{Combine like terms.}$$
$$\frac{30}{2} = \frac{2W}{2} \qquad \text{Divide by 2.}$$
$$15 = W$$

Step 5 **State the answer.** The width is 15 m.

Step 6 **Check.** If the width is 15 m and the length is 25 m, the distance around the rectangular lot (perimeter) is $2(25) + 2(15) = 50 + 30 = 80$ m, as required.

Now Try Exercise 37.

FIGURE 7

EXAMPLE 3 Finding the Height of a Triangular Sail

The area of a triangular sail of a sailboat is 126 ft². (Recall that ft² means "square feet.") The base of the sail is 12 ft. Find the height of the sail.

Step 1 **Read.** We must find the height of the triangular sail.

Step 2 **Assign a variable.** Let h = the height of the sail in feet. See Figure 7.

Step 3 **Write an equation.** The formula for the area of a triangle is $A = \frac{1}{2}bh$, where A is the area, b is the base, and h is the height. Using the information given in the problem, we substitute 126 for A and 12 for b in the formula.

$$A = \frac{1}{2}bh$$

$$126 = \frac{1}{2}(12)h \qquad A = 126,\ b = 12$$

Step 4 **Solve.**
$$126 = 6h \qquad \text{Multiply.}$$
$$21 = h \qquad \text{Divide by 6.}$$

Step 5 **State the answer.** The height of the sail is 21 ft.

Step 6 **Check** to see that the values $A = 126$, $b = 12$, and $h = 21$ satisfy the formula for the area of a triangle.

Now Try Exercise 39.

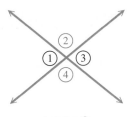

FIGURE 8

OBJECTIVE 3 Solve problems involving vertical angles and straight angles. Figure 8 shows two intersecting lines forming angles that are numbered ①, ②, ③, and ④. Angles ① and ③ lie "opposite" each other. They are called **vertical angles.** Another pair of vertical angles is ② and ④. In geometry, it is shown that vertical angles have equal measures.

Now look at angles ① and ②. When their measures are added, we get 180°, the measure of a straight angle. There are three other such pairs of angles: ② and ③, ③ and ④, and ① and ④.

The next example uses these ideas.

EXAMPLE 4 Finding Angle Measures

Refer to the appropriate figure in each part.

(a) Find the measure of each marked angle in Figure 9.

Since the marked angles are vertical angles, they have equal measures. Set $4x + 19$ equal to $6x - 5$ and solve.

$$4x + 19 = 6x - 5$$
$$19 = 2x - 5 \qquad \text{Subtract } 4x.$$
$$24 = 2x \qquad \text{Add 5.}$$
$$12 = x \qquad \text{Divide by 2.}$$

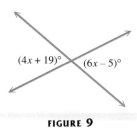

FIGURE 9

Since $x = 12$, one angle has measure $4(12) + 19 = 67$ degrees. The other has the same measure, since $6(12) - 5 = 67$ as required. Each angle measures 67°.

(b) Find the measure of each marked angle in Figure 10.

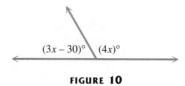

FIGURE 10

The measures of the marked angles must add to $180°$ because together they form a straight angle. The equation to solve is

$$(3x - 30) + 4x = 180.$$
$$7x - 30 = 180 \qquad \text{Combine like terms.}$$
$$7x = 210 \qquad \text{Add 30.}$$
$$x = 30 \qquad \text{Divide by 7.}$$

To find the measures of the angles, replace x with 30 in the two expressions.

$$3x - 30 = 3(30) - 30 = 90 - 30 = 60$$
$$4x = 4(30) = 120$$

The two angle measures are $60°$ and $120°$.

Now Try Exercises 45 and 47.

CAUTION   In Example 4, the answer is *not* the value of x. Remember to substitute the value of the variable into the expression given for each angle.

OBJECTIVE 4 Solve a formula for a specified variable.   Sometimes it is necessary to solve a number of problems that use the same formula. For example, a surveying class might need to solve several problems that involve the formula for the area of a rectangle, $A = LW$. Suppose that in each problem the area (A) and the length (L) of a rectangle are given, and the width (W) must be found. Rather than solving for W each time the formula is used, it would be simpler to *rewrite the formula* so that it is solved for W. This process is called **solving for a specified variable.**

In solving a formula for a specified variable, we treat the specified variable as if it were the *only* variable in the equation, and treat the other variables as if they were numbers. We use the same steps to solve the equation for the specified variable that we use to solve equations with just one variable.

EXAMPLE 5 Solving for a Specified Variable

Solve $A = LW$ for W.

Think of undoing what has been done to W. Since W is multiplied by L, undo the multiplication by dividing each side of $A = LW$ by L.

$$A = LW$$
$$\frac{A}{L} = \frac{LW}{L} \qquad \text{Divide by } L.$$
$$\frac{A}{L} = W \quad \text{or} \quad W = \frac{A}{L} \qquad \tfrac{L}{L} = 1;\ 1W = W$$

Now Try Exercise 51.

EXAMPLE 6 Solving for a Specified Variable

Solve $P = 2L + 2W$ for L.

We want to get L alone on one side of the equation. We begin by subtracting $2W$ from each side.

$$P = 2L + 2W$$

$$P - 2W = 2L + 2W - 2W \qquad \text{Subtract } 2W.$$

$$P - 2W = 2L \qquad \text{Combine like terms.}$$

$$\frac{P - 2W}{2} = \frac{2L}{2} \qquad \text{Divide by 2.}$$

$$\frac{P - 2W}{2} = L \qquad \frac{2}{2} = 1; \ 1L = L$$

or
$$L = \frac{P - 2W}{2}$$

The last step gives the formula solved for L, as required.

Now Try Exercise 67.

EXAMPLE 7 Solving for a Specified Variable

Solve $F = \dfrac{9}{5}C + 32$ for C. (This is the formula for converting from Celsius to Fahrenheit.)

We need to isolate C on one side of the equation. First we undo the addition of 32 to $\frac{9}{5}C$ by subtracting 32 from each side.

$$F = \frac{9}{5}C + 32$$

$$F - 32 = \frac{9}{5}C + 32 - 32 \qquad \text{Subtract 32.}$$

$$F - 32 = \frac{9}{5}C$$

Now we multiply each side by $\frac{5}{9}$, using parentheses on the left.

$$\frac{5}{9}(F - 32) = \frac{5}{9} \cdot \frac{9}{5}C \qquad \text{Multiply by } \tfrac{5}{9}.$$

$$\frac{5}{9}(F - 32) = C \quad \text{or} \quad C = \frac{5}{9}(F - 32)$$

This last result is the formula for converting temperatures from Fahrenheit to Celsius.

Now Try Exercise 69.

EXAMPLE 8 Solving for a Specified Variable

Solve $A = \dfrac{1}{2}(b + B)h$ for B.

To get B alone, begin by multiplying each side by 2 to clear the fraction.

$$A = \frac{1}{2}(b + B)h$$

$$2A = 2 \cdot \frac{1}{2}(b + B)h$$ Multiply by 2. (*Do not distribute the 2 on the right side.*)

$$2A = (b + B)h$$ $2 \cdot \frac{1}{2} = \frac{2}{2} = 1$

$$2A = bh + Bh$$ Distributive property

$$2A - bh = bh + Bh - bh$$ Subtract *bh*.

$$2A - bh = Bh$$ Combine like terms.

$$\frac{2A - bh}{h} = \frac{Bh}{h}$$ Divide by *h*.

$$\frac{2A - bh}{h} = B \quad \text{or} \quad B = \frac{2A - bh}{h}$$

Now Try Exercise 73.

NOTE The result in Example 8 can be written in a different form as follows.

$$B = \frac{2A - bh}{h} = \frac{2A}{h} - \frac{bh}{h} = \frac{2A}{h} - b$$

Either form is correct.

2.5 EXERCISES

For Extra Help

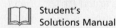
Student's
Solutions Manual

MyMathLab

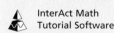
InterAct Math
Tutorial Software

AW Math
Tutor Center

Math XP MathXL

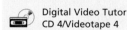
Digital Video Tutor
CD 4/Videotape 4

1. In your own words, explain what is meant by each term.

    (a) Perimeter of a plane geometric figure

    (b) Area of a plane geometric figure

2. Perimeter is to a polygon as _____ is to a circle.

3. If a formula has exactly five variables, how many values would you need to be given in order to find the value of any one variable?

4. The formula for changing Celsius to Fahrenheit is given in Example 7 as $F = \frac{9}{5}C + 32$. Sometimes it is seen as $F = \frac{9C}{5} + 32$. These are both correct. Why is it true that $\frac{9}{5}C$ is equal to $\frac{9C}{5}$?

Decide whether perimeter or area would be used to solve a problem concerning the measure of the quantity.

    5. Sod for a lawn

    6. Carpeting for a bedroom

    7. Baseboards for a living room

    8. Fencing for a yard

    9. Fertilizer for a garden

    10. Tile for a bathroom

    11. Determining the cost of planting rye grass in a lawn for the winter

    12. Determining the cost of replacing a linoleum floor with a wood floor

In the following exercises a formula is given, along with the values of all but one of the variables in the formula. Find the value of the variable that is not given. (When necessary, use 3.14 as an approximation for π.) See Example 1.

13. $P = 2L + 2W$ (perimeter of a rectangle); $L = 8$, $W = 5$

14. $P = 2L + 2W$; $L = 6$, $W = 4$

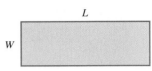

15. $A = \frac{1}{2}bh$ (area of a triangle); $b = 8$, $h = 16$

16. $A = \frac{1}{2}bh$; $b = 10$, $h = 14$

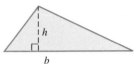

17. $P = a + b + c$ (perimeter of a triangle); $P = 12$, $a = 3$, $c = 5$

18. $P = a + b + c$; $P = 15$, $a = 3$, $b = 7$

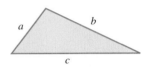

19. $d = rt$ (distance formula); $d = 252$, $r = 45$

20. $d = rt$; $d = 100$, $t = 2.5$

21. $I = prt$ (simple interest); $p = 7500$, $r = .035$, $t = 6$

22. $I = prt$; $p = 5000$, $r = .025$, $t = 7$

23. $A = \frac{1}{2}h(b + B)$ (area of a trapezoid); $A = 91$, $h = 7$, $b = 12$

24. $A = \frac{1}{2}h(b + B)$; $A = 75$, $b = 19$, $B = 31$

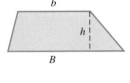

25. $C = 2\pi r$ (circumference of a circle); $C = 16.328$

26. $C = 2\pi r$; $C = 8.164$

27. $A = \pi r^2$ (area of a circle); $r = 4$

28. $A = \pi r^2$; $r = 12$

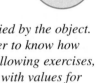

*The **volume** of a three-dimensional object is a measure of the space occupied by the object. For example, we would need to know the volume of a gasoline tank in order to know how many gallons of gasoline it would take to completely fill the tank. In the following exercises, a formula for the volume (V) of a three-dimensional object is given, along with values for the other variables. Evaluate V. (Use 3.14 as an approximation for π.) See Example 1.*

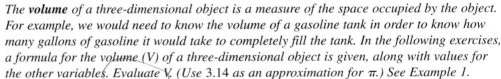

29. $V = LWH$ (volume of a rectangular box); $L = 10$, $W = 5$, $H = 3$

30. $V = LWH$; $L = 12$, $W = 8$, $H = 4$

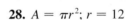

31. $V = \frac{1}{3}Bh$ (volume of a pyramid); $B = 12$, $h = 13$

32. $V = \frac{1}{3}Bh$; $B = 36$, $h = 4$

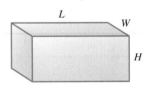

33. $V = \frac{4}{3}\pi r^3$ (volume of a sphere); $r = 12$

34. $V = \frac{4}{3}\pi r^3$; $r = 6$

Use a formula to write an equation for each application, and then use the problem-solving method of Section 2.4 to solve. (Use 3.14 as an approximation for π.) Formulas are found on the inside covers of this book. See Examples 2 and 3.

35. Recently, a prehistoric ceremonial site dating to about 3000 B.C. was discovered at Stanton Drew in southwestern England. The site, which is larger than Stonehenge, is a nearly perfect circle, consisting of nine concentric rings that probably held upright wooden posts. Around this timber temple is a wide, encircling ditch enclosing an area with a diameter of 443 ft. Find this enclosed area to the nearest thousand square feet. (*Hint:* Find the radius. Then use $A = \pi r^2$.) (*Source: Archaeology,* vol. 51, no. 1, Jan./Feb. 1998.)

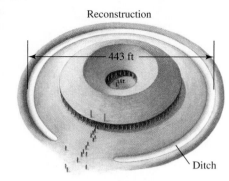

Reconstruction

443 ft

Ditch

36. The Skydome in Toronto, Canada, is the first stadium with a hard-shell, retractable roof. The steel dome is 630 ft in diameter. To the nearest foot, what is the circumference of this dome? (*Source:* www.4ballparks.com)

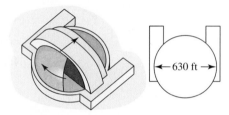

630 ft

37. The *Daily Banner,* published in Roseberg, Oregon, in the 19th century, had page size 3 in. by 3.5 in. What was the perimeter? What was the area? (*Source: Guinness Book of Records.*)

38. The newspaper *The Constellation,* printed in 1859 in New York City as part of the Fourth of July celebration, had length 51 in. and width 35 in. What was the perimeter? What was the area? (*Source: Guinness Book of Records.*)

39. The largest drum ever constructed was played at the Royal Festival Hall in London in 1987. It had a diameter of 13 ft. What was the area of the circular face of the drum? (*Hint:* $A = \pi r^2$.) (*Source: Guinness Book of Records.*)

40. What was the circumference of the drum described in Exercise 39? (*Hint:* Use $C = 2\pi r$.)

41. The survey plat depicted here shows two lots that form a trapezoid. The measures of the parallel sides are 115.80 ft and 171.00 ft. The height of the trapezoid is 165.97 ft. Find the combined area of the two lots. Round your answer to the nearest hundredth of a square foot.

42. Lot A in the figure is in the shape of a trapezoid. The parallel sides measure 26.84 ft and 82.05 ft. The height of the trapezoid is 165.97 ft. Find the area of Lot A. Round your answer to the nearest hundredth of a square foot.

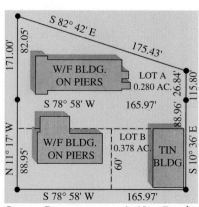

S 82° 42' E

171.00'

82.05'

175.43'

W/F BLDG. ON PIERS

LOT A 0.280 AC.

26.84'

115.80

S 78° 58' W   165.97'

N 11° 17' W

88.95'

W/F BLDG. ON PIERS

LOT B 0.378 AC.

60'

TIN BLDG

88.96

S 10° 36' E

S 78° 58' W   165.97'

Source: Property survey in New Roads, Louisiana.

43. The U.S. Postal Service requires that any box sent through the mail have length plus girth (distance around) totaling no more than 108 in. The maximum volume that meets this condition is contained by a box with a square end 18 in. on each side. What is the length of the box? What is the maximum volume?

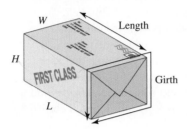

44. The largest box of popcorn was filled by students in Jacksonville, Florida. The box was approximately 40 ft long, $20\frac{2}{3}$ ft wide, and 8 ft high. To the nearest cubic foot, what was the volume of the box? (*Source: Guinness Book of Records.*)

Find the measure of each marked angle. See Example 4.

45.
$(x + 1)°$ $(4x - 56)°$

46.
$(10x + 7)°$ $(7x + 3)°$

47.
$(5x - 129)°$ $(2x - 21)°$

48.
$(3x + 45)°$ $(7x + 5)°$

49.
$(10x + 15)°$
$(12x - 3)°$

50.
$(11x - 37)°$ $(7x + 27)°$

Solve each formula for the specified variable. See Examples 5–8.

51. $d = rt$ for t

52. $d = rt$ for r

53. $A = bh$ for b

54. $A = LW$ for L

55. $C = \pi d$ for d

56. $P = 4s$ for s

57. $V = LWH$ for H

58. $V = LWH$ for W

59. $I = prt$ for r

60. $I = prt$ for p

61. $A = \frac{1}{2}bh$ for h

62. $A = \frac{1}{2}bh$ for b

63. $V = \frac{1}{3}\pi r^2 h$ for h

64. $V = \pi r^2 h$ for h

65. $P = a + b + c$ for b

66. $P = a + b + c$ for a

67. $P = 2L + 2W$ for W

68. $A = p + prt$ for r

69. $y = mx + b$ for m

70. $y = mx + b$ for x

71. $Ax + By = C$ for y

72. $Ax + By = C$ for x

73. $M = C(1 + r)$ for r

74. $C = \frac{5}{9}(F - 32)$ for F

2.6   Ratios and Proportions

OBJECTIVES

1 Write ratios.

2 Solve proportions.

3 Solve applied problems using proportions.

OBJECTIVE 1   Write ratios.   A **ratio** is a comparison of two quantities using a quotient.

> **Ratio**
>
> The ratio of the number a to the number b ($b \neq 0$) is written
>
> $$a \text{ to } b, \qquad a\!:\!b, \qquad \text{or} \qquad \frac{a}{b}.$$

The last way of writing a ratio is most common in algebra.

Percents are ratios where the second number is always 100. For example, 50% represents the ratio of 50 to 100, 27% represents the ratio of 27 to 100, and so on.

EXAMPLE 1   Writing Word Phrases as Ratios

Write a ratio for each word phrase.

(a) The ratio of 5 hr to 3 hr is

$$\frac{5 \text{ hr}}{3 \text{ hr}} = \frac{5}{3}.$$

(b) To find the ratio of 6 hr to 3 days, first convert 3 days to hours.

$$3 \text{ days} = 3 \cdot 24$$
$$= 72 \text{ hr}$$

The ratio of 6 hr to 3 days is thus

$$\frac{6 \text{ hr}}{3 \text{ days}} = \frac{6 \text{ hr}}{72 \text{ hr}} = \frac{6}{72} = \frac{1}{12}.$$

Now Try Exercises 3 and 7.

An application of ratios is in unit pricing, to see which size of an item offered in different sizes produces the best price per unit. To do this, set up the ratio of the price of the item to the number of units on the label. Then divide to obtain the price per unit.

EXAMPLE 2   Finding Price per Unit

The Winn-Dixie supermarket in Mandeville, Louisiana, charges the prices shown in the table for a box of trash bags. Which size is the best buy? That is, which size has the lowest unit price?

Size	Price
10-count	$1.28
20-count	$2.68
30-count	$3.88

To find the best buy, write ratios comparing the price for each box size to the number of units (bags) per box. The results in the table on the next page are rounded to the nearest thousandth.

Size	Unit Cost (dollars per bag)
10-count	$\dfrac{\$1.28}{10} = \$.128 \longleftarrow$ The best buy
20-count	$\dfrac{\$2.68}{20} = \$.134$
30-count	$\dfrac{\$3.88}{30} = \$.129$

Because the 10-count size produces the lowest unit cost, it is the best buy. This example shows that buying the largest size does not always provide the best buy, although this is often true.

Now Try Exercise 15.

OBJECTIVE 2 Solve proportions. A ratio is used to compare two numbers or amounts. A **proportion** says that two ratios are equal, so it is a special type of equation. For example,

$$\frac{3}{4} = \frac{15}{20}$$

is a proportion that says that the ratios $\frac{3}{4}$ and $\frac{15}{20}$ are equal. In the proportion

$$\frac{a}{b} = \frac{c}{d} \quad (b, d \neq 0),$$

a, b, c, and d are the **terms** of the proportion. The a and d terms are called the **extremes,** and the b and c terms are called the **means.** We read the proportion $\frac{a}{b} = \frac{c}{d}$ as "a is to b as c is to d." Beginning with this proportion and multiplying each side by the common denominator, bd, gives

$$\frac{a}{b} = \frac{c}{d}$$

$$bd \cdot \frac{a}{b} = bd \cdot \frac{c}{d}$$

$$\frac{b}{b}(d \cdot a) = \frac{d}{d}(b \cdot c) \qquad \text{Associative and commutative properties}$$

$$ad = bc. \qquad \text{Commutative and identity properties}$$

We can also find the products ad and bc by multiplying diagonally.

$$\frac{a}{b} = \frac{c}{d}$$

bc

ad

For this reason, ad and bc are called **cross products.**

Cross Products

If $\dfrac{a}{b} = \dfrac{c}{d}$, then the cross products ad and bc are equal.

Also, if $ad = bc$, then $\dfrac{a}{b} = \dfrac{c}{d}$ $(b, d \neq 0)$.

From this rule, if $\frac{a}{b} = \frac{c}{d}$ then $ad = bc$; that is, the product of the extremes equals the product of the means.

NOTE If $\frac{a}{c} = \frac{b}{d}$, then $ad = cb$, or $ad = bc$. This means that the two proportions are equivalent, and

$$\text{the proportion } \frac{a}{b} = \frac{c}{d} \text{ can also be written as } \frac{a}{c} = \frac{b}{d} \quad (c \neq 0).$$

Sometimes one form is more convenient to work with than the other.

EXAMPLE 3 Deciding Whether Proportions Are True

Decide whether each proportion is true or false.

(a) $\dfrac{3}{4} = \dfrac{15}{20}$

Check to see whether the cross products are equal.

$$4 \cdot 15 = 60$$
$$\frac{3}{4} = \frac{15}{20}$$
$$3 \cdot 20 = 60$$

The cross products are equal, so the proportion is true.

(b) $\dfrac{6}{7} = \dfrac{30}{32}$

The cross products are $6 \cdot 32 = 192$ and $7 \cdot 30 = 210$. The cross products are not equal, so the proportion is false.

Now Try Exercises 23 and 25.

Four numbers are used in a proportion. If any three of these numbers are known, the fourth can be found.

EXAMPLE 4 Finding an Unknown in a Proportion

Solve the proportion

$$\frac{5}{9} = \frac{x}{63}.$$

The cross products must be equal.

$$5 \cdot 63 = 9 \cdot x \qquad \text{Cross products}$$
$$315 = 9x \qquad \text{Multiply.}$$
$$35 = x \qquad \text{Divide by 9.}$$

Check by substituting 35 for x in the proportion. The solution set is $\{35\}$.

Now Try Exercise 29.

CAUTION The cross product method cannot be used directly if there is more than one term on either side of the equals sign.

EXAMPLE 5 Solving an Equation Using Cross Products

Solve the equation

$$\frac{m-2}{5} = \frac{m+1}{3}.$$

Find the cross products.

$3(m-2) = 5(m+1)$	Be sure to use parentheses.
$3m - 6 = 5m + 5$	Distributive property
$3m = 5m + 11$	Add 6.
$-2m = 11$	Subtract 5m.
$m = -\dfrac{11}{2}$	Divide by −2.

The solution set is $\left\{-\frac{11}{2}\right\}$.

Now Try Exercise 37.

NOTE When you set cross products equal to each other, you are really multiplying each ratio in the proportion by a common denominator.

OBJECTIVE 3 Solve applied problems using proportions. Proportions are useful in many practical applications. We continue to use the six-step method, although the steps are not numbered here.

EXAMPLE 6 Applying Proportions

After Lee Ann Spahr pumped 5.0 gal of gasoline, the display showing the price read $7.90. When she finished pumping the gasoline, the price display read $21.33. How many gallons did she pump?

To solve this problem, set up a proportion, with prices in the numerators and gallons in the denominators. Make sure that the corresponding numbers appear together.

Let x = the number of gallons she pumped. Then

Price $\longrightarrow \dfrac{\$7.90}{5.0} = \dfrac{\$21.33}{x} \longleftarrow$ Price
Gallons $\longrightarrow \qquad\qquad\quad \longleftarrow$ Gallons

$7.90x = 5.0(21.33)$	Cross products
$7.90x = 106.65$	Multiply.
$x = 13.5.$	Divide by 7.90.

She pumped 13.5 gal. Check this answer. Using a calculator to perform the arithmetic reduces the possibility of errors. Notice that the way the proportion was set up uses the fact that the unit price is the same, no matter how many gallons are purchased.

Now Try Exercise 43.

2.6 EXERCISES

1. Match each ratio in Column I with the ratio equivalent to it in Column II.

I	II
(a) 75 to 100	**A.** 80 to 100
(b) 5 to 4	**B.** 50 to 100
(c) $\dfrac{1}{2}$	**C.** 3 to 4
	D. 15 to 12
(d) 4 to 5	

2. Give three different, equivalent forms of the ratio $\frac{4}{3}$.

Write a ratio for each word phrase. In Exercises 7–12, first write the amounts with the same units. Write fractions in lowest terms. See Example 1.

3. 60 ft to 70 ft

4. 40 mi to 30 mi

5. 72 dollars to 220 dollars

6. 120 people to 90 people

7. 30 in. to 8 ft

8. 20 yd to 8 ft

9. 16 min to 1 hr

10. 24 min to 2 hr

11. 5 days to 40 hr

12. 60 in. to 2 yd

A supermarket was surveyed to find the prices charged for items in various sizes. Find the best buy (based on price per unit) for each item. See Example 2.

13. Seasoning mix

8-oz size: $1.75
17-oz size: $2.88

14. Red beans

1-lb package: $.89
2-lb package: $1.79

15. Prune juice

32-oz can: $1.95
48-oz can: $2.89
64-oz can: $3.29

16. Corn oil

24-oz bottle: $2.08
64-oz bottle: $3.94
128-oz bottle: $7.65

17. Artificial sweetener packets

50-count: $1.19
100-count: $1.85
250-count: $3.79
500-count: $6.38

18. Chili (no beans)

7.5-oz can: $1.19
10.5-oz can: $1.29
15-oz can: $1.78
25-oz can: $2.59

19. Extra crunchy peanut butter

12-oz size: $1.49
28-oz size: $1.99
40-oz size: $3.99

20. Tomato ketchup

14-oz size: $.93
32-oz size: $1.19
44-oz size: $2.19

21. Explain the distinction between *ratio* and *proportion*. Give examples.

22. Suppose that someone told you to use cross products to multiply fractions. How would you explain to the person what is wrong with his or her thinking?

Decide whether each proportion is true *or* false. *See Example 3.*

23. $\dfrac{5}{35} = \dfrac{8}{56}$

24. $\dfrac{4}{12} = \dfrac{7}{21}$

25. $\dfrac{120}{82} = \dfrac{7}{10}$

26. $\dfrac{27}{160} = \dfrac{18}{110}$

27. $\dfrac{\frac{1}{2}}{5} = \dfrac{1}{10}$

28. $\dfrac{\frac{1}{3}}{6} = \dfrac{1}{18}$

Solve each equation. See Examples 4 and 5.

29. $\dfrac{k}{4} = \dfrac{175}{20}$

30. $\dfrac{x}{6} = \dfrac{18}{4}$

31. $\dfrac{49}{56} = \dfrac{z}{8}$

32. $\dfrac{20}{100} = \dfrac{z}{80}$

33. $\dfrac{a}{24} = \dfrac{15}{16}$

34. $\dfrac{x}{4} = \dfrac{12}{30}$

35. $\dfrac{z}{2} = \dfrac{z+1}{3}$

36. $\dfrac{m}{5} = \dfrac{m-2}{2}$

37. $\dfrac{3y-2}{5} = \dfrac{6y-5}{11}$

38. $\dfrac{2r+8}{4} = \dfrac{3r-9}{3}$

39. $\dfrac{5k+1}{6} = \dfrac{3k-2}{3}$

40. $\dfrac{x+4}{6} = \dfrac{x+10}{8}$

41. $\dfrac{2p+7}{3} = \dfrac{p-1}{4}$

42. $\dfrac{3m-2}{5} = \dfrac{4-m}{3}$

Solve each problem. See Example 6.

43. If nine pairs of jeans cost $121.50, find the cost of five pairs. (Assume all are equally priced.)

44. If 7 shirts cost $87.50, find the cost of 11 shirts. (Assume all are equally priced.)

45. If 6 gal of premium unleaded gasoline cost $11.34, how much would it cost to completely fill a 15-gal tank?

46. If sales tax on a $16.00 compact disc is $1.32, how much would the sales tax be on a $120.00 compact disc player?

47. The distance between Kansas City, Missouri, and Denver is 600 mi. On a certain wall map, this is represented by a length of 2.4 ft. On the map, how many feet would there be between Memphis and Philadelphia, two cities that are actually 1000 mi apart?

48. The distance between Singapore and Tokyo is 3300 mi. On a certain wall map, this distance is represented by 11 in. The actual distance between Mexico City and Cairo is 7700 mi. How far apart are they on the same map?

49. A chain saw requires a mixture of 2-cycle engine oil and gasoline. According to the directions on a bottle of Oregon 2-cycle Engine Oil, for a 50 to 1 ratio requirement, approximately 2.5 fluid oz of oil are required for 1 gal of gasoline. For 2.75 gal, how many fluid ounces of oil are required?

50. The directions on the bottle mentioned in Exercise 49 indicate that if the ratio requirement is 24 to 1, approximately 5.5 oz of oil are required for 1 gal of gasoline. If gasoline is to be mixed with 22 oz of oil, how much gasoline is to be used?

51. In a recent year, the average exchange rate between British pounds and U.S. dollars was 1 pound to $1.6762. Margaret went to London and exchanged her U.S. currency for British pounds, and received 400 pounds. How much in U.S. dollars did Margaret exchange? (*Note:* Great Britain and Switzerland (see Exercise 52) are among the European nations not using euros.)

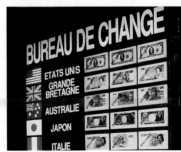

52. If 3 U.S. dollars can be exchanged for 4.5204 Swiss francs, how many Swiss francs can be obtained for $49.20? (Round to the nearest hundredth.)

53. Biologists tagged 500 fish in Willow Lake on October 5. At a later date they found 7 tagged fish in a sample of 700. Estimate the total number of fish in Willow Lake to the nearest hundred.

54. On May 13 researchers at Argyle Lake tagged 840 fish. When they returned a few weeks later, their sample of 1000 fish contained 18 that were tagged. Give an approximation of the fish population in Argyle Lake to the nearest hundred.

The International Olympic Committee has come to rely more and more on television rights and major corporate sponsors to finance the games. The circle graphs show funding plans for the first Olympics in Athens and the 1996 Olympics in Atlanta 100 yr later. Use proportions and the graphs to work Exercises 55 and 56.

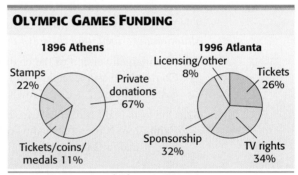

OLYMPIC GAMES FUNDING

Source: International Olympic Committee.

55. In the 1996 Olympics, total revenue of $350 million was raised. There were 10 major sponsors.

  (a) Write a proportion to find the amount of revenue provided by tickets. Solve it.

  (b) What amount was provided by sponsors? Assuming the sponsors contributed equally, how much was provided per sponsor?

  (c) What amount was raised by TV rights?

56. Suppose the amount of revenue raised in the 1896 Olympics was equivalent to the $350 million in 1996.

  (a) Write a proportion for the amount of revenue provided by stamps and solve it.

  (b) What amount (in dollars) would have been provided by private donations?

Two triangles are **similar** if they have the same shape (but not necessarily the same size). Similar triangles have sides that are proportional. The figure shows two similar triangles. Notice that the ratios of the corresponding sides all equal $\frac{3}{2}$:

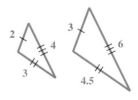

$$\frac{3}{2} = \frac{3}{2} \qquad \frac{4.5}{3} = \frac{3}{2} \qquad \frac{6}{4} = \frac{3}{2}.$$

If we know that two triangles are similar, we can set up a proportion to solve for the length of an unknown side.

(continued)

Use a proportion to find the length x, given that each pair of triangles is similar.

57.

58.

59.

60.

For Exercises 61 and 62, (a) draw a sketch consisting of two right triangles, depicting the situation described, and (b) solve the problem. (Source: Guinness Book of Records.)

61. An enlarged version of the chair used by George Washington at the Constitutional Convention casts a shadow 18 ft long at the same time a vertical pole 12 ft high casts a shadow 4 ft long. How tall is the chair?

62. One of the tallest candles ever constructed was exhibited at the 1897 Stockholm Exhibition. If it cast a shadow 5 ft long at the same time a vertical pole 32 ft high cast a shadow 2 ft long, how tall was the candle?

The Consumer Price Index provides a means of determining the purchasing power of the U.S. dollar from one year to the next. Using the period from 1982 to 1984 as a measure of 100.0, the Consumer Price Index for selected years from 1990 through 2000 is shown in the table. To use the Consumer Price Index to predict a price in a particular year, we set up a proportion and compare it with a known price in another year, as follows:

$$\frac{\text{price in year } A}{\text{index in year } A} = \frac{\text{price in year } B}{\text{index in year } B}.$$

Year	Consumer Price Index
1990	130.7
1992	140.3
1994	148.2
1996	156.9
1998	163.0
1999	166.6
2000	172.2

Source: U.S. Bureau of Labor Statistics.

Use the Consumer Price Index figures in the table to find the amount that would be charged for the use of the same amount of electricity that cost $225 in 1990. Give your answer to the nearest dollar.

63. in 1996   **64.** in 1998   **65.** in 1999   **66.** in 2000

RELATING CONCEPTS (EXERCISES 67–70)

For Individual or Group Work

*In Section 2.3 we solved equations with fractions by first multiplying each side of the equation by the common denominator. A proportion with a variable is this kind of equation. **Work Exercises 67–70 in order.** The steps justify the method of solving a proportion by cross products.*

67. What is the LCD of the fractions in the equation $\dfrac{x}{6} = \dfrac{2}{5}$?

68. Solve the equation in Exercise 67 as follows.

(a) Multiply each side by the LCD. What equation do you get?

(b) Solve the equation from part (a) by dividing each side by the coefficient of x.

69. Solve the equation in Exercise 67 using cross products.

70. Compare your solutions from Exercises 68 and 69. What do you notice?

2.7 More about Problem Solving

OBJECTIVES

1 Use percent in problems involving rates.

2 Solve problems involving mixtures.

3 Solve problems involving simple interest.

4 Solve problems involving denominations of money.

5 Solve problems involving distance, rate, and time.

OBJECTIVE 1 Use percent in problems involving rates. Recall that percent means "per hundred." Thus, percents are ratios where the second number is always 100. For example, 50% represents the ratio of 50 to 100 and 27% represents the ratio of 27 to 100.

| **PROBLEM SOLVING** |

Percents are often used in problems involving mixing different concentrations of a substance or different interest rates. In each case, to get the amount of pure substance or the interest, we multiply.

Mixture Problems	**Interest Problems (annual)**
base × rate (%) = percentage	principal × rate (%) = interest
$b \quad \times \quad r \quad = \quad p$	$p \quad \times \quad r \quad = \quad I$

In an equation, *percent is always written as a decimal.* For example, 35% is written .35, *not* 35, and 7% is written .07, *not* 7.

■ **EXAMPLE 1** Using Percents to Find Percentages

(a) If a chemist has 40 L of a 35% acid solution, then the amount of pure acid in the solution is

$$40 \qquad \times \qquad .35 \qquad = \qquad 14 \text{ L.}$$

Amount of solution　　Rate of concentration　　Amount of pure acid

(b) If \$1300 is invested for one year at 7% simple interest, the amount of interest earned in the year is

$$\$1300 \quad \times \quad .07 \quad = \quad \$91.$$

Principal　　Interest rate　　Interest earned

Now Try Exercises 1 and 3.

PROBLEM SOLVING

In the examples that follow, we use tables to organize the information in the problems. A table enables us to more easily set up an equation, which is usually the most difficult step.

OBJECTIVE 2 Solve problems involving mixtures. In the next example, we use percent to solve a mixture problem.

EXAMPLE 2   Solving a Mixture Problem

A chemist needs to mix 20 L of 40% acid solution with some 70% acid solution to get a mixture that is 50% acid. How many liters of the 70% acid solution should be used?

Step 1 **Read** the problem. Note the percent of each solution and of the mixture.

Step 2 **Assign a variable.**

Let x = the number of liters of 70% acid solution needed.

Recall from Example 1(a) that the amount of pure acid in this solution will be given by the product of the percent of strength and the number of liters of solution, or

liters of pure acid in x L of 70% solution = .70x.

The amount of pure acid in the 20 L of 40% solution is

liters of pure acid in the 40% solution = .40(20) = 8.

The new solution will contain $(x + 20)$ L of 50% solution. The amount of pure acid in this solution is

liters of pure acid in the 50% solution = .50$(x + 20)$.

Figure 11 illustrates this information, which is summarized in the table.

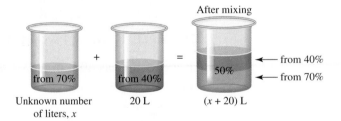

FIGURE 11

Liters of Solution	Rate (as a decimal)	Liters of Pure Acid
x	.70	.70x
20	.40	.40(20) = 8
$x + 20$	.50	.50$(x + 20)$

Sum must equal

Step 3 **Write an equation.** The number of liters of pure acid in the 70% solution added to the number of liters of pure acid in the 40% solution will equal the number of liters of pure acid in the final mixture, so the equation is

<table>
<tr><td>Pure acid
in 70%</td><td>plus</td><td>pure acid
in 40%</td><td>is</td><td>pure acid
in 50%.</td></tr>
<tr><td>↓</td><td>↓</td><td>↓</td><td>↓</td><td>↓</td></tr>
<tr><td>.70x</td><td>+</td><td>.40(20)</td><td>=</td><td>.50(x + 20).</td></tr>
</table>

Step 4 **Solve** the equation. Clear parentheses on the right side, then multiply by 100 to clear decimals.

$$.70x + .40(20) = .50x + .50(20) \qquad \text{Distributive property}$$
$$70x + 40(20) = 50x + 50(20) \qquad \text{Multiply by 100.}$$
$$70x + 800 = 50x + 1000$$
$$20x + 800 = 1000 \qquad \text{Subtract } 50x.$$
$$20x = 200 \qquad \text{Subtract 800.}$$
$$x = 10 \qquad \text{Divide by 20.}$$

Step 5 **State the answer.** The chemist needs to use 10 L of 70% solution.

Step 6 **Check.** Since

$$.70(10) + .40(20) = 7 + 8 = 15$$

and

$$.50(10 + 20) = .50(30) = 15,$$

the answer checks.

Now Try Exercise 15.

NOTE In a problem such as Example 2, the concentration of the final mixture must be *between* the concentrations of the two solutions making up the mixture.

OBJECTIVE 3 Solve problems involving simple interest. The next example uses the formula for simple interest, $I = prt$. Remember that when $t = 1$, the formula becomes $I = pr$, as shown in the Problem-Solving box at the beginning of this section. Once again, the idea of multiplying the total amount (principal) by the rate (rate of interest) gives the percentage (amount of interest).

EXAMPLE 3 Solving a Simple Interest Problem

Elizabeth Suco receives an inheritance. She plans to invest part of it at 9% and $2000 more than this amount at 10%. To earn $1150 per year in interest, how much should she invest at each rate?

Step 1 **Read** the problem again.

Step 2 **Assign a variable.**

Let $\qquad x =$ the amount invested at 9% (in dollars).

Then $\qquad x + 2000 =$ the amount invested at 10% (in dollars).

Use a table to arrange the information given in this problem.

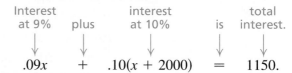

Amount Invested in Dollars	Rate of Interest	Interest for One Year
x	.09	.09x
$x + 2000$	.10	.10($x + 2000$)

Step 3 **Write an equation.** Multiply amount by rate to get the interest earned. Since the total interest is to be $1150, the equation is

Interest at 9%	plus	interest at 10%	is	total interest.
↓	↓	↓	↓	↓
.09x	+	.10($x + 2000$)	=	1150.

Step 4 **Solve** the equation. Clear parentheses; then clear decimals.

$$.09x + .10x + .10(2000) = 1150 \qquad \text{Distributive property}$$
$$9x + 10x + 10(2000) = 115{,}000 \qquad \text{Multiply by 100.}$$
$$9x + 10x + 20{,}000 = 115{,}000$$
$$19x + 20{,}000 = 115{,}000 \qquad \text{Combine like terms.}$$
$$19x = 95{,}000 \qquad \text{Subtract 20,000.}$$
$$x = 5000 \qquad \text{Divide by 19.}$$

Step 5 **State the answer.** She should invest $5000 at 9% and $5000 + $2000 = $7000 at 10%.

Step 6 **Check.** Investing $5000 at 9% and $7000 at 10% gives total interest of .09($5000) + .10($7000) = $450 + $700 = $1150, as required in the original problem.

Now Try Exercise 25.

OBJECTIVE 4 Solve problems involving denominations of money.

PROBLEM SOLVING

Problems that involve different denominations of money or items with different monetary values are very similar to mixture and interest problems. To get the total value, we multiply.

Money Problems

number × value of one item = total value

For example, 30 dimes have a monetary value of 30($.10) = $3. Fifteen $5 bills have a value of 15($5) = $75. A table is helpful for these problems, too.

EXAMPLE 4 Solving a Money Problem

A bank teller has 25 more $5 bills than $10 bills. The total value of the money is $200. How many of each denomination of bill does she have?

Step 1 **Read** the problem. We must find the number of each denomination of bill that the teller has.

Step 2 **Assign a variable.**

Let x = the number of $10 bills.

Then $x + 25$ = the number of $5 bills.

Organize the given information in a table.

Number of Bills	Denomination	Total Value
x	10	10x
$x + 25$	5	5(x + 25)

Step 3 **Write an equation.** Multiplying the number of bills by the denomination gives the monetary value. The value of the tens added to the value of the fives must be $200.

$$
\begin{array}{ccccc}
\text{Value of} & & \text{value of} & & \\
\text{tens} & \text{plus} & \text{fives} & \text{is} & \$200. \\
\downarrow & \downarrow & \downarrow & \downarrow & \downarrow \\
10x & + & 5(x + 25) & = & 200
\end{array}
$$

Step 4 **Solve.**

$$10x + 5x + 125 = 200 \qquad \text{Distributive property}$$
$$15x + 125 = 200 \qquad \text{Combine like terms.}$$
$$15x = 75 \qquad \text{Subtract 125.}$$
$$x = 5 \qquad \text{Divide by 15.}$$

Step 5 **State the answer.** The teller has 5 tens and $5 + 25 = 30$ fives.

Step 6 **Check.** The teller has $30 - 5 = 25$ more fives, and the value of the money is $5(\$10) + 30(\$5) = \$200$, as required.

Now Try Exercise 29.

OBJECTIVE 5 **Solve problems involving distance, rate, and time.** If your car travels at an average rate of 50 mph for 2 hr, then it travels $50 \times 2 = 100$ mi. This is an example of the basic relationship between distance, rate, and time,

$$\text{distance} = \text{rate} \times \text{time},$$

given by the formula $d = rt$. By solving, in turn, for r and t in the formula, we obtain two other equivalent forms of the formula. The three forms are given here.

Distance, Rate, and Time Relationship

$$d = rt \qquad r = \frac{d}{t} \qquad t = \frac{d}{r}$$

The next examples illustrate the uses of these formulas.

▧ EXAMPLE 5 Finding Distance, Rate, or Time

(a) The speed of sound is 1088 ft per sec at sea level at 32°F. In 5 sec under these conditions, sound travels

$$1088 \times 5 = 5440 \text{ ft.}$$

$$\text{Rate} \times \text{Time} = \text{Distance}$$

Here, we found distance given rate and time, using $d = rt$.

(b) The winner of the first Indianapolis 500 race (in 1911) was Ray Harroun, driving a Marmon Wasp at an average speed of 74.59 mph. (*Source: Universal Almanac,* 1997.) To complete the 500 mi, it took him

$$\text{Distance} \to \frac{500}{74.59} = 6.70 \text{ hr} \quad \text{(rounded).} \quad \leftarrow \text{Time}$$
$$\text{Rate} \to$$

Here, we found time given rate and distance using $t = \frac{d}{r}$. To convert .70 hr to minutes, we multiply by 60 to get .70(60) = 42. It took Harroun about 6 hr, 42 min to complete the race.

(c) At the 2000 Olympic Games in Sydney, Australia, Dutch swimmer Inge de Bruijn set a world record in the women's 50-m freestyle swimming event of 24.13 sec. (*Source: World Almanac and Book of Facts,* 2002.) Her rate was

$$\text{Rate} = \frac{\text{Distance} \to}{\text{Time} \to} \frac{50}{24.13} = 2.07 \text{ m per sec (rounded).}$$

Now Try Exercises 37, 39, and 41.

▧ EXAMPLE 6 Solving a Motion Problem

Two cars leave Baton Rouge, Louisiana, at the same time and travel east on Interstate 10. One travels at a constant speed of 55 mph, and the other travels at a constant speed of 63 mph. In how many hours will the distance between them be 24 mi?

Step 1 **Read** the problem. We must find the time it will take for the distance between the cars to be 24 mi.

Step 2 **Assign a variable.** Since we are looking for time, we let t = the number of hours until the distance between them is 24 mi. The sketch in Figure 12 shows what is happening in the problem.

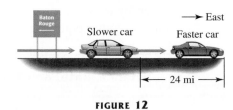

FIGURE 12

Now we construct a table using the information given in the problem and t for the time traveled by each car. We multiply rate by time to get the expressions for distances traveled.

	Rate	$\times$ Time	$=$ Distance	
Faster Car	63	t	$63t$	Difference is 24 mi.
Slower Car	55	t	$55t$	

The quantities $63t$ and $55t$ represent the two distances. Refer to Figure 12, and notice that the *difference* between the larger distance and the smaller distance is 24 mi.

Step 3 **Write an equation.**

$$63t - 55t = 24$$

Step 4 **Solve.** $8t = 24$ Combine like terms.

$t = 3$ Divide by 8.

Step 5 **State the answer.** It will take the cars 3 hr to be 24 mi apart.

Step 6 **Check.** After 3 hr the faster car will have traveled $63 \times 3 = 189$ mi, and the slower car will have traveled $55 \times 3 = 165$ mi. Since $189 - 165 = 24$, the conditions of the problem are satisfied.

Now Try Exercise 45.

NOTE In motion problems like the one in Example 6, once you have filled in two pieces of information in each row of the table, you should automatically fill in the third piece of information, using the appropriate form of the formula relating distance, rate, and time. Set up the equation based on your sketch and the information in the table.

EXAMPLE 7 Solving a Motion Problem

Two planes leave Los Angeles at the same time. One heads south to San Diego; the other heads north to San Francisco. The San Francisco plane flies 50 mph faster. In $\frac{1}{2}$ hr, the planes are 275 mi apart. What are their speeds?

Step 1 **Read** the problem carefully.

Step 2 **Assign a variable.**

Let $r =$ the speed of the slower plane.

Then $r + 50 =$ the speed of the faster plane.

Fill in a table.

	Rate	Time	Distance	
Slower Plane	r	$\frac{1}{2}$	$\frac{1}{2}r$	Sum is 275 mi.
Faster Plane	$r + 50$	$\frac{1}{2}$	$\frac{1}{2}(r + 50)$	

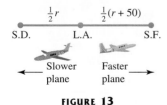

$\frac{1}{2}r$ $\frac{1}{2}(r + 50)$

S.D.   L.A.   S.F.

Slower   Faster
plane   plane

FIGURE 13

Step 3 **Write an equation.** As Figure 13 shows, the planes are headed in *opposite* directions. The *sum* of their distances equals 275 mi, so

$$\frac{1}{2}r + \frac{1}{2}(r + 50) = 275.$$

Step 4 **Solve.** $\dfrac{1}{2}r + \dfrac{1}{2}(r + 50) = 275$

$$r + (r + 50) = 550 \qquad \text{Multiply by 2.}$$
$$2r + 50 = 550 \qquad \text{Combine like terms.}$$
$$2r = 500 \qquad \text{Subtract 50.}$$
$$r = 250 \qquad \text{Divide by 2.}$$

Step 5 **State the answer.** The slower plane (headed south) has a speed of 250 mph. The speed of the faster plane is $250 + 50 = 300$ mph.

Step 6 **Check.** Verify that $\frac{1}{2}(250) + \frac{1}{2}(300) = 275$ mi.

Now Try Exercise 51.

Another way to solve the problems in this section is given in Chapter 8.

2.7 EXERCISES

For Extra Help

 Student's Solutions Manual

MyMathLab

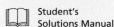 InterAct Math Tutorial Software

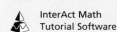 AW Math Tutor Center

 MathXL

 Digital Video Tutor CD 4/Videotape 4

Use the concepts of this section to answer each question. See Example 1 and the Problem-Solving box before Example 4.

1. How much pure acid is in 250 mL of a 14% acid solution?

2. How much pure alcohol is in 150 L of a 30% alcohol solution?

3. If $10,000 is invested for 1 yr at 3.5% simple interest, how much interest is earned?

4. If $25,000 is invested at 3% simple interest for 2 yr, how much interest is earned?

5. What is the monetary amount of 283 nickels?

6. What is the monetary amount of 35 half-dollars?

Solve each percent problem. Remember that base × rate = percentage.

7. The 2000 U.S. Census showed that the population of Alabama was 4,447,000, with 26.0% represented by African-Americans. What is the best estimate of the African-American population in Alabama? (*Source:* U.S. Bureau of the Census.)

   A. 500,000   **B.** 750,000   **C.** 1,100,000   **D.** 1,500,000

8. The 2000 U.S. Census showed that the population of New Mexico was 1,819,000, with 42.1% being Hispanic. What is the best estimate of the Hispanic population in New Mexico? (*Source:* U.S. Bureau of the Census.)

   A. 720,000   **B.** 72,000   **C.** 650,000
   D. 36,000

9. The graph shows the breakdown, by approximate percents, of colors chosen for new compact/sports cars in the 2000 model year. If about 2.5 million compact/sports cars were sold in 2000, about how many were each color? (*Source:* Ward's Communications.)

   (a) Red   **(b)** White   **(c)** Black

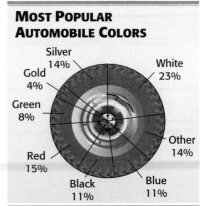

MOST POPULAR AUTOMOBILE COLORS

Silver 14%
White 23%
Gold 4%
Green 8%
Other 14%
Red 15%
Black 11%
Blue 11%

Source: DuPont Automotive Products.

10. An average middle-income family will spend $160,140 to raise a child born in 1999 from birth to age 17. The graph shows the breakdown, by approximate percents, for various expense categories. To the nearest dollar, about how much will be spent to provide the following?

(a) Housing (b) Food
(c) Health care

THE COST OF PARENTHOOD

Housing 33%
Miscellaneous 11%
Child care/education 10%
Health care 7%
Clothing 7%
Transportation 14%
Food 18%

Source: U.S. Department of Agriculture.

11. In 1998, the U.S. civilian labor force consisted of 137,673,000 persons. Of this total, 6,210,000 were unemployed. To the nearest tenth, what was the percent of unemployment? (*Source:* U.S. Bureau of Labor Statistics.)

12. In 1998, the U.S. labor force (excluding agricultural employees, self-employed persons, and the unemployed) consisted of 116,730,000 persons. Of this total, 16,211,000 were union members. To the nearest tenth, what percent of this labor force belonged to unions? (*Source:* U.S. Bureau of Labor Statistics.)

Use the concepts of this section to answer each question.

13. Suppose that a chemist is mixing two acid solutions, one of 20% concentration and the other of 30% concentration. Which one of the following concentrations could *not* be obtained?

    A. 22%   **B.** 24%   **C.** 28%   **D.** 32%

14. Suppose that pure alcohol is added to a 24% alcohol mixture. Which one of the following concentrations could *not* be obtained?

    A. 22%   **B.** 26%   **C.** 28%   **D.** 30%

Work each mixture problem. See Example 2.

15. How many gallons of 50% antifreeze must be mixed with 80 gal of 20% antifreeze to get a mixture that is 40% antifreeze?

Gallons of Mixture	Rate	Gallons of Antifreeze
x	.50	.50x
80	.20	.20(80)
$x + 80$	.40	.40(x + 80)

16. How many liters of 25% acid solution must be added to 80 L of 40% solution to get a solution that is 30% acid?

Liters of Solution	Rate	Liters of Acid
x	.25	.25x
80	.40	.40(80)
$x + 80$	.30	.30(x + 80)

17. A certain metal is 20% tin. How many kilograms of this metal must be mixed with 80 kg of a metal that is 70% tin to get a metal that is 50% tin?

Kilograms of Metal	Rate	Kilograms of Pure Tin
x	.20	
	.70	
	.50	

18. A pharmacist has 20 L of a 10% drug solution. How many liters of 5% solution must be added to get a mixture that is 8%?

Liters of Solution	Rate	Liters of Pure Drug
20		20(.10)
	.05	
	.08	

19. In a chemistry class, 12 L of a 12% alcohol solution must be mixed with a 20% solution to get a 14% solution. How many liters of the 20% solution are needed?

20. How many liters of a 60% acid solution must be mixed with a 75% acid solution to get 20 L of a 72% solution?

21. How many liters of a 10% alcohol solution must be mixed with 40 L of a 50% solution to get a 40% solution?

22. How many gallons of a 12% indicator solution must be mixed with a 20% indicator solution to get 10 gal of a 14% solution?

23. Minoxidil is a drug that has recently proven to be effective in treating male pattern baldness. A pharmacist wishes to mix a solution that is 2% minoxidil. She has on hand 50 mL of a 1% solution, and she wishes to add some 4% solution to it to obtain the desired 2% solution. How much 4% solution should she add?

24. Water must be added to 20 mL of a 4% minoxidil solution to dilute it to a 2% solution. How many milliliters of water should be used? (*Hint:* Water is 0% minoxidil.)

Work each investment problem using simple interest. See Example 3.

25. Li Nguyen invested some money at 3% and $4000 less than that amount at 5%. The two investments produced a total of $200 interest in 1 yr. How much was invested at each rate?

26. LaShondra Williams inherited some money from her uncle. She deposited part of the money in a savings account paying 2%, and $3000 more than that amount in a different account paying 3%. Her annual interest income was $690. How much did she deposit at each rate?

27. With income earned by selling the rights to his life story, an actor invests some of the money at 3% and $30,000 more than twice as much at 4%. The total annual interest earned from the investments is $5600. How much is invested at each rate?

28. An artist invests her earnings in two ways. Some goes into a tax-free bond paying 6%, and $6000 more than three times as much goes into mutual funds paying 5%. Her total annual interest income from the investments is $825. How much does she invest at each rate?

Work each problem involving monetary values. See Example 4.

29. A bank teller has some $5 bills and some $20 bills. The teller has 5 more twenties than fives. The total value of the money is $725. Find the number of $5 bills that the teller has.

Number of Bills	Denomination	Total Value
x	5	
x + 5	20	

30. A coin collector has $1.70 in dimes and nickels. She has 2 more dimes than nickels. How many nickels does she have?

Number of Coins	Denomination	Total Value
x	.05	.05x
	.10	

31. A cashier has a total of 126 bills, made up of fives and tens. The total value of the money is $840. How many of each kind does he have?

32. A convention manager finds that she has $1290, made up of twenties and fifties. She has a total of 42 bills. How many of each kind does she have?

33. A merchant wishes to mix candy worth $5 per lb with 40 lb of candy worth $2 per lb to get a mixture that can be sold for $3 per lb. How many pounds of $5 candy should be used?

34. At Vern's Grill, hamburgers cost $2.70 each, and a bag of french fries costs $1.20. How many hamburgers and how many bags of french fries can a customer buy with $26.40 if he wants twice as many hamburgers as bags of french fries?

Vern's Grill

Hamburgers	$2.70 each
French fries	$1.20 a bag
Soda pop	$ .75 each
Shakes	$2.70 each
Cherry pie	$2.70 a slice
Ice cream	$1.20 a scoop

Solve each problem involving distance, rate, and time. See Example 5.

35. Which choice is the best estimate for the average speed of a trip of 405 mi that lasted 8.2 hr?

   A. 50 mph   **B.** 30 mph   **C.** 60 mph   **D.** 40 mph

36. Suppose that an automobile averages 45 mph, and travels for 30 min. Is the distance traveled $45 \times 30 = 1350$ mi? If not, explain why not, and give the correct distance.

37. A driver averaged 53 mph and took 10 hr to travel from Memphis to Chicago. What is the distance between Memphis and Chicago?

38. A small plane traveled from Warsaw to Rome, averaging 164 mph. The trip took 2 hr. What is the distance from Warsaw to Rome?

39. The winner of the 1998 Charlotte 500 (mile) race was Mark Martin, who drove his Ford to victory with a rate of 123.188 mph. What was his time? (*Source: Sports Illustrated 2000 Sports Almanac.*)

40. In 1998, Jeff Gordon drove his Chevrolet to victory in the North Carolina 400 (mile) race. His rate was 128.423 mph. What was his time? (*Source: Sports Illustrated 2000 Sports Almanac.*)

In Exercises 41–44, find the rate based on the information provided. Use a calculator and round your answers to the nearest hundredth. All events were at the 2000 Summer Olympics. (Source: http://espn.go.com/oly/summer00)

	Event	Participant	Distance	Time
41.	100-m hurdles, Women	Olga Shishigina, Kazakhstan	100 m	12.65 sec
42.	400-m hurdles, Women	Irina Privalova, Russia	400 m	53.02 sec
43.	400-m hurdles, Men	Angelo Taylor, USA	400 m	47.50 sec
44.	400-m dash, Men	Michael Johnson, USA	400 m	43.84 sec

Solve each motion problem. See Examples 6 and 7.

45. St. Louis and Portland are 2060 mi apart. A small plane leaves Portland, traveling toward St. Louis at an average speed of 90 mph. Another plane leaves St. Louis at the same time, traveling toward Portland, averaging 116 mph. How long will it take them to meet?

	r	t	d
Plane Leaving Portland	90	t	$90t$
Plane Leaving St. Louis	116	t	$116t$

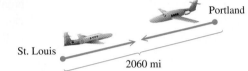

46. Atlanta and Cincinnati are 440 mi apart. John leaves Cincinnati, driving toward Atlanta at an average speed of 60 mph. Pat leaves Atlanta at the same time, driving toward Cincinnati in her antique auto, averaging 28 mph. How long will it take them to meet?

	r	t	d
John	60	t	$60t$
Pat	28	t	$28t$

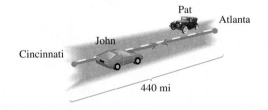

47. Two steamers leave a port on a river at the same time, traveling in opposite directions. Each is traveling 22 mph. How long will it take for them to be 110 mi apart?

48. A train leaves Kansas City, Kansas, and travels north at 85 km per hr. Another train leaves at the same time and travels south at 95 km per hour. How long will it take before they are 315 km apart?

49. At a given hour two steamboats leave a city in the same direction on a straight canal. One travels at 18 mph and the other travels at 25 mph. In how many hours will the boats be 35 mi apart?

50. From a point on a straight road, Lupe and Maria ride bicycles in the same direction. Lupe rides 10 mph and Maria rides 12 mph. In how many hours will they be 5 mi apart?

51. Two trains leave a city at the same time. One travels north and the other travels south at 20 mph faster. In 2 hr the trains are 280 mi apart. Find their speeds.

	r	t	d
Northbound	x	2	
Southbound	$x + 20$	2	

52. Two planes leave an airport at the same time, one flying east, the other flying west. The eastbound plane travels 150 mph slower. They are 2250 mi apart after 3 hr. Find the speed of each plane.

	r	t	d
Eastbound	$x - 150$	3	
Westbound	x	3	

53. Two cars leave towns 230 km apart at the same time, traveling directly toward one another. One car travels 15 km per hr slower than the other. They pass one another 2 hr later. What are their speeds?

54. Two cars start from towns 400 mi apart and travel toward each other. They meet after 4 hr. Find the speed of each car if one travels 20 mph faster than the other.

2.8 Solving Linear Inequalities

OBJECTIVES

1 Graph intervals on a number line.

2 Use the addition property of inequality.

3 Use the multiplication property of inequality.

4 Solve linear inequalities using both properties of inequality.

5 Solve linear inequalities with three parts.

6 Solve applied problems using inequalities.

Inequalities are algebraic expressions related by

$<$ "is less than," $\leq$ "is less than or equal to,"

$>$ "is greater than," $\geq$ "is greater than or equal to."

We solve an inequality by finding all real number solutions for it. For example, the solution set of $x \leq 2$ includes *all real numbers* that are less than or equal to 2, not just the *integers* less than or equal to 2.

OBJECTIVE 1 Graph intervals on a number line. A good way to show the solution set of an inequality is by graphing. We graph all the real numbers satisfying $x \leq 2$ by placing a square bracket at 2 on a number line and drawing an arrow extending from the bracket to the left (to represent the fact that all numbers less than 2 are also part of the graph). The graph is shown in Figure 14.

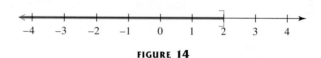

FIGURE 14

The set of numbers less than or equal to 2 is an example of an **interval** on the number line. To write intervals, we use **interval notation.** For example, using this notation, the interval of all numbers less than or equal to 2 is written as $(-\infty, 2]$. The **negative infinity** symbol $-\infty$ does not indicate a number. It is used to show that the interval includes all real numbers less than 2. As on the number line, the square bracket indicates that 2 is part of the solution. A parenthesis is always used next to the infinity symbol. The set of real numbers is written in interval notation as $(-\infty, \infty)$.

EXAMPLE 1 Graphing Intervals Written in Interval Notation on a Number Line

Write each inequality in interval notation and graph the interval.

(a) $x > -5$

The statement $x > -5$ says that x can represent any number greater than -5 but cannot equal -5. The interval is written $(-5, \infty)$. We show this on a graph by placing a parenthesis at −5 and drawing an arrow to the right, as shown in Figure 15. The parenthesis at −5 indicates that −5 is not part of the graph.

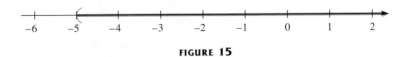

FIGURE 15

(b) $-1 \le x < 3$

The statement is read "−1 is less than or equal to x *and* x is less than 3." Thus, we want the set of numbers that are *between* −1 and 3, with −1 included and 3 excluded. In interval notation, we write $[-1, 3)$, using a square bracket at −1 because −1 is part of the graph, and a parenthesis at 3 because 3 is not part of the graph. The graph is shown in Figure 16.

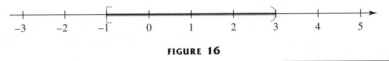

FIGURE 16

Now Try Exercises 7 and 15.

We summarize the various types of intervals here.

Interval Notation

Type of Interval	Set	Interval Notation	Graph
Open interval	$\{x \mid a < x\}$	(a, ∞)	
	$\{x \mid a < x < b\}$	(a, b)	
	$\{x \mid x < b\}$	$(-\infty, b)$	
	$\{x \mid x \text{ is a real number}\}$	$(-\infty, \infty)$	
Half-open interval	$\{x \mid a \le x\}$	$[a, \infty)$	
	$\{x \mid a < x \le b\}$	$(a, b]$	
	$\{x \mid a \le x < b\}$	$[a, b)$	
	$\{x \mid x \le b\}$	$(-\infty, b]$	
Closed interval	$\{x \mid a \le x \le b\}$	$[a, b]$	

OBJECTIVE 2 Use the addition property of inequality. Solving inequalities is similar to solving equations.

Linear Inequality in One Variable

A **linear inequality in one variable** can be written in the form

$$Ax + B < C,$$

where A, B, and C are real numbers, with $A \neq 0$.

(Throughout this section we give definitions and rules only for $<$, but they are also valid for $>$, $\leq$, and $\geq$.)

Examples of linear inequalities in one variable include

$$x + 5 < 2, \qquad y - 3 \geq 5, \qquad \text{and} \qquad 2k + 5 \leq 10.$$

Consider the inequality $2 < 5$. If 4 is added to each side of this inequality, the result is

$$2 + 4 < 5 + 4$$
$$6 < 9,$$

a true sentence. Now subtract 8 from each side:

$$2 - 8 < 5 - 8$$
$$-6 < -3.$$

The result is again a true sentence. These examples suggest the **addition property of inequality,** which states that the same real number can be added to each side of an inequality without changing the solutions.

Addition Property of Inequality

For any real numbers A, B, and C, the inequalities

$$A < B \qquad \text{and} \qquad A + C < B + C$$

have exactly the same solutions.

    That is, the same number may be added to each side of an inequality without changing the solutions.

As with the addition property of equality, the same number may also be *subtracted* from each side of an inequality.

EXAMPLE 2 Using the Addition Property of Inequality

Solve $7 + 3k > 2k - 5$.

    Use the addition property of inequality twice, once to get the terms containing k alone on one side of the inequality and a second time to get the integers together on the other side. (These steps can be done in either order.)

$$7 + 3k > 2k - 5$$

$$7 + 3k - 2k > 2k - 5 - 2k \qquad \text{Subtract } 2k.$$

$$7 + k > -5 \qquad \text{Combine like terms.}$$

$$7 + k - 7 > -5 - 7 \qquad \text{Subtract } 7.$$

$$k > -12 \qquad \text{Combine like terms.}$$

The solution set is $(-12, \infty)$. Its graph is shown in Figure 17.

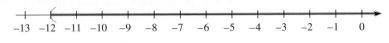

FIGURE 17

Now Try Exercise 21.

OBJECTIVE 3   Use the multiplication property of inequality.   The addition property of inequality cannot be used to solve inequalities such as $4y \geq 28$. These inequalities require the *multiplication property of inequality*. To see how this property works, we look at some examples.

First, start with the inequality $3 < 7$ and multiply each side by the positive number 2.

$$3 < 7$$

$$2(3) < 2(7) \qquad \text{Multiply each side by 2.}$$

$$6 < 14 \qquad \text{True}$$

Now multiply each side of $3 < 7$ by the negative number -5.

$$3 < 7$$

$$-5(3) < -5(7) \qquad \text{Multiply each side by } -5.$$

$$-15 < -35 \qquad \text{False}$$

To get a true statement when multiplying each side by -5, we must reverse the direction of the inequality symbol.

$$3 < 7$$

$$-5(3) > -5(7) \qquad \text{Multiply by } -5; \text{ reverse the symbol.}$$

$$-15 > -35 \qquad \text{True}$$

Take the inequality $-6 < 2$ as another example. Multiply each side by the positive number 4.

$$-6 < 2$$

$$4(-6) < 4(2) \qquad \text{Multiply by 4.}$$

$$-24 < 8 \qquad \text{True}$$

Multiplying each side of $-6 < 2$ by -5 *and at the same time reversing the direction of the inequality symbol* gives

$$-6 < 2$$

$$-5(-6) > -5(2) \qquad \text{Multiply by } -5; \text{ reverse the symbol.}$$

$$30 > -10. \qquad \text{True}$$

In summary, the **multiplication property of inequality** has two parts.

Multiplication Property of Inequality

For any real numbers A, B, and C, with $C \neq 0$,

1. if C is *positive,* then the inequalities

$$A < B \qquad \text{and} \qquad AC < BC$$

have exactly the same solutions;

2. if C is *negative,* then the inequalities

$$A < B \qquad \text{and} \qquad AC > BC$$

have exactly the same solutions.

That is, each side of an inequality may be multiplied by the same positive number without changing the solutions. If the multiplier is negative, we must reverse the direction of the inequality symbol.

The multiplication property of inequality also permits *division* of each side of an inequality by the same nonzero number.

It is important to remember the differences in the multiplication property for positive and negative numbers.

1. When each side of an inequality is multiplied or divided by a positive number, the direction of the inequality symbol *does not change.* (Adding or subtracting terms on each side also does not change the symbol.)

2. When each side of an inequality is multiplied or divided by a negative number, the direction of the symbol *does change.* ***Reverse the direction of the inequality symbol only when multiplying or dividing each side by a negative number.***

EXAMPLE 3   Using the Multiplication Property of Inequality

Solve each inequality and graph the solution set.

(a) $3r < -18$

Using the multiplication property of inequality, we divide each side by 3. Since 3 is a positive number, the direction of the inequality symbol *does not* change.

$$3r < -18$$

$$\frac{3r}{3} < \frac{-18}{3} \qquad \text{Divide by 3.}$$

$$r < -6$$

The solution set is $(-\infty, -6)$. The graph is shown in Figure 18.

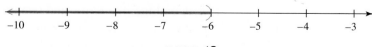

FIGURE 18

(b) $-4t \geq 8$

Here each side of the inequality must be divided by -4, a negative number, which *does* change the direction of the inequality symbol.

$$-4t \geq 8$$

$$\frac{-4t}{-4} \leq \frac{8}{-4} \qquad \text{Divide by } -4; \text{ reverse the symbol.}$$

$$t \leq -2$$

The solution set $(-\infty, -2]$ is graphed in Figure 19.

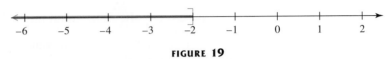

FIGURE 19

Now Try Exercises 29 and 31.

CAUTION Even though the number on the right side of the inequality in Example 3(a) is negative (-18), *do not reverse the direction of the inequality symbol.* Reverse the symbol only when multiplying or dividing by a negative number, as shown in Example 3(b).

OBJECTIVE 4 Solve linear inequalities using both properties of inequality. To solve a linear inequality, follow these steps.

Solving a Linear Inequality

Step 1 **Simplify each side separately.** Use the distributive property to clear parentheses and combine like terms on each side as needed.

Step 2 **Isolate the variable terms on one side.** Use the addition property of inequality to get all terms with variables on one side of the inequality and all numbers on the other side.

Step 3 **Isolate the variable.** Use the multiplication property of inequality to change the inequality to the form $x < k$ or $x > k$.

Remember: Reverse the direction of the inequality symbol *only when multiplying or dividing each side of an inequality by a negative number.*

EXAMPLE 4 Solving a Linear Inequality

Solve $3z + 2 - 5 > -z + 7 + 2z$ and graph the solution set.

Step 1 Combine like terms and simplify.

$$3z + 2 - 5 > -z + 7 + 2z$$

$$3z - 3 > z + 7$$

Step 2   Use the addition property of inequality.

$$3z - 3 + 3 > z + 7 + 3 \qquad \text{Add 3.}$$
$$3z > z + 10$$
$$3z - z > z + 10 - z \qquad \text{Subtract } z.$$
$$2z > 10$$

Step 3   Use the multiplication property of inequality.

$$\frac{2z}{2} > \frac{10}{2} \qquad \text{Divide by 2.}$$
$$z > 5$$

Since 2 is positive, the direction of the inequality symbol is not changed in Step 3. The solution set is $(5, \infty)$. Its graph is shown in Figure 20.

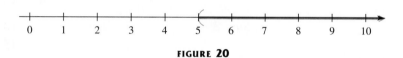

FIGURE 20

Now Try Exercise 41.

EXAMPLE 5   Solving a Linear Inequality

Solve $5(k - 3) - 7k \geq 4(k - 3) + 9$ and graph the solution set.

Step 1   Simplify and combine like terms.

$$5(k - 3) - 7k \geq 4(k - 3) + 9$$
$$5k - 15 - 7k \geq 4k - 12 + 9 \qquad \text{Distributive property}$$
$$-2k - 15 \geq 4k - 3 \qquad \text{Combine like terms.}$$

Step 2   Use the addition property.

$$-2k - 15 - 4k \geq 4k - 3 - 4k \qquad \text{Subtract } 4k.$$
$$-6k - 15 \geq -3$$
$$-6k - 15 + 15 \geq -3 + 15 \qquad \text{Add 15.}$$
$$-6k \geq 12$$

Step 3   Divide each side by -6, a negative number. Change the direction of the inequality symbol.

$$\frac{-6k}{-6} \leq \frac{12}{-6} \qquad \text{Divide by } -6; \text{ reverse the symbol.}$$
$$k \leq -2$$

The solution set is $(-\infty, -2]$. Its graph is shown in Figure 21.

FIGURE 21

Now Try Exercise 45.

■ **EXAMPLE 6** Solving a Linear Inequality with Fractions

Solve $-\dfrac{2}{3}(r - 3) - \dfrac{1}{2} < \dfrac{1}{2}(5 - r)$ and graph the solution set.

To clear fractions, multiply each side by the least common denominator, 6.

$$6\left[-\dfrac{2}{3}(r - 3) - \dfrac{1}{2}\right] < 6\left[\dfrac{1}{2}(5 - r)\right] \qquad \text{Multiply by 6.}$$

$$6\left[-\dfrac{2}{3}(r - 3)\right] - 6\left(\dfrac{1}{2}\right) < 6\left[\dfrac{1}{2}(5 - r)\right] \qquad \text{Distributive property}$$

$$-4(r - 3) - 3 < 3(5 - r)$$

Step 1 $\qquad -4r + 12 - 3 < 15 - 3r \qquad$ Distributive property

$$-4r + 9 < 15 - 3r$$

Step 2 $\qquad -4r + 9 + 3r < 15 - 3r + 3r \qquad$ Add 3r.

$$-r + 9 < 15$$

$$-r + 9 - 9 < 15 - 9 \qquad \text{Subtract 9.}$$

$$-r < 6$$

Step 3   To solve for r, multiply each side of the inequality by -1. Since -1 is negative, change the direction of the inequality symbol.

$$-1(-r) > -1(6) \qquad \text{Multiply by } -1, \text{ change } < \text{ to } >.$$

$$r > -6$$

The solution set is $(-6, \infty)$. See the graph in Figure 22.

FIGURE 22

Now Try Exercise 47.

OBJECTIVE 5 **Solve linear inequalities with three parts.**   Inequalities that say that one number is *between* two other numbers are **three-part inequalities.** For example,

$$-3 < 5 < 7$$

says that 5 is between -3 and 7. For some applications, it is necessary to work with an inequality such as

$$3 < x + 2 < 8,$$

where $x + 2$ is between 3 and 8. To solve this inequality, we subtract 2 from each of the three parts of the inequality, giving

$$3 - 2 < x + 2 - 2 < 8 - 2$$

$$1 < x < 6.$$

The idea is to get the inequality in the form

a number $< x <$ another number,

using "is less than." The solution set (in this case the interval $(1, 6)$) can then easily be graphed.

CAUTION When inequalities have three parts, the order of the parts is important. It would be *wrong* to write an inequality as $8 < x + 2 < 3$, since this would imply that $8 < 3$, a false statement. In general, three-part inequalities are written so that the symbols point in the same direction, and both point toward the smaller number.

▨ EXAMPLE 7 Solving Three-Part Inequalities

Solve each inequality and graph the solution set.

(a) $\qquad 4 \le 3x - 5 < 6$

$\qquad 4 + 5 \le 3x - 5 + 5 < 6 + 5 \qquad$ Add 5 to each part.

$\qquad\qquad 9 \le 3x < 11$

$\qquad\qquad \dfrac{9}{3} \le \dfrac{3x}{3} < \dfrac{11}{3} \qquad\qquad$ Divide each part by 3.

$\qquad\qquad 3 \le x < \dfrac{11}{3}$

The solution set is $\left[3, \frac{11}{3}\right)$. Its graph is shown in Figure 23.

FIGURE 23

(b) $-4 \le \dfrac{2}{3}m - 1 < 8$

To clear the fraction, multiply each part by 3.

$\qquad 3(-4) \le 3\left(\dfrac{2}{3}m - 1\right) < 3(8) \qquad$ Multiply each part by 3.

$\qquad\qquad -12 \le 2m - 3 < 24 \qquad$ Distributive property

$\qquad -12 + 3 \le 2m - 3 + 3 < 24 + 3 \qquad$ Add 3 to each part.

$\qquad\qquad -9 \le 2m < 27$

$\qquad\qquad \dfrac{-9}{2} \le \dfrac{2m}{2} < \dfrac{27}{2} \qquad$ Divide each part by 2.

$\qquad\qquad -\dfrac{9}{2} \le m < \dfrac{27}{2}$

The solution set is $\left[-\frac{9}{2}, \frac{27}{2}\right)$. Its graph is shown in Figure 24.

FIGURE 24

Now Try Exercises 59 and 65.

NOTE The inequality in Example 7(b) can also be solved by first adding 1 to each part, and then multiplying each part by $\frac{3}{2}$. Do this and confirm that the same solution set results.

Examples of the types of solution sets to be expected from solving linear equations or linear inequalities are shown below.

Solution Sets of Linear Equations and Inequalities

Equation or Inequality	Typical Solution Set	Graph of Solution Set
Linear equation $5x + 4 = 14$	$\{2\}$	(graph: point at 2)
Linear inequality $5x + 4 < 14$	$(-\infty, 2)$	(graph: ray left, open at 2)
Linear inequality $5x + 4 > 14$	$(2, \infty)$	(graph: ray right, open at 2)
Three-part inequality $-1 \le 5x + 4 \le 14$	$[-1, 2]$	(graph: segment from −1 to 2, closed)

OBJECTIVE 6 Solve applied problems using inequalities.   Until now, the applied problems that we have studied have all led to equations.

PROBLEM SOLVING

Inequalities can be used to solve applied problems involving phrases that suggest inequality. The table gives some of the more common such phrases along with examples and translations.

Phrase	Example	Inequality
Is more than	A number *is more than* 4.	$x > 4$
Is less than	A number *is less than* −12.	$x < -12$
Is at least	A number *is at least* 6.	$x \ge 6$
Is at most	A number *is at most* 8.	$x \le 8$

We use the same six problem-solving steps from Section 2.4, changing Step 3 to "Write an inequality" instead of "Write an equation."

CAUTION   Do not confuse statements like "5 is more than a number" with the phrase "5 more than a number." The first of these is expressed as $5 > x$ while the second is expressed as $x + 5$ or $5 + x$.

The next example shows an application of algebra that is important to anyone who has ever asked, "What score can I make on my next test and have a (particular grade) in this course?" It uses the idea of finding the average of a number of grades. In general, to find the average of n numbers, add the numbers, then divide by n.

EXAMPLE 8 Finding an Average Test Score

Brent has test grades of 86, 88, and 78 on his first three tests in geometry. If he wants an average of at least 80 after his fourth test, what are the possible scores he can make on his fourth test?

Step 1 **Read** the problem again.

Step 2 **Assign a variable.** Let x = Brent's score on his fourth test.

Step 3 **Write an inequality.** To find his average after 4 tests, add the test scores and divide by 4.

$$\underset{\text{Average}}{\underbrace{\frac{86 + 88 + 78 + x}{4}}} \underset{\substack{\text{is at} \\ \text{least } 80.}}{\ge} 80$$

Step 4 **Solve.**

$$\frac{252 + x}{4} \ge 80 \qquad \text{Add the known scores.}$$

$$4\left(\frac{252 + x}{4}\right) \ge 4(80) \qquad \text{Multiply by 4.}$$

$$252 + x \ge 320$$

$$252 + x - 252 \ge 320 - 252 \qquad \text{Subtract 252.}$$

$$x \ge 68 \qquad \text{Combine like terms.}$$

Step 5 **State the answer.** He must score 68 or more on the fourth test to have an average of *at least* 80.

Step 6 **Check.**

$$\frac{86 + 88 + 78 + 68}{4} = \frac{320}{4} = 80$$

Now Try Exercise 77.

CAUTION Errors often occur when the phrases "at least" and "at most" appear in applied problems. Remember that

| at least | **translates as** | greater than or equal to |

and

| at most | **translates as** | less than or equal to. |

EXAMPLE 9 Using a Linear Inequality to Solve a Rental Problem

A rental company charges $15 to rent a chain saw, plus $2 per hr. Al Ghandi can spend no more than $35 to clear some logs from his yard. What is the *maximum* amount of time he can use the rented saw?

Step 1 **Read** the problem again.

Step 2 **Assign a variable.** Let h = the number of hours he can rent the saw.

Step 3 **Write an inequality.** He must pay $15, plus $2h, to rent the saw for h hours, and this amount must be *no more than* $35.

Cost of renting	is no more than	35 dollars.
$15 + 2h$	$\leq$	35

Step 4 **Solve.**
$$2h \leq 20 \qquad \text{Subtract 15.}$$
$$h \leq 10 \qquad \text{Divide by 2.}$$

Step 5 **State the answer.** He can use the saw for a maximum of 10 hr. (Of course, he may use it for less time, as indicated by the inequality $h \leq 10$.)

Step 6 **Check.** If Al uses the saw for 10 hr, he will spend $15 + 2(10) = 35$ dollars, the maximum amount.

Now Try Exercise 87.

CONNECTIONS

Many applications from economics involve inequalities rather than equations. For example, a company that produces videocassettes has found that revenue from the sales of the cassettes is $5 per cassette less sales costs of $100. Production costs are $125 plus $4 per cassette. Profit (*P*) is given by revenue (*R*) less cost (*C*), so the company must find the production level *x* that makes

$$P = R - C > 0.$$

For Discussion or Writing

Write an expression for revenue, letting *x* represent the production level (number of cassettes to be produced). Write an expression for production costs using *x*. Write an expression for profit and solve the inequality shown above. Describe the solution in terms of the problem.

2.8 EXERCISES

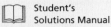

1. Explain how to determine whether to use a parenthesis or a square bracket at the endpoint when graphing an inequality on a number line.

2. How does the graph of $t \geq -7$ differ from the graph of $t > -7$?

Write an inequality involving the variable x that describes each set of numbers graphed. See Example 1.

3.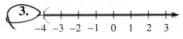

4.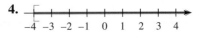

5.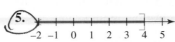

6.

Write each inequality in interval notation and graph the interval. See Example 1.

7. $k \leq 4$ **8.** $r \leq -11$ **9.** $x < -3$ **10.** $y < 3$

11. $t > 4$ **12.** $m > 5$ **13.** $8 \leq x \leq 10$ **14.** $3 \leq x \leq 5$

15. $0 < y \leq 10$ **16.** $-3 \leq x < 5$

17. Why is it wrong to write $3 < x < -2$ to indicate that x is between -2 and 3?

18. If $p < q$ and $r < 0$, which one of the following statements is false?

 A. $pr < qr$ **B.** $pr > qr$ **C.** $p + r < q + r$ **D.** $p - r < q - r$

Solve each inequality. Write the solution set in interval notation and graph it. See Example 2.

19. $z - 8 \geq -7$ **20.** $p - 3 \geq -11$ **21.** $2k + 3 \geq k + 8$

22. $3x + 7 \geq 2x + 11$ **23.** $3n + 5 < 2n - 6$ **24.** $5x - 2 < 4x - 5$

25. Under what conditions must the inequality symbol be reversed when solving an inequality?

26. Your friend tells you that when solving the inequality $6x < -42$ he reversed the direction of the inequality symbol because of the -42. How would you respond?

Solve each inequality. Write the solution set in interval notation and graph it. See Example 3.

27. $3x < 18$ **28.** $5x < 35$ **29.** $2y \geq -20$ **30.** $6m \geq -24$

31. $-8t > 24$ **32.** $-7x > 49$ **33.** $-x \geq 0$ **34.** $-k < 0$

35. $-\dfrac{3}{4}r < -15$ **36.** $-\dfrac{7}{8}t < -14$ **37.** $-.02x \leq .06$ **38.** $-.03v \geq -.12$

Solve each inequality. Write the solution set in interval notation and graph it. See Examples 4–6.

39. $5r + 1 \geq 3r - 9$ **40.** $6t + 3 < 3t + 12$

41. $6x + 3 + x < 2 + 4x + 4$ **42.** $-4w + 12 + 9w \geq w + 9 + w$

43. $-x + 4 + 7x \leq -2 + 3x + 6$ **44.** $14y - 6 + 7y > 4 + 10y - 10$

45. $5(x + 3) - 6x \leq 3(2x + 1) - 4x$ **46.** $2(x - 5) + 3x < 4(x - 6) + 1$

47. $\dfrac{2}{3}(p + 3) > \dfrac{5}{6}(p - 4)$ **48.** $\dfrac{7}{9}(y - 4) \leq \dfrac{4}{3}(y + 5)$

49. $-\dfrac{1}{4}(p + 6) + \dfrac{3}{2}(2p - 5) < 10$ **50.** $\dfrac{3}{5}(k - 2) - \dfrac{1}{4}(2k - 7) \leq 3$

51. $4x - (6x + 1) \leq 8x + 2(x - 3)$ **52.** $2y - (4y + 3) > 6y + 3(y + 4)$

53. $5(2k + 3) - 2(k - 8) > 3(2k + 4) + k - 2$

54. $2(3z - 5) + 4(z + 6) \geq 2(3z + 2) + 3z - 15$

Write a three-part inequality involving the variable x that describes each set of numbers graphed. See Example 1(b).

55.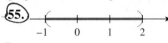

56.

57.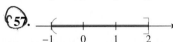

58.

Solve each inequality. Write the solution set in interval notation and graph it. See Example 7.

59. $-5 \le 2x - 3 \le 9$

60. $-7 \le 3x - 4 \le 8$

61. $5 < 1 - 6m < 12$

62. $-1 \le 1 - 5q \le 16$

63. $10 < 7p + 3 < 24$

64. $-8 \le 3r - 1 \le -1$

65. $-12 \le \dfrac{1}{2}z + 1 \le 4$

66. $-6 \le 3 + \dfrac{1}{3}a \le 5$

67. $1 \le 3 + \dfrac{2}{3}p \le 7$

68. $2 < 6 + \dfrac{3}{4}y < 12$

69. $-7 \le \dfrac{5}{4}r - 1 \le -1$

70. $-12 \le \dfrac{3}{7}a + 2 \le -4$

RELATING CONCEPTS (EXERCISES 71–76)

For Individual or Group Work

The methods for solving linear equations and linear inequalities are quite similar. **Work Exercises 71–76 in order,** *to see how the solutions of an inequality are closely connected to the solution of the corresponding equation.*

71. Solve the equation $3x + 2 = 14$ and graph the solution set as a single point on the number line.

72. Solve the inequality $3x + 2 > 14$ and graph the solution set as an interval on the number line.

73. Solve the inequality $3x + 2 < 14$ and graph the solution set as an interval on the number line.

74. If you were to graph all the solution sets from Exercises 71–73 on the same number line, what would the graph be? (This is called the *union* of all the solution sets.)

75. Based on your results from Exercises 71–74, if you were to graph the union of the solution sets of

$$-4x + 3 = -1, \qquad -4x + 3 > -1, \qquad \text{and} \qquad -4x + 3 < -1,$$

what do you think the graph would be?

76. Comment on the following statement: *Equality is the boundary between less than and greater than.*

Solve each problem. See Examples 8 and 9.

77. Inkie Landry has scores of 76 and 81 on her first two algebra tests. If she wants an average of at least 80 after her third test, what possible scores can she make on her third test?

78. Mabimi Pampo has scores of 96 and 86 on his first two geometry tests. What possible scores can he make on his third test so that his average is at least 90?

79. When 2 is added to the difference between six times a number and 5, the result is greater than 13 added to five times the number. Find all such numbers.

80. When 8 is subtracted from the sum of three times a number and 6, the result is less than 4 more than the number. Find all such numbers.

81. The formula for converting Fahrenheit temperature to Celsius is

$$C = \frac{5}{9}(F - 32).$$

If the Celsius temperature on a certain summer day in Toledo is never more than 30°, how would you describe the corresponding Fahrenheit temperatures?

82. The formula for converting Celsius temperature to Fahrenheit is

$$F = \frac{9}{5}C + 32.$$

The Fahrenheit temperature of Key West, Florida, has never exceeded 95°. How would you describe this using Celsius temperature?

83. For what values of x would the rectangle have perimeter of at least 400?

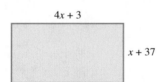

4x + 3

x + 37

84. For what values of x would the triangle have perimeter of at least 72?

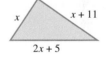

x   x + 11

2x + 5

85. A long-distance phone call costs $2.00 for the first three minutes plus $.30 per minute for each minute or fractional part of a minute after the first three minutes. If x represents the number of minutes of the length of the call after the first three minutes, then $2 + .30x$ represents the cost of the call. If Jorge has $5.60 to spend on a call, what is the maximum total time he can use the phone?

86. If the call described in Exercise 85 costs between $5.60 and $6.50, what are the possible total time lengths for the call?

87. At the Speedy Gas'n Go, a car wash costs $3.00, and gasoline is selling for $1.50 per gal. Lee Ann Spahr has $17.25 to spend, and her car is so dirty that she must have it washed. What is the maximum number of gallons of gasoline that she can purchase?

88. A taxicab driver charges $1.50 for the first $\frac{1}{5}$ mi and $.25 for each additional $\frac{1}{5}$ mi. Dantrell Jackson has only $3.75. What is the maximum distance he can travel (not including a tip for the cabbie)?

89. A product will break even or produce a profit if the revenue R from selling the product is at least equal to the cost C of producing it. Suppose that the cost C (in dollars) to produce x units of bicycle helmets is $C = 50x + 5000$, while the revenue R (in dollars) collected from the sale of x units is $R = 60x$. For what values of x does the product break even or produce a profit?

90. (See Exercise 89.) If the cost to produce x units of baseball cards is $C = 100x + 6000$ (in dollars), and the revenue collected from selling x units is $R = 500x$ (in dollars), for what values of x does the product break even or produce a profit?

91. A BMI (body mass index) between 19 and 25 is considered healthy. Use the formula

$$\text{BMI} = \frac{704 \times (\text{weight in pounds})}{(\text{height in inches})^2}$$

to find the weight range w, to the nearest pound, that gives a healthy BMI for each height. (*Source: Washington Post.*)

(a) 72 in.   **(b)** Your height in inches

92. To achieve the maximum benefit from exercising, the heart rate in beats per minute should be in the target heart rate zone (THR). For a person aged A, the formula is

$$.7(220 - A) \le \text{THR} \le .85(220 - A).$$

Find the THR to the nearest whole number for each age. (*Source:* Hockey, Robert V., *Physical Fitness: The Pathway to Healthful Living,* Times Mirror/Mosby College Publishing, 1989.)

(a) 35   **(b)** Your age

The July 14th weather forecast by time of day for the 2000 U.S. Olympic Track and Field Trials, held July 14–23, 2000, in Sacramento, California, is shown in the figure. Use this graph to work Exercises 93–96.

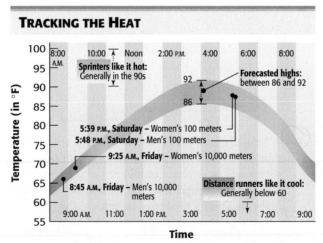

Source: Accuweather, Bee research.

93. Sprinters prefer Fahrenheit temperatures in the 90s. Using the upper boundary of the forecast, in what time period is the temperature expected to be at least 90°F?

94. Distance runners prefer cool temperatures. During what time period are temperatures predicted to be no more than 70°F? Use the lower forecast boundary.

95. What range of temperatures is predicted for the Women's 100-m event?

96. What range of temperatures is forecast for the Men's 10,000-m event?

Chapter **2** **Group Activity**

Are You a Race-Walker?

OBJECTIVE   Use proportions to calculate walking speeds.

MATERIALS   Students will need a large area for walking. Each group will need a stopwatch.

Race-walking at speeds exceeding 8 mph is a high fitness, long-distance competitive sport. The table contains gold medal winners in Olympic race-walking competition. Complete the table by applying the proportion given below to find the race-walker's steps per minute.

Use 10 km ≈ 6.21 mi. (Round all answers except those for steps per minute to the nearest thousandth. Round steps per minute to the nearest whole number.)

$$\frac{70 \text{ steps per min}}{2 \text{ mph}} = \frac{x \text{ steps per min}}{y \text{ mph}}$$

Event	Gold Medal Winner	Country	Time in Hours: Minutes: Seconds	Time in Minutes	Time in Hours	y Miles per Hour	x Steps per Minute
10-km Walk, Women	Yelena Nikolayeva	Russia	0:41:49				
20-km Walk, Men	Jefferson Perez	Ecuador	1:20:07				
50-km Walk, Men	Robert Korzeniowski	Poland	3:43:30				

Source: World Almanac and Book of Facts, 1999.

A. Using a stopwatch, take turns counting how many steps each member of the group takes in one minute while walking at a normal pace. Record the results in the following table. Then do it again at a fast pace. Record these results.

	Normal Pace		Fast Pace	
Name	x Steps per Minute	y Miles per Hour	x Steps per Minute	y Miles per Hour

B. Use the given proportion to convert the numbers from part A to miles per hour and complete the table.

1. Find the average speed for the group at a normal pace and at a fast pace.

2. What is the minimum number of steps per minute you would have to take to be a race-walker?

3. At a fast pace did anyone in the group walk fast enough to be a race-walker? Explain how you decided.

CHAPTER 2 SUMMARY

KEY TERMS

2.1	linear equation in one variable solution set equivalent equations	2.4	degree complementary angles right angle		vertical angles volume
2.3	conditional equation identity contradiction empty (null) set	2.5	supplementary angles straight angle consecutive integers area perimeter	2.6	ratio proportion terms of a proportion extremes means cross products

2.8 interval on the number line
interval notation
linear inequality in one variable
three-part inequality

NEW SYMBOLS

$\emptyset$	empty set	a to b, $a{:}b$, or $\dfrac{a}{b}$		(a, b)	interval notation for $a < x < b$
$1°$	one degree	the ratio of a to b		$[a, b]$	interval notation for $a \leq x \leq b$

∞	infinity
$-\infty$	negative infinity
$(-\infty, \infty)$	set of all real numbers

TEST YOUR WORD POWER

See how well you have learned the vocabulary in this chapter. Answers, with examples, follow the Quick Review.

1. A **solution set** is the set of numbers that
   A. make an expression undefined
   B. make an equation false
   C. make an equation true
   D. make an expression equal to 0.

2. The **empty set** is a set
   A. with 0 as its only element
   B. with an infinite number of elements
   C. with no elements
   D. of ideas.

3. **Complementary angles** are angles
   A. formed by two parallel lines
   B. whose sum is 90°
   C. whose sum is 180°
   D. formed by perpendicular lines.

4. **Supplementary angles** are angles
   A. formed by two parallel lines
   B. whose sum is 90°
   C. whose sum is 180°
   D. formed by perpendicular lines.

5. A **ratio**
   A. compares two quantities using a quotient
   B. says that two quotients are equal
   C. is a product of two quantities
   D. is a difference between two quantities.

6. A **proportion**
   A. compares two quantities using a quotient
   B. says that two quotients are equal
   C. is a product of two quantities
   D. is a difference between two quantities.

7. An **inequality** is
   A. a statement that two algebraic expressions are equal
   B. a point on a number line
   C. an equation with no solutions
   D. a statement with algebraic expressions related by $<$, $\leq$, $>$, or $\geq$.

8. **Interval notation** is
   A. a portion of a number line
   B. a special notation for describing a point on a number line
   C. a way to use symbols to describe an interval on a number line
   D. a notation to describe unequal quantities.

QUICK REVIEW

CONCEPTS	EXAMPLES

2.1 THE ADDITION PROPERTY OF EQUALITY

The same number may be added to (or subtracted from) each side of an equation without changing the solution.

Solve.

$$x - 6 = 12$$
$$x - 6 + 6 = 12 + 6 \qquad \text{Add 6.}$$
$$x = 18 \qquad \text{Combine like terms.}$$

Solution set: $\{18\}$

2.2 THE MULTIPLICATION PROPERTY OF EQUALITY

Each side of an equation may be multiplied (or divided) by the same nonzero number without changing the solution.

Solve.

$$\frac{3}{4}x = -9$$
$$\frac{4}{3} \cdot \frac{3}{4}x = \frac{4}{3}(-9) \qquad \text{Multiply by } \frac{4}{3}.$$
$$x = -12$$

Solution set: $\{-12\}$

2.3 MORE ON SOLVING LINEAR EQUATIONS

Step 1 Simplify each side separately.

Solve.
$$2x + 2(x + 1) = 14 + x$$
$$2x + 2x + 2 = 14 + x \qquad \text{Distributive property}$$
$$4x + 2 = 14 + x \qquad \text{Combine like terms.}$$

Step 2 Isolate the variable term on one side.

$$4x + 2 - x - 2 = 14 + x - x - 2 \qquad \text{Subtract } x; \text{ subtract 2.}$$
$$3x = 12 \qquad \text{Combine like terms.}$$

Step 3 Isolate the variable.

$$\frac{3x}{3} = \frac{12}{3} \qquad \text{Divide by 3.}$$
$$x = 4$$

Step 4 Check.

Check:
$$2(4) + 2(4 + 1) = 14 + 4 \qquad ? \qquad \text{Let } x = 4.$$
$$18 = 18 \qquad \text{True}$$

Solution set: $\{4\}$

2.4 AN INTRODUCTION TO APPLICATIONS OF LINEAR EQUATIONS

One number is 5 more than another. Their sum is 21. What are the numbers?

Step 1 Read.

We are looking for two numbers.

Step 2 Assign a variable.

Let x represent the smaller number. Then $x + 5$ represents the larger number.

(continued)

CONCEPTS	EXAMPLES

Step 3 Write an equation.

Step 4 Solve the equation.

$$x + (x + 5) = 21$$

$$2x + 5 = 21 \qquad \text{Combine like terms.}$$

$$2x + 5 - 5 = 21 - 5 \qquad \text{Subtract 5.}$$

$$2x = 16 \qquad \text{Combine like terms.}$$

$$\frac{2x}{2} = \frac{16}{2} \qquad \text{Divide by 2.}$$

$$x = 8$$

Step 5 State the answer.

Step 6 Check.

The numbers are 8 and 13.

13 is 5 more than 8, and $8 + 13 = 21$. It checks.

2.5 FORMULAS AND APPLICATIONS FROM GEOMETRY

To find the value of one of the variables in a formula, given values for the others, substitute the known values into the formula.

Find L if $A = LW$, given that $A = 24$ and $W = 3$.

$$24 = L \cdot 3 \qquad A = 24, W = 3$$

$$\frac{24}{3} = \frac{L \cdot 3}{3} \qquad \text{Divide by 3.}$$

$$8 = L$$

To solve a formula for one of the variables, isolate that variable by treating the other variables as numbers and using the steps for solving equations.

Solve $P = 2L + 2W$ for W.

$$P - 2L = 2L + 2W - 2L \qquad \text{Subtract } 2L.$$

$$P - 2L = 2W \qquad \text{Combine like terms.}$$

$$\frac{P - 2L}{2} = \frac{2W}{2} \qquad \text{Divide by 2.}$$

$$\frac{P - 2L}{2} = W \quad \text{or} \quad W = \frac{P - 2L}{2}$$

2.6 RATIOS AND PROPORTIONS

To write a ratio, express quantities in the same units.

$$4 \text{ ft to 8 in.} = 48 \text{ in. to 8 in.} = \frac{48}{8} = \frac{6}{1}$$

To solve a proportion, use the method of cross products.

Solve $\dfrac{x}{12} = \dfrac{35}{60}$.

$$60x = 12 \cdot 35 \qquad \text{Cross products}$$

$$60x = 420 \qquad \text{Multiply.}$$

$$\frac{60x}{60} = \frac{420}{60} \qquad \text{Divide by 60.}$$

$$x = 7$$

Solution set: $\{7\}$

CONCEPTS	EXAMPLES

2.7 MORE ABOUT PROBLEM SOLVING

Step 1 Read.

A sum of money is invested at simple interest in two ways. Part is invested at 12%, and $20,000 less than that amount is invested at 10%. If the total interest for 1 yr is $9000, find the amount invested at each rate.

Step 2 Assign a variable. Make a table to help solve the problems in this section.

Let $x =$ amount invested at 12%.
Then $x - 20,000 =$ amount invested at 10%.

Dollars Invested	Rate of Interest	Interest for One Year
x	.12	.12x
$x - 20,000$	.10	.10(x − 20,000)

Step 3 Write an equation.
Step 4 Solve the equation.

$$.12x + .10(x - 20,000) = 9000$$
$$.12x + .10x + .10(-20,000) = 9000 \quad \text{Distributive property}$$
$$12x + 10x + 10(-20,000) = 900,000 \quad \text{Multiply by 100.}$$
$$12x + 10x - 200,000 = 900,000$$
$$22x - 200,000 = 900,000$$
$$22x = 1,100,000 \quad \text{Add 200,000.}$$
$$x = 50,000 \quad \text{Divide by 22.}$$

Steps 5 and 6 State the answer and check the solution.

$50,000 is invested at 12% and $30,000 is invested at 10%.

The three forms of the formula relating distance, rate, and time are

$$d = rt, \quad r = \frac{d}{t}, \quad \text{and} \quad t = \frac{d}{r}.$$

Two cars leave from the same point, traveling in opposite directions. One travels at 45 mph and the other at 60 mph. How long will it take them to be 210 mi apart?

Let $t =$ time it takes for them to be 210 mi apart.

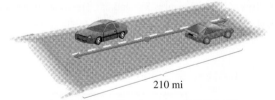

210 mi

The table gives the information from the problem.

	Rate	Time	Distance
One Car	45	t	45t
Other Car	60	t	60t

The sum of the distances, 45t and 60t, must be 210 mi.

$$45t + 60t = 210$$
$$105t = 210 \quad \text{Combine like terms.}$$
$$t = 2 \quad \text{Divide by 105.}$$

It will take them 2 hr to be 210 mi apart. (continued)

CONCEPTS	EXAMPLES

2.8 SOLVING LINEAR INEQUALITIES

Step 1 Simplify each side separately.

Solve and graph the solution set.

$$3(1 - x) + 5 - 2x > 9 - 6$$

$3 - 3x + 5 - 2x > 9 - 6$ Clear parentheses.

$8 - 5x > 3$ Combine like terms.

Step 2 Isolate the variable term on one side.

$8 - 5x - 8 > 3 - 8$ Subtract 8.

$-5x > -5$ Combine like terms.

Step 3 Isolate the variable.
Be sure to reverse the direction of the inequality symbol when multiplying or dividing by a negative number.

$$\frac{-5x}{-5} < \frac{-5}{-5}$$ Divide by -5; change $>$ to $<$.

$$x < 1$$

Solution set: $(-\infty, 1)$

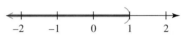

To solve an inequality such as

$$4 < 2x + 6 < 8$$

work with all three expressions at the same time.

Solve.

$$4 < 2x + 6 < 8$$

$4 - 6 < 2x + 6 - 6 < 8 - 6$ Subtract 6.

$-2 < 2x < 2$ Combine like terms.

$$\frac{-2}{2} < \frac{2x}{2} < \frac{2}{2}$$ Divide by 2.

$$-1 < x < 1$$

Solution set: $(-1, 1)$

Answers to Test Your Word Power
1. C; *Example:* {8} is the solution set of $2x + 5 = 21$.   **2.** C; *Example:* The empty set $\emptyset$ is the solution set of $5x + 3 = 5x + 4$.
3. B; *Example:* Angles with measures 35° and 55° are complementary angles.   **4.** C; *Example:* Angles with measures 112° and 68°
are supplementary angles.   **5.** A; *Example:* $\frac{7 \text{ in.}}{12 \text{ in.}}$ or $\frac{7}{12}$ **6.** B; *Example:* $\frac{2}{3} = \frac{8}{12}$ **7.** D; *Examples:* $x < 5, 7 + 2y \geq 11$,
$-5 < 2z - 1 \leq 3$ **8.** C; *Examples:* $(-\infty, 5], (1, \infty), [-3, 3)$

CHAPTER 2 REVIEW EXERCISES

[2.1–2.3] *Solve each equation.*

1. $m - 5 = 1$ **2.** $y + 8 = -4$ **3.** $3k + 1 = 2k + 8$

4. $5k = 4k + \dfrac{2}{3}$ **5.** $(4r - 2) - (3r + 1) = 8$ **6.** $3(2y - 5) = 2 + 5y$

7. $7k = 35$ **8.** $12r = -48$ **9.** $2p - 7p + 8p = 15$

10. $\dfrac{m}{12} = -1$ **11.** $\dfrac{5}{8}k = 8$ **12.** $12m + 11 = 59$

13. $3(2x + 6) - 5(x + 8) = x - 22$ **14.** $5x + 9 - (2x - 3) = 2x - 7$

15. $\dfrac{1}{2}r - \dfrac{r}{3} = \dfrac{r}{6}$

16. $.10(x + 80) + .20x = 14$

17. $3x - (-2x + 6) = 4(x - 4) + x$

18. $2(y - 3) - 4(y + 12) = -2(y + 27)$

[2.4] *Use the six-step method to solve each problem.*

19. In a recent year, the state of Florida had a total of 120 members in its House of Representatives, consisting of only Democrats and Republicans. There were 30 more Democrats than Republicans. How many representatives from each party were there?

20. The land area of Hawaii is 5213 mi^2 greater than the area of Rhode Island. Together, the areas total 7637 mi^2. What is the area of each of the two states?

21. The height of Seven Falls in Colorado is $\frac{5}{2}$ the height of Twin Falls in Idaho. The sum of the heights is 420 ft. Find the height of each. (*Source: World Almanac and Book of Facts.*)

22. The supplement of an angle measures 10 times the measure of its complement. What is the measure of the angle?

23. Find two consecutive odd integers such that when the smaller is added to twice the larger, the result is 24 more than the larger integer.

[2.5] *A formula is given along with the values for all but one of the variables. Find the value of the variable that is not given.*

24. $A = \dfrac{1}{2}bh$; $A = 44$, $b = 8$

25. $A = \dfrac{1}{2}h(b + B)$; $b = 3$, $B = 4$, $h = 8$

26. $C = 2\pi r$; $C = 29.83$, $\pi = 3.14$

27. $V = \dfrac{4}{3}\pi r^3$; $r = 6$, $\pi = 3.14$

Solve each formula for the specified variable.

28. $A = bh$ for h

29. $A = \dfrac{1}{2}h(b + B)$ for h

Find the measure of each marked angle.

30.

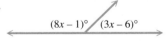

$(8x - 1)°$ $(3x - 6)°$

31.

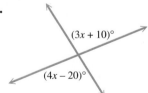

$(3x + 10)°$

$(4x - 20)°$

Solve each problem.

32. The perimeter of a certain rectangle is 16 times the width. The length is 12 cm more than the width. Find the width of the rectangle.

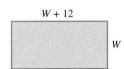

$W + 12$

W

33. The Ziegfield Room in Reno, Nevada, has a circular turntable on which its showgirls dance. The circumference of the table is 62.5 ft. What is the diameter? What is the radius? What is the area? (Use $\pi = 3.14$.) (*Source: Guinness Book of Records.*)

34. A baseball diamond is a square with a side of 90 ft. The pitcher's rubber is located 60.5 ft from home plate, as shown in the figure. Find the measures of the angles marked in the figure. (*Hint:* Recall that the sum of the measures of the angles of any triangle is 180°.)

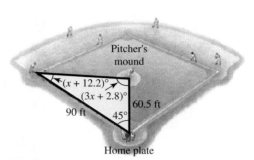

Pitcher's mound

$(x + 12.2)°$

$(3x + 2.8)°$

60.5 ft

90 ft

45°

Home plate

35. What is wrong with the following problem? "The formula for the area of a trapezoid is $A = \frac{1}{2}h(b + B)$. If $h = 12$ and $b = 14$, find the value of A."

[2.6] *Give a ratio for each word phrase, writing fractions in lowest terms.*

36. 60 cm to 40 cm

37. 5 days to 2 weeks

38. 90 in. to 10 ft

39. 3 mo to 3 yr

Solve each equation.

40. $\dfrac{p}{21} = \dfrac{5}{30}$

41. $\dfrac{5 + x}{3} = \dfrac{2 - x}{6}$

42. $\dfrac{y}{5} = \dfrac{6y - 5}{11}$

Use proportions to solve each problem.

43. If 2 lb of fertilizer will cover 150 ft^2 of lawn, how many pounds would be needed to cover 500 ft^2?

44. The tax on a $24.00 item is $2.04. How much tax would be paid on a $36.00 item?

45. The distance between two cities on a road map is 32 cm. The two cities are actually 150 km apart. The distance on the map between two other cities is 80 cm. How far apart are these cities?

46. In the 2000 Olympics held in Sydney, Australia, Russian athletes earned 88 medals. Four of every 11 medals were gold. How many gold medals did Russia earn? (*Source: Times Picayune,* October 2, 2000.)

47. Which is the best buy for a popular breakfast cereal?

15-oz size: $2.69
20-oz size: $3.29
25.5-oz size: $3.49

[2.7] *Solve each problem.*

48. The San Francisco Giants baseball team borrowed $160 million of the $290 million needed to build Pacific Bell Park. Corporate sponsors provided the rest of the cost. What percent of the cost of the park was borrowed? (*Source:* Michael K. Ozanian, "Fields of Debt," *Forbes,* December 15, 1997.)

49. A nurse must mix 15 L of a 10% solution of a drug with some 60% solution to get a 20% mixture. How many liters of the 60% solution will be needed?

50. Todd Cardella invested $10,000 from which he earns an annual income of $550 per yr. He invested part of it at 5% annual interest and the remainder in bonds paying 6% interest. How much did he invest at each rate?

51. In 1846, the vessel *Yorkshire* traveled from Liverpool to New York, a distance of 3150 mi, in 384 hr. What was the *Yorkshire's* average speed? Round your answer to the nearest tenth.

52. Honey Kirk drove from Louisville to Dallas, a distance of 819 mi, averaging 63 mph. What was her driving time?

53. Two planes leave St. Louis at the same time. One flies north at 350 mph and the other flies south at 420 mph. In how many hours will they be 1925 mi apart?

[2.8] *Write each inequality in interval notation and graph it.*

54. $p \geq -4$ **55.** $x < 7$ **56.** $-5 \leq y < 6$

Solve each inequality and graph the solution set.

57. $y + 6 \geq 3$ **58.** $5t < 4t + 2$

59. $-6x \leq -18$ **60.** $8(k - 5) - (2 + 7k) \geq 4$

61. $4x - 3x > 10 - 4x + 7x$

62. $3(2w + 5) + 4(8 + 3w) < 5(3w + 2) + 2w$

63. $-3 \leq 2m + 1 \leq 4$ **64.** $9 < 3m + 5 \leq 20$

Solve each problem by writing an inequality.

65. Carlotta Valdes has grades of 94 and 88 on her first two calculus tests. What possible scores on a third test will give her an average of at least 90?

66. If nine times a number is added to 6, the result is at most 3. Find all such numbers.

■ MIXED REVIEW EXERCISES*

Solve.

67. $\dfrac{x}{7} = \dfrac{x - 5}{2}$ **68.** $I = prt$ for r

69. $-6 \leq \dfrac{4}{3}x - 2 \leq 2$ **70.** $2k - 5 = 4k + 13$

71. $.05x + .02x = 4.9$ **72.** $\dfrac{5}{3}(m - 2) + \dfrac{2}{5}(m + 1) > 1$

73. $9x - (7x + 2) = 3x + (2 - x)$ **74.** $\dfrac{1}{3}s + \dfrac{1}{2}s + 7 = \dfrac{5}{6}s + 5 + 2$

75. A student solved $3 - (8 + 4x) = 2x + 7$ and gave the answer as $\{6\}$. Verify that this answer is incorrect by checking it in the equation. Then explain the error and give the correct solution set. (*Hint:* The error involves the subtraction sign.)

---

*The order of the exercises in this final group does not correspond to the order in which topics occur in the chapter. This random ordering should help you prepare for the chapter test.

76. Athletes in vigorous training programs can eat 50 calories per day for every 2.2 lb of body weight. To the nearest hundred, how many calories can a 175 lb athlete consume per day? (*Source: The Gazette,* March 23, 2002.)

77. The Golden Gate Bridge in San Francisco is 2605 ft longer than the Brooklyn Bridge. Together, their spans total 5795 ft. How long is each bridge? (*Source: World Almanac and Book of Facts.*)

78. Which is the best buy for apple juice?

32-oz size: $1.19
48-oz size: $1.79
64-oz size: $1.99

79. If 1 qt of oil must be mixed with 24 qt of gasoline, how much oil would be needed for 192 qt of gasoline?

80. Two trains are 390 mi apart. They start at the same time and travel toward one another, meeting 3 hr later. If the speed of one train is 30 mph more than the speed of the other train, find the speed of each train.

81. The perimeter of a triangle is 96 m. One side is twice as long as another, and the third side is 30 m long. What is the length of the longest side?

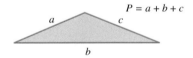

82. The perimeter of a certain square cannot be greater than 200 m. Find the possible values for the length of a side.

CHAPTER **2** TEST

Solve each equation.

1. $5x + 9 = 7x + 21$

2. $-\dfrac{4}{7}x = -12$

3. $7 - (m - 4) = -3m + 2(m + 1)$

4. $.06(x + 20) + .08(x - 10) = 4.6$

5. $-8(2x + 4) = -4(4x + 8)$

Solve each problem.

6. In the 2000–2001 baseball season, the Seattle Mariners tied a league record set by the 1906 Chicago Cubs for most wins in a season. The Mariners won 24 more than twice as many games as they lost. They played 162 regular season games. How many wins and losses did the Mariners have?

7. The three largest islands in the Hawaiian island chain are Hawaii (the Big Island), Maui, and Kauai. Together, their areas total 5300 mi². The island of Hawaii is 3293 mi² larger than the island of Maui, and Maui is 177 mi² larger than Kauai. What is the area of each island?

8. Find the measure of an angle if its supplement measures 10° more than three times its complement.

9. The formula for the perimeter of a rectangle is $P = 2L + 2W$.

   (a) Solve for W.
   (b) If $P = 116$ and $L = 40$, find the value of W.

10. Find the measure of each marked angle.

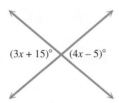

$(3x + 15)°$ $(4x - 5)°$

Solve each proportion.

11. $\dfrac{z}{8} = \dfrac{12}{16}$

12. $\dfrac{x + 5}{3} = \dfrac{x - 3}{4}$

Solve each problem.

13. Which is the better buy for processed cheese slices: 8 slices for $2.19 or 12 slices for $3.30?

14. The distance between Milwaukee and Boston is 1050 mi. On a certain map, this distance is represented by 42 in. On the same map, Seattle and Cincinnati are 92 in. apart. What is the actual distance between Seattle and Cincinnati?

15. Steven Pusztai invested some money at 3% simple interest and $6000 more than that amount at 4.5% simple interest. After 1 yr his total interest from the two accounts was $870. How much did he invest at each rate?

16. Two cars leave from the same point, traveling in opposite directions. One travels at a constant rate of 50 mph while the other travels at a constant rate of 65 mph. How long will it take for them to be 460 mi apart?

Solve each inequality and graph the solution set.

17. $-4x + 2(x - 3) \geq 4x - (3 + 5x) - 7$ **18.** $-10 < 3k - 4 \leq 14$

19. Twylene Johnson has grades of 76 and 81 on her first two algebra tests. If she wants an average of at least 80 after her third test, what score must she make on her third test?

20. Write a short explanation of the additional (extra) rule that must be remembered when solving an inequality (as opposed to solving an equation).

CUMULATIVE REVIEW EXERCISES CHAPTERS 1–2

1. Write $\frac{108}{144}$ in lowest terms.

Perform each indicated operation.

2. $\dfrac{5}{6} + \dfrac{1}{4} - \dfrac{7}{15}$

3. $\dfrac{9}{8} \cdot \dfrac{16}{3} \div \dfrac{5}{8}$

Translate from words to symbols. Use x as the variable, if necessary.

4. The difference between half a number and 18

5. The quotient of 6 and 12 more than a number is 2.

6. True or false? $\dfrac{8(7) - 5(6 + 2)}{3 \cdot 5 + 1} \geq 1$

Perform each indicated operation.

7. $9 - (-4) + (-2)$

8. $\dfrac{-4(9)(-2)}{-3^2}$

9. $(-7 - 1)(-4) + (-4)$

10. Find the value of $\dfrac{3x^2 - y^3}{-4z}$ when $x = -2$, $y = -4$, and $z = 3$.

Name each property illustrated.

11. $7(k + m) = 7k + 7m$

12. $3 + (5 + 2) = 3 + (2 + 5)$

13. Simplify $-4(k + 2) + 3(2k - 1)$ by combining like terms.

Solve each equation, then check the solution.

14. $2r - 6 = 8r$

15. $4 - 5(a + 2) = 3(a + 1) - 1$

16. $\dfrac{2}{3}x + \dfrac{3}{4}x = -17$

17. $\dfrac{2x + 3}{5} = \dfrac{x - 4}{2}$

Solve each formula for the indicated variable.

18. $3x + 4y = 24$ for y

19. $A = P(1 + ni)$ for n

Solve each inequality. Graph the solution set.

20. $6(r - 1) + 2(3r - 5) \leq -4$

21. $-18 \leq -9z < 9$

Solve each problem.

22. For a woven hanging, Miguel Hidalgo needs three pieces of yarn, which he will cut from a 40-cm piece. The longest piece is to be 3 times as long as the middle-sized piece, and the shortest piece is to be 5 cm shorter than the middle-sized piece. What lengths should he cut?

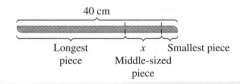

23. A fully inflated professional basketball has a circumference of 78 cm. What is the radius of a circular cross section through the center of the ball? (Use 3.14 as the approximation for π.) Round your answer to the nearest hundredth.

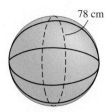

24. A cook wants to increase a recipe that serves 6 to make enough for 20 people. The recipe calls for $1\frac{1}{4}$ cups of grated cheese. How much cheese will be needed to serve 20?

25. Two cars are 400 mi apart. Both start at the same time and travel toward one another. They meet 4 hr later. If the speed of one car is 20 mph faster than the other, what is the speed of each car?

Linear Equations in Two Variables

3

U.S. debt from credit cards continues to increase. In recent years, college campuses have become fertile territory as credit card companies pitch their plastic to students at bookstores, student unions, and sporting events. As a result, three out of four undergrads now have at least one credit card and carry an average balance of $2748. (*Source:* Nellie Mae.) In Example 6 of Section 3.2, we use the concepts of this chapter to investigate credit card debt.

3.1 Reading Graphs; Linear Equations in Two Variables

Pie charts (circle graphs) and bar graphs were introduced in Chapter 1, and we have seen many examples of them in the first two chapters of this book. It is important to be able to interpret graphs correctly.

OBJECTIVE 1 Interpret graphs. A *bar graph* is used to show comparisons. We begin with a bar graph where we must estimate the heights of the bars.

■ EXAMPLE 1 Interpreting a Bar Graph

The bar graph in Figure 1 shows sales in millions of dollars for CarMax Auto Superstores, Inc., which has emerged as the most successful used-car megastore. The graph compares sales for 5 years.

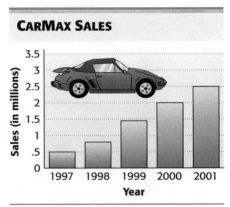

CarMax Sales

Source: Circuit City CarMax Group.

FIGURE 1

(a) Estimate sales in 1998.

Move horizontally from the top of the bar for 1998 to the scale on the left to see that sales in 1998 were about $.8 million.

(b) In what years were sales greater than $1 million?

Locate 1 on the vertical scale and follow the line across to the right. Three years—1999, 2000, and 2001—have bars that extend above the line for 1, so sales were greater than $1 million in those years.

(c) As the years progress, describe the change in sales.

Sales increase steadily as the years progress, from about $.5 million to $2.5 million.

Now Try Exercises 1 and 3.

A *line graph* is used to show changes or trends in data over time. To form a line graph, we connect a series of points representing data with line segments.

EXAMPLE 2 Interpreting a Line Graph

The line graph in Figure 2 shows average prices for personal computers (PCs) for the years 1993 through 1999.

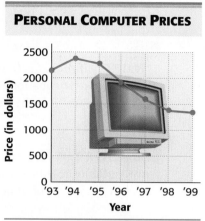

PERSONAL COMPUTER PRICES

Source: CNW Marketing/Research; *USA Today.*

FIGURE 2

(a) Between which years did the average price of a PC increase?

The line between 1993 and 1994 rises, so PC prices increased from 1993 to 1994.

(b) What has been the general trend in average PC prices since 1994?

The line graph falls from 1994 to 1999, so PC prices have been decreasing over these years.

(c) Estimate average PC prices in 1996 and 1999. About how much did PC prices decline between 1996 and 1999?

Move up from 1996 on the horizontal scale to the point plotted for 1996. Then move across to the vertical scale. The average price of a PC in 1996 was about $2000.

The point for 1999 is a little more than halfway between the lines for $1000 and $1500, so estimate the average price in 1999 at about $1300.

Between 1996 and 1999, PC prices declined about

$$\$2000 - \$1300 = \$700.$$

Now Try Exercises 5 and 7.

Many everyday situations, such as those illustrated in Examples 1 and 2, involve two quantities that are related. The equations and applications we discussed in Chapter 2 had only one variable. In this chapter, we extend those ideas to *linear equations in two variables.*

Linear Equation in Two Variables

A **linear equation in two variables** is an equation that can be written in the form

$$Ax + By = C,$$

where A, B, and C are real numbers and A and B are not both 0.

NOTE If A or B is 0, we have equations such as $y = -1$ or $x = 3$. We think of these as linear equations in two variables by writing $y = -1$ as $0x + y = -1$. Similarly, $x = 3$ is the same as $x + 0y = 3$.

OBJECTIVE 2 Write a solution as an ordered pair. A solution of a linear equation in *two* variables requires *two* numbers, one for each variable. For example, the equation $y = 4x + 5$ is a true statement if x is replaced with 2 and y is replaced with 13, since

$$13 = 4(2) + 5. \qquad \text{Let } x = 2; y = 13.$$

The pair of numbers $x = 2$ and $y = 13$ gives a solution of the equation $y = 4x + 5$. The phrase "$x = 2$ and $y = 13$" is abbreviated

x-value $\qquad\qquad$ y-value

$$(2, 13)$$

Ordered pair

with the x-value, 2, and the y-value, 13, given as a pair of numbers written inside parentheses. *The x-value is always given first.* A pair of numbers such as $(2, 13)$ is called an **ordered pair.** As the name indicates, the order in which the numbers are written is important. The ordered pairs $(2, 13)$ and $(13, 2)$ are not the same. The second pair indicates that $x = 13$ and $y = 2$.

OBJECTIVE 3 Decide whether a given ordered pair is a solution of a given equation. An ordered pair that is a solution of an equation is said to *satisfy* the equation.

■ **EXAMPLE 3 Deciding Whether an Ordered Pair Satisfies an Equation**

Decide whether the given ordered pair is a solution of the given equation.

(a) $(3, 2)$; $2x + 3y = 12$

To see whether $(3, 2)$ is a solution of the equation $2x + 3y = 12$, we substitute 3 for x and 2 for y in the given equation.

$$2x + 3y = 12$$
$$2(3) + 3(2) = 12 \qquad ? \qquad \text{Let } x = 3; \text{ let } y = 2.$$
$$6 + 6 = 12 \qquad ?$$
$$12 = 12 \qquad\qquad \text{True}$$

This result is true, so $(3, 2)$ satisfies $2x + 3y = 12$.

(b) $(-2, -7)$; $m + 5n = 33$

$$m + 5n = 33$$
$$-2 + 5(-7) = 33 \qquad ? \qquad \text{Let } m = -2; \text{ let } n = -7.$$
$$-2 + (-35) = 33 \qquad ?$$
$$-37 = 33 \qquad\qquad \text{False}$$

This result is false, so $(-2, -7)$ is *not* a solution of $m + 5n = 33$.

Now Try Exercises 17 and 21.

OBJECTIVE 4 Complete ordered pairs for a given equation.   Choosing a number for one variable in a linear equation makes it possible to find the value of the other variable.

EXAMPLE 4   Completing an Ordered Pair

Complete the ordered pair $(7, \quad)$ for the equation $y = 4x + 5$.

In this ordered pair, $x = 7$. (Remember that x always comes first.) To find the corresponding value of y, replace x with 7 in the equation $y = 4x + 5$.

$$y = 4(7) + 5 = 28 + 5 = 33$$

The ordered pair is $(7, 33)$.

Now Try Exercise 31.

Ordered pairs often are displayed in a **table of values.** Although we usually write tables of values vertically, they may be written horizontally as well.

EXAMPLE 5   Completing Tables of Values

Complete the given table of values for each equation. Then write the results as ordered pairs.

(a) $x - 2y = 8$

x	y
2	
10	
	0
	-2

To complete the first two ordered pairs, let $x = 2$ and $x = 10$, respectively.

	If $\quad x = 2,$		If $\quad x = 10,$
then	$x - 2y = 8$	then	$x - 2y = 8$
becomes	$2 - 2y = 8$	becomes	$10 - 2y = 8$
	$-2y = 6$		$-2y = -2$
	$y = -3.$		$y = 1.$

Now complete the last two ordered pairs by letting $y = 0$ and $y = -2$, respectively.

If	$y = 0,$		If	$y = -2,$
then	$x - 2y = 8$		then	$x - 2y = 8$
becomes	$x - 2(0) = 8$		becomes	$x - 2(-2) = 8$
	$x - 0 = 8$			$x + 4 = 8$
	$x = 8.$			$x = 4.$

The completed table of values is as follows.

x	y
2	-3
10	1
8	0
4	-2

The corresponding ordered pairs are $(2, -3)$, $(10, 1)$, $(8, 0)$, and $(4, -2)$.

(b) $x = 5$

x	y
	-2
	6
	3

The given equation is $x = 5$. No matter which value of y might be chosen, the value of x is always the same, 5.

x	y
5	-2
5	6
5	3

The ordered pairs are $(5, -2)$, $(5, 6)$, and $(5, 3)$.

Now Try Exercises 41 and 45.

OBJECTIVE 5 Plot ordered pairs. Every linear equation in two variables has an infinite number of ordered pairs as solutions. Each choice of a number for one variable leads to a particular real number for the other variable. To graph these solutions, represented as ordered pairs (x, y), we need *two* number lines, one for each variable. These two number lines are drawn as shown in Figure 3. The horizontal number line is called the **x-axis.** The vertical line is called the **y-axis.** Together, the x-axis and y-axis form a **rectangular coordinate system,** also called the **Cartesian coordinate system,** in honor of René Descartes.

The coordinate system is divided into four regions, called **quadrants.** These quadrants are numbered counterclockwise, as shown in Figure 3. Points on the axes themselves are not in any quadrant. The point at which the *x*-axis and *y*-axis meet is called the **origin.** The origin, labeled 0 in Figure 3, is the point corresponding to (0, 0).

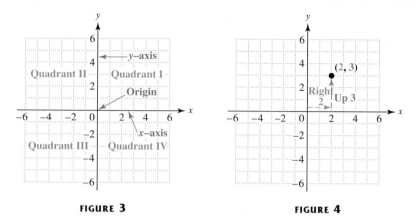

FIGURE 3   FIGURE 4

As mentioned previously, the coordinate system that we use to plot points is credited to René Descartes (1596–1650), a French mathematician of the seventeenth century. It has been said that he developed the coordinate system while lying in bed, watching an insect move across the ceiling. He realized that he could locate the position of the insect at any given time by finding its distance from each of two perpendicular walls.

The *x*-axis and *y*-axis in a coordinate system determine a **plane,** a flat surface similar to a sheet of paper. By referring to the two axes, every point on the plane can be associated with an ordered pair. The numbers in the ordered pair are called the **coordinates** of the point. For example, locate the point associated with the ordered pair (2, 3) by starting at the origin. Since the *x*-coordinate is 2, go 2 units to the right along the *x*-axis. Then, since the *y*-coordinate is 3, turn and go up 3 units on a line parallel to the *y*-axis. This is called **plotting** the point (2, 3). (See Figure 4.) From now on we refer to the point with *x*-coordinate 2 and *y*-coordinate 3 as the point (2, 3).

> **NOTE** On a plane, both numbers in the ordered pair are needed to locate a point. The ordered pair is a name for the point.

EXAMPLE 6 Plotting Ordered Pairs

Plot the given points on a coordinate system.

(a) (1, 5)   **(b)** (−2, 3)   **(c)** (−1, −4)

(d) (7, −2)   **(e)** $\left(\dfrac{3}{2}, 2\right)$ **(f)** (5, 0)

Figure 5 on the next page shows the graphs of the points. Locate the point (−1, −4) in part (c) by first going 1 unit to the left along the *x*-axis. Then turn and go 4 units down, parallel to the *y*-axis. Plot the point $\left(\frac{3}{2}, 2\right)$ in part (e) by going $\frac{3}{2}$ (or $1\frac{1}{2}$) units to the right along the *x*-axis. Then turn and go 2 units up, parallel to the *y*-axis.

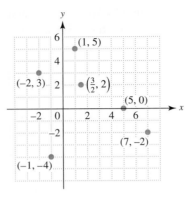

FIGURE 5

Now Try Exercises 49 and 53.

Sometimes we can use a linear equation to mathematically describe, or *model,* a real-life situation, as shown in the next example.

EXAMPLE 7 Completing Ordered Pairs to Estimate Annual Costs of Doctors' Visits

The amount Americans pay annually for doctors' visits has increased steadily from 1990 through 2000. This amount can be closely approximated by the linear equation

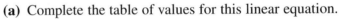

Cost ⌐↓   ⌐ Year

$$y = 34.3x - 67{,}693,$$

which relates x, the year, and y, the cost in dollars. (*Source:* U.S. Health Care Financing Administration.)

(a) Complete the table of values for this linear equation.

x (Year)	y (Cost)
1990	
1996	
2000	

To find y when $x = 1990$, we substitute into the equation.

$$y = 34.3(1990) - 67{,}693 \qquad \text{Let } x = 1990.$$
$$y = 564 \qquad\qquad\qquad \text{Use a calculator.}$$

This means that in 1990, Americans each spent about $564 on doctors' visits.

We substitute the years 1996 and 2000 in the same way to complete the table as follows.

x (Year)	y (Cost)
1990	564
1996	770
2000	907

We can write the results from the table of values as ordered pairs (x, y). Each year x is paired with its cost y:

$$(1990, 564), \quad (1996, 770), \quad \text{and} \quad (2000, 907).$$

(b) Graph the ordered pairs found in part (a).

The ordered pairs are graphed in Figure 6. This graph of ordered pairs of data is called a **scatter diagram.** Notice how the axes are labeled: x represents the year, and y represents the cost in dollars. Different scales are used on the two axes. Here, each square represents two units in the horizontal direction and 100 units in the vertical direction. Because the numbers in the first ordered pair are so large, we show a break in the axes near the origin.

COSTS OF DOCTORS' VISITS

FIGURE 6

A scatter diagram enables us to tell whether two quantities are related to each other. In Figure 6, the plotted points could be connected to form a straight *line,* so the variables x (year) and y (cost) have a *line*ar relationship. The increase in costs is also reflected.

Now Try Exercise 71.

CAUTION   The equation in Example 7 is valid only for the years 1990 through 2000 because it was based on data for those years. Do not assume that this equation would provide reliable data for other years since the data for those years may not follow the same pattern.

We can think of ordered pairs as representing an input value x and an output value y. If we input x into the equation, the output is y. We encounter many examples of this type of relationship every day.

• The cost to fill a tank with gasoline depends on how many gallons are needed; the number of gallons is the input, and the cost is the output.

• The distance traveled depends on the traveling time; input a time, and the output is a distance.

• The growth of a plant depends on the amount of sun it gets; the input is the amount of sun, and the output is the growth.

This idea is illustrated in Figure 7 on the next page with an input-output "machine."

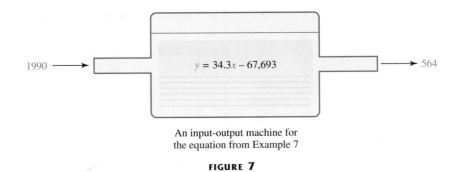

1990 $\longrightarrow$ | $y = 34.3x - 67{,}693$ | $\longrightarrow$ 564

An input-output machine for
the equation from Example 7

FIGURE 7

In Section 7.3, we extend this input/output idea to the concept of a *function.*

3.1 EXERCISES

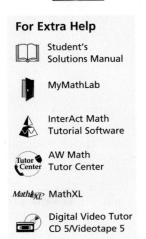
The bar graph compares egg production in millions of eggs for six states in June 1999. Use the bar graph to work Exercises 1–4. See Example 1.

1. Name the top two egg-producing states in June 1999. Estimate their production.

2. Which states had egg production less than 400 million eggs?

3. Which states appear to have had equal production? Estimate this production.

4. How does egg production in Ohio compare to egg production in North Carolina?

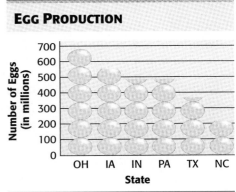

EGG PRODUCTION

Source: Iowa Agricultural Statistics.

The line graph shows the average price, adjusted for inflation, that Americans have paid for a gallon of gasoline for selected years since 1970. Use the line graph to work Exercises 5–8. See Example 2.

5. Over which period of years did the greatest increase in the price of a gallon of gas occur? About how much was this increase?

6. Estimate the price of a gallon of gas during 1985, 1990, 1995, and 2000.

7. Describe the trend in gas prices from 1980 to 1995.

8. During which year(s) did a gallon of gas cost $1.50?

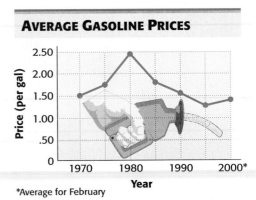

AVERAGE GASOLINE PRICES

*Average for February

Source: American Petroleum Institute;
AP research.

Fill in each blank with the correct response.

9. The symbol (x, y) _____ represent an ordered pair, while the symbols $[x, y]$ and
(does/does not)
$\{x, y\}$ _____ represent ordered pairs.
(do/do not)

10. The ordered pair $(3, 2)$ is a solution of the equation $2x - 5y =$ _____ .

11. The point whose graph has coordinates $(-4, 2)$ is in quadrant _____ .
x y

12. The point whose graph has coordinates $(0, 5)$ lies along the _____-axis.

13. The ordered pair $(4, \underline{})$ is a solution of the equation $y = 3$.

14. The ordered pair $(\underline{}, -2)$ is a solution of the equation $x = 6$.

15. Define a linear equation in one variable and a linear equation in two variables, and give examples of each.

Decide whether the given ordered pair is a solution of the given equation. See Example 3.

16. $x + y = 9$; $(0, 9)$ **17.** $x + y = 8$; $(0, 8)$ **18.** $2p - q = 6$; $(4, 2)$

19. $2v + w = 5$; $(3, -1)$ **20.** $4x - 3y = 6$; $(2, 1)$ **21.** $5x - 3y = 15$; $(5, 2)$

22. $y = 3x$; $(2, 6)$ **23.** $x = -4y$; $(-8, 2)$ **24.** $x = -6$; $(-6, 5)$

25. $y = 2$; $(4, 2)$ **26.** $x + 4 = 0$; $(-6, 2)$ **27.** $x - 6 = 0$; $(4, 2)$

28. Do $(4, -1)$ and $(-1, 4)$ represent the same ordered pair? Explain.

29. Do the ordered pairs $(3, 4)$ and $(4, 3)$ correspond to the same point on the plane? Explain.

Complete each ordered pair for the equation $y = 2x + 7$. See Example 4.

30. $(2, \quad)$ **31.** $(5, \quad)$ **32.** $(\quad, 0)$ **33.** $(\quad, -3)$

Complete each ordered pair for the equation $y = -4x - 4$. See Example 4.

34. $(0, \quad)$ **35.** $(\quad, 0)$ **36.** $(\quad, 16)$ **37.** $(\quad, 24)$

38. Explain why it would be easier to find the corresponding y-value for $x = \frac{1}{3}$ than for $x = \frac{1}{7}$ in the equation $y = 6x + 2$.

39. For the equation $y = mx + b$, what is the y-value corresponding to $x = 0$ for *any* value of m?

Complete each table of values. See Example 5.

40. $2x + 3y = 12$

x	y
0	
	0
	8

41. $4x + 3y = 24$

x	y
0	
	0
	4

42. $3x - 5y = -15$

x	y
0	
	0
	-6

43. $4x - 9y = -36$

x	y
	0
0	
	8

44. $x = -9$

x	y
	6
	2
	-3

45. $x = 12$

x	y
	3
	8
	0

46. $y = -6$

x	y
8	
4	
-2	

47. $y = -10$

x	y
4	
0	
-4	

48. Give the ordered pairs that correspond to the points labeled in the figure. (All coordinates are integers.)

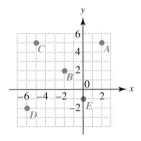

Plot each ordered pair in a rectangular coordinate system. See Example 6.

49. $(6, 2)$ **50.** $(5, 3)$ **51.** $(-4, 2)$ **52.** $(-3, 5)$

53. $\left(-\dfrac{4}{5}, -1 \right)$ **54.** $\left(-\dfrac{3}{2}, -4 \right)$ **55.** $(0, 4)$ **56.** $(-3, 0)$

Fill in each blank with the word positive *or the word* negative.

The point with coordinates (x, y) is in

57. quadrant III if x is _____ and y is _____.

58. quadrant II if x is _____ and y is _____.

59. quadrant IV if x is _____ and y is _____.

60. quadrant I if x is _____ and y is _____.

Complete each table of values and then plot the ordered pairs. See Examples 5 and 6.

61. $x - 2y = 6$

x	y
0	
	0
2	
	-1

62. $2x - y = 4$

x	y
0	
	0
1	
	-6

63. $3x - 4y = 12$

x	y
0	
	0
-4	
	-4

64. $2x - 5y = 10$

x	y
0	
	0
-5	
	-3

65. $y + 4 = 0$

x	y
0	
5	
-2	
-3	

66. $x - 5 = 0$

x	y
	1
	0
	6
	-4

67. Look at your graphs of the ordered pairs in Exercises 61–66. Describe the pattern indicated by the plotted points.

Solve each problem. See Example 7.

68. Suppose that it costs $5000 to start up a business selling snow cones. Furthermore, it costs $.50 per cone in labor, ice, syrup, and overhead. Then the cost to make x snow cones is given by y dollars, where $y = .50x + 5000$. Express as an ordered pair each of the following.

(a) When 100 snow cones are made, the cost is $5050. (*Hint:* What does x represent? What does y represent?)

(b) When the cost is $6000, the number of snow cones made is 2000.

69. It costs a flat fee of $20 plus $5 per day to rent a pressure washer. Therefore, the cost to rent the pressure washer for x days is given by $y = 5x + 20$, where y is in dollars. Express as an ordered pair each of the following.

(a) When the washer is rented for 5 days, the cost is $45.

(b) I paid $50 when I returned the washer, so I must have rented it for 6 days.

Work each problem. See Example 7.

70. The table shows on-line retail spending in billions of dollars.

Year	Spending (in billions)
1998	7.8
1999	14.9
2000	23.1
2001	34.6
2002	53.0

Source: Jupiter Communications.

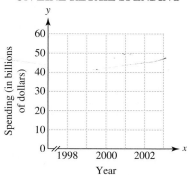

ON-LINE RETAIL SPENDING

(a) Write the data from the table as ordered pairs (x, y), where x represents the year and y represents on-line spending in billions of dollars.

(b) What does the ordered pair $(2003, 78.0)$ mean in the context of this problem?

(c) Make a scatter diagram of the data using the ordered pairs from part (a) and the given grid.

(d) Describe the pattern indicated by the points on the scatter diagram. What is the trend in on-line spending?

71. The table shows the rate (in percent) at which 4-yr college students graduate within 5 yr.

Year	Rate (%)
1996	53.3
1997	52.8
1998	52.1
1999	51.6

Source: ACT.

4-YEAR COLLEGE STUDENTS GRADUATING WITHIN 5 YEARS

(a) Write the data from the table as ordered pairs (x, y), where x represents the year and y represents graduation rate.

(b) What does the ordered pair $(1995, 54.0)$ mean in the context of this problem?

(c) Make a scatter diagram of the data using the ordered pairs from part (a) and the given grid.

(d) Describe the pattern indicated by the points on the scatter diagram. What is happening to graduation rates for 4-yr college students within 5 yr?

72. The maximum benefit for the heart from exercising occurs if the heart rate is in the target heart rate zone. The lower limit of this target zone can be approximated by the linear equation

$$y = -.7x + 154,$$

where x represents age and y represents heartbeats per minute. (*Source:* Hockey, R. V., *Physical Fitness: The Pathway to Healthy Living,* Times Mirror/Mosby College Publishing, 1989.)

Age	Heartbeats (per minute)
20	
40	
60	
80	

(a) Complete the table of values for this linear equation.

(b) Write the data from the table of values as ordered pairs.

(c) Make a scatter diagram of the data. Do the points lie in an approximately linear pattern?

73. (See Exercise 72.) The upper limit of the target heart rate zone can be approximated by the linear equation

$$y = -.8x + 186,$$

where x represents age and y represents heartbeats per minute. (*Source:* Hockey, R. V., *Physical Fitness: The Pathway to Healthy Living,* Times Mirror/Mosby College Publishing, 1989.)

Age	Heartbeats (per minute)
20	
40	
60	
80	

(a) Complete the table of values for this linear equation.

(b) Write the data from the table of values as ordered pairs.

(c) Make a scatter diagram of the data. Describe the pattern indicated by the data.

74. Refer to Exercises 72 and 73. What is the target heart rate zone for age 20? age 40?

3.2 Graphing Linear Equations in Two Variables

OBJECTIVES

1 Graph linear equations.

2 Find intercepts.

3 Graph linear equations of the form $Ax + By = 0$.

4 Graph linear equations of the form $y = b$ or $x = a$.

5 Use a linear equation to model data.

In this section we use a few ordered pairs that satisfy a linear equation to graph the equation.

OBJECTIVE 1 Graph linear equations. We know that infinitely many ordered pairs satisfy a linear equation. Some ordered pairs that are solutions of $x + 2y = 7$ are graphed in Figure 8. Notice that the points plotted in this figure all appear to lie on a straight line as shown in Figure 9. In fact,

every point on the line represents a solution of the equation $x + 2y = 7$, and each solution of the equation corresponds to a point on the line.

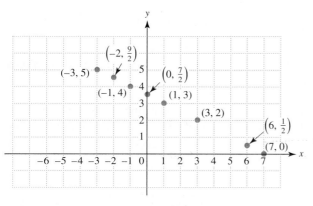

FIGURE 8

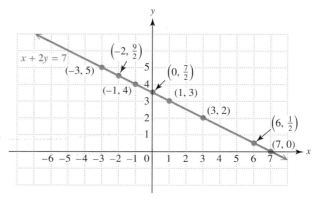

FIGURE 9

The line is a "picture" of all the solutions of the equation $x + 2y = 7$. Only a portion of the line is shown here, but it extends indefinitely in both directions, as suggested by the arrowhead on each end of the line. The line is called the **graph** of the equation, and the process of plotting the ordered pairs and drawing the line through the corresponding points is called **graphing.**

The preceding discussion can be generalized.

Graph of a Linear Equation

The graph of any linear equation in two variables is a straight line.

Notice that the word *line* appears in the name "*line*ar equation."

Since two distinct points determine a line, we can graph a straight line by finding any two different points on the line. However, it is a good idea to plot a third point as a check.

■ EXAMPLE 1 Graphing a Linear Equation

Graph the linear equation $2y = -3x + 6$.

Although this equation is not in the form $Ax + By = C$, it *could* be put in that form, and so is a linear equation. For most linear equations, we can find two different points on the graph by first letting $x = 0$, and then letting $y = 0$. Doing this gives the ordered pairs $(0, 3)$ and $(2, 0)$. We get a third ordered pair (as a check) by letting x or y equal some other number. For example, if $x = -2$, we find that $y = 6$, giving the ordered pair $(-2, 6)$. These three ordered pairs are shown in the table of values with Figure 10. The table of values is an alternative way of listing the ordered pairs $(0, 3)$, $(2, 0)$, and $(-2, 6)$.

x	y
0	3
2	0
-2	6

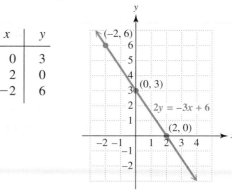

FIGURE 10

We plot the corresponding points, then draw a line through them. This line, shown in Figure 10, is the graph of $2y = -3x + 6$.

Now Try Exercise 5.

OBJECTIVE 2 **Find intercepts.** In Figure 10 the graph intersects (crosses) the y-axis at $(0, 3)$ and the x-axis at $(2, 0)$. For this reason $(0, 3)$ is called the **y-intercept** and $(2, 0)$ is called the **x-intercept** of the graph. The intercepts are particularly useful for graphing linear equations.

Finding Intercepts

To find the x-intercept, let $y = 0$ in the given equation and solve for x. Then $(x, 0)$ is the x-intercept.

To find the y-intercept, let $x = 0$ in the given equation and solve for y. Then $(0, y)$ is the y-intercept.

EXAMPLE 2   Finding Intercepts

Find the intercepts for the graph of $2x + y = 4$. Draw the graph.

Find the y-intercept by letting $x = 0$; find the x-intercept by letting $y = 0$.

$$2x + y = 4$$
$$2(0) + y = 4 \quad \text{Let } x = 0.$$
$$0 + y = 4$$
$$y = 4$$

$$2x + y = 4$$
$$2x + 0 = 4 \quad \text{Let } y = 0.$$
$$2x = 4$$
$$x = 2$$

The y-intercept is $(0, 4)$. The x-intercept is $(2, 0)$. The graph, with the two intercepts shown in red, is given in Figure 11. Get a third point as a check. For example, choosing $x = 4$ gives $y = -4$. These three ordered pairs are shown in the table with Figure 11. Plot $(0, 4)$, $(2, 0)$, and $(4, -4)$ and draw a line through them. This line, shown in Figure 11, is the graph of $2x + y = 4$.

x	y
0	4
2	0
4	−4

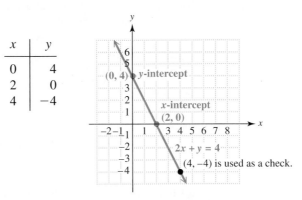

FIGURE 11

Now Try Exercise 13.

CAUTION When choosing *x*- or *y*-values to find ordered pairs to plot, be careful to choose so that the resulting points are not too close together. For example, using $(-1, -1)$, $(0, 0)$, and $(1, 1)$ may result in an inaccurate line. It is better to choose points where the *x*-values differ by at least 2.

OBJECTIVE 3 Graph linear equations of the form *Ax + By* = 0. In earlier examples, the *x*- and *y*-intercepts were used to help draw the graphs. This is not always possible. Example 3 shows what to do when the *x*- and *y*-intercepts are the same point.

■ EXAMPLE 3   Graphing an Equation of the Form *Ax + By* = 0

Graph $x - 3y = 0$.

If we let $x = 0$, then $y = 0$, giving the ordered pair $(0, 0)$. Letting $y = 0$ also gives $(0, 0)$. This is the same ordered pair, so we choose two *other* values for x or y. Choosing 2 for y gives $x - 3 \cdot 2 = 0$, or $x = 6$, giving the ordered pair $(6, 2)$. For a check point, we choose -6 for x getting -2 for y. We use the ordered pairs $(-6, -2)$, $(0, 0)$, and $(6, 2)$ to get the graph shown in Figure 12.

x	y
0	0
6	2
-6	-2

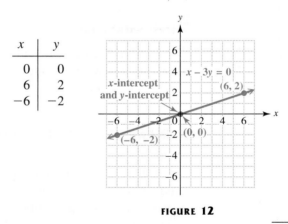

FIGURE 12

Now Try Exercise 31.

Example 3 can be generalized as follows.

Line through the Origin

If A and B are nonzero real numbers, the graph of a linear equation of the form

$$Ax + By = 0$$

goes through the origin $(0, 0)$.

OBJECTIVE 4 Graph linear equations of the form *y = b* or *x = a*. The equation $y = -4$ is a linear equation in which the coefficient of x is 0. (Write $y = -4$ as $0x + y = -4$ to see this.) Also, $x = 3$ is a linear equation in which the coefficient of y is 0. These equations lead to horizontal or vertical straight lines.

■ **EXAMPLE 4**   **Graphing an Equation of the Form $y = b$**

Graph $y = -4$.

As the equation states, for any value of x, y is always equal to -4. To get ordered pairs that are solutions of this equation, we choose any numbers for x, always using -4 for y. Three ordered pairs that satisfy the equation are shown in the table of values with Figure 13. Drawing a line through these points gives the horizontal line shown in Figure 13. The y-intercept is $(0, -4)$; there is no x-intercept.

x	y
-2	-4
0	-4
3	-4

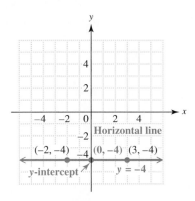

FIGURE 13

Now Try Exercise 35.

Horizontal Line

The graph of a linear equation $y = b$, where b is a real number, is the horizontal line with y-intercept $(0, b)$ and no x-intercept.

■ **EXAMPLE 5**   **Graphing an Equation of the Form $x = a$**

Graph $x - 3 = 0$.

First we add 3 to each side of the equation $x - 3 = 0$ to get $x = 3$. All the ordered pairs that are solutions of this equation have an x-value of 3. Any number can be used for y. We show three ordered pairs that satisfy the equation in the table of values with Figure 14. Drawing a line through these points gives the vertical line shown in Figure 14. The x-intercept is $(3, 0)$; there is no y-intercept.

x	y
3	3
3	0
3	-2

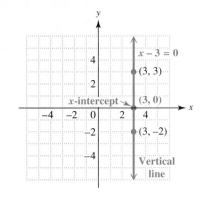

FIGURE 14

Now Try Exercise 37.

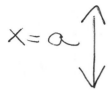

Vertical Line

The graph of a linear equation $x = a$, where a is a real number, is the vertical line with x-intercept $(a, 0)$ and no y-intercept.

In particular, notice that the horizontal line $y = 0$ is the x-axis and the vertical line $x = 0$ is the y-axis.

CAUTION The equations of horizontal and vertical lines are often confused with each other. Remember that the graph of $y = b$ is parallel to the x-axis and that of $x = a$ is parallel to the y-axis.

The different forms of straight-line equations and the methods of graphing them are given in the following summary.

Graphing Straight Lines

Equation	To Graph	Example
$y = b$	Draw a horizontal line, through $(0, b)$.	
$x = a$	Draw a vertical line, through $(a, 0)$.	
$Ax + By = 0$	Graph goes through $(0, 0)$. Get additional points that lie on the graph by choosing any value of x or y, except 0.	

(continued)

$Ax + By = C$ but not of the types above	Find any two points the line goes through. A good choice is to find the intercepts: let $x = 0$, and find the corresponding value of y; then let $y = 0$, and find x. As a check, get a third point by choosing a value of x or y that has not yet been used.	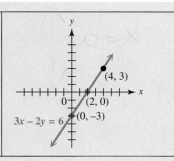

CONNECTIONS

Beginning in this chapter we include information on the basic features of graphing calculators. The most obvious feature is their ability to graph equations. We must solve the equation for y in order to enter it into the calculator. Also, we must select an appropriate "window" for the graph. The window is determined by the minimum and maximum values of x and y. Graphing calculators have a standard window, often from $x = -10$ to $x = 10$ and from $y = -10$ to $y = 10$. Sometimes this is written $[-10, 10]$, $[-10, 10]$, with the x-interval shown first.

For example, to graph the equation $2x + y = 4$, discussed in Example 2, we first solve for y.

$$2x + y = 4$$
$$y = -2x + 4 \qquad \text{Subtract } 2x.$$

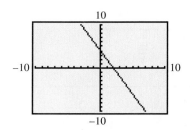

If we enter this equation as $y = -2x + 4$ and choose the standard window, the calculator displays the graph shown in the margin.

The x-value of the point on the graph where $y = 0$ (the x-intercept) gives the solution of the equation

$$y = 0,$$
$$\text{or} \qquad -2x + 4 = 0. \qquad \text{Substitute } -2x + 4 \text{ for } y.$$

Since each tick mark on the x-axis represents 1, the graph shows that $(2, 0)$ is the x-intercept.

For Discussion or Writing

How would you rewrite these equations, with one side equal to 0, to enter them into a graphing calculator for solution? (It is not necessary to clear parentheses or combine terms.)

1. $3x + 4 - 2x - 7 = 4x + 3$ **2.** $5x - 15 = 3(x - 2)$

OBJECTIVE 5 Use a linear equation to model data.

EXAMPLE 6 Using a Linear Equation to Model Credit Card Debt

Credit card debt in the United States increased steadily from 1992 through 1999. The amount of debt y in billions of dollars can be modeled by the linear equation

$$y = 47.3x + 281,$$

where $x = 0$ represents 1992, $x = 1$ represents 1993, and so on. (*Source:* Board of Governors of the Federal Reserve System.)

(a) Use the equation to approximate credit card debt in the years 1992, 1993, and 1999.

For 1992:	$y = 47.3(0) + 281$	Replace *x* with 0.
	$y = 281$ billion dollars	
For 1993:	$y = 47.3(1) + 281$	Replace *x* with 1.
	$y = 328.3$ billion dollars	
For 1999:	$y = 47.3(7) + 281$	$1999 - 1992 = 7$;
	$y = 612.1$ billion dollars	replace *x* with 7.

(b) Write the information from part (a) as three ordered pairs, and use them to graph the given linear equation.

Since x represents the year and y represents the debt in billions of dollars, the ordered pairs are (0, 281), (1, 328.3), and (7, 612.1). Figure 15 shows a graph of these ordered pairs and the line through them. (Note that arrowheads are not included with the graphed line since the data are for the years 1992 to 1999 only, that is, from $x = 0$ to $x = 7$.)

U.S. CREDIT CARD DEBT

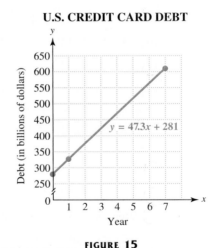

FIGURE 15

(c) Use the graph and then the equation to approximate credit card debt in 1996.

For 1996, $x = 4$. On the graph, find 4 on the horizontal axis and move up to the graphed line, then across to the vertical axis. It appears that credit card debt in 1996 was about 470 billion dollars.

To use the equation, substitute 4 for x.

$$y = 47.3(4) + 281 \qquad \text{Let } x = 4.$$

$$y = 470.2 \text{ billion dollars}$$

This result is quite similar to our estimate using the graph.

Now Try Exercise 45.

3.2 EXERCISES

Complete the given ordered pairs using the given equation. Then graph each equation by plotting the points and drawing a line through them. See Example 1.

1. $y = -x + 5$
(0,   ), (,0), (2,   )

2. $y = x - 2$
(0,   ), (,0), (5,   )

3. $y = \frac{2}{3}x + 1$
(0,   ), (3,   ), (−3,   )

4. $y = -\frac{3}{4}x + 2$
(0,   ), (4,   ), (−4,   )

5. $3x = -y - 6$
(0,   ), (,0), $\left(-\frac{1}{3},\quad\right)$

6. $x = 2y + 3$
(,0), (0,   ), $\left(\quad,\frac{1}{2}\right)$

In Exercises 7–12, match the information about each graph in Column I with the correct linear equation in Column II.

I

7. The graph of the equation has x-intercept $(4, 0)$.

8. The graph of the equation has y-intercept $(0, -4)$.

9. The graph of the equation goes through the origin.

10. The graph of the equation is a vertical line.

11. The graph of the equation is a horizontal line.

12. The graph of the equation goes through $(9, 2)$.

II

A. $x = 5$

B. $y = -3$

C. $2x - 5y = 8$

D. $x + 4y = 0$

E. $3x + y = -4$

Find the x-intercept and the y-intercept for the graph of each equation. See Examples 2, 4, and 5.

13. $2x - 3y = 24$
14. $-3x + 8y = 48$
15. $x + 6y = 0$
16. $3x - y = 0$

17. $5x - 2y = 20$
18. $-3x + 2y = 12$
19. $x - 4 = 0$
20. $y = 5$

21. What is the equation of the x-axis? What is the equation of the y-axis?

22. A student attempted to graph $4x + 5y = 0$ by finding intercepts. She first let $x = 0$ and found y; then she let $y = 0$ and found x. In both cases, the resulting point was $(0, 0)$. She knew that she needed at least two different points to graph the line, but was unsure what to do next since finding intercepts gave her only one point. How would you explain to her what to do next?

Graph each linear equation. See Examples 1–5.

23. $x = y + 2$
24. $x = -y + 6$
25. $x - y = 4$
26. $x - y = 5$

27. $2x + y = 6$
28. $-3x + y = -6$
29. $3x + 7y = 14$
30. $6x - 5y = 18$

31. $y - 2x = 0$
32. $y + 3x = 0$
33. $y = -6x$
34. $y = 4x$

35. $y + 1 = 0$
36. $y - 3 = 0$
37. $x = -2$
38. $x = 4$

39. Write a paragraph summarizing how to graph a linear equation in two variables.

| **TECHNOLOGY INSIGHTS** (EXERCISES 40–44) |

In Exercises 40–43, we give the graph of a linear equation in two variables solved for y and a corresponding linear equation in one variable, with y replaced by 0. Solve the equation in one variable using the methods of Section 2.3. Is the solution you get the same as the x-intercept (labeled "Zero") on the calculator screen?

40. $8 - 2(3x - 4) - 2x = 0$

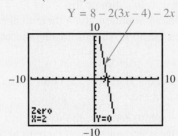

41. $5(2x - 1) - 4(2x + 1) - 7 = 0$

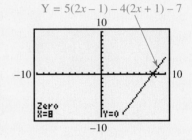

42. $.6x - .1x - x + 2.5 = 0$

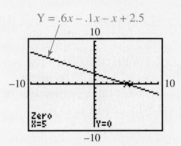

43. $-\dfrac{2}{7}x + 2x - \dfrac{1}{2}x - \dfrac{17}{2} = 0$

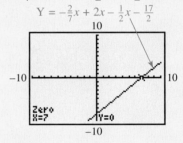

44. From the results in Exercises 40–43, what can you conclude?

Solve each problem. See Example 6.

45. The height y (in centimeters) of a woman is related to the length of her radius bone x (from the wrist to the elbow) and is approximated by the linear equation

$$y = 3.9x + 73.5.$$

(a) Use the equation to find the approximate heights of women with radius bones of lengths 20 cm, 26 cm, and 22 cm.

(b) Graph the equation using the data from part (a).

(c) Use the graph to estimate the length of the radius bone in a woman who is 167 cm tall. Then use the equation to find the length of this radius bone to the nearest centimeter. (*Hint:* Substitute for y in the equation.)

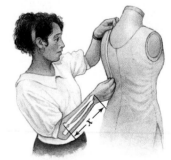

46. The weight y (in pounds) of a man taller than 60 in. can be roughly approximated by the linear equation

$$y = 5.5x - 220,$$

where x is the height of the man in inches.

(a) Use the equation to approximate the weights of men whose heights are 62 in., 66 in., and 72 in.

(b) Graph the equation using the data from part (a).

(c) Use the graph to estimate the height of a man who weighs 155 lb. Then use the equation to find the height of this man to the nearest inch. (*Hint:* Substitute for y in the equation.)

47. Refer to Section 3.1 Exercise 72. Draw a line through the points you plotted in the scatter diagram there.

(a) Use the graph to estimate the lower limit of the target heart rate zone for age 30.

(b) Use the linear equation given there to approximate the lower limit for age 30.

(c) How does the approximation using the equation compare to the estimate from the graph?

48. Refer to Section 3.1 Exercise 73. Draw a line through the points you plotted in the scatter diagram there.

(a) Use the graph to estimate the upper limit of the target heart rate zone for age 30.

(b) Use the linear equation given there to approximate the upper limit for age 30.

(c) How does the approximation using the equation compare to the estimate from the graph?

49. Use the results of Exercises 47(b) and 48(b) to determine the target heart rate zone for age 30.

50. Should the graphs of the target heart rate zone in Section 3.1 Exercises 72 and 73 be used to estimate the target heart rate zone for ages below 20 or above 80? Why or why not?

51. Per capita consumption of carbonated soft drinks increased for the years 1992 through 1997 as shown in the graph. If $x = 0$ represents 1992, $x = 1$ represents 1993, and so on, per capita consumption can be modeled by the linear equation

$$y = .8x + 49,$$

where y is in gallons.

(a) Use the equation to approximate consumption in 1993, 1995, and 1997.

(b) Use the graph to estimate consumption for the same years.

(c) How do the approximations using the equation compare to the estimates from the graph?

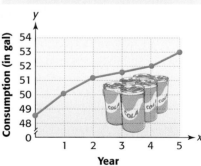

SOFT DRINK CONSUMPTION

Source: U.S. Department of Agriculture.

52. Sporting goods sales y (in billions of dollars) from 1995 through 2000 are modeled by the linear equation

$$y = 3.606x + 41.86,$$

where $x = 0$ corresponds to 1990, $x = 5$ corresponds to 1995, and so on.

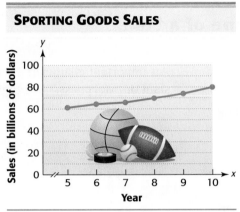

SPORTING GOODS SALES

Source: U.S. Bureau of the Census.

(a) Use the equation to approximate sporting goods sales in 1995, 1997, and 2000. Round your answers to the nearest billion dollars.

(b) Use the graph to estimate sales for the same years.

(c) How do the approximations using the equation compare to the estimates using the graph?

53. The graph shows the value of a certain sport utility vehicle over the first 5 yr of ownership.

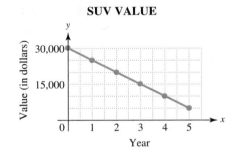

SUV VALUE

Use the graph to do the following.

(a) Determine the initial value of the SUV.

(b) Find the *depreciation* (loss in value) from the original value after the first 3 yr.

(c) What is the annual or yearly depreciation in each of the first 5 yr?

(d) What does the ordered pair $(5, 5000)$ mean in the context of this problem?

54. Demand for an item is often closely related to its price. As price increases, demand decreases, and as price decreases, demand increases. Suppose demand for a video game is 2000 units when the price is $40, and demand is 2500 units when the price is $30.

(a) Let x be the price and y be the demand for the game. Graph the two given pairs of prices and demands.

(b) Assume the relationship is linear. Draw a line through the two points from part (a). From your graph, estimate the demand if the price drops to $20.

(c) Use the graph to estimate the price if the demand is 3500 units.

3.3 The Slope of a Line

An important characteristic of the lines we graphed in the previous section is their slant or "steepness." See Figure 16.

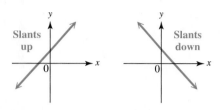

FIGURE 16

One way to measure the steepness of a line is to compare the vertical change in the line to the horizontal change while moving along the line from one fixed point to another. This measure of steepness is called the *slope* of the line.

OBJECTIVE 1 Find the slope of a line given two points. Figure 17 shows a line through two nonspecific points (x_1, y_1) and (x_2, y_2). (This notation is called **subscript notation.** Read x_1 as "x-sub-one" and x_2 as "x-sub-two.")

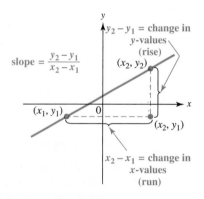

FIGURE 17

Moving along the line from the point (x_1, y_1) to the point (x_2, y_2) causes y to change by $y_2 - y_1$ units. This is the vertical change or **rise.** Similarly, x changes by $x_2 - x_1$ units, which is the horizontal change or **run.** (In both cases, the change is expressed as a *difference*.) Remember from Section 2.6 that one way to compare two numbers is by using a ratio. **Slope** is the ratio of the vertical change in y to the horizontal change in x.

EXAMPLE 1 Comparing Rise to Run

Figure 18 shows two lines. Find the slope ratio of the rise to the run for each line.

We use the two points shown on each line. For the line in Figure 18(a), we find the rise from point Q to point P by determining the vertical change from -4 to 4, the

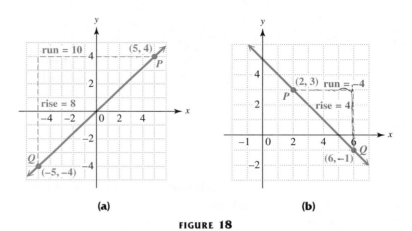

(a)   **(b)**

FIGURE 18

difference $4 - (-4) = 8$. Similarly, we find the run by determining the horizontal change from Q to P, $5 - (-5) = 10$. Slope is the ratio

$$\frac{\text{rise}}{\text{run}} = \frac{8}{10} \text{ or } \frac{4}{5}.$$

From Q to P, the line in Figure 18(b) has rise $3 - (-1) = 4$ and run $2 - 6 = -4$, so the slope is the ratio

$$\frac{\text{rise}}{\text{run}} = \frac{4}{-4} \text{ or } -1.$$

To confirm our slope ratios, count grid squares from one point on the line to another. For example, starting at the origin in Figure 18(a), count up 4 squares $\left(\text{the rise in the slope ratio } \frac{4}{5}\right)$ and then 5 squares to the right (the run) to arrive at the point $(5, 4)$ on the line.

> **Now Try Exercise 3.**

The slopes we found for the two lines in Figure 18 suggest that a line with positive slope slants upward from left to right, and a line with negative slope slants downward. These facts can be generalized.

Positive and Negative Slopes

A line with positive slope rises from left to right.

A line with negative slope falls from left to right.

Hill

The idea of slope is used in many everyday situations. For example, because $10\% = \frac{1}{10}$, a highway with a 10% grade (or slope) rises 1 meter for every 10 horizontal meters. A highway sign is used to warn of a downgrade ahead that may be long or steep. Architects specify the pitch of a roof using slope; a $\frac{5}{12}$ roof means that the roof rises 5 ft for every 12 ft in the horizontal direction. See the figure on the next page. The slope of a stairwell also indicates the ratio of the vertical rise to the horizontal run. The slope of the stairs in the figure is $\frac{8}{14}$.

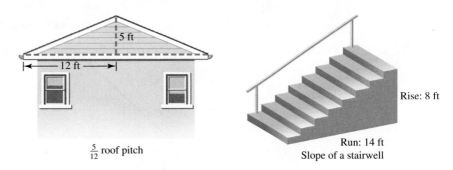

$\frac{5}{12}$ roof pitch

Slope of a stairwell

Traditionally, the letter m represents slope. The slope m of a line is defined as follows.

Slope Formula

The **slope** of the line through the points (x_1, y_1) and (x_2, y_2) is

$$m = \frac{\textbf{change in } y}{\textbf{change in } x} = \frac{y_2 - y_1}{x_2 - x_1}, \quad \text{if } x_1 \neq x_2.$$

The slope of a line tells how fast y changes for each unit of change in x; that is, the slope gives the rate of change in y for each unit of change in x.

EXAMPLE 2 Finding Slopes of Lines

Find the slope of each line.

(a) The line through $(1, -2)$ and $(-4, 7)$

Use the slope formula. Let $(-4, 7) = (x_2, y_2)$ and $(1, -2) = (x_1, y_1)$. Then

$$\text{slope } m = \frac{\text{change in } y}{\text{change in } x} = \frac{y_2 - y_1}{x_2 - x_1} = \frac{7 - (-2)}{-4 - 1} = \frac{9}{-5} = -\frac{9}{5}.$$

$$\frac{y_2 - y_1}{x_2 - x_1}$$

See Figure 19.

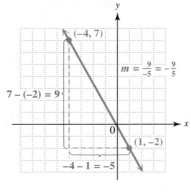

FIGURE 19

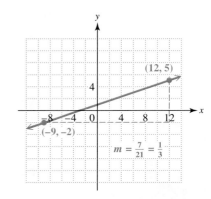

FIGURE 20

(b) The line through $(-9, -2)$ and $(12, 5)$

$$m = \frac{y_2 - y_1}{x_2 - x_1} = \frac{5 - (-2)}{12 - (-9)} = \frac{7}{21} = \frac{1}{3}$$

See Figure 20. Note that the same slope is obtained by subtracting in reverse order.

$$m = \frac{-2-5}{-9-12} = \frac{-7}{-21} = \frac{1}{3}$$

Now Try Exercises 17 and 21.

CAUTION   It makes no difference which point is (x_1, y_1) or (x_2, y_2); however, be consistent. Start with the x- and y-values of one point (either one) and subtract the corresponding values of the other point. Also, the slope of a line is the same for *any* two points on the line.

EXAMPLE 3   Finding the Slope of a Horizontal Line

Find the slope of the line through $(-8, 4)$ and $(2, 4)$.

Use the slope formula.

$$m = \frac{4-4}{-8-2} = \frac{0}{-10} = 0 \qquad \text{Zero slope}$$

As shown in Figure 21, the line through these two points is horizontal, with equation $y = 4$. *All horizontal lines have slope 0,* since the difference in y-values is always 0.

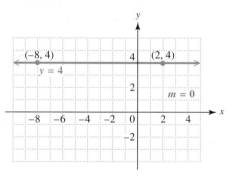

FIGURE 21

Now Try Exercise 23.

EXAMPLE 4   Finding the Slope of a Vertical Line

Find the slope of the line through $(6, 2)$ and $(6, -9)$.

$$m = \frac{2-(-9)}{6-6} = \frac{11}{0} \qquad \text{Undefined slope}$$

Since division by 0 is undefined, the slope is undefined. The graph in Figure 22 shows that the line through these two points is vertical with equation $x = 6$. All points on a vertical line have the same x-value, so *the slope of any vertical line is undefined.*

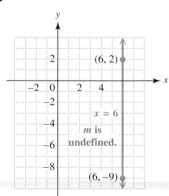

FIGURE 22

Now Try Exercise 25.

> **Slopes of Horizontal and Vertical Lines**
>
> **Horizontal lines,** with equations of the form $y = b$, have **slope 0.**
>
> **Vertical lines,** with equations of the form $x = a$, have **undefined slope.**

OBJECTIVE 2 Find the slope from the equation of a line. The slope of a line can be found directly from its equation. For example, the slope of the line

$$y = -3x + 5$$

is found using any two points on the line. We get these two points by first choosing two different values of x and then finding the corresponding values of y. We choose $x = -2$ and $x = 4$.

$y = -3x + 5$	$y = -3x + 5$
$y = -3(-2) + 5$ Let $x = -2$.	$y = -3(4) + 5$ Let $x = 4$.
$y = 6 + 5$	$y = -12 + 5$
$y = 11$	$y = -7$

The ordered pairs are $(-2, 11)$ and $(4, -7)$. Now we use the slope formula.

$$m = \frac{11 - (-7)}{-2 - 4} = \frac{18}{-6} = -3$$

The slope, -3, is the same number as the coefficient of x in the equation $y = -3x + 5$. It can be shown that this always happens, *as long as the equation is solved for y.* This fact is used to find the slope of a line from its equation.

> **Finding the Slope of a Line from Its Equation**
>
> *Step 1*   Solve the equation for y.
>
> *Step 2*   The slope is given by the coefficient of x.

▌ **EXAMPLE 5 Finding Slopes from Equations**

Find the slope of each line.

(a) $2x - 5y = 4$

Solve the equation for y.

$$2x - 5y = 4$$
$$-5y = -2x + 4 \qquad \text{Subtract } 2x \text{ from each side.}$$
$$y = \frac{2}{5}x - \frac{4}{5} \qquad \text{Divide by } -5.$$

The slope is given by the coefficient of x, so the slope is $\frac{2}{5}$.

(b) $8x + 4y = 1$

Solve the equation for y.

$$8x + 4y = 1$$

$$4y = -8x + 1 \qquad \text{Subtract } 8x.$$

$$y = -2x + \frac{1}{4} \qquad \text{Divide by 4.}$$

The slope of this line is given by the coefficient of x, -2.

Now Try Exercise 33.

OBJECTIVE 3 **Use slope to determine whether two lines are parallel, perpendicular, or neither.** Two lines in a plane that never intersect are **parallel.** We use slopes to tell whether two lines are parallel. For example, Figure 23 shows the graphs of $x + 2y = 4$ and $x + 2y = -6$. These lines appear to be parallel. Solve for y to find that both $x + 2y = 4$ and $x + 2y = -6$ have slope $-\frac{1}{2}$. Nonvertical parallel lines always have equal slopes.

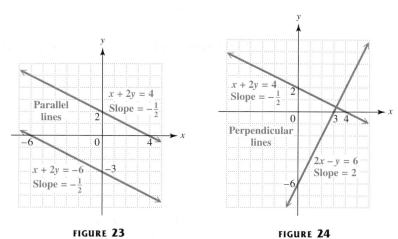

FIGURE 23   FIGURE 24

Figure 24 shows the graphs of $x + 2y = 4$ and $2x - y = 6$. These lines appear to be **perpendicular** (that is, they intersect at a 90° angle). Solving for y shows that the slope of $x + 2y = 4$ is $-\frac{1}{2}$, while the slope of $2x - y = 6$ is 2. The product of $-\frac{1}{2}$ and 2 is

$$-\frac{1}{2}(2) = -1.$$

This is true in general; the product of the slopes of two perpendicular lines, neither of which is vertical, is always -1. This means that the slopes of perpendicular lines are negative reciprocals; if one slope is the nonzero number a, the other is $-\frac{1}{a}$.

Slopes of Parallel and Perpendicular Lines

Two lines with the same slope are parallel.

Two lines whose slopes have a product of -1 are perpendicular.

EXAMPLE 6 Deciding Whether Two Lines Are Parallel or Perpendicular

Decide whether each pair of lines is *parallel, perpendicular,* or *neither.*

(a) $x + 2y = 7$

$-2x + \ y = 3$

Find the slope of each line by first solving each equation for y.

$x + 2y = 7$	$-2x + y = 3$
$2y = -x + 7$	$y = 2x + 3$
$y = -\dfrac{1}{2}x + \dfrac{7}{2}$	
Slope: $-\dfrac{1}{2}$	Slope: 2

Since the slopes are not equal, the lines are not parallel. Check the product of the slopes: $-\frac{1}{2}(2) = -1$. The two lines are perpendicular because the product of their slopes is -1, indicating that the slopes are negative reciprocals.

(b) $3x - \ y = 4$ *Solve for y.* $y = 3x - 4$

$6x - 2y = -12$ ⟶ $y = 3x + 6$

Both lines have slope 3, so the lines are parallel.

(c) $4x + 3y = 6$ *Solve for y.* $y = -\dfrac{4}{3}x + 2$

$2x - \ y = 5$ ⟶ $y = 2x - 5$

Here the slopes are $-\frac{4}{3}$ and 2. These lines are neither parallel nor perpendicular because $-\frac{4}{3} \neq 2$ and $-\frac{4}{3} \cdot 2 \neq -1$.

(d) $5x - y = 1$ $y = 5x - 1$

                        Solve for y.

$x - 5y = -10$ ⟶ $y = \dfrac{1}{5}x + 2$

The slopes are 5 and $\frac{1}{5}$. The lines are not parallel, nor are they perpendicular. $\left(\textit{Be careful! } 5\left(\frac{1}{5}\right) = 1, \textit{ not } -1.\right)$

Now Try Exercises 51, 55, and 57.

CONNECTIONS

Because the viewing window of a graphing calculator is a rectangle, the graphs of perpendicular lines will not appear perpendicular unless appropriate intervals are used for x and y. Graphing calculators usually have a key to select a "square" window automatically. In a square window, the x-interval is about 1.5 times the y-interval because the screen is about 1.5 times as wide as it is high. The equations from Figure 24 are graphed with the standard (nonsquare) window and then with a square window on the next page.

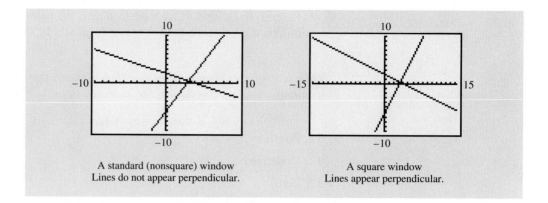

A standard (nonsquare) window
Lines do not appear perpendicular.

A square window
Lines appear perpendicular.

3.3 EXERCISES

 1. In the context of the graph of a straight line, what is meant by "rise"? What is meant by "run"?

Use the coordinates of the indicated points to find the ratio of rise to run for each line. See Example 1.

2.

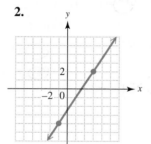

3.

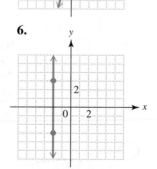

4.

5.

6.

7.

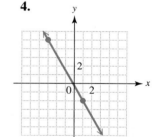

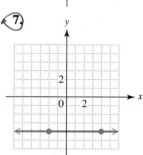

8. Look at the graph in Exercise 2 and answer the following.

(a) Start at the point $(-1, -4)$ and count vertically up to the horizontal line that goes through the other plotted point. What is this vertical change? (Remember: "up" means positive, "down" means negative.)

(b) From this new position, count horizontally to the other plotted point. What is this horizontal change? (Remember: "right" means positive, "left" means negative.)

(c) What is the quotient of the numbers found in parts (a) and (b)? What do we call this number?

9. Refer to Exercise 8. If we were to *start* at the point $(3, 2)$ and *end* at the point $(-1, -4)$, do you think that the answer to part (c) would be the same? Explain why or why not.

On a pair of axes similar to the one shown, sketch the graph of a straight line having the indicated slope.

10. Negative

11. Positive

12. Undefined

13. Zero

14. Explain in your own words what is meant by *slope* of a line.

15. A student was asked to find the slope of the line through the points $(2, 5)$ and $(-1, 3)$. His answer, $-\frac{2}{3}$, was incorrect. He showed his work as

$$\frac{3-5}{2-(-1)} = \frac{-2}{3} = -\frac{2}{3}.$$

$-1 \ -2$

What was his error? Give the correct slope.

Find the slope of the line through each pair of points. See Examples 2–4.

16. $(4, -1)$ and $(-2, -8)$ **17.** $(1, -2)$ and $(-3, -7)$ **18.** $(-8, 0)$ and $(0, -5)$

19. $(0, 3)$ and $(-2, 0)$ **20.** $(-4, -5)$ and $(-5, -8)$ **21.** $(-2, 4)$ and $(-3, 7)$

22. $(6, -5)$ and $(-12, -5)$ **23.** $(4, 3)$ and $(-6, 3)$

24. $(-8, 6)$ and $(-8, -1)$ **25.** $(-12, 3)$ and $(-12, -7)$

26. $(3.1, 2.6)$ and $(1.6, 2.1)$ **27.** $\left(-\frac{7}{5}, \frac{3}{10}\right)$ and $\left(\frac{1}{5}, -\frac{1}{2}\right)$

Find the slope of each line. See Example 5.

28. $y = 2x - 3$ **29.** $y = 5x + 12$ **30.** $2y = -x + 4$ **31.** $4y = x + 1$

32. $-6x + 4y = 4$ **33.** $3x - 2y = 3$ **34.** $2x + 4y = 5$ **35.** $-3x + 2y = 5$

36. $x = -2$ **37.** $y = -5$ **38.** $y = 4$ **39.** $x = 6$

40. What is the slope of a line whose graph is parallel to the graph of $-5x + y = -3$? Perpendicular to the graph of $-5x + y = -3$?

41. What is the slope of a line whose graph is parallel to the graph of $3x + y = 7$? Perpendicular to the graph of $3x + y = 7$?

The figure shows a line that has a positive slope (because it rises from left to right) and a positive y-value for the y-intercept (because it intersects the y-axis above the origin).

*For each line in Exercises 42–47, decide whether **(a)** the slope is positive, negative, or zero and **(b)** the y-value of the y-intercept is positive, negative, or zero.*

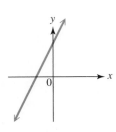

42.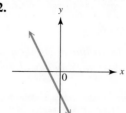

43.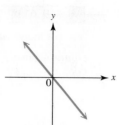

44.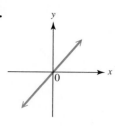

45.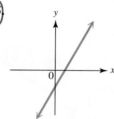

46.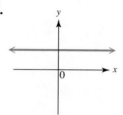

47.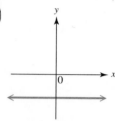

48. If two nonvertical lines are parallel, what do we know about their slopes? If two lines are perpendicular and neither is parallel to an axis, what do we know about their slopes? Why must the lines be nonvertical?

49. If two lines are both vertical or both horizontal, which of the following are they?

   A. Parallel   **B.** Perpendicular   **C.** Neither parallel nor perpendicular

50. If a line is vertical, what is true of any line that is perpendicular to it?

For each pair of equations, give the slopes of the lines and then determine whether the two lines are parallel, perpendicular, *or* neither parallel nor perpendicular. *See Example 6.*

51. $2x + 5y = 4$
   $4x + 10y = 1$

52. $-4x + 3y = 4$
   $-8x + 6y = 0$

53. $8x - 9y = 6$
   $8x + 6y = -5$

54. $5x - 3y = -2$
   $3x - 5y = -8$

55. $3x - 2y = 6$
   $2x + 3y = 3$

56. $3x - 5y = -1$
   $5x + 3y = 2$

57. $5x - y = 1$
   $x - 5y = -10$

58. $3x - 4y = 12$
   $4x + 3y = 12$

59. What is the slope (or pitch) of this roof?

60. What is the slope (or grade) of this hill?

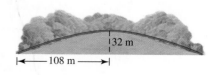

For Individual or Group Work

Figure A gives public school enrollment (in thousands) in grades 9–12 in the United States. Figure B gives the (average) number of public school students per computer.

PUBLIC SCHOOL ENROLLMENT

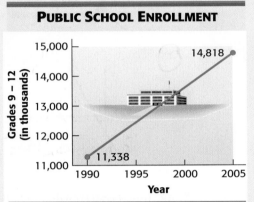

Source: *Digest of Educational Statistics*, annual, and *Projections of Educational Statistics*, annual.

FIGURE A

STUDENTS PER COMPUTER

Source: Quality Education Data.

FIGURE B

Work Exercises 61–66 in order.

61. Use the ordered pairs (1990, 11,338) and (2005, 14,818) to find the slope of the line in Figure A.

62. The slope of the line in Figure A is _____. This means that
(positive/negative)
during the period represented, enrollment _____.
(increased/decreased)

63. The slope of a line represents its *rate of change*. Based on Figure A, what was the increase in students *per year* during the period shown?

64. Use the given information to find the slope of the line in Figure B.

65. The slope of the line in Figure B is _____. This means that
(positive/negative)
during the period represented, the number of students per computer

_____.
(increased/decreased)

66. Based on Figure B, what was the decrease in students per computer *per year*
during the period shown?

67. The growth in retail square footage, in billions, is shown in the line graph. This graph
looks like a straight line. If the change in square footage each year is the same, then it
is a straight line. Find the change in square footage for the years shown in the graph.
(*Hint:* To find the change in square footage from 1991 to 1992, subtract the *y*-value for
1991 from the *y*-value for 1992.) Is the graph a straight line?

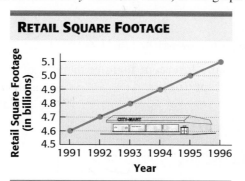

RETAIL SQUARE FOOTAGE

Source: International Council of Shopping
Centers.

68. Find the slope of the line in Exercise 67 by using any two of the points shown on the
line. How does the slope compare with the yearly change in square footage?

TECHNOLOGY INSIGHTS (EXERCISES 69–72)

*Some graphing calculators have the capability of
displaying a table of points for a graph. The table
shown here gives several points that lie on a line
designated* Y_1.

X	Y_1
-12	-.8
-10	0
-8	.8
-6	1.6
-4	2.4
-2	3.2
0	4

X=-12

69. Use any two of the displayed ordered pairs to
find the slope of the line.

70. What is the *x*-intercept of the line?

71. What is the *y*-intercept of the line?

72. Which one of the two lines shown is the graph of Y_1?

A.

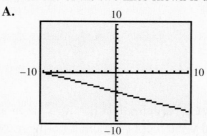

B.

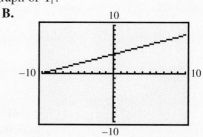

3.4 Equations of a Line

OBJECTIVES

1 Write an equation of a line given its slope and *y*-intercept.

2 Graph a line given its slope and a point on the line.

3 Write an equation of a line given its slope and any point on the line.

4 Write an equation of a line given two points on the line.

In the previous section we found the slope (steepness) of a line from the equation of the line by solving the equation for *y*. In that form, the slope is the coefficient of *x*. For example, the slope of the line with equation $y = 2x + 3$ is 2, the coefficient of *x*. What does the number 3 represent? If $x = 0$, the equation becomes

$$y = 2(0) + 3 = 0 + 3 = 3.$$

Since $y = 3$ corresponds to $x = 0$, $(0, 3)$ is the *y*-intercept of the graph of $y = 2x + 3$. An equation like $y = 2x + 3$ that is solved for *y* is said to be in *slope-intercept form* because both the slope and the *y*-intercept of the line can be read directly from the equation.

Slope-Intercept Form

The **slope-intercept form** of the equation of a line with slope *m* and *y*-intercept $(0, b)$ is

$$y = mx + b.$$

Remember that the intercept in the slope-intercept form is the *y-intercept*.

NOTE The slope-intercept form is the most useful form for a linear equation because of the information we can determine from it. It is also the form used by graphing calculators and the one that describes a *linear function,* an important concept in mathematics.

OBJECTIVE 1 Write an equation of a line given its slope and *y*-intercept. Given the slope and *y*-intercept of a line, we can use the slope-intercept form to find an equation of the line.

EXAMPLE 1 Finding an Equation of a Line

Find an equation of the line with slope $\frac{2}{3}$ and *y*-intercept $(0, -1)$.
    Here $m = \frac{2}{3}$ and $b = -1$, so the equation is

$$\underset{\text{Slope}}{y = mx} + \underset{y\text{-intercept}}{b}$$

$$y = \frac{2}{3}x - 1.$$

Now Try Exercise 11.

OBJECTIVE 2 Graph a line given its slope and a point on the line. We can use the slope and *y*-intercept to graph a line. For example, to graph $y = \frac{2}{3}x - 1$, we first locate the *y*-intercept, $(0, -1)$, on the *y*-axis. From the definition of slope and the fact

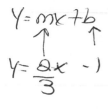

that the slope of the line is $\frac{2}{3}$,

$$m = \frac{\text{rise}}{\text{run}} = \frac{2}{3}.$$

Another point P on the graph of the line can be found by counting from the y-intercept 2 units up and then counting 3 units to the right. We then draw the line through point P and the y-intercept, as shown in Figure 25.

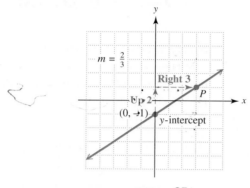

FIGURE 25

This method can be extended to graph a line given its slope and any point on the line, not just the y-intercept.

EXAMPLE 2 Graphing a Line Given a Point and the Slope

Graph the line through $(-2, 3)$ with slope -4.

First, locate the point $(-2, 3)$. Write the slope as

$$m = \frac{\text{rise}}{\text{run}} = -4 = \frac{-4}{1}.$$

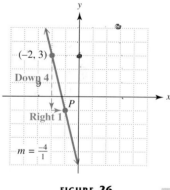

Locate another point on the line by counting 4 units down (because of the negative sign) and then 1 unit to the right. Finally, draw the line through this new point P and the given point $(-2, 3)$. See Figure 26.

FIGURE 26

Now Try Exercise 19.

N O T E In Example 2, we could have written the slope as $\frac{4}{-1}$ instead. In this case, we would move 4 units up from $(-2, 3)$ and then 1 unit to the left (because of the negative sign). Verify that this produces the same line.

OBJECTIVE 3 Write an equation of a line given its slope and any point on the line. Let m represent the slope of a line and let (x_1, y_1) represent a given point on the line. Let (x, y) represent any other point on the line. Then by the definition of slope,

$$\frac{y - y_1}{x - x_1} = m \quad \text{or} \quad y - y_1 = m(x - x_1).$$

This result is the *point-slope form* of the equation of a line.

Point-Slope Form

The **point-slope form** of the equation of a line with slope m going through (x_1, y_1) is

$$y - y_1 = m(x - x_1).$$

EXAMPLE 3 Using the Point-Slope Form to Write Equations

Find an equation of each line. Write the equation in slope-intercept form.

(a) Through $(-2, 4)$, with slope -3

The given point is $(-2, 4)$ so $x_1 = -2$ and $y_1 = 4$. Also, $m = -3$. First substitute these values into the point-slope form to get an equation of the line. Then solve for y to write the equation in slope-intercept form.

$$
\begin{array}{ll}
y - y_1 = m(x - x_1) & \text{Point-slope form} \\
y - 4 = -3[x - (-2)] & \text{Let } y_1 = 4, m = -3, x_1 = -2. \\
y - 4 = -3(x + 2) & \\
y - 4 = -3x - 6 & \text{Distributive property} \\
y = -3x - 2 & \text{Add 4.}
\end{array}
$$

The last equation is in slope-intercept form.

(b) Through $(4, 2)$, with slope $\frac{3}{5}$

$$
\begin{array}{ll}
y - y_1 = m(x - x_1) & \\
y - 2 = \dfrac{3}{5}(x - 4) & \text{Let } y_1 = 2, m = \frac{3}{5}, x_1 = 4. \\[2mm]
y - 2 = \dfrac{3}{5}x - \dfrac{12}{5} & \text{Distributive property} \\[2mm]
y = \dfrac{3}{5}x - \dfrac{12}{5} + \dfrac{10}{5} & \text{Add } 2 = \frac{10}{5} \text{ to both sides.} \\[2mm]
y = \dfrac{3}{5}x - \dfrac{2}{5} & \text{Combine terms.}
\end{array}
$$

We did not clear fractions after the substitution step because we want the equation in slope-intercept form; that is, solved for y.

Now Try Exercises 29 and 31.

OBJECTIVE 4 Write an equation of a line given two points on the line. We can also use the point-slope form to find an equation of a line when two points on the line are known.

EXAMPLE 4 Finding the Equation of a Line Given Two Points

Find an equation of the line through the points $(-2, 5)$ and $(3, 4)$. Write the equation in slope-intercept form.

First, find the slope of the line, using the slope formula.

$$\text{slope } m = \frac{y_2 - y_1}{x_2 - x_1} = \frac{5 - 4}{-2 - 3} = \frac{1}{-5} = -\frac{1}{5}$$

Now use either $(-2, 5)$ or $(3, 4)$ and the point-slope form. We choose $(3, 4)$.

$$y - y_1 = m(x - x_1)$$

$$y - 4 = -\frac{1}{5}(x - 3) \qquad \text{Let } y_1 = 4, \ m = -\tfrac{1}{5}, \ x_1 = 3.$$

$$y - 4 = -\frac{1}{5}x + \frac{3}{5} \qquad \text{Distributive property}$$

$$y = -\frac{1}{5}x + \frac{3}{5} + \frac{20}{5} \qquad \text{Add } 4 = \tfrac{20}{5} \text{ to both sides.}$$

$$y = -\frac{1}{5}x + \frac{23}{5} \qquad \text{Combine terms.}$$

The same result would be found by using $(-2, 5)$ for (x_1, y_1).

Now Try Exercise 39.

Many of the linear equations in Sections 3.1 through 3.3 were given in the form $Ax + By = C$, which is called *standard form*.

Standard Form

A linear equation is in **standard form** if it is written as

$$Ax + By = C,$$

where A, B, and C are integers and $A > 0$, $B \neq 0$.

NOTE   The definition of standard form is not the same in all texts. A linear equation can be written in this form in many different, equally correct, ways. For example, $3x + 4y = 12$, $6x + 8y = 24$, and $9x + 12y = 36$ represent the same set of ordered pairs. Let us agree that $3x + 4y = 12$ is preferable to the other forms because the greatest common factor of 3, 4, and 12 is 1.

A summary of the forms of linear equations follows.

Linear Equations

$x = a$	**Vertical line**
	Slope is undefined; x-intercept is $(a, 0)$.
$y = b$	**Horizontal line**
	Slope is 0; y-intercept is $(0, b)$.
$y = mx + b$	**Slope-intercept form**
	Slope is m; y-intercept is $(0, b)$.
$y - y_1 = m(x - x_1)$	**Point-slope form**
	Slope is m; line goes through (x_1, y_1).
$Ax + By = C$	**Standard form**
	Slope is $-\frac{A}{B}$; x-intercept is $\left(\frac{C}{A}, 0\right)$; y-intercept is $\left(0, \frac{C}{B}\right)$.

3.4 EXERCISES

Match the correct equation in Column II with the description in Column I.

I	II
1. Slope $= -2$, through the point $(4, 1)$	**A.** $y = 4x$
2. Slope $= -2$, y-intercept $(0, 1)$	**B.** $y = \dfrac{1}{4}x$
3. Passing through the points $(0, 0)$ and $(4, 1)$	**C.** $y = -2x + 1$
4. Passing through the points $(0, 0)$ and $(1, 4)$	**D.** $y - 1 = -2(x - 4)$

5. Explain why the equation of a vertical line cannot be written in the form $y = mx + b$.

6. Match each equation with the graph that would most closely resemble its graph.

(a) $y = x + 3$
(b) $y = -x + 3$
(c) $y = x - 3$
(d) $y = -x - 3$

A.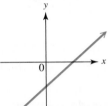

B.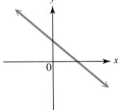

C.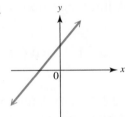

D.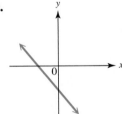

Use the geometric interpretation of slope (rise divided by run, from Section 3.3) to find the slope of each line. Then, by identifying the y-intercept from the graph, write the slope-intercept form of the equation of the line.

7.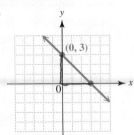
$(0, -3)$

8.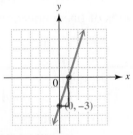
$(0, -4)$

9.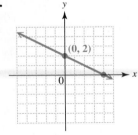
$(0, 3)$

10.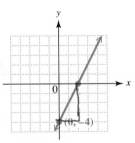
$(0, 2)$

$y = mx + b$

$y = mx + x + 3$

Write the equation of each line with the given slope and y-intercept. See Example 1.

11. $m = 4, (0, -3)$

12. $m = -5, (0, 6)$

13. $m = 0, (0, 3)$

14. $m = 0, (0, -4)$

15. Undefined slope, $(0, -2)$

16. Undefined slope, $(0, 5)$

Graph each line passing through the given point and having the given slope. (In Exercises 22–25, recall the types of lines having slope 0 and undefined slope.) See Example 2.

17. $(0, 2), m = 3$

18. $(0, -5), m = -2$

19. $(1, -5), m = -\dfrac{2}{5}$

20. $(2, -1), m = -\dfrac{1}{3}$

21. $(-1, 4), m = \dfrac{2}{5}$

22. $(3, 2), m = 0$

23. $(-2, 3), m = 0$

24. $(3, -2),$ undefined slope

25. $(2, 4),$ undefined slope

26. What is the common name given to a vertical line whose *x*-intercept is the origin?

27. What is the common name given to a line with slope 0 whose *y*-intercept is the origin?

Write an equation for each line passing through the given point and having the given slope. Write the equation in slope-intercept form. See Example 3.

28. $(4, 1), m = 2$

29. $(-1, 3), m = -4$

30. $(2, 7), m = 3$

31. $(-2, 5), m = \dfrac{2}{3}$

32. $(4, 2), m = -\dfrac{1}{3}$

33. $(-4, 1), m = \dfrac{3}{4}$

34. $(6, -3), m = -\dfrac{4}{5}$

35. $(2, 1), m = \dfrac{5}{2}$

36. $(7, -2), m = -\dfrac{7}{2}$

37. If a line passes through the origin and a second point whose *x*- and *y*-coordinates are equal, what is an equation of the line?

Write an equation for the line passing through the given pair of points. Write each equation in slope-intercept form. See Example 4.

38. $(8, 5)$ and $(9, 6)$

39. $(4, 10)$ and $(6, 12)$

40. $(-1, -7)$ and $(-8, -2)$

41. $(-2, -1)$ and $(3, -4)$

42. $(0, -2)$ and $(-3, 0)$

43. $(-4, 0)$ and $(0, 2)$

44. $\left(\dfrac{1}{2}, \dfrac{3}{2}\right)$ and $\left(-\dfrac{1}{4}, \dfrac{5}{4}\right)$

45. $\left(-\dfrac{2}{3}, \dfrac{8}{3}\right)$ and $\left(\dfrac{1}{3}, \dfrac{7}{3}\right)$

46. Describe in your own words the slope-intercept and point-slope forms of the equation of a line. Tell what information must be given to use each form to write an equation. Include examples.

RELATING CONCEPTS (EXERCISES 47–54)

For Individual or Group Work

If we think of ordered pairs of the form (C, F), *then the two most common methods of measuring temperature, Celsius and Fahrenheit, can be related as follows: When* $C = 0$, $F = 32$, *and when* $C = 100$, $F = 212$. **Work Exercises 47–54 in order.**

47. Write two ordered pairs relating these two temperature scales.

48. Find the slope of the line through the two points.

49. Use the point-slope form to find an equation of the line. (Your variables should be C and F rather than x and y.)

50. Write an equation for F in terms of C.

51. Use the equation from Exercise 50 to write an equation for C in terms of F.

52. Use the equation from Exercise 50 to find the Fahrenheit temperature when $C = 30$.

53. Use the equation from Exercise 51 to find the Celsius temperature when $F = 50$.

54. For what temperature is $F = C$?

Write an equation of the line satisfying the given conditions. Write the equation in slope-intercept form.

55. Through $(2, -3)$, parallel to $3x = 4y + 5$

56. Through $(-1, 4)$, perpendicular to $2x + 3y = 8$

57. Perpendicular to $x - 2y = 7$, y-intercept $(0, -3)$

58. Parallel to $5x = 2y + 10$, y-intercept $(0, 4)$

The cost to produce x items is, in some cases, expressed as $y = mx + b$. The number b gives the fixed cost *(the cost that is the same no matter how many items are produced), and the number m is the* variable cost *(the cost to produce an additional item). Use this information to work Exercises 59 and 60.*

59. It costs $400 to start up a business selling snow cones. Each snow cone costs $.25 to produce.

  (a) What is the fixed cost?
  (b) What is the variable cost?
  (c) Write the cost equation.
  (d) What will be the cost to produce 100 snow cones, based on the cost equation?
  (e) How many snow cones will be produced if total cost is $775?

60. It costs $2000 to purchase a copier, and each copy costs $.02 to make.

  (a) What is the fixed cost?
  (b) What is the variable cost?
  (c) Write the cost equation.
  (d) What will be the cost to produce 10,000 copies, based on the cost equation?
  (e) How many copies will be produced if total cost is $2600?

TECHNOLOGY INSIGHTS (EXERCISES 61–64)

In Exercises 61 and 62, two graphing calculator views of the same line are shown. Use the displays at the bottom of the screen to find an equation of the form $y = mx + b$ for each line.

61.

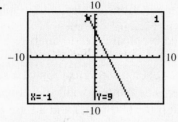

62.

In Exercises 63 and 64, a table of points, generated by a graphing calculator, is shown for a line Y_1. Use any two points to find the equation of each line. Write the equation in slope-intercept form.*

63.

X	Y1	
-2	-.5	
-1	.25	
0	1	
1	1.75	
2	2.5	
3	3.25	
4	4	
X= -2		

64.

X	Y1	
-4	14	
-3	10	
-2	6	
-1	2	
0	-2	
1	-6	
2	-10	
X= -4		

*With graphing calculators, we use capital Y_1 and X like the calculator does.

Chapter **3** **Group Activity**

Determining Business Profit

OBJECTIVE Use equations, tables, and graphs to make business decisions.

Graphs are used by businesses to analyze and make decisions. This activity will explore such a process.

The Parent Teacher Organization of a school decides to sell bags of popcorn as a fund-raiser. The parents want to estimate their profit. Costs include $14 for popcorn (enough for 680 bags) and a total of $7 for bags to hold the popcorn.

A. Develop a profit formula.

1. From the information above, determine the total costs. (Assume the organization is buying enough supplies for 680 bags of popcorn.)

2. If the filled bags sell for $.25 each, write an expression for the total revenue from n bags of popcorn.

3. Since Profit = Revenue − Cost, write an equation that represents the profit P for n bags of popcorn sold.

4. Work in pairs. One person should use the profit equation from part 3 to complete the table below. The other person should decide on an appropriate scale and graph the profit equation.

n	P
0	
	0
100	

B. Choose a different price for a bag of popcorn (between $.20 and $.75).

1. Write the profit equation for this cost.

2. Switch roles, that is, if you drew the graph in part A, now make a table of values for this equation. Have your partner graph the profit equation on the same coordinate system you used in part A.

C. Compare your findings and answer the following.

1. What is the break-even point (that is, when profits are 0 or revenue equals costs) for each equation?

2. What does it mean if you end up with a negative value for P?

3. If you sell all 680 bags, what will your profits be for the two different prices? Explain how you would estimate this from your graph.

4. What are the advantages and/or disadvantages of charging a higher price?

5. Together, decide on the price you would charge for a bag of popcorn. Explain why you chose this price.

CHAPTER 3 SUMMARY

KEY TERMS

3.1 linear equation in two variables
ordered pair
table of values
x-axis
y-axis

rectangular (Cartesian) coordinate system
quadrant
origin
plane

coordinates
plot
scatter diagram
3.2 graph, graphing
y-intercept
x-intercept

3.3 subscript notation
rise
run
slope
parallel lines
perpendicular lines

NEW SYMBOLS

(a, b) an ordered pair

(x_1, y_1) x-sub-one, y-sub-one

m slope

TEST YOUR WORD POWER

See how well you have learned the vocabulary in this chapter. Answers, with examples, follow the Quick Review.

1. An **ordered pair** is a pair of numbers written
   A. in numerical order between brackets
   B. between parentheses or brackets
   C. between parentheses in which order is important
   D. between parentheses in which order does not matter.

2. An **intercept** is
   A. the point where the x-axis and y-axis intersect
   B. a pair of numbers written in parentheses in which order matters

   C. one of the four regions determined by a rectangular coordinate system
   D. the point where a graph intersects the x-axis or the y-axis.

3. The **slope** of a line is
   A. the measure of the run over the rise of the line
   B. the distance between two points on the line
   C. the ratio of the change in y to the change in x along the line
   D. the horizontal change compared to the vertical change of two points on the line.

4. Two lines in a plane are **parallel** if
   A. they represent the same line
   B. they never intersect
   C. they intersect at a 90° angle
   D. one has a positive slope and one has a negative slope.

5. Two lines in a plane are **perpendicular** if
   A. they represent the same line
   B. they never intersect
   C. they intersect at a 90° angle
   D. one has a positive slope and one has a negative slope.

QUICK REVIEW

CONCEPTS	EXAMPLES

3.1 READING GRAPHS; LINEAR EQUATIONS IN TWO VARIABLES

An ordered pair is a solution of an equation if it satisfies the equation.

Is $(2, -5)$ or $(0, -6)$ a solution of $4x - 3y = 18$?

$$4(2) - 3(-5) = 23 \neq 18 \qquad 4(0) - 3(-6) = 18$$

$(2, -5)$ is not a solution. $(0, -6)$ is a solution.

If a value of either variable in an equation is given, the other variable can be found by substitution.

Complete the ordered pair $(0, \quad)$ for $3x = y + 4$.

$$3(0) = y + 4$$
$$0 = y + 4$$
$$-4 = y$$

The ordered pair is $(0, -4)$.

Plot the ordered pair $(-3, 4)$ by starting at the origin, going 3 units to the left, then going 4 units up.

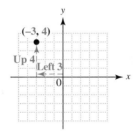

3.2 GRAPHING LINEAR EQUATIONS IN TWO VARIABLES

The graph of $y = b$ is a horizontal line through $(0, b)$.

The graph of $x = a$ is a vertical line through $(a, 0)$.

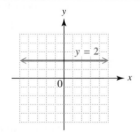 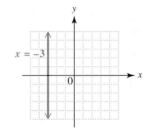

The graph of $Ax + By = 0$ goes through the origin. Find and plot another point that satisfies the equation. Then draw the line through the two points.

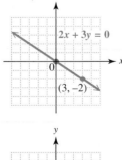

To graph a linear equation:

Step 1 Find at least two ordered pairs that satisfy the equation.

Step 2 Plot the corresponding points.

Step 3 Draw a straight line through the points.

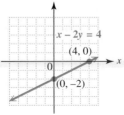

CONCEPTS	EXAMPLES

3.3 THE SLOPE OF A LINE

The slope of the line through (x_1, y_1) and (x_2, y_2) is

$$m = \frac{y_2 - y_1}{x_2 - x_1} \quad (x_1 \neq x_2).$$

Horizontal lines have slope 0.

Vertical lines have undefined slope.

To find the slope of a line from its equation, solve for y. The slope is the coefficient of x.

The line through $(-2, 3)$ and $(4, -5)$ has slope

$$m = \frac{-5 - 3}{4 - (-2)} = \frac{-8}{6} = -\frac{4}{3}.$$

The line $y = -2$ has slope 0.

The line $x = 4$ has undefined slope.

Find the slope of $3x - 4y = 12$.

$$-4y = -3x + 12$$

$$y = \frac{3}{4}x - 3$$

The slope is $\frac{3}{4}$.

Parallel lines have the same slope.

The lines $y = 3x - 1$ and $y = 3x + 4$ are parallel because both have slope 3.

The slopes of perpendicular lines are negative reciprocals (that is, their product is -1).

The lines $y = -3x - 1$ and $y = \frac{1}{3}x + 4$ are perpendicular because their slopes are -3 and $\frac{1}{3}$, and $-3\left(\frac{1}{3}\right) = -1$.

3.4 EQUATIONS OF A LINE

Slope-Intercept Form
$$y = mx + b$$
m is the slope.
$(0, b)$ is the y-intercept.

Find an equation of the line with slope 2 and y-intercept $(0, -5)$.

The equation is $y = 2x - 5$.

Point-Slope Form
$$y - y_1 = m(x - x_1)$$
m is the slope.
(x_1, y_1) is a point on the line.

Find an equation of the line with slope $-\frac{1}{2}$ through $(-4, 5)$.

$$y - 5 = -\frac{1}{2}[x - (-4)]$$

$$y - 5 = -\frac{1}{2}(x + 4)$$

$$y - 5 = -\frac{1}{2}x - 2$$

$$y = -\frac{1}{2}x + 3$$

Standard Form
$$Ax + By = C$$
A, B, and C are integers and $A > 0$, $B \neq 0$.

This equation is written in standard form as

$$x + 2y = 6,$$

with $A = 1$, $B = 2$, and $C = 6$.

Answers to Test Your Word Power

1. C; *Examples:* $(0, 3)$, $(-3, 8)$, $(4, 0)$ **2.** D; *Example:* The graph of the equation $4x - 3y = 12$ has x-intercept $(3, 0)$ and y-intercept $(0, -4)$.   **3.** C; *Example:* The line through $(3, 6)$ and $(5, 4)$ has slope $\frac{4 - 6}{5 - 3} = \frac{-2}{2} = -1$.   **4.** B; *Example:* See Figure 23 in Section 3.3.   **5.** C; *Example:* See Figure 24 in Section 3.3.

CHAPTER **3** REVIEW EXERCISES

[3.1]

1. The line graph shows average prices for a gallon of gasoline in Cedar Rapids, Iowa, from June 1999 through June 2000. Use the graph to work parts (a)–(d).

(a) About how much did a gallon of gas cost in June 1999? in June 2000?

(b) How much did the price of a gallon of gas increase over this 1-yr period? What percent increase is this?

(c) During which months did the biggest increase in the price of a gallon of gas occur? About how much did the price increase during this time?

(d) Between which months did the price of a gallon of gas decrease?

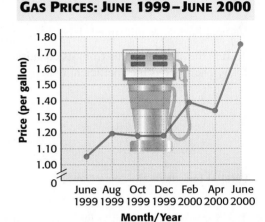

GAS PRICES: JUNE 1999 – JUNE 2000

Source: Iowa DNR.

Complete the given ordered pairs for each equation.

2. $y = 3x + 2$;　$(-1,\ )$, $(0,\ )$, $(\ ,5)$　　　**3.** $4x + 3y = 6$;　$(0,\ )$, $(\ ,0)$, $(-2,\ )$

4. $x = 3y$;　$(0,\ )$, $(8,\ )$, $(\ ,-3)$　　　**5.** $x - 7 = 0$;　$(\ ,-3)$, $(\ ,0)$, $(\ ,5)$

Determine whether the given ordered pair is a solution of the given equation.

6. $x + y = 7$;　$(2, 5)$　　　　　　　　　**7.** $2x + y = 5$;　$(-1, 3)$

8. $3x - y = 4$;　$\left(\dfrac{1}{3}, -3\right)$

Name the quadrant in which each pair lies. Then plot each ordered pair in a rectangular coordinate system.

9. $(2, 3)$　　　　**10.** $(-4, 2)$　　　　**11.** $(3, 0)$　　　　**12.** $(0, -6)$

13. If $xy > 0$, in what quadrant or quadrants must (x, y) lie?

[3.2]　*Find the x- and y-intercepts for the line that is the graph of each equation, and graph the line.*

14. $y = 2x + 5$　　　　　**15.** $3x + 2y = 8$　　　　　**16.** $x + 2y = -4$

[3.3]　*Find the slope of each line.*

17. Through $(2, 3)$ and $(-4, 6)$　　　　　**18.** Through $(2, 5)$ and $(2, 8)$

19. $y = 3x - 4$　　　　　　　　　　　　　**20.** $y = 5$

21.

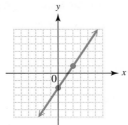

22.

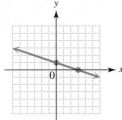

23. The line having these points

x	y
0	1
2	4
6	10

24. (a) A line parallel to the graph of
$y = 2x + 3$

(b) A line perpendicular to the graph
of $y = -3x + 3$

Decide whether each pair of lines is parallel, perpendicular, *or* neither.

25. $3x + 2y = 6$
$6x + 4y = 8$

26. $x - 3y = 1$
$3x + y = 4$

27. $x - 2y = 8$
$x + 2y = 8$

[3.4] *Write an equation for each line in the form* $y = mx + b$, *if possible.*

28. $m = -1, b = \dfrac{2}{3}$

29. Through $(2, 3)$ and $(-4, 6)$

30. Through $(4, -3), m = 1$

31. Through $(-1, 4), m = \dfrac{2}{3}$

32. Through $(1, -1), m = -\dfrac{3}{4}$

33. $m = -\dfrac{1}{4}, b = \dfrac{3}{2}$

34. Slope 0, through $(-4, 1)$

35. Through $\left(\dfrac{1}{3}, -\dfrac{5}{4}\right)$ with undefined slope

▰ MIXED REVIEW EXERCISES

In Exercises 36–41, match each statement to the appropriate graph or graphs in A–D.
Graphs may be used more than once.

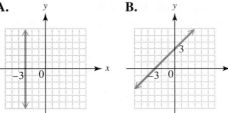

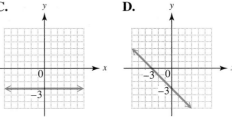

36. The line shown in the graph has undefined slope.

37. The graph of the equation has y-intercept $(0, -3)$.

38. The graph of the equation has x-intercept $(-3, 0)$.

39. The line shown in the graph has negative slope.

40. The graph is that of the equation $y = -3$.

41. The line shown in the graph has slope 1.

Find the intercepts and the slope of each line. Then graph the line.

42. $y = -2x - 5$ **43.** $x + 3y = 0$ **44.** $y - 5 = 0$

Write an equation for each line. Write the equation in slope-intercept form.

45. $m = -\dfrac{1}{4}, b = -\dfrac{5}{4}$ **46.** Through $(8, 6)$, $m = -3$ **47.** Through $(3, -5)$ and $(-4, -1)$

RELATING CONCEPTS (EXERCISES 48–54)

For Individual or Group Work

The total amount spent (in billions of dollars) on video rentals in the United States from 1996 through 2000 is shown in the graph. Use the graph to **work Exercises 48–54 in order.**

48. About how much did the amount spent on video rentals decrease during the years shown in the graph?

49. Since the points of the graph lie approximately in a linear pattern, a straight line can be used to model the data. Will this line have positive or negative slope? Explain.

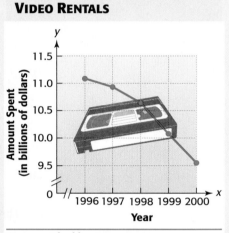

VIDEO RENTALS

Source: Blockbuster.

50. The table gives the actual amounts spent on video rentals in 1996 and 2000.
   (a) Write two ordered pairs for the data.
   (b) Use the ordered pairs to find the slope of the line through them.

Year x	Amount y (in billions of dollars)
1996	11.1
2000	9.6

51. The amount spent on video rentals y in billions of dollars from 1996 through 2000 can be modeled by the linear equation

$$y = -.375x + 759.6,$$

where x represents the year. Based on this equation, what is the slope of the line? Does it agree with your answers in Exercises 49 and 50?

52. Use the equation from Exercise 51 to approximate the amount spent on video rentals from 1997 through 1999, and complete the table. Round your answers to the nearest tenth.

53. The actual amounts spent on video rentals are given in the following ordered pairs.

$(1996, 11.1), (1997, 10.9), (1998, 10.6), (1999, 10.1), (2000, 9.6)$

How do the actual amounts compare to those you found in Exercise 52 using the linear equation?

x	y
1996	11.1
1997	
1998	
1999	
2000	9.6

54. Since the equation in Exercise 51 models the data fairly well, use it to estimate the amount spent on video rentals in 2002.

CHAPTER **3** TEST

1. Complete these ordered pairs for the equation $3x + 5y = -30$: $(0, \quad)$, $(\quad, 0)$, $(\quad, -3)$.

2. Is $(4, -1)$ a solution of $4x - 7y = 9$?

 3. How do you find the *x*-intercept of the graph of a linear equation in two variables? How do you find the *y*-intercept?

Graph each linear equation. Give the x- and y-intercepts.

4. $3x + y = 6$ 5. $y - 2x = 0$ 6. $x + 3 = 0$

7. $y = 1$ 8. $x - y = 4$

Find the slope of each line.

9. Through $(-4, 6)$ and $(-1, -2)$ 10. $2x + y = 10$

11. $x + 12 = 0$ 12.

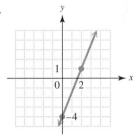

13. A line parallel to the graph of $y - 4 = 6$

Write an equation for each line. Write the equation in slope-intercept form.

14. Through $(-1, 4)$; $m = 2$ 15. The line in Exercise 12

16. Through $(2, -6)$ and $(1, 3)$

The graph shows total food and drink sales at U.S. restaurants from 1970 through 2000, where 1970 corresponds to $x = 0$. Use the graph to work Exercises 17–20 on the next page.

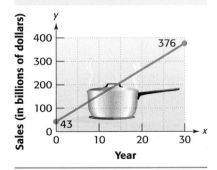

WHAT'S FOR DINNER?

Source: National Restaurant Association.

17. Is the slope of the line in the graph positive or negative? Explain.

18. Write two ordered pairs for the data points shown in the graph. Use them to find the slope of the line.

19. The linear equation

$$y = 11.1x + 43$$

approximates food and drink sales y in billions of dollars, where $x = 0$ again represents 1970. Use the equation to approximate food and drink sales for 1990 and 1995.

20. What does the ordered pair $(30, 376)$ mean in the context of this situation?

CUMULATIVE REVIEW EXERCISES CHAPTERS 1–3

Perform each indicated operation.

1. $10\dfrac{5}{8} - 3\dfrac{1}{10}$
  2. $\dfrac{3}{4} \div \dfrac{1}{8}$
  3. $5 - (-4) + (-2)$

4. $\dfrac{(-3)^2 - (-4)(2^4)}{5(2) - (-2)^3}$
  5. True or false? $\dfrac{4(3 - 9)}{2 - 6} \geq 6$

6. Find the value of $xz^3 - 5y^2$ when $x = -2$, $y = -3$, and $z = -1$.

7. What property does $3(-2 + x) = -6 + 3x$ illustrate?

8. Simplify $-4p - 6 + 3p + 8$ by combining terms.

Solve.

9. $V = \dfrac{1}{3}\pi r^2 h$ for h
  10. $6 - 3(1 + a) = 2(a + 5) - 2$

11. $-(m - 3) = 5 - 2m$
  12. $\dfrac{x - 2}{3} = \dfrac{2x + 1}{5}$

Solve each inequality, and graph the solutions.

13. $-2.5x < 6.5$
  14. $4(x + 3) - 5x < 12$
  15. $\dfrac{2}{3}x - \dfrac{1}{6}x \leq -2$

Solve each problem.

16. The gap in average annual earnings by level of education continues to increase. Based on the most recent statistics available, a person with a bachelor's degree can expect to earn $17,583 more each year than someone with a high school diploma. Together the individuals would earn $63,373. How much can a person at each level of education expect to earn? (*Source:* U.S. Bureau of the Census.)

17. Mount Mayon in the Philippines is the most perfectly shaped conical volcano in the world. Its base is a perfect circle with circumference 80 mi, and it has a height of about 8200 ft. (One mile is 5280 ft.) Find the radius of the circular base to the nearest mile. (*Hint:* This problem has some unneeded information.) (*Source: Microsoft Encarta Encyclopedia 2000.*)

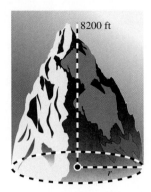

Circumference = 80 mi

18. The winning times in seconds for the women's 1000-m speed skating event in the Winter Olympics for the years 1960 through 1998 can be closely approximated by the linear equation

$$y = -.4685x + 95.07,$$

where x is the number of years since 1960. That is, $x = 4$ represents 1964, $x = 8$ represents 1968, and so on. (*Source: The Universal Almanac,* 1998.)

(a) Use this equation to complete the table of values. Round times to the nearest hundredth of a second.

(b) What does the ordered pair $(20, 85.7)$ mean in the context of the problem?

x	y
12	
28	
36	

19. Baby boomers are expected to inherit $10.4 trillion from their parents over the next 45 yr, an average of $50,000 each. The circle graph shows how they plan to spend their inheritances.

(a) How much of the $50,000 is expected to go toward home purchase?

(b) How much is expected to go toward retirement?

(c) Use the answer from part (b) to estimate the amount expected to go toward paying off debts or funding children's education.

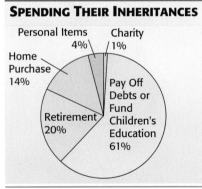

SPENDING THEIR INHERITANCES

Personal Items 4%
Charity 1%
Home Purchase 14%
Pay Off Debts or Fund Children's Education 61%
Retirement 20%

Source: First Interstate Bank Trust and Private Banking Group.

Consider the linear equation $-3x + 4y = 12$. *Find the following.*

20. The x- and y-intercepts   **21.** The slope   **22.** The graph

23. Are the lines with equations $x + 5y = -6$ and $y = 5x - 8$ *parallel, perpendicular,* or *neither*?

Write an equation for each line. Write the equation in slope-intercept form, if possible.

24. Through $(2, -5)$ with slope 3   **25.** Through $(0, 4)$ and $(2, 4)$

Exponents and Polynomials

4

The expression $100x - 13x^2$ gives the distance in feet a car going approximately 68 mph will skid in x sec. This expression in x is an example of a *polynomial,* one topic of this chapter. Accident investigators use polynomials like this to determine the length of a skid or the elapsed time during a skid. In Exercises 95 and 96 of Section 4.4, we use this polynomial to approximate skidding distance.

243

4.1 The Product Rule and Power Rules for Exponents

OBJECTIVES

1 Identify bases and exponents.

2 Use the product rule for exponents.

3 Use the rule $(a^m)^n = a^{mn}$.

4 Use the rule $(ab)^m = a^m b^m$.

5 Use the rule $\left(\dfrac{a}{b}\right)^m = \dfrac{a^m}{b^m}$.

6 Use combinations of rules.

OBJECTIVE 1 Identify bases and exponents. Recall from Section 1.2 that in the expression 5^2, the number 5 is the *base* and 2 is the *exponent* or *power*. The expression 5^2 is called an *exponential expression*. Usually we do not write the exponent when it is 1; however, sometimes it is convenient to do so. In general, for any quantity a,

$$a^1 = a.$$

EXAMPLE 1 Determining the Base and Exponent in Exponential Expressions

Evaluate each exponential expression. Name the base and the exponent.

	Base	Exponent
(a) $5^4 = 5 \cdot 5 \cdot 5 \cdot 5 = 625$	5	4
(b) $-5^4 = -1 \cdot 5^4 = -1 \cdot (5 \cdot 5 \cdot 5 \cdot 5) = -625$	5	4
(c) $(-5)^4 = (-5)(-5)(-5)(-5) = 625$	-5	4

Now Try Exercises 15 and 17.

CAUTION Note the differences between parts (b) and (c) of Example 1. In -5^4 the lack of parentheses shows that the exponent 4 refers only to the base 5, not -5; in $(-5)^4$ the parentheses show that the exponent 4 refers to the base -5. In summary, $-a^n$ and $(-a)^n$ are not necessarily the same.

Expression	Base	Exponent	Example
$-a^n$	a	n	$-3^2 = -(3 \cdot 3) = -9$
$(-a)^n$	$-a$	n	$(-3)^2 = (-3)(-3) = 9$

OBJECTIVE 2 Use the product rule for exponents. By the definition of exponents,

$$2^4 \cdot 2^3 = \overbrace{(2 \cdot 2 \cdot 2 \cdot 2)}^{4 \text{ factors}}\overbrace{(2 \cdot 2 \cdot 2)}^{3 \text{ factors}}$$
$$= \underbrace{2 \cdot 2 \cdot 2 \cdot 2 \cdot 2 \cdot 2 \cdot 2}_{4 + 3 = 7 \text{ factors}}$$
$$= 2^7.$$

Also,

$$6^2 \cdot 6^3 = (6 \cdot 6)(6 \cdot 6 \cdot 6)$$
$$= 6 \cdot 6 \cdot 6 \cdot 6 \cdot 6$$
$$= 6^5.$$

Generalizing from these examples, $2^4 \cdot 2^3 = 2^{4+3} = 2^7$ and $6^2 \cdot 6^3 = 6^{2+3} = 6^5$, suggests the **product rule for exponents.**

Product Rule for Exponents

For any positive integers m and n, $a^m \cdot a^n = a^{m+n}$.
(Keep the same base; add the exponents.)

Example: $6^2 \cdot 6^5 = 6^{2+5} = 6^7$

CAUTION Avoid the common error of multiplying the bases when using the product rule:

$$6^2 \cdot 6^5 \neq 36^7.$$

Keep the same base and add the exponents.

◼ EXAMPLE 2 Using the Product Rule

Use the product rule for exponents to find each result when possible.

(a) $6^3 \cdot 6^5 = 6^{3+5} = 6^8$ **(b)** $(-4)^7(-4)^2 = (-4)^{7+2} = (-4)^9$

(c) $x^2 \cdot x = x^2 \cdot x^1 = x^{2+1} = x^3$ **(d)** $m^4 m^3 m^5 = m^{4+3+5} = m^{12}$

(e) $2^3 \cdot 3^2$

The product rule does not apply to the product $2^3 \cdot 3^2$, since the bases are different.

$$2^3 \cdot 3^2 = 8 \cdot 9 = 72$$

(f) $2^3 + 2^4$

The product rule does not apply to $2^3 + 2^4$, since this is a *sum,* not a *product.*

$$2^3 + 2^4 = 8 + 16 = 24$$

(g) $(2x^3)(3x^7)$

Since $2x^3$ means $2 \cdot x^3$ and $3x^7$ means $3 \cdot x^7$, we use the associative and commutative properties to get

$$(2x^3)(3x^7) = (2 \cdot 3) \cdot (x^3 \cdot x^7) = 6x^{10}.$$

Now Try Exercises 25, 29, 31, 35, and 39.

CAUTION Be sure you understand the difference between *adding* and *multiplying* exponential expressions. For example,

$$8x^3 + 5x^3 = (8 + 5)x^3 = 13x^3,$$

but $$(8x^3)(5x^3) = (8 \cdot 5)x^{3+3} = 40x^6.$$

OBJECTIVE 3 Use the rule $(a^m)^n = a^{mn}$. We simplify an expression such as $(8^3)^2$ with the product rule for exponents.

$$(8^3)^2 = (8^3)(8^3) = 8^{3+3} = 8^6$$

The exponents in $(8^3)^2$ are multiplied to give the exponent in 8^6. As another example,

$$
\begin{aligned}
(5^2)^4 &= 5^2 \cdot 5^2 \cdot 5^2 \cdot 5^2 \qquad &\text{Definition of exponent} \\
&= 5^{2+2+2+2} &\text{Product rule} \\
&= 5^8
\end{aligned}
$$

and $2 \cdot 4 = 8$. These examples suggest **power rule (a) for exponents.**

Power Rule (a) for Exponents

For any positive integers m and n, $\quad (a^m)^n = a^{mn}$.
(Raise a power to a power by multiplying exponents.)
Example: $(3^2)^4 = 3^{2 \cdot 4} = 3^8$

EXAMPLE 3 Using Power Rule (a)

Use power rule (a) for exponents to simplify each expression.

(a) $(2^5)^3 = 2^{5 \cdot 3} = 2^{15}$ **(b)** $(5^7)^2 = 5^{7(2)} = 5^{14}$ **(c)** $(x^2)^5 = x^{2(5)} = x^{10}$

Now Try Exercises 43 and 45.

OBJECTIVE 4 Use the rule $(ab)^m = a^m b^m$. We can use the properties studied in Chapter 1 to develop two more rules for exponents. Using the definition of an exponential expression and the commutative and associative properties, we can rewrite the expression $(4x)^3$ as follows.

$$(4x)^3 = (4x)(4x)(4x) \qquad \text{Definition of exponent}$$
$$= (4 \cdot 4 \cdot 4)(x \cdot x \cdot x) \qquad \text{Commutative and associative properties}$$
$$= 4^3 \cdot x^3 \qquad \text{Definition of exponent}$$

This example suggests **power rule (b) for exponents.**

Power Rule (b) for Exponents

For any positive integer m, $\quad (ab)^m = a^m b^m$.
(Raise a product to a power by raising each factor to the power.)
Example: $(2p)^5 = 2^5 p^5$

EXAMPLE 4 Using Power Rule (b)

Use power rule (b) for exponents to simplify each expression.

(a) $(3xy)^2 = 3^2 x^2 y^2 = 9x^2 y^2$ Power rule (a)

(b) $5(pq)^2 = 5(p^2 q^2)$ Power rule (b)
$\qquad\quad = 5p^2 q^2$ Multiply.

(c) $3(2m^2 p^3)^4 = 3[2^4 (m^2)^4 (p^3)^4]$ Power rule (b)
$\qquad\qquad\quad = 3 \cdot 2^4 m^8 p^{12}$ Power rule (a)
$\qquad\qquad\quad = 48 m^8 p^{12}$ Multiply.

(d) $(-5^6)^3 = (-1 \cdot 5^6)^3$
$\qquad\qquad = (-1)^3 \cdot (5^6)^3$
$\qquad\qquad = -1 \cdot 5^{18}$
$\qquad\qquad = -5^{18}$

Now Try Exercises 49 and 53.

CAUTION Power rule (b) *does not* apply to a *sum*.
$$(x + 4)^2 \neq x^2 + 4^2$$

OBJECTIVE 5 Use the rule $\left(\dfrac{a}{b}\right)^m = \dfrac{a^m}{b^m}$. Since the quotient $\dfrac{a}{b}$ can be written as $a\left(\dfrac{1}{b}\right)$, we use power rule (b), together with some of the properties of real numbers, to get **power rule (c) for exponents.**

Power Rule (c) for Exponents

For any positive integer m,

$$\left(\frac{a}{b}\right)^m = \frac{a^m}{b^m} \quad (b \neq 0).$$

(Raise a quotient to a power by raising both numerator and denominator to the power.)

Example: $\left(\dfrac{5}{3}\right)^2 = \dfrac{5^2}{3^2}$

EXAMPLE 5 Using Power Rule (c)

Use power rule (c) for exponents to simplify each expression.

(a) $\left(\dfrac{2}{3}\right)^5 = \dfrac{2^5}{3^5}$ **(b)** $\left(\dfrac{m}{n}\right)^3 = \dfrac{m^3}{n^3} \quad (n \neq 0)$

Now Try Exercises 59 and 61.

In the following box, we list the rules for exponents discussed in this section. These rules are basic to the study of algebra.

Rules for Exponents

For positive integers m and n:

		Examples
Product rule	$a^m \cdot a^n = a^{m+n}$	$6^2 \cdot 6^5 = 6^{2+5} = 6^7$
Power rules (a)	$(a^m)^n = a^{mn}$	$(3^2)^4 = 3^{2\cdot4} = 3^8$
(b)	$(ab)^m = a^m b^m$	$(2p)^5 = 2^5 p^5$
(c)	$\left(\dfrac{a}{b}\right)^m = \dfrac{a^m}{b^m} \quad (b \neq 0)$	$\left(\dfrac{5}{3}\right)^2 = \dfrac{5^2}{3^2}$

OBJECTIVE 6 Use combinations of rules. As shown in the next example, more than one rule may be needed to simplify an expression with exponents.

▨ EXAMPLE 6 Using Combinations of Rules

Use the rules for exponents to simplify each expression.

(a) $\left(\dfrac{2}{3}\right)^2 \cdot 2^3 = \dfrac{2^2}{3^2} \cdot \dfrac{2^3}{1}$ Power rule (c)

$= \dfrac{2^2 \cdot 2^3}{3^2 \cdot 1}$ Multiply fractions.

$= \dfrac{2^5}{3^2}$ Product rule

(b) $(5x)^3(5x)^4 = (5x)^7$ Product rule

$= 5^7x^7$ Power rule (b)

(c) $(2x^2y^3)^4(3xy^2)^3 = 2^4(x^2)^4(y^3)^4 \cdot 3^3x^3(y^2)^3$ Power rule (b)

$= 2^4x^8y^{12} \cdot 3^3x^3y^6$ Power rule (a)

$= 2^4 \cdot 3^3x^8x^3y^{12}y^6$ Commutative and associative properties

$= 16 \cdot 27x^{11}y^{18}$ Product rule

$= 432x^{11}y^{18}$

(d) $(-x^3y)^2(-x^5y^4)^3 = (-1 \cdot x^3y)^2(-1 \cdot x^5y^4)^3$

$= (-1)^2(x^3)^2y^2 \cdot (-1)^3(x^5)^3(y^4)^3$ Power rule (b)

$= (-1)^2(x^6)(y^2)(-1)^3(x^{15})(y^{12})$ Power rule (a)

$= (-1)^5(x^{21})(y^{14})$ Product rule

$= -x^{21}y^{14}$

Now Try Exercises 63, 67, 75, and 77.

CAUTION   Refer to Example 6(c). Notice that
$$(2x^2y^3)^4 = 2^4x^{2 \cdot 4}y^{3 \cdot 4}, \quad \text{not} \quad (2 \cdot 4)x^{2 \cdot 4}y^{3 \cdot 4}.$$
Do not multiply the coefficient 2 and the exponent 4.

4.1 EXERCISES

Decide whether each statement is true *or* false.

1. $3^3 = 9$ **2.** $(-2)^4 = 2^4$ **3.** $(a^2)^3 = a^5$ **4.** $\left(\dfrac{1}{4}\right)^2 = \dfrac{1}{4^2}$

Write each expression using exponents.

5. $w \cdot w \cdot w \cdot w \cdot w \cdot w$ **6.** $t \cdot t \cdot t \cdot t \cdot t \cdot t \cdot t \cdot t$

7. $\dfrac{1}{4 \cdot 4 \cdot 4 \cdot 4}$ **8.** $\dfrac{1}{3 \cdot 3 \cdot 3}$

9. $(-7x)(-7x)(-7x)(-7x)$ **10.** $(-8p)(-8p)$

11. $\left(\dfrac{1}{2}\right)\left(\dfrac{1}{2}\right)\left(\dfrac{1}{2}\right)\left(\dfrac{1}{2}\right)\left(\dfrac{1}{2}\right)\left(\dfrac{1}{2}\right)$ **12.** $\left(-\dfrac{1}{4}\right)\left(-\dfrac{1}{4}\right)\left(-\dfrac{1}{4}\right)\left(-\dfrac{1}{4}\right)\left(-\dfrac{1}{4}\right)$

13. Explain how the expressions $(-3)^4$ and -3^4 are different.

14. Explain how the expressions $(5x)^3$ and $5x^3$ are different.

Identify the base and the exponent for each exponential expression. In Exercises 15–18, also evaluate each expression. See Example 1.

15. 3^5 **16.** 2^7 **17.** $(-3)^5$ **18.** $(-2)^7$

19. $(-6x)^4$ **20.** $(-8x)^4$ **21.** $-6x^4$ **22.** $-8x^4$

23. Explain why the product rule does not apply to the expression $5^2 + 5^3$. Then evaluate the expression by finding the individual powers and adding the results.

24. Repeat Exercise 23 for the expression $(-4)^3 + (-4)^4$.

Use the product rule, if possible, to simplify each expression. Write each answer in exponential form. See Example 2.

25. $5^2 \cdot 5^6$ **26.** $3^6 \cdot 3^7$ **27.** $4^2 \cdot 4^7 \cdot 4^3$

28. $5^3 \cdot 5^8 \cdot 5^2$ **29.** $(-7)^3(-7)^6$ **30.** $(-9)^8(-9)^5$

31. $t^3 \cdot t^8 \cdot t^{13}$ **32.** $n^5 \cdot n^6 \cdot n^9$ **33.** $(-8r^4)(7r^3)$

34. $(10a^7)(-4a^3)$ **35.** $(-6p^5)(-7p^5)$ **36.** $(-5w^8)(-9w^8)$

37. $(5x^2)(-2x^3)(3x^4)$ **38.** $(12y^3)(4y)(-3y^5)$ **39.** $3^8 + 3^9$

40. $4^{12} + 4^5$ **41.** $5^8 \cdot 3^9$ **42.** $6^3 \cdot 8^9$

Use the power rules for exponents to simplify each expression. Write each answer in exponential form. See Examples 3–5.

43. $(4^3)^2$ **44.** $(8^3)^6$ **45.** $(t^4)^5$

46. $(y^6)^5$ **47.** $(7r)^3$ **48.** $(11x)^4$

49. $(5xy)^5$ **50.** $(9pq)^6$ **51.** $(-5^2)^6$

52. $(-9^4)^8$ **53.** $(-8^3)^5$ **54.** $(-7^5)^7$

55. $8(qr)^3$ **56.** $4(vw)^5$ **57.** $\left(\dfrac{1}{2}\right)^3$

58. $\left(\dfrac{1}{3}\right)^5$ **59.** $\left(\dfrac{a}{b}\right)^3$ $(b \neq 0)$ **60.** $\left(\dfrac{r}{t}\right)^4$ $(t \neq 0)$

61. $\left(\dfrac{9}{5}\right)^8$ **62.** $\left(\dfrac{12}{7}\right)^3$

Use a combination of the rules of exponents to simplify each expression. See Example 6.

63. $\left(\dfrac{5}{2}\right)^3 \cdot \left(\dfrac{5}{2}\right)^2$ **64.** $\left(\dfrac{3}{4}\right)^5 \cdot \left(\dfrac{3}{4}\right)^6$ **65.** $\left(\dfrac{9}{8}\right)^3 \cdot 9^2$

66. $\left(\dfrac{8}{5}\right)^4 \cdot 8^3$ **67.** $(2x)^9(2x)^3$ **68.** $(6y)^5(6y)^8$

69. $(-6p)^4(-6p)^5$ **70.** $(-13q)^3(-13q)$ **71.** $(6x^2y^3)^5$

72. $(5r^5t^6)^7$ **73.** $(x^2)^3(x^3)^5$ **74.** $(y^4)^5(y^3)^5$

75. $(2w^2x^3y)^2(x^4y)^5$ **76.** $(3x^4y^2z)^3(yz^4)^5$ **77.** $(-r^4s)^2(-r^2s^3)^5$

78. $(-ts^6)^4(-t^3s^5)^3$ **79.** $\left(\dfrac{5a^2b^5}{c^6}\right)^3$ $(c \neq 0)$ **80.** $\left(\dfrac{6x^3y^9}{z^5}\right)^4$ $(z \neq 0)$

81. A student tried to simplify $(10^2)^3$ as 1000^6. Is this correct? If not, how is it simplified using the product rule for exponents?

82. Explain why $(3x^2y^3)^4$ is *not* equivalent to $(3 \cdot 4)x^8y^{12}$.

Find the area of each figure. Use the formulas found on the inside covers. (The small squares in the figures indicate $90°$ right angles.)

83.

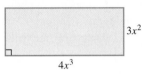

$3x^2$

$4x^3$

84.

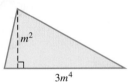

m^2

$3m^4$

85.

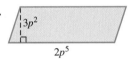

$3p^2$

$2p^5$

86.

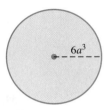

$6a^3$

Find the volume of each figure. Use the formulas found on the inside covers.

87.

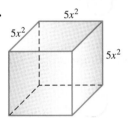

$5x^2$

$5x^2$

$5x^2$

88.

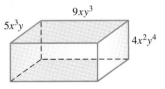

$9xy^3$

$5x^3y$

$4x^2y^4$

89. Assume a is a positive number greater than 1. Arrange the following terms in order from smallest to largest: $-(-a)^3$, $-a^3$, $(-a)^4$, $-a^4$. Explain how you decided on the order.

90. Devise a rule to tell whether an exponential expression with a negative base is positive or negative.

In Chapter 2 we used the formula for simple interest, $I = prt$, which deals with interest paid only on the principal. With **compound interest,** interest is paid on the principal and the interest earned earlier. The formula for compound interest, which involves an exponential expression, is

$$A = P(1 + r)^n.$$

Here A is the amount accumulated from a principal of P dollars left untouched for n years with an annual interest rate r (expressed as a decimal).

In Exercises 91–94, use this formula and a calculator to find A to the nearest cent.

91. $P = \$250, r = .04, n = 5$ **92.** $P = \$400, r = .04, n = 3$

93. $P = \$1500, r = .035, n = 6$ **94.** $P = \$2000, r = .025, n = 4$

4.2 Integer Exponents and the Quotient Rule

In Section 4.1 we studied the product rule for exponents. In all our earlier work, exponents were positive integers. Now we want to develop meaning for exponents that are not positive integers.

Consider the following list of exponential expressions.

$$2^4 = 16$$
$$2^3 = 8$$
$$2^2 = 4$$

Do you see the pattern in the values? Each time we reduce the exponent by 1, the value is divided by 2 (the base). Using this pattern, we can continue the list to smaller and smaller integer exponents.

$$2^1 = 2$$
$$2^0 = 1$$
$$2^{-1} = \frac{1}{2}$$
$$2^{-2} = \frac{1}{4}$$
$$2^{-3} = \frac{1}{8}$$

From the preceding list, it appears that we should define 2^0 as 1 and negative exponents as reciprocals.

OBJECTIVE 1 Use 0 as an exponent. We want the definitions of 0 and negative exponents to satisfy the rules for exponents from Section 4.1. For example, if $6^0 = 1$,

$$6^0 \cdot 6^2 = 1 \cdot 6^2 = 6^2 \qquad \text{and} \qquad 6^0 \cdot 6^2 = 6^{0+2} = 6^2,$$

so the product rule is satisfied. Check that the power rules are also valid for a 0 exponent. Thus, we define a 0 exponent as follows.

Zero Exponent

For any nonzero real number a, $\qquad a^0 = 1.$

Example: $17^0 = 1$

EXAMPLE 1 Using Zero Exponents

Evaluate each exponential expression.

(a) $60^0 = 1$

(b) $(-60)^0 = 1$

(c) $-60^0 = -(1) = -1$

(d) $y^0 = 1 \quad (y \neq 0)$

(e) $6y^0 = 6(1) = 6 \quad (y \neq 0)$

(f) $(6y)^0 = 1 \quad (y \neq 0)$

Now Try Exercises 1 and 5.

CAUTION Notice the difference between parts (b) and (c) of Example 1. In $(-60)^0$, the base is -60 and the exponent is 0. Any nonzero base raised to the 0 exponent is 1. But in -60^0, the base is 60. Then $60^0 = 1$, and $-60^0 = -1$.

OBJECTIVE 2 Use negative numbers as exponents. From the lists at the beginning of this section, since $2^{-2} = \frac{1}{4}$ and $2^{-3} = \frac{1}{8}$, we can deduce that 2^{-n} should equal $\frac{1}{2^n}$. Is the product rule valid in such cases? For example, if we multiply 6^{-2} by 6^2, we get

$$6^{-2} \cdot 6^2 = 6^{-2+2} = 6^0 = 1.$$

The expression 6^{-2} behaves as if it were the reciprocal of 6^2 because their product is 1. The reciprocal of 6^2 is also $\frac{1}{6^2}$, leading us to define 6^{-2} as $\frac{1}{6^2}$. This is a particular case of the definition of negative exponents.

Negative Exponents

For any nonzero real number a and any integer n, $\qquad a^{-n} = \dfrac{1}{a^n}.$

Example: $3^{-2} = \dfrac{1}{3^2}$

By definition, a^{-n} and a^n are reciprocals, since

$$a^n \cdot a^{-n} = a^n \cdot \frac{1}{a^n} = 1.$$

Since $1^n = 1$, the definition of a^{-n} can also be written

$$a^{-n} = \frac{1}{a^n} = \frac{1^n}{a^n} = \left(\frac{1}{a}\right)^n.$$

For example,

$$6^{-3} = \left(\frac{1}{6}\right)^3 \qquad \text{and} \qquad \left(\frac{1}{3}\right)^{-2} = 3^2.$$

EXAMPLE 2 Using Negative Exponents

Simplify by writing each expression with positive exponents.

(a) $3^{-2} = \dfrac{1}{3^2} = \dfrac{1}{9}$
  (b) $5^{-3} = \dfrac{1}{5^3} = \dfrac{1}{125}$

(c) $\left(\dfrac{1}{2}\right)^{-3} = 2^3 = 8 \qquad \frac{1}{2}$ and 2 are reciprocals.

Notice that we can change the base to its reciprocal if we also change the sign of the exponent.

(d) $\left(\dfrac{2}{5}\right)^{-4} = \left(\dfrac{5}{2}\right)^4 \qquad \frac{2}{5}$ and $\frac{5}{2}$ are reciprocals.
  (e) $\left(\dfrac{4}{3}\right)^{-5} = \left(\dfrac{3}{4}\right)^5$

(f) $4^{-1} - 2^{-1} = \dfrac{1}{4} - \dfrac{1}{2} = \dfrac{1}{4} - \dfrac{2}{4} = -\dfrac{1}{4}$

We apply the exponents first, then subtract.

(g) $p^{-2} = \dfrac{1}{p^2}$ $(p \neq 0)$

(h) $\dfrac{1}{x^{-4}}$ $(x \neq 0)$

$$\dfrac{1}{x^{-4}} = \dfrac{1^{-4}}{x^{-4}} \qquad 1^{-4} = 1$$

$$= \left(\dfrac{1}{x}\right)^{-4} \qquad \text{Power rule (c)}$$

$$= x^4 \qquad \tfrac{1}{x} \text{ and } x \text{ are reciprocals.}$$

Now Try Exercises 19, 21, 23, and 27.

CAUTION A negative exponent does not indicate a negative number; negative exponents lead to reciprocals.

Expression	Example	
a^{-n}	$3^{-2} = \dfrac{1}{3^2} = \dfrac{1}{9}$	Not negative
$-a^{-n}$	$-3^{-2} = -\dfrac{1}{3^2} = -\dfrac{1}{9}$	Negative

The definition of negative exponents allows us to move factors across a fraction bar if we also change the signs of the exponents. For example,

$$\dfrac{2^{-3}}{3^{-4}} = \dfrac{\tfrac{1}{2^3}}{\tfrac{1}{3^4}} = \dfrac{1}{2^3} \cdot \dfrac{3^4}{1} = \dfrac{3^4}{2^3},$$

so

$$\dfrac{2^{-3}}{3^{-4}} = \dfrac{3^4}{2^3}.$$

Changing from Negative to Positive Exponents

For any nonzero numbers a and b, and any integers m and n,

$$\dfrac{a^{-m}}{b^{-n}} = \dfrac{b^n}{a^m} \qquad \text{and} \qquad \left(\dfrac{a}{b}\right)^{-m} = \left(\dfrac{b}{a}\right)^m.$$

Examples: $\dfrac{3^{-5}}{2^{-4}} = \dfrac{2^4}{3^5}$ and $\left(\dfrac{4}{5}\right)^{-3} = \left(\dfrac{5}{4}\right)^3$

EXAMPLE 3 Changing from Negative to Positive Exponents

Write with only positive exponents. Assume all variables represent nonzero real numbers.

(a) $\dfrac{4^{-2}}{5^{-3}} = \dfrac{5^3}{4^2}$ or $\dfrac{125}{16}$

(b) $\dfrac{m^{-5}}{p^{-1}} = \dfrac{p^1}{m^5} = \dfrac{p}{m^5}$

(c) $\dfrac{a^{-2}b}{3d^{-3}} = \dfrac{bd^3}{3a^2}$

Notice that b in the numerator and 3 in the denominator were not affected.

(d) $x^3y^{-4} = \dfrac{x^3y^{-4}}{1} = \dfrac{x^3}{y^4}$

(e) $\left(\dfrac{x}{2y}\right)^{-4} = \left(\dfrac{2y}{x}\right)^4 = \dfrac{2^4y^4}{x^4}$

Now Try Exercises 31, 35, and 47.

CAUTION Be careful. We cannot change negative exponents to positive exponents using this rule if the exponents occur in a sum of terms. For example,

$$\frac{5^{-2} + 3^{-1}}{7 - 2^{-3}}$$

cannot be written with positive exponents using the rule given here. We would have to use the definition of a negative exponent to rewrite this expression with positive exponents, as

$$\frac{\dfrac{1}{5^2} + \dfrac{1}{3}}{7 - \dfrac{1}{2^3}}.$$

OBJECTIVE 3 Use the quotient rule for exponents. How should we handle the quotient of two exponential expressions with the same base? We know that

$$\frac{6^5}{6^3} = \frac{6 \cdot 6 \cdot 6 \cdot 6 \cdot 6}{6 \cdot 6 \cdot 6} = 6^2.$$

Notice that the difference between the exponents, $5 - 3 = 2$, is the exponent in the quotient. Also,

$$\frac{6^2}{6^4} = \frac{6 \cdot 6}{6 \cdot 6 \cdot 6 \cdot 6} = \frac{1}{6^2} = 6^{-2}.$$

Here, $2 - 4 = -2$. These examples suggest the **quotient rule for exponents.**

Quotient Rule for Exponents

For any nonzero real number a and any integers m and n,

$$\frac{a^m}{a^n} = a^{m-n}.$$

(Keep the base and subtract the exponents.)

Example: $\dfrac{5^8}{5^4} = 5^{8-4} = 5^4$

CAUTION   A common **error** is to write $\dfrac{5^8}{5^4} = 1^{8-4} = 1^4$. Notice that by the quotient rule, the quotient should have the *same base*, 5. That is,

$$\frac{5^8}{5^4} = 5^{8-4} = 5^4.$$

If you are not sure, use the definition of an exponent to write out the factors:
$$5^8 = 5 \cdot 5 \cdot 5 \cdot 5 \cdot 5 \cdot 5 \cdot 5 \cdot 5 \quad \text{and} \quad 5^4 = 5 \cdot 5 \cdot 5 \cdot 5.$$

Then it is clear that the quotient is 5^4.

EXAMPLE 4   Using the Quotient Rule

Simplify, using the quotient rule for exponents. Write answers with positive exponents.

(a) $\dfrac{5^8}{5^6} = 5^{8-6} = 5^2$ or 25

(b) $\dfrac{4^2}{4^9} = 4^{2-9} = 4^{-7} = \dfrac{1}{4^7}$

(c) $\dfrac{5^{-3}}{5^{-7}} = 5^{-3-(-7)} = 5^4$ or 625

(d) $\dfrac{q^5}{q^{-3}} = q^{5-(-3)} = q^8$ $(q \neq 0)$

(e) $\dfrac{3^2 x^5}{3^4 x^3} = \dfrac{3^2}{3^4} \cdot \dfrac{x^5}{x^3} = 3^{2-4} \cdot x^{5-3} = 3^{-2} x^2 = \dfrac{x^2}{3^2}$ $(x \neq 0)$

(f) $\dfrac{(m+n)^{-2}}{(m+n)^{-4}} = (m+n)^{-2-(-4)} = (m+n)^{-2+4} = (m+n)^2$ $(m \neq -n)$

(g) $\dfrac{7x^{-3} y^2}{2^{-1} x^2 y^{-5}} = \dfrac{7 \cdot 2^1 y^2 y^5}{x^2 x^3} = \dfrac{14 y^7}{x^5}$ $(x, y \neq 0)$

Now Try Exercises 29, 41, 45, and 49.

The definitions and rules for exponents given in this section and Section 4.1 are summarized here.

Definitions and Rules for Exponents

For any integers m and n: **Examples**

Product rule	$a^m \cdot a^n = a^{m+n}$	$7^4 \cdot 7^5 = 7^9$
Zero exponent	$a^0 = 1$ $(a \neq 0)$	$(-3)^0 = 1$
Negative exponent	$a^{-n} = \dfrac{1}{a^n}$ $(a \neq 0)$	$5^{-3} = \dfrac{1}{5^3}$
Quotient rule	$\dfrac{a^m}{a^n} = a^{m-n}$ $(a \neq 0)$	$\dfrac{2^2}{2^5} = 2^{2-5} = 2^{-3} = \dfrac{1}{2^3}$
Power rules (a)	$(a^m)^n = a^{mn}$	$(4^2)^3 = 4^6$
(b)	$(ab)^m = a^m b^m$	$(3k)^4 = 3^4 k^4$
(c)	$\left(\dfrac{a}{b}\right)^m = \dfrac{a^m}{b^m}$ $(b \neq 0)$	$\left(\dfrac{2}{3}\right)^2 = \dfrac{2^2}{3^2}$

(continued)

Negative to positive rules	$\dfrac{a^{-m}}{b^{-n}} = \dfrac{b^n}{a^m}$ $(a \neq 0, b \neq 0)$	$\dfrac{2^{-4}}{5^{-3}} = \dfrac{5^3}{2^4}$
	$\left(\dfrac{a}{b}\right)^{-m} = \left(\dfrac{b}{a}\right)^{m}$	$\left(\dfrac{4}{7}\right)^{-2} = \left(\dfrac{7}{4}\right)^{2}$

OBJECTIVE 4 Use combinations of rules. We may sometimes need to use more than one rule to simplify an expression.

EXAMPLE 5 Using Combinations of Rules

Use the rules for exponents to simplify each expression. Assume all variables represent nonzero real numbers.

(a) $\dfrac{(4^2)^3}{4^5} = \dfrac{4^6}{4^5}$ Power rule (a)

$\qquad = 4^{6-5}$ Quotient rule

$\qquad = 4^1 = 4$

(b) $(2x)^3(2x)^2 = (2x)^5$ Product rule

$\qquad\qquad = 2^5x^5 \text{ or } 32x^5$ Power rule (b)

(c) $\left(\dfrac{2x^3}{5}\right)^{-4} = \left(\dfrac{5}{2x^3}\right)^{4}$ Negative to positive rule

$\qquad\qquad = \dfrac{5^4}{2^4x^{12}} \text{ or } \dfrac{625}{16x^{12}}$ Power rules (a)–(c)

(d) $\left(\dfrac{3x^{-2}}{4^{-1}y^3}\right)^{-3} = \dfrac{3^{-3}x^6}{4^3y^{-9}}$ Power rules (a)–(c)

$\qquad\qquad = \dfrac{x^6y^9}{4^3 \cdot 3^3} \text{ or } \dfrac{x^6y^9}{1728}$ Negative to positive rule

(e) $\dfrac{(4m)^{-3}}{(3m)^{-4}} = \dfrac{4^{-3}m^{-3}}{3^{-4}m^{-4}}$ Power rule (b)

$\qquad\qquad = \dfrac{3^4m^4}{4^3m^3}$ Negative to positive rule

$\qquad\qquad = \dfrac{3^4m^{4-3}}{4^3}$ Quotient rule

$\qquad\qquad = \dfrac{3^4m}{4^3} \text{ or } \dfrac{81m}{64}$

Now Try Exercises 57, 63, 65, and 67.

NOTE Since the steps can be done in several different orders, there are many equally correct ways to simplify expressions like those in Examples 5(d) and 5(e).

4.2 EXERCISES

Each expression is either equal to 0, 1, or −1. Decide which is correct. See Example 1.

1. 9^0

2. 5^0

3. $(-4)^0$

4. $(-10)^0$

5. -9^0

6. -5^0

7. $(-2)^0 - 2^0$

8. $(-8)^0 - 8^0$

9. $\dfrac{0^{10}}{10^0}$

10. $\dfrac{0^5}{5^0}$

Match each expression in Column I with the equivalent expression in Column II. Choices in Column II may be used once, more than once, or not at all.

I	II
11. -2^{-4}	**A.** 8
12. $(-2)^{-4}$	**B.** 16
13. 2^{-4}	**C.** $-\dfrac{1}{16}$
14. $\dfrac{1}{2^{-4}}$	**D.** -8
15. $\dfrac{1}{-2^{-4}}$	**E.** -16
16. $\dfrac{1}{(-2)^{-4}}$	**F.** $\dfrac{1}{16}$

Evaluate each expression. See Examples 1 and 2.

17. $7^0 + 9^0$

18. $8^0 + 6^0$

19. 4^{-3}

20. 5^{-4}

21. $\left(\dfrac{1}{2}\right)^{-4}$

22. $\left(\dfrac{1}{3}\right)^{-3}$

23. $\left(\dfrac{6}{7}\right)^{-2}$

24. $\left(\dfrac{2}{3}\right)^{-3}$

25. $(-3)^{-4}$

26. $(-4)^{-3}$

27. $5^{-1} + 3^{-1}$

28. $6^{-1} + 2^{-1}$

Use the quotient rule to simplify each expression. Write the expression with positive exponents. Assume that all variables represent nonzero real numbers. See Examples 2–4.

29. $\dfrac{5^8}{5^5}$

30. $\dfrac{11^6}{11^3}$

31. $\dfrac{3^{-2}}{5^{-3}}$

32. $\dfrac{4^{-3}}{3^{-2}}$

33. $\dfrac{5}{5^{-1}}$

34. $\dfrac{6}{6^{-2}}$

35. $\dfrac{x^{12}}{x^{-3}}$

36. $\dfrac{y^4}{y^{-6}}$

37. $\dfrac{1}{6^{-3}}$

38. $\dfrac{1}{5^{-2}}$

39. $\dfrac{2}{r^{-4}}$

40. $\dfrac{3}{s^{-8}}$

41. $\dfrac{4^{-3}}{5^{-2}}$

42. $\dfrac{6^{-2}}{5^{-4}}$

43. $p^5 q^{-8}$

44. $x^{-8} y^4$

45. $\dfrac{r^5}{r^{-4}}$

46. $\dfrac{a^6}{a^{-4}}$

47. $\dfrac{x^{-3} y}{4z^{-2}}$

48. $\dfrac{p^{-5} q^{-4}}{9r^{-3}}$

49. $\dfrac{(a+b)^{-3}}{(a+b)^{-4}}$

50. $\dfrac{(x+y)^{-8}}{(x+y)^{-9}}$

51. $\dfrac{(x+2y)^{-3}}{(x+2y)^{-5}}$

52. $\dfrac{(p-3q)^{-2}}{(p-3q)^{-4}}$

| **RELATING CONCEPTS** (EXERCISES 53–56) |

For Individual or Group Work

In Objective 1, we showed how 6^0 acts as 1 when it is applied to the product rule, thus motivating the definition for 0 as an exponent. We can also use the quotient rule to motivate this definition. **Work Exercises 53–56 in order.**

53. Consider the expression $\frac{25}{25}$. What is its simplest form?

54. Because $25 = 5^2$, the expression $\frac{25}{25}$ can be written as the quotient of powers of 5. Write the expression in this way.

55. Apply the quotient rule for exponents to the expression you wrote in Exercise 54. Give the answer as a power of 5.

56. Your answers in Exercises 53 and 55 must be equal because they both represent $\frac{25}{25}$. Write this equality. What definition does this result support?

Use a combination of the rules for exponents to simplify each expression. Write answers with only positive exponents. Assume that all variables represent nonzero real numbers. See Example 5.

57. $\dfrac{(7^4)^3}{7^9}$

58. $\dfrac{(5^3)^2}{5^2}$

59. $x^{-3} \cdot x^5 \cdot x^{-4}$

60. $y^{-8} \cdot y^5 \cdot y^{-2}$

61. $\dfrac{(3x)^{-2}}{(4x)^{-3}}$

62. $\dfrac{(2y)^{-3}}{(5y)^{-4}}$

63. $\left(\dfrac{x^{-1}y}{z^2}\right)^{-2}$

64. $\left(\dfrac{p^{-4}q}{r^{-3}}\right)^{-3}$

65. $(6x)^4(6x)^{-3}$

66. $(10y)^9(10y)^{-8}$

67. $\dfrac{(m^7n)^{-2}}{m^{-4}n^3}$

68. $\dfrac{(m^8n^{-4})^2}{m^{-2}n^5}$

69. $\dfrac{(x^{-1}y^2z)^{-2}}{(x^{-3}y^3z)^{-1}}$

70. $\dfrac{(a^{-2}b^{-3}c^{-4})^{-5}}{(a^2b^3c^4)^5}$

71. $\left(\dfrac{xy^{-2}}{x^2y}\right)^{-3}$

72. $\left(\dfrac{wz^{-5}}{w^{-3}z}\right)^{-2}$

73. $\dfrac{(4a^2b^3)^{-2}(2ab^{-1})^3}{(a^3b)^{-4}}$

74. $\dfrac{(m^6n)^{-2}(m^2n^{-2})^3}{m^{-1}n^{-2}}$

75. $\dfrac{(2y^{-1}z^2)^2(3y^{-2}z^{-3})^3}{(y^3z^2)^{-1}}$

76. $\dfrac{(3p^{-2}q^3)^2(5p^{-1}q^{-4})^{-1}}{(p^2q^{-2})^{-3}}$

77. $\dfrac{(9^{-1}z^{-2}x)^{-1}(4z^2x^4)^{-2}}{(5z^{-2}x^{-3})^2}$

78. $\dfrac{(4^{-1}a^{-1}b^{-2})^{-2}(5a^{-3}b^4)^{-2}}{(3a^{-3}b^{-5})^2}$

79. Consider the following typical student **error:**

$$\frac{16^3}{2^2} = \left(\frac{16}{2}\right)^{3-2} = 8^1 = 8.$$

Explain what the student did incorrectly, and then give the correct answer.

80. Consider the following typical student **error:**

$$-5^4 = (-5)^4 = 625.$$

Explain what the student did incorrectly, and then give the correct answer.

SUMMARY EXERCISES ON THE RULES FOR EXPONENTS

Use the rules for exponents to simplify each expression. Use only positive exponents in your answers. Assume all variable expressions represent positive numbers.

1. $\left(\dfrac{6x^2}{5}\right)^{12}$

2. $\left(\dfrac{rs^2t^3}{3t^4}\right)^6$

3. $(10x^2y^4)^2(10xy^2)^3$

4. $(-2ab^3c)^4(-2a^2b)^3$

5. $\left(\dfrac{9wx^3}{y^4}\right)^3$

6. $(4x^{-2}y^{-3})^{-2}$

7. $\dfrac{c^{11}(c^2)^4}{(c^3)^3(c^2)^{-6}}$

8. $\left(\dfrac{k^4t^2}{k^2t^{-4}}\right)^{-2}$

9. $5^{-1} + 6^{-1}$

10. $\dfrac{(3y^{-1}z^3)^{-1}(3y^2)}{(y^3z^2)^{-3}}$

11. $\dfrac{(2xy^{-1})^3}{2^3x^{-3}y^2}$

12. $-8^0 + (-8)^0$

13. $(z^4)^{-3}(z^{-2})^{-5}$

14. $\left(\dfrac{r^2st^5}{3r}\right)^{-2}$

15. $\dfrac{(3^{-1}x^{-3}y)^{-1}(2x^2y^{-3})^2}{(5x^{-2}y^2)^{-2}}$

16. $\left(\dfrac{5x^2}{3x^{-4}}\right)^{-1}$

17. $\left(\dfrac{-2x^{-2}}{2x^2}\right)^{-2}$

18. $\dfrac{(x^{-4}y^2)^3(x^2y)^{-1}}{(xy^2)^{-3}}$

19. $\dfrac{(a^{-2}b^3)^{-4}}{(a^{-3}b^2)^{-2}(ab)^{-4}}$

20. $(2a^{-30}b^{-29})(3a^{31}b^{30})$

21. $5^{-2} + 6^{-2}$

22. $\left[\dfrac{(x^{47}y^{23})^2}{x^{-26}y^{-42}}\right]^0$

23. $\left(\dfrac{7a^2b^3}{2}\right)^3$

24. $-(-12^0)$

25. $-(-12)^0$

26. $\dfrac{0^{12}}{12^0}$

27. $\dfrac{(2xy^{-3})^{-2}}{(3x^{-2}y^4)^{-3}}$

28. $\left(\dfrac{a^2b^3c^4}{a^{-2}b^{-3}c^{-4}}\right)^{-2}$

29. $(6x^{-5}z^3)^{-3}$

30. $(2p^{-2}qr^{-3})(2p)^{-4}$

31. $\dfrac{(xy)^{-3}(xy)^5}{(xy)^{-4}}$

32. $42^0 - (-12)^0$

33. $\dfrac{(7^{-1}x^{-3})^{-2}(x^4)^{-6}}{7^{-1}x^{-3}}$

34. $\left(\dfrac{3^{-4}x^{-3}}{3^{-3}x^{-6}}\right)^{-2}$

35. $(5p^{-2}q)^{-3}(5pq^3)^4$

36. $8^{-1} + 6^{-1}$

37. $\left[\dfrac{4r^{-6}s^{-2}t}{2r^8s^{-4}t^2}\right]^{-1}$

38. $(13x^{-6}y)(13x^{-6}y)^{-1}$

39. $\dfrac{(8pq^{-2})^4}{(8p^{-2}q^{-3})^3}$

40. $\left(\dfrac{mn^{-2}p}{m^2np^4}\right)^{-2}\left(\dfrac{mn^{-2}p}{m^2np^4}\right)^3$

41. $-(-3^0)^0$

42. $5^{-1} - 8^{-1}$

4.3 An Application of Exponents: Scientific Notation

OBJECTIVE 1 Express numbers in scientific notation. One example of the use of exponents comes from science. The numbers occurring in science are often extremely large (such as the distance from Earth to the sun, 93,000,000 mi) or extremely small (the wavelength of yellow-green light, approximately .0000006 m). Because of the difficulty of working with many zeros, scientists often express such numbers with exponents. Each number is written as $a \times 10^n$, where $1 \le |a| < 10$ and n is an integer. This form is called **scientific notation.** There is always one nonzero digit before the decimal point. This is shown in the following examples. (In work with scientific notation, the times symbol, $\times$, is commonly used.)

$3.19 \times 10^1 = 3.19 \times 10 = 31.9$ Decimal point moves 1 place to the right.

$3.19 \times 10^2 = 3.19 \times 100 = 319.$ Decimal point moves 2 places to the right.

$3.19 \times 10^3 = 3.19 \times 1000 = 3190.$ Decimal point moves 3 places to the right.

$3.19 \times 10^{-1} = 3.19 \times .1 = .319$ Decimal point moves 1 place to the left.

$3.19 \times 10^{-2} = 3.19 \times .01 = .0319$ Decimal point moves 2 places to the left.

$3.19 \times 10^{-3} = 3.19 \times .001 = .00319$ Decimal point moves 3 places to the left.

A number in scientific notation is always written with the decimal point after the first nonzero digit and then multiplied by the appropriate power of 10. For example, 35 is written 3.5×10^1, or 3.5×10; 56,200 is written 5.62×10^4, since

$$56{,}200 = 5.62 \times 10{,}000 = 5.62 \times 10^4.$$

To write a number in scientific notation, follow these steps.

Writing a Number in Scientific Notation

Step 1   Move the decimal point to the right of the first nonzero digit.

Step 2   Count the number of places you moved the decimal point.

Step 3   The number of places in Step 2 is the absolute value of the exponent on 10.

Step 4   The exponent on 10 is positive if the original number is larger than the number in Step 1; the exponent is negative if the original number is smaller than the number in Step 1. If the decimal point is not moved, the exponent is 0.

EXAMPLE 1 Using Scientific Notation

Write each number in scientific notation.

(a) 93,000,000

The number will be written in scientific notation as 9.3×10^n. To find the value of n, first compare the original number, 93,000,000, with 9.3. Here 93,000,000 is *larger* than 9.3. Therefore, multiply by a *positive* power of 10 so the product 9.3×10^n will equal the larger number.

Move the decimal point to follow the first nonzero digit (the 9). Count the number of places the decimal point was moved.

$$93,\!000,\!000 \qquad \text{7 places}$$

Since the decimal point was moved 7 places, and since n is positive,

$$93,\!000,\!000 = 9.3 \times 10^7.$$

(b) $63,\!200,\!000,\!000 = 6.3200000000 = 6.32 \times 10^{10}$

10 places

(c) .00462

Move the decimal point to the right of the first nonzero digit, and count the number of places the decimal point was moved.

$$.00462 \qquad \text{3 places}$$

Since .00462 is *smaller* than 4.62, the exponent must be *negative*.

$$.00462 = 4.62 \times 10^{-3}$$

(d) $.0000762 = 7.62 \times 10^{-5}$

Now Try Exercises 15 and 19.

NOTE To choose the exponent when writing a number in scientific notation, think: If the original number is "large," like 93,000,000, use a *positive* exponent on 10, since positive is larger than negative. However, if the original number is "small," like .00462, use a *negative* exponent on 10, since negative is smaller than positive.

OBJECTIVE 2 Convert numbers in scientific notation to numbers without exponents. To convert a number written in scientific notation to a number without exponents, work in reverse. Multiplying by a positive power of 10 will make the number larger; multiplying by a negative power of 10 will make the number smaller.

EXAMPLE 2 Writing Numbers without Exponents

Write each number without exponents.

(a) 6.2×10^3

Since the exponent is positive, make 6.2 larger by moving the decimal point 3 places to the right, inserting zeros as needed.

$$6.2 \times 10^3 = 6.200 = 6200$$

(b) $4.283 \times 10^5 = 4.28300 = 428,\!300$ Move 5 places to the right.

(c) $7.04 \times 10^{-3} = .00704$ Move 3 places to the left.

The exponent tells the number of places and the direction that the decimal point is moved.

Now Try Exercises 23 and 29.

OBJECTIVE 3 Use scientific notation in calculations. The next example uses scientific notation with products and quotients.

EXAMPLE 3 Multiplying and Dividing with Scientific Notation

Write each product or quotient without exponents.

(a) $(6 \times 10^3)(5 \times 10^{-4}) = (6 \times 5)(10^3 \times 10^{-4})$ Commutative and associative properties

$= 30 \times 10^{-1}$ Product rule

$= 3$ Write without exponents.

(b) $\dfrac{4 \times 10^{-5}}{2 \times 10^3} = \dfrac{4}{2} \times \dfrac{10^{-5}}{10^3} = 2 \times 10^{-8} = .00000002$

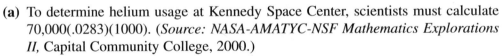

Now Try Exercises 33 and 43.

NOTE Multiplying or dividing numbers written in scientific notation may produce an answer in the form $a \times 10^0$. Since $10^0 = 1$, $a \times 10^0 = a$. For example,

$$(8 \times 10^{-4})(5 \times 10^4) = 40 \times 10^0 = 40.$$

EXAMPLE 4 Calculating Using Scientific Notation

Convert to scientific notation, perform each computation, then give the result without scientific notation.

(a) To determine helium usage at Kennedy Space Center, scientists must calculate 70,000(.0283)(1000). (*Source: NASA-AMATYC-NSF Mathematics Explorations II,* Capital Community College, 2000.)

$$70{,}000(.0283)(1000) = (7 \times 10^4)(2.83 \times 10^{-2})(1 \times 10^3)$$
$$= (7 \times 2.83 \times 1)(10^{4-2+3})$$
$$= 19.81 \times 10^5$$
$$= 1{,}981{,}000$$

(b) The ratio of the tidal force exerted by the moon compared to that exerted by the sun is given by

$$\frac{73.5 \times 10^{21} \times (1.5 \times 10^8)^3}{1.99 \times 10^{30} \times (3.84 \times 10^5)^3}.$$

(*Source:* Kastner, Bernice, *Space Mathematics,* NASA.)

$$\frac{7.35 \times 10^1 \times 10^{21} \times 1.5^3 \times 10^{24}}{1.99 \times 10^{30} \times 3.84^3 \times 10^{15}} \approx .22 \times 10^{1+21+24-30-15}$$
$$= .22 \times 10^1$$
$$= 2.2$$

Now Try Exercises 61 and 65.

| **CONNECTIONS** |

Charles F. Richter devised a scale in 1935 to compare the intensities, or relative power, of earthquakes. The *intensity* of an earthquake is measured relative to the intensity of a standard *zero-level* earthquake of intensity I_0. The relationship is equivalent to $I = I_0 \times 10^R$, where R is the *Richter scale* measure. For example, if an earthquake has magnitude 5.0 on the Richter scale, then its intensity is calculated as $I = I_0 \times 10^{5.0} = I_0 \times 100{,}000$, which is 100,000 times as intense as a zero-level earthquake. The following diagram illustrates the intensities of earthquakes and their Richter scale magnitudes.

Intensity $\quad I_0 \times 10^0 \quad I_0 \times 10^1 \quad I_0 \times 10^2 \quad I_0 \times 10^3 \quad I_0 \times 10^4 \quad I_0 \times 10^5 \quad I_0 \times 10^6 \quad I_0 \times 10^7 \quad I_0 \times 10^8$

Richter Scale $\quad\quad 0 \quad\quad 1 \quad\quad 2 \quad\quad 3 \quad\quad 4 \quad\quad 5 \quad\quad 6 \quad\quad 7 \quad\quad 8$

To compare two earthquakes to each other, a ratio of the intensities is calculated. For example, to compare an earthquake that measures 8.0 on the Richter scale to one that measures 5.0, simply find the ratio of the intensities:

$$\frac{\text{intensity } 8.0}{\text{intensity } 5.0} = \frac{I_0 \times 10^{8.0}}{I_0 \times 10^{5.0}} = \frac{10^8}{10^5} = 10^{8-5} = 10^3 = 1000.$$

Therefore an earthquake that measures 8.0 on the Richter Scale is 1000 times as intense as one that measures 5.0.

For Discussion or Writing

The table gives Richter scale measurements for several earthquakes.

Earthquake	Richter Scale Measurement
1960 Concepción, Chile	9.5
1906 San Francisco, California	8.3
1939 Erzincan, Turkey	8.0
1998 Sumatra, Indonesia	7.0
1998 Adana, Turkey	6.3

Source: World Almanac and Books of Facts, 2000.

1. Compare the intensity of the 1939 Erzincan earthquake to the 1998 Sumatra earthquake.
2. Compare the intensity of the 1998 Adana earthquake to the 1906 San Francisco earthquake.
3. Compare the intensity of the 1939 Erzincan earthquake to the 1998 Adana earthquake.
4. Suppose an earthquake measures 7.2 on the Richter scale. How would the intensity of a second earthquake compare if its Richter scale measure differed by $+3.0$? By -1.0?

4.3 EXERCISES

Match each number written in scientific notation in Column I with the correct choice from Column II.

I	II
1. 4.6×10^{-4}	**A.** .00046
2. 4.6×10^{4}	**B.** 46,000
3. 4.6×10^{5}	**C.** 460,000
4. 4.6×10^{-5}	**D.** .000046

Determine whether or not each number is written in scientific notation as defined in Objective 1. If it is not, write it as such.

5. 4.56×10^{3} **6.** 7.34×10^{5} **7.** 5,600,000 **8.** 34,000

9. $.8 \times 10^{2}$ **10.** $.9 \times 10^{3}$ **11.** .004 **12.** .0007

13. Explain in your own words what it means for a number to be written in scientific notation. Give examples.

14. Explain how to multiply a number by a positive power of ten. Then explain how to multiply a number by a negative power of ten.

Write each number in scientific notation. See Example 1.

15. 5,876,000,000 **16.** 9,994,000,000 **17.** 82,350 **18.** 78,330

19. .000007 **20.** .0000004 **21.** .00203 **22.** .0000578

Write each number without exponents. See Example 2.

23. 7.5×10^{5} **24.** 8.8×10^{6} **25.** 5.677×10^{12} **26.** 8.766×10^{9}

27. 6.21×10^{0} **28.** 8.56×10^{0} **29.** 7.8×10^{-4} **30.** 8.9×10^{-5}

31. 5.134×10^{-9} **32.** 7.123×10^{-10}

Use properties and rules for exponents to perform the indicated operations, and write each answer without exponents. See Example 3.

33. $(2 \times 10^{8}) \times (3 \times 10^{3})$ **34.** $(4 \times 10^{7}) \times (3 \times 10^{3})$

35. $(5 \times 10^{4}) \times (3 \times 10^{2})$ **36.** $(8 \times 10^{5}) \times (2 \times 10^{3})$

37. $(3 \times 10^{-4}) \times (2 \times 10^{8})$ **38.** $(4 \times 10^{-3}) \times (2 \times 10^{7})$

39. $\dfrac{9 \times 10^{-5}}{3 \times 10^{-1}}$ **40.** $\dfrac{12 \times 10^{-4}}{4 \times 10^{-3}}$ **41.** $\dfrac{8 \times 10^{3}}{2 \times 10^{2}}$

42. $\dfrac{5 \times 10^{4}}{1 \times 10^{3}}$ **43.** $\dfrac{2.6 \times 10^{-3}}{2 \times 10^{2}}$ **44.** $\dfrac{9.5 \times 10^{-1}}{5 \times 10^{3}}$

TECHNOLOGY INSIGHTS (EXERCISES 45–50)

Graphing calculators such as the TI-83 Plus can display numbers in scientific notation (when in scientific mode), using the format shown in the screen at the top left on the next page. For example, the calculator displays 5.4E3 to represent 5.4×10^{3}, *the scientific notation form for 5400. The display* 5.4E−4 *means* 5.4 × 10⁻⁴. *It will also perform operations with numbers entered in scientific notation, as shown in the screen on the right. Notice how the rules for exponents are applied.*

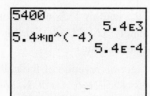

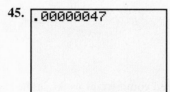

Predict the display the calculator would give for the expression shown in each screen.

45.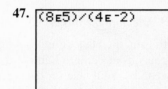

.00000047

46.

.000021

47.

(8E5)/(4E⁻2)

48.

(9E⁻4)/(3E3)

49.

(2E6)*(2E⁻3)/(4E
2)

50.

(5E⁻3)*(1E9)/(5E
3)

Each statement comes from Astronomy! A Brief Edition *by James B. Kaler (Addison-Wesley, 1997). If the number in italics is in scientific notation, write it without exponents. If the number is written without exponents, write it in scientific notation.*

51. Multiplying this view over the whole sky yields a galaxy count of more than *10 billion.* (page 496)

52. The circumference of the solar orbit is . . . about *4.7 million* km . . . . (in reference to the orbit of Jupiter, page 395)

53. The solar luminosity requires that *2 × 10⁹* kg of mass be converted into energy every second. (page 327)

54. At maximum, a cosmic ray particle—a mere atomic nucleus of only *10⁻¹³* cm across— can carry the energy of a professionally pitched baseball. (page 445)

Each of the following statements contains a number in italics. Write the number in scientific notation.

55. During the 1999–2000 Broadway season, gross receipts were *$603,000,000*. (*Source:* The League of American Theatres and Producers, Inc.)

56. In 1999, the leading U.S. advertiser was the General Motors Corporation, which spent approximately *$2,900,000,000*. (*Source:* Competitive Media Reporting and Publishers Information Bureau.)

57. In 2000, service revenue of the cellular telephone industry was *$52,466,000,000*. (*Source:* Cellular Telecommunications & Internet Association.)

58. Assets of the insured commercial banks in the state of New York totaled *$1,304,300,000*. (*Source:* U.S. Federal Deposit Insurance Corporation.)

Use scientific notation to calculate the answer to each problem. See Example 4.

59. The distance to Earth from the planet Pluto is 4.58×10^9 km. In April 1983, Pioneer 10 transmitted radio signals from Pluto to Earth at the speed of light, 3.00×10^5 km per sec. How long (in seconds) did it take for the signals to reach Earth?

60. In Exercise 59, how many hours did it take for the signals to reach Earth?

61. In a recent year, the state of Texas had about 1.3×10^6 farms with an average of 7.1×10^2 acres per farm. What was the total number of acres devoted to farmland in Texas that year? (*Source:* National Agricultural Statistics Service, U.S. Department of Agriculture.)

62. The graph depicts aerospace industry sales. The figures at the tops of the bars represent billions of dollars.

   (a) For each year, write the figure in scientific notation.

   (b) If a line segment is drawn between the tops of the bars for 1997 and 1999, what is its slope?

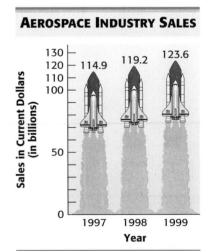

AEROSPACE INDUSTRY SALES

Source: U.S. Bureau of the Census, *Current Industrial Reports,*

63. There are 10^9 Social Security numbers. The population of the United States is about 3×10^8. How many Social Security numbers are available for each person? (*Source:* U.S. Bureau of the Census.)

64. The top-grossing movie of 1997 was *Titanic*, with box office receipts of about 6×10^8 dollars. That amount represented a fraction of about 9.5×10^{-3} of the total receipts for motion pictures in that year. What were the total receipts? (*Source:* U.S. Bureau of the Census.)

65. The body of a 150-lb person contains about 2.3×10^{-4} lb of copper. How much copper is contained in the bodies of 1200 such people?

66. There were 6.3×10^{10} dollars spent to attend motion pictures in a recent year. Approximately 1.3×10^8 adults attended a motion picture theater at least once. What was the average amount spent per person that year? (*Source:* U.S. National Endowment for the Arts.)

4.4 Adding and Subtracting Polynomials; Graphing Simple Polynomials

OBJECTIVES

1 Identify terms and coefficients.

2 Add like terms.

3 Know the vocabulary for polynomials.

4 Evaluate polynomials.

5 Add and subtract polynomials.

6 Graph equations defined by polynomials of degree 2.

OBJECTIVE 1 Identify terms and coefficients. In Chapter 1 we saw that in an expression such as

$$4x^3 + 6x^2 + 5x + 8,$$

the quantities $4x^3$, $6x^2$, $5x$, and 8 are called *terms*. As mentioned earlier, in the term $4x^3$, the number 4 is called the *numerical coefficient,* or simply the *coefficient,* of x^3. In the same way, 6 is the coefficient of x^2 in the term $6x^2$, 5 is the coefficient of x in the term $5x$, and 8 is the coefficient in the term 8. A constant term, like 8 in the expression above, can be thought of as $8 \cdot 1 = 8x^0$, since $x^0 = 1$.

■ EXAMPLE 1 Identifying Coefficients

Name the (numerical) coefficient of each term in these expressions.

(a) $4x^3$

The coefficient is 4.

(b) $x - 6x^4$

The coefficient of x is 1 because $x = 1 \cdot x$ or $1x$. The coefficient of x^4 is -6 since we can write $x - 6x^4$ as the sum $x + (-6x^4)$.

(c) $5 - v^3$

The coefficient of the term 5 is 5 because $5 = 5v^0$. By writing $5 - v^3$ as a sum, $5 + (-v^3)$, or $5 + (-1v^3)$, we can identify the coefficient of v^3 as -1. ■

Now Try Exercises 9 and 13.

OBJECTIVE 2 Add like terms. Recall from Section 1.8 that *like terms* have exactly the same combination of variables with the same exponents on the variables. Only the coefficients may differ. Examples of like terms are

$$19m^5 \quad \text{and} \quad 14m^5,$$
$$6y^9, \quad -37y^9, \quad \text{and} \quad y^9,$$
$$3pq \quad \text{and} \quad -2pq,$$
$$2xy^2 \quad \text{and} \quad -xy^2.$$

Using the distributive property, we combine, or add, like terms by adding their coefficients.

EXAMPLE 2 Adding Like Terms

Simplify each expression by adding like terms.

(a) $-4x^3 + 6x^3 = (-4 + 6)x^3 = 2x^3$ Distributive property

(b) $9x^6 - 14x^6 + x^6 = (9 - 14 + 1)x^6 = -4x^6$

(c) $12m^2 + 5m + 4m^2 = (12 + 4)m^2 + 5m = 16m^2 + 5m$

(d) $3x^2y + 4x^2y - x^2y = (3 + 4 - 1)x^2y = 6x^2y$

Now Try Exercises 17, 23, and 27.

In Example 2(c), we cannot combine $16m^2$ and $5m$. These two terms are unlike because the exponents on the variables are different. *Unlike terms* have different variables or different exponents on the same variables.

OBJECTIVE 3 Know the vocabulary for polynomials. A **polynomial in** x is a term or the sum of a finite number of terms of the form ax^n, for any real number a and any whole number n. For example,

$$16x^8 - 7x^6 + 5x^4 - 3x^2 + 4$$

is a polynomial in x. (The 4 can be written as $4x^0$.) This polynomial is written in **descending powers** of the variable, since the exponents on x decrease from left to right. On the other hand,

$$2x^3 - x^2 + \frac{4}{x}$$

is not a polynomial in x, since a variable appears in a denominator. Of course, we could define *polynomial* using any variable and not just x, as in Example 2(c). In fact, polynomials may have terms with more than one variable, as in Example 2(d).

The **degree of a term** is the sum of the exponents on the variables. For example, $3x^4$ has degree 4, while $6x^{17}$ has degree 17. The term $5x$ has degree 1, -7 has degree 0 (since -7 can be written as $-7x^0$), and $2x^2y$ has degree $2 + 1 = 3$. (y has an exponent of 1.) The **degree of a polynomial** is the greatest degree of any nonzero term of the polynomial. For example, $3x^4 - 5x^2 + 6$ is of degree 4, the polynomial $5x + 7$ is of degree 1, 3 (or $3x^0$) is of degree 0, and $x^2y + xy - 5xy^2$ is of degree 3.

Three types of polynomials are very common and are given special names. A polynomial with exactly three terms is called a **trinomial.** (*Tri-* means "three," as in *tri*angle.) Examples are

$$9m^3 - 4m^2 + 6, \qquad 19y^2 + 8y + 5, \qquad \text{and} \qquad -3m^5n^2 + 2n^3 - m^4.$$

A polynomial with exactly two terms is called a **binomial.** (*Bi-* means "two," as in *bi*cycle.) Examples are

$$-9x^4 + 9x^3, \qquad 8m^2 + 6m, \qquad \text{and} \qquad 3m^5n^2 - 9m^2n^4.$$

A polynomial with only one term is called a **monomial.** (*Mon(o)-* means "one," as in *mono*rail.) Examples are

$$9m, \qquad -6y^5, \qquad a^2b^2, \qquad \text{and} \qquad 6.$$

EXAMPLE 3   Classifying Polynomials

For each polynomial, first simplify if possible by combining like terms. Then give the degree and tell whether it is a monomial, a binomial, a trinomial, or none of these.

(a) $2x^3 + 5$

The polynomial cannot be simplified. The degree is 3. The polynomial is a binomial.

(b) $4xy - 5xy + 2xy$

Add like terms to simplify: $4xy - 5xy + 2xy = xy$, which is a monomial of degree 2.

Now Try Exercises 29 and 31.

OBJECTIVE 4 Evaluate polynomials.   A polynomial usually represents different numbers for different values of the variable, as shown in the next example.

EXAMPLE 4   Evaluating a Polynomial

Find the value of $3x^4 + 5x^3 - 4x - 4$ when $x = -2$ and when $x = 3$.

First, substitute -2 for x.

$$\begin{aligned}
3x^4 + 5x^3 - 4x - 4 &= 3(-2)^4 + 5(-2)^3 - 4(-2) - 4 \\
&= 3 \cdot 16 + 5 \cdot (-8) - 4(-2) - 4 \qquad \text{Apply exponents.} \\
&= 48 - 40 + 8 - 4 \qquad \text{Multiply.} \\
&= 12 \qquad \text{Add and subtract.}
\end{aligned}$$

Next, replace x with 3.

$$\begin{aligned}
3x^4 + 5x^3 - 4x - 4 &= 3(3)^4 + 5(3)^3 - 4(3) - 4 \\
&= 3 \cdot 81 + 5 \cdot 27 - 12 - 4 \\
&= 362
\end{aligned}$$

Now Try Exercise 37.

CAUTION   Notice the use of parentheses around the numbers that are substituted for the variable in Example 4. This is particularly important when substituting a negative number for a variable that is raised to a power, so that the sign of the product is correct.

OBJECTIVE 5 Add and subtract polynomials.   Polynomials may be added, subtracted, multiplied, and divided.

Adding Polynomials

To add two polynomials, add like terms.

EXAMPLE 5 Adding Polynomials Vertically

Add $6x^3 - 4x^2 + 3$ and $-2x^3 + 7x^2 - 5$.

Write like terms in columns.

$$6x^3 - 4x^2 + 3$$
$$-2x^3 + 7x^2 - 5$$

Now add, column by column.

$$
\begin{array}{ccc}
6x^3 & -4x^2 & 3 \\
-2x^3 & 7x^2 & -5 \\
\hline
4x^3 & 3x^2 & -2
\end{array}
$$

Add the three sums together.

$$4x^3 + 3x^2 + (-2) = 4x^3 + 3x^2 - 2$$

Now Try Exercise 45.

The polynomials in Example 5 also can be added horizontally.

EXAMPLE 6 Adding Polynomials Horizontally

Add $6x^3 - 4x^2 + 3$ and $-2x^3 + 7x^2 - 5$.

Write the sum as

$$(6x^3 - 4x^2 + 3) + (-2x^3 + 7x^2 - 5).$$

Use the associative and commutative properties to rewrite this sum with the parentheses removed and with the subtractions changed to additions of inverses.

$$6x^3 + (-4x^2) + 3 + (-2x^3) + 7x^2 + (-5)$$

Place like terms together.

$$6x^3 + (-2x^3) + (-4x^2) + 7x^2 + 3 + (-5)$$

Combine like terms to get

$$4x^3 + 3x^2 + (-2), \qquad \text{or} \qquad 4x^3 + 3x^2 - 2,$$

the same answer found in Example 5.

Now Try Exercise 61.

Earlier, we defined the difference $x - y$ as $x + (-y)$. (We find the difference $x - y$ by adding x and the opposite of y.) For example,

$$7 - 2 = 7 + (-2) = 5 \qquad \text{and} \qquad -8 - (-2) = -8 + 2 = -6.$$

A similar method is used to subtract polynomials.

Subtracting Polynomials

To subtract two polynomials, change all the signs in the second polynomial and add the result to the first polynomial.

EXAMPLE 7 Subtracting Polynomials

(a) Perform the subtraction $(5x - 2) - (3x - 8)$.
By the definition of subtraction,
$$(5x - 2) - (3x - 8) = (5x - 2) + [-(3x - 8)].$$

As shown in Chapter 1, the distributive property gives
$$-(3x - 8) = -1(3x - 8) = -3x + 8,$$

so
$$(5x - 2) - (3x - 8) = (5x - 2) + (-3x + 8)$$
$$= 2x + 6.$$

(b) Subtract $6x^3 - 4x^2 + 2$ from $11x^3 + 2x^2 - 8$.
Write the problem.
$$(11x^3 + 2x^2 - 8) - (6x^3 - 4x^2 + 2)$$

Change all signs in the second polynomial and add.   *Really addition Problem),*
$$(11x^3 + 2x^2 - 8) + (-6x^3 + 4x^2 - 2) = 5x^3 + 6x^2 - 10$$

To check a subtraction problem, use the fact that if $a - b = c$, then $a = b + c$. For example, $6 - 2 = 4$, so check by writing $6 = 2 + 4$, which is correct. Check the polynomial subtraction above by adding $6x^3 - 4x^2 + 2$ and $5x^3 + 6x^2 - 10$. Since the sum is $11x^3 + 2x^2 - 8$, the subtraction was performed correctly.

Now Try Exercise 59.

Subtraction also can be done in columns (vertically). We use vertical subtraction in Section 4.7 when we divide polynomials.

EXAMPLE 8 Subtracting Polynomials Vertically

Use the method of subtracting by columns to find
$$(14y^3 - 6y^2 + 2y - 5) - (2y^3 - 7y^2 - 4y + 6).$$

Arrange like terms in columns.
$$14y^3 - 6y^2 + 2y - 5$$
$$2y^3 - 7y^2 - 4y + 6$$

Change all signs in the second row, and then add.

$$\begin{array}{l} 14y^3 - 6y^2 + 2y - 5 \\ \underline{-2y^3 + 7y^2 + 4y - 6} \quad \text{Change all signs.} \\ 12y^3 + y^2 + 6y - 11 \quad \text{Add.} \end{array}$$

Now Try Exercise 55.

Either the horizontal or the vertical method may be used to add and subtract polynomials.

We add and subtract polynomials in more than one variable by combining like terms, just as with single variable polynomials.

EXAMPLE 9   Adding and Subtracting Polynomials with More Than One Variable

Add or subtract as indicated.

(a) $(4a + 2ab - b) + (3a - ab + b) = 4a + 2ab - b + 3a - ab + b$

$\qquad\qquad\qquad\qquad\qquad\qquad\qquad = 7a + ab$ Combine like terms.

(b) $(2x^2y + 3xy + y^2) - (3x^2y - xy - 2y^2)$

$\qquad = 2x^2y + 3xy + y^2 - 3x^2y + xy + 2y^2$

$\qquad = -x^2y + 4xy + 3y^2$

Now Try Exercises 73 and 75.

OBJECTIVE 6 **Graph equations defined by polynomials of degree 2.** In Chapter 3 we introduced graphs of straight lines. These graphs were defined by linear equations (which are actually polynomial equations of degree 1). By selective point-plotting, we can graph polynomial equations of degree 2.

EXAMPLE 10   Graphing Equations Defined by Polynomials with Degree 2

Graph each equation.

(a) $y = x^2$

Select several values for x; then find the corresponding y-values. For example, selecting $x = 2$ gives

$$y = 2^2 = 4,$$

and so the point $(2, 4)$ is on the graph of $y = x^2$. (Recall that in an ordered pair such as $(2, 4)$, the x-value comes first and the y-value second.) We show some ordered pairs that satisfy $y = x^2$ in the table next to Figure 1. If we plot the ordered pairs from the table on a coordinate system and draw a smooth curve through them, we obtain the graph shown in Figure 1.

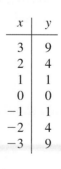

x	y
3	9
2	4
1	1
0	0
-1	1
-2	4
-3	9

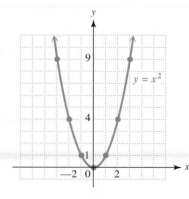

FIGURE 1

The graph of $y = x^2$ is the graph of a function, since each input x is related to just one output y. The curve in Figure 1 is called a **parabola.** The point $(0, 0)$, the lowest point on this graph, is called the **vertex** of the parabola. The vertical line through the vertex (the y-axis here) is called the **axis** of the parabola. The axis of a parabola is a **line of symmetry** for the graph. If the graph is folded on this line, the two halves will match.

(b) $y = -x^2 + 3$

Once again plot points to obtain the graph. For example, if $x = -2$,

$$y = -(-2)^2 + 3 = -4 + 3 = -1.$$

This point and several others are shown in the table that accompanies the graph in Figure 2. The vertex of this parabola is $(0, 3)$. This time the vertex is the *highest* point on the graph. The graph opens downward because x^2 has a negative coefficient.

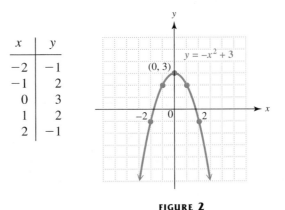

x	y
-2	-1
-1	2
0	3
1	2
2	-1

FIGURE 2

Now Try Exercises 87 and 91.

NOTE   All polynomials of degree 2 have parabolas as their graphs. When graphing by plotting points, it is necessary to continue finding points until the vertex and points on either side of it are located. (In this section, all parabolas have their vertices on the x-axis or the y-axis.)

CONNECTIONS

In Section 3.1 we introduced the idea of a *function:* for every input x, there is one output y. For example, according to the U.S. National Aeronautics and Space Administration (NASA), the budget in millions of dollars for space station research for 1996–2001 can be approximated by the polynomial equation

$$y = -10.25x^2 - 126.04x + 5730.21,$$

where $x = 0$ represents 1996, $x = 1$ represents 1997, and so on, up to $x = 5$ representing 2001. The actual budget for 1998 was 5327 million dollars; an input of $x = 2$ (for 1998) in the equation gives approximately $y = 5437$. Considering the magnitude of the numbers, this is a good approximation.

(continued)

For Discussion or Writing

Use the given polynomial equation to approximate the budget in other years between 1996 and 2001. Compare to the actual figures given here.

Year	Budget (in millions of dollars)
1996	5710
1997	5675
1998	5327
1999	5306
2000	5077
2001	4832

4.4 EXERCISES

For Extra Help

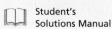 Student's Solutions Manual

MyMathLab

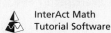 InterAct Math Tutorial Software

 AW Math Tutor Center

 MathXL

 Digital Video Tutor CD 6/Videotape 6

Fill in each blank with the correct response.

1. In the term $7x^5$, the coefficient is _____ and the exponent is _____.

2. The expression $5x^3 - 4x^2$ has _____ term(s).
(how many?)

3. The degree of the term $-4x^8$ is _____.

4. The polynomial $4x^2 - y^2$ _____ an example of a trinomial.
(is/is not)

5. When $x^2 + 10$ is evaluated for $x = 4$, the result is _____.

6. $5x^{\underline{}} + 3x^3 - 7x$ is a trinomial of degree 4.

7. $3xy + 2xy - 5xy =$ _____.

8. _____ is an example of a monomial with coefficient 5, in the variable x, having degree 9.

For each polynomial, determine the number of terms and name the coefficients of the terms. See Example 1.

9. $6x^4$ **10.** $-9y^5$ **11.** t^4 **12.** s^7

13. $-19r^2 - r$ **14.** $2y^3 - y$ **15.** $x + 8x^2 + 5x^3$ **16.** $v - 2v^3 - v^7$

In each polynomial, add like terms whenever possible. Write the result in descending powers of the variable. See Example 2.

17. $-3m^5 + 5m^5$ **18.** $-4y^3 + 3y^3$

19. $2r^5 + (-3r^5)$ **20.** $-19y^2 + 9y^2$

21. $.2m^5 - .5m^2$ **22.** $-.9y + .9y^2$

23. $-3x^5 + 2x^5 - 4x^5$ **24.** $6x^3 - 8x^3 + 9x^3$

25. $-4p^7 + 8p^7 + 5p^9$ **26.** $-3a^8 + 4a^8 - 3a^2$

27. $-4xy^2 + 3xy^2 - 2xy^2 + xy^2$ **28.** $3pr^5 - 8pr^5 + pr^5 + 2pr^5$

For each polynomial, first simplify, if possible, and write it in descending powers of the variable. Then give the degree of the resulting polynomial and tell whether it is a monomial, binomial, trinomial, or none of these. See Example 3.

29. $6x^4 - 9x$ **30.** $7t^3 - 3t$

31. $5m^4 - 3m^2 + 6m^4 - 7m^3$ **32.** $6p^5 + 4p^3 - 8p^5 + 10p^2$

33. $\dfrac{5}{3}x^4 - \dfrac{2}{3}x^4$ **34.** $\dfrac{4}{5}r^6 + \dfrac{1}{5}r^6$

35. $.8x^4 - .3x^4 - .5x^4 + 7$ **36.** $1.2t^3 - .9t^3 - .3t^3 + 9$

*Find the value of each polynomial when (**a**) x = 2 and when (**b**) x = −1. See Example 4.*

37. $2x^5 - 4x^4 + 5x^3 - x^2$ **38.** $2x^2 + 5x + 1$

39. $-3x^2 + 14x - 2$ **40.** $-2x^2 + 3$

41. $2x^2 - 3x - 5$ **42.** $x^2 + 5x - 10$

Add. See Example 5.

43. $2x^2 - 4x$ **44.** $-5y^3 + 3y$ **45.** $3m^2 + 5m + 6$
     $\underline{3x^2 + 2x}$ $\underline{8y^3 - 4y}$ $\underline{2m^2 - 2m - 4}$

46. $4a^3 - 4a^2 - 4$ **47.** $\dfrac{2}{3}x^2 + \dfrac{1}{5}x + \dfrac{1}{6}$ **48.** $\dfrac{4}{7}y^2 - \dfrac{1}{5}y + \dfrac{7}{9}$
     $\underline{6a^3 + 5a^2 - 8}$ $\underline{\dfrac{1}{2}x^2 - \dfrac{1}{3}x + \dfrac{2}{3}}$ $\underline{\dfrac{1}{3}y^2 - \dfrac{1}{3}y + \dfrac{2}{5}}$

49. $9m^3 - 5m^2 + 4m - 8$ **50.** $12r^5 + 11r^4 - 7r^3 - 2r^2$
     $\underline{-3m^3 + 6m^2 + 8m - 6}$ $\underline{-8r^5 + 10r^4 + 3r^3 + 2r^2}$

Subtract. See Example 8.

51. $5y^3 - 3y^2$ **52.** $-6t^3 + 4t^2$
     $\underline{2y^3 + 8y^2}$ $\underline{8t^3 - 6t^2}$

53. $12x^4 - x^2 + x$ **54.** $13y^5 - y^3 - 8y^2$
     $\underline{8x^4 + 3x^2 - 3x}$ $\underline{7y^5 + 5y^3 + y^2}$

55. $12m^3 - 8m^2 + 6m + 7$ **56.** $5a^4 - 3a^3 + 2a^2 - a + 6$
     $\underline{-3m^3 + 5m^2 - 2m - 4}$ $\underline{-6a^4 + a^3 - a^2 + a - 1}$

57. After reading Examples 5–8, do you have a preference regarding horizontal or vertical addition and subtraction of polynomials? Explain your answer.

58. Write a paragraph explaining how to add and subtract polynomials. Give an example using addition.

Perform each indicated operation. See Examples 6 and 7.

59. $(8m^2 - 7m) - (3m^2 + 7m - 6)$ **60.** $(x^2 + x) - (3x^2 + 2x - 1)$

61. $(16x^3 - x^2 + 3x) + (-12x^3 + 3x^2 + 2x)$ **62.** $(-2b^6 + 3b^4 - b^2) + (b^6 + 2b^4 + 2b^2)$

63. $(7y^4 + 3y^2 + 2y) - (18y^4 - 5y^2 + y)$ **64.** $(8t^5 + 3t^3 + 5t) - (19t^5 - 6t^3 + t)$

65. $(9a^4 - 3a^2 + 2) + (4a^4 - 4a^2 + 2) + (-12a^4 + 6a^2 - 3)$

66. $(4m^2 - 3m + 2) + (5m^2 + 13m - 4) - (16m^2 + 4m - 3)$

67. $[(8m^2 + 4m - 7) - (2m^2 - 5m + 2)] - (m^2 + m + 1)$

68. $[(9b^3 - 4b^2 + 3b + 2) - (-2b^3 - 3b^2 + b)] - (8b^3 + 6b + 4)$

69. $[(3x^2 - 2x + 7) - (4x^2 + 2x - 3)] - [(9x^2 + 4x - 6) + (-4x^2 + 4x + 4)]$

70. $[(6t^2 - 3t + 1) - (12t^2 + 2t - 6)] - [(4t^2 - 3t - 8) + (-6t^2 + 10t - 12)]$

71. Without actually performing the operations, determine mentally the coefficient of x^2 in the simplified form of $(-4x^2 + 2x - 3) - (-2x^2 + x - 1) + (-8x^2 + 3x - 4)$.

72. Without actually performing the operations, determine mentally the coefficient of x in the simplified form of $(-8x^2 - 3x + 2) - (4x^2 - 3x + 8) - (-2x^2 - x + 7)$.

Add or subtract as indicated. See Example 9.

73. $(6b + 3c) + (-2b - 8c)$

74. $(-5t + 13s) + (8t - 3s)$

75. $(4x + 2xy - 3) - (-2x + 3xy + 4)$

76. $(8ab + 2a - 3b) - (6ab - 2a + 3b)$

77. $(5x^2y - 2xy + 9xy^2) - (8x^2y + 13xy + 12xy^2)$

78. $(16t^3s^2 + 8t^2s^3 + 9ts^4) - (-24t^3s^2 + 3t^2s^3 - 18ts^4)$

For Exercises 79–82, use the formulas found on the inside covers. Find the perimeter of each rectangle.

79.
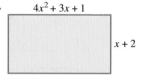
$4x^2 + 3x + 1$

$x + 2$

80.
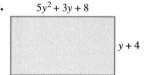
$5y^2 + 3y + 8$

$y + 4$

*Find **(a)** a polynomial representing the perimeter of each triangle and **(b)** the measures of the angles of the triangle.*

81.
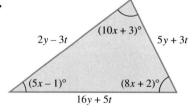
$2y - 3t$ $(10x + 3)°$ $5y + 3t$

$(5x - 1)°$ $(8x + 2)°$

$16y + 5t$

82.
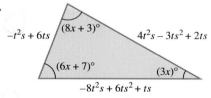
$-t^2s + 6ts$ $(8x + 3)°$ $4t^2s - 3ts^2 + 2ts$

$(6x + 7)°$ $(3x)°$

$-8t^2s + 6ts^2 + ts$

The concepts required to work Exercises 83–86 have been covered, but the usual wording of the problem has been changed. Perform the indicated operations.

83. Subtract $9x^2 - 6x + 5$ from $3x^2 - 2$.

84. Find the difference when $9x^4 + 3x^2 + 5$ is subtracted from $8x^4 - 2x^3 + x - 1$.

85. Find the difference between the sum of $5x^2 + 2x - 3$ and $x^2 - 8x + 2$ and the sum of $7x^2 - 3x + 6$ and $-x^2 + 4x - 6$.

86. Subtract the sum of $9t^3 - 3t + 8$ and $t^2 - 8t + 4$ from the sum of $12t + 8$ and $t^2 - 10t + 3$.

Graph each equation by completing the table of values. See Example 10.

87. $y = x^2 - 4$

x	y
-2	
-1	
0	
1	
2	

88. $y = x^2 - 6$

x	y
-2	
-1	
0	
1	
2	

89. $y = 2x^2 - 1$

x	y
-2	
-1	
0	
1	
2	

90. $y = 2x^2 + 2$

x	y
-2	
-1	
0	
1	
2	

91. $y = -x^2 + 4$

x	y
-2	
-1	
0	
1	
2	

92. $y = -x^2 + 2$

x	y
-2	
-1	
0	
1	
2	

93. $y = (x + 3)^2$

x	-5	-4	-3	-2	-1
y					

94. $y = (x - 4)^2$

x	2	3	4	5	6
y					

In the introduction to this chapter, we gave a polynomial that models the distance in feet that a car going approximately 68 mph will skid in x seconds. If we let D represent this distance, then

$$D = 100x - 13x^2.$$

Each time we evaluate this polynomial for a value of x, we get one and only one output value D. This idea is basic to the concept of a function, introduced in Section 3.1. Exercises 95 and 96 illustrate the idea with this polynomial.

95. Use the given polynomial equation to approximate the skidding distance in feet if $x = 5$ sec.

96. Use the given polynomial equation to find the distance the car will skid in 1 sec. Write an ordered pair of the form (x, D).

RELATING CONCEPTS (EXERCISES 97–100)

For Individual or Group Work

As explained earlier in this section, the polynomial equation

$$y = -10.25x^2 - 126.04x + 5730.21$$

gives a good approximation of NASA's budget for space station research, in millions of dollars, for 1996 through 2001, where $x = 0$ represents 1996, $x = 1$ represents

(continued)

1997, and so on. If we evaluate the polynomial for a specific input value x, we will get one and only one output value y as a result. This idea is basic to the study of functions, one of the most important concepts in mathematics. **Work Exercises 97–100 in order.**

97. If gasoline costs \$1.25 per gallon, then the monomial $1.25x$ gives the cost of x gallons. Evaluate this monomial for 4, and then use the result to fill in the blanks: If _____ gallons are purchased, then the cost is _____.

98. If it costs \$15 to rent a chain saw plus \$2 per day, the binomial $2x + 15$ gives the cost to rent the chain saw for x days. Evaluate this polynomial for 6 and then use the result to fill in the blanks: If the saw is rented for _____ days, then the cost is _____.

99. If an object is thrown upward under certain conditions, its height in feet is given by the trinomial $-16x^2 + 60x + 80$, where x is in seconds. Evaluate this polynomial for 2.5 and then use the result to fill in the blanks: If _____ seconds have elapsed, then the height of the object is _____ feet.

100. The polynomial $2.69x^2 + 4.75x + 452.43$ gives a good approximation for the number of revenue passenger miles, in billions, for the U.S. airline industry during the period from 1990 to 1995, where $x = 0$ represents 1990. Use this polynomial to approximate the number of revenue passenger miles in 1991. (*Hint:* Any power of 1 is equal to 1, so simply add the coefficients and the constant.) (*Source:* Air Transportation Association of America.)

4.5 Multiplying Polynomials

OBJECTIVES

1 Multiply a monomial and a polynomial.

2 Multiply two polynomials.

3 Multiply binomials by the FOIL method.

OBJECTIVE 1 Multiply a monomial and a polynomial.   As shown earlier, we find the product of two monomials by using the rules for exponents and the commutative and associative properties. For example,

$$(-8m^6)(-9n^6) = (-8)(-9)(m^6)(n^6) = 72m^6n^6.$$

CAUTION Do not confuse *addition* of terms with *multiplication* of terms. For example,

$$7q^5 + 2q^5 = 9q^5, \quad \text{but} \quad (7q^5)(2q^5) = 7 \cdot 2q^{5+5} = 14q^{10}.$$

To find the product of a monomial and a polynomial with more than one term, we use the distributive property and multiplication of monomials.

EXAMPLE 1 Multiplying Monomials and Polynomials

Use the distributive property to find each product.

(a) $4x^2(3x + 5)$

$$4x^2(3x + 5) = 4x^2(3x) + 4x^2(5) \qquad \text{Distributive property}$$
$$= 12x^3 + 20x^2 \qquad \text{Multiply monomials.}$$

(b) $-8m^3(4m^3 + 3m^2 + 2m - 1)$
$$= -8m^3(4m^3) + (-8m^3)(3m^2)$$
$$+ (-8m^3)(2m) + (-8m^3)(-1) \qquad \text{Distributive property}$$
$$= -32m^6 - 24m^5 - 16m^4 + 8m^3 \qquad \text{Multiply monomials.}$$

Now Try Exercises 11 and 19.

OBJECTIVE 2 Multiply two polynomials. We use the distributive property repeatedly to find the product of any two polynomials. For example, to find the product of the polynomials $x^2 + 3x + 5$ and $x - 4$, think of $x - 4$ as a single quantity and use the distributive property as follows.

$$(x^2 + 3x + 5)(x - 4) = x^2(x - 4) + 3x(x - 4) + 5(x - 4)$$

Now use the distributive property three times to find $x^2(x - 4)$, $3x(x - 4)$, and $5(x - 4)$.

$$x^2(x - 4) + 3x(x - 4) + 5(x - 4)$$
$$= x^2(x) + x^2(-4) + 3x(x) + 3x(-4) + 5(x) + 5(-4)$$
$$= x^3 - 4x^2 + 3x^2 - 12x + 5x - 20 \qquad \text{Multiply monomials.}$$
$$= x^3 - x^2 - 7x - 20 \qquad \text{Combine like terms.}$$

This example suggests the following rule.

Multiplying Polynomials

To multiply two polynomials, multiply each term of the second polynomial by each term of the first polynomial and add the products.

EXAMPLE 2 Multiplying Two Polynomials

Multiply $(m^2 + 5)(4m^3 - 2m^2 + 4m)$.

Multiply each term of the second polynomial by each term of the first.
$$(m^2 + 5)(4m^3 - 2m^2 + 4m)$$
$$= m^2(4m^3) + m^2(-2m^2) + m^2(4m) + 5(4m^3) + 5(-2m^2) + 5(4m)$$
$$= 4m^5 - 2m^4 + 4m^3 + 20m^3 - 10m^2 + 20m$$
$$= 4m^5 - 2m^4 + 24m^3 - 10m^2 + 20m \qquad \text{Combine like terms.}$$

Now Try Exercise 25.

When at least one of the factors in a product of polynomials has three or more terms, the multiplication can be simplified by writing one polynomial above the other vertically.

EXAMPLE 3   Multiplying Polynomials Vertically

Multiply $(x^3 + 2x^2 + 4x + 1)(3x + 5)$ using the vertical method.

Write the polynomials as follows.

$$\begin{array}{r} x^3 + 2x^2 + 4x + 1 \\ 3x + 5 \\ \hline \end{array}$$

It is not necessary to line up terms in columns, because any terms may be multiplied (not just like terms). Begin by multiplying each of the terms in the top row by 5.

$$\begin{array}{r} x^3 + 2x^2 + 4x + 1 \\ 3x + 5 \\ \hline 5x^3 + 10x^2 + 20x + 5 \end{array} \qquad 5(x^3 + 2x^2 + 4x + 1)$$

Notice how this process is similar to multiplication of whole numbers. Now multiply each term in the top row by $3x$. Be careful to place like terms in columns, since the final step will involve addition (as in multiplying two whole numbers).

$$\begin{array}{r} x^3 + 2x^2 + 4x + 1 \\ 3x + 5 \\ \hline 5x^3 + 10x^2 + 20x + 5 \\ 3x^4 + 6x^3 + 12x^2 + 3x \end{array} \qquad 3x(x^3 + 2x^2 + 4x + 1)$$

Add like terms.

$$\begin{array}{r} x^3 + 2x^2 + 4x + 1 \\ 3x + 5 \\ \hline 5x^3 + 10x^2 + 20x + 5 \\ 3x^4 + 6x^3 + 12x^2 + 3x \\ \hline 3x^4 + 11x^3 + 22x^2 + 23x + 5 \end{array}$$

The product is $3x^4 + 11x^3 + 22x^2 + 23x + 5$.

Now Try Exercise 29.

EXAMPLE 4   Multiplying Polynomials with Fractional Coefficients Vertically

Find the product of $4m^3 - 2m^2 + 4m$ and $\frac{1}{2}m^2 + \frac{5}{2}$.

$$\begin{array}{r} 4m^3 - 2m^2 + 4m \\ \frac{1}{2}m^2 + \frac{5}{2} \\ \hline 10m^3 - 5m^2 + 10m \\ 2m^5 - m^4 + 2m^3 \\ \hline 2m^5 - m^4 + 12m^3 - 5m^2 + 10m \end{array}$$

Terms of top row multiplied by $\frac{5}{2}$
Terms of top row multiplied by $\frac{1}{2}m^2$
Add.

Now Try Exercise 35.

We can use a rectangle to model polynomial multiplication. For example, to find the product

$$(2x + 1)(3x + 2),$$

label a rectangle with each term as shown here.

	$3x$	2
$2x$		
1		

Now put the product of each pair of monomials in the appropriate box.

	$3x$	2
$2x$	$6x^2$	$4x$
1	$3x$	2

The product of the original binomials is the sum of these four monomial products.

$$(2x + 1)(3x + 2) = 6x^2 + 4x + 3x + 2$$
$$= 6x^2 + 7x + 2$$

This approach can be extended to polynomials with any number of terms.

OBJECTIVE 3   Multiply binomials by the FOIL method.   In algebra, many of the polynomials to be multiplied are both binomials (with just two terms). For these products, the **FOIL method** reduces the rectangle method to a systematic approach without the rectangle. To develop the FOIL method, we use the distributive property to find $(x + 3)(x + 5)$.

$$(x + 3)(x + 5) = (x + 3)x + (x + 3)5$$
$$= x(x) + 3(x) + x(5) + 3(5)$$
$$= x^2 + 3x + 5x + 15$$
$$= x^2 + 8x + 15$$

Here is where the letters of the word FOIL originate.

$(x + 3)(x + 5)$ Multiply the **First** terms: $x(x)$.   F

$(x + 3)(x + 5)$ Multiply the **Outer** terms: $x(5)$.   O
                        This is the **outer product.**

$(x + 3)(x + 5)$ Multiply the **Inner** terms: $3(x)$.   I
                        This is the **inner product.**

$(x + 3)(x + 5)$ Multiply the **Last** terms: $3(5)$.   L

The inner product and the outer product should be added mentally so that the three terms of the answer can be written without extra steps as

$$(x + 3)(x + 5) = x^2 + 8x + 15.$$

A summary of the steps in the FOIL method follows.

Multiplying Binomials by the FOIL Method

Step 1 Multiply the two **F**irst terms of the binomials to get the first term of the answer.

Step 2 Find the **O**uter product and the **I**nner product and add them (when possible) to get the middle term of the answer.

Step 3 Multiply the two **L**ast terms of the binomials to get the last term of the answer.

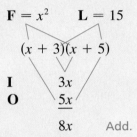

▨ EXAMPLE 5 Using the FOIL Method

Use the FOIL method to find the product $(x + 8)(x - 6)$.

Step 1 **F** Multiply the first terms.

$$x(x) = x^2$$

Step 2 **O** Find the outer product.

$$x(-6) = -6x$$

     I Find the inner product.

$$8(x) = 8x$$

Add the outer and inner products mentally.

$$-6x + 8x = 2x$$

Step 3 **L** Multiply the last terms.

$$8(-6) = -48$$

The product of $x + 8$ and $x - 6$ is the sum of the four terms found in three steps above, so

$$(x + 8)(x - 6) = x^2 + 2x - 48.$$

As a shortcut, this product can be found in the following manner.

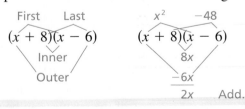

Now Try Exercise 37.

It is not possible to add the inner and outer products of the FOIL method if unlike terms result, as shown in the next example.

EXAMPLE 6   Using the FOIL Method

Multiply $(9x - 2)(3y + 1)$.

First	$(9x - 2)(3y + 1)$	$27xy$
Outer	$(9x - 2)(3y + 1)$	$9x$ ←
Inner	$(9x - 2)(3y + 1)$	$-6y$ ←
Last	$(9x - 2)(3y + 1)$	-2

Unlike terms

$$\begin{array}{cccc} \text{F} & \text{O} & \text{I} & \text{L} \end{array}$$

$$(9x - 2)(3y + 1) = 27xy + 9x - 6y - 2$$

Now Try Exercise 51.

EXAMPLE 7   Using the FOIL Method

Find each product.

$$\begin{array}{cccc} \text{F} & \text{O} & \text{I} & \text{L} \end{array}$$

(a) $(2k + 5y)(k + 3y) = 2k(k) + 2k(3y) + 5y(k) + 5y(3y)$
$$= 2k^2 + 6ky + 5ky + 15y^2$$
$$= 2k^2 + 11ky + 15y^2$$

(b) $(7p + 2q)(3p - q) = 21p^2 - pq - 2q^2$ FOIL

(c) $2x^2(x - 3)(3x + 4) = 2x^2(3x^2 - 5x - 12)$ FOIL
$$= 6x^4 - 10x^3 - 24x^2$$ Distributive property

Now Try Exercises 53 and 55.

NOTE   Example 7(c) showed one way to multiply three polynomials. We could have multiplied $2x^2$ and $x - 3$ first, then multiplied that product and $3x + 4$ as follows.

$$2x^2(x - 3)(3x + 4) = (2x^3 - 6x^2)(3x + 4)$$
$$= 6x^4 - 10x^3 - 24x^2$$ FOIL

4.5   EXERCISES

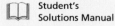

1. Match each product in Column I with the correct monomial in Column II.

I	II
(a) $(5x^3)(6x^5)$	**A.** $125x^{15}$
(b) $(-5x^5)(6x^3)$	**B.** $30x^8$
(c) $(5x^5)^3$	**C.** $-216x^9$
(d) $(-6x^3)^3$	**D.** $-30x^8$

2. Match each product in Column I with the correct polynomial in Column II.

I	II
(a) $(x - 5)(x + 3)$	**A.** $x^2 + 8x + 15$
(b) $(x + 5)(x + 3)$	**B.** $x^2 - 8x + 15$
(c) $(x - 5)(x - 3)$	**C.** $x^2 - 2x - 15$
(d) $(x + 5)(x - 3)$	**D.** $x^2 + 2x - 15$

Find each product. Use the rules for exponents discussed earlier in the chapter.

3. $(5y^4)(3y^7)$

4. $(10p^2)(5p^3)$

5. $(-15a^4)(-2a^5)$

6. $(-3m^6)(-5m^4)$

7. $(5p)(3q^2)$

8. $(4a^3)(3b^2)$

9. $(-6m^3)(3n^2)$

10. $(9r^3)(-2s^2)$

Find each product. See Example 1.

11. $2m(3m + 2)$

12. $4x(5x + 3)$

13. $3p(-2p^3 + 4p^2)$

14. $4x(3 + 2x + 5x^3)$

15. $-8z(2z + 3z^2 + 3z^3)$

16. $-7y(3 + 5y^2 - 2y^3)$

17. $2y^3(3 + 2y + 5y^4)$

18. $2m^4(3m^2 + 5m + 6)$

19. $-4r^3(-7r^2 + 8r - 9)$

20. $-9a^5(-3a^6 - 2a^4 + 8a^2)$

21. $3a^2(2a^2 - 4ab + 5b^2)$

22. $4z^3(8z^2 + 5zy - 3y^2)$

23. $7m^3n^2(3m^2 + 2mn - n^3)$

24. $2p^2q(3p^2q^2 - 5p + 2q^2)$

Find each product. See Examples 2–4.

25. $(6x + 1)(2x^2 + 4x + 1)$

26. $(9y - 2)(8y^2 - 6y + 1)$

27. $(9a + 2)(9a^2 + a + 1)$

28. $(2r - 1)(3r^2 + 4r - 4)$

29. $(4m + 3)(5m^3 - 4m^2 + m - 5)$

30. $(y + 4)(3y^4 - 2y^2 + 1)$

31. $(2x - 1)(3x^5 - 2x^3 + x^2 - 2x + 3)$

32. $(2a + 3)(a^4 - a^3 + a^2 - a + 1)$

33. $(5x^2 + 2x + 1)(x^2 - 3x + 5)$

34. $(2m^2 + m - 3)(m^2 - 4m + 5)$

35. $(6x^4 - 4x^2 + 8x)\left(\dfrac{1}{2}x + 3\right)$

36. $(8y^6 + 4y^4 - 12y^2)\left(\dfrac{3}{4}y^2 + 2\right)$

Find each product. Use the FOIL method. See Examples 5–7.

37. $(m + 7)(m + 5)$

38. $(n - 1)(n + 4)$

39. $(x + 5)(x - 5)$

40. $(y + 8)(y - 8)$

41. $(2x + 3)(6x - 4)$

42. $(4m + 3)(4m + 3)$

43. $(3x - 2)(3x - 2)$

44. $(b + 8)(6b - 2)$

45. $(5a + 1)(2a + 7)$

46. $(8 - 3a)(2 + a)$

47. $(6 - 5m)(2 + 3m)$

48. $(-4 + k)(2 - k)$

49. $(5 - 3x)(4 + x)$

50. $(2m - 3n)(m + 5n)$

51. $(4x + 3)(2y - 1)$

52. $(5x + 7)(3y - 8)$

53. $(3x + 2y)(5x - 3y)$

54. $x(2x - 5)(x + 3)$

55. $3y^3(2y + 3)(y - 5)$

56. $5t^4(t + 3)(3t - 1)$

57. $-8r^3(5r^2 + 2)(5r^2 - 2)$

58. Find a polynomial that represents the area of this square.

$6x + 2$

59. Find a polynomial that represents the area of this rectangle.

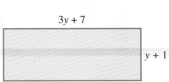

$3y + 7$

$y + 1$

60. Perform the indicated multiplications, and then describe the pattern that you observe in the products.

(a) $(x + 4)(x - 4)$; $(y + 2)(y - 2)$; $(r + 7)(r - 7)$
(b) $(x + 4)(x + 4)$; $(y - 2)(y - 2)$; $(r + 7)(r + 7)$

Find each product. In Exercises 71–74 and 78–80, apply the meaning of exponents.

61. $\left(3p + \dfrac{5}{4}q\right)\left(2p - \dfrac{5}{3}q\right)$

62. $\left(-x + \dfrac{2}{3}y\right)\left(3x - \dfrac{3}{4}y\right)$

63. $(x + 7)^2$

64. $(m + 6)^2$

65. $(a - 4)(a + 4)$

66. $(b - 10)(b + 10)$

67. $(2p - 5)^2$

68. $(3m + 1)^2$

69. $(5k + 3q)^2$

70. $(8m - 3n)^2$

71. $(m - 5)^3$

72. $(p + 3)^3$

73. $(2a + 1)^3$

74. $(3m - 1)^3$

75. $7(4m - 3)(2m + 1)$

76. $-4r(3r + 2)(2r - 5)$

77. $-3a(3a + 1)(a - 4)$

78. $(k + 1)^4$

79. $(3r - 2s)^4$

80. $(2z + 5y)^4$

81. $3p^3(2p^2 + 5p)(p^3 + 2p + 1)$

82. $5k^2(k^2 - k + 4)(k^3 - 3)$

83. $-2x^5(3x^2 + 2x - 5)(4x + 2)$

84. $-4x^3(3x^4 + 2x^2 - x)(-2x + 1)$

Find a polynomial that represents the area of each shaded region. In Exercises 87 and 88 leave π in your answer. Use the formulas found on the inside covers.

85.

86.

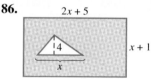

87.

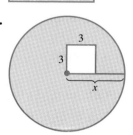

88.

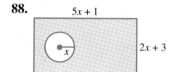

RELATING CONCEPTS (EXERCISES 89–96)

For Individual or Group Work

Work Exercises 89–96 in order. *Refer to the figure as necessary.*

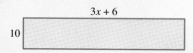

89. Find a polynomial that represents the area of the rectangle.

(continued)

90. Suppose you know that the area of the rectangle is 600 yd^2. Use this information and the polynomial from Exercise 89 to write an equation that allows you to solve for x.

91. Solve for x.

92. What are the dimensions of the rectangle (assume units are all in yards)?

93. Suppose the rectangle represents a lawn and it costs $3.50 per yd^2 to lay sod on the lawn. How much will it cost to sod the entire lawn?

94. Use the result of Exercise 92 to find the perimeter of the lawn.

95. Again, suppose the rectangle represents a lawn and it costs $9.00 per yd to fence the lawn. How much will it cost to fence the lawn?

96. (a) Suppose that it costs k dollars per yd^2 to sod the lawn. Determine a polynomial in the variables x and k that represents the cost to sod the entire lawn.

(b) Suppose that it costs r dollars per yd to fence the lawn. Determine a polynomial in the variables x and r that represents the cost to fence the lawn.

97. Explain the FOIL method for multiplying two binomials. Give an example.

98. Why does the FOIL method not apply to the product of a binomial and a trinomial? Give an example.

4.6 Special Products

In this section, we develop shortcuts to find certain binomial products that occur frequently.

OBJECTIVES

1 Square binomials.

2 Find the product of the sum and difference of two terms.

3 Find higher powers of binomials.

OBJECTIVE 1 Square binomials. The square of a binomial can be found quickly by using the method suggested by Example 1.

EXAMPLE 1 Squaring a Binomial

Find $(m + 3)^2$.

Squaring $m + 3$ by the FOIL method gives

$$(m + 3)(m + 3) = m^2 + 3m + 3m + 9$$
$$= m^2 + 6m + 9.$$

Now Try Exercise 1.

The result above has the squares of the first and the last terms of the binomial:

$$m^2 = m^2 \quad \text{and} \quad 3^2 = 9.$$

The middle term is twice the product of the two terms of the binomial, since the outer and inner products are $m(3)$ and $3(m)$, and

$$m(3) + 3(m) = 2(m)(3) = 6m.$$

This example suggests the following rules.

Square of a Binomial

The square of a binomial is a trinomial consisting of the square of the first term of the binomial, plus twice the product of the two terms, plus the square of the last term of the binomial. For x and y,

$$(x + y)^2 = x^2 + 2xy + y^2.$$

Also,
$$(x - y)^2 = x^2 - 2xy + y^2.$$

EXAMPLE 2 Squaring Binomials

Use the rules to square each binomial.

$$(x - y)^2 = x^2 - 2 \cdot x \cdot y + y^2$$

(a) $(5z - 1)^2 = (5z)^2 - 2(5z)(1) + (1)^2$
$$= 25z^2 - 10z + 1 \qquad (5z)^2 = 5^2z^2 = 25z^2$$

(b) $(3b + 5r)^2 = (3b)^2 + 2(3b)(5r) + (5r)^2$
$$= 9b^2 + 30br + 25r^2$$

(c) $(2a - 9x)^2 = 4a^2 - 36ax + 81x^2$

(d) $\left(4m + \dfrac{1}{2}\right)^2 = (4m)^2 + 2(4m)\left(\dfrac{1}{2}\right) + \left(\dfrac{1}{2}\right)^2$

$$= 16m^2 + 4m + \dfrac{1}{4}$$

Now Try Exercises 5, 7, 9, and 15.

Notice that in the square of a sum all of the terms are positive, as in Examples 2(b) and (d). In the square of a difference, the middle term is negative, as in Examples 2(a) and (c).

CAUTION A common error when squaring a binomial is to forget the middle term of the product. In general,

$$(x + y)^2 \neq x^2 + y^2.$$

OBJECTIVE 2 Find the product of the sum and difference of two terms. Binomial products of the form $(x + y)(x - y)$ also occur frequently. In these products, one binomial is the sum of two terms, and the other is the difference of the same two terms. For example, the product of $x + 2$ and $x - 2$ is

$$(x + 2)(x - 2) = x^2 - 2x + 2x - 4$$
$$= x^2 - 4.$$

As this example suggests, the product of $x + y$ and $x - y$ is the difference between two squares.

Product of the Sum and Difference of Two Terms

$$(x + y)(x - y) = x^2 - y^2$$

EXAMPLE 3 Finding the Product of the Sum and Difference of Two Terms

Find each product.

(a) $(x + 4)(x - 4)$

Use the rule for the product of the sum and difference of two terms.

$$(x + 4)(x - 4) = x^2 - 4^2$$
$$= x^2 - 16$$

(b) $\left(\dfrac{2}{3} - w\right)\left(\dfrac{2}{3} + w\right)$

By the commutative property, this product is the same as $\left(\frac{2}{3} + w\right)\left(\frac{2}{3} - w\right)$.

$$\left(\frac{2}{3} - w\right)\left(\frac{2}{3} + w\right) = \left(\frac{2}{3} + w\right)\left(\frac{2}{3} - w\right)$$

$$= \left(\frac{2}{3}\right)^2 - w^2$$

$$= \frac{4}{9} - w^2$$

Now Try Exercises 23 and 37.

EXAMPLE 4 Finding the Product of the Sum and Difference of Two Terms

Find each product.

$$\begin{array}{cccc} (x & + y) & (x & - y) \\ \downarrow & \downarrow & \downarrow & \downarrow \end{array}$$

(a) $(5m + 3)(5m - 3)$

Use the rule for the product of the sum and difference of two terms.

$$(5m + 3)(5m - 3) = (5m)^2 - 3^2$$
$$= 25m^2 - 9$$

(b) $(4x + y)(4x - y) = (4x)^2 - y^2$
$$= 16x^2 - y^2$$

(c) $\left(z - \dfrac{1}{4}\right)\left(z + \dfrac{1}{4}\right) = z^2 - \dfrac{1}{16}$

Now Try Exercises 31 and 33.

The product rules of this section will be important later, particularly in Chapters 5 and 6. Therefore, it is essential to learn these rules and practice using them.

 OBJECTIVE 3 Find higher powers of binomials. The methods used in the previous section and this section can be combined to find higher powers of binomials.

EXAMPLE 5 Finding Higher Powers of Binomials

Find each product.

(a) $(x + 5)^3 = (x + 5)^2(x + 5)$ $a^3 = a^2 \cdot a$

$= (x^2 + 10x + 25)(x + 5)$ Square the binomial.

$= x^3 + 10x^2 + 25x + 5x^2 + 50x + 125$ Multiply polynomials.

$= x^3 + 15x^2 + 75x + 125$ Combine like terms.

(b) $(2y - 3)^4 = (2y - 3)^2(2y - 3)^2$ $a^4 = a^2 \cdot a^2$

$= (4y^2 - 12y + 9)(4y^2 - 12y + 9)$ Square each binomial.

$= 16y^4 - 48y^3 + 36y^2 - 48y^3 + 144y^2$ Multiply polynomials.

$- 108y + 36y^2 - 108y + 81$

$= 16y^4 - 96y^3 + 216y^2 - 216y + 81$ Combine like terms.

Now Try Exercises 41 and 49.

4.6 EXERCISES

For Extra Help

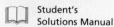
Student's
Solutions Manual

MyMathLab

InterAct Math
Tutorial Software

AW Math
Tutor Center

MathXL

Digital Video Tutor
CD 7/Videotape 7

1. Consider the square $(2x + 3)^2$.

    (a) What is the square of the first term, $(2x)^2$?

    (b) What is twice the product of the two terms, $2(2x)(3)$?

    (c) What is the square of the last term, 3^2?

    (d) Write the final product, which is a trinomial, using your results in parts (a)–(c).

2. Explain in your own words how to square a binomial. Give an example.

Find each product. See Examples 1 and 2.

3. $(m + 2)^2$ **4.** $(x + 8)^2$ **5.** $(r - 3)^2$

6. $(z - 5)^2$ **7.** $(x + 2y)^2$ **8.** $(3m - p)^2$

9. $(5p + 2q)^2$ **10.** $(8a - 3b)^2$ **11.** $(4a + 5b)^2$

12. $(9y + z)^2$ **13.** $(7t + s)^2$ **14.** $\left(5x + \dfrac{2}{5}y \right)^2$

15. $\left(6m - \dfrac{4}{5}n \right)^2$ **16.** $x(2x + 5)^2$ **17.** $t(3t - 1)^2$

18. $-(4r - 2)^2$ **19.** $-(3y - 8)^2$

20. Consider the product $(7x + 3y)(7x - 3y)$.

    (a) What is the product of the first terms, $(7x)(7x)$?

    (b) Multiply the outer terms, $(7x)(-3y)$. Then multiply the inner terms, $(3y)(7x)$. Add the results. What is this sum?

    (c) What is the product of the last terms, $(3y)(-3y)$?

    (d) Write the complete product using your answers in parts (a) and (c). Why is the sum found in part (b) omitted here?

21. Explain in your own words how to find the product of the sum and the difference of two terms. Give an example.

22. The square of a binomial leads to a polynomial with how many terms? The product of the sum and difference of two terms leads to a polynomial with how many terms?

Find each product. See Examples 3 and 4.

23. $(a + 8)(a - 8)$

24. $(k + 5)(k - 5)$

25. $(2 + p)(2 - p)$

26. $(4 - 3t)(4 + 3t)$

27. $(2m + 5)(2m - 5)$

28. $(5x + 2)(5x - 2)$

29. $(3x + 4y)(3x - 4y)$

30. $(6a - p)(6a + p)$

31. $(5y + 3x)(5y - 3x)$

32. $(10x + 3y)(10x - 3y)$

33. $(13r + 2z)(13r - 2z)$

34. $(2x^2 - 5)(2x^2 + 5)$

35. $(9y^2 - 2)(9y^2 + 2)$

36. $\left(7x + \dfrac{3}{7}\right)\left(7x - \dfrac{3}{7}\right)$

37. $\left(9y + \dfrac{2}{3}\right)\left(9y - \dfrac{2}{3}\right)$

38. $p(3p + 7)(3p - 7)$

39. $q(5q - 1)(5q + 1)$

40. Does $(a + b)^3$ equal $a^3 + b^3$ in general? Explain.

Find each product. See Example 5.

41. $(x + 1)^3$

42. $(y + 2)^3$

43. $(t - 3)^3$

44. $(m - 5)^3$

45. $(r + 5)^3$

46. $(p + 3)^3$

47. $(2a + 1)^3$

48. $(3m - 1)^3$

49. $(3r - 2t)^4$

50. $(2z + 5y)^4$

RELATING CONCEPTS (EXERCISES 51–60)

For Individual or Group Work

Special products can be illustrated by using areas of rectangles. Use the figure, and **work Exercises 51–56 in order** *to justify the special product*

$$(a + b)^2 = a^2 + 2ab + b^2.$$

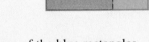

51. Express the area of the large square as the square of a binomial.

52. Give the monomial that represents the area of the red square.

53. Give the monomial that represents the sum of the areas of the blue rectangles.

54. Give the monomial that represents the area of the yellow square.

55. What is the sum of the monomials you obtained in Exercises 52–54?

56. Explain why the binomial square you found in Exercise 51 must equal the polynomial you found in Exercise 55.

To understand how the special product $(a + b)^2 = a^2 + 2ab + b^2$ *can be applied to a purely numerical problem,* **work Exercises 57–60 in order.**

57. Evaluate 35^2 using either traditional paper-and-pencil methods or a calculator.

58. The number 35 can be written as $30 + 5$. Therefore, $35^2 = (30 + 5)^2$. Use the special product for squaring a binomial with $a = 30$ and $b = 5$ to write an expression for $(30 + 5)^2$. Do not simplify at this time.

59. Use the order of operations to simplify the expression you found in Exercise 58.

60. How do the answers in Exercises 57 and 59 compare?

The special product

$$(a + b)(a - b) = a^2 - b^2$$

can be used to perform some multiplication problems. For example,

$$
\begin{aligned}
51 \times 49 &= (50 + 1)(50 - 1) \\
&= 50^2 - 1^2 \\
&= 2500 - 1 \\
&= 2499.
\end{aligned}
\qquad
\begin{aligned}
102 \times 98 &= (100 + 2)(100 - 2) \\
&= 100^2 - 2^2 \\
&= 10{,}000 - 4 \\
&= 9996.
\end{aligned}
$$

Once these patterns are recognized, multiplications of this type can be done mentally. Use this method to calculate each product mentally.

61. 101×99

62. 103×97

63. 201×199

64. 301×299

65. $20\dfrac{1}{2} \times 19\dfrac{1}{2}$

66. $30\dfrac{1}{3} \times 29\dfrac{2}{3}$

Determine a polynomial that represents the area of each figure. Use the formulas found on the inside covers.

67.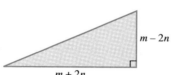

$m - 2n$

$m + 2n$

68.

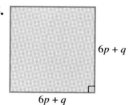

$6p + q$

$6p + q$

69.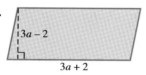

$3a - 2$

$3a + 2$

70.

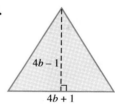

$4b - 1$

$4b + 1$

71.

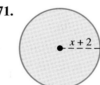

$x + 2$

72.

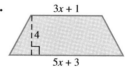

$3x + 1$

4

$5x + 3$

In Exercises 73 and 74, refer to the figure shown here.

73. Find a polynomial that represents the volume of the cube.

74. If the value of x is 6, what is the volume of the cube?

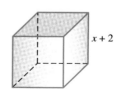

$x + 2$

4.7   Dividing Polynomials

OBJECTIVES

1 Divide a polynomial by a monomial.

2 Divide a polynomial by a polynomial.

OBJECTIVE 1   Divide a polynomial by a monomial.   We add two fractions with a common denominator as follows.

$$\frac{a}{c} + \frac{b}{c} = \frac{a + b}{c}$$

Looking at this statement in reverse gives us a rule for dividing a polynomial by a monomial.

Dividing a Polynomial by a Monomial

To divide a polynomial by a monomial, divide each term of the polynomial by the monomial:

$$\frac{a + b}{c} = \frac{a}{c} + \frac{b}{c} \quad (c \neq 0).$$

Examples: $\dfrac{2 + 5}{3} = \dfrac{2}{3} + \dfrac{5}{3}$ and $\dfrac{x + 3z}{2y} = \dfrac{x}{2y} + \dfrac{3z}{2y}$

The parts of a division problem are named here.

$$\text{Dividend} \rightarrow \quad \frac{12x^2 + 6x}{6x} = 2x + 1 \quad \leftarrow \text{Quotient}$$
$$\text{Divisor} \rightarrow$$

EXAMPLE 1   Dividing a Polynomial by a Monomial

Divide $5m^5 - 10m^3$ by $5m^2$.

Use the preceding rule, with $+$ replaced by $-$. Then use the quotient rule.

$$\frac{5m^5 - 10m^3}{5m^2} = \frac{5m^5}{5m^2} - \frac{10m^3}{5m^2} = m^3 - 2m$$

Check by multiplying: $5m^2(m^3 - 2m) = 5m^5 - 10m^3$.

Because division by 0 is undefined, the quotient

$$\frac{5m^5 - 10m^3}{5m^2}$$

is undefined if $m = 0$. From now on, we assume that no denominators are 0.

Now Try Exercise 15.

EXAMPLE 2   Dividing a Polynomial by a Monomial

Divide $\dfrac{16a^5 - 12a^4 + 8a^2}{4a^3}$.

Divide each term of $16a^5 - 12a^4 + 8a^2$ by $4a^3$.

$$\frac{16a^5 - 12a^4 + 8a^2}{4a^3} = \frac{16a^5}{4a^3} - \frac{12a^4}{4a^3} + \frac{8a^2}{4a^3}$$

$$= 4a^2 - 3a + \frac{2}{a} \qquad \text{Quotient rule}$$

The quotient is not a polynomial because of the expression $\frac{2}{a}$, which has a variable in the denominator. While the sum, difference, and product of two polynomials are always polynomials, the quotient of two polynomials may not be. Again, check by multiplying.

$$4a^3\left(4a^2 - 3a + \frac{2}{a}\right) = 4a^3(4a^2) + 4a^3(-3a) + 4a^3\left(\frac{2}{a}\right)$$

$$= 16a^5 - 12a^4 + 8a^2$$

Now Try Exercise 13.

◼ EXAMPLE 3   Dividing a Polynomial by a Monomial with a Negative Coefficient

Divide $-7x^3 + 12x^4 - 4x$ by $-4x$.

The polynomial should be written in descending powers before dividing. Write it as $12x^4 - 7x^3 - 4x$; then divide by $-4x$.

$$\frac{12x^4 - 7x^3 - 4x}{-4x} = \frac{12x^4}{-4x} + \frac{-7x^3}{-4x} + \frac{-4x}{-4x}$$

$$= -3x^3 + \frac{7x^2}{4} + 1 \quad \text{or} \quad -3x^3 + \frac{7}{4}x^2 + 1$$

Check by multiplying.

Now Try Exercise 23.

CAUTION   In Example 3, notice the quotient $\frac{-4x}{-4x} = 1$. It is a common error to leave this term out of the answer. A check by multiplying will show that the answer $-3x^3 + \frac{7}{4}x^2$ is not correct.

◼ EXAMPLE 4   Dividing a Polynomial by a Monomial

Divide the polynomial

$$180x^4y^{10} - 150x^3y^8 + 120x^2y^6 - 90xy^4 + 100y$$

by the monomial $30xy^2$.

$$\frac{180x^4y^{10} - 150x^3y^8 + 120x^2y^6 - 90xy^4 + 100y}{30xy^2}$$

$$= \frac{180x^4y^{10}}{30xy^2} - \frac{150x^3y^8}{30xy^2} + \frac{120x^2y^6}{30xy^2} - \frac{90xy^4}{30xy^2} + \frac{100y}{30xy^2}$$

$$= 6x^3y^8 - 5x^2y^6 + 4xy^4 - 3y^2 + \frac{10}{3xy}$$

Now Try Exercise 31.

OBJECTIVE 2 Divide a polynomial by a polynomial. We use a method of "long division" to divide a polynomial by a polynomial (other than a monomial). This method is similar to the method of long division used for two whole numbers. For comparison, the division of whole numbers is shown alongside the division of polynomials. Both polynomials must first be written in descending powers.

Dividing Whole Numbers	Dividing Polynomials
Step 1	
Divide 6696 by 27.	Divide $8x^3 - 4x^2 - 14x + 15$ by $2x + 3$.
$27\overline{)6696}$	$2x + 3\overline{)8x^3 - 4x^2 - 14x + 15}$
Step 2	
66 divided by $27 = 2$; $2 \cdot 27 = 54$.	$8x^3$ divided by $2x = 4x^2$; $4x^2(2x + 3) = 8x^3 + 12x^2$.
$\begin{array}{r} 2 \\ 27\overline{)6696} \\ 54 \end{array}$	$\begin{array}{r} 4x^2 \\ 2x + 3\overline{)8x^3 - 4x^2 - 14x + 15} \\ 8x^3 + 12x^2 \end{array}$
Step 3	
Subtract; then bring down the next digit.	Subtract; then bring down the next term.
$\begin{array}{r} 2 \\ 27\overline{)6696} \\ 54\downarrow \\ \overline{129} \end{array}$	$\begin{array}{r} 4x^2 \\ 2x + 3\overline{)8x^3 - 4x^2 - 14x + 15} \\ 8x^3 + 12x^2 \quad\downarrow \\ \overline{-16x^2 - 14x} \end{array}$
	(To subtract two polynomials, change the signs of the second and then add.)
Step 4	
129 divided by $27 = 4$; $4 \cdot 27 = 108$.	$-16x^2$ divided by $2x = -8x$; $-8x(2x + 3) = -16x^2 - 24x$.
$\begin{array}{r} 24 \\ 27\overline{)6696} \\ 54 \\ \overline{129} \\ 108 \end{array}$	$\begin{array}{r} 4x^2 - 8x \\ 2x + 3\overline{)8x^3 - 4x^2 - 14x + 15} \\ 8x^3 + 12x^2 \\ \overline{-16x^2 - 14x} \\ -16x^2 - 24x \end{array}$
Step 5	
Subtract; then bring down the next digit.	Subtract; then bring down the next term.
$\begin{array}{r} 24 \\ 27\overline{)6696} \\ 54 \\ \overline{129} \\ 108\downarrow \\ \overline{216} \end{array}$	$\begin{array}{r} 4x^2 - 8x \\ 2x + 3\overline{)8x^3 - 4x^2 - 14x + 15} \\ 8x^3 + 12x^2 \\ \overline{-16x^2 - 14x} \\ -16x^2 - 24x \quad\downarrow \\ \overline{10x + 15} \end{array}$

Step 6

216 divided by $27 = \mathbf{8}$;
$8 \cdot 27 = 216$.

$$
\begin{array}{r}
248 \\
27\overline{)6696} \\
54 \\
\hline
129 \\
108 \\
\hline
216 \\
216 \\
\hline
0
\end{array}
$$

6696 divided by 27 is 248.
There is no remainder.

Step 7

Check by multiplying.
$$27 \cdot 248 = 6696$$

$10x$ divided by $2x = \mathbf{5}$;
$5(2x + 3) = 10x + 15$.

$$
\begin{array}{r}
4x^2 - 8x + 5 \\
2x + 3\overline{)8x^3 - 4x^2 - 14x + 15} \\
8x^3 + 12x^2 \\
\hline
-16x^2 - 14x \\
-16x^2 - 24x \\
\hline
10x + 15 \\
10x + 15 \\
\hline
0
\end{array}
$$

$8x^3 - 4x^2 - 14x + 15$ divided by
$2x + 3$ is $4x^2 - 8x + 5$. There is no
remainder.

Check by multiplying.
$$(2x + 3)(4x^2 - 8x + 5)$$
$$= 8x^3 - 4x^2 - 14x + 15$$

Now Try Exercise 53.

EXAMPLE 5 Dividing a Polynomial by a Polynomial

Divide $\dfrac{5x + 4x^3 - 8 - 4x^2}{2x - 1}$.

 The first polynomial must be written with the exponents in descending order as
$4x^3 - 4x^2 + 5x - 8$. Then begin the division process.
 Divide $4x^3 - 4x^2 + 5x - 8$ by $2x - 1$.

$$
\begin{array}{r}
2x^2 - x + 2 \\
2x - 1\overline{)4x^3 - 4x^2 + 5x - 8} \\
4x^3 - 2x^2 \\
\hline
-2x^2 + 5x \\
-2x^2 + x \\
\hline
4x - 8 \\
4x - 2 \\
\hline
-6 \quad \leftarrow \text{Remainder}
\end{array}
$$

Step 1 $4x^3$ divided by $2x = 2x^2$; $2x^2(2x - 1) = 4x^3 - 2x^2$.

Step 2 Subtract; bring down the next term.

Step 3 $-2x^2$ divided by $2x = -x$; $-x(2x - 1) = -2x^2 + x$.

Step 4 Subtract; bring down the next term.

Step 5 $4x$ divided by $2x = 2$; $2(2x - 1) = 4x - 2$.

Step 6   Subtract. The remainder is -6. Thus $4x^3 - 4x^2 + 5x - 8$ divided by $2x - 1$ has a quotient of $2x^2 - x + 2$ and a remainder of -6. Write the remainder as the numerator of a fraction that has $2x - 1$ as its denominator. The answer is not a polynomial because of the remainder.

$$\frac{4x^3 - 4x^2 + 5x - 8}{2x - 1} = 2x^2 - x + 2 + \frac{-6}{2x - 1}$$

Step 7   Check by multiplying.

$$(2x - 1)\left(2x^2 - x + 2 + \frac{-6}{2x - 1}\right)$$

$$= (2x - 1)(2x^2) + (2x - 1)(-x) + (2x - 1)(2) + (2x - 1)\left(\frac{-6}{2x - 1}\right)$$

$$= 4x^3 - 2x^2 - 2x^2 + x + 4x - 2 - 6$$

$$= 4x^3 - 4x^2 + 5x - 8$$

Now Try Exercise 59.

EXAMPLE 6   Dividing into a Polynomial with Missing Terms

Divide $x^3 - 1$ by $x - 1$.

Here the polynomial $x^3 - 1$ is missing the x^2-term and the x-term. When terms are missing, use 0 as the coefficient for each missing term. (Zero acts as a placeholder here, just as it does in our number system.)

$$x^3 - 1 = x^3 + 0x^2 + 0x - 1$$

Now divide.

$$
\begin{array}{r}
x^2 + x + 1 \\
x - 1{\overline{\smash{\big)}\,x^3 + 0x^2 + 0x - 1}} \\
\underline{x^3 - x^2} \\
x^2 + 0x \\
\underline{x^2 - x} \\
x - 1 \\
\underline{x - 1} \\
0
\end{array}
$$

The remainder is 0. The quotient is $x^2 + x + 1$. Check by multiplying.

$$(x^2 + x + 1)(x - 1) = x^3 - 1$$

Now Try Exercise 65.

EXAMPLE 7   Dividing by a Polynomial with Missing Terms

Divide $x^4 + 2x^3 + 2x^2 - x - 1$ by $x^2 + 1$.

Since $x^2 + 1$ has a missing x-term, write it as $x^2 + 0x + 1$. Then go through the division process as follows.

$$
\begin{array}{r}
x^2 + 2x + 1 \\
x^2 + 0x + 1{\overline{\smash{\big)}\,x^4 + 2x^3 + 2x^2 - x - 1}} \\
\underline{x^4 + 0x^3 + x^2} \\
2x^3 + x^2 - x \\
\underline{2x^3 + 0x^2 + 2x} \\
x^2 - 3x - 1 \\
\underline{x^2 + 0x + 1} \\
-3x - 2 \;\; \leftarrow \text{Remainder}
\end{array}
$$

When the result of subtracting ($-3x - 2$, in this case) is a polynomial of smaller degree than the divisor ($x^2 + 0x + 1$), that polynomial is the remainder. Write the answer as

$$
x^2 + 2x + 1 + \frac{-3x - 2}{x^2 + 1}.
$$

Multiply to check that this is correct.

Now Try Exercise 69.

EXAMPLE 8 Dividing a Polynomial When the Quotient Has Fractional Coefficients

Divide $4x^3 + 2x^2 + 3x + 1$ by $4x - 4$.

$$
\begin{array}{r}
x^2 + \dfrac{3}{2}x + \dfrac{9}{4} \\
4x - 4{\overline{\smash{\big)}\,4x^3 + 2x^2 + 3x + 1}} \\
\underline{4x^3 - 4x^2} \\
6x^2 + 3x \\
\underline{6x^2 - 6x} \\
9x + 1 \\
\underline{9x - 9} \\
10
\end{array}
$$

The answer is $x^2 + \dfrac{3}{2}x + \dfrac{9}{4} + \dfrac{10}{4x - 4}.$

Now Try Exercise 73.

| **CONNECTIONS** |

In Section 4.4, we found the value of a polynomial in x for a given value of x by substituting that number for x. Surprisingly, we can accomplish the same thing by division. Suppose we want to find the value of $2x^3 - 4x^2 + 3x - 5$ for $x = -3$. Instead of substituting -3 for x in the polynomial, we divide the polynomial by $x - (-3) = x + 3$. The remainder will give the value of the

(continued)

polynomial for $x = -3$. In general, when a polynomial P is divided by $x - r$, the remainder is equal to P evaluated at $x = r$.

For Discussion or Writing

1. Evaluate $2x^3 - 4x^2 + 3x - 5$ for $x = -3$.
2. Divide $2x^3 - 4x^2 + 3x - 5$ by $x + 3$. Give the remainder.
3. Compare the answers to Exercises 1 and 2. What do you notice?
4. Choose another polynomial and evaluate it both ways at some value of the variable. Do the answers agree?

4.7 EXERCISES

For Extra Help

 Student's Solutions Manual

 MyMathLab

 InterAct Math Tutorial Software

Tutor Center AW Math Tutor Center

Math*XP* MathXL

 Digital Video Tutor CD 7/Videotape 7

Fill in each blank with the correct response.

1. In the statement $\dfrac{6x^2 + 8}{2} = 3x^2 + 4$, _____ is the dividend, _____ is the divisor, and _____ is the quotient.

2. The expression $\dfrac{3x + 12}{x}$ is undefined if $x =$ _____.

3. To check the division shown in Exercise 1, multiply _____ by _____ and show that the product is _____.

4. The expression $5x^2 - 3x + 6 + \dfrac{2}{x}$ _____ a polynomial.
   (is/is not)

5. Explain why the division problem $\dfrac{16m^3 - 12m^2}{4m}$ can be performed using the methods of this section, while the division problem $\dfrac{4m}{16m^3 - 12m^2}$ cannot.

6. Suppose that a polynomial in the variable x has degree 5 and it is divided by a monomial in the variable x having degree 3. What is the degree of the quotient?

Perform each division. See Examples 1–3.

7. $\dfrac{60x^4 - 20x^2 + 10x}{2x}$

8. $\dfrac{120x^6 - 60x^3 + 80x^2}{2x}$

9. $\dfrac{20m^5 - 10m^4 + 5m^2}{5m^2}$

10. $\dfrac{12t^5 - 6t^3 + 6t^2}{6t^2}$

11. $\dfrac{8t^5 - 4t^3 + 4t^2}{2t}$

12. $\dfrac{8r^4 - 4r^3 + 6r^2}{2r}$

13. $\dfrac{4a^5 - 4a^2 + 8}{4a}$

14. $\dfrac{5t^8 + 5t^7 + 15}{5t}$

Divide each polynomial by $3x^2$. See Examples 1–3.

15. $12x^5 - 9x^4 + 6x^3$

16. $24x^6 - 12x^5 + 30x^4$

17. $3x^2 + 15x^3 - 27x^4$

18. $3x^2 - 18x^4 + 30x^5$

19. $36x + 24x^2 + 6x^3$

20. $9x - 12x^2 + 9x^3$

21. $4x^4 + 3x^3 + 2x$

22. $5x^4 - 6x^3 + 8x$

Perform each division. See Examples 1–4.

23. $\dfrac{-27r^4 + 36r^3 - 6r^2 - 26r + 2}{-3r}$

24. $\dfrac{-8k^4 + 12k^3 + 2k^2 - 7k + 3}{-2k}$

25. $\dfrac{2m^5 - 6m^4 + 8m^2}{-2m^3}$

26. $\dfrac{6r^5 - 8r^4 + 10r^2}{-2r^4}$

27. $(20a^4 - 15a^5 + 25a^3) \div (5a^4)$

28. $(16y^5 - 8y^2 + 12y) \div (4y^2)$

29. $(120x^{11} - 60x^{10} + 140x^9 - 100x^8) \div (10x^{12})$

30. $(120x^{12} - 84x^9 + 60x^8 - 36x^7) \div (12x^9)$

31. $(120x^5y^4 - 80x^2y^3 + 40x^2y^4 - 20x^5y^3) \div (20xy^2)$

32. $(200a^5b^6 - 160a^4b^7 - 120a^3b^9 + 40a^2b^2) \div (40a^2b)$

33. The area of the rectangle is given by the polynomial $15x^3 + 12x^2 - 9x + 3$. What polynomial expresses the length?

34. The area of the triangle is given by the polynomial $24m^3 + 48m^2 + 12m$. What polynomial expresses the length of the base?

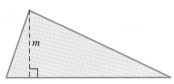

35. The quotient of a certain polynomial and $-7m^2$ is $9m^2 + 3m + 5 - \dfrac{2}{m}$. Find the polynomial.

36. Suppose that a polynomial of degree n is divided by a monomial of degree m to get a *polynomial* quotient.

 (a) How do m and n compare in value?

 (b) What is the expression that gives the degree of the quotient?

RELATING CONCEPTS (EXERCISES 37–40)

For Individual or Group Work

Our system of numeration is called a decimal system. It is based on powers of ten. In a whole number such as 2846, each digit is understood to represent the number of powers of ten for its place value. The 2 represents two thousands (2×10^3), the 8 represents eight hundreds (8×10^2), the 4 represents four tens (4×10^1), and the 6 represents six ones (or units) (6×10^0). In expanded form we write

$$2846 = (2 \times 10^3) + (8 \times 10^2) + (4 \times 10^1) + (6 \times 10^0).$$

*Keeping this information in mind, **work Exercises 37–40 in order.***

37. Divide 2846 by 2, using paper-and-pencil methods: $2\overline{)2846}$.

38. Write your answer from Exercise 37 in expanded form.

39. Use the methods of this section to divide the polynomial $2x^3 + 8x^2 + 4x + 6$ by 2.

40. Compare your answers in Exercises 38 and 39. How are they similar? How are they different? For what value of x does the answer in Exercise 39 equal the answer in Exercise 38?

Perform each division. See Example 5.

41. $\dfrac{x^2 - x - 6}{x - 3}$

42. $\dfrac{m^2 - 2m - 24}{m - 6}$

43. $\dfrac{2y^2 + 9y - 35}{y + 7}$

44. $\dfrac{2y^2 + 9y + 7}{y + 1}$

45. $\dfrac{p^2 + 2p + 20}{p + 6}$

46. $\dfrac{x^2 + 11x + 16}{x + 8}$

47. $(r^2 - 8r + 15) \div (r - 3)$

48. $(t^2 + 2t - 35) \div (t - 5)$

49. $\dfrac{12m^2 - 20m + 3}{2m - 3}$

50. $\dfrac{12y^2 + 20y + 7}{2y + 1}$

51. $\dfrac{4a^2 - 22a + 32}{2a + 3}$

52. $\dfrac{9w^2 + 6w + 10}{3w - 2}$

53. $\dfrac{8x^3 - 10x^2 - x + 3}{2x + 1}$

54. $\dfrac{12t^3 - 11t^2 + 9t + 18}{4t + 3}$

55. $\dfrac{8k^4 - 12k^3 - 2k^2 + 7k - 6}{2k - 3}$

56. $\dfrac{27r^4 - 36r^3 - 6r^2 + 26r - 24}{3r - 4}$

57. $\dfrac{5y^4 + 5y^3 + 2y^2 - y - 3}{y + 1}$

58. $\dfrac{2r^3 - 5r^2 - 6r + 15}{r - 3}$

59. $\dfrac{3k^3 - 4k^2 - 6k + 10}{k - 2}$

60. $\dfrac{5z^3 - z^2 + 10z + 2}{z + 2}$

61. $\dfrac{6p^4 - 16p^3 + 15p^2 - 5p + 10}{3p + 1}$

62. $\dfrac{6r^4 - 11r^3 - r^2 + 16r - 8}{2r - 3}$

Perform each division. See Examples 5–8.

63. $\dfrac{5 - 2r^2 + r^4}{r^2 - 1}$

64. $\dfrac{4t^2 + t^4 + 7}{t^2 + 1}$

65. $\dfrac{y^3 + 1}{y + 1}$

66. $\dfrac{y^3 - 1}{y - 1}$

67. $\dfrac{a^4 - 1}{a^2 - 1}$

68. $\dfrac{a^4 - 1}{a^2 + 1}$

69. $\dfrac{x^4 - 4x^3 + 5x^2 - 3x + 2}{x^2 + 3}$

70. $\dfrac{3t^4 + 5t^3 - 8t^2 - 13t + 2}{t^2 - 5}$

71. $\dfrac{2x^5 + 9x^4 + 8x^3 + 10x^2 + 14x + 5}{2x^2 + 3x + 1}$

72. $\dfrac{4t^5 - 11t^4 - 6t^3 + 5t^2 - t + 3}{4t^2 + t - 3}$

73. $(3a^2 - 11a + 17) \div (2a + 6)$

74. $(4x^2 + 11x - 8) \div (3x + 6)$

75. Suppose that one of your classmates asks you the following question: "How do I know when to stop the division process in a problem like the one in Exercise 69?" How would you respond?

76. Suppose that someone asks you if the following division problem is correct:

$$(6x^3 + 4x^2 - 3x + 9) \div (2x - 3) = 4x^2 + 9x - 3.$$

Tell how, by looking only at the *first term* of the quotient, you immediately know that the problem has been worked incorrectly.

Find a polynomial that describes each quantity required. Use the formulas found on the inside covers.

77. Give the length of the rectangle.

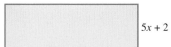

The area is $5x^3 + 7x^2 - 13x - 6$ sq. units.

78. Find the measure of the base of the parallelogram.

The area is $2x^3 + 2x^2 - 3x - 1$ sq. units.

79. If the distance traveled is $5x^3 - 6x^2 + 3x + 14$ mi and the rate is $x + 1$ mph, what is the time traveled?

80. If it costs $4x^5 + 3x^4 + 2x^3 + 9x^2 - 29x + 2$ dollars to fertilize a garden, and fertilizer costs $x + 2$ dollars per yd^2, what is the area of the garden?

RELATING CONCEPTS (EXERCISES 81–84)

For Individual or Group Work

Students often would like to know quickly whether the quotient obtained in a polynomial division problem is actually correct or whether an error was made. While the method described here is not 100% foolproof, it will at least give a fairly good idea as to the accuracy of the result. To illustrate, suppose that $4x^4 + 2x^3 - 14x^2 + 19x + 10$ is divided by $2x + 5$. Lakeisha and Stan obtain the following answers:

Lakeisha	*Stan*
$2x^3 - 4x^2 + 3x + 2$	$2x^3 - 4x^2 - 3x + 2.$

As a "quick check" we can evaluate Lakeisha's answer for $x = 1$ and Stan's answer for $x = 1$:

Lakeisha: When $x = 1$, her answer gives $2(1)^3 - 4(1)^2 + 3(1) + 2 = 3$.
Stan: When $x = 1$, his answer gives $2(1)^3 - 4(1)^2 - 3(1) + 2 = -3$.

Now, if the original quotient of the two polynomials is evaluated for $x = 1$, we get

$$\frac{4x^4 + 2x^3 - 14x^2 + 19x + 10}{2x + 5} = \frac{4(1)^4 + 2(1)^3 - 14(1)^2 + 19(1) + 10}{2(1) + 5}$$

$$= \frac{21}{7}$$

$$= 3.$$

Because Stan's answer, -3, is different from the quotient 3 just obtained, Stan can conclude his answer is incorrect. Lakeisha's answer, 3, agrees with the quotient just obtained, and while this does not *guarantee* that she is correct, at least she can feel better and go on to the next problem.

(continued)

In Exercises 81–84, a division problem is given, along with two possible answers.
One is correct and one is incorrect. Use the method just described to determine which
one is correct and which one is not.

81. Problem   Possible answers

$$\frac{2x^2 + 3x - 14}{x - 2}$$

A. $2x + 7$ **B.** $2x - 7$

82. Problem   Possible answers

$$\frac{x^4 + 4x^3 - 5x^2 - 12x + 6}{x^2 - 3}$$

A. $x^2 - 4x - 2$ **B.** $x^2 + 4x - 2$

83. Problem   Possible answers

$$\frac{2y^3 + 17y^2 + 37y + 7}{2y + 7}$$

A. $y^2 + 5y + 1$ **B.** $y^2 - 5y + 1$

84. In the explanation preceding Exercise 81 we used 1 for the value of x to check
our work. This is because a polynomial is easy to evaluate for 1. Why is this so?
Why would we not be able to use 1 if the divisor is $x - 1$?

 Chapter **4** **Group Activity**

Measuring the Flight of a Rocket

OBJECTIVE Graph quadratic equations to solve an application problem.

A physics class observes the launch of two rockets that are slightly different. One
has an initial velocity of 48 ft per sec. The second one has an initial velocity of 64 ft
per sec. Follow the steps below to determine the maximum height each rocket will
achieve and how long it will take it to reach that height.

The following formula is used to find the height of a projectile when shot
vertically into the air:

$$H = -16t^2 + Vt,$$

where $H =$ height above the launch pad in feet, $t =$ time in seconds, and $V =$
initial velocity.

A. One student should complete the table on the next page for the first rocket. The
other student should do the same for the second rocket. Use the formula for the
height of a projectile and the values provided.

Initial Velocity = 48 Feet per Second		Initial Velocity = 64 Feet per Second	
Time (seconds)	Height (feet)	Time (seconds)	Height (feet)
0		0	
.5		.5	
1		1	
1.5		1.5	
2		2	
2.5		2.5	
3		3	
		3.5	
		4	

B. Graph the results. You must determine the scales for the two axes on the coordinate system. Clearly label each graph and its scale.

C. Answer the following questions for each rocket.

   1. Find the time in seconds when the rocket is at height 0 ft. Why does this happen twice?

   2. What is the maximum height the rocket reached?

   3. At what time (in seconds) did the rocket reach its maximum height?

D. Compare the data from both rockets and answer the following questions.

   1. Which rocket had the greater maximum height? Explain how you decided.

   2. Which rocket reached its maximum height first? Explain how you decided.

   3. Consider the answers for Exercises D1 and D2 and discuss why this occurred.

   4. Do both rockets reach a height of 48 ft? If yes, at what times? What observations about projectiles might this lead to?

   5. What are some other possible uses for the formula of the height of a projectile?

CHAPTER 4 SUMMARY

KEY TERMS

4.3 scientific notation
4.4 polynomial
     descending powers
     degree of a term

degree of a
    polynomial
trinomial
binomial

monomial
parabola
vertex
axis

line of symmetry
4.5 outer product
     inner product
     FOIL

NEW SYMBOLS

x^{-n} x to the negative n
           power

TEST YOUR WORD POWER

See how well you have learned the vocabulary in this chapter. Answers, with examples, follow the Quick Review.

1. A **polynomial** is an algebraic
   expression made up of
   A. a term or a finite product of terms
      with positive coefficients and
      exponents
   B. a term or a finite sum of terms
      with real coefficients and whole
      number exponents
   C. the product of two or more terms
      with positive exponents
   D. the sum of two or more terms
      with whole number coefficients
      and exponents.

2. The **degree of a term** is
   A. the number of variables in the
      term

   B. the product of the exponents on
      the variables
   C. the smallest exponent on the
      variables
   D. the sum of the exponents on the
      variables.

3. A **trinomial** is a polynomial with
   A. only one term
   B. exactly two terms
   C. exactly three terms
   D. more than three terms.

4. A **binomial** is a polynomial with
   A. only one term
   B. exactly two terms
   C. exactly three terms

   D. more than three terms.

5. A **monomial** is a polynomial with
   A. only one term
   B. exactly two terms
   C. exactly three terms
   D. more than three terms.

6. **FOIL** is a method for
   A. adding two binomials
   B. adding two trinomials
   C. multiplying two binomials
   D. multiplying two trinomials.

QUICK REVIEW

CONCEPTS	EXAMPLES

4.1 THE PRODUCT RULE AND POWER RULES FOR EXPONENTS

For any integers m and n with no denominators
zero:

Product Rule $a^m \cdot a^n = a^{m+n}$

Power Rules (a) $(a^m)^n = a^{mn}$

(b) $(ab)^m = a^m b^m$

(c) $\left(\dfrac{a}{b}\right)^m = \dfrac{a^m}{b^m}$ $(b \neq 0)$

Perform the operations by using rules for exponents.

$$2^4 \cdot 2^5 = 2^9$$

$$(3^4)^2 = 3^8$$

$$(6a)^5 = 6^5 a^5$$

$$\left(\frac{2}{3}\right)^4 = \frac{2^4}{3^4}$$

CONCEPTS	EXAMPLES

4.2 INTEGER EXPONENTS AND THE QUOTIENT RULE

If $a \neq 0$, for integers m and n:

Zero Exponent $a^0 = 1$

Negative Exponent $a^{-n} = \dfrac{1}{a^n}$

Quotient Rule $\dfrac{a^m}{a^n} = a^{m-n}$

Negative to Positive Rules

$$\dfrac{a^{-m}}{b^{-n}} = \dfrac{b^n}{a^m} \quad (b \neq 0)$$

$$\left(\dfrac{a}{b}\right)^{-m} = \left(\dfrac{b}{a}\right)^{m} \quad (b \neq 0)$$

Simplify by using the rules for exponents.

$$15^0 = 1$$

$$5^{-2} = \dfrac{1}{5^2} = \dfrac{1}{25}$$

$$\dfrac{4^8}{4^3} = 4^5$$

$$\dfrac{4^{-2}}{3^{-5}} = \dfrac{3^5}{4^2}$$

$$\left(\dfrac{6}{5}\right)^{-3} = \left(\dfrac{5}{6}\right)^{3}$$

4.3 AN APPLICATION OF EXPONENTS: SCIENTIFIC NOTATION

To write a number in scientific notation (as $a \times 10^n$), move the decimal point to follow the first nonzero digit. If moving the decimal point makes the number smaller, n is positive. If it makes the number larger, n is negative. If the decimal point is not moved, n is 0.

Write in scientific notation.

$$247 = 2.47 \times 10^2$$
$$.0051 = 5.1 \times 10^{-3}$$
$$4.8 = 4.8 \times 10^0$$

Write without exponents.

$$3.25 \times 10^5 = 325,000$$
$$8.44 \times 10^{-6} = .00000844$$

4.4 ADDING AND SUBTRACTING POLYNOMIALS; GRAPHING SIMPLE POLYNOMIALS

Adding Polynomials
Add like terms.

Add.

$$\begin{array}{r} 2x^2 + 5x - 3 \\ 5x^2 - 2x + 7 \\ \hline 7x^2 + 3x + 4 \end{array}$$

Subtracting Polynomials
Change the signs of the terms in the second polynomial and add to the first polynomial.

Subtract.

$$(2x^2 + 5x - 3) - (5x^2 - 2x + 7)$$
$$= (2x^2 + 5x - 3) + (-5x^2 + 2x - 7)$$
$$= -3x^2 + 7x - 10$$

Graphing Simple Polynomials
To graph a simple polynomial equation such as $y = x^2 - 2$, plot points near the vertex. (In this chapter, all parabolas have a vertex on the x-axis or the y-axis.)

Graph $y = x^2 - 2$.

x	y
-2	2
-1	-1
0	-2
1	-1
2	2

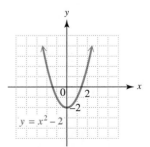

$y = x^2 - 2$

(continued)

CONCEPTS	EXAMPLES

4.5 MULTIPLYING POLYNOMIALS

General Method for Multiplying Polynomials
Multiply each term of the first polynomial by each term of the second polynomial. Then add like terms.

Multiply.

$$
\begin{array}{r}
3x^3 - 4x^2 + 2x - 7 \\
4x + 3 \\
\hline
9x^3 - 12x^2 + 6x - 21 \\
12x^4 - 16x^3 + 8x^2 - 28x \\
\hline
12x^4 - 7x^3 - 4x^2 - 22x - 21
\end{array}
$$

FOIL Method for Multiplying Binomials

Step 1 Multiply the two first terms to get the first term of the answer.

Step 2 Find the outer product and the inner product and mentally add them, when possible, to get the middle term of the answer.

Step 3 Multiply the two last terms to get the last term of the answer.

Multiply. $(2x + 3)(5x - 4)$

$$2x(5x) = 10x^2$$

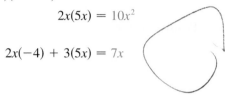

$$2x(-4) + 3(5x) = 7x$$

$$3(-4) = -12$$

The product of $(2x + 3)$ and $(5x - 4)$ is $10x^2 + 7x - 12$.

4.6 SPECIAL PRODUCTS

Square of a Binomial

$$(x + y)^2 = x^2 + 2xy + y^2$$
$$(x - y)^2 = x^2 - 2xy + y^2$$

Product of the Sum and Difference of Two Terms

$$(x + y)(x - y) = x^2 - y^2$$

Multiply.

$$(3x + 1)^2 = 9x^2 + 6x + 1$$
$$(2m - 5n)^2 = 4m^2 - 20mn + 25n^2$$

$$(4a + 3)(4a - 3) = 16a^2 - 9$$

4.7 DIVIDING POLYNOMIALS

Dividing a Polynomial by a Monomial
Divide each term of the polynomial by the monomial.

$$\frac{a + b}{c} = \frac{a}{c} + \frac{b}{c}$$

Divide.

$$\frac{4x^3 - 2x^2 + 6x - 8}{2x} = 2x^2 - x + 3 - \frac{4}{x}$$

Dividing a Polynomial by a Polynomial
Use "long division."

Divide.

$$
\begin{array}{r}
2x - 5 \\
3x + 4 \overline{)6x^2 - 7x - 21} \\
\underline{6x^2 + 8x} \\
-15x - 21 \\
\underline{-15x - 20} \\
-1 \leftarrow \text{Remainder}
\end{array}
$$

The final answer is $2x - 5 + \dfrac{-1}{3x + 4}$.

Answers to Test Your Word Power

1. B; *Example:* $5x^3 + 2x^2 - 7$ **2.** D; *Examples:* The term 6 has degree 0, $3x$ has degree 1, $-2x^8$ has degree 8, and $5x^2y^4$ has degree 6.

3. C; *Example:* $2a^2 - 3ab + b^2$ **4.** B; *Example:* $3t^3 + 5t$ **5.** A; *Examples:* -5 and $4xy^5$

        F   O   I   L

6. C; *Example:* $(m + 4)(m - 3) = m(m) - 3m + 4m + 4(-3) = m^2 + m - 12$

CHAPTER **4** REVIEW EXERCISES

[4.1]   *Use the product rule, power rules, or both to simplify each expression. Write the answers in exponential form.*

1. $4^3 \cdot 4^8$

2. $(-5)^6(-5)^5$

3. $(-8x^4)(9x^3)$

4. $(2x^2)(5x^3)(x^9)$

5. $(19x)^5$

6. $(-4y)^7$

7. $5(pt)^4$

8. $\left(\dfrac{7}{5}\right)^6$

9. $(6x^2z^4)^2(x^3yz^2)^4$

10. $\left(\dfrac{2m^3n}{p^2}\right)^3$

11. Why does the product rule for exponents not apply to the expression $7^2 + 7^4$?

[4.2]   *Evaluate each expression.*

12. $6^0 + (-6)^0$

13. $(-23)^0 - (-23)^0$

14. -10^0

Write each expression using only positive exponents.

15. -7^{-2}

16. $\left(\dfrac{5}{8}\right)^{-2}$

17. $(5^{-2})^{-4}$

18. $9^3 \cdot 9^{-5}$

19. $2^{-1} + 4^{-1}$

20. $\dfrac{6^{-5}}{6^{-3}}$

Simplify. Write each answer with only positive exponents. Assume that all variables represent nonzero real numbers.

21. $\dfrac{x^{-7}}{x^{-9}}$

22. $\dfrac{y^4 \cdot y^{-2}}{y^{-5}}$

23. $(3r^{-2})^{-4}$

24. $(3p)^4(3p^{-7})$

25. $\dfrac{ab^{-3}}{a^4b^2}$

26. $\dfrac{(6r^{-1})^2(2r^{-4})}{r^{-5}(r^2)^{-3}}$

[4.3]   *Write each number in scientific notation.*

27. 48,000,000

28. 28,988,000,000

29. .0000000824

Write each number without exponents.

30. 2.4×10^4

31. 7.83×10^7

32. 8.97×10^{-7}

Perform each indicated operation and write the answer without exponents.

33. $(2 \times 10^{-3}) \times (4 \times 10^5)$

34. $\dfrac{8 \times 10^4}{2 \times 10^{-2}}$

35. $\dfrac{12 \times 10^{-5} \times 5 \times 10^4}{4 \times 10^3 \times 6 \times 10^{-2}}$

Each quote is taken from the source cited. Write each number that is in scientific notation in the quote without exponents.

36. The muon, a close relative of the electron produced by the bombardment of cosmic rays against the upper atmosphere, has a half-life of 2 millionths of a second (2×10^{-6} s). (Excerpt from *Conceptual Physics,* 6th edition, by Paul G. Hewitt. Copyright © by Paul G. Hewitt. Published by HarperCollins College Publishers.)

37. There are 13 red balls and 39 black balls in a box. Mix them up and draw 13 out one at a time without returning any ball . . . the probability that the 13 drawings each will produce a red ball is . . . 1.6×10^{-12}. (Weaver, Warren, *Lady Luck,* Doubleday, 1963, pp. 298–299.)

Each quote is taken from the source cited. Write each number in scientific notation.

38. An electron and a positron attract each other in two ways: the electromagnetic attraction of their opposite electric charges, and the gravitational attraction of their two masses. The electromagnetic attraction is

$$4,200,000,000,000,000,000,000,000,000,000,000,000,000$$

times as strong as the gravitational. (Asimov, Isaac, *Isaac Asimov's Book of Facts,* Bell Publishing Company, 1981, p. 106.)

39. The aircraft carrier USS John Stennis is a 97,000-ton nuclear powered floating city with a crew of 5000. (*Source:* Seelye, Katharine Q., "Staunch Allies Hard to Beat: Defense Dept., Hollywood," *New York Times,* in *Plain Dealer,* June 10, 2002.)

40. A googol is

$$10,000,000,000,000,000,000,000,000,000,000,000,000,000,000,000,$$
$$000,000,000,000,000,000,000,000,000,000,000,000,000,000,000.$$

The Web search engine Google is named after a googol. Sergey Brin, president and co-founder of Google, Inc., was a math major. He chose the name Google to describe the vast reach of this search engine. (*Source: The Gazette,* March 2, 2001.)

41. According to Campbell, Mitchell, and Reece in *Biology Concepts and Connections* (Benjamin Cummings, 1994, p. 230), "The amount of DNA in a human cell is about 1000 times greater than the DNA in *E. coli.* Does this mean humans have 1000 times as many genes as the 2000 in *E. coli?* The answer is probably no; the human genome is thought to carry between 50,000 and 100,000 genes, which code for various proteins (as well as for tRNA and rRNA)."

Write each number from this quote using scientific notation.

(a) 1000　　**(b)** 2000　　**(c)** 50,000　　**(d)** 100,000

[4.4] *Combine like terms where possible in each polynomial. Write the answer in descending powers of the variable. Give the degree of the answer. Identify the polynomial as a* monomial, binomial, trinomial, *or* none of these.

42. $9m^2 + 11m^2 + 2m^2$

43. $-4p + p^3 - p^2 + 8p + 2$

44. $12a^5 - 9a^4 + 8a^3 + 2a^2 - a + 3$

45. $-7y^5 - 8y^4 - y^5 + y^4 + 9y$

46. $(12r^4 - 7r^3 + 2r^2) - (5r^4 - 3r^3 + 2r^2 - 1)$

47. Simplify $(5x^3y^2 - 3xy^5 + 12x^2) - (-9x^2 - 8x^3y^2 + 2xy^5)$.

Add or subtract as indicated.

48. Add.

$$-2a^3 + 5a^2$$
$$3a^3 - a^2$$

49. Subtract.

$$6y^2 - 8y + 2$$
$$5y^2 + 2y - 7$$

50. Subtract.

$$-12k^4 - 8k^2 + 7k$$
$$k^4 + 7k^2 - 11k$$

Graph each equation by completing the table of values.

51. $y = -x^2 + 5$

x	-2	-1	0	1	2
y					

52. $y = 3x^2 - 2$

x	-2	-1	0	1	2
y					

[4.5] *Find each product.*

53. $(a + 2)(a^2 - 4a + 1)$

54. $(3r - 2)(2r^2 + 4r - 3)$

55. $(5p^2 + 3p)(p^3 - p^2 + 5)$

56. $(m - 9)(m + 2)$

57. $(3k - 6)(2k + 1)$

58. $(a + 3b)(2a - b)$

59. $(6k + 5q)(2k - 7q)$

60. $(s - 1)^3$

61. Find a polynomial that represents the area of the rectangle shown.

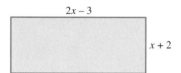

62. If the side of a square has a measure represented by $5x^4 + 2x^2$, what polynomial represents its area?

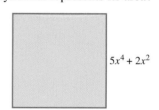

[4.6] *Find each product.*

63. $(a + 4)^2$

64. $(2r + 5t)^2$

65. $(6m - 5)(6m + 5)$

66. $(5a + 6b)(5a - 6b)$

67. $(r + 2)^3$

68. $t(5t - 3)^2$

69. Choose values for x and y to show that, in general, the following hold true.

   (a) $(x + y)^2 \neq x^2 + y^2$ **(b)** $(x + y)^3 \neq x^3 + y^3$

70. Write an explanation on how to raise a binomial to the third power. Give an example.

71. Refer to Exercise 69. Suppose that you happened to let $x = 0$ and $y = 1$. Would your results be sufficient to illustrate the truth, in general, of the inequalities shown? If not, what would you need to do as your next step in working the exercise?

72. What is the volume of a cube with one side having length $x^2 + 2$ cm?

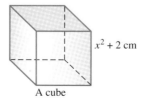

A cube

73. What is the volume of a sphere with radius $x + 1$ in.?

A sphere

[4.7] *Perform each division.*

74. $\dfrac{-15y^4}{9y^2}$

75. $\dfrac{6y^4 - 12y^2 + 18y}{6y}$

76. $(-10m^4n^2 + 5m^3n^2 + 6m^2n^4) \div (5m^2n)$

77. What polynomial, when multiplied by $6m^2n$, gives the product $12m^3n^2 + 18m^6n^3 - 24m^2n^2$?

78. One of your friends in class simplified $\dfrac{6x^2 - 12x}{6}$ as $x^2 - 12x$. Is this correct? If not, what error did your friend make and how would you explain the correct method of performing the division?

Perform each division.

79. $\dfrac{2r^2 + 3r - 14}{r - 2}$

80. $\dfrac{10a^3 + 9a^2 - 14a + 9}{5a - 3}$

81. $\dfrac{x^4 - 5x^2 + 3x^3 - 3x + 4}{x^2 - 1}$

82. $\dfrac{m^4 + 4m^3 - 12m - 5m^2 + 6}{m^2 - 3}$

83. $\dfrac{16x^2 - 25}{4x + 5}$

84. $\dfrac{25y^2 - 100}{5y + 10}$

85. $\dfrac{y^3 - 8}{y - 2}$

86. $\dfrac{1000x^6 + 1}{10x^2 + 1}$

87. $\dfrac{6y^4 - 15y^3 + 14y^2 - 5y - 1}{3y^2 + 1}$

88. $\dfrac{4x^5 - 8x^4 - 3x^3 + 22x^2 - 15}{4x^2 - 3}$

MIXED REVIEW EXERCISES

Perform each indicated operation. Write with positive exponents only. Assume all variables represent nonzero real numbers.

89. $5^0 + 7^0$

90. $\left(\dfrac{6r^2p}{5}\right)^3$

91. $(12a + 1)(12a - 1)$

92. 2^{-4}

93. $(8^{-3})^4$

94. $\dfrac{2p^3 - 6p^2 + 5p}{2p^2}$

95. $\dfrac{(2m^{-5})(3m^2)^{-1}}{m^{-2}(m^{-1})^2}$

96. $(3k - 6)(2k^2 + 4k + 1)$

97. $\dfrac{r^9 \cdot r^{-5}}{r^{-2} \cdot r^{-7}}$

98. $(2r + 5s)^2$

99. $(-5y^2 + 3y - 11) + (4y^2 - 7y + 15)$

100. $(2r + 5)(5r - 2)$

101. $\dfrac{2y^3 + 17y^2 + 37y + 7}{2y + 7}$

102. $(25x^2y^3 - 8xy^2 + 15x^3y) \div (10x^2y^3)$

103. $(6p^2 - p - 8) - (-4p^2 + 2p - 3)$

104. $\dfrac{5^8}{5^{19}}$

105. $(-7 + 2k)^2$

106. $\left(\dfrac{x}{y^{-3}}\right)^{-4}$

CHAPTER **4** TEST

Evaluate each expression.

1. 5^{-4}

2. $(-3)^0 + 4^0$

3. $4^{-1} + 3^{-1}$

4. Use the rules for exponents to simplify $\dfrac{(3x^2y)^2(xy^3)^2}{(xy)^3}$. Assume x and y are nonzero.

Simplify, and write the answer using only positive exponents. Assume that variables represent nonzero numbers.

5. $\dfrac{8^{-1} \cdot 8^4}{8^{-2}}$

6. $\dfrac{(x^{-3})^{-2}(x^{-1}y)^2}{(xy^{-2})^2}$

7. Determine whether each expression represents a number that is *positive, negative,* or *zero.*

    (a) 3^{-4} **(b)** $(-3)^4$ **(c)** -3^4 **(d)** 3^0 **(e)** $(-3)^0 - 3^0$ **(f)** $(-3)^{-3}$

8. (a) Write 45,000,000,000 using scientific notation.

    (b) Write 3.6×10^{-6} without using exponents.

    (c) Write the quotient without using exponents: $\dfrac{9.5 \times 10^{-1}}{5 \times 10^3}$.

9. A satellite galaxy of our own Milky Way, known as the Large Magellanic Cloud, is 1000 light-years across. A *light-year* is equal to 5,890,000,000,000 mi. (*Source:* "Images of Brightest Nebula Unveiled," *USA Today,* June 12, 2002.)

    (a) Write these two numbers using scientific notation.

    (b) How many miles across is the Large Magellanic Cloud?

For each polynomial, combine like terms when possible and write the polynomial in descending powers of the variable. Give the degree of the simplified polynomial. Decide whether the simplified polynomial is a monomial, binomial, trinomial, *or* none of these.

10. $5x^2 + 8x - 12x^2$

11. $13n^3 - n^2 + n^4 + 3n^4 - 9n^2$

12. Use the table to complete a set of ordered pairs that lie on the graph of $y = 2x^2 - 4$. Then graph the equation.

x	-2	-1	0	1	2
y					

Perform each indicated operation.

13. $(2y^2 - 8y + 8) + (-3y^2 + 2y + 3) - (y^2 + 3y - 6)$

14. $(-9a^3b^2 + 13ab^5 + 5a^2b^2) - (6ab^5 + 12a^3b^2 + 10a^2b^2)$

15. Subtract.

    $9t^3 - 4t^2 + 2t + 2$
    $\underline{9t^3 + 8t^2 - 3t - 6}$

16. $3x^2(-9x^3 + 6x^2 - 2x + 1)$

17. $(t - 8)(t + 3)$

18. $(4x + 3y)(2x - y)$

19. $(5x - 2y)^2$

20. $(10v + 3w)(10v - 3w)$

21. $(2r - 3)(r^2 + 2r - 5)$

22. What polynomial expression represents the area of this square?

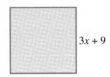

$3x + 9$

Perform each division.

23. $\dfrac{8y^3 - 6y^2 + 4y + 10}{2y}$

24. $(-9x^2y^3 + 6x^4y^3 + 12xy^3) \div (3xy)$

25. $(3x^3 - x + 4) \div (x - 2)$

CUMULATIVE REVIEW EXERCISES CHAPTERS 1–4

Write each fraction in lowest terms.

1. $\dfrac{28}{16}$

2. $\dfrac{55}{11}$

Perform each operation.

3. $\dfrac{2}{3} + \dfrac{1}{8}$

4. $\dfrac{7}{4} - \dfrac{9}{5}$

5. A contractor installs toolsheds. Each requires $1\frac{1}{4}$ yd^3 of concrete. How much concrete would be needed for 25 sheds?

6. A retailer has \$34,000 invested in her business. She finds that last year she earned 5.4% on this investment. How much did she earn?

7. List all positive integer factors of 45.

8. Find the value of

$$\frac{4x - 2y}{x + y}$$

if $x = -2$ and $y = 4$.

Perform each indicated operation.

9. $\dfrac{(-13 + 15) - (3 + 2)}{6 - 12}$

10. $-7 - 3[2 + (5 - 8)]$

Decide which property justifies each statement.

11. $(9 + 2) + 3 = 9 + (2 + 3)$

12. $6(4 + 2) = 6(4) + 6(2)$

13. Simplify the expression $-3(2x^2 - 8x + 9) - (4x^2 + 3x + 2)$.

Solve each equation.

14. $2 - 3(t - 5) = 4 + t$

15. $2(5h + 1) = 10h + 4$

16. $d = rt$ for r

17. $\dfrac{x}{5} = \dfrac{x - 2}{7}$

18. $\dfrac{1}{3}p - \dfrac{1}{6}p = -2$

19. $.05x + .15(50 - x) = 5.50$

20. $4 - (3x + 12) = (2x - 9) - (5x - 1)$

Solve each problem.

21. A 1-oz mouse takes about 16 times as many breaths as does a 3-ton elephant. If the two animals take a combined total of 170 breaths per minute, how many breaths does each take during that time period? (*Source:* McGowan, Christopher, *Dinosaurs, Spitfires, and Sea Dragons,* Harvard University Press, 1991.)

22. If a number is subtracted from 8 and this difference is tripled, the result is 3 times the number. Find this number, and you will learn how many times a dolphin rests during a 24-hr period.

23. One side of a triangle is twice as long as a second side. The third side of the triangle is 17 ft long. The perimeter of the triangle cannot be more than 50 ft. Find the longest possible values for the other two sides of the triangle.

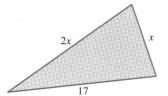

Solve each inequality.

24. $-8x \le -80$

25. $-2(x + 4) > 3x + 6$

26. $-3 \le 2x + 5 < 9$

27. Graph $y = -3x + 6$.

28. Consider the two points $(-1, 5)$ and $(2, 8)$.

    (a) Find the slope of the line joining them.
    (b) Find the equation of the line joining them.

Evaluate each expression.

29. $4^{-1} + 3^0$

30. $2^{-4} \cdot 2^5$

31. $\dfrac{8^{-5} \cdot 8^7}{8^2}$

32. Write with positive exponents only: $\dfrac{(a^{-3}b^2)^2}{(2a^{-4}b^{-3})^{-1}}$.

33. Write in scientific notation: 34,500.

34. Write without exponents: 5.36×10^{-7}.

35. It takes about 3.6×10^1 sec at a speed of 3.0×10^5 km per sec for light from the sun to reach Venus. How far is Venus from the sun? (*Source: World Almanac and Book of Facts,* 2000.)

36. Graph $y = (x + 4)^2$, using the x-values $-6, -5, -4, -3,$ and -2 to obtain a set of points.

Perform each indicated operation.

37. $(7x^3 - 12x^2 - 3x + 8) + (6x^2 + 4) - (-4x^3 + 8x^2 - 2x - 2)$

38. $6x^5(3x^2 - 9x + 10)$ **39.** $(7x + 4)(9x + 3)$

40. $(5x + 8)^2$ **41.** $\dfrac{14x^3 - 21x^2 + 7x}{7x}$

42. $\dfrac{y^3 - 3y^2 + 8y - 6}{y - 1}$

Factoring and Applications

5

Galileo Galilei, born near Pisa, Italy, in 1564, became a professor of mathematics at the University of Pisa at age 25. He and his students conducted experiments involving the famous Leaning Tower to investigate the relationship between an object's speed of fall and its weight. (*Source: Microsoft Encarta Encyclopedia 2002.*) We use the concepts of this chapter and the formula Galileo developed from his experiments in Section 5.6.

5.1 The Greatest Common Factor; Factoring by Grouping

OBJECTIVES

1 Find the greatest common factor of a list of terms.

2 Factor out the greatest common factor.

3 Factor by grouping.

Recall from Chapter 1 that to **factor** means to write a quantity as a product. That is, factoring is the opposite of multiplying. For example,

Multiplying   *Factoring*

$6 \cdot 2 = 12,$ $12 = 6 \cdot 2.$

Factors Product   Product Factors

Other **factored forms** of 12 are

$$-6(-2), \quad 3 \cdot 4, \quad -3(-4), \quad 12 \cdot 1, \quad \text{and} \quad -12(-1).$$

More than two factors may be used, so another factored form of 12 is $2 \cdot 2 \cdot 3$. The positive integer factors of 12 are

$$1, 2, 3, 4, 6, 12.$$

OBJECTIVE 1 **Find the greatest common factor of a list of terms.** An integer that is a factor of two or more integers is called a **common factor** of those integers. For example, 6 is a common factor of 18 and 24 since 6 is a factor of both 18 and 24. Other common factors of 18 and 24 are 1, 2, and 3. The **greatest common factor (GCF)** of a list of integers is the largest common factor of those integers. Thus, 6 is the greatest common factor of 18 and 24, since it is the largest of their common factors.

NOTE Factors of a number are also divisors of the number. The greatest common factor is actually the same as the greatest common divisor. There are many rules for deciding what numbers divide into a given number. Here are some especially useful divisibility rules for small numbers. It is surprising how many people do not know them.

A Whole Number Divisible by:	Must Have the Following Property:
2	Ends in 0, 2, 4, 6, or 8
3	Sum of its digits is divisible by 3.
4	Last two digits form a number divisible by 4
5	Ends in 0 or 5
6	Divisible by both 2 and 3
8	Last three digits form a number divisible by 8
9	Sum of its digits is divisible by 9.
10	Ends in 0

Recall from Chapter 1 that a prime number has only itself and 1 as factors. In Section 1.1 we factored numbers into prime factors. This is the first step in finding the greatest common factor of a list of numbers. We find the greatest common factor (GCF) of a list of numbers as follows.

Finding the Greatest Common Factor (GCF)

Step 1 **Factor.** Write each number in prime factored form.

Step 2 **List common factors.** List each prime number that is a factor of every number in the list. (If a prime does not appear in one of the prime factored forms, it cannot appear in the greatest common factor.)

Step 3 **Choose smallest exponents.** Use as exponents on the common prime factors the *smallest* exponent from the prime factored forms.

Step 4 **Multiply.** Multiply the primes from Step 3. If there are no primes left after Step 3, the greatest common factor is 1.

EXAMPLE 1 Finding the Greatest Common Factor for Numbers

Find the greatest common factor for each list of numbers.

(a) 30, 45

First write each number in prime factored form.

$$30 = 2 \cdot 3 \cdot 5$$
$$45 = 3 \cdot 3 \cdot 5$$

Use each prime the *least* number of times it appears in *all* the factored forms. There is no 2 in the prime factored form of 45, so there will be no 2 in the greatest common factor. The least number of times 3 appears in all the factored forms is 1, and the least number of times 5 appears is also 1. From this, the

$$\text{GCF} = 3^1 \cdot 5^1 = 15.$$

(b) 72, 120, 432

Find the prime factored form of each number.

$$72 = 2 \cdot 2 \cdot 2 \cdot 3 \cdot 3$$
$$120 = 2 \cdot 2 \cdot 2 \cdot 3 \cdot 5$$
$$432 = 2 \cdot 2 \cdot 2 \cdot 2 \cdot 3 \cdot 3 \cdot 3$$

The least number of times 2 appears in all the factored forms is 3, and the least number of times 3 appears is 1. There is no 5 in the prime factored form of either 72 or 432, so the

$$\text{GCF} = 2^3 \cdot 3^1 = 24.$$

(c) 10, 11, 14

Write the prime factored form of each number.

$$10 = 2 \cdot 5$$
$$11 = 11$$
$$14 = 2 \cdot 7$$

There are no primes common to all three numbers, so the GCF is 1.

Now Try Exercise 1.

The greatest common factor can also be found for a list of variable terms. For example, the terms x^4, x^5, x^6, and x^7 have x^4 as the greatest common factor because each of these terms can be written with x^4 as a factor.

$$x^4 = 1 \cdot x^4, \quad x^5 = x \cdot x^4, \quad x^6 = x^2 \cdot x^4, \quad x^7 = x^3 \cdot x^4$$

NOTE The exponent on a variable in the GCF is the *smallest* exponent that appears in *all* the common factors.

EXAMPLE 2 Finding the Greatest Common Factor for Variable Terms

Find the greatest common factor for each list of terms.

(a) $21m^7$, $-18m^6$, $45m^8$, $-24m^5$

$$21m^7 = 3 \cdot 7 \cdot m^7$$
$$-18m^6 = -1 \cdot 2 \cdot 3^2 \cdot m^6$$
$$45m^8 = 3^2 \cdot 5 \cdot m^8$$
$$-24m^5 = -1 \cdot 2^3 \cdot 3 \cdot m^5$$

First, 3 is the greatest common factor of the coefficients 21, -18, 45, and -24. The smallest exponent on m is 5, so the GCF of the terms is $3m^5$.

(b) x^4y^2, x^7y^5, x^3y^7, y^{15}

$$x^4y^2 = x^4 \cdot y^2$$
$$x^7y^5 = x^7 \cdot y^5$$
$$x^3y^7 = x^3 \cdot y^7$$
$$y^{15} = y^{15}$$

There is no x in the last term, y^{15}, so x will not appear in the greatest common factor. There is a y in each term, however, and 2 is the smallest exponent on y. The GCF is y^2.

(c) $-a^2b$, $-ab^2$

$$-a^2b = -1a^2b = -1 \cdot 1 \cdot a^2b$$
$$-ab^2 = -1ab^2 = -1 \cdot 1 \cdot ab^2$$

The factors of -1 are -1 and 1. Since $1 > -1$, the GCF is $1ab$ or ab.

Now Try Exercises 9 and 13.

NOTE In a list of negative terms, sometimes a negative common factor is preferable (even though it is not the greatest common factor). In Example 2(c), for instance, we might prefer $-ab$ as the common factor. In factoring exercises, either answer will be acceptable.

OBJECTIVE 2 **Factor out the greatest common factor.** Writing a polynomial (a sum) in factored form as a product is called **factoring.** For example, the polynomial

$$3m + 12$$

has two terms, $3m$ and 12. The greatest common factor for these two terms is 3. We can write $3m + 12$ so that each term is a product with 3 as one factor.

$$3m + 12 = 3 \cdot m + 3 \cdot 4$$

Now we use the distributive property.

$$3m + 12 = 3 \cdot m + 3 \cdot 4 = 3(m + 4)$$

The factored form of $3m + 12$ is $3(m + 4)$. This process is called **factoring out the greatest common factor.**

CAUTION The polynomial $3m + 12$ is *not* in factored form when written as

$$3 \cdot m + 3 \cdot 4.$$

The *terms* are factored, but the polynomial is not. The factored form of $3m + 12$ is the *product*

$$3(m + 4).$$

EXAMPLE 3 Factoring Out the Greatest Common Factor

Factor out the greatest common factor.

(a) $5y^2 + 10y = 5y(y) + 5y(2)$ GCF = 5y

$\qquad\qquad = 5y(y + 2)$ Distributive property

To check, multiply out the factored form: $5y(y + 2) = 5y^2 + 10y$, which is the original polynomial.

(b) $20m^5 + 10m^4 + 15m^3$

The GCF for the terms of this polynomial is $5m^3$.

$$20m^5 + 10m^4 + 15m^3 = 5m^3(4m^2) + 5m^3(2m) + 5m^3(3)$$
$$= 5m^3(4m^2 + 2m + 3)$$

Check: $5m^3(4m^2 + 2m + 3) = 20m^5 + 10m^4 + 15m^3$, which is the original polynomial.

(c) $x^5 + x^3 = x^3(x^2) + x^3(1) = x^3(x^2 + 1)$ Don't forget the 1.

(d) $20m^7p^2 - 36m^3p^4 = 4m^3p^2(5m^4) - 4m^3p^2(9p^2)$ GCF = $4m^3p^2$

$\qquad\qquad\qquad\qquad = 4m^3p^2(5m^4 - 9p^2)$

(e) $\dfrac{1}{6}n^2 + \dfrac{5}{6}n = \dfrac{1}{6}n(n) + \dfrac{1}{6}n(5) = \dfrac{1}{6}n(n + 5)$ GCF = $\frac{1}{6}n$

Now Try Exercises 37, 41, and 47.

CAUTION Be sure to include the 1 in a problem like Example 3(c). *Always* check that the factored form can be multiplied out to give the original polynomial.

■ **EXAMPLE 4** **Factoring Out the Greatest Common Factor**

Factor out the greatest common factor.

(a) $a(a + 3) + 4(a + 3)$

The binomial $a + 3$ is the greatest common factor here.

Same

$$a(a + 3) + 4(a + 3) = (a + 3)(a + 4)$$

(b) $x^2(x + 1) - 5(x + 1) = (x + 1)(x^2 - 5)$ Factor out $x + 1$.

> **Now Try Exercise 55.**

OBJECTIVE 3 **Factor by grouping.**   When a polynomial has four terms, common factors can sometimes be used to **factor by grouping.**

■ **EXAMPLE 5** **Factoring by Grouping**

Factor by grouping.

(a) $(2x + 6) + (ax + 3a)$

The first two terms have a common factor of 2, and the last two terms have a common factor of a.

$$2x + 6 + ax + 3a = 2(x + 3) + a(x + 3)$$

The expression is still not in factored form because it is the *sum* of two terms. Now, however, $x + 3$ is a common factor and can be factored out.

$$2x + 6 + ax + 3a = 2(x + 3) + a(x + 3)$$
$$= (x + 3)(2 + a)$$

The final result is in factored form because it is a *product*. Note that the goal in factoring by grouping is to get a common factor, $x + 3$ here, so that the last step is possible.

Check: $(x + 3)(2 + a) = 2x + 6 + ax + 3a$, which is the original polynomial.

(b) $2x^2 - 10x + 3xy - 15y = (2x^2 - 10x) + (3xy - 15y)$ Group terms.

$$= 2x(x - 5) + 3y(x - 5)$$ Factor each group.

$$= (x - 5)(2x + 3y)$$ Factor out the common factor, $x - 5$.

Check: $(x - 5)(2x + 3y) = 2x^2 + 3xy - 10x - 15y$ FOIL

$$= 2x^2 - 10x + 3xy - 15y$$ Original polynomial

(c) $t^3 + 2t^2 - 3t - 6 = (t^3 + 2t^2) + (-3t - 6)$ Group terms.

$$= t^2(t + 2) - 3(t + 2)$$ Factor out -3 so there is a common factor, $t + 2$; $-3(t + 2) = -3t - 6$.

$$= (t + 2)(t^2 - 3)$$ Factor out $t + 2$.

Check by multiplying.

> **Now Try Exercises 69, 73, and 77.**

CAUTION   Use negative signs carefully when grouping, as in Example 5(c). Otherwise, sign errors may result. Always check by multiplying.

Use these steps when factoring four terms by grouping.

Factoring by Grouping

Step 1   **Group terms.**   Collect the terms into two groups so that each group has a common factor.

Step 2   **Factor within groups.**   Factor out the greatest common factor from each group.

Step 3   **Factor the entire polynomial.**   Factor a common binomial factor from the results of Step 2.

Step 4   **If necessary, rearrange terms.**   If Step 2 does not result in a common binomial factor, try a different grouping.

■ **EXAMPLE 6   Rearranging Terms Before Factoring by Grouping**

Factor by grouping.

(a) $10x^2 - 12y + 15x - 8xy$

Factoring out the common factor of 2 from the first two terms and the common factor of x from the last two terms gives

$$10x^2 - 12y + 15x - 8xy = 2(5x^2 - 6y) + x(15 - 8y).$$

This did not lead to a common factor, so we try rearranging the terms. There is usually more than one way to do this. We try

$$10x^2 - 8xy - 12y + 15x,$$

and group the first two terms and the last two terms as follows.

$$10x^2 - 8xy - 12y + 15x = 2x(5x - 4y) + 3(-4y + 5x)$$
$$= 2x(5x - 4y) + 3(5x - 4y)$$
$$= (5x - 4y)(2x + 3)$$

Check: $(5x - 4y)(2x + 3) = 10x^2 + 15x - 8xy - 12y$ FOIL
$$= 10x^2 - 12y + 15x - 8xy$$ Original polynomial

(b) $2xy + 3 - 3y - 2x$

We need to rearrange these terms to get two groups that each have a common factor. Trial and error suggests the following grouping.

$$2xy + 3 - 3y - 2x = (2xy - 3y) + (-2x + 3)$$ Group terms.
$$= y(2x - 3) - 1(2x - 3)$$ Factor each group.
$$= (2x - 3)(y - 1)$$ Factor out the common binomial factor.

Since the quantities in parentheses in the second step must be the same, we factored out -1 rather than 1. Check by multiplying. ■

Now Try Exercise 81.

5.1 EXERCISES

Find the greatest common factor for each list of numbers. See Example 1.

1. 40, 20, 4 **2.** 50, 30, 5 **3.** 18, 24, 36, 48 **4.** 15, 30, 45, 75

5. How can you check your answer when you factor a polynomial?

6. Explain how to find the greatest common factor of a list of terms. Use examples.

Find the greatest common factor for each list of terms. See Examples 1 and 2.

7. $16y$, 24

8. $18w$, 27

9. $30x^3$, $40x^6$, $50x^7$

10. $60z^4$, $70z^8$, $90z^9$

11. $12m^3n^2$, $18m^5n^4$, $36m^8n^3$

12. $25p^5r^7$, $30p^7r^8$, $50p^5r^3$

13. $-x^4y^3$, $-xy^2$

14. $-a^4b^5$, $-a^3b$

15. $42ab^3$, $-36a$, $90b$, $-48ab$

16. $45c^3d$, $75c$, $90d$, $-105cd$

An expression is factored when it is written as a product, not a sum. Which of the following are not factored?

17. $2k^2(5k)$

18. $2k^2(5k + 1)$

19. $2k^2 + (5k + 1)$

20. $(2k^2 + 1)(5k + 1)$

21. Is $-xy$ a common factor of $-x^4y^3$ and $-xy^2$? If so, what is the other factor that when multiplied by $-xy$ gives $-x^4y^3$?

22. Is $-a^5b^2$ a common factor of $-a^4b^5$ and $-a^3b$?

Complete each factoring.

23. $9m^4 = 3m^2($ $)$

24. $12p^5 = 6p^3($ $)$

25. $-8z^9 = -4z^5($ $)$

26. $-15k^{11} = -5k^8($ $)$

27. $6m^4n^5 = 3m^3n($ $)$

28. $27a^3b^2 = 9a^2b($ $)$

29. $12y + 24 = 12($ $)$

30. $18p + 36 = 18($ $)$

31. $10a^2 - 20a = 10a($ $)$

32. $15x^2 - 30x = 15x($ $)$

33. $8x^2y + 12x^3y^2 = 4x^2y($ $)$

34. $18s^3t^2 + 10st = 2st($ $)$

Factor out the greatest common factor. See Examples 3 and 4.

35. $27m^3 - 9m$

36. $36p^3 + 24p$

37. $16z^4 + 22z^2$

38. $25k^4 - 15k^2$

39. $\frac{1}{4}d^2 - \frac{3}{4}d$

40. $\frac{1}{5}z^2 + \frac{3}{5}z$

41. $12x^3 + 6x^2$

42. $21b^3 - 7b^2$

43. $65y^{10} + 35y^6$

44. $100a^5 + 16a^3$

45. $11w^3 - 100$

46. $13z^5 - 80$

47. $8m^2n^3 + 24m^2n^2$

48. $19p^2y - 38p^2y^3$

49. $13y^8 + 26y^4 - 39y^2$

50. $5x^5 + 25x^4 - 20x^3$

51. $45q^4p^5 + 36qp^6 + 81q^2p^3$

52. $125a^3z^5 + 60a^4z^4 - 85a^5z^2$

53. $a^5 + 2a^3b^2 - 3a^5b^2 + 4a^4b^3$

54. $x^6 + 5x^4y^3 - 6xy^4 + 10xy$

55. $c(x + 2) - d(x + 2)$

56. $r(5 - x) + t(5 - x)$

57. $m(m + 2n) + n(m + 2n)$

58. $3p(1 - 4p) - 2q(1 - 4p)$

Students often have difficulty when factoring by grouping because they are not able to tell when the polynomial is completely factored. For example,

$$5y(2x - 3) + 8t(2x - 3)$$

is not in factored form, because it is the *sum* of two terms, $5y(2x - 3)$ and $8t(2x - 3)$. However, because $2x - 3$ is a common factor of these two terms, the expression can now be factored as

$$(2x - 3)(5y + 8t).$$

The factored form is a *product* of two factors, $2x - 3$ and $5y + 8t$.

Determine whether each expression is in factored form or is not in factored form. If it is not in factored form, factor it if possible.

59. $8(7t + 4) + x(7t + 4)$

60. $3r(5x - 1) + 7(5x - 1)$

61. $(8 + x)(7t + 4)$

62. $(3r + 7)(5x - 1)$

63. $18x^2(y + 4) + 7(y + 4)$

64. $12k^3(s - 3) + 7(s + 3)$

65. Tell why it is not possible to factor the expression in Exercise 64.

66. Summarize the method of factoring a polynomial with four terms by grouping. Give an example.

Factor by grouping. See Examples 5 and 6.

67. $p^2 + 4p + pq + 4q$

68. $m^2 + 2m + mn + 2n$

69. $a^2 - 2a + ab - 2b$

70. $y^2 - 6y + yw - 6w$

71. $7z^2 + 14z - az - 2a$

72. $5m^2 + 15mp - 2mr - 6pr$

73. $18r^2 + 12ry - 3xr - 2xy$

74. $8s^2 - 4st + 6sy - 3yt$

75. $3a^3 + 3ab^2 + 2a^2b + 2b^3$

76. $4x^3 + 3x^2y + 4xy^2 + 3y^3$

77. $1 - a + ab - b$

78. $6 - 3x - 2y + xy$

79. $16m^3 - 4m^2p^2 - 4mp + p^3$

80. $10t^3 - 2t^2s^2 - 5ts + s^3$

81. $5m - 6p - 2mp + 15$

82. $y^2 + 3x - 3y - xy$

83. $18r^2 - 2ty + 12ry - 3rt$

84. $2b^3 + 3a^3 + 3ab^2 + 2a^2b$

85. $a^5 - 3 + 2a^5b - 6b$

86. $4b^3 + a^2b - 4a - ab^4$

RELATING CONCEPTS (EXERCISES 87–90)

For Individual or Group Work

*In many cases, the choice of which pairs of terms to group when factoring by grouping can be made in different ways. To see this for Example 6(b), **work Exercises 87–90 in order.***

87. Start with the polynomial from Example 6(b), $2xy + 3 - 3y - 2x$, and rearrange the terms as follows: $2xy - 2x - 3y + 3$. What property from Section 1.7 allows this?

88. Group the first two terms and the last two terms of the rearranged polynomial in Exercise 87. Then factor each group.

89. Is your result from Exercise 88 in factored form? Explain your answer.

90. If your answer to Exercise 89 is *no*, factor the polynomial. Is the result the same as the one shown for Example 6(b)?

91. Refer to Exercise 77. The answer given in the back of the book is $(1 - a)(1 - b)$. A student factored this same polynomial and got the result $(a - 1)(b - 1)$.

   (a) Is the student's answer correct?

   (b) If your answer to part (a) is *yes*, explain why these two seemingly different answers are both acceptable.

92. A student factored $18x^3y^2 + 9xy$ as $9xy(2x^2y)$. Is this correct? If not, explain the error and factor correctly.

5.2   Factoring Trinomials

OBJECTIVES

1 Factor trinomials with a coefficient of 1 for the squared term.

2 Factor such trinomials after factoring out the greatest common factor.

Using FOIL, the product of the binomials $k - 3$ and $k + 1$ is

$$(k - 3)(k + 1) = k^2 - 2k - 3. \qquad \text{Multiplying}$$

Suppose instead that we are given the polynomial $k^2 - 2k - 3$ and want to rewrite it as the product $(k - 3)(k + 1)$. That is,

$$k^2 - 2k - 3 = (k - 3)(k + 1). \qquad \text{Factoring}$$

Recall from the previous section that this process is called factoring the polynomial. Factoring reverses or "undoes" multiplying.

OBJECTIVE 1   Factor trinomials with a coefficient of 1 for the squared term.   When factoring polynomials with integer coefficients, we use only integers in the factors. For example, we can factor $x^2 + 5x + 6$ by finding integers m and n such that

$$x^2 + 5x + 6 = (x + m)(x + n).$$

To find these integers m and n, we first use FOIL to multiply the two binomials on the right side of the equation:

$$(x + m)(x + n) = x^2 + nx + mx + mn.$$

By the distributive property,

$$x^2 + nx + mx + mn = x^2 + (n + m)x + mn.$$

Comparing this result with $x^2 + 5x + 6$ shows that we must find integers m and n having a sum of 5 and a product of 6.

Product of m and n is 6.

$$x^2 + 5x + 6 = x^2 + (n + m)x + mn$$

Sum of m and n is 5.

Since many pairs of integers have a sum of 5, it is best to begin by listing those pairs of integers whose product is 6. Both 5 and 6 are positive, so we consider only pairs in which both integers are positive.

Pair	Product	Sum	
1, 6	$1 \cdot 6 = 6$	$1 + 6 = 7$	
2, 3	$2 \cdot 3 = 6$	$2 + 3 = 5$	Sum is 5.

Both pairs have a product of 6, but only the pair 2 and 3 has a sum of 5. So 2 and 3 are the required integers, and

$$x^2 + 5x + 6 = (x + 2)(x + 3).$$

Check by multiplying the binomials using FOIL. Make sure that the sum of the outer and inner products produces the correct middle term.

$$(x + 2)(x + 3) = x^2 + 5x + 6$$

$$2x$$
$$\underline{3x}$$
$$5x \qquad \text{Add.}$$

We can use this method of factoring only for trinomials that have 1 as the coefficient of the squared term. We give methods for factoring other trinomials in the next section.

EXAMPLE 1 Factoring a Trinomial with All Positive Terms

Factor $m^2 + 9m + 14$.

Look for two integers whose product is 14 and whose sum is 9. List the pairs of integers whose product is 14. Then examine the sums. Again, only positive integers are needed because all signs in $m^2 + 9m + 14$ are positive.

Factors of 14	Sums of Factors	
14, 1	$14 + 1 = 15$	
7, 2	$7 + 2 = 9$	Sum is 9.

From the list, 7 and 2 are the required integers, since $7 \cdot 2 = 14$ and $7 + 2 = 9$. Thus

$$m^2 + 9m + 14 = (m + 7)(m + 2).$$

Check by multiplying on the right side to be sure the original trinomial results. ∎

Now Try Exercise 25.

NOTE In Example 1, the answer also could have been written $(m + 2)(m + 7)$. Because of the commutative property of multiplication, the order of the factors does not matter. Always check by multiplying.

EXAMPLE 2 Factoring a Trinomial with a Negative Middle Term

Factor $x^2 - 9x + 20$.

We must find two integers whose product is 20 and whose sum is -9. Since the numbers we are looking for have a positive product and a negative sum, we consider only pairs of negative integers.

Factors of 20	Sums of Factors	
$-20, -1$	$-20 + (-1) = -21$	
$-10, -2$	$-10 + (-2) = -12$	
$-5, -4$	$-5 + (-4) = -9$	Sum is -9.

The required integers are -5 and -4, so
$$x^2 - 9x + 20 = (x - 5)(x - 4).$$

Check: $(x - 5)(x - 4) = x^2 - 4x - 5x + 20$
$$= x^2 - 9x + 20$$

Now Try Exercise 29.

■ **EXAMPLE 3** Factoring a Trinomial with Two Negative Terms

Factor $p^2 - 2p - 15$.

Find two integers whose product is -15 and whose sum is -2. If these numbers do not come to mind right away, find them (if they exist) by listing all the pairs of integers whose product is -15. Because the last term, -15, is negative, list pairs of integers with different signs.

Factors of -15	Sums of Factors	
$15, -1$	$15 + (-1) = 14$	
$-15, 1$	$-15 + 1 = -14$	
$5, -3$	$5 + (-3) = 2$	
$-5, 3$	$-5 + 3 = -2$	Sum is -2.

The required integers are -5 and 3, so
$$p^2 - 2p - 15 = (p - 5)(p + 3).$$

Check: Multiply $(p - 5)(p + 3)$ to get $p^2 - 2p - 15$.

Now Try Exercise 35.

NOTE In Examples 1–3, notice that we listed factors in descending order (disregarding sign) when we were looking for the required pair of integers. This helps avoid skipping the correct combination.

As shown in the next example, some trinomials cannot be factored using only integers. We call such trinomials **prime polynomials.**

■ **EXAMPLE 4** Deciding Whether Polynomials Are Prime

Factor each trinomial.

(a) $x^2 - 5x + 12$

As in Example 2, both factors must be negative to give a positive product and a negative sum. First, list all pairs of negative integers whose product is 12. Then examine the sums.

Factors of 12	Sums of Factors
$-12, -1$	$-12 + (-1) = -13$
$-6, -2$	$-6 + (-2) = -8$
$-4, -3$	$-4 + (-3) = -7$

None of the pairs of integers has a sum of -5. Therefore, the trinomial $x^2 - 5x + 12$ *cannot be factored using only integers; it is a prime polynomial.*

(b) $k^2 - 8k + 11$

There is no pair of integers whose product is 11 and whose sum is -8, so $k^2 - 8k + 11$ is a prime polynomial.

Now Try Exercise 31.

The procedure for factoring a trinomial of the form $x^2 + bx + c$ is summarized here.

Factoring $x^2 + bx + c$

Find two integers whose product is c and whose sum is b.

1. Both integers must be positive if b and c are positive.
2. Both integers must be negative if c is positive and b is negative.
3. One integer must be positive and one must be negative if c is negative.

EXAMPLE 5   Factoring a Trinomial with Two Variables

Factor $z^2 - 2bz - 3b^2$.

The coefficient of z in the middle term is $-2b$, so we need to find two expressions whose product is $-3b^2$ and whose sum is $-2b$. The expressions are $-3b$ and b, so

$$z^2 - 2bz - 3b^2 = (z - 3b)(z + b).$$

Check by multiplying.

Now Try Exercise 45.

OBJECTIVE 2   Factor such trinomials after factoring out the greatest common factor. The trinomial in the next example does not have a coefficient of 1 for the squared term. (In fact, there is no squared term.) However, there may be a common factor.

EXAMPLE 6   Factoring a Trinomial with a Common Factor

Factor $4x^5 - 28x^4 + 40x^3$.

First, factor out the greatest common factor, $4x^3$.

$$4x^5 - 28x^4 + 40x^3 = 4x^3(x^2 - 7x + 10)$$

Now factor $x^2 - 7x + 10$. The integers -5 and -2 have a product of 10 and a sum of -7. The complete factored form is

$$4x^5 - 28x^4 + 40x^3 = 4x^3(x - 5)(x - 2).$$

Remember to include $4x^3$.

Check: $4x^3(x - 5)(x - 2) = 4x^3(x^2 - 7x + 10)$
$$= 4x^5 - 28x^4 + 40x^3$$

Now Try Exercise 53.

CAUTION When factoring, always look for a common factor first. Remember to include the common factor as part of the answer. As a check, multiplying out the factored form should always give the original polynomial.

5.2 EXERCISES

In Exercises 1–4, list all pairs of integers with the given product. Then find the pair whose sum is given.

1. Product: 48;   Sum: −19

2. Product: 48;   Sum: 14

3. Product: −24;   Sum: −5

4. Product: −36;   Sum: −16

5. In factoring a trinomial in x as $(x + a)(x + b)$, what must be true of a and b if the coefficient of the last term of the trinomial is negative?

6. In Exercise 5, what must be true of a and b if the coefficient of the last term is positive?

7. What is meant by a *prime polynomial*?

8. How can you check your work when factoring a trinomial? Does the check ensure that the trinomial is completely factored?

9. Which one of the following is the correct factored form of $x^2 - 12x + 32$?

    A. $(x - 8)(x + 4)$ **B.** $(x + 8)(x - 4)$

    C. $(x - 8)(x - 4)$ **D.** $(x + 8)(x + 4)$

10. What would be the first step in factoring $2x^3 + 8x^2 - 10x$?

Complete each factoring. See Examples 1–4.

11. $p^2 + 11p + 30 = (p + 5)(\quad\quad)$

12. $x^2 + 10x + 21 = (x + 7)(\quad\quad)$

13. $x^2 + 15x + 44 = (x + 4)(\quad\quad)$

14. $r^2 + 15r + 56 = (r + 7)(\quad\quad)$

15. $x^2 - 9x + 8 = (x - 1)(\quad\quad)$

16. $t^2 - 14t + 24 = (t - 2)(\quad\quad)$

17. $y^2 - 2y - 15 = (y + 3)(\quad\quad)$

18. $t^2 - t - 42 = (t + 6)(\quad\quad)$

19. $x^2 + 9x - 22 = (x - 2)(\quad\quad)$

20. $x^2 + 6x - 27 = (x - 3)(\quad\quad)$

21. $y^2 - 7y - 18 = (y + 2)(\quad\quad)$

22. $y^2 - 2y - 24 = (y + 4)(\quad\quad)$

Factor completely. If the polynomial cannot be factored, write prime. *See Examples 1–4.*

23. $y^2 + 9y + 8$

24. $a^2 + 9a + 20$

25. $b^2 + 8b + 15$

26. $x^2 + 6x + 8$

27. $m^2 + m - 20$

28. $p^2 + 4p - 5$

29. $y^2 - 8y + 15$

30. $y^2 - 6y + 8$

31. $x^2 + 4x + 5$

32. $t^2 + 11t + 12$

33. $z^2 - 15z + 56$

34. $x^2 - 13x + 36$

35. $r^2 - r - 30$

36. $q^2 - q - 42$

37. $a^2 - 8a - 48$

38. $m^2 - 10m - 25$

39. $x^2 + 3x - 39$

40. $d^2 + 4d - 45$

41. Explain how you would factor $8 + 6x + x^2$.

42. Use your answer to Exercise 41 to factor $5 - 4x - x^2$.

Factor completely. See Examples 5 and 6.

43. $r^2 + 3ra + 2a^2$ **44.** $x^2 + 5xa + 4a^2$ **45.** $t^2 - tz - 6z^2$

46. $a^2 - ab - 12b^2$ **47.** $x^2 + 4xy + 3y^2$ **48.** $p^2 + 9pq + 8q^2$

49. $v^2 - 11vw + 30w^2$ **50.** $v^2 - 11vx + 24x^2$ **51.** $4x^2 + 12x - 40$

52. $5y^2 - 5y - 30$ **53.** $2t^3 + 8t^2 + 6t$ **54.** $3t^3 + 27t^2 + 24t$

55. $2x^6 + 8x^5 - 42x^4$ **56.** $4y^5 + 12y^4 - 40y^3$ **57.** $5m^5 + 25m^4 - 40m^2$

58. $12k^5 - 6k^3 + 10k^2$ **59.** $m^3n - 10m^2n^2 + 24mn^3$ **60.** $y^3z + 3y^2z^2 - 54yz^3$

61. Use the FOIL method from Section 4.5 to show that $(2x + 4)(x - 3) = 2x^2 - 2x - 12$. If you are asked to completely factor $2x^2 - 2x - 12$, why would it be incorrect to give $(2x + 4)(x - 3)$ as your answer?

62. If you are asked to completely factor the polynomial $3x^2 + 9x - 12$, why would it be incorrect to give $(x - 1)(3x + 12)$ as your answer?

Use a combination of the factoring methods discussed in this section to factor each polynomial.

63. $a^5 + 3a^4b - 4a^3b^2$ **64.** $m^3n - 2m^2n^2 - 3mn^3$ **65.** $y^3z + y^2z^2 - 6yz^3$

66. $k^7 - 2k^6m - 15k^5m^2$ **67.** $z^{10} - 4z^9y - 21z^8y^2$ **68.** $x^9 + 5x^8w - 24x^7w^2$

69. $(a + b)x^2 + (a + b)x - 12(a + b)$

70. $(x + y)n^2 + (x + y)n + 16(x + y)$

71. $(2p + q)r^2 - 12(2p + q)r + 27(2p + q)$

72. $(3m - n)k^2 - 13(3m - n)k + 40(3m - n)$

73. What polynomial can be factored as $(a + 9)(a + 4)$?

74. What polynomial can be factored as $(y - 7)(y + 3)$?

5.3 | More on Factoring Trinomials

OBJECTIVES

1 Factor trinomials by grouping when the coefficient of the squared term is not 1.

2 Factor trinomials using FOIL.

Trinomials like $2x^2 + 7x + 6$, in which the coefficient of the squared term is *not* 1, are factored with extensions of the methods from the previous sections. One such method uses factoring by grouping from Section 5.1.

OBJECTIVE 1 Factor trinomials by grouping when the coefficient of the squared term is not 1.   Recall that a trinomial such as $m^2 + 3m + 2$ is factored by finding two numbers whose product is 2 and whose sum is 3. To factor $2x^2 + 7x + 6$, we look for two integers whose product is $2 \cdot 6 = 12$ and whose sum is 7.

Sum is 7.

$$2x^2 + 7x + 6$$

Product is $2 \cdot 6 = 12$.

By considering pairs of positive integers whose product is 12, we find the necessary integers to be 3 and 4. We use these integers to write the middle term, $7x$, as $7x = 3x + 4x$. The trinomial $2x^2 + 7x + 6$ becomes

$$2x^2 + 7x + 6 = 2x^2 + \underbrace{3x + 4x}_{7x} + 6.$$

$$= (2x^2 + 3x) + (4x + 6) \qquad \text{Group terms.}$$
$$= x(2x + 3) + 2(2x + 3) \qquad \text{Factor each group.}$$

Must be the same factor

$$2x^2 + 7x + 6 = (2x + 3)(x + 2) \qquad \text{Factor out } 2x + 3.$$

Check: $(2x + 3)(x + 2) = 2x^2 + 7x + 6$

In this example, we could have written $7x$ as $4x + 3x$. Factoring by grouping this way would give the same answer.

■ **EXAMPLE 1** Factoring Trinomials by Grouping

Factor each trinomial.

(a) $6r^2 + r - 1$

We must find two integers with a product of $6(-1) = -6$ and a sum of 1.

Sum is 1.

$$6r^2 + r - 1 = 6r^2 + 1r - 1$$

Product is $6(-1) = -6$.

The integers are -2 and 3. We write the middle term, r, as $-2r + 3r$.

$$6r^2 + r - 1 = 6r^2 - 2r + 3r - 1 \qquad r = -2r + 3r$$
$$= (6r^2 - 2r) + (3r - 1) \qquad \text{Group terms.}$$
$$= 2r(3r - 1) + 1(3r - 1) \qquad \text{The binomials must be the same.}$$
$$= (3r - 1)(2r + 1) \qquad \text{Factor out } 3r - 1.$$

Check: $(3r - 1)(2r + 1) = 6r^2 + r - 1$

(b) $12z^2 - 5z - 2$

Look for two integers whose product is $12(-2) = -24$ and whose sum is -5. The required integers are 3 and -8, so

$$12z^2 - 5z - 2 = 12z^2 + 3z - 8z - 2 \qquad -5z = 3z - 8z$$
$$= (12z^2 + 3z) + (-8z - 2) \qquad \text{Group terms.}$$
$$= 3z(4z + 1) - 2(4z + 1) \qquad \text{Factor each group; be careful with signs.}$$
$$= (4z + 1)(3z - 2). \qquad \text{Factor out } 4z + 1.$$

Check: $(4z + 1)(3z - 2) = 12z^2 - 5z - 2$

(c) $10m^2 + mn - 3n^2$

Two integers whose product is $10(-3) = -30$ and whose sum is 1 are -5 and 6. Rewrite the trinomial with four terms.

$$10m^2 + mn - 3n^2 = 10m^2 - 5mn + 6mn - 3n^2 \qquad \textit{mn} = -5mn + 6mn$$
$$= 5m(2m - n) + 3n(2m - n) \qquad \text{Group terms;} \\ \text{factor each group.}$$
$$= (2m - n)(5m + 3n) \qquad \text{Factor out } 2m - n.$$

Check by multiplying.

Now Try Exercises 21, 27, and 47.

▨ **EXAMPLE 2** Factoring a Trinomial with a Common Factor by Grouping

Factor $28x^5 - 58x^4 - 30x^3$.

First factor out the greatest common factor, $2x^3$.

$$28x^5 - 58x^4 - 30x^3 = 2x^3(14x^2 - 29x - 15)$$

To factor $14x^2 - 29x - 15$, find two integers whose product is $14(-15) = -210$ and whose sum is -29. Factoring 210 into prime factors gives

$$210 = 2 \cdot 3 \cdot 5 \cdot 7.$$

Combine these prime factors in pairs in different ways, using one positive and one negative (to get -210). The factors 6 and -35 have the correct sum. Now rewrite the given trinomial and factor it.

$$28x^5 - 58x^4 - 30x^3 = 2x^3(14x^2 + 6x - 35x - 15)$$
$$= 2x^3[(14x^2 + 6x) + (-35x - 15)]$$
$$= 2x^3[2x(7x + 3) - 5(7x + 3)]$$
$$= 2x^3[(7x + 3)(2x - 5)]$$
$$= 2x^3(7x + 3)(2x - 5)$$

Check by multiplying.

Now Try Exercise 43.

CAUTION Remember to include the common factor in the final result.

OBJECTIVE 2 Factor trinomials using FOIL. In the rest of this section we show an alternative method of factoring trinomials in which the coefficient of the squared term is not 1. This method generalizes the factoring procedure explained in Section 5.2.

To factor $2x^2 + 7x + 6$ (the trinomial factored at the beginning of this section) by the method in Section 5.2, use FOIL backwards. We want to write $2x^2 + 7x + 6$ as the product of two binomials.

$$2x^2 + 7x + 6 = (\qquad)(\qquad)$$

The product of the two first terms of the binomials is $2x^2$. The possible factors of $2x^2$ are $2x$ and x or $-2x$ and $-x$. Since all terms of the trinomial are positive, we consider only positive factors. Thus, we have

$$2x^2 + 7x + 6 = (2x \qquad)(x \qquad).$$

The product of the two last terms, 6, can be factored as $1 \cdot 6, 6 \cdot 1, 2 \cdot 3,$ or $3 \cdot 2$. Try each pair to find the pair that gives the correct middle term, $7x$.

$(2x + 1)(x + 6)$ Incorrect | $(2x + 6)(x + 1)$ Incorrect

x $6x$

$\underline{12x}$ $\underline{2x}$

$13x$ Add. $8x$ Add.

Since $2x + 6 = 2(x + 3)$, the binomial $2x + 6$ has a common factor of 2, while $2x^2 + 7x + 6$ has no common factor other than 1. The product $(2x + 6)(x + 1)$ cannot be correct.

NOTE If the original polynomial has no common factor, then none of its binomial factors will either.

Now try the numbers 2 and 3 as factors of 6. Because of the common factor of 2 in $2x + 2$, $(2x + 2)(x + 3)$ will not work, so we try $(2x + 3)(x + 2)$.

$$(2x + 3)(x + 2) = 2x^2 + 7x + 6 \qquad \text{Correct}$$

$3x$

$\underline{4x}$

$7x$ Add.

Thus, $2x^2 + 7x + 6$ factors as

$$2x^2 + 7x + 6 = (2x + 3)(x + 2).$$

Check by multiplying $2x + 3$ and $x + 2$.

EXAMPLE 3 Factoring a Trinomial with All Positive Terms Using FOIL

Factor $8p^2 + 14p + 5$.

The number 8 has several possible pairs of factors, but 5 has only 1 and 5 or -1 and -5. For this reason, it is easier to begin by considering the factors of 5. Ignore the negative factors since all coefficients in the trinomial are positive. If $8p^2 + 14p + 5$ can be factored, the factors will have the form

$$(\quad + 5)(\quad + 1).$$

The possible pairs of factors of $8p^2$ are $8p$ and p, or $4p$ and $2p$. Try various combinations, checking in each case to see if the middle term is $14p$.

$(8p + 5)(p + 1)$ Incorrect | $(p + 5)(8p + 1)$ Incorrect

$5p$ $40p$

$\underline{8p}$ $\underline{p}$

$13p$ Add. $41p$ Add.

$(4p + 5)(2p + 1)$ Correct

$10p$

$\underline{4p}$

$14p$ Add.

Since $14p$ is the correct middle term,

$$8p^2 + 14p + 5 = (4p + 5)(2p + 1).$$

Check: $(4p + 5)(2p + 1) = 8p^2 + 14p + 5$

Now Try Exercise 23.

▒ EXAMPLE 4 Factoring a Trinomial with a Negative Middle Term Using FOIL

Factor $6x^2 - 11x + 3$.

Since 3 has only 1 and 3 or -1 and -3 as factors, it is better here to begin by factoring 3. The last term of the trinomial $6x^2 - 11x + 3$ is positive and the middle term has a negative coefficient, so we consider only negative factors. We need two negative factors because the *product* of two negative factors is positive and their *sum* is negative, as required. Use -3 and -1 as factors of 3:

$$(\quad - 3)(\quad - 1).$$

The factors of $6x^2$ may be either $6x$ and x, or $2x$ and $3x$. We try $2x$ and $3x$.

$$(2x - 3)(3x - 1) \qquad \text{Correct}$$
$$-9x$$
$$-2x$$
$$\overline{-11x} \qquad \text{Add.}$$

These factors give the correct middle term, so

$$6x^2 - 11x + 3 = (2x - 3)(3x - 1).$$

Check by multiplying.

Now Try Exercise 29.

▒ EXAMPLE 5 Factoring a Trinomial with a Negative Last Term Using FOIL

Factor $8x^2 + 6x - 9$.

The integer 8 has several possible pairs of factors, as does -9. Since the last term is negative, one positive factor and one negative factor of -9 are needed. Since the coefficient of the middle term is small, it is wise to avoid large factors such as 8 or 9. We try $4x$ and $2x$ as factors of $8x^2$, and 3 and -3 as factors of -9, and check the middle term.

$$(4x + 3)(2x - 3) \qquad \text{Incorrect}$$
$$6x$$
$$-12x$$
$$\overline{-6x} \qquad \text{Add.}$$

Now, we try interchanging 3 and -3, since only the sign of the middle term is incorrect.

$$(4x - 3)(2x + 3) \qquad \text{Correct}$$
$$-6x$$
$$12x$$
$$\overline{6x} \qquad \text{Add.}$$

This combination produces the correct middle term, so

$$8x^2 + 6x - 9 = (4x - 3)(2x + 3).$$

Now Try Exercise 33.

▒ EXAMPLE 6 Factoring a Trinomial with Two Variables

Factor $12a^2 - ab - 20b^2$.

There are several pairs of factors of $12a^2$, including $12a$ and a, $6a$ and $2a$, and $3a$ and $4a$, just as there are many pairs of factors of $-20b^2$, including $20b$ and $-b$,

$-20b$ and b, $10b$ and $-2b$, $-10b$ and $2b$, $4b$ and $-5b$, and $-4b$ and $5b$. Once again, since the desired middle term is small, avoid the larger factors. Try the factors $6a$ and $2a$, and $4b$ and $-5b$.

$$(6a + 4b)(2a - 5b)$$

This cannot be correct, as mentioned before, since $6a + 4b$ has a common factor while the given trinomial has none. Try $3a$ and $4a$ with $4b$ and $-5b$.

$$(3a + 4b)(4a - 5b) = 12a^2 + ab - 20b^2 \qquad \text{Incorrect}$$

Here the middle term has the wrong sign, so we interchange the signs in the factors.

$$(3a - 4b)(4a + 5b) = 12a^2 - ab - 20b^2 \qquad \text{Correct}$$

Now Try Exercise 41.

EXAMPLE 7 Factoring Trinomials with Common Factors

Factor each trinomial.

(a) $15y^3 + 55y^2 + 30y$

First factor out the greatest common factor, $5y$.

$$15y^3 + 55y^2 + 30y = 5y(3y^2 + 11y + 6)$$

Now factor $3y^2 + 11y + 6$. Try $3y$ and y as factors of $3y^2$, and 2 and 3 as factors of 6.

$$(3y + 2)(y + 3) = 3y^2 + 11y + 6 \qquad \text{Correct}$$

The complete factored form of $15y^3 + 55y^2 + 30y$ is

$$15y^3 + 55y^2 + 30y = 5y(3y + 2)(y + 3).$$

Check by multiplying.

(b) $-24a^3 - 42a^2 + 45a$

The common factor could be $3a$ or $-3a$. If we factor out $-3a$, the first term of the trinomial will be positive, which makes it easier to factor.

$$-24a^3 - 42a^2 + 45a = -3a(8a^2 + 14a - 15) \qquad \text{Factor out } -3a.$$
$$= -3a(4a - 3)(2a + 5) \qquad \text{Use FOIL.}$$

Check by multiplying.

Now Try Exercise 45.

CAUTION This caution bears repeating: Remember to include the common factor in the final factored form.

The two methods for factoring trinomials are summarized here.

Factoring Trinomials of the Form $ax^2 + bx + c$

Write the middle term of the trinomial as the sum of two terms to get a polynomial with four terms. Use factoring by grouping as shown in Section 5.1.

Factor the trinomial using FOIL backwards as explained in Section 5.2 and this section.

5.3 EXERCISES

Factor each polynomial by grouping. (The middle term of an equivalent trinomial has already been rewritten.) See Example 1.

1. $10t^2 + 5t + 4t + 2$

2. $6x^2 + 9x + 4x + 6$

3. $15z^2 - 10z - 9z + 6$

4. $12p^2 - 9p - 8p + 6$

5. $8s^2 - 4st + 6st - 3t^2$

6. $3x^2 - 7xy + 6xy - 14y^2$

7. Which pair of integers would be used to rewrite the middle term when factoring $12y^2 + 5y - 2$ by grouping?

A. $-8, 3$ **B.** $8, -3$
C. $-6, 4$ **D.** $6, -4$

8. Which pair of integers would be used to rewrite the middle term when factoring $20b^2 - 13b + 2$ by grouping?

A. $10, 3$ **B.** $-10, -3$
C. $8, 5$ **D.** $-8, -5$

Complete the steps to factor each trinomial by grouping.

9. $2m^2 + 11m + 12$

(a) Find two integers whose product is
_____ · _____ = _____
and whose sum is _____.

(b) The required integers are _____ and _____.

(c) Write the middle term $11m$ as
_____ + _____.

(d) Rewrite the given trinomial as
_____.

(e) Factor the polynomial in part (d) by grouping.

(f) Check by multiplying.

10. $6y^2 - 19y + 10$

(a) Find two integers whose product is
_____ · _____ = _____
and whose sum is _____.

(b) The required integers are _____ and _____.

(c) Write the middle term $-19y$ as
_____ + _____.

(d) Rewrite the given trinomial as
_____.

(e) Factor the polynomial in part (d) by grouping.

(f) Check by multiplying.

Decide which is the correct factored form of the given polynomial.

11. $2x^2 - x - 1$

A. $(2x - 1)(x + 1)$
B. $(2x + 1)(x - 1)$

12. $3a^2 - 5a - 2$

A. $(3a + 1)(a - 2)$
B. $(3a - 1)(a + 2)$

13. $4y^2 + 17y - 15$

A. $(y + 5)(4y - 3)$
B. $(2y - 5)(2y + 3)$

14. $12c^2 - 7c - 12$

A. $(6c - 2)(2c + 6)$
B. $(4c + 3)(3c - 4)$

Complete each factoring.

15. $6a^2 + 7ab - 20b^2 = (3a - 4b)(\qquad)$

16. $9m^2 - 3mn - 2n^2 = (3m + n)(\qquad)$

17. $2x^2 + 6x - 8 = 2(\qquad)$
$\qquad\qquad = 2(\qquad)(\qquad)$

18. $3x^2 - 9x - 30 = 3(\qquad)$
$\qquad\qquad = 3(\qquad)(\qquad)$

19. For the polynomial $12x^2 + 7x - 12$, 2 is not a common factor. Explain why the binomial $2x - 6$, then, cannot be a factor of the polynomial.

✐ **20.** Factor $4k^2 + 7k - 15$ twice, using the two methods discussed in the text. Do your answers agree? Which method do you prefer?

Factor each trinomial completely. See Examples 1–7.

21. $3a^2 + 10a + 7$ **22.** $7r^2 + 8r + 1$ **23.** $2y^2 + 7y + 6$

24. $5z^2 + 12z + 4$ **25.** $15m^2 + m - 2$ **26.** $6x^2 + x - 1$

27. $12s^2 + 11s - 5$ **28.** $20x^2 + 11x - 3$ **29.** $10m^2 - 23m + 12$

30. $6x^2 - 17x + 12$ **31.** $8w^2 - 14w + 3$ **32.** $9p^2 - 18p + 8$

33. $20y^2 - 39y - 11$ **34.** $10x^2 - 11x - 6$ **35.** $3x^2 - 15x + 16$

36. $2t^2 + 13t - 18$ **37.** $20x^2 + 22x + 6$ **38.** $36y^2 + 81y + 45$

39. $24x^2 - 42x + 9$ **40.** $48b^2 - 74b - 10$ **41.** $40m^2q + mq - 6q$

42. $15a^2b + 22ab + 8b$ **43.** $15n^4 - 39n^3 + 18n^2$ **44.** $24a^4 + 10a^3 - 4a^2$

45. $15x^2y^2 - 7xy^2 - 4y^2$ **46.** $14a^2b^3 + 15ab^3 - 9b^3$ **47.** $5a^2 - 7ab - 6b^2$

48. $6x^2 - 5xy - y^2$ **49.** $12s^2 + 11st - 5t^2$ **50.** $25a^2 + 25ab + 6b^2$

51. $6m^6n + 7m^5n^2 + 2m^4n^3$ **52.** $12k^3q^4 - 4k^2q^5 - kq^6$ **53.** $5 - 6x + x^2$

54. $7 + 8x + x^2$ **55.** $16 + 16x + 3x^2$ **56.** $18 + 65x + 7x^2$

57. $-10x^3 + 5x^2 + 140x$ **58.** $-18k^3 - 48k^2 + 66k$

✐ **59.** On a quiz, a student factored $16x^2 - 24x + 5$ by grouping as follows.

$$16x^2 - 24x + 5 = 16x^2 - 4x - 20x + 5$$
$$= 4x(4x - 1) - 5(4x - 1) \qquad \text{His answer}$$

He thought his answer was correct since it checked by multiplying. Why was the answer marked wrong? What is the correct factored form?

✐ **60.** On the same quiz, another student factored $3k^3 - 12k^2 - 15k$ by first factoring out the common factor $3k$ to get $3k(k^2 - 4k - 5)$. Then she wrote

$$k^2 - 4k - 5 = k^2 - 5k + k - 5$$
$$= k(k - 5) + 1(k - 5)$$
$$= (k - 5)(k + 1). \qquad \text{Her answer}$$

Why was the answer marked wrong? What is the correct factored form?

If a trinomial has a negative coefficient for the squared term, such as $-2x^2 + 11x - 12$, it is usually easier to factor by first factoring out the common factor -1:

$$-2x^2 + 11x - 12 = -1(2x^2 - 11x + 12)$$
$$= -1(2x - 3)(x - 4).$$

Use this method to factor each trinomial. See Example 7(b).

61. $-x^2 - 4x + 21$ **62.** $-x^2 + x + 72$

63. $-3x^2 - x + 4$ **64.** $-5x^2 + 2x + 16$

65. $-2a^2 - 5ab - 2b^2$ **66.** $-3p^2 + 13pq - 4q^2$

✐ **67.** The answer given in the back of the book for Exercise 61 is $-1(x + 7)(x - 3)$. Is $(x + 7)(3 - x)$ also a correct answer? Explain.

✐ **68.** One answer for Exercise 62 is $-1(x + 8)(x - 9)$. Is $(-x - 8)(-x + 9)$ also a correct answer? Explain.

Factor each polynomial. Remember to factor out the greatest common factor as the first step.

69. $25q^2(m + 1)^3 - 5q(m + 1)^3 - 2(m + 1)^3$

70. $18x^2(y - 3)^2 - 21x(y - 3)^2 - 4(y - 3)^2$

71. $15x^2(r + 3)^3 - 34xy(r + 3)^3 - 16y^2(r + 3)^3$

72. $4t^2(k + 9)^7 + 20ts(k + 9)^7 + 25s^2(k + 9)^7$

Find all integers k so that the trinomial can be factored using the methods of this section. (Hint: Try all possible factored forms with the given first and last terms. The coefficient of the variable will give the values of k.)

73. $5x^2 + kx - 1$ **74.** $2c^2 + kc - 3$

75. $2m^2 + km + 5$ **76.** $3y^2 + ky + 3$

5.4 Special Factoring Rules

OBJECTIVES

1 Factor a difference of squares.

2 Factor a perfect square trinomial.

3 Factor a difference of cubes.

4 Factor a sum of cubes.

By reversing the rules for multiplication of binomials from the last chapter, we get rules for factoring polynomials in certain forms.

OBJECTIVE 1 Factor a difference of squares. The formula for the product of the sum and difference of the same two terms is

$$(x + y)(x - y) = x^2 - y^2.$$

Reversing this rule leads to the following special factoring rule.

Factoring a Difference of Squares

$$x^2 - y^2 = (x + y)(x - y)$$

For example,

$$m^2 - 16 = m^2 - 4^2 = (m + 4)(m - 4).$$

As the next examples show, the following conditions must be true for a binomial to be a difference of squares.

1. Both terms of the binomial must be squares, such as

$$x^2, \quad 9y^2, \quad 25, \quad 1, \quad m^4.$$

2. The second terms of the binomials must have different signs (one positive and one negative).

EXAMPLE 1 Factoring Differences of Squares

Factor each binomial, if possible.

$$x^2 - y^2 = (x + y)(x - y)$$

(a) $a^2 - 49 = a^2 - 7^2 = (a + 7)(a - 7)$

(b) $y^2 - m^2 = (y + m)(y - m)$

(c) $z^2 - \dfrac{9}{16} = z^2 - \left(\dfrac{3}{4}\right)^2 = \left(z + \dfrac{3}{4}\right)\left(z - \dfrac{3}{4}\right)$

(d) $x^2 - 8$

Because 8 is not the square of an integer, this binomial is not a difference of squares. It is a prime polynomial.

(e) $p^2 + 16$

Since $p^2 + 16$ is a *sum* of squares, it is not equal to $(p + 4)(p - 4)$. Also, using FOIL,

$$(p - 4)(p - 4) = p^2 - 8p + 16 \neq p^2 + 16$$

and

$$(p + 4)(p + 4) = p^2 + 8p + 16 \neq p^2 + 16,$$

so $p^2 + 16$ is a prime polynomial.

Now Try Exercises 7, 9, and 11.

CAUTION As Example 1(e) suggests, after any common factor is removed, a *sum* of squares cannot be factored.

EXAMPLE 2 Factoring Differences of Squares

Factor each difference of squares.

$$x^2 \quad - \quad y^2 = (x \quad + \quad y)(x \quad - \quad y)$$

(a) $25m^2 - 16 = (5m)^2 - 4^2 = (5m + 4)(5m - 4)$

(b) $49z^2 - 64 = (7z)^2 - 8^2 = (7z + 8)(7z - 8)$

Now Try Exercise 13.

NOTE As in previous sections, you should always check a factored form by multiplying.

EXAMPLE 3 Factoring More Complex Differences of Squares

Factor completely.

(a) $81y^2 - 36$

First factor out the common factor, 9.

$$\begin{aligned}
81y^2 - 36 &= 9(9y^2 - 4) &&\text{Factor out 9.}\\
&= 9[(3y)^2 - 2^2]\\
&= 9(3y + 2)(3y - 2) &&\text{Difference of squares}
\end{aligned}$$

(b) $9x^2 - 4z^2 = (3x)^2 - (2z)^2 = (3x + 2z)(3x - 2z)$

(c) $p^4 - 36 = (p^2)^2 - 6^2 = (p^2 + 6)(p^2 - 6)$

Neither $p^2 + 6$ nor $p^2 - 6$ can be factored further.

(d) $m^4 - 16 = (m^2)^2 - 4^2$

$\qquad\qquad = (m^2 + 4)(m^2 - 4)$ — Difference of squares

$\qquad\qquad = (m^2 + 4)(m + 2)(m - 2)$ — Difference of squares again

Now Try Exercises 17, 21, and 25.

CAUTION Remember to factor again when any of the factors is a difference of squares, as in Example 3(d). Check by multiplying.

OBJECTIVE 2 Factor a perfect square trinomial. The expressions 144, $4x^2$, and $81m^6$ are called *perfect squares* because

$$144 = 12^2, \qquad 4x^2 = (2x)^2, \qquad \text{and} \qquad 81m^6 = (9m^3)^2.$$

A **perfect square trinomial** is a trinomial that is the square of a binomial. For example, $x^2 + 8x + 16$ is a perfect square trinomial because it is the square of the binomial $x + 4$:

$$x^2 + 8x + 16 = (x + 4)(x + 4) = (x + 4)^2.$$

For a trinomial to be a perfect square, *two of its terms must be perfect squares.* For this reason, $16x^2 + 4x + 15$ is not a perfect square trinomial because only the term $16x^2$ is a perfect square.

On the other hand, even if two of the terms are perfect squares, the trinomial may not be a perfect square trinomial. For example, $x^2 + 6x + 36$ has two perfect square terms, x^2 and 36, but it is not a perfect square trinomial. (Try to find a binomial that can be squared to give $x^2 + 6x + 36$.)

We can multiply to see that the square of a binomial gives one of the following perfect square trinomials.

Factoring Perfect Square Trinomials

$$x^2 + 2xy + y^2 = (x + y)^2$$
$$x^2 - 2xy + y^2 = (x - y)^2$$

The middle term of a perfect square trinomial is always twice the product of the two terms in the squared binomial (as shown in Section 4.6). Use this rule to check any attempt to factor a trinomial that appears to be a perfect square.

EXAMPLE 4 Factoring a Perfect Square Trinomial

Factor $x^2 + 10x + 25$.

The term x^2 is a perfect square, and so is 25. Try to factor the trinomial as

$$x^2 + 10x + 25 = (x + 5)^2.$$

To check, take twice the product of the two terms in the squared binomial.

$$2 \cdot x \cdot 5 = 10x$$

Twice · First term of binomial · Last term of binomial

Since $10x$ is the middle term of the trinomial, the trinomial is a perfect square and can be factored as $(x + 5)^2$. Thus,

$$x^2 + 10x + 25 = (x + 5)^2.$$

Now Try Exercise 33.

EXAMPLE 5 Factoring Perfect Square Trinomials

Factor each trinomial.

(a) $x^2 - 22x + 121$

The first and last terms are perfect squares ($121 = 11^2$ or $(-11)^2$). Check to see whether the middle term of $x^2 - 22x + 121$ is twice the product of the first and last terms of the binomial $x - 11$.

$$2 \cdot x \cdot (-11) = -22x$$

Twice ⎯ First term ⎯ Last term

Since twice the product of the first and last terms of the binomial is the middle term, $x^2 - 22x + 121$ is a perfect square trinomial and

$$x^2 - 22x + 121 = (x - 11)^2.$$

Notice that the sign of the second term in the squared binomial is the same as the sign of the middle term in the trinomial.

(b) $9m^2 - 24m + 16 = (3m)^2 + 2(3m)(-4) + (-4)^2 = (3m - 4)^2$

Twice ⎯ First term ⎯ Last term

(c) $25y^2 + 20y + 16$

The first and last terms are perfect squares.

$$25y^2 = (5y)^2 \qquad \text{and} \qquad 16 = 4^2$$

Twice the product of the first and last terms of the binomial $5y + 4$ is

$$2 \cdot 5y \cdot 4 = 40y,$$

which is not the middle term of $25y^2 + 20y + 16$. This trinomial is not a perfect square. In fact, the trinomial cannot be factored even with the methods of the previous sections; it is a prime polynomial.

(d) $12z^3 + 60z^2 + 75z$

Factor out the common factor, $3z$, first.

$$12z^3 + 60z^2 + 75z = 3z(4z^2 + 20z + 25)$$
$$= 3z[(2z)^2 + 2(2z)(5) + 5^2]$$
$$= 3z(2z + 5)^2$$

Now Try Exercises 35, 43, and 51.

N O T E As mentioned in Example 5(a), the sign of the second term in the squared binomial is always the same as the sign of the middle term in the trinomial. Also, the first and last terms of a perfect square trinomial must be *positive*, because they are squares. For example, the polynomial $x^2 - 2x - 1$ cannot be a perfect square because the last term is negative.

Perfect square trinomials can also be factored using grouping or FOIL, although using the method of this section is often easier.

OBJECTIVE 3 Factor a difference of cubes. Just as we factored the difference of squares in Objective 1, we can also factor the **difference of cubes** using the following pattern.

Factoring a Difference of Cubes
$$x^3 - y^3 = (x - y)(x^2 + xy + y^2)$$

This pattern *should be memorized*. Multiply on the right to see that the pattern gives the correct factors.

$$
\begin{array}{r}
x^2 + xy + y^2 \\
x - y \\
\hline
-x^2y - xy^2 - y^3 \\
x^3 + x^2y + xy^2 \\
\hline
x^3 \qquad\qquad - y^3
\end{array}
$$

Notice the pattern of the terms in the factored form of $x^3 - y^3$.

- $x^3 - y^3 = $ (a binomial factor)(a trinomial factor)
- The binomial factor has the difference of the cube roots of the given terms.
- The terms in the trinomial factor are all positive.
- What you write in the binomial factor determines the trinomial factor:

$$
x^3 - y^3 = (x - y)(\underset{\substack{\text{First term}\\\text{squared}}}{x^2} + \underset{\substack{\text{positive}\\\text{product of}\\\text{the terms}}}{xy} + \underset{\substack{\text{second term}\\\text{squared}}}{y^2}).
$$

CAUTION The polynomial $x^3 - y^3$ is not equivalent to $(x - y)^3$, because $(x - y)^3$ can also be written as

$$
\begin{aligned}
(x - y)^3 &= (x - y)(x - y)(x - y) \\
&= (x - y)(x^2 - 2xy + y^2)
\end{aligned}
$$

but

$$x^3 - y^3 = (x - y)(x^2 + xy + y^2).$$

▨ **EXAMPLE 6** Factoring Differences of Cubes

Factor the following.

(a) $m^3 - 125$

Let $x = m$ and $y = 5$ in the pattern for the difference of cubes.

$$x^3 - y^3 = (x - y)(x^2 + xy + y^2)$$

$$m^3 - 125 = m^3 - 5^3 = (m - 5)(m^2 + 5m + 5^2) \qquad \text{Let } x = m, y = 5.$$
$$= (m - 5)(m^2 + 5m + 25)$$

(b) $8p^3 - 27$

Since $8p^3 = (2p)^3$ and $27 = 3^3$,

$$8p^3 - 27 = (2p)^3 - 3^3$$
$$= (2p - 3)[(2p)^2 + (2p)3 + 3^2]$$
$$= (2p - 3)(4p^2 + 6p + 9).$$

(c) $4m^3 - 32 = 4(m^3 - 8)$
$$= 4(m^3 - 2^3)$$
$$= 4(m - 2)(m^2 + 2m + 4)$$

(d) $125t^3 - 216s^6 = (5t)^3 - (6s^2)^3$
$$= (5t - 6s^2)[(5t)^2 + 5t(6s^2) + (6s^2)^2]$$
$$= (5t - 6s^2)(25t^2 + 30ts^2 + 36s^4)$$

▨

Now Try Exercises 59, 63, and 69.

CAUTION A common error in factoring a difference of cubes, such as $x^3 - y^3 = (x - y)(x^2 + xy + y^2)$, is to try to factor $x^2 + xy + y^2$. It is easy to confuse this factor with a perfect square trinomial, $x^2 + 2xy + y^2$. Because there is no 2 in $x^2 + xy + y^2$, it is very unusual to be able to further factor an expression of the form $x^2 + xy + y^2$.

OBJECTIVE 4 Factor a sum of cubes. A sum of squares, such as $m^2 + 25$, cannot be factored using real numbers, but a **sum of cubes** can be factored by the following pattern, *which should be memorized.*

Factoring a Sum of Cubes

$$x^3 + y^3 = (x + y)(x^2 - xy + y^2)$$

Compare the pattern for the *sum* of cubes with the pattern for the *difference* of cubes. The only difference between them is the positive and negative signs.

$$x^3 - y^3 = (x - y)(x^2 + xy + y^2) \qquad \text{Difference of cubes}$$

Positive
Same sign
Opposite sign

$$x^3 + y^3 = (x + y)(x^2 - xy + y^2) \qquad \text{Sum of cubes}$$

Positive
Same sign
Opposite sign

Observing these relationships should help you to remember these patterns.

EXAMPLE 7   Factoring Sums of Cubes

Factor.

(a) $k^3 + 27 = k^3 + 3^3$

$\qquad = (k + 3)(k^2 - 3k + 3^2)$

$\qquad = (k + 3)(k^2 - 3k + 9)$

(b) $8m^3 + 125 = (2m)^3 + 5^3$

$\qquad = (2m + 5)[(2m)^2 - 2m(5) + 5^2]$

$\qquad = (2m + 5)(4m^2 - 10m + 25)$

(c) $1000a^6 + 27b^3 = (10a^2)^3 + (3b)^3$

$\qquad = (10a^2 + 3b)[(10a^2)^2 - (10a^2)(3b) + (3b)^2]$

$\qquad = (10a^2 + 3b)(100a^4 - 30a^2b + 9b^2)$

Now Try Exercises 61 and 71.

The methods of factoring discussed in this section are summarized here.

Special Factorizations

Difference of squares	$x^2 - y^2 = (x + y)(x - y)$
Perfect square trinomials	$x^2 + 2xy + y^2 = (x + y)^2$
	$x^2 - 2xy + y^2 = (x - y)^2$
Difference of cubes	$x^3 - y^3 = (x - y)(x^2 + xy + y^2)$
Sum of cubes	$x^3 + y^3 = (x + y)(x^2 - xy + y^2)$

Remember the *sum* of *squares* can be factored only if the terms have a common factor.

EXERCISES

1. To help you factor the difference of squares, complete the following list of squares.

$1^2 = $ _____ $2^2 = $ _____ $3^2 = $ _____ $4^2 = $ _____ $5^2 = $ _____

$6^2 = $ _____ $7^2 = $ _____ $8^2 = $ _____ $9^2 = $ _____ $10^2 = $ _____

$11^2 = $ _____ $12^2 = $ _____ $13^2 = $ _____ $14^2 = $ _____ $15^2 = $ _____

$16^2 = $ _____ $17^2 = $ _____ $18^2 = $ _____ $19^2 = $ _____ $20^2 = $ _____

2. The following powers of x are all perfect squares: x^2, x^4, x^6, x^8, x^{10}. Based on this observation, we may make a conjecture (an educated guess) that if the power of a variable is divisible by _____ (with 0 remainder), then we have a perfect square.

3. To help you factor the sum or difference of cubes, complete the following list of cubes.

$1^3 = $ _____ $2^3 = $ _____ $3^3 = $ _____ $4^3 = $ _____ $5^3 = $ _____

$6^3 = $ _____ $7^3 = $ _____ $8^3 = $ _____ $9^3 = $ _____ $10^3 = $ _____

4. The following powers of x are all perfect cubes: x^3, x^6, x^9, x^{12}, x^{15}. Based on this observation, we may make a conjecture that if the power of a variable is divisible by _____ (with 0 remainder), then we have a perfect cube.

5. Identify each monomial as a perfect square, a perfect cube, both of these, or neither of these.

(a) $64x^6y^{12}$ **(b)** $125t^6$ **(c)** $49x^{12}$ **(d)** $81r^{10}$

6. What must be true for x^n to be both a perfect square and a perfect cube?

Factor each binomial completely. Use your answers from Exercises 1 and 2 as necessary. See Examples 1–3.

7. $y^2 - 25$ **8.** $t^2 - 16$ **9.** $p^2 - \dfrac{1}{9}$ **10.** $q^2 - \dfrac{1}{4}$

11. $m^2 + 64$ **12.** $k^2 + 49$ **13.** $9r^2 - 4$ **14.** $4x^2 - 9$

15. $36m^2 - \dfrac{16}{25}$ **16.** $100b^2 - \dfrac{4}{49}$ **17.** $36x^2 - 16$ **18.** $32a^2 - 8$

19. $196p^2 - 225$ **20.** $361q^2 - 400$ **21.** $16r^2 - 25a^2$ **22.** $49m^2 - 100p^2$

23. $100x^2 + 49$ **24.** $81w^2 + 16$ **25.** $p^4 - 49$ **26.** $r^4 - 25$

27. $x^4 - 1$ **28.** $y^4 - 16$ **29.** $p^4 - 256$ **30.** $16k^4 - 1$

31. When a student was directed to factor $x^4 - 81$ completely, his teacher did not give him full credit for the answer $(x^2 + 9)(x^2 - 9)$. The student argued that since his answer does indeed give $x^4 - 81$ when multiplied out, he should be given full credit. Was the teacher justified in her grading of this item? Why or why not?

32. The binomial $4x^2 + 16$ is a sum of squares that *can* be factored. How is this binomial factored? When can the sum of squares be factored?

Factor each trinomial completely. It may be necessary to factor out the greatest common factor first. See Examples 4 and 5.

33. $w^2 + 2w + 1$ **34.** $p^2 + 4p + 4$ **35.** $x^2 - 8x + 16$

36. $x^2 - 10x + 25$ **37.** $t^2 + t + \dfrac{1}{4}$ **38.** $m^2 + \dfrac{2}{3}m + \dfrac{1}{9}$

39. $x^2 - 1.0x + .25$

40. $y^2 - 1.4y + .49$

41. $2x^2 + 24x + 72$

42. $3y^2 - 48y + 192$

43. $16x^2 - 40x + 25$

44. $36y^2 - 60y + 25$

45. $49x^2 - 28xy + 4y^2$

46. $4z^2 - 12zw + 9w^2$

47. $64x^2 + 48xy + 9y^2$

48. $9t^2 + 24tr + 16r^2$

49. $-50h^2 + 40hy - 8y^2$

50. $-18x^2 - 48xy - 32y^2$

51. $4k^3 - 4k^2 + 9k$

52. $9r^3 + 6r^2 + 16r$

53. $25z^4 + 5z^3 + z^2$

54. In the polynomial $9y^2 + 14y + 25$, the first and last terms are perfect squares. Can the polynomial be factored? If it can, factor it. If it cannot, explain why it is not a perfect square trinomial.

Find the value of the indicated variable. (Hint: *Perform the squaring on the right side, then solve the equation using the steps given in Chapter 2.*)

55. Find a value of b so that $x^2 + bx + 25 = (x + 5)^2$.

56. For what value of c is $4m^2 - 12m + c = (2m - 3)^2$?

57. Find a so that $ay^2 - 12y + 4 = (3y - 2)^2$.

58. Find b so that $100a^2 + ba + 9 = (10a + 3)^2$.

Factor each binomial completely. Use your answers from Exercises 3 and 4 as necessary. See Examples 6 and 7.

59. $a^3 - 1$

60. $m^3 - 8$

61. $m^3 + 8$

62. $b^3 + 1$

63. $27x^3 - 64$

64. $64y^3 - 27$

65. $6p^3 + 6$

66. $81x^3 + 3$

67. $5x^3 + 40$

68. $128y^3 - 54$

69. $2y^3 - 16x^3$

70. $27w^3 - 216z^3$

71. $8p^3 + 729q^3$

72. $64x^3 + 125y^3$

73. $27a^3 + 64b^3$

74. $125m^3 - 8p^3$

75. $125t^3 + 8s^3$

76. $27r^3 + 1000s^3$

77. State and give the name of each of the five special factoring rules. For each rule, give the numbers of three exercises from this exercise set that are examples.

RELATING CONCEPTS (EXERCISES 78–81)

For Individual or Group Work

We have seen that multiplication and factoring are reverse processes. We know that multiplication and division are also related: To check a division problem, we multiply the quotient by the divisor to get the dividend. To see how factoring and division are related, **work Exercises 78–81 in order.**

78. Factor $10x^2 + 11x - 6$.

79. Use long division to divide $10x^2 + 11x - 6$ by $2x + 3$.

80. Could we have predicted the result in Exercise 79 from the result in Exercise 78? Explain.

81. Divide $x^3 - 1$ by $x - 1$. Use your answer to factor $x^3 - 1$.

Extend the methods of factoring presented so far in this chapter to factor each polynomial completely.

82. $(m + n)^2 - (m - n)^2$

83. $(a - b)^3 - (a + b)^3$

84. $m^2 - p^2 + 2m + 2p$

85. $3r - 3k + 3r^2 - 3k^2$

▬▬ **SUMMARY EXERCISES ON FACTORING**
_____

These mixed exercises are included to give you practice in selecting an appropriate method for factoring a particular polynomial. As you factor a polynomial, ask yourself these questions to decide on a suitable factoring technique.

Factoring a Polynomial

1. Is there a common factor? If so, factor it out.

2. How many terms are in the polynomial?

   Two terms: Check to see whether it is a difference of squares or the sum or difference of cubes. If so, factor as in Section 5.4.

   Three terms: Is it a perfect square trinomial? If the trinomial is not a perfect square, check to see whether the coefficient of the squared term is 1. If so, use the method of Section 5.2. If the coefficient of the squared term of the trinomial is not 1, use the general factoring methods of Section 5.3.

   Four terms: Try to factor the polynomial by grouping as in Section 5.1.

3. Can any factors be factored further? If so, factor them.

Factor each polynomial completely.

1. $a^2 - 4a - 12$

2. $a^2 + 17a + 72$

3. $6y^2 - 6y - 12$

4. $7y^6 + 14y^5 - 168y^4$

5. $6a + 12b + 18c$

6. $m^2 - 3mn - 4n^2$

7. $p^2 - 17p + 66$

8. $z^2 - 6z + 7z - 42$

9. $10z^2 - 7z - 6$

10. $2m^2 - 10m - 48$

11. $m^2 - n^2 + 5m - 5n$

12. $15y + 5$

13. $8a^5 - 8a^4 - 48a^3$

14. $8k^2 - 10k - 3$

15. $z^2 - 3za - 10a^2$

16. $50z^2 - 100$

17. $x^2 - 4x - 5x + 20$

18. $100n^2r^2 + 30nr^3 - 50n^2r$

19. $6n^2 - 19n + 10$

20. $9y^2 + 12y - 5$

21. $16x + 20$

22. $m^2 + 2m - 15$

23. $6y^2 - 5y - 4$

24. $m^2 - 81$

25. $6z^2 + 31z + 5$

26. $5z^2 + 24z - 5 + 3z + 15$

27. $4k^2 - 12k + 9$

28. $8p^2 + 23p - 3$

29. $54m^2 - 24z^2$

30. $8m^2 - 2m - 3$

31. $3k^2 + 4k - 4$

32. $45a^3b^5 - 60a^4b^2 + 75a^6b^4$

33. $14k^3 + 7k^2 - 70k$

34. $5 + r - 5s - rs$

35. $y^4 - 16$

36. $20y^5 - 30y^4$

37. $8m - 16m^2$

38. $k^2 - 16$

39. $z^3 - 8$

40. $y^2 - y - 56$

41. $k^2 + 9$

42. $27p^{10} - 45p^9 - 252p^8$

43. $32m^9 + 16m^5 + 24m^3$

44. $8m^3 + 125$

45. $16r^2 + 24rm + 9m^2$

46. $z^2 - 12z + 36$

47. $15h^2 + 11hg - 14g^2$

48. $5z^3 - 45z^2 + 70z$

49. $k^2 - 11k + 30$

50. $64p^2 - 100m^2$

51. $3k^3 - 12k^2 - 15k$

52. $y^2 - 4yk - 12k^2$

53. $1000p^3 + 27$

54. $64r^3 - 343$

55. $6 + 3m + 2p + mp$

56. $2m^2 + 7mn - 15n^2$

57. $16z^2 - 8z + 1$

58. $125m^4 - 400m^3n + 195m^2n^2$

59. $108m^2 - 36m + 3$

60. $100a^2 - 81y^2$

61. $64m^2 - 40mn + 25n^2$

62. $4y^2 - 25$

63. $32z^3 + 56z^2 - 16z$

64. $10m^2 + 25m - 60$

65. $20 + 5m + 12n + 3mn$

66. $4 - 2q - 6p + 3pq$

67. $6a^2 + 10a - 4$

68. $36y^6 - 42y^5 - 120y^4$

69. $a^3 - b^3 + 2a - 2b$

70. $16k^2 - 48k + 36$

71. $64m^2 - 80mn + 25n^2$

72. $72y^3z^2 + 12y^2 - 24y^4z^2$

73. $8k^2 - 2kh - 3h^2$

74. $2a^2 - 7a - 30$

75. $(m + 1)^3 + 1$

76. $8a^3 - 27$

77. $10y^2 - 7yz - 6z^2$

78. $m^2 - 4m + 4$

79. $8a^2 + 23ab - 3b^2$

80. $a^4 - 625$

RELATING CONCEPTS (EXERCISES 81–88)

For Individual or Group Work

A binomial may be *both* a difference of squares *and* a difference of cubes. One example of such a binomial is $x^6 - 1$. Using the techniques of Section 5.4, one factoring method will give the complete factored form, while the other will not. ***Work Exercises 81–88 in order*** *to determine the method to use if you have to make such a decision.*

81. Factor $x^6 - 1$ as the difference of squares.

82. The factored form obtained in Exercise 81 consists of a difference of cubes multiplied by a sum of cubes. Factor each binomial further.

83. Now start over and factor $x^6 - 1$ as the difference of cubes.

84. The factored form obtained in Exercise 83 consists of a binomial which is a difference of squares and a trinomial. Factor the binomial further.

85. Compare your results in Exercises 82 and 84. Which one of these is the completely factored form?

86. Verify that the trinomial in the factored form in Exercise 84 is the product of the two trinomials in the factored form in Exercise 82.

87. Use the results of Exercises 81–86 to complete the following statement: In general, if I must choose between factoring first using the method for difference of squares or the method for difference of cubes, I should choose the _____ method to eventually obtain the complete factored form.

88. Find the *complete* factored form of $x^6 - 729$ using the knowledge you have gained in Exercises 81–87.

5.5 Solving Quadratic Equations by Factoring

OBJECTIVES

1 Solve quadratic equations by factoring.

2 Solve other equations by factoring.

Galileo Galilei (1564–1642) developed theories to explain physical phenomena and set up experiments to test his ideas. According to legend, Galileo dropped objects of different weights from the Leaning Tower of Pisa to disprove the belief that heavier objects fall faster than lighter objects. He developed a formula for freely falling objects described by

$$d = 16t^2,$$

where d is the distance in feet that an object falls (disregarding air resistance) in t seconds, regardless of weight.

The equation $d = 16t^2$ is a *quadratic equation,* the subject of this section. A quadratic equation contains a squared term and no terms of higher degree.

> **Quadratic Equation**
>
> A **quadratic equation** is an equation that can be written in the form
>
> $$ax^2 + bx + c = 0,$$
>
> where a, b, and c are real numbers, with $a \neq 0$.

The form $ax^2 + bx + c = 0$ is the **standard form** of a quadratic equation. For example,

$$x^2 + 5x + 6 = 0, \qquad 2a^2 - 5a = 3, \qquad \text{and} \qquad y^2 = 4$$

are all quadratic equations, but only $x^2 + 5x + 6 = 0$ is in standard form.

Up to now, we have factored *expressions,* including many quadratic expressions of the form $ax^2 + bx + c$. In this section, we see how we can use factored quadratic expressions to solve quadratic *equations.*

OBJECTIVE 1 Solve quadratic equations by factoring. We use the **zero-factor property** to solve a quadratic equation by factoring.

> **Zero-Factor Property**
>
> If a and b are real numbers and if $ab = 0$, then $a = 0$ or $b = 0$.
>
> That is, if the product of two numbers is 0, then at least one of the numbers must be 0. One number *must* be 0, but both *may* be 0.

EXAMPLE 1 Using the Zero-Factor Property

Solve each equation.

(a) $(x + 3)(2x - 1) = 0$

The product $(x + 3)(2x - 1)$ is equal to 0. By the zero-factor property, the only way that the product of these two factors can be 0 is if at least one of the factors equals 0. Therefore, either $x + 3 = 0$ or $2x - 1 = 0$. Solve each of these two linear equations as in Chapter 2.

$$x + 3 = 0 \quad \text{or} \quad 2x - 1 = 0 \qquad \text{Zero-factor property}$$
$$x = -3 \quad \text{or} \qquad 2x = 1 \qquad \text{Add 1 to each side.}$$
$$x = \frac{1}{2} \qquad \text{Divide each side by 2.}$$

The given equation, $(x + 3)(2x - 1) = 0$, has two solutions, -3 and $\frac{1}{2}$. Check these solutions by substituting -3 for x in the original equation, $(x + 3)(2x - 1) = 0$. Then start over and substitute $\frac{1}{2}$ for x.

If $x = -3$, then

$$(x + 3)(2x - 1) = 0$$
$$(-3 + 3)[2(-3) - 1] = 0 \quad ?$$
$$0(-7) = 0. \qquad \text{True}$$

If $x = \frac{1}{2}$, then

$$(x + 3)(2x - 1) = 0$$
$$\left(\frac{1}{2} + 3\right)\left(2 \cdot \frac{1}{2} - 1\right) = 0 \quad ?$$
$$\frac{7}{2}(1 - 1) = 0 \quad ?$$
$$\frac{7}{2} \cdot 0 = 0. \qquad \text{True}$$

Both -3 and $\frac{1}{2}$ result in true equations, so the solution set is $\left\{-3, \frac{1}{2}\right\}$.

(b) $y(3y - 4) = 0$

$$y(3y - 4) = 0$$
$$y = 0 \quad \text{or} \quad 3y - 4 = 0 \qquad \text{Zero-factor property}$$
$$3y = 4$$
$$y = \frac{4}{3}$$

Check these solutions by substituting each one in the original equation. The solution set is $\left\{0, \frac{4}{3}\right\}$.

Now Try Exercises 3 and 5.

NOTE The word *or* as used in Example 1 means "one or the other or both."

In Example 1, each equation to be solved was given with the polynomial in factored form. If the polynomial in an equation is not already factored, first make sure that the equation is in standard form. Then factor.

EXAMPLE 2 Solving Quadratic Equations

Solve each equation.

(a) $x^2 - 5x = -6$

First, rewrite the equation in standard form by adding 6 to each side.

$$x^2 - 5x = -6$$
$$x^2 - 5x + 6 = 0 \qquad \text{Add 6.}$$

Now factor $x^2 - 5x + 6$. Find two numbers whose product is 6 and whose sum is -5. These two numbers are -2 and -3, so the equation becomes

$$(x - 2)(x - 3) = 0. \qquad \text{Factor.}$$
$$x - 2 = 0 \quad \text{or} \quad x - 3 = 0 \qquad \text{Zero-factor property}$$
$$x = 2 \quad \text{or} \qquad x = 3 \qquad \text{Solve each equation.}$$

Check: If $x = 2$, then   If $x = 3$, then

$x^2 - 5x = -6$	$x^2 - 5x = -6$
$2^2 - 5(2) = -6$?	$3^2 - 5(3) = -6$?
$4 - 10 = -6$?	$9 - 15 = -6$?
$-6 = -6.$ True	$-6 = -6.$ True

Both solutions check, so the solution set is $\{2, 3\}$.

(b) $y^2 = y + 20$

Rewrite the equation in standard form.

$$y^2 = y + 20$$
$$y^2 - y - 20 = 0 \qquad \text{Subtract } y \text{ and } 20.$$
$$(y - 5)(y + 4) = 0 \qquad \text{Factor.}$$
$$y - 5 = 0 \quad \text{or} \quad y + 4 = 0 \qquad \text{Zero-factor property}$$
$$y = 5 \quad \text{or} \qquad y = -4 \qquad \text{Solve each equation.}$$

Check these solutions by substituting each one in the original equation. The solution set is $\{-4, 5\}$.

Now Try Exercise 21.

In summary, follow these steps to solve quadratic equations by factoring.

Solving a Quadratic Equation by Factoring

Step 1   **Write the equation in standard form,** that is, with all terms on one side of the equals sign in descending powers of the variable and 0 on the other side.

Step 2   **Factor** completely.

Step 3   **Use the zero-factor property** to set each factor with a variable equal to 0, and solve the resulting equations.

Step 4   **Check** each solution in the original equation.

NOTE   Not all quadratic equations can be solved by factoring. A more general method for solving such equations is given in Chapter 11.

EXAMPLE 3   **Solving a Quadratic Equation with a Common Factor**

Solve $4p^2 + 40 = 26p$.

Subtract $26p$ from each side and write the equation in standard form to get

$$4p^2 - 26p + 40 = 0.$$

$$2(2p^2 - 13p + 20) = 0 \qquad \text{Factor out 2.}$$
$$2p^2 - 13p + 20 = 0 \qquad \text{Divide each side by 2.}$$
$$(2p - 5)(p - 4) = 0 \qquad \text{Factor.}$$
$$2p - 5 = 0 \quad \text{or} \quad p - 4 = 0 \qquad \text{Zero-factor property}$$
$$2p = 5 \quad \text{or} \qquad p = 4$$
$$p = \frac{5}{2}$$

Check that the solution set is $\left\{\frac{5}{2}, 4\right\}$ by substituting each solution in the original equation.

Now Try Exercise 31.

EXAMPLE 4   Solving Quadratic Equations

Solve each equation.

(a) $16m^2 - 25 = 0$

Factor the left side of the equation as the difference of squares.

$$(4m + 5)(4m - 5) = 0$$
$$4m + 5 = 0 \qquad \text{or} \quad 4m - 5 = 0 \qquad \text{Zero-factor property}$$
$$4m = -5 \quad \text{or} \qquad 4m = 5 \qquad \text{Solve each equation.}$$
$$m = -\frac{5}{4} \quad \text{or} \qquad m = \frac{5}{4}$$

Check the two solutions, $-\frac{5}{4}$ and $\frac{5}{4}$, in the original equation. The solution set is $\left\{-\frac{5}{4}, \frac{5}{4}\right\}$.

(b) $y^2 = 2y$

First write the equation in standard form.

$$y^2 - 2y = 0 \qquad \text{Standard form}$$
$$y(y - 2) = 0 \qquad \text{Factor.}$$
$$y = 0 \quad \text{or} \quad y - 2 = 0 \qquad \text{Zero-factor property}$$
$$y = 2 \qquad \text{Solve.}$$

The solution set is $\{0, 2\}$.

(c) $k(2k + 1) = 3$

Write the equation in standard form.

$$k(2k + 1) = 3$$
$$2k^2 + k = 3 \qquad \text{Distributive property}$$
$$2k^2 + k - 3 = 0 \qquad \text{Subtract 3.}$$
$$(k - 1)(2k + 3) = 0 \qquad \text{Factor.}$$
$$k - 1 = 0 \quad \text{or} \quad 2k + 3 = 0 \qquad \text{Zero-factor property}$$
$$k = 1 \quad \text{or} \qquad 2k = -3$$
$$k = -\frac{3}{2}$$

The solution set is $\left\{1, -\frac{3}{2}\right\}$.

Now Try Exercises 37, 41, and 45.

CAUTION In Example 4(b) it is tempting to begin by dividing both sides of the equation $y^2 = 2y$ by y to get $y = 2$. Note that we do not get the other solution, 0, if we divide by a variable. (We may divide each side of an equation by a *nonzero* real number, however. For instance, in Example 3 we divided each side by 2.)

In Example 4(c) we could not use the zero-factor property to solve the equation $k(2k + 1) = 3$ in its given form because of the 3 on the right. Remember that the zero-factor property applies only to a product that equals 0.

OBJECTIVE 2 Solve other equations by factoring. We can also use the zero-factor property to solve equations that involve more than two factors with variables, as shown in Example 5. (These equations are *not* quadratic equations. Why not?)

> ### EXAMPLE 5 Solving Equations with More Than Two Variable Factors
> Solve each equation.
> **(a)**
> $$6z^3 - 6z = 0$$
> $$6z(z^2 - 1) = 0 \qquad \text{Factor out } 6z.$$
> $$6z(z + 1)(z - 1) = 0 \qquad \text{Factor } z^2 - 1.$$
>
> By an extension of the zero-factor property, this product can equal 0 only if at least one of the factors is 0. Write and solve three equations, one for each factor with a variable.
> $$6z = 0 \quad \text{or} \quad z + 1 = 0 \quad \text{or} \quad z - 1 = 0$$
> $$z = 0 \quad \text{or} \qquad z = -1 \quad \text{or} \qquad z = 1$$
>
> Check by substituting, in turn, 0, -1, and 1 in the original equation. The solution set is $\{-1, 0, 1\}$.
>
> **(b)**
> $$(3x - 1)(x^2 - 9x + 20) = 0$$
> $$(3x - 1)(x - 5)(x - 4) = 0 \qquad \text{Factor } x^2 - 9x + 20.$$
> $$3x - 1 = 0 \quad \text{or} \quad x - 5 = 0 \quad \text{or} \quad x - 4 = 0 \qquad \text{Zero-factor property}$$
> $$x = \frac{1}{3} \quad \text{or} \qquad x = 5 \quad \text{or} \qquad x = 4$$
>
> The solutions of the original equation are $\frac{1}{3}$, 4, and 5. Check each solution to verify that the solution set is $\left\{\frac{1}{3}, 4, 5\right\}$.

Now Try Exercises 51 and 55.

CAUTION In Example 5(b), it would be unproductive to begin by multiplying the two factors together. Keep in mind that the zero-factor property requires the *product* of two or more factors; this product must equal 0. Always consider first whether an equation is given in an appropriate form for the zero-factor property.

EXAMPLE 6 Solving an Equation Requiring Multiplication Before Factoring

Solve $(3x + 1)x = (x + 1)^2 + 5$.

The zero-factor property requires the *product* of two or more factors to equal 0. To write this equation in the required form, we must first multiply on both sides and collect terms on one side.

$$(3x + 1)x = (x + 1)^2 + 5$$
$$3x^2 + x = x^2 + 2x + 1 + 5 \qquad \text{Multiply.}$$
$$3x^2 + x = x^2 + 2x + 6 \qquad \text{Combine like terms.}$$
$$2x^2 - x - 6 = 0 \qquad \text{Standard form}$$
$$(2x + 3)(x - 2) = 0 \qquad \text{Factor.}$$
$$2x + 3 = 0 \quad \text{or} \quad x - 2 = 0 \qquad \text{Zero-factor property}$$
$$x = -\frac{3}{2} \quad \text{or} \qquad x = 2$$

Check that the solution set is $\left\{-\frac{3}{2}, 2\right\}$.

Now Try Exercise 65.

5.5 EXERCISES

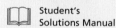

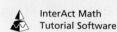

Solve each equation and check your solutions. See Example 1.

1. $(x + 5)(x - 2) = 0$ **2.** $(x - 1)(x + 8) = 0$ **3.** $(2m - 7)(m - 3) = 0$

4. $(6k + 5)(k + 4) = 0$ **5.** $t(6t + 5) = 0$ **6.** $w(4w + 1) = 0$

7. $2x(3x - 4) = 0$ **8.** $6y(4y + 9) = 0$

9. $\left(x + \dfrac{1}{2}\right)\left(2x - \dfrac{1}{3}\right) = 0$ **10.** $\left(a + \dfrac{2}{3}\right)\left(5a - \dfrac{1}{2}\right) = 0$

11. $(.5z - 1)(2.5z + 2) = 0$ **12.** $(.25x + 1)(x - .5) = 0$

13. $(x - 9)(x - 9) = 0$ **14.** $(2y + 1)(2y + 1) = 0$

15. What is wrong with this "solution"?

$$2x(3x - 4) = 0$$
$$x = 2 \quad \text{or} \quad x = 0 \quad \text{or} \quad 3x - 4 = 0$$
$$x = \frac{4}{3}$$

The solution set is $\left\{2, 0, \frac{4}{3}\right\}$.

16. What is wrong with this "solution"?

$$x(7x - 1) = 0$$
$$7x - 1 = 0 \qquad \text{Zero-factor property}$$
$$x = \frac{1}{7}$$

The solution set is $\left\{\frac{1}{7}\right\}$.

Solve each equation and check your solutions. See Examples 2–4.

17. $y^2 + 3y + 2 = 0$ **18.** $p^2 + 8p + 7 = 0$ **19.** $y^2 - 3y + 2 = 0$

20. $r^2 - 4r + 3 = 0$ **21.** $x^2 = 24 - 5x$ **22.** $t^2 = 2t + 15$

23. $x^2 = 3 + 2x$ **24.** $m^2 = 4 + 3m$ **25.** $z^2 + 3z = -2$

26. $p^2 - 2p = 3$ **27.** $m^2 + 8m + 16 = 0$ **28.** $b^2 - 6b + 9 = 0$

29. $3x^2 + 5x - 2 = 0$ **30.** $6r^2 - r - 2 = 0$ **31.** $12p^2 = 8 - 10p$

32. $18x^2 = 12 + 15x$ **33.** $9s^2 + 12s = -4$ **34.** $36x^2 + 60x = -25$

35. $y^2 - 9 = 0$ **36.** $m^2 - 100 = 0$ **37.** $16k^2 - 49 = 0$

38. $4w^2 - 9 = 0$ **39.** $n^2 = 121$ **40.** $x^2 = 400$

Solve each equation and check your solutions. See Examples 4–6.

41. $x^2 = 7x$ **42.** $t^2 = 9t$ **43.** $6r^2 = 3r$

44. $10y^2 = -5y$ **45.** $g(g - 7) = -10$ **46.** $r(r - 5) = -6$

47. $3z(2z + 7) = 12$ **48.** $4b(2b + 3) = 36$

49. $2(y^2 - 66) = -13y$ **50.** $3(t^2 + 4) = 20t$

51. $(2r + 5)(3r^2 - 16r + 5) = 0$ **52.** $(3m + 4)(6m^2 + m - 2) = 0$

53. $(2x + 7)(x^2 + 2x - 3) = 0$ **54.** $(x + 1)(6x^2 + x - 12) = 0$

55. $9y^3 - 49y = 0$ **56.** $16r^3 - 9r = 0$

57. $r^3 - 2r^2 - 8r = 0$ **58.** $x^3 - x^2 - 6x = 0$

59. $a^3 + a^2 - 20a = 0$ **60.** $y^3 - 6y^2 + 8y = 0$

61. $r^4 = 2r^3 + 15r^2$ **62.** $x^3 = 3x + 2x^2$

63. $3x(x + 1) = (2x + 3)(x + 1)$ **64.** $2k(k + 3) = (3k + 1)(k + 3)$

65. $x^2 + (x + 1)^2 = (x + 2)^2$ **66.** $(x - 7)^2 + x^2 = (x + 1)^2$

67. $(2x)^2 = (2x + 4)^2 - (x + 5)^2$ **68.** $5 - (x - 1)^2 = (x - 2)^2$

69. $6p^2(p + 1) = 4(p + 1) - 5p(p + 1)$

70. $6x^2(2x + 3) - 5x(2x + 3) = 4(2x + 3)$

71. $(k + 3)^2 - (2k - 1)^2 = 0$

72. $(4y - 3)^3 - 9(4y - 3) = 0$

73. Galileo's formula for freely falling objects, $d = 16t^2$, was given at the beginning of this section. The distance d in feet an object falls depends on the time elapsed t in seconds. (This is an example of an important mathematical concept, the *function.*)

  (a) Use Galileo's formula and complete the following table. (*Hint:* Substitute each given value into the formula and solve for the unknown value.)

t in seconds	0	1	2	3		
d in feet	0	16			256	576

  (b) When $t = 0$, $d = 0$. Explain this in the context of the problem.

  (c) When you substituted 256 for d and solved for t, you should have found two solutions, 4 and -4. Why doesn't -4 make sense as an answer?

TECHNOLOGY INSIGHTS (EXERCISES 74–77)

In Section 3.2 we showed how an equation in one variable can be solved with a graphing calculator by getting 0 on one side, then replacing 0 with y to get a corresponding equation in two variables. The x-values of the x-intercepts of the graph of the two-variable equation then give the solutions of the original equation.

Use the calculator screens to determine the solution set of each quadratic equation. Verify your answers by substitution.

74. $x^2 + .4x - .05 = 0$

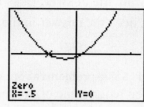
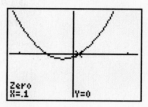

75. $2x^2 - 7.2x + 6.3 = 0$

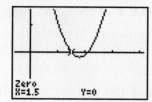
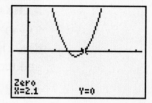

76. $2x^2 + 7.2x + 5.5 = 0$

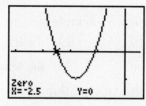
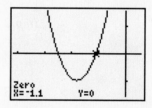

77. $4x^2 - x - 33 = 0$

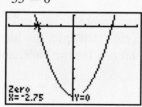
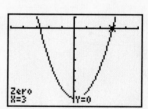

5.6 Applications of Quadratic Equations

OBJECTIVES

1 Solve problems about geometric figures.

2 Solve problems about consecutive integers.

3 Solve problems using the Pythagorean formula.

4 Solve problems using given quadratic models.

We can now use factoring to solve quadratic equations that arise in application problems. We follow the same six problem-solving steps given in Section 2.4.

Solving an Applied Problem

Step 1 **Read** the problem carefully until you understand what is given and what is to be found.

Step 2 **Assign a variable** to represent the unknown value, using diagrams or tables as needed. Write down what the variable represents. If necessary, express any other unknown values in terms of the variable.

(continued)

Step 3 **Write an equation** using the variable expression(s).

Step 4 **Solve** the equation.

Step 5 **State the answer.** Does it seem reasonable?

Step 6 **Check** the answer in the words of the original problem.

OBJECTIVE 1 Solve problems about geometric figures. Some of the applied problems in this section require one of the formulas given on the inside covers of the text.

EXAMPLE 1 Solving an Area Problem

The O'Connors want to plant a flower bed in a triangular area in a corner of their garden. One leg of the right-triangular flower bed will be 2 m shorter than the other leg, and they want it to have an area of 24 m². See Figure 1. Find the lengths of the legs.

Step 1 **Read** the problem carefully. We need to find the lengths of the legs of a right triangle with area 24 m².

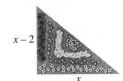

FIGURE 1

Step 2 **Assign a variable.**

Let x = the length of one leg.

Then $x - 2$ = the length of the other leg.

Step 3 **Write an equation.** The area of a right triangle is given by the formula

$$\text{area} = \frac{1}{2} \times \text{base} \times \text{height} = \frac{1}{2}bh.$$

In a right triangle, the legs are the base and height, so we substitute 24 for the area, x for the base, and $x - 2$ for the height in the formula.

$$A = \frac{1}{2}bh$$

$$24 = \frac{1}{2}x(x - 2) \qquad \text{Let } A = 24, b = x, h = x - 2.$$

Step 4 **Solve.**
$$48 = x(x - 2) \qquad \text{Multiply by 2.}$$
$$48 = x^2 - 2x \qquad \text{Distributive property}$$
$$x^2 - 2x - 48 = 0 \qquad \text{Standard form}$$
$$(x + 6)(x - 8) = 0 \qquad \text{Factor.}$$
$$x + 6 = 0 \quad \text{or} \quad x - 8 = 0 \qquad \text{Zero-factor property}$$
$$x = -6 \quad \text{or} \qquad x = 8$$

Step 5 **State the answer.** The solutions are -6 and 8. Because a triangle cannot have a side of negative length, we discard the solution -6. Then the lengths of the legs will be 8 m and $8 - 2 = 6$ m.

Step 6 **Check.** The length of one leg is 2 m less than the length of the other leg, and the area is $\frac{1}{2}(8)(6) = 24$ m², as required.

Now Try Exercise 7.

CAUTION When solving applied problems, *always* check solutions against physical facts and discard any answers that are not appropriate.

OBJECTIVE 2 Solve problems about consecutive integers. Recall from our work in Section 2.4 that consecutive integers are integers that are next to each other on a number line, such as 5 and 6, or -11 and -10. Consecutive odd integers are *odd* integers that are next to each other, such as 5 and 7, or -13 and -11. Consecutive even integers are defined similarly; for example, 4 and 6 are consecutive even integers, as are -10 and -8. The following list may be helpful.

Consecutive Integers

Let x represent the first of the integers.

Two consecutive integers	$x, x + 1$
Three consecutive integers	$x, x + 1, x + 2$
Two consecutive even or odd integers	$x, x + 2$
Three consecutive even or odd integers	$x, x + 2, x + 4$

▇ **EXAMPLE 2 Solving a Consecutive Integer Problem**

The product of the second and third of three consecutive integers is 2 more than 7 times the first integer. Find the integers.

Step 1 **Read** the problem. Note that the integers are consecutive.

Step 2 **Assign a variable.**

Let $x =$ the first integer.

Then $x + 1 =$ the second integer,

and $x + 2 =$ the third integer.

Step 3 **Write an equation.**

The product of the second and third is 2 more than 7 times the first.

$$(x + 1)(x + 2) = 7x + 2$$

Step 4 **Solve.**

$$x^2 + 3x + 2 = 7x + 2 \qquad \text{Multiply.}$$
$$x^2 - 4x = 0 \qquad \text{Standard form}$$
$$x(x - 4) = 0 \qquad \text{Factor.}$$
$$x = 0 \quad \text{or} \quad x = 4 \qquad \text{Zero-factor property}$$

Step 5 **State the answer.** The solutions 0 and 4 each lead to a correct answer:

$$0, 1, 2 \quad \text{or} \quad 4, 5, 6.$$

Step 6 **Check.** Since $1 \cdot 2 = 7 \cdot 0 + 2$ and $5 \cdot 6 = 7 \cdot 4 + 2$, both sets of consecutive integers satisfy the statement of the problem.

Now Try Exercise 17.

OBJECTIVE 3 Solve problems using the Pythagorean formula. The next example requires the Pythagorean formula from geometry.

Pythagorean Formula

If a right triangle (a triangle with a 90° angle) has longest side of length c and two other sides of lengths a and b, then

$$a^2 + b^2 = c^2.$$

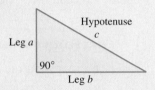

The longest side, the **hypotenuse,** is opposite the right angle. The two shorter sides are the **legs** of the triangle.

▮ **EXAMPLE 3** Using the Pythagorean Formula

Ed and Mark leave their office, with Ed traveling north and Mark traveling east. When Mark is 1 mi farther than Ed from the office, the distance between them is 2 mi more than Ed's distance from the office. Find their distances from the office and the distance between them.

Step 1 **Read** the problem again. There will be three answers to this problem.

Step 2 **Assign a variable.** Let x represent Ed's distance from the office, $x + 1$ represent Mark's distance from the office, and $x + 2$ represent the distance between them. Place these on a right triangle, as in Figure 2.

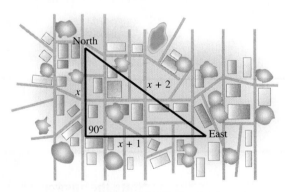

FIGURE 2

Step 3 **Write an equation.** Substitute into the Pythagorean formula.

$$a^2 + b^2 = c^2$$
$$x^2 + (x + 1)^2 = (x + 2)^2$$

Step 4 **Solve.**

$$x^2 + x^2 + 2x + 1 = x^2 + 4x + 4$$

$$x^2 - 2x - 3 = 0 \qquad \text{Standard form}$$

$$(x - 3)(x + 1) = 0 \qquad \text{Factor.}$$

$$x - 3 = 0 \quad \text{or} \quad x + 1 = 0 \qquad \text{Zero-factor property}$$

$$x = 3 \quad \text{or} \qquad x = -1$$

Step 5 **State the answer.** Since -1 cannot represent a distance, 3 is the only possible answer. Ed's distance is 3 mi, Mark's distance is $3 + 1 = 4$ mi, and the distance between them is $3 + 2 = 5$ mi.

Step 6 **Check.** Since $3^2 + 4^2 = 5^2$, the answers are correct.

Now Try Exercise 25.

CAUTION When solving a problem involving the Pythagorean formula, be sure that the expressions for the sides are properly placed.

$$\text{leg}^2 + \text{leg}^2 = \text{hypotenuse}^2$$

OBJECTIVE 4 **Solve problems using given quadratic models.** In Examples 1–3, we wrote quadratic equations to model, or mathematically describe, various situations and then solved the equations. In the final examples, we are given the quadratic models and must use them to determine data.

EXAMPLE 4 **Finding the Height of a Ball**

A tennis player's serve travels 180 ft per sec (125 mph). If she hits the ball directly upward, the height h of the ball in feet at time t in seconds is modeled by the quadratic equation

$$h = -16t^2 + 180t + 6.$$

How long will it take for the ball to reach a height of 206 ft?

A height of 206 ft means $h = 206$, so we substitute 206 for h in the equation.

$$206 = -16t^2 + 180t + 6 \qquad \text{Let } h = 206.$$

To solve the equation, we first write it in standard form. For convenience, we reverse the sides of the equation.

$$-16t^2 + 180t + 6 = 206$$

$$-16t^2 + 180t - 200 = 0 \qquad \text{Standard form}$$

$$4t^2 - 45t + 50 = 0 \qquad \text{Divide by } -4.$$

$$(4t - 5)(t - 10) = 0 \qquad \text{Factor.}$$

$$4t - 5 = 0 \quad \text{or} \quad t - 10 = 0 \qquad \text{Zero-factor property}$$

$$t = \frac{5}{4} \quad \text{or} \qquad t = 10$$

Since we found two acceptable answers, the ball will be 206 ft above the ground twice (once on its way up and once on its way down)—at $\frac{5}{4}$ sec and at 10 sec. See Figure 3.

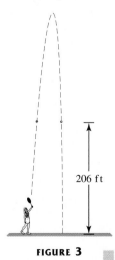

FIGURE 3

Now Try Exercise 29.

EXAMPLE 5   Modeling Increases in Drug Prices

The annual percent increase y in the amount pharmacies paid wholesalers for drugs in the years 1990–1999 can be modeled by the quadratic equation

$$y = .23x^2 - 2.6x + 9,$$

where $x = 0$ represents 1990, $x = 1$ represents 1991, and so on. (*Source: IMS Health,* Retail and Provider Perspective.)

(a) Use the model to find the annual percent increase to the nearest tenth in 1997.

In 1997, $x = 1997 - 1990 = 7$. Substitute 7 for x in the equation.

$$y = .23(7)^2 - 2.6(7) + 9 \qquad \text{Let } x = 7.$$
$$y = 2.07 \qquad\qquad\qquad \text{Use a calculator.}$$

To the nearest tenth, pharmacies paid about 2.1% more for drugs in 1997.

(b) Repeat part (a) for 1999.

For 1999, $x = 9$.

$$y = .23(9)^2 - 2.6(9) + 9 \qquad \text{Let } x = 1999 - 1990 = 9.$$
$$y = 4.23$$

In 1999, pharmacies paid about 4.2% more for drugs.

(c) The model used in parts (a) and (b) was developed using the data shown in the table. How do the results in parts (a) and (b) compare to the actual data from the table?

Year	Percent Increase
1990	8.4
1991	7.2
1992	5.5
1993	3.0
1994	1.7
1995	1.9
1996	1.6
1997	2.5
1998	3.2
1999	4.2

From the table, the actual data for 1997 is 2.5%. Our answer, 2.1%, is a little low. For 1999, the actual data is 4.2%, which is the same as our answer in part (b).

Now Try Exercise 31.

NOTE A graph of the quadratic equation from Example 5 is shown in Figure 4. Notice the basic shape of this graph, which follows the general pattern of the data in the table—it decreases from 1990 to 1996 (with the exception of the data for 1995) and then increases from 1997 to 1999. We consider such graphs of quadratic equations, called *parabolas,* in more detail in Chapter 11.

$y = .23x^2 - 2.6x + 9$

FIGURE 4

5.6 EXERCISES

1. To review the six problem-solving steps first introduced in Section 2.4, complete each statement.

Step 1: _____ the problem carefully until you understand what is given and what must be found.

Step 2: Assign a _____ to represent the unknown value.

Step 3: Write a(n) _____ using the variable expression(s).

Step 4: _____ the equation.

Step 5: State the _____.

Step 6: _____ the answer in the words of the _____ problem.

2. A student solves an applied problem and gets 6 or -3 for the length of the side of a square. Which of these answers is reasonable? Explain.

In Exercises 3–6, a figure and a corresponding geometric formula are given. Using x as the variable, complete Steps 3–6 for each problem. (Refer to the steps in Exercise 1 as needed.)

3.

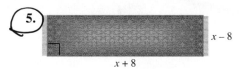

Area of a parallelogram: $A = bh$

The area of this parallelogram is 45 sq. units. Find its base and height.

4.

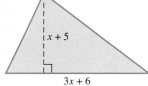

Area of a triangle: $A = \dfrac{1}{2}bh$

The area of this triangle is 60 sq. units. Find its base and height.

5.

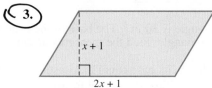

Area of a rectangular rug: $A = LW$

The area of this rug is 80 sq. units. Find its length and width.

6.

Volume of a rectangular Chinese box: $V = LWH$

The volume of this box is 192 cu. units. Find its length and width.

Solve each problem. Check your answers to be sure they are reasonable. Refer to the formulas on the inside covers. See Example 1.

7. The length of a VHS videocassette shell is 3 in. more than its width. The area of the rectangular top side of the shell is 28 in.^{2}. Find the length and width of the videocassette shell.

8. A plastic box that holds a standard audiocassette has length 4 cm longer than its width. The area of the rectangular top of the box is 77 cm^2. Find the length and width of the box.

9. The dimensions of a Gateway EV700 computer monitor screen are such that its length is 3 in. more than its width. If the length is increased by 1 in. while the width remains the same, the area is increased by 10 in.². What are the dimensions of the screen? (*Source:* Author's computer.)

10. The keyboard of the computer in Exercise 9 is 11 in. longer than it is wide. If both its length and width are increased by 2 in., the area of the top of the keyboard is increased by 54 in.². Find the length and width of the keyboard. (*Source:* Author's computer.)

11. The area of a triangle is 30 in.². The base of the triangle measures 2 in. more than twice the height of the triangle. Find the measures of the base and the height.

12. A certain triangle has its base equal in measure to its height. The area of the triangle is 72 m². Find the equal base and height measure.

13. A 10-gal aquarium is 3 in. higher than it is wide. Its length is 21 in., and its volume is 2730 in.³. What are the height and width of the aquarium?

14. A toolbox is 2 ft high, and its width is 3 ft less than its length. If its volume is 80 ft³, find the length and width of the box.

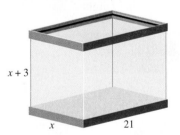

15. A square mirror has sides measuring 2 ft less than the sides of a square painting. If the difference between their areas is 32 ft², find the lengths of the sides of the mirror and the painting.

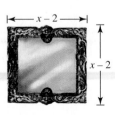

16. The sides of one square have length 3 m more than the sides of a second square. If the area of the larger square is subtracted from 4 times the area of the smaller square, the result is 36 m². What are the lengths of the sides of each square?

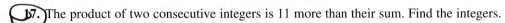

Solve each problem about consecutive integers. See Example 2.

17. The product of two consecutive integers is 11 more than their sum. Find the integers.

18. The product of two consecutive integers is 4 less than 4 times their sum. Find the integers.

19. Find three consecutive odd integers such that 3 times the sum of all three is 18 more than the product of the smaller two.

20. Find three consecutive odd integers such that the sum of all three is 42 less than the product of the larger two.

21. Find three consecutive even integers such that the sum of the squares of the smaller two is equal to the square of the largest.

22. Find three consecutive even integers such that the square of the sum of the smaller two is equal to twice the largest.

Use the Pythagorean formula to solve each problem. See Example 3.

23. The hypotenuse of a right triangle is 1 cm longer than the longer leg. The shorter leg is 7 cm shorter than the longer leg. Find the length of the longer leg of the triangle.

24. The longer leg of a right triangle is 1 m longer than the shorter leg. The hypotenuse is 1 m shorter than twice the shorter leg. Find the length of the shorter leg of the triangle.

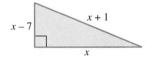

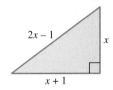

25. Tram works due north of home. Her husband Alan works due east. They leave for work at the same time. By the time Tram is 5 mi from home, the distance between them is 1 mi more than Alan's distance from home. How far from home is Alan?

26. Two cars left an intersection at the same time. One traveled north. The other traveled 14 mi farther, but to the east. How far apart were they then, if the distance between them was 4 mi more than the distance traveled east?

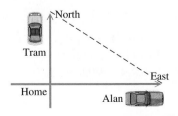

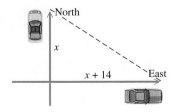

27. A ladder is leaning against a building. The distance from the bottom of the ladder to the building is 4 ft less than the length of the ladder. How high up the side of the building is the top of the ladder if that distance is 2 ft less than the length of the ladder?

28. A lot has the shape of a right triangle with one leg 2 m longer than the other. The hypotenuse is 2 m less than twice the length of the shorter leg. Find the length of the shorter leg.

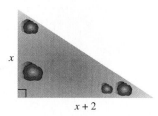

Solve each problem. See Examples 4 and 5.

29. An object propelled from a height of 48 ft with an initial velocity of 32 ft per sec after *t* sec has height

$$h = -16t^2 + 32t + 48.$$

48 ft

(a) After how many seconds is the height 64 ft? (*Hint:* Let $h = 64$ and solve.)

(b) After how many seconds is the height 60 ft?

(c) After how many seconds does the object hit the ground? (*Hint:* When the object hits the ground, $h = 0$.)

🖉 **(d)** The quadratic equation from part (c) has two solutions, yet only one of them is appropriate for answering the question. Why is this so?

30. If an object is propelled upward from ground level with an initial velocity of 64 ft per sec, its height *h* in feet *t* sec later is

$$h = -16t^2 + 64t.$$

(a) After how many seconds is the height 48 ft?

(b) The object reaches its maximum height 2 sec after it is propelled. What is this maximum height?

(c) After how many seconds does the object hit the ground?

🖉 **(d)** The quadratic equation from part (c) has two solutions, yet only one of them is appropriate for answering the question. Why is this so?

31. The table shows the number of cellular phones (in millions) owned by Americans.

Year	Cellular Phones (in millions)
1990	5
1992	11
1994	24
1996	44
1998	62
1999	86

Source: Cellular Telecommunications Industry Association.

The quadratic equation

$$y = .813x^2 + 1.27x + 5.28$$

models the number of cellular phones y (in millions) in the year x, where $x = 0$ represents 1990, $x = 2$ represents 1992, and so on.

(a) Use the model to find the number of cellular phones to the nearest tenth of a million in 1994. How does the result compare to the actual data in the table?

(b) What value of x corresponds to 1999?

(c) Use the model to find the number of cellular phones to the nearest tenth of a million in 1999. How does the result compare to the actual data in the table?

(d) Assuming that the trend in the data continues, use the quadratic equation to approximate the number of cellular phones to the nearest tenth of a million in 2002.

RELATING CONCEPTS (EXERCISES 32–40)

For Individual or Group Work

The U.S. trade deficit represents the amount by which exports are less than imports. It provides not only a sign of economic prosperity but also a warning of potential decline. The data in the table shows the U.S. trade deficit for 1995 through 1999.

Year	Deficit (in billions of dollars)
1995	97.5
1996	104.3
1997	104.7
1998	164.3
1999	271.3

Source: U.S. Department of Commerce.

Use the data to **work Exercises 32–40 in order.**

32. How much did the trade deficit increase from 1998 to 1999? What percent increase is this (to the nearest percent)?

33. The U.S. trade deficit for the years shown in the table can be approximated by the linear equation

$$y = 40.8x + 66.9,$$

(continued)

where y is the deficit in billions of dollars. Here $x = 0$ represents 1995, $x = 1$ represents 1996, and so on. Use this equation to approximate the trade deficits in 1995, 1997, and 1999.

34. How do your answers from Exercise 33 compare to the actual data in the table?

35. The trade deficit y (in billions of dollars) can also be approximated by the quadratic equation

$$y = 18.5x^2 - 33.4x + 104,$$

where $x = 0$ again represents 1995, $x = 1$ represents 1996, and so on. Use this equation to approximate the trade deficits in 1995, 1997, and 1999.

36. Compare your answers from Exercise 35 to the actual data in the table. Which equation, the linear or quadratic one, models the data better?

37. We can also see graphically why the linear equation is not a very good model for the data. To do so, write the data from the table as a set of ordered pairs (x, y), where x represents the number of years since 1995 and y represents the trade deficit in billions of dollars.

38. Plot the ordered pairs from Exercise 37 on a graph. Recall from Chapter 3 that a linear equation has a straight line for its graph. Do the ordered pairs you plotted lie in a linear pattern?

39. Assuming that the trend in the data continues and since the quadratic equation models the data fairly well, use the quadratic equation to predict the trade deficit for the year 2000.

40. The actual trade deficit for the year 2000 was 369.7 billion dollars. (*Source:* www.census.gov)

(a) How does the actual deficit for 2000 compare to your prediction from Exercise 39?

(b) Should the quadratic equation be used to predict the U.S. trade deficit for years after 2000? Explain.

Chapter **5** **Group Activity**

Factoring Trinomials Made Easy

OBJECTIVE Use an organized approach to factoring by grouping.

To factor a trinomial using FOIL, we must find the outer and inner coefficients that sum to give the coefficient of the middle term. Our approach begins with a **key number,** found by multiplying the coefficients of the first and last terms. For the trinomial $6x^2 - x - 2$, for instance, the key number is -12 since $6(-2) = -12$.

Step 1 Display the factors of -12 by entering $Y_1 = -12/X$ in a graphing calculator (Screen 1) and using an automatic table (Screen 2). Factors of -12 are automatically displayed in pairs as $1, -12$; $2, -6$; $3, -4$; $4, -3$; and $6, -2$ (Screen 3). You could scroll up or down to find other factors. Note that $5, -2.4$ and $7, -1.714$ are not factor pairs since -2.4 and -1.714 are not integers.

| SCREEN 1 | SCREEN 2 | SCREEN 3 |

Step 2   Find the pair of factors that sum to the *middle* term coefficient, -1. We can let the calculator do this, too. Enter $Y_2 = X + -12/X$. In this case, X is one of the factors and $-12/X$ is the other, so Y_2 will give the sum (Screen 4). Look for -1 in the Y_2 column in Screen 5. (You may have to scroll up or down to find it.)

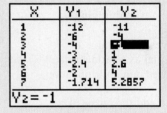

| SCREEN 4 | SCREEN 5 |

Step 3   Screen 5 shows that the coefficients of the outer and inner products are 3 and -4. Write $6x^2 - x - 2$ as $6x^2 + 3x - 4x - 2$. Using factoring by grouping,

$$(6x^2 + 3x) + (-4x - 2) = 3x(2x + 1) - 2(2x + 1) = (3x - 2)(2x + 1).$$

A. To factor each trinomial given in the column heads of the following table, first find the key number, and then use a calculator to help you find the coefficients of the outer and inner products of FOIL.

Trinomial	$3x^2 - 2x - 8$	$2x^2 - 11x + 15$	$10x^2 + 11x - 6$	$4x^2 + 5x + 3$
Key Number				
Outer, Inner Coefficients				
Factor by Grouping				(*Hint:* What does it mean if the middle term coefficient is *not* listed in the Y_2 column?)

B. Factor each trinomial by grouping.

CHAPTER **5** SUMMARY

KEY TERMS

5.1 factor factored form common factor greatest common factor (GCF)	**5.2** prime polynomial **5.4** perfect square trinomial	**5.5** quadratic equation standard form	**5.6** hypotenuse legs

TEST YOUR WORD POWER

See how well you have learned the vocabulary in this chapter. Answers, with examples, follow the Quick Review.

1. **Factoring** is
   A. a method of multiplying polynomials
   B. the process of writing a polynomial as a product
   C. the answer in a multiplication problem
   D. a way to add the terms of a polynomial.

2. A polynomial is in **factored form** when
   A. it is prime
   B. it is written as a sum

   C. the squared term has a coefficient of 1
   D. it is written as a product.

3. A **perfect square trinomial** is a trinomial
   A. that can be factored as the square of a binomial
   B. that cannot be factored
   C. that is multiplied by a binomial
   D. where all terms are perfect squares.

4. A **quadratic equation** is a polynomial equation of
   A. degree one
   B. degree two
   C. degree three
   D. degree four.

5. A **hypotenuse** is
   A. either of the two shorter sides of a triangle
   B. the shortest side of a triangle
   C. the side opposite the right angle in a triangle
   D. the longest side in any triangle.

QUICK REVIEW

CONCEPTS	EXAMPLES

5.1 THE GREATEST COMMON FACTOR; FACTORING BY GROUPING

Finding the Greatest Common Factor (GCF)
1. Include the largest numerical factor of every term.
2. Include each variable that is a factor of every term raised to the smallest exponent that appears in a term.

Find the greatest common factor of
$$4x^2y, \qquad -6x^2y^3, \qquad 2xy^2.$$
$$4x^2y = 2^2 \cdot x^2 \cdot y$$
$$-6x^2y^3 = -1 \cdot 2 \cdot 3 \cdot x^2 \cdot y^3$$
$$2xy^2 = 2 \cdot x \cdot y^2$$

The greatest common factor is $2xy$.

Factoring by Grouping

Step 1 Group the terms.

Step 2 Factor out the greatest common factor in each group.

Step 3 Factor a common binomial factor from the result of Step 2.

Step 4 If necessary try a different grouping.

Factor by grouping.
$$3x^2 + 5x - 24xy - 40y = (3x^2 + 5x) + (-24xy - 40y)$$
$$= x(3x + 5) - 8y(3x + 5)$$
$$= (3x + 5)(x - 8y)$$

CONCEPTS	EXAMPLES

5.2 FACTORING TRINOMIALS

To factor $x^2 + bx + c$, find m and n such that $mn = c$ and $m + n = b$.

$$\begin{array}{c} mn = c \\ \downarrow \\ x^2 + bx + c \\ \uparrow \\ m + n = b \end{array}$$

Then $x^2 + bx + c = (x + m)(x + n)$.

Check by multiplying.

Factor $x^2 + 6x + 8$.

$$\begin{array}{c} mn = 8 \\ \downarrow \\ x^2 + 6x + 8 \\ \uparrow \\ m + n = 6 \end{array}$$

$m = 2$ and $n = 4$

$$x^2 + 6x + 8 = (x + 2)(x + 4)$$

Check: $(x + 2)(x + 4) = x^2 + 4x + 2x + 8$
$$= x^2 + 6x + 8$$

5.3 MORE ON FACTORING TRINOMIALS

To factor $ax^2 + bx + c$:

By Grouping
Find m and n.

$$\begin{array}{c} mn = ac \\ \downarrow \qquad \downarrow \\ ax^2 + bx + c \\ \uparrow \\ m + n = b \end{array}$$

By Trial and Error
Use FOIL backwards.

Factor $3x^2 + 14x - 5$.

$$\overbrace{3x^2 + 14x - 5}^{\;}$$
$$-15$$

$mn = -15, m + n = 14$

By trial and error or by grouping,
$$3x^2 + 14x - 5 = (3x - 1)(x + 5).$$

5.4 SPECIAL FACTORING RULES

Difference of Squares
$x^2 - y^2 = (x + y)(x - y)$

Perfect Square Trinomials
$x^2 + 2xy + y^2 = (x + y)^2$
$x^2 - 2xy + y^2 = (x - y)^2$

Difference of Cubes
$x^3 - y^3 = (x - y)(x^2 + xy + y^2)$

Sum of Cubes
$x^3 + y^3 = (x + y)(x^2 - xy + y^2)$

Factor.
$$4x^2 - 9 = (2x + 3)(2x - 3)$$

$$9x^2 + 6x + 1 = (3x + 1)^2$$
$$4x^2 - 20x + 25 = (2x - 5)^2$$

$$m^3 - 8 = m^3 - 2^3 = (m - 2)(m^2 + 2m + 4)$$

$$27 + z^3 = 3^3 + z^3 = (3 + z)(9 - 3z + z^2)$$

CONCEPTS	EXAMPLES

5.5 SOLVING QUADRATIC EQUATIONS BY FACTORING

Zero-Factor Property

If a and b are real numbers and if $ab = 0$, then $a = 0$ or $b = 0$.

If $(x - 2)(x + 3) = 0$, then $x - 2 = 0$ or $x + 3 = 0$.

Solving a Quadratic Equation by Factoring

Solve $2x^2 = 7x + 15$.

Step 1 Write in standard form.

$$2x^2 - 7x - 15 = 0$$

Step 2 Factor.

$$(2x + 3)(x - 5) = 0$$

Step 3 Use the zero-factor property.

$$2x + 3 = 0 \quad \text{or} \quad x - 5 = 0$$
$$2x = -3 \qquad\qquad x = 5$$
$$x = -\frac{3}{2}$$

Step 4 Check.

Both solutions satisfy the original equation; the solution set is $\left\{-\frac{3}{2}, 5\right\}$.

5.6 APPLICATIONS OF QUADRATIC EQUATIONS

Pythagorean Formula

In a right triangle, the square of the hypotenuse equals the sum of the squares of the legs.

$$a^2 + b^2 = c^2$$

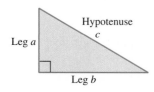

Leg a

Hypotenuse c

Leg b

In a right triangle, one leg measures 2 ft longer than the other. The hypotenuse measures 4 ft longer than the shorter leg. Find the lengths of the three sides of the triangle.

Let $x =$ the length of the shorter leg. Then

$$x^2 + (x + 2)^2 = (x + 4)^2.$$

Verify that the solutions of this equation are -2 and 6. Discard -2 as a solution. Check that the sides are 6, $6 + 2 = 8$, and $6 + 4 = 10$ ft in length.

Answers to Test Your Word Power

1. B; *Example:* $x^2 - 5x - 14 = (x - 7)(x + 2)$ **2.** D; *Example:* The factored form of $x^2 - 5x - 14$ is $(x - 7)(x + 2)$.
3. A; *Example:* $a^2 + 2a + 1$ is a perfect square trinomial; its factored form is $(a + 1)^2$.   **4.** B; *Examples:* $y^2 - 3y + 2 = 0, x^2 - 9 = 0$, $2m^2 = 6m + 8$ **5.** C; *Example:* In the triangle above, the hypotenuse is the side labeled c.

CHAPTER 5 REVIEW EXERCISES

[5.1] *Factor out the greatest common factor or factor by grouping.*

1. $7t + 14$

2. $60z^3 + 30z$

3. $2xy - 8y + 3x - 12$

4. $6y^2 + 9y + 4xy + 6x$

[5.2] *Factor completely.*

5. $x^2 + 5x + 6$

6. $y^2 - 13y + 40$

7. $q^2 + 6q - 27$

8. $r^2 - r - 56$

9. $r^2 - 4rs - 96s^2$

10. $p^2 + 2pq - 120q^2$

11. $8p^3 - 24p^2 - 80p$

12. $3x^4 + 30x^3 + 48x^2$

13. $p^7 - p^6q - 2p^5q^2$

14. $3r^5 - 6r^4s - 45r^3s^2$

[5.3]

15. To begin factoring $6r^2 - 5r - 6$, what are the possible first terms of the two binomial factors if we consider only positive integer coefficients?

16. What is the first step you would use to factor $2z^3 + 9z^2 - 5z$?

Factor completely.

17. $2k^2 - 5k + 2$ **18.** $3r^2 + 11r - 4$ **19.** $6r^2 - 5r - 6$

20. $10z^2 - 3z - 1$ **21.** $8v^2 + 17v - 21$ **22.** $24x^5 - 20x^4 + 4x^3$

23. $-6x^2 + 3x + 30$ **24.** $10r^3s + 17r^2s^2 + 6rs^3$

[5.4]

25. Which one of the following is the difference of squares?

    A. $32x^2 - 1$ **B.** $4x^2y^2 - 25z^2$ **C.** $x^2 + 36$ **D.** $25y^3 - 1$

26. Which one of the following is a perfect square trinomial?

    A. $x^2 + x + 1$ **B.** $y^2 - 4y + 9$ **C.** $4x^2 + 10x + 25$ **D.** $x^2 - 20x + 100$

Factor completely.

27. $n^2 - 49$ **28.** $25b^2 - 121$ **29.** $49y^2 - 25w^2$

30. $144p^2 - 36q^2$ **31.** $x^2 + 100$ **32.** $r^2 - 12r + 36$

33. $9t^2 - 42t + 49$ **34.** $m^3 + 1000$ **35.** $125k^3 + 64x^3$

36. $343x^3 - 64$

[5.5] *Solve each equation and check your solutions.*

37. $(4t + 3)(t - 1) = 0$ **38.** $(x + 7)(x - 4)(x + 3) = 0$

39. $x(2x - 5) = 0$ **40.** $z^2 + 4z + 3 = 0$

41. $m^2 - 5m + 4 = 0$ **42.** $x^2 = -15 + 8x$

43. $3z^2 - 11z - 20 = 0$ **44.** $81t^2 - 64 = 0$

45. $y^2 = 8y$ **46.** $n(n - 5) = 6$

47. $t^2 - 14t + 49 = 0$ **48.** $t^2 = 12(t - 3)$

49. $(5z + 2)(z^2 + 3z + 2) = 0$ **50.** $x^2 = 9$

[5.6] *Solve each problem.*

51. The length of a rug is 6 ft more than the width. The area is 40 ft^2. Find the length and width of the rug.

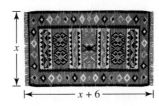

52. The surface area S of a box is given by

$$S = 2WH + 2WL + 2LH.$$

A treasure chest from a sunken galleon has dimensions as shown in the figure. Its surface area is 650 ft^2. Find its width.

53. The length of a rectangle is three times the width. If the width were increased by 3 m while the length remained the same, the new rectangle would have an area of 30 m². Find the length and width of the original rectangle.

54. The volume of a rectangular box is 120 m³. The width of the box is 4 m, and the height is 1 m less than the length. Find the length and height of the box.

55. The product of two consecutive integers is 29 more than their sum. What are the integers?

56. Two cars left an intersection at the same time. One traveled west, and the other traveled 14 mi less, but to the south. How far apart were they then, if the distance between them was 16 mi more than the distance traveled south?

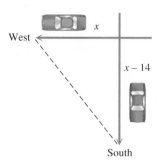

If an object is propelled upward with an initial velocity of 128 ft per sec, its height h after t sec is

$$h = 128t - 16t^2.$$

Find the height of the object after each period of time.

57. 1 sec   **58.** 2 sec   **59.** 4 sec

60. For the object described above, when does it return to the ground?

61. Annual revenue in millions of dollars for eBay is shown in the table.

Year	Annual Revenue (in millions of dollars)
1997	5.1
1998	47.4
1999	224.7

Source: eBay.

Using the data, we developed the quadratic equation

$$y = 67.5x^2 - 25.2x + 5.1$$

to model eBay revenues y in year x, where $x = 0$ represents 1997, $x = 1$ represents 1998, and so on. Because only three years of data were used to determine the model, we must be careful about using it to predict revenue for years beyond 1999.

(a) Use the model to predict annual revenue for eBay in 2000.

(b) The revenue for eBay through the first half of 2000 was $183.2 million. Given this information, do you think your prediction in part (a) is reliable? Explain.

▰ MIXED REVIEW EXERCISES

62. Which of the following is *not* factored completely?

    A. $3(7t)$ **B.** $3x(7t + 4)$ **C.** $(3 + x)(7t + 4)$ **D.** $3(7t + 4) + x(7t + 4)$

63. Although $(2x + 8)(3x - 4) = 6x^2 + 16x - 32$ is a true statement, the polynomial is not factored completely. Explain why and give the complete factored form.

Factor completely.

64. $z^2 - 11zx + 10x^2$

65. $3k^2 + 11k + 10$

66. $15m^2 + 20m - 12mp - 16p$

67. $y^4 - 625$

68. $6m^3 - 21m^2 - 45m$

69. $24ab^3c^2 - 56a^2bc^3 + 72a^2b^2c$

70. $25a^2 + 15ab + 9b^2$

71. $12x^2yz^3 + 12xy^2z - 30x^3y^2z^4$

72. $2a^5 - 8a^4 - 24a^3$

73. $12r^2 + 18rq - 10r - 15q$

74. $1000a^3 + 27$

75. $49t^2 + 56t + 16$

Solve.

76. $t(t - 7) = 0$

77. $x^2 + 3x = 10$

78. $25x^2 + 20x + 4 = 0$

79. The numbers of alternative-fueled vehicles, in thousands, in use for the years 1998–2001 are given in the table. Using statistical methods, we constructed the quadratic equation

$$y = .25x^2 - 25.65x + 496.6$$

to model the number of vehicles y in year x. Here we used $x = 98$ for 1998, $x = 99$ for 1999, and so on. Because only four years of data were used to determine the model, we must be particularly careful about using it to estimate for years before 1998 or after 2001.

(a) What prediction for 2002 is given by the equation?

(b) Why might the prediction for 2002 be unreliable?

ALTERNATIVE-FUELED VEHICLES

Year	Number (in thousands)
1998	384
1999	407
2000	432
2001	456

Source: Energy Information Administration, Alternatives to Traditional Fuels, 2001.

80. A lot is shaped like a right triangle. The hypotenuse is 3 m longer than the longer leg. The longer leg is 6 m longer than twice the length of the shorter leg. Find the lengths of the sides of the lot.

81. A pyramid has a rectangular base with a length that is 2 m more than the width. The height of the pyramid is 6 m, and its volume is 48 m³. Find the length and width of the base.

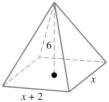

82. The product of the smaller two of three consecutive integers is equal to 23 plus the largest. Find the integers.

83. If an object is dropped, the distance d in feet it falls in t sec (disregarding air resistance) is given by the quadratic equation

$$d = 16t^2.$$

Find the distance an object would fall in the following times.

(a) 4 sec   (b) 8 sec

84. The floor plan for a house is a rectangle with length 7 m more than its width. The area is 170 m². Find the width and length of the house.

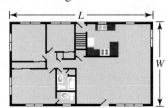

85. The triangular sail of a schooner has an area of 30 m². The height of the sail is 4 m more than the base. Find the base of the sail.

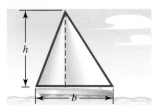

CHAPTER 5 TEST

1. Which one of the following is the correct, completely factored form of $2x^2 - 2x - 24$?

 A. $(2x + 6)(x - 4)$ **B.** $(x + 3)(2x - 8)$

 C. $2(x + 4)(x - 3)$ **D.** $2(x + 3)(x - 4)$

Factor each polynomial completely.

2. $12x^2 - 30x$

3. $2m^3n^2 + 3m^3n - 5m^2n^2$

4. $2ax - 2bx + ay - by$

5. $x^2 - 5x - 24$

6. $2x^2 + x - 3$

7. $10z^2 - 17z + 3$

8. $t^2 + 2t + 3$

9. $x^2 + 36$

10. $12 - 6a + 2b - ab$

11. $9y^2 - 64$

12. $4x^2 - 28xy + 49y^2$

13. $-2x^2 - 4x - 2$

14. $6t^4 + 3t^3 - 108t^2$

15. $r^3 - 125$

16. $8k^3 + 64$

17. $x^4 - 81$

✍ **18.** Why is $(p + 3)(p + 3)$ not the correct factored form of $p^2 + 9$?

Solve each equation.

19. $2r^2 - 13r + 6 = 0$

20. $25x^2 - 4 = 0$

21. $x(x - 20) = -100$

22. $t^3 = 9t$

Solve each problem.

23. The length of a rectangular flower bed is 3 ft less than twice its width. The area of the bed is 54 ft². Find the dimensions of the flower bed.

24. Find two consecutive integers such that the square of the sum of the two integers is 11 more than the smaller integer.

25. A carpenter needs to cut a brace to support a wall stud, as shown in the figure. The brace should be 7 ft less than three times the length of the stud. If the brace will be anchored on the floor 15 ft away from the stud, how long should the brace be?

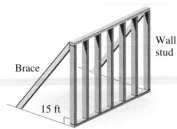

26. TV viewers have more choices than ever. The number of cable TV channels y from 1984 through 1999 can be approximated by the quadratic equation

$$y = .57x^2 + .31x + 48,$$

where $x = 0$ represents 1984, $x = 1$ represents 1985, and so on. (*Source:* National Cable Television Association.) Use the model to estimate the number of cable TV channels in 1999. Round your answer to the nearest whole number.

CUMULATIVE REVIEW EXERCISES CHAPTERS **1–5**

Solve each equation.

1. $3x + 2(x - 4) = 4(x - 2)$

2. $.3x + .9x = .06$

3. $\dfrac{2}{3}y - \dfrac{1}{2}(y - 4) = 3$

4. Solve for P: $A = P + Prt.$

5. From a list of "everyday items" often taken for granted, adults were recently surveyed as to those items they wouldn't want to live without. Complete the results shown in the table if 500 adults were surveyed.

Item	Percent That Wouldn't Want to Live Without	Number That Wouldn't Want to Live Without
Toilet paper	69%	
Zipper	42%	
Frozen foods		190
Self-stick note pads		75

(Other items included tape, hairspray, pantyhose, paper clips, and Velcro.)
Source: Market Facts for Kleenex Cottonelle.

Solve each problem.

6. At the 1998 Winter Olympics in Nagano, Japan, the top medal winner was Germany with 29. Germany won 1 more silver medal than bronze and 3 more gold medals than silver. Find the number of each type of medal won. (*Source: World Almanac and Book of Facts,* 2000.)

7. In July 2000, roughly 144 million people surfed the Web from home. This was a 35% increase from the same month the previous year. How many people, to the nearest million, surfed the Web from home in July 1999? (*Source: The Gazette,* September 3, 2000.)

8. Find the measures of the marked angles.

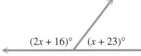

$(2x + 16)°$ $(x + 23)°$

9. Fill in each blank with *positive* or *negative*. The point with coordinates (a, b) is in

 (a) quadrant II if a is _____ and b is _____.
 (b) quadrant III if a is _____ and b is _____.

Consider the equation $y = 12x + 3$. Find the following.

10. The x- and y-intercepts

11. The slope

12. The graph

13. The points on the graph show the number of U.S. radio stations in the years 1993 through 1999, along with the graph of a linear equation that models the data. Use the ordered pairs shown on the graph to find the slope of the line to the nearest whole number. Interpret the slope.

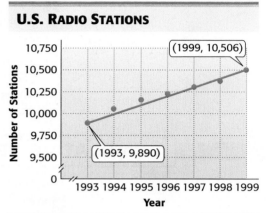

U.S. RADIO STATIONS

(1999, 10,506)

(1993, 9,890)

Source: M Street Corporation.

Write an equation for each line. Write the equation in slope-intercept form if possible.

14. Slope $-\dfrac{1}{2}$, y-intercept $(0, 5)$

15. Through $\left(\dfrac{1}{4}, -\dfrac{2}{3}\right)$, slope 0

Evaluate each expression.

16. $2^{-3} \cdot 2^5$

17. $\left(\dfrac{3}{4}\right)^{-2}$

18. $\left(\dfrac{4^{-3} \cdot 4^4}{4^5}\right)^{-1}$

Simplify each expression and write the answer using only positive exponents. Assume no denominators are 0.

19. $\dfrac{(p^2)^3 p^{-4}}{(p^{-3})^{-1} p}$

20. $\dfrac{(m^{-2})^3 m}{m^5 m^{-4}}$

Perform each indicated operation.

21. $(2k^2 + 4k) - (5k^2 - 2) - (k^2 + 8k - 6)$ **22.** $(9x + 6)(5x - 3)$

23. $(3p + 2)^2$

24. $\dfrac{8x^4 + 12x^3 - 6x^2 + 20x}{2x}$

25. To make a pound of honey, bees may travel 55,000 mi and visit more than 2,000,000 flowers. (*Source: Home & Garden.*) Write the two given numbers in scientific notation.

Factor completely.

26. $2a^2 + 7a - 4$ **27.** $10m^2 + 19m + 6$ **28.** $8t^2 + 10tv + 3v^2$

29. $4p^2 - 12p + 9$ **30.** $25r^2 - 81t^2$ **31.** $2pq + 6p^3q + 8p^2q$

Solve each equation.

32. $6m^2 + m - 2 = 0$ **33.** $8x^2 = 64x$

34. The length of the hypotenuse of a right triangle is twice the length of the shorter leg, plus 3 m. The longer leg is 7 m longer than the shorter leg. Find the lengths of the sides.

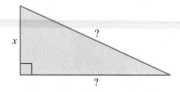

Rational Expressions and Applications

6

Americans have been car crazy ever since the first automobiles hit the road early in the twentieth century. Today there are about 213.5 million vehicles in the United States driving on 3.4 million mi of paved roadways. There is even a new museum devoted exclusively to our four-wheeled passion and its influence on our lives and culture. The Museum of Automobile History in Syracuse, N.Y., features some 200 years of automobile memorabilia, including rare advertising pieces, designer drawings, and Hollywood movie posters. (*Source: Home and Away,* May/June 2002.) In Exercises 75 and 76 of Section 6.4, we use a rational expression to determine the cost of restoring a vintage automobile.

6.1 The Fundamental Property of Rational Expressions

The quotient of two integers (with denominator not 0), such as $\frac{2}{3}$ or $-\frac{3}{4}$, is called a *rational number*. In the same way, the quotient of two polynomials with denominator not equal to 0 is called a *rational expression*.

Rational Expression

A **rational expression** is an expression of the form

$$\frac{P}{Q},$$

where P and Q are polynomials, with $Q \neq 0$.

Examples of rational expressions include

$$\frac{-6x}{x^3 + 8}, \qquad \frac{9x}{y + 3}, \qquad \text{and} \qquad \frac{2m^3}{8}.$$

Our work with rational expressions requires much of what we learned in Chapters 4 and 5 on polynomials and factoring, as well as the rules for fractions from Section 1.1.

OBJECTIVE 1 Find the numerical value of a rational expression. We use substitution to evaluate a rational expression for a given value of the variable.

▦ EXAMPLE 1 Evaluating Rational Expressions

Find the numerical value of $\dfrac{3x + 6}{2x - 4}$ for each value of x.

(a) $x = 1$

$$\frac{3x + 6}{2x - 4} = \frac{3(1) + 6}{2(1) - 4} \qquad \text{Let } x = 1.$$

$$= \frac{9}{-2} \quad \text{or} \quad -\frac{9}{2}$$

(b) $x = -2$

$$\frac{3x + 6}{2x - 4} = \frac{3(-2) + 6}{2(-2) - 4} \qquad \text{Let } x = -2.$$

$$= \frac{0}{-8} \quad \text{or} \quad 0$$

> **Now Try Exercise 3.**

OBJECTIVE 2 Find the values of the variable for which a rational expression is undefined. In the definition of a rational expression $\frac{P}{Q}$, Q cannot equal 0. The denominator of a rational expression cannot equal 0 because division by 0 is undefined.

For instance, in the rational expression

$$\frac{3x + 6}{2x - 4} \leftarrow \text{Denominator cannot equal 0.}$$

from Example 1, the variable x can take on any real number value except 2. If x is 2, then the denominator becomes $2(2) - 4 = 0$, making the expression undefined. Thus, x cannot equal 2. We indicate this restriction by writing $x \neq 2$.

NOTE The *numerator* of a rational expression may be *any* real number. If the numerator equals 0 and the denominator does not equal 0, then the rational expression equals 0. See Example 1(b).

To determine the values for which a rational expression is undefined, use the following procedure.

Determining When a Rational Expression is Undefined

Step 1 Set the denominator of the rational expression equal to 0.

Step 2 Solve this equation.

Step 3 The solutions of the equation are the values that make the rational expression undefined.

EXAMPLE 2 Finding Values That Make Rational Expressions Undefined

Find any values of the variable for which each rational expression is undefined.

(a) $\dfrac{p + 5}{3p + 2}$

Remember that the *numerator* may be any real number; you must find any value of p that makes the *denominator* equal to 0 since division by 0 is undefined.

Step 1 Set the denominator equal to 0.

$$3p + 2 = 0$$

Step 2 Solve this equation.

$$3p = -2$$

$$p = -\frac{2}{3}$$

Step 3 The given expression is undefined for $-\frac{2}{3}$, so $p \neq -\frac{2}{3}$.

(b) $\dfrac{8x^2 + 1}{x - 3}$

The denominator $x - 3 = 0$ when x is 3. The given expression is undefined for 3, so $x \neq 3$.

(c) $\dfrac{9m^2}{m^2 - 5m + 6}$

Set the denominator equal to 0, and then find the solutions of the equation $m^2 - 5m + 6 = 0$.

$$(m - 2)(m - 3) = 0 \qquad \text{Factor.}$$
$$m - 2 = 0 \quad \text{or} \quad m - 3 = 0 \qquad \text{Zero-factor property}$$
$$m = 2 \quad \text{or} \qquad m = 3$$

The given expression is undefined for 2 and 3, so $m \neq 2$ and $m \neq 3$.

(d) $\dfrac{2r}{r^2 + 1}$

This denominator will not equal 0 for any value of r because r^2 is always greater than or equal to 0, and adding 1 makes the sum greater than 0. Thus, there are no values for which this rational expression is undefined.

Now Try Exercises 17 and 21.

OBJECTIVE 3 **Write rational expressions in lowest terms.** A fraction such as $\frac{2}{3}$ is said to be in *lowest terms*. How can "lowest terms" be defined? We use the idea of greatest common factor for this definition, which applies to all rational expressions.

Lowest Terms

A rational expression $\frac{P}{Q}$ $(Q \neq 0)$ is in **lowest terms** if the greatest common factor of its numerator and denominator is 1.

The properties of rational numbers also apply to rational expressions. We use the **fundamental property of rational expressions** to write a rational expression in lowest terms.

Fundamental Property of Rational Expressions

If $\frac{P}{Q}$ $(Q \neq 0)$ is a rational expression and if K represents any polynomial, where $K \neq 0$, then

$$\frac{PK}{QK} = \frac{P}{Q}.$$

This property is based on the identity property of multiplication, since

$$\frac{PK}{QK} = \frac{P}{Q} \cdot \frac{K}{K} = \frac{P}{Q} \cdot 1 = \frac{P}{Q}.$$

The next example shows how to write both a rational number and a rational expression in lowest terms. Notice the similarity in the procedures. In both cases, we factor and then divide out the greatest common factor.

EXAMPLE 3 Writing in Lowest Terms

Write each expression in lowest terms.

(a) $\dfrac{30}{72}$

Begin by factoring.

$$\frac{30}{72} = \frac{2 \cdot 3 \cdot 5}{2 \cdot 2 \cdot 2 \cdot 3 \cdot 3}$$

Group any factors common to the numerator and denominator.

$$\frac{30}{72} = \frac{5 \cdot (2 \cdot 3)}{2 \cdot 2 \cdot 3 \cdot (2 \cdot 3)}$$

Use the fundamental property.

$$\frac{30}{72} = \frac{5}{2 \cdot 2 \cdot 3} = \frac{5}{12}$$

(b) $\dfrac{14k^2}{2k^3}$

Write k^2 as $k \cdot k$ and k^3 as $k \cdot k \cdot k$.

$$\frac{14k^2}{2k^3} = \frac{2 \cdot 7 \cdot k \cdot k}{2 \cdot k \cdot k \cdot k}$$

$$\frac{14k^2}{2k^3} = \frac{7(2 \cdot k \cdot k)}{k(2 \cdot k \cdot k)}$$

$$\frac{14k^2}{2k^3} = \frac{7}{k}$$

Now Try Exercise 27.

To write a rational expression in lowest terms, follow these steps.

Writing a Rational Expression in Lowest Terms

Step 1 **Factor** the numerator and denominator completely.

Step 2 **Use the fundamental property** to divide out any common factors.

EXAMPLE 4 Writing in Lowest Terms

Write each rational expression in lowest terms.

(a) $\dfrac{3x - 12}{5x - 20}$

Step 1 Factor both numerator and denominator.

$$\frac{3x - 12}{5x - 20} = \frac{3(x - 4)}{5(x - 4)}$$

Step 2 Use the fundamental property.

$$= \frac{3}{5} \qquad \text{Divide out the common factor.}$$

The given expression is equal to $\frac{3}{5}$ for all values of x, where $x \neq 4$ (since the denominator of the original rational expression is 0 when x is 4).

(b) $\dfrac{2y^2 - 8}{2y + 4} = \dfrac{2(y^2 - 4)}{2(y + 2)}$ Factor. (Step 1)

$$= \frac{2(y + 2)(y - 2)}{2(y + 2)} \qquad \text{Be sure to factor completely.}$$

$$= y - 2 \qquad \text{Fundamental property (Step 2)}$$

Here, $y \neq -2$ since the denominator of the original expression is 0 for this value.

(c) $\dfrac{m^2 + 2m - 8}{2m^2 - m - 6} = \dfrac{(m + 4)(m - 2)}{(2m + 3)(m - 2)}$ Factor.

$\qquad\qquad\quad = \dfrac{m + 4}{2m + 3}$ Fundamental property

Here, $m \neq -\frac{3}{2}$ and $m \neq 2$, since the denominator of the original expression is 0 for these values of m.

Now Try Exercises 33, 41, and 49.

From now on, we write statements of equality of rational expressions with the understanding that they apply only to those real numbers that make neither denominator equal to 0.

CAUTION   *Rational expressions cannot be written in lowest terms until after the numerator and denominator have been factored. Only common factors can be divided out,* not common terms. For example,

$$\dfrac{6x + 9}{4x + 6} = \dfrac{3(2x + 3)}{2(2x + 3)} = \dfrac{3}{2} \qquad \dfrac{6 + x}{4x} \leftarrow \text{Numerator cannot be factored.}$$

Divide out the common factor.   Already in lowest terms

EXAMPLE 5   Writing in Lowest Terms (Factors Are Opposites)

Write $\dfrac{x - y}{y - x}$ in lowest terms.

At first glance, there does not seem to be any way in which $x - y$ and $y - x$ can be factored to get a common factor. However, the denominator $y - x$ can be factored as

$$y - x = -1(-y + x) = -1(x - y).$$

Now, use the fundamental property to simplify.

$$\dfrac{x - y}{y - x} = \dfrac{1(x - y)}{-1(x - y)} = \dfrac{1}{-1} = -1$$

Now Try Exercise 63.

NOTE   Either the numerator or the denominator could have been factored in the first step in Example 5. Factor -1 from the numerator and confirm that the result is the same.

In Example 5, notice that $y - x$ is the opposite of $x - y$. A general rule for this situation follows.

If the numerator and the denominator of a rational expression are opposites, as in $\frac{x - y}{y - x}$, then the rational expression is equal to -1.

■ **EXAMPLE 6** Writing in Lowest Terms (Factors Are Opposites)

Write each rational expression in lowest terms.

(a) $\dfrac{2 - m}{m - 2}$

Each term in the numerator differs only in sign from the same term in the denominator, so we can factor -1 from either the numerator or the denominator.

$$\frac{2 - m}{m - 2} = \frac{-1(m - 2)}{m - 2} = -1$$

(b) $\dfrac{4x^2 - 9}{6 - 4x}$

Factor the numerator and denominator.

$$\begin{aligned}\frac{4x^2 - 9}{6 - 4x} &= \frac{(2x + 3)(2x - 3)}{2(3 - 2x)} \\[2mm] &= \frac{(2x + 3)(2x - 3)}{2(-1)(2x - 3)} \qquad \text{Write } 3 - 2x \text{ as } -1(2x - 3). \\[2mm] &= \frac{2x + 3}{2(-1)} \\[2mm] &= \frac{2x + 3}{-2} \quad \text{or} \quad -\frac{2x + 3}{2} \qquad \frac{a}{-b} = -\frac{a}{b}\end{aligned}$$

(c) $\dfrac{3 + r}{3 - r}$

The quantity $3 - r$ is *not* the opposite of $3 + r$ because the 3 is positive in both cases. This rational expression is already in lowest terms. ■

Now Try Exercises 65 and 69.

OBJECTIVE 4 **Recognize equivalent forms of rational expressions.** When working with rational expressions, it is important to be able to recognize equivalent forms of an expression. For example, the common fraction $-\frac{5}{6}$ can also be written $\frac{-5}{6}$ and $\frac{5}{-6}$. Consider also the rational expression

$$-\frac{2x + 3}{2}.$$

The $-$ sign representing the -1 factor is in front of the expression. The -1 factor may instead be placed in the numerator or in the denominator. Thus, some other equivalent forms of this rational expression are

Use parentheses.

$$\frac{-(2x + 3)}{2} \qquad \text{and} \qquad \frac{2x + 3}{-2}.$$

By the distributive property,

$$\frac{-(2x + 3)}{2} \qquad \text{can also be written} \qquad \frac{-2x - 3}{2}.$$

CAUTION $\frac{-2x + 3}{2}$ is *not* an equivalent form of $\frac{-(2x + 3)}{2}$. The sign preceding 3 in the numerator of $\frac{-2x + 3}{2}$ should be $-$ rather than $+$. Be careful to apply the distributive property correctly.

EXAMPLE 7   Writing Equivalent Forms of a Rational Expression

Write four equivalent forms of the rational expression

$$-\frac{3x + 2}{x - 6}.$$

If we apply the negative sign to the numerator, we obtain the equivalent forms

$$\frac{-(3x + 2)}{x - 6} \quad \text{or, by the distributive property,} \quad \frac{-3x - 2}{x - 6}.$$

If we apply the negative sign to the denominator, we obtain

$$\frac{3x + 2}{-(x - 6)} \quad \text{or, distributing once again,} \quad \frac{3x + 2}{-x + 6}.$$

Now Try Exercise 73.

CAUTION   Recall that $-\frac{5}{6} \neq \frac{-5}{-6}$. Thus, in Example 7, it would be incorrect to distribute the negative sign to *both* the numerator *and* the denominator. (Doing this would actually lead to the *opposite* of the original expression.)

| CONNECTIONS |

In Chapter 4 we used long division to find the quotient of two polynomials. For example, we found $(2x^2 + 5x - 12) \div (2x - 3)$ as follows:

$$\begin{array}{r} x + 4 \\ 2x - 3 \overline{) 2x^2 + 5x - 12} \\ \underline{2x^2 - 3x} \\ 8x - 12 \\ \underline{8x - 12} \\ 0 \end{array}$$

The quotient is $x + 4$. We also get the same quotient by expressing the division problem as a rational expression (fraction) and writing this rational expression in lowest terms.

$$\frac{2x^2 + 5x - 12}{2x - 3} = \frac{(2x - 3)(x + 4)}{2x - 3}$$

$$= x + 4$$

For Discussion or Writing

What kind of division problem has a quotient that cannot be found by writing a fraction in lowest terms? Try using rational expressions to solve each division problem. Then use long division and compare.

1. $(3x^2 + 11x + 8) \div (x + 2)$ **2.** $(x^3 - 8) \div (x^2 + 2x + 4)$

6.1 EXERCISES

For Extra Help

Student's
Solutions Manual

MyMathLab

InterAct Math
Tutorial Software

Tutor
Center
AW Math
Tutor Center

Math MathXL

Digital Video Tutor
CD 9/Videotape 9

 1. Define *rational expression* in your own words, and give an example.

*Find the numerical value of each rational expression when (**a**) x = 2 and (**b**) x = −3. See Example 1.*

2. $\dfrac{5x - 2}{4x}$

3. $\dfrac{3x + 1}{5x}$

4. $\dfrac{2x^2 - 4x}{3x - 1}$

5. $\dfrac{x^2 - 4}{2x + 1}$

6. $\dfrac{(-3x)^2}{4x + 12}$

7. $\dfrac{(-2x)^3}{3x + 9}$

8. $\dfrac{5x + 2}{2x^2 + 11x + 12}$

9. $\dfrac{7 - 3x}{3x^2 - 7x + 2}$

10. Fill in each blank with the correct response: The rational expression $\frac{x + 5}{x - 3}$ is undefined when x is _____, so $x \neq$ _____. This rational expression is equal to 0 when $x =$ _____.

 11. Why can't the denominator of a rational expression equal 0?

 12. If 2 is substituted for x in the rational expression $\dfrac{x - 2}{x^2 - 4}$, the result is $\dfrac{0}{0}$. An often-heard statement is "Any number divided by itself is 1." Does this mean that this expression is equal to 1 for $x = 2$? If not, explain.

Find any values for which each rational expression is undefined. See Example 2.

13. $\dfrac{12}{5y}$

14. $\dfrac{-7}{3z}$

15. $\dfrac{x + 1}{x - 6}$

16. $\dfrac{m - 2}{m - 5}$

17. $\dfrac{4x^2}{3x + 5}$

18. $\dfrac{2x^3}{3x + 4}$

19. $\dfrac{5m + 2}{m^2 + m - 6}$

20. $\dfrac{2r - 5}{r^2 - 5r + 4}$

21. $\dfrac{x^2 + 3x}{4}$

22. $\dfrac{x^2 - 4x}{6}$

23. $\dfrac{3x - 1}{x^2 + 2}$

24. $\dfrac{4q + 2}{q^2 + 9}$

25. (a) Identify the two *terms* in the numerator and the two *terms* in the denominator of the rational expression $\dfrac{x^2 + 4x}{x + 4}$.

 (b) Describe the steps you would use to write this rational expression in lowest terms. (*Hint:* It simplifies to x.)

26. Only one of the following rational expressions can be simplified. Which one is it?

A. $\dfrac{x^2 + 2}{x^2}$ **B.** $\dfrac{x^2 + 2}{2}$ **C.** $\dfrac{x^2 + y^2}{y^2}$ **D.** $\dfrac{x^2 - 5x}{x}$

Write each rational expression in lowest terms. See Examples 3 and 4.

27. $\dfrac{18r^3}{6r}$

28. $\dfrac{27p^2}{3p}$

29. $\dfrac{4(y-2)}{10(y-2)}$

30. $\dfrac{15(m-1)}{9(m-1)}$

31. $\dfrac{(x+1)(x-1)}{(x+1)^2}$

32. $\dfrac{(t+5)(t-3)}{(t-1)(t+5)}$

33. $\dfrac{7m+14}{5m+10}$

34. $\dfrac{8z-24}{4z-12}$

35. $\dfrac{6m-18}{7m-21}$

36. $\dfrac{5r+20}{3r+12}$

37. $\dfrac{m^2-n^2}{m+n}$

38. $\dfrac{a^2-b^2}{a-b}$

39. $\dfrac{2t+6}{t^2-9}$

40. $\dfrac{5s-25}{s^2-25}$

41. $\dfrac{12m^2-3}{8m-4}$

42. $\dfrac{20p^2-45}{6p-9}$

43. $\dfrac{3m^2-3m}{5m-5}$

44. $\dfrac{6t^2-6t}{2t-2}$

45. $\dfrac{9r^2-4s^2}{9r+6s}$

46. $\dfrac{16x^2-9y^2}{12x-9y}$

47. $\dfrac{5k^2-13k-6}{5k+2}$

48. $\dfrac{7t^2-31t-20}{7t+4}$

49. $\dfrac{x^2+2x-15}{x^2+6x+5}$

50. $\dfrac{y^2-5y-14}{y^2+y-2}$

51. $\dfrac{2x^2-3x-5}{2x^2-7x+5}$

52. $\dfrac{3x^2+8x+4}{3x^2-4x-4}$

These exercises involve factoring by grouping (Section 5.1) and factoring sums and differences of cubes (Section 5.4). Write each rational expression in lowest terms.

53. $\dfrac{zw+4z-3w-12}{zw+4z+5w+20}$

54. $\dfrac{km+4k+4m+16}{km+4k+5m+20}$

55. $\dfrac{m^2-n^2-4m-4n}{2m-2n-8}$

56. $\dfrac{x^2y+y+x^2z+z}{xy+xz}$

57. $\dfrac{b^3-a^3}{a^2-b^2}$

58. $\dfrac{k^3+8}{k^2-4}$

59. $\dfrac{z^3+27}{z^3-3z^2+9z}$

60. $\dfrac{1-8r^3}{8r^2+4r+2}$

61. Which two of the following rational expressions equal -1?

  A. $\dfrac{2x+3}{2x-3}$ **B.** $\dfrac{2x-3}{3-2x}$ **C.** $\dfrac{2x+3}{3+2x}$ **D.** $\dfrac{2x+3}{-2x-3}$

62. Make the correct choice for the blank: $\dfrac{4-r^2}{4+r^2}$ _____ equal to -1.
                                          (is/is not)

Write each rational expression in lowest terms. See Examples 5 and 6.

63. $\dfrac{6-t}{t-6}$

64. $\dfrac{2-k}{k-2}$

65. $\dfrac{m^2-1}{1-m}$

66. $\dfrac{a^2-b^2}{b-a}$

67. $\dfrac{q^2-4q}{4q-q^2}$

68. $\dfrac{z^2-5z}{5z-z^2}$

69. $\dfrac{p+6}{p-6}$

70. $\dfrac{5-x}{5+x}$

71. Only one of the following rational expressions is *not* equivalent to $\frac{x-3}{4-x}$. Which one is it?

    A. $\dfrac{3-x}{x-4}$ **B.** $\dfrac{x+3}{4+x}$ **C.** $-\dfrac{3-x}{4-x}$ **D.** $-\dfrac{x-3}{x-4}$

72. Make the correct choice for the blank: $\frac{5+2x}{3-x}$ and $\frac{-5-2x}{x-3}$ _____ equivalent rational expressions. (are/are not)

Write four equivalent forms for each rational expression. See Example 7.

73. $-\dfrac{x+4}{x-3}$ **74.** $-\dfrac{x+6}{x-1}$ **75.** $-\dfrac{2x-3}{x+3}$

76. $-\dfrac{5x-6}{x+4}$ **77.** $\dfrac{-3x+1}{5x-6}$ **78.** $\dfrac{-2x-9}{3x+1}$

79. The area of the rectangle is represented by

$$x^4 + 10x^2 + 21.$$

What is the width? $\left(Hint:\text{ Use }W = \frac{A}{L}.\right)$

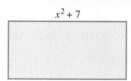

$x^2 + 7$

80. The volume of the box is represented by

$$(x^2 + 8x + 15)(x + 4).$$

Find the polynomial that represents the area of the bottom of the box.

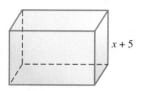

$x + 5$

Solve each problem.

81. The average number of vehicles waiting in line to enter a sports arena parking area is approximated by the rational expression

$$\frac{x^2}{2(1-x)},$$

where x is a quantity between 0 and 1 known as the *traffic intensity*. (*Source:* Mannering, F. and W. Kilareski, *Principles of Highway Engineering and Traffic Control,* John Wiley and Sons, 1990.) To the nearest tenth, find the average number of vehicles waiting if the traffic intensity is

    (a) .1   **(b)** .8   **(c)** .9.

    (d) What happens to waiting time as traffic intensity increases?

82. The percent of deaths caused by smoking is modeled by the rational expression

$$\frac{x-1}{x},$$

where x is the number of times a smoker is more likely to die of lung cancer than a nonsmoker. This is called the *incidence rate*. (*Source:* Walker, A., *Observation and Inference: An Introduction to the Methods of Epidemiology,* Epidemiology Resources Inc., 1991.) For example, $x = 10$ means that a smoker is 10 times more likely than a nonsmoker to die of lung cancer. Find the percent of deaths if the incidence rate is

    (a) 5   **(b)** 10   **(c)** 20.

    (d) Can the incidence rate equal 0? Explain.

6.2 Multiplying and Dividing Rational Expressions

OBJECTIVES

1 Multiply rational expressions.

2 Divide rational expressions.

OBJECTIVE 1 Multiply rational expressions. The product of two fractions is found by multiplying the numerators and multiplying the denominators. Rational expressions are multiplied in the same way.

Multiplying Rational Expressions

The product of the rational expressions $\frac{P}{Q}$ and $\frac{R}{S}$ is

$$\frac{P}{Q} \cdot \frac{R}{S} = \frac{PR}{QS}.$$

That is, to multiply rational expressions, multiply the numerators and multiply the denominators.

In the following example, the parallel discussion with rational numbers and rational expressions lets you compare the steps.

EXAMPLE 1 Multiplying Rational Expressions

Multiply. Write each answer in lowest terms.

(a) $\dfrac{3}{10} \cdot \dfrac{5}{9}$ **(b)** $\dfrac{6}{x} \cdot \dfrac{x^2}{12}$

Indicate the product of the numerators and the product of the denominators.

$$\frac{3}{10} \cdot \frac{5}{9} = \frac{3 \cdot 5}{10 \cdot 9} \qquad\qquad \frac{6}{x} \cdot \frac{x^2}{12} = \frac{6 \cdot x^2}{x \cdot 12}$$

Leave the products in factored form because common factors are needed to write the product in lowest terms. Factor the numerator and denominator to further identify any common factors. Then use the fundamental property to write each product in lowest terms.

$$\frac{3}{10} \cdot \frac{5}{9} = \frac{3 \cdot 5}{2 \cdot 5 \cdot 3 \cdot 3} = \frac{1}{6} \qquad\qquad \frac{6}{x} \cdot \frac{x^2}{12} = \frac{6 \cdot x \cdot x}{2 \cdot 6 \cdot x} = \frac{x}{2}$$

Now Try Exercise 3.

NOTE It is also possible to divide out common factors in the numerator and denominator *before* multiplying the rational expressions. For example,

$$\frac{6}{5} \cdot \frac{35}{22} = \frac{2 \cdot 3}{5} \cdot \frac{5 \cdot 7}{2 \cdot 11} \qquad \text{Identify common factors.}$$

$$= \frac{3 \cdot 7}{11} \qquad \text{Lowest terms}$$

$$= \frac{21}{11}. \qquad \text{Multiply in numerator.}$$

EXAMPLE 2 Multiplying Rational Expressions

Multiply. Write the answer in lowest terms.

$$\frac{x + y}{2x} \cdot \frac{x^2}{(x + y)^2} = \frac{(x + y)x^2}{2x(x + y)^2} \quad \text{Multiply numerators.}$$
$$\text{Multiply denominators.}$$

$$= \frac{(x + y)x \cdot x}{2x(x + y)(x + y)} \quad \text{Factor; identify common factors.}$$

$$= \frac{x}{2(x + y)} \quad \text{Lowest terms}$$

Notice the quotient of factors $\frac{(x + y)x}{x(x + y)}$ in the second line of the solution. Since it is equal to 1, the final product is $\frac{x}{2(x + y)}$.

Now Try Exercise 9.

EXAMPLE 3 Multiplying Rational Expressions

Multiply. Write the answer in lowest terms.

$$\frac{x^2 + 3x}{x^2 - 3x - 4} \cdot \frac{x^2 - 5x + 4}{x^2 + 2x - 3} = \frac{(x^2 + 3x)(x^2 - 5x + 4)}{(x^2 - 3x - 4)(x^2 + 2x - 3)} \quad \begin{array}{l}\text{Definition of}\\\text{multiplication}\end{array}$$

$$= \frac{x(x + 3)(x - 4)(x - 1)}{(x - 4)(x + 1)(x + 3)(x - 1)} \quad \text{Factor.}$$

$$= \frac{x}{x + 1} \quad \text{Lowest terms}$$

The quotients $\frac{x + 3}{x + 3}$, $\frac{x - 4}{x - 4}$, and $\frac{x - 1}{x - 1}$ all equal 1, justifying the final product $\frac{x}{x + 1}$.

Now Try Exercise 43.

OBJECTIVE 2 Divide rational expressions. To develop a method for dividing rational numbers and rational expressions, consider the following problem. Suppose you have $\frac{7}{8}$ gal of milk and want to find how many quarts you have. Since 1 qt is $\frac{1}{4}$ gal, you ask yourself, "How many $\frac{1}{4}$s are there in $\frac{7}{8}$?" This would be interpreted as

$$\frac{7}{8} \div \frac{1}{4} \quad \text{or} \quad \frac{\frac{7}{8}}{\frac{1}{4}}$$

since the fraction bar means division.

The fundamental property of rational expressions discussed earlier can be applied to rational number values of P, Q, and K. With $P = \frac{7}{8}$, $Q = \frac{1}{4}$, and $K = 4$ $\left(\text{since 4 is the reciprocal of } Q = \frac{1}{4}\right)$,

$$\frac{P}{Q} = \frac{P \cdot K}{Q \cdot K} = \frac{\frac{7}{8} \cdot 4}{\frac{1}{4} \cdot 4} = \frac{\frac{7}{8} \cdot 4}{1} = \frac{7}{8} \cdot \frac{4}{1}.$$

So, to divide $\frac{7}{8}$ by $\frac{1}{4}$, we multiply $\frac{7}{8}$ by the reciprocal of $\frac{1}{4}$, namely 4. Since $\frac{7}{8}(4) = \frac{7}{2}$, there are $\frac{7}{2}$ or $3\frac{1}{2}$ qt in $\frac{7}{8}$ gal.

The preceding discussion illustrates the rule for dividing common fractions. To divide $\frac{a}{b}$ by $\frac{c}{d}$, multiply $\frac{a}{b}$ by the reciprocal of $\frac{c}{d}$. Division of rational expressions is defined in the same way.

Dividing Rational Expressions

If $\frac{P}{Q}$ and $\frac{R}{S}$ are any two rational expressions, with $\frac{R}{S} \neq 0$, then

$$\frac{P}{Q} \div \frac{R}{S} = \frac{P}{Q} \cdot \frac{S}{R} = \frac{PS}{QR}.$$

That is, to divide one rational expression by another rational expression, multiply the first rational expression by the reciprocal of the second rational expression.

The next example shows the division of two rational numbers and the division of two rational expressions.

EXAMPLE 4 Dividing Rational Expressions

Divide. Write each answer in lowest terms.

(a) $\dfrac{5}{8} \div \dfrac{7}{16}$

(b) $\dfrac{y}{y+3} \div \dfrac{4y}{y+5}$

Multiply the first expression by the reciprocal of the second.

$$\frac{5}{8} \div \frac{7}{16} = \frac{5}{8} \cdot \frac{16}{7} \qquad \text{Reciprocal of } \frac{7}{16}$$

$$= \frac{5 \cdot 16}{8 \cdot 7}$$

$$= \frac{5 \cdot 8 \cdot 2}{8 \cdot 7}$$

$$= \frac{10}{7}$$

$$\frac{y}{y+3} \div \frac{4y}{y+5}$$

$$= \frac{y}{y+3} \cdot \frac{y+5}{4y} \qquad \text{Reciprocal of } \frac{4y}{y+5}$$

$$= \frac{y(y+5)}{(y+3)(4y)}$$

$$= \frac{y+5}{4(y+3)}$$

Now Try Exercise 19.

EXAMPLE 5 Dividing Rational Expressions

Divide. Write the answer in lowest terms.

$$\frac{(3m)^2}{(2p)^3} \div \frac{6m^3}{16p^2} = \frac{(3m)^2}{(2p)^3} \cdot \frac{16p^2}{6m^3} \qquad \text{Multiply by the reciprocal.}$$

$$= \frac{9m^2}{8p^3} \cdot \frac{16p^2}{6m^3} \qquad \text{Power rule for exponents}$$

$$= \frac{9 \cdot 16m^2p^2}{8 \cdot 6p^3m^3} \qquad \text{Multiply numerators.}$$
$$\qquad\qquad\qquad \text{Multiply denominators.}$$

$$= \frac{3}{mp} \qquad\qquad \text{Lowest terms}$$

Now Try Exercise 17.

▦ EXAMPLE 6 Dividing Rational Expressions

Divide. Write the answer in lowest terms.

$$\frac{x^2 - 4}{(x + 3)(x - 2)} \div \frac{(x + 2)(x + 3)}{-2x}$$

$$= \frac{x^2 - 4}{(x + 3)(x - 2)} \cdot \frac{-2x}{(x + 2)(x + 3)} \qquad \text{Multiply by the reciprocal.}$$

$$= \frac{-2x(x^2 - 4)}{(x + 3)(x - 2)(x + 2)(x + 3)} \qquad \begin{array}{l}\text{Multiply numerators.}\\ \text{Multiply denominators.}\end{array}$$

$$= \frac{-2x(x + 2)(x - 2)}{(x + 3)(x - 2)(x + 2)(x + 3)} \qquad \text{Factor.}$$

$$= \frac{-2x}{(x + 3)^2} \qquad\qquad \text{Lowest terms}$$

$$= -\frac{2x}{(x + 3)^2} \qquad\qquad \frac{-a}{b} = -\frac{a}{b}$$

Now Try Exercise 35.

▦ EXAMPLE 7 Dividing Rational Expressions (Factors Are Opposites)

Divide. Write the answer in lowest terms.

$$\frac{m^2 - 4}{m^2 - 1} \div \frac{2m^2 + 4m}{1 - m}$$

$$= \frac{m^2 - 4}{m^2 - 1} \cdot \frac{1 - m}{2m^2 + 4m} \qquad \text{Multiply by the reciprocal.}$$

$$= \frac{(m^2 - 4)(1 - m)}{(m^2 - 1)(2m^2 + 4m)} \qquad \begin{array}{l}\text{Multiply numerators.}\\ \text{Multiply denominators.}\end{array}$$

$$= \frac{(m + 2)(m - 2)(1 - m)}{(m + 1)(m - 1)(2m)(m + 2)} \qquad \text{Factor; } 1 - m \text{ and } m - 1 \text{ differ only in sign.}$$

$$= \frac{-1(m - 2)}{2m(m + 1)} \qquad \text{From Section 6.1, } \frac{1 - m}{m - 1} = -1.$$

$$= \frac{-m + 2}{2m(m + 1)} \quad \text{or} \quad \frac{2 - m}{2m(m + 1)} \qquad \begin{array}{l}\text{Distribute the negative sign in the}\\ \text{numerator.}\end{array}$$

Now Try Exercise 37.

In summary, follow these steps to multiply or divide rational expressions.

> **Multiplying or Dividing Rational Expressions**
>
> *Step 1* **Note the operation.** If the operation is division, use the definition of division to rewrite as multiplication.
>
> *Step 2* **Multiply** numerators and multiply denominators.
>
> *Step 3* **Factor** all numerators and denominators completely.
>
> *Step 4* **Write in lowest terms** using the fundamental property.
>
> *Note:* Steps 2 and 3 may be interchanged based on personal preference.

6.2 EXERCISES

For Extra Help

 Student's Solutions Manual

 MyMathLab

 InterAct Math Tutorial Software

 AW Math Tutor Center

Math XP MathXL

Digital Video Tutor CD 9/Videotape 9

1. Match each multiplication problem in Column I with the correct product in Column II.

I	II
(a) $\dfrac{5x^3}{10x^4} \cdot \dfrac{10x^7}{2x}$	A. $\dfrac{2}{5x^5}$
(b) $\dfrac{10x^4}{5x^3} \cdot \dfrac{10x^7}{2x}$	B. $\dfrac{5x^5}{2}$
(c) $\dfrac{5x^3}{10x^4} \cdot \dfrac{2x}{10x^7}$	C. $\dfrac{1}{10x^7}$
(d) $\dfrac{10x^4}{5x^3} \cdot \dfrac{2x}{10x^7}$	D. $10x^7$

2. Match each division problem in Column I with the correct quotient in Column II.

I	II
(a) $\dfrac{5x^3}{10x^4} \div \dfrac{10x^7}{2x}$	A. $\dfrac{5x^5}{2}$
(b) $\dfrac{10x^4}{5x^3} \div \dfrac{10x^7}{2x}$	B. $10x^7$
(c) $\dfrac{5x^3}{10x^4} \div \dfrac{2x}{10x^7}$	C. $\dfrac{2}{5x^5}$
(d) $\dfrac{10x^4}{5x^3} \div \dfrac{2x}{10x^7}$	D. $\dfrac{1}{10x^7}$

Multiply. Write each answer in lowest terms. See Examples 1 and 2.

3. $\dfrac{15a^2}{14} \cdot \dfrac{7}{5a}$

4. $\dfrac{27k^3}{9k} \cdot \dfrac{24}{9k^2}$

5. $\dfrac{12x^4}{18x^3} \cdot \dfrac{-8x^5}{4x^2}$

6. $\dfrac{12m^5}{-2m^2} \cdot \dfrac{6m^6}{28m^3}$

7. $\dfrac{2(c+d)}{3} \cdot \dfrac{18}{6(c+d)^2}$

8. $\dfrac{4(y-2)}{x} \cdot \dfrac{3x}{6(y-2)^2}$

9. $\dfrac{(x-y)^2}{2} \cdot \dfrac{24}{3(x-y)}$

10. $\dfrac{(a+b)^2}{5} \cdot \dfrac{30}{2(a+b)}$

11. $\dfrac{t-4}{8} \cdot \dfrac{4t^2}{t-4}$

12. $\dfrac{z+9}{12} \cdot \dfrac{3z^2}{z+9}$

13. $\dfrac{3x}{x+3} \cdot \dfrac{(x+3)^2}{6x^2}$

14. $\dfrac{(t-2)^2}{4t^2} \cdot \dfrac{2t}{t-2}$

Divide. Write each answer in lowest terms. See Examples 4 and 5.

15. $\dfrac{9z^4}{3z^5} \div \dfrac{3z^2}{5z^3}$

16. $\dfrac{35q^8}{9q^5} \div \dfrac{25q^6}{10q^5}$

17. $\dfrac{4t^4}{2t^5} \div \dfrac{(2t)^3}{-6}$

18. $\dfrac{-12a^6}{3a^2} \div \dfrac{(2a)^3}{27a}$

19. $\dfrac{3}{2y-6} \div \dfrac{6}{y-3}$

20. $\dfrac{4m+16}{10} \div \dfrac{3m+12}{18}$

21. $\dfrac{7t + 7}{-6} \div \dfrac{4t + 4}{15}$ **22.** $\dfrac{8z - 16}{-20} \div \dfrac{3z - 6}{40}$ **23.** $\dfrac{2x}{x - 1} \div \dfrac{x^2}{x + 2}$

24. $\dfrac{y^2}{y + 1} \div \dfrac{3y}{y - 3}$ **25.** $\dfrac{(x - 3)^2}{6x} \div \dfrac{x - 3}{x^2}$ **26.** $\dfrac{2a}{a + 4} \div \dfrac{a^2}{(a + 4)^2}$

✍ **27.** Use an example to explain how to multiply rational expressions.

✍ **28.** Use an example to explain how to divide rational expressions.

Multiply or divide. Write each answer in lowest terms. See Examples 3, 6, and 7.

29. $\dfrac{5x - 15}{3x + 9} \cdot \dfrac{4x + 12}{6x - 18}$ **30.** $\dfrac{8r + 16}{24r - 24} \cdot \dfrac{6r - 6}{3r + 6}$ **31.** $\dfrac{2 - t}{8} \div \dfrac{t - 2}{6}$

32. $\dfrac{4}{m - 2} \div \dfrac{16}{2 - m}$ **33.** $\dfrac{27 - 3z}{4} \cdot \dfrac{12}{2z - 18}$ **34.** $\dfrac{5 - x}{5 + x} \cdot \dfrac{x + 5}{x - 5}$

35. $\dfrac{p^2 + 4p - 5}{p^2 + 7p + 10} \div \dfrac{p - 1}{p + 4}$ **36.** $\dfrac{z^2 - 3z + 2}{z^2 + 4z + 3} \div \dfrac{z - 1}{z + 1}$ **37.** $\dfrac{m^2 - 4}{16 - 8m} \div \dfrac{m + 2}{8}$

38. $\dfrac{2}{3 - x} \div \dfrac{2x + 6}{x^2 - 9}$ **39.** $\dfrac{2x^2 - 7x + 3}{x - 3} \cdot \dfrac{x + 2}{x - 1}$ **40.** $\dfrac{3x^2 - 5x - 2}{x - 2} \cdot \dfrac{x - 3}{x + 1}$

41. $\dfrac{2k^2 - k - 1}{2k^2 + 5k + 3} \div \dfrac{4k^2 - 1}{2k^2 + k - 3}$ **42.** $\dfrac{2m^2 - 5m - 12}{m^2 + m - 20} \div \dfrac{4m^2 - 9}{m^2 + 4m - 5}$

43. $\dfrac{2k^2 + 3k - 2}{6k^2 - 7k + 2} \cdot \dfrac{4k^2 - 5k + 1}{k^2 + k - 2}$ **44.** $\dfrac{2m^2 - 5m - 12}{m^2 - 10m + 24} \div \dfrac{4m^2 - 9}{m^2 - 9m + 18}$

45. $\dfrac{m^2 + 2mp - 3p^2}{m^2 - 3mp + 2p^2} \div \dfrac{m^2 + 4mp + 3p^2}{m^2 + 2mp - 8p^2}$ **46.** $\dfrac{r^2 + rs - 12s^2}{r^2 - rs - 20s^2} \div \dfrac{r^2 - 2rs - 3s^2}{r^2 + rs - 30s^2}$

47. $\dfrac{m^2 + 3m + 2}{m^2 + 5m + 4} \cdot \dfrac{m^2 + 10m + 24}{m^2 + 5m + 6}$ **48.** $\dfrac{z^2 - z - 6}{z^2 - 2z - 8} \cdot \dfrac{z^2 + 7z + 12}{z^2 - 9}$

49. $\dfrac{y^2 + y - 2}{y^2 + 3y - 4} \div \dfrac{y + 2}{y + 3}$ **50.** $\dfrac{r^2 + r - 6}{r^2 + 4r - 12} \div \dfrac{r + 3}{r - 1}$

51. $\dfrac{2m^2 + 7m + 3}{m^2 - 9} \cdot \dfrac{m^2 - 3m}{2m^2 + 11m + 5}$ **52.** $\dfrac{m^2 + 2mp - 3p^2}{m^2 - 3mp + 2p^2} \div \dfrac{m^2 + 4mp + 3p^2}{m^2 + 2mp - 8p^2}$

53. $\dfrac{r^2 + rs - 12s^2}{r^2 - rs - 20s^2} \div \dfrac{r^2 - 2rs - 3s^2}{r^2 + rs - 30s^2}$ **54.** $\dfrac{(x + 1)^3(x + 4)}{x^2 + 5x + 4} \div \dfrac{x^2 + 2x + 1}{x^2 + 3x + 2}$

55. $\dfrac{(q - 3)^4(q + 2)}{q^2 + 3q + 2} \div \dfrac{q^2 - 6q + 9}{q^2 + 4q + 4}$ **56.** $\dfrac{(x + 4)^3(x - 3)}{x^2 - 9} \div \dfrac{x^2 + 8x + 16}{x^2 + 6x + 9}$

These exercises involve grouping symbols (Section 1.2), factoring by grouping (Section 5.1), and factoring sums and differences of cubes (Section 5.4). Multiply or divide as indicated. Write each answer in lowest terms.

57. $\dfrac{x + 5}{x + 10} \div \left(\dfrac{x^2 + 10x + 25}{x^2 + 10x} \cdot \dfrac{10x}{x^2 + 15x + 50} \right)$

58. $\dfrac{m - 8}{m - 4} \div \left(\dfrac{m^2 - 12m + 32}{8m} \cdot \dfrac{m^2 - 8m}{m^2 - 8m + 16} \right)$

59. $\dfrac{3a - 3b - a^2 + b^2}{4a^2 - 4ab + b^2} \cdot \dfrac{4a^2 - b^2}{2a^2 - ab - b^2}$

60. $\dfrac{4r^2 - t^2 + 10r - 5t}{2r^2 + rt + 5r} \cdot \dfrac{4r^3 + 4r^2t + rt^2}{2r + t}$

61. $\dfrac{-x^3 - y^3}{x^2 - 2xy + y^2} \div \dfrac{3y^2 - 3xy}{x^2 - y^2}$

62. $\dfrac{b^3 - 8a^3}{4a^3 + 4a^2b + ab^2} \div \dfrac{4a^2 + 2ab + b^2}{-a^3 - ab^3}$

63. If the rational expression $\dfrac{5x^2y^3}{2pq}$ represents the area of a rectangle and $\dfrac{2xy}{p}$ represents the length, what rational expression represents the width?

Width

Length $= \dfrac{2xy}{p}$

The area is $\dfrac{5x^2y^3}{2pq}$.

64. If you are given the problem $\dfrac{4y + 12}{2y - 10} \div \dfrac{?}{y^2 - y - 20} = \dfrac{2(y + 4)}{y - 3}$, what must be the polynomial that is represented by the question mark?

6.3 Least Common Denominators

OBJECTIVES

1 Find the least common denominator for a group of fractions.

2 Rewrite rational expressions with given denominators.

OBJECTIVE 1 Find the least common denominator for a group of fractions. Adding or subtracting rational expressions (to be discussed in the next section) often requires a **least common denominator (LCD),** the simplest expression that is divisible by all denominators. For example, the least common denominator for the fractions $\frac{2}{9}$ and $\frac{5}{12}$ is 36 because 36 is the smallest positive number divisible by both 9 and 12.

We can often find least common denominators by inspection. For example, the LCD for $\frac{1}{6}$ and $\frac{2}{3m}$ is $6m$. In other cases, we find the LCD by a procedure similar to that used in Chapter 5 for finding the greatest common factor.

> **Finding the Least Common Denominator (LCD)**
>
> *Step 1* **Factor** each denominator into prime factors.
>
> *Step 2* **List each different denominator factor** the *greatest* number of times it appears in any of the denominators.
>
> *Step 3* **Multiply** the denominator factors from Step 2 to get the LCD.

When each denominator is factored into prime factors, every prime factor must be a factor of the least common denominator.

In Example 1, we find the LCD for both numerical and algebraic denominators.

EXAMPLE 1 Finding the LCD

Find the LCD for each pair of fractions.

(a) $\dfrac{1}{24}, \dfrac{7}{15}$

(b) $\dfrac{1}{8x}, \dfrac{3}{10x}$

Step 1   Write each denominator in factored form with numerical coefficients in prime factored form.

$$24 = 2 \cdot 2 \cdot 2 \cdot 3 = 2^3 \cdot 3 \qquad\qquad 8x = 2 \cdot 2 \cdot 2 \cdot x = 2^3 \cdot x$$
$$15 = 3 \cdot 5 \qquad\qquad\qquad\qquad\qquad 10x = 2 \cdot 5 \cdot x$$

Step 2   Find the LCD by taking each different factor the *greatest* number of times it appears as a factor in any of the denominators.

The factor 2 appears three times in one product and not at all in the other, so the greatest number of times 2 appears is three. The greatest number of times both 3 and 5 appear is one.

Here 2 appears three times in one product and once in the other, so the greatest number of times 2 appears is three. The greatest number of times 5 appears is one, and the greatest number of times x appears in either product is one.

Step 3
$$\begin{aligned} \text{LCD} &= 2 \cdot 2 \cdot 2 \cdot 3 \cdot 5 \\ &= 2^3 \cdot 3 \cdot 5 \\ &= 120 \end{aligned}$$

$$\begin{aligned} \text{LCD} &= 2 \cdot 2 \cdot 2 \cdot 5 \cdot x \\ &= 2^3 \cdot 5 \cdot x \\ &= 40x \end{aligned}$$

Now Try Exercises 5 and 11.

EXAMPLE 2   Finding the LCD

Find the LCD for $\dfrac{5}{6r^2}$ and $\dfrac{3}{4r^3}$.

Step 1   Factor each denominator.

$$6r^2 = 2 \cdot 3 \cdot r^2$$
$$4r^3 = 2 \cdot 2 \cdot r^3 = 2^2 \cdot r^3$$

Step 2   The greatest number of times 2 appears is two, the greatest number of times 3 appears is one, and the greatest number of times r appears is three; therefore,

Step 3 $\text{LCD} = 2^2 \cdot 3 \cdot r^3 = 12r^3.$

Now Try Exercise 13.

EXAMPLE 3   Finding the LCD

Find each LCD.

(a) $\dfrac{6}{5m}, \dfrac{4}{m^2 - 3m}$

Factor each denominator.

$$5m = 5 \cdot m$$
$$m^2 - 3m = m(m - 3)$$

Use each different factor the greatest number of times it appears.

$$\text{LCD} = 5 \cdot m \cdot (m - 3) = 5m(m - 3)$$

Because m is not a *factor* of $m - 3$, both factors, m and $m - 3$, must appear in the LCD.

(b) $\dfrac{1}{r^2 - 4r - 5}, \dfrac{3}{r^2 - r - 20}, \dfrac{1}{r^2 - 10r + 25}$

Factor each denominator.

$$r^2 - 4r - 5 = (r - 5)(r + 1)$$
$$r^2 - r - 20 = (r - 5)(r + 4)$$
$$r^2 - 10r + 25 = (r - 5)^2$$

Use each different factor the greatest number of times it appears as a factor. The LCD is

$$(r - 5)^2(r + 1)(r + 4).$$

(c) $\dfrac{1}{q - 5}, \dfrac{3}{5 - q}$

The expressions $q - 5$ and $5 - q$ are opposites of each other because

$$-(q - 5) = -q + 5 = 5 - q.$$

Therefore, either $q - 5$ or $5 - q$ can be used as the LCD.

Now Try Exercises 19, 35, and 45.

OBJECTIVE 2 Rewrite rational expressions with given denominators.   Once the LCD has been found, the next step in preparing to add or subtract two rational expressions is to use the fundamental property to write equivalent rational expressions. The next example shows how to do this with both numerical and algebraic fractions.

EXAMPLE 4   Writing Rational Expressions with Given Denominators

Rewrite each rational expression with the indicated denominator.

(a) $\dfrac{3}{8} = \dfrac{}{40}$

(b) $\dfrac{9k}{25} = \dfrac{}{50k}$

For each example, first factor the denominator on the right. Then compare the denominator on the left with the one on the right to decide what factors are missing. (It may sometimes be necessary to factor both denominators.)

$$\dfrac{3}{8} = \dfrac{}{5 \cdot 8} \qquad\qquad \dfrac{9k}{25} = \dfrac{}{25 \cdot 2k}$$

A factor of 5 is missing. Using the fundamental property, multiply $\frac{3}{8}$ by $\frac{5}{5}$.

$$\dfrac{3}{8} = \dfrac{3}{8} \cdot \dfrac{5}{5} = \dfrac{15}{40}$$
$$\dfrac{5}{5} = 1 \longrightarrow$$

Factors of 2 and k are missing. Multiply by $\frac{2k}{2k}$.

$$\dfrac{9k}{25} = \dfrac{9k}{25} \cdot \dfrac{2k}{2k} = \dfrac{18k^2}{50k}$$
$$\dfrac{2k}{2k} = 1 \longrightarrow$$

Now Try Exercises 57 and 59.

EXAMPLE 5   Writing Rational Expressions with Given Denominators

Rewrite each rational expression with the indicated denominator.

(a) $\dfrac{8}{3x + 1} = \dfrac{}{12x + 4}$

Factor the denominator on the right.

$$\frac{8}{3x + 1} = \frac{}{4(3x + 1)} \qquad \text{Factor.}$$

The missing factor is 4, so multiply the fraction on the left by $\frac{4}{4}$.

$$\frac{8}{3x + 1} \cdot \frac{4}{4} = \frac{32}{12x + 4} \qquad \text{Fundamental property}$$

(b) $\dfrac{12p}{p^2 + 8p} = \dfrac{}{p^3 + 4p^2 - 32p}$

Factor $p^2 + 8p$ as $p(p + 8)$. Compare with the denominator on the right, which factors as $p(p + 8)(p - 4)$. The factor $p - 4$ is missing, so multiply $\frac{12p}{p(p + 8)}$ by $\frac{p - 4}{p - 4}$.

$$\frac{12p}{p^2 + 8p} = \frac{12p}{p(p + 8)} \cdot \frac{p - 4}{p - 4} \qquad \text{Fundamental property}$$

$$= \frac{12p(p - 4)}{p(p + 8)(p - 4)} \qquad \begin{array}{l}\text{Multiply numerators.}\\ \text{Multiply denominators.}\end{array}$$

$$= \frac{12p^2 - 48p}{p^3 + 4p^2 - 32p} \qquad \text{Multiply the factors.}$$

Now Try Exercises 63 and 67.

NOTE In the next section we add and subtract rational expressions, which sometimes requires the steps illustrated in Examples 4 and 5. While it is beneficial to leave the denominator in factored form, we multiplied the factors in the denominator in Example 5 to give the answer in the same form as the original problem.

6.3 EXERCISES

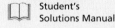

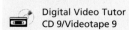
Choose the correct response in Exercises 1–4.

1. Suppose that the greatest common factor of a and b is 1. Then the least common denominator for $\frac{1}{a}$ and $\frac{1}{b}$ is

   A. a **B.** b **C.** ab **D.** 1.

2. If a is a factor of b, then the least common denominator for $\frac{1}{a}$ and $\frac{1}{b}$ is

   A. a **B.** b **C.** ab **D.** 1.

3. The least common denominator for $\frac{11}{20}$ and $\frac{1}{2}$ is

   A. 40   **B.** 2   **C.** 20   **D.** none of these.

4. Suppose that we wish to write the fraction $\dfrac{1}{(x - 4)^2(y - 3)}$ with denominator $(x - 4)^3(y - 3)^2$. We must multiply both the numerator and the denominator by

   A. $(x - 4)(y - 3)$ **B.** $(x - 4)^2$ **C.** $x - 4$ **D.** $(x - 4)^2(y - 3)$.

Find the LCD for the fractions in each list. See Examples 1–3.

5. $\dfrac{7}{15}, \dfrac{21}{20}$

6. $\dfrac{9}{10}, \dfrac{12}{25}$

7. $\dfrac{17}{100}, \dfrac{23}{120}, \dfrac{43}{180}$

8. $\dfrac{17}{250}, \dfrac{-21}{300}, \dfrac{127}{360}$

9. $\dfrac{9}{x^2}, \dfrac{8}{x^5}$

10. $\dfrac{12}{m^7}, \dfrac{13}{m^8}$

11. $\dfrac{-2}{5p}, \dfrac{15}{6p}$

12. $\dfrac{14}{15k}, \dfrac{9}{4k}$

13. $\dfrac{17}{15y^2}, \dfrac{55}{36y^4}$

14. $\dfrac{4}{25m^3}, \dfrac{-9}{10m^4}$

15. $\dfrac{6}{21r^3}, \dfrac{8}{12r^5}$

16. $\dfrac{9}{35t^2}, \dfrac{5}{49t^6}$

17. $\dfrac{13}{5a^2b^3}, \dfrac{29}{15a^5b}$

18. $\dfrac{-7}{3r^4s^5}, \dfrac{-22}{9r^6s^8}$

19. $\dfrac{7}{6p}, \dfrac{15}{4p - 8}$

20. $\dfrac{7}{8k}, \dfrac{-23}{12k - 24}$

21. $\dfrac{9}{28m^2}, \dfrac{3}{12m - 20}$

22. $\dfrac{15}{27a^3}, \dfrac{8}{9a - 45}$

23. $\dfrac{7}{5b - 10}, \dfrac{11}{6b - 12}$

24. $\dfrac{3}{7x^2 + 21x}, \dfrac{1}{5x^2 + 15x}$

RELATING CONCEPTS (EXERCISES 25–28)

For Individual or Group Work

Suppose we want to find the LCD for the two common fractions

$$\frac{1}{24} \quad \text{and} \quad \frac{1}{20}.$$

In their prime factored forms, the denominators are

$$24 = 2^3 \cdot 3 \quad \text{and} \quad 20 = 2^2 \cdot 5.$$

Refer to this information as necessary, and **work Exercises 25–28 in order.**

25. What is the prime factored form of the LCD of the two fractions?

26. Suppose that two algebraic fractions have denominators $(t + 4)^3(t - 3)$ and $(t + 4)^2(t + 8)$. What is the factored form of the LCD of these?

27. What is the similarity between your answers in Exercises 25 and 26?

28. Comment on the following statement: The method for finding the LCD for two algebraic fractions is the same as the method for finding the LCD for two common fractions.

Find the LCD for the fractions in each list. See Examples 1–3.

29. $\dfrac{37}{6r - 12}, \dfrac{25}{9r - 18}$

30. $\dfrac{-14}{5p - 30}, \dfrac{5}{6p - 36}$

31. $\dfrac{5}{12p + 60}, \dfrac{17}{p^2 + 5p}, \dfrac{16}{p^2 + 10p + 25}$

32. $\dfrac{13}{r^2 + 7r}, \dfrac{-3}{5r + 35}, \dfrac{-7}{r^2 + 14r + 49}$

33. $\dfrac{3}{8y + 16}, \dfrac{22}{y^2 + 3y + 2}$

34. $\dfrac{-2}{9m - 18}, \dfrac{-9}{m^2 - 7m + 10}$

35. $\dfrac{5}{c - d}, \dfrac{8}{d - c}$

36. $\dfrac{4}{y - x}, \dfrac{7}{x - y}$

37. $\dfrac{12}{m - 3}, \dfrac{-4}{3 - m}$

38. $\dfrac{-17}{8 - a}, \dfrac{2}{a - 8}$

39. $\dfrac{29}{p - q}, \dfrac{18}{q - p}$

40. $\dfrac{16}{z - x}, \dfrac{8}{x - z}$

41. $\dfrac{3}{k^2 + 5k}, \dfrac{2}{k^2 + 3k - 10}$

42. $\dfrac{1}{z^2 - 4z}, \dfrac{4}{z^2 - 3z - 4}$

43. $\dfrac{6}{a^2 + 6a}, \dfrac{-5}{a^2 + 3a - 18}$

44. $\dfrac{8}{y^2 - 5y}, \dfrac{-2}{y^2 - 2y - 15}$

45. $\dfrac{5}{p^2 + 8p + 15}, \dfrac{3}{p^2 - 3p - 18}, \dfrac{2}{p^2 - p - 30}$

46. $\dfrac{10}{y^2 - 10y + 21}, \dfrac{2}{y^2 - 2y - 3}, \dfrac{5}{y^2 - 6y - 7}$

47. $\dfrac{-5}{k^2 + 2k - 35}, \dfrac{-8}{k^2 + 3k - 40}, \dfrac{9}{k^2 - 2k - 15}$

48. $\dfrac{19}{z^2 + 4z - 12}, \dfrac{16}{z^2 + z - 30}, \dfrac{6}{z^2 + 2z - 24}$

49. Suppose that $(2x - 5)^2$ is the LCD for two fractions. Is $(5 - 2x)^2$ also acceptable as an LCD? Why or why not?

50. Suppose that $(4t - 3)(5t - 6)$ is the LCD for two fractions. Is $(3 - 4t)(6 - 5t)$ also acceptable as an LCD? Why or why not?

RELATING CONCEPTS (EXERCISES 51–56)

For Individual or Group Work

Work Exercises 51–56 in order.

51. Suppose that you want to write $\frac{3}{4}$ as an equivalent fraction with denominator 28. By what number must you multiply both the numerator and the denominator?

52. If you write $\frac{3}{4}$ as an equivalent fraction with denominator 28, by what number are you actually multiplying the fraction?

53. What property of multiplication is being used when we write a common fraction as an equivalent one with a larger denominator? (See Section 1.7.)

54. Suppose that you want to write $\frac{2x + 5}{x - 4}$ as an equivalent fraction with denominator $7x - 28$. By what number must you multiply both the numerator and the denominator?

55. If you write $\frac{2x + 5}{x - 4}$ as an equivalent fraction with denominator $7x - 28$, by what number are you actually multiplying the fraction?

56. Repeat Exercise 53, changing "a common" to "an algebraic."

Rewrite each rational expression with the indicated denominator. See Examples 4 and 5.

57. $\dfrac{4}{11} = \dfrac{}{55}$

58. $\dfrac{6}{7} = \dfrac{}{42}$

59. $\dfrac{-5}{k} = \dfrac{}{9k}$

60. $\dfrac{-3}{q} = \dfrac{}{6q}$

61. $\dfrac{15m^2}{8k} = \dfrac{}{32k^4}$

62. $\dfrac{5t^2}{3y} = \dfrac{}{9y^2}$

63. $\dfrac{19z}{2z - 6} = \dfrac{}{6z - 18}$

64. $\dfrac{2r}{5r - 5} = \dfrac{}{15r - 15}$

65. $\dfrac{-2a}{9a - 18} = \dfrac{}{18a - 36}$

66. $\dfrac{-5y}{6y + 18} = \dfrac{}{24y + 72}$

67. $\dfrac{6}{k^2 - 4k} = \dfrac{}{k(k - 4)(k + 1)}$

68. $\dfrac{15}{m^2 - 9m} = \dfrac{}{m(m - 9)(m + 8)}$

69. $\dfrac{36r}{r^2 - r - 6} = \dfrac{}{(r - 3)(r + 2)(r + 1)}$

70. $\dfrac{4m}{m^2 - 8m + 15} = \dfrac{}{(m - 5)(m - 3)(m + 2)}$

71. $\dfrac{a + 2b}{2a^2 + ab - b^2} = \dfrac{}{2a^3b + a^2b^2 - ab^3}$

72. $\dfrac{m - 4}{6m^2 + 7m - 3} = \dfrac{}{12m^3 + 14m^2 - 6m}$

73. $\dfrac{4r - t}{r^2 + rt + t^2} = \dfrac{}{t^3 - r^3}$

74. $\dfrac{3x - 1}{x^2 + 2x + 4} = \dfrac{}{x^3 - 8}$

75. $\dfrac{2(z - y)}{y^2 + yz + z^2} = \dfrac{}{y^4 - z^3y}$

76. $\dfrac{2p + 3q}{p^2 + 2pq + q^2} = \dfrac{}{(p + q)(p^3 + q^3)}$

✒ **77.** Write an explanation of how to find the least common denominator for a group of denominators. Give an example.

✒ **78.** Write an explanation of how to write a rational expression as an equivalent rational expression with a given denominator. Give an example.

6.4 Adding and Subtracting Rational Expressions

OBJECTIVES

1 Add rational expressions having the same denominator.

2 Add rational expressions having different denominators.

To add and subtract rational expressions, we need the skills developed in the previous section to find least common denominators and to write equivalent fractions with the LCD.

OBJECTIVE 1 Add rational expressions having the same denominator. We find the sum of two rational expressions with the same procedure that we used for adding two fractions in Section 1.1.

3 Subtract rational expressions.

> **Adding Rational Expressions**
>
> If $\frac{P}{Q}$ and $\frac{R}{Q}$ ($Q \neq 0$) are rational expressions, then
>
> $$\frac{P}{Q} + \frac{R}{Q} = \frac{P + R}{Q}.$$
>
> That is, to add rational expressions with the same denominator, add the numerators and keep the same denominator.

The first example shows how addition of rational expressions compares with that of rational numbers.

EXAMPLE 1   Adding Rational Expressions with the Same Denominator

Add. Write each answer in lowest terms.

(a) $\dfrac{4}{9} + \dfrac{2}{9}$

(b) $\dfrac{3x}{x + 1} + \dfrac{3}{x + 1}$

The denominators are the same, so the sum is found by adding the two numerators and keeping the same (common) denominator.

$$\frac{4}{9} + \frac{2}{9} = \frac{4 + 2}{9}$$
$$= \frac{6}{9}$$
$$= \frac{2}{3}$$

$$\frac{3x}{x + 1} + \frac{3}{x + 1} = \frac{3x + 3}{x + 1}$$
$$= \frac{3(x + 1)}{x + 1}$$
$$= 3$$

Now Try Exercises 9 and 19.

OBJECTIVE 2 Add rational expressions having different denominators. We use the following steps to add two rational expressions with different denominators. These are the same steps we used to add fractions with different denominators in Section 1.1.

> **Adding with Different Denominators**
>
> *Step 1*   **Find the least common denominator (LCD).**
>
> *Step 2*   **Rewrite each rational expression** as an equivalent rational expression with the LCD as the denominator.
>
> *Step 3*   **Add** the numerators to get the numerator of the sum. The LCD is the denominator of the sum.
>
> *Step 4*   **Write in lowest terms** using the fundamental property.

▓ **EXAMPLE 2** Adding Rational Expressions with Different Denominators

Add. Write each answer in lowest terms.

(a) $\dfrac{1}{12} + \dfrac{7}{15}$ **(b)** $\dfrac{2}{3y} + \dfrac{1}{4y}$

Step 1 First find the LCD using the methods of the previous section.

$$\text{LCD} = 2^2 \cdot 3 \cdot 5 = 60 \qquad\qquad \text{LCD} = 2^2 \cdot 3 \cdot y = 12y$$

Step 2 Now rewrite each rational expression as a fraction with the LCD (either 60 or 12y) as the denominator.

$$\frac{1}{12} + \frac{7}{15} = \frac{1(5)}{12(5)} + \frac{7(4)}{15(4)} \qquad\qquad \frac{2}{3y} + \frac{1}{4y} = \frac{2(4)}{3y(4)} + \frac{1(3)}{4y(3)}$$

$$= \frac{5}{60} + \frac{28}{60} \qquad\qquad\qquad = \frac{8}{12y} + \frac{3}{12y}$$

Step 3 Since the fractions now have common denominators, add the numerators.

Step 4 Write in lowest terms if necessary.

$$\frac{5}{60} + \frac{28}{60} = \frac{5 + 28}{60} \qquad\qquad \frac{8}{12y} + \frac{3}{12y} = \frac{8 + 3}{12y}$$

$$= \frac{33}{60} = \frac{11}{20} \qquad\qquad\qquad = \frac{11}{12y}$$

▓

Now Try Exercises 25 and 31.

▓ **EXAMPLE 3** Adding Rational Expressions

Add. Write the answer in lowest terms.

$$\frac{2x}{x^2 - 1} + \frac{-1}{x + 1}$$

Step 1 Since the denominators are different, find the LCD.

$$x^2 - 1 = (x + 1)(x - 1)$$

$$x + 1 \text{ is prime.}$$

The LCD is $(x + 1)(x - 1)$.

Step 2 Rewrite each rational expression as a fraction with common denominator $(x + 1)(x - 1)$.

$$\frac{2x}{x^2 - 1} + \frac{-1}{x + 1} = \frac{2x}{(x + 1)(x - 1)} + \frac{-1(x - 1)}{(x + 1)(x - 1)} \qquad \text{Multiply the second fraction by } \frac{x - 1}{x - 1}.$$

$$= \frac{2x}{(x + 1)(x - 1)} + \frac{-x + 1}{(x + 1)(x - 1)} \qquad \text{Distributive property}$$

Step 3
$$= \frac{2x - x + 1}{(x + 1)(x - 1)} \qquad \text{Add numerators; keep the same denominator.}$$

$$= \frac{x + 1}{(x + 1)(x - 1)} \qquad \text{Combine like terms.}$$

Step 4 $= \dfrac{1(x + 1)}{(x + 1)(x - 1)}$ Identity property of multiplication

$= \dfrac{1}{x - 1}$ Fundamental property

Now Try Exercise 41.

EXAMPLE 4   Adding Rational Expressions

Add. Write the answer in lowest terms.

$$\dfrac{2x}{x^2 + 5x + 6} + \dfrac{x + 1}{x^2 + 2x - 3}$$

$= \dfrac{2x}{(x + 2)(x + 3)} + \dfrac{x + 1}{(x + 3)(x - 1)}$ Factor the denominators.

The LCD is $(x + 2)(x + 3)(x - 1)$. Use the fundamental property.

$= \dfrac{2x(x - 1)}{(x + 2)(x + 3)(x - 1)} + \dfrac{(x + 1)(x + 2)}{(x + 2)(x + 3)(x - 1)}$

$= \dfrac{2x(x - 1) + (x + 1)(x + 2)}{(x + 2)(x + 3)(x - 1)}$ Add numerators; keep the same denominator.

$= \dfrac{2x^2 - 2x + x^2 + 3x + 2}{(x + 2)(x + 3)(x - 1)}$ Multiply.

$= \dfrac{3x^2 + x + 2}{(x + 2)(x + 3)(x - 1)}$ Combine like terms.

It is usually more convenient to leave the denominator in factored form. The numerator cannot be factored here, so the expression is in lowest terms.

Now Try Exercise 43.

Rational expressions to be added or subtracted may have denominators that are opposites of each other.

EXAMPLE 5   Adding Rational Expressions with Denominators That Are Opposites

Add. Write the answer in lowest terms.

$$\dfrac{y}{y - 2} + \dfrac{8}{2 - y}$$

One way to get a common denominator is to multiply the second expression by -1 in both the numerator and the denominator, giving $y - 2$ as a common denominator.

$\dfrac{y}{y - 2} + \dfrac{8}{2 - y} = \dfrac{y}{y - 2} + \dfrac{8(-1)}{(2 - y)(-1)}$ Fundamental property

$= \dfrac{y}{y - 2} + \dfrac{-8}{y - 2}$ Distributive property

$= \dfrac{y - 8}{y - 2}$ Add numerators; keep the same denominator.

If we had chosen to use $2 - y$ as the common denominator, the final answer would be $\frac{8 - y}{2 - y}$, which is equivalent to $\frac{y - 8}{y - 2}$.

Now Try Exercise 51.

OBJECTIVE 3 **Subtract rational expressions.** To subtract rational expressions, use the following rule.

Subtracting Rational Expressions

If $\frac{P}{Q}$ and $\frac{R}{Q}$ $(Q \neq 0)$ are rational expressions, then

$$\frac{P}{Q} - \frac{R}{Q} = \frac{P - R}{Q}.$$

That is, to subtract rational expressions with the same denominator, subtract the numerators and keep the same denominator.

EXAMPLE 6 **Subtracting Rational Expressions with the Same Denominator**

Subtract. Write the answer in lowest terms.

Use parentheses around the quantity being subtracted.

$$\frac{2m}{m - 1} - \frac{m + 3}{m - 1} = \frac{2m - (m + 3)}{m - 1}$$ Subtract numerators; keep the same denominator.

$$= \frac{2m - m - 3}{m - 1}$$ Distributive property

$$= \frac{m - 3}{m - 1}$$ Combine like terms.

Now Try Exercise 15.

CAUTION Sign errors often occur in subtraction problems like the one in Example 6. Remember that the numerator of the fraction being subtracted must be treated as a single quantity. Be sure to use parentheses after the subtraction sign.

EXAMPLE 7 **Subtracting Rational Expressions with Different Denominators**

Subtract. Write the answer in lowest terms.

$$\frac{9}{x - 2} - \frac{3}{x} = \frac{9x}{x(x - 2)} - \frac{3(x - 2)}{x(x - 2)}$$ The LCD is $x(x - 2)$.

$$= \frac{9x - 3(x - 2)}{x(x - 2)}$$ Subtract numerators; keep the same denominator.

$$= \frac{9x - 3x + 6}{x(x - 2)}$$ Distributive property; Be careful with signs.

$$= \frac{6x + 6}{x(x - 2)}$$ Combine like terms.

$$= \frac{6(x + 1)}{x(x - 2)}$$ Factor the numerator.

Now Try Exercise 39.

NOTE We factored the final numerator in Example 7 to get $\frac{6(x + 1)}{x(x - 2)}$; however, the fundamental property does not apply since there are no common factors that allow us to write the answer in lower terms.

EXAMPLE 8 Subtracting Rational Expressions with Denominators That Are Opposites

Subtract. Write the answer in lowest terms.

$$\frac{3x}{x - 5} - \frac{2x - 25}{5 - x}$$

The denominators are opposites, so either may be used as the common denominator. We choose $x - 5$.

$$\frac{3x}{x - 5} - \frac{2x - 25}{5 - x} = \frac{3x}{x - 5} - \frac{(2x - 25)(-1)}{(5 - x)(-1)}$$ Fundamental property

$$= \frac{3x}{x - 5} - \frac{-2x + 25}{x - 5}$$ Multiply.

$$= \frac{3x - (-2x + 25)}{x - 5}$$ Subtract numerators; use parentheses.

$$= \frac{3x + 2x - 25}{x - 5}$$ Distributive property; Be careful with signs.

$$= \frac{5x - 25}{x - 5}$$ Combine like terms.

$$= \frac{5(x - 5)}{x - 5}$$ Factor.

$$= 5$$ Lowest terms

Now Try Exercise 53.

EXAMPLE 9 Subtracting Rational Expressions

Subtract. Write the answer in lowest terms.

$$\frac{6x}{x^2 - 2x + 1} - \frac{1}{x^2 - 1}$$

We begin by factoring the denominators.

$$x^2 - 2x + 1 = (x - 1)(x - 1) \quad \text{and} \quad x^2 - 1 = (x - 1)(x + 1)$$

From the factored denominators, we identify the LCD, $(x - 1)(x - 1)(x + 1)$. We use the factor $x - 1$ twice, since it appears twice in the first denominator.

$$\frac{6x}{(x - 1)(x - 1)} - \frac{1}{(x - 1)(x + 1)}$$

$$= \frac{6x(x + 1)}{(x - 1)(x - 1)(x + 1)} - \frac{1(x - 1)}{(x - 1)(x - 1)(x + 1)} \qquad \text{Fundamental property}$$

$$= \frac{6x(x + 1) - 1(x - 1)}{(x - 1)(x - 1)(x + 1)} \qquad \text{Subtract numerators.}$$

$$= \frac{6x^2 + 6x - x + 1}{(x - 1)(x - 1)(x + 1)} \qquad \text{Distributive property}$$

$$= \frac{6x^2 + 5x + 1}{(x - 1)(x - 1)(x + 1)} \qquad \text{Combine like terms.}$$

$$= \frac{(2x + 1)(3x + 1)}{(x - 1)^2(x + 1)} \qquad \text{Factor the numerator.}$$

Verify that the final expression is in lowest terms.

Now Try Exercise 63.

6.4 EXERCISES

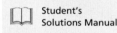

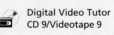
Match the problem in Column I with the correct sum or difference in Column II.

I

1. $\dfrac{x}{x + 6} + \dfrac{6}{x + 6}$

2. $\dfrac{2x}{x - 6} - \dfrac{12}{x - 6}$

3. $\dfrac{6}{x - 6} - \dfrac{x}{x - 6}$

4. $\dfrac{6}{x + 6} - \dfrac{x}{x + 6}$

5. $\dfrac{x}{x + 6} - \dfrac{6}{x + 6}$

6. $\dfrac{1}{x} + \dfrac{1}{6}$

7. $\dfrac{1}{6} - \dfrac{1}{x}$

8. $\dfrac{1}{6x} - \dfrac{1}{6x}$

II

A. 2

B. $\dfrac{x - 6}{x + 6}$

C. -1

D. $\dfrac{6 + x}{6x}$

E. 1

F. 0

G. $\dfrac{x - 6}{6x}$

H. $\dfrac{6 - x}{x + 6}$

Note: When adding and subtracting rational expressions, several different equivalent forms of the answer often exist. If your answer does not look exactly like the one given in the back of the book, check to see whether you have written an equivalent form. (See Example 7 of Section 6.1.)

Add or subtract. Write each answer in lowest terms. See Examples 1 and 6.

9. $\dfrac{4}{m} + \dfrac{7}{m}$

10. $\dfrac{5}{p} + \dfrac{11}{p}$

11. $\dfrac{5}{y+4} - \dfrac{1}{y+4}$

12. $\dfrac{4}{y+3} - \dfrac{1}{y+3}$

13. $\dfrac{x}{x+y} + \dfrac{y}{x+y}$

14. $\dfrac{a}{a+b} + \dfrac{b}{a+b}$

15. $\dfrac{5m}{m+1} - \dfrac{1+4m}{m+1}$

16. $\dfrac{4x}{x+2} - \dfrac{2+3x}{x+2}$

17. $\dfrac{a+b}{2} - \dfrac{a-b}{2}$

18. $\dfrac{x-y}{2} - \dfrac{x+y}{2}$

19. $\dfrac{x^2}{x+5} + \dfrac{5x}{x+5}$

20. $\dfrac{t^2}{t-3} + \dfrac{-3t}{t-3}$

21. $\dfrac{y^2-3y}{y+3} + \dfrac{-18}{y+3}$

22. $\dfrac{r^2-8r}{r-5} + \dfrac{15}{r-5}$

23. Explain with an example how to add or subtract rational expressions with the same denominator.

24. Explain with an example how to add or subtract rational expressions with different denominators.

Add or subtract. Write each answer in lowest terms. See Examples 2, 3, 4, and 7.

25. $\dfrac{z}{5} + \dfrac{1}{3}$

26. $\dfrac{p}{8} + \dfrac{3}{5}$

27. $\dfrac{5}{7} - \dfrac{r}{2}$

28. $\dfrac{10}{9} - \dfrac{z}{3}$

29. $-\dfrac{3}{4} - \dfrac{1}{2x}$

30. $-\dfrac{5}{8} - \dfrac{3}{2a}$

31. $\dfrac{6}{5x} + \dfrac{9}{2x}$

32. $\dfrac{3}{2x} + \dfrac{4}{7x}$

33. $\dfrac{x+1}{6} + \dfrac{3x+3}{9}$

34. $\dfrac{2x-6}{4} + \dfrac{x+5}{6}$

35. $\dfrac{x+3}{3x} + \dfrac{2x+2}{4x}$

36. $\dfrac{x+2}{5x} + \dfrac{6x+3}{3x}$

37. $\dfrac{7}{3p^2} - \dfrac{2}{p}$

38. $\dfrac{12}{5m^2} - \dfrac{5}{m}$

39. $\dfrac{1}{k+4} - \dfrac{2}{k}$

40. $\dfrac{3}{m+1} - \dfrac{4}{m}$

41. $\dfrac{x}{x-2} + \dfrac{-8}{x^2-4}$

42. $\dfrac{2x}{x-1} + \dfrac{-4}{x^2-1}$

43. $\dfrac{4m}{m^2+3m+2} + \dfrac{2m-1}{m^2+6m+5}$

44. $\dfrac{a}{a^2+3a-4} + \dfrac{4a}{a^2+7a+12}$

45. $\dfrac{4y}{y^2-1} - \dfrac{5}{y^2+2y+1}$

46. $\dfrac{2x}{x^2-16} - \dfrac{3}{x^2+8x+16}$

47. $\dfrac{t}{t+2} + \dfrac{5-t}{t} - \dfrac{4}{t^2+2t}$

48. $\dfrac{2p}{p-3} + \dfrac{2+p}{p} - \dfrac{-6}{p^2-3p}$

49. What are the two possible LCDs that could be used for the sum $\dfrac{10}{m-2} + \dfrac{5}{2-m}$?

50. If one form of the correct answer to a sum or difference of rational expressions is $\dfrac{4}{k-3}$, what would an alternative form of the answer be if the denominator is $3-k$?

Add or subtract. Write each answer in lowest terms. See Examples 5 and 8.

51. $\dfrac{4}{x-5} + \dfrac{6}{5-x}$

52. $\dfrac{10}{m-2} + \dfrac{5}{2-m}$

53. $\dfrac{-1}{1-y} - \dfrac{4y-3}{y-1}$

54. $\dfrac{-4}{p-3} - \dfrac{p+1}{3-p}$

55. $\dfrac{2}{x-y^2} + \dfrac{7}{y^2-x}$

56. $\dfrac{-8}{p-q^2} + \dfrac{3}{q^2-p}$

57. $\dfrac{x}{5x-3y} - \dfrac{y}{3y-5x}$

58. $\dfrac{t}{8t-9s} - \dfrac{s}{9s-8t}$

59. $\dfrac{3}{4p-5} + \dfrac{9}{5-4p}$

60. $\dfrac{8}{3-7y} - \dfrac{2}{7y-3}$

In these subtraction problems, the rational expression that follows the subtraction sign has a numerator with more than one term. Be very careful with signs and find each difference. See Example 9.

61. $\dfrac{2m}{m-n} - \dfrac{5m+n}{2m-2n}$

62. $\dfrac{5p}{p-q} - \dfrac{3p+1}{4p-4q}$

63. $\dfrac{5}{x^2-9} - \dfrac{x+2}{x^2+4x+3}$

64. $\dfrac{1}{a^2-1} - \dfrac{a-1}{a^2+3a-4}$

65. $\dfrac{2q+1}{3q^2+10q-8} - \dfrac{3q+5}{2q^2+5q-12}$

66. $\dfrac{4y-1}{2y^2+5y-3} - \dfrac{y+3}{6y^2+y-2}$

Perform each indicated operation. See Examples 1–9.

67. $\dfrac{4}{r^2-r} + \dfrac{6}{r^2+2r} - \dfrac{1}{r^2+r-2}$

68. $\dfrac{6}{k^2+3k} - \dfrac{1}{k^2-k} + \dfrac{2}{k^2+2k-3}$

69. $\dfrac{x+3y}{x^2+2xy+y^2} + \dfrac{x-y}{x^2+4xy+3y^2}$

70. $\dfrac{m}{m^2-1} + \dfrac{m-1}{m^2+2m+1}$

71. $\dfrac{r+y}{18r^2+12ry-3ry-2y^2} + \dfrac{3r-y}{36r^2-y^2}$

72. $\dfrac{2x-z}{2x^2-4xz+5xz-10z^2} - \dfrac{x+z}{x^2-4z^2}$

73. Refer to the rectangle in the figure.

   (a) Find an expression that represents its perimeter. Give the simplified form.

   (b) Find an expression that represents its area. Give the simplified form.

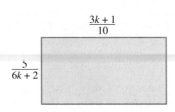

74. Refer to the triangle in the figure. Find an expression that represents its perimeter.

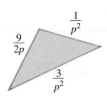

A concours d'elegance is a competition in which a maximum of 100 points is awarded to a car based on its general attractiveness. The rational expression

$$\frac{1010}{49(101 - x)} - \frac{10}{49}$$

approximates the cost, in thousands of dollars, of restoring a car so that it will win x points.
    Use this information to work Exercises 75 and 76.

75. Simplify the expression by performing the indicated subtraction.

76. Use the simplified expression from Exercise 75 to determine how much it would cost to win 95 points.

6.5 Complex Fractions

OBJECTIVES

1 Simplify a complex fraction by writing it as a division problem (Method 1).

2 Simplify a complex fraction by multiplying by the least common denominator (Method 2).

The quotient of two mixed numbers in arithmetic, such as $2\frac{1}{2} \div 3\frac{1}{4}$, can be written as a fraction:

$$2\frac{1}{2} \div 3\frac{1}{4} = \frac{2\frac{1}{2}}{3\frac{1}{4}} = \frac{2 + \frac{1}{2}}{3 + \frac{1}{4}}.$$

The last expression is the quotient of expressions that involve fractions. In algebra, some rational expressions also have fractions in the numerator, or denominator, or both.

> **Complex Fraction**
>
> A rational expression with one or more fractions in the numerator, denominator, or both, is called a **complex fraction.**

Examples of complex fractions include

$$\frac{2 + \frac{1}{2}}{3 + \frac{1}{4}}, \qquad \frac{\frac{3x^2 - 5x}{6x^2}}{2x - \frac{1}{x}}, \qquad \text{and} \qquad \frac{3 + x}{5 - \frac{2}{x}}.$$

The parts of a complex fraction are named as follows.

$$\left.\begin{array}{c} \dfrac{2}{p} - \dfrac{1}{q} \\ \hline \dfrac{3}{p} + \dfrac{5}{q} \end{array}\right.$$

    ←Numerator of complex fraction
    ←Main fraction bar
    ←Denominator of complex fraction

OBJECTIVE 1 Simplify a complex fraction by writing it as a division problem (Method 1).
Since the main fraction bar represents division in a complex fraction, one method of simplifying a complex fraction involves division.

Method 1

To simplify a complex fraction:

Step 1 Write both the numerator and denominator as single fractions.

Step 2 Change the complex fraction to a division problem.

Step 3 Perform the indicated division.

Once again, in this section the first example shows complex fractions from both arithmetic and algebra.

EXAMPLE 1 Simplifying Complex Fractions (Method 1)

Simplify each complex fraction.

(a) $\dfrac{\dfrac{2}{3} + \dfrac{5}{9}}{\dfrac{1}{4} + \dfrac{1}{12}}$

(b) $\dfrac{6 + \dfrac{3}{x}}{\dfrac{x}{4} + \dfrac{1}{8}}$

Step 1 First, write each numerator as a single fraction.

$$\frac{2}{3} + \frac{5}{9} = \frac{2(3)}{3(3)} + \frac{5}{9}$$

$$= \frac{6}{9} + \frac{5}{9} = \frac{11}{9}$$

$$6 + \frac{3}{x} = \frac{6}{1} + \frac{3}{x}$$

$$= \frac{6x}{x} + \frac{3}{x} = \frac{6x + 3}{x}$$

Do the same thing with each denominator.

$$\frac{1}{4} + \frac{1}{12} = \frac{1(3)}{4(3)} + \frac{1}{12}$$

$$= \frac{3}{12} + \frac{1}{12} = \frac{4}{12}$$

$$\frac{x}{4} + \frac{1}{8} = \frac{x(2)}{4(2)} + \frac{1}{8}$$

$$= \frac{2x}{8} + \frac{1}{8} = \frac{2x + 1}{8}$$

Step 2 The original complex fraction can now be written as follows.

$$\frac{\dfrac{11}{9}}{\dfrac{4}{12}} = \frac{11}{9} \div \frac{4}{12}$$

$$\frac{\dfrac{6x + 3}{x}}{\dfrac{2x + 1}{8}} = \frac{6x + 3}{x} \div \frac{2x + 1}{8}$$

Step 3 Now use the rule for division and the fundamental property.

Multiply by the reciprocal.

$$\frac{11}{9} \div \frac{4}{12} = \frac{11}{9} \cdot \frac{12}{4}$$

Multiply by the reciprocal.

$$\frac{6x + 3}{x} \div \frac{2x + 1}{8} = \frac{6x + 3}{x} \cdot \frac{8}{2x + 1}$$

$$= \frac{11 \cdot 3 \cdot 4}{3 \cdot 3 \cdot 4}$$

$$= \frac{11}{3}$$

$$= \frac{3(2x + 1)}{x} \cdot \frac{8}{2x + 1}$$

$$= \frac{24}{x}$$

Now Try Exercises 1 and 15.

EXAMPLE 2 Simplifying a Complex Fraction (Method 1)

Simplify the complex fraction.

$$\frac{\dfrac{xp}{q^3}}{\dfrac{p^2}{qx^2}} = \frac{xp}{q^3} \div \frac{p^2}{qx^2} = \frac{xp}{q^3} \cdot \frac{qx^2}{p^2} = \frac{x^3}{q^2 p}$$

Now Try Exercise 9.

EXAMPLE 3 Simplifying a Complex Fraction (Method 1)

Simplify the complex fraction.

$$\frac{\dfrac{3}{x+2} - 4}{\dfrac{2}{x+2} + 1} = \frac{\dfrac{3}{x+2} - \dfrac{4(x+2)}{x+2}}{\dfrac{2}{x+2} + \dfrac{1(x+2)}{x+2}}$$

Write both second terms with a denominator of $x + 2$.

$$= \frac{\dfrac{3 - 4(x+2)}{x+2}}{\dfrac{2 + 1(x+2)}{x+2}}$$

Subtract in the numerator.

Add in the denominator.

$$= \frac{\dfrac{3 - 4x - 8}{x+2}}{\dfrac{2 + x + 2}{x+2}}$$

Distributive property

$$= \frac{\dfrac{-5 - 4x}{x+2}}{\dfrac{4 + x}{x+2}}$$

Combine like terms.

$$= \frac{-5 - 4x}{x+2} \cdot \frac{x+2}{4+x}$$

Multiply by the reciprocal.

$$= \frac{-5 - 4x}{4 + x}$$

Lowest terms

Now Try Exercise 33.

OBJECTIVE 2 Simplify a complex fraction by multiplying by the least common denominator (Method 2). Since any expression can be multiplied by a form of 1 to get an equivalent expression, we can multiply both the numerator and the denominator of a complex fraction by the same nonzero expression to get an equivalent complex fraction. If we choose the expression to be the LCD of all the fractions within the complex fraction, the complex fraction will be simplified. This is Method 2.

Method 2

To simplify a complex fraction:

Step 1 Find the LCD of all fractions within the complex fraction.

Step 2 Multiply both the numerator and the denominator of the complex fraction by this LCD using the distributive property as necessary. Write in lowest terms.

In the next example, Method 2 is used to simplify the complex fractions from Example 1.

EXAMPLE 4   Simplifying Complex Fractions (Method 2)

Simplify each complex fraction.

(a) $\dfrac{\dfrac{2}{3} + \dfrac{5}{9}}{\dfrac{1}{4} + \dfrac{1}{12}}$
(b) $\dfrac{6 + \dfrac{3}{x}}{\dfrac{x}{4} + \dfrac{1}{8}}$

Step 1 Find the LCD for all denominators in the complex fraction.

The LCD for 3, 9, 4, and 12 is 36. | The LCD for x, 4, and 8 is $8x$.

Step 2 Multiply numerator and denominator of the complex fraction by the LCD.

$$\frac{\dfrac{2}{3} + \dfrac{5}{9}}{\dfrac{1}{4} + \dfrac{1}{12}} = \frac{36\left(\dfrac{2}{3} + \dfrac{5}{9}\right)}{36\left(\dfrac{1}{4} + \dfrac{1}{12}\right)}$$

$$\frac{6 + \dfrac{3}{x}}{\dfrac{x}{4} + \dfrac{1}{8}} = \frac{8x\left(6 + \dfrac{3}{x}\right)}{8x\left(\dfrac{x}{4} + \dfrac{1}{8}\right)}$$

$$= \frac{36\left(\dfrac{2}{3}\right) + 36\left(\dfrac{5}{9}\right)}{36\left(\dfrac{1}{4}\right) + 36\left(\dfrac{1}{12}\right)}$$

$$= \frac{8x(6) + 8x\left(\dfrac{3}{x}\right)}{8x\left(\dfrac{x}{4}\right) + 8x\left(\dfrac{1}{8}\right)} \qquad \text{Distributive property}$$

$$= \frac{24 + 20}{9 + 3}$$

$$= \frac{48x + 24}{2x^2 + x}$$

$$= \frac{44}{12} = \frac{4 \cdot 11}{4 \cdot 3}$$

$$= \frac{24(2x + 1)}{x(2x + 1)} \qquad \text{Factor.}$$

$$= \frac{11}{3}$$

$$= \frac{24}{x} \qquad \text{Lowest terms}$$

Now Try Exercises 2 and 19.

■ **EXAMPLE 5** Simplifying a Complex Fraction (Method 2)

Simplify the complex fraction.

$$
\frac{\dfrac{3}{5m} - \dfrac{2}{m^2}}{\dfrac{9}{2m} + \dfrac{3}{4m^2}} = \frac{20m^2\left(\dfrac{3}{5m} - \dfrac{2}{m^2}\right)}{20m^2\left(\dfrac{9}{2m} + \dfrac{3}{4m^2}\right)}
$$

The LCD for $5m$, m^2, $2m$, and $4m^2$ is $20m^2$.

$$
= \frac{20m^2\left(\dfrac{3}{5m}\right) - 20m^2\left(\dfrac{2}{m^2}\right)}{20m^2\left(\dfrac{9}{2m}\right) + 20m^2\left(\dfrac{3}{4m^2}\right)}
$$

Distributive property

$$
= \frac{12m - 40}{90m + 15}
$$

Now Try Exercise 25.

Either of the two methods can be used to simplify a complex fraction. You may want to choose one method and stick with it to eliminate confusion. However, some students prefer to use Method 1 for problems like Example 2, which is the quotient of two fractions. They prefer Method 2 for problems like Examples 1, 3, 4, and 5, which have sums or differences in the numerators or denominators or both.

■ **EXAMPLE 6** Deciding on a Method and Simplifying Complex Fractions

Simplify each complex fraction.

(a) $\dfrac{\dfrac{1}{y} + \dfrac{2}{y+2}}{\dfrac{4}{y} - \dfrac{3}{y+2}}$

Although either method will work, we use Method 2 here since there are sums and differences in the numerator and denominator. The LCD is $y(y+2)$. Multiply the numerator and denominator by the LCD.

$$
\frac{\dfrac{1}{y} + \dfrac{2}{y+2}}{\dfrac{4}{y} - \dfrac{3}{y+2}} \cdot \frac{y(y+2)}{y(y+2)} = \frac{1(y+2) + 2y}{4(y+2) - 3y}
$$

Fundamental property

$$
= \frac{y + 2 + 2y}{4y + 8 - 3y}
$$

Distributive property

$$
= \frac{3y + 2}{y + 8}
$$

Combine like terms.

(b) $\dfrac{\dfrac{x+2}{x-3}}{\dfrac{x^2-4}{x^2-9}}$

Since this is simply a quotient of two rational expressions, we use Method 1.

$$\dfrac{\dfrac{x+2}{x-3}}{\dfrac{x^2-4}{x^2-9}} = \dfrac{x+2}{x-3} \div \dfrac{x^2-4}{x^2-9}$$

$$= \dfrac{x+2}{x-3} \cdot \dfrac{x^2-9}{x^2-4} \qquad \text{Definition of division}$$

$$= \dfrac{x+2}{x-3} \cdot \dfrac{(x+3)(x-3)}{(x+2)(x-2)} \qquad \text{Factor.}$$

$$= \dfrac{x+3}{x-2} \qquad \text{Lowest terms}$$

Now Try Exercises 29 and 31.

6.5   EXERCISES

For Extra Help

Student's
Solutions Manual

MyMathLab

InterAct Math
Tutorial Software

AW Math
Tutor Center

MathXL

Digital Video Tutor
CD 10/Videotape 10

Note: In many problems involving complex fractions, several different equivalent forms of the answer exist. If your answer does not look exactly like the one given in the back of the book, check to see whether you have written an equivalent form.

1. Consider the complex fraction $\dfrac{\frac{1}{2}-\frac{1}{3}}{\frac{5}{6}-\frac{1}{12}}$. Answer each part, outlining Method 1 for simplifying this complex fraction.

(a) To combine the terms in the numerator, we must find the LCD of $\frac{1}{2}$ and $\frac{1}{3}$. What is this LCD? Determine the simplified form of the numerator of the complex fraction.

(b) To combine the terms in the denominator, we must find the LCD of $\frac{5}{6}$ and $\frac{1}{12}$. What is this LCD? Determine the simplified form of the denominator of the complex fraction.

(c) Now use the results from parts (a) and (b) to write the complex fraction as a division problem using the symbol $\div$.

(d) Perform the operation from part (c) to obtain the final simplification.

2. Consider the same complex fraction given in Exercise 1, $\dfrac{\frac{1}{2}-\frac{1}{3}}{\frac{5}{6}-\frac{1}{12}}$. Answer each part, outlining Method 2 for simplifying this complex fraction.

(a) We must determine the LCD of all the fractions within the complex fraction. What is this LCD?

(b) Multiply every term in the complex fraction by the LCD found in part (a), but do not combine the terms in the numerator and the denominator yet.

(c) Combine the terms from part (b) to obtain the simplified form of the complex fraction.

✎ **3.** Which complex fraction is equivalent to $\dfrac{3-\frac{1}{2}}{2-\frac{1}{4}}$? Answer this question without showing any work, and explain your reasoning.

A. $\dfrac{3+\frac{1}{2}}{2+\frac{1}{4}}$ **B.** $\dfrac{-3+\frac{1}{2}}{2-\frac{1}{4}}$ **C.** $\dfrac{-3-\frac{1}{2}}{-2-\frac{1}{4}}$ **D.** $\dfrac{-3+\frac{1}{2}}{-2+\frac{1}{4}}$

✎ **4.** Only one of these choices is equal to $\dfrac{\frac{1}{2}+\frac{1}{4}}{\frac{1}{3}+\frac{1}{12}}$. Which one is it? Answer this question without showing any work, and explain your reasoning.

A. $\dfrac{9}{5}$ **B.** $-\dfrac{9}{5}$ **C.** $-\dfrac{5}{9}$ **D.** -12

✎ **5.** Describe Method 1 for simplifying complex fractions. Illustrate with the example $\dfrac{\frac{1}{2}}{\frac{2}{3}}$.

✎ **6.** Describe Method 2 for simplifying complex fractions. Illustrate with the example $\dfrac{\frac{1}{2}}{\frac{2}{3}}$.

Simplify each complex fraction. Use either method. See Examples 1–6.

7. $\dfrac{-\frac{4}{3}}{\frac{2}{9}}$ **8.** $\dfrac{-\frac{5}{6}}{\frac{5}{4}}$ **9.** $\dfrac{\frac{x}{y^2}}{\frac{x^2}{y}}$ **10.** $\dfrac{\frac{p^4}{r}}{\frac{p^2}{r^2}}$

11. $\dfrac{\frac{4a^4b^3}{3a}}{\frac{2ab^4}{b^2}}$ **12.** $\dfrac{\frac{2r^4t^2}{3t}}{\frac{5r^2t^5}{3r}}$ **13.** $\dfrac{\frac{m+2}{3}}{\frac{m-4}{m}}$ **14.** $\dfrac{\frac{q-5}{q}}{\frac{q+5}{3}}$

15. $\dfrac{\frac{2}{x}-3}{\frac{2-3x}{2}}$ **16.** $\dfrac{6+\frac{2}{r}}{\frac{3r+1}{4}}$ **17.** $\dfrac{\frac{1}{x}+x}{\frac{x^2+1}{8}}$ **18.** $\dfrac{\frac{3}{m}-m}{\frac{3-m^2}{4}}$

19. $\dfrac{a-\frac{5}{a}}{a+\frac{1}{a}}$ **20.** $\dfrac{q+\frac{1}{q}}{q+\frac{4}{q}}$ **21.** $\dfrac{\frac{5}{8}+\frac{2}{3}}{\frac{7}{3}-\frac{1}{4}}$ **22.** $\dfrac{\frac{6}{5}-\frac{1}{9}}{\frac{2}{5}+\frac{5}{3}}$

23. $\dfrac{\frac{1}{x^2}+\frac{1}{y^2}}{\frac{1}{x}-\frac{1}{y}}$ **24.** $\dfrac{\frac{1}{a^2}-\frac{1}{b^2}}{\frac{1}{a}-\frac{1}{b}}$ **25.** $\dfrac{\frac{2}{p^2}-\frac{3}{5p}}{\frac{4}{p}+\frac{1}{4p}}$ **26.** $\dfrac{\frac{2}{m^2}-\frac{3}{m}}{\frac{2}{5m^2}+\frac{1}{3m}}$

27. $\dfrac{\frac{5}{x^2y}-\frac{2}{xy^2}}{\frac{3}{x^2y^2}+\frac{4}{xy}}$ **28.** $\dfrac{\frac{1}{m^3p}+\frac{2}{mp^2}}{\frac{4}{mp}+\frac{1}{m^2p}}$ **29.** $\dfrac{\frac{1}{4}-\frac{1}{a^2}}{\frac{1}{2}+\frac{1}{a}}$ **30.** $\dfrac{\frac{1}{9}-\frac{1}{m^2}}{\frac{1}{3}+\frac{1}{m}}$

31. $\dfrac{\dfrac{1}{z+5}}{\dfrac{4}{z^2-25}}$

32. $\dfrac{\dfrac{1}{a+1}}{\dfrac{2}{a^2-1}}$

33. $\dfrac{\dfrac{1}{m+1}-1}{\dfrac{1}{m+1}+1}$

34. $\dfrac{\dfrac{2}{x-1}+2}{\dfrac{2}{x-1}-2}$

35. $\dfrac{\dfrac{1}{m-1}+\dfrac{2}{m+2}}{\dfrac{2}{m+2}-\dfrac{1}{m-3}}$

36. $\dfrac{\dfrac{5}{r+3}-\dfrac{1}{r-1}}{\dfrac{2}{r+2}+\dfrac{3}{r+3}}$

37. In a fraction, what operation does the fraction bar represent?

38. What property of real numbers justifies Method 2 of simplifying complex fractions?

RELATING CONCEPTS (EXERCISES 39–42)

For Individual or Group Work

To find the average of two numbers, we add them and divide by 2. Suppose that we wish to find the average of $\frac{3}{8}$ and $\frac{5}{6}$. **Work Exercises 39–42 in order,** *to see how a complex fraction occurs in a problem like this.*

39. Write in symbols: the sum of $\frac{3}{8}$ and $\frac{5}{6}$, divided by 2. Your result should be a complex fraction.

40. Simplify the complex fraction from Exercise 39 using Method 1.

41. Simplify the complex fraction from Exercise 39 using Method 2.

42. Your answers in Exercises 40 and 41 should be the same. Which method did you prefer? Why?

6.6 Solving Equations with Rational Expressions

OBJECTIVES

1 Distinguish between operations with rational expressions and equations with terms that are rational expressions.

2 Solve equations with rational expressions.

3 Solve a formula for a specified variable.

In Section 2.3 we solved equations with fractions as coefficients. By using the multiplication property of equality, we cleared the fractions by multiplying by the LCD. We continue this work here.

OBJECTIVE 1 **Distinguish between operations with rational expressions and equations with terms that are rational expressions.** Before solving equations with rational expressions, you must understand the difference between *sums* and *differences* of terms with rational coefficients, and *equations* with terms that are rational expressions. **Sums and differences are operations to perform, while equations are solved.**

EXAMPLE 1 **Distinguishing between Operations and Equations**

Identify each of the following as an operation or an equation. Then perform the operation or solve the equation.

(a) $\dfrac{3}{4}x - \dfrac{2}{3}x$

This is a difference of two terms, so it is an operation. (There is no equals sign.) Find the LCD, write each coefficient with this LCD, and combine like terms.

$$\frac{3}{4}x - \frac{2}{3}x = \frac{9}{12}x - \frac{8}{12}x \qquad \text{Get a common denominator.}$$

$$= \frac{1}{12}x \qquad \text{Combine like terms.}$$

(b) $\dfrac{3}{4}x - \dfrac{2}{3}x = \dfrac{1}{2}$

Because of the equals sign, this is an equation to be solved. Proceed as in Section 2.3, using the multiplication property of equality to clear fractions. The LCD is 12.

$$\frac{3}{4}x - \frac{2}{3}x = \frac{1}{2}$$

$$12\left(\frac{3}{4}x - \frac{2}{3}x\right) = 12\left(\frac{1}{2}\right) \qquad \text{Multiply by 12.}$$

$$12\left(\frac{3}{4}x\right) - 12\left(\frac{2}{3}x\right) = 12\left(\frac{1}{2}\right) \qquad \text{Distributive property; Multiply } each \text{ term by 12.}$$

$$9x - 8x = 6 \qquad \text{Multiply.}$$

$$x = 6 \qquad \text{Combine like terms.}$$

Check:

$$\frac{3}{4}x - \frac{2}{3}x = \frac{1}{2} \qquad \text{Original equation}$$

$$\frac{3}{4}(6) - \frac{2}{3}(6) = \frac{1}{2} \qquad ? \qquad \text{Let } x = 6.$$

$$\frac{9}{2} - 4 = \frac{1}{2} \qquad ? \qquad \text{Multiply.}$$

$$\frac{1}{2} = \frac{1}{2} \qquad \text{True}$$

Since a true statement results, {6} is the solution set of the equation.

Now Try Exercises 1 and 3.

The ideas of Example 1 can be summarized as follows.

When adding or subtracting, the LCD must be kept throughout the simplification.

When solving an equation, the LCD is used to multiply each side so that denominators are eliminated.

OBJECTIVE 2 Solve equations with rational expressions. When an equation involves fractions as in Example 1(b), we use the multiplication property of equality to clear the fractions. Choose as multiplier the LCD of all denominators in the fractions of the equation.

▮ **EXAMPLE 2** Solving an Equation with Rational Expressions

Solve $\dfrac{p}{2} - \dfrac{p-1}{3} = 1$.

$$6\left(\dfrac{p}{2} - \dfrac{p-1}{3}\right) = 6(1) \qquad \text{Multiply by the LCD, 6.}$$

$$6\left(\dfrac{p}{2}\right) - 6\left(\dfrac{p-1}{3}\right) = 6 \qquad \text{Distributive property}$$

$$3p - 2(p-1) = 6 \qquad \text{Multiply; use parentheses around } p-1.$$

$$3p - 2(p) - 2(-1) = 6 \qquad \text{Distributive property}$$

$$3p - 2p + 2 = 6 \qquad \text{Be careful with signs.}$$

$$p + 2 = 6 \qquad \text{Combine like terms.}$$

$$p = 4 \qquad \text{Subtract 2.}$$

Check to see that {4} is the solution set by replacing p with 4 in the original equation.

Now Try Exercise 37.

CAUTION Note that the use of the LCD here is different from its use in the previous section. Here, we use the multiplication property of equality to multiply each side of an *equation* by the LCD. Earlier, we used the fundamental property to multiply a *fraction* by another fraction that had the LCD as both its numerator and denominator. Be careful not to confuse these two methods.

Recall from Section 6.1 that the denominator of a rational expression cannot equal 0 since division by 0 is undefined. Therefore, when solving an equation with rational expressions that have variables in the denominator, *the solution cannot be a number that makes the denominator equal 0.* If it does, the proposed solution must be rejected.

▮ **EXAMPLE 3** Solving an Equation with Rational Expressions

Solve $\dfrac{x}{x-2} = \dfrac{2}{x-2} + 2$. Check the proposed solution.

Note that $x \neq 2$ here since both denominators in the given expression equal 0 if x is 2. The common denominator is $x - 2$. Solve the equation by multiplying each side by $x - 2$.

$$(x-2)\left(\dfrac{x}{x-2}\right) = (x-2)\left(\dfrac{2}{x-2} + 2\right)$$

$$(x-2)\left(\dfrac{x}{x-2}\right) = (x-2)\left(\dfrac{2}{x-2}\right) + (x-2)(2) \qquad \text{Distributive property}$$

$$x = 2 + 2x - 4$$

$$x = -2 + 2x \qquad \text{Combine like terms.}$$

$$-x = -2 \qquad \text{Subtract } 2x.$$

$$x = 2 \qquad \text{Multiply by } -1.$$

As noted, x cannot equal 2 since replacing x with 2 in the original equation causes the denominators to equal 0.

Check:
$$\frac{2}{2-2} = \frac{2}{2-2} + 2 \quad ? \qquad \text{Let } x = 2 \text{ in the original equation.}$$

$$\frac{2}{0} = \frac{2}{0} + 2 \quad ?$$

Thus, 2 must be rejected as a solution, and the solution set is $\emptyset$.

Now Try Exercise 31.

While it is always a good idea to check solutions to guard against arithmetic and algebraic errors, it is *essential* to check proposed solutions when variables appear in denominators in the original equation. Some students like to determine which numbers cannot be solutions *before* solving the equation, as we did in Example 3.

The steps used to solve an equation with rational expressions follow.

Solving an Equation with Rational Expressions

Step 1 **Multiply each side of the equation by the LCD** to clear the equation of fractions.

Step 2 **Solve** the resulting equation.

Step 3 **Check** each proposed solution by substituting it in the original equation. Reject any that cause a denominator to equal 0.

EXAMPLE 4   Solving an Equation with Rational Expressions

Solve $\dfrac{2}{x^2 - x} = \dfrac{1}{x^2 - 1}$. Check the proposed solution.

Step 1   Factor the denominators to find the LCD.

$$\frac{2}{x(x-1)} = \frac{1}{(x+1)(x-1)}$$

The LCD is $x(x+1)(x-1)$. Notice that $x \neq 0, -1,$ or 1; otherwise a denominator will equal 0. Thus, 0, -1, and 1 cannot be solutions of this equation. Multiply each side of the equation by $x(x+1)(x-1)$.

$$x(x+1)(x-1)\frac{2}{x(x-1)} = x(x+1)(x-1)\frac{1}{(x+1)(x-1)}$$

Step 2
$$2(x+1) = x$$
$$2x + 2 = x \qquad \text{Distributive property}$$
$$2 = -x \qquad \text{Subtract 2x.}$$
$$x = -2 \qquad \text{Multiply by } -1.$$

Step 3 The proposed solution is -2, which does not make any denominator equal 0.

Check:

$$\frac{2}{x^2 - x} = \frac{1}{x^2 - 1} \qquad \text{Original equation}$$

$$\frac{2}{(-2)^2 - (-2)} = \frac{1}{(-2)^2 - 1} \qquad ? \qquad \text{Let } x = -2.$$

$$\frac{2}{4 + 2} = \frac{1}{4 - 1} \qquad ?$$

$$\frac{1}{3} = \frac{1}{3} \qquad \text{True}$$

Thus, the solution set is $\{-2\}$.

Now Try Exercise 43.

■ **EXAMPLE 5** Solving an Equation with Rational Expressions

Solve $\dfrac{2m}{m^2 - 4} + \dfrac{1}{m - 2} = \dfrac{2}{m + 2}$.

Factor the first denominator on the left to find the LCD.

$$\frac{2m}{(m + 2)(m - 2)} + \frac{1}{m - 2} = \frac{2}{m + 2}$$

The LCD is $(m + 2)(m - 2)$. Here $m \neq -2$ or 2, so these numbers cannot be solutions of this equation. Multiply each side by the LCD.

$$(m + 2)(m - 2)\left(\frac{2m}{(m + 2)(m - 2)} + \frac{1}{m - 2} \right)$$

$$= (m + 2)(m - 2)\frac{2}{m + 2}$$

$$(m + 2)(m - 2)\frac{2m}{(m + 2)(m - 2)} + (m + 2)(m - 2)\frac{1}{m - 2}$$

$$= (m + 2)(m - 2)\frac{2}{m + 2}$$

$$2m + m + 2 = 2(m - 2)$$

$$3m + 2 = 2m - 4 \qquad \text{Combine like terms; distributive property}$$

$$m + 2 = -4 \qquad \text{Subtract } 2m.$$

$$m = -6 \qquad \text{Subtract 2.}$$

A check verifies that $\{-6\}$ is the solution set.

Now Try Exercise 53.

■ **EXAMPLE 6** Solving an Equation with Rational Expressions

Solve $\dfrac{1}{x - 1} + \dfrac{1}{2} = \dfrac{2}{x^2 - 1}$.

Factor the denominator on the right.

$$\frac{1}{x - 1} + \frac{1}{2} = \frac{2}{(x + 1)(x - 1)}$$

Notice that $x \neq 1$ or -1. Multiply each side of the equation by the LCD, $2(x + 1)(x - 1)$.

$$2(x + 1)(x - 1)\left(\frac{1}{x - 1} + \frac{1}{2}\right) = 2(x + 1)(x - 1)\frac{2}{(x + 1)(x - 1)}$$

$$2(x + 1)(x - 1)\frac{1}{x - 1} + 2(x + 1)(x - 1)\frac{1}{2} = 2(x + 1)(x - 1)\frac{2}{(x + 1)(x - 1)}$$

$$2(x + 1) + (x + 1)(x - 1) = 4$$

$2x + 2 + x^2 - 1 = 4$	Distributive property
$x^2 + 2x + 1 = 4$	Combine like terms.
$x^2 + 2x - 3 = 0$	Subtract 4.
$(x + 3)(x - 1) = 0$	Factor.
$x + 3 = 0 \quad$ or $\quad x - 1 = 0$	Zero-factor property
$x = -3 \quad$ or $\quad\quad x = 1$	

-3 and 1 are proposed solutions. However, as noted, 1 makes an original denominator equal 0, so 1 is not a solution. Check that -3 is a solution.

Check:

$$\frac{1}{x - 1} + \frac{1}{2} = \frac{2}{x^2 - 1} \qquad \text{Original equation}$$

$$\frac{1}{-3 - 1} + \frac{1}{2} = \frac{2}{(-3)^2 - 1} \quad ? \qquad \text{Let } x = -3.$$

$$\frac{1}{-4} + \frac{1}{2} = \frac{2}{9 - 1} \quad ? \qquad \text{Simplify.}$$

$$\frac{1}{4} = \frac{1}{4} \qquad\qquad \text{True}$$

The solution set is $\{-3\}$.

Now Try Exercise 63.

EXAMPLE 7 Solving an Equation with Rational Expressions

Solve $\dfrac{1}{k^2 + 4k + 3} + \dfrac{1}{2k + 2} = \dfrac{3}{4k + 12}$.

Factoring each denominator gives the equation

$$\frac{1}{(k + 1)(k + 3)} + \frac{1}{2(k + 1)} = \frac{3}{4(k + 3)}.$$

The LCD is $4(k + 1)(k + 3)$, indicating that $k \neq -1$ or -3. Multiply each side by this LCD.

$$4(k + 1)(k + 3)\left(\frac{1}{(k + 1)(k + 3)} + \frac{1}{2(k + 1)}\right)$$

$$= 4(k + 1)(k + 3)\frac{3}{4(k + 3)}$$

$$4(k + 1)(k + 3)\frac{1}{(k + 1)(k + 3)} + 2 \cdot 2(k + 1)(k + 3)\frac{1}{2(k + 1)}$$

$$= 4(k + 1)(k + 3)\frac{3}{4(k + 3)}$$

$$4 + 2(k + 3) = 3(k + 1)$$
$$4 + 2k + 6 = 3k + 3 \qquad \text{Distributive property}$$
$$2k + 10 = 3k + 3 \qquad \text{Combine like terms.}$$
$$7 = k \qquad \text{Subtract } 2k \text{ and } 3.$$

The proposed solution, 7, does not make an original denominator equal 0. A check shows that the algebra is correct, so {7} is the solution set.

Now Try Exercise 65.

OBJECTIVE 3 **Solve a formula for a specified variable.** Solving a formula for a specified variable was first discussed in Chapter 2. Remember to treat the variable for which you are solving as if it were the only variable, and all others as if they were constants.

EXAMPLE 8 **Solving for a Specified Variable**

Solve each formula for the specified variable.

(a) $F = \dfrac{k}{d - D}$ for d

We want to get d alone on one side of the equation. We begin by multiplying each side by $d - D$ to clear the fraction.

$$F = \frac{k}{d - D} \qquad \text{Given equation}$$

$$F(d - D) = \frac{k}{d - D}(d - D) \qquad \text{Clear the fraction.}$$

$$F(d - D) = k \qquad \text{Simplify.}$$

$$Fd - FD = k \qquad \text{Distributive property}$$

$$Fd = k + FD \qquad \text{Add } FD.$$

$$d = \frac{k + FD}{F} \quad \text{or} \quad d = \frac{k}{F} + D \qquad \text{Divide by } F.$$

(b) $\dfrac{1}{a} = \dfrac{1}{b} + \dfrac{1}{c}$ for c

The LCD of all the fractions in the equation is abc, so we multiply each side by abc.

$$abc\left(\frac{1}{a}\right) = abc\left(\frac{1}{b} + \frac{1}{c}\right)$$

$$abc\left(\frac{1}{a}\right) = abc\left(\frac{1}{b}\right) + abc\left(\frac{1}{c}\right) \qquad \text{Distributive property}$$

$$bc = ac + ab$$

Since we are solving for c, we need to get all terms with c on one side of the equation. We do this by subtracting ac from each side.

$$bc - ac = ab \qquad \text{Subtract } ac.$$

$$c(b - a) = ab \qquad \text{Factor out } c.$$

Finally, we divide each side by the coefficient of c, which is $b - a$.

$$c = \frac{ab}{b - a}$$

Now Try Exercises 77 and 83.

CAUTION Students often have trouble in the step that involves factoring out the variable for which they are solving. In Example 8(b), we had to factor out c on the left side so that we could divide both sides by $b - a$.

When solving an equation for a specified variable, *be sure that the specified variable appears alone on only one side of the equals sign in the final equation.*

6.6 EXERCISES

Identify as an expression or an equation. Then perform the operation or solve the equation. See Example 1.

1. $\dfrac{7}{8}x + \dfrac{1}{5}x$

2. $\dfrac{4}{7}x + \dfrac{3}{5}x$

3. $\dfrac{7}{8}x + \dfrac{1}{5}x = 1$

4. $\dfrac{4}{7}x + \dfrac{3}{5}x = 1$

5. $\dfrac{3}{5}x - \dfrac{7}{10}x$

6. $\dfrac{2}{3}x - \dfrac{7}{4}x$

7. $\dfrac{3}{5}x - \dfrac{7}{10}x = 1$

8. $\dfrac{2}{3}x - \dfrac{7}{4}x = -13$

When solving an equation with variables in denominators, we must determine the values that cause these denominators to equal 0, so we can reject these values if they appear as possible solutions. Find all values for which at least one denominator is equal to 0. Do not solve. See Examples 3–7.

9. $\dfrac{3}{x + 2} - \dfrac{5}{x} = 1$

10. $\dfrac{7}{x} + \dfrac{9}{x - 3} = 5$

11. $\dfrac{-1}{(x + 3)(x - 4)} = \dfrac{1}{2x + 1}$

12. $\dfrac{8}{(x - 8)(x + 2)} = \dfrac{7}{3x - 10}$

13. $\dfrac{4}{x^2 + 8x - 9} + \dfrac{1}{x^2 - 4} = 0$

14. $\dfrac{-3}{x^2 + 9x - 10} - \dfrac{12}{x^2 - 16} = 0$

 15. Explain how the LCD is used in a different way when adding and subtracting rational expressions compared to solving equations with rational expressions.

 16. If we multiply each side of the equation $\frac{6}{x + 5} = \frac{6}{x + 5}$ by $x + 5$, we get $6 = 6$. Are all real numbers solutions of this equation? Explain.

Solve each equation, and check your answers. See Examples 1(b), 2, and 3.

17. $\dfrac{5}{m} - \dfrac{3}{m} = 8$

18. $\dfrac{4}{y} + \dfrac{1}{y} = 2$

19. $\dfrac{5}{y} + 4 = \dfrac{2}{y}$

20. $\dfrac{11}{q} = 3 - \dfrac{1}{q}$

21. $\dfrac{3x}{5} - 6 = x$

22. $\dfrac{5t}{4} + t = 9$

23. $\dfrac{4m}{7} + m = 11$

24. $a - \dfrac{3a}{2} = 1$

25. $\dfrac{z-1}{4} = \dfrac{z+3}{3}$

26. $\dfrac{r-5}{2} = \dfrac{r+2}{3}$

27. $\dfrac{3p+6}{8} = \dfrac{3p-3}{16}$

28. $\dfrac{2z+1}{5} = \dfrac{7z+5}{15}$

29. $\dfrac{2x+3}{x} = \dfrac{3}{2}$

30. $\dfrac{5-2x}{x} = \dfrac{1}{4}$

31. $\dfrac{k}{k-4} - 5 = \dfrac{4}{k-4}$

32. $\dfrac{-5}{a+5} = \dfrac{a}{a+5} + 2$

33. $\dfrac{q+2}{3} + \dfrac{q-5}{5} = \dfrac{7}{3}$

34. $\dfrac{t}{6} + \dfrac{4}{3} = \dfrac{t-2}{3}$

35. $\dfrac{x}{2} = \dfrac{5}{4} + \dfrac{x-1}{4}$

36. $\dfrac{8p}{5} = \dfrac{3p-4}{2} + \dfrac{5}{2}$

Solve each equation, and check your answers. Be careful with signs. See Example 2.

37. $\dfrac{a+7}{8} - \dfrac{a-2}{3} = \dfrac{4}{3}$

38. $\dfrac{x+3}{7} - \dfrac{x+2}{6} = \dfrac{1}{6}$

39. $\dfrac{p}{2} - \dfrac{p-1}{4} = \dfrac{5}{4}$

40. $\dfrac{r}{6} - \dfrac{r-2}{3} = -\dfrac{4}{3}$

41. $\dfrac{3x}{5} - \dfrac{x-5}{7} = 3$

42. $\dfrac{8k}{5} - \dfrac{3k-4}{2} = \dfrac{5}{2}$

Solve each equation, and check your answers. See Examples 3–7.

43. $\dfrac{4}{x^2-3x} = \dfrac{1}{x^2-9}$

44. $\dfrac{2}{t^2-4} = \dfrac{3}{t^2-2t}$

45. $\dfrac{2}{m} = \dfrac{m}{5m+12}$

46. $\dfrac{x}{4-x} = \dfrac{2}{x}$

47. $\dfrac{-2}{z+5} + \dfrac{3}{z-5} = \dfrac{20}{z^2-25}$

48. $\dfrac{3}{r+3} - \dfrac{2}{r-3} = \dfrac{-12}{r^2-9}$

49. $\dfrac{3}{x-1} + \dfrac{2}{4x-4} = \dfrac{7}{4}$

50. $\dfrac{2}{p+3} + \dfrac{3}{8} = \dfrac{5}{4p+12}$

51. $\dfrac{x}{3x+3} = \dfrac{2x-3}{x+1} - \dfrac{2x}{3x+3}$

52. $\dfrac{2k+3}{k+1} - \dfrac{3k}{2k+2} = \dfrac{-2k}{2k+2}$

53. $\dfrac{2p}{p^2-1} = \dfrac{2}{p+1} - \dfrac{1}{p-1}$

54. $\dfrac{2x}{x^2-16} - \dfrac{2}{x-4} = \dfrac{4}{x+4}$

55. $\dfrac{5x}{14x+3} = \dfrac{1}{x}$

56. $\dfrac{m}{8m+3} = \dfrac{1}{3m}$

57. $\dfrac{2}{x-1} - \dfrac{2}{3} = \dfrac{-1}{x+1}$

58. $\dfrac{5}{p-2} = 7 - \dfrac{10}{p+2}$

59. $\dfrac{x}{2x+2} = \dfrac{-2x}{4x+4} + \dfrac{2x-3}{x+1}$

60. $\dfrac{5t+1}{3t+3} = \dfrac{5t-5}{5t+5} + \dfrac{3t-1}{t+1}$

61. $\dfrac{8x+3}{x} = 3x$

62. $\dfrac{2}{x} = \dfrac{x}{5x-12}$

63. $\dfrac{1}{x+4} + \dfrac{x}{x-4} = \dfrac{-8}{x^2-16}$

64. $\dfrac{x}{x-3} + \dfrac{4}{x+3} = \dfrac{18}{x^2-9}$

65. $\dfrac{4}{3x + 6} - \dfrac{3}{x + 3} = \dfrac{8}{x^2 + 5x + 6}$

66. $\dfrac{-13}{t^2 + 6t + 8} + \dfrac{4}{t + 2} = \dfrac{3}{2t + 8}$

67. $\dfrac{3x}{x^2 + 5x + 6} = \dfrac{5x}{x^2 + 2x - 3} - \dfrac{2}{x^2 + x - 2}$

68. $\dfrac{m}{m^2 + m - 2} + \dfrac{m}{m^2 - 1} = \dfrac{m}{m^2 + 3m + 2}$

69. $\dfrac{x + 4}{x^2 - 3x + 2} - \dfrac{5}{x^2 - 4x + 3} = \dfrac{x - 4}{x^2 - 5x + 6}$

70. $\dfrac{3}{r^2 + r - 2} - \dfrac{1}{r^2 - 1} = \dfrac{7}{2(r^2 + 3r + 2)}$

71. If you are solving a formula for the letter k, and your steps lead to the equation $kr - mr = km$, what would be your next step?

72. If you are solving a formula for the letter k, and your steps lead to the equation $kr - km = mr$, what would be your next step?

Solve each formula for the specified variable. See Example 8.

73. $m = \dfrac{kF}{a}$ for F

74. $I = \dfrac{kE}{R}$ for E

75. $m = \dfrac{kF}{a}$ for a

76. $I = \dfrac{kE}{R}$ for R

77. $I = \dfrac{E}{R + r}$ for R

78. $I = \dfrac{E}{R + r}$ for r

79. $h = \dfrac{2A}{B + b}$ for A

80. $d = \dfrac{2S}{n(a + L)}$ for S

81. $d = \dfrac{2S}{n(a + L)}$ for a

82. $h = \dfrac{2A}{B + b}$ for B

83. $\dfrac{1}{x} = \dfrac{1}{y} - \dfrac{1}{z}$ for y

84. $\dfrac{3}{k} = \dfrac{1}{p} + \dfrac{1}{q}$ for q

85. $9x + \dfrac{3}{z} = \dfrac{5}{y}$ for z

86. $\dfrac{1}{a} = \dfrac{1}{b} + \dfrac{1}{c}$ for a

▨ SUMMARY EXERCISES ON OPERATIONS AND EQUATIONS WITH RATIONAL EXPRESSIONS

We have performed the four operations of arithmetic with rational expressions and solved equations with rational expressions. The exercises in this summary include a mixed variety of problems of these types. Since students often confuse *operations* on rational expressions with *solving equations* with rational expressions, we review the four operations using the rational expressions $\frac{1}{x}$ and $\frac{1}{x - 2}$ as follows.

Add:

$$\frac{1}{x} + \frac{1}{x - 2} = \frac{1(x - 2)}{x(x - 2)} + \frac{x(1)}{x(x - 2)} \qquad \text{Write with a common denominator.}$$

$$= \frac{x - 2 + x}{x(x - 2)} \qquad \text{Add numerators; keep the same denominator.}$$

$$= \frac{2x - 2}{x(x - 2)} \qquad \text{Combine like terms.}$$

Subtract:

$$\frac{1}{x} - \frac{1}{x-2} = \frac{1(x-2)}{x(x-2)} - \frac{x(1)}{x(x-2)} \qquad \text{Write with a common denominator.}$$

$$= \frac{x-2-x}{x(x-2)} \qquad \text{Subtract numerators; keep the same denominator.}$$

$$= \frac{-2}{x(x-2)} \qquad \text{Combine like terms.}$$

Multiply:

$$\frac{1}{x} \cdot \frac{1}{x-2} = \frac{1}{x(x-2)} \qquad \text{Multiply numerators and multiply denominators.}$$

Divide:

$$\frac{1}{x} \div \frac{1}{x-2} = \frac{1}{x} \cdot \frac{x-2}{1} = \frac{x-2}{x} \qquad \text{Multiply by the reciprocal of the second fraction.}$$

On the other hand, consider the *equation*

$$\frac{1}{x} + \frac{1}{x-2} = \frac{3}{4}.$$

Neither 0 nor 2 can be a solution of this equation, since each will cause a denominator to equal 0. We use the multiplication property of equality and multiply each side by the LCD, $4x(x-2)$, to clear fractions.

$$4x(x-2)\frac{1}{x} + 4x(x-2)\frac{1}{x-2} = 4x(x-2)\frac{3}{4}$$

$$4(x-2) + 4x = 3x(x-2)$$

$$4x - 8 + 4x = 3x^2 - 6x \qquad \text{Distributive property}$$

$$0 = 3x^2 - 14x + 8 \qquad \text{Get 0 on one side.}$$

$$0 = (3x-2)(x-4) \qquad \text{Factor.}$$

$$3x - 2 = 0 \quad \text{or} \quad x - 4 = 0 \qquad \text{Zero-factor property}$$

$$x = \frac{2}{3} \quad \text{or} \qquad x = 4$$

Both $\frac{2}{3}$ and 4 are solutions since neither makes a denominator equal 0; the solution set is $\left\{\frac{2}{3}, 4\right\}$.

In conclusion, remember the following points when working exercises involving rational expressions.

Points to Remember When Working with Rational Expressions

1. The fundamental property is applied only after numerators and denominators have been *factored*.

2. When adding and subtracting rational expressions, the common denominator must be kept throughout the problem and in the final result.

3. Always look to see if the answer is in lowest terms; if it is not, use the fundamental property.

4. When solving equations, the LCD is used to clear the equation of fractions. Multiply each side by the LCD. (Notice how this differs from the use of the LCD in Point 2.)

5. When solving equations with rational expressions, reject any proposed solution that causes an original denominator to equal 0.

For each exercise, indicate "operation" if an operation is to be performed or "equation" if an equation is to be solved. Then perform the operation or solve the equation.

1. $\dfrac{4}{p} + \dfrac{6}{p}$

2. $\dfrac{x^3y^2}{x^2y^4} \cdot \dfrac{y^5}{x^4}$

3. $\dfrac{1}{x^2 + x - 2} \div \dfrac{4x^2}{2x - 2}$

4. $\dfrac{8}{m - 5} = 2$

5. $\dfrac{2y^2 + y - 6}{2y^2 - 9y + 9} \cdot \dfrac{y^2 - 2y - 3}{y^2 - 1}$

6. $\dfrac{2}{k^2 - 4k} + \dfrac{3}{k^2 - 16}$

7. $\dfrac{x - 4}{5} = \dfrac{x + 3}{6}$

8. $\dfrac{3t^2 - t}{6t^2 + 15t} \div \dfrac{6t^2 + t - 1}{2t^2 - 5t - 25}$

9. $\dfrac{4}{p + 2} + \dfrac{1}{3p + 6}$

10. $\dfrac{1}{x} + \dfrac{1}{x - 3} = -\dfrac{5}{4}$

11. $\dfrac{3}{t - 1} + \dfrac{1}{t} = \dfrac{7}{2}$

12. $\dfrac{6}{y} - \dfrac{2}{3y}$

13. $\dfrac{5}{4z} - \dfrac{2}{3z}$

14. $\dfrac{k + 2}{3} = \dfrac{2k - 1}{5}$

15. $\dfrac{1}{m^2 + 5m + 6} + \dfrac{2}{m^2 + 4m + 3}$

16. $\dfrac{2k^2 - 3k}{20k^2 - 5k} \div \dfrac{2k^2 - 5k + 3}{4k^2 + 11k - 3}$

17. $\dfrac{2}{x + 1} + \dfrac{5}{x - 1} = \dfrac{10}{x^2 - 1}$

18. $\dfrac{3}{x + 3} + \dfrac{4}{x + 6} = \dfrac{9}{x^2 + 9x + 18}$

19. $\dfrac{4t^2 - t}{6t^2 + 10t} \div \dfrac{8t^2 + 2t - 1}{3t^2 + 11t + 10}$

20. $\dfrac{x}{x - 2} + \dfrac{3}{x + 2} = \dfrac{8}{x^2 - 4}$

6.7 Applications of Rational Expressions

OBJECTIVES

1 Solve problems about numbers.

2 Solve problems about distance, rate, and time.

3 Solve problems about work.

When we learn how to solve a new type of equation, we are able to apply our knowledge to solving new types of applications. In Section 6.6 we solved equations with rational expressions; now we can solve applications that involve this type of equation. The six-step problem solving method of Chapter 2 still applies.

OBJECTIVE 1 Solve problems about numbers. We begin with an example about an unknown number.

EXAMPLE 1 Solving a Problem about an Unknown Number

If the same number is added to both the numerator and the denominator of the fraction $\frac{2}{5}$, the result is equivalent to $\frac{2}{3}$. Find the number.

Step 1 **Read** the problem carefully. We are trying to find a number.

Step 2 **Assign a variable.** Here, we let $x =$ the number added to the numerator and the denominator.

Step 3 **Write an equation.** The fraction

$$\frac{2 + x}{5 + x}$$

represents the result of adding the same number to both the numerator and the denominator. Since this result is equivalent to $\frac{2}{3}$, the equation is

$$\frac{2 + x}{5 + x} = \frac{2}{3}.$$

Step 4 **Solve** this equation by multiplying each side by the LCD, $3(5 + x)$.

$$3(5 + x)\frac{2 + x}{5 + x} = 3(5 + x)\frac{2}{3}$$

$$3(2 + x) = 2(5 + x)$$

$$6 + 3x = 10 + 2x \qquad \text{Distributive property}$$

$$x = 4 \qquad \text{Subtract } 2x; \text{ subtract } 6.$$

Step 5 **State the answer.** The number is 4.

Step 6 **Check** the solution in the words of the original problem. If 4 is added to both the numerator and the denominator of $\frac{2}{5}$, the result is $\frac{6}{9} = \frac{2}{3}$, as required. ∎

Now Try Exercise 3.

OBJECTIVE 2 Solve problems about distance, rate, and time. Recall from Chapter 2 the following formulas relating distance, rate, and time.

Distance, Rate, and Time Relationship

$$d = rt \qquad r = \frac{d}{t} \qquad t = \frac{d}{r}$$

You may wish to refer to Example 5 in Section 2.7 to review the basic use of these formulas. We continue our work with motion problems here.

EXAMPLE 2   Solving a Problem about Distance, Rate, and Time

The Tickfaw River has a current of 3 mph. A motorboat takes as long to go 12 mi downstream as to go 8 mi upstream. What is the speed of the boat in still water?

Step 1   **Read** the problem again. We are looking for the speed of the boat in still water.

Step 2   **Assign a variable.** Let x = the speed of the boat in still water. Because the current pushes the boat when the boat is going downstream, the speed of the boat downstream will be the sum of the speed of the boat and the speed of the current, $x + 3$ mph. Because the current slows down the boat when the boat is going upstream, the boat's speed going upstream is given by the difference between the speed of the boat and the speed of the current, $x - 3$ mph. See Figure 1.

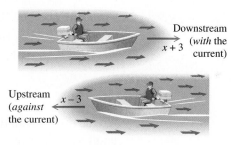

FIGURE 1

This information is summarized in the table.

	d	r	t
Downstream	12	x + 3	
Upstream	8	x − 3	

We fill in the column representing time by using the formula $t = \frac{d}{r}$. Then the time downstream is the distance divided by the rate, or

$$t = \frac{d}{r} = \frac{12}{x + 3},$$

and the time upstream is also the distance divided by the rate, or

$$t = \frac{d}{r} = \frac{8}{x - 3}.$$

The completed table follows.

	d	r	t	
Downstream	12	$x + 3$	$\dfrac{12}{x + 3}$	←
Upstream	8	$x - 3$	$\dfrac{8}{x - 3}$	←

Times are equal.

Step 3 **Write an equation.** According to the original problem, the time downstream equals the time upstream. The two times from the table must therefore be equal, giving the equation

$$\frac{12}{x + 3} = \frac{8}{x - 3}.$$

Step 4 **Solve.** We begin by multiplying each side by the LCD, $(x + 3)(x - 3)$.

$$(x + 3)(x - 3)\frac{12}{x + 3} = (x + 3)(x - 3)\frac{8}{x - 3}$$

$$12(x - 3) = 8(x + 3)$$

$$12x - 36 = 8x + 24 \qquad \text{Distributive property}$$

$$4x = 60 \qquad \text{Subtract } 8x; \text{ add } 36.$$

$$x = 15 \qquad \text{Divide by } 4.$$

Step 5 **State the answer.** The speed of the boat in still water is 15 mph.

Step 6 **Check.** First we find the speed of the boat going downstream, which is $15 + 3 = 18$ mph. Traveling 12 mi would take

$$t = \frac{d}{r} = \frac{12}{18} = \frac{2}{3} \text{ hr.}$$

On the other hand, the speed of the boat going upstream is $15 - 3 = 12$ mph, and traveling 8 mi would take

$$t = \frac{d}{r} = \frac{8}{12} = \frac{2}{3} \text{ hr.}$$

The time upstream equals the time downstream, as required.

Now Try Exercise 21.

OBJECTIVE 3 Solve problems about work. Suppose that you can mow your lawn in 4 hr. Then after 1 hr, you will have mowed $\frac{1}{4}$ of the lawn. After 2 hr, you will have mowed $\frac{2}{4}$ or $\frac{1}{2}$ of the lawn, and so on. This idea is generalized as follows.

Rate of Work

If a job can be completed in t units of time, then the rate of work is

$$\frac{1}{t} \text{ job per unit of time.}$$

◼◼ PROBLEM SOLVING ◼◼

The relationship between problems involving work and problems involving distance is a close one. Recall that the formula $d = rt$ says that distance traveled is equal to rate of travel multiplied by time traveled. Similarly, the fractional part of a job accomplished is equal to the rate of work multiplied by the time worked. In the lawn mowing example, after 3 hr, the fractional part of the job done is

$$\underbrace{\frac{1}{4}}_{\substack{\text{Rate of} \\ \text{work}}} \cdot \underbrace{3}_{\substack{\text{Time} \\ \text{worked}}} = \underbrace{\frac{3}{4}}_{\substack{\text{Fractional part} \\ \text{of job done}}}.$$

After 4 hr, $\frac{1}{4}(4) = 1$ whole job has been done.

◼ **EXAMPLE 3** Solving a Problem about Work Rates

With spraying equipment, Mateo can paint the woodwork in a small house in 8 hr. His assistant, Chet, needs 14 hr to complete the same job painting by hand. If Mateo and Chet work together, how long will it take them to paint the woodwork?

Step 1 **Read** the problem again. We are looking for time working together.

Step 2 **Assign a variable.** Let $x =$ the number of hours it will take for Mateo and Chet to paint the woodwork, working together.

Certainly, x will be less than 8, since Mateo alone can complete the job in 8 hr. We begin by making a table as shown. Remember that based on the previous discussion, Mateo's rate alone is $\frac{1}{8}$ job per hour, and Chet's rate is $\frac{1}{14}$ job per hour.

	Rate	Time Working Together	Fractional Part of the Job Done When Working Together	
Mateo	$\frac{1}{8}$	x	$\frac{1}{8}x$	← Sum is 1 whole job.
Chet	$\frac{1}{14}$	x	$\frac{1}{14}x$	←

Step 3 **Write an equation.** Since together Mateo and Chet complete 1 whole job, we must add their individual fractional parts and set the sum equal to 1.

$$\underbrace{\frac{1}{8}x}_{\substack{\text{Fractional part} \\ \text{done by Mateo}}} + \underbrace{\frac{1}{14}x}_{\substack{\text{Fractional part} \\ \text{done by Chet}}} = \underbrace{1}_{\text{1 whole job.}}$$

Step 4 **Solve.** $\qquad 56\left(\frac{1}{8}x + \frac{1}{14}x\right) = 56(1) \qquad$ Multiply by the LCD, 56.

$$56\left(\frac{1}{8}x\right) + 56\left(\frac{1}{14}x\right) = 56(1) \qquad \text{Distributive property}$$

$$7x + 4x = 56$$

$$11x = 56 \qquad \text{Combine like terms.}$$

$$x = \frac{56}{11} \qquad \text{Divide by 11.}$$

Step 5   **State the answer.** Working together, Mateo and Chet can paint the wood-work in $\frac{56}{11}$ hr, or $5\frac{1}{11}$ hr.

Step 6   **Check** to be sure the answer is correct.

Now Try Exercise 33.

NOTE   An alternative approach in work problems is to consider the part of the job that can be done in 1 hr. For instance, in Example 3 Mateo can do the entire job in 8 hr, and Chet can do it in 14 hr. Thus, their work rates, as we saw in Example 3, are $\frac{1}{8}$ and $\frac{1}{14}$, respectively. Since it takes them x hr to complete the job when working together, in 1 hr they can paint $\frac{1}{x}$ of the woodwork. The amount painted by Mateo in 1 hr plus the amount painted by Chet in 1 hr must equal the amount they can do together. This leads to the equation

Amount by Chet
$$\text{Amount by Mateo} \rightarrow \ \frac{1}{8} + \frac{1}{14} = \frac{1}{x}. \ \leftarrow \text{Amount together}$$

Compare this with the equation in Example 3. Multiplying each side by $56x$ leads to

$$7x + 4x = 56,$$

the same equation found in the third line of Step 4 in the example. The same solution results.

6.7   EXERCISES

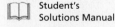

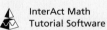
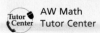

Use Steps 2 and 3 of the six-step method to set up the equation you would use to solve each problem. (Remember that Step 1 is to read the problem carefully.) Do not actually solve the equation. See Example 1.

1. The numerator of the fraction $\frac{5}{6}$ is increased by an amount so that the value of the resulting fraction is equivalent to $\frac{13}{3}$. By what amount was the numerator increased?

(a) Let $x = $ _____ . (*Step 2*)

(b) Write an expression for "the numerator of the fraction $\frac{5}{6}$ is increased by an amount."

(c) Set up an equation to solve the problem. (*Step 3*)

2. If the same number is added to the numerator and subtracted from the denominator of $\frac{23}{12}$, the resulting fraction is equivalent to $\frac{3}{2}$. What is the number?

(a) Let $x =$ _____ . (*Step 2*)

(b) Write an expression for "a number is added to the numerator of $\frac{23}{12}$." Then write an expression for "the same number is subtracted from the denominator of $\frac{23}{12}$."

(c) Set up an equation to solve the problem. (*Step 3*)

Use the six-step method to solve each problem. See Example 1.

3. In a certain fraction, the denominator is 6 more than the numerator. If 3 is added to both the numerator and the denominator, the resulting fraction is equivalent to $\frac{5}{7}$. What was the original fraction?

4. In a certain fraction, the denominator is 4 less than the numerator. If 3 is added to both the numerator and the denominator, the resulting fraction is equivalent to $\frac{3}{2}$. What was the original fraction?

5. The numerator of a certain fraction is four times the denominator. If 6 is added to both the numerator and the denominator, the resulting fraction is equivalent to 2. What was the original fraction?

6. The denominator of a certain fraction is three times the numerator. If 2 is added to the numerator and subtracted from the denominator, the resulting fraction is equivalent to 1. What was the original fraction?

7. One-third of a number is 2 more than one-sixth of the same number. What is the number?

8. One-sixth of a number is 5 more than the same number. What is the number?

9. A quantity, its $\frac{2}{3}$, its $\frac{1}{2}$, and its $\frac{1}{7}$, added together, become 33. What is the quantity? (*Source:* Rhind Mathematical Papyrus.)

10. A quantity, its $\frac{3}{4}$, its $\frac{1}{2}$, and its $\frac{1}{3}$, added together, become 93. What is the quantity? (*Source:* Rhind Mathematical Papyrus.)

Solve each problem. See Example 5 in Section 2.7.

11. In the 2002 Winter Olympics, Catriona LeMay Doan of Canada won the 500-m speed skating event for women. Her rate was 6.6889 m per sec. What was her time (to the nearest hundredth of a second)? (*Source:* www.espn.com)

12. In the 1998 World Championships, Amy Van Dyken of the United States won the 50-m freestyle swimming event for women. Her rate was 1.9881 m per sec. What was her time (to the nearest hundredth of a second)? (*Source: Sports Illustrated 2000 Sports Almanac.*)

13. The winner of the women's 1500-m race in the 2000 Olympics was Nouria Merah-Benida of Algeria with a time of 4.085 min. What was her rate (to three decimal places)? (*Source: World Almanac and Book of Facts, 2002.*)

14. Gabriela Szabo of Romania won the women's 5000-m race in the 2000 Olympics with a time of 14.680 min. What was her rate (to three decimal places)? (*Source: World Almanac and Book of Facts, 2002.*)

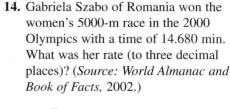

15. The winner of the 2001 Daytona 500 (mile) race was Michael Waltrip, who drove his Chevrolet to victory with a rate of 161.794 mph. What was his time (to the nearest thousandth of an hour)? (*Source: World Almanac and Book of Facts, 2002.*)

16. In 2001, Helio Castroneves drove his Reynard-Honda to victory in the Indianapolis 500 (mile) race. His rate was 131.294 mph. What was his time (to the nearest thousandth of an hour)? (*Source: World Almanac and Book of Facts, 2002.*)

Use the concepts of this section to solve each problem.

17. Suppose Stephanie walks D mi at R mph in the same time that Wally walks d mi at r mph. Give an equation relating D, R, d, and r.

18. If a migrating hawk travels m mph in still air, what is its rate when it flies into a steady headwind of 5 mph? What is its rate with a tailwind of 5 mph?

Set up the equation you would use to solve each problem. Do not actually solve the equation. See Example 2.

19. Julio flew his airplane 500 mi against the wind in the same time it took him to fly 600 mi with the wind. If the speed of the wind was 10 mph, what was the average speed of his plane? (Let x = speed of the plane in still air.)

	d	r	t
Against the Wind	500	$x - 10$	
With the Wind	600	$x + 10$	

20. Luvenia can row 4 mph in still water. She takes as long to row 8 mi upstream as 24 mi downstream. How fast is the current? (Let x = speed of the current.)

	d	r	t
Upstream	8	$4 - x$	
Downstream	24	$4 + x$	

Solve each problem. See Example 2.

21. A boat can go 20 mi against a current in the same time that it can go 60 mi with the current. The current is 4 mph. Find the speed of the boat in still water.

22. Lucinda can fly her plane 200 mi against the wind in the same time it takes her to fly 300 mi with the wind. The wind blows at 30 mph. Find the speed of her plane in still air.

23. An airplane, maintaining a constant airspeed, takes as long to go 450 mi with the wind as it does to go 375 mi against the wind. If the wind is blowing at 15 mph, what is the speed of the plane?

24. A river has a current of 4 km per hr. Find the speed of Lynn McTernan's boat in still water if it goes 40 km downstream in the same time that it takes to go 24 km upstream.

25. Kellen's boat goes 12 mph. Find the rate of the current of the river if she can go 6 mi upstream in the same amount of time she can go 10 mi downstream.

26. Kasey can travel 8 mi upstream in the same time it takes her to go 12 mi downstream. Her boat goes 15 mph in still water. What is the rate of the current?

27. The distance from Seattle, Washington, to Victoria, British Columbia, is about 148 mi by ferry. It takes about 4 hr less to travel by the same ferry from Victoria to Vancouver, British Columbia, a distance of about 74 mi. What is the average speed of the ferry?

28. Driving from Tulsa to Detroit, Jeff averaged 50 mph. He figured that if he had averaged 60 mph, his driving time would have decreased 3 hr. How far is it from Tulsa to Detroit?

Use the concepts of this section to solve each problem.

29. If it takes Elayn 10 hr to do a job, what is her rate?

30. If it takes Clay 12 hr to do a job, how much of the job does he do in 8 hr?

In Exercises 31 and 32, set up the equation you would use to solve each problem. Do not actually solve the equation. See Example 3.

31. Working alone, Jorge can paint a room in 8 hr. Caterina can paint the same room working alone in 6 hr. How long will it take them if they work together? (Let *x* represent the time working together.)

	r	t	w
Jorge		x	
Caterina		x	

32. Edwin Bedford can tune up his Chevy in 2 hr working alone. His son, Beau, can do the job in 3 hr working alone. How long would it take them if they worked together? (Let *t* represent the time working together.)

	r	t	w
Edwin		t	
Beau		t	

Solve each problem. See Example 3.

33. Geraldo and Luisa Hernandez operate a small laundry. Luisa, working alone, can clean a day's laundry in 9 hr. Geraldo can clean a day's laundry in 8 hr. How long would it take them if they work together?

34. Lea can groom the horses in her boarding stable in 5 hr, while Tran needs 4 hr. How long will it take them to groom the horses if they work together?

35. A pump can pump the water out of a flooded basement in 10 hr. A smaller pump takes 12 hr. How long would it take to pump the water from the basement using both pumps?

36. Doug Todd's copier can do a printing job in 7 hr. Scott's copier can do the same job in 12 hr. How long would it take to do the job using both copiers?

37. An experienced employee can enter tax data into a computer twice as fast as a new employee. Working together, it takes the employees 2 hr. How long would it take the experienced employee working alone?

38. One roofer can put a new roof on a house three times faster than another. Working together they can roof a house in 4 days. How long would it take the faster roofer working alone?

39. One pipe can fill a swimming pool in 6 hr, and another pipe can do it in 9 hr. How long will it take the two pipes working together to fill the pool $\frac{3}{4}$ full?

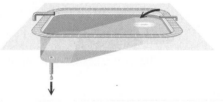

40. An inlet pipe can fill a swimming pool in 9 hr, and an outlet pipe can empty the pool in 12 hr. Through an error, both pipes are left open. How long will it take to fill the pool?

41. A cold water faucet can fill a sink in 12 min, and a hot water faucet can fill it in 15 min. The drain can empty the sink in 25 min. If both faucets are on and the drain is open, how long will it take to fill the sink?

42. Refer to Exercise 40. Assume the error was discovered after both pipes had been running for 3 hr, and the outlet pipe was then closed. How much more time would then be required to fill the pool? (*Hint:* Consider how much of the job had been done when the error was discovered.)

Chapter **Group Activity**

Buying a Car

OBJECTIVE Use a complex fraction to calculate monthly car payments.

You are shopping for a sports car and have put aside a certain amount of money each month for a car payment. Your instructor will assign this amount to you. After looking through a variety of resources, you have narrowed your choices to the cars listed in the table.

Year/Make/Model	Retail Price	Fuel Tank Size (in gallons)	Miles per Gallon (city)	Miles per Gallon (highway)
2002 Mitsubishi Eclipse	$24,197	16.4	21	28
2002 Chevy Camaro	18,415	16.8	19	31
2002 Ford Mustang	18,080	15.7	20	29
2002 Pontiac Firebird	20,050	16.8	19	31
2002 BMW-Z3	37,700	13.5	21	29
2002 Toyota Celica GTS	21,555	14.5	23	32

Source: http://aolsvc.carguides.aol.com

As a group, work through the following steps to determine which car you can afford to buy.

A. Decide which cars you think are within your budget.

B. Select one of the cars you identified in part A. Have each member of the group calculate the monthly payment for this car using a different financing option. Use the formula given below, where P is principal, r is interest rate, and m is the number of monthly payments, along with the financing options table.

Financing Options

Time (in years)	Interest Rate
4	7.0%
5	8.5%
6	10.0%

$$\text{Monthly Payment} = \frac{\dfrac{Pr}{12}}{1 - \left(\dfrac{12}{12 + r}\right)^{m}}$$

C. Have each group member determine the amount of money paid in interest over the duration of the loan for their financing option.

D. Consider fuel expenses.

   1. Assume you will travel an average of 75 mi in the city and 400 mi on the highway each week. How many gallons of gas will you need to buy each month?

   2. Using typical prices for gas in your area at this time, how much money will you need to have available for buying gas?

E. Repeat parts B–D as necessary until your group can reach a consensus on the car you will buy and the financing option you will use. Write a paragraph to explain your choices.

CHAPTER 6 SUMMARY

KEY TERMS

6.1 rational expression
lowest terms

6.3 least common denominator (LCD)

6.5 complex fraction

TEST YOUR WORD POWER

See how well you have learned the vocabulary in this chapter. Answers, with examples, follow the Quick Review.

1. A **rational expression** is
   A. an algebraic expression made up of a term or the sum of a finite number of terms with real coefficients and whole number exponents
   B. a polynomial equation of degree 2
   C. an expression with one or more fractions in the numerator, denominator, or both

   D. the quotient of two polynomials with denominator not 0.

2. A **complex fraction** is
   A. an algebraic expression made up of a term or the sum of a finite number of terms with real coefficients and whole number exponents

   B. a polynomial equation of degree 2
   C. a rational expression with one or more fractions in the numerator, denominator, or both
   D. the quotient of two polynomials with denominator not 0.

QUICK REVIEW

CONCEPTS	EXAMPLES

6.1 THE FUNDAMENTAL PROPERTY OF RATIONAL EXPRESSIONS

To find the value(s) for which a rational expression is undefined, set the denominator equal to 0 and solve the equation.

Find the values for which the expression $\dfrac{x-4}{x^2-16}$ is undefined.

$$x^2 - 16 = 0$$
$$(x-4)(x+4) = 0$$
$$x - 4 = 0 \quad \text{or} \quad x + 4 = 0$$
$$x = 4 \quad \text{or} \qquad x = -4$$

The rational expression is undefined for 4 and -4, so $x \neq 4$ and $x \neq -4$.

Writing a Rational Expression in Lowest Terms

Step 1 Factor the numerator and denominator.

Step 2 Use the fundamental property to divide out common factors.

Write $\dfrac{x^2-1}{(x-1)^2}$ in lowest terms.

$$\frac{x^2-1}{(x-1)^2} = \frac{(x-1)(x+1)}{(x-1)(x-1)}$$

$$= \frac{x+1}{x-1}$$

There are often several different equivalent forms of a rational expression.

Give four equivalent forms of $-\dfrac{x-1}{x+2}$.

Distribute the $-$ sign in the numerator to get $\dfrac{-(x-1)}{x+2}$ or

$\dfrac{-x+1}{x+2}$; do so in the denominator to get $\dfrac{x-1}{-(x+2)}$ or $\dfrac{x-1}{-x-2}$.

CONCEPTS	EXAMPLES

6.2 MULTIPLYING AND DIVIDING RATIONAL EXPRESSIONS

Multiplying Rational Expressions

Step 1 Multiply numerators and multiply denominators.

Step 2 Factor.

Step 3 Write in lowest terms.

Multiply. $\dfrac{3x + 9}{x - 5} \cdot \dfrac{x^2 - 3x - 10}{x^2 - 9}$

$$= \frac{(3x + 9)(x^2 - 3x - 10)}{(x - 5)(x^2 - 9)}$$

$$= \frac{3(x + 3)(x - 5)(x + 2)}{(x - 5)(x + 3)(x - 3)}$$

$$= \frac{3(x + 2)}{x - 3}$$

Dividing Rational Expressions

Step 1 Multiply the first rational expression by the reciprocal of the second.

Step 2 Multiply numerators and multiply denominators.

Step 3 Factor.

Step 4 Write in lowest terms.

Divide. $\dfrac{2x + 1}{x + 5} \div \dfrac{6x^2 - x - 2}{x^2 - 25}$

$$= \frac{2x + 1}{x + 5} \cdot \frac{x^2 - 25}{6x^2 - x - 2}$$

$$= \frac{(2x + 1)(x^2 - 25)}{(x + 5)(6x^2 - x - 2)}$$

$$= \frac{(2x + 1)(x + 5)(x - 5)}{(x + 5)(2x + 1)(3x - 2)}$$

$$= \frac{x - 5}{3x - 2}$$

6.3 LEAST COMMON DENOMINATORS

Finding the LCD

Step 1 Factor each denominator into prime factors.

Step 2 List each different factor the greatest number of times it appears.

Step 3 Multiply the factors from Step 2 to get the LCD.

Find the LCD for $\dfrac{3}{k^2 - 8k + 16}$ and $\dfrac{1}{4k^2 - 16k}$.

$$k^2 - 8k + 16 = (k - 4)^2$$
$$4k^2 - 16k = 4k(k - 4)$$
$$\text{LCD} = (k - 4)^2 \cdot 4 \cdot k$$

$$= 4k(k - 4)^2$$

Writing a Rational Expression with a Specified Denominator

Step 1 Factor both denominators.

Step 2 Decide what factors the denominator must be multiplied by to equal the specified denominator.

Step 3 Multiply the rational expression by that factor divided by itself (multiply by 1).

Find the numerator: $\dfrac{5}{2z^2 - 6z} = \dfrac{}{4z^3 - 12z^2}$.

$$\frac{5}{2z(z - 3)} = \frac{}{4z^2(z - 3)}$$

$2z(z - 3)$ must be multiplied by $2z$.

$$\frac{5}{2z(z - 3)} \cdot \frac{2z}{2z} = \frac{10z}{4z^2(z - 3)} = \frac{10z}{4z^3 - 12z^2}$$

(continued)

CONCEPTS	EXAMPLES

6.4 ADDING AND SUBTRACTING RATIONAL EXPRESSIONS

Adding Rational Expressions

Step 1 Find the LCD.

Add. $\dfrac{2}{3m + 6} + \dfrac{m}{m^2 - 4}$

$$3m + 6 = 3(m + 2)$$
$$m^2 - 4 = (m + 2)(m - 2)$$

The LCD is $3(m + 2)(m - 2)$.

Step 2 Rewrite each rational expression with the LCD as denominator.

$$= \dfrac{2(m - 2)}{3(m + 2)(m - 2)} + \dfrac{3m}{3(m + 2)(m - 2)}$$

Step 3 Add the numerators to get the numerator of the sum. The LCD is the denominator of the sum.

$$= \dfrac{2m - 4 + 3m}{3(m + 2)(m - 2)}$$

Step 4 Write in lowest terms.

$$= \dfrac{5m - 4}{3(m + 2)(m - 2)}$$

Subtracting Rational Expressions

Follow the same steps as for addition, but subtract in Step 3.

Subtract. $\dfrac{6}{k + 4} - \dfrac{2}{k}$

The LCD is $k(k + 4)$.

$$\dfrac{6k}{(k + 4)k} - \dfrac{2(k + 4)}{k(k + 4)} = \dfrac{6k - 2(k + 4)}{k(k + 4)}$$

Be careful with signs when subtracting the numerators.

$$= \dfrac{6k - 2k - 8}{k(k + 4)}$$

$$= \dfrac{4k - 8}{k(k + 4)} \quad \text{or} \quad \dfrac{4(k - 2)}{k(k + 4)}$$

6.5 COMPLEX FRACTIONS

Simplifying Complex Fractions

Simplify.

Method 1 Simplify the numerator and denominator separately. Then divide the simplified numerator by the simplified denominator.

Method 1 $\dfrac{\dfrac{1}{a} - a}{1 - a} = \dfrac{\dfrac{1}{a} - \dfrac{a^2}{a}}{1 - a} = \dfrac{\dfrac{1 - a^2}{a}}{1 - a}$

$$= \dfrac{1 - a^2}{a} \div (1 - a)$$

$$= \dfrac{1 - a^2}{a} \cdot \dfrac{1}{1 - a}$$

$$= \dfrac{(1 - a)(1 + a)}{a(1 - a)} = \dfrac{1 + a}{a}$$

Method 2 Multiply the numerator and denominator of the complex fraction by the LCD of all the denominators in the complex fraction. Write in lowest terms.

Method 2 $\dfrac{\dfrac{1}{a} - a}{1 - a} = \dfrac{\dfrac{1}{a} - a}{1 - a} \cdot \dfrac{a}{a} = \dfrac{\dfrac{a}{a} - a^2}{(1 - a)a}$

$$= \dfrac{1 - a^2}{(1 - a)a} = \dfrac{(1 + a)(1 - a)}{(1 - a)a}$$

$$= \dfrac{1 + a}{a}$$

CONCEPTS	EXAMPLES

6.6 SOLVING EQUATIONS WITH RATIONAL EXPRESSIONS

Solving Equations with Rational Expressions

Step 1 Find the LCD of all denominators in the equation.

Step 2 Multiply each side of the equation by the LCD.

Step 3 Solve the resulting equation, which should have no fractions.

Step 4 Check each proposed solution.

Solve $\dfrac{2}{x-1} + \dfrac{3}{4} = \dfrac{5}{x-1}$.

The LCD is $4(x-1)$. Note that 1 cannot be a solution.

$$4(x-1)\left(\frac{2}{x-1} + \frac{3}{4}\right) = 4(x-1)\left(\frac{5}{x-1}\right)$$

$$4(x-1)\left(\frac{2}{x-1}\right) + 4(x-1)\left(\frac{3}{4}\right) = 4(x-1)\left(\frac{5}{x-1}\right)$$

$$8 + 3(x-1) = 20$$
$$8 + 3x - 3 = 20$$
$$3x = 15$$
$$x = 5$$

The proposed solution, 5, checks. The solution set is $\{5\}$.

6.7 APPLICATIONS OF RATIONAL EXPRESSIONS

Solving Problems about Distance, Rate, Time
Use the six-step method.

Step 1 **Read** the problem carefully.

Step 2 **Assign a variable.** Use a table to identify distance, rate, and time. Solve $d = rt$ for the unknown quantity in the table.

On a trip from Sacramento to Monterey, Marge traveled at an average speed of 60 mph. The return trip, at an average speed of 64 mph, took $\frac{1}{4}$ hr less. How far did she travel between the two cities?

Let x = the unknown distance.

	d	r	$t = \dfrac{d}{r}$
Going	x	60	$\dfrac{x}{60}$
Returning	x	64	$\dfrac{x}{64}$

Step 3 **Write an equation.** From the wording in the problem, decide the relationship between the quantities. Use those expressions to write an equation.

Step 4 **Solve** the equation.

Since the time for the return trip was $\frac{1}{4}$ hr less, the time going equals the time returning plus $\frac{1}{4}$.

$$\frac{x}{60} = \frac{x}{64} + \frac{1}{4}$$
$$16x = 15x + 240 \qquad \text{Multiply by 960.}$$
$$x = 240 \qquad \text{Subtract } 15x.$$

Step 5 **State the answer.**

Step 6 **Check** the solution.

She traveled 240 mi.

The trip there took $\frac{240}{60} = 4$ hr, while the return trip took $\frac{240}{64} = 3\frac{3}{4}$ hr, which is $\frac{1}{4}$ hr less time. The solution checks.

(continued)

CONCEPTS	EXAMPLES

Solving Problems about Work

Step 1 **Read** the problem carefully.

Step 2 **Assign a variable.** State what the variable represents. Put the information from the problem in a table. If a job is done in t units of time, the rate is $\frac{1}{t}$.

It takes the regular mail carrier 6 hr to cover her route. A substitute takes 8 hr to cover the same route. How long would it take them to cover the route together?

Let x = the number of hours to cover the route together.

The rate of the regular carrier is $\frac{1}{6}$ job per hr; the rate of the substitute is $\frac{1}{8}$ job per hr. Multiply rate by time to get the fractional part of the job done.

	Rate	Time	Part of the Job Done
Regular	$\frac{1}{6}$	x	$\frac{1}{6}x$
Substitute	$\frac{1}{8}$	x	$\frac{1}{8}x$

Step 3 **Write an equation.** The sum of the fractional parts should equal 1 (whole job).

The equation is $\frac{1}{6}x + \frac{1}{8}x = 1$.

Step 4 **Solve** the equation.

The solution of the equation is $\frac{24}{7}$.

Steps 5 and 6 **State the answer** and **check** the solution.

It would take them $\frac{24}{7}$ or $3\frac{3}{7}$ hr to cover the route together. The solution checks because $\frac{1}{6}\left(\frac{24}{7}\right) + \frac{1}{8}\left(\frac{24}{7}\right) = 1$ is true.

Answers to Test Your Word Power

1. D; *Examples:* $-\frac{3}{4y}, \frac{5x^3}{x+2}, \frac{a+3}{a^2-4a-5}$ 2. C; *Examples:* $\dfrac{\frac{2}{3}}{\frac{4}{7}}, \dfrac{x-\frac{1}{y}}{x+\frac{1}{y}}, \dfrac{\frac{2}{a+1}}{a^2-1}$

CHAPTER 6 REVIEW EXERCISES

[6.1] *Find the numerical value of each rational expression when (**a**) $x = -2$ and (**b**) $x = 4$.*

1. $\dfrac{4x-3}{5x+2}$

2. $\dfrac{3x}{x^2-4}$

Find the value(s) of the variable for which each rational expression is undefined.

3. $\dfrac{4}{x-3}$

4. $\dfrac{y+3}{2y}$

5. $\dfrac{2k+1}{3k^2+17k+10}$

✍ **6.** How would you determine the values of the variable for which a rational expression is undefined?

Write each rational expression in lowest terms.

7. $\dfrac{5a^3b^3}{15a^4b^2}$

8. $\dfrac{m-4}{4-m}$

9. $\dfrac{4x^2-9}{6-4x}$

10. $\dfrac{4p^2+8pq-5q^2}{10p^2-3pq-q^2}$

Write four equivalent forms for each rational expression.

11. $-\dfrac{4x-9}{2x+3}$

12. $\dfrac{8-3x}{3+6x}$

[6.2] *Multiply or divide, and write each answer in lowest terms.*

13. $\dfrac{18p^3}{6}\cdot\dfrac{24}{p^4}$

14. $\dfrac{8x^2}{12x^5}\cdot\dfrac{6x^4}{2x}$

15. $\dfrac{x-3}{4}\cdot\dfrac{5}{2x-6}$

16. $\dfrac{2r+3}{r-4}\cdot\dfrac{r^2-16}{6r+9}$

17. $\dfrac{6a^2+7a-3}{2a^2-a-6}\div\dfrac{a+5}{a-2}$

18. $\dfrac{y^2-6y+8}{y^2+3y-18}\div\dfrac{y-4}{y+6}$

19. $\dfrac{2p^2+13p+20}{p^2+p-12}\cdot\dfrac{p^2+2p-15}{2p^2+7p+5}$

20. $\dfrac{3z^2+5z-2}{9z^2-1}\cdot\dfrac{9z^2+6z+1}{z^2+5z+6}$

[6.3] *Find the least common denominator for each list of fractions.*

21. $\dfrac{4}{9y},\dfrac{7}{12y^2},\dfrac{5}{27y^4}$

22. $\dfrac{3}{x^2+4x+3},\dfrac{5}{x^2+5x+4}$

Rewrite each rational expression with the given denominator.

23. $\dfrac{3}{2a^3}=\dfrac{}{10a^4}$

24. $\dfrac{9}{x-3}=\dfrac{}{18-6x}$

25. $\dfrac{-3y}{2y-10}=\dfrac{}{50-10y}$

26. $\dfrac{4b}{b^2+2b-3}=\dfrac{}{(b+3)(b-1)(b+2)}$

[6.4] *Add or subtract, and write each answer in lowest terms.*

27. $\dfrac{10}{x}+\dfrac{5}{x}$

28. $\dfrac{6}{3p}-\dfrac{12}{3p}$

29. $\dfrac{9}{k}-\dfrac{5}{k-5}$

30. $\dfrac{4}{y}+\dfrac{7}{7+y}$

31. $\dfrac{m}{3}-\dfrac{2+5m}{6}$

32. $\dfrac{12}{x^2}-\dfrac{3}{4x}$

33. $\dfrac{5}{a-2b}+\dfrac{2}{a+2b}$

34. $\dfrac{4}{k^2-9}-\dfrac{k+3}{3k-9}$

35. $\dfrac{8}{z^2+6z}-\dfrac{3}{z^2+4z-12}$

36. $\dfrac{11}{2p-p^2}-\dfrac{2}{p^2-5p+6}$

[6.5]

37. Simplify the complex fraction $\dfrac{\dfrac{a^4}{b^2}}{\dfrac{a^3}{b}}$ by

    (a) Method 1 as described in Section 6.5.
    (b) Method 2 as described in Section 6.5.
    (c) Explain which method you prefer, and why.

Simplify each complex fraction.

38. $\dfrac{\dfrac{2}{3} - \dfrac{1}{6}}{\dfrac{1}{4} + \dfrac{2}{5}}$

39. $\dfrac{\dfrac{y - 3}{y}}{\dfrac{y + 3}{4y}}$

40. $\dfrac{\dfrac{1}{p} - \dfrac{1}{q}}{\dfrac{1}{q - p}}$

41. $\dfrac{x + \dfrac{1}{w}}{x - \dfrac{1}{w}}$

42. $\dfrac{\dfrac{1}{r + t} - 1}{\dfrac{1}{r + t} + 1}$

[6.6]

43. Before even beginning the solution process, how do you know that 2 cannot be a solution to the equation found in Exercise 46?

Solve each equation, and check your solutions.

44. $\dfrac{4 - z}{z} + \dfrac{3}{2} = \dfrac{-4}{z}$

45. $\dfrac{3x - 1}{x - 2} = \dfrac{5}{x - 2} + 1$

46. $\dfrac{3}{m - 2} + \dfrac{1}{m - 1} = \dfrac{7}{m^2 - 3m + 2}$

Solve each formula for the specified variable.

47. $m = \dfrac{Ry}{t}$ for t

48. $x = \dfrac{3y - 5}{4}$ for y

49. $p^2 = \dfrac{4}{3m - q}$ for m

[6.7] *Solve each problem.*

50. In a certain fraction, the denominator is 4 less than the numerator. If 3 is added to both the numerator and the denominator, the resulting fraction is equal to $\frac{3}{2}$. Find the original fraction.

51. The denominator of a certain fraction is 3 times the numerator. If 2 is added to the numerator and subtracted from the denominator, the resulting fraction is equal to 1. Find the original fraction.

52. A plane flies 350 mi with the wind in the same time that it can fly 310 mi against the wind. The plane has a still-air speed of 165 mph. Find the speed of the wind.

53. A man can plant his garden in 5 hr, working alone. His daughter can do the same job in 8 hr. How long would it take them if they worked together?

54. The head gardener can mow the lawns in the city park twice as fast as his assistant. Working together, they can complete the job in $1\frac{1}{3}$ hr. How long would it take the head gardener working alone?

███   **MIXED REVIEW EXERCISES**

Perform each indicated operation.

55. $\dfrac{4}{m-1} - \dfrac{3}{m+1}$

56. $\dfrac{8p^5}{5} \div \dfrac{2p^3}{10}$

57. $\dfrac{r-3}{8} \div \dfrac{3r-9}{4}$

58. $\dfrac{\dfrac{5}{x} - 1}{\dfrac{5-x}{3x}}$

59. $\dfrac{4}{z^2 - 2z + 1} - \dfrac{3}{z^2 - 1}$

Solve.

60. $a = \dfrac{v - w}{t}$ for v

61. $\dfrac{2}{z} - \dfrac{z}{z+3} = \dfrac{1}{z+3}$

62. $\dfrac{5+m}{m} + \dfrac{3}{4} = -\dfrac{2}{m}$

63. "If Joe can paint a house in 3 hours, and Sam can paint the same house in 5 hours, how long does it take for them to do it together?" (From the movie *Little Big League*)

64. Anne Kelly flew her plane 400 km with the wind in the same time it took her to go 200 km against the wind. The speed of the wind is 50 km per hr. Find the speed of the plane in still air.

███   **RELATING CONCEPTS**   (EXERCISES 65–74)

For Individual or Group Work

In these exercises, we summarize the various concepts involving rational expressions. **Work Exercises 65–74 in order.**

Let P, Q, and R be rational expressions defined as follows.

$$P = \dfrac{6}{x+3} \qquad Q = \dfrac{5}{x+1} \qquad R = \dfrac{4x}{x^2 + 4x + 3}$$

65. Find the value or values for which the expression is undefined.

    (a) P **(b)** Q **(c)** R

66. Find and express in lowest terms: $(P \cdot Q) \div R$.

67. Why is $(P \cdot Q) \div R$ not defined if $x = 0$?

68. Find the LCD for P, Q, and R.

69. Perform the operations and express in lowest terms: $P + Q - R$.

70. Simplify the complex fraction $\dfrac{P+Q}{R}$.

(continued)

71. Solve the equation $P + Q = R$.

72. How does your answer to Exercise 65 help you work Exercise 71?

✍ **73.** Suppose that a car travels 6 mi in $x + 3$ min. Explain why P represents the rate of the car (in miles per minute).

74. For what value or values of x is $R = \frac{40}{77}$?

CHAPTER **6** TEST

1. Find the numerical value of $\dfrac{6r + 1}{2r^2 - 3r - 20}$ when

   (a) $r = -2$ **(b)** $r = 4$.

2. Find any values for which $\dfrac{3x - 1}{x^2 - 2x - 8}$ is undefined.

3. Write four rational expressions equivalent to $-\dfrac{6x - 5}{2x + 3}$.

Write each rational expression in lowest terms.

4. $\dfrac{-15x^6 y^4}{5x^4 y}$

5. $\dfrac{6a^2 + a - 2}{2a^2 - 3a + 1}$

Multiply or divide. Write each answer in lowest terms.

6. $\dfrac{5(d - 2)}{9} \div \dfrac{3(d - 2)}{5}$

7. $\dfrac{6k^2 - k - 2}{8k^2 + 10k + 3} \cdot \dfrac{4k^2 + 7k + 3}{3k^2 + 5k + 2}$

8. $\dfrac{4a^2 + 9a + 2}{3a^2 + 11a + 10} \div \dfrac{4a^2 + 17a + 4}{3a^2 + 2a - 5}$

Find the least common denominator for each list of fractions.

9. $\dfrac{-3}{10p^2}, \dfrac{21}{25p^3}, \dfrac{-7}{30p^5}$

10. $\dfrac{r + 1}{2r^2 + 7r + 6}, \dfrac{-2r + 1}{2r^2 - 7r - 15}$

Rewrite each rational expression with the given denominator.

11. $\dfrac{15}{4p} = \dfrac{}{64p^3}$

12. $\dfrac{3}{6m - 12} = \dfrac{}{42m - 84}$

Add or subtract. Write each answer in lowest terms.

13. $\dfrac{4x + 2}{x + 5} + \dfrac{-2x + 8}{x + 5}$

14. $\dfrac{-4}{y + 2} + \dfrac{6}{5y + 10}$

15. $\dfrac{x + 1}{3 - x} + \dfrac{x^2}{x - 3}$

16. $\dfrac{3}{2m^2 - 9m - 5} - \dfrac{m + 1}{2m^2 - m - 1}$

Simplify each complex fraction.

17. $\dfrac{\dfrac{2p}{k^2}}{\dfrac{3p^2}{k^3}}$

18. $\dfrac{\dfrac{1}{x+3}-1}{1+\dfrac{1}{x+3}}$

Solve.

19. $\dfrac{2x}{x-3}+\dfrac{1}{x+3}=\dfrac{-6}{x^2-9}$

20. $F=\dfrac{k}{d-D}$ for D

Solve each problem.

21. A boat goes 7 mph in still water. It takes as long to go 20 mi upstream as 50 mi downstream. Find the speed of the current.

22. A man can paint a room in his house, working alone, in 5 hr. His wife can do the job in 4 hr. How long will it take them to paint the room if they work together?

CUMULATIVE REVIEW EXERCISES CHAPTERS **1–6**

1. Use the order of operations to evaluate $3+4\left(\frac{1}{2}-\frac{3}{4}\right)$.

Solve.

2. $3(2y-5)=2+5y$　　　**3.** $A=\dfrac{1}{2}bh$ for b　　　**4.** $\dfrac{2+m}{2-m}=\dfrac{3}{4}$

5. $5y\le 6y+8$　　　**6.** $5m-9>2m+3$

7. For the graph of $4x+3y=-12$,

　　(**a**) what is the x-intercept?　　(**b**) what is the y-intercept?

Sketch each graph.

8. $y=-3x+2$　　　　　　**9.** $y=-x^2+1$

Find the slope of each line described in Exercises 10 and 11.

10. Through $(-5,8)$ and $(-1,2)$　　　**11.** Perpendicular to $4x-3y=12$

Simplify each expression. Write with only positive exponents.

12. $\dfrac{(2x^3)^{-1}\cdot x}{2^3x^5}$　　　**13.** $\dfrac{(m^{-2})^3m}{m^5m^{-4}}$　　　**14.** $\dfrac{2p^3q^4}{8p^5q^3}$

Perform each indicated operation.

15. $(2k^2+3k)-(k^2+k-1)$　　　**16.** $8x^2y^2(9x^4y^5)$

17. $(2a-b)^2$　　　**18.** $(y^2+3y+5)(3y-1)$

19. $\dfrac{12p^3+2p^2-12p+4}{2p-2}$

20. A computer can do one operation in 1.4×10^{-7} sec. How long would it take for the computer to do one trillion (10^{12}) operations?

Factor completely.

21. $8t^2 + 10tv + 3v^2$ **22.** $8r^2 - 9rs + 12s^2$ **23.** $16x^4 - 1$

Solve each equation.

24. $r^2 = 2r + 15$ **25.** $(r - 5)(2r + 1)(3r - 2) = 0$

Solve each problem.

26. One number is 4 more than another. The product of the numbers is 2 less than the smaller number. Find the smaller number.

27. The length of a rectangle is 2 m less than twice the width. The area is 60 m². Find the width of the rectangle.

28. One of the following is equal to 1 for *all* real numbers. Which one is it?

A. $\dfrac{k^2 + 2}{k^2 + 2}$ **B.** $\dfrac{4 - m}{4 - m}$ **C.** $\dfrac{2x + 9}{2x + 9}$ **D.** $\dfrac{x^2 - 1}{x^2 - 1}$

29. Which one of the following rational expressions is *not* equivalent to $\dfrac{4 - 3x}{7}$?

A. $-\dfrac{-4 + 3x}{7}$ **B.** $-\dfrac{4 - 3x}{-7}$ **C.** $\dfrac{-4 + 3x}{-7}$ **D.** $\dfrac{-(3x + 4)}{7}$

Perform each operation and write the answer in lowest terms.

30. $\dfrac{5}{q} - \dfrac{1}{q}$ **31.** $\dfrac{3}{7} + \dfrac{4}{r}$

32. $\dfrac{4}{5q - 20} - \dfrac{1}{3q - 12}$ **33.** $\dfrac{2}{k^2 + k} - \dfrac{3}{k^2 - k}$

34. $\dfrac{7z^2 + 49z + 70}{16z^2 + 72z - 40} \div \dfrac{3z + 6}{4z^2 - 1}$ **35.** $\dfrac{\dfrac{4}{a} + \dfrac{5}{2a}}{\dfrac{7}{6a} - \dfrac{1}{5a}}$

36. What values of x cannot possibly be solutions of the equation $\dfrac{1}{x - 4} = \dfrac{3}{2x}$?

Solve each equation. Check your solutions.

37. $\dfrac{r + 2}{5} = \dfrac{r - 3}{3}$ **38.** $\dfrac{1}{x} = \dfrac{1}{x + 1} + \dfrac{1}{2}$

Solve each problem.

39. Juanita can weed the yard in 3 hr. Benito can weed the yard in 2 hr. How long would it take them if they worked together?

40. A canal has a current of 2 mph. Find the speed of Amy's boat in still water if it goes 11 mi downstream in the same time that it goes 8 mi upstream.

Equations of Lines; Functions

7

Graphs are widely used in the media because they present a lot of information in an easy-to-understand form. As the saying goes, "A picture is worth a thousand words." It is important to be able to read graphs correctly and understand how to use the data they provide. In Section 7.1, Example 10, we use a graph to find the average rate of change each year from 1997 to 2001 in the number of U.S. households owning more than one personal computer.

7.1 Review of Graphs and Slopes of Lines

This section and the next review some of the main topics of linear equations in two variables, first introduced in Chapter 3.

OBJECTIVE 1 Plot ordered pairs. Each of the pairs of numbers $(3, 1)$, $(-5, 6)$, and $(4, -1)$ is an example of an **ordered pair;** that is, a pair of numbers written within parentheses in which the order of the numbers is important. We graph an ordered pair using two perpendicular number lines that intersect at their 0 points, as shown in Figure 1. The common 0 point is called the **origin.** The position of any point in this plane is determined by referring to the horizontal number line, the **x-axis,** and the vertical number line, the **y-axis.** The first number in the ordered pair indicates the position relative to the x-axis, and the second number indicates the position relative to the y-axis. The x-axis and the y-axis make up a **rectangular** (or **Cartesian,** for Descartes) **coordinate system.**

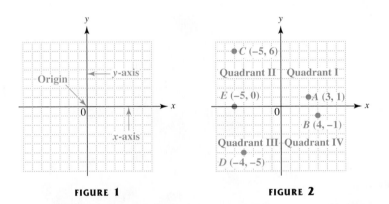

FIGURE 1   **FIGURE 2**

To locate, or **plot,** the point on the graph that corresponds to the ordered pair $(3, 1)$, we move three units from 0 to the right along the x-axis, and then one unit up parallel to the y-axis. The point corresponding to the ordered pair $(3, 1)$ is labeled A in Figure 2. Additional points are labeled $B-E$. The phrase "the point corresponding to the ordered pair $(3, 1)$" is often abbreviated as "the point $(3, 1)$." The numbers in an ordered pair are called the **coordinates** of the corresponding point.

The four regions of the graph, shown in Figure 2, are called **quadrants I, II, III,** and **IV,** reading counterclockwise from the upper right quadrant. The points on the x-axis and y-axis do not belong to any quadrant. For example, point E in Figure 2 belongs to no quadrant.

Now Try Exercises 3, 5, 7, and 9.

OBJECTIVE 2 Graph lines and find intercepts. Each solution to an equation with two variables, such as $2x + 3y = 6$, includes two numbers, one for each variable. To keep track of which number goes with which variable, we write the solutions as ordered pairs. (If x and y are used as the variables, the x-value is given first.) For example, we can show that $(6, -2)$ is a solution of $2x + 3y = 6$ by substitution.

$$2x + 3y = 6$$
$$2(6) + 3(-2) = 6 \qquad ? \qquad \text{Let } x = 6, y = -2.$$
$$12 - 6 = 6 \qquad ?$$
$$6 = 6 \qquad \text{True}$$

Because the ordered pair $(6, -2)$ makes the equation true, it is a solution. On the other hand, $(5, 1)$ is *not* a solution of the equation $2x + 3y = 6$ because

$$2(5) + 3(1) = 10 + 3 = 13 \neq 6.$$

To find ordered pairs that satisfy an equation, select any number for one of the variables, substitute it into the equation for that variable, and then solve for the other variable. Two other ordered pairs satisfying $2x + 3y = 6$ are $(0, 2)$ and $(3, 0)$. Since any real number could be selected for one variable and would lead to a real number for the other variable, linear equations in two variables have an infinite number of ordered-pair solutions.

How might we express the solution set of an equation like $2x + 3y = 6$? The graph of an equation is the set of points corresponding to *all* ordered pairs that satisfy the equation. It gives a "picture" of the equation. The graph of the equation $2x + 3y = 6$ is shown in Figure 3 along with a table of ordered pairs.

x	y
-3	4
0	2
3	0
6	-2
9	-4

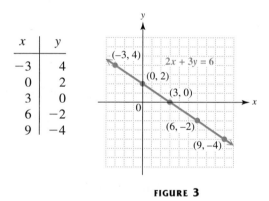

FIGURE 3

Now Try Exercise 11.

The equation $2x + 3y = 6$ is called a **first-degree equation** because it has no term with a variable to a power greater than 1.

The graph of any first-degree equation in two variables is a straight line.

Since first-degree equations with two variables have straight-line graphs, they are called *linear equations in two variables.*

Linear Equation in Two Variables

A **linear equation in two variables** can be written in the form

$$Ax + By = C,$$

where A, B, and C are real numbers (A and B not both 0).

A straight line is determined if any two different points on the line are known, so finding two different points is enough to graph the line. Two useful points for graphing are the *x*- and *y*-intercepts. The **x-intercept** is the point (if any) where the line intersects the *x*-axis; likewise, the **y-intercept** is the point (if any) where the line intersects the *y*-axis.* In Figure 3, the *y*-value of the point where the line intersects the *x*-axis is 0. Similarly, the *x*-value of the point where the line intersects the *y*-axis is 0. This suggests a method for finding the *x*- and *y*-intercepts.

Finding Intercepts

In the equation of a line, let $y = 0$ to find the *x*-intercept; let $x = 0$ to find the *y*-intercept.

EXAMPLE 1 Finding Intercepts

Find the *x*- and *y*-intercepts of $4x - y = -3$ and graph the equation.

We find the *x*-intercept by letting $y = 0$.

$$4x - 0 = -3 \qquad \text{Let } y = 0.$$

$$4x = -3$$

$$x = -\frac{3}{4} \qquad \text{\textit{x}-intercept is } \left(-\tfrac{3}{4}, 0\right).$$

For the *y*-intercept, we let $x = 0$.

$$4(0) - y = -3 \qquad \text{Let } x = 0.$$

$$-y = -3$$

$$y = 3 \qquad \text{\textit{y}-intercept is } (0, 3).$$

The intercepts are the two points $\left(-\tfrac{3}{4}, 0\right)$ and $(0, 3)$. We show these ordered pairs in the table next to Figure 4 and use them to draw the graph.

x	y
$-\frac{3}{4}$	0
0	3

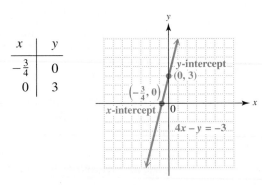

FIGURE 4

Now Try Exercise 17.

---

*Some texts define an intercept as a number, not a point. For example, "*y*-intercept (0, 4)" would be given as "*y*-intercept 4."

NOTE   While two points, such as the two intercepts in Figure 4, are sufficient to graph a straight line, it is a good idea to use a third point to guard against errors. Verify by substitution that $(-2, -5)$ also lies on the graph of $4x - y = -3$.

Some lines have both the x- and y-intercepts at the origin.

▨ EXAMPLE 2   Graphing a Line That Passes through the Origin

Graph $x + 2y = 0$.

Find the intercepts.

$x + 2y = 0$	$x + 2y = 0$
$x + 2(0) = 0$ Let $y = 0$.	$0 + 2y = 0$ Let $x = 0$.
$x + 0 = 0$	$2y = 0$
$x = 0$ x-intercept is $(0, 0)$.	$y = 0$ y-intercept is $(0, 0)$.

Both intercepts are the same point, $(0, 0)$, which means that the graph passes through the origin. To find another point so that we can graph the line, we choose any nonzero number for x or y. If we choose $x = 4$ and solve for y, then

$$x + 2y = 0$$
$$4 + 2y = 0 \qquad \text{Let } x = 4.$$
$$2y = -4$$
$$y = -2.$$

This gives the ordered pair $(4, -2)$. These two points lead to the graph shown in Figure 5. As a check, verify that $(-2, 1)$ also lies on the line.

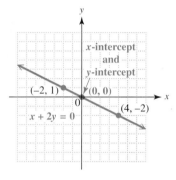

FIGURE 5

Now Try Exercise 23.

OBJECTIVE 3   Recognize equations of vertical and horizontal lines.   Two special cases of straight-line graphs are horizontal and vertical lines.

▨ EXAMPLE 3   Graphing a Horizontal Line

Graph $y = 2$.

Writing $y = 2$ as $0x + 1y = 2$ shows that any value of x, including $x = 0$, gives $y = 2$, making the y-intercept $(0, 2)$. Since y is always 2, there is no value of x correspon-ding to $y = 0$, so the graph has no x-intercept. The graph, shown with a table of ordered pairs in Figure 6, is a horizontal line.

x	y
-1	2
0	2
3	2

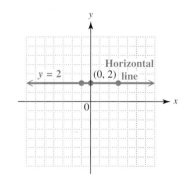

FIGURE 6

Now Try Exercise 27.

EXAMPLE 4   Graphing a Vertical Line

Graph $x + 1 = 0$.

The form $1x + 0y = -1$ shows that every value of y leads to $x = -1$, making the x-intercept $(-1, 0)$. No value of y makes $x = 0$, so the graph has no y-intercept. The only way a straight line can have no y-intercept is to be vertical, as shown in Figure 7.

x	y
-1	-4
-1	0
-1	5

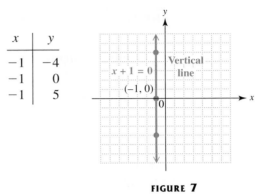

FIGURE 7

Now Try Exercise 29.

CAUTION   To avoid confusing equations of horizontal and vertical lines remember that

1. An equation with only the variable x will always intersect the *x-axis* and thus will be *vertical*.

2. An equation with only the variable y will always intersect the *y-axis* and thus will be *horizontal*.

| **CONNECTIONS** |

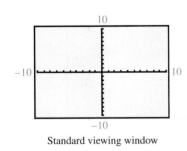

Standard viewing window

FIGURE 8

When graphing by hand, we first set up a rectangular coordinate system, then plot points and draw the graph. Similarly, when graphing with a graphing calculator, we first tell the calculator how to set up a rectangular coordinate system. This involves choosing the minimum and maximum x- and y-values that will determine the viewing screen. In the screen shown in Figure 8, we chose minimum x- and y-values of -10 and maximum x- and y-values of 10. The *scale* on each axis determines the distance between the tick marks; in the screen shown, the scale is 1 for both axes. We refer to this as the *standard viewing window*.

For example, to graph $4x - y = 3$ with a graphing calculator, we use the intercepts to determine an appropriate window. Here, the x-intercept is $(.75, 0)$ and the y-intercept is $(0, -3)$. Although many choices are possible, we choose the standard viewing window. We must solve the equation for y to enter it into the calculator.

$$4x - y = 3$$

$$-y = -4x + 3 \qquad \text{Subtract } 4x.$$

$$y = 4x - 3 \qquad \text{Multiply by } -1.$$

The graph is shown in Figures 9 and 10, which also give the intercepts at the bottoms of the screens. Some calculators have the capability of locating the *x*-intercept (called "Root" or "Zero"). Consult your owner's manual.

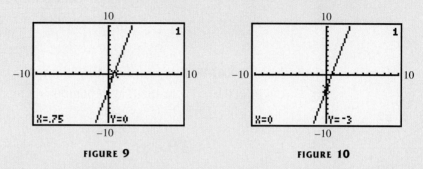

FIGURE 9 **FIGURE 10**

For Discussion or Writing

Write each equation in the form needed to enter it into a graphing calculator. Then graph the equation in the standard viewing window of a graphing calculator, and locate the intercepts.

1. $5x + 2y = -10$ **2.** $3x - 4y = -6$

3. $3.2x - y = -5.8$ **4.** $y - 4.2 = -1.5x$

OBJECTIVE 4 Find the slope of a line. Slope (steepness) is used in many practical ways. The slope of a highway (sometimes called the *grade*) is often given as a percent. For example, a 10% $\left(\text{or } \frac{10}{100} = \frac{1}{10}\right)$ slope means the highway rises 1 unit for every 10 horizontal units. Stairs and roofs have slopes too, as shown in Figure 11.

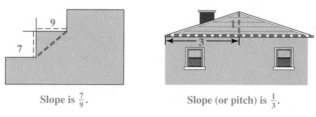

Slope is $\frac{7}{9}$.   Slope (or pitch) is $\frac{1}{3}$.

FIGURE 11

In each example mentioned, slope is the ratio of vertical change, or **rise,** to horizontal change, or **run.** A simple way to remember this is to think "slope is rise over run."

To get a formal definition of the slope of a line, we designate two different points on the line. To differentiate between the points, we write them as (x_1, y_1) and (x_2, y_2). See Figure 12. (The small numbers 1 and 2 in these ordered pairs are called *subscripts*. Read (x_1, y_1) as "*x*-sub-one, *y*-sub-one.")

As we move along the line in Figure 12 from (x_1, y_1) to (x_2, y_2), the *y*-value changes (vertically) from y_1 to y_2, an amount equal to $y_2 - y_1$. As *y* changes from y_1 to y_2, the value of *x* changes (horizontally) from x_1 to x_2 by the amount $x_2 - x_1$. The ratio of the change in *y* to the change in *x* (the rise over the run) is called the *slope* of the line, with the letter *m* traditionally used for slope.

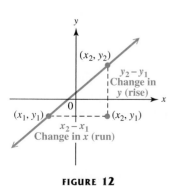

FIGURE 12

Slope Formula

The **slope** of the line through the distinct points (x_1, y_1) and (x_2, y_2) is

$$m = \frac{\text{rise}}{\text{run}} = \frac{\text{change in } y}{\text{change in } x} = \frac{y_2 - y_1}{x_2 - x_1} \quad (x_1 \neq x_2).$$

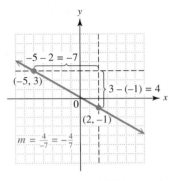

FIGURE 13

EXAMPLE 5 Finding the Slope of a Line

Find the slope of the line through the points $(2, -1)$ and $(-5, 3)$.

If $(2, -1) = (x_1, y_1)$ and $(-5, 3) = (x_2, y_2)$, then

$$m = \frac{y_2 - y_1}{x_2 - x_1} = \frac{3 - (-1)}{-5 - 2} = \frac{4}{-7} = -\frac{4}{7}.$$

See Figure 13. If the pairs are reversed so that $(2, -1) = (x_2, y_2)$ and $(-5, 3) = (x_1, y_1)$, the slope is the same.

$$m = \frac{-1 - 3}{2 - (-5)} = \frac{-4}{7} = -\frac{4}{7}$$

Now Try Exercise 61.

Example 5 suggests that the slope is the same no matter which point we consider first. Also, using similar triangles from geometry, we can show that the slope is the same no matter which two different points on the line we choose.

CAUTION In calculating slope, be careful to subtract the y-values and the x-values in the *same order*.

Correct		Incorrect	
$\dfrac{y_2 - y_1}{x_2 - x_1}$	or $\dfrac{y_1 - y_2}{x_1 - x_2}$	$\dfrac{y_2 - y_1}{x_1 - x_2}$ or	$\dfrac{y_1 - y_2}{x_2 - x_1}$

Also, remember that the change in y is the *numerator* and the change in x is the *denominator*.

When an equation of a line is given, one way to find the slope is to use the definition of slope by first finding two different points on the line.

EXAMPLE 6 Finding the Slope of a Line

Find the slope of the line $4x - y = -8$.

The intercepts can be used as the two different points needed to find the slope. Let $y = 0$ to find that the x-intercept is $(-2, 0)$. Then let $x = 0$ to find that the y-intercept is $(0, 8)$. Use these two points in the slope formula. The slope is

$$m = \frac{\text{rise}}{\text{run}} = \frac{8 - 0}{0 - (-2)} = \frac{8}{2} = 4.$$

Now Try Exercise 71.

We review the following special cases of slope.

Slopes of Horizontal and Vertical Lines

The slope of a horizontal line is 0; the slope of a vertical line is undefined.

Now Try Exercises 77 and 79.

The slope of a line can also be found directly from its equation. Look again at the equation $4x - y = -8$ from Example 6. Solve this equation for y.

$$4x - y = -8 \qquad \text{Equation from Example 6}$$
$$-y = -4x - 8 \qquad \text{Subtract } 4x.$$
$$y = 4x + 8 \qquad \text{Multiply by } -1.$$

Notice that the slope, 4, found using the slope formula in Example 6 is the same number as the coefficient of x in the equation $y = 4x + 8$. This always happens, *as long as the equation is solved for y.*

EXAMPLE 7 Finding the Slope from an Equation

Find the slope of the graph of $3x - 5y = 8$.
    Solve the equation for y.

$$3x - 5y = 8$$
$$-5y = -3x + 8 \qquad \text{Subtract } 3x.$$
$$y = \frac{3}{5}x - \frac{8}{5} \qquad \text{Divide by } -5.$$

The slope is given by the coefficient of x, so the slope is $\frac{3}{5}$.

Now Try Exercise 73.

OBJECTIVE 5 Graph a line given its slope and a point on the line.

EXAMPLE 8 Using the Slope and a Point to Graph Lines

Graph each line.

(a) With slope $\frac{2}{3}$ passing through the point $(-1, 4)$
    First locate the point $P(-1, 4)$ on a graph as shown in Figure 14. Then use the slope to find a second point. From the slope formula,

$$m = \frac{\text{change in } y}{\text{change in } x} = \frac{2}{3},$$

so move *up* 2 units and then 3 units to the *right* to locate another point on the graph (labeled R). The line through $(-1, 4)$ and R is the required graph.

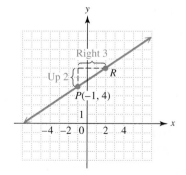

FIGURE 14

(b) Through $(3, 1)$ with slope -4

Start by locating the point $P(3, 1)$ on a graph. Find a second point R on the line by writing the slope -4 as $\frac{-4}{1}$ and using the slope formula.

$$m = \frac{\text{change in } y}{\text{change in } x} = \frac{-4}{1}$$

Move *down* 4 units from $(3, 1)$, and then move 1 unit to the *right*. Draw a line through this second point R and $(3, 1)$, as shown in Figure 15.

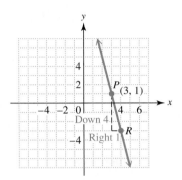

FIGURE 15

The slope also could be written as

$$m = \frac{\text{change in } y}{\text{change in } x} = \frac{4}{-1}.$$

In this case the second point R is located *up* 4 units and 1 unit to the *left*. Verify that this approach also produces the line in Figure 15.

Now Try Exercises 83 and 85.

In Example 8(a), the slope of the line is the *positive* number $\frac{2}{3}$. The graph of the line in Figure 14 goes up (rises) from left to right. The line in Example 8(b) has *negative* slope, -4. As Figure 15 shows, its graph goes down (falls) from left to right. These facts suggest the following generalization.

A positive slope indicates that the line goes *up* from left to right;

a negative slope indicates that the line goes *down* from left to right.

Figure 16 shows lines of positive, 0, negative, and undefined slopes.

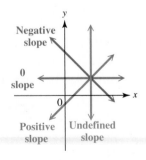

FIGURE 16

OBJECTIVE 6 Use slopes to determine whether two lines are parallel, perpendicular, or neither. Recall that the slopes of a pair of parallel or perpendicular lines are related in a special way.

Slopes of Parallel or Perpendicular Lines

Two nonvertical lines with the same slope are parallel; two nonvertical parallel lines have the same slope.

If neither is vertical, perpendicular lines have slopes that are negative reciprocals; that is, their product is -1. Also, lines with slopes that are negative reciprocals are perpendicular.

EXAMPLE 9 Determining Whether Two Lines Are Parallel, Perpendicular, or Neither

Determine whether the two lines described are *parallel, perpendicular,* or *neither.*

(a) The lines L_1, through $(-2, 1)$ and $(4, 5)$, and L_2, through $(3, 0)$ and $(0, -2)$

The slope of L_1 is

$$m_1 = \frac{5 - 1}{4 - (-2)} = \frac{4}{6} = \frac{2}{3}.$$

The slope of L_2 is

$$m_2 = \frac{-2 - 0}{0 - 3} = \frac{-2}{-3} = \frac{2}{3}.$$

Because the slopes are equal, the two lines are parallel.

(b) The lines with equations $y = \frac{2}{5}x + 3$ and $y = -\frac{2}{5}x - 4$

The slopes of the lines are $\frac{2}{5}$ and $-\frac{2}{5}$, which are neither equal nor negative reciprocals. Therefore, the lines are neither parallel nor perpendicular.

(c) The lines with equations $2y = 3x - 6$ and $2x + 3y = -6$

Find the slope of each line by first solving each equation for y.

$$2y = 3x - 6$$
$$y = \frac{3}{2}x - 3$$
$$\uparrow$$
$$\text{Slope}$$

$$2x + 3y = -6$$
$$3y = -2x - 6$$
$$y = -\frac{2}{3}x - 2$$
$$\uparrow$$
$$\text{Slope}$$

Since the product of the slopes of the two lines is $\frac{3}{2}\left(-\frac{2}{3}\right) = -1$, the lines are perpendicular.

Now Try Exercises 89, 91, and 93.

OBJECTIVE 7 Solve problems involving average rate of change. We know that the slope of a line is the ratio of the vertical change in y to the horizontal change in x. Thus, slope gives the *average rate of change* in y per unit of change in x, where the value of y depends on the value of x. The next examples illustrate this idea. We assume a linear relationship between x and y.

EXAMPLE 10 Interpreting Slope as Average Rate of Change

The graph in Figure 17 approximates the percent of U.S. households owning multiple personal computers in the years 1997 through 2001. Find the average rate of change in percent per year.

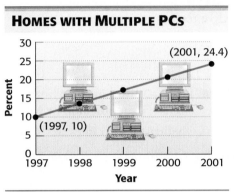

Source: The Yankee Group.

FIGURE 17

To determine the average rate of change, we need two pairs of data. From the graph, if $x = 1997$, then $y = 10$ and if $x = 2001$, then $y = 24.4$, so we have the ordered pairs (1997, 10) and (2001, 24.4). By the slope formula,

$$\text{average rate of change} = \frac{\text{change in } y}{\text{change in } x} = \frac{24.4 - 10}{2001 - 1997} = \frac{14.4}{4} = 3.6.$$

This means that the number of U.S. households owning multiple computers *increased* by 3.6% each year from 1997 to 2001.

Now Try Exercise 105.

EXAMPLE 11 Interpreting Slope as Average Rate of Change

In 1997, sales of VCRs numbered 16.7 million. In 2002, estimated sales of VCRs were 13.3 million. Find the average rate of change, in millions, per year. (*Source: The Gazette,* June 22, 2002.)

To use the slope formula, we need two ordered pairs. Here, if $x = 1997$, then $y = 16.7$ and if $x = 2002$, then $y = 13.3$, which gives the ordered pairs (1997, 16.7) and (2002, 13.3). (Note that y is in millions.)

$$\text{average rate of change} = \frac{13.3 - 16.7}{2002 - 1997} = \frac{-3.4}{5} = -.68$$

The graph in Figure 18 confirms that the line through the ordered pairs falls from left to right and therefore has negative slope. Thus, sales of VCRs *decreased* by .68 million each year from 1997 to 2002.

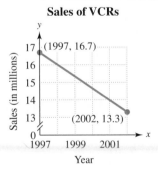

FIGURE 18

Now Try Exercise 107.

7.1 EXERCISES

In Exercises 1 and 2, answer each question by locating ordered pairs on the graphs.

1. The graph shows the percent of women in math or computer science professions.

   (a) If (x, y) represents a point on the graph, what does x represent? What does y represent?

   (b) In what decade (10-yr period) did the percent of women in math or computer science professions decrease?

   (c) Write an ordered pair (x, y) that gives the approximate percent of women in math or computer science professions in 1990.

   (d) What does the ordered pair (2000, 30) mean in the context of this graph?

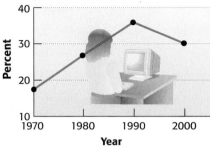

WOMEN IN MATH OR COMPUTER SCIENCE PROFESSIONS

Source: U.S. Bureau of the Census and Bureau of Labor Statistics.

2. The graph indicates federal government tax revenues in billions of dollars.

   (a) If (x, y) represents a point on the graph, what does x represent? What does y represent?

   (b) Estimate revenue in 1996.

   (c) Write an ordered pair (x, y) that gives approximate federal tax revenues in 1995.

   (d) What does the ordered pair (1998, 1720) mean in the context of this graph?

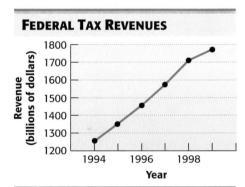

FEDERAL TAX REVENUES

Source: U.S. Office of Management and Budget.

3. Name the quadrant, if any, in which each point is located.

   (a) $(1, 6)$ **(b)** $(-4, -2)$ **(c)** $(-3, 6)$
   (d) $(7, -5)$ **(e)** $(-3, 0)$ **(f)** $(0, -8)$

4. Use the given information to determine the possible quadrants in which the point (x, y) must lie.

   (a) $xy > 0$ **(b)** $xy < 0$ **(c)** $\dfrac{x}{y} < 0$ **(d)** $\dfrac{x}{y} > 0$

Plot each point on a rectangular coordinate system.

   5. $(2, 3)$ **6.** $(1, -4)$ **7.** $(0, 5)$

   8. $(-2, -4)$ **9.** $(-2, 4)$ **10.** $(3, 0)$

*In Exercises 11–13, **(a)** complete the given table for each equation, and then **(b)** graph the equation. See Figure 3.*

11. $x - y = 3$

x	y
0	
	0
5	
2	

12. $x + 3y = -5$

x	y
0	
	0
1	
	-1

13. $4x - 5y = 20$

x	y
0	
	0
2	
	-3

✐ 14. Explain how to determine the intercepts and graph of the linear equation $4x - 3y = 12$.

✐ 15. Explain why the graph of $x + y = k$ cannot pass through quadrant III if $k > 0$.

✐ 16. A student attempted to graph $4x + 5y = 0$ by finding intercepts. She first let $x = 0$ and found y; then she let $y = 0$ and found x. In both cases, the resulting point was $(0, 0)$. She knew that she needed at least two points to graph the line, but was unsure what to do next because finding intercepts gave her only one point. Explain to her what to do next.

Find the x- and y-intercepts. Then graph each equation. See Examples 1–4.

17. $2x + 3y = 12$

18. $5x + 2y = 10$

19. $x - 3y = 6$

20. $x - 2y = -4$

21. $\frac{2}{3}x - 3y = 7$

22. $\frac{5}{7}x + \frac{6}{7}y = -2$

23. $x + 5y = 0$

24. $x - 3y = 0$

25. $2x = 3y$

26. $4y = 3x$

27. $y = 5$

28. $x = -3$

29. $x + 4 = 0$

30. $y + 2 = 0$

TECHNOLOGY INSIGHTS (EXERCISES 31–33)

31. The screens show the graph of one of the equations in A–D. Which equation is it?

    A. $3x + 2y = 6$ **B.** $-3x + 2y = 6$ **C.** $-3x - 2y = 6$ **D.** $3x - 2y = 6$

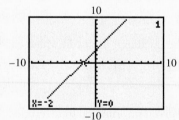
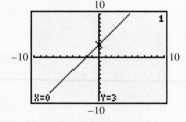

32. The table of ordered pairs was generated by a graphing calculator with a TABLE feature.

  (a) What is the x-intercept?

  (b) What is the y-intercept?

  (c) Which equation corresponds to this table of values?

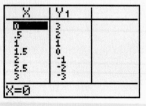

      A. $Y_1 = 2X - 3$ **B.** $Y_1 = -2X - 3$

      C. $Y_1 = 2X + 3$ **D.** $Y_1 = -2X + 3$

33. The screens each show the graph of $x + y = 15$ (which was entered as $y = -x + 15$). However, different viewing windows are used. Which window would be more useful for this graph? Why?

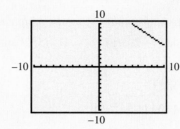 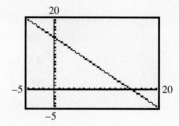

RELATING CONCEPTS (EXERCISES 34–39)

For Individual or Group Work

If the endpoints of a line segment are known, then the coordinates of the midpoint of the segment can be found. The figure shows the coordinates of the points P and Q. Let $\overline{PQ}$ represent the line segment with endpoints at P and Q. To derive a formula for the midpoint of $\overline{PQ}$, **work Exercises 34–39 in order.**

34. In the figure, R is the point with the same x-coordinate as Q and the same y-coordinate as P. Write the ordered pair that corresponds to R.

35. From the graph, determine the coordinates of the midpoint of $\overline{PR}$.

36. From the graph, determine the coordinates of the midpoint of $\overline{QR}$.

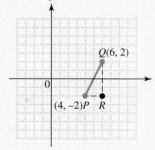

37. The x-coordinate of the midpoint M of $\overline{PQ}$ is the x-coordinate of the midpoint of $\overline{PR}$ and the y-coordinate is the y-coordinate of the midpoint of $\overline{QR}$. Write the ordered pair that corresponds to M.

38. The average of two numbers is found by dividing their sum by 2. Find the average of the x-coordinates of points P and Q. Find the average of the y-coordinates of points P and Q.

39. Comparing your answers to Exercises 37 and 38, what connection is there between the coordinates of P and Q and the coordinates of M?

The result of the preceding Relating Concepts exercises leads to the **midpoint formula.**

Midpoint Formula

If the endpoints of a line segment PQ are (x_1, y_1) and (x_2, y_2), then its midpoint M is

$$\left(\frac{x_1 + x_2}{2}, \frac{y_1 + y_2}{2} \right).$$

For example, the midpoint of the segment with endpoints $(4, -3)$ *and* $(6, -1)$ *is*

$$\left(\frac{4 + 6}{2}, \frac{-3 + (-1)}{2}\right) = \left(\frac{10}{2}, \frac{-4}{2}\right) = (5, -2).$$

Use the midpoint formula to find the midpoint of each segment with the given endpoints.

40. $(-8, 4)$ and $(-2, -6)$ **41.** $(5, 2)$ and $(-1, 8)$ **42.** $(3, -6)$ and $(6, 3)$

43. $(-10, 4)$ and $(7, 1)$ **44.** $(-9, 3)$ and $(9, 8)$

45. $(4, -3)$ and $(-1, 3)$ **46.** $(2.5, 3.1)$ and $(1.7, -1.3)$

47. $(6.2, 5.8)$ and $(1.4, -.6)$ **48.** $(-.4, -.9)$ and $(-.6, -.1)$

Use the concept of slope to solve each problem.

49. A ski slope drops 30 ft for every horizontal 100 ft. Which of the following express its slope? (There are several correct choices.)

A. $-.3$ **B.** $-\frac{3}{10}$ **C.** $-3\frac{1}{3}$

D. $-\frac{30}{100}$ **E.** $-\frac{10}{3}$

50. A hill has slope $-.05$. How many feet in the vertical direction correspond to a run of 50 ft?

51. Match each situation in (a)–(c) with the most appropriate graph in A–C.

  (a) Sales rose sharply during the first quarter, leveled off during the second quarter, and then rose slowly for the rest of the year.

  (b) Sales rose sharply during the first quarter, and then fell to the original level during the second quarter before rising steadily for the rest of the year.

  (c) Sales fell during the first two quarters of the year, leveled off during the third quarter, and rose during the fourth quarter.

    A.   **B.**   **C.**

Determine the slope of each line segment in the given figure.

52. *AB*   **53.** *BC*   **54.** *CD*

55. *DE*   **56.** *EF*

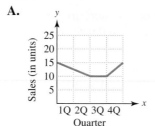

57. Which of the following forms of the slope formula are correct? Explain.

A. $\dfrac{y_1 - y_2}{x_2 - x_1}$ **B.** $\dfrac{y_1 - y_2}{x_1 - x_2}$ **C.** $\dfrac{x_2 - x_1}{y_2 - y_1}$ **D.** $\dfrac{y_2 - y_1}{x_2 - x_1}$

Find the slope of the line through each pair of points. See Example 5.

58. $(-2, -3)$ and $(-1, 5)$ **59.** $(-4, 3)$ and $(-3, 4)$ **60.** $(-4, 1)$ and $(2, 6)$

61. $(-3, -3)$ and $(5, 6)$ **62.** $(2, 4)$ and $(-4, 4)$ **63.** $(-6, 3)$ and $(2, 3)$

Find the slope of each line.

64. 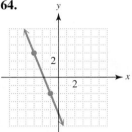 **65.** 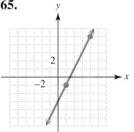 **66.**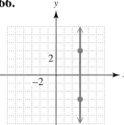

Based on the figure shown here, determine which line satisfies the given description.

67. The line has positive slope.

68. The line has negative slope.

69. The line has slope 0.

70. The line has undefined slope.

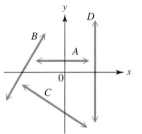

Find the slope of the line and sketch the graph. See Examples 5–7.

71. $x + 2y = 4$ **72.** $x + 3y = -6$ **73.** $5x - 2y = 10$

74. $4x - y = 4$ **75.** $y = 4x$ **76.** $y = -3x$

77. $x - 3 = 0$ **78.** $y + 5 = 0$ **79.** $y = -4$

Graph the line described. See Example 8.

80. Through $(-4, 2)$; $m = \dfrac{1}{2}$ **81.** Through $(-2, -3)$; $m = \dfrac{5}{4}$

82. Through $(0, -2)$; $m = -\dfrac{2}{3}$ **83.** Through $(0, -4)$; $m = -\dfrac{3}{2}$

84. Through $(-1, -2)$; $m = 3$ **85.** Through $(-2, -4)$; $m = 4$

86. $m = 0$; through $(2, -5)$ **87.** Undefined slope; through $(-3, 1)$

88. If a line has slope $-\frac{4}{9}$, then any line parallel to it has slope _____, and any line perpendicular to it has slope _____.

Decide whether each pair of lines is parallel, perpendicular, *or* neither. *See Example 9.*

89. The line through $(4, 6)$ and $(-8, 7)$ and the line through $(-5, 5)$ and $(7, 4)$

90. The line through $(15, 9)$ and $(12, -7)$ and the line through $(8, -4)$ and $(5, -20)$

91. $2x + 5y = -7$ and $5x - 2y = 1$

92. $x + 4y = 7$ and $4x - y = 3$

93. $2x + y = 6$ and $x - y = 4$

94. $4x - 3y = 6$ and $3x - 4y = 2$

95. $2x + 5y = -8$ and $6 + 2x = 5y$

96. $4x + y = 0$ and $5x - 8 = 2y$

97. $4x - 3y = 8$ and $4y + 3x = 12$

98. $2x = y + 3$ and $2y + x = 3$

Find and interpret the average rate of change illustrated in each graph.

99.

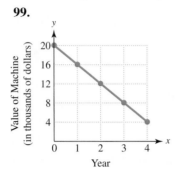

100.

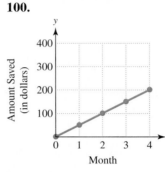

101.

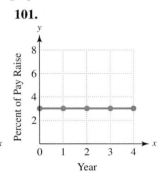

102. If the graph of a linear equation rises from left to right, then the average rate of change is _____. If the graph of a linear equation falls from left to right, then
 (positive/negative)
the average rate of change is _____.
 (positive/negative)

Solve each problem. See Examples 10 and 11.

103. The table gives book publishers' approximate net dollar sales (in millions) from 1995 through 2000.

    (a) Find the average rate of change for 1995–1996, 1995–1999, and 1998–2000.

    (b) What do you notice about your answers in part (a)? What does this tell you?

Book Publishers' Sales

Year	Sales (in millions)
1995	19,000
1996	20,000
1997	21,000
1998	22,000
1999	23,000
2000	24,000

Source: Book Industry Study Group.

104. The table gives the number of cellular telephone subscribers (in thousands) from 1994 through 1999.

    (a) Find the average rate of change in subscribers for 1994–1995, 1995–1996, and so on.

    (b) Is the average rate of change in successive years approximately the same? If the ordered pairs in the table were plotted, could an approximately straight line be drawn through them?

Cellular Telephone Subscribers

Year	Subscribers (in thousands)
1994	24,134
1995	33,786
1996	44,043
1997	55,312
1998	69,209
1999	86,047

Source: Cellular Telecommunications Industry Association, Washington, D.C., *State of the Cellular Industry* (Annual).

105. Merck pharmaceutical company research and development expenditures (in millions of dollars) in recent years are closely approximated by the graph.

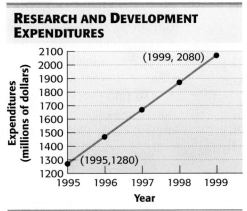

RESEARCH AND DEVELOPMENT EXPENDITURES

Source: Merck & Co., Inc. 1999 Annual Report.

(a) Use the given ordered pairs to determine the average rate of change in these expenditures per year.

(b) Explain how a positive rate of change is interpreted in this situation.

106. The graph provides a good approximation of the number of food stamp recipients (in millions) from 1994 through 1998.

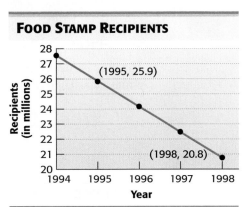

FOOD STAMP RECIPIENTS

Source: U.S. Bureau of the Census.

(a) Use the given ordered pairs to find the average rate of change in food stamp recipients per year during this period.

(b) Interpret what a negative slope means in this situation.

107. When introduced in 1997, a DVD player sold for about $500. In 2002, the average price was $155. Find and interpret the average rate of change in price per year. (*Source: The Gazette,* June 22, 2002.)

108. In 1997 when DVD players entered the market, .349 million (that is, 349,000) were sold. In 2002, sales of DVD players reached 15.5 million (estimated). Find and interpret the average rate of change in sales, in millions, per year. Round your answer to the nearest hundredth. (*Source: The Gazette,* June 22, 2002.)

7.2 Review of Equations of Lines; Linear Models

OBJECTIVE 1 Write an equation of a line given its slope and y-intercept. Recall that we can find the slope of a line from the equation of the line by solving the equation for y. For example, the slope of the line with equation $y = 4x + 8$ is 4, the coefficient of x. What does the number 8 represent?

To find out, suppose a line has slope m and y-intercept $(0, b)$. We can find an equation of this line by choosing another point (x, y) on the line, as shown in Figure 19. Using the slope formula,

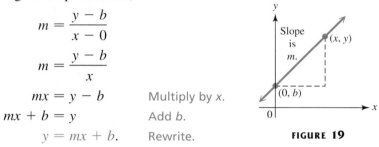

$$m = \frac{y - b}{x - 0}$$

$$m = \frac{y - b}{x}$$

$$mx = y - b \qquad \text{Multiply by } x.$$

$$mx + b = y \qquad \text{Add } b.$$

$$y = mx + b. \qquad \text{Rewrite.}$$

FIGURE 19

This last equation is the *slope-intercept form* of the equation of a line, because we can identify the slope and y-intercept at a glance. Thus, in the line with equation $y = 4x + 8$, the number 8 indicates that the y-intercept is $(0, 8)$.

Slope-Intercept Form

The **slope-intercept form** of the equation of a line with slope m and y-intercept $(0, b)$ is

$$y = mx + b.$$

Slope y-intercept is $(0, b)$.

■ EXAMPLE 1 Using the Slope-Intercept Form to Find an Equation of a Line

Find an equation of the line with slope $-\frac{4}{5}$ and y-intercept $(0, -2)$.

Here $m = -\frac{4}{5}$ and $b = -2$. Substitute these values into the slope-intercept form.

$$y = mx + b \qquad \text{Slope-intercept form}$$

$$y = -\frac{4}{5}x - 2 \qquad m = -\frac{4}{5}; b = -2$$

Now Try Exercise 19.

OBJECTIVE 2 Graph a line using its slope and y-intercept. If the equation of a line is written in slope-intercept form, we can use the slope and y-intercept to obtain its graph.

■ **EXAMPLE 2** Graphing Lines Using Slope and *y*-Intercept

Graph each line using the slope and *y*-intercept.

(a) $y = 3x - 6$

Here $m = 3$ and $b = -6$. Plot the *y*-intercept $(0, -6)$. The slope 3 can be interpreted as

$$m = \frac{\text{rise}}{\text{run}} = \frac{\text{change in } y}{\text{change in } x} = \frac{3}{1}.$$

From $(0, -6)$, move *up* 3 units and to the *right* 1 unit, and plot a second point at $(1, -3)$. Join the two points with a straight line to obtain the graph in Figure 20.

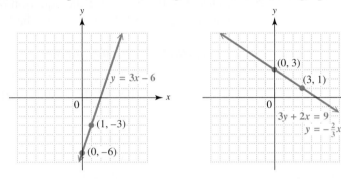

FIGURE 20   **FIGURE 21**

(b) $3y + 2x = 9$

Write the equation in slope-intercept form by solving for *y*.

$$3y + 2x = 9$$
$$3y = -2x + 9 \qquad \text{Subtract } 2x.$$
$$y = -\frac{2}{3}x + 3 \qquad \text{Slope-intercept form}$$

Slope ⟶   ⟵ *y*-intercept is $(0, 3)$.

To graph this equation, plot the *y*-intercept $(0, 3)$. The slope can be interpreted as either $\frac{-2}{3}$ or $\frac{2}{-3}$. Using $\frac{-2}{3}$, move from $(0, 3)$ *down* 2 units and to the *right* 3 units to locate the point $(3, 1)$. The line through these two points is the required graph. See Figure 21. $\left(\text{Verify that the point obtained using } \frac{2}{-3} \text{ as the slope is also on this line.}\right)$

■

Now Try Exercise 25.

OBJECTIVE 3 Write an equation of a line given its slope and a point on the line. Let *m* represent the slope of a line and (x_1, y_1) represent a given point on the line. Let (x, y) represent any other point on the line. See Figure 22. Then by the slope formula,

$$m = \frac{y - y_1}{x - x_1}$$
$$m(x - x_1) = y - y_1 \qquad \text{Multiply each side by } x - x_1.$$
$$y - y_1 = m(x - x_1). \qquad \text{Rewrite.}$$

This last equation is the *point-slope form* of the equation of a line.

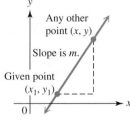

FIGURE 22

Point-Slope Form

The **point-slope form** of the equation of a line with slope m passing through the point (x_1, y_1) is

Slope
↓
$$y - y_1 = m(x - x_1).$$
↑ Given point ↑

To use this form to write the equation of a line, we need to know the coordinates of a point (x_1, y_1) and the slope m of the line.

EXAMPLE 3   Using the Point-Slope Form

Find an equation of the line with slope $\frac{1}{3}$ passing through the point $(-2, 5)$.

Use the point-slope form of the equation of a line, with $(x_1, y_1) = (-2, 5)$ and $m = \frac{1}{3}$.

$$y - y_1 = m(x - x_1) \qquad \text{Point-slope form}$$

$$y - 5 = \frac{1}{3}[x - (-2)] \qquad y_1 = 5,\ m = \tfrac{1}{3},\ x_1 = -2$$

$$y - 5 = \frac{1}{3}(x + 2)$$

$$3y - 15 = x + 2 \qquad \text{Multiply by 3.}$$
$$-x + 3y = 17 \qquad \text{Subtract } x;\ \text{add 15.}$$

Recall that a linear equation is in **standard form** if it is written as

$$Ax + By = C,$$

where A, B, and C are integers, with $A > 0$, $B \neq 0$. Let us also agree that integers A, B, and C have no common factor (except 1). For example, the final equation in Example 3, $-x + 3y = 17$, is written in standard form as $x - 3y = -17$.

Now Try Exercise 31.

OBJECTIVE 4 Write an equation of a line given two points on the line. To find an equation of a line when two points on the line are known, first use the slope formula to find the slope of the line. Then use the slope with either of the given points and the point-slope form of the equation of a line.

EXAMPLE 4   Finding an Equation of a Line Given Two Points

Find an equation of the line passing through the points $(-4, 3)$ and $(5, -7)$. Write the equation in standard form.

First find the slope by using the slope formula.

$$m = \frac{-7 - 3}{5 - (-4)} = -\frac{10}{9}$$

Use either $(-4, 3)$ or $(5, -7)$ as (x_1, y_1) in the point-slope form of the equation of a line. If you choose $(-4, 3)$, then $-4 = x_1$ and $3 = y_1$.

$$y - y_1 = m(x - x_1) \qquad \text{Point-slope form}$$

$$y - 3 = -\frac{10}{9}[x - (-4)] \qquad y_1 = 3,\ m = -\tfrac{10}{9},\ x_1 = -4$$

$$y - 3 = -\frac{10}{9}(x + 4)$$

$$9y - 27 = -10x - 40 \qquad \text{Multiply by 9; distributive property.}$$

$$10x + 9y = -13 \qquad \text{Standard form}$$

Verify that if $(5, -7)$ were used, the same equation would result.

> **Now Try Exercise 49.**

A horizontal line has slope 0. Using point-slope form, the equation of a horizontal line through the point (a, b) is

$$y - y_1 = m(x - x_1)$$

$$y - b = 0(x - a) \qquad y_1 = b,\ m = 0,\ x_1 = a$$

$$y - b = 0$$

$$y = b.$$

Notice that point-slope form does not apply to a vertical line, since the slope of a vertical line is undefined. A vertical line through the point (a, b) has equation $x = a$.

In summary, horizontal and vertical lines have the following special equations.

Equations of Horizontal and Vertical Lines

The horizontal line through the point (a, b) has equation $y = b.$
The vertical line through the point (a, b) has equation $x = a.$

> **Now Try Exercises 41 and 43.**

OBJECTIVE 5 Write an equation of a line parallel or perpendicular to a given line. Recall that parallel lines have the same slope and perpendicular lines have slopes that are negative reciprocals of each other.

EXAMPLE 5 Finding Equations of Parallel or Perpendicular Lines

Find an equation of the line passing through the point $(-4, 5)$ and **(a)** parallel to the line $2x + 3y = 6$; **(b)** perpendicular to the line $2x + 3y = 6$. Write each equation in slope-intercept form.

(a) We find the slope of the line $2x + 3y = 6$ by solving for y.

$$2x + 3y = 6$$

$$3y = -2x + 6 \qquad \text{Subtract } 2x.$$

$$y = -\frac{2}{3}x + 2 \qquad \text{Divide by 3.}$$

$$\underset{\text{Slope}}{\uparrow}$$

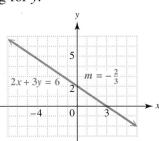

The slope is given by the coefficient of x, so $m = -\frac{2}{3}$.
See the figure.

The required equation of the line through $(-4, 5)$ and parallel to $2x + 3y = 6$ must also have slope $-\frac{2}{3}$. To find this equation, we use the point-slope form, with $(x_1, y_1) = (-4, 5)$ and $m = -\frac{2}{3}$.

$$y - 5 = -\frac{2}{3}[x - (-4)] \qquad y_1 = 5,\ m = -\frac{2}{3},\ x_1 = -4$$

$$y - 5 = -\frac{2}{3}(x + 4)$$

$$y - 5 = -\frac{2}{3}x - \frac{8}{3} \qquad \text{Distributive property}$$

$$y = -\frac{2}{3}x - \frac{8}{3} + \frac{15}{3} \qquad \text{Add } 5 = \frac{15}{3}.$$

$$y = -\frac{2}{3}x + \frac{7}{3} \qquad \text{Combine like terms.}$$

We did not clear fractions after the substitution step here because we want the equation in slope-intercept form—that is, solved for y. Both lines are shown in the figure.

(b) To be perpendicular to the line $2x + 3y = 6$, a line must have a slope that is the negative reciprocal of $-\frac{2}{3}$, which is $\frac{3}{2}$. We use $(-4, 5)$ and slope $\frac{3}{2}$ in the point-slope form to get the equation of the perpendicular line shown in the figure.

$$y - 5 = \frac{3}{2}[x - (-4)] \qquad y_1 = 5,\ m = \frac{3}{2},\ x_1 = -4$$

$$y - 5 = \frac{3}{2}(x + 4)$$

$$y - 5 = \frac{3}{2}x + 6 \qquad \text{Distributive property}$$

$$y = \frac{3}{2}x + 11 \qquad \text{Add 5.}$$

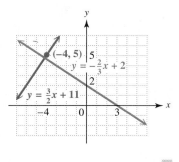

Now Try Exercises 61 and 65.

A summary of the various forms of linear equations follows.

Forms of Linear Equations

Equation	Description	When to Use
$y = mx + b$	**Slope-Intercept Form** Slope is m. y-intercept is $(0, b)$.	The slope and y-intercept can be easily identified and used to quickly graph the equation.
$y - y_1 = m(x - x_1)$	**Point-Slope Form** Slope is m. Line passes through (x_1, y_1).	This form is ideal for finding the equation of a line if the slope and a point on the line or two points on the line are known.

Equation	Description	When to Use
Ax + By = C	**Standard Form** (A, B, and C integers, $A > 0$) Slope is $-\frac{A}{B}$ ($B \neq 0$). x-intercept is $\left(\frac{C}{A}, 0\right)$ ($A \neq 0$). y-intercept is $\left(0, \frac{C}{B}\right)$ ($B \neq 0$).	The x- and y-intercepts can be found quickly and used to graph the equation. Slope must be calculated.
y = b	**Horizontal Line** Slope is 0. y-intercept is $(0, b)$.	If the graph intersects only the y-axis, then y is the only variable in the equation.
x = a	**Vertical Line** Slope is undefined. x-intercept is $(a, 0)$.	If the graph intersects only the x-axis, then x is the only variable in the equation.

OBJECTIVE 6 Write an equation of a line that models real data. We can use the information presented in this section to write equations of lines that mathematically describe, or *model,* real data if the given set of data changes at a fairly constant rate. In this case, the data fit a linear pattern, and the rate of change is the slope of the line.

■ **EXAMPLE 6** Determining a Linear Equation to Describe Real Data

Suppose it is time to fill your car with gasoline. At your local station, 89-octane gas is selling for $1.60 per gal.

(a) Write an equation that describes the cost y to buy x gal of gas.

Experience has taught you that the total price you pay is determined by the number of gallons you buy multiplied by the price per gallon (in this case, $1.60). As you pump the gas, two sets of numbers spin by: the number of gallons pumped and the price for that number of gallons.

The table uses ordered pairs to illustrate this situation.

Number of Gallons Pumped	Price of This Number of Gallons
0	0($1.60) = $0.00
1	1($1.60) = $1.60
2	2($1.60) = $3.20
3	3($1.60) = $4.80
4	4($1.60) = $6.40

If we let x denote the number of gallons pumped, then the total price y in dollars can be found by the linear equation

Total price ——┐   ┌—— Number of gallons

$$y = 1.60x.$$

Theoretically, there are infinitely many ordered pairs (x, y) that satisfy this equation, but here we are limited to nonnegative values for x, since we cannot have a negative number of gallons. There is also a practical maximum value for x in this situation, which varies from one car to another. What determines this maximum value?

(b) You can also get a car wash at the gas station if you pay an additional $3.00. Write an equation that defines the price for gas and a car wash.

Since an additional $3.00 will be charged, you pay $1.60x + 3.00$ dollars for x gallons of gas and a car wash, or

$$y = 1.6x + 3. \qquad \text{Delete unnecessary 0s.}$$

(c) Interpret the ordered pairs $(5, 11)$ and $(10, 19)$ in relation to the equation from part (b).

The ordered pair $(5, 11)$ indicates that the price of 5 gal of gas and a car wash is $11.00. Similarly, $(10, 19)$ indicates that the price of 10 gal of gas and a car wash is $19.00.

Now Try Exercises 69 and 73.

NOTE In Example 6(a), the ordered pair $(0, 0)$ satisfied the equation, so the linear equation has the form $y = mx$, where $b = 0$. If a realistic situation involves an initial charge plus a charge per unit as in Example 6(b), the equation has the form $y = mx + b$, where $b \neq 0$.

EXAMPLE 7 **Finding an Equation of a Line That Models Data**

Average annual tuition and fees for in-state students at public 4-year colleges are shown in the table for selected years and graphed as ordered pairs of points in the *scatter diagram* in Figure 23, where $x = 0$ represents 1990, $x = 4$ represents 1994, and so on, and y represents the cost in dollars.

Year	Cost (in dollars)
1990	2035
1994	2820
1996	3151
1998	3486
2000	3774

Source: U.S. National Center for Education Statistics.

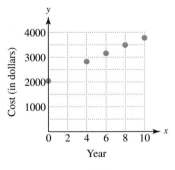

FIGURE 23

(a) Find an equation that models the data.

Since the points in Figure 23 lie approximately on a straight line, we can write a linear equation that models the relationship between year x and cost y. We choose two data points, $(0, 2035)$ and $(10, 3774)$, to find the slope of the line.

$$m = \frac{3774 - 2035}{10 - 0} = \frac{1739}{10} = 173.9$$

The slope 173.9 indicates that the cost of tuition and fees for in-state students at public 4-year colleges increased by about $174 per year from 1990 to 2000. We use this slope, the y-intercept $(0, 2035)$, and the slope-intercept form to write an equation of the line. Thus,

$$y = 173.9x + 2035.$$

(b) Use the equation from part (a) to approximate the cost of tuition and fees at public 4-year colleges in 2002.

The value $x = 12$ corresponds to the year 2002, so we substitute 12 for x in the equation.

$$y = 173.9x + 2035$$
$$y = 173.9(12) + 2035$$
$$y = 4121.8$$

According to the model, average tuition and fees for in-state students at public 4-year colleges in 2002 were about $4122.

Now Try Exercise 79.

NOTE In Example 7, if we had chosen different data points, we would have gotten a slightly different equation. However, all such equations should be similar.

EXAMPLE 8 Finding an Equation of a Line That Models Data

Retail spending (in billions of dollars) on prescription drugs in the United States is shown in the graph in Figure 24.

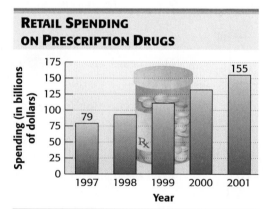

RETAIL SPENDING ON PRESCRIPTION DRUGS

Source: American Institute for Research analysis of Scott-Levin data.

FIGURE 24

(a) Write an equation that models the data.

The data shown in the bar graph increase linearly; that is, we could draw a straight line through the tops of any two bars that would be close to the top of each bar. We can use the data and the point-slope form of the equation of a line to get an equation that models the relationship between year x and spending on prescription drugs y. If we let $x = 7$ represent 1997, $x = 8$ represent 1998, and so on, the given data for 1997 and 2001 can be written as the ordered pairs (7, 79) and (11, 155). The slope of the line through these two points is

$$m = \frac{155 - 79}{11 - 7} = \frac{76}{4} = 19.$$

Thus, retail spending on prescription drugs increased by about $19 billion per year. Using this slope, one of the points, say $(7, 79)$, and the point-slope form, we obtain

$$y - y_1 = m(x - x_1) \qquad \text{Point-slope form}$$
$$y - 79 = 19(x - 7) \qquad (x_1, y_1) = (7, 79); \ m = 19$$
$$y - 79 = 19x - 133 \qquad \text{Distributive property}$$
$$y = 19x - 54. \qquad \text{Slope-intercept form}$$

Thus, retail spending y (in billions of dollars) on prescription drugs in the United States in year x can be approximated by the equation $y = 19x - 54$.

(b) Use the equation from part (a) to predict retail spending on prescription drugs in the United States in 2004. (Assume a constant rate of change.)

Since $x = 7$ represents 1997 and 2004 is 7 yr after 1997, $x = 14$ represents 2004. We substitute 14 for x in the equation.

$$y = 19x - 54 = 19(14) - 54 = 212$$

According to the model, $212 billion will be spent on prescription drugs in 2004.

Now Try Exercise 81.

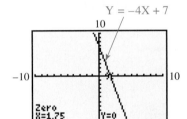

$Y = -4X + 7$

$Y = -2X - 4(2 - X) - 3X - 4$

CONNECTIONS

In the Connections box in Section 7.1, we graphed linear equations in two variables and located their intercepts using a graphing calculator. The top screen in the margin shows the graph of

$$y = -4x + 7.$$

From the values at the bottom of the screen, we see that when $x = 1.75$, $y = 0$. This means that $x = 1.75$ satisfies the equation

$$-4x + 7 = 0,$$

a linear equation in *one* variable. Therefore, the solution set of $-4x + 7 = 0$ is $\{1.75\}$. We can verify this algebraically by substitution. (Recall that the word "Zero" indicates that the x-intercept has been located.)

To solve $-2x - 4(2 - x) = 3x + 4$ using a graphing calculator, we must write the equation as an equivalent equation with 0 on one side.

$$-2x - 4(2 - x) - 3x - 4 = 0 \qquad \text{Subtract } 3x \text{ and } 4.$$

Then we graph

$$y = -2x - 4(2 - x) - 3x - 4$$

to find the x-intercept. The standard viewing window cannot be used because the x-intercept does not lie in the interval $[-10, 10]$. As seen in the bottom screen in the margin, the x-intercept of the graph is $(-12, 0)$, and thus the solution (or zero) of the equation is -12. The solution set is $\{-12\}$. Check this algebraically by substitution.

For Discussion or Writing

Use a graphing calculator to solve each linear equation in one variable. Check solutions algebraically by substitution.

1. $2x + 7 - x = 4x - 2$ **2.** $7x - 2x + 4 - 5 = 3x + 1$
3. $4(x - 3) - x = x - 6$ **4.** $3(2x + 1) - 2(x - 2) = 5$

7.2 EXERCISES

1. The following equations all represent the same line. Which one is in standard form as defined in the text?

   A. $3x - 2y = 5$ **B.** $2y = 3x - 5$ **C.** $\dfrac{3}{5}x - \dfrac{2}{5}y = 1$ **D.** $3x = 2y + 5$

2. Which equation is in point-slope form?

   A. $y = 6x + 2$ **B.** $4x + y = 9$ **C.** $y - 3 = 2(x - 1)$ **D.** $2y = 3x - 7$

3. Which equation in Exercise 2 is in slope-intercept form?

4. Write the equation $y + 2 = -3(x - 4)$ in slope-intercept form.

5. Write the equation from Exercise 4 in standard form.

6. Write the equation $10x - 7y = 70$ in slope-intercept form.

Match each equation with the graph that it most closely resembles. (Hint: Determine the signs of m and b to help you make your decision.)

7. $y = 2x + 3$

8. $y = -2x + 3$

9. $y = -2x - 3$

10. $y = 2x - 3$

11. $y = 2x$

12. $y = -2x$

13. $y = 3$

14. $y = -3$

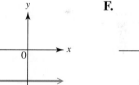

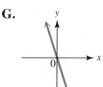

Find the equation in slope-intercept form of the line satisfying the given conditions. See Example 1.

15. $m = 5;\ b = 15$ **16.** $m = -2;\ b = 12$

17. $m = -\dfrac{2}{3};\ b = \dfrac{4}{5}$ **18.** $m = -\dfrac{5}{8};\ b = -\dfrac{1}{3}$

19. Slope $\dfrac{2}{5}$; *y*-intercept $(0, 5)$ **20.** Slope $-\dfrac{3}{4}$; *y*-intercept $(0, 7)$

Write an equation in slope-intercept form of the line shown in each graph. (Hint: Use the indicated points to find the slope.)

21.

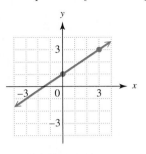

22.

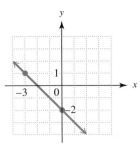

For each equation, (a) write it in slope-intercept form, (b) give the slope of the line, (c) give the y-intercept, and (d) graph the line. See Example 2.

23. $-x + y = 4$ **24.** $-x + y = 6$ **25.** $6x + 5y = 30$

26. $3x + 4y = 12$ **27.** $4x - 5y = 20$ **28.** $7x - 3y = 3$

29. $x + 2y = -4$ **30.** $x + 3y = -9$

Find an equation of the line that satisfies the given conditions. Write the equation in standard form. See Example 3.

31. Through $(-2, 4)$; slope $-\dfrac{3}{4}$ **32.** Through $(-1, 6)$; slope $-\dfrac{5}{6}$

33. Through $(5, 8)$; slope -2 **34.** Through $(12, 10)$; slope 1

35. Through $(-5, 4)$; slope $\dfrac{1}{2}$ **36.** Through $(7, -2)$; slope $\dfrac{1}{4}$

37. x-intercept $(3, 0)$; slope 4 **38.** x-intercept $(-2, 0)$; slope -5

39. In your own words, list all the forms of linear equations in two variables and describe when each form should be used.

40. Explain why the point-slope form of an equation cannot be used to find the equation of a vertical line.

Write an equation of the line that satisfies the given conditions.

41. Through $(9, 5)$; slope 0 **42.** Through $(-4, -2)$; slope 0

43. Through $(9, 10)$; undefined slope **44.** Through $(-2, 8)$; undefined slope

45. Through $(.5, .2)$; vertical **46.** Through $\left(\dfrac{5}{8}, \dfrac{2}{9}\right)$; vertical

47. Through $(-7, 8)$; horizontal **48.** Through $(2, 7)$; horizontal

Find an equation of the line passing through the given points. Write the equation in standard form. See Example 4.

49. $(3, 4)$ and $(5, 8)$ **50.** $(5, -2)$ and $(-3, 14)$

51. $(6, 1)$ and $(-2, 5)$ **52.** $(-2, 5)$ and $(-8, 1)$

53. $\left(-\dfrac{2}{5}, \dfrac{2}{5}\right)$ and $\left(\dfrac{4}{3}, \dfrac{2}{3}\right)$ **54.** $\left(\dfrac{3}{4}, \dfrac{8}{3}\right)$ and $\left(\dfrac{2}{5}, \dfrac{2}{3}\right)$

55. $(2, 5)$ and $(1, 5)$ **56.** $(-2, 2)$ and $(4, 2)$

57. $(7, 6)$ and $(7, -8)$

58. $(13, 5)$ and $(13, -1)$

59. $(1, -3)$ and $(-1, -3)$

60. $(-4, -6)$ and $(5, -6)$

Find an equation of the line satisfying the given conditions. Write the equation in slope-intercept form. See Example 5.

61. Through $(7, 2)$; parallel to $3x - y = 8$

62. Through $(4, 1)$; parallel to $2x + 5y = 10$

63. Through $(-2, -2)$; parallel to $-x + 2y = 10$

64. Through $(-1, 3)$; parallel to $-x + 3y = 12$

65. Through $(8, 5)$; perpendicular to $2x - y = 7$

66. Through $(2, -7)$; perpendicular to $5x + 2y = 18$

67. Through $(-2, 7)$; perpendicular to $x = 9$

68. Through $(8, 4)$; perpendicular to $x = -3$

Write an equation in the form $y = mx$ for each situation. Then give the three ordered pairs associated with the equation for x-values 0, 5, and 10. See Example 6(a).

69. x represents the number of hours traveling at 45 mph, and y represents the distance traveled (in miles).

70. x represents the number of compact discs sold at \$16 each, and y represents the total cost of the discs (in dollars).

71. x represents the number of gallons of gas sold at \$1.50 per gal, and y represents the total cost of the gasoline (in dollars).

72. x represents the number of days a videocassette is rented at \$3.50 per day, and y represents the total charge for the rental (in dollars).

For each situation, (a) write an equation in the form $y = mx + b$; (b) find and interpret the ordered pair associated with the equation for $x = 5$; and (c) answer the question. See Examples 6(b) and 6(c).

73. A membership to the Midwest Athletic Club costs \$99 plus \$39 per month. (*Source:* Midwest Athletic Club.) Let x represent the number of months selected. How much does the first year's membership cost?

74. For a family membership, the athletic club in Exercise 73 charges a membership fee of \$159 plus \$60 for each additional family member after the first. Let x represent the number of additional family members. What is the membership fee for a four-person family?

75. A cell phone plan includes 900 anytime minutes for \$50 per month, plus a one-time activation fee of \$25. A Nokia 5165 cell phone is included at no additional charge. (*Source:* U.S. Cellular.) Let x represent the number of months of service. If you sign a 2-yr contract, how much will this cell phone plan cost? (Assume that you never use more than the allotted number of minutes.)

76. Another cell phone plan includes 450 anytime minutes for $35 per month, plus $19.95 for a Nokia 5165 cell phone and $25 for a one-time activation fee. (*Source:* U.S. Cellular.) Let x represent the number of months of service. If you sign a 1-yr contract, how much will this cell phone package cost? (Assume that you never use more than the allotted number of minutes.)

77. A rental car costs $50 plus $.20 per mile. Let x represent the number of miles driven, and y represent the total charge to the renter. How many miles was the car driven if the renter paid $84.60?

78. There is a $30 fee to rent a chain saw, plus $6 per day. Let x represent the number of days the saw is rented and y represent the charge to the user in dollars. If the total charge is $138, for how many days is the saw rented?

Solve each problem. In part (a), give equations in slope-intercept form. See Examples 7 and 8. (Source for Exercises 79 and 80: Jupiter Media Metrix.)

79. The percent of households that access the Internet by high-speed broadband is shown in the graph, where the year 2000 corresponds to $x = 0$.

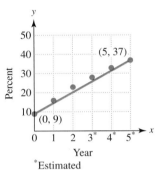

    (a) Use the ordered pairs from the graph to write an equation that models the data. What does the slope tell us in the context of this problem?

    (b) Use the equation from part (a) to predict the percent of U.S. households that will access the Internet by broadband in 2006. Round your answer to the nearest percent.

80. The percent of U.S. households that access the Internet by dial-up is shown in the graph, where the year 2000 corresponds to $x = 0$.

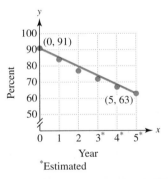

    (a) Use the ordered pairs from the graph to write an equation that models the data. What does the slope tell us in the context of this problem?

    (b) Use the equation from part (a) to predict the percent of U.S. households that will access the Internet by dial-up in 2006. Round your answer to the nearest percent.

81. The number of post offices in the United States is shown in the bar graph.

    (a) Use the information given for the years 1995 and 2000, letting $x = 5$ represent 1995, $x = 10$ represent 2000, and y represent the number of post offices, to write an equation that models the data.

    (b) Use the equation to approximate the number of post offices in 1998. How does this result compare to the actual value, 27,952?

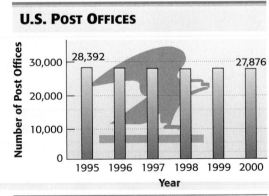

Source: U.S. Postal Service, *Annual Report of the Postmaster General.*

82. Median household income of African-Americans is shown in the bar graph.

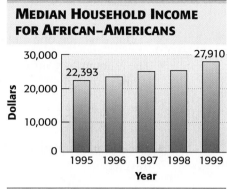

MEDIAN HOUSEHOLD INCOME FOR AFRICAN-AMERICANS

Source: U.S. Bureau of the Census.

(a) Use the information given for the years 1995 and 1999, letting $x = 5$ represent 1995, $x = 9$ represent 1999, and y represent the median income, to write an equation that models median household income.

(b) Use the equation to approximate the median income for 1997. How does your result compare to the actual value, $25,050?

RELATING CONCEPTS (EXERCISES 83–88)

For Individual or Group Work

In Section 2.5 we learned how formulas can be applied to problem solving. **Work Exercises 83–88 in order,** *to see how the formula that relates Celsius and Fahrenheit temperatures is derived.*

83. There is a linear relationship between Celsius and Fahrenheit temperatures. When $C = 0°$, $F = $ _____°, and when $C = 100°$, $F = $ _____°.

84. Think of ordered pairs of temperatures (C, F), where C and F represent corresponding Celsius and Fahrenheit temperatures. The equation that relates the two scales has a straight-line graph that contains the two points determined in Exercise 83. What are these two points?

85. Find the slope of the line described in Exercise 84.

86. Now think of the point-slope form of the equation in terms of C and F, where C replaces x and F replaces y. Use the slope you found in Exercise 85 and one of the two points determined earlier, and find the equation that gives F in terms of C.

87. To obtain another form of the formula, use the equation you found in Exercise 86 and solve for C in terms of F.

88. The equation found in Exercise 86 is graphed on the graphing calculator screen shown here. Interpret the display at the bottom, in the context of this group of exercises.

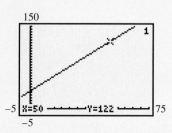

7.3 Functions

OBJECTIVES

1 Define and identify relations and functions.

2 Find domain and range.

3 Identify functions defined by graphs and equations.

4 Use function notation.

5 Identify linear functions.

We often describe one quantity in terms of another. Consider the following.

- The amount of your paycheck if you are paid hourly depends on the number of hours you worked.

- The cost at the gas station depends on the number of gallons of gas you pumped into your car.

- The distance traveled by a car moving at a constant speed depends on the time traveled.

We can use ordered pairs to represent these corresponding quantities. For example, we indicate the relationship between the amount of your paycheck and hours worked by writing ordered pairs in which the first number represents hours worked and the second number represents paycheck amount in dollars. Then the ordered pair $(5, 40)$ indicates that when you work 5 hr, your paycheck is $40. Similarly, the ordered pairs $(10, 80)$ and $(20, 160)$ show that working 10 hr results in an $80 paycheck and working 20 hr results in a $160 paycheck. In this example, what would the ordered pair $(40, 320)$ indicate?

Since the amount of your paycheck *depends* on the number of hours worked, your paycheck amount is called the *dependent variable,* and the number of hours worked is called the *independent variable.* Generalizing, if the value of the variable y depends on the value of the variable x, then y is the **dependent variable** and x is the **independent variable.**

Independent variable ⌐ ⌐Dependent variable
$$(x, y)$$

OBJECTIVE 1 Define and identify relations and functions. Since we can write related quantities using ordered pairs, a set of ordered pairs such as

$$\{(5, 40), (10, 80), (20, 160), (40, 320)\}$$

is called a *relation.*

Relation

A **relation** is a set of ordered pairs.

A special kind of relation, called a *function,* is very important in mathematics and its applications.

Function

A **function** is a relation in which, for each value of the first component of the ordered pairs, there is *exactly one value* of the second component.

EXAMPLE 1 Determining Whether Relations Are Functions

Tell whether each relation defines a function.

$$F = \{(1, 2), (-2, 4), (3, -1)\}$$
$$G = \{(-2, -1), (-1, 0), (0, 1), (1, 2), (2, 2)\}$$
$$H = \{(-4, 1), (-2, 1), (-2, 0)\}$$

Relations F and G are functions, because for each different x-value there is exactly one y-value. Notice that in G, the last two ordered pairs have the same y-value (1 is paired with 2, and 2 is paired with 2). This does not violate the definition of function, since the first components (x-values) are different and each is paired with only one second component (y-value).

In relation H, however, the last two ordered pairs have the *same* x-value paired with *two different* y-values (-2 is paired with both 1 and 0), so H is a relation but not a function. ***In a function, no two ordered pairs can have the same first component and different second components.***

Different y-values

$$H = \{(-4, 1), (-2, 1), (-2, 0)\} \qquad \text{Not a function}$$

Same x-value

Now Try Exercises 5 and 7.

In a function, there is *exactly one* value of the dependent variable, the second component, for each value of the independent variable, the first component. This is what makes functions so important in applications.

> **NOTE** The relation from the beginning of this section representing hours worked and corresponding paycheck amount is a function since each x-value is paired with exactly one y-value. You would not be happy, for example, if you and a coworker each worked 20 hr at the same hourly rate and your paycheck was $160 while his was $200.

Relations and functions can also be expressed as a correspondence or *mapping* from one set to another, as shown in Figure 25 for function F and relation H from Example 1. The arrow from 1 to 2 indicates that the ordered pair (1, 2) belongs to F—each first component is paired with exactly one second component. In the mapping for set H, which is not a function, the first component -2 is paired with two different second components, 1 and 0.

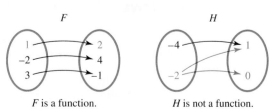

F is a function. H is not a function.

FIGURE 25

Since relations and functions are sets of ordered pairs, we can represent them using tables and graphs. A table and graph for function F from Example 1 is shown in Figure 26.

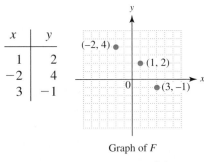

x	y
1	2
−2	4
3	−1

Graph of F

FIGURE 26

Finally, we can describe a relation or function using a rule that tells how to determine the dependent variable for a specific value of the independent variable. The rule may be given in words: the dependent variable is twice the independent variable. Usually the rule is an equation:

$$y = 2x.$$

Dependent variable   Independent variable

This is the most efficient way to define a relation or function.

NOTE Another way to think of a function relationship is to think of the independent variable as an input and the dependent variable as an output. This is illustrated by the input-output (function) machine for the function defined by $y = 2x$.

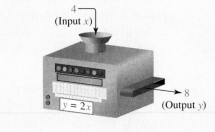

4
(Input x)

8
(Output y)

$y = 2x$

Function machine

OBJECTIVE 2 Find domain and range.

Domain and Range

In a relation, the set of all values of the independent variable (x) is the **domain;** the set of all values of the dependent variable (y) is the **range.**

EXAMPLE 2 Finding Domains and Ranges of Relations

Give the domain and range of each relation. Tell whether the relation defines a function.

(a) $\{(3, -1), (4, 2), (4, 5), (6, 8)\}$

The domain, the set of x-values, is $\{3, 4, 6\}$; the range, the set of y-values, is $\{-1, 2, 5, 8\}$. This relation is not a function because the same x-value 4 is paired with two different y-values, 2 and 5.

(b)

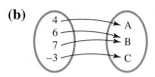

The domain of this relation is
$$\{4, 6, 7, -3\};$$
the range is
$$\{A, B, C\}.$$
This mapping defines a function—each x-value corresponds to exactly one y-value.

(c)

x	y
-5	2
0	2
5	2

This is a table of ordered pairs, so the domain is the set of x-values $\{-5, 0, 5\}$ and the range is the set of y-values $\{2\}$. The table defines a function because each different x-value corresponds to exactly one y-value (even though it is the same y-value).

Now Try Exercises 11, 13, and 15.

As mentioned previously, the graph of a relation is the graph of its ordered pairs. The graph gives a picture of the relation, which can be used to determine its domain and range.

EXAMPLE 3 Finding Domains and Ranges from Graphs

Give the domain and range of each relation.

(a)

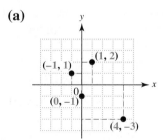

The domain is the set of x-values,
$$\{-1, 0, 1, 4\}.$$
The range is the set of y-values,
$$\{-3, -1, 1, 2\}.$$

(b)

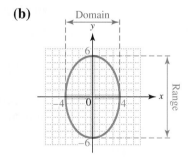

The x-values of the points on the graph include all numbers between -4 and 4, inclusive. The y-values include all numbers between -6 and 6, inclusive. Using interval notation,

the domain is $[-4, 4]$;
the range is $[-6, 6]$.

(c)

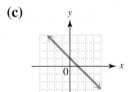

(d)

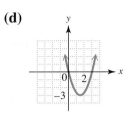

The arrowheads indicate that the line extends indefinitely left and right, as well as up and down. Therefore, both the domain and the range include all real numbers, written $(-\infty, \infty)$.

The arrowheads indicate that the graph extends indefinitely left and right, as well as upward. The domain is $(-\infty, \infty)$. Because there is a least y-value, -3, the range includes all numbers greater than or equal to -3, written $[-3, \infty)$.

Now Try Exercises 17 and 19.

Since relations are often defined by equations, such as $y = 2x + 3$ and $y^2 = x$, we must sometimes determine the domain of a relation from its equation. In this book, we assume the following agreement on the domain of a relation.

Agreement on Domain

Unless specified otherwise, the domain of a relation is assumed to be all real numbers that produce real numbers when substituted for the independent variable.

To illustrate this agreement, since any real number can be used as a replacement for x in $y = 2x + 3$, the domain of this function is the set of all real numbers. As another example, the function defined by $y = \frac{1}{x}$ has all real numbers except 0 as domain, since y is undefined if $x = 0$. In general, the domain of a function defined by an algebraic expression is all real numbers, except those numbers that lead to division by 0 or an even root of a negative number.

OBJECTIVE 3 Identify functions defined by graphs and equations. Most of the relations we have seen in the examples are functions—that is, each x-value corresponds to exactly one y-value. Since each value of x leads to only one value of y in a function, any vertical line drawn through the graph of a function must intersect the graph in at most one point. This is the *vertical line test* for a function.

Vertical Line Test

If every vertical line intersects the graph of a relation in no more than one point, then the relation represents a function.

For example, the graph shown in Figure 27(a) is not the graph of a function since a vertical line intersects the graph in more than one point. The graph in Figure 27(b) does represent a function.

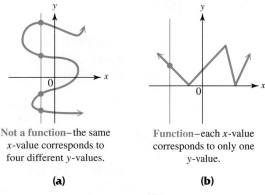

Not a function–the same *x*-value corresponds to four different *y*-values.

Function–each *x*-value corresponds to only one *y*-value.

(a)

(b)

FIGURE 27

▓ EXAMPLE 4 Using the Vertical Line Test

Use the vertical line test to determine whether each relation graphed in Example 3 is a function.

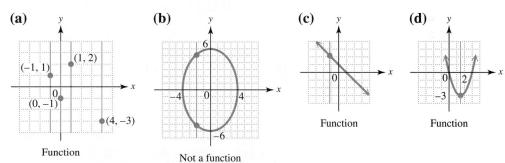

The graphs in (a), (c), and (d) represent functions. The graph of the relation in (b) fails the vertical line test, since the same *x*-value corresponds to two different *y*-values; therefore, it is not the graph of a function.

Now Try Exercise 21.

NOTE Graphs that do not represent functions are still relations. Remember that all equations and graphs represent relations and that all relations have a domain and range.

The vertical line test is a simple method for identifying a function defined by a graph. It is more difficult to decide whether a relation defined by an equation is a function. The next example gives some hints that may help.

▓ EXAMPLE 5 Identifying Functions from Their Equations

Decide whether each relation defines a function and give the domain.

(a) $y = x + 4$

In the defining equation (or rule), $y = x + 4$, y is always found by adding 4 to x. Thus, each value of x corresponds to just one value of y and the relation defines a function; x can be any real number, so the domain is $\{x \mid x \text{ is a real number}\}$ or $(-\infty, \infty)$.

(b) $y^2 = x$

The ordered pairs $(16, 4)$ and $(16, -4)$ both satisfy this equation. Since one value of x, 16, corresponds to two values of y, 4 and -4, this equation does not define a function. Because x is equal to the square of y, the values of x must always be nonnegative. The domain of the relation is $[0, \infty)$.

(c) $y = \dfrac{5}{x - 1}$

Given any value of x in the domain, we find y by subtracting 1, then dividing the result into 5. This process produces exactly one value of y for each value in the domain, so this equation defines a function. The domain includes all real numbers except those that make the denominator 0. We find these numbers by setting the denominator equal to 0 and solving for x.

$$x - 1 = 0$$
$$x = 1$$

Thus, the domain includes all real numbers except 1. In interval notation this is written as $(-\infty, 1) \cup (1, \infty)$.

Now Try Exercises 25, 29, and 33.

In summary, three variations of the definition of function are given here.

Variations of the Definition of Function

1. A **function** is a relation in which, for each value of the first component of the ordered pairs, there is exactly one value of the second component.

2. A **function** is a set of ordered pairs in which no first component is repeated.

3. A **function** is a rule or correspondence that assigns exactly one range value to each domain value.

OBJECTIVE 4 Use function notation.   When a function f is defined with a rule or an equation using x and y for the independent and dependent variables, we say "y is a function of x" to emphasize that y *depends on* x. We use the notation

$$y = f(x),$$

called **function notation,** to express this and read $f(x)$ as "f of x." (In this special notation the parentheses do not indicate multiplication.) The letter f stands for *function*. For example, if $y = 9x - 5$, we can name this function f and write

$$f(x) = 9x - 5.$$

Note that $f(x)$ *is just another name for the dependent variable y.* For example, if $y = f(x) = 9x - 5$ and $x = 2$, then we find y, or $f(2)$, by replacing x with 2.

$$y = f(2)$$
$$= 9 \cdot 2 - 5$$
$$= 18 - 5$$
$$= 13.$$

The statement "if $x = 2$, then $y = 13$" represents the ordered pair $(2, 13)$ and is abbreviated with function notation as

$$f(2) = 13.$$

Read $f(2)$ as "f of 2" or "f at 2." Also,

$$f(0) = 9 \cdot 0 - 5 = -5 \qquad \text{and} \qquad f(-3) = 9(-3) - 5 = -32.$$

These ideas and the symbols used to represent them can be illustrated as follows.

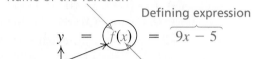

Name of the function

Defining expression

$$y \; = \; f(x) \; = \; 9x - 5$$

Value of the function   Name of the independent variable

CAUTION   The symbol $f(x)$ *does not* indicate "f times x," but represents the y-value for the indicated x-value. As just shown, $f(2)$ is the y-value that corresponds to the x-value 2.

EXAMPLE 6   Using Function Notation

Let $f(x) = -x^2 + 5x - 3$. Find the following.

(a) $f(2)$

Replace x with 2.

$$f(x) = -x^2 + 5x - 3$$
$$f(2) = -2^2 + 5 \cdot 2 - 3$$
$$= -4 + 10 - 3$$
$$= 3$$

Thus, $f(2) = 3$; the ordered pair $(2, 3)$ belongs to f.

(b) $f(q)$

$$f(x) = -x^2 + 5x - 3$$
$$f(q) = -q^2 + 5q - 3 \qquad \text{Replace } x \text{ with } q.$$

The replacement of one variable with another is important in later courses.

Now Try Exercises 41 and 45.

Sometimes letters other than f, such as g, h, or capital letters F, G, and H are used to name functions.

EXAMPLE 7   Using Function Notation

Let $g(x) = 2x + 3$. Find and simplify $g(a + 1)$.

$$g(x) = 2x + 3$$
$$g(a + 1) = 2(a + 1) + 3 \qquad \text{Replace } x \text{ with } a + 1.$$
$$= 2a + 2 + 3$$
$$= 2a + 5$$

Now Try Exercise 49.

Functions can be evaluated in a variety of ways, as shown in Example 8.

▌ **EXAMPLE 8** Using Function Notation

For each function, find $f(3)$.

(a) $f(x) = 3x - 7$

$f(3) = 3(3) - 7$ Replace x with 3.

$f(3) = 2$

(b) $f = \{(-3, 5), (0, 3), (3, 1), (6, -1)\}$

We want $f(3)$, the y-value of the ordered pair where $x = 3$. As indicated by the ordered pair $(3, 1)$, when $x = 3, y = 1$, so $f(3) = 1$.

(c)

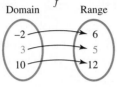

The domain element 3 is paired with 5 in the range, so $f(3) = 5$.

(d)

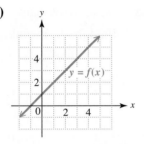

To evaluate $f(3)$, find 3 on the x-axis. See Figure 28. Then move up until the graph of f is reached. Moving horizontally to the y-axis gives 4 for the corresponding y-value. Thus, $f(3) = 4$.

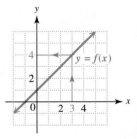

FIGURE 28

Now Try Exercises 53, 55, and 57.

If a function f is defined by an equation with x and y, not with function notation, use the following steps to find $f(x)$.

Finding an Expression for *f(x)*

Step 1   Solve the equation for y.

Step 2   Replace y with $f(x)$.

EXAMPLE 9   Writing Equations Using Function Notation

Rewrite each equation using function notation. Then find $f(-2)$ and $f(a)$.

(a) $y = x^2 + 1$

This equation is already solved for y. Since $y = f(x)$,

$$f(x) = x^2 + 1.$$

To find $f(-2)$, let $x = -2$.

$$f(-2) = (-2)^2 + 1$$
$$= 4 + 1$$
$$= 5$$

Find $f(a)$ by letting $x = a$: $f(a) = a^2 + 1$.

(b) $x - 4y = 5$

First solve $x - 4y = 5$ for y. Then replace y with $f(x)$.

$$x - 4y = 5$$
$$x - 5 = 4y$$
$$y = \frac{x - 5}{4} \quad \text{so} \quad f(x) = \frac{1}{4}x - \frac{5}{4}$$

Now find $f(-2)$ and $f(a)$.

$$f(-2) = \frac{1}{4}(-2) - \frac{5}{4} = -\frac{7}{4} \qquad \text{Let } x = -2.$$

$$f(a) = \frac{1}{4}a - \frac{5}{4} \qquad \text{Let } x = a.$$

Now Try Exercise 59.

OBJECTIVE 5 Identify linear functions.   Our first two-dimensional graphing was of straight lines. Linear equations (except for vertical lines with equations $x = a$) define *linear functions.*

Linear Function

A function that can be defined by

$$f(x) = mx + b$$

for real numbers m and b is a **linear function.**

Recall that m is the slope of the line and $(0, b)$ is the y-intercept. In Example 9(b), we wrote the equation $x - 4y = 5$ as the linear function defined by

$$f(x) = \frac{1}{4}x - \frac{5}{4}.$$

Slope ↗   ↑ y-intercept is $\left(0, -\frac{5}{4}\right)$.

To graph this function, plot the y-intercept and use the definition of slope as $\frac{\text{rise}}{\text{run}}$ to find a second point on the line. Draw the straight line through the points to obtain the graph. See Figure 29.

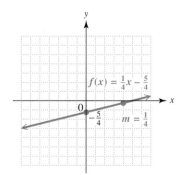

FIGURE 29

A linear function defined by $f(x) = b$ (whose graph is a horizontal line) is sometimes called a **constant function.** The domain of any linear function is $(-\infty, \infty)$. The range of a nonconstant linear function is $(-\infty, \infty)$, while the range of the constant function defined by $f(x) = b$ is $\{b\}$.

Now Try Exercise 67.

7.3 EXERCISES

For Extra Help

 Student's
Solutions Manual

MyMathLab

InterAct Math
Tutorial Software

AW Math
Tutor Center

MathXL MathXL

Digital Video Tutor
CD 11/Videotape 11

 1. In your own words, define a function and give an example.

 2. In your own words, define the domain of a function and give an example.

3. In an ordered pair of a relation, is the first element the independent or the dependent variable?

4. Give an example of a relation that is not a function, having domain $\{-3, 2, 6\}$ and range $\{4, 6\}$. (There are many possible correct answers.)

Tell whether each relation defines a function. See Example 1.

5. $\{(5, 1), (3, 2), (4, 9), (7, 6)\}$

6. $\{(8, 0), (5, 4), (9, 3), (3, 8)\}$

7. $\{(2, 4), (0, 2), (2, 5)\}$

8. $\{(9, -2), (-3, 5), (9, 2)\}$

9. $\{(-3, 1), (4, 1), (-2, 7)\}$

10. $\{(-12, 5), (-10, 3), (8, 3)\}$

Decide whether each relation defines a function and give the domain and range. See Examples 1–4.

11. $\{(1, 1), (1, -1), (0, 0), (2, 4), (2, -4)\}$

12. $\{(2, 5), (3, 7), (4, 9), (5, 11)\}$

13.

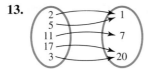

14.

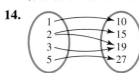

15.

x	y
1	5
1	2
1	-1
1	-4

16.

x	y
4	-3
2	-3
0	-3
-2	-3

17.

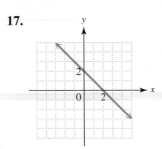

18.

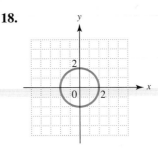

19.
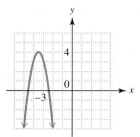

20.
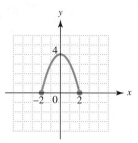

21.
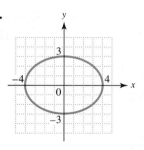

22.
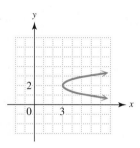

Decide whether each relation defines y as a function of x. Give the domain. See Example 5.

23. $y = 10x$ **24.** $y = -5x$ **25.** $y = 2x - 6$ **26.** $y = -6x + 8$

27. $y = x^2$ **28.** $y = x^3$ **29.** $x = y^6$ **30.** $x = y^4$

31. $y = \dfrac{1}{x}$ **32.** $y = -\dfrac{3}{x}$ **33.** $y = \dfrac{2}{x - 9}$ **34.** $y = \dfrac{-7}{x - 16}$

35. $y = \dfrac{1}{4x + 2}$ **36.** $y = \dfrac{1}{9 - 2x}$ **37.** $y = |x|$ **38.** $y = -|x|$

39. Choose the correct response: The notation $f(3)$ means

     A. the variable f times 3 or $3f$.
     B. the value of the dependent variable when the independent variable is 3.
     C. the value of the independent variable when the dependent variable is 3.
     D. f equals 3.

40. Give an example of a function from everyday life. (*Hint:* Fill in the blanks: _____ depends on _____, so _____ is a function of _____.)

Let $f(x) = -3x + 4$ and $g(x) = -x^2 + 4x + 1$. Find the following. See Examples 6 and 7.

41. $f(0)$ **42.** $f(-3)$ **43.** $g(-2)$ **44.** $g(10)$

45. $f(p)$ **46.** $g(k)$ **47.** $f(-x)$ **48.** $g(-x)$

49. $f(x + 2)$ **50.** $f(a + 4)$ **51.** $f(2m - 3)$ **52.** $f(3t - 2)$

*For each function, find (**a**) $f(2)$ and (**b**) $f(-1)$. See Example 8.*

53. $f = \{(-1, 3), (4, 7), (0, 6), (2, 2)\}$ **54.** $f = \{(2, 5), (3, 9), (-1, 11), (5, 3)\}$

55.

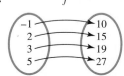

56.

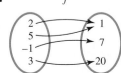

57.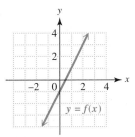

58.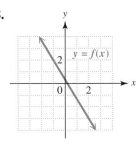

An equation that defines y as a function of x is given. (a) Solve for y in terms of x and replace y with the function notation f(x). (b) Find f(3). See Example 9.

59. $x + 3y = 12$ **60.** $x - 4y = 8$ **61.** $y + 2x^2 = 3$

62. $y - 3x^2 = 2$ **63.** $4x - 3y = 8$ **64.** $-2x + 5y = 9$

65. Fill in each blank with the correct response.

The equation $2x + y = 4$ has a straight _____ as its graph. One point that lies on the graph is (3, _____). If we solve the equation for y and use function notation, we obtain $f(x) =$ _____. For this function, $f(3) =$ _____, meaning that the point (_____, _____) lies on the graph of the function.

66. Which of the following defines a linear function?

    A. $y = \dfrac{x - 5}{4}$ **B.** $y = \dfrac{1}{x}$ **C.** $y = x^2$ **D.** $y = \sqrt{x}$

Graph each linear function. Give the domain and range. See Objective 5 and Figure 29.

67. $f(x) = -2x + 5$ **68.** $g(x) = 4x - 1$ **69.** $h(x) = \dfrac{1}{2}x + 2$

70. $F(x) = -\dfrac{1}{4}x + 1$ **71.** $G(x) = 2x$ **72.** $H(x) = -3x$

73. $g(x) = -4$ **74.** $f(x) = 5$

75. Suppose that a package weighing x lb costs $f(x)$ dollars to mail to a given location, where

$$f(x) = 2.75x.$$

    (a) What is the value of $f(3)$?

   (b) In your own words, describe what 3 and the value $f(3)$ mean in part (a), using the terminology *independent variable* and *dependent variable*.

    (c) How much would it cost to mail a 5-lb package? Interpret this question and its answer using function notation.

76. Suppose that a Yellow Cab driver charges $1.50 per mile.

    (a) Fill in the table with the correct response for the price $f(x)$ he charges for a trip of x mi.

x	$f(x)$
0	
1	
2	
3	

    (b) The linear function that gives a rule for the amount charged is $f(x) =$ _____.

    (c) Graph this function for the domain {0, 1, 2, 3}.

Forensic scientists use the lengths of certain bones to calculate the height of a person. Two bones often used are the tibia (t), the bone from the ankle to the knee, and the femur (r), the bone from the knee to the hip socket. A person's height (h) is determined from the lengths of these bones using functions defined by the following formulas. All measurements are in centimeters.

$$\text{For men:} \quad h(r) = 69.09 + 2.24r \quad \text{or} \quad h(t) = 81.69 + 2.39t$$
$$\text{For women:} \quad h(r) = 61.41 + 2.32r \quad \text{or} \quad h(t) = 72.57 + 2.53t$$

77. Find the height of a man with a femur measuring 56 cm.

78. Find the height of a man with a tibia measuring 40 cm.

79. Find the height of a woman with a femur measuring 50 cm.

80. Find the height of a woman with a tibia measuring 36 cm.

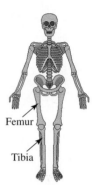

Femur

Tibia

Federal regulations set standards for the size of the quarters of marine mammals. A pool to house sea otters must have a volume of "the square of the sea otter's average adult length (in meters) multiplied by 3.14 and by .91 meter." If x represents the sea otter's average adult length and f(x) represents the volume (in cubic meters) of the corresponding pool size, this formula can be written as

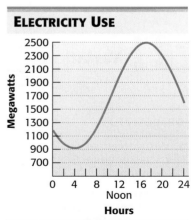

$$f(x) = .91(3.14)x^2.$$

Find the volume of the pool for each adult sea otter length (in meters). Round answers to the nearest hundredth.

81. .8   **82.** 1.0   **83.** 1.2   **84.** 1.5

85. The graph shows the daily megawatts of electricity used on a record-breaking summer day in Sacramento, California.

  (a) Is this the graph of a function?

  (b) What is the domain?

  (c) Estimate the number of megawatts used at 8 A.M.

  (d) At what time was the most electricity used? the least electricity?

  (e) Call this function f. What is $f(12)$? What does it mean?

ELECTRICITY USE

Source: Sacramento Municipal Utility District.

86. Refer to the graph to answer the questions.

  (a) What numbers are possible values of the independent variable? the dependent variable?

  (b) For how long is the water level increasing? decreasing?

  (c) How many gallons of water are in the pool after 90 hr?

  (d) Call this function f. What is $f(0)$? What does it mean?

  (e) What is $f(25)$? What does it mean?

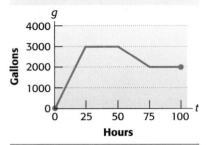

GALLONS OF WATER IN A POOL AT TIME t

TECHNOLOGY INSIGHTS (EXERCISES 87 AND 88)

87. The calculator screen shows the graph of a linear function $y = f(x)$, along with the display of coordinates of a point on the graph. Use function notation to write what the display indicates.

88. The table was generated by a graphing calculator for a linear function $Y_1 = f(X)$. Use the table to work parts (a)–(e).

  (a) What is $f(2)$?

  (b) If $f(X) = -3.7$, what is the value of X?

  (c) What is the slope of the line?

  (d) What is the y-intercept of the line?

  (e) Find the expression for $f(X)$.

X	Y₁
0	3.5
1	2.3
2	1.1
3	-.1
4	-1.3
5	-2.5
6	-3.7

X=0

7.4 Variation

OBJECTIVES

1 Write an equation expressing direct variation.

2 Find the constant of variation, and solve direct variation problems.

Certain types of functions are very common, especially in business and the physical sciences. These are functions where y depends on a multiple of x, or y depends on a number divided by x. In such situations, y is said to *vary directly as x* (in the first case) or *vary inversely as x* (in the second case). For example, by the distance formula, the distance traveled varies directly as the rate (or speed) and the time. Formulas for area and volume are other familiar examples of *direct variation*.

On the other hand, the force required to keep a car from skidding on a curve varies inversely as the radius of the curve. Another example of *inverse variation* is how travel time is inversely proportional to rate or speed.

3 Solve inverse variation problems.

4 Solve joint variation problems.

5 Solve combined variation problems.

$C = 2\pi r$

OBJECTIVE 1 Write an equation expressing direct variation. The circumference of a circle is given by the formula $C = 2\pi r$, where r is the radius of the circle. See the figure. Circumference is always a constant multiple of the radius. (C is always found by multiplying r by the constant 2π.) Thus,

As the *radius increases,* the *circumference increases.*

The reverse is also true.

As the *radius decreases,* the *circumference decreases.*

Because of this, the circumference is said to *vary directly* as the radius.

Direct Variation

y **varies directly as** *x* if there exists a real number k such that

$$y = kx.$$

Also, y is said to be **proportional to** x. The number k is called the **constant of variation.** In direct variation, for $k > 0$, as the value of x increases, the value of y also increases. Similarly, as x decreases, y decreases.

OBJECTIVE 2 Find the constant of variation, and solve direct variation problems. The direct variation equation $y = kx$ defines a linear function, where the constant of variation k is the slope of the line. For example, we wrote the equation

$$y = 1.60x$$

to describe the cost y to buy x gal of gas in Example 6 of Section 7.2. The cost varies directly as, or is proportional to, the number of gallons of gas purchased. That is, as the number of gallons of gas increases, cost increases; also, as the number of gallons of gas decreases, cost decreases. The constant of variation k is 1.60, the cost of 1 gal of gas.

EXAMPLE 1 Finding the Constant of Variation and the Variation Equation

Steven Pusztai is paid an hourly wage. One week he worked 43 hr and was paid $795.50. How much does he earn per hour?

Let h represent the number of hours he works and P represent his corresponding pay. Then, P varies directly as h, so

$$P = kh.$$

Here, k represents Steven's hourly wage. Since $P = 795.50$ when $h = 43$,

$$795.50 = 43k$$
$$k = 18.50. \qquad \text{Use a calculator.}$$

His hourly wage is $18.50, and P and h are related by

$$P = 18.50h.$$

Now Try Exercise 31.

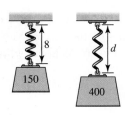

FIGURE 30

EXAMPLE 2 Solving a Direct Variation Problem

Hooke's law for an elastic spring states that the distance a spring stretches is proportional to the force applied. If a force of 150 newtons* stretches a certain spring 8 cm, how much will a force of 400 newtons stretch the spring? See Figure 30.

If d is the distance the spring stretches and f is the force applied, then $d = kf$ for some constant k. Since a force of 150 newtons stretches the spring 8 cm, use these values to find k.

$$d = kf \qquad \text{Variation equation}$$
$$8 = k \cdot 150 \qquad \text{Let } d = 8 \text{ and } f = 150.$$
$$k = \frac{8}{150} \qquad \text{Find } k.$$
$$k = \frac{4}{75}$$

Substitute $\frac{4}{75}$ for k in the variation equation $d = kf$ to get

$$d = \frac{4}{75}f.$$

For a force of 400 newtons,

$$d = \frac{4}{75}(400) \qquad \text{Let } f = 400.$$
$$= \frac{64}{3}.$$

The spring will stretch $\frac{64}{3}$ cm if a force of 400 newtons is applied.

Now Try Exercise 35.

In summary, use the following steps to solve a variation problem.

Solving a Variation Problem

Step 1 Write the variation equation.

Step 2 Substitute the initial values and solve for k.

Step 3 Rewrite the variation equation with the value of k from Step 2.

Step 4 Substitute the remaining values, solve for the unknown, and find the required answer.

The direct variation equation $y = kx$ is a linear equation. However, other kinds of variation involve other types of equations. For example, one variable can be proportional to a power of another variable.

---

*A newton is a unit of measure of force used in physics.

Direct Variation as a Power

y **varies directly as the** *n***th power of** *x* if there exists a real number *k* such that

$$y = kx^n.$$

$A = \pi r^2$

An example of direct variation as a power is the formula for the area of a circle, $A = \pi r^2$. Here, π is the constant of variation, and the area varies directly as the *square* of the radius.

EXAMPLE 3 Solving a Direct Variation Problem

The distance a body falls from rest varies directly as the square of the time it falls (disregarding air resistance). If a skydiver falls 64 ft in 2 sec, how far will she fall in 8 sec?

Step 1 If *d* represents the distance the skydiver falls and *t* the time it takes to fall, then *d* is a function of *t*, and, for some constant *k*,

$$d = kt^2.$$

Step 2 To find the value of *k*, use the fact that the skydiver falls 64 ft in 2 sec.

$$d = kt^2 \qquad \text{Variation equation}$$
$$64 = k(2)^2 \qquad \text{Let } d = 64 \text{ and } t = 2.$$
$$k = 16 \qquad \text{Find } k.$$

Step 3 Using 16 for *k*, the variation equation becomes

$$d = 16t^2.$$

Step 4 Now let $t = 8$ to find the number of feet the skydiver will fall in 8 sec.

$$d = 16(8)^2 \qquad \text{Let } t = 8.$$
$$= 1024$$

The skydiver will fall 1024 ft in 8 sec.

Now Try Exercise 37.

OBJECTIVE 3 Solve inverse variation problems.   In direct variation, where $k > 0$, as *x* increases, *y* increases. Similarly, as *x* decreases, *y* decreases. Another type of variation is *inverse variation*. With inverse variation, where $k > 0$, as one variable increases, the other variable decreases. For example, in a closed space, volume decreases as pressure increases, as illustrated by a trash compactor. See Figure 31. As the compactor presses down, the pressure on the trash increases; in turn, the trash occupies a smaller space.

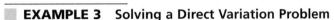

As pressure on trash increases, volume of trash decreases.

FIGURE 31

Inverse Variation

y **varies inversely as** *x* if there exists a real number *k* such that

$$y = \frac{k}{x}.$$

(continued)

Also, **y varies inversely as the nth power of x** if there exists a real number k such that

$$y = \frac{k}{x^n}.$$

The inverse variation equation also defines a function. Since x is in the denominator, these functions are *rational functions*. (See Chapter 6 and Section 11.1.) Another example of inverse variation comes from the distance formula. In its usual form, the formula is

$$d = rt.$$

Dividing each side by r gives

$$t = \frac{d}{r}.$$

Here, t (time) varies inversely as r (rate or speed), with d (distance) serving as the constant of variation. For example, if the distance between Chicago and Des Moines is 300 mi, then

$$t = \frac{300}{r}$$

and the values of r and t might be any of the following.

$$\left.\begin{array}{l} r = 50, t = 6 \\ r = 60, t = 5 \\ r = 75, t = 4 \end{array}\right\} \begin{array}{l}\text{As } r \text{ increases,} \\ t \text{ decreases.}\end{array} \qquad \left.\begin{array}{l} r = 30, t = 10 \\ r = 25, t = 12 \\ r = 20, t = 15 \end{array}\right\} \begin{array}{l}\text{As } r \text{ decreases,} \\ t \text{ increases.}\end{array}$$

If we *increase* the rate (speed) we drive, time *decreases*. If we *decrease* the rate (speed) we drive, what happens to time?

EXAMPLE 4 Solving an Inverse Variation Problem

In the manufacturing of a certain medical syringe, the cost of producing the syringe varies inversely as the number produced. If 10,000 syringes are produced, the cost is $2 per syringe. Find the cost per syringe to produce 25,000 syringes.

$$\begin{array}{ll} \text{Let} & x = \text{the number of syringes produced,} \\ \text{and} & c = \text{the cost per syringe.} \end{array}$$

Here, as production increases, cost decreases and as production decreases, cost increases. Since c varies inversely as x, there is a constant k such that

$$c = \frac{k}{x}.$$

Find k by replacing c with 2 and x with 10,000.

$$2 = \frac{k}{10,000}$$

$$20,000 = k \qquad \text{Multiply by 10,000.}$$

Since $c = \frac{k}{x}$,

$$c = \frac{20{,}000}{25{,}000} = .80. \qquad \text{Let } k = 20{,}000 \text{ and } x = 25{,}000.$$

The cost per syringe to make 25,000 syringes is $.80.

Now Try Exercise 39.

EXAMPLE 5 Solving an Inverse Variation Problem

The weight of an object above Earth varies inversely as the square of its distance from the center of Earth. A space shuttle in an elliptical orbit has a maximum distance from the center of Earth (*apogee*) of 6700 mi. Its minimum distance from the center of Earth (*perigee*) is 4090 mi. See Figure 32. If an astronaut in the shuttle weighs 57 lb at its apogee, what does the astronaut weigh at its perigee?

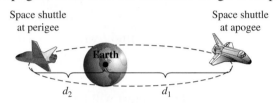

FIGURE 32

If w is the weight and d is the distance from the center of Earth, then

$$w = \frac{k}{d^2}$$

for some constant k. At the apogee the astronaut weighs 57 lb, and the distance from the center of Earth is 6700 mi. Use these values to find k.

$$57 = \frac{k}{(6700)^2} \qquad \text{Let } w = 57 \text{ and } d = 6700.$$

$$k = 57(6700)^2$$

Then the weight at the perigee with $d = 4090$ mi is

$$w = \frac{57(6700)^2}{(4090)^2} \approx 153 \text{ lb.} \qquad \text{Use a calculator.}$$

Now Try Exercise 43.

OBJECTIVE 4 Solve joint variation problems. It is common for one variable to depend on several others. If one variable varies directly as the *product* of several other variables (perhaps raised to powers), the first variable is said to *vary jointly* as the others.

Joint Variation

y varies jointly as x and z if there exists a real number k such that

$$y = kxz.$$

CAUTION   Note that *and* in the expression "y varies directly as x *and* z" translates as the product $y = kxz$. The word *and* does not indicate addition here.

▨ EXAMPLE 6 Solving a Joint Variation Problem

The interest on a loan or an investment is given by the formula $I = prt$. Here, for a given principal p, the interest earned I varies jointly as the interest rate r and the time t the principal is left at interest. If an investment earns \$100 interest at 5% for 2 yr, how much interest will the same principal earn at 4.5% for 3 yr?

We use the formula $I = prt$, where p is the constant of variation because it is the same for both investments. For the first investment, $I = 100$, $r = .05$, and $t = 2$, so

$$I = prt$$
$$100 = p(.05)(2) \qquad \text{Let } I = 100, r = .05, \text{ and } t = 2.$$
$$100 = .1p$$
$$\frac{100}{.1} = p$$
$$p = 1000.$$

Now we find I when $p = 1000$, $r = .045$, and $t = 3$.

$$I = 1000(.045)(3) \qquad \text{Let } p = 1000, r = .045, \text{ and } t = 3.$$
$$I = 135$$

The interest will be \$135.

Now Try Exercise 45.

OBJECTIVE 5 Solve combined variation problems. There are many combinations of direct and inverse variation. Example 7 shows a typical **combined variation** problem.

▨ EXAMPLE 7 Solving a Combined Variation Problem

Body mass index, or BMI, is used by physicians to assess a person's level of fatness. A BMI from 19 through 25 is considered desirable. BMI varies directly as an individual's weight in pounds and inversely as the square of the individual's height in inches. A person who weighs 118 lb and is 64 in. tall has a BMI of 20. (The BMI is rounded to the nearest whole number.) Find the BMI of a person who weighs 165 lb with a height of 70 in. (*Source: Washington Post.*)

Let B represent the BMI, w the weight, and h the height. Then

$$B = \frac{kw}{h^2}. \quad \begin{array}{l} \longleftarrow \text{ BMI varies directly as the weight.} \\ \longleftarrow \text{ BMI varies inversely as the square of the height.} \end{array}$$

To find k, let $B = 20$, $w = 118$, and $h = 64$.

$$20 = \frac{k(118)}{64^2}$$
$$k = \frac{20(64^2)}{118} \qquad \text{Multiply by } 64^2; \text{ divide by 118.}$$
$$k \approx 694 \qquad \text{Use a calculator.}$$

Now find B when $k = 694$, $w = 165$, and $h = 70$.

$$B = \frac{694(165)}{70^2} \approx 23 \qquad \text{Nearest whole number}$$

The person's BMI is 23.

Now Try Exercise 47.

7.4 EXERCISES

Use personal experience or intuition to determine whether the situation suggests direct *or* inverse *variation.*

1. The number of different lottery tickets you buy and your probability of winning that lottery

2. The rate and the distance traveled by a pickup truck in 3 hr

3. The amount of pressure put on the accelerator of a car and the speed of the car

4. The number of days from now until December 25 and the magnitude of the frenzy of Christmas shopping

5. Your age and the probability that you believe in Santa Claus

6. The surface area of a balloon and its diameter

7. The number of days until the end of the baseball season and the number of home runs that Barry Bonds has

8. The amount of gasoline you pump and the amount you pay

Determine whether each equation represents direct, inverse, joint, *or* combined *variation.*

9. $y = \dfrac{3}{x}$

10. $y = \dfrac{8}{x}$

11. $y = 10x^2$

12. $y = 2x^3$

13. $y = 3xz^4$

14. $y = 6x^3z^2$

15. $y = \dfrac{4x}{wz}$

16. $y = \dfrac{6x}{st}$

17. For $k > 0$, if y varies directly as x, when x increases, y _____, and when x decreases, y _____.

18. For $k > 0$, if y varies inversely as x, when x increases, y _____, and when x decreases, y _____.

Solve each problem.

19. If x varies directly as y, and $x = 9$ when $y = 3$, find x when $y = 12$.

20. If x varies directly as y, and $x = 10$ when $y = 7$, find y when $x = 50$.

21. If a varies directly as the square of b, and $a = 4$ when $b = 3$, find a when $b = 2$.

22. If h varies directly as the square of m, and $h = 15$ when $m = 5$, find h when $m = 7$.

23. If z varies inversely as w, and $z = 10$ when $w = .5$, find z when $w = 8$.

24. If t varies inversely as s, and $t = 3$ when $s = 5$, find s when $t = 5$.

25. If m varies inversely as p^2, and $m = 20$ when $p = 2$, find m when $p = 5$.

26. If a varies inversely as b^2, and $a = 48$ when $b = 4$, find a when $b = 7$.

27. p varies jointly as q and r^2, and $p = 200$ when $q = 2$ and $r = 3$. Find p when $q = 5$ and $r = 2$.

28. f varies jointly as g^2 and h, and $f = 50$ when $g = 4$ and $h = 2$. Find f when $g = 3$ and $h = 6$.

29. Explain the difference between inverse variation and direct variation.

30. What is meant by the constant of variation in a direct variation problem? If you were to graph the linear equation $y = kx$ for some nonnegative constant k, what role would the value of k play in the graph?

Solve each problem involving variation. See Examples 1–7.

31. Todd bought 8 gal of gasoline and paid $13.59. To the nearest tenth of a cent, what is the price of gasoline per gallon?

32. Melissa gives horseback rides at Shadow Mountain Ranch. A 2.5-hr ride costs $50.00. What is the price per hour?

33. The volume of a can of tomatoes is proportional to the height of the can. If the volume of the can is 300 cm³ when its height is 10.62 cm, find the volume of a can with height 15.92 cm.

34. The force required to compress a spring is proportional to the change in length of the spring. If a force of 20 newtons is required to compress a certain spring 2 cm, how much force is required to compress the spring from 20 cm to 8 cm?

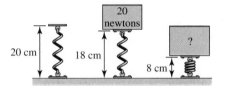

35. The weight of an object on Earth is directly proportional to the weight of that same object on the moon. A 200-lb astronaut would weigh 32 lb on the moon. How much would a 50-lb dog weigh on the moon?

36. The pressure exerted by a certain liquid at a given point varies directly as the depth of the point beneath the surface of the liquid. The pressure at 30 m is 80 newtons per m². What pressure is exerted at 50 m?

37. For a body falling freely from rest (disregarding air resistance), the distance the body falls varies directly as the square of the time. If an object is dropped from the top of a tower 576 ft high and hits the ground in 6 sec, how far did it fall in the first 4 sec?

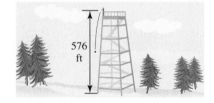

38. The amount of water emptied by a pipe varies directly as the square of the diameter of the pipe. For a certain constant water flow, a pipe emptying into a canal will allow 200 gal of water to escape in an hour. The diameter of the pipe is 6 in. How much water would a 12-in. pipe empty into the canal in an hour, assuming the same water flow?

39. Over a specified distance, speed varies inversely with time. If a Dodge Viper on a test track goes a certain distance in one-half minute at 160 mph, what speed is needed to go the same distance in three-fourths minute?

40. For a constant area, the length of a rectangle varies inversely as the width. The length of a rectangle is 27 ft when the width is 10 ft. Find the width of a rectangle with the same area if the length is 18 ft.

41. The frequency of a vibrating string varies inversely as its length. That is, a longer string vibrates fewer times in a second than a shorter string. Suppose a piano string 2 ft long vibrates 250 cycles per sec. What frequency would a string 5 ft long have?

42. The current in a simple electrical circuit varies inversely as the resistance. If the current is 20 amps when the resistance is 5 ohms, find the current when the resistance is 7.5 ohms.

43. The amount of light (measured in foot-candles) produced by a light source varies inversely as the square of the distance from the source. If the illumination produced 1 m from a light source is 768 foot-candles, find the illumination produced 6 m from the same source.

44. The force with which Earth attracts an object above Earth's surface varies inversely with the square of the distance of the object from the center of Earth. If an object 4000 mi from the center of Earth is attracted with a force of 160 lb, find the force of attraction if the object were 6000 mi from the center of Earth.

45. For a given interest rate, simple interest varies jointly as principal and time. If $2000 left in an account for 4 yr earned interest of $280, how much interest would be earned in 6 yr?

46. The collision impact of an automobile varies jointly as its mass and the square of its speed. Suppose a 2000-lb car traveling at 55 mph has a collision impact of 6.1. What is the collision impact of the same car at 65 mph?

47. The force needed to keep a car from skidding on a curve varies inversely as the radius of the curve and jointly as the weight of the car and the square of the speed. If 242 lb of force keep a 2000-lb car from skidding on a curve of radius 500 ft at 30 mph, what force would keep the same car from skidding on a curve of radius 750 ft at 50 mph?

48. The maximum load that a cylindrical column with a circular cross section can hold varies directly as the fourth power of the diameter of the cross section and inversely as the square of the height. A 9-m column 1 m in diameter will support 8 metric tons. How many metric tons can be supported by a column 12 m high and $\frac{2}{3}$ m in diameter?

Load = 8 metric tons

49. The number of long-distance phone calls between two cities in a certain time period varies jointly as the populations of the cities, p_1 and p_2, and inversely as the distance between them. If 80,000 calls are made between two cities 400 mi apart, with populations of 70,000 and 100,000, how many calls are made between cities with populations of 50,000 and 75,000 that are 250 mi apart?

50. A body mass index from 27 through 29 carries a slight risk of weight-related health problems, while one of 30 or more indicates a great increase in risk. Use your own height and weight and the information in Example 7 to determine your BMI and whether you are at risk.

51. Natural gas provides 35.8% of U.S. energy. (*Source:* U.S. Energy Department.) The volume of gas varies inversely as the pressure and directly as the temperature. (Temperature must be measured in *Kelvin* (K), a unit of measurement used in physics.) If a certain gas occupies a volume of 1.3 L at 300 K and a pressure of 18 newtons per cm^2, find the volume at 340 K and a pressure of 24 newtons per cm^2.

52. The maximum load of a horizontal beam that is supported at both ends varies directly as the width and the square of the height and inversely as the length between the supports. A beam 6 m long, .1 m wide, and .06 m high supports a load of 360 kg. What is the maximum load supported by a beam 16 m long, .2 m wide, and .08 m high?

Exercises 53 and 54 describe weight-estimation formulas that fishermen have used over the years. Girth *is the distance around the body of the fish.* (*Source: Sacramento Bee,* November 9, 2000.)

53. The weight of a bass varies jointly as its girth and the square of its length. A prize-winning bass weighed in at 22.7 lb and measured 36 in. long with 21 in. girth. How much would a bass 28 in. long with 18 in. girth weigh?

54. The weight of a trout varies jointly as its length and the square of its girth. One angler caught a trout that weighed 10.5 lb and measured 26 in. long with 18 in. girth. Find the weight of a trout that is 22 in. long with 15 in. girth.

RELATING CONCEPTS (EXERCISES 55–62)

For Individual or Group Work

A routine activity such as pumping gasoline can be related to many of the concepts studied in this chapter. Suppose that premium unleaded costs $1.25 per gallon. **Work Exercises 55–62 in order.**

55. Zero gallons of gasoline cost $0.00, while 1 gallon costs $1.25. Represent these two pieces of information as ordered pairs of the form (gallons, price).

56. Use the information from Exercise 55 to find the slope of the line on which the two points lie.

57. Write the slope-intercept form of the equation of the line on which the two points lie.

58. Using function notation, if $f(x) = ax + b$ represents the line from Exercise 57, what are the values of a and b?

59. How does the value of a from Exercise 58 relate to gasoline in this situation? With relationship to the line, what do we call this number?

60. Why does the equation from Exercise 57 satisfy the conditions for direct variation? In the context of variation, what do we call the value of a?

61. The graph of the equation from Exercise 57 is shown in the calculator screen. How is the display at the bottom of the screen interpreted in the context of these exercises?

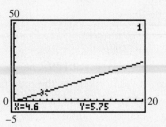

62. The table was generated by a graphing calculator, with Y_1 entered as the equation from Exercise 57. Interpret the entry for X = 12 in the context of these exercises.

X	Y_1	
7	8.75	
8	10	
9	11.25	
10	12.5	
11	13.75	
12	15	
13	16.25	

X=12

| Chapter **7** | **Group Activity** |

Choosing an Energy Source (or How to Get Hot Water)

OBJECTIVE Write and graph linear functions that model given data.

There are many different ways to heat water. In this activity you will look at three different energy sources that may be used to provide heat for a 40-gal home water tank.

Have each student in your group choose one of the three types of water heaters listed in the table.

Type of Hot Water Heater	Size (in gallons)	Price	Operating Cost per Month (manufacturer's estimate)	Hot Water Temperature
Kenmore Economizer 6 —Electric	40	$139.99	$35.00	120°–130°
Kenmore Economizer 6 —Natural Gas	40	$139.99	$13.25	120°–130°
Sunbather Water Heater —Solar	40	$950.00	$0.00	*

Source: Jade Mountain 1999.

A. Using data for the water heater you selected, find a linear equation that represents total cost y of heating water with respect to time. Let x represent number of months. Write the equation in slope-intercept form.

B. Graph your equation using domain [0, 60] and range [0, 1000].

C. As a group, compare the graphs of your equations.

 1. What are the y-intercepts?

 2. How do the slopes compare? Which is the steepest? Which has 0 slope?

D. Discuss other factors to consider when choosing each type of water heater. Which of these water heaters would you choose to heat your home?

---

*No temperature listed but the ad says "Best for warm climates, preheating water, summer only in cold places, or when hot water needed only in afternoons and evenings."

CHAPTER 7 SUMMARY

◼ KEY TERMS

7.1 ordered pair
origin
x-axis
y-axis
rectangular
(Cartesian)
coordinate system
plot
coordinate
quadrant

graph of an equation
first-degree equation
linear equation in two
variables
x-intercept
y-intercept
rise
run
slope

7.2 slope-intercept form
point-slope form
standard form
7.3 dependent variable
independent variable
relation
function
domain
range

function notation
linear function
constant function
7.4 vary directly
proportional
constant of variation
vary inversely
vary jointly
combined variation

◼ NEW SYMBOLS

(a, b) ordered pair

x_1 a specific value of the variable x (read "x-sub-one")

m slope

$f(x)$ function of x (read "f of x")

◼ TEST YOUR WORD POWER

See how well you have learned the vocabulary in this chapter. Answers, with examples, follow the Quick Review.

1. A **linear equation in two variables**
is an equation that can be written in
the form
A. $Ax + By < C$
B. $ax = b$
C. $y = x^2$
D. $Ax + By = C.$

2. The **slope** of a line is
A. the measure of the run over the
rise of the line
B. the distance between two points
on the line
C. the ratio of the change in y to the
change in x along the line
D. the horizontal change compared
to the vertical change between
two points on the line.

3. In a relationship between two
variables x and y, the **independent
variable** is
A. x, if x depends on y
B. x, if y depends on x

C. either x or y
D. the larger of x and y.

4. In a relationship between two
variables x and y, the **dependent
variable** is
A. y, if y depends on x
B. y, if x depends on y
C. either x or y
D. the smaller of x and y.

5. A **relation** is
A. a set of ordered pairs
B. the ratio of the change in y to the
change in x along a line
C. the set of all possible values of
the independent variable
D. all the second components of a
set of ordered pairs.

6. A **function** is
A. the numbers in an ordered pair
B. a set of ordered pairs in which
each x-value corresponds to
exactly one y-value

C. a pair of numbers written
between parentheses in which
order matters
D. the set of all ordered pairs that
satisfy an equation.

7. The **domain** of a function is
A. the set of all possible values of
the dependent variable y
B. a set of ordered pairs
C. the difference between the
x-values
D. the set of all possible values of
the independent variable x.

8. The **range** of a function is
A. the set of all possible values of
the dependent variable y
B. a set of ordered pairs
C. the difference between the
y-values
D. the set of all possible values of
the independent variable x.

QUICK REVIEW

CONCEPTS	EXAMPLES

7.1 REVIEW OF GRAPHS AND SLOPES OF LINES

Finding Intercepts

To find the x-intercept, let $y = 0$.

To find the y-intercept, let $x = 0$.

The graph of $2x + 3y = 12$ has

$$x\text{-intercept} \quad (6, 0)$$
and
$$y\text{-intercept} \quad (0, 4).$$

Finding Slope

If $x_2 \neq x_1$, then

$$m = \frac{\text{rise}}{\text{run}} = \frac{\text{change in } y}{\text{change in } x} = \frac{y_2 - y_1}{x_2 - x_1}.$$

A vertical line has undefined slope.

A horizontal line has 0 slope.

Parallel lines have equal slopes.

For $2x + 3y = 12$,

$$m = \frac{4 - 0}{0 - 6} = -\frac{2}{3}.$$

$x = 3$ has undefined slope.

$y = -5$ has $m = 0$.

$$
\begin{array}{l|l}
y = 2x + 3 & 4x - 2y = 6 \\
m = 2 & -2y = -4x + 6 \\
 & y = 2x - 3 \\
 & m = 2
\end{array}
$$

These lines are parallel.

The slopes of perpendicular lines are negative reciprocals with a product of -1.

$$
\begin{array}{l|l}
y = 3x - 1 & x + 3y = 4 \\
m = 3 & 3y = -x + 4 \\
 & y = -\frac{1}{3}x + \frac{4}{3} \\
 & m = -\frac{1}{3}
\end{array}
$$

These lines are perpendicular.

7.2 REVIEW OF EQUATIONS OF LINES; LINEAR MODELS

Slope-Intercept Form

$y = mx + b$

$y = 2x + 3 \qquad m = 2,\ y\text{-intercept is } (0, 3).$

Point-Slope Form

$y - y_1 = m(x - x_1)$

$y - 3 = 4(x - 5) \qquad (5, 3)$ is on the line, $m = 4$.

Standard Form

$Ax + By = C \quad (A, B, C \text{ integers}, A > 0, B \neq 0)$

$2x - 5y = 8$

Horizontal Line

$y = b$

$y = 4$

Vertical Line

$x = a$

$x = -1$

(continued)

CONCEPTS	EXAMPLES

7.3 FUNCTIONS

A **function** is a set of ordered pairs such that for each first component there is one and only one second component. The set of first components is called the **domain,** and the set of second components is called the **range.**

$y = f(x) = x^2$ defines a function f, with domain $(-\infty, \infty)$ and range $[0, \infty)$.

To evaluate a function using function notation (that is, $f(x)$ notation) for a given value of x, substitute the value wherever x appears.

If $f(x) = x^2 - 7x + 12$, then
$$f(1) = 1^2 - 7(1) + 12 = 6.$$

To write an equation that defines a function in function notation,

Write $2x + 3y = 12$ using function notation.

$$3y = -2x + 12 \qquad \text{Subtract } 2x.$$

Step 1   Solve the equation for y.

$$y = -\frac{2}{3}x + 4 \qquad \text{Divide by 3.}$$

Step 2   Replace y with $f(x)$.

$$f(x) = -\frac{2}{3}x + 4$$

7.4 VARIATION

If there is some constant k such that:

$y = kx^n$,   then y varies directly as x^n.

$y = \dfrac{k}{x^n}$,   then y varies inversely as x^n.

The area of a circle varies directly as the square of the radius.
$$A = kr^2 \qquad \text{Here, } k = \pi.$$

Pressure varies inversely as volume.
$$p = \frac{k}{V}$$

$y = kxz$,   then y varies jointly as x and z.

For a given principal, interest varies jointly as interest rate and time.
$$I = krt \qquad k \text{ is the given principal.}$$

Answers to Test Your Word Power

1. D; *Examples:* $3x + 2y = 6$, $x = y - 7$, $4x = y$ **2.** C; *Example:* The line through (3, 6) and (5, 4) has slope $\frac{4-6}{5-3} = \frac{-2}{2} = -1$.
3. B; *Example:* See Answer 4, which follows.   **4.** A; *Example:* When borrowing money, the amount you borrow (independent variable) determines the size of your payments (dependent variable).   **5.** A; *Example:* The set $\{(2, 0), (4, 3), (6, 6), (8, 9)\}$ defines a relation.
6. B; The relation given in Answer 5 is a function since the x-value of each ordered pair corresponds to exactly one y-value.
7. D; *Example:* In the function in Answer 5, the domain is the set of x-values, $\{2, 4, 6, 8\}$.   **8.** A; *Example:* In the function in Answer 5, the range is the set of y-values, $\{0, 3, 6, 9\}$.

CHAPTER 7 REVIEW EXERCISES

[7.1] *Complete the table of ordered pairs for each equation. Then graph the equation.*

1. $3x + 2y = 10$

x	y
0	
	0
2	
	−2

2. $x − y = 8$

x	y
2	
	−3
3	
	−2

Find the x- and y-intercepts and then graph each equation.

3. $4x − 3y = 12$

4. $5x + 7y = 28$

5. $2x + 5y = 20$

6. $x − 4y = 8$

 7. Explain how the signs of the *x*- and *y*-coordinates of a point determine the quadrant in which the point lies.

Find the slope of each line.

8. Through $(−1, 2)$ and $(4, −5)$

9. Through $(0, 3)$ and $(−2, 4)$

10. $y = 2x + 3$

11. $3x − 4y = 5$

12. $x = 5$

13. Parallel to $3y = 2x + 5$

14. Perpendicular to $3x − y = 4$

15. Through $(−1, 5)$ and $(−1, −4)$

16. $y + 6 = 0$

17.

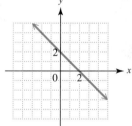

Tell whether each line has positive, negative, 0, *or* undefined *slope.*

18.

19.

20.

21.

22. If a walkway rises 2 ft for every 10 ft on the horizontal, which of the following express its slope (or grade)? (There are several correct choices.)

A. .2 **B.** $\frac{2}{10}$ **C.** $\frac{1}{5}$ **D.** 20%

E. 5 **F.** $\frac{20}{100}$ **G.** 500% **H.** $\frac{10}{2}$

23. If the pitch of a roof is $\frac{1}{4}$, how many feet in the horizontal direction correspond to a rise of 3 ft?

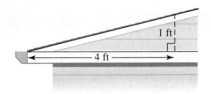

24. Family income in the United States has steadily increased for many years (primarily due to inflation). In 1970 the median family income was about $10,000 a year. In 1999 it was about $49,000 a year. Find the average rate of change of median family income to the nearest dollar over that period. (*Source:* U.S. Bureau of the Census.)

[7.2] *Find an equation for each line, if possible. Write the equation in slope-intercept form.*

25. Slope $-\frac{1}{3}$; *y*-intercept $(0, -1)$

26. Slope 0; *y*-intercept $(0, -2)$

27. Slope $-\frac{4}{3}$; through $(2, 7)$

28. Slope 3; through $(-1, 4)$

29. Vertical; through $(2, 5)$

30. Through $(2, -5)$ and $(1, 4)$

31. Through $(-3, -1)$ and $(2, 6)$

32. The line pictured in Exercise 17

33. Parallel to $4x - y = 3$ and through $(7, -1)$

34. Perpendicular to $2x - 5y = 7$ and through $(4, 3)$

35. The Midwest Athletic Club (Section 7.2, Exercises 73 and 74) offers two special membership plans. (*Source:* Midwest Athletic Club.) For each plan, write a linear equation in slope-intercept form and give the cost *y* in dollars of a 1-yr membership. Let *x* represent the number of months.

  (a) Executive VIP/Gold membership: $159 fee plus $57 per month

  (b) Executive Regular/Silver membership: $159 fee plus $47 per month

36. The percent of tax returns filed electronically for the years 1996–2001 is shown in the graph.

  (a) Use the information given for the years 1996 and 2001, letting $x = 6$ represent 1996, $x = 11$ represent 2001, and *y* represent the percent of returns filed electronically to find a linear equation that models the data. Write the equation in slope-intercept form. Interpret the slope of this equation.

  (b) Use your equation from part (a) to predict the percent of tax returns that will be filed electronically in 2005. (Assume a constant rate of change.)

E-FILING TAXPAYERS

Source: Internal Revenue Service.

[7.3] *In Exercises 37–40, give the domain and range of each relation. Identify any functions.*

37. $\{(-4, 2), (-4, -2), (1, 5), (1, -5)\}$

38.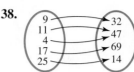

39.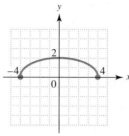

40.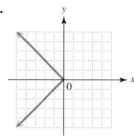

Determine whether each equation defines y as a function of x. Give the domain in each case. Identify any linear functions.

41. $y = 3x - 3$

42. $x = y^2$

43. $y = \dfrac{7}{x - 6}$

44. Explain the test that allows us to determine whether a graph is that of a function.

Given $f(x) = -2x^2 + 3x - 6$, find each function value or expression.

45. $f(0)$

46. $f(2.1)$

47. $f\left(-\dfrac{1}{2}\right)$

48. $f(k)$

49. The table shows profits for cable television station CNBC from 1994 through 1999.

(a) Does the table define a function?
(b) What are the domain and range?
(c) Call this function f. Give two ordered pairs that belong to f.
(d) Find $f(1994)$. What does it mean?
(e) If $f(x) = 180$, what does x equal?

Year	Profit (in millions of dollars)
1994	40
1995	60
1996	80
1997	130
1998	180
1999	200

Source: Fortune, May 24, 1999, p. 142.

50. The equation $2x^2 - y = 0$ defines y as a function of x. Rewrite it using $f(x)$ notation, and find $f(3)$.

51. Suppose that $2x - 5y = 7$ defines a function. If $y = f(x)$, which one of the following defines the same function?

A. $f(x) = \dfrac{7 - 2x}{5}$

B. $f(x) = \dfrac{-7 - 2x}{5}$

C. $f(x) = \dfrac{-7 + 2x}{5}$

D. $f(x) = \dfrac{7 + 2x}{5}$

52. Can the graph of a linear function have undefined slope? Explain.

RELATING CONCEPTS (EXERCISES 53–64)

For Individual or Group Work

Refer to the straight-line graph and **work Exercises 53–64**
in order.

53. By just looking at the graph, how can you tell
whether the slope is positive, negative, 0,
or undefined?

54. Use the slope formula to find the slope of the line.

55. What is the slope of any line parallel to the
line shown? perpendicular to the line shown?

56. Find the *x*-intercept of the graph.

57. Find the *y*-intercept of the graph.

58. Use function notation to write the equation of the line. Use *f* to designate the
function.

59. Find $f(8)$.

60. If $f(x) = -8$, what is the value of *x*?

61. Graph the solution set of $f(x) \geq 0$.

62. What is the solution set of $f(x) = 0$?

63. What is the solution set of $f(x) < 0$? (Use the graph and the result of
Exercise 62.)

64. What is the solution set of $f(x) > 0$? (Use the graph and the result of
Exercise 62.)

The graph shows a line labeled $2y = -3x + 7$ passing through points $(-1, 5)$ and $(3, -1)$.

[7.4]

65. In which one of the following does *y* vary inversely as *x*?

A. $y = 2x$ **B.** $y = \dfrac{x}{3}$ **C.** $y = \dfrac{3}{x}$ **D.** $y = x^2$

Solve each problem.

66. For the subject in a photograph to appear in the same perspective in the photograph as
in real life, the viewing distance must be properly related to the amount of enlargement.
For a particular camera, the viewing distance varies directly as the amount of
enlargement. A picture taken with this camera that is enlarged 5 times should be viewed
from a distance of 250 mm. Suppose a print 8.6 times the size of the negative is made.
From what distance should it be viewed?

67. The frequency (number of vibrations per second) of a vibrating guitar string varies
inversely as its length. That is, a longer string vibrates fewer times in a second than a
shorter string. Suppose a guitar string .65 m long vibrates 4.3 times per sec. What
frequency would a string .5 m long have?

68. The volume of a rectangular box of a given height is proportional to its width and
length. A box with width 2 ft and length 4 ft has volume 12 ft^3. Find the volume of a
box with the same height that is 3 ft wide and 5 ft long.

CHAPTER 7 TEST

1. Complete the table of ordered pairs for the equation $2x - 3y = 12$.

x	y
1	
3	
	-4

Find the x- and y-intercepts, and graph each equation.

2. $3x - 2y = 20$ **3.** $y = 5$ **4.** $x = 2$

5. Find the slope of the line through the points $(6, 4)$ and $(-4, -1)$.

6. Describe how the graph of a line with undefined slope is situated in a rectangular coordinate system.

Determine whether each pair of lines is parallel, perpendicular, or neither.

7. $5x - y = 8$ and $5y = -x + 3$ **8.** $2y = 3x + 12$ and $3y = 2x - 5$

 9. In 1980, there were 119,000 farms in Iowa. As of 2001, there were 93,500. Find and interpret the average rate of change in the number of farms per year. Round your answer to the nearest whole number. (*Source:* Iowa Agricultural Statistics Service.)

Find an equation of each line, and write it in slope-intercept form.

10. Through $(4, -1)$; $m = -5$ **11.** Through $(-3, 14)$; horizontal

12. Through $(-7, 2)$ and parallel to $3x + 5y = 6$

13. Through $(-7, 2)$ and perpendicular to $y = 2x$

14. Through $(-2, 3)$ and $(6, -1)$

15. Which one of the following has positive slope and negative y-coordinate for its y-intercept?

A.   **B.**   **C.**   **D.**

16. The bar graph shows median household income for Hispanics.

(a) Use the information for the years 1995 and 1999 to find an equation that models the data. Let $x = 5$ represent 1995, $x = 9$ represent 1999, and y represent the median income. Write the equation in slope-intercept form.

(b) Use the equation from part (a) to approximate median household income for 1997 to the nearest dollar. How does your result compare to the actual value, $26,628?

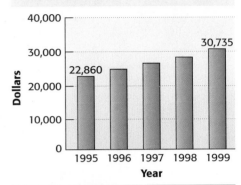

MEDIAN HOUSEHOLD INCOME FOR HISPANICS

Source: U.S. Bureau of the Census.

17. Which one of the following is the graph of a function?

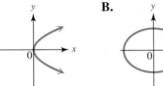

A.   **B.**   **C.**   **D.**

18. Which of the following does not define a function?

A. $\{(0, 1), (-2, 3), (4, 8)\}$ **B.** $y = 2x - 6$ **C.** $y = \dfrac{1}{x + 2}$ **D.**

x	y
0	1
3	2
0	2
6	3

19. Give the domain and range of the relation shown in **(a)** choice A of Problem 17 and **(b)** choice A of Problem 18.

20. If $f(x) = -x^2 + 2x - 1$, find $f(1)$ and $f(a)$.

21. Graph the linear function defined by $f(x) = \frac{2}{3}x - 1$. What is its domain and range?

Solve each problem.

22. The current in a simple electrical circuit is inversely proportional to the resistance. If the current is 80 amps when the resistance is 30 ohms, find the current when the resistance is 12 ohms.

23. The force of the wind blowing on a vertical surface varies jointly as the area of the surface and the square of the velocity. If a wind blowing at 40 mph exerts a force of 50 lb on a surface of 500 ft², how much force will a wind of 80 mph place on a surface of 2 ft²?

CUMULATIVE REVIEW EXERCISES CHAPTERS **1–7**

Decide whether each statement is always true, sometimes true, *or* never true. *If the statement is* sometimes true, *give examples where it is true and where it is false.*

1. The absolute value of a negative number equals the additive inverse of the number.

2. The quotient of two integers with nonzero denominator is a rational number.

3. The sum of two negative numbers is positive.

4. The sum of a positive number and a negative number is 0.

Simplify.

5. $-|-2| - 4 + |-3| + 7$

6. $3x^2 - 4x + 4 + 9x - x^2$

Evaluate each expression if $p = -4$, $q = -2$, *and* $r = 5$.

7. $-3(2q - 3p)$

8. $\dfrac{r}{-p + 2q}$

Solve.

9. $2z - 5 + 3z = 4 - (z + 2)$

10. $\dfrac{3a - 1}{5} + \dfrac{a + 2}{2} = -\dfrac{3}{10}$

11. $V = \dfrac{1}{3}\pi r^2 h$ for h

Solve each problem.

12. If each side of a square were increased by 4 in., the perimeter would be 8 in. less than twice the perimeter of the original square. Find the length of a side of the original square.

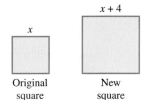

Original square   New square

13. Two planes leave the Dallas–Fort Worth airport at the same time. One travels east at 550 mph, and the other travels west at 500 mph. Assuming no wind, how long will it take for the planes to be 2100 mi apart?

West ⟵ ✈ Airport ✈ ⟶ East

Solve. Write each solution set in interval notation and graph it.

14. $-4 < 3 - 2k < 9$

15. $-.3x + 2.1(x - 4) \le -6.6$

16. Find the x- and y-intercepts of the line with equation $3x + 5y = 12$ and graph the line.

17. Consider the points $A(-2, 1)$ and $B(3, -5)$.

  (a) Find the slope of the line AB.
  (b) Find the slope of a line perpendicular to line AB.

Write an equation for each line. Express the equation in slope-intercept form.

18. Slope $-\dfrac{3}{4}$; y-intercept $(0, -1)$

19. Horizontal; through $(2, -2)$

20. Through $(4, -3)$ and $(1, 1)$

Perform the indicated operations. In Exercises 21 and 22, assume that variables represent nonzero real numbers.

21. $(3x^2y^{-1})^{-2}(2x^{-3}y)^{-1}$

22. $\dfrac{5m^{-2}y^3}{3m^{-3}y^{-1}}$

23. $(3x^3 + 4x^2 - 7) - (2x^3 - 8x^2 + 3x)$

24. $(7x + 3y)^2$

25. $(2p + 3)(5p^2 - 4p - 8)$

26. $\dfrac{m^3 - 3m^2 + 5m - 3}{m - 1}$

Factor.

27. $16w^2 + 50wz - 21z^2$

28. $4x^2 - 4x + 1 - y^2$

29. $4y^2 - 36y + 81$

30. $100x^4 - 81$

31. $8p^3 + 27$

Solve.

32. $(p - 1)(2p + 3)(p + 4) = 0$

33. $9q^2 = 6q - 1$

34. A sign is to have the shape of a triangle with a height 3 ft greater than the length of the base. How long should the base be if the area is to be 14 ft²?

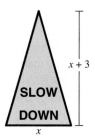

35. A game board has the shape of a rectangle. The longer sides are each 2 in. longer than the distance between them. The area of the board is 288 in.². Find the length of the longer sides and the distance between them.

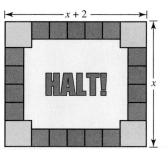

Perform each indicated operation. Write the answer in lowest terms.

36. $\dfrac{8}{x + 1} - \dfrac{2}{x + 3}$

37. $\dfrac{x^2 - 25}{3x + 6} \cdot \dfrac{4x + 8}{x^2 + 10x + 25}$

38. $\dfrac{x^2 + 5x + 6}{3x} \div \dfrac{x^2 - 4}{x^2 + x - 6}$

39. $\dfrac{\dfrac{12}{x + 6}}{\dfrac{4}{2x + 12}}$

40. Solve $\dfrac{2}{x - 1} = \dfrac{5}{x - 1} - \dfrac{3}{4}$.

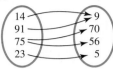 **41.** Give the domain and range of the relation. Does it define a function? Explain.

```
   14 ───────► 9
   91 ───────► 70
   75 ═══════► 56
   23 ───────► 5
```

42. For the function defined by

$$f(x) = -4x + 10,$$

    (a) find the domain and range.
    (b) what is $f(-3)$?

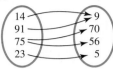 **43.** Use the information in the graph to find and interpret the average rate of change in millions of U.S. cell phone subscribers per year from 1992 to 2000.

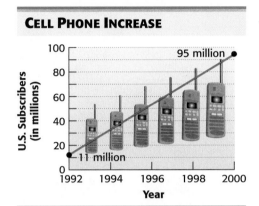

Source: Cellular Telecommunications Industry Association, Intel Corp.

44. The cost of a pizza varies directly as the square of its radius. If a pizza with a 7-in. radius costs $6.00, how much should a pizza with a 9-in. radius cost?

Systems of Linear Equations

8

Americans continue to flock to movie theaters in record numbers. Box office grosses reached $8.14 billion in 2001, the first time they exceeded $8 billion, as more than 1.4 billion movie tickets were sold. Surprisingly, the top three movies of the year were family films—*Harry Potter and the Sorcerer's Stone, Shrek,* and *Monsters, Inc.* not only attracted scores of adults with kids, but also unaccompanied adults, perhaps wishing to get away from it all for a few hours. (*Source:* ACNielsen EDI.) In Exercise 29 of the Chapter 8 Review Exercises, we use systems of linear equations to find out just how much money these top films earned.

8.1 Solving Systems of Linear Equations by Graphing

8.2 Solving Systems of Linear Equations by Substitution

8.3 Solving Systems of Linear Equations by Elimination

Summary Exercises on Solving Systems of Linear Equations

8.4 Systems of Linear Equations in Three Variables

8.5 Applications of Systems of Linear Equations

8.6 Solving Systems of Linear Equations by Matrix Methods

8.1   Solving Systems of Linear Equations by Graphing

OBJECTIVES

1 Decide whether a given ordered pair is a solution of a system.

2 Solve linear systems by graphing.

3 Solve special systems by graphing.

4 Identify special systems without graphing.

5 Use a graphing calculator to solve a linear system.

A **system of linear equations,** often called a **linear system,** consists of two or more linear equations with the same variables. Examples of systems of two linear equations include

$$2x + 3y = 4 \qquad x + 3y = 1 \qquad x - y = 1$$
$$3x - y = -5 \qquad -y = 4 - 2x \qquad y = 3.$$

In the system on the right, think of $y = 3$ as an equation in two variables by writing it as $0x + y = 3$.

OBJECTIVE 1   Decide whether a given ordered pair is a solution of a system.   A **solution of a system** of linear equations is an ordered pair that makes both equations true at the same time. A solution of an equation is said to *satisfy* the equation.

▓ **EXAMPLE 1   Determining Whether an Ordered Pair Is a Solution**

Is $(4, -3)$ a solution of each system?

(a) $x + 4y = -8$
   $3x + 2y = 6$
   To decide whether $(4, -3)$ is a solution of the system, substitute 4 for x and -3 for y in each equation.

$x + 4y = -8$	$3x + 2y = 6$
$4 + 4(-3) = -8$?	$3(4) + 2(-3) = 6$?
$4 + (-12) = -8$? Multiply.	$12 + (-6) = 6$? Multiply.
$-8 = -8$ True	$6 = 6$ True

Because $(4, -3)$ satisfies both equations, it is a solution of the system.

(b) $2x + 5y = -7$
   $3x + 4y = 2$
   Again, substitute 4 for x and -3 for y in both equations.

$2x + 5y = -7$	$3x + 4y = 2$
$2(4) + 5(-3) = -7$?	$3(4) + 4(-3) = 2$?
$8 + (-15) = -7$? Multiply.	$12 + (-12) = 2$? Multiply.
$-7 = -7$ True	$0 = 2$ False

The ordered pair $(4, -3)$ is not a solution of this system because it does not satisfy the second equation.

Now Try Exercises 3 and 5.

We discuss three methods of solving a system of two linear equations in two variables in this chapter.

OBJECTIVE 2   Solve linear systems by graphing.   The set of all ordered pairs that are solutions of a system is its **solution set.** One way to find the solution set of a system of two linear equations is to graph both equations on the same axes. The graph of each line shows points whose coordinates satisfy the equation of that line. Any intersection point would be on both lines and would therefore be a solution of both equations. Thus, the coordinates of any point where the lines intersect give a solution of the system. Because two *different* straight lines can intersect at no more than one point, there can never be more than one solution for such a system. The graph in Figure 1 shows that the solution of the system in Example 1(a) is the intersection point $(4, -3)$.

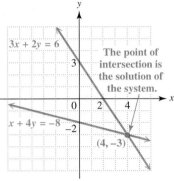

FIGURE 1

![] **EXAMPLE 2** Solving a System by Graphing

Solve the system of equations by graphing both equations on the same axes.

$$2x + 3y = 4$$
$$3x - y = -5$$

To graph these two equations, we choose *any* number for either x or y to get an ordered pair. The intercepts are often convenient choices. It is a good idea to use a third ordered pair as a check.

$2x + 3y = 4$

x	y
0	$\frac{4}{3}$
2	0
-2	$\frac{8}{3}$

$3x - y = -5$

x	y
0	5
$-\frac{5}{3}$	0
-2	-1

The lines in Figure 2 suggest that the graphs intersect at the point $(-1, 2)$. We check this by substituting -1 for x and 2 for y in both equations.

$$2x + 3y = 4$$
$$2(-1) + 3(2) = 4 \quad ?$$
$$4 = 4 \qquad \text{True}$$

$$3x - y = -5$$
$$3(-1) - 2 = -5 \quad ?$$
$$-5 = -5 \qquad \text{True}$$

Because $(-1, 2)$ satisfies both equations, the solution set of this system is $\{(-1, 2)\}$.

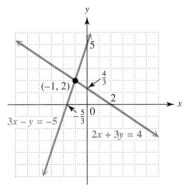

FIGURE 2

Now Try Exercise 13.

To solve a system by graphing, follow these steps.

Solving a Linear System by Graphing

Step 1 **Graph each equation** of the system on the same coordinate axes.

Step 2 **Find the coordinates of the point of intersection** of the graphs, if possible. This is the solution of the system.

Step 3 **Check** the solution in both of the original equations. Then write the solution set.

CAUTION A difficulty with the graphing method of solution is that it may not be possible to determine from the graph the exact coordinates of the point that represents the solution, particularly if these coordinates are not integers. The graphing method does, however, show geometrically how solutions are found and is useful when approximate answers will do.

OBJECTIVE 3 Solve special systems by graphing. Sometimes the graphs of the two equations in a system either do not intersect at all or are the same line.

▮ EXAMPLE 3 Solving Special Systems by Graphing

Solve each system by graphing.

(a) $2x + y = 2$
$2x + y = 8$

The graphs of these lines are shown in Figure 3. The two lines are parallel and have no points in common. For such a system, there is no solution; we write the solution set as $\emptyset$.

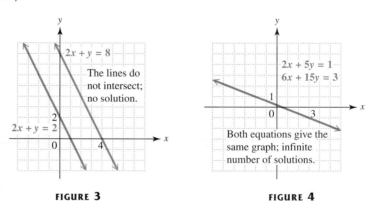

FIGURE 3

FIGURE 4

(b) $2x + 5y = 1$
$6x + 15y = 3$

The graphs of these two equations are the same line. See Figure 4. We can obtain the second equation by multiplying each side of the first equation by 3. In this case, every point on the line is a solution of the system, and the solution set contains an infinite number of ordered pairs. We write the solution set using set-builder notation as $\{(x, y) \mid 2x + 5y = 1\}$.

NOTE In Example 3(b), either equation of the system could be used to write the solution set. We prefer to use the equation (in standard form) with coefficients that are integers ($A > 0$) having no common factor (except 1). Other texts may express such solutions differently.

The system in Example 2 has exactly one solution. A system with at least one solution is called a **consistent system.** A system with no solution, such as the one in Example 3(a), is called an **inconsistent system.** The equations in Example 2 are **independent equations** with different graphs. The equations of the system in Example 3(b) have the same graph and are equivalent. Because they are different forms of the same equation, these equations are called **dependent equations.** Examples 2 and 3 show the three cases that may occur when solving a system of equations with two variables.

Possible Types of Solutions

1. The graphs intersect at exactly one point, which gives the (single) ordered pair solution of the system. The **system is consistent** and the **equations are independent.** See Figure 5(a).

2. The graphs are parallel lines, so there is no solution and the solution set is ∅. The **system is inconsistent.** See Figure 5(b).

3. The graphs are the same line. There are an infinite number of solutions. The **equations are dependent.** See Figure 5(c).

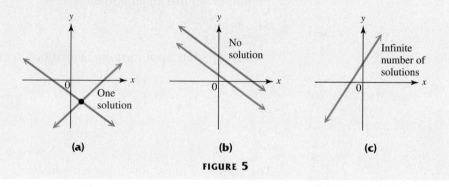

FIGURE 5

Now Try Exercises 23 and 25.

OBJECTIVE 4 Identify special systems without graphing. Example 3 showed that the graphs of an inconsistent system are parallel lines and the graphs of a system of dependent equations are the same line. We can recognize these special kinds of systems without graphing by using slopes.

EXAMPLE 4 Identifying the Three Cases Using Slopes

Describe each system without graphing.

(a) $3x + 2y = 6$
    $-2y = 3x - 5$

Write each equation in slope-intercept form by solving for y.

$$3x + 2y = 6 \qquad\qquad -2y = 3x - 5$$
$$2y = -3x + 6 \qquad\qquad y = -\frac{3}{2}x + \frac{5}{2}$$
$$y = -\frac{3}{2}x + 3$$

Both equations have slope $-\frac{3}{2}$ but they have different y-intercepts, 3 and $\frac{5}{2}$. In Chapter 3 we found that lines with the same slope are parallel, so these equations have graphs that are parallel lines. The system has no solution.

(b) $2x - y = 4$

$$x = \frac{y}{2} + 2$$

Again, write the equations in slope-intercept form.

$$2x - y = 4 \qquad\qquad x = \frac{y}{2} + 2$$
$$-y = -2x + 4 \qquad\qquad \frac{y}{2} + 2 = x$$
$$y = 2x - 4 \qquad\qquad \frac{y}{2} = x - 2$$
$$\qquad\qquad\qquad\qquad y = 2x - 4$$

The equations are exactly the same; their graphs are the same line. The system has an infinite number of solutions.

(c) $x - 3y = 5$
$2x + \ y = 8$

In slope-intercept form, the equations are as follows.

$$x - 3y = 5 \qquad\qquad 2x + y = 8$$
$$-3y = -x + 5 \qquad\qquad y = -2x + 8$$
$$y = \frac{1}{3}x - \frac{5}{3}$$

The graphs of these equations are neither parallel nor the same line since the slopes are different. This system has exactly one solution.

Now Try Exercises 31, 33, and 35.

OBJECTIVE 5 Use a graphing calculator to solve a linear system. In the next example we use a graphing calculator to find the solution of the system in Example 1(a).

EXAMPLE 5 Finding the Solution Set of a System with a Graphing Calculator

Solve the system by graphing with a calculator.

$$x + 4y = -8$$
$$3x + 2y = 6$$

To enter the equations in a graphing calculator, first solve each equation for y.

$$x + 4y = -8$$
$$4y = -x - 8$$
$$y = -\frac{1}{4}x - 2$$

$$3x + 2y = 6$$
$$2y = -3x + 6$$
$$y = -\frac{3}{2}x + 3$$

The calculator allows us to enter several equations to be graphed at the same time. We designate the first one Y_1 and the second one Y_2. See Figure 6(a). Notice the careful use of parentheses with the fractions. We graph the two equations using a standard window and then use the capability of the calculator to find the coordinates of the point of intersection of the graphs. The display at the bottom of Figure 6(b) indicates that the solution set is $\{(4, -3)\}$.

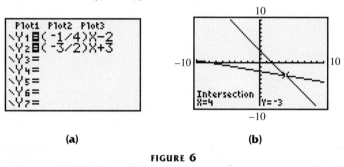

(a) (b)

FIGURE 6

Now Try Exercise 51.

8.1 EXERCISES

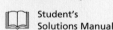

1. Each ordered pair in (a)–(d) is a solution of one of the systems graphed in A–D. Because of the location of the point of intersection, you should be able to determine the correct system for each solution. Match each system from A–D with its solution from (a)–(d).

(a) $(3, 4)$

(b) $(-2, 3)$

(c) $(-4, -1)$

(d) $(5, -2)$

A.

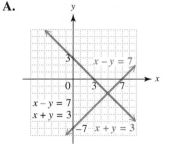

B.

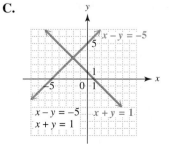

C.

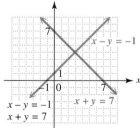

D.

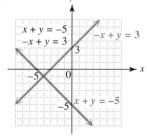

✎ **2.** When a student was asked to determine whether the ordered pair $(1, -2)$ is a solution of the system

$$x + y = -1$$
$$2x + y = 4,$$

he answered "yes." His reasoning was that the ordered pair satisfies the equation $x + y = -1$ since $1 + (-2) = -1$ is true. Why is the student's answer wrong?

Decide whether the given ordered pair is a solution of the given system. See Example 1.

3. $(2, -3)$
$$x + y = -1$$
$$2x + 5y = 19$$

4. $(4, 3)$
$$x + 2y = 10$$
$$3x + 5y = 3$$

5. $(-1, -3)$
$$3x + 5y = -18$$
$$4x + 2y = -10$$

6. $(-9, -2)$
$$2x - 5y = -8$$
$$3x + 6y = -39$$

7. $(7, -2)$
$$4x = 26 - y$$
$$3x = 29 + 4y$$

8. $(9, 1)$
$$2x = 23 - 5y$$
$$3x = 24 + 3y$$

9. $(6, -8)$
$$-2y = x + 10$$
$$3y = 2x + 30$$

10. $(-5, 2)$
$$5y = 3x + 20$$
$$3y = -2x - 4$$

11. Which ordered pair could not possibly be a solution of the system graphed? Why is it the only valid choice?

A. $(-4, -4)$ **B.** $(-2, 2)$

C. $(-4, 4)$ **D.** $(-3, 3)$

12. Which ordered pair could possibly be a solution of the system graphed? Why is it the only valid choice?

A. $(2, 0)$ **B.** $(0, 2)$

C. $(-2, 0)$ **D.** $(0, -2)$

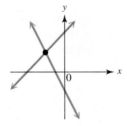

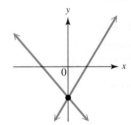

Solve each system of equations by graphing. If the system is inconsistent or the equations are dependent, say so. See Examples 2 and 3.

13. $x - y = 2$
$$x + y = 6$$

14. $x - y = 3$
$$x + y = -1$$

15. $x + y = 4$
$$y - x = 4$$

16. $x + y = -5$
$$x - y = 5$$

17. $x - 2y = 6$
$$x + 2y = 2$$

18. $2x - y = 4$
$$4x + y = 2$$

19. $3x - 2y = -3$
$$-3x - y = -6$$

20. $2x - y = 4$
$$2x + 3y = 12$$

21. $2x - 3y = -6$
$$y = -3x + 2$$

22. $x + 2y = 4$
$$2x + 4y = 12$$

23. $2x - y = 6$
$$4x - 2y = 8$$

24. $2x - y = 4$
$$4x = 2y + 8$$

25. $3x = 5 - y$
   $6x + 2y = 10$

26. $-3x + y = -3$
   $y = x - 3$

27. $3x - 4y = 24$
   $y = -\dfrac{3}{2}x + 3$

✎ **28.** Solve the system

$$2x + 3y = 6$$
$$x - 3y = 5$$

by graphing. Can you check your solution? Why or why not?

✎ **29.** Explain one of the drawbacks of solving a system of equations graphically.

✎ **30.** Explain the three situations that may occur (regarding the number of solutions) when solving a system of two linear equations in two variables by graphing.

Without graphing, answer the following questions for each linear system. See Example 4.

(a) *Is the system inconsistent, are the equations dependent, or neither?*
(b) *Is the graph a pair of intersecting lines, a pair of parallel lines, or one line?*
(c) *Does the system have one solution, no solution, or an infinite number of solutions?*

31. $y - x = -5$
   $x + y = 1$

32. $2x + y = 6$
   $x - 3y = -4$

33. $x + 2y = 0$
   $4y = -2x$

34. $y = 3x$
   $y + 3 = 3x$

35. $5x + 4y = 7$
   $10x + 8y = 4$

36. $2x + 3y = 12$
   $2x - y = 4$

37. $x - 3y = 5$
   $2x + y = 8$

38. $2x - y = 4$
   $x = \dfrac{1}{2}y + 2$

*An application of mathematics in economics deals with **supply and demand.** Typically, as the price of an item increases, the demand for the item decreases, while the supply increases. (There are exceptions to this, however.) If supply and demand can be described by straight-line equations, the point at which the lines intersect determines the **equilibrium supply** and **equilibrium demand.***

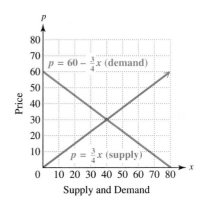

Supply and Demand

Suppose that an economist has studied the supply and demand for aluminum siding and has concluded that the price per unit, p, and the demand, x, are related by the demand equation $p = 60 - \frac{3}{4}x$, while the supply is given by the equation $p = \frac{3}{4}x$. The graphs of these two equations are shown in the figure.

Use the graph to answer Exercises 39–42.

39. At what value of x does supply equal demand?

40. At what value of p does supply equal demand?

41. What are the coordinates of the point of intersection of the two lines?

42. When $x > 40$, does demand exceed supply or does supply exceed demand?

The personal computer market share for two manufacturers is shown in the graph.

43. For which years was Hewlett-Packard's share less than Packard Bell, NEC's share?

44. Estimate the year in which market share for Hewlett-Packard and Packard Bell, NEC was the same. About what was this share?

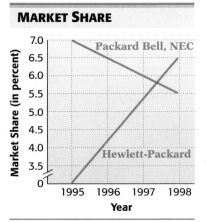

MARKET SHARE

Source: Intelliquest; IDC.

45. The graph shows network share (the percentage of TV sets in use) for the early evening news programs for the three major broadcast networks from 1986 through 2000.

(a) Between what years did the ABC early evening news dominate?

(b) During what year did ABC's dominance end? Which network equaled ABC's share that year? What was that share?

(c) During what years did ABC and CBS have equal network share? What was the share for each of these years?

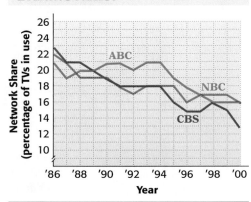

WHO'S WATCHING THE EVENING NEWS?

Source: Nielsen Media Research.

(d) Which networks most recently had equal share? Write their share as an ordered pair of the form (year, share).

 (e) Describe the general trend in viewership for the three major networks during these years.

46. The graph shows how the production of vinyl LPs, audiocassettes, and compact discs (CDs) changed over the years from 1986 through 1998.

(a) In what year did cassette production and CD production reach equal levels? What was that level?

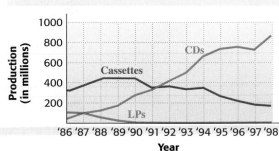

THE SOUNDS OF MUSIC

Source: Recording Industry Association of America.

(b) Express the point of intersection of the graphs of LP production and CD production as an ordered pair of the form (year, production level).

(c) Between what years did cassette production first stabilize and remain fairly constant?

 (d) Describe the trend in CD production from 1986 through 1998. If a straight line were used to approximate its graph, would the line have positive, negative, or 0 slope?

 (e) If a straight line were used to approximate the graph of cassette production from 1990 through 1998, would the line have positive, negative, or 0 slope? Explain.

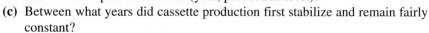

TECHNOLOGY INSIGHTS (EXERCISES 47–50)

Match the graphing calculator screens in choices A–D with the appropriate system in Exercises 47–50. See Example 5.

A.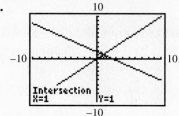

B.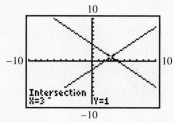

C.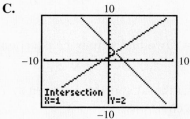

D.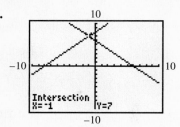

47. $x + y = 4$
$x - y = 2$

48. $x + y = 6$
$x - y = -8$

49. $2x + 3y = 5$
$x - y = 0$

50. $3x + 2y = 7$
$-x + y = 1$

Use a graphing calculator to solve each system. See Example 5.

51. $3x + y = 2$
$2x - y = -7$

52. $x + 2y = -2$
$2x - y = 11$

53. $8x + 4y = 0$
$4x - 2y = 2$

54. $3x + 3y = 0$
$4x + 2y = 3$

8.2 Solving Systems of Linear Equations by Substitution

OBJECTIVES

1 Solve linear systems by substitution.

2 Solve special systems by substitution.

3 Solve linear systems with fractions.

OBJECTIVE 1 Solve linear systems by substitution. Graphing to solve a system of equations has a serious drawback: It is difficult to accurately find a solution such as $\left(\frac{1}{3}, -\frac{5}{6}\right)$ from a graph. One algebraic method for solving a system of equations is the **substitution method.** This method is particularly useful for solving systems where one equation is already solved, or can be solved quickly, for one of the variables.

EXAMPLE 1 Using the Substitution Method

Solve the system by the substitution method.

$$3x + 5y = 26$$
$$y = 2x$$

The second equation is already solved for y. This equation says that $y = 2x$. Substituting $2x$ for y in the first equation gives

$$3x + 5y = 26$$
$$3x + 5(2x) = 26 \quad \text{Let } y = 2x.$$
$$3x + 10x = 26 \quad \text{Multiply.}$$
$$13x = 26 \quad \text{Combine like terms.}$$
$$x = 2. \quad \text{Divide by 13.}$$

Because $x = 2$, we find y from the equation $y = 2x$ by substituting 2 for x.

$$y = 2(2) = 4 \quad \text{Let } x = 2.$$

We check that the solution of the given system is $(2, 4)$ by substituting 2 for x and 4 for y in *both* equations.

Check:

$3x + 5y = 26$	$y = 2x$
$3(2) + 5(4) = 26$?	$4 = 2(2)$?
$6 + 20 = 26$?	$4 = 4$ True
$26 = 26$ True	

Since $(2, 4)$ satisfies both equations, the solution set is $\{(2, 4)\}$.

Now Try Exercise 3.

EXAMPLE 2 Using the Substitution Method

Solve the system

$$2x + 5y = 7$$
$$x = -1 - y.$$

The second equation gives x in terms of y. Substitute $-1 - y$ for x in the first equation.

$$2x + 5y = 7$$
$$2(-1 - y) + 5y = 7 \quad \text{Let } x = -1 - y.$$
$$-2 - 2y + 5y = 7 \quad \text{Distributive property}$$

$$-2 + 3y = 7 \qquad \text{Combine like terms.}$$
$$3y = 9 \qquad \text{Add 2.}$$
$$y = 3 \qquad \text{Divide by 3.}$$

To find x, substitute 3 for y in the equation $x = -1 - y$ to get

$$x = -1 - 3 = -4.$$

Check that the solution set of the given system is $\{(-4, 3)\}$.

Now Try Exercise 5.

CAUTION Be careful when you write the ordered pair solution of a system. Even though we found y first in Example 2, the x-coordinate is *always* written first in the ordered pair.

To solve a system by substitution, follow these steps.

Solving a Linear System by Substitution

Step 1 **Solve one equation for either variable.** If one of the variables has coefficient 1 or -1, choose it, since the substitution method is usually easier this way.

Step 2 **Substitute** for that variable in the other equation. The result should be an equation with just one variable.

Step 3 **Solve** the equation from Step 2.

Step 4 **Substitute** the result from Step 3 into the equation from Step 1 to find the value of the other variable.

Step 5 **Check** the solution in both of the original equations. Then write the solution set.

▨ **EXAMPLE 3** Using the Substitution Method

Use substitution to solve the system

$$2x = 4 - y \qquad (1)$$
$$5x + 3y = 10. \qquad (2)$$

Step 1 We must solve one of the equations for either x or y. Because the coefficient of y in equation (1) is -1, we avoid fractions by solving this equation for y.

$$2x = 4 - y \qquad (1)$$
$$2x - 4 = -y \qquad \text{Subtract 4.}$$
$$-2x + 4 = y \qquad \text{Multiply by } -1.$$

Step 2 Now substitute $-2x + 4$ for y in equation (2).

$$5x + 3y = 10 \qquad (2)$$
$$5x + 3(-2x + 4) = 10 \qquad \text{Let } y = -2x + 4.$$

Step 3 Solve the equation from Step 2.

$$5x - 6x + 12 = 10 \qquad \text{Distributive property}$$
$$-x + 12 = 10 \qquad \text{Combine like terms.}$$
$$-x = -2 \qquad \text{Subtract 12.}$$
$$x = 2 \qquad \text{Multiply by } -1.$$

Step 4 Since $y = -2x + 4$ and $x = 2$,

$$y = -2(2) + 4 = 0,$$

and the solution is $(2, 0)$.

Step 5 *Check:*

$$
\begin{array}{lll|lll}
2x = 4 - y & (1) & & 5x + 3y = 10 & (2) \\
2(2) = 4 - 0 & \text{?} & & 5(2) + 3(0) = 10 & \text{?} \\
4 = 4 & \text{True} & & 10 = 10 & \text{True}
\end{array}
$$

Since both results are true, the solution set of the system is $\{(2, 0)\}$.

Now Try Exercise 7.

OBJECTIVE 2 Solve special systems by substitution. In the previous section we solved inconsistent systems with graphs that are parallel lines and systems of dependent equations with graphs that are the same line. We can also solve these special systems with the substitution method.

EXAMPLE 4 Solving an Inconsistent System by Substitution

Use substitution to solve the system

$$x = 5 - 2y \qquad (1)$$
$$2x + 4y = 6. \qquad (2)$$

Substitute $5 - 2y$ for x in equation (2).

$$2x + 4y = 6 \qquad (2)$$
$$2(5 - 2y) + 4y = 6 \qquad \text{Let } x = 5 - 2y.$$
$$10 - 4y + 4y = 6 \qquad \text{Distributive property}$$
$$10 = 6 \qquad \text{False}$$

This false result means that the equations in the system have graphs that are parallel lines. The system is inconsistent and has no solution, so the solution set is $\emptyset$. See Figure 7.

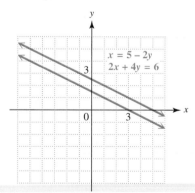

FIGURE 7

Now Try Exercise 17.

CAUTION It is a common error to give "false" as the solution of an inconsistent system. The correct response is ∅.

EXAMPLE 5 Solving a System with Dependent Equations by Substitution

Solve the system by the substitution method.

$$3x - y = 4 \qquad (1)$$

$$-9x + 3y = -12 \qquad (2)$$

Begin by solving equation (1) for y to get $y = 3x - 4$. Substitute $3x - 4$ for y in equation (2) and solve the resulting equation.

$$-9x + 3y = -12 \qquad (2)$$

$$-9x + 3(3x - 4) = -12 \qquad \text{Let } y = 3x - 4.$$

$$-9x + 9x - 12 = -12 \qquad \text{Distributive property}$$

$$0 = 0 \qquad \text{Add 12; combine like terms.}$$

This true result means that every solution of one equation is also a solution of the other, so the system has an infinite number of solutions: all the ordered pairs corresponding to points that lie on the common graph. The solution set is $\{(x, y) \mid 3x - y = 4\}$. A graph of the equations of this system is shown in Figure 8.

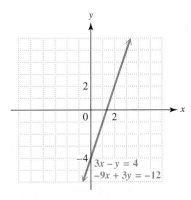

FIGURE 8

Now Try Exercise 19.

CAUTION It is a common error to give "true" as the solution of a system of dependent equations. The correct response is "infinite number of solutions."

OBJECTIVE 3 Solve linear systems with fractions. When a system includes an equation with fractions as coefficients, eliminate the fractions by multiplying each side of the equation by a common denominator. Then solve the resulting system.

▨ **EXAMPLE 6** **Using the Substitution Method with Fractions as Coefficients**

Solve the system by the substitution method.

$$3x + \frac{1}{4}y = 2 \qquad (1)$$

$$\frac{1}{2}x + \frac{3}{4}y = -\frac{5}{2} \qquad (2)$$

Clear equation (1) of fractions by multiplying each side by 4.

$$4\left(3x + \frac{1}{4}y\right) = 4(2) \qquad \text{Multiply by 4.}$$

$$4(3x) + 4\left(\frac{1}{4}y\right) = 4(2) \qquad \text{Distributive property}$$

$$12x + y = 8 \qquad (3)$$

Now clear equation (2) of fractions by multiplying each side by the common denominator 4.

$$4\left(\frac{1}{2}x + \frac{3}{4}y\right) = 4\left(-\frac{5}{2}\right) \qquad \text{Multiply by 4.}$$

$$4\left(\frac{1}{2}x\right) + 4\left(\frac{3}{4}y\right) = 4\left(-\frac{5}{2}\right) \qquad \text{Distributive property}$$

$$2x + 3y = -10 \qquad (4)$$

The given system of equations has been simplified to the equivalent system

$$12x + y = 8 \qquad (3)$$

$$2x + 3y = -10. \qquad (4)$$

Solve this system by the substitution method. Equation (3) can be solved for y by subtracting $12x$ from each side.

$$12x + y = 8 \qquad (3)$$

$$y = -12x + 8 \qquad \text{Subtract } 12x.$$

Now substitute this result for y in equation (4).

$$2x + 3y = -10 \qquad (4)$$

$$2x + 3(-12x + 8) = -10 \qquad \text{Let } y = -12x + 8.$$

$$2x - 36x + 24 = -10 \qquad \text{Distributive property}$$

$$-34x = -34 \qquad \text{Combine like terms; subtract 24.}$$

$$x = 1 \qquad \text{Divide by } -34.$$

Substitute 1 for x in $y = -12x + 8$ to get

$$y = -12(1) + 8 = -4.$$

Check by substituting 1 for x and -4 for y in both of the original equations. The solution set is $\{(1, -4)\}$.

▨

Now Try Exercise 25.

8.2 EXERCISES

1. A student solves the system

$$5x - y = 15$$
$$7x + y = 21$$

and finds that $x = 3$, which is the correct value for x. The student gives the solution set as $\{3\}$. Is this correct? Explain.

2. A student solves the system

$$x + y = 4$$
$$2x + 2y = 8$$

and obtains the equation $0 = 0$. The student gives the solution set as $\{(0, 0)\}$. Is this correct? Explain.

Solve each system by the substitution method. Check each solution. See Examples 1–5.

3. $x + y = 12$
$y = 3x$

4. $x + 3y = -28$
$y = -5x$

5. $3x + 2y = 27$
$x = y + 4$

6. $4x + 3y = -5$
$x = y - 3$

7. $3x + 5y = 25$
$x - 2y = -10$

8. $5x + 2y = -15$
$2x - y = -6$

9. $3x + 4 = -y$
$2x + y = 0$

10. $2x - 5 = -y$
$x + 3y = 0$

11. $7x + 4y = 13$
$x + y = 1$

12. $3x - 2y = 19$
$x + y = 8$

13. $3x - y = 5$
$y = 3x - 5$

14. $4x - y = -3$
$y = 4x + 3$

15. $2x + y = 0$
$4x - 2y = 2$

16. $x + y = 0$
$4x + 2y = 3$

17. $2x + 8y = 3$
$x = 8 - 4y$

18. $2x + 10y = 3$
$x = 1 - 5y$

19. $2x = -12 + y$
$2y = 4x + 24$

20. $3x = 7 - y$
$2y = 14 - 6x$

21. When you use the substitution method, how can you tell that a system has

(a) no solution?
(b) an infinite number of solutions?

22. Solve each system.

(a) $5x - 4y = 7$
$x = 3$

(b) $5x - 4y = 7$
$y = -3$

Why are these systems easier to solve than the examples in this section?

Solve each system by the substitution method. First clear all fractions. Check each solution. See Example 6.

23. $\dfrac{1}{2}x + \dfrac{1}{3}y = 3$
$y = 3x$

24. $\dfrac{1}{4}x - \dfrac{1}{5}y = 9$
$y = 5x$

25. $\dfrac{1}{2}x + \dfrac{1}{3}y = -\dfrac{1}{3}$
$\dfrac{1}{2}x + 2y = -7$

26. $\dfrac{1}{6}x + \dfrac{1}{6}y = 1$
$-\dfrac{1}{2}x - \dfrac{1}{3}y = -5$

27. $\dfrac{x}{5} + 2y = \dfrac{8}{5}$
$\dfrac{3x}{5} + \dfrac{y}{2} = -\dfrac{7}{10}$

28. $\dfrac{x}{2} + \dfrac{y}{3} = \dfrac{7}{6}$
$\dfrac{x}{4} - \dfrac{3y}{2} = \dfrac{9}{4}$

29. $\dfrac{x}{5} + y = \dfrac{6}{5}$

$\dfrac{x}{10} + \dfrac{y}{3} = \dfrac{5}{6}$

30. $\dfrac{1}{2}x - \dfrac{1}{8}y = -\dfrac{1}{4}$

$-4x + y = 2$

31. $\dfrac{1}{6}x + \dfrac{1}{3}y = 8$

$\dfrac{1}{4}x + \dfrac{1}{2}y = 12$

32. One student solved the system

$$\dfrac{1}{3}x - \dfrac{1}{2}y = 7$$

$$\dfrac{1}{6}x + \dfrac{1}{3}y = 0$$

and wrote as his answer "$x = 12$," while another solved it and wrote as her answer "$y = -6$." Who, if either, was correct? Why?

RELATING CONCEPTS (EXERCISES 33–36)

For Individual or Group Work

A system of linear equations can be used to model the cost and the revenue of a business. **Work Exercises 33–36 in order.**

33. Suppose that you start a business manufacturing and selling bicycles, and it costs you $5000 to get started. You determine that each bicycle will cost $400 to manufacture. Explain why the linear equation $y_1 = 400x + 5000$ gives your *total* cost to manufacture x bicycles (y_1 in dollars).

34. You decide to sell each bike for $600. What expression in x represents the revenue you will take in if you sell x bikes? Write an equation using y_2 to express your revenue when you sell x bikes (y_2 in dollars).

35. Form a system from the two equations in Exercises 33 and 34 and then solve the system.

36. The value of x from Exercise 35 is the number of bikes it takes to *break even*. Fill in the blanks: When _____ bikes are sold, the break-even point is reached. At that point, you have spent _____ dollars and taken in _____ dollars.

Solve each system by substitution. Then graph both lines in the standard viewing window of a graphing calculator and use the intersection feature to support your answer. See Example 5 in Section 8.1. (In Exercises 41 and 42, you will need to solve each equation for y first before graphing.)

37. $y = 6 - x$

$y = 2x$

38. $y = 4x - 4$

$y = -3x - 11$

39. $y = -\dfrac{4}{3}x + \dfrac{19}{3}$

$y = \dfrac{15}{2}x - \dfrac{5}{2}$

40. $y = -\dfrac{15}{2}x + 10$

$y = \dfrac{25}{3}x - \dfrac{65}{3}$

41. $4x + 5y = 5$

$2x + 3y = 1$

42. $6x + 5y = 13$

$3x + 3y = 4$

43. If the point of intersection does not appear on your screen when solving a linear system using a graphing calculator, how can you find the point of intersection?

44. Suppose that you were asked to solve the system

$$y = 1.73x + 5.28$$
$$y = -2.94x - 3.85.$$

Why would it probably be easier to solve this system using a graphing calculator than using the substitution method?

8.3 Solving Systems of Linear Equations by Elimination

OBJECTIVES

1 Solve linear systems by elimination.

2 Multiply when using the elimination method.

3 Use an alternative method to find the second value in a solution.

4 Use the elimination method to solve special systems.

OBJECTIVE 1 Solve linear systems by elimination.   An algebraic method that depends on the addition property of equality can also be used to solve systems. As mentioned earlier, adding the same quantity to each side of an equation results in equal sums.

$$\text{If} \quad A = B, \quad \text{then} \quad A + C = B + C.$$

We can take this addition a step further. Adding *equal* quantities, rather than the *same* quantity, to each side of an equation also results in equal sums.

$$\text{If} \quad A = B \quad \text{and} \quad C = D, \quad \text{then} \quad A + C = B + D.$$

Using the addition property to solve systems is called the **elimination method.** When using this method, the idea is to *eliminate* one of the variables. To do this, one pair of variable terms in the two equations must have coefficients that are opposites.

■ **EXAMPLE 1   Using the Elimination Method**

Use the elimination method to solve the system

$$x + y = 5$$
$$x - y = 3.$$

Each equation in this system is a statement of equality, so the sum of the right sides equals the sum of the left sides. Adding in this way gives

$$(x + y) + (x - y) = 5 + 3.$$

Combine terms and simplify to get

$$2x = 8$$
$$x = 4. \qquad \text{Divide by 2.}$$

Notice that y has been eliminated. The result, $x = 4$, gives the x-value of the solution of the given system. To find the y-value of the solution, substitute 4 for x in either of the two equations of the system. Choosing the first equation, $x + y = 5$, gives

$$x + y = 5$$
$$4 + y = 5 \qquad \text{Let } x = 4.$$
$$y = 1. \qquad \text{Subtract 4.}$$

Check the solution, (4, 1), by substituting 4 for x and 1 for y in both equations of the given system.

$x + y = 5$		$x - y = 3$	
$4 + 1 = 5$	?	$4 - 1 = 3$	?
$5 = 5$	True	$3 = 3$	True

Since both results are true, the solution set of the system is $\{(4, 1)\}$.

Now Try Exercise 5.

CAUTION A system is not completely solved until values for *both* x and y are found. Do not stop after finding the value of only one variable. Remember to write the solution set as a set containing an ordered pair.

In general, use the following steps to solve a linear system of equations by the elimination method.

Solving a Linear System by Elimination

Step 1 **Write both equations in standard form** $Ax + By = C$.

Step 2 **Transform so that the coefficients of one pair of variable terms are opposites.** Multiply one or both equations by appropriate numbers so that the sum of the coefficients of either the x- or y-terms is 0.

Step 3 **Add** the new equations to eliminate a variable. The sum should be an equation with just one variable.

Step 4 **Solve** the equation from Step 3 for the remaining variable.

Step 5 **Substitute** the result from Step 4 into either of the original equations and solve for the other variable.

Step 6 **Check** the solution in both of the original equations. Then write the solution set.

It does not matter which variable is eliminated first. Usually we choose the one that is more convenient to work with.

EXAMPLE 2 Using the Elimination Method

Solve the system

$$y + 11 = 2x$$
$$5x = y + 26.$$

Step 1 Rewrite both equations in standard form $Ax + By = C$ to get the system

$$-2x + y = -11 \quad \text{Subtract } 2x \text{ and } 11.$$
$$5x - y = 26. \quad \text{Subtract } y.$$

Step 2 Because the coefficients of y are 1 and -1, adding will eliminate y. It is not necessary to multiply either equation by a number.

Step 3 Add the two equations. This time we use vertical addition.

$$\begin{aligned} -2x + y &= -11 \\ \underline{5x - y} &= \underline{26} \\ 3x &= 15 \end{aligned}$$ Add in columns.

Step 4 Solve the equation.

$$3x = 15$$
$$x = 5 \qquad \text{Divide by 3.}$$

Step 5 Find the value of y by substituting 5 for x in either of the original equations. Choosing the first gives

$$\begin{aligned} y + 11 &= 2x \\ y + 11 &= 2(5) \\ y + 11 &= 10 \qquad \text{Let } x = 5. \\ y &= -1. \qquad \text{Subtract 11.} \end{aligned}$$

Step 6 Check by substituting $x = 5$ and $y = -1$ in both of the original equations.

Check:

$y + 11 = 2x$	$5x = y + 26$
$(-1) + 11 = 2(5)$?	$5(5) = -1 + 26$?
$10 = 10$ True	$25 = 25$ True

Since $(5, -1)$ is a solution of *both* equations, the solution set is $\{(5, -1)\}$.

Now Try Exercise 9.

OBJECTIVE 2 Multiply when using the elimination method. In both of the preceding examples, a variable was eliminated by adding the equations. Sometimes we need to multiply each side of one or both equations in a system by some number before adding will eliminate a variable.

EXAMPLE 3 Multiplying Both Equations When Using the Elimination Method

Solve the system

$$\begin{aligned} 2x + 3y &= -15 \qquad (1) \\ 5x + 2y &= 1. \qquad (2) \end{aligned}$$

Adding the two equations gives $7x + 5y = -14$, which does not eliminate either variable. However, we can multiply each equation by a suitable number so that the coefficients of one of the two variables are opposites. For example, to eliminate x, we multiply each side of equation (1) by 5, and each side of equation (2) by -2.

$$\begin{aligned} 10x + 15y &= -75 \qquad \text{Multiply equation (1) by 5.} \\ \underline{-10x - 4y} &= \underline{-2} \qquad \text{Multiply equation (2) by } -2. \\ 11y &= -77 \qquad \text{Add.} \\ y &= -7 \qquad \text{Divide by 11.} \end{aligned}$$

Substituting -7 for y in either equation (1) or (2) gives $x = 3$. Check that the solution set of the system is $\{(3, -7)\}$.

Now Try Exercise 19.

CAUTION   When using the elimination method, remember to *multiply both sides* of an equation by the same nonzero number.

OBJECTIVE 3   Use an alternative method to find the second value in a solution.   Sometimes it is easier to find the value of the second variable in a solution by using the elimination method twice. The next example shows this approach.

■ EXAMPLE 4   Finding the Second Value Using an Alternative Method

Solve the system

$$4x = 9 - 3y \qquad (1)$$
$$5x - 2y = 8. \qquad (2)$$

Rearrange the terms in equation (1) so that like terms are aligned in columns. Add $3y$ to each side to get the system

$$4x + 3y = 9 \qquad (3)$$
$$5x - 2y = 8. \qquad (2)$$

One way to proceed is to eliminate y by multiplying each side of equation (3) by 2 and each side of equation (2) by 3, and then adding.

$8x + 6y = 18$	Multiply equation (3) by 2.
$15x - 6y = 24$	Multiply equation (2) by 3.
$23x \qquad = 42$	Add.
$x = \dfrac{42}{23}$	Divide by 23.

Substituting $\frac{42}{23}$ for x in one of the given equations would give y, but the arithmetic involved would be messy. Instead, solve for y by starting again with the original equations and eliminating x. Multiply each side of equation (3) by 5 and each side of equation (2) by -4, and then add.

$20x + 15y = 45$	Multiply equation (3) by 5.
$-20x + 8y = -32$	Multiply equation (2) by -4.
$23y = 13$	Add.
$y = \dfrac{13}{23}$	Divide by 23.

Check that the solution set is $\left\{\left(\frac{42}{23}, \frac{13}{23}\right)\right\}$.

Now Try Exercise 27.

NOTE   When the value of the first variable is a fraction, the method used in Example 4 helps avoid arithmetic errors. Of course, this method could be used to solve any system of equations.

OBJECTIVE 4   Use the elimination method to solve special systems.   The next example shows the elimination method when a system is inconsistent or the equations of the system are dependent. To contrast the elimination method with the substitution method, in part (b) we use the same system solved in Example 5 of the previous section.

> **EXAMPLE 5   Using the Elimination Method for an Inconsistent System or Dependent Equations**

Solve each system by the elimination method.

(a) $2x + 4y = 5$
$4x + 8y = -9$

Multiply each side of $2x + 4y = 5$ by -2; then add to $4x + 8y = -9$.

$$-4x - 8y = -10$$
$$\underline{4x + 8y = \quad -9}$$
$$0 = -19 \qquad \text{False}$$

The false statement $0 = -19$ indicates that the given system has solution set $\emptyset$.

(b) $\quad 3x - \quad y = 4$
$-9x + 3y = -12$

Multiply each side of the first equation by 3; then add the two equations.

$$9x - 3y = \quad 12$$
$$\underline{-9x + 3y = -12}$$
$$0 = \quad 0 \qquad \text{True}$$

A true statement occurs when the equations are equivalent. As before, this result indicates that every solution of one equation is also a solution of the other; there are an infinite number of solutions. The solution set is $\{(x, y) \mid 3x - y = 4\}$.

> **Now Try Exercises 35 and 37.**

8.3 EXERCISES

For Extra Help

Student's
Solutions Manual

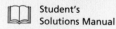
MyMathLab

InterAct Math
Tutorial Software

AW Math
Tutor Center

*Math*XP MathXL

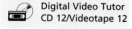
Digital Video Tutor
CD 12/Videotape 12

Answer true *or* false *for each statement. If false, tell why.*

1. To eliminate the y-terms in the system

$$2x + 12y = 7$$
$$3x + \quad 4y = 1,$$

we should multiply the bottom equation by 3 and then add.

2. The ordered pair $(0, 0)$ *must* be a solution of a system of the form

$$Ax + By = 0$$
$$Cx + Dy = 0.$$

3. The system

$$x + y = 1$$
$$x + y = 2$$

has $\emptyset$ as its solution set.

4. The ordered pair $(4, -5)$ cannot be a solution of a system that contains the equation $5x - 4y = 0$.

Solve each system by the elimination method. Check each solution. See Examples 1 and 2.

5. $x - y = -2$
  $x + y = 10$

6. $x + y = 10$
  $x - y = -6$

7. $2x + y = -5$
  $x - y = 2$

8. $2x + y = -15$
  $-x - y = 10$

9. $2y = -3x$
  $-3x - y = 3$

10. $5x = y + 5$
  $-5x + 2y = 0$

11. $6x - y = -1$
  $5y = 17 + 6x$

12. $y = 9 - 6x$
  $-6x + 3y = 15$

Solve each system by the elimination method. (Hint: In Exercises 33 and 34, first clear all fractions.) Check each solution. See Examples 3–5.*

13. $2x - y = 12$
  $3x + 2y = -3$

14. $x + y = 3$
  $-3x + 2y = -19$

15. $x + 4y = 16$
  $3x + 5y = 20$

16. $2x + y = 8$
  $5x - 2y = -16$

17. $2x - 8y = 0$
  $4x + 5y = 0$

18. $3x - 15y = 0$
  $6x + 10y = 0$

19. $3x + 3y = 33$
  $5x - 2y = 27$

20. $4x - 3y = -19$
  $3x + 2y = 24$

21. $5x + 4y = 12$
  $3x + 5y = 15$

22. $2x + 3y = 21$
  $5x - 2y = -14$

23. $5x - 4y = 15$
  $-3x + 6y = -9$

24. $4x + 5y = -16$
  $5x - 6y = -20$

25. $6x - 2y = -21$
  $-3x + 4y = 36$

26. $6x - 2y = -21$
  $3x + 4y = 36$

27. $3x - 7y = 1$
  $-5x + 4y = 4$

28. $-4x + 3y = 2$
  $5x - 2y = -3$

29. $2x + 3y = 0$
  $4x + 12 = 9y$

30. $-4x + 3y = 2$
  $5x + 3 = -2y$

31. $24x + 12y = -7$
  $16x - 17 = 18y$

32. $9x + 4y = -3$
  $6x + 7 = -6y$

33. $3x = 3 + 2y$
  $-\dfrac{4}{3}x + y = \dfrac{1}{3}$

34. $3x = 27 + 2y$
  $x - \dfrac{7}{2}y = -25$

35. $-x + 3y = 4$
  $-2x + 6y = 8$

36. $6x - 2y = 24$
  $-3x + y = -12$

37. $5x - 2y = 3$
  $10x - 4y = 5$

38. $3x - 5y = 1$
  $6x - 10y = 4$

RELATING CONCEPTS (EXERCISES 39–44)

For Individual or Group Work

The graph on the next page shows movie attendance from 1991 through 1999. In 1991, attendance was 1141 million, as represented by the point $P(1991, 1141)$. In 1999, attendance was 1465 million, as represented by the point $Q(1999, 1465)$. We can find an equation of line segment PQ using a system of equations and then use the equation to approximate the attendance in any of the years between 1991 and 1999. **Work Exercises 39–44 in order.**

---

*The authors thank Mitchel Levy of Broward Community College for his suggestions for this group of exercises.

MOVIE BOX OFFICE ATTENDANCE/ ADMISSIONS

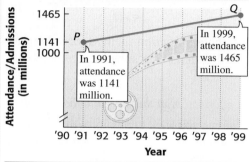

In 1991, attendance was 1141 million.

In 1999, attendance was 1465 million.

Source: Motion Picture Association of America.

39. The line segment has an equation that can be written in the form $y = ax + b$. Using the coordinates of point P with $x = 1991$ and $y = 1141$, write an equation in the variables a and b.

40. Using the coordinates of point Q with $x = 1999$ and $y = 1465$, write a second equation in the variables a and b.

41. Write the system of equations formed from the two equations in Exercises 39 and 40, and solve the system using the elimination method.

42. What is the equation of the segment PQ?

43. Let $x = 1998$ in the equation of Exercise 42, and solve for y. How does the result compare with the actual figure of 1481 million?

44. The actual data points for the years 1991 through 1999 do not lie in a perfectly straight line. Explain the pitfalls of relying too heavily on using the equation in Exercise 42 to approximate attendance.

SUMMARY EXERCISES ON SOLVING SYSTEMS OF LINEAR EQUATIONS

The exercises in this summary include a variety of problems on solving systems of linear equations. Since we do not usually specify the method of solution, use the following guidelines to help you decide whether to use substitution or elimination.

Choosing a Method When Solving a System of Linear Equations

1. If one of the equations of the system is already solved for one of the variables, as in the systems

$$3x + 4y = 9 \qquad -5x + 3y = 9$$
$$y = 2x - 6 \quad \text{or} \quad x = 3y - 7,$$

the substitution method is the better choice.

2. If both equations are in standard $Ax + By = C$ form, as in

$$4x - 11y = 3$$
$$-2x + 3y = 4,$$

and none of the variables has coefficient -1 or 1, the elimination method is the better choice.

(continued)

3. If one or both of the equations are in standard form and the coefficient of one of the variables is -1 or 1, as in the systems

$$\begin{array}{ll} 3x + y = -2 \\ -5x + 2y = 4 \end{array} \quad \text{or} \quad \begin{array}{ll} -x + 3y = -4 \\ 3x - 2y = 8, \end{array}$$

use the elimination method, or solve for the variable with coefficient -1 or 1 and then use the substitution method.

✒ *Use the information in the preceding box to solve each problem.*

1. Assuming you want to minimize the amount of work required, tell whether you would use the substitution or elimination method to solve each system. Explain your answers. *Do not actually solve.*

(a) $3x + 5y = 69$
$y = 4x$

(b) $3x + y = -7$
$x - y = -5$

(c) $3x - 2y = 0$
$9x + 8y = 7$

2. Which one of the following systems would be easier to solve using the substitution method? Why?

$$\begin{array}{ll} 5x - 3y = 7 \\ 2x + 8y = 3 \end{array} \qquad \begin{array}{ll} 7x + 2y = 4 \\ y = -3x + 1 \end{array}$$

In Exercises 3 and 4, (a) solve the system by the elimination method, (b) solve the system by the substitution method, and (c) tell which method you prefer for that particular system and why.

3. $4x - 3y = -8$
$x + 3y = 13$

4. $2x + 5y = 0$
$x = -3y + 1$

Solve each system by the method of your choice. (For Exercises 5–7, see your answers for Exercise 1.)

5. $3x + 5y = 69$
$y = 4x$

6. $3x + y = -7$
$x - y = -5$

7. $3x - 2y = 0$
$9x + 8y = 7$

8. $x + y = 7$
$x = -3 - y$

9. $6x + 7y = 4$
$5x + 8y = -1$

10. $6x - y = 5$
$y = 11x$

11. $4x - 6y = 10$
$-10x + 15y = -25$

12. $3x - 5y = 7$
$2x + 3y = 30$

13. $5x = 7 + 2y$
$5y = 5 - 3x$

14. $4x + 3y = 1$
$3x + 2y = 2$

15. $2x - 3y = 7$
$-4x + 6y = 14$

16. $2x + 3y = 10$
$-3x + y = 18$

17. $2x + 5y = 4$
$x + y = -1$

18. $x - 3y = 7$
$4x + y = 5$

Solve each system by any method. First clear all fractions.

19. $\dfrac{1}{3}x - \dfrac{1}{2}y = \dfrac{1}{6}$
$3x - 2y = 9$

20. $\dfrac{1}{5}x + \dfrac{2}{3}y = -\dfrac{8}{5}$
$3x - y = 9$

21. $\dfrac{1}{6}x + \dfrac{1}{6}y = 2$
$-\dfrac{1}{2}x - \dfrac{1}{3}y = -8$

22. $\dfrac{x}{2} - \dfrac{y}{3} = 9$
$\dfrac{x}{5} - \dfrac{y}{4} = 5$

23. $\dfrac{x}{3} - \dfrac{3y}{4} = -\dfrac{1}{2}$
$\dfrac{x}{6} + \dfrac{y}{8} = \dfrac{3}{4}$

24. $\dfrac{x}{5} + 2y = \dfrac{16}{5}$
$\dfrac{3x}{5} + \dfrac{y}{2} = -\dfrac{7}{5}$

8.4 Systems of Linear Equations in Three Variables

A solution of an equation in three variables, such as

$$2x + 3y - z = 4,$$

is called an **ordered triple** and is written (x, y, z). For example, the ordered triple $(0, 1, -1)$ is a solution of the equation, because

$$2(0) + 3(1) - (-1) = 0 + 3 + 1 = 4.$$

Verify that another solution of this equation is $(10, -3, 7)$.

The term *linear equation* can be extended to equations of the form

$$Ax + By + Cz + \cdots + Dw = K,$$

where not all the coefficients $A, B, C, \ldots, D$ equal 0. For example,

$$2x + 3y - 5z = 7 \qquad \text{and} \qquad x - 2y - z + 3u - 2w = 8$$

are linear equations, the first with three variables and the second with five variables.

OBJECTIVE 1 Understand the geometry of systems of three equations in three variables. Consider the solution of a system of linear equations in three variables, such as

$$4x + 8y + \ z = 2$$
$$x + 7y - 3z = -14$$
$$2x - 3y + 2z = 3.$$

Theoretically, a system of this type can be solved by graphing. However, the graph of a linear equation with three variables is a *plane,* not a line. Since the graph of each equation of the system is a plane, which requires three-dimensional graphing, this method is not practical. However, it does illustrate the number of solutions possible for such systems, as shown in Figure 9.

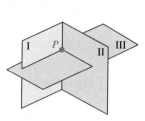

A single solution

(a)

Points of a line in common

(b)

All points in common

(c)

No points in common

(d)

No points in common

(e)

No points in common

(f)

FIGURE 9

Figure 9 illustrates the following cases.

Graphs of Linear Systems in Three Variables

1. The three planes may meet at a single, common point that is the solution of the system. See Figure 9(a).
2. The three planes may have the points of a line in common so that the infinite set of points that satisfy the equation of the line is the solution of the system. See Figure 9(b).
3. The three planes may coincide so that the solution of the system is the set of all points on a plane. See Figure 9(c).
4. The planes may have no points common to all three so that there is no solution of the system. See Figures 9(d), (e), and (f).

OBJECTIVE 2 **Solve linear systems (with three equations and three variables) by elimination.** Since graphing to find the solution set of a system of three equations in three variables is impractical, these systems are solved with an extension of the elimination method, summarized as follows.

Solving a Linear System in Three Variables

Step 1 **Eliminate a variable.** Use the elimination method to eliminate any variable from any two of the original equations. The result is an equation in two variables.

Step 2 **Eliminate the same variable again.** Eliminate the *same* variable from any *other* two equations. The result is an equation in the same two variables as in Step 1.

Step 3 **Eliminate a different variable and solve.** Use the elimination method to eliminate a second variable from the two equations in two variables that result from Steps 1 and 2. The result is an equation in one variable that gives the value of that variable.

Step 4 **Find a second value.** Substitute the value of the variable found in Step 3 into either of the equations in two variables to find the value of the second variable.

Step 5 **Find a third value.** Use the values of the two variables from Steps 3 and 4 to find the value of the third variable by substituting into an appropriate equation.

Step 6 **Check** the solution in all of the original equations. Then write the solution set.

EXAMPLE 1 Solving a System in Three Variables

Solve the system

$$
\begin{array}{ll}
4x + 8y + z = 2 & (1) \\
x + 7y - 3z = -14 & (2) \\
2x - 3y + 2z = 3. & (3)
\end{array}
$$

Step 1 As before, the elimination method involves eliminating a variable from the sum of two equations. The choice of which variable to eliminate is arbitrary. Suppose we decide to begin by eliminating z. We multiply equation (1) by 3 and then add the result to equation (2).

$$
\begin{array}{ll}
12x + 24y + 3z = 6 & \text{Multiply each side of (1) by 3.} \\
\underline{x + 7y - 3z = -14} & \text{(2)} \\
13x + 31y = -8 & \text{Add. \quad (4)}
\end{array}
$$

Step 2 Equation (4) has only two variables. To get another equation without z, we multiply equation (1) by -2 and add the result to equation (3). It is essential at this point to *eliminate the same variable, z*.

$$
\begin{array}{ll}
-8x - 16y - 2z = -4 & \text{Multiply each side of (1) by } -2. \\
\underline{2x - 3y + 2z = 3} & \text{(3)} \\
-6x - 19y = -1 & \text{Add. \quad (5)}
\end{array}
$$

Step 3 Now we solve the system of equations (4) and (5) for x and y. This step is possible only if the *same* variable is eliminated in Steps 1 and 2.

$$
\begin{array}{ll}
78x + 186y = -48 & \text{Multiply each side of (4) by 6.} \\
\underline{-78x - 247y = -13} & \text{Multiply each side of (5) by 13.} \\
-61y = -61 & \text{Add.} \\
y = 1
\end{array}
$$

Step 4 Now we substitute 1 for y in either equation (4) or (5). Choosing (5) gives

$$
\begin{array}{ll}
-6x - 19y = -1 & \text{(5)} \\
-6x - 19(1) = -1 & \text{Let } y = 1. \\
-6x - 19 = -1 & \\
-6x = 18 & \\
x = -3. &
\end{array}
$$

Step 5 We substitute -3 for x and 1 for y in any one of the three original equations to find z. Choosing (1) gives

$$
\begin{array}{ll}
4x + 8y + z = 2 & \text{(1)} \\
4(-3) + 8(1) + z = 2 & \text{Let } x = -3 \text{ and } y = 1. \\
-4 + z = 2 & \\
z = 6. &
\end{array}
$$

Step 6 It appears that the ordered triple $(-3, 1, 6)$ is the only solution of the system. We must check that the solution satisfies all three equations of the system. For equation (1),

$$
\begin{array}{ll}
4x + 8y + z = 2 & \text{(1)} \\
4(-3) + 8(1) + 6 = 2 & \text{?} \\
-12 + 8 + 6 = 2 & \text{?} \\
2 = 2. & \text{True}
\end{array}
$$

Because $(-3, 1, 6)$ also satisfies equations (2) and (3), the solution set is $\{(-3, 1, 6)\}$.

Now Try Exercise 3.

OBJECTIVE 3 Solve linear systems (with three equations and three variables) where some of the equations have missing terms. When this happens, one elimination step can be omitted.

EXAMPLE 2 Solving a System of Equations with Missing Terms

Solve the system

$$6x - 12y = -5 \quad (1)$$
$$8y + z = 0 \quad (2)$$
$$9x - z = 12. \quad (3)$$

Since equation (3) is missing the variable y, eliminate y using equations (1) and (2).

$$
\begin{array}{ll}
12x - 24y \qquad = -10 & \text{Multiply each side of (1) by 2.} \\
\underline{\qquad 24y + 3z = \quad 0} & \text{Multiply each side of (2) by 3.} \\
12x \qquad + 3z = -10 & \text{Add. \quad (4)}
\end{array}
$$

Use this result, together with equation (3), to eliminate z. Multiply equation (3) by 3.

$$
\begin{array}{ll}
27x - 3z = \quad 36 & \text{Multiply each side of (3) by 3.} \\
\underline{12x + 3z = -10} & (4) \\
39x \qquad = \quad 26 & \text{Add.}
\end{array}
$$

$$x = \frac{26}{39} = \frac{2}{3}$$

Substituting into equation (3) gives

$$
\begin{array}{ll}
9x - z = 12 & (3) \\
9\left(\dfrac{2}{3}\right) - z = 12 & \text{Let } x = \tfrac{2}{3}. \\
6 - z = 12 & \\
z = -6. &
\end{array}
$$

Substituting -6 for z in equation (2) gives

$$
\begin{array}{ll}
8y + z = 0 & (2) \\
8y - 6 = 0 & \text{Let } z = -6. \\
8y = 6 & \\
y = \dfrac{3}{4}. &
\end{array}
$$

Check in each of the original equations of the system to verify that the solution set of the system is $\left\{\left(\tfrac{2}{3}, \tfrac{3}{4}, -6\right)\right\}$.

Now Try Exercise 21.

OBJECTIVE 4 Solve special systems (with three equations and three variables). Linear systems with three variables may be inconsistent or may include dependent equations. The next examples illustrate these cases.

EXAMPLE 3   **Solving an Inconsistent System with Three Variables**

Solve the system

$$2x - 4y + 6z = 5 \qquad (1)$$
$$-x + 3y - 2z = -1 \qquad (2)$$
$$x - 2y + 3z = 1. \qquad (3)$$

Eliminate x by adding equations (2) and (3) to get the equation

$$y + z = 0.$$

Now, *eliminate x again,* using equations (1) and (3).

$$
\begin{array}{ll}
-2x + 4y - 6z = -2 & \text{Multiply each side of (3) by } -2. \\
\underline{2x - 4y + 6z = 5} & (1) \\
 0 = 3 & \text{False}
\end{array}
$$

The resulting false statement indicates that equations (1) and (3) have no common solution. Thus, the system is inconsistent and the solution set is $\emptyset$. The graph of this system would show these two planes parallel to one another.

Now Try Exercise 29.

NOTE   If you get a false statement when adding as in Example 3, you do not need to go any further with the solution. Since two of the three planes are parallel, it is not possible for the three planes to have any common points.

EXAMPLE 4   **Solving a System of Dependent Equations with Three Variables**

Solve the system

$$2x - 3y + 4z = 8 \qquad (1)$$
$$-x + \frac{3}{2}y - 2z = -4 \qquad (2)$$
$$6x - 9y + 12z = 24. \qquad (3)$$

Multiplying each side of equation (1) by 3 gives equation (3). Multiplying each side of equation (2) by -6 also gives equation (3). Because of this, the equations are dependent. All three equations have the same graph, as illustrated in Figure 9(c). The solution set is written

$$\{(x, y, z) \mid 2x - 3y + 4z = 8\}.$$

Although any one of the three equations could be used to write the solution set, we use the equation in standard form with coefficients that are integers with no common factor (except 1), as we did in Section 8.1.

Now Try Exercise 33.

We can extend the method discussed in this section to solve larger systems. For example, to solve a system of four equations in four variables, eliminate a variable from three pairs of equations to get a system of three equations in three unknowns. Then proceed as shown above.

 **EXERCISES**

1. Explain what the following statement means: The solution set of the system

$$2x + y + z = 3$$
$$3x - y + z = -2$$
$$4x - y + 2z = 0$$

is $\{(-1, 2, 3)\}$.

2. The two equations

$$x + y + z = 6$$
$$2x - y + z = 3$$

have a common solution of $(1, 2, 3)$. Which equation would complete a system of three linear equations in three variables having solution set $\{(1, 2, 3)\}$?

A. $3x + 2y - z = 1$ **B.** $3x + 2y - z = 4$
C. $3x + 2y - z = 5$ **D.** $3x + 2y - z = 6$

Solve each system of equations. See Example 1.

3. $2x - 5y + 3z = -1$
$x + 4y - 2z = 9$
$x - 2y - 4z = -5$

4. $x + 3y - 6z = 7$
$2x - y + z = 1$
$x + 2y + 2z = -1$

5. $3x + 2y + z = 8$
$2x - 3y + 2z = -16$
$x + 4y - z = 20$

6. $-3x + y - z = -10$
$-4x + 2y + 3z = -1$
$2x + 3y - 2z = -5$

7. $2x + 5y + 2z = 0$
$4x - 7y - 3z = 1$
$3x - 8y - 2z = -6$

8. $5x - 2y + 3z = -9$
$4x + 3y + 5z = 4$
$2x + 4y - 2z = 14$

9. $x + 2y + z = 4$
$2x + y - z = -1$
$x - y - z = -2$

10. $x - 2y + 5z = -7$
$-2x - 3y + 4z = -14$
$-3x + 5y - z = -7$

11. $\dfrac{1}{3}x + \dfrac{1}{6}y - \dfrac{2}{3}z = -1$

$-\dfrac{3}{4}x - \dfrac{1}{3}y - \dfrac{1}{4}z = 3$

$\dfrac{1}{2}x + \dfrac{3}{2}y + \dfrac{3}{4}z = 21$

12. $\dfrac{2}{3}x - \dfrac{1}{4}y + \dfrac{5}{8}z = 0$

$\dfrac{1}{5}x + \dfrac{2}{3}y - \dfrac{1}{4}z = -7$

$-\dfrac{3}{5}x + \dfrac{4}{3}y - \dfrac{7}{8}z = -5$

13. $-x + 2y + 6z = 2$
$3x + 2y + 6z = 6$
$x + 4y - 3z = 1$

14. $2x + y + 2z = 1$
$x + 2y + z = 2$
$x - y - z = 0$

15. $x + y - z = -2$
$2x - y + z = -5$
$-x + 2y - 3z = -4$

16. $x + 2y + 3z = 1$
$-x - y + 3z = 2$
$-6x + y + z = -2$

Solve each system of equations. See Example 2.

17. $2x - 3y + 2z = -1$
$x + 2y + z = 17$
$2y - z = 7$

18. $2x - y + 3z = 6$
$x + 2y - z = 8$
$2y + z = 1$

19. $4x + 2y - 3z = 6$
$x - 4y + z = -4$
$-x + 2z = 2$

20. $2x + 3y - 4z = 4$
$x - 6y + z = -16$
$-x + 3z = 8$

21. $2x + y = 6$
$3y - 2z = -4$
$3x - 5z = -7$

22. $4x - 8y = -7$
$4y + z = 7$
$-8x + z = -4$

23. $-5x + 2y + z = 5$
$-3x - 2y - z = 3$
$-x + 6y = 1$

24. $x + y - z = 0$
$2y - z = 1$
$2x + 3y - 4z = -4$

25. $4x - z = -6$
$\dfrac{3}{5}y + \dfrac{1}{2}z = 0$
$\dfrac{1}{3}x + \dfrac{2}{3}z = -5$

26. $5x - 2z = 8$
$4y + 3z = -9$
$\dfrac{1}{2}x + \dfrac{2}{3}y = -1$

✎ 27. Using your immediate surroundings, give an example of three planes that

    (a) intersect in a single point;
    (b) do not intersect;
    (c) intersect in infinitely many points.

28. Suppose that a system has infinitely many ordered triple solutions of the form (x, y, z) such that

$$x + y + 2z = 1.$$

Give three specific ordered triples that are solutions of the system.

Solve each system of equations. If the system is inconsistent or has dependent equations, say so. See Examples 1, 3, and 4.

29. $2x + 2y - 6z = 5$
$-3x + y - z = -2$
$-x - y + 3z = 4$

30. $-2x + 5y + z = -3$
$5x + 14y - z = -11$
$7x + 9y - 2z = -5$

31. $-5x + 5y - 20z = -40$
$x - y + 4z = 8$
$3x - 3y + 12z = 24$

32. $x + 4y - z = 3$
$-2x - 8y + 2z = -6$
$3x + 12y - 3z = 9$

33. $2x + y - z = 6$
$4x + 2y - 2z = 12$
$-x - \dfrac{1}{2}y + \dfrac{1}{2}z = -3$

34. $2x - 8y + 2z = -10$
$-x + 4y - z = 5$
$\dfrac{1}{8}x - \dfrac{1}{2}y + \dfrac{1}{8}z = -\dfrac{5}{8}$

35. $x + y - 2z = 0$
$3x - y + z = 0$
$4x + 2y - z = 0$

36. $2x + 3y - z = 0$
$x - 4y + 2z = 0$
$3x - 5y - z = 0$

Extend the method of this section to solve each system. Express the solution in the form (x, y, z, w).

37. $x + y + z - w = 5$
$2x + y - z + w = 3$
$x - 2y + 3z + w = 18$
$-x - y + z + 2w = 8$

38. $3x + y - z + 2w = 9$
$x + y + 2z - w = 10$
$x - y - z + 3w = -2$
$-x + y - z + w = -6$

39.
$$3x + y - z + w = -3$$
$$2x + 4y + z - w = -7$$
$$-2x + 3y - 5z + w = 3$$
$$5x + 4y - 5z + 2w = -7$$

40.
$$x - 3y + 7z + w = 11$$
$$2x + 4y + 6z - 3w = -3$$
$$3x + 2y + z + 2w = 19$$
$$4x + y - 3z + w = 22$$

RELATING CONCEPTS (EXERCISES 41–50)

For Individual or Group Work

Suppose that on a distant planet a function of the form

$$f(x) = ax^2 + bx + c \quad (a \neq 0)$$

describes the height in feet of a projectile x sec after it has been projected upward. **Work Exercises 41–50 in order,** *to see how this can be related to a system of three equations in three variables a, b, and c.*

41. After 1 sec, the height of a certain projectile is 128 ft. Thus, $f(1) = 128$. Use this information to find one equation in the variables a, b, and c. (*Hint:* Substitute 1 for x and 128 for $f(x)$.)

42. After 1.5 sec, the height is 140 ft. Find a second equation in a, b, and c.

43. After 3 sec, the height is 80 ft. Find a third equation in a, b, and c.

44. Write a system of three equations in a, b, and c, based on your answers in Exercises 41–43. Solve the system.

45. What is the function f for this particular projectile?

46. In the function f written in Exercise 45, the _____ of the projectile is a function of the _____ elapsed since it was projected.

47. What was the initial height of the projectile? (*Hint:* Find $f(0)$.)

48. The projectile reaches its maximum height in 1.625 sec. Find its maximum height.

49. In Chapter 11 we discuss graphs of functions of the form $f(x) = ax^2 + bx + c$ ($a \neq 0$). Use a system of equations to find the values of a, b, and c for the function of this form that satisfies $f(1) = 2$, $f(-1) = 0$, and $f(-2) = 8$. Then write the expression for $f(x)$.

50. The accompanying table was generated by a graphing calculator for a function $Y_1 = aX^2 + bX + c$. Use any three points shown to find the values of a, b, and c. Then write the expression for Y_1.

X	Y₁	
1	8	
2	15	
3	24	
4	35	
5	48	
6	63	
7	80	

X=1

51. Discuss why it is necessary to eliminate the same variable in the first two steps of the elimination method with three equations and three variables.

52. In Step 3 of the elimination method for solving systems in three variables, does it matter which variable is eliminated? Explain.

8.5 Applications of Systems of Linear Equations

Many applied problems involve more than one unknown quantity. Although some problems with two unknowns can be solved using just one variable, it is often easier to use two variables. To solve a problem with two unknowns, we must write two equations that relate the unknown quantities. The system formed by the pair of equations can then be solved using the methods of this chapter.

The following steps, based on the six-step problem-solving method first introduced in Chapter 2, give a strategy for solving applied problems using more than one variable.

Solving an Applied Problem by Writing a System of Equations

Step 1 **Read** the problem carefully until you understand what is given and what is to be found.

Step 2 **Assign variables** to represent the unknown values, using diagrams or tables as needed. *Write down* what each variable represents.

Step 3 **Write a system of equations** that relates the unknowns.

Step 4 **Solve** the system of equations.

Step 5 **State the answer** to the problem. Does it seem reasonable?

Step 6 **Check** the answer in the words of the original problem.

OBJECTIVE 1 Solve geometry problems using two variables. Problems about the perimeter of a geometric figure often involve two unknowns and can be solved using systems of equations.

EXAMPLE 1 Finding the Dimensions of a Soccer Field

Unlike football, where the dimensions of a playing field cannot vary, a rectangular soccer field may have a width between 50 and 100 yd and a length between 50 and 100 yd. Suppose that one particular field has a perimeter of 320 yd. Its length measures 40 yd more than its width. What are the dimensions of this field? (*Source: Microsoft Encarta Encyclopedia 2000.*)

Step 1 **Read** the problem again. We are asked to find the dimensions of the field.

Step 2 **Assign variables.** Let L = the length and W = the width. Figure 10 shows a soccer field with the length labeled L and the width labeled W.

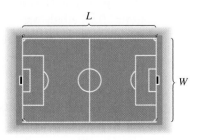

FIGURE 10

Step 3 **Write a system of equations.** Because the perimeter is 320 yd, we find one equation by using the perimeter formula:

$$2L + 2W = 320.$$

Because the length is 40 yd more than the width, we have

$$L = W + 40.$$

The system is

$$2L + 2W = 320 \qquad (1)$$
$$L = W + 40. \qquad (2)$$

Step 4 **Solve** the system of equations. Since equation (2) is solved for *L*, we can use the substitution method. We substitute $W + 40$ for *L* in equation (1), and solve for *W*.

$$2L + 2W = 320 \qquad \text{(1)}$$
$$2(W + 40) + 2W = 320 \qquad \text{Let } L = W + 40.$$
$$2W + 80 + 2W = 320 \qquad \text{Distributive property}$$
$$4W + 80 = 320 \qquad \text{Combine terms.}$$
$$4W = 240 \qquad \text{Subtract 80.}$$
$$W = 60 \qquad \text{Divide by 4.}$$

Let $W = 60$ in the equation $L = W + 40$ to find *L*.

$$L = 60 + 40 = 100$$

Step 5 **State the answer.** The length is 100 yd, and the width is 60 yd. The answer is reasonable, since both dimensions are within the ranges given in the problem.

Step 6 **Check.** The perimeter of this soccer field is

$$2(100) + 2(60) = 320 \text{ yd,}$$

and the length, 100 yd, is 40 yd more than the width, since

$$100 - 40 = 60.$$

The answer is correct.

Now Try Exercise 3.

OBJECTIVE 2 Solve money problems using two variables. Professional sport ticket prices increase annually. Average per-ticket prices in three of the four major sports (football, basketball, and hockey) now exceed $30.00.

EXAMPLE 2 Solving a Problem about Ticket Prices

During recent National Hockey League and National Basketball Association seasons, two hockey tickets and one basketball ticket purchased at their average prices would have cost $110.40. One hockey ticket and two basketball tickets would have cost $106.32. What were the average ticket prices for the two sports? (*Source:* Team Marketing Report, Chicago.)

Step 1 **Read** the problem again. There are two unknowns.

Step 2 **Assign variables.** Let h represent the average price for a hockey ticket and b represent the average price for a basketball ticket.

Step 3 **Write a system of equations.** Because two hockey tickets and one basketball ticket cost a total of $110.40, one equation for the system is

$$2h + b = 110.40.$$

By similar reasoning, the second equation is

$$h + 2b = 106.32.$$

Therefore, the system is

$$2h + b = 110.40 \qquad (1)$$
$$h + 2b = 106.32. \qquad (2)$$

Step 4 **Solve** the system of equations. To eliminate h, multiply equation (2) by -2 and add.

$$2h + b = 110.40 \qquad (1)$$
$$\underline{-2h - 4b = -212.64} \qquad \text{Multiply each side of (2) by } -2.$$
$$-3b = -102.24 \qquad \text{Add.}$$
$$b = 34.08 \qquad \text{Divide by } -3.$$

To find the value of h, let $b = 34.08$ in equation (2).

$$h + 2b = 106.32 \qquad (2)$$
$$h + 2(34.08) = 106.32 \qquad \text{Let } b = 34.08.$$
$$h + 68.16 = 106.32 \qquad \text{Multiply.}$$
$$h = 38.16 \qquad \text{Subtract 68.16.}$$

Step 5 **State the answer.** The average price for one basketball ticket was $34.08. For one hockey ticket, the average price was $38.16.

Step 6 **Check** that these values satisfy the conditions stated in the problem.

Now Try Exercise 11.

OBJECTIVE 3 Solve mixture problems using two variables. We solved mixture problems earlier using one variable. For many mixture problems it seems more natural to use more than one variable and a system of equations.

EXAMPLE 3 Solving a Mixture Problem

How many ounces each of 5% hydrochloric acid and 20% hydrochloric acid must be combined to get 10 oz of solution that is 12.5% hydrochloric acid?

Step 1 **Read** the problem. Two solutions of different strengths are being mixed together to get a specific amount of a solution with an "in-between" strength.

Step 2 **Assign variables.** Let x represent the number of ounces of 5% solution and y represent the number of ounces of 20% solution. Use a table to summarize the information from the problem.

Ounces of Solution	Percent (as a decimal)	Ounces of Pure Acid
x	5% = .05	.05x
y	20% = .20	.20y
10	12.5% = .125	(.125)10

Figure 11 illustrates what is happening in the problem.

FIGURE 11

Step 3 **Write a system of equations.** When the *x* oz of 5% solution and the *y* oz of 20% solution are combined, the total number of ounces is 10, so

$$x + y = 10. \quad (1)$$

The ounces of acid in the 5% solution (.05*x*) plus the ounces of acid in the 20% solution (.20*y*) should equal the total ounces of acid in the mixture, which is (.125)10, or 1.25. That is,

$$.05x + .20y = 1.25. \quad (2)$$

Notice that these equations can be quickly determined by reading down in the table or using the labels in Figure 9.

Step 4 **Solve** the system of equations (1) and (2). Eliminate *x* by first multiplying equation (2) by 100 to clear it of decimals and then multiplying equation (1) by −5.

$$
\begin{aligned}
5x + 20y &= 125 \qquad \text{Multiply each side of (2) by 100.} \\
-5x - 5y &= -50 \qquad \text{Multiply each side of (1) by } -5. \\
\hline
15y &= 75 \qquad \text{Add.} \\
y &= 5
\end{aligned}
$$

Because $y = 5$ and $x + y = 10$, *x* is also 5.

Step 5 **State the answer.** The desired mixture will require 5 oz of the 5% solution and 5 oz of the 20% solution.

Step 6 **Check** that these values satisfy both equations of the system.

Now Try Exercise 17.

| **CONNECTIONS** |

Problems that can be solved by writing a system of equations have been of interest historically. The following problem appeared in a Hindu work that dates back to about 850 A.D.

> The mixed price of 9 citrons (a lemonlike fruit shown in the photo) and 7 fragrant wood apples is 107; again, the mixed price of 7 citrons and 9 fragrant wood apples is 101. O you arithmetician, tell me quickly the price of a citron and the price of a wood apple here, having distinctly separated those prices well.

For Discussion or Writing

What do you think is meant by "the mixed price" in the problem quoted above? Write a system of equations for this problem. (You will be asked to solve it in Exercise 35.)

OBJECTIVE 4 Solve distance-rate-time problems using two variables. Motion problems require the distance formula, $d = rt$, where d is distance, r is rate (or speed), and t is time. These applications often lead to systems of equations, as in the next example.

EXAMPLE 4   Solving a Motion Problem

A car travels 250 km in the same time that a truck travels 225 km. If the speed of the car is 8 km per hr faster than the speed of the truck, find both speeds.

Step 1 **Read** the problem again. Given the distances traveled, we need to find the speed of each vehicle.

Step 2 **Assign variables.**

$$\text{Let} \quad x = \text{the speed of the car,}$$
$$\text{and} \quad y = \text{the speed of the truck.}$$

As in Example 3, a table helps organize the information. Fill in the given information for each vehicle (in this case, distance) and use the assigned variables for the unknown speeds (rates).

	d	r	t
Car	250	x	
Truck	225	y	

The table shows nothing about time. To get an expression for time, solve the distance formula, $d = rt$, for t.

$$\frac{d}{r} = t$$

The two times can be written as $\frac{250}{x}$ and $\frac{225}{y}$.

Step 3 **Write a system of equations.** The problem states that the car travels 8 km per hr faster than the truck. Since the two speeds are x and y,

$$x = y + 8. \qquad (1)$$

Both vehicles travel for the same time, so from the table,

$$\frac{250}{x} = \frac{225}{y}.$$

This is not a linear equation. However, multiplying each side by xy gives

$$250y = 225x,$$

which is linear. The system is

$$x = y + 8$$
$$250y = 225x. \qquad (2)$$

Step 4 **Solve** the system of equations by substitution. Replace x with $y + 8$ in equation (2).

$$
\begin{array}{lll}
250y = 225x & & (2) \\
250y = 225(y + 8) & & \text{Let } x = y + 8. \\
250y = 225y + 1800 & & \text{Distributive property} \\
25y = 1800 & & \text{Subtract } 225y. \\
y = 72 & & \text{Divide by 25.}
\end{array}
$$

Because $x = y + 8$, the value of x is $72 + 8 = 80$.

Step 5   **State the answer.** The car's speed is 80 km per hr, and the truck's speed is 72 km per hr.

Step 6   **Check.** This is especially important since one of the equations had variable denominators.

$$\text{Car:} \quad t = \frac{d}{r} = \frac{250}{80} = 3.125$$

$$\text{Truck:} \quad t = \frac{d}{r} = \frac{225}{72} = 3.125$$

Times are equal.

Since $80 - 72 = 8$, the conditions of the problem are satisfied.

Now Try Exercise 27.

OBJECTIVE 5   Solve problems with three variables using a system of three equations. To solve such problems, we extend the method used for two unknowns. Since three variables are used, three equations are necessary to find a solution.

EXAMPLE 5   Solving a Problem Involving Prices

At Panera Bread, a loaf of honey wheat bread costs \$2.40, a loaf of pumpernickel bread costs \$3.35, and a loaf of French bread costs \$2.10. On a recent day, three times as many loaves of honey wheat were sold as pumpernickel. The number of loaves of French bread sold was 5 less than the number of loaves of honey wheat sold. Total receipts for these breads were \$56.90. How many loaves of each type of bread were sold? (*Source:* Panera Bread menu.)

Step 1   **Read** the problem again. There are three unknowns in this problem.

Step 2   **Assign variables** to represent the three unknowns.

Let $x =$ the number of loaves of honey wheat,

$y =$ the number of loaves of pumpernickel,

and $z =$ the number of loaves of French bread.

Step 3   **Write a system of three equations** using the information in the problem. Since three times as many loaves of honey wheat were sold as pumpernickel,

$$x = 3y, \quad \text{or} \quad x - 3y = 0. \qquad (1)$$

Also,

Number of loaves of French bread	equals	5 less than the number of loaves of honey wheat.
↓	↓	↓
z	$=$	$x - 5,$

so $x - z = 5.$ (2)

Multiplying the cost of a loaf of each kind of bread by the number of loaves of that kind sold and adding gives the total receipts.

$$2.40x + 3.35y + 2.10z = 56.90$$

Multiply each side of this equation by 100 to clear it of decimals.

$$240x + 335y + 210z = 5690 \qquad (3)$$

Step 4 **Solve** the system of three equations using the method shown in Section 8.4. Solving the system

$$x - 3y = 0 \qquad \text{(1)}$$
$$x - z = 5 \qquad \text{(2)}$$
$$240x + 335y + 210z = 5690 \qquad \text{(3)}$$

leads to

$$x = 12, \quad y = 4, \quad \text{and} \quad z = 7.$$

Step 5 **State the answer.** The solution is (12, 4, 7), so 12 loaves of honey wheat, 4 loaves of pumpernickel, and 7 loaves of French bread were sold.

Step 6 **Check.** Since $12 = 3 \cdot 4$, the number of loaves of honey wheat is three times the number of loaves of pumpernickel. Also, $12 - 7 = 5$, so the number of loaves of French bread is 5 less than the number of loaves of honey wheat. Multiply the appropriate cost per loaf by the number of loaves sold and add the results to check that total receipts were $56.90.

Now Try Exercise 45.

EXAMPLE 6 **Solving a Business Production Problem**

A company produces three color television sets, models X, Y, and Z. Each model X set requires 2 hr of electronics work, 2 hr of assembly time, and 1 hr of finishing time. Each model Y requires 1, 3, and 1 hr of electronics, assembly, and finishing time, respectively. Each model Z requires 3, 2, and 2 hr of the same work, respectively. There are 100 hr available for electronics, 100 hr available for assembly, and 65 hr available for finishing per week. How many of each model should be produced each week if all available time must be used?

Step 1 **Read** the problem again. There are three unknowns.

Step 2 **Assign variables.**

Let $x =$ the number of model X produced per week,

$y =$ the number of model Y produced per week,

and $z =$ the number of model Z produced per week.

We organize the information in a table.

	Each Model X	Each Model Y	Each Model Z	Totals
Hours of Electronics Work	2	1	3	100
Hours of Assembly Time	2	3	2	100
Hours of Finishing Time	1	1	2	65

Step 3 **Write a system of three equations.** The x model X sets require $2x$ hr of electronics, the y model Y sets require $1y$ (or y) hr of electronics, and the

z model Z sets require $3z$ hr of electronics. Since 100 hr are available for electronics,

$$2x + y + 3z = 100. \quad (1)$$

Similarly, from the fact that 100 hr are available for assembly,

$$2x + 3y + 2z = 100, \quad (2)$$

and the fact that 65 hr are available for finishing leads to the equation

$$x + y + 2z = 65. \quad (3)$$

Again, notice the advantage of setting up a table. By reading across, we can easily determine the coefficients and constants in the equations of the system.

Step 4 **Solve** the system

$$2x + y + 3z = 100$$
$$2x + 3y + 2z = 100$$
$$x + y + 2z = 65$$

to find $x = 15$, $y = 10$, and $z = 20$.

Step 5 **State the answer.** The company should produce 15 model X, 10 model Y, and 20 model Z sets per week.

Step 6 **Check** that these values satisfy the conditions of the problem.

Now Try Exercise 47.

8.5 EXERCISES

For Extra Help

 Student's Solutions Manual

 MyMathLab

 InterAct Math Tutorial Software

 AW Math Tutor Center

Math XP MathXL

 Digital Video Tutor CD 12/Videotape 12

Solve each problem. See Example 1.

1. During the 2000 Major League Baseball season, the St. Louis Cardinals played 162 games. They won 28 more games than they lost. What was their win–loss record that year?

2. Refer to Exercise 1. During the same 162-game season, the Chicago Cubs lost 32 more games than they won. What was the team's win–loss record?

2000 MLB FINAL STANDINGS NATIONAL LEAGUE CENTRAL

Team	W	L
St. Louis	___	___
Cincinnati	85	77
Milwaukee	73	89
Houston	72	90
Pittsburgh	69	93
Chicago	___	___

Source: www.mlb.com

3. Venus and Serena measured a tennis court and found that it was 42 ft longer than it was wide and had a perimeter of 228 ft. What were the length and the width of the tennis court?

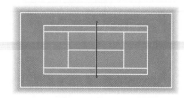

4. Shaq and Kobe found that the width of their basketball court was 44 ft less than the length. If the perimeter was 288 ft, what were the length and the width of their court?

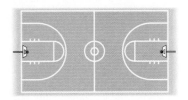

5. The two biggest U.S. companies in terms of revenue in 2000 were ExxonMobil and General Motors. ExxonMobil's revenue was $29 billion more than that of General Motors. Total revenue for the two companies was $399 billion. What was the revenue for each company? (*Source:* Bridge News, MarketGuide.com)

6. The top two U.S. trading partners during the first four months of 2000 were Canada and Mexico. Exports and imports with Mexico were $57 billion less than those with Canada. Total exports and imports involving these two countries were $211 billion. How much were U.S. exports and imports with each country? (*Source:* U.S. Bureau of the Census.)

In Exercises 7 and 8, find the measures of the angles marked x and y. Remember that (1) *the sum of the measures of the angles of a triangle is* 180°, (2) *supplementary angles have a sum of* 180°, *and* (3) *vertical angles have equal measures.*

7.

$x°$

$y°$ $(3x + 10)°$

8.

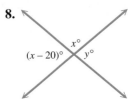

$(x - 20)°$ $x°$ $y°$

The Fan Cost Index (FCI) represents the cost of four average-price tickets, four small soft drinks, two small beers, four hot dogs, parking for one car, two game programs, and two souvenir caps to a sporting event. For example, in a recent year, the FCI for Major League Baseball was $105.63. This was by far the least for the four major professional sports. (Source: Team Marketing Report, Chicago.)

Use the concept of FCI in Exercises 9 and 10. See Example 2.

9. The FCI prices for the National Hockey League and the National Basketball Association totaled $423.12. The hockey FCI was $16.36 more than that of basketball. What were the FCIs for these sports?

10. The FCI prices for Major League Baseball and the National Football League totaled $311.03. The football FCI was $105.87 more than that of baseball. What were the FCIs for these sports?

Solve each problem. See Example 2.

11. Andrew McGinnis works at Wendy's Old Fashioned Hamburgers. During one particular lunch hour, he sold 15 single hamburgers and 10 double hamburgers, totaling $63.25. Another lunch hour, he sold 30 singles and 5 doubles, totaling $78.65. How much did each type of burger cost? (*Source:* Wendy's Old Fashioned Hamburgers menu.)

12. Tokyo and New York are among the most expensive cities worldwide for business travelers. Using average costs per day for each city (which includes room, meals, laundry, and two taxi fares), 2 days in Tokyo and 3 days in New York cost $2015. Four days in Tokyo and 2 days in New York cost $2490. What is the average cost per day for each city? (*Source:* ECA International.)

The formulas p = br (percentage = base × rate) and I = prt (simple interest = principal × rate × time) are used in the applications in Exercises 17–24. In general, we are using

$$\text{portion} = \text{whole} \times \text{percent}.$$

To prepare to use these formulas, answer the questions in Exercises 13 and 14.

13. If a container of liquid contains 60 oz of solution, what is the number of ounces of pure acid if the given solution contains the following acid concentrations?

(a) 10%   **(b)** 25%   **(c)** 40%   **(d)** 50%

14. If $5000 is invested in an account paying simple annual interest, how much interest will be earned during the first year at the following rates?

(a) 2%   **(b)** 3%   **(c)** 4%   **(d)** 3.5%

15. If a pound of turkey costs $.99, how much will x pounds cost?

16. If a ticket to the movie *Eight Legged Freaks* costs $8 and y tickets are sold, how much is collected from the sale?

Solve each problem. See Example 3.

17. How many gallons each of 25% alcohol and 35% alcohol should be mixed to get 20 gal of 32% alcohol?

Gallons of Solution	Percent (as a decimal)	Gallons of Pure Alcohol
x	25% = .25	
y	35% = .35	
20	32% =	

18. How many liters each of 15% acid and 33% acid should be mixed to get 120 L of 21% acid?

Liters of Solution	Percent (as a decimal)	Liters of Pure Acid
x	15% = .15	
y	33% =	
120	21% =	

19. Pure acid is to be added to a 10% acid solution to obtain 54 L of a 20% acid solution. What amounts of each should be used?

20. A truck radiator holds 36 L of fluid. How much pure antifreeze must be added to a mixture that is 4% antifreeze to fill the radiator with a mixture that is 20% antifreeze?

21. A party mix is made by adding nuts that sell for $2.50 per kg to a cereal mixture that sells for $1 per kg. How much of each should be added to get 30 kg of a mix that will sell for $1.70 per kg?

	Number of Kilograms	Price per Kilogram	Value
Nuts	x	2.50	
Cereal	y	1.00	
Mixture		1.70	

22. A popular fruit drink is made by mixing fruit juices. Such a drink with 50% juice is to be mixed with another drink that is 30% juice to get 200 L of a drink that is 45% juice. How much of each should be used?

	Liters of Drink	Percent (as a decimal)	Liters of Pure Juice
50% Juice	x	.50	
30% Juice	y	.30	
Mixture		.45	

23. A total of $3000 is invested, part at 2% simple interest and part at 4%. If the total annual return from the two investments is $100, how much is invested at each rate?

Principal	Rate (as a decimal)	Interest
x	.02	.02x
y	.04	.04y
3000		100

24. An investor will invest a total of $15,000 in two accounts, one paying 4% annual simple interest, and the other 3%. If he wants to earn $550 annual interest, how much should he invest at each rate?

Principal	Rate (as a decimal)	Interest
x	.04	
y	.03	
15,000		

The formula d = rt (distance = rate × time) is used in the applications in Exercises 27–30. To prepare to use this formula, answer the questions in Exercises 25 and 26.

25. If the speed of a killer whale is 25 mph and the whale swims for *y* hr, how many miles does the whale travel?

26. If the speed of a boat in still water is 10 mph, and the speed of the current of a river is *x* mph, what is the speed of the boat

Upstream (against the current)

Downstream (with the current)

   (a) going upstream (that is, against the current, which slows the boat down);
   (b) going downstream (that is, with the current, which speeds the boat up)?

Solve each problem. See Example 4.

27. A train travels 150 km in the same time that a plane covers 400 km. If the speed of the plane is 20 km per hr less than 3 times the speed of the train, find both speeds.

	r	t	d
Train	x		150
Plane	y		400

28. A freight train and an express train leave towns 390 km apart, traveling toward one another. The freight train travels 30 km per hr slower than the express train. They pass one another 3 hr later. What are their speeds?

	r	t	d
Freight Train	x	3	
Express Train	y	3	

29. In his motorboat, Bill Ruhberg travels upstream at top speed to his favorite fishing spot, a distance of 36 mi, in 2 hr. Returning, he finds that the trip downstream, still at top speed, takes only 1.5 hr. Find the speed of Bill's boat and the speed of the current.

	r	t	d
Upstream	x − y	2	
Downstream	x + y		

30. Traveling for 3 hr into a steady headwind, a plane flies 1650 mi. The pilot determines that flying *with* the same wind for 2 hr, he could make a trip of 1300 mi. Find the speed of the plane and the speed of the wind.

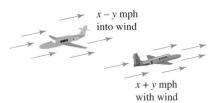

x − y mph
into wind

x + y mph
with wind

Use the problem-solving techniques of this section to solve each problem with two variables. See Examples 1–4.

31. At age 61, rock icon Tina Turner generated the most revenue on the concert circuit in 2000. Turner and second-place 'N Sync together took in $157 million from ticket sales. If 'N Sync took in $3.8 million less than Turner, how much did each generate? (*Source:* Pollstar.)

32. Carol Britz plans to mix pecan clusters that sell for $3.60 per lb with chocolate truffles that sell for $7.20 per lb to get a mixture that she can sell in Valentine boxes for $4.95 per lb. How much of the $3.60 clusters and the $7.20 truffles should she use to create 80 lb of the mix?

	Number of Pounds	Price per Pound	Value
Pecan Clusters	x		
Chocolate Truffles	y		
Valentine Mixture	80		

33. Tickets to a production of *King Lear* at the College of DuPage cost $5 for general admission or $4 with a student ID. If 184 people paid to see a performance and $812 was collected, how many of each type of ticket were sold?

34. At a business meeting at Panera Bread, the bill for two cappuccinos and three house lattes was $10.95. At another table, the bill for one cappuccino and two house lattes was $6.65. How much did each type of beverage cost? (*Source:* Panera Bread menu.)

35. The mixed price of 9 citrons and 7 fragrant wood apples is 107; again, the mixed price of 7 citrons and 9 fragrant wood apples is 101. O you arithmetician, tell me quickly the price of a citron and the price of a wood apple here, having distinctly separated those prices well. (*Source:* Hindu work, A.D. 850.)

36. Braving blizzard conditions on the planet Hoth, Luke Skywalker sets out at top speed in his snow speeder for a rebel base 4800 mi away. He travels into a steady headwind and makes the trip in 3 hr. Returning, he finds that the trip back, still at top speed but now with a tailwind, takes only 2 hr. Find the top speed of Luke's snow speeder and the speed of the wind.

	r	t	d
Into Headwind			
With Tailwind			

Solve each problem involving three variables. See Examples 5 and 6. (In Exercises 37–40, remember that the sum of the measures of the angles of a triangle is $180°$.)

37. In the figure, $z = x + 10$ and $x + y = 100$. Determine a third equation involving x, y, and z, and then find the measures of the three angles.

38. In the figure, x is 10 less than y and 20 less than z. Write a system of equations and find the measures of the three angles.

39. In a certain triangle, the measure of the second angle is $10°$ more than three times the first. The third angle measure is equal to the sum of the measures of the other two. Find the measures of the three angles.

40. The measure of the largest angle of a triangle is $12°$ less than the sum of the measures of the other two. The smallest angle measures $58°$ less than the largest. Find the measures of the angles.

41. The perimeter of a triangle is 70 cm. The longest side is 4 cm less than the sum of the other two sides. Twice the shortest side is 9 cm less than the longest side. Find the length of each side of the triangle.

42. The perimeter of a triangle is 56 in. The longest side measures 4 in. less than the sum of the other two sides. Three times the shortest side is 4 in. more than the longest side. Find the lengths of the three sides.

43. In a random sample of 100 Americans of voting age, 10 more Americans identify themselves as Independents than Republicans. Six fewer Americans identify themselves as Republicans than Democrats. Assuming that all of those sampled are Republican, Democrat, or Independent, how many of those in the sample identify themselves with each political affiliation? (*Source:* The Gallup Organization.)

44. In the 2000 Summer Olympics in Sydney, Australia, the United States earned 14 more gold medals than silver. The number of bronze medals earned was 17 less than twice the number of silver medals. The United States earned a total of 97 medals. How many of each kind of medal did the United States earn? (*Source: The Gazette,* October 2, 2000.)

45. Tickets for one show on the Harlem Globetrotters' 75th Anniversary Tour cost $10, $18, or, for VIP seats, $30. So far, five times as many $18 tickets have been sold as VIP tickets. The number of $10 tickets equals the number of $18 tickets plus twice the number of VIP tickets. Sales of these tickets total $9500. How many of each kind of ticket have been sold? (*Source:* www.ticketmaster.com)

46. Three kinds of tickets are available for a *Prosthetic Forehead* concert: "up close," "in the middle," and "far out." "Up close" tickets cost $10 more than "in the middle" tickets, while "in the middle" tickets cost $10 more than "far out" tickets. Twice the cost of an "up close" ticket is $20 more than 3 times the cost of a "far out" ticket. Find the price of each kind of ticket.

47. A hardware supplier manufactures three kinds of clamps, types A, B, and C. Production restrictions require it to make 10 units more type C clamps than the total of the other types and twice as many type B clamps as type A. The shop must produce a total of 490 units of clamps per day. How many units of each type can be made per day?

48. A Mardi Gras trinket manufacturer supplies three wholesalers, A, B, and C. The output from a day's production is 320 cases of trinkets. She must send wholesaler A three times as many cases as she sends B, and she must send wholesaler C 160 cases less than she provides A and B together. How many cases should she send to each wholesaler to distribute the entire day's production to them?

49. A plant food is to be made from three chemicals. The mix must include 60% of the first and second chemicals. The second and third chemicals must be in a ratio of 4 to 3 by weight. How much of each chemical is needed to make 750 kg of the plant food?

50. How many ounces of 5% hydrochloric acid, 20% hydrochloric acid, and water must be combined to get 10 oz of solution that is 8.5% hydrochloric acid, if the amount of water used must equal the total amount of the other two solutions?

51. During a recent National Hockey League regular season, the Dallas Stars played 82 games. Together, their wins and losses totaled 74. They tied 18 fewer games than they lost. How many wins, losses, and ties did they have that year?

Team	GP	W	L	T	GF	GA	Pts
Dallas	82	__	__	__	252	198	104
Detroit	82	38	26	18	253	197	94
Phoenix	82	38	37	7	240	243	83
St. Louis	82	36	35	11	236	239	83
Chicago	82	34	35	13	223	210	81
Toronto	82	30	44	8	230	273	68

Source: Sports Illustrated Sports Almanac.

52. During a recent National Hockey League season, the Boston Bruins played 82 games. Their losses and ties totaled 56, and they had 21 fewer wins than losses. How many wins, losses, and ties did they have that year?

Team	GP	W	L	T	GF	GA	Pts
Buffalo	82	40	30	12	237	208	92
Pittsburgh	82	38	36	8	285	280	84
Ottawa	82	31	36	15	226	234	77
Montreal	82	31	36	15	249	276	77
Hartford	82	32	39	11	226	256	75
Boston	82	__	__	__	234	300	61

Source: Sports Illustrated Sports Almanac.

8.6 Solving Systems of Linear Equations by Matrix Methods

FIGURE 12

OBJECTIVE 1 Define a matrix. An ordered array of numbers such as

$$\text{Rows} \begin{bmatrix} 2 & 3 & 5 \\ 7 & 1 & 2 \end{bmatrix} \quad \text{Columns}$$

is called a **matrix.** The numbers are called **elements** of the matrix. Matrices (the plural of *matrix*) are named according to the number of **rows** and **columns** they contain. The rows are read horizontally, and the columns are read vertically. For example, the first row in the preceding matrix is 2 3 5 and the first column is $\dfrac{2}{7}$. This matrix is a 2 × 3 (read "two by three") matrix because it has 2 rows and 3 columns. The number of rows is given first, and then the number of columns. Two other examples follow.

$$\begin{bmatrix} -1 & 0 \\ 1 & -2 \end{bmatrix} \quad \begin{matrix} 2 \times 2 \\ \text{matrix} \end{matrix} \qquad \begin{bmatrix} 8 & -1 & -3 \\ 2 & 1 & 6 \\ 0 & 5 & -3 \\ 5 & 9 & 7 \end{bmatrix} \quad \begin{matrix} 4 \times 3 \\ \text{matrix} \end{matrix}$$

A **square matrix** is one that has the same number of rows as columns. The 2 × 2 matrix is a square matrix.

Figure 12 shows how a graphing calculator displays the preceding two matrices. Work with matrices is made much easier by using technology when available. Consult your owner's manual for details.

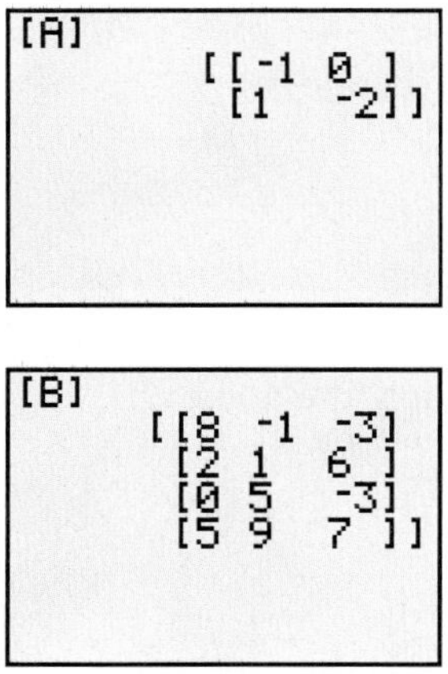

In this section, we discuss a matrix method of solving linear systems that is really just a very structured way of using the elimination method. The advantage of this new method is that it can be done by a graphing calculator or a computer, allowing large systems of equations to be solved easily.

OBJECTIVE 2 Write the augmented matrix for a system. To begin, we write an *augmented matrix* for the system. An **augmented matrix** has a vertical bar that separates the columns of the matrix into two groups. For example, to solve the system

$$x - 3y = 1$$
$$2x + y = -5,$$

start with the augmented matrix

$$\left[\begin{array}{cc|c} 1 & -3 & 1 \\ 2 & 1 & -5 \end{array} \right].$$

Place the coefficients of the variables to the left of the bar, and the constants to the right. The bar separates the coefficients from the constants. The matrix is just a shorthand way of writing the system of equations, so the rows of the augmented matrix can be treated the same as the equations of a system of equations.

We know that exchanging the position of two equations in a system does not change the system. Also, multiplying any equation in a system by a nonzero number does not change the system. Comparable changes to the augmented matrix of a system of equations produce new matrices that correspond to systems with the same solutions as the original system.

The following **row operations** produce new matrices that lead to systems having the same solutions as the original system.

Matrix Row Operations

1. Any two rows of the matrix may be interchanged.
2. The elements in any row may be multiplied by any nonzero real number.
3. Any row may be changed by adding to the elements of the row the product of a real number and the corresponding elements of another row.

Examples of these row operations follow.

Row operation 1:

$$\begin{bmatrix} 2 & 3 & 9 \\ 4 & 8 & -3 \\ 1 & 0 & 7 \end{bmatrix} \quad \text{becomes} \quad \begin{bmatrix} 1 & 0 & 7 \\ 4 & 8 & -3 \\ 2 & 3 & 9 \end{bmatrix}.$$

Interchange row 1 and row 3.

Row operation 2:

$$\begin{bmatrix} 2 & 3 & 9 \\ 4 & 8 & -3 \\ 1 & 0 & 7 \end{bmatrix} \quad \text{becomes} \quad \begin{bmatrix} 6 & 9 & 27 \\ 4 & 8 & -3 \\ 1 & 0 & 7 \end{bmatrix}.$$

Multiply the numbers in row 1 by 3.

Row operation 3:

$$\begin{bmatrix} 2 & 3 & 9 \\ 4 & 8 & -3 \\ 1 & 0 & 7 \end{bmatrix} \quad \text{becomes} \quad \begin{bmatrix} 0 & 3 & -5 \\ 4 & 8 & -3 \\ 1 & 0 & 7 \end{bmatrix}.$$

Multiply the numbers in row 3 by -2; add them to the corresponding numbers in row 1.

The third row operation corresponds to the way we eliminated a variable from a pair of equations in the previous sections.

OBJECTIVE 3 **Use row operations to solve a system with two equations.** Row operations can be used to rewrite a matrix until it is the matrix of a system where the solution is easy to find. The goal is a matrix in the form

$$\left[\begin{array}{cc|c} 1 & a & b \\ 0 & 1 & c \end{array}\right] \quad \text{or} \quad \left[\begin{array}{ccc|c} 1 & a & b & c \\ 0 & 1 & d & e \\ 0 & 0 & 1 & f \end{array}\right]$$

for systems with two or three equations, respectively. Notice that there are 1s down the diagonal from upper left to lower right and 0s below the 1s. A matrix written this way is said to be in **row echelon form.** When these matrices are rewritten as systems of equations, the value of one variable is known, and the rest can be found by substitution. The following examples illustrate this method.

EXAMPLE 1 Using Row Operations to Solve a System with Two Variables

Use row operations to solve the system

$$x - 3y = 1$$
$$2x + y = -5.$$

We start with the augmented matrix of the system.

$$\begin{bmatrix} 1 & -3 & | & 1 \\ 2 & 1 & | & -5 \end{bmatrix}$$

Now we use the various row operations to change this matrix into one that leads to a system that is easier to solve.

It is best to work by columns. We start with the first column and make sure that there is a 1 in the first row, first column position. There is already a 1 in this position. Next, we get 0 in every position below the first. To get a 0 in row two, column one, we use the third row operation and add to the numbers in row two the result of multiplying each number in row one by -2. (We abbreviate this as $-2R_1 + R_2$.) Row one remains unchanged.

$$\begin{bmatrix} 1 & -3 & | & 1 \\ 2 + 1(-2) & 1 + -3(-2) & | & -5 + 1(-2) \end{bmatrix}$$

Original number from row two -2 times number from row one

$$\begin{bmatrix} 1 & -3 & | & 1 \\ 0 & 7 & | & -7 \end{bmatrix} \quad -2R_1 + R_2$$

The matrix now has a 1 in the first position of column one, with 0 in every position below the first.

Now we go to column two. A 1 is needed in row two, column two. We get this 1 by using the second row operation, multiplying each number of row two by $\frac{1}{7}$.

$$\begin{bmatrix} 1 & -3 & | & 1 \\ 0 & 1 & | & -1 \end{bmatrix} \quad \frac{1}{7}R_2$$

This augmented matrix leads to the system of equations

$$\begin{array}{lll} 1x - 3y = 1 & \text{or} & x - 3y = 1 \\ 0x + 1y = -1 & & y = -1. \end{array}$$

From the second equation, $y = -1$. We substitute -1 for y in the first equation to get

$$x - 3y = 1$$
$$x - 3(-1) = 1$$
$$x + 3 = 1$$
$$x = -2.$$

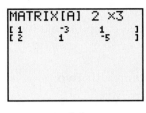

(a)

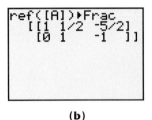

(b)

FIGURE 13

The solution set of the system is $\{(-2, -1)\}$. Check this solution by substitution in both equations of the system.

Now Try Exercise 3.

If the augmented matrix of the system in Example 1 is entered as matrix [A] in a graphing calculator (Figure 13(a)) and the row echelon form of the matrix is found (Figure 13(b)), then the system becomes

$$x + \frac{1}{2}y = -\frac{5}{2}$$
$$y = -1.$$

While this system looks different from the one we obtained in Example 1, it is equivalent, since its solution set is also $\{(-2, -1)\}$.

OBJECTIVE 4 Use row operations to solve a system with three equations. A linear system with three equations is solved in a similar way. We use row operations to get 1s down the diagonal from left to right and all 0s below each 1.

EXAMPLE 2 Using Row Operations to Solve a System with Three Variables

Use row operations to solve the system

$$\begin{aligned} x - y + 5z &= -6 \\ 3x + 3y - z &= 10 \\ x + 3y + 2z &= 5. \end{aligned}$$

Start by writing the augmented matrix of the system.

$$\left[\begin{array}{ccc|c} 1 & -1 & 5 & -6 \\ 3 & 3 & -1 & 10 \\ 1 & 3 & 2 & 5 \end{array}\right]$$

This matrix already has 1 in row one, column one. Next get 0s in the rest of column one. First, add to row two the results of multiplying each number of row one by -3. This gives the matrix

$$\left[\begin{array}{ccc|c} 1 & -1 & 5 & -6 \\ 0 & 6 & -16 & 28 \\ 1 & 3 & 2 & 5 \end{array}\right]. \qquad -3R_1 + R_2$$

Now add to the numbers in row three the results of multiplying each number of row one by -1.

$$\left[\begin{array}{ccc|c} 1 & -1 & 5 & -6 \\ 0 & 6 & -16 & 28 \\ 0 & 4 & -3 & 11 \end{array}\right] \qquad -1R_1 + R_3$$

Introduce 1 in row two, column two by multiplying each number in row two by $\frac{1}{6}$.

$$\begin{bmatrix} 1 & -1 & 5 & | & -6 \\ 0 & 1 & -\frac{8}{3} & | & \frac{14}{3} \\ 0 & 4 & -3 & | & 11 \end{bmatrix} \qquad \frac{1}{6}R_2$$

To obtain 0 in row three, column two, add to row three the results of multiplying each number in row two by -4.

$$\begin{bmatrix} 1 & -1 & 5 & | & -6 \\ 0 & 1 & -\frac{8}{3} & | & \frac{14}{3} \\ 0 & 0 & \frac{23}{3} & | & -\frac{23}{3} \end{bmatrix} \qquad -4R_2 + R_3$$

Finally, obtain 1 in row three, column three by multiplying each number in row three by $\frac{3}{23}$.

$$\begin{bmatrix} 1 & -1 & 5 & | & -6 \\ 0 & 1 & -\frac{8}{3} & | & \frac{14}{3} \\ 0 & 0 & 1 & | & -1 \end{bmatrix} \qquad \frac{3}{23}R_3$$

This final matrix gives the system of equations

$$x - y + 5z = -6$$
$$y - \frac{8}{3}z = \frac{14}{3}$$
$$z = -1.$$

Substitute -1 for z in the second equation, $y - \frac{8}{3}z = \frac{14}{3}$, to find that $y = 2$. Finally, substitute 2 for y and -1 for z in the first equation, $x - y + 5z = -6$, to determine that $x = 1$. The solution set of the original system is $\{(1, 2, -1)\}$. Check by substitution.

Now Try Exercise 15.

OBJECTIVE 5 Use row operations to solve special systems. In the final example we show how to recognize inconsistent systems or systems with dependent equations when solving these systems with row operations.

EXAMPLE 3 Recognizing Inconsistent Systems or Dependent Equations

Use row operations to solve each system.

(a) $2x - 3y = 8$
$-6x + 9y = 4$

$$\begin{bmatrix} 2 & -3 & | & 8 \\ -6 & 9 & | & 4 \end{bmatrix} \qquad \text{Write the augmented matrix.}$$

$$\begin{bmatrix} 1 & -\frac{3}{2} & | & 4 \\ -6 & 9 & | & 4 \end{bmatrix} \qquad \frac{1}{2}R_1$$

$$\begin{bmatrix} 1 & -\frac{3}{2} & | & 4 \\ 0 & 0 & | & 28 \end{bmatrix} \qquad 6R_1 + R_2$$

The corresponding system of equations is

$$x - \frac{3}{2}y = 4$$

$$0 = 28, \qquad \text{False}$$

which has no solution and is inconsistent. The solution set is $\emptyset$.

(b) $-10x + 12y = 30$

$\qquad 5x - 6y = -15$

$$\begin{bmatrix} -10 & 12 & | & 30 \\ 5 & -6 & | & -15 \end{bmatrix} \qquad \text{Write the augmented matrix.}$$

$$\begin{bmatrix} 1 & -\frac{6}{5} & | & -3 \\ 5 & -6 & | & -15 \end{bmatrix} \qquad -\frac{1}{10}R_1$$

$$\begin{bmatrix} 1 & -\frac{6}{5} & | & -3 \\ 0 & 0 & | & 0 \end{bmatrix} \qquad -5R_1 + R_2$$

The corresponding system is

$$x - \frac{6}{5}y = -3$$

$$0 = 0, \qquad \text{True}$$

which has dependent equations. Using the second equation of the original system, we write the solution set as

$$\{(x, y) \mid 5x - 6y = -15\}.$$

Now Try Exercises 11 and 13.

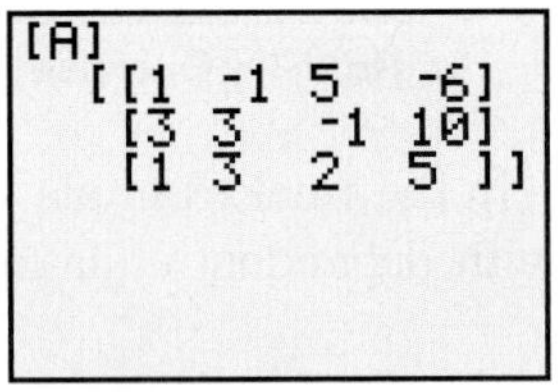

[A]
```
[[1 -1 5 -6]
 [3 3 -1 10]
 [1 3 2 5 ]]
```

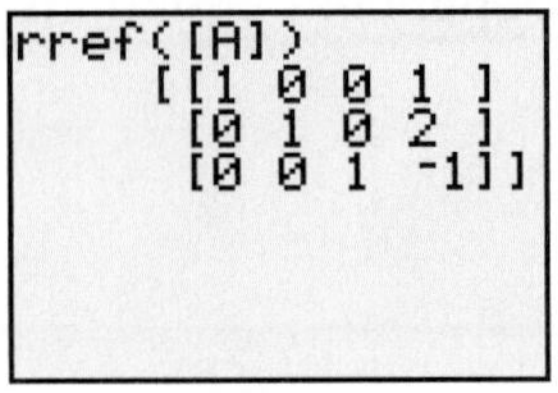

rref([A])
```
[[1 0 0 1 ]
 [0 1 0 2 ]
 [0 0 1 -1]]
```

FIGURE 14

| **CONNECTIONS** |

An extension of the matrix method described in this section involves transforming an augmented matrix into **reduced row echelon form.** This form has 1s down the main diagonal and 0s above and below this diagonal. For example, the matrix for the system in Example 2 could be transformed into the matrix

$$\begin{bmatrix} 1 & 0 & 0 & | & 1 \\ 0 & 1 & 0 & | & 2 \\ 0 & 0 & 1 & | & -1 \end{bmatrix} \qquad \text{which gives the equivalent system} \qquad \begin{array}{l} x = 1 \\ y = 2 \\ z = -1. \end{array}$$

The calculator screens in Figure 14 indicate how easily this transformation can be obtained using technology.

For Discussion or Writing

1. Write the reduced row echelon form for the matrix of the system in Example 1.
2. If transforming to reduced row echelon form leads to all 0s in the final row, what kind of system is represented?

8.6 EXERCISES

For Extra Help

 Student's
Solutions Manual

 MyMathLab

 InterAct Math
Tutorial Software

 AW Math
Tutor Center

$Math_{XP}$ MathXL

 Digital Video Tutor
CD 12/Videotape 12

1. Consider the matrix $\begin{bmatrix} -2 & 3 & 1 \\ 0 & 5 & -3 \\ 1 & 4 & 8 \end{bmatrix}$ and answer the following.

 (a) What are the elements of the second row?

 (b) What are the elements of the third column?

 (c) Is this a square matrix? Explain why or why not.

 (d) Give the matrix obtained by interchanging the first and third rows.

 (e) Give the matrix obtained by multiplying the first row by $-\frac{1}{2}$.

 (f) Give the matrix obtained by multiplying the third row by 3 and adding to the first row.

2. Give the dimensions of each matrix.

 (a) $\begin{bmatrix} 3 & -7 \\ 4 & 5 \\ -1 & 0 \end{bmatrix}$ **(b)** $\begin{bmatrix} 4 & 9 & 0 \\ -1 & 2 & -4 \end{bmatrix}$

 (c)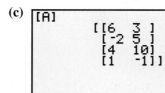

 (d)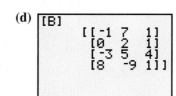

Complete the steps in the matrix solution of each system by filling in the boxes. Give the final system and the solution set. See Example 1.

3. $4x + 8y = 44$
  $2x - y = -3$

$\begin{bmatrix} 4 & 8 & | & 44 \\ 2 & -1 & | & -3 \end{bmatrix}$

$\begin{bmatrix} 1 & \blacksquare & | & \blacksquare \\ 2 & -1 & | & -3 \end{bmatrix}$ $\frac{1}{4}R_1$

$\begin{bmatrix} 1 & 2 & | & 11 \\ 0 & \blacksquare & | & \blacksquare \end{bmatrix}$ $-2R_1 + R_2$

$\begin{bmatrix} 1 & 2 & | & 11 \\ 0 & 1 & | & \blacksquare \end{bmatrix}$ $-\frac{1}{5}R_2$

4. $2x - 5y = -1$
  $3x + y = 7$

$\begin{bmatrix} 2 & -5 & | & -1 \\ 3 & 1 & | & 7 \end{bmatrix}$

$\begin{bmatrix} 1 & -\dfrac{5}{2} & | & \blacksquare \\ 3 & 1 & | & 7 \end{bmatrix}$ $\frac{1}{2}R_1$

$\begin{bmatrix} 1 & -\dfrac{5}{2} & | & -\dfrac{1}{2} \\ 0 & \blacksquare & | & \blacksquare \end{bmatrix}$ $-3R_1 + R_2$

$\begin{bmatrix} 1 & -\dfrac{5}{2} & | & -\dfrac{1}{2} \\ 0 & 1 & | & \blacksquare \end{bmatrix}$ $\frac{2}{17}R_2$

Use row operations to solve each system. See Examples 1 and 3.

5. $x + y = 5$
  $x - y = 3$

6. $x + 2y = 7$
  $x - y = -2$

7. $2x + 4y = 6$
  $3x - y = 2$

8. $4x + 5y = -7$
  $x - y = 5$

9. $3x + 4y = 13$
$2x - 3y = -14$

10. $5x + 2y = 8$
$3x - y = 7$

11. $-4x + 12y = 36$
$x - 3y = 9$

12. $2x - 4y = 8$
$-3x + 6y = 5$

13. $2x + y = 4$
$4x + 2y = 8$

14. $-3x - 4y = 1$
$6x + 8y = -2$

Complete the steps in the matrix solution of each system by filling in the boxes. Give the final system and the solution set. See Example 2.

15. $x + y - z = -3$
$2x + y + z = 4$
$5x - y + 2z = 23$

$$\begin{bmatrix} 1 & 1 & -1 & \bigm| & -3 \\ 2 & 1 & 1 & \bigm| & 4 \\ 5 & -1 & 2 & \bigm| & 23 \end{bmatrix}$$

$$\begin{bmatrix} 1 & 1 & -1 & \bigm| & -3 \\ 0 & \blacksquare & \blacksquare & \bigm| & \blacksquare \\ 0 & \blacksquare & \blacksquare & \bigm| & \blacksquare \end{bmatrix} \begin{array}{l} -2R_1 + R_2 \\ -5R_1 + R_3 \end{array}$$

$$\begin{bmatrix} 1 & 1 & -1 & \bigm| & -3 \\ 0 & 1 & \blacksquare & \bigm| & \blacksquare \\ 0 & -6 & 7 & \bigm| & 38 \end{bmatrix} \quad -1R_2$$

$$\begin{bmatrix} 1 & 1 & -1 & \bigm| & -3 \\ 0 & 1 & -3 & \bigm| & -10 \\ 0 & 0 & \blacksquare & \bigm| & \blacksquare \end{bmatrix} \quad 6R_2 + R_3$$

$$\begin{bmatrix} 1 & 1 & -1 & \bigm| & -3 \\ 0 & 1 & -3 & \bigm| & -10 \\ 0 & 0 & 1 & \bigm| & \blacksquare \end{bmatrix} \quad -\tfrac{1}{11}R_3$$

16. $2x + y + 2z = 11$
$2x - y - z = -3$
$3x + 2y + z = 9$

$$\begin{bmatrix} 2 & 1 & 2 & \bigm| & 11 \\ 2 & -1 & -1 & \bigm| & -3 \\ 3 & 2 & 1 & \bigm| & 9 \end{bmatrix}$$

$$\begin{bmatrix} 1 & \blacksquare & \blacksquare & \bigm| & \blacksquare \\ 2 & -1 & -1 & \bigm| & -3 \\ 3 & 2 & 1 & \bigm| & 9 \end{bmatrix} \quad \tfrac{1}{2}R_1$$

$$\begin{bmatrix} 1 & \tfrac{1}{2} & 1 & \bigm| & \tfrac{11}{2} \\ 0 & \blacksquare & \blacksquare & \bigm| & \blacksquare \\ 0 & \blacksquare & \blacksquare & \bigm| & \blacksquare \end{bmatrix} \begin{array}{l} -2R_1 + R_2 \\ -3R_1 + R_3 \end{array}$$

$$\begin{bmatrix} 1 & \tfrac{1}{2} & 1 & \bigm| & \tfrac{11}{2} \\ 0 & 1 & \blacksquare & \bigm| & \blacksquare \\ 0 & \tfrac{1}{2} & -2 & \bigm| & -\tfrac{15}{2} \end{bmatrix} \quad -\tfrac{1}{2}R_2$$

$$\begin{bmatrix} 1 & \tfrac{1}{2} & 1 & \bigm| & \tfrac{11}{2} \\ 0 & 1 & \tfrac{3}{2} & \bigm| & 7 \\ 0 & 0 & \blacksquare & \bigm| & \blacksquare \end{bmatrix} \quad -\tfrac{1}{2}R_2 + R_3$$

$$\begin{bmatrix} 1 & \tfrac{1}{2} & 1 & \bigm| & \tfrac{11}{2} \\ 0 & 1 & \tfrac{3}{2} & \bigm| & 7 \\ 0 & 0 & 1 & \bigm| & \blacksquare \end{bmatrix} \quad -\tfrac{4}{11}R_3$$

Use row operations to solve each system. See Examples 2 and 3.

17. $x + y - 3z = 1$
$2x - y + z = 9$
$3x + y - 4z = 8$

18. $2x + 4y - 3z = -18$
$3x + y - z = -5$
$x - 2y + 4z = 14$

19. $x + y - z = 6$
$2x - y + z = -9$
$x - 2y + 3z = 1$

20. $x + 3y - 6z = 7$
$2x - y + 2z = 0$
$x + y + 2z = -1$

CHAPTER 8 SUMMARY

KEY TERMS

8.1 system of linear equations (linear system)
solution of a system
solution set of a system

consistent system
inconsistent system
independent equations
dependent equations
8.4 ordered triple

8.6 matrix
element of a matrix
row
column
square matrix

augmented matrix
row operations
row echelon form
reduced row echelon form

NEW SYMBOLS

(x, y, z) ordered triple

$$\begin{bmatrix} a & b & c \\ d & e & f \end{bmatrix}$$ matrix with two rows, three columns

TEST YOUR WORD POWER

See how well you have learned the vocabulary in this chapter. Answers, with examples, follow the Quick Review.

1. A **system of linear equations** consists of
   A. at least two linear equations with different variables
   B. two or more linear equations that have an infinite number of solutions
   C. two or more linear equations with the same variables
   D. two or more linear inequalities.

2. A **solution of a system** of linear equations is
   A. an ordered pair that makes one equation of the system true
   B. an ordered pair that makes all the equations of the system true at the same time

   C. any ordered pair that makes one or the other or both equations of the system true
   D. the set of values that make all the equations of the system false.

3. A **consistent system** is a system of equations
   A. with one solution
   B. with no solution
   C. with an infinite number of solutions
   D. that have the same graph.

4. An **inconsistent system** is a system of equations
   A. with one solution
   B. with no solution

   C. with an infinite number of solutions
   D. that have the same graph.

5. **Dependent equations**
   A. have different graphs
   B. have no solution
   C. have one solution
   D. are different forms of the same equation.

6. A **matrix** is
   A. an ordered pair of numbers
   B. an array of numbers with the same number of rows and columns
   C. a pair of numbers written between brackets
   D. a rectangular array of numbers.

QUICK REVIEW

CONCEPTS	EXAMPLES

8.1 SOLVING SYSTEMS OF LINEAR EQUATIONS BY GRAPHING

An ordered pair is a solution of a system if it makes all equations of the system true at the same time.

Is $(4, -1)$ a solution of the system $\begin{array}{l} x + y = 3 \\ 2x - y = 9? \end{array}$

Yes, because $4 + (-1) = 3$, and $2(4) - (-1) = 9$ are both true.

To solve a linear system by graphing,

Step 1 Graph each equation of the system on the same axes.

Step 2 Find the coordinates of the point of intersection.

Step 3 Check. Write the solution set.

Solve the system by graphing: $\begin{array}{l} x + y = 5 \\ 2x - y = 4. \end{array}$

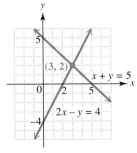
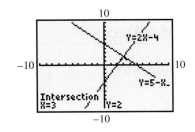

A graphing calculator can find the solution of a system by locating the point of intersection of the graphs.

The solution $(3, 2)$ checks, so $\{(3, 2)\}$ is the solution set.

8.2 SOLVING SYSTEMS OF LINEAR EQUATIONS BY SUBSTITUTION

Step 1 Solve one equation for either variable.

Solve by substitution.

$$x + 2y = -5 \quad (1)$$
$$y = -2x - 1 \quad (2)$$

Equation (2) is already solved for y.

Step 2 Substitute for that variable in the other equation to get an equation in one variable.

Substitute $-2x - 1$ for y in equation (1).

$$x + 2(-2x - 1) = -5$$

Step 3 Solve the equation from Step 2.

Solve to get $x = 1$.

Step 4 Substitute the result into the equation from Step 1 to get the value of the other variable.

To find y, let $x = 1$ in equation (2).

$$y = -2(1) - 1 = -3$$

Step 5 Check. Write the solution set.

The solution, $(1, -3)$, checks so $\{(1, -3)\}$ is the solution set.

CONCEPTS	EXAMPLES

8.3 SOLVING SYSTEMS OF LINEAR EQUATIONS BY ELIMINATION

Step 1 Write both equations in standard form $Ax + By = C$.

Solve by elimination.

$$x + 3y = 7 \quad (1)$$
$$3x - y = 1 \quad (2)$$

Step 2 Multiply to make the coefficients of one pair of variable terms opposites.

Multiply equation (1) by -3 to eliminate the x-terms.

Step 3 Add the equations to get an equation with only one variable.

$$-3x - 9y = -21$$
$$\underline{3x - y = 1}$$

Step 4 Solve the equation from Step 3.

$$-10y = -20 \qquad \text{Add.}$$
$$y = 2 \qquad \text{Divide by } -10.$$

Step 5 Substitute the solution from Step 4 into either of the original equations to find the value of the remaining variable.

Substitute to get the value of x.

$$x + 3y = 7 \qquad (1)$$
$$x + 3(2) = 7 \qquad \text{Let } y = 2.$$
$$x + 6 = 7 \qquad \text{Multiply.}$$
$$x = 1 \qquad \text{Subtract 6.}$$

Step 6 Check. Write the solution set.

Since $1 + 3(2) = 7$ and $3(1) - 2 = 1$, the solution set is $\{(1, 2)\}$.

If the result of the addition step (Step 3) is a false statement, such as $0 = 4$, the graphs are parallel lines and *the solution set is $\emptyset$*.

$$x - 2y = 6$$
$$\underline{-x + 2y = -2}$$
$$0 = 4 \qquad \text{Solution set: } \emptyset$$

If the result is a true statement, such as $0 = 0$, the graphs are the same line, and an *infinite number of ordered pairs are solutions.*

$$x - 2y = 6$$
$$\underline{-x + 2y = -6}$$
$$0 = 0 \qquad \text{Solution set: } \{(x, y) \mid x - 2y = 6\}$$

8.4 SYSTEMS OF LINEAR EQUATIONS IN THREE VARIABLES

Solving a Linear System in Three Variables

Solve the system

Step 1 Use the elimination method to eliminate any variable from any two of the original equations.

$$x + 2y - z = 6 \qquad (1)$$
$$x + y + z = 6 \qquad (2)$$
$$2x + y - z = 7. \qquad (3)$$

Add equations (1) and (2); z is eliminated and the result is $2x + 3y = 12$.

Step 2 Eliminate the *same* variable from any *other* two equations.

Eliminate z again by adding equations (2) and (3) to get $3x + 2y = 13$. Now solve the system

$$2x + 3y = 12 \qquad (4)$$
$$3x + 2y = 13. \qquad (5) \qquad \text{(continued)}$$

CONCEPTS	EXAMPLES

Step 3 Eliminate a second variable from the two equations in two variables that result from Steps 1 and 2. The result is an equation in one variable that gives the value of that variable.

To eliminate x, multiply equation (4) by -3 and equation (5) by 2.

$$\begin{array}{r} -6x - 9y = -36 \\ \underline{6x + 4y = 26} \\ -5y = -10 \\ y = 2 \end{array}$$

Step 4 Substitute the value of the variable found in Step 3 into either of the equations in two variables to find the value of the second variable.

Let $y = 2$ in equation (4).

$$2x + 3(2) = 12$$
$$2x + 6 = 12$$
$$2x = 6$$
$$x = 3$$

Step 5 Use the values of the two variables from Steps 3 and 4 to find the value of the third variable by substituting into an appropriate equation.

Let $y = 2$ and $x = 3$ in any of the original equations to find $z = 1$.

Step 6 Check the solution in all of the original equations. Then write the solution set.

Check. The solution set is $\{(3, 2, 1)\}$.

8.5 APPLICATIONS OF SYSTEMS OF LINEAR EQUATIONS

Use the six-step problem-solving method.

Step 1 Read the problem carefully.

The perimeter of a rectangle is 18 ft. The length is 3 ft more than twice the width. What are the dimensions of the rectangle?

Step 2 Assign variables.

Let x represent the length and y represent the width.

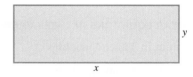

Step 3 Write a system of equations that relates the unknowns.

From the perimeter formula, one equation is $2x + 2y = 18$. From the problem, another equation is $x = 3 + 2y$.

Step 4 Solve the system.

Solve the system

$$2x + 2y = 18$$
$$x = 3 + 2y$$

to find that $x = 7$ and $y = 2$.

Steps 5 and 6 State the answer and check.

The length is 7 ft, and the width is 2 ft. Since the perimeter is

$$2(7) + 2(2) = 18, \quad \text{and} \quad 3 + 2(2) = 7,$$

the answer checks.

CONCEPTS	EXAMPLES

8.6 SOLVING SYSTEMS OF LINEAR EQUATIONS BY MATRIX METHODS

Matrix Row Operations

1. Any two rows of the matrix may be interchanged.

$$\begin{bmatrix} 1 & 5 & 7 \\ 3 & 9 & -2 \\ 0 & 6 & 4 \end{bmatrix} \text{ becomes } \begin{bmatrix} 3 & 9 & -2 \\ 1 & 5 & 7 \\ 0 & 6 & 4 \end{bmatrix} \quad \text{Interchange } R_1 \text{ and } R_2.$$

2. The elements in any row may be multiplied by any nonzero real number.

$$\begin{bmatrix} 1 & 5 & 7 \\ 3 & 9 & -2 \\ 0 & 6 & 4 \end{bmatrix} \text{ becomes } \begin{bmatrix} 1 & 5 & 7 \\ 1 & 3 & -\frac{2}{3} \\ 0 & 6 & 4 \end{bmatrix} \quad \frac{1}{3}R_2$$

3. Any row may be changed by adding to the elements of the row the product of a real number and the elements of another row.

$$\begin{bmatrix} 1 & 5 & 7 \\ 3 & 9 & -2 \\ 0 & 6 & 4 \end{bmatrix} \text{ becomes } \begin{bmatrix} 1 & 5 & 7 \\ 0 & -6 & -23 \\ 0 & 6 & 4 \end{bmatrix} \quad -3R_1 + R_2$$

A system can be solved by matrix methods. Write the augmented matrix and use row operations to obtain a matrix in row echelon form.

Solve using row operations:
$$\begin{aligned} x + 3y &= 7 \\ 2x + y &= 4. \end{aligned}$$

$$\begin{bmatrix} 1 & 3 & | & 7 \\ 2 & 1 & | & 4 \end{bmatrix} \quad \text{Augmented matrix}$$

$$\begin{bmatrix} 1 & 3 & | & 7 \\ 0 & -5 & | & -10 \end{bmatrix} \quad -2R_1 + R_2$$

$$\begin{bmatrix} 1 & 3 & | & 7 \\ 0 & 1 & | & 2 \end{bmatrix} \quad -\frac{1}{5}R_2$$

$$\begin{aligned} x + 3y &= 7 \\ y &= 2 \end{aligned}$$

When $y = 2$, $x + 3(2) = 7$, so $x = 1$. The solution set is $\{(1, 2)\}$.

Answers to Test Your Word Power

1. C; *Example:* $2x + y = 7$, $3x - y = 3$ **2.** B; *Example:* The ordered pair $(2, 3)$ satisfies both equations of the system in Answer 1, so it is a solution of the system.   **3.** A; *Example:* The system in Answer 1 is consistent. The graphs of the equations intersect at exactly one point, in this case the solution $(2, 3)$.   **4.** B; *Example:* The equations of two parallel lines make up an inconsistent system; their graphs never intersect, so there is no solution to the system.   **5.** D; *Example:* The equations $4x - y = 8$ and $8x - 2y = 16$ are dependent because their graphs are the same line.   **6.** D; *Examples:* $\begin{bmatrix} 3 & -1 & 0 \\ 4 & 2 & 1 \end{bmatrix}, \begin{bmatrix} 1 & 2 \\ 4 & 3 \end{bmatrix}$

CHAPTER **8** REVIEW EXERCISES

[8.1] *Decide whether the given ordered pair is a solution of the given system.*

1. $(3, 4)$
$$4x - 2y = 4$$
$$5x + y = 19$$

2. $(-5, 2)$
$$x - 4y = -13$$
$$2x + 3y = 4$$

Solve each system by graphing.

3. $x + y = 4$
$2x - y = 5$

4. $x - 2y = 4$
$2x + y = -2$

5. $2x + 4 = 2y$
$y - x = -3$

6. The graph shows the trends during the years 1974 through 1996 relating to bachelor's degrees awarded in the United States.

(a) Between what years shown on the horizontal axis did the number of degrees for men and women reach equal numbers?

(b) When the number of degrees for men and women reached equal numbers, what was that number (approximately)?

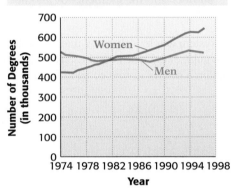

BACHELOR'S DEGREES IN THE U.S.

Source: U.S. National Center for Education Statistics, *Digest of Education Statistics*, annual.

[8.2]

 **7.** A student solves the system $\begin{array}{l} 2x + y = 6 \\ -2x - y = 4 \end{array}$ and gets the equation $0 = 10$. The student gives the solution set as $\{(0, 10)\}$. Is this correct? Explain.

Solve each system by the substitution method.

8. $3x + y = 7$
$x = 2y$

9. $2x - 5y = -19$
$y = x + 2$

10. $5x + 15y = 30$
$x + 3y = 6$

 **11.** After solving a system of linear equations by the substitution method, a student obtained the equation "$0 = 0$." He gave the solution set of the system as $\{(0, 0)\}$. Was his answer correct? Why or why not?

 **12.** Suppose that you were asked to solve the system $\begin{array}{l} 5x - 3y = 7 \\ -x + 2y = 4 \end{array}$ by substitution. Which variable in which equation would be easiest to solve for in your first step? Why?

[8.3]

13. Only one of the following systems does not require that we multiply one or both equations by a constant to solve the system by the elimination method. Which one is it?

A. $-4x + 3y = 7$
$3x - 4y = 4$

B. $5x + 8y = 13$
$12x + 24y = 36$

C. $2x + 3y = 5$
$x - 3y = 12$

D. $x + 2y = 9$
$3x - y = 6$

14. For the system

$$2x + 12y = 7$$
$$3x + 4y = 1,$$

if we were to multiply the first equation by -3, by what number would we have to multiply the second equation to

(a) eliminate the x-terms when solving by the elimination method?

(b) eliminate the y-terms when solving by the elimination method?

Solve each system by the elimination method.

15. $2x - y = 13$
$\quad\;\; x + y = 8$

16. $-4x + 3y = 25$
$\quad\;\;\;\; 6x - 5y = -39$

17. $3x - 4y = 9$
$\quad\;\; 6x - 8y = 18$

18. $\quad 2x + y = 3$
$\quad -4x - 2y = 6$

[8.1–8.3] *Solve each system by any method.*

19. $2x + 3y = -5$
$\quad\; 3x + 4y = -8$

20. $6x - 9y = 0$
$\quad\; 2x - 3y = 0$

21. $x - 2y = 5$
$\quad\; y = x - 7$

22. $\dfrac{x}{2} + \dfrac{y}{3} = 7$

$\quad\;\; \dfrac{x}{4} + \dfrac{2y}{3} = 8$

23. List the methods of solving systems discussed in this chapter. Choose one and discuss its advantages and disadvantages.

24. Why would it be easier to solve system B by the substitution method than system A?

$$\textbf{A:}\; -5x + 6y = 7 \qquad \textbf{B:}\; 2x + 9y = 13$$
$$\quad\;\;\; 2x + 5y = -5 \qquad\qquad\;\; y = 3x - 2$$

[8.4] *Solve each system. If a system is inconsistent or has dependent equations, say so.*

25. $\quad 2x + 3y - z = -16$
$\quad\quad x + 2y + 2z = -3$
$\quad -3x + y + z = -5$

26. $4x - y = 2$
$\quad\; 3y + z = 9$
$\quad\;\; x + 2z = 7$

27. $\quad 3x - y - z = -8$
$\quad\quad 4x + 2y + 3z = 15$
$\quad -6x + 2y + 2z = 10$

[8.5] *Solve each problem using a system of equations.*

28. A regulation National Hockey League ice rink has perimeter 570 ft. The length is 30 ft longer than twice the width. What are the dimensions of an NHL ice rink? (*Source: Microsoft Encarta Encyclopedia 2000.*)

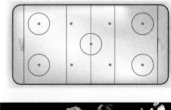

29. The two top-grossing movies of 2001 were *Harry Potter and the Sorcerer's Stone* and *Shrek*. *Shrek* grossed $26 million less than *Harry Potter and the Sorcerer's Stone*, and together the two films took in $562 million. (*Source:* ACNielsen EDI.)

(a) How much did each of these movies earn?

(b) If *Shrek* earned $27 million more than *Monsters, Inc.*, how much did *Monsters, Inc.* earn? (*Hint:* Use your answer from part (a).)

(c) What is the total amount these top three films earned?

30. A plane flies 560 mi in 1.75 hr traveling with the wind. The return trip later against the same wind takes the plane 2 hr. Find the speed of the plane and the speed of the wind.

	r	t	d
With Wind	x + y	1.75	
Against Wind		2	

31. Sweet's Candy Store is offering a special mix for Valentine's Day. Ms. Sweet will mix some $2-per-lb nuts with some $1-per-lb chocolate candy to get 100 lb of mix, which she will sell at $1.30 per lb. How many pounds of each should she use?

	Number of Pounds	Price per Pound	Value
Nuts	x		
Chocolate	y		
Mixture	100		

32. The sum of the measures of the angles of a triangle is 180°. The largest angle measures 10° less than the sum of the other two. The measure of the middle-sized angle is the average of the other two. Find the measures of the three angles.

33. Maria Gonzales sells real estate. On three recent sales, she made 10% commission, 6% commission, and 5% commission. Her total commissions on these sales were $17,000, and she sold property worth $280,000. If the 5% sale amounted to the sum of the other two, what were the three sales prices?

34. How many liters each of 8%, 10%, and 20% hydrogen peroxide should be mixed together to get 8 L of 12.5% solution, if the amount of 8% solution used must be 2 L more than the amount of 20% solution used?

35. In the great baseball year of 1961, Yankee teammates Mickey Mantle, Roger Maris, and John Blanchard combined for 136 home runs. Mantle hit 7 fewer than Maris. Maris hit 40 more than Blanchard. What were the home run totals for each player? (*Source:* Neft, David S. and Richard M. Cohen, *The Sports Encyclopedia: Baseball 1997.*)

[8.6] *Solve each system of equations using row operations.*

36. $2x + 5y = -4$
$4x - y = 14$

37. $6x + 3y = 9$
$-7x + 2y = 17$

38. $x + 2y - z = 1$
$3x + 4y + 2z = -2$
$-2x - y + z = -1$

39. $x + 3y = 7$
$3x + z = 2$
$y - 2z = 4$

MIXED REVIEW EXERCISES

Solve by any method.

40. $\dfrac{2}{3}x + \dfrac{1}{6}y = \dfrac{19}{2}$

$\dfrac{1}{3}x - \dfrac{2}{9}y = 2$

41. $2x + 5y - z = 12$
$-x + y - 4z = -10$
$-8x - 20y + 4z = 31$

42. $x = 7y + 10$
$2x + 3y = 3$

43. $x + 4y = 17$
$-3x + 2y = -9$

44. $-7x + 3y = 12$
$5x + 2y = 8$

45. $2x - 5y = 8$
$3x + 4y = 10$

46. To make a 10% acid solution for chemistry class, Xavier wants to mix some 5% solution with 10 L of 20% solution. How many liters of 5% solution should he use?

Liters of Solution	Percent (as a decimal)	Liters of Pure Acid

47. At the end of 2001, Subway topped McDonald's as the largest restaurant chain in the United States. Subway operated 148 more restaurants than McDonald's, and together the two chains had 26,346 restaurants. How many restaurants did each company operate? (*Source: USA Today,* February 4, 2002.)

48. Candy that sells for $1.30 per lb is to be mixed with candy selling for $.90 per lb to get 100 lb of a mix that will sell for $1 per lb. How much of each type should be used?

49. In the 2000 Summer Olympics in Sydney, Australia, the top medal-winning countries were the United States, Russia, and China, with a combined total of 244 medals. The United States won 9 more medals than Russia, while China won 29 fewer medals than Russia. How many medals did each country win? (*Source: The Gazette,* October 2, 2000.)

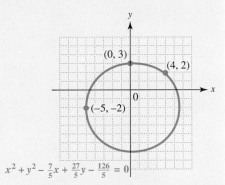

RELATING CONCEPTS (EXERCISES 50–54)

For Individual or Group Work

Thus far in this text we have studied only linear *equations. In later chapters we will study the graphs of other kinds of equations. One such graph is a* circle, *which has an equation of the form*

$$x^2 + y^2 + ax + by + c = 0.$$

It is a fact from geometry that given three noncollinear points (that is, points that do not all lie on the same straight line), there will be a circle that contains them. For example, the points $(4, 2)$, $(-5, -2)$, *and* $(0, 3)$ *lie on the circle whose equation is shown in the figure.* **Work Exercises 50–54 in order,** *to find an equation of the circle passing through the points* $(2, 1)$, $(-1, 0)$, *and* $(3, 3)$.

$$x^2 + y^2 - \tfrac{7}{5}x + \tfrac{27}{5}y - \tfrac{126}{5} = 0$$

50. Let $x = 2$ and $y = 1$ in the equation $x^2 + y^2 + ax + by + c = 0$ to find an equation in a, b, and c.

(continued)

51. Let $x = -1$ and $y = 0$ to find a second equation in a, b, and c.

52. Let $x = 3$ and $y = 3$ to find a third equation in a, b, and c.

53. Solve the system of equations formed by your answers in Exercises 50–52 to find the values of a, b, and c. What is the equation of the circle?

54. Explain why the relation whose graph is a circle is not a function.

CHAPTER 8 TEST

The graph shows a company's costs to produce computer parts and the revenue from the sale of those parts.

1. At what production level does the cost equal the revenue?

2. What is the revenue at that point?

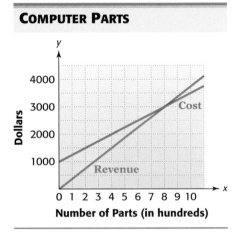

COMPUTER PARTS

3. Decide whether each ordered pair is a solution of the system $\begin{aligned} 2x + y &= -3 \\ x - y &= -9. \end{aligned}$

    (a) $(1, -5)$ **(b)** $(1, 10)$ **(c)** $(-4, 5)$

4. Use a graph to solve the system $\begin{aligned} x + y &= 7 \\ x - y &= 5. \end{aligned}$

Solve each system by substitution. If a system is inconsistent or has dependent equations, say so.

5. $2x - 3y = 24$
    $y = -\dfrac{2}{3}x$

6. $12x - 5y = 8$
    $3x = \dfrac{5}{4}y + 2$

Solve each system by elimination. If a system is inconsistent or has dependent equations, say so.

7. $3x + y = 12$
    $2x - y = 3$

8. $-5x + 2y = -4$
    $6x + 3y = -6$

9. $3x + 4y = 8$
    $8y = 7 - 6x$

10. $\dfrac{6}{5}x - \dfrac{1}{3}y = -20$
    $-\dfrac{2}{3}x + \dfrac{1}{6}y = 11$

11. $3x + 5y + 3z = 2$
    $6x + 5y + z = 0$
    $3x + 10y - 2z = 6$

12. $4x + y + z = 11$
    $x - y - z = 4$
    $y + 2z = 0$

Solve each problem using a system of equations.

13. Julia Roberts is one of the biggest box office stars in Hollywood. As of July 2001, her two top-grossing domestic films, *Pretty Woman* and *Runaway Bride,* together earned $330.7 million. If *Runaway Bride* grossed $26.1 million less than *Pretty Woman,* how much did each film gross? (*Source:* ACNielsen EDI.)

14. Two cars start from points 420 mi apart and travel toward each other. They meet after 3.5 hr. Find the average speed of each car if one travels 30 mph slower than the other.

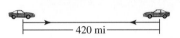

420 mi

15. A chemist needs 12 L of a 40% alcohol solution. She must mix a 20% solution and a 50% solution. How many liters of each will be required to obtain what she needs?

16. A local electronics store will sell 7 AC adaptors and 2 rechargeable flashlights for $86, or 3 AC adaptors and 4 rechargeable flashlights for $84. What is the price of a single AC adaptor and a single rechargeable flashlight?

17. The owner of a tea shop wants to mix three kinds of tea to make 100 oz of a mixture that will sell for $.83 per oz. He uses Orange Pekoe, which sells for $.80 per oz, Irish Breakfast, for $.85 per oz, and Earl Grey, for $.95 per oz. If he wants to use twice as much Orange Pekoe as Irish Breakfast, how much of each kind of tea should he use?

Solve each system using row operations.

18. $3x + 2y = 4$
$5x + 5y = 9$

19. $x + 3y + 2z = 11$
$3x + 7y + 4z = 23$
$5x + 3y - 5z = -14$

20. Use any method described in this chapter to solve the system

$4x - 2y = -8$
$3y - 5z = 14$
$2x + z = -10.$

CUMULATIVE REVIEW EXERCISES CHAPTERS **1–8**

1. Find the value of the expression $\dfrac{3x^2 + 2y^2}{10y + 3}$ if $x = 1$ and $y = 5$.

Name the property that justifies each statement.

2. $5 + (-4) = (-4) + 5$

3. $r(s - k) = rs - rk$

4. Evaluate $-2 + 6[3 - (4 - 9)]$.

Solve each linear equation.

5. $2 - 3(6x + 2) = 4(x + 1) + 18$

6. $\dfrac{3}{2}\left(\dfrac{1}{3}x + 4\right) = 6\left(\dfrac{1}{4} + x\right)$

7. Solve the formula $P = \dfrac{kT}{V}$ for T.

Solve each linear inequality.

8. $-\dfrac{5}{6}x < 15$

9. $-8 < 2x + 3$

10. A recent survey measured public recognition of the most popular contemporary advertising slogans. Complete the results shown in the table if 2500 people were surveyed.

Slogan (product or company)	Percent Recognition (nearest tenth of a percent)	Actual Number Who Recognized Slogan (nearest whole number)
Please Don't Squeeze the . . . (Charmin)	80.4%	
The Breakfast of Champions (Wheaties)	72.5%	
The King of Beers (Budweiser)		1570
Like a Good Neighbor (State Farm)		1430

(Other slogans included "You're in Good Hands" (Allstate), "Snap, Crackle, Pop" (Rice Krispies), and "The Un-Cola" (7-Up).)
Source: Department of Integrated Marketing Communications, Northwestern University.

Solve each problem.

11. On February 12, 1999, the U.S. Senate voted to acquit William Jefferson Clinton on both counts of impeachment (perjury and obstruction of justice). Of the 200 "guilty" or "not guilty" votes cast that day, there were 10 more "not guilty" votes than "guilty" votes. How many of each vote were there? (*Source:* MSNBC Web site, February 13, 1999.)

12. Two angles of a triangle have the same measure. The measure of the third angle is 4° less than twice the measure of each of the equal angles. Find the measures of the three angles.

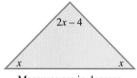

Measures are in degrees.

13. No baseball fan should be without a copy of *The Sports Encyclopedia: Baseball 2000* by David S. Neft and Richard M. Cohen. The 20th edition provides exhaustive statistics for professional baseball dating back to 1876. This book has a perimeter of 38 in., and its width measures 2.5 in. less than its length. What are the dimensions of the book?

Graph each linear equation.

14. $x - y = 4$

15. $3x + y = 6$

Find the slope of each line.

16. Through $(-5, 6)$ and $(1, -2)$

17. Perpendicular to the line $y = 4x - 3$

Find an equation for each line. Write it in slope-intercept form.

18. Through $(-4, 1)$ with slope $\frac{1}{2}$

19. Through the points $(1, 3)$ and $(-2, -3)$

20. (a) Write an equation of the vertical line through $(9, -2)$.
   (b) Write an equation of the horizontal line through $(4, -1)$.

Simplify. Write each answer with only positive exponents. Assume all variables represent nonzero real numbers.

21. $\left(\dfrac{a^{-3}b^4}{a^2b^{-1}}\right)^{-2}$

22. $\left(\dfrac{m^{-4}n^2}{m^2n^{-3}}\right) \cdot \left(\dfrac{m^5n^{-1}}{m^{-2}n^5}\right)$

Perform the indicated operations.

23. $(3y^2 - 2y + 6) - (-y^2 + 5y + 12)$

24. $(4f + 3)(3f - 1)$

25. $\left(\dfrac{1}{4}x + 5\right)^2$

26. $(3x^3 + 13x^2 - 17x - 7) \div (3x + 1)$

Factor each polynomial completely.

27. $2x^2 - 13x - 45$

28. $100t^4 - 25$

29. $8p^3 + 125$

30. Solve the equation $3x^2 + 4x = 7$.

31. Write $\dfrac{y^2 - 16}{y^2 - 8y + 16}$ in lowest terms.

Perform the indicated operations. Express the answer in lowest terms.

32. $\dfrac{2a^2}{a + b} \cdot \dfrac{a - b}{4a}$

33. $\dfrac{2x}{2x - 1} + \dfrac{4}{2x + 1} + \dfrac{8}{4x^2 - 1}$

34. Solve the equation $\dfrac{-3x}{x + 1} + \dfrac{4x + 1}{x} = \dfrac{-3}{x^2 + x}$.

Decide whether each relation defined in Exercises 35 and 36 is a function, and give its domain and range.

35. Average Hourly Wages in Mexico

Year	Wage (in dollars)
1990	1.25
1992	1.61
1994	1.80
1996	1.21
1998	1.94
2000	2.26

Source: John Christman, CIEMEX-WEFA.

36.

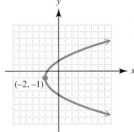

37. Given the equation $5x - 3y = 8$,

(a) write it with function notation $f(x)$;

(b) find $f(1)$.

38. If y varies directly as x and $y = 5$ when $x = 12$, find y when $x = 42$.

Solve by any method.

39. $-2x + 3y = -15$
$4x - y = 15$

40. $x - 3y = 7$
$2x - 6y = 14$

41. $x + y + z = 10$
$x - y - z = 0$
$-x + y - z = -4$

Solve each problem using a system of equations.

42. Two of the best-selling toys in a recent year were Tickle Me Elmo and Snacktime Kid. Based on their average retail prices, Elmo cost $8.63 less than Kid, and together they cost $63.89. What was the average retail price for each toy? (*Source:* NPD Group, Inc.)

43. At the Chalmette Nut Shop, 6 lb of peanuts and 12 lb of cashews cost $60, while 3 lb of peanuts and 4 lb of cashews cost $22. Find the cost of each type of nut.

44. Eboni Perkins compared the monthly payments she would incur for two types of mortgages: fixed-rate and variable-rate. Her observations led to the graph at the right. Use it to answer the following questions.

   (a) For which years would the monthly payment be more for the fixed-rate mortgage than for the variable-rate mortgage?

   (b) In what year would the payments be the same, and what would those payments be?

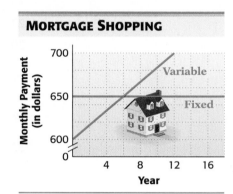

Inequalities and Absolute Value

The cost of a college education has risen rapidly in the last decade. Average higher education tuition and fees increased by 78.8% from the 1990–1991 school year to the 2000–2001 school year. (*Source:* National Center for Education Statistics, U.S. Department of Education.) In Exercises 69–72 of Section 9.1, we apply the concepts of this chapter to college student expenses.

9.1 Set Operations and Compound Inequalities

OBJECTIVES

1 Find the intersection of two sets.

2 Solve compound inequalities with the word *and*.

3 Find the union of two sets.

4 Solve compound inequalities with the word *or*.

The table shows symptoms of an underactive thyroid and an overactive thyroid.

Underactive Thyroid	*Overactive Thyroid*
Sleepiness, *s*	Insomnia, *i*
Dry hands, *d*	Moist hands, *m*
Intolerance of cold, *c*	Intolerance of heat, *h*
Goiter, *g*	Goiter, *g*

Source: The Merck Manual of Diagnosis and Therapy,
16th Edition, Merck Research Laboratories, 1992.

Let *N* be the set of symptoms for an underactive thyroid, and let *O* be the set of symptoms for an overactive thyroid. Suppose we are interested in the set of symptoms that are found in *both* sets *N and O*. In this section we discuss the use of the words *and* and *or* as they relate to sets and inequalities.

OBJECTIVE 1 Find the intersection of two sets. The intersection of two sets is defined using the word *and*.

Intersection of Sets

For any two sets *A* and *B*, the **intersection** of *A* and *B*, symbolized *A* ∩ *B*, is defined as follows:

$$A \cap B = \{x \mid x \text{ is an element of } A \text{ and } x \text{ is an element of } B\}.$$

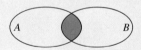

▉ **EXAMPLE 1 Finding the Intersection of Two Sets**

Let $A = \{1, 2, 3, 4\}$ and $B = \{2, 4, 6\}$. Find $A \cap B$.

The set $A \cap B$ contains those elements that belong to both *A and B*: the numbers 2 and 4. Therefore,

$$A \cap B = \{1, 2, 3, 4\} \cap \{2, 4, 6\}$$
$$= \{2, 4\}.$$

Now Try Exercise 7.

A **compound inequality** consists of two inequalities linked by a connective word such as *and* or *or*. Examples of compound inequalities are

$$x + 1 \leq 9 \quad \text{and} \quad x - 2 \geq 3$$

and
$$2x > 4 \quad \text{or} \quad 3x - 6 < 5.$$

OBJECTIVE 2 Solve compound inequalities with the word *and*. Use the following steps.

Solving a Compound Inequality with *and*

Step 1   Solve each inequality in the compound inequality individually.

Step 2   Since the inequalities are joined with *and*, the solution set of the compound inequality will include all numbers that satisfy both inequalities in Step 1 (the intersection of the solution sets).

EXAMPLE 2   Solving a Compound Inequality with *and*

Solve the compound inequality

$$x + 1 \leq 9 \quad \text{and} \quad x - 2 \geq 3.$$

Step 1   Solve each inequality in the compound inequality individually.

$$x + 1 \leq 9 \qquad \text{and} \qquad x - 2 \geq 3$$
$$x + 1 - 1 \leq 9 - 1 \quad \text{and} \quad x - 2 + 2 \geq 3 + 2$$
$$x \leq 8 \qquad \text{and} \qquad x \geq 5$$

Step 2   Because the inequalities are joined with the word *and*, the solution set will include all numbers that satisfy both inequalities in Step 1 at the same time. Thus, the compound inequality is true whenever $x \leq 8$ and $x \geq 5$ are both true. The top graph in Figure 1 shows $x \leq 8$, and the bottom graph shows $x \geq 5$.

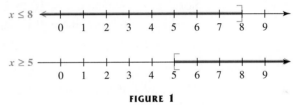

FIGURE 1

Find the intersection of the two graphs in Figure 1 to get the solution set of the compound inequality. The intersection of the two graphs in Figure 2 shows that the solution set in interval notation is $[5, 8]$.

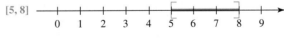

FIGURE 2

> **Now Try Exercise 27.**

EXAMPLE 3   Solving a Compound Inequality with *and*

Solve the compound inequality

$$-3x - 2 > 5 \quad \text{and} \quad 5x - 1 \leq -21.$$

Step 1   Solve each inequality separately.

$$-3x - 2 > 5 \qquad \text{and} \quad 5x - 1 \leq -21$$
$$-3x > 7 \qquad \text{and} \qquad 5x \leq -20$$
$$x < -\frac{7}{3} \quad \text{and} \qquad x \leq -4$$

The graphs of $x < -\frac{7}{3}$ and $x \leq -4$ are shown in Figure 3.

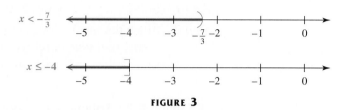

FIGURE 3

Step 2 Now find all values of x that satisfy both conditions; that is, the real numbers that are less than $-\frac{7}{3}$ and also less than or equal to -4. As shown by the graph in Figure 4, the solution set is $(-\infty, -4]$.

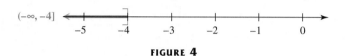

FIGURE 4

Now Try Exercise 31.

EXAMPLE 4 Solving a Compound Inequality with *and*

Solve $x + 2 < 5$ and $x - 10 > 2$.

First solve each inequality separately.

$$x + 2 < 5 \quad \text{and} \quad x - 10 > 2$$
$$x < 3 \quad \text{and} \quad x > 12$$

The graphs of $x < 3$ and $x > 12$ are shown in Figure 5.

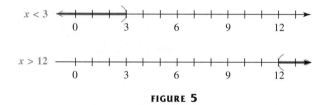

FIGURE 5

There is no number that is both less than 3 *and* greater than 12, so the given compound inequality has no solution. The solution set is $\emptyset$. See Figure 6.

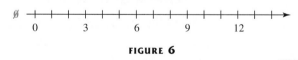

FIGURE 6

Now Try Exercise 25.

OBJECTIVE 3 **Find the union of two sets.** The union of two sets is defined using the word *or*.

Union of Sets

For any two sets A and B, the **union** of A and B, symbolized $A \cup B$, is defined as follows:

$$A \cup B = \{x \,|\, x \text{ is an element of } A \textbf{ or } x \text{ is an element of } B\}.$$

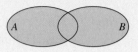

▨ EXAMPLE 5 Finding the Union of Two Sets

Let $A = \{1, 2, 3, 4\}$ and $B = \{2, 4, 6\}$. Find $A \cup B$.

Begin by listing all the elements of set A: 1, 2, 3, 4. Then list any additional elements from set B. In this case the elements 2 and 4 are already listed, so the only additional element is 6. Therefore,

$$\begin{aligned} A \cup B &= \{1, 2, 3, 4\} \cup \{2, 4, 6\} \\ &= \{1, 2, 3, 4, 6\}. \end{aligned}$$

The union consists of all elements in either A *or* B (or both).

Now Try Exercise 13.

In Example 5, notice that although the elements 2 and 4 appeared in both sets A and B, they are written only once in $A \cup B$.

OBJECTIVE 4 Solve compound inequalities with the word *or*. Use the following steps.

Solving a Compound Inequality with *or*

Step 1 Solve each inequality in the compound inequality individually.

Step 2 Since the inequalities are joined with *or*, the solution set of the compound inequality includes all numbers that satisfy either one of the two inequalities in Step 1 (the union of the solution sets).

▨ EXAMPLE 6 Solving a Compound Inequality with *or*

Solve $6x - 4 < 2x$ or $-3x \le -9$.

Step 1 Solve each inequality separately.

$$6x - 4 < 2x \quad \text{or} \quad -3x \le -9$$
$$4x < 4$$
$$x < 1 \quad \text{or} \quad x \ge 3$$

The graphs of these two inequalities are shown in Figure 7 on the next page.

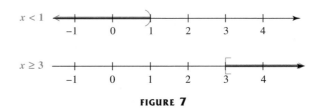

FIGURE 7

Step 2 Since the inequalities are joined with *or,* find the union of the two solution sets. The union is shown in Figure 8 and is written

$$(-\infty, 1) \cup [3, \infty).$$

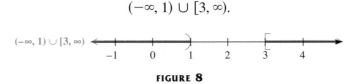

FIGURE 8

Now Try Exercise 43.

CAUTION When inequalities are used to write the solution set in Example 6, it *should* be written as

$$x < 1 \quad \text{or} \quad x \geq 3,$$

which keeps the numbers 1 and 3 in their order on the number line. Writing $3 \leq x < 1$ would imply that $3 \leq 1$, which is **FALSE.** There is no other way to write the solution set of such a union.

EXAMPLE 7 Solving a Compound Inequality with *or*

Solve $-4x + 1 \geq 9$ or $5x + 3 \leq -12$.

First we solve each inequality separately.

$$-4x + 1 \geq 9 \qquad \text{or} \quad 5x + 3 \leq -12$$
$$-4x \geq 8 \qquad \text{or} \qquad 5x \leq -15$$
$$x \leq -2 \quad \text{or} \qquad x \leq -3$$

The graphs of these two inequalities are shown in Figure 9.

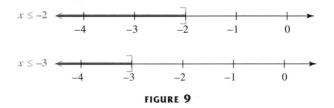

FIGURE 9

By taking the union, we obtain the interval $(-\infty, -2]$. It is graphed in Figure 10.

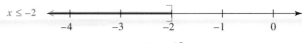

FIGURE 10

Now Try Exercise 37.

EXAMPLE 8 Applying Intersection and Union

The five highest domestic grossing films (adjusted for inflation) are listed in the table.

Five All-Time Highest Grossing Films

Film	Admissions	Gross Income
Gone with the Wind	200,605,313	$972,900,000
Star Wars	178,119,595	$863,900,000
The Sound of Music	142,415,376	$690,700,000
E.T.	135,987,938	$659,500,000
The Ten Commandments	131,000,000	$635,400,000

Source: *New York Times Almanac*, 2001.

List the elements of the following sets.

(a) The set of top five films with admissions greater than 180,000,000 *and* gross greater than $800,000,000

The only film that satisfies both conditions is *Gone with the Wind,* so the set is

$$\{Gone\ with\ the\ Wind\}.$$

(b) The set of top five films with admissions less than 170,000,000 *or* gross greater than $700,000,000

Here, a film that satisfies at least one of the conditions is in the set. This set includes all five films:

$$\{Gone\ with\ the\ Wind,\ Star\ Wars,\ The\ Sound\ of\ Music,\ E.T.,\ The\ Ten\ Commandments\}.$$

Now Try Exercises 69 and 71.

9.1 EXERCISES

Decide whether each statement is true *or* false. *If it is false, explain why.*

1. The union of the solution sets of $x + 1 = 5$, $x + 1 < 5$, and $x + 1 > 5$ is $(-\infty, \infty)$.

2. The intersection of the sets $\{x \mid x \geq 7\}$ and $\{x \mid x \leq 7\}$ is $\emptyset$.

3. The union of the sets $(-\infty, 8)$ and $(8, \infty)$ is $\{8\}$.

4. The intersection of the sets $(-\infty, 8]$ and $[8, \infty)$ is $\{8\}$.

5. The intersection of the set of rational numbers and the set of irrational numbers is $\{0\}$.

6. The union of the set of rational numbers and the set of irrational numbers is the set of real numbers.

Let $A = \{1, 2, 3, 4, 5, 6\}$, $B = \{1, 3, 5\}$, $C = \{1, 6\}$, and $D = \{4\}$. Specify each set. See Examples 1 and 5.

7. $B \cap A$ **8.** $A \cap B$ **9.** $A \cap D$ **10.** $B \cap C$

11. $B \cap \emptyset$ **12.** $A \cap \emptyset$ **13.** $A \cup B$ **14.** $B \cup D$

✎ **15.** Give an example of intersection applied to a real-life situation.

✎ **16.** A compound inequality uses one of the words *and* or *or*. Explain how you will determine whether to use *intersection* or *union* when graphing the solution set.

Two sets are specified by graphs. Graph the intersection of the two sets.

17.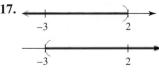

18.

19.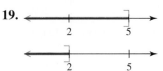

20.

For each compound inequality, give the solution set in both interval and graph forms. See Examples 2–4.

21. $x < 2$ and $x > -3$ **22.** $x < 5$ and $x > 0$ **23.** $x \leq 2$ and $x \leq 5$

24. $x \geq 3$ and $x \geq 6$ **25.** $x \leq 3$ and $x \geq 6$ **26.** $x \leq -1$ and $x \geq 3$

27. $x - 3 \leq 6$ and $x + 2 \geq 7$ **28.** $x + 5 \leq 11$ and $x - 3 \geq -1$

29. $-3x > 3$ and $x + 3 > 0$ **30.** $-3x < 3$ and $x + 2 < 6$

31. $3x - 4 \leq 8$ and $-4x + 1 \geq -15$ **32.** $7x + 6 \leq 48$ and $-4x \geq -24$

Two sets are specified by graphs. Graph the union of the two sets.

33.

34.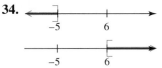

35.

36.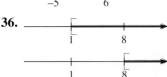

For each compound inequality, give the solution set in both interval and graph forms. See Examples 6 and 7.

37. $x \leq 1$ or $x \leq 8$ **38.** $x \geq 1$ or $x \geq 8$

39. $x \geq -2$ or $x \geq 5$ **40.** $x \leq -2$ or $x \leq 6$

41. $x \geq -2$ or $x \leq 4$ **42.** $x \geq 5$ or $x \leq 7$

43. $x + 2 > 7$ or $1 - x > 6$ **44.** $x + 1 > 3$ or $x + 4 < 2$

45. $x + 1 > 3$ or $-4x + 1 > 5$ **46.** $3x < x + 12$ or $x + 1 > 10$

Express each set in the simplest interval form. (Hint: Graph each set and look for the intersection or union.)

47. $(-\infty, -1] \cap [-4, \infty)$ **48.** $[-1, \infty) \cap (-\infty, 9]$

49. $(-\infty, -6] \cap [-9, \infty)$ **50.** $(5, 11] \cap [6, \infty)$

51. $(-\infty, 3) \cup (-\infty, -2)$ **52.** $[-9, 1] \cup (-\infty, -3)$

53. $[3, 6] \cup (4, 9)$ **54.** $[-1, 2] \cup (0, 5)$

For each compound inequality, decide whether intersection *or* union *should be used. Then give the solution set in both interval and graph forms. See Examples 2, 3, 4, 6, and 7.*

55. $x < -1$ and $x > -5$

56. $x > -1$ and $x < 7$

57. $x < 4$ or $x < -2$

58. $x < 5$ or $x < -3$

59. $-3x \leq -6$ or $-3x \geq 0$

60. $2x - 6 \leq -18$ and $2x \geq -18$

61. $x + 1 \geq 5$ and $x - 2 \leq 10$

62. $-8x \leq -24$ or $-5x \geq 15$

RELATING CONCEPTS (EXERCISES 63–68)

For Individual or Group Work

The figures represent the backyards of neighbors Luigi, Maria, Than, and Joe. Find the area and the perimeter of each yard. Suppose that each resident has 150 ft of fencing and enough sod to cover 1400 ft² of lawn. Give the name or names of the residents whose yards satisfy each description. **Work Exercises 63–68 in order.**

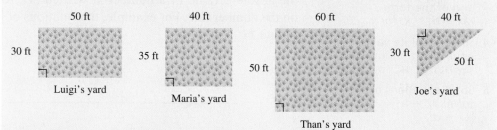

63. The yard can be fenced *and* the yard can be sodded.

64. The yard can be fenced *and* the yard cannot be sodded.

65. The yard cannot be fenced *and* the yard can be sodded.

66. The yard cannot be fenced *and* the yard cannot be sodded.

67. The yard can be fenced *or* the yard can be sodded.

68. The yard cannot be fenced *or* the yard can be sodded.

Average expenses for full-time college students during the 2000–2001 academic year are shown in the table.

College Expenses in 2000–2001 (in dollars)

Type of Expense	Public Schools	Private Schools
Tuition and fees	2600	14,690
Board rates	2454	2989
Dormitory charges	2566	3370

Source: U.S. National Center for Education Statistics, U.S. Department of Education.

Use the table to list the elements of each set.

69. The set of expenses less than $2700 for public schools *and* greater than $3500 for private schools

70. The set of expenses less than $2600 for public schools *and* less than $3500 for private schools

71. The set of expenses less than $2500 for public schools *or* greater than $3500 for private schools

72. The set of expenses greater than $10,000 for private schools *or* less than $2000 for public schools

9.2 Absolute Value Equations and Inequalities

In a production line, quality is controlled by randomly choosing items from the line and checking to see how selected measurements vary from the optimum measure. These differences are sometimes positive and sometimes negative, so they are expressed with absolute value. For example, a machine that fills quart milk cartons might be set to release 1 qt (32 oz) plus or minus 2 oz per carton. Then the number of ounces in each carton should satisfy the *absolute value inequality* $|x - 32| \leq 2$, where x is the number of ounces.

OBJECTIVE 1 Use the distance definition of absolute value. In Chapter 1 we saw that the absolute value of a number x, written $|x|$, represents the distance from x to 0 on the number line. For example, the solutions of $|x| = 4$ are 4 and -4, as shown in Figure 11.

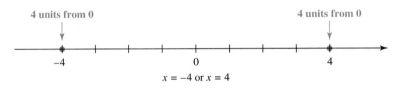

$$x = -4 \text{ or } x = 4$$

FIGURE 11

Because absolute value represents distance from 0, it is reasonable to interpret the solutions of $|x| > 4$ to be all numbers that are *more* than 4 units from 0. The set $(-\infty, -4) \cup (4, \infty)$ fits this description. Figure 12 shows the graph of the solution set of $|x| > 4$. Because the graph consists of two separate intervals, the solution set is described using *or* as

$$x < -4 \quad \text{or} \quad x > 4.$$

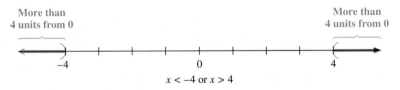

$$x < -4 \text{ or } x > 4$$

FIGURE 12

The solution set of $|x| < 4$ consists of all numbers that are *less* than 4 units from 0 on the number line. Another way of thinking of this is to think of all numbers *between* -4 and 4. This set of numbers is given by $(-4, 4)$, as shown in Figure 13. Here, the graph shows that $-4 < x < 4$, which means $x > -4$ *and* $x < 4$.

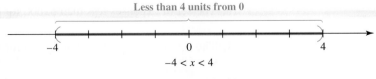

$$-4 < x < 4$$

FIGURE 13

The equation and inequalities just described are examples of **absolute value equations and inequalities.** They involve the absolute value of a variable expression and generally take the form

$$|ax + b| = k, \qquad |ax + b| > k, \qquad \text{or} \qquad |ax + b| < k,$$

where k is a positive number. From Figures 11–13, we see that

$$|x| = 4 \text{ has the same solution set as } x = -4 \text{ or } x = 4,$$
$$|x| > 4 \text{ has the same solution set as } x < -4 \text{ or } x > 4,$$
$$|x| < 4 \text{ has the same solution set as } x > -4 \text{ and } x < 4.$$

Thus, we solve an absolute value equation or inequality by first rewriting it as an equivalent statement without absolute value bars. Notice that, except for special cases, the solution set of an absolute value *equation* includes exactly *two points.* However, the solution set of an absolute value *inequality* includes one or more *intervals.* We summarize these facts in the next box.

Solving Absolute Value Equations and Inequalities

Let k be a positive real number, and p and q be real numbers.

1. To solve $|ax + b| = k,$ solve the compound equation

$$ax + b = k \quad \text{or} \quad ax + b = -k.$$

The solution set is usually of the form $\{p, q\}$, which includes two numbers.

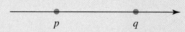

2. To solve $|ax + b| > k,$ solve the compound inequality

$$ax + b > k \quad \text{or} \quad ax + b < -k.$$

The solution set is of the form $(-\infty, p) \cup (q, \infty)$, which consists of two separate intervals.

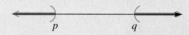

3. To solve $|ax + b| < k,$ solve the compound inequality

$$-k < ax + b < k.$$

The solution set is of the form (p, q), a single interval.

NOTE Some people prefer to write the compound statements in parts 1 and 2 of the preceding summary as

$$ax + b = k \quad \text{or} \quad -(ax + b) = k$$

and

$$ax + b > k \quad \text{or} \quad -(ax + b) > k.$$

These forms are equivalent to those we give in the summary and produce the same results.

OBJECTIVE 2 Solve equations of the form $|ax + b| = k$, for $k > 0$. The next example shows how we use a compound equation to solve a typical absolute value equation. Remember that because absolute value refers to distance from the origin, each absolute value equation will have two parts.

■ EXAMPLE 1   Solving an Absolute Value Equation

Solve $|2x + 1| = 7$.

For $|2x + 1|$ to equal 7, $2x + 1$ must be 7 units from 0 on the number line. This can happen only when $2x + 1 = 7$ or $2x + 1 = -7$. This is the first case in the preceding summary. Solve this compound equation as follows.

$$2x + 1 = 7 \quad \text{or} \quad 2x + 1 = -7$$
$$2x = 6 \quad \text{or} \quad 2x = -8$$
$$x = 3 \quad \text{or} \quad x = -4$$

Check by substitution in the original absolute value equation to verify that the solution set is $\{-4, 3\}$. The graph is shown in Figure 14.

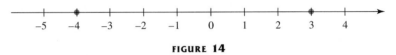

FIGURE 14

Now Try Exercise 11.

OBJECTIVE 3 Solve inequalities of the form $|ax + b| < k$ and of the form $|ax + b| > k$, for $k > 0$.

■ EXAMPLE 2   Solving an Absolute Value Inequality with >

Solve $|2x + 1| > 7$.

By part 2 of the summary, this absolute value inequality is rewritten as

$$2x + 1 > 7 \quad \text{or} \quad 2x + 1 < -7,$$

because $2x + 1$ must represent a number that is *more* than 7 units from 0 on either side of the number line. Now, solve the compound inequality.

$$2x + 1 > 7 \quad \text{or} \quad 2x + 1 < -7$$
$$2x > 6 \quad \text{or} \quad 2x < -8$$
$$x > 3 \quad \text{or} \quad x < -4$$

Check these solutions. The solution set is $(-\infty, -4) \cup (3, \infty)$. See Figure 15. Notice that the graph consists of two intervals.

FIGURE 15

Now Try Exercise 25.

EXAMPLE 3 Solving an Absolute Value Inequality with <

Solve $|2x + 1| < 7$.

The expression $2x + 1$ must represent a number that is less than 7 units from 0 on either side of the number line. Another way of thinking of this is to realize that $2x + 1$ must be between -7 and 7. As part 3 of the summary shows, this is written as the three-part inequality

$$-7 < 2x + 1 < 7.$$

We solved such inequalities in Section 2.8 by working with all three parts at the same time.

$$-7 < 2x + 1 < 7$$
$$-8 < 2x < 6 \qquad \text{Subtract 1 from each part.}$$
$$-4 < x < 3 \qquad \text{Divide each part by 2.}$$

Check that the solution set is $(-4, 3)$, so the graph consists of the single interval shown in Figure 16.

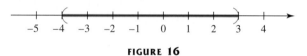

FIGURE 16

Now Try Exercise 39.

Look back at Figures 14, 15, and 16, with the graphs of $|2x + 1| = 7$, $|2x + 1| > 7$, and $|2x + 1| < 7$. If we find the union of the three sets, we get the set of all real numbers. This is because for any value of x, $|2x + 1|$ will satisfy one and only one of the following: it is equal to 7, greater than 7, or less than 7.

CAUTION When solving absolute value equations and inequalities of the types in Examples 1, 2, and 3, remember the following.

1. The methods described apply when the constant is alone on one side of the equation or inequality and is *positive*.

2. Absolute value equations and absolute value inequalities in the form $|ax + b| > k$ translate into "or" compound statements.

3. Absolute value inequalities in the form $|ax + b| < k$ translate into "and" compound statements, which may be written as three-part inequalities.

4. An "or" statement *cannot* be written in three parts. It would be incorrect to write

$$-7 > 2x + 1 > 7$$

in Example 2, because this would imply that $-7 > 7$, which is *false*.

OBJECTIVE 4 Solve absolute value equations that involve rewriting. Sometimes an absolute value equation or inequality requires some rewriting before it can be set up as a compound statement, as shown in the next example.

▓ **EXAMPLE 4** Solving an Absolute Value Equation That Requires Rewriting

Solve $|x + 3| + 5 = 12$.

First get the absolute value alone on one side of the equals sign by subtracting 5 from each side.

$$|x + 3| + 5 - 5 = 12 - 5 \qquad \text{Subtract 5.}$$
$$|x + 3| = 7$$

Now use the method shown in Example 1.

$$x + 3 = 7 \quad \text{or} \quad x + 3 = -7$$
$$x = 4 \quad \text{or} \qquad x = -10$$

Check that the solution set is $\{4, -10\}$ by substituting into the original equation. ▓

Now Try Exercise 63.

We use a similar method to solve an absolute value *inequality* that requires rewriting.

OBJECTIVE 5 Solve equations of the form $|ax + b| = |cx + d|$. By definition, for two expressions to have the same absolute value, they must either be equal or be negatives of each other.

Solving $|ax + b| = |cx + d|$

To solve an absolute value equation of the form

$$|ax + b| = |cx + d|,$$

solve the compound equation

$$ax + b = cx + d \quad \text{or} \quad ax + b = -(cx + d).$$

▓ **EXAMPLE 5** Solving an Equation with Two Absolute Values

Solve $|z + 6| = |2z - 3|$.

This equation is satisfied either if $z + 6$ and $2z - 3$ are equal to each other or if $z + 6$ and $2z - 3$ are negatives of each other.

$$z + 6 = 2z - 3 \quad \text{or} \quad z + 6 = -(2z - 3)$$

Solve each equation.

$$z + 6 = 2z - 3 \quad \text{or} \quad z + 6 = -2z + 3$$
$$6 + 3 = 2z - z \qquad\qquad 3z = -3$$
$$9 = z \qquad \text{or} \qquad z = -1$$

The solution set is $\{9, -1\}$.

Now Try Exercise 71.

OBJECTIVE 6 Solve special cases of absolute value equations and inequalities. When a typical absolute value equation or inequality involves a *negative constant or 0* alone on one side, use the properties of absolute value to solve. Keep in mind the following.

1. The absolute value of an expression can never be negative: $|a| \geq 0$ for all real numbers a.

2. The absolute value of an expression equals 0 only when the expression is equal to 0.

The next two examples illustrate these special cases.

EXAMPLE 6 Solving Special Cases of Absolute Value Equations

Solve each equation.

(a) $|5r - 3| = -4$

Since the absolute value of an expression can never be negative, there are no solutions for this equation. The solution set is $\emptyset$.

(b) $|7x - 3| = 0$

The expression $7x - 3$ will equal 0 *only* if

$$7x - 3 = 0.$$

The solution of this equation is $\frac{3}{7}$. The solution set is $\left\{\frac{3}{7}\right\}$. It consists of only one element that checks by substitution in the original equation.

Now Try Exercises 79 and 81.

EXAMPLE 7 Solving Special Cases of Absolute Value Inequalities

Solve each inequality.

(a) $|x| \geq -4$

The absolute value of a number is never negative. For this reason, $|x| \geq -4$ is true for *all* real numbers. The solution set is $(-\infty, \infty)$.

(b) $|k + 6| - 3 < -5$

Add 3 to both sides to get the absolute value expression alone on one side.

$$|k + 6| < -2$$

There is no number whose absolute value is less than -2, so this inequality has no solution. The solution set is $\emptyset$.

(c) $|m - 7| + 4 \leq 4$

Adding -4 to both sides gives

$$|m - 7| \leq 0.$$

The value of $|m - 7|$ will never be less than 0. However, $|m - 7|$ will equal 0 when $m = 7$. Therefore, the solution set is $\{7\}$.

Now Try Exercises 85 and 91.

CONNECTIONS

Absolute value is used to find the relative error of a measurement in science, engineering, manufacturing, and other fields. If x_t represents the expected value of a measurement and x represents the actual measurement, then the *relative error in x* equals the absolute value of the difference between x_t and x divided by x_t. That is,

$$\text{relative error in } x = \left| \frac{x_t - x}{x_t} \right|.$$

In many situations in the work world, the relative error must be less than some predetermined amount. For example, suppose a machine filling *quart* milk cartons is set for a relative error no greater than .05. Here $x_t = 32$ oz, the relative error = .05 oz, and we must find x, given

$$\left| \frac{32 - x}{32} \right| = \left| 1 - \frac{x}{32} \right| \le .05.$$

For Discussion or Writing

With this tolerance level, how many ounces may a carton contain?

 EXERCISES

For Extra Help

 Student's
Solutions Manual

MyMathLab

 InterAct Math
Tutorial Software

 AW Math
Tutor Center

 MathXL

Digital Video Tutor
CD 13/Videotape 13

Match each absolute value equation or inequality in Column I with the graph of its solution set in Column II.

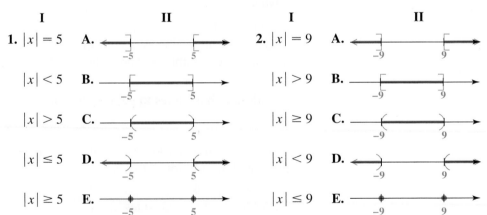

1. $|x| = 5$ **A.**

$|x| < 5$ **B.**

$|x| > 5$ **C.**

$|x| \le 5$ **D.**

$|x| \ge 5$ **E.**

2. $|x| = 9$ **A.**

$|x| > 9$ **B.**

$|x| \ge 9$ **C.**

$|x| < 9$ **D.**

$|x| \le 9$ **E.**

3. Explain when to use *and* and when to use *or* if you are solving an absolute value equation or inequality of the form $|ax + b| = k$, $|ax + b| < k$, or $|ax + b| > k$, where k is a positive number.

4. How many solutions will $|ax + b| = k$ have if
   (a) $k = 0$;   **(b)** $k > 0$;   **(c)** $k < 0$?

Solve each equation. See Example 1.

5. $|x| = 12$ **6.** $|k| = 14$ **7.** $|4x| = 20$ **8.** $|5x| = 30$

9. $|y - 3| = 9$ **10.** $|p - 5| = 13$ **11.** $|2x - 1| = 11$ **12.** $|2y + 3| = 19$

13. $|4r - 5| = 17$ **14.** $|5t - 1| = 21$ **15.** $|2y + 5| = 14$ **16.** $|2x - 9| = 18$

17. $\left|\dfrac{1}{2}x + 3\right| = 2$ **18.** $\left|\dfrac{2}{3}q - 1\right| = 5$

19. $\left|1 + \dfrac{3}{4}k\right| = 7$ **20.** $\left|2 - \dfrac{5}{2}m\right| = 14$

Solve each inequality and graph the solution set. See Example 2.

21. $|x| > 3$ **22.** $|y| > 5$ **23.** $|k| \geq 4$ **24.** $|r| \geq 6$

25. $|r + 5| \geq 20$ **26.** $|3x - 1| \geq 8$ **27.** $|t + 2| > 10$ **28.** $|4x + 1| \geq 21$

29. $|3 - x| > 5$ **30.** $|5 - x| > 3$ **31.** $|-5x + 3| \geq 12$ **32.** $|-2x - 4| \geq 5$

33. The graph of the solution set of $|2x + 1| = 9$ is given here.

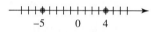

$$-5 \quad 0 \quad 4$$

Without actually doing the algebraic work, graph the solution set of each inequality, referring to the graph above.

(a) $|2x + 1| < 9$ **(b)** $|2x + 1| > 9$

34. The graph of the solution set of $|3x - 4| < 5$ is given here.

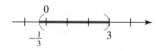

$$-\dfrac{1}{3} \qquad 3$$

Without actually doing the algebraic work, graph the solution set of the following, referring to the graph above.

(a) $|3x - 4| = 5$ **(b)** $|3x - 4| > 5$

Solve each inequality and graph the solution set. See Example 3. (Hint: Compare your answers to those in Exercises 21–32.)

35. $|x| \leq 3$ **36.** $|y| \leq 5$ **37.** $|k| < 4$ **38.** $|r| < 6$

39. $|r + 5| \leq 20$ **40.** $|3x - 1| < 8$ **41.** $|t + 2| \leq 10$ **42.** $|4x + 1| < 21$

43. $|3 - x| \leq 5$ **44.** $|5 - x| \leq 3$ **45.** $|-5x + 3| \leq 12$ **46.** $|-2x - 4| \leq 5$

Decide which method you should use to solve each absolute value equation or inequality. Find the solution set and in Exercises 47–58, graph the solution set. See Examples 1–3.

47. $|-4 + k| > 9$ **48.** $|-3 + t| > 8$ **49.** $|r + 5| > 20$ **50.** $|2x - 1| < 7$

51. $|7 + 2z| = 5$ **52.** $|9 - 3p| = 3$ **53.** $|3r - 1| \leq 11$ **54.** $|2s - 6| \leq 6$

55. $|-6x - 6| \leq 1$ **56.** $|-2x - 6| \leq 5$ **57.** $|2x - 1| \geq 7$ **58.** $|-4 + k| \leq 9$

59. $|x| - 1 = 4$ **60.** $|x + 3| = 10$ **61.** $|x + 2| = 3$ **62.** $|x - 4| = 1$

Solve each equation or inequality. Give the solution set in set notation for equations and in interval notation for inequalities. See Example 4.

63. $|x + 4| + 1 = 2$

64. $|x + 5| - 2 = 12$

65. $|2x + 1| + 3 > 8$

66. $|6x - 1| - 2 > 6$

67. $|x + 5| - 6 \le -1$

68. $|r - 2| - 3 \le 4$

69. $|2 - x| > 3$

70. $|4 - x| < 1$

Solve each equation. See Example 5.

71. $|3x + 1| = |2x + 4|$

72. $|7x + 12| = |x - 8|$

73. $\left| m - \dfrac{1}{2} \right| = \left| \dfrac{1}{2}m - 2 \right|$

74. $\left| \dfrac{2}{3}r - 2 \right| = \left| \dfrac{1}{3}r + 3 \right|$

75. $|6x| = |9x + 1|$

76. $|13x| = |2x + 1|$

77. $|2p - 6| = |2p + 11|$

78. $|3x - 1| = |3x + 9|$

Solve each equation or inequality. See Examples 6 and 7.

79. $|12t - 3| = -8$

80. $|13w + 1| = -3$

81. $|4x + 1| = 0$

82. $|6r - 2| = 0$

83. $|2q - 1| = -6$

84. $|8n + 4| = -4$

85. $|x + 5| > -9$

86. $|x + 9| > -3$

87. $|7x + 3| \le 0$

88. $|4x - 1| \le 0$

89. $|5x - 2| = 0$

90. $|4 + 7x| = 0$

91. $|10z + 7| + 3 < 1$

92. $|4x + 1| - 2 < -5$

93. The 1998 recommended daily intake (RDI) of calcium for females aged 19–50 is 1000 mg. (*Source: World Almanac and Book of Facts,* 2000.) Actual vitamin needs vary from person to person. Write an absolute value inequality, with x representing the RDI, to express the RDI plus or minus 100 mg and solve it.

94. The average clotting time of blood is 7.45 sec with a variation of plus or minus 3.6 sec. Write this statement as an absolute value inequality with x representing the time and solve it.

RELATING CONCEPTS (EXERCISES 95–98)

For Individual or Group Work

The ten tallest buildings in Kansas City, Missouri, are listed along with their heights.

Building	Height (in feet)
One Kansas City Place	632
AT&T Town Pavilion	590
Hyatt Regency	504
Kansas City Power and Light	476
City Hall	443
Fidelity Bank and Trust Building	433
1201 Walnut	427
Federal Office Building	413
Commerce Tower	407
City Center Square	404

Source: World Almanac and Book of Facts, 2001.

*Use this information to **work Exercises 95–98 in order.***

95. To find the average of a group of numbers, we add the numbers and then divide by the number of items added. Use a calculator to find the average of the heights.

96. Let k represent the average height of these buildings. If a height x satisfies the inequality

$$|x - k| < t,$$

then the height is said to be within t ft of the average. Using your result from Exercise 95, list the buildings that are within 50 ft of the average.

97. Repeat Exercise 96, but list the buildings that are within 75 ft of the average.

98. (a) Write an absolute value inequality that describes the height of a building that is *not* within 75 ft of the average.
(b) Solve the inequality you wrote in part (a).
(c) Use the result of part (b) to list the buildings that are not within 75 ft of the average.
(d) Confirm that your answer to part (c) makes sense by comparing it with your answer to Exercise 97.

9.3 Linear Inequalities in Two Variables

OBJECTIVES

1 Graph linear inequalities in two variables.

2 Graph the intersection of two linear inequalities.

3 Graph the union of two linear inequalities.

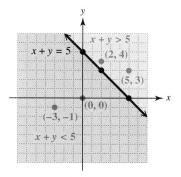

FIGURE 17

OBJECTIVE 1 Graph linear inequalities in two variables. In earlier sections we graphed linear inequalities in one variable on the number line. In this section we graph linear inequalities in two variables on a rectangular coordinate system.

Linear Inequality in Two Variables

An inequality that can be written as

$$Ax + By < C \qquad \text{or} \qquad Ax + By > C,$$

where A, B, and C are real numbers and A and B are not both 0, is a **linear inequality in two variables.**

The symbols $\leq$ and $\geq$ may replace $<$ and $>$ in the definition.

Consider the graph in Figure 17. The graph of the line $x + y = 5$ divides the points in the rectangular coordinate system into three sets: those points that lie on the line itself and satisfy the equation $x + y = 5$ [like $(0, 5)$, $(2, 3)$, and $(5, 0)$], those that lie in the half-plane above the line and satisfy the inequality $x + y > 5$ [like $(5, 3)$ and $(2, 4)$], and those that lie in the half-plane below the line and satisfy the inequality $x + y < 5$ [like $(0, 0)$ and $(-3, -1)$]. The graph of the line $x + y = 5$ is called the **boundary line** for the inequalities $x + y > 5$ and $x + y < 5$. Graphs of linear inequalities in two variables are *regions* in the real number plane that may or may not include boundary lines.

To graph a linear inequality in two variables, follow these steps.

Graphing a Linear Inequality

Step 1 **Draw the graph of the straight line that is the boundary.** Make the line solid if the inequality involves $\leq$ or $\geq$; make the line dashed if the inequality involves $<$ or $>$.

(continued)

Step 2   **Choose a test point.** Choose any point not on the line and substitute the coordinates of this point in the inequality.

Step 3   **Shade the appropriate region.** Shade the region that includes the test point if it satisfies the original inequality; otherwise, shade the region on the other side of the boundary line.

■ EXAMPLE 1   Graphing a Linear Inequality

Graph $3x + 2y \geq 6$.

Step 1   First graph the line $3x + 2y = 6$. The graph of this line, the boundary of the graph of the inequality, is shown in Figure 18.

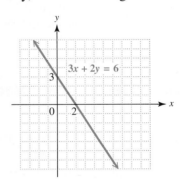

FIGURE 18

Step 2   The graph of the inequality $3x + 2y \geq 6$ includes the points of the line $3x + 2y = 6$ and either the points *above* the line $3x + 2y = 6$ or the points *below* that line. To decide which, select any point not on the boundary line $3x + 2y = 6$ as a test point. The origin, $(0, 0)$, is often a good choice because the substitution is easy. Substitute the values from the test point $(0, 0)$ for x and y in the inequality $3x + 2y > 6$.

$$3(0) + 2(0) > 6 \qquad ?$$

$$0 > 6 \qquad \text{False}$$

Step 3   Because the result is false, $(0, 0)$ does *not* satisfy the inequality, and so the solution set includes all points on the other side of the line. This region is shaded in Figure 19.

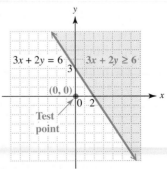

FIGURE 19

Now Try Exercise 7.

If the inequality is written in the form $y > mx + b$ or $y < mx + b$, then the inequality symbol indicates which half-plane to shade.

If $y > mx + b$, then shade above the boundary line;

if $y < mx + b$, then shade below the boundary line.

This method works *only* if the inequality is solved for y.

▮ EXAMPLE 2 Graphing a Linear Inequality

Graph $x - 3y < 4$.

First graph the boundary line, shown in Figure 20. The points of the boundary line do not belong to the inequality $x - 3y < 4$ (because the inequality symbol is $<$, not $\leq$). For this reason, the line is dashed. Now solve the inequality for y.

$$x - 3y < 4$$
$$-3y < -x + 4 \qquad \text{Subtract } x.$$
$$y > \frac{1}{3}x - \frac{4}{3} \qquad \text{Multiply by } -\tfrac{1}{3}; \text{ change } < \text{ to } >.$$

Because of the *is greater than* symbol, shade *above* the line. As a check, choose a test point not on the line, say $(0, 0)$, and substitute for x and y in the original inequality.

$$0 - 3(0) < 4 \qquad ?$$
$$0 < 4 \qquad \text{True}$$

This result agrees with the decision to shade above the line. The solution set, graphed in Figure 20, includes only those points in the shaded half-plane (not those on the line).

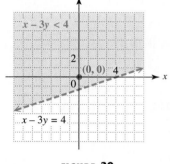

FIGURE 20

Now Try Exercise 9.

▮ EXAMPLE 3 Graphing a Linear Inequality with a Boundary Line through the Origin

Graph $x - 2y \leq 0$.

Begin by graphing $x - 2y = 0$ as a solid line because of the $\leq$ symbol. Since $(0, 0)$ is on the line $x - 2y = 0$, it cannot be used as a test point. Instead, we choose a test point off the line, such as $(1, 3)$.

$$x - 2y \leq 0 \qquad \text{Original inequality}$$
$$1 - 2(3) \leq 0 \qquad ? \qquad \text{Let } x = 1 \text{ and } y = 3.$$
$$-5 \leq 0 \qquad \text{True}$$

Since $-5 \leq 0$ is true, shade the side of the graph containing the test point $(1, 3)$. See Figure 21.

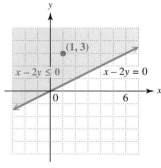

FIGURE 21

Now Try Exercise 13.

| | **CONNECTIONS** | |

Recall from Section 7.2 that the x-intercept of the graph of the line $y = mx + b$ indicates the solution of the equation $mx + b = 0$. We can extend this observation to find solutions of the associated inequalities

$$mx + b > 0 \quad \text{and} \quad mx + b < 0.$$

The solution set of $mx + b > 0$ is the set of all x-values for which the graph of $y = mx + b$ is *above* the x-axis. (We consider points above because the symbol is $>$.) On the other hand, the solution set of $mx + b < 0$ is the set of all x-values for which the graph of $y = mx + b$ is *below* the x-axis. (We consider points below because the symbol is $<$.)

For example, in Figure 22 the x-intercept of $y = 3x - 9$ is $(3, 0)$. Therefore,

$$\text{the solution set of } 3x - 9 = 0 \text{ is } \{3\}.$$

Because the graph of y lies above the x-axis for x-values greater than 3,

$$\text{the solution set of } 3x - 9 > 0 \text{ is } (3, \infty).$$

Because the graph lies below the x-axis for x-values less than 3,

$$\text{the solution set of } 3x - 9 < 0 \text{ is } (-\infty, 3).$$

To solve the equation $-2(3x + 1) = -2x + 18$ and the associated inequalities $-2(3x + 1) > -2x + 18$ and $-2(3x + 1) < -2x + 18$, we must rewrite the equation so that the right side equals 0:

$$-2(3x + 1) + 2x - 18 = 0.$$

Graphing

$$y = -2(3x + 1) + 2x - 18$$

yields the x-intercept $(-5, 0)$, as shown in Figure 23. Because the graph of y lies *above* the x-axis for x-values less than -5,

$$\text{the solution set of } -2(3x + 1) > -2x + 18 \text{ is } (-\infty, -5).$$

Because the graph of y lies *below* the x-axis for x-values greater than -5,

$$\text{the solution set of } -2(3x + 1) < -2x + 18 \text{ is } (-5, \infty).$$

For Discussion or Writing

1. Discuss the pros and cons of using a calculator to solve a linear inequality in one variable.
2. Use a graphing calculator to solve each inequality.
   (a) $-2x + 4 \geq 0$
   (b) $-2x + 4 \leq 0$ (Use the result from part (a).)
   (c) $3x + 2 - 5 > -x + 7 + 2x$
   (d) $3x + 2 - 5 < -x + 7 + 2x$ (Use the result from part (c).)

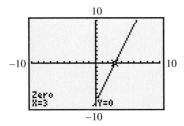

FIGURE 22

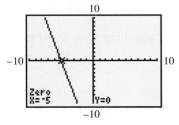

FIGURE 23

OBJECTIVE 2 Graph the intersection of two linear inequalities. In Section 9.1, we used the words *and* and *or* to solve compound inequalities. In that section, the inequalities had one variable. We can extend those ideas to include inequalities in two variables. A pair of inequalities joined with the word *and* is interpreted as the intersection of the solution sets of the inequalities. The graph of the intersection of two or more inequalities is the region of the plane where all points satisfy all of the inequalities at the same time.

EXAMPLE 4 Graphing the Intersection of Two Inequalities

Graph $2x + 4y \geq 5$ and $x \geq 1$.

To begin, we graph each of the two inequalities $2x + 4y \geq 5$ and $x \geq 1$ separately. The graph of $2x + 4y \geq 5$ is shown in Figure 24(a), and the graph of $x \geq 1$ is shown in Figure 24(b).

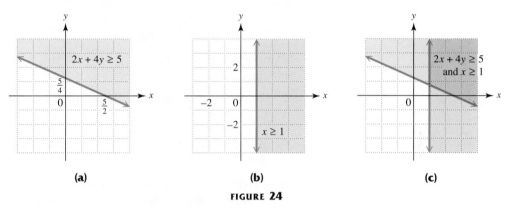

(a)   (b)   (c)

FIGURE 24

In practice, the graphs in Figures 24(a) and (b) are graphed on the same axes. Then we use heavy shading to identify the intersection of the graphs, as shown in Figure 24(c).

To check, we use a test point from each of the four regions formed by the intersection of the boundary lines. Verify that only ordered pairs in the heavily shaded region satisfy both inequalities.

Now Try Exercise 19.

OBJECTIVE 3 Graph the union of two linear inequalities. When two inequalities are joined by the word *or,* we must find the union of the graphs of the inequalities. The graph of the union of two inequalities includes all of the points that satisfy either inequality.

EXAMPLE 5 Graphing the Union of Two Inequalities

Graph $2x + 4y \geq 5$ or $x \geq 1$.

The graphs of the two inequalities are shown in Figures 24(a) and (b) in Example 3. The graph of the union is shown in Figure 25.

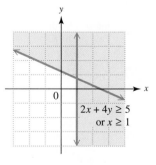

FIGURE 25

Now Try Exercise 29.

CONNECTIONS

Suppose a factory can have *no more than* 200 workers on a shift, but must have *at least* 100 and must manufacture *at least* 3000 units at minimum cost. The managers need to know how many workers should be on a shift in order to produce the required units at minimal cost. *Linear programming* is a method for finding the optimal (best possible) solution that meets all the conditions for such problems. The first step in solving linear programming problems with two variables is to express the conditions (constraints) as inequalities, graph the system of inequalities, and identify the region that satisfies all the inequalities at once.

For Discussion or Writing

Let x represent the number of workers and y represent the number of units manufactured.

1. Write three inequalities expressing the conditions given in the problem.

2. Graph the inequalities from Item 1 and shade the intersection.

3. The cost per worker is $50 per day and the cost to manufacture 1 unit is $100. Write an expression representing the total daily cost, C.

4. Find values of x and y for several points in or on the boundary of the shaded region. Include any "corner points."

5. Of the values of x and y that you chose in Item 4, which gives the least cost when substituted in the cost equation from Item 3? What does your answer mean in terms of the given problem? Is your answer reasonable? Explain.

9.3 EXERCISES

For Extra Help

- 📖 Student's Solutions Manual
- 🚪 MyMathLab
- 🔺 InterAct Math Tutorial Software
- Tutor Center AW Math Tutor Center
- MathXL MathXL
- 📼 Digital Video Tutor CD 13/Videotape 13

In Exercises 1–4, fill in the first blank with either solid *or* dashed. *Fill in the second blank with* either above *or* below.

1. The boundary of the graph of $y \leq -x + 2$ will be a _____ line, and the shading will be _____ the line.

2. The boundary of the graph of $y < -x + 2$ will be a _____ line, and the shading will be _____ the line.

3. The boundary of the graph of $y > -x + 2$ will be a _____ line, and the shading will be _____ the line.

4. The boundary of the graph of $y \geq -x + 2$ will be a _____ line, and the shading will be _____ the line.

📝 5. How is the boundary line $Ax + By = C$ used in graphing either $Ax + By < C$ or $Ax + By > C$?

📝 6. Describe the two methods discussed in the text for deciding which region is the solution set of a linear inequality in two variables.

Graph each linear inequality in two variables. See Examples 1–3.

7. $x + y \leq 2$ **8.** $x + y \leq -3$ **9.** $4x - y < 4$

10. $3x - y < 3$ **11.** $x + 3y \geq -2$ **12.** $x + 4y \geq -3$

13. $x + y > 0$ **14.** $x + 2y > 0$ **15.** $x - 3y \leq 0$

16. $x - 5y \leq 0$ **17.** $y < x$ **18.** $y \leq 4x$

Graph each compound inequality. See Example 4.

19. $x + y \leq 1$ and $x \geq 1$ **20.** $x - y \geq 2$ and $x \geq 3$

21. $2x - y \geq 2$ and $y < 4$ **22.** $3x - y \geq 3$ and $y < 3$

23. $x + y > -5$ and $y < -2$ **24.** $6x - 4y < 10$ and $y > 2$

Use the method described in Section 9.2 to write each inequality as a compound inequality, and graph its solution set in the rectangular coordinate plane.

25. $|x| < 3$ **26.** $|y| < 5$ **27.** $|x + 1| < 2$ **28.** $|y - 3| < 2$

Graph each compound inequality. See Example 5.

29. $x - y \geq 1$ or $y \geq 2$ **30.** $x + y \leq 2$ or $y \geq 3$

31. $x - 2 > y$ or $x < 1$ **32.** $x + 3 < y$ or $x > 3$

33. $3x + 2y < 6$ or $x - 2y > 2$ **34.** $x - y \geq 1$ or $x + y \leq 4$

TECHNOLOGY INSIGHTS (EXERCISES 35–42)

Match each inequality with its calculator graph. (Hint: Use the slope, y-intercept, and inequality symbol in making your choice.)

35. $y \leq 3x - 6$ **36.** $y \geq 3x - 6$

37. $y \leq -3x - 6$ **38.** $y \geq -3x - 6$

A.

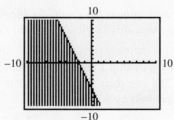

B.

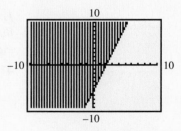

C.

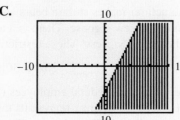

D.

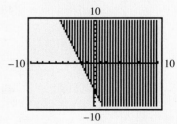

(continued)

The graph of a linear equation $y = mx + b$ is shown on a graphing calculator screen, along with the x-value of the x-intercept of the line. Use the screen to solve **(a)** $y = 0$, **(b)** $y < 0$, *and* **(c)** $y > 0$. *See the Connections box after Example 3.*

39.

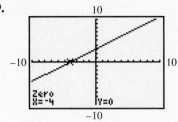

40.

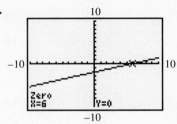

41.

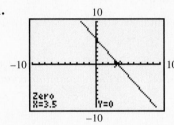

42.

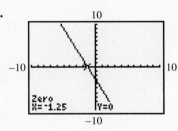

Solve the equation in part (a) and the associated inequalities in parts (b) and (c) using the methods of Chapter 2. Then graph the left side as y in the standard viewing window of a graphing calculator and explain how the graph supports your answers in parts (a)–(c).

43. **(a)** $5x + 3 = 0$
     (b) $5x + 3 > 0$
     (c) $5x + 3 < 0$

44. **(a)** $6x + 3 = 0$
     (b) $6x + 3 > 0$
     (c) $6x + 3 < 0$

45. **(a)** $-8x - (2x + 12) = 0$
     (b) $-8x - (2x + 12) \geq 0$
     (c) $-8x - (2x + 12) \leq 0$

46. **(a)** $-4x - (2x + 18) = 0$
     (b) $-4x - (2x + 18) \geq 0$
     (c) $-4x - (2x + 18) \leq 0$

For the given information **(a)** *graph the inequality. Here $x \geq 0$ and $y \geq 0$, so graph only the part of the inequality in quadrant I.* **(b)** *Give three ordered pairs that satisfy the inequality.*

47. A company will ship x units of merchandise to outlet I and y units of merchandise to outlet II. The company must ship a total of at least 500 units to these two outlets. This can be expressed by writing

$$x + y \geq 500.$$

48. A toy manufacturer makes stuffed bears and geese. It takes 20 minutes to sew a bear and 30 minutes to sew a goose. There is a total of 480 minutes of sewing time available to make x bears and y geese. These restrictions lead to the inequality

$$20x + 30y \leq 480.$$

49. The number of civilian federal employees (in thousands) is approximated by $y = -51.1x + 3153$, where x represents the year. Here, $x = 1$ corresponds to 1991, $x = 2$ corresponds to 1992, and so on. Thus, the total number of federal employees (including noncivilians) in each year is described by the inequality

$$y \geq -51.1x + 3153.$$

(*Hint:* Use tick marks of 1 to 10 on the *x*-axis and 2500 to 3500, at intervals of 200, on the *y*-axis.) (*Source:* U.S. Bureau of the Census.)

50. The equation $y = 57.1x + 1798$ approximates the number of free lunches served in public schools each year in millions. The year 1992 is represented by $x = 2$, 1993 is represented by $x = 3$, and so on. Free school breakfasts are also served in some (but not all) of the same schools. The number of children receiving only breakfast is described by the inequality

$$y \leq 57.1x + 1798.$$

(*Hint:* Use tick marks of 1 to 10 on the *x*-axis and 1800, 1900, 2000, 2100, 2200, 2300, and 2400 on the *y*-axis.) (*Source:* U.S. Department of Agriculture, Food and Consumer Service.)

Chapter **Group Activity**

Comparing Vacation Options

OBJECTIVE   Use tables, linear equations, and compound inequalities to compare admission costs to Universal Studios theme parks.

The North family, which includes parents Wayne and Sharon and daughters Marissa age 7 and Ce-Ce age 5, plans to visit Universal Studios Escape on their vacation to Orlando. The Universal Studios complex consists of Universal Studios Florida and the Islands of Adventure theme park.

A. A one-day pass to the Islands of Adventure is the same cost as a pass to Universal Studios Florida: $46.64 for adults, $37.10 for children ages 3–9, and free for children under age 3.

1. Complete the following table.

	Ticket Cost		
Number of Days	Adults	Children	Total Cost (in dollars)
1			
2			
x			

2. Find a linear equation that gives the total cost of the family's admission based on the number of days spent at the parks.

3. Assume the North family plans to visit the parks as many days as possible but wishes to keep the total admission cost between $400 and $500. Use the results found in Problem 2 to set up a compound inequality. Solve this inequality to find the number of days the family can visit the parks based on their budget.

B. Universal Studios also offers two package admission plans.

Plan 1   A 2-day Escape pass, which allows guests to visit both parks in the same day, costs $84.75 for adults and $68.85 for children ages 3–9.

Plan 2   A 3-day Escape pass costs $105.95 for adults and $84.75 for children ages 3–9.

1. How much can the North family save on a two-day visit by purchasing 2-day Escape passes? Express this range of admission costs using a compound inequality.

2. How much can the North family save on a three-day visit by purchasing 3-day Escape passes? Express this range of admission costs using a compound inequality.

3. If the family decides to visit the parks for four days, what is the most economical way to do so? What is the least economical way to do so? Express this range of admission costs using a compound inequality.

Source: St. Louis Post Dispatch; www.uescape.com

CHAPTER **9** SUMMARY

 KEY TERMS

9.1 intersection compound inequality union	**9.2** absolute value equation absolute value inequality	**9.3** linear inequality in two variables boundary line

NEW SYMBOLS

$\cap$ set intersection	$\cup$ set union

TEST YOUR WORD POWER

See how well you have learned the vocabulary in this chapter. Answers, with examples, follow the Quick Review.

1. The **intersection** of two sets A and B is the set of elements that belong
  A. to both A and B
  B. to either A or B, or both
  C. to either A or B, but not both
  D. to just A.

2. The **union** of two sets A and B is the set of elements that belong
  A. to both A and B
  B. to either A or B, or both
  C. to either A or B, but not both
  D. to just B.

3. A **linear inequality in two variables** is an inequality that can be written in the form
  A. $Ax + By < C$ or $Ax + By > C$
   ($\leq$ or $\geq$ can be used)
  B. $ax < b$
  C. $y \geq x^2$
  D. $Ax + By = C$.

QUICK REVIEW

CONCEPTS	EXAMPLES

9.1 SET OPERATIONS AND COMPOUND INEQUALITIES

Solving a Compound Inequality

Step 1 Solve each inequality in the compound inequality individually.

Step 2 If the inequalities are joined with *and,* then the solution set is the intersection of the two individual solution sets.

If the inequalities are joined with *or,* then the solution set is the union of the two individual solution sets.

Solve $x + 1 > 2$ and $2x < 6$.

$$x + 1 > 2 \quad \text{and} \quad 2x < 6$$
$$x > 1 \quad \text{and} \quad x < 3$$

The solution set is $(1, 3)$.

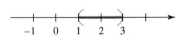

Solve $x \geq 4$ or $x \leq 0$.
The solution set is $(-\infty, 0] \cup [4, \infty)$.

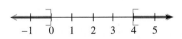

(continued)

CONCEPTS	EXAMPLES

9.2 ABSOLUTE VALUE EQUATIONS AND INEQUALITIES

Solving Absolute Value Equations and Inequalities

Let k be a positive number.
To solve $|ax + b| = k,$ solve the compound equation

$$ax + b = k \quad \text{or} \quad ax + b = -k.$$

Solve $|x - 7| = 3.$

$$x - 7 = 3 \quad \text{or} \quad x - 7 = -3$$
$$x = 10 \quad \text{or} \quad x = 4$$

The solution set is $\{4, 10\}.$

To solve $|ax + b| > k,$ solve the compound inequality

$$ax + b > k \quad \text{or} \quad ax + b < -k.$$

Solve $|x - 7| > 3.$

$$x - 7 > 3 \quad \text{or} \quad x - 7 < -3$$
$$x > 10 \quad \text{or} \quad x < 4$$

The solution set is $(-\infty, 4) \cup (10, \infty).$

To solve $|ax + b| < k,$ solve the compound inequality

$$-k < ax + b < k.$$

Solve $|x - 7| < 3.$

$$-3 < x - 7 < 3$$
$$4 < x < 10 \qquad \text{Add 7.}$$

The solution set is $(4, 10).$

To solve an absolute value equation of the form

$$|ax + b| = |cx + d|,$$

solve the compound equation

$$ax + b = cx + d \quad \text{or} \quad ax + b = -(cx + d).$$

Solve $|x + 2| = |2x - 6|.$

$$x + 2 = 2x - 6 \quad \text{or} \quad x + 2 = -(2x - 6)$$
$$x = 8 \qquad \text{or} \qquad x = \frac{4}{3}$$

The solution set is $\left\{\frac{4}{3}, 8\right\}.$

9.3 LINEAR INEQUALITIES IN TWO VARIABLES

Graphing a Linear Inequality

Step 1 Draw the graph of the line that is the boundary. Make the line solid if the inequality involves $\leq$ or $\geq$; make the line dashed if the inequality involves $<$ or $>$.

Step 2 Choose any point not on the line as a test point. Substitute the coordinates in the inequality.

Graph $2x - 3y \leq 6.$

Draw the graph of $2x - 3y = 6.$ Use a solid line because of $\leq$.

Choose $(1, 2).$

$$2(1) - 3(2) = 2 - 6 \leq 6 \qquad \text{True}$$

CONCEPTS	EXAMPLES
Step 3 Shade the region that includes the test point if the test point satisfies the original inequality; otherwise, shade the region on the other side of the boundary line.	Shade the side of the line that includes (1, 2). 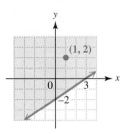

Answers to Test Your Word Power

1. A; *Example:* If $A = \{2, 4, 6, 8\}$ and $B = \{1, 2, 3\}$, $A \cap B = \{2\}$.   **2.** B; *Example:* Using the sets A and B from Answer 1, $A \cup B = \{1, 2, 3, 4, 6, 8\}$.   **3.** A; *Examples:* $4x + 3y < 12, x > 6y, 2x \geq 4y + 5$

CHAPTER 9 REVIEW EXERCISES

[9.1]   *Let $A = \{a, b, c, d\}$, $B = \{a, c, e, f\}$, and $C = \{a, e, f, g\}$. Find each set.*

1. $A \cap B$ **2.** $A \cap C$ **3.** $B \cup C$ **4.** $A \cup C$

Solve each compound inequality. Give the solution set in both interval and graph forms.

5. $x > 6$ and $x < 9$ **6.** $x + 4 > 12$ and $x - 2 < 12$

7. $x > 5$ or $x \leq -3$ **8.** $x \geq -2$ or $x < 2$

9. $x - 4 > 6$ and $x + 3 \leq 10$ **10.** $-5x + 1 \geq 11$ or $3x + 5 \geq 26$

Express each union or intersection in simplest interval form.

11. $(-3, \infty) \cap (-\infty, 4)$ **12.** $(-\infty, 6) \cap (-\infty, 2)$

13. $(4, \infty) \cup (9, \infty)$ **14.** $(1, 2) \cup (1, \infty)$

[9.2]   *Solve each absolute value equation.*

15. $|x| = 7$ **16.** $|x + 2| = 9$ **17.** $|3k - 7| = 8$

18. $|z - 4| = -12$ **19.** $|2k - 7| + 4 = 11$ **20.** $|4a + 2| - 7 = -3$

21. $|3p + 1| = |p + 2|$ **22.** $|2m - 1| = |2m + 3|$

Solve each absolute value inequality. Give the solution set in interval form.

23. $|p| < 14$ **24.** $|-t + 6| \leq 7$ **25.** $|2p + 5| \leq 1$

26. $|x + 1| \geq -3$ **27.** $|5r - 1| > 9$ **28.** $|11x - 3| \leq -2$

29. $|11x - 3| \geq -2$ **30.** $|11x - 3| \leq 0$

[9.3]   *Graph the solution set of each inequality or compound inequality.*

31. $3x - 2y \leq 12$ **32.** $5x - y > 6$

33. $x \geq 2$ **34.** $y < -4x$

35. $x \geq 2$ or $y \geq 2$ **36.** $2x + y \leq 1$ and $x \geq 2y$

MIXED REVIEW EXERCISES

Solve.

37. $x < 3$ and $x \geq -2$ **38.** $|3k + 6| \geq 0$ **39.** $|3x + 2| + 4 = 9$

40. $|m + 3| \leq 13$ **41.** $|m - 1| = |2m + 3|$ **42.** $|5r - 1| > 14$

43. $x \geq -2$ or $x < 4$ **44.** $|3k - 7| = 4$ **45.** $|m + 3| \leq 1$

In Exercises 46–49, graph the solution set of each inequality or compound inequality.

46. $-5x + 1 \geq 11$ or $3x + 5 \geq 26$ **47.** $x > 6$ and $x < 8$

48. $3x + 5y > 9$ **49.** $2x - 3y > -6$

The 2000 median weekly earnings of full-time workers by occupation were as shown in the table.

Weekly Earnings of Full-time Workers (in dollars)

Occupation	Men	Women
Managerial/Professional	994	709
Technical/Sales/Administrative Support	655	452
Service	357	316
Operators/Fabricators/Laborers	487	351

Source: U.S. Bureau of Labor Statistics.

List the elements of each set.

50. The set of occupations with median earnings for men less than \$900 and for women greater than \$500

51. The set of occupations with median earnings for men greater than \$600 or for women less than \$400

52. The solution set of $|3x + 4| = 7$ is shown on the number line.

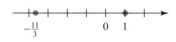

(a) What is the solution set of $|3x + 4| \geq 7$?

(b) What is the solution set of $|3x + 4| \leq 7$?

CHAPTER **9** TEST

Use the graphs to answer Exercises 1 and 2.

1. In which years did the number of players with 30–39 home runs exceed 20 *and* the number of players with 40 or more home runs exceed 45?

2. In which years were the number of players with 20–29 home runs less than 20 *or* the number of players with 30–39 home runs at least 20?

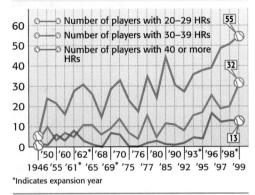

GOING, GOING, GONE
Home runs have been flying out of major-league ballparks at an increasing rate. A breakdown:

○——○ Number of players with 20–29 HRs 55
○——○ Number of players with 30–39 HRs
○——○ Number of players with 40 or more HRs 32

13

'50 '60 '62* '68 '70 '76 '80 '90 '93* '96 '98*
1946 '55 '61* '65 '69* '75 '77 '85 '92 '95 '97 '99
*Indicates expansion year

Source: Bee research.

Let $A = \{1, 2, 5, 7\}$ and $B = \{1, 5, 9, 12\}$. Find each set.

3. $A \cap B$

4. $A \cup B$

Solve each compound inequality. Give the solution set in both interval and graph forms.

5. $3k \geq 6$ and $k - 4 < 5$

6. $-4x \leq -24$ or $4x - 2 < 10$

Solve each absolute value equation or inequality. Give the solution set in interval form.

7. $|4x - 3| = 7$

8. $|5 - 6x| > 12$

9. $|7 - x| \leq -1$

10. $|3 - 5x| = |2x + 8|$

11. $|-3x + 4| - 4 < -1$

12. $|12t + 7| \geq 0$

13. If $k < 0$, what is the solution set of

   (a) $|5x + 3| < k$ (b) $|5x + 3| > k$ (c) $|5x + 3| = k$?

Graph the solution set of each inequality or compound inequality.

14. $3x - 2y > 6$

15. $3x - y > 0$

16. $y < 2x - 1$ and $x - y < 3$

17. $x - 2 \geq y$ or $y \geq 3$

CUMULATIVE REVIEW EXERCISES CHAPTERS 1–9

1. Match each number in Column I with the choice or choices of sets of numbers in Column II to which the number belongs.

I	II
(a) 34	**A.** Natural numbers
(b) 0	**B.** Whole numbers
(c) 2.16	**C.** Integers
(d) $-\pi$	**D.** Rational numbers
(e) $\sqrt{3}$	**E.** Irrational numbers
(f) $-\dfrac{4}{5}$	**F.** Real numbers

Evaluate.

2. $9 \cdot 4 - 16 \div 4$

3. $-|8 - 13| - |-4| + |-9|$

Solve.

4. $-5(8 - 2z) + 4(7 - z) = 7(8 + z) - 3$

5. $3(x + 2) - 5(x + 2) = -2x - 4$

6. $A = p + prt$ for t

7. $2(m + 5) - 3m + 1 > 5$

8. A recent survey polled teens about the most important inventions of the twentieth century. Complete the results shown in the table if 1500 teens were surveyed.

Most Important Invention	Percent	Actual Number
Personal computer		480
Pacemaker	26%	
Wireless communication	18%	
Television		150

Source: Lemelson-MIT Program.

9. Find the measure of each angle of the triangle. (*Hint:* The sum of the measures of any triangle is 180°.)

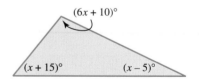

Find the slope of each line described.

10. Through $(-4, 5)$ and $(2, -3)$

11. Horizontal, through $(4, 5)$

Find an equation of each line. Write the equation in slope-intercept form.

12. Through $(4, -1)$, $m = -4$

13. Through $(0, 0)$ and $(1, 4)$

14. Graph $-3x + 4y = 12$.

Simplify. Write answers with only positive exponents. Assume all variables represent positive real numbers.

15. $\left(\dfrac{2m^3n}{p^2}\right)^3$

16. $\dfrac{x^{-6}y^3z^{-1}}{x^7y^{-4}z}$

Perform the indicated operations.

17. $2(3x^2 - 8x + 1) - 4(x^2 - 3x - 9)$

18. $(3x + 2y)(5x - y)$

19. $(8m + 5n)(8m - 5n)$

20. $(x + 2y)(x^2 - 2xy + 4y^2)$

21. $\dfrac{m^3 - 3m^2 + 5m - 3}{m - 1}$

Factor each polynomial completely.

22. $m^2 + 12m + 32$

23. $25t^4 - 36$

24. $81z^2 + 72z + 16$

25. Solve the equation $(x + 4)(x - 1) = -6$.

26. For what real number(s) is the expression $\dfrac{3}{x^2 + 5x - 14}$ undefined?

Perform each indicated operation. Express answers in lowest terms.

27. $\dfrac{x^2 - 3x - 4}{x^2 + 3x} \cdot \dfrac{x^2 + 2x - 3}{x^2 - 5x + 4}$

28. $\dfrac{t^2 + 4t - 5}{t + 5} \div \dfrac{t - 1}{t^2 + 8t + 15}$

29. $\dfrac{2}{x + 3} - \dfrac{4}{x - 1}$

30. $\dfrac{\dfrac{2}{3} + \dfrac{1}{2}}{\dfrac{1}{9} - \dfrac{1}{6}}$

31. Solve the equation $\dfrac{x}{x + 8} - \dfrac{3}{x - 8} = \dfrac{128}{x^2 - 64}$.

32. The graph shows the annual number of twin births in the United States for selected years.

   ✍ **(a)** Use the information given on the graph to find and interpret the average rate of change in the number of twin births per year.

   (b) If $x = 0$ represents 1990, use your answer from part (a) to write an equation of the line in slope-intercept form that models the annual number of twin births for the years 1990 through 2000.

   (c) Use the equation from part (b) to approximate the number of twin births in 2002.

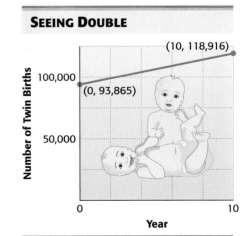

SEEING DOUBLE

(10, 118,916)

(0, 93,865)

Number of Twin Births

100,000

50,000

0   Year   10

Source: National Center for Health Statistics.

33. Give the domain and range of the relation $\{(-4, -2), (-1, 0), (2, 0), (5, 2)\}$. Does this relation define a function?

34. One source of renewable energy is wind, although as of 2000, it provided less than 5 trillion BTUs in the United States. (*Source: U.S. Energy Information Administration, Annual Energy Review.*) The force of the wind blowing on a vertical surface varies jointly as the area of the surface and the square of the velocity. If a wind of 40 mph exerts a force of 50 lb on a surface of $\frac{1}{2}$ ft^2, how much force will a wind of 80 mph place on a surface of 2 ft^2?

Solve each system.

35. $3x - 4y = 1$
$\quad\;\; 2x + 3y = 12$

36. $\quad 3x - 2y = 4$
$\quad -6x + 4y = 7$

37. $\;\; x + 3y - 6z = 7$
$\quad 2x - \;\; y + \;\; z = 1$
$\quad\;\; x + 2y + 2z = -1$

Solve each problem using a system of equations.

38. The Star-Spangled Banner that flew over Fort McHenry during the War of 1812 had a perimeter of 144 ft. Its length measured 12 ft more than its width. Find the dimensions of this flag, which is displayed in the Smithsonian Institution's Museum of American History in Washington, D.C. (*Source:* National Park Service brochure.)

39. A chemist needs 9 L of a 20% solution of alcohol. She has a 15% solution on hand, as well as a 30% solution. How many liters of the 15% solution and the 30% solution should she mix to get the 20% solution she needs?

Solve each equation or inequality.

40. $x > -4$ and $x < 4$

41. $2x + 1 > 5$ or $2 - x \geq 2$

42. $|3x - 1| = 2$

43. $|3z + 1| \geq 7$

Graph the solution set of each inequality or compound inequality.

44. $y \leq 2x - 6$

45. $3x + 2y < 0$

46. $x - y \geq 3$ and $3x + 4y \leq 12$

Roots, Radicals, and Root Functions

10

Tom Skilling is the chief meteorologist for the *Chicago Tribune*. He writes a column titled "Ask Tom Why," where readers question him on a variety of topics. In the Saturday, August 17, 2002 issue, reader Ted Fleischaker wrote: "I cannot remember the formula to calculate the distance to the horizon. I have a stunning view from my 14th floor condo, 150 feet above the ground. How far can I see?" (See Exercise 118 in Section 10.3.)

In Skilling's answer, he explained the formula for finding the distance d to the horizon in miles,

$$d = 1.224\sqrt{h},$$

where h is the height in feet. Square roots such as this one are often found in formulas. This chapter deals with roots and radicals.

10.1 Radical Expressions and Graphs

In Section 1.2, we discussed the idea of the *square* of a number. Recall that squaring a number means multiplying the number by itself.

$$\text{If } a = 8, \qquad \text{then} \qquad a^2 = 8 \cdot 8 = 64.$$
$$\text{If } a = -4, \qquad \text{then} \qquad a^2 = (-4)(-4) = 16.$$
$$\text{If } a = -\frac{1}{2}, \qquad \text{then} \qquad a^2 = \left(-\frac{1}{2}\right)\left(-\frac{1}{2}\right) = \frac{1}{4}.$$

In this chapter, we consider the opposite process.

$$\text{If } a^2 = 49, \qquad \text{then} \qquad a = \,?.$$
$$\text{If } a^2 = 100, \qquad \text{then} \qquad a = \,?.$$
$$\text{If } a^2 = 25, \qquad \text{then} \qquad a = \,?.$$

OBJECTIVE 1 Find square roots. To find a in the three preceding statements, we must find a number that when multiplied by itself results in the given number. The number a is called a **square root** of the number a^2.

EXAMPLE 1 Finding All Square Roots of a Number

Find all square roots of 49.

To find a square root of 49, think of a number that when multiplied by itself gives 49. One square root is 7 because $7 \cdot 7 = 49$. Another square root of 49 is -7 because $(-7)(-7) = 49$. The number 49 has two square roots, 7 and -7; one is positive, and one is negative.

Now Try Exercise 7.

The **positive** or **principal square root** of a number is written with the symbol $\sqrt{}$. For example, the positive square root of 121 is 11, written

$$\sqrt{121} = 11.$$

The symbol $-\sqrt{}$ is used for the **negative square root** of a number. For example, the negative square root of 121 is -11, written

$$-\sqrt{121} = -11.$$

The symbol $\sqrt{}$ called a **radical sign,** always represents the positive square root $\left(\text{except that } \sqrt{0} = 0\right)$. The number inside the radical sign is called the **radicand,** and the entire expression—radical sign and radicand—is called a **radical.**

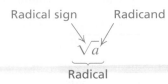

Radical sign   Radicand

$$\sqrt{a}$$

Radical

Early radical symbol

Radicals have a long mathematical history. The radical sign $\sqrt{}$ has been used since sixteenth-century Germany and was probably derived from the letter R. The radical symbol in the margin on the preceding page comes from the Latin word for root, *radix*. It was first used by Leonardo of Pisa (Fibonacci) in 1220.

We summarize our discussion of square roots as follows.

Square Roots of *a*

If *a* is a positive real number, then

$$\sqrt{a} \text{ is the positive or principal square root of } a,$$

and $-\sqrt{a}$ is the negative square root of *a*.

For nonnegative *a*,

$$\sqrt{a} \cdot \sqrt{a} = \left(\sqrt{a}\right)^2 = a \quad \text{and} \quad -\sqrt{a} \cdot \left(-\sqrt{a}\right) = \left(-\sqrt{a}\right)^2 = a.$$

Also, $\sqrt{0} = 0$.

EXAMPLE 2 Finding Square Roots

Find each square root.

(a) $\sqrt{144}$

The radical $\sqrt{144}$ represents the positive or principal square root of 144. Think of a positive number whose square is 144.

$$12^2 = 144, \quad \text{so} \quad \sqrt{144} = 12.$$

(b) $-\sqrt{1024}$

This symbol represents the negative square root of 1024. A calculator with a square root key can be used to find $\sqrt{1024} = 32$. Then, $-\sqrt{1024} = -32$.

(c) $\sqrt{\dfrac{4}{9}} = \dfrac{2}{3}$ **(d)** $-\sqrt{\dfrac{16}{49}} = -\dfrac{4}{7}$

Now Try Exercises 19, 21, and 23.

As shown in the preceding definition, when the square root of a positive real number is squared, the result is that positive real number. $\left(\text{Also, } \left(\sqrt{0}\right)^2 = 0.\right)$

EXAMPLE 3 Squaring Radical Expressions

Find the *square* of each radical expression.

(a) $\sqrt{13}$

$$\left(\sqrt{13}\right)^2 = 13 \qquad \text{Definition of square root}$$

(b) $-\sqrt{29}$

$$\left(-\sqrt{29}\right)^2 = 29 \qquad \text{The square of a } \textit{negative} \text{ number is positive.}$$

(c) $\sqrt{p^2 + 1}$

$$\left(\sqrt{p^2 + 1}\right)^2 = p^2 + 1$$

Now Try Exercises 27, 29, and 33.

OBJECTIVE 2 Decide whether a given root is rational, irrational, or not a real number.
All numbers with square roots that are rational are called **perfect squares**. For example, 144 and $\frac{4}{9}$ are perfect squares since their respective square roots, 12 and $\frac{2}{3}$, are rational numbers.

A number that is not a perfect square has a square root that is not a rational number. For example, $\sqrt{5}$ is not a rational number because it cannot be written as the ratio of two integers. Its decimal neither terminates nor repeats. However, $\sqrt{5}$ is a real number and corresponds to a point on the number line. As mentioned in Chapter 1, a real number that is not rational is called an *irrational number*. The number $\sqrt{5}$ is irrational. Many square roots of integers are irrational.

If a is a positive real number that is not a perfect square, then $\sqrt{a}$ is irrational.

Not every number has a *real number* square root. For example, there is no real number that can be squared to get -36. (The square of a real number can never be negative.) Because of this, $\sqrt{-36}$ is not a real number.

If a is a negative real number, then $\sqrt{a}$ is not a real number.

CAUTION Be careful not to confuse $\sqrt{-36}$ and $-\sqrt{36}$. $\sqrt{-36}$ is not a real number since there is no real number that can be squared to get -36. However, $-\sqrt{36}$ is the negative square root of 36, which is -6.

EXAMPLE 4 Identifying Types of Square Roots

Tell whether each square root is *rational, irrational,* or *not a real number.*

(a) $\sqrt{17}$

Because 17 is not a perfect square, $\sqrt{17}$ is irrational.

(b) $\sqrt{64}$

The number 64 is a perfect square, 8^2, so $\sqrt{64} = 8$, a rational number.

(c) $\sqrt{-25}$

There is no real number whose square is -25. Therefore, $\sqrt{-25}$ is not a real number.

Now Try Exercises 39, 41, and 47.

NOTE Not all irrational numbers are square roots of integers. For example, π (approximately 3.14159) is an irrational number that is not a square root of any integer.

OBJECTIVE 3 Find higher roots. Finding the square root of a number is the inverse (reverse) of squaring a number. In a similar way, there are inverses to finding the cube of a number, or finding the fourth or higher power of a number. These inverses are the **cube root,** written $\sqrt[3]{a}$, and the **fourth root,** written $\sqrt[4]{a}$. Similar symbols are used for higher roots. In general, we have the following.

$\sqrt[n]{a}$

The *n*th root of *a*, written $\sqrt[n]{a}$, is a number whose *n*th power equals *a*. That is,

$$\sqrt[n]{a} = b \qquad \text{means} \qquad b^n = a.$$

In $\sqrt[n]{a}$, the number *n* is the **index** or **order** of the radical. It is possible to write $\sqrt[2]{a}$ instead of $\sqrt{a}$, but the simpler symbol $\sqrt{a}$ is customary since the square root is the most commonly used root.

When working with cube roots or fourth roots, it is helpful to memorize the first few *perfect cubes* ($2^3 = 8$, $3^3 = 27$, and so on) and the first few perfect fourth powers ($2^4 = 16$, $3^4 = 81$, and so on).

EXAMPLE 5 Finding Cube Roots

Find each cube root.

(a) $\sqrt[3]{8}$

Look for a number that can be cubed to give 8. Because $2^3 = 8$, $\sqrt[3]{8} = 2$.

(b) $\sqrt[3]{-8} = -2$ because $(-2)^3 = -8$.

(c) $-\sqrt[3]{216} = -6$ because $\sqrt[3]{216} = 6$, and thus $-\sqrt[3]{216} = -6$.

Now Try Exercises 65, 67, and 69.

Notice in Example 5(b) that we can find the cube root of a negative number. (Contrast this with the square root of a negative number, which is not real.) In fact, the cube root of a positive number is positive, and the cube root of a negative number is negative. *There is only one real number cube root for each real number.*

When the index of the radical is even (square root, fourth root, and so on), *the radicand must be nonnegative* to get a real number root. Also, for even indexes, the symbols $\sqrt{}$, $\sqrt[4]{}$, $\sqrt[6]{}$, and so on are used for the positive (principal) roots. The symbols $-\sqrt{}$, $-\sqrt[4]{}$, $-\sqrt[6]{}$, and so on are used for the negative roots.

EXAMPLE 6 Finding Higher Roots

Find each root.

(a) $\sqrt[4]{16} = 2$ because 2 is positive and $2^4 = 16$.

(b) $-\sqrt[4]{16}$

From part (a), $\sqrt[4]{16} = 2$, so the negative root $-\sqrt[4]{16} = -2$.

(c) $\sqrt[4]{-16}$

For a real number fourth root, the radicand must be nonnegative. There is no real number that equals $\sqrt[4]{-16}$.

(d) $-\sqrt[5]{32}$

First find $\sqrt[5]{32}$. Because 2 is the number whose fifth power is 32, $\sqrt[5]{32} = 2$. If $\sqrt[5]{32} = 2$, then

$$-\sqrt[5]{32} = -2.$$

(e) $\sqrt[5]{-32} = -2$, because $(-2)^5 = -32$.

Now Try Exercises 71, 73, 75, and 77.

OBJECTIVE 4   Graph functions defined by radical expressions.   A **radical expression** is an algebraic expression that contains radicals. For example,

$$3 - \sqrt{x}, \qquad \sqrt[3]{x}, \qquad \text{and} \qquad \sqrt{2x - 1}$$

are radical expressions.

In Section 7.3 we graphed functions defined by linear expressions. Now we examine the graphs of functions defined by the basic radical expressions $f(x) = \sqrt{x}$ and $f(x) = \sqrt[3]{x}$.

Figure 1 shows the graph of the **square root function** with a table of selected points. Only nonnegative values can be used for x, so the domain is $[0, \infty)$. Because $\sqrt{x}$ is the principal square root of x, it always has a nonnegative value, so the range is also $[0, \infty)$.

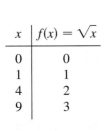

x	$f(x) = \sqrt{x}$
0	0
1	1
4	2
9	3

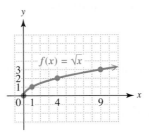

x	$f(x) = \sqrt[3]{x}$
-8	-2
-1	-1
0	0
1	1
8	2

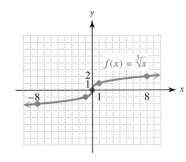

FIGURE 1

FIGURE 2

Figure 2 shows the graph of the **cube root function** and a table of selected points. Since any real number (positive, negative, or 0) can be used for x in the cube root function, $\sqrt[3]{x}$ can be positive, negative, or 0. Thus both the domain and the range of the cube root function are $(-\infty, \infty)$.

EXAMPLE 7   Graphing Functions Defined with Radicals

Graph each function by creating a table of values. Give the domain and range.

(a) $f(x) = \sqrt{x - 3}$

A table of values is given below. The x-values were chosen in such a way that the function values are all integers. For the radicand to be nonnegative, we must have $x - 3 \geq 0$, or $x \geq 3$. Therefore, the domain is $[3, \infty)$. Function values are positive or 0, so the range is $[0, \infty)$. The graph is shown in Figure 3.

x	$f(x) = \sqrt{x - 3}$
3	$\sqrt{3 - 3} = 0$
4	$\sqrt{4 - 3} = 1$
7	$\sqrt{7 - 3} = 2$

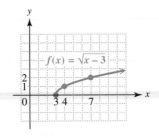

FIGURE 3

(b) $f(x) = \sqrt[3]{x} + 2$

See the table and Figure 4. Both the domain and range are $(-\infty, \infty)$.

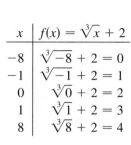

x	$f(x) = \sqrt[3]{x} + 2$
-8	$\sqrt[3]{-8} + 2 = 0$
-1	$\sqrt[3]{-1} + 2 = 1$
0	$\sqrt[3]{0} + 2 = 2$
1	$\sqrt[3]{1} + 2 = 3$
8	$\sqrt[3]{8} + 2 = 4$

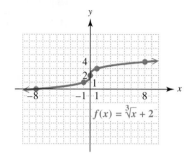

FIGURE 4

Now Try Exercises 87 and 91.

OBJECTIVE 5 Find *n*th roots of *n*th powers. A square root of a^2 (where $a \neq 0$) is a number that can be squared to give a^2. This number is either a or $-a$. Since the symbol $\sqrt{a^2}$ represents the *nonnegative* square root, we express $\sqrt{a^2}$ with absolute value bars as $|a|$, because a may be a negative number.

$\sqrt{\boldsymbol{a^2}}$

For any real number a,

$$\sqrt{a^2} = |a|.$$

EXAMPLE 8 Simplifying Square Roots Using Absolute Value

Find each square root.

(a) $\sqrt{7^2} = |7| = 7$ **(b)** $\sqrt{(-7)^2} = |-7| = 7$

(c) $\sqrt{k^2} = |k|$ **(d)** $\sqrt{(-k)^2} = |-k| = |k|$

Now Try Exercises 95, 97, and 103.

We can generalize this idea to any *n*th root.

$\sqrt[n]{\boldsymbol{a^n}}$

If n is an *even* positive integer, then $\sqrt[n]{a^n} = |a|,$

and if n is an *odd* positive integer, then $\sqrt[n]{a^n} = a.$

That is, use absolute value when n is even; absolute value is not necessary when n is odd.

■ EXAMPLE 9 Simplifying Higher Roots Using Absolute Value

Simplify each root.

(a) $\sqrt[6]{(-3)^6} = |-3| = 3$ *n* is even; use absolute value.

(b) $\sqrt[5]{(-4)^5} = -4$ *n* is odd.

(c) $-\sqrt[4]{(-9)^4} = -|-9| = -9$ **(d)** $\sqrt[3]{\dfrac{8}{27}} = \sqrt[3]{\left(\dfrac{2}{3}\right)^3} = \dfrac{2}{3}$

(e) $-\sqrt{m^4} = -|m^2| = -m^2$

No absolute value bars are needed here because m^2 is nonnegative for any real number value of *m*.

(f) $\sqrt[3]{a^{12}} = a^4$ because $a^{12} = (a^4)^3$.

(g) $\sqrt[4]{x^{12}} = |x^3|$

We use absolute value bars to guarantee that the result is not negative (because x^3 can be either positive or negative, depending on *x*). Also, $|x^3|$ can be written as $x^2 \cdot |x|$.

Now Try Exercises 99, 101, 105, and 107.

OBJECTIVE 6 Use a calculator to find roots. While numbers such as $\sqrt{9}$ and $\sqrt[3]{-8}$ are rational, radicals are often irrational numbers. To find approximations of roots such as $\sqrt{15}$, $\sqrt[3]{10}$, and $\sqrt[4]{2}$, we usually use scientific or graphing calculators. Using a calculator, we find

$$\sqrt{15} \approx 3.872983346, \quad \sqrt[3]{10} \approx 2.15443469, \quad \text{and} \quad \sqrt[4]{2} \approx 1.189207115,$$

where the symbol $\approx$ means "is approximately equal to." In this book we usually show approximations rounded to three decimal places. Thus, we would write

$$\sqrt{15} \approx 3.873, \quad \sqrt[3]{10} \approx 2.154, \quad \text{and} \quad \sqrt[4]{2} \approx 1.189.$$

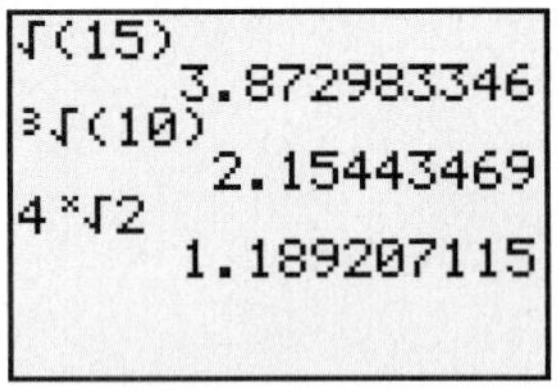

Figure 5 shows how the preceding approximations are displayed on a TI-83 Plus graphing calculator. In Figure 5(a), eight or nine decimal places are shown, while in Figure 5(b), the number of decimal places is fixed at three.

There is a simple way to check that a calculator approximation is "in the ballpark." Because 16 is a little larger than 15, $\sqrt{16} = 4$ should be a little larger than $\sqrt{15}$. Thus, 3.873 is a reasonable approximation for $\sqrt{15}$.

```
√(15)
        3.872983346
³√(10)
        2.15443469
4 ˣ√2
        1.189207115
```

(a)

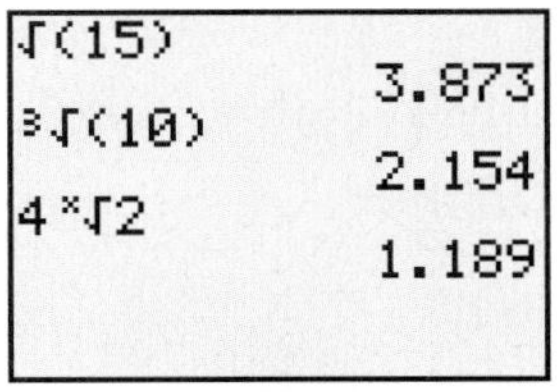

```
√(15)
              3.873
³√(10)
              2.154
4 ˣ√2
              1.189
```

(b)

FIGURE 5

NOTE The methods for finding approximations differ among makes and models of calculators. You should always consult your owner's manual for keystroke instructions. Be aware that graphing calculators often differ from scientific calculators in the order in which keystrokes are made.

■ EXAMPLE 10 Finding Approximations for Roots

Use a calculator to verify that each approximation is correct.

(a) $\sqrt{39} \approx 6.245$ **(b)** $-\sqrt{72} \approx -8.485$

(c) $\sqrt[3]{93} \approx 4.531$ **(d)** $\sqrt[4]{39} \approx 2.499$

Now Try Exercises 109, 111, and 113.

■ **EXAMPLE 11** Using Roots to Calculate Resonant Frequency

In electronics, the resonant frequency f of a circuit may be found by the formula

$$f = \frac{1}{2\pi\sqrt{LC}},$$

where f is in cycles per second, L is in henrys, and C is in farads.* Find the resonant frequency f if $L = 5 \times 10^{-4}$ henrys and $C = 3 \times 10^{-10}$ farads. Give your answer to the nearest thousand.

Find the value of f when $L = 5 \times 10^{-4}$ and $C = 3 \times 10^{-10}$.

$$f = \frac{1}{2\pi\sqrt{LC}} \qquad \text{Given formula}$$

$$= \frac{1}{2\pi\sqrt{(5 \times 10^{-4})(3 \times 10^{-10})}} \qquad \text{Substitute for } L \text{ and } C.$$

$$\approx 411{,}000 \qquad \text{Use a calculator.}$$

The resonant frequency f is approximately 411,000 cycles per sec.

Now Try Exercise 117.

10.1 EXERCISES

For Extra Help

Student's
Solutions Manual

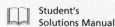
MyMathLab

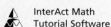
InterAct Math
Tutorial Software

AW Math
Tutor Center

MathXL MathXL

Digital Video Tutor
CD 14/Videotape 14

▨ *Decide whether each statement is* true *or* false. *If false, tell why.*

1. Every positive number has two real square roots.

2. A negative number has negative square roots.

3. Every nonnegative number has two real square roots.

4. The positive square root of a positive number is its principal square root.

5. The cube root of every real number has the same sign as the number itself.

6. Every positive number has three real cube roots.

Find all square roots of each number. See Example 1.

7. 9   **8.** 16   **9.** 64   **10.** 100   **11.** 144

12. 225   **13.** $\dfrac{25}{196}$ **14.** $\dfrac{81}{400}$ **15.** 900   **16.** 1600

Find each square root. See Examples 2 and 4(c).

17. $\sqrt{1}$ **18.** $\sqrt{4}$ **19.** $\sqrt{49}$ **20.** $\sqrt{81}$ **21.** $-\sqrt{121}$

22. $-\sqrt{196}$ **23.** $-\sqrt{\dfrac{144}{121}}$ **24.** $-\sqrt{\dfrac{49}{36}}$ **25.** $\sqrt{-121}$ **26.** $\sqrt{-64}$

Find the square of each radical expression. See Example 3.

27. $\sqrt{19}$ **28.** $\sqrt{59}$ **29.** $-\sqrt{19}$ **30.** $-\sqrt{99}$

31. $\sqrt{\dfrac{2}{3}}$ **32.** $\sqrt{\dfrac{5}{7}}$ **33.** $\sqrt{3x^2 + 4}$ **34.** $\sqrt{9y^2 + 3}$

*Henrys and farads are units of measure in electronics.

What must be true about the variable a for each statement in Exercises 35–38 to be true?

35. $\sqrt{a}$ represents a positive number.

36. $-\sqrt{a}$ represents a negative number.

37. $\sqrt{a}$ is not a real number.

38. $-\sqrt{a}$ is not a real number.

Write rational, irrational, *or* not a real number *for each number. If a number is rational, give its exact value. If a number is irrational, give a decimal approximation to the nearest thousandth. Use a calculator as necessary. See Examples 4 and 10.*

39. $\sqrt{25}$

40. $\sqrt{169}$

41. $\sqrt{29}$

42. $\sqrt{33}$

43. $-\sqrt{64}$

44. $-\sqrt{81}$

45. $-\sqrt{300}$

46. $-\sqrt{500}$

47. $\sqrt{-29}$

48. $\sqrt{-47}$

49. $\sqrt{1200}$

50. $\sqrt{1500}$

Match each expression from Column I with the equivalent choice from Column II. Answers may be used more than once.

I	II
51. $-\sqrt{16}$	**A.** 3
52. $\sqrt{-16}$	**B.** -2
53. $\sqrt[3]{-27}$	**C.** 2
54. $\sqrt[5]{-32}$	**D.** -3
55. $\sqrt[4]{81}$	**E.** -4
56. $\sqrt[3]{8}$	**F.** Not a real number

Choose the closest approximation of each square root.

57. $\sqrt{123.5}$

    A. 9   **B.** 10   **C.** 11   **D.** 12

58. $\sqrt{67.8}$

    A. 7   **B.** 8   **C.** 9   **D.** 10

Refer to the rectangle to answer the questions in Exercises 59 and 60.

59. Which one of the following is the best estimate of its area?

    A. 2500   **B.** 250   **C.** 50   **D.** 100

60. Which one of the following is the best estimate of its perimeter?

    A. 15   **B.** 250   **C.** 100   **D.** 30

61. Consider the expression $-\sqrt{-a}$. Decide whether it is *positive, negative,* 0, or *not a real number* if

    (a) $a > 0$,   **(b)** $a < 0$,   **(c)** $a = 0$.

62. If n is odd, under what conditions is $\sqrt[n]{a}$

    (a) positive,   **(b)** negative,   **(c)** 0?

Find each root that is a real number. Use a calculator as necessary. See Examples 5 and 6.

63. $\sqrt[3]{1}$

64. $\sqrt[3]{27}$

65. $-\sqrt[3]{125}$

66. $-\sqrt[3]{343}$

67. $\sqrt[3]{-64}$

68. $\sqrt[3]{-125}$

69. $\sqrt[3]{512}$

70. $\sqrt[3]{1000}$

71. $\sqrt[4]{625}$ **72.** $\sqrt[4]{1296}$ **73.** $-\sqrt[4]{625}$ **74.** $-\sqrt[4]{256}$

75. $\sqrt[4]{-625}$ **76.** $\sqrt[4]{-256}$ **77.** $\sqrt[6]{64}$ **78.** $\sqrt[8]{256}$

79. $\sqrt[6]{-32}$ **80.** $\sqrt[8]{-1}$ **81.** $\sqrt[4]{\dfrac{256}{81}}$ **82.** $\sqrt[3]{\dfrac{1000}{27}}$

83. $\sqrt[3]{\dfrac{8}{27}}$ **84.** $\sqrt[4]{\dfrac{81}{16}}$ **85.** $\sqrt[6]{\dfrac{1}{64}}$ **86.** $\sqrt[5]{\dfrac{1}{32}}$

Graph each function and give its domain and range. See Example 7.

87. $f(x) = \sqrt{x + 3}$ **88.** $f(x) = \sqrt{x - 5}$ **89.** $f(x) = \sqrt{x} - 2$

90. $f(x) = \sqrt{x} + 4$ **91.** $f(x) = \sqrt[3]{x} - 3$ **92.** $f(x) = \sqrt[3]{x} + 1$

93. $f(x) = \sqrt[3]{x - 3}$ **94.** $f(x) = \sqrt[3]{x + 1}$

Simplify each root. See Examples 8 and 9.

95. $\sqrt{12^2}$ **96.** $\sqrt{19^2}$ **97.** $\sqrt{(-10)^2}$ **98.** $-\sqrt{13^2}$

99. $\sqrt[6]{(-2)^6}$ **100.** $\sqrt[6]{(-4)^6}$ **101.** $\sqrt[5]{(-9)^5}$ **102.** $\sqrt[5]{(-8)^5}$

103. $\sqrt{x^2}$ **104.** $-\sqrt{x^2}$ **105.** $\sqrt[3]{x^3}$ **106.** $-\sqrt[3]{x^3}$

107. $\sqrt[3]{x^{15}}$ **108.** $\sqrt[4]{k^{20}}$

Find a decimal approximation for each radical. Round the answer to three decimal places. See Example 10.

109. $-\sqrt{82}$ **110.** $-\sqrt{91}$ **111.** $\sqrt[3]{423}$ **112.** $\sqrt[3]{555}$

113. $\sqrt[4]{100}$ **114.** $\sqrt[4]{250}$ **115.** $\sqrt[5]{23.8}$ **116.** $\sqrt[5]{98.4}$

Solve each problem. See Example 11.

117. Use the formula in Example 11 to calculate the resonant frequency of a circuit to the nearest thousand if $L = 7.237 \times 10^{-5}$ henrys and $C = 2.5 \times 10^{-10}$ farads.

118. The time for one complete swing of a simple pendulum is

$$t = 2\pi\sqrt{\dfrac{L}{g}},$$

2 ft

where t is time in seconds, L is the length of the pendulum in feet, and g, the force due to gravity, is about 32 ft per sec². Find the time of a complete swing of a 2-ft pendulum to the nearest tenth of a second.

119. Heron's formula gives a method of finding the area of a triangle if the lengths of its sides are known. Suppose that a, b, and c are the lengths of the sides. Let s denote one-half of the perimeter of the triangle (called the *semiperimeter*); that is,

$$s = \dfrac{1}{2}(a + b + c).$$

Then the area of the triangle is

$$A = \sqrt{s(s - a)(s - b)(s - c)}.$$

Find the area of the Bermuda Triangle, if the "sides" of this triangle measure approximately 850 mi, 925 mi, and 1300 mi. Give your answer to the nearest thousand square miles.

120. The Vietnam Veterans' Memorial in Washington, D.C., is in the shape of an unenclosed isosceles triangle with equal sides of length 246.75 ft. If the triangle were enclosed, the third side would have length 438.14 ft. Use Heron's formula from the previous exercise to find the area of this enclosure to the nearest hundred square feet. (*Source:* Information pamphlet obtained at the Vietnam Veterans' Memorial.)

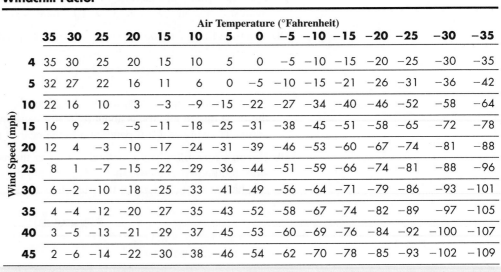

246.75 ft
246.75 ft
438.14 ft
Not to scale

121. The formula

$$I = \sqrt{\dfrac{2P}{L}}$$

relates the coefficient of self-induction L (in henrys), the energy P stored in an electronic circuit (in joules), and the current I (in amps). Find I if $P = 120$ and $L = 80$.

122. When the wind blows, the air feels much colder than the actual temperature. The **windchill factor** measures the cooling effect that the wind has on one's skin. Until recently, the formula that the National Weather Service used to compute windchill was

$$T_{wc} = .0817\left(3.71\sqrt{V} + 5.81 - .25V\right)(T - 91.4) + 91.4,$$

where T_{wc} is windchill, V is wind speed in miles per hour (mph), and T is air temperature in degrees Fahrenheit. The windchill for various wind speeds and temperatures is shown in the table.

Windchill Factor

Wind Speed (mph)	Air Temperature (°Fahrenheit)														
	35	30	25	20	15	10	5	0	−5	−10	−15	−20	−25	−30	−35
4	35	30	25	20	15	10	5	0	−5	−10	−15	−20	−25	−30	−35
5	32	27	22	16	11	6	0	−5	−10	−15	−21	−26	−31	−36	−42
10	22	16	10	3	−3	−9	−15	−22	−27	−34	−40	−46	−52	−58	−64
15	16	9	2	−5	−11	−18	−25	−31	−38	−45	−51	−58	−65	−72	−78
20	12	4	−3	−10	−17	−24	−31	−39	−46	−53	−60	−67	−74	−81	−88
25	8	1	−7	−15	−22	−29	−36	−44	−51	−59	−66	−74	−81	−88	−96
30	6	−2	−10	−18	−25	−33	−41	−49	−56	−64	−71	−79	−86	−93	−101
35	4	−4	−12	−20	−27	−35	−43	−52	−58	−67	−74	−82	−89	−97	−105
40	3	−5	−13	−21	−29	−37	−45	−53	−60	−69	−76	−84	−92	−100	−107
45	2	−6	−14	−22	−30	−38	−46	−54	−62	−70	−78	−85	−93	−102	−109

Source: USA Today.

Choose a temperature of 10°F. Use the formula to calculate the windchill for wind speeds of 4, 10, 25, and 40 mph. Round the results to the nearest degree. Do your results match those in the tables?

10.2 Rational Exponents

OBJECTIVE 1 Use exponential notation for *n*th roots. In mathematics we often formulate definitions so that previous rules remain valid. In Chapter 4 we defined 0 as an exponent in such a way that the rules for products, quotients, and powers would still be valid. Now we look at exponents that are rational numbers of the form $\frac{1}{n}$, where *n* is a natural number.

For the rules of exponents to remain valid, the product $(3^{1/2})^2 = 3^{1/2} \cdot 3^{1/2}$ should be found by adding exponents.

$$(3^{1/2})^2 = 3^{1/2} \cdot 3^{1/2}$$
$$= 3^{1/2+1/2}$$
$$= 3^1$$
$$= 3$$

However, by definition $(\sqrt{3})^2 = \sqrt{3} \cdot \sqrt{3} = 3$. Since both $(3^{1/2})^2$ and $(\sqrt{3})^2$ are equal to 3, it is reasonable to have

$$3^{1/2} = \sqrt{3}.$$

This suggests the following generalization.

$a^{1/n}$

If $\sqrt[n]{a}$ is a real number, then $\quad a^{1/n} = \sqrt[n]{a}.$

▮ EXAMPLE 1 Evaluating Exponentials of the Form $a^{1/n}$

Evaluate each expression.

(a) $64^{1/3} = \sqrt[3]{64} = 4$　　　　　　　**(b)** $100^{1/2} = \sqrt{100} = 10$

(c) $-256^{1/4} = -\sqrt[4]{256} = -4$

(d) $(-256)^{1/4} = \sqrt[4]{-256}$ is not a real number because the radicand, -256, is negative and the index is even.

(e) $(-32)^{1/5} = \sqrt[5]{-32} = -2$　　　　**(f)** $\left(\dfrac{1}{8}\right)^{1/3} = \sqrt[3]{\dfrac{1}{8}} = \dfrac{1}{2}$

Now Try Exercises 11, 13, 19, and 25.

CAUTION Notice the difference between parts (c) and (d) in Example 1. The radical in part (c) is the *negative fourth root* of a positive number, while the radical in part (d) is the *principal fourth root of a negative number,* which is not a real number.

OBJECTIVE 2 Define and use expressions of the form $a^{m/n}$. We know that $8^{1/3} = \sqrt[3]{8}$. How should we define a number like $8^{2/3}$? For past rules of exponents to be valid,

$$8^{2/3} = 8^{(1/3)2} = (8^{1/3})^2.$$

Since $8^{1/3} = \sqrt[3]{8}$,

$$8^{2/3} = \left(\sqrt[3]{8}\right)^2 = 2^2 = 4.$$

Generalizing from this example, we define $a^{m/n}$ as follows.

$a^{m/n}$

If m and n are positive integers with m/n in lowest terms, then

$$a^{m/n} = (a^{1/n})^m,$$

provided that $a^{1/n}$ is a real number. If $a^{1/n}$ is not a real number, then $a^{m/n}$ is not a real number.

▨ EXAMPLE 2 Evaluating Exponentials of the Form $a^{m/n}$

Evaluate each exponential.

(a) $36^{3/2} = (36^{1/2})^3 = 6^3 = 216$ **(b)** $125^{2/3} = (125^{1/3})^2 = 5^2 = 25$

(c) $-4^{5/2} = -(4^{5/2}) = -(4^{1/2})^5 = -(2)^5 = -32$

(d) $(-27)^{2/3} = [(-27)^{1/3}]^2 = (-3)^2 = 9$

Notice how the $-$ sign is used in parts (c) and (d). In part (c), we first evaluate the exponential and then find its negative. In part (d), the $-$ sign is part of the base, -27.

(e) $(-100)^{3/2}$ is not a real number since $(-100)^{1/2}$ is not a real number. ▨

Now Try Exercises 21 and 23.

When a rational exponent is negative, the earlier interpretation of negative exponents is applied.

$a^{-m/n}$

If $a^{m/n}$ is a real number, then

$$a^{-m/n} = \frac{1}{a^{m/n}} \quad (a \neq 0).$$

▨ EXAMPLE 3 Evaluating Exponentials with Negative Rational Exponents

Evaluate each exponential.

(a) $16^{-3/4}$

By the definition of a negative exponent,

$$16^{-3/4} = \frac{1}{16^{3/4}}.$$

Since $16^{3/4} = \left(\sqrt[4]{16}\right)^3 = 2^3 = 8$,

$$16^{-3/4} = \frac{1}{16^{3/4}} = \frac{1}{8}.$$

(b) $25^{-3/2} = \dfrac{1}{25^{3/2}} = \dfrac{1}{\left(\sqrt{25}\right)^3} = \dfrac{1}{5^3} = \dfrac{1}{125}$

(c) $\left(\dfrac{8}{27}\right)^{-2/3} = \dfrac{1}{\left(\dfrac{8}{27}\right)^{2/3}} = \dfrac{1}{\left(\sqrt[3]{\dfrac{8}{27}}\right)^2} = \dfrac{1}{\left(\dfrac{2}{3}\right)^2} = \dfrac{1}{\dfrac{4}{9}} = \dfrac{9}{4}$

We could also use the rule $\left(\dfrac{b}{a}\right)^{-m} = \left(\dfrac{a}{b}\right)^m$ here, as follows.

$$\left(\frac{8}{27}\right)^{-2/3} = \left(\frac{27}{8}\right)^{2/3} = \left(\sqrt[3]{\frac{27}{8}}\right)^2 = \left(\frac{3}{2}\right)^2 = \frac{9}{4}$$

Now Try Exercises 27 and 29.

> **CAUTION** When using the rule in Example 3(c), we take the reciprocal only of the base, *not* the exponent. Also, be careful to distinguish between exponential expressions like $-16^{1/4}$, $16^{-1/4}$, and $-16^{-1/4}$.
>
> $$-16^{1/4} = -2, \qquad 16^{-1/4} = \frac{1}{2}, \qquad \text{and} \qquad -16^{-1/4} = -\frac{1}{2}.$$

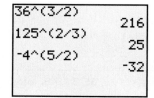

FIGURE 6

The screens in Figures 6 and 7 illustrate how a graphing calculator performs some of the evaluations seen in Examples 2 and 3. (All results on the screens are rational numbers.)

We obtain an alternative definition of $a^{m/n}$ by using the power rule for exponents a little differently than in the earlier definition. If all indicated roots are real numbers,

then $\qquad\qquad a^{m/n} = a^{m(1/n)} = (a^m)^{1/n},$

so $\qquad\qquad a^{m/n} = (a^m)^{1/n}.$

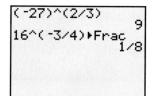

FIGURE 7

> ### $a^{m/n}$
>
> If all indicated roots are real numbers, then
>
> $$a^{m/n} = (a^{1/n})^m = (a^m)^{1/n}.$$

We can now evaluate an expression such as $27^{2/3}$ in two ways:

$$27^{2/3} = (27^{1/3})^2 = 3^2 = 9$$

or $\qquad\qquad 27^{2/3} = (27^2)^{1/3} = 729^{1/3} = 9.$

In most cases, it is easier to use $(a^{1/n})^m$.

This rule can also be expressed with radicals as follows.

> ### Radical Form of $a^{m/n}$
>
> If all indicated roots are real numbers, then
>
> $$a^{m/n} = \sqrt[n]{a^m} = \left(\sqrt[n]{a}\right)^m.$$

That is, raise a to the power and then take the root, or take the root and then raise a to the power.

For example,

$$8^{2/3} = \sqrt[3]{8^2} = \sqrt[3]{64} = 4, \qquad \text{and} \qquad 8^{2/3} = \left(\sqrt[3]{8}\right)^2 = 2^2 = 4,$$

so

$$8^{2/3} = \sqrt[3]{8^2} = \left(\sqrt[3]{8}\right)^2.$$

OBJECTIVE 3 Convert between radicals and rational exponents. Using the definition of rational exponents, we can simplify many problems involving radicals by converting the radicals to numbers with rational exponents. After simplifying, we convert the answer back to radical form.

EXAMPLE 4 Converting between Rational Exponents and Radicals

Write each exponential as a radical. Assume all variables represent positive real numbers. Use the definition that takes the root first.

(a) $13^{1/2} = \sqrt{13}$

(b) $6^{3/4} = \left(\sqrt[4]{6}\right)^3$

(c) $9m^{5/8} = 9\left(\sqrt[8]{m}\right)^5$

(d) $6x^{2/3} - (4x)^{3/5} = 6\left(\sqrt[3]{x}\right)^2 - \left(\sqrt[5]{4x}\right)^3$

(e) $r^{-2/3} = \dfrac{1}{r^{2/3}} = \dfrac{1}{\left(\sqrt[3]{r}\right)^2}$

(f) $(a^2 + b^2)^{1/2} = \sqrt{a^2 + b^2}$ Note that $\sqrt{a^2 + b^2} \neq a + b$.

In (g)–(i), write each radical as an exponential. Simplify. Assume all variables represent positive real numbers.

(g) $\sqrt{10} = 10^{1/2}$

(h) $\sqrt[4]{3^8} = 3^{8/4} = 3^2 = 9$

(i) $\sqrt[6]{z^6} = z$ since z is positive.

Now Try Exercises 33, 35, 37, 49, and 51.

NOTE In Example 4(i), it was not necessary to use absolute value bars since the directions specifically stated that the variable represents a positive real number. Since the absolute value of the positive real number z is z itself, the answer is simply z. When working exercises with radicals, we often assume that variables represent positive real numbers, which will eliminate the need for absolute value.

OBJECTIVE 4 Use the rules for exponents with rational exponents. The definition of rational exponents allows us to apply the rules for exponents first introduced in Chapter 4.

Rules for Rational Exponents

Let r and s be rational numbers. For all real numbers a and b for which the indicated expressions exist:

$$a^r \cdot a^s = a^{r+s} \qquad a^{-r} = \frac{1}{a^r} \qquad \frac{a^r}{a^s} = a^{r-s} \qquad \left(\frac{a}{b}\right)^{-r} = \frac{b^r}{a^r}$$

$$(a^r)^s = a^{rs} \qquad (ab)^r = a^r b^r \qquad \left(\frac{a}{b}\right)^r = \frac{a^r}{b^r} \qquad a^{-r} = \left(\frac{1}{a}\right)^r.$$

▮ **EXAMPLE 5** Applying Rules for Rational Exponents

Write with only positive exponents. Assume all variables represent positive real numbers.

(a) $2^{1/2} \cdot 2^{1/4} = 2^{1/2+1/4} = 2^{3/4}$ Product rule

(b) $\dfrac{5^{2/3}}{5^{7/3}} = 5^{2/3-7/3} = 5^{-5/3} = \dfrac{1}{5^{5/3}}$ Quotient rule

(c) $\dfrac{(x^{1/2}y^{2/3})^4}{y} = \dfrac{(x^{1/2})^4(y^{2/3})^4}{y}$ Power rule

$\quad\quad = \dfrac{x^2 y^{8/3}}{y^1}$ Power rule

$\quad\quad = x^2 y^{8/3-1}$ Quotient rule

$\quad\quad = x^2 y^{5/3}$

(d) $\left(\dfrac{x^4 y^{-6}}{x^{-2} y^{1/3}}\right)^{-2/3} = \dfrac{(x^4)^{-2/3}(y^{-6})^{-2/3}}{(x^{-2})^{-2/3}(y^{1/3})^{-2/3}}$

$\quad\quad = \dfrac{x^{-8/3} y^4}{x^{4/3} y^{-2/9}}$ Power rule

$\quad\quad = x^{-8/3-4/3} y^{4-(-2/9)}$ Quotient rule

$\quad\quad = x^{-4} y^{38/9}$

$\quad\quad = \dfrac{y^{38/9}}{x^4}$ Definition of negative exponent

The same result is obtained if we simplify within the parentheses first, leading to

$$(x^6 y^{-19/3})^{-2/3}.$$

Then, apply the power rule. (Show that the result is the same.)

(e) $m^{3/4}(m^{5/4} - m^{1/4}) = m^{3/4} \cdot m^{5/4} - m^{3/4} \cdot m^{1/4}$ Distributive property

$\quad\quad = m^{3/4+5/4} - m^{3/4+1/4}$ Product rule

$\quad\quad = m^{8/4} - m^{4/4}$

$\quad\quad = m^2 - m$

Do not make the common mistake of multiplying exponents in the first step.   ▮

Now Try Exercises 57, 59, 65, 75, and 77.

CAUTION Use the rules of exponents in problems like those in Example 5. Do not convert the expressions to radical form.

▮ **EXAMPLE 6** Applying Rules for Rational Exponents

Rewrite all radicals as exponentials, and then apply the rules for rational exponents. Leave answers in exponential form. Assume all variables represent positive real numbers.

(a) $\sqrt[3]{x^2} \cdot \sqrt[4]{x} = x^{2/3} \cdot x^{1/4}$ Convert to rational exponents.

$\qquad\qquad\qquad = x^{2/3+1/4}$ Product rule

$\qquad\qquad\qquad = x^{8/12+3/12}$ Write exponents with a common denominator.

$\qquad\qquad\qquad = x^{11/12}$

(b) $\dfrac{\sqrt{x^3}}{\sqrt[3]{x^2}} = \dfrac{x^{3/2}}{x^{2/3}} = x^{3/2-2/3} = x^{5/6}$

(c) $\sqrt{\sqrt[4]{z}} = \sqrt{z^{1/4}} = (z^{1/4})^{1/2} = z^{1/8}$

Now Try Exercises 83, 85, and 89.

10.2   EXERCISES

Match each expression from Column I with the equivalent choice from Column II.

I	II
1. $2^{1/2}$	**A.** -4
2. $(-27)^{1/3}$	**B.** 8
3. $-16^{1/2}$	**C.** $\sqrt{2}$
4. $(-16)^{1/2}$	**D.** $-\sqrt{6}$
5. $(-32)^{1/5}$	**E.** -3
6. $(-32)^{2/5}$	**F.** $\sqrt{6}$
7. $4^{3/2}$	**G.** 4
8. $6^{2/4}$	**H.** -2
9. $-6^{2/4}$	**I.** 6
10. $36^{.5}$	**J.** Not a real number

Simplify each expression involving rational exponents. See Examples 1–3.

11. $169^{1/2}$ **12.** $121^{1/2}$ **13.** $729^{1/3}$ **14.** $512^{1/3}$

15. $16^{1/4}$ **16.** $625^{1/4}$ **17.** $\left(\dfrac{64}{81}\right)^{1/2}$ **18.** $\left(\dfrac{8}{27}\right)^{1/3}$

19. $(-27)^{1/3}$ **20.** $(-32)^{1/5}$ **21.** $100^{3/2}$ **22.** $64^{3/2}$

23. $-16^{5/2}$ **24.** $-32^{3/5}$ **25.** $(-144)^{1/2}$ **26.** $(-36)^{1/2}$

27. $64^{-3/2}$ **28.** $81^{-3/2}$ **29.** $\left(-\dfrac{8}{27}\right)^{-2/3}$ **30.** $\left(-\dfrac{64}{125}\right)^{-2/3}$

 31. Explain why $(-64)^{1/2}$ is not a real number, while $-64^{1/2}$ is a real number.

32. Explain why $a^{1/n}$ is defined to be equal to $\sqrt[n]{a}$ when $\sqrt[n]{a}$ is real.

Write with radicals. Assume all variables represent positive real numbers. See Example 4.

33. $12^{1/2}$ **34.** $3^{1/2}$ **35.** $8^{3/4}$

36. $7^{2/3}$ **37.** $(9q)^{5/8} - (2x)^{2/3}$ **38.** $(3p)^{3/4} + (4x)^{1/3}$

39. $(2m)^{-3/2}$ **40.** $(5y)^{-3/5}$ **41.** $(2y + x)^{2/3}$

42. $(r + 2z)^{3/2}$ **43.** $(3m^4 + 2k^2)^{-2/3}$ **44.** $(5x^2 + 3z^3)^{-5/6}$

45. Show that, in general, $\sqrt{a^2 + b^2} \neq a + b$ by replacing a with 3 and b with 4.

46. Suppose someone claims that $\sqrt[n]{a^n + b^n}$ must equal $a + b$, since when $a = 1$ and $b = 0$, a true statement results:

$$\sqrt[n]{a^n + b^n} = \sqrt[n]{1^n + 0^n} = \sqrt[n]{1^n} = 1 = 1 + 0 = a + b.$$

Explain why this is faulty reasoning.

Simplify by first converting to rational exponents. Assume all variables represent positive real numbers. See Example 4.

47. $\sqrt{2^{12}}$ **48.** $\sqrt{5^{10}}$ **49.** $\sqrt[3]{4^9}$ **50.** $\sqrt[4]{6^8}$ **51.** $\sqrt{x^{20}}$

52. $\sqrt{r^{50}}$ **53.** $\sqrt[3]{x} \cdot \sqrt{x}$ **54.** $\sqrt[4]{y} \cdot \sqrt[5]{y^2}$ **55.** $\dfrac{\sqrt[3]{t^4}}{\sqrt[5]{t^4}}$ **56.** $\dfrac{\sqrt[4]{w^3}}{\sqrt[6]{w}}$

Use the rules of exponents to simplify each expression. Write all answers with positive exponents. Assume all variables represent positive real numbers. See Example 5.

57. $3^{1/2} \cdot 3^{3/2}$ **58.** $6^{4/3} \cdot 6^{2/3}$ **59.** $\dfrac{64^{5/3}}{64^{4/3}}$

60. $\dfrac{125^{7/3}}{125^{5/3}}$ **61.** $y^{7/3} \cdot y^{-4/3}$ **62.** $r^{-8/9} \cdot r^{17/9}$

63. $\dfrac{k^{1/3}}{k^{2/3} \cdot k^{-1}}$ **64.** $\dfrac{z^{3/4}}{z^{5/4} \cdot z^{-2}}$ **65.** $\dfrac{(x^{1/4}y^{2/5})^{20}}{x^2}$

66. $\dfrac{(r^{1/5}s^{2/3})^{15}}{r^2}$ **67.** $\dfrac{(x^{2/3})^2}{(x^2)^{7/3}}$ **68.** $\dfrac{(p^3)^{1/4}}{(p^{5/4})^2}$

69. $\dfrac{m^{3/4}n^{-1/4}}{(m^2n)^{1/2}}$ **70.** $\dfrac{(a^2b^5)^{-1/4}}{(a^{-3}b^2)^{1/6}}$ **71.** $\dfrac{p^{1/5}p^{7/10}p^{1/2}}{(p^3)^{-1/5}}$

72. $\dfrac{z^{1/3}z^{-2/3}z^{1/6}}{(z^{-1/6})^3}$ **73.** $\left(\dfrac{b^{-3/2}}{c^{-5/3}}\right)^2 (b^{-1/4}c^{-1/3})^{-1}$ **74.** $\left(\dfrac{m^{-2/3}}{a^{-3/4}}\right)^4 (m^{-3/8}a^{1/4})^{-2}$

75. $\left(\dfrac{p^{-1/4}q^{-3/2}}{3^{-1}p^{-2}q^{-2/3}}\right)^{-2}$ **76.** $\left(\dfrac{2^{-2}w^{-3/4}x^{-5/8}}{w^{3/4}x^{-1/2}}\right)^{-3}$ **77.** $p^{2/3}(p^{1/3} + 2p^{4/3})$

78. $z^{5/8}(3z^{5/8} + 5z^{11/8})$ **79.** $k^{1/4}(k^{3/2} - k^{1/2})$ **80.** $r^{3/5}(r^{1/2} + r^{3/4})$

81. $6a^{7/4}(a^{-7/4} + 3a^{-3/4})$ **82.** $4m^{5/3}(m^{-2/3} - 4m^{-5/3})$

Write with rational exponents, and then apply the properties of exponents. Assume all radicands represent positive real numbers. Give answers in exponential form. See Example 6.

83. $\sqrt[5]{x^3} \cdot \sqrt[4]{x}$ **84.** $\sqrt[6]{y^5} \cdot \sqrt[3]{y^2}$ **85.** $\dfrac{\sqrt{x^5}}{\sqrt{x^8}}$ **86.** $\dfrac{\sqrt[3]{k^5}}{\sqrt[3]{k^7}}$

87. $\sqrt{y} \cdot \sqrt[3]{yz}$ **88.** $\sqrt[3]{xz} \cdot \sqrt{z}$ **89.** $\sqrt[4]{\sqrt[3]{m}}$ **90.** $\sqrt[3]{\sqrt{k}}$

91. $\sqrt{\sqrt[3]{\sqrt[4]{x}}}$ **92.** $\sqrt[3]{\sqrt[5]{\sqrt{y}}}$

Solve each problem.

93. Meteorologists can determine the duration of a storm by using the function defined by

$$T(D) = .07D^{3/2},$$

where D is the diameter of the storm in miles and T is the time in hours. Find the duration of a storm with a diameter of 16 mi. Round your answer to the nearest tenth of an hour.

94. The threshold weight T, in pounds, for a person is the weight above which the risk of death increases greatly. The threshold weight in pounds for men aged 40–49 is related to height in inches by the function defined by

$$h(T) = (1860.867T)^{1/3}.$$

What height corresponds to a threshold weight of 200 lb for a 46-yr-old man? Round your answer to the nearest inch, and then to the nearest tenth of a foot.

RELATING CONCEPTS (EXERCISES 95–102)

For Individual or Group Work

Earlier, we factored expressions like $x^4 - x^5$ by factoring out the greatest common factor to get

$$x^4 - x^5 = x^4(1 - x).$$

We can adapt this approach to factor expressions with rational exponents. When one or more of the exponents is negative or a fraction, we use order on the number line discussed in Chapter 1 to decide on the common factor. In this type of factoring, we want the binomial factor to have only positive exponents, so we always factor out the variable with the least *exponent. A positive exponent is greater than a negative exponent, so in $7z^{5/8} + z^{-3/4}$, we factor out $z^{-3/4}$, because $-\frac{3}{4}$ is less than $\frac{5}{8}$.*

Factor out the given common factor from each expression. Assume all variables represent positive real numbers.

95. $3x^{-1/2} - 4x^{1/2}$; $x^{-1/2}$

96. $m^3 - 3m^{5/2}$; $m^{5/2}$

97. $4t^{-1/2} + 7t^{3/2}$; $t^{-1/2}$

98. $8x^{2/3} + 5x^{-1/3}$; $x^{-1/3}$

99. $4p - p^{3/4}$; $p^{3/4}$

100. $2m^{1/8} - m^{5/8}$; $m^{1/8}$

101. $9k^{-3/4} - 2k^{-1/4}$; $k^{-3/4}$

102. $7z^{-5/8} - z^{-3/4}$; $z^{-3/4}$

10.3 Simplifying Radical Expressions

OBJECTIVE 1 Use the product rule for radicals.   We now develop rules for multiplying and dividing radicals that have the same index. For example, is the product of two nth-root radicals equal to the nth root of the product of the radicands? Are the expressions $\sqrt{36 \cdot 4}$ and $\sqrt{36} \cdot \sqrt{4}$ equal? To find out, we do the computations:

$$\sqrt{36 \cdot 4} = \sqrt{144} = 12$$
$$\sqrt{36} \cdot \sqrt{4} = 6 \cdot 2 = 12.$$

Notice that in both cases the result is the same. This is an example of the **product rule for radicals.**

Product Rule for Radicals

If $\sqrt[n]{a}$ and $\sqrt[n]{b}$ are real numbers and n is a natural number, then

$$\sqrt[n]{a} \cdot \sqrt[n]{b} = \sqrt[n]{ab}.$$

That is, the product of two radicals is the radical of the product.

We justify the product rule using the rules for rational exponents. Since $\sqrt[n]{a} = a^{1/n}$ and $\sqrt[n]{b} = b^{1/n}$,

$$\sqrt[n]{a} \cdot \sqrt[n]{b} = a^{1/n} \cdot b^{1/n} = (ab)^{1/n} = \sqrt[n]{ab}.$$

CAUTION   Use the product rule only when the radicals have the *same* index.

■ EXAMPLE 1 Using the Product Rule

Multiply. Assume all variables represent positive real numbers.

(a) $\sqrt{5} \cdot \sqrt{7} = \sqrt{5 \cdot 7} = \sqrt{35}$ **(b)** $\sqrt{2} \cdot \sqrt{19} = \sqrt{2 \cdot 19} = \sqrt{38}$

(c) $\sqrt{11} \cdot \sqrt{p} = \sqrt{11p}$ **(d)** $\sqrt{7} \cdot \sqrt{11xyz} = \sqrt{77xyz}$

Now Try Exercises 7, 9, and 11.

■ EXAMPLE 2 Using the Product Rule

Multiply. Assume all variables represent positive real numbers.

(a) $\sqrt[3]{3} \cdot \sqrt[3]{12} = \sqrt[3]{3 \cdot 12} = \sqrt[3]{36}$ **(b)** $\sqrt[4]{8y} \cdot \sqrt[4]{3r^2} = \sqrt[4]{24yr^2}$

(c) $\sqrt[6]{10m^4} \cdot \sqrt[6]{5m} = \sqrt[6]{50m^5}$

(d) $\sqrt[4]{2} \cdot \sqrt[5]{2}$ cannot be simplified using the product rule for radicals because the indexes (4 and 5) are different.

Now Try Exercises 13, 15, 17, and 19.

OBJECTIVE 2 Use the quotient rule for radicals.   The **quotient rule for radicals** is similar to the product rule.

Quotient Rule for Radicals

If $\sqrt[n]{a}$ and $\sqrt[n]{b}$ are real numbers, $b \neq 0$, and n is a natural number, then

$$\sqrt[n]{\frac{a}{b}} = \frac{\sqrt[n]{a}}{\sqrt[n]{b}}.$$

That is, the radical of a quotient is the quotient of the radicals.

EXAMPLE 3   Using the Quotient Rule

Simplify. Assume all variables represent positive real numbers.

(a) $\sqrt{\dfrac{16}{25}} = \dfrac{\sqrt{16}}{\sqrt{25}} = \dfrac{4}{5}$

(b) $\sqrt{\dfrac{7}{36}} = \dfrac{\sqrt{7}}{\sqrt{36}} = \dfrac{\sqrt{7}}{6}$

(c) $\sqrt[3]{-\dfrac{8}{125}} = \sqrt[3]{\dfrac{-8}{125}} = \dfrac{\sqrt[3]{-8}}{\sqrt[3]{125}} = \dfrac{-2}{5} = -\dfrac{2}{5}$

(d) $\sqrt[3]{\dfrac{7}{216}} = \dfrac{\sqrt[3]{7}}{\sqrt[3]{216}} = \dfrac{\sqrt[3]{7}}{6}$

(e) $\sqrt[5]{\dfrac{x}{32}} = \dfrac{\sqrt[5]{x}}{\sqrt[5]{32}} = \dfrac{\sqrt[5]{x}}{2}$

(f) $-\sqrt[3]{\dfrac{m^6}{125}} = -\dfrac{\sqrt[3]{m^6}}{\sqrt[3]{125}} = -\dfrac{m^2}{5}$

Now Try Exercises 23, 25, 31, 33, and 35.

OBJECTIVE 3 Simplify radicals.   We use the product and quotient rules to simplify radicals. A radical is **simplified** if the following four conditions are met.

Simplified Radical

1. The radicand has no factor raised to a power greater than or equal to the index.
2. The radicand has no fractions.
3. No denominator contains a radical.
4. Exponents in the radicand and the index of the radical have no common factor (except 1).

EXAMPLE 4   Simplifying Roots of Numbers

Simplify.

(a) $\sqrt{24}$

Check to see whether 24 is divisible by a perfect square (the square of a natural number) such as 4, 9, ... Choose the largest perfect square that divides into 24. The largest such number is 4. Write 24 as the product of 4 and 6, and then use the product rule.

$$\sqrt{24} = \sqrt{4 \cdot 6} = \sqrt{4} \cdot \sqrt{6} = 2\sqrt{6}$$

(b) $\sqrt{108}$

The number 108 is divisible by the perfect square 36: $\sqrt{108} = \sqrt{36 \cdot 3}$. If this is not obvious, try factoring 108 into its prime factors.

$$\sqrt{108} = \sqrt{2^2 \cdot 3^3}$$
$$= \sqrt{2^2 \cdot 3^2 \cdot 3}$$
$$= 2 \cdot 3 \cdot \sqrt{3} \qquad \text{Product rule}$$
$$= 6\sqrt{3}$$

(c) $\sqrt{10}$

No perfect square (other than 1) divides into 10, so $\sqrt{10}$ cannot be simplified further.

(d) $\sqrt[3]{16}$

Look for the largest perfect *cube* that divides into 16. The number 8 satisfies this condition, so write 16 as $8 \cdot 2$ (or factor 16 into prime factors).

$$\sqrt[3]{16} = \sqrt[3]{8 \cdot 2} = \sqrt[3]{8} \cdot \sqrt[3]{2} = 2\sqrt[3]{2}$$

(e) $-\sqrt[4]{162} = -\sqrt[4]{81 \cdot 2} \qquad$ 81 is a perfect 4th power.

$$= -\sqrt[4]{81} \cdot \sqrt[4]{2} \qquad \text{Product rule}$$
$$= -3\sqrt[4]{2}$$

Now Try Exercises 39, 41, 49, and 55.

CAUTION In simplifying an expression like that in Example 4(b), be careful with which factors belong *outside* the radical sign and which belong *inside*. Note how $2 \cdot 3$ is written outside because $\sqrt{2^2} = 2$ and $\sqrt{3^2} = 3$, while the remaining 3 is left inside the radical.

EXAMPLE 5 Simplifying Radicals Involving Variables

Simplify. Assume all variables represent positive real numbers.

(a) $\sqrt{16m^3} = \sqrt{16m^2 \cdot m}$

$$= \sqrt{16m^2} \cdot \sqrt{m}$$
$$= 4m\sqrt{m}$$

No absolute value bars are needed around the m in color because of the assumption that all the variables represent *positive* real numbers.

(b) $\sqrt{200k^7q^8} = \sqrt{10^2 \cdot 2 \cdot (k^3)^2 \cdot k \cdot (q^4)^2} \qquad$ Factor.

$$= 10k^3q^4\sqrt{2k} \qquad \text{Remove perfect square factors.}$$

(c) $\sqrt[3]{8x^4y^5} = \sqrt[3]{(8x^3y^3)(xy^2)} \qquad 8x^3y^3$ is the largest perfect cube that divides $8x^4y^5$.

$$= \sqrt[3]{8x^3y^3} \cdot \sqrt[3]{xy^2}$$
$$= 2xy\sqrt[3]{xy^2}$$

(d) $-\sqrt[4]{32y^9} = -\sqrt[4]{(16y^8)(2y)}$ $16y^8$ is the largest 4th power that divides $32y^9$.

$$= -\sqrt[4]{16y^8} \cdot \sqrt[4]{2y}$$

$$= -2y^2\sqrt[4]{2y}$$

Now Try Exercises 75, 79, 83, and 87.

NOTE From Example 5 we see that if a variable is raised to a power with an exponent divisible by 2, it is a perfect square. If it is raised to a power with an exponent divisible by 3, it is a perfect cube. In general, if it is raised to a power with an exponent divisible by n, it is a perfect nth power.

The conditions for a simplified radical given earlier state that an exponent in the radicand and the index of the radical should have no common factor (except 1). The next example shows how to simplify radicals with such common factors.

EXAMPLE 6   Simplifying Radicals by Using Smaller Indexes

Simplify. Assume all variables represent positive real numbers.

(a) $\sqrt[9]{5^6}$

We can write this radical using rational exponents and then write the exponent in lowest terms. We then express the answer as a radical.

$$\sqrt[9]{5^6} = 5^{6/9} = 5^{2/3} = \sqrt[3]{5^2} \quad \text{or} \quad \sqrt[3]{25}$$

(b) $\sqrt[4]{p^2} = p^{2/4} = p^{1/2} = \sqrt{p}$ (Recall the assumption that $p > 0$.)

Now Try Exercises 93 and 97.

These examples suggest the following rule.

If m is an integer, n and k are natural numbers, and all indicated roots exist, then

$$\sqrt[kn]{a^{km}} = \sqrt[n]{a^m}.$$

OBJECTIVE 4 Simplify products and quotients of radicals with different indexes. Since the product and quotient rules for radicals apply only when they have the same index, we multiply and divide radicals with different indexes by using rational exponents.

EXAMPLE 7   Multiplying Radicals with Different Indexes

Simplify $\sqrt{7} \cdot \sqrt[3]{2}$.

Because the different indexes, 2 and 3, have a least common index of 6, use rational exponents to write each radical as a sixth root.

$$\sqrt{7} = 7^{1/2} = 7^{3/6} = \sqrt[6]{7^3} = \sqrt[6]{343}$$

$$\sqrt[3]{2} = 2^{1/3} = 2^{2/6} = \sqrt[6]{2^2} = \sqrt[6]{4}$$

Therefore,

$$\sqrt{7} \cdot \sqrt[3]{2} = \sqrt[6]{343} \cdot \sqrt[6]{4} = \sqrt[6]{1372}. \quad \text{Product rule}$$

Now Try Exercise 99.

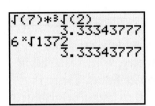

FIGURE 8

Results such as the one in Example 7 can be supported using a calculator, as shown in Figure 8. Notice that the calculator gives the same approximation for the initial product and the final radical that we obtained.

> **CAUTION** The computation in Figure 8 is not *proof* that the two expressions are equal. The algebra in Example 7, however, is valid proof of their equality.

OBJECTIVE 5 Use the Pythagorean formula. Recall from Section 5.6 that the **Pythagorean formula** relates the lengths of the three sides of a right triangle.

Pythagorean Formula

If c is the length of the longest side of a right triangle and a and b are the lengths of the shorter sides, then

$$c^2 = a^2 + b^2.$$

The longest side is the **hypotenuse** and the two shorter sides are the **legs** of the triangle. The hypotenuse is the side opposite the right angle.

▨ **EXAMPLE 8 Using the Pythagorean Formula**

Use the Pythagorean formula to find the length of the hypotenuse in the triangle in Figure 9.

To find the length of the hypotenuse c, let $a = 4$ and $b = 6$. Then use the formula.

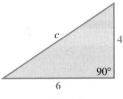

FIGURE 9

$$c^2 = a^2 + b^2$$
$$c^2 = 4^2 + 6^2 \qquad \text{Let } a = 4 \text{ and } b = 6.$$
$$c^2 = 52$$
$$c = \sqrt{52} \qquad \text{Choose the principal root.}$$
$$c = \sqrt{4 \cdot 13} \qquad \text{Factor.}$$
$$c = \sqrt{4} \cdot \sqrt{13} \qquad \text{Product rule}$$
$$c = 2\sqrt{13}$$

The length of the hypotenuse is $2\sqrt{13}$.

▨

Now Try Exercise 109.

> **CAUTION** When using the equation $c^2 = a^2 + b^2$, be sure that the length of the hypotenuse is substituted for c, and that the lengths of the legs are substituted for a and b. Errors often occur because values are substituted incorrectly.

| | **CONNECTIONS** | |

The Pythagorean formula is undoubtedly one of the most widely used and oldest formulas we have. It is very important in trigonometry, which is used in surveying, drafting, engineering, navigation, and many other fields. There is evidence that the Babylonians knew the concept quite well. Although attributed to Pythagoras, it was known to every surveyor from Egypt to China for a thousand years before Pythagoras. In the 1939 movie *The Wizard of Oz,* the Scarecrow asks the Wizard for a brain. When the Wizard presents him with a diploma granting him a Th.D. (Doctor of Thinkology), the Scarecrow recites the following:

> The sum of the square roots of any two sides of an isosceles triangle is equal to the square root of the remaining side. . . .
> Oh joy! Rapture! I've got a brain.

For Discussion or Writing

Did the Scarecrow recite the Pythagorean formula? (An *isosceles triangle* is a triangle with two equal sides.) Is his statement true? Explain.

■ **EXAMPLE 9** Using a Formula from Electronics

The impedance Z of an alternating series circuit is given by the formula

$$Z = \sqrt{R^2 + X^2},$$

where R is the resistance and X is the reactance, both in ohms. Find the value of the impedance if $R = 40$ ohms and $X = 30$ ohms.

Substitute 40 for R and 30 for X in the formula.

$$Z = \sqrt{R^2 + X^2} \qquad \text{Given formula}$$
$$= \sqrt{40^2 + 30^2} \qquad \text{Let } R = 40 \text{ and } X = 30.$$
$$= \sqrt{1600 + 900}$$
$$= \sqrt{2500}$$
$$= 50$$

The impedance is 50 ohms.

Now Try Exercise 113.

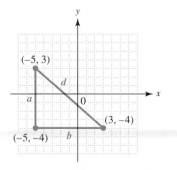

FIGURE 10

OBJECTIVE 6 Use the distance formula. An important result in algebra is derived by using the Pythagorean formula. The *distance formula* allows us to find the distance between two points in the coordinate plane, or the length of the line segment joining those two points. Figure 10 shows the points $(3, -4)$ and $(-5, 3)$. The vertical line through $(-5, 3)$ and the horizontal line through $(3, -4)$ intersect at the point $(-5, -4)$. Thus, the point $(-5, -4)$ becomes the vertex of the right angle in a right triangle. By the Pythagorean formula, the square of the length of the hypotenuse, d, of the right triangle in Figure 10 is equal to the sum of the squares of the lengths of the two legs a and b:

$$d^2 = a^2 + b^2.$$

The length a is the difference between the y-coordinates of the endpoints. Since the x-coordinate of both points in Figure 10 is -5, the side is vertical, and we can find a by finding the difference between the y-coordinates. We subtract -4 from 3 to get a positive value for a.

$$a = 3 - (-4) = 7$$

Similarly, we find b by subtracting -5 from 3.

$$b = 3 - (-5) = 8$$

Substituting these values into the formula, we obtain

$$d^2 = a^2 + b^2$$
$$d^2 = 7^2 + 8^2 \qquad \text{Let } a = 7 \text{ and } b = 8.$$
$$d^2 = 49 + 64$$
$$d^2 = 113$$
$$d = \sqrt{113}.$$

We choose the principal root since distance cannot be negative. Therefore, the distance between $(-5, 3)$ and $(3, -4)$ is $\sqrt{113}$.

NOTE It is customary to leave the distance in radical form. Do not use a calculator to get an approximation unless you are specifically directed to do so.

This result can be generalized. Figure 11 shows the two points (x_1, y_1) and (x_2, y_2). To find a formula for the distance d between these two points, notice that the distance between (x_1, y_1) and (x_2, y_1) is given by

$$a = |x_2 - x_1|,$$

and the distance between (x_2, y_2) and (x_2, y_1) is given by

$$b = |y_2 - y_1|.$$

From the Pythagorean formula,

$$d^2 = a^2 + b^2$$
$$d^2 = (x_2 - x_1)^2 + (y_2 - y_1)^2. \qquad |p - q|^2 = (p - q)^2$$

Choosing the principal square root gives the **distance formula.**

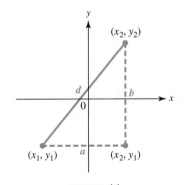

FIGURE 11

Distance Formula

The distance between the points (x_1, y_1) and (x_2, y_2) is

$$d = \sqrt{(x_2 - x_1)^2 + (y_2 - y_1)^2}.$$

EXAMPLE 10 Using the Distance Formula

Find the distance between the points $(-3, 5)$ and $(6, 4)$.

When using the distance formula to find the distance between two points, designating the points as (x_1, y_1) and (x_2, y_2) is arbitrary. We choose $(x_1, y_1) = (-3, 5)$ and $(x_2, y_2) = (6, 4)$.

$$d = \sqrt{(x_2 - x_1)^2 + (y_2 - y_1)^2}$$
$$= \sqrt{[6 - (-3)]^2 + (4 - 5)^2} \qquad x_2 = 6, \ y_2 = 4, \ x_1 = -3, \ y_1 = 5$$
$$= \sqrt{9^2 + (-1)^2}$$
$$= \sqrt{82}$$

Now Try Exercise 121.

10.3 EXERCISES

For Extra Help

 Student's
Solutions Manual

 MyMathLab

 InterAct Math
Tutorial Software

 AW Math
Tutor Center

MathXL MathXL

 Digital Video Tutor
CD 14/Videotape 14

Decide whether each statement is true *or* false *by using the product rule explained in this section. Then support your answer by finding a calculator approximation for each expression.*

1. $2\sqrt{12} = \sqrt{48}$ **2.** $\sqrt{72} = 2\sqrt{18}$

3. $3\sqrt{8} = 2\sqrt{18}$ **4.** $5\sqrt{72} = 6\sqrt{50}$

5. Which one of the following is *not* equal to $\sqrt{\frac{1}{2}}$? (Do not use calculator approximations.)

     A. $\sqrt{.5}$ **B.** $\sqrt{\frac{2}{4}}$ **C.** $\sqrt{\frac{3}{6}}$ **D.** $\frac{\sqrt{4}}{\sqrt{16}}$

6. Use the π key on your calculator to get a value for π. Now find an approximation for $\sqrt[4]{\frac{2143}{22}}$. Does the result mean that π is actually equal to $\sqrt[4]{\frac{2143}{22}}$? Why or why not?

Multiply using the product rule. Assume all variables represent positive real numbers. See Examples 1 and 2.

7. $\sqrt{5} \cdot \sqrt{6}$ **8.** $\sqrt{10} \cdot \sqrt{3}$ **9.** $\sqrt{14} \cdot \sqrt{x}$ **10.** $\sqrt{23} \cdot \sqrt{t}$

11. $\sqrt{14} \cdot \sqrt{3pqr}$ **12.** $\sqrt{7} \cdot \sqrt{5xt}$ **13.** $\sqrt[3]{7x} \cdot \sqrt[3]{2y}$ **14.** $\sqrt[3]{9x} \cdot \sqrt[3]{4y}$

15. $\sqrt[4]{11} \cdot \sqrt[4]{3}$ **16.** $\sqrt[4]{6} \cdot \sqrt[4]{9}$ **17.** $\sqrt[4]{2x} \cdot \sqrt[4]{3y^2}$ **18.** $\sqrt[4]{3y^2} \cdot \sqrt[4]{6yz}$

19. $\sqrt[3]{7} \cdot \sqrt[4]{3}$ **20.** $\sqrt[5]{8} \cdot \sqrt[6]{12}$

21. Explain the product rule for radicals in your own words. Give examples.

22. Explain the quotient rule for radicals in your own words. Give examples.

Simplify each radical. Assume all variables represent positive real numbers. See Example 3.

23. $\sqrt{\frac{64}{121}}$ **24.** $\sqrt{\frac{16}{49}}$ **25.** $\sqrt{\frac{3}{25}}$ **26.** $\sqrt{\frac{13}{49}}$

27. $\sqrt{\frac{x}{25}}$ **28.** $\sqrt{\frac{k}{100}}$ **29.** $\sqrt{\frac{p^6}{81}}$ **30.** $\sqrt{\frac{w^{10}}{36}}$

31. $\sqrt[3]{\frac{27}{64}}$ **32.** $\sqrt[3]{\frac{216}{125}}$ **33.** $\sqrt[3]{-\frac{r^2}{8}}$ **34.** $\sqrt[3]{-\frac{t}{125}}$

35. $-\sqrt[4]{\frac{81}{x^4}}$ **36.** $-\sqrt[4]{\frac{625}{y^4}}$ **37.** $\sqrt[5]{\frac{1}{x^{15}}}$ **38.** $\sqrt[5]{\frac{32}{y^{20}}}$

Express each radical in simplified form. See Example 4.

39. $\sqrt{12}$ **40.** $\sqrt{18}$ **41.** $\sqrt{288}$ **42.** $\sqrt{72}$ **43.** $-\sqrt{32}$

44. $-\sqrt{48}$ **45.** $-\sqrt{28}$ **46.** $-\sqrt{24}$ **47.** $\sqrt{-300}$ **48.** $\sqrt{-150}$

49. $\sqrt[3]{128}$ **50.** $\sqrt[3]{24}$ **51.** $\sqrt[3]{-16}$ **52.** $\sqrt[3]{-250}$ **53.** $\sqrt[3]{40}$

54. $\sqrt[3]{375}$ **55.** $-\sqrt[4]{512}$ **56.** $-\sqrt[4]{1250}$ **57.** $\sqrt[5]{64}$ **58.** $\sqrt[5]{128}$

59. A student claimed that $\sqrt[3]{14}$ is not in simplified form, since $14 = 8 + 6$, and 8 is a perfect cube. Was his reasoning correct? Why or why not?

60. Explain in your own words why $\sqrt[3]{k^4}$ is not a simplified radical.

Express each radical in simplified form. Assume all variables represent positive real numbers. See Example 5.

61. $\sqrt{72k^2}$ **62.** $\sqrt{18m^2}$ **63.** $\sqrt{144x^3y^9}$

64. $\sqrt{169s^5t^{10}}$ **65.** $\sqrt{121x^6}$ **66.** $\sqrt{256z^{12}}$

67. $-\sqrt[3]{27t^{12}}$ **68.** $-\sqrt[3]{64y^{18}}$ **69.** $-\sqrt{100m^8z^4}$

70. $-\sqrt{25t^6s^{20}}$ **71.** $-\sqrt[3]{-125a^6b^9c^{12}}$ **72.** $-\sqrt[3]{-216y^{15}x^6z^3}$

73. $\sqrt[4]{\dfrac{1}{16}r^8t^{20}}$ **74.** $\sqrt[4]{\dfrac{81}{256}t^{12}u^8}$ **75.** $\sqrt{50x^3}$ **76.** $\sqrt{300z^3}$

77. $-\sqrt{500r^{11}}$ **78.** $-\sqrt{200p^{13}}$ **79.** $\sqrt{13x^7y^8}$ **80.** $\sqrt{23k^9p^{14}}$

81. $\sqrt[3]{8z^6w^9}$ **82.** $\sqrt[3]{64a^{15}b^{12}}$ **83.** $\sqrt[3]{-16z^5t^7}$ **84.** $\sqrt[3]{-81m^4n^{10}}$

85. $\sqrt[4]{81x^{12}y^{16}}$ **86.** $\sqrt[4]{81t^8u^{28}}$ **87.** $-\sqrt[4]{162r^{15}s^{10}}$ **88.** $-\sqrt[4]{32k^5m^{10}}$

89. $\sqrt{\dfrac{y^{11}}{36}}$ **90.** $\sqrt{\dfrac{v^{13}}{49}}$ **91.** $\sqrt[3]{\dfrac{x^{16}}{27}}$ **92.** $\sqrt[3]{\dfrac{y^{17}}{125}}$

Simplify each radical. Assume that $x \geq 0$. See Example 6.

93. $\sqrt[4]{48^2}$ **94.** $\sqrt[4]{50^2}$ **95.** $\sqrt[4]{25}$

96. $\sqrt[6]{8}$ **97.** $\sqrt[10]{x^{25}}$ **98.** $\sqrt[12]{x^{44}}$

Simplify by first writing the radicals as radicals with the same index. Then multiply. Assume all variables represent positive real numbers. See Example 7.

99. $\sqrt[3]{4} \cdot \sqrt{3}$ **100.** $\sqrt[3]{5} \cdot \sqrt{6}$ **101.** $\sqrt[4]{3} \cdot \sqrt[3]{4}$

102. $\sqrt[5]{7} \cdot \sqrt[7]{5}$ **103.** $\sqrt{x} \cdot \sqrt[3]{x}$ **104.** $\sqrt[3]{y} \cdot \sqrt[4]{y}$

TECHNOLOGY INSIGHTS (EXERCISES 105–108)

A graphing calculator can be used to test whether two quantities are equal. In the screen shown here, the first two lines of entries both represent true statements, and thus the calculator returns a 1 to indicate true. The third entry is false, and the calculator returns a 0. These can be verified algebraically using the rules for radicals found in this section.

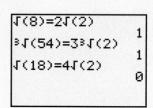

Determine whether the calculator should return a 1 or a 0 for each screen.

105.

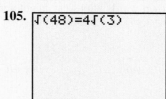

106.

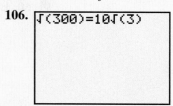

107.

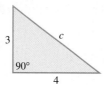

108.

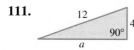

Find the unknown length in each right triangle. Simplify the answer if necessary. See Example 8.

109.

110.

111.

112.

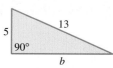

Solve each problem. See Example 9.

113. The length of the diagonal of a box is given by

$$D = \sqrt{L^2 + W^2 + H^2},$$

where L, W, and H are the length, width, and height of the box. Find the length of the diagonal, D, of a box that is 4 ft long, 3 ft high, and 2 ft wide. Give the exact value, then round to the nearest tenth of a foot.

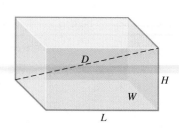

114. A Sanyo color television, model AVM-2755, has a rectangular screen with a 21.7-in. width. Its height is 16 in. What is the diagonal of the screen to the nearest tenth of an inch? (*Source:* Actual measurements of the author's television.)

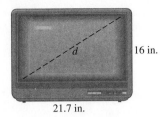

16 in.

21.7 in.

115. A formula from electronics dealing with impedance of parallel resonant circuits is

$$I = \frac{E}{\sqrt{R^2 + \omega^2 L^2}},$$

where the variables are in appropriate units. Find I if $E = 282$, $R = 100$, $L = 264$, and $\omega = 120\pi$. Give your answer to the nearest thousandth.

116. In the study of sound, one version of the law of tensions is

$$f_1 = f_2 \sqrt{\frac{F_1}{F_2}}.$$

If $F_1 = 300$, $F_2 = 60$, and $f_2 = 260$, find f_1 to the nearest unit.

117. The illumination I, in foot-candles, produced by a light source is related to the distance d, in feet, from the light source by the equation

$$d = \sqrt{\frac{k}{I}},$$

where k is a constant. If $k = 640$, how far from the light source will the illumination be 2 foot-candles? Give the exact value, and then round to the nearest tenth of a foot.

118. The following letter appeared in the column "Ask Tom Why," written by Tom Skilling of the *Chicago Tribune*.

> *Dear Tom,*
> *I cannot remember the formula to calculate the distance to the horizon. I have a stunning view from my 14th floor condo, 150 feet above the ground. How far can I see?*
> *Ted Fleischaker; Indianapolis, Ind.*

Skilling's answer was as follows.

> To find the distance to the horizon in miles, take the square root of the height of your view in feet and multiply that result by 1.224. Your answer will be the number of miles to the horizon. (*Source: Chicago Tribune*, August 17, 2002.)

Assuming Ted's eyes are 6 ft above the ground, the total height from the ground is $150 + 6 = 156$ ft. To the nearest tenth of a mile, how far can he see to the horizon?

Find the distance between each pair of points. See Example 10.

119. (6, 13) and (1, 1)

120. (8, 13) and (2, 5)

121. (−6, 5) and (3, −4)

122. (−1, 5) and (−7, 7)

123. (−8, 2) and (−4, 1)

124. (−1, 2) and (5, 3)

125. $(4.7, 2.3)$ and $(1.7, -1.7)$

126. $(-2.9, 18.2)$ and $(2.1, 6.2)$

127. $\left(\sqrt{2}, \sqrt{6}\right)$ and $\left(-2\sqrt{2}, 4\sqrt{6}\right)$

128. $\left(\sqrt{7}, 9\sqrt{3}\right)$ and $\left(-\sqrt{7}, 4\sqrt{3}\right)$

129. $(x + y, y)$ and $(x - y, x)$

130. $(c, c - d)$ and $(d, c + d)$

131. As given in the text, the distance formula is expressed with a radical. Write the distance formula using rational exponents.

132. An alternative form of the distance formula is

$$d = \sqrt{(x_1 - x_2)^2 + (y_1 - y_2)^2}.$$

Compare this to the form given in this section, and explain why the two forms are equivalent.

Find the perimeter of each triangle. $\left(\textit{Hint: For Exercise 133, } \sqrt{k} + \sqrt{k} = 2\sqrt{k}.\right)$

133.

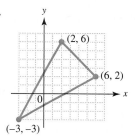

134.

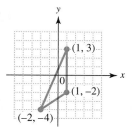

10.4 Adding and Subtracting Radical Expressions

OBJECTIVE 1 Define a radical expression. Recall from Section 10.1 that a **radical expression** is an algebraic expression that contains radicals. For example,

$$\sqrt[4]{3} + \sqrt{6}, \qquad \sqrt{x + 2y} - 1, \qquad \text{and} \qquad \sqrt{8} - \sqrt{2r}$$

are radical expressions. The examples in the previous section discussed simplifying radical expressions that involve multiplication and division. Now we show how to simplify radical expressions that involve addition and subtraction.

OBJECTIVE 2 Simplify radical expressions involving addition and subtraction. An expression such as $4\sqrt{2} + 3\sqrt{2}$ can be simplified by using the distributive property.

$$4\sqrt{2} + 3\sqrt{2} = (4 + 3)\sqrt{2} = 7\sqrt{2}$$

As another example, $2\sqrt{3} - 5\sqrt{3} = (2 - 5)\sqrt{3} = -3\sqrt{3}$. This is similar to simplifying $4x + 3x$ to $7x$ or $2y - 5y$ to $-3y$.

CAUTION Only radical expressions with the *same index* and the *same radicand* may be combined. Expressions such as $5\sqrt{3} + 2\sqrt{2}$ or $3\sqrt{3} + 2\sqrt[3]{3}$ cannot be simplified by combining terms.

EXAMPLE 1 Adding and Subtracting Radicals

Add or subtract to simplify each radical expression.

(a) $3\sqrt{24} + \sqrt{54}$

Begin by simplifying each radical; then use the distributive property to combine terms.

$$
\begin{aligned}
3\sqrt{24} + \sqrt{54} &= 3\sqrt{4} \cdot \sqrt{6} + \sqrt{9} \cdot \sqrt{6} \qquad &\text{Product rule} \\
&= 3 \cdot 2\sqrt{6} + 3\sqrt{6} \\
&= 6\sqrt{6} + 3\sqrt{6} \\
&= 9\sqrt{6} &\text{Combine terms.}
\end{aligned}
$$

(b) $\begin{aligned}[t]
2\sqrt{20x} - \sqrt{45x} &= 2\sqrt{4} \cdot \sqrt{5x} - \sqrt{9} \cdot \sqrt{5x} \qquad &\text{Product rule} \\
&= 2 \cdot 2\sqrt{5x} - 3\sqrt{5x} \\
&= 4\sqrt{5x} - 3\sqrt{5x} \\
&= \sqrt{5x}, \quad x \geq 0 &\text{Combine terms.}
\end{aligned}$

(c) $2\sqrt{3} - 4\sqrt{5}$

Here the radicands differ and are already simplified, so $2\sqrt{3} - 4\sqrt{5}$ cannot be simplified further.

Now Try Exercises 7, 15, and 19.

CAUTION Do not confuse the product rule with combining like terms. The root of a sum *does not equal* the sum of the roots. For example,

$$\sqrt{9 + 16} \neq \sqrt{9} + \sqrt{16}$$

since $\quad \sqrt{9 + 16} = \sqrt{25} = 5, \quad$ but $\quad \sqrt{9} + \sqrt{16} = 3 + 4 = 7.$

EXAMPLE 2 Adding and Subtracting Radicals with Higher Indexes

Add or subtract to simplify each radical expression. Assume all variables represent positive real numbers.

(a) $\begin{aligned}[t]
2\sqrt[3]{16} - 5\sqrt[3]{54} &= 2\sqrt[3]{8 \cdot 2} - 5\sqrt[3]{27 \cdot 2} \qquad &\text{Factor.} \\
&= 2\sqrt[3]{8} \cdot \sqrt[3]{2} - 5\sqrt[3]{27} \cdot \sqrt[3]{2} &\text{Product rule} \\
&= 2 \cdot 2 \cdot \sqrt[3]{2} - 5 \cdot 3 \cdot \sqrt[3]{2} \\
&= 4\sqrt[3]{2} - 15\sqrt[3]{2} \\
&= -11\sqrt[3]{2} &\text{Combine terms.}
\end{aligned}$

(b) $\begin{aligned}[t]
2\sqrt[3]{x^2y} + \sqrt[3]{8x^5y^4} &= 2\sqrt[3]{x^2y} + \sqrt[3]{(8x^3y^3)x^2y} \qquad &\text{Factor.} \\
&= 2\sqrt[3]{x^2y} + 2xy\sqrt[3]{x^2y} &\text{Product rule} \\
&= (2 + 2xy)\sqrt[3]{x^2y} &\text{Distributive property}
\end{aligned}$

Now Try Exercises 23 and 29.

CAUTION Remember to write the index when working with cube roots, fourth roots, and so on.

EXAMPLE 3 Adding and Subtracting Radicals with Fractions

Perform the indicated operations. Assume all variables represent positive real numbers.

(a) $2\sqrt{\dfrac{75}{16}} + 4\dfrac{\sqrt{8}}{\sqrt{32}} = 2\dfrac{\sqrt{25 \cdot 3}}{\sqrt{16}} + 4\dfrac{\sqrt{4 \cdot 2}}{\sqrt{16 \cdot 2}}$ Quotient rule

$\qquad\qquad\qquad\;\; = 2\left(\dfrac{5\sqrt{3}}{4}\right) + 4\left(\dfrac{2\sqrt{2}}{4\sqrt{2}}\right)$ Product rule

$\qquad\qquad\qquad\;\; = \dfrac{5\sqrt{3}}{2} + 2$ Multiply; $\dfrac{\sqrt{2}}{\sqrt{2}} = 1.$

$\qquad\qquad\qquad\;\; = \dfrac{5\sqrt{3}}{2} + \dfrac{4}{2}$ Write with a common denominator.

$\qquad\qquad\qquad\;\; = \dfrac{5\sqrt{3} + 4}{2}$

(b) $10\sqrt[3]{\dfrac{5}{x^6}} - 3\sqrt[3]{\dfrac{4}{x^9}} = 10\dfrac{\sqrt[3]{5}}{\sqrt[3]{x^6}} - 3\dfrac{\sqrt[3]{4}}{\sqrt[3]{x^9}}$ Quotient rule

$\qquad\qquad\qquad\;\; = \dfrac{10\sqrt[3]{5}}{x^2} - \dfrac{3\sqrt[3]{4}}{x^3}$

$\qquad\qquad\qquad\;\; = \dfrac{10x\sqrt[3]{5}}{x^3} - \dfrac{3\sqrt[3]{4}}{x^3}$ Write with a common denominator.

$\qquad\qquad\qquad\;\; = \dfrac{10x\sqrt[3]{5} - 3\sqrt[3]{4}}{x^3}$

Now Try Exercises 47 and 53.

A calculator can support some of the results obtained in the examples of this section. In Example 1(a), we simplified $3\sqrt{24} + \sqrt{54}$ to obtain $9\sqrt{6}$. The screen in Figure 12(a) shows that the approximations are the same, suggesting that our simplification was correct. Figure 12(b) shows support for the result of Example 2(a): $2\sqrt[3]{16} - 5\sqrt[3]{54} = -11\sqrt[3]{2}$. Figure 12(c) supports the result of Example 3(a).

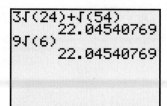

```
3√(24)+√(54)
        22.04540769
9√(6)
        22.04540769
```

(a)

```
2³√(16)-5³√(54)
       -13.85913155
-11³√(2)
       -13.85913155
```

(b)

```
2√(75/16)+4√(8)/
√(32)
       6.330127019
5√(3)/2+2
       6.330127019
```

(c)

FIGURE 12

CONNECTIONS

A triangle that has whole number measures for the lengths of two sides may have an irrational number as the measure of the third side. For example, a right triangle with the two shorter sides measuring 1 and 2 units will have a longest side measuring $\sqrt{5}$ units. The ratio of the dimensions of the *golden rectangle,* considered to have the most pleasing dimensions of any rectangle, is irrational. To sketch a golden rectangle, begin with the square *ONRS.* Divide it into two equal parts by segment *MK,* as shown in the figure. Let *M* be the center of a circle with radius *MN.* Sketch the rectangle *PQRS.* This is a golden rectangle, with the property that if the original square is taken away, *PQNO* is still a golden rectangle. If the square with side *OP* is taken away, another golden rectangle results, and so on.

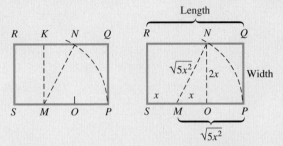

If the sides of the generating square have measure $2x$, then by the Pythagorean formula,
$$MN = \sqrt{x^2 + (2x)^2} = \sqrt{x^2 + 4x^2} = \sqrt{5x^2}.$$
Since *NP* is an arc of the circle with radius *MN,*
$$MP = MN = \sqrt{5x^2}.$$
The ratio of length to width is
$$\frac{\text{length}}{\text{width}} = \frac{x + \sqrt{5x^2}}{2x} = \frac{x + x\sqrt{5}}{2x} = \frac{x(1 + \sqrt{5})}{2x} = \frac{1 + \sqrt{5}}{2},$$
which is an irrational number.

For Discussion or Writing

1. The golden rectangle has been widely used in art and architecture. See whether you can find some examples of its use. Use a calculator to approximate the ratio found above, called the *golden ratio.*

2. The sequence 1, 1, 2, 3, 5, 8, 13, 21, 34, 55, . . . is called the *Fibonacci sequence.* After the first two terms, both 1, every term is found by adding the two preceding terms. Form a sequence of ratios of the successive terms:
$$\frac{1}{1}, \frac{2}{1}, \frac{3}{2}, \frac{5}{3}, \frac{8}{5}, \frac{13}{8}, \frac{21}{13}, \frac{34}{21}, \frac{55}{34}, \cdots$$

Now use a calculator to find approximations of these ratios. What seems to be happening?

10.4 EXERCISES

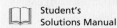

1. Which one of the following sums could be simplified without first simplifying the individual radical expressions?

A. $\sqrt{50} + \sqrt{32}$ **B.** $3\sqrt{6} + 9\sqrt{6}$ **C.** $\sqrt[3]{32} - \sqrt[3]{108}$ **D.** $\sqrt[5]{6} - \sqrt[5]{192}$

2. Let $a = 1$ and let $b = 64$.

  (a) Evaluate $\sqrt{a} + \sqrt{b}$. Then find $\sqrt{a + b}$. Are they equal?

  (b) Evaluate $\sqrt[3]{a} + \sqrt[3]{b}$. Then find $\sqrt[3]{a + b}$. Are they equal?

  (c) Complete the following: In general, $\sqrt[n]{a} + \sqrt[n]{b} \neq$ _____, based on the observations in parts (a) and (b) of this exercise.

3. Even though the root indexes of the terms are not equal, the sum $\sqrt{64} + \sqrt[3]{125} + \sqrt[4]{16}$ can be simplified quite easily. What is this sum? Why can we add these terms so easily?

4. Explain why $28 - 4\sqrt{2}$ is not equal to $24\sqrt{2}$. (This is a common error among algebra students.)

Simplify. Assume all variables represent positive real numbers. See Examples 1 and 2.

5. $\sqrt{36} - \sqrt{100}$ **6.** $\sqrt{25} - \sqrt{81}$ **7.** $-2\sqrt{48} + 3\sqrt{75}$

8. $4\sqrt{32} - 2\sqrt{8}$ **9.** $\sqrt[3]{16} + 4\sqrt[3]{54}$ **10.** $3\sqrt[3]{24} - 2\sqrt[3]{192}$

11. $\sqrt[4]{32} + 3\sqrt[4]{2}$ **12.** $\sqrt[4]{405} - 2\sqrt[4]{5}$

13. $6\sqrt{18} - \sqrt{32} + 2\sqrt{50}$ **14.** $5\sqrt{8} + 3\sqrt{72} - 3\sqrt{50}$

15. $5\sqrt{6} + 2\sqrt{10}$ **16.** $3\sqrt{11} - 5\sqrt{13}$

17. $2\sqrt{5} + 3\sqrt{20} + 4\sqrt{45}$ **18.** $5\sqrt{54} - 2\sqrt{24} - 2\sqrt{96}$

19. $8\sqrt{2x} - \sqrt{8x} + \sqrt{72x}$ **20.** $4\sqrt{18k} - \sqrt{72k} + \sqrt{50k}$

21. $3\sqrt{72m^2} - 5\sqrt{32m^2} - 3\sqrt{18m^2}$ **22.** $9\sqrt{27p^2} - 14\sqrt{108p^2} + 2\sqrt{48p^2}$

23. $-\sqrt[3]{54} + 2\sqrt[3]{16}$ **24.** $15\sqrt[3]{81} - 4\sqrt[3]{24}$

25. $2\sqrt[3]{27x} - 2\sqrt[3]{8x}$ **26.** $6\sqrt[3]{128m} + 3\sqrt[3]{16m}$

27. $\sqrt[3]{x^2y} - \sqrt[3]{8x^2y}$ **28.** $3\sqrt[3]{x^2y^2} - 2\sqrt[3]{64x^2y^2}$

29. $3x\sqrt[3]{xy^2} - 2\sqrt[3]{8x^4y^2}$ **30.** $6q^2\sqrt[3]{5q} - 2q\sqrt[3]{40q^4}$

31. $5\sqrt[4]{32} + 3\sqrt[4]{162}$ **32.** $2\sqrt[4]{512} + 4\sqrt[4]{32}$

33. $3\sqrt[4]{x^5y} - 2x\sqrt[4]{xy}$ **34.** $2\sqrt[4]{m^9p^6} - 3m^2p\sqrt[4]{mp^2}$

35. $2\sqrt[4]{32a^3} + 5\sqrt[4]{2a^3}$ **36.** $-\sqrt[4]{16r} + 5\sqrt[4]{r}$

37. $\sqrt[3]{64xy^2} + \sqrt[3]{27x^4y^5}$ **38.** $\sqrt[4]{625s^3t} - \sqrt[4]{81s^7t^5}$

Simplify. Assume all variables represent positive real numbers. See Example 3.

39. $\sqrt{8} - \dfrac{\sqrt{64}}{\sqrt{16}}$ **40.** $\sqrt{48} - \dfrac{\sqrt{81}}{\sqrt{9}}$ **41.** $\dfrac{2\sqrt{5}}{3} + \dfrac{\sqrt{5}}{6}$

42. $\dfrac{4\sqrt{3}}{3} + \dfrac{2\sqrt{3}}{9}$

43. $\sqrt{\dfrac{8}{9}} + \sqrt{\dfrac{18}{36}}$

44. $\sqrt{\dfrac{12}{16}} + \sqrt{\dfrac{48}{64}}$

45. $\dfrac{\sqrt{32}}{3} + \dfrac{2\sqrt{2}}{3} - \dfrac{\sqrt{2}}{\sqrt{9}}$

46. $\dfrac{\sqrt{27}}{2} - \dfrac{3\sqrt{3}}{2} + \dfrac{\sqrt{3}}{\sqrt{4}}$

47. $3\sqrt{\dfrac{50}{9}} + 8\dfrac{\sqrt{2}}{\sqrt{8}}$

48. $9\sqrt{\dfrac{48}{25}} - 2\dfrac{\sqrt{2}}{\sqrt{98}}$

49. $\sqrt{\dfrac{25}{x^8}} - \sqrt{\dfrac{9}{x^6}}$

50. $\sqrt{\dfrac{100}{y^4}} + \sqrt{\dfrac{81}{y^{10}}}$

51. $3\sqrt[3]{\dfrac{m^5}{27}} - 2m\sqrt[3]{\dfrac{m^2}{64}}$

52. $2a\sqrt[4]{\dfrac{a}{16}} - 5a\sqrt[4]{\dfrac{a}{81}}$

53. $3\sqrt[3]{\dfrac{2}{x^6}} - 4\sqrt[3]{\dfrac{5}{x^9}}$

54. $-4\sqrt[3]{\dfrac{4}{t^9}} + 3\sqrt[3]{\dfrac{9}{t^{12}}}$

In Example 1(a) we showed that $3\sqrt{24} + \sqrt{54} = 9\sqrt{6}$. *To support this result, we can find a calculator approximation of* $3\sqrt{24}$, *then find a calculator approximation of* $\sqrt{54}$, *and add these two approximations. Then, we find a calculator approximation of* $9\sqrt{6}$. *It should correspond to the sum that we just found. (For this example, both approximations are 22.04540769. Due to rounding procedures, there may be a discrepancy in the final digit if you try to duplicate this work.) Follow this procedure to support the statements in Exercises 55 and 56.*

55. $3\sqrt{32} - 2\sqrt{8} = 8\sqrt{2}$

56. $2\sqrt{40} + 6\sqrt{90} - 3\sqrt{160} = 10\sqrt{10}$

57. A rectangular yard has a length of $\sqrt{192}$ m and a width of $\sqrt{48}$ m. Choose the best estimate of its dimensions. Then estimate the perimeter.

  A. 14 m by 7 m   **B.** 5 m by 7 m   **C.** 14 m by 8 m   **D.** 15 m by 8 m

58. If the sides of a triangle are $\sqrt{65}$ in., $\sqrt{35}$ in., and $\sqrt{26}$ in., which one of the following is the best estimate of its perimeter?

  A. 20 in.   **B.** 26 in.   **C.** 19 in.   **D.** 24 in.

Solve each problem. Give answers as simplified radical expressions.

59. Find the perimeter of the triangle.

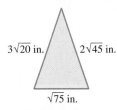

$3\sqrt{20}$ in. $2\sqrt{45}$ in.

$\sqrt{75}$ in.

60. Find the perimeter of the rectangle.

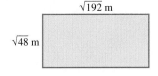

$\sqrt{192}$ m

$\sqrt{48}$ m

61. What is the perimeter of the computer graphic?

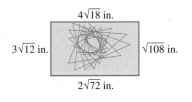

$4\sqrt{18}$ in.

$3\sqrt{12}$ in. $\sqrt{108}$ in.

$2\sqrt{72}$ in.

62. Find the area of the trapezoid.

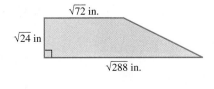

$\sqrt{72}$ in.

$\sqrt{24}$ in

$\sqrt{288}$ in.

10.5 Multiplying and Dividing Radical Expressions

OBJECTIVES

1 Multiply radical expressions.

2 Rationalize denominators with one radical term.

3 Rationalize denominators with binomials involving radicals.

4 Write radical quotients in lowest terms.

OBJECTIVE 1 Multiply radical expressions. We multiply binomial expressions involving radicals by using the FOIL (First, Outer, Inner, Last) method. For example, we find the product of the binomials $\sqrt{5} + 3$ and $\sqrt{6} + 1$ as follows.

$$
(\sqrt{5} + 3)(\sqrt{6} + 1) = \overbrace{\sqrt{5} \cdot \sqrt{6}}^{\text{First}} + \overbrace{\sqrt{5} \cdot 1}^{\text{Outer}} + \overbrace{3 \cdot \sqrt{6}}^{\text{Inner}} + \overbrace{3 \cdot 1}^{\text{Last}}
$$
$$
= \sqrt{30} + \sqrt{5} + 3\sqrt{6} + 3
$$

This result cannot be simplified further.

EXAMPLE 1 Multiplying Binomials Involving Radical Expressions

Multiply using FOIL.

(a) $(7 - \sqrt{3})(\sqrt{5} + \sqrt{2}) = 7\sqrt{5} + 7\sqrt{2} - \sqrt{3} \cdot \sqrt{5} - \sqrt{3} \cdot \sqrt{2}$

$$= 7\sqrt{5} + 7\sqrt{2} - \sqrt{15} - \sqrt{6}$$

(b) $(\sqrt{10} + \sqrt{3})(\sqrt{10} - \sqrt{3})$

$$= \sqrt{10} \cdot \sqrt{10} - \sqrt{10} \cdot \sqrt{3} + \sqrt{10} \cdot \sqrt{3} - \sqrt{3} \cdot \sqrt{3}$$
$$= 10 - 3$$
$$= 7$$

Notice that this is the kind of product that results in the difference of squares:

$$(x + y)(x - y) = x^2 - y^2.$$

Here, $x = \sqrt{10}$ and $y = \sqrt{3}$.

(c) $(\sqrt{7} - 3)^2 = (\sqrt{7} - 3)(\sqrt{7} - 3)$

$$= \sqrt{7} \cdot \sqrt{7} - 3\sqrt{7} - 3\sqrt{7} + 3 \cdot 3$$
$$= 7 - 6\sqrt{7} + 9$$
$$= 16 - 6\sqrt{7}$$

(d) $(5 - \sqrt[3]{3})(5 + \sqrt[3]{3}) = 5 \cdot 5 + 5\sqrt[3]{3} - 5\sqrt[3]{3} - \sqrt[3]{3} \cdot \sqrt[3]{3}$

$$= 25 - \sqrt[3]{3^2}$$
$$= 25 - \sqrt[3]{9}$$

(e) $(\sqrt{k} + \sqrt{y})(\sqrt{k} - \sqrt{y}) = (\sqrt{k})^2 - (\sqrt{y})^2$

$$= k - y, \quad k \geq 0 \text{ and } y \geq 0$$

Now Try Exercises 13, 17, 23, 27, and 39.

> **NOTE** In Example 1(c) we could have used the formula for the square of a binomial,
> $$(x - y)^2 = x^2 - 2xy + y^2,$$
> to obtain the same result:
> $$\left(\sqrt{7} - 3\right)^2 = \left(\sqrt{7}\right)^2 - 2\left(\sqrt{7}\right)(3) + 3^2$$
> $$= 7 - 6\sqrt{7} + 9$$
> $$= 16 - 6\sqrt{7}.$$

OBJECTIVE 2 Rationalize denominators with one radical term. As defined earlier, a simplified radical expression will have no radical in the denominator. The origin of this agreement no doubt occurred before the days of high-speed calculation, when computation was a tedious process performed by hand. To see this, consider the radical expression $\frac{1}{\sqrt{2}}$. To find a decimal approximation by hand, it would be necessary to divide 1 by a decimal approximation for $\sqrt{2}$, such as 1.414. It would be much easier if the divisor were a whole number. This can be accomplished by multiplying $\frac{1}{\sqrt{2}}$ by 1 in the form $\frac{\sqrt{2}}{\sqrt{2}}$:

$$\frac{1}{\sqrt{2}} \cdot \frac{\sqrt{2}}{\sqrt{2}} = \frac{\sqrt{2}}{2}.$$

Now the computation would require dividing 1.414 by 2 to obtain .707, a much easier task.

With current technology, either form of this fraction can be approximated with the same number of keystrokes. See Figure 13, which shows how a calculator gives the same approximation for both forms of the expression.

A common way of "standardizing" the form of a radical expression is to have the denominator contain no radicals. The process of removing radicals from a denominator so that the denominator contains only rational numbers is called **rationalizing the denominator.**

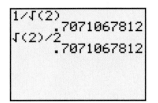

FIGURE 13

EXAMPLE 2 Rationalizing Denominators with Square Roots

Rationalize each denominator.

(a) $\dfrac{3}{\sqrt{7}}$

Multiply the numerator and denominator by $\sqrt{7}$. This is, in effect, multiplying by 1.

$$\frac{3}{\sqrt{7}} = \frac{3 \cdot \sqrt{7}}{\sqrt{7} \cdot \sqrt{7}}$$

In the denominator, since $\sqrt{7} \cdot \sqrt{7} = \sqrt{7 \cdot 7} = \sqrt{49} = 7$,

$$\frac{3}{\sqrt{7}} = \frac{3\sqrt{7}}{7}.$$

The denominator is now a rational number.

(b) $\dfrac{5\sqrt{2}}{\sqrt{5}} = \dfrac{5\sqrt{2} \cdot \sqrt{5}}{\sqrt{5} \cdot \sqrt{5}} = \dfrac{5\sqrt{10}}{5} = \sqrt{10}$

(c) $\dfrac{6}{\sqrt{12}}$

Less work is involved if we simplify the radical in the denominator first.

$$\dfrac{6}{\sqrt{12}} = \dfrac{6}{\sqrt{4 \cdot 3}} = \dfrac{6}{2\sqrt{3}} = \dfrac{3}{\sqrt{3}}$$

Now we rationalize the denominator by multiplying the numerator and denominator by $\sqrt{3}$.

$$\dfrac{3 \cdot \sqrt{3}}{\sqrt{3} \cdot \sqrt{3}} = \dfrac{3\sqrt{3}}{3} = \sqrt{3}$$

Now Try Exercises 43, 49, and 51.

EXAMPLE 3 Rationalizing Denominators in Roots of Fractions

Simplify each radical. In part (b), $p > 0$.

(a) $\sqrt{\dfrac{18}{125}} = \dfrac{\sqrt{18}}{\sqrt{125}}$ Quotient rule

$= \dfrac{\sqrt{9 \cdot 2}}{\sqrt{25 \cdot 5}}$ Factor.

$= \dfrac{3\sqrt{2}}{5\sqrt{5}}$ Product rule

$= \dfrac{3\sqrt{2} \cdot \sqrt{5}}{5\sqrt{5} \cdot \sqrt{5}}$ Multiply by $\dfrac{\sqrt{5}}{\sqrt{5}}$.

$= \dfrac{3\sqrt{10}}{5 \cdot 5}$ Product rule

$= \dfrac{3\sqrt{10}}{25}$

(b) $\sqrt{\dfrac{50m^4}{p^5}} = \dfrac{\sqrt{50m^4}}{\sqrt{p^5}}$ Quotient rule

$= \dfrac{5m^2\sqrt{2}}{p^2\sqrt{p}}$ Product rule

$= \dfrac{5m^2\sqrt{2} \cdot \sqrt{p}}{p^2\sqrt{p} \cdot \sqrt{p}}$ Multiply by $\dfrac{\sqrt{p}}{\sqrt{p}}$.

$= \dfrac{5m^2\sqrt{2p}}{p^2 \cdot p}$ Product rule

$= \dfrac{5m^2\sqrt{2p}}{p^3}$

Now Try Exercises 55 and 63.

▨ **EXAMPLE 4** Rationalizing Denominators with Cube and Fourth Roots

Simplify.

(a) $\sqrt[3]{\dfrac{27}{16}}$

Use the quotient rule and simplify the numerator and denominator.

$$\sqrt[3]{\frac{27}{16}} = \frac{\sqrt[3]{27}}{\sqrt[3]{16}} = \frac{3}{\sqrt[3]{8} \cdot \sqrt[3]{2}} = \frac{3}{2\sqrt[3]{2}}$$

To get a rational denominator, multiply the numerator and denominator by a number that will result in a perfect cube in the radicand in the denominator. Since $2 \cdot 4 = 8$, a perfect cube, multiply the numerator and denominator by $\sqrt[3]{4}$.

$$\sqrt[3]{\frac{27}{16}} = \frac{3}{2\sqrt[3]{2}} = \frac{3 \cdot \sqrt[3]{4}}{2\sqrt[3]{2} \cdot \sqrt[3]{4}} = \frac{3\sqrt[3]{4}}{2\sqrt[3]{8}} = \frac{3\sqrt[3]{4}}{2 \cdot 2} = \frac{3\sqrt[3]{4}}{4}$$

(b) $\sqrt[4]{\dfrac{5x}{z}} = \dfrac{\sqrt[4]{5x}}{\sqrt[4]{z}} \cdot \dfrac{\sqrt[4]{z^3}}{\sqrt[4]{z^3}} = \dfrac{\sqrt[4]{5xz^3}}{\sqrt[4]{z^4}} = \dfrac{\sqrt[4]{5xz^3}}{z}, \quad x \geq 0, z > 0$

▨

Now Try Exercises 71 and 81.

CAUTION It is easy to make mistakes in problems like the one in Example 4(a). A typical error is to multiply the numerator and denominator by $\sqrt[3]{2}$, forgetting that

$$\sqrt[3]{2} \cdot \sqrt[3]{2} \neq 2.$$

You need *three* factors of 2 to obtain 2^3 under the radical. As implied in Example 4(a),

$$\sqrt[3]{2} \cdot \sqrt[3]{2} \cdot \sqrt[3]{2} = 2.$$

OBJECTIVE 3 Rationalize denominators with binomials involving radicals. Recall the special product

$$(x + y)(x - y) = x^2 - y^2.$$

To rationalize a denominator that contains a binomial expression (one that contains exactly two terms) involving radicals, such as

$$\frac{3}{1 + \sqrt{2}},$$

we must use *conjugates*. The conjugate of $1 + \sqrt{2}$ is $1 - \sqrt{2}$. In general, $x + y$ and $x - y$ are **conjugates.**

Rationalizing a Binomial Denominator

Whenever a radical expression has a sum or difference with square root radicals in the denominator, rationalize the denominator by multiplying both the numerator and denominator by the conjugate of the denominator.

For the expression $\dfrac{3}{1+\sqrt{2}}$, we rationalize the denominator by multiplying both the numerator and denominator by $1-\sqrt{2}$, the conjugate of the denominator.

$$\frac{3}{1+\sqrt{2}} = \frac{3(1-\sqrt{2})}{(1+\sqrt{2})(1-\sqrt{2})}$$

Then $(1+\sqrt{2})(1-\sqrt{2}) = 1^2 - (\sqrt{2})^2 = 1 - 2 = -1$. Placing -1 in the denominator gives

$$= \frac{3(1-\sqrt{2})}{-1}$$

$$= \frac{3}{-1}(1-\sqrt{2})$$

$$= -3(1-\sqrt{2}) \quad \text{or} \quad -3 + 3\sqrt{2}.$$

EXAMPLE 5   Rationalizing Binomial Denominators

Rationalize each denominator.

(a) $\dfrac{5}{4-\sqrt{3}}$

To rationalize the denominator, multiply both the numerator and denominator by the conjugate of the denominator, $4+\sqrt{3}$.

$$\frac{5}{4-\sqrt{3}} = \frac{5(4+\sqrt{3})}{(4-\sqrt{3})(4+\sqrt{3})}$$

$$= \frac{5(4+\sqrt{3})}{16-3}$$

$$= \frac{5(4+\sqrt{3})}{13}$$

Notice that the numerator is left in factored form. This makes it easier to determine whether the expression is written in lowest terms.

(b) $\dfrac{\sqrt{2}-\sqrt{3}}{\sqrt{5}+\sqrt{3}}$

Multiply the numerator and denominator by $\sqrt{5}-\sqrt{3}$ to rationalize the denominator.

$$\frac{\sqrt{2}-\sqrt{3}}{\sqrt{5}+\sqrt{3}} = \frac{(\sqrt{2}-\sqrt{3})(\sqrt{5}-\sqrt{3})}{(\sqrt{5}+\sqrt{3})(\sqrt{5}-\sqrt{3})}$$

$$= \frac{\sqrt{10}-\sqrt{6}-\sqrt{15}+3}{5-3}$$

$$= \frac{\sqrt{10}-\sqrt{6}-\sqrt{15}+3}{2}$$

(c) $\dfrac{3}{\sqrt{5m} - \sqrt{p}} = \dfrac{3(\sqrt{5m} + \sqrt{p})}{(\sqrt{5m} - \sqrt{p})(\sqrt{5m} + \sqrt{p})}$

$\qquad\qquad\quad = \dfrac{3(\sqrt{5m} + \sqrt{p})}{5m - p}, \quad 5m \neq p,\, m > 0,\, p > 0$

Now Try Exercises 85, 91, and 95.

OBJECTIVE 4 Write radical quotients in lowest terms.

EXAMPLE 6 Writing Radical Quotients in Lowest Terms

Write each quotient in lowest terms.

(a) $\dfrac{6 + 2\sqrt{5}}{4}$

Factor the numerator and denominator, then write in lowest terms.

$$\frac{6 + 2\sqrt{5}}{4} = \frac{2(3 + \sqrt{5})}{2 \cdot 2} = \frac{3 + \sqrt{5}}{2}$$

Here is an alternative method for writing this expression in lowest terms.

$$\frac{6 + 2\sqrt{5}}{4} = \frac{6}{4} + \frac{2\sqrt{5}}{4} = \frac{3}{2} + \frac{\sqrt{5}}{2} = \frac{3 + \sqrt{5}}{2}$$

(b) $\dfrac{5y - \sqrt{8y^2}}{6y} = \dfrac{5y - 2y\sqrt{2}}{6y}, \quad y > 0$ $\qquad$ Product rule

$\qquad\qquad = \dfrac{y(5 - 2\sqrt{2})}{6y}$ $\qquad\qquad$ Factor the numerator.

$\qquad\qquad = \dfrac{5 - 2\sqrt{2}}{6}$

Note that the final fraction cannot be simplified further because there is no common factor of 2 in the numerator.

Now Try Exercises 107 and 109.

CAUTION Be careful to factor *before* writing a quotient in lowest terms.

CONNECTIONS

In calculus, it is sometimes desirable to rationalize the *numerator* in an expression. The procedure is similar to rationalizing the denominator. For example, to rationalize the numerator of

$$\frac{6 - \sqrt{2}}{4},$$

we multiply the numerator and the denominator by the conjugate of the numerator.

(continued)

$$\frac{6 - \sqrt{2}}{4} = \frac{(6 - \sqrt{2})(6 + \sqrt{2})}{4(6 + \sqrt{2})} = \frac{36 - 2}{4(6 + \sqrt{2})} = \frac{34}{4(6 + \sqrt{2})} = \frac{17}{2(6 + \sqrt{2})}$$

In the final expression, the numerator is rationalized and is in lowest terms.

For Discussion or Writing

Rationalize the numerators of the following expressions, assuming a and b are nonnegative real numbers.

1. $\dfrac{8\sqrt{5} - 1}{6}$ **2.** $\dfrac{3\sqrt{a} + \sqrt{b}}{b}$ **3.** $\dfrac{3\sqrt{a} + \sqrt{b}}{\sqrt{b} - \sqrt{a}}$

4. Rationalize the denominator of the expression in Exercise 3, and then describe the difference in the procedure you used from what you did in Exercise 3.

10.5 EXERCISES

For Extra Help

 Student's Solutions Manual

 MyMathLab

 InterAct Math Tutorial Software

 AW Math Tutor Center

MathXL MathXL

 Digital Video Tutor CD 15/Videotape 15

Match each part of a rule for a special product in Column I with the other part in Column II.

I	II
1. $(x + \sqrt{y})(x - \sqrt{y})$	**A.** $x - y$
2. $(\sqrt{x} + y)(\sqrt{x} - y)$	**B.** $x + 2y\sqrt{x} + y^2$
3. $(\sqrt{x} + \sqrt{y})(\sqrt{x} - \sqrt{y})$	**C.** $x - y^2$
4. $(\sqrt{x} + \sqrt{y})^2$	**D.** $x - 2\sqrt{xy} + y$
5. $(\sqrt{x} - \sqrt{y})^2$	**E.** $x^2 - y$
6. $(\sqrt{x} + y)^2$	**F.** $x + 2\sqrt{xy} + y$

Multiply, then simplify each product. Assume all variables represent positive real numbers. See Example 1.

7. $\sqrt{6}(3 + \sqrt{2})$ **8.** $\sqrt{2}(\sqrt{32} - \sqrt{9})$ **9.** $5(\sqrt{72} - \sqrt{8})$

10. $\sqrt{3}(\sqrt{12} + 2)$ **11.** $(\sqrt{7} + 3)(\sqrt{7} - 3)$ **12.** $(\sqrt{3} - 5)(\sqrt{3} + 5)$

13. $(\sqrt{2} - \sqrt{3})(\sqrt{2} + \sqrt{3})$ **14.** $(\sqrt{7} + \sqrt{3})(\sqrt{7} - \sqrt{3})$

15. $(\sqrt{8} - \sqrt{2})(\sqrt{8} + \sqrt{2})$ **16.** $(\sqrt{20} - \sqrt{5})(\sqrt{20} + \sqrt{5})$

17. $(\sqrt{2} + 1)(\sqrt{3} - 1)$ **18.** $(\sqrt{3} + 3)(\sqrt{5} - 2)$

19. $(\sqrt{11} - \sqrt{7})(\sqrt{2} + \sqrt{5})$ **20.** $(\sqrt{6} + \sqrt{2})(\sqrt{3} + \sqrt{2})$

21. $(2\sqrt{3} + \sqrt{5})(3\sqrt{3} - 2\sqrt{5})$ **22.** $(\sqrt{7} - \sqrt{11})(2\sqrt{7} + 3\sqrt{11})$

23. $(\sqrt{5} + 2)^2$ **24.** $(\sqrt{11} - 1)^2$

25. $(\sqrt{21} - \sqrt{5})^2$ **26.** $(\sqrt{6} - \sqrt{2})^2$

27. $(2 + \sqrt[3]{6})(2 - \sqrt[3]{6})$ **28.** $(\sqrt[3]{3} + 6)(\sqrt[3]{3} - 6)$

29. $\left(2 + \sqrt[3]{2}\right)\left(4 - 2\sqrt[3]{2} + \sqrt[3]{4}\right)$ **30.** $\left(\sqrt[3]{3} - 1\right)\left(\sqrt[3]{9} + \sqrt[3]{3} + 1\right)$

31. $\left(3\sqrt{x} - \sqrt{5}\right)\left(2\sqrt{x} + 1\right)$ **32.** $\left(4\sqrt{p} + \sqrt{7}\right)\left(\sqrt{p} - 9\right)$

33. $\left(3\sqrt{r} - \sqrt{s}\right)\left(3\sqrt{r} + \sqrt{s}\right)$ **34.** $\left(\sqrt{k} + 4\sqrt{m}\right)\left(\sqrt{k} - 4\sqrt{m}\right)$

35. $\left(\sqrt[3]{2y} - 5\right)\left(4\sqrt[3]{2y} + 1\right)$ **36.** $\left(\sqrt[3]{9z} - 2\right)\left(5\sqrt[3]{9z} + 7\right)$

37. $\left(\sqrt{3x} + 2\right)\left(\sqrt{3x} - 2\right)$ **38.** $\left(\sqrt{6y} - 4\right)\left(\sqrt{6y} + 4\right)$

39. $\left(2\sqrt{x} + \sqrt{y}\right)\left(2\sqrt{x} - \sqrt{y}\right)$ **40.** $\left(\sqrt{p} + 5\sqrt{s}\right)\left(\sqrt{p} - 5\sqrt{s}\right)$

41. $\left[\left(\sqrt{2} + \sqrt{3}\right) - \sqrt{6}\right]\left[\left(\sqrt{2} + \sqrt{3}\right) + \sqrt{6}\right]$

42. $\left[\left(\sqrt{5} - \sqrt{2}\right) - \sqrt{3}\right]\left[\left(\sqrt{5} - \sqrt{2}\right) + \sqrt{3}\right]$

Rationalize the denominator in each expression. Assume all variables represent positive real numbers. See Examples 2 and 3.

43. $\dfrac{7}{\sqrt{7}}$ **44.** $\dfrac{11}{\sqrt{11}}$ **45.** $\dfrac{15}{\sqrt{3}}$ **46.** $\dfrac{12}{\sqrt{6}}$ **47.** $\dfrac{\sqrt{3}}{\sqrt{2}}$

48. $\dfrac{\sqrt{7}}{\sqrt{6}}$ **49.** $\dfrac{9\sqrt{3}}{\sqrt{5}}$ **50.** $\dfrac{3\sqrt{2}}{\sqrt{11}}$ **51.** $\dfrac{-6}{\sqrt{18}}$ **52.** $\dfrac{-5}{\sqrt{24}}$

53. $\sqrt{\dfrac{7}{2}}$ **54.** $\sqrt{\dfrac{10}{3}}$ **55.** $-\sqrt{\dfrac{7}{50}}$ **56.** $-\sqrt{\dfrac{13}{75}}$ **57.** $\sqrt{\dfrac{24}{x}}$

58. $\sqrt{\dfrac{52}{y}}$ **59.** $\dfrac{-8\sqrt{3}}{\sqrt{k}}$ **60.** $\dfrac{-4\sqrt{13}}{\sqrt{m}}$ **61.** $-\sqrt{\dfrac{150m^5}{n^3}}$ **62.** $-\sqrt{\dfrac{98r^3}{s^5}}$

63. $\sqrt{\dfrac{288x^7}{y^9}}$ **64.** $\sqrt{\dfrac{242t^9}{u^{11}}}$ **65.** $\dfrac{5\sqrt{2m}}{\sqrt{y^3}}$

66. $\dfrac{2\sqrt{5r}}{\sqrt{m^3}}$ **67.** $-\sqrt{\dfrac{48k^2}{z}}$ **68.** $-\sqrt{\dfrac{75m^3}{p}}$

Simplify. Assume all variables represent positive real numbers. See Example 4.

69. $\sqrt[3]{\dfrac{2}{3}}$ **70.** $\sqrt[3]{\dfrac{4}{5}}$ **71.** $\sqrt[3]{\dfrac{4}{9}}$ **72.** $\sqrt[3]{\dfrac{5}{16}}$ **73.** $\sqrt[3]{\dfrac{9}{32}}$

74. $\sqrt[3]{\dfrac{10}{9}}$ **75.** $-\sqrt[3]{\dfrac{2p}{r^2}}$ **76.** $-\sqrt[3]{\dfrac{6x}{y^2}}$ **77.** $\sqrt[3]{\dfrac{x^6}{y}}$ **78.** $\sqrt[3]{\dfrac{m^9}{q}}$

79. $\sqrt[4]{\dfrac{16}{x}}$ **80.** $\sqrt[4]{\dfrac{81}{y}}$ **81.** $\sqrt[4]{\dfrac{2y}{z}}$ **82.** $\sqrt[4]{\dfrac{7t}{s^2}}$

✐ **83.** Explain the procedure you will use to rationalize the denominator of the expression in Exercise 85: $\dfrac{3}{4 + \sqrt{5}}$. Would multiplying both the numerator and the denominator of this fraction by $4 + \sqrt{5}$ lead to a rationalized denominator? Why or why not?

✐ **84.** Show, in two ways, that the reciprocal of $\sqrt{6} - \sqrt{5}$ is $\sqrt{6} + \sqrt{5}$. (In general, however, the conjugate is not equal to the reciprocal.)

Rationalize the denominator in each expression. Assume all variables represent positive real numbers and no denominators are 0. See Example 5.

85. $\dfrac{3}{4 + \sqrt{5}}$

86. $\dfrac{4}{3 - \sqrt{7}}$

87. $\dfrac{\sqrt{8}}{3 - \sqrt{2}}$

88. $\dfrac{\sqrt{27}}{2 + \sqrt{3}}$

89. $\dfrac{2}{3\sqrt{5} + 2\sqrt{3}}$

90. $\dfrac{-1}{3\sqrt{2} - 2\sqrt{7}}$

91. $\dfrac{\sqrt{2} - \sqrt{3}}{\sqrt{6} - \sqrt{5}}$

92. $\dfrac{\sqrt{5} + \sqrt{6}}{\sqrt{3} - \sqrt{2}}$

93. $\dfrac{m - 4}{\sqrt{m} + 2}$

94. $\dfrac{r - 9}{\sqrt{r} - 3}$

95. $\dfrac{4\sqrt{x}}{\sqrt{x} - 2\sqrt{y}}$

96. $\dfrac{5\sqrt{r}}{3\sqrt{r} + \sqrt{s}}$

97. $\dfrac{\sqrt{x} - \sqrt{y}}{\sqrt{x} + \sqrt{y}}$

98. $\dfrac{\sqrt{a} + \sqrt{b}}{\sqrt{a} - \sqrt{b}}$

99. $\dfrac{5\sqrt{k}}{2\sqrt{k} + \sqrt{q}}$

100. $\dfrac{3\sqrt{x}}{\sqrt{x} - 2\sqrt{y}}$

101. If a and b are both positive numbers and $a^2 = b^2$, then $a = b$. Use this fact to show that $\dfrac{\sqrt{6} - \sqrt{2}}{4} = \dfrac{\sqrt{2} - \sqrt{3}}{2}$.

102. Use a calculator approximation to support your result in Exercise 101.

Write each expression in lowest terms. Assume all variables represent positive real numbers. See Example 6.

103. $\dfrac{30 - 20\sqrt{6}}{10}$

104. $\dfrac{24 + 12\sqrt{5}}{12}$

105. $\dfrac{3 - 3\sqrt{5}}{3}$

106. $\dfrac{-5 + 5\sqrt{2}}{5}$

107. $\dfrac{16 - 4\sqrt{8}}{12}$

108. $\dfrac{12 - 9\sqrt{72}}{18}$

109. $\dfrac{6p + \sqrt{24p^3}}{3p}$

110. $\dfrac{11y - \sqrt{242y^5}}{22y}$

Rationalize each denominator. Assume all radicals represent real numbers and no denominators are 0.

111. $\dfrac{1}{\sqrt{x + y}}$

112. $\dfrac{5}{\sqrt{m - n}}$

113. $\dfrac{p}{\sqrt{p + 2}}$

114. $\dfrac{3q}{\sqrt{5 + q}}$

115. The following expression occurs in a certain standard problem in trigonometry:

$$\frac{1}{\sqrt{2}} \cdot \frac{\sqrt{3}}{2} - \frac{1}{\sqrt{2}} \cdot \frac{1}{2}.$$

Show that it simplifies to $\frac{\sqrt{6} - \sqrt{2}}{4}$. Then verify using a calculator approximation.

116. The following expression occurs in a certain standard problem in trigonometry:

$$\frac{\sqrt{3} + 1}{1 - \sqrt{3}}.$$

Show that it simplifies to $-2 - \sqrt{3}$. Then verify using a calculator approximation.

═══ **RELATING CONCEPTS** (EXERCISES 117–124) ═══

For Individual or Group Work

In Chapter 5 we presented methods of factoring, where the terms in the factors were integers. For example, the binomial $x^2 - 9$ is a difference of squares and factors as $(x + 3)(x - 3)$. However, we can also use this pattern to factor any binomial if we allow square root radicals in the terms of the factors. For example, $t - 5$ can be factored as $(\sqrt{t} + \sqrt{5})(\sqrt{t} - \sqrt{5})$.

Similarly, we can factor any binomial as the sum or difference of cubes, using the patterns $x^3 + y^3 = (x + y)(x^2 - xy + y^2)$ and $x^3 - y^3 = (x - y)(x^2 + xy + y^2)$. For example, we can factor $y + 2$ and $y - 2$ as follows:

$$y + 2 = \left(\sqrt[3]{y} + \sqrt[3]{2}\right)\left(\sqrt[3]{y^2} - \sqrt[3]{2y} + \sqrt[3]{4}\right)$$
$$y - 2 = \left(\sqrt[3]{y} - \sqrt[3]{2}\right)\left(\sqrt[3]{y^2} + \sqrt[3]{2y} + \sqrt[3]{4}\right).$$

Use these ideas to **work Exercises 117–124 in order.**

117. Factor $x - 7$ as the difference of squares.

118. Factor $x - 7$ as the difference of cubes.

119. Factor $x + 7$ as the sum of cubes.

120. Use the result of Exercise 117 to rationalize the denominator of $\dfrac{x + 3}{\sqrt{x} - \sqrt{7}}$.

121. Use the result of Exercise 118 to rationalize the denominator of $\dfrac{x + 3}{\sqrt[3]{x} - \sqrt[3]{7}}$.

122. Use the result of Exercise 119 to rationalize the denominator of
$\dfrac{x + 3}{\sqrt[3]{x^2} - \sqrt[3]{7x} + \sqrt[3]{49}}$.

123. Factor the integer 2 as a difference of cubes by first writing it as $5 - 3$.

124. Use the result of Exercise 123 to rationalize the denominator of $\dfrac{2}{\sqrt[3]{5} - \sqrt[3]{3}}$.

Rationalize the numerator in each expression. Assume all variables represent positive real numbers. (Hint: See the Connections box following Example 6.)

125. $\dfrac{6 - \sqrt{2}}{4}$ **126.** $\dfrac{8\sqrt{5} - 1}{6}$ **127.** $\dfrac{3\sqrt{a} + \sqrt{b}}{b}$ **128.** $\dfrac{\sqrt{p} - 3\sqrt{q}}{4q}$

███ **SUMMARY EXERCISES ON OPERATIONS WITH RADICALS**

Recall that a simplified radical satisfies the following conditions.

1. The radicand has no factor raised to a power greater than or equal to the index.

2. The radicand has no fractions.

3. No denominator contains a radical.

4. Exponents in the radicand and the index of the radical have no common factor (except 1).

Perform all indicated operations and express each answer in simplest form. Assume all variables represent positive real numbers.

1. $6\sqrt{10} - 12\sqrt{10}$
2. $\sqrt{7}(\sqrt{7} - \sqrt{2})$
3. $(1 - \sqrt{3})(2 + \sqrt{6})$

4. $\sqrt{50} - \sqrt{98} + \sqrt{72}$
5. $(3\sqrt{5} + 2\sqrt{7})^2$
6. $\dfrac{-3}{\sqrt{6}}$

7. $\dfrac{8}{\sqrt{7} + \sqrt{5}}$
8. $\sqrt[3]{16x^2} - \sqrt[3]{54x^2} + \sqrt[3]{128x^2}$

9. $\dfrac{1 - \sqrt{2}}{1 + \sqrt{2}}$
10. $(1 - \sqrt[3]{3})(1 + \sqrt[3]{3} + \sqrt[3]{9})$

11. $(\sqrt{5} + 7)(\sqrt{5} - 7)$
12. $\dfrac{1}{\sqrt{x} - \sqrt{5}}, \quad x \neq 5$
13. $\sqrt[3]{8a^3b^5c^9}$

14. $\dfrac{15}{\sqrt[3]{9}}$
15. $\dfrac{3}{\sqrt{5} + 2}$
16. $\sqrt{\dfrac{3}{5x}}$

17. $\dfrac{16\sqrt{3}}{5\sqrt{12}}$
18. $\dfrac{2\sqrt{25}}{8\sqrt{50}}$
19. $\dfrac{-10}{\sqrt[3]{10}}$

20. $\dfrac{\sqrt{6} + \sqrt{5}}{\sqrt{6} - \sqrt{5}}$
21. $\sqrt{12x} - \sqrt{75x}$
22. $(5 - 3\sqrt{3})^2$

23. $(\sqrt{74} - \sqrt{73})(\sqrt{74} + \sqrt{73})$
24. $\sqrt[3]{\dfrac{13}{81}}$

25. $-t^2\sqrt[4]{t} + 3\sqrt[4]{t^9} - t\sqrt[4]{t^5}$

10.6 Solving Equations with Radicals

OBJECTIVES

1 Solve radical equations using the power rule.

2 Solve radical equations that require additional steps.

3 Solve radical equations with indexes greater than 2.

4 Solve radical equations using a graphing calculator.

5 Use the power rule to solve a formula for a specified variable.

OBJECTIVE 1 Solve radical equations using the power rule. An equation that includes one or more radical expressions with a variable is called a **radical equation.** Some examples of radical equations are

$$\sqrt{x - 4} = 8, \qquad \sqrt{5x + 12} = 3\sqrt{2x - 1}, \qquad \text{and} \qquad \sqrt[3]{6 + x} = 27.$$

The equation $x = 1$ has only one solution. Its solution set is $\{1\}$. If we square both sides of this equation, we get $x^2 = 1$. This new equation has two solutions: -1 and 1. Notice that the solution of the original equation is also a solution of the squared equation. However, the squared equation has another solution, -1, that is *not* a solution of the original equation. When solving equations with radicals, we use this idea of raising both sides to a power. It is an application of the **power rule.**

Power Rule for Solving an Equation with Radicals

If both sides of an equation are raised to the same power, all solutions of the original equation are also solutions of the new equation.

Read the power rule carefully; it does *not* say that all solutions of the new equation are solutions of the original equation. They may or may not be. Solutions that do not satisfy the original equation are called **extraneous solutions;** they must be discarded.

CAUTION When the power rule is used to solve an equation, *every solution of the new equation* **must** *be checked in the original equation.*

▨ EXAMPLE 1 Using the Power Rule

Solve $\sqrt{3x + 4} = 8$.

Use the power rule and square both sides to obtain

$$\left(\sqrt{3x + 4}\right)^2 = 8^2$$
$$3x + 4 = 64$$
$$3x = 60$$
$$x = 20.$$

To check, substitute the potential solution in the *original* equation.

Check:
$$\sqrt{3x + 4} = 8$$
$$\sqrt{3 \cdot 20 + 4} = 8 \quad ? \qquad \text{Let } x = 20.$$
$$\sqrt{64} = 8 \quad ?$$
$$8 = 8 \qquad \text{True}$$

Since 20 satisfies the *original* equation, the solution set is $\{20\}$. ▨

Now Try Exercise 9.

The solution of the equation in Example 1 can be generalized to give a method for solving equations with radicals.

Solving an Equation with Radicals

Step 1 **Isolate the radical.** Make sure that one radical term is alone on one side of the equation.

Step 2 **Apply the power rule.** Raise both sides of the equation to a power that is the same as the index of the radical.

Step 3 **Solve** the resulting equation; if it still contains a radical, repeat Steps 1 and 2.

Step 4 **Check** all potential solutions in the original equation.

CAUTION Remember Step 4 or you may get an incorrect solution set.

▨ EXAMPLE 2 Using the Power Rule

Solve $\sqrt{5q - 1} + 3 = 0$.

Step 1 To get the radical alone on one side, subtract 3 from each side.

$$\sqrt{5q - 1} = -3$$

Step 2 Now square both sides.

$$\left(\sqrt{5q-1}\right)^2 = (-3)^2$$

Step 3

$$5q - 1 = 9$$
$$5q = 10$$
$$q = 2$$

Step 4 Check the potential solution, 2, by substituting it in the original equation.

Check:

$$\sqrt{5q-1} + 3 = 0$$
$$\sqrt{5 \cdot 2 - 1} + 3 = 0 \qquad ? \qquad \text{Let } q = 2.$$
$$3 + 3 = 0 \qquad \qquad \text{False}$$

This false result shows that 2 is *not* a solution of the original equation; it is extraneous. The solution set is $\varnothing$.

Now Try Exercise 11.

NOTE We could have determined after Step 1 that the equation in Example 2 has no solution because the expression on the left cannot be negative.

OBJECTIVE 2 Solve radical equations that require additional steps. The next examples involve finding the square of a binomial. Recall that

$$(x + y)^2 = x^2 + 2xy + y^2.$$

■ **EXAMPLE 3 Using the Power Rule; Squaring a Binomial**

Solve $\sqrt{4 - x} = x + 2$.

Step 1 The radical is alone on the left side of the equation.

Step 2 Square both sides; the square of $x + 2$ is $(x + 2)^2 = x^2 + 4x + 4$.

$$\left(\sqrt{4-x}\right)^2 = (x+2)^2$$
$$4 - x = x^2 + 4x + 4$$

⌐Twice the product of 2 and x

Step 3 The new equation is quadratic, so get 0 on one side.

$$0 = x^2 + 5x \qquad \text{Subtract 4 and add } x.$$
$$0 = x(x + 5) \qquad \text{Factor.}$$
$$x = 0 \quad \text{or} \quad x + 5 = 0 \qquad \text{Zero-factor property}$$
$$x = -5$$

Step 4 Check each potential solution in the original equation.

Check: If $x = 0$, then

$$\sqrt{4 - x} = x + 2$$
$$\sqrt{4 - 0} = 0 + 2 \quad ?$$
$$\sqrt{4} = 2 \qquad ?$$
$$2 = 2. \qquad \text{True}$$

If $x = -5$, then

$$\sqrt{4 - x} = x + 2$$
$$\sqrt{4 - (-5)} = -5 + 2 \quad ?$$
$$\sqrt{9} = -3 \qquad ?$$
$$3 = -3. \qquad \text{False}$$

The solution set is $\{0\}$. The other potential solution, -5, is extraneous.

Now Try Exercise 27.

CAUTION   When a radical equation requires squaring a binomial as in Example 3, remember to include the middle term.

$$(x + 2)^2 \neq x^2 + 4 \qquad\qquad (x + 2)^2 = x^2 + 4x + 4$$
INCORRECT   **CORRECT**

EXAMPLE 4   Using the Power Rule; Squaring a Binomial

Solve $\sqrt{x^2 - 4x + 9} = x - 1$.

Squaring both sides gives $(x - 1)^2 = x^2 - 2(x)(1) + 1^2$ on the right.

$$\left(\sqrt{x^2 - 4x + 9}\right)^2 = (x - 1)^2$$
$$x^2 - 4x + 9 = x^2 - 2x + 1$$

                    └─ Twice the product of x and -1

Subtract x^2 and 1 from each side; then add $4x$ to each side to obtain

$$8 = 2x$$
$$4 = x.$$

Check this potential solution in the original equation.

Check: $\sqrt{x^2 - 4x + 9} = x - 1$

            $\sqrt{4^2 - 4 \cdot 4 + 9} = 4 - 1$?   Let $x = 4$.

                          $3 = 3$ True

The solution set of the original equation is $\{4\}$.

Now Try Exercise 29.

EXAMPLE 5   Using the Power Rule; Squaring Twice

Solve $\sqrt{5x + 6} + \sqrt{3x + 4} = 2$.

Start by getting one radical alone on one side of the equation by subtracting $\sqrt{3x + 4}$ from each side.

$$\sqrt{5x + 6} = 2 - \sqrt{3x + 4}$$
$$\left(\sqrt{5x + 6}\right)^2 = \left(2 - \sqrt{3x + 4}\right)^2 \qquad\qquad \text{Square both sides.}$$
$$5x + 6 = 4 - 4\sqrt{3x + 4} + (3x + 4)$$

                     └─── Twice the product of 2 and $-\sqrt{3x + 4}$

This equation still contains a radical, so square both sides again. Before doing this, isolate the radical term on the right.

$$5x + 6 = 8 + 3x - 4\sqrt{3x + 4}$$
$$2x - 2 = -4\sqrt{3x + 4} \qquad\qquad \text{Subtract 8 and } 3x.$$
$$x - 1 = -2\sqrt{3x + 4} \qquad\qquad \text{Divide by 2.}$$
$$(x - 1)^2 = \left(-2\sqrt{3x + 4}\right)^2 \qquad\qquad \text{Square both sides again.}$$
$$x^2 - 2x + 1 = (-2)^2\left(\sqrt{3x + 4}\right)^2 \qquad\qquad (ab)^2 = a^2 b^2$$
$$x^2 - 2x + 1 = 4(3x + 4)$$
$$x^2 - 2x + 1 = 12x + 16 \qquad\qquad \text{Distributive property}$$

$$x^2 - 14x - 15 = 0 \qquad \text{Standard form}$$
$$(x - 15)(x + 1) = 0 \qquad \text{Factor.}$$
$$x - 15 = 0 \quad \text{or} \quad x + 1 = 0 \qquad \text{Zero-factor property}$$
$$x = 15 \quad \text{or} \qquad x = -1$$

Check each of these potential solutions in the original equation. Only -1 checks, so the solution set, $\{-1\}$, has only one element.

Now Try Exercise 51.

OBJECTIVE 3 Solve radical equations with indexes greater than 2.   The power rule also works for powers greater than 2.

 EXAMPLE 6   Using the Power Rule for a Power Greater than 2

Solve $\sqrt[3]{z + 5} = \sqrt[3]{2z - 6}$.

Raise both sides to the third power.

$$\left(\sqrt[3]{z + 5}\right)^3 = \left(\sqrt[3]{2z - 6}\right)^3$$
$$z + 5 = 2z - 6$$
$$11 = z$$

Check: $\qquad \sqrt[3]{z + 5} = \sqrt[3]{2z - 6} \qquad$ Original equation

$\qquad\qquad \sqrt[3]{11 + 5} = \sqrt[3]{2 \cdot 11 - 6} \qquad ? \qquad$ Let $z = 11$.

$\qquad\qquad\qquad \sqrt[3]{16} = \sqrt[3]{16} \qquad\qquad$ True

The solution set is $\{11\}$.

Now Try Exercise 37.

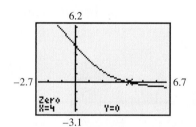

FIGURE 14

OBJECTIVE 4 Solve radical equations using a graphing calculator.   In Example 4 we solved the equation $\sqrt{x^2 - 4x + 9} = x - 1$ using algebraic methods. If we write this equation with one side equal to 0, we get

$$\sqrt{x^2 - 4x + 9} - x + 1 = 0.$$

Using a graphing calculator to graph the function defined by

$$f(x) = \sqrt{x^2 - 4x + 9} - x + 1,$$

we obtain the graph shown in Figure 14. Notice that its zero (x-value of the x-intercept) is 4, which is the solution we found in Example 4.

In Example 3, we found that the single solution of $\sqrt{4 - x} = x + 2$ is 0, with an extraneous value of -5. If we graph $f(x) = \sqrt{4 - x}$ and $g(x) = x + 2$ in the same window, we find that the x-coordinate of the point of intersection of the two graphs is 0, which is the solution of the equation. See Figure 15.

We solved the equation in Example 3 by squaring both sides, obtaining $4 - x = x^2 + 4x + 4$. In Figure 16 on the next page, we show that the two functions defined by $f(x) = 4 - x$ and $g(x) = x^2 + 4x + 4$ have two points of intersection. The extraneous value -5 that we found in Example 3 shows up as an x-value of one of these points of intersection. However, our check showed that -5 was not a solution of the *original* equation (before the squaring step). Here we see a graphical interpretation of the extraneous value.

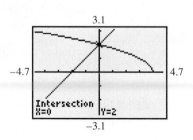

FIGURE 15

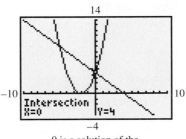
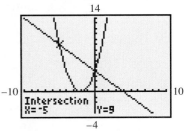

0 is a solution of the
original equation.

−5 is *not* a solution of the
original equation.

FIGURE 16

OBJECTIVE 5 Use the power rule to solve a formula for a specified variable.

EXAMPLE 7 Solving a Formula from Electronics for a Variable

An important property of a radio frequency transmission line is its *characteristic impedance,* represented by Z and measured in ohms. If L and C are the inductance and capacitance, respectively, per unit of length of the line, then these quantities are related by the formula $Z = \sqrt{\dfrac{L}{C}}$. Solve this formula for C.

$$Z = \sqrt{\frac{L}{C}}$$ Given formula

$$Z^2 = \frac{L}{C}$$ Square both sides.

$$CZ^2 = L$$ Multiply by C.

$$C = \frac{L}{Z^2}$$ Divide by Z^2.

Now Try Exercise 67.

10.6 EXERCISES

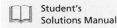

Check each equation to see if the given value for x is a solution.

1. $\sqrt{3x + 18} = x$

   (a) 6   (b) −3

2. $\sqrt{3x - 3} = x - 1$

   (a) 1   (b) 4

3. $\sqrt{x + 2} = \sqrt{9x - 2} - 2\sqrt{x - 1}$

   (a) 2   (b) 7

4. $\sqrt{8x - 3} = 2x$

   (a) $\dfrac{3}{2}$ (b) $\dfrac{1}{2}$

5. Is 9 a solution of the equation $\sqrt{x} = -3$? If not, what is the solution of this equation? Explain.

6. Before even attempting to solve $\sqrt{3x + 18} = x$, how can you be sure that the equation cannot have a negative solution?

Solve each equation. See Examples 1–4.

7. $\sqrt{r-2}=3$

8. $\sqrt{q+1}=7$

9. $\sqrt{6k-1}=1$

10. $\sqrt{7m-3}=5$

11. $\sqrt{4r+3}+1=0$

12. $\sqrt{5k-3}+2=0$

13. $\sqrt{3k+1}-4=0$

14. $\sqrt{5z+1}-11=0$

15. $4-\sqrt{x-2}=0$

16. $9-\sqrt{4k+1}=0$

17. $\sqrt{9a-4}=\sqrt{8a+1}$

18. $\sqrt{4p-2}=\sqrt{3p+5}$

19. $2\sqrt{x}=\sqrt{3x+4}$

20. $2\sqrt{m}=\sqrt{5m-16}$

21. $3\sqrt{z-1}=2\sqrt{2z+2}$

22. $5\sqrt{4a+1}=3\sqrt{10a+25}$

23. $k=\sqrt{k^2+4k-20}$

24. $p=\sqrt{p^2-3p+18}$

25. $a=\sqrt{a^2+3a+9}$

26. $z=\sqrt{z^2-4z-8}$

27. $\sqrt{9-x}=x+3$

28. $\sqrt{5-x}=x+1$

29. $\sqrt{k^2+2k+9}=k+3$

30. $\sqrt{a^2-3a+3}=a-1$

31. $\sqrt{r^2+9r+3}=-r$

32. $\sqrt{p^2-15p+15}=p-5$

33. $\sqrt{z^2+12z-4}+4-z=0$

34. $\sqrt{m^2+3m+12}-m-2=0$

35. What is *wrong* with this first step in the solution process for $\sqrt{3x+4}=8-x$. Solve it correctly.

$$3x+4=64+x^2$$

36. Explain what is *wrong* with this first step in the solution process for the equation $\sqrt{5x+6}-\sqrt{x+3}=3$. Then solve it correctly.

$$(5x+6)+(x+3)=9$$

Solve each equation. See Examples 5 and 6.

37. $\sqrt[3]{2x+5}=\sqrt[3]{6x+1}$

38. $\sqrt[3]{p-1}=2$

39. $\sqrt[3]{a^2+5a+1}=\sqrt[3]{a^2+4a}$

40. $\sqrt[3]{r^2+2r+8}=\sqrt[3]{r^2}$

41. $\sqrt[3]{2m-1}=\sqrt[3]{m+13}$

42. $\sqrt[3]{2k-11}-\sqrt[3]{5k+1}=0$

43. $\sqrt[4]{a+8}=\sqrt[4]{2a}$

44. $\sqrt[4]{z+11}=\sqrt[4]{2z+6}$

45. $\sqrt[3]{x-8}+2=0$

46. $\sqrt[3]{r+1}+1=0$

47. $\sqrt[4]{2k-5}+4=0$

48. $\sqrt[4]{8z-3}+2=0$

49. $\sqrt{k+2}-\sqrt{k-3}=1$

50. $\sqrt{r+6}-\sqrt{r-2}=2$

51. $\sqrt{2r+11}-\sqrt{5r+1}=-1$

52. $\sqrt{3x-2}-\sqrt{x+3}=1$

53. $\sqrt{3p+4}-\sqrt{2p-4}=2$

54. $\sqrt{4x+5}-\sqrt{2x+2}=1$

55. $\sqrt{3-3p}-3=\sqrt{3p+2}$

56. $\sqrt{4x+7}-4=\sqrt{4x-1}$

57. $\sqrt{2\sqrt{x+11}}=\sqrt{4x+2}$

58. $\sqrt{1+\sqrt{24-10x}}=\sqrt{3x+5}$

59. What is the smallest power to which you can raise both sides of the radical equation $\sqrt{x+3}=\sqrt[3]{5+4x}$ so that the radicals are eliminated?

60. What is the smallest power to which you can raise both sides of the radical equation $\sqrt[4]{x+3}=\sqrt[3]{10x+14}$ so that the radicals are eliminated?

61. Use a graphing calculator to solve $\sqrt{3 - 3x} = 3 + \sqrt{3x + 2}$. What is the domain of $y = \sqrt{3 - 3x} - 3 - \sqrt{3x + 2}$?

62. Use a graphing calculator with a window of $[-1, 4]$ by $[-1, 3]$ to solve $\sqrt{2\sqrt{7x + 2}} = \sqrt{3x + 2}$. What is the domain of $f(x) = \sqrt{2\sqrt{7x + 2}} - \sqrt{3x + 2}$?

For each equation, rewrite the expressions with rational exponents as radical expressions, and then solve using the procedures explained in this section.

63. $(2x - 9)^{1/2} = 2 + (x - 8)^{1/2}$

64. $(3w + 7)^{1/2} = 1 + (w + 2)^{1/2}$

65. $(2w - 1)^{2/3} - w^{1/3} = 0$

66. $(x^2 - 2x)^{1/3} - x^{1/3} = 0$

Solve each formula from electricity and radio for the indicated variable. See Example 7. (Source: Cooke, Nelson M., and Joseph B. Orleans, Mathematics Essential to Electricity and Radio, *McGraw-Hill, 1943.)*

67. $V = \sqrt{\dfrac{2K}{m}}$ for K

68. $V = \sqrt{\dfrac{2K}{m}}$ for m

69. $f = \dfrac{1}{2\pi\sqrt{LC}}$ for L

70. $r = \sqrt{\dfrac{Mm}{F}}$ for F

71. A number of useful formulas involve radicals or radical expressions. Many occur in the mathematics needed for working with objects in space. The formula

$$N = \frac{1}{2\pi}\sqrt{\frac{a}{r}}$$

is used to find the rotational rate N of a space station. Here a is the acceleration and r represents the radius of the space station in meters. To find the value of r that will make N simulate the effect of gravity on Earth, the equation must be solved for r, using the required value of N. (*Source:* Kastner, Bernice, *Space Mathematics,* NASA, 1972.)

(a) Solve the equation for r.

(b) Find the value of r that makes $N = .063$ rotation per sec, if $a = 9.8$ m per sec^2.

(c) Find the value of r that makes $N = .04$ rotation per sec, if $a = 9.8$ m per sec^2.

If x is the number of years since 1900, the equation $y = x^{.7}$ approximates the timber grown in the United States in billions of cubic feet. Let $x = 20$ represent 1920, $x = 52$ represent 1952, and so on.

72. Replace x in the equation for each year shown in the graph and use a calculator to find the value of y. (Round answers to the nearest billion.)

73. Use the graph to estimate the amount of timber grown for each year shown.

74. Compare the values found from the equation with your estimates from the graph. Does the equation give a good approximation of the data from the graph? In which year is the approximation best?

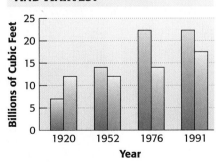

U.S. TIMBER GROWTH AND HARVEST

Billions of Cubic Feet

- Timber grown
- Timber harvested

Source: Figures from U.S. Forest Service. Adapted from *"The Truth about America's Forests,"* Evergreen, 4025 Crater Lake Hwy., Medford, Ore. 97504.

75. From the graph, estimate the amount of timber harvested in each year shown.

76. Use the equation $y = x^{.62}$ and a calculator to approximate the amount of timber harvested in each of the given years. (Round answers to the nearest billion.)

77. Compare your answers from Exercises 75 and 76. Does the equation give a good approximation? For which year is it poorest?

10.7 Complex Numbers

OBJECTIVES

1 Simplify numbers of the form $\sqrt{-b}$, where $b > 0$.

2 Recognize imaginary complex numbers.

3 Add and subtract complex numbers.

4 Multiply complex numbers.

5 Divide complex numbers.

6 Find powers of i.

As we saw in Chapter 1, the set of real numbers includes many other number sets (the rational numbers, integers, and natural numbers, for example). In this section a new set of numbers is introduced that includes the set of real numbers, as well as numbers that are even roots of negative numbers, like $\sqrt{-2}$.

OBJECTIVE 1 Simplify numbers of the form $\sqrt{-b}$, where $b > 0$. The equation $x^2 + 1 = 0$ has no real number solution since any solution must be a number whose square is -1. In the set of real numbers, all squares are nonnegative numbers because the product of two positive numbers or two negative numbers is positive and $0^2 = 0$. To provide a solution for the equation $x^2 + 1 = 0$, a new number i is defined so that

$$i^2 = -1.$$

That is, i is a number whose square is -1, so $i = \sqrt{-1}$. This definition of i makes it possible to define any square root of a negative real number as follows.

$\sqrt{-b}$

For any positive real number b,

$$\sqrt{-b} = i\sqrt{b}.$$

■ **EXAMPLE 1 Simplifying Square Roots of Negative Numbers**

Write each number as a product of a real number and i.

(a) $\sqrt{-100} = i\sqrt{100} = 10i$ **(b)** $-\sqrt{-36} = -i\sqrt{36} = -6i$

(c) $\sqrt{-2} = i\sqrt{2}$

Now Try Exercises 7, 9, and 11.

CAUTION It is easy to mistake $\sqrt{2}i$ for $\sqrt{2i}$, with the i under the radical. For this reason, we usually write $\sqrt{2}i$ as $i\sqrt{2}$, as in the definition of $\sqrt{-b}$.

When finding a product such as $\sqrt{-4} \cdot \sqrt{-9}$, we cannot use the product rule for radicals because it applies only to nonnegative radicands. For this reason, we change $\sqrt{-b}$ to the form $i\sqrt{b}$ before performing any multiplications or divisions. For example,

$$\sqrt{-4} \cdot \sqrt{-9} = i\sqrt{4} \cdot i\sqrt{9}$$
$$= i \cdot 2 \cdot i \cdot 3$$
$$= 6i^2$$
$$= 6(-1) \qquad \text{Substitute: } i^2 = -1.$$
$$= -6.$$

CAUTION Using the product rule for radicals *before* using the definition of $\sqrt{-b}$ gives a *wrong* answer. The preceding example shows that

$$\sqrt{-4} \cdot \sqrt{-9} = -6,$$

but

$$\sqrt{-4(-9)} = \sqrt{36} = 6,$$

so

$$\sqrt{-4} \cdot \sqrt{-9} \neq \sqrt{-4(-9)}.$$

▧ **EXAMPLE 2** Multiplying Square Roots of Negative Numbers

Multiply.

(a) $\sqrt{-3} \cdot \sqrt{-7} = i\sqrt{3} \cdot i\sqrt{7}$
$$= i^2\sqrt{3 \cdot 7}$$
$$= (-1)\sqrt{21} \qquad \text{Substitute: } i^2 = -1.$$
$$= -\sqrt{21}$$

(b) $\sqrt{-2} \cdot \sqrt{-8} = i\sqrt{2} \cdot i\sqrt{8}$
$$= i^2\sqrt{2 \cdot 8}$$
$$= (-1)\sqrt{16}$$
$$= (-1)4$$
$$= -4$$

(c) $\sqrt{-5} \cdot \sqrt{6} = i\sqrt{5} \cdot \sqrt{6} = i\sqrt{30}$

Now Try Exercises 15, 17, and 19.

The methods used to find products also apply to quotients.

▧ **EXAMPLE 3** Dividing Square Roots of Negative Numbers

Divide.

(a) $\dfrac{\sqrt{-75}}{\sqrt{-3}} = \dfrac{i\sqrt{75}}{i\sqrt{3}} = \sqrt{\dfrac{75}{3}} = \sqrt{25} = 5$

(b) $\dfrac{\sqrt{-32}}{\sqrt{8}} = \dfrac{i\sqrt{32}}{\sqrt{8}} = i\sqrt{\dfrac{32}{8}} = i\sqrt{4} = 2i$

Now Try Exercises 21 and 23.

OBJECTIVE 2 **Recognize imaginary complex numbers.** With the imaginary number i and the real numbers, a new set of numbers can be formed that includes the real numbers as a subset. The *complex numbers* are defined as follows.

> **Complex Number**
>
> If a and b are real numbers, then any number of the form $a + bi$ is called a **complex number.**

In the complex number $a + bi$, the number a is called the **real part** and b is called the **imaginary part.** When $b = 0$, $a + bi$ is a real number, so the real numbers are a subset of the complex numbers. Complex numbers with $b \neq 0$ are called **imaginary numbers.*** In spite of their name, imaginary numbers are very useful in applications, particularly in work with electricity.

The relationships among the various sets of numbers discussed in this book are shown in Figure 17.

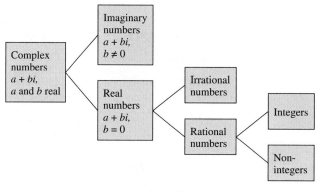

FIGURE 17

OBJECTIVE 3 **Add and subtract complex numbers.** The commutative, associative, and distributive properties for real numbers are also valid for complex numbers. Thus, to add complex numbers, we add their real parts and add their imaginary parts.

▨ **EXAMPLE 4** Adding Complex Numbers

Add.

(a) $(2 + 3i) + (6 + 4i)$
$$= (2 + 6) + (3 + 4)i \qquad \text{Commutative, associative, and distributive properties}$$
$$= 8 + 7i$$

(b) $(4 + 2i) + (3 - i) + (-6 + 3i) = [4 + 3 + (-6)] + [2 + (-1) + 3]i$
$$= 1 + 4i$$

Now Try Exercises 27 and 35.

---

*Some texts define bi as the imaginary part of the complex number $a + bi$. Also, imaginary numbers are sometimes defined as complex numbers with $a = 0$ and $b \neq 0$.

We subtract complex numbers by subtracting their real parts and subtracting their imaginary parts.

_____

EXAMPLE 5 Subtracting Complex Numbers

Subtract.

(a) $(6 + 5i) - (3 + 2i) = (6 - 3) + (5 - 2)i$
$$= 3 + 3i$$

(b) $(7 - 3i) - (8 - 6i) = (7 - 8) + [-3 - (-6)]i$
$$= -1 + 3i$$

(c) $(-9 + 4i) - (-9 + 8i) = (-9 + 9) + (4 - 8)i$
$$= 0 - 4i$$
$$= -4i$$

Now Try Exercises 31 and 33.

In Example 5(c), the answer was written as $0 - 4i$ and then as just $-4i$. A complex number written in the form $a + bi$, like $0 - 4i$, is in **standard form.** In this section, most answers will be given in standard form, but if a or b is 0, we consider answers such as a or bi to be in standard form.

OBJECTIVE 4 Multiply complex numbers. We multiply complex numbers as we multiply polynomials. Complex numbers of the form $a + bi$ have the same form as binomials, so we multiply two complex numbers in standard form by using the FOIL method for multiplying binomials. (Recall that FOIL stands for *First, Outer, Inner, Last.*)

_____

EXAMPLE 6 Multiplying Complex Numbers

Multiply.

(a) $4i(2 + 3i) = 4i(2) + 4i(3i)$ Distributive property
$$= 8i + 12i^2$$
$$= 8i + 12(-1) \qquad \text{Substitute: } i^2 = -1.$$
$$= -12 + 8i$$

(b) $(3 + 5i)(4 - 2i)$

Use the FOIL method.

$$(3 + 5i)(4 - 2i) = \underbrace{3(4)}_{\text{First}} + \underbrace{3(-2i)}_{\text{Outer}} + \underbrace{5i(4)}_{\text{Inner}} + \underbrace{5i(-2i)}_{\text{Last}}$$

$$= 12 - 6i + 20i - 10i^2$$
$$= 12 + 14i - 10(-1) \qquad \text{Substitute: } i^2 = -1.$$
$$= 12 + 14i + 10$$
$$= 22 + 14i$$

(c) $(2 + 3i)(1 - 5i) = 2(1) + 2(-5i) + 3i(1) + 3i(-5i)$ FOIL

$$= 2 - 10i + 3i - 15i^2$$

$$= 2 - 7i - 15(-1)$$

$$= 2 - 7i + 15$$

$$= 17 - 7i$$

Now Try Exercises 45 and 47.

The two complex numbers $a + bi$ and $a - bi$ are called **complex conjugates,** or simply *conjugates,* of each other. The product of a complex number and its conjugate is always a real number, as shown here.

$$(a + bi)(a - bi) = a^2 - abi + abi - b^2i^2$$

$$= a^2 - b^2(-1)$$

$$(a + bi)(a - bi) = a^2 + b^2$$

For example, $(3 + 7i)(3 - 7i) = 3^2 + 7^2 = 9 + 49 = 58$.

OBJECTIVE 5 **Divide complex numbers.** The quotient of two complex numbers should be a complex number. To write the quotient as a complex number, we need to eliminate i in the denominator. We use conjugates to do this.

EXAMPLE 7 **Dividing Complex Numbers**

Find each quotient.

(a) $\dfrac{8 + 9i}{5 + 2i}$

Multiply both the numerator and denominator by the conjugate of the denominator. The conjugate of $5 + 2i$ is $5 - 2i$.

$$\frac{8 + 9i}{5 + 2i} = \frac{(8 + 9i)(5 - 2i)}{(5 + 2i)(5 - 2i)} \qquad \tfrac{5 - 2i}{5 - 2i} = 1$$

$$= \frac{40 - 16i + 45i - 18i^2}{5^2 + 2^2}$$

$$= \frac{58 + 29i}{29} \qquad \text{Substitute: } i^2 = -1; \text{ combine terms.}$$

$$= \frac{29(2 + i)}{29} \qquad \text{Factor the numerator.}$$

$$= 2 + i \qquad \text{Lowest terms}$$

Notice that this is just like rationalizing a denominator. The final result is in standard form.

(b) $\dfrac{1 + i}{i}$

The conjugate of i is $-i$. Multiply both the numerator and denominator by $-i$.

$$\begin{aligned}
\frac{1 + i}{i} &= \frac{(1 + i)(-i)}{i(-i)} \\
&= \frac{-i - i^2}{-i^2} \\
&= \frac{-i - (-1)}{-(-1)} \qquad \text{Substitute: } i^2 = -1. \\
&= \frac{-i + 1}{1} \\
&= 1 - i
\end{aligned}$$

Now Try Exercises 61 and 67.

In Examples 4–7, we showed how complex numbers can be added, subtracted, multiplied, and divided using algebraic methods. Many current models of graphing calculators can perform these operations. Figure 18 shows how the computations in parts of Examples 4–7 are carried out by a TI-83 Plus calculator. It is important to use parentheses as shown.

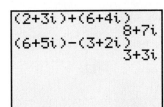
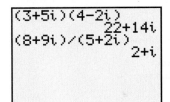

FIGURE 18

OBJECTIVE 6 Find powers of i. Because i^2 is defined to be -1, we can find higher powers of i as shown in the following examples.

$$\begin{array}{ll}
i^3 = i \cdot i^2 = i(-1) = -i & i^6 = i^2 \cdot i^4 = (-1) \cdot 1 = -1 \\
i^4 = i^2 \cdot i^2 = (-1)(-1) = 1 & i^7 = i^3 \cdot i^4 = (-i) \cdot 1 = -i \\
i^5 = i \cdot i^4 = i \cdot 1 = i & i^8 = i^4 \cdot i^4 = 1 \cdot 1 = 1
\end{array}$$

As these examples suggest, the powers of i rotate through the four numbers i, -1, $-i$, and 1. Larger powers of i can be simplified by using the fact that $i^4 = 1$. For example,

$$i^{75} = (i^4)^{18} \cdot i^3 = 1^{18} \cdot i^3 = 1 \cdot i^3 = i^3 = -i.$$

This example suggests a quick method for simplifying larger powers of i.

EXAMPLE 8 Simplifying Powers of i

Find each power of i.

(a) $i^{12} = (i^4)^3 = 1^3 = 1$

(b) $i^{39} = i^{36} \cdot i^3 = (i^4)^9 \cdot i^3 = 1^9 \cdot (-i) = -i$

(c) $i^{-2} = \dfrac{1}{i^2} = \dfrac{1}{-1} = -1$

(d) $i^{-1} = \dfrac{1}{i}$

To simplify this quotient, multiply both the numerator and denominator by $-i$, the conjugate of i.

$$\frac{1}{i} = \frac{1(-i)}{i(-i)} = \frac{-i}{-i^2} = \frac{-i}{-(-1)} = \frac{-i}{1} = -i$$

Now Try Exercises 73 and 81.

10.7 EXERCISES

Decide whether each expression is equal to 1, -1, i, or $-i$.

1. $\sqrt{-1}$ **2.** $-\sqrt{-1}$ **3.** i^2 **4.** $-i^2$ **5.** $\dfrac{1}{i}$ **6.** $(-i)^2$

Write each number as a product of a real number and i. Simplify all radical expressions. See Example 1.

7. $\sqrt{-169}$ **8.** $\sqrt{-225}$ **9.** $-\sqrt{-144}$ **10.** $-\sqrt{-196}$

11. $\sqrt{-5}$ **12.** $\sqrt{-21}$ **13.** $\sqrt{-48}$ **14.** $\sqrt{-96}$

Multiply or divide as indicated. See Examples 2 and 3.

15. $\sqrt{-7} \cdot \sqrt{-15}$ **16.** $\sqrt{-3} \cdot \sqrt{-19}$ **17.** $\sqrt{-4} \cdot \sqrt{-25}$ **18.** $\sqrt{-9} \cdot \sqrt{-81}$

19. $\sqrt{-3} \cdot \sqrt{11}$ **20.** $\sqrt{-10} \cdot \sqrt{2}$ **21.** $\dfrac{\sqrt{-300}}{\sqrt{-100}}$ **22.** $\dfrac{\sqrt{-40}}{\sqrt{-10}}$

23. $\dfrac{\sqrt{-75}}{\sqrt{3}}$ **24.** $\dfrac{\sqrt{-160}}{\sqrt{10}}$

25. **(a)** Every real number is a complex number. Explain why this is so.

    (b) Not every complex number is a real number. Give an example of this and explain why this statement is true.

26. Explain how to add, subtract, multiply, and divide complex numbers. Give examples.

Add or subtract as indicated. Write your answers in the form a + bi. See Examples 4 and 5.

27. $(3 + 2i) + (-4 + 5i)$ **28.** $(7 + 15i) + (-11 + 14i)$

29. $(5 - i) + (-5 + i)$ **30.** $(-2 + 6i) + (2 - 6i)$

31. $(4 + i) - (-3 - 2i)$ **32.** $(9 + i) - (3 + 2i)$

33. $(-3 - 4i) - (-1 - 4i)$ **34.** $(-2 - 3i) - (-5 - 3i)$

35. $(-4 + 11i) + (-2 - 4i) + (7 + 6i)$ **36.** $(-1 + i) + (2 + 5i) + (3 + 2i)$

37. $[(7 + 3i) - (4 - 2i)] + (3 + i)$ **38.** $[(7 + 2i) + (-4 - i)] - (2 + 5i)$

39. Fill in the blank with the correct response:

Because $(4 + 2i) - (3 + i) = 1 + i$, using the definition of subtraction, we can check this to find that $(1 + i) + (3 + i) = $ _____.

40. Fill in the blank with the correct response:

Because $\dfrac{-5}{2 - i} = -2 - i$, using the definition of division, we can check this to find that $(-2 - i)(2 - i) = $ _____.

Multiply. See Example 6.

41. $(3i)(27i)$ **42.** $(5i)(125i)$ **43.** $(-8i)(-2i)$

44. $(-32i)(-2i)$ **45.** $5i(-6 + 2i)$ **46.** $3i(4 + 9i)$

47. $(4 + 3i)(1 - 2i)$ **48.** $(7 - 2i)(3 + i)$ **49.** $(4 + 5i)^2$

50. $(3 + 2i)^2$ **51.** $2i(-4 - i)^2$ **52.** $3i(-3 - i)^2$

53. $(12 + 3i)(12 - 3i)$ **54.** $(6 + 7i)(6 - 7i)$ **55.** $(4 + 9i)(4 - 9i)$

56. $(7 + 2i)(7 - 2i)$

57. What is the conjugate of $a + bi$?

58. If we multiply $a + bi$ by its conjugate, we get _____, which is always a real number.

Write each expression in standard form $a + bi$. See Example 7.

59. $\dfrac{2}{1 - i}$ **60.** $\dfrac{29}{5 + 2i}$ **61.** $\dfrac{-7 + 4i}{3 + 2i}$ **62.** $\dfrac{-38 - 8i}{7 + 3i}$

63. $\dfrac{8i}{2 + 2i}$ **64.** $\dfrac{-8i}{1 + i}$ **65.** $\dfrac{2 - 3i}{2 + 3i}$ **66.** $\dfrac{-1 + 5i}{3 + 2i}$

67. $\dfrac{3 + i}{i}$ **68.** $\dfrac{5 - i}{-i}$

TECHNOLOGY INSIGHTS (EXERCISES 69–70)

Predict the answer that the calculator screen will provide for the given complex number operation entry.

69.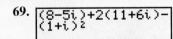

70.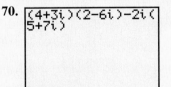

71. Recall that if $a \neq 0$, then $\frac{1}{a}$ is called the reciprocal of a. Use this definition to express the reciprocal of $5 - 4i$ in the form $a + bi$.

72. Recall that if $a \neq 0$, then a^{-1} is defined to be $\frac{1}{a}$. Use this definition to express $(4 - 3i)^{-1}$ in the form $a + bi$.

Find each power of i. See Example 8.

73. i^{18} **74.** i^{26} **75.** i^{89} **76.** i^{48} **77.** i^{38}

78. i^{102} **79.** i^{43} **80.** i^{83} **81.** i^{-5} **82.** i^{-17}

83. A student simplified i^{-18} as follows:

$$i^{-18} = i^{-18} \cdot i^{20} = i^{-18+20} = i^2 = -1.$$

Explain the mathematical justification for this correct work.

84. Explain why

$$(46 + 25i)(3 - 6i) \quad \text{and} \quad (46 + 25i)(3 - 6i)i^{12}$$

must be equal. (Do not actually perform the computation.)

Ohm's law for the current I in a circuit with voltage E, resistance R, capacitance reactance X_c, and inductive reactance X_L is

$$I = \frac{E}{R + (X_L - X_c)i}.$$

Use this law to work Exercises 85 and 86.

85. Find I if $E = 2 + 3i$, $R = 5$, $X_L = 4$, and $X_c = 3$.

86. Find E if $I = 1 - i$, $R = 2$, $X_L = 3$, and $X_c = 1$.

Complex numbers will appear again in this book in Chapter 11, when we study quadratic equations. The following exercises examine how a complex number can be a solution of a quadratic equation.

87. Show that $1 + 5i$ is a solution of $x^2 - 2x + 26 = 0$. Then show that its conjugate is also a solution.

88. Show that $3 + 2i$ is a solution of $x^2 - 6x + 13 = 0$. Then show that its conjugate is also a solution.

RELATING CONCEPTS (EXERCISES 89–94)

For Individual or Group Work

Consider the following expressions:

Binomials	**Complex Numbers**
$x + 2, \quad 3x - 1$	$1 + 2i, \quad 3 - i$

When we add, subtract, or multiply complex numbers in standard form, the rules are the same as those for the corresponding operations on binomials. That is, we add or subtract like terms, and we use FOIL to multiply. Division, however, is comparable to division by the sum or difference of radicals, where we multiply by the conjugate to get a rational denominator. To express the quotient of two complex numbers in standard form, we also multiply by the conjugate of the denominator. **Work Exercises 89–94 in order,** *to better understand these ideas.*

89. (a) Add the two binomials.   **(b)** Add the two complex numbers.

90. (a) Subtract the second binomial from the first.
   (b) Subtract the second complex number from the first.

91. (a) Multiply the two binomials.
   (b) Multiply the two complex numbers.

92. (a) Rationalize the denominator: $\frac{\sqrt{3}-1}{1+\sqrt{2}}$.

   (b) Write in standard form: $\frac{3-i}{1+2i}$.

93. Explain why the answers for (a) and (b) in Exercise 91 do not correspond as the answers in Exercises 89–90 do.

94. Explain why the answers for (a) and (b) in Exercise 92 do not correspond as the answers in Exercises 89–90 do.

Perform the indicated operations. Give answers in standard form.

95. $\dfrac{3}{2-i} + \dfrac{5}{1+i}$

96. $\dfrac{2}{3+4i} + \dfrac{4}{1-i}$

97. $\left(\dfrac{2+i}{2-i} + \dfrac{i}{1+i}\right)i$

98. $\left(\dfrac{4-i}{1+i} - \dfrac{2i}{2+i}\right)4i$

| Chapter **10** | Group Activity |

Solar Electricity

OBJECTIVE Apply the Pythagorean formula.

In this activity you will determine the sizes of frames needed to support solar electric panels on a flat roof.

A. The following table gives three different solar modules by Solarex. Have each member of the group choose one of the solar panels.

Model	Watts	Volts	Amps	Size (in inches)	Cost (in dollars)
MSX-77	77	16.9	4.56	44 × 26	475
MSX-83	83	17.1	4.85	44 × 24	490
MSX-60	60	17.1	3.5	44 × 20	382

Source: Solarex table in *Jade Mountain* catalog.

B. To use your solar panel, you must make a wooden frame to support it. The sides of this frame will form a right triangle. The hypotenuse of the triangle will be the width of the solar panel you chose. Make a sketch and use the Pythagorean formula to find the dimensions of the legs for each frame given the following conditions. Round answers to the nearest tenth.

1. The legs have equal length.
2. One leg is twice the length of the other.
3. One leg is 3 times the length of the other.

C. Compare the different frame sizes for each panel. What factors might determine which of the triangles you would use in your frame?

CHAPTER **10** SUMMARY

KEY TERMS

10.1 square root	fourth root	**10.5** rationalizing the	real part
principal square root	index (order)	denominator	imaginary part
radicand	radical expression	conjugates	imaginary numbers
radical	square root function	**10.6** radical equation	standard form
perfect square	cube root function	extraneous solution	complex conjugates
cube root	**10.3** simplified radical	**10.7** complex number	

NEW SYMBOLS

$\sqrt{}$ radical sign	$\sqrt[n]{a}$ principal nth root of a	$a^{1/n}$ a to the power $\dfrac{1}{n}$	i a number whose square is -1
$\sqrt[3]{a}$ cube root of a			
$\sqrt[4]{a}$ principal fourth root of a	$\approx$ is approximately equal to	$a^{m/n}$ a to the power $\dfrac{m}{n}$	

TEST YOUR WORD POWER

See how well you have learned the vocabulary in this chapter. Answers, with examples, follow the Quick Review.

1. A **radicand** is
   A. the index of a radical
   B. the number or expression under the radical sign
   C. the positive root of a number
   D. the radical sign.

2. The **Pythagorean formula** states that, in a right triangle,
   A. the sum of the measures of the angles is 180°
   B. the sum of the lengths of the two shorter sides equals the length of the longest side
   C. the longest side is opposite the right angle
   D. the square of the length of the longest side equals the sum of the squares of the lengths of the two shorter sides.

3. A **hypotenuse** is
   A. either of the two shorter sides of a triangle
   B. the shortest side of a triangle
   C. the side opposite the right angle in a right triangle
   D. the longest side in any triangle.

4. **Rationalizing the denominator** is the process of
   A. eliminating fractions from a radical expression
   B. changing the denominator of a fraction from a radical to a rational number

   C. clearing a radical expression of radicals
   D. multiplying radical expressions.

5. An **extraneous solution** is a solution
   A. that does not satisfy the original equation
   B. that makes an equation true
   C. that makes an expression equal 0
   D. that checks in the original equation.

6. A **complex number** is
   A. a real number that includes a complex fraction
   B. a zero multiple of i
   C. a number of the form $a + bi$, where a and b are real numbers
   D. the square root of -1.

QUICK REVIEW

CONCEPTS	EXAMPLES

10.1   RADICAL EXPRESSIONS AND GRAPHS

$\sqrt[n]{a} = b$ means $b^n = a$.

$\sqrt[n]{a}$ is the principal nth root of a.

$\sqrt[n]{a^n} = |a|$ if n is even.

$\sqrt[n]{a^n} = a$ if n is odd.

The two square roots of 64 are $\sqrt{64} = 8$, the principal square root, and $-\sqrt{64} = -8$.

$$\sqrt[4]{(-2)^4} = |-2| = 2$$
$$\sqrt[3]{-27} = -3$$

(continued)

CONCEPTS	EXAMPLES

Functions Defined by Radical Expressions

The square root function with $f(x) = \sqrt{x}$ and the cube root function with $f(x) = \sqrt[3]{x}$ are two important functions defined by radical expressions.

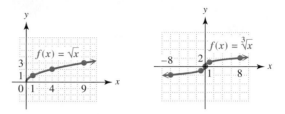

10.2 RATIONAL EXPONENTS

$a^{1/n} = \sqrt[n]{a}$ whenever $\sqrt[n]{a}$ exists.

If m and n are positive integers with $\frac{m}{n}$ in lowest terms, then $a^{m/n} = (a^{1/n})^m$, provided that $a^{1/n}$ is a real number.

All of the usual definitions and rules for exponents are valid for rational exponents.

$81^{1/2} = \sqrt{81} = 9 \qquad -64^{1/3} = -\sqrt[3]{64} = -4$

$8^{5/3} = (8^{1/3})^5 = 2^5 = 32$

$5^{-1/2} \cdot 5^{1/4} = 5^{-1/2+1/4} = 5^{-1/4} = \dfrac{1}{5^{1/4}} \qquad (y^{2/5})^{10} = y^4$

$\dfrac{x^{-1/3}}{x^{-1/2}} = x^{-1/3-(-1/2)} = x^{-1/3+1/2} = x^{1/6}, \quad x > 0$

10.3 SIMPLIFYING RADICAL EXPRESSIONS

Product and Quotient Rules for Radicals

If $\sqrt[n]{a}$ and $\sqrt[n]{b}$ are real numbers and n is a natural number, then

$$\sqrt[n]{a} \cdot \sqrt[n]{b} = \sqrt[n]{ab}$$

and

$$\sqrt[n]{\dfrac{a}{b}} = \dfrac{\sqrt[n]{a}}{\sqrt[n]{b}}, \qquad b \neq 0.$$

$$\sqrt{3} \cdot \sqrt{7} = \sqrt{21}$$

$$\sqrt[5]{x^3 y} \cdot \sqrt[5]{xy^2} = \sqrt[5]{x^4 y^3}$$

$$\dfrac{\sqrt{x^5}}{\sqrt{x^4}} = \sqrt{\dfrac{x^5}{x^4}} = \sqrt{x}, \quad x > 0$$

Simplified Radical

1. The radicand has no factor raised to a power greater than or equal to the index.
2. The radicand has no fractions.
3. No denominator contains a radical.
4. Exponents in the radicand and the index of the radical have no common factor (except 1).

$$\sqrt{18} = \sqrt{9 \cdot 2} = 3\sqrt{2}$$

$$\sqrt[3]{54x^5y^3} = \sqrt[3]{27x^3y^3 \cdot 2x^2} = 3xy\sqrt[3]{2x^2}$$

$$\sqrt{\dfrac{7}{4}} = \dfrac{\sqrt{7}}{\sqrt{4}} = \dfrac{\sqrt{7}}{2}$$

$$\sqrt[9]{x^3} = x^{3/9} = x^{1/3} \quad \text{or} \quad \sqrt[3]{x}$$

Pythagorean Formula

If c is the length of the longest side of a right triangle and a and b are the lengths of the shorter sides, then $c^2 = a^2 + b^2$. The longest side is the hypotenuse and the two shorter sides are the legs of the triangle. The hypotenuse is opposite the right angle.

Find b for the triangle in the figure.

$$10^2 + b^2 = (2\sqrt{61})^2$$

$$b^2 = 4(61) - 100$$

$$b^2 = 144$$

$$b = 12$$

Distance Formula

The distance between (x_1, y_1) and (x_2, y_2) is

$$d = \sqrt{(x_2 - x_1)^2 + (y_2 - y_1)^2}.$$

The distance between $(3, -2)$ and $(-1, 1)$ is

$$\sqrt{(-1-3)^2 + [1-(-2)]^2} = \sqrt{(-4)^2 + 3^2}$$

$$= \sqrt{16 + 9} = \sqrt{25} = 5.$$

CONCEPTS	EXAMPLES

10.4   ADDING AND SUBTRACTING RADICAL EXPRESSIONS

Only radical expressions with the same index and the same radicand may be combined.

$$3\sqrt{17} + 2\sqrt{17} - 8\sqrt{17} = (3 + 2 - 8)\sqrt{17}$$
$$= -3\sqrt{17}$$

$$\sqrt[3]{2} - \sqrt[3]{250} = \sqrt[3]{2} - 5\sqrt[3]{2}$$
$$= -4\sqrt[3]{2}$$

$$\left.\begin{array}{l}\sqrt{15} + \sqrt{30} \\ \sqrt{3} + \sqrt[3]{9}\end{array}\right\} \quad \begin{array}{l}\text{cannot be} \\ \text{simplified further}\end{array}$$

10.5   MULTIPLYING AND DIVIDING RADICAL EXPRESSIONS

Multiply binomial radical expressions by using the FOIL method. Special products from Section 4.6 may apply.

$$\left(\sqrt{2} + \sqrt{7}\right)\left(\sqrt{3} - \sqrt{6}\right)$$
$$= \sqrt{6} - 2\sqrt{3} + \sqrt{21} - \sqrt{42} \qquad \sqrt{12} = 2\sqrt{3}$$

$$\left(\sqrt{5} - \sqrt{10}\right)\left(\sqrt{5} + \sqrt{10}\right) = 5 - 10 = -5$$

$$\left(\sqrt{3} - \sqrt{2}\right)^2 = 3 - 2\sqrt{3}\cdot\sqrt{2} + 2 = 5 - 2\sqrt{6}$$

Rationalize the denominator by multiplying both the numerator and denominator by the same expression.

$$\frac{\sqrt{7}}{\sqrt{5}} = \frac{\sqrt{7}}{\sqrt{5}}\cdot\frac{\sqrt{5}}{\sqrt{5}} = \frac{\sqrt{35}}{5}$$

$$\frac{\sqrt[3]{2}}{\sqrt[3]{4}} = \frac{\sqrt[3]{2}}{\sqrt[3]{4}}\cdot\frac{\sqrt[3]{2}}{\sqrt[3]{2}} = \frac{\sqrt[3]{4}}{\sqrt[3]{8}} = \frac{\sqrt[3]{4}}{2}$$

$$\frac{4}{\sqrt{5} - \sqrt{2}} = \frac{4}{\sqrt{5} - \sqrt{2}}\cdot\frac{\sqrt{5} + \sqrt{2}}{\sqrt{5} + \sqrt{2}}$$
$$= \frac{4\left(\sqrt{5} + \sqrt{2}\right)}{5 - 2} = \frac{4\left(\sqrt{5} + \sqrt{2}\right)}{3}$$

$$\frac{5 + 15\sqrt{6}}{10} = \frac{5\left(1 + 3\sqrt{6}\right)}{5\cdot 2} = \frac{1 + 3\sqrt{6}}{2}$$

10.6   SOLVING EQUATIONS WITH RADICALS

Solving an Equation with Radicals

Step 1   Isolate one radical on one side of the equation.

Step 2   Raise both sides of the equation to a power that is the same as the index of the radical.

Step 3   Solve the resulting equation; if it still contains a radical, repeat Steps 1 and 2.

Step 4   Check all potential solutions in the *original* equation.

Potential solutions that do not check are extraneous; they are not part of the solution set.

Solve $\sqrt{2x + 3} - x = 0$.

$$\sqrt{2x + 3} = x$$
$$\left(\sqrt{2x + 3}\right)^2 = x^2$$
$$2x + 3 = x^2$$
$$x^2 - 2x - 3 = 0$$
$$(x - 3)(x + 1) = 0$$
$$x - 3 = 0 \quad \text{or} \quad x + 1 = 0$$
$$x = 3 \quad \text{or} \qquad x = -1$$

A check shows that 3 is a solution, but -1 is extraneous. The solution set is $\{3\}$.

(continued)

CONCEPTS	EXAMPLES

10.7 COMPLEX NUMBERS

$i^2 = -1$, so $i = \sqrt{-1}$.

For any positive number b, $\sqrt{-b} = i\sqrt{b}$.

To multiply radicals with negative radicands, first change each factor to the form $i\sqrt{b}$, then multiply. The same procedure applies to quotients.

$$\sqrt{-25} = i\sqrt{25} = 5i$$

$$\sqrt{-3} \cdot \sqrt{-27} = i\sqrt{3} \cdot i\sqrt{27}$$
$$= i^2\sqrt{81}$$
$$= -1 \cdot 9$$
$$= -9$$

$$\frac{\sqrt{-18}}{\sqrt{-2}} = \frac{i\sqrt{18}}{i\sqrt{2}} = \sqrt{\frac{18}{2}} = \sqrt{9} = 3$$

Adding and Subtracting Complex Numbers

Add (or subtract) the real parts and add (or subtract) the imaginary parts.

$$(5 + 3i) + (8 - 7i) = 13 - 4i$$
$$(5 + 3i) - (8 - 7i) = -3 + 10i$$

Multiplying and Dividing Complex Numbers

Multiply complex numbers by using the FOIL method.

$$(2 + i)(5 - 3i) = 10 - 6i + 5i - 3i^2$$
$$= 10 - i - 3(-1)$$
$$= 10 - i + 3$$
$$= 13 - i$$

Divide complex numbers by multiplying the numerator and the denominator by the conjugate of the denominator.

$$\frac{20}{3 + i} = \frac{20(3 - i)}{(3 + i)(3 - i)} = \frac{20(3 - i)}{9 - i^2}$$
$$= \frac{20(3 - i)}{10} = 2(3 - i) = 6 - 2i$$

Answers to Test Your Word Power

1. B; *Example:* In $\sqrt{3xy}$, $3xy$ is the radicand.   **2.** D; *Example:* In a right triangle where $a = 6$, $b = 8$, and $c = 10$, $6^2 + 8^2 = 10^2$.

3. C; *Example:* In a right triangle where the sides measure 9, 12, and 15 units, the hypotenuse is the side opposite the right angle, with

measure 15 units.   **4.** B; *Example:* To rationalize the denominator of $\dfrac{5}{\sqrt{3} + 1}$, multiply both the numerator and denominator by $\sqrt{3} - 1$ to

get $\dfrac{5(\sqrt{3} - 1)}{2}$.   **5.** A; *Example:* The potential solution 2 is extraneous in $\sqrt{5q - 1} + 3 = 0$.   **6.** C; *Examples:* -5 (or $-5 + 0i$), $7i$

(or $0 + 7i$), $\sqrt{2} - 4i$

CHAPTER 10 REVIEW EXERCISES

[10.1] *Find each root.*

1. $\sqrt{1764}$ **2.** $-\sqrt{289}$ **3.** $\sqrt[3]{216}$

4. $\sqrt[3]{-125}$ **5.** $-\sqrt[3]{27}$ **6.** $\sqrt[5]{-32}$

7. Under what conditions is $\sqrt[n]{a}$ not a real number?

8. Simplify each radical so that no radicals appear. Assume x represents any real number.

     (a) $\sqrt{x^2}$ **(b)** $-\sqrt{x^2}$ **(c)** $\sqrt[3]{x^3}$

Use a calculator to find a decimal approximation for each number. Give the answer to the nearest thousandth.

9. $-\sqrt{47}$

10. $\sqrt[3]{-129}$

11. $\sqrt[4]{605}$

12. $500^{-3/4}$

13. $-500^{4/3}$

14. $-28^{-1/2}$

Graph each function. Give the domain and the range.

15. $f(x) = \sqrt{x-1}$

16. $f(x) = \sqrt[3]{x} + 4$

17. What is the best estimate of the area of the triangle shown here?

A. 3600 **B.** 30 **C.** 60 **D.** 360

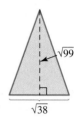

[10.2]

18. Fill in the blanks with the correct responses: One way to evaluate $8^{2/3}$ is to first find the _____ root of _____, which is _____. Then raise that result to the _____ power, to get an answer of _____. Therefore, $8^{2/3} =$ _____.

19. Which one of the following is a positive number?

A. $(-27)^{2/3}$ **B.** $(-64)^{5/3}$ **C.** $(-100)^{1/2}$ **D.** $(-32)^{1/5}$

20. If a is a negative number and n is odd, then what must be true about m for $a^{m/n}$ to be

(a) positive **(b)** negative?

21. If a is negative and n is even, then what can be said about $a^{1/n}$?

Simplify. If the expression does not represent a real number, say so.

22. $49^{1/2}$

23. $-121^{1/2}$

24. $16^{5/4}$

25. $-8^{2/3}$

26. $-\left(\dfrac{36}{25}\right)^{3/2}$

27. $\left(-\dfrac{1}{8}\right)^{-5/3}$

28. $\left(\dfrac{81}{10,000}\right)^{-3/4}$

29. $(-16)^{3/4}$

30. Solve the Pythagorean formula $a^2 + b^2 = c^2$ for b, where $b > 0$.

31. Explain the relationship between the expressions $a^{m/n}$ and $\sqrt[n]{a^m}$. Give an example.

Write each expression as a radical.

32. $(m + 3n)^{1/2}$

33. $(3a + b)^{-5/3}$

Write each expression with a rational exponent.

34. $\sqrt{7^9}$

35. $\sqrt[5]{p^4}$

Use the rules for exponents to simplify each expression. Write the answer with only positive exponents. Assume all variables represent positive real numbers.

36. $5^{1/4} \cdot 5^{7/4}$

37. $\dfrac{96^{2/3}}{96^{-1/3}}$

38. $\dfrac{(a^{1/3})^4}{a^{2/3}}$

39. $\dfrac{y^{-1/3} \cdot y^{5/6}}{y}$

40. $\left(\dfrac{z^{-1}x^{-3/5}}{2^{-2}z^{-1/2}x}\right)^{-1}$

41. $r^{-1/2}(r + r^{3/2})$

Simplify by first writing each radical in exponential form. Leave the answer in exponential form. Assume all variables represent positive real numbers.

42. $\sqrt[8]{s^4}$

43. $\sqrt[6]{r^9}$

44. $\dfrac{\sqrt{p^5}}{p^2}$

45. $\sqrt[4]{k^3} \cdot \sqrt{k^3}$

46. $\sqrt[3]{m^5} \cdot \sqrt[3]{m^8}$

47. $\sqrt[4]{\sqrt[3]{z}}$

48. $\sqrt{\sqrt{\sqrt{x}}}$

49. $\sqrt[3]{\sqrt[5]{x}}$

50. $\sqrt{\sqrt[6]{\sqrt[3]{x}}}$

51. By the product rule for exponents, we know that $2^{1/4} \cdot 2^{1/5} = 2^{9/20}$. However, there is no exponent rule to simplify $3^{1/4} \cdot 2^{1/5}$. Why?

[10.3] *Simplify each radical. Assume all variables represent positive real numbers.*

52. $\sqrt{6} \cdot \sqrt{11}$

53. $\sqrt{5} \cdot \sqrt{r}$

54. $\sqrt[3]{6} \cdot \sqrt[3]{5}$

55. $\sqrt[4]{7} \cdot \sqrt[4]{3}$

56. $\sqrt{20}$

57. $\sqrt{75}$

58. $-\sqrt{125}$

59. $\sqrt[3]{-108}$

60. $\sqrt{100y^7}$

61. $\sqrt[3]{64p^4q^6}$

62. $\sqrt[3]{108a^8b^5}$

63. $\sqrt[3]{632r^8t^4}$

64. $\sqrt{\dfrac{y^3}{144}}$

65. $\sqrt[3]{\dfrac{m^{15}}{27}}$

66. $\sqrt[3]{\dfrac{r^2}{8}}$

67. $\sqrt[4]{\dfrac{a^9}{81}}$

Simplify each radical expression.

68. $\sqrt[6]{15^3}$

69. $\sqrt[4]{p^6}$

70. $\sqrt[3]{2} \cdot \sqrt[4]{5}$

71. $\sqrt{x} \cdot \sqrt[5]{x}$

72. Find the missing length in the right triangle. Simplify the answer if applicable.

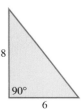

73. Find the distance between the points $(-4, 7)$ and $(10, 6)$.

[10.4] *Perform the indicated operations. Assume all variables represent positive real numbers.*

74. $2\sqrt{8} - 3\sqrt{50}$

75. $8\sqrt{80} - 3\sqrt{45}$

76. $-\sqrt{27y} + 2\sqrt{75y}$

77. $2\sqrt{54m^3} + 5\sqrt{96m^3}$

78. $3\sqrt[3]{54} + 5\sqrt[3]{16}$

79. $-6\sqrt[4]{32} + \sqrt[4]{512}$

80. $\dfrac{3}{\sqrt{16}} - \dfrac{\sqrt{5}}{2}$

81. $\dfrac{4}{\sqrt{25}} + \dfrac{\sqrt{5}}{4}$

In Exercises 82 and 83, leave answers as simplified radicals.

82. Find the perimeter of a rectangular electronic billboard having sides of lengths shown in the figure.

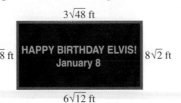

83. Find the perimeter of a triangular electronic highway road sign having the dimensions shown in the figure.

[10.5] *Multiply.*

84. $\left(\sqrt{3} + 1\right)\left(\sqrt{3} - 2\right)$

85. $\left(\sqrt{7} + \sqrt{5}\right)\left(\sqrt{7} - \sqrt{5}\right)$

86. $\left(3\sqrt{2} + 1\right)\left(2\sqrt{2} - 3\right)$

87. $\left(\sqrt{13} - \sqrt{2}\right)^2$

88. $\left(\sqrt[3]{2} + 3\right)\left(\sqrt[3]{4} - 3\sqrt[3]{2} + 9\right)$

89. $\left(\sqrt[3]{4y} - 1\right)\left(\sqrt[3]{4y} + 3\right)$

90. Use a calculator to show that the answer to Exercise 87, $15 - 2\sqrt{26}$, is not equal to $13\sqrt{26}$.

91. A friend wants to rationalize the denominator of the fraction $\dfrac{5}{\sqrt[3]{6}}$, and she decides to multiply the numerator and denominator by $\sqrt[3]{6}$. Why will her plan *not* work?

Rationalize each denominator. Assume all variables represent positive real numbers.

92. $\dfrac{\sqrt{6}}{\sqrt{5}}$

93. $\dfrac{-6\sqrt{3}}{\sqrt{2}}$

94. $\dfrac{3\sqrt{7p}}{\sqrt{y}}$

95. $\sqrt{\dfrac{11}{8}}$

96. $-\sqrt[3]{\dfrac{9}{25}}$

97. $\sqrt[3]{\dfrac{108m^3}{n^5}}$

98. $\dfrac{1}{\sqrt{2} + \sqrt{7}}$

99. $\dfrac{-5}{\sqrt{6} - 3}$

Write in lowest terms.

100. $\dfrac{2 - 2\sqrt{5}}{8}$

101. $\dfrac{4 - 8\sqrt{8}}{12}$

102. $\dfrac{-18 + \sqrt{27}}{6}$

[10.6] *Solve each equation.*

103. $\sqrt{8x + 9} = 5$

104. $\sqrt{2z - 3} - 3 = 0$

105. $\sqrt{3m + 1} - 2 = -3$

106. $\sqrt{7z + 1} = z + 1$

107. $3\sqrt{m} = \sqrt{10m - 9}$

108. $\sqrt{p^2 + 3p + 7} = p + 2$

109. $\sqrt{a + 2} - \sqrt{a - 3} = 1$

110. $\sqrt[3]{5m - 1} = \sqrt[3]{3m - 2}$

111. $\sqrt[3]{2x^2 + 3x - 7} = \sqrt[3]{2x^2 + 4x + 6}$

112. $\sqrt[3]{3y^2 - 4y + 6} = \sqrt[3]{3y^2 - 2y + 8}$

113. $\sqrt[3]{1 - 2k} - \sqrt[3]{-k - 13} = 0$

114. $\sqrt[3]{11 - 2t} - \sqrt[3]{-1 - 5t} = 0$

115. $\sqrt[4]{x - 1} + 2 = 0$

116. $\sqrt[4]{2k + 3} + 1 = 0$

117. $\sqrt[4]{x + 7} = \sqrt[4]{2x}$

118. $\sqrt[4]{x + 8} = \sqrt[4]{3x}$

RELATING CONCEPTS (EXERCISES 119–125)

For Individual or Group Work

Solve the equations in Exercises 119–124 in order, and then use a generalization to fill in the blanks in Exercise 125.

119. $x = 3$

120. $x = -3$

121. $x^2 = 9$

122. $x^3 = 27$

123. $x^4 = 81$

124. $x^5 = -243$

125. Suppose both sides of $x = k$ are raised to the nth power.

(a) If n is even, the number of solutions of the new equation is
_____ the number of solutions of the original
(more than/the same as/fewer than)
equation.

(b) If n is odd, the number of solutions of the new equation is
_____ the number of solutions of the original
(more than/the same as/fewer than)
equation.

126. Carpenters stabilize wall frames with a diagonal brace, as shown in the figure. The length of the brace is given by $L = \sqrt{H^2 + W^2}$.

    (a) Solve this formula for H.

    (b) If the bottom of the brace is attached 9 ft from the corner and the brace is 12 ft long, how far up the corner post should it be nailed? Give your answer to the nearest tenth of a foot.

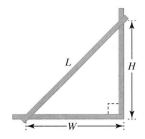

[10.7] *Write each expression as a product of a real number and i.*

127. $\sqrt{-25}$ **128.** $\sqrt{-200}$

129. If a is a positive real number, is $-\sqrt{-a}$ a real number?

Perform the indicated operations. Write each imaginary number answer in standard form $a + bi$.

130. $(-2 + 5i) + (-8 - 7i)$ **131.** $(5 + 4i) - (-9 - 3i)$ **132.** $\sqrt{-5} \cdot \sqrt{-7}$

133. $\sqrt{-25} \cdot \sqrt{-81}$ **134.** $\dfrac{\sqrt{-72}}{\sqrt{-8}}$ **135.** $(2 + 3i)(1 - i)$

136. $(6 - 2i)^2$ **137.** $\dfrac{3 - i}{2 + i}$ **138.** $\dfrac{5 + 14i}{2 + 3i}$

Find each power of i.

139. i^{11} **140.** i^{36} **141.** i^{-10} **142.** i^{-8}

▮ MIXED REVIEW EXERCISES

Simplify. Assume all variables represent positive real numbers.

143. $-\sqrt[4]{256}$ **144.** $1000^{-2/3}$ **145.** $\dfrac{z^{-1/5} \cdot z^{3/10}}{z^{7/10}}$

146. $\sqrt[4]{k^{24}}$ **147.** $\sqrt[3]{54z^9t^8}$ **148.** $-5\sqrt{18} + 12\sqrt{72}$

149. $8\sqrt[3]{x^3y^2} - 2x\sqrt[3]{y^2}$ **150.** $\left(\sqrt{5} - \sqrt{3}\right)\left(\sqrt{7} + \sqrt{3}\right)$

151. $\dfrac{-1}{\sqrt{12}}$ **152.** $\sqrt[3]{\dfrac{12}{25}}$ **153.** i^{-1000}

154. $\sqrt{-49}$ **155.** $(4 - 9i) + (-1 + 2i)$ **156.** $\dfrac{\sqrt{50}}{\sqrt{-2}}$

157. $\dfrac{3 + \sqrt{54}}{6}$ **158.** $(3 + 2i)^2$

Solve each equation.

159. $\sqrt{x + 4} = x - 2$ **160.** $\sqrt[3]{2x - 9} = \sqrt[3]{5x + 3}$

161. $\sqrt{6 + 2x} - 1 = \sqrt{7 - 2x}$ **162.** $\sqrt{7x + 11} - 5 = 0$

163. $\sqrt{6x + 2} - \sqrt{5x + 3} = 0$ **164.** $\sqrt{3 + 5x} - \sqrt{x + 11} = 0$

165. $3\sqrt{x} = \sqrt{8x + 9}$ **166.** $6\sqrt{p} = \sqrt{30p + 24}$

167. $\sqrt{11 + 2x} + 1 = \sqrt{5x + 1}$ **168.** $\sqrt{5x + 6} - \sqrt{x + 3} = 3$

CHAPTER **10** TEST

Evaluate.

1. $-\sqrt{841}$ **2.** $\sqrt[3]{-512}$ **3.** $125^{1/3}$

4. For $\sqrt{146.25}$, which choice gives the best estimate?
   A. 10   **B.** 11   **C.** 12   **D.** 13

Use a calculator to approximate each root to the nearest thousandth.

5. $\sqrt{478}$ **6.** $\sqrt[3]{-832}$

7. Graph the function defined by $f(x) = \sqrt{x + 6}$, and give the domain and range.

Simplify each expression. Assume all variables represent positive real numbers.

8. $\left(\dfrac{16}{25}\right)^{-3/2}$ **9.** $(-64)^{-4/3}$ **10.** $\dfrac{3^{2/5}x^{-1/4}y^{2/5}}{3^{-8/5}x^{7/4}y^{1/10}}$

11. $\left(\dfrac{x^{-4}y^{-6}}{x^{-2}y^3}\right)^{-2/3}$ **12.** $7^{3/4} \cdot 7^{-1/4}$ **13.** $\sqrt[3]{a^4} \cdot \sqrt[3]{a^7}$

14. Use the Pythagorean formula to find the exact length of side b in the figure.

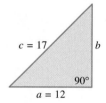

$c = 17$ b $90°$ $a = 12$

15. Find the distance between the points $(-4, 2)$ and $(2, 10)$.

Simplify each expression. Assume all variables represent positive real numbers.

16. $\sqrt{54x^5y^6}$ **17.** $\sqrt[4]{32a^7b^{13}}$

18. $\sqrt{2} \cdot \sqrt[3]{5}$ (Express as a radical.)   **19.** $3\sqrt{20} - 5\sqrt{80} + 4\sqrt{500}$

20. $\sqrt[3]{16t^3s^5} - \sqrt[3]{54t^6s^2}$ **21.** $(7\sqrt{5} + 4)(2\sqrt{5} - 1)$

22. $(\sqrt{3} - 2\sqrt{5})^2$ **23.** $\dfrac{-5}{\sqrt{40}}$

24. $\dfrac{2}{\sqrt[3]{5}}$ **25.** $\dfrac{-4}{\sqrt{7} + \sqrt{5}}$

26. Write $\dfrac{6 + \sqrt{24}}{2}$ in lowest terms.

27. The following formula is used in physics, relating the velocity of sound V to the temperature T.

$$V = \frac{V_0}{\sqrt{1 - kT}}$$

   (a) Find an approximation of V to the nearest tenth if $V_0 = 50$, $k = .01$, and $T = 30$. Use a calculator.

   (b) Solve the formula for T.

Solve each equation.

28. $\sqrt[3]{5x} = \sqrt[3]{2x - 3}$

29. $x + \sqrt{x + 6} = 9 - x$

30. $\sqrt{x + 4} - \sqrt{1 - x} = -1$

Perform the indicated operations. Express the answers in standard form $a + bi$.

31. $(-2 + 5i) - (3 + 6i) - 7i$

32. $(1 + 5i)(3 + i)$

33. $\dfrac{7 + i}{1 - i}$

34. Simplify i^{37}.

35. Answer *true* or *false* to each of the following.

    (a) $i^2 = -1$ **(b)** $i = \sqrt{-1}$ **(c)** $i = -1$ **(d)** $\sqrt{-3} = i\sqrt{3}$

CUMULATIVE REVIEW EXERCISES CHAPTERS **1–10**

Evaluate each expression if $a = -3$, $b = 5$, and $c = -4$.

1. $|2a^2 - 3b + c|$

2. $\dfrac{(a + b)(a + c)}{3b - 6}$

Solve each equation or inequality.

3. $3(x + 2) - 4(2x + 3) = -3x + 2$

4. $\dfrac{1}{3}x + \dfrac{1}{4}(x + 8) = x + 7$

5. $.04x + .06(100 - x) = 5.88$

6. $-5 - 3(m - 2) < 11 - 2(m + 2)$

7. $2k + 4 < 10$ and $3k - 1 > 5$

8. $2k + 4 > 10$ or $3k - 1 < 5$

9. $|6x + 7| = 13$

10. $|-2x + 4| = |-2x - 3|$

11. $|2p - 5| \geq 9$

12. Find the measures of the marked angles.

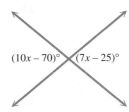

$(10x - 70)°$ $(7x - 25)°$

Solve each problem.

13. A piggy bank has 50 coins, all of which are nickels and quarters. The total value of the money is $8.90. How many of each denomination are there in the bank?

14. How many liters of pure alcohol must be mixed with 40 L of 18% alcohol to obtain a 22% alcohol solution?

15. Graph the equation $4x - 3y = 12$.

16. Find the slope of the line passing through the points $(-4, 6)$ and $(2, -3)$. Then find the equation of the line and write it in the form $y = mx + b$.

Perform the indicated operations.

17. $(3k^3 - 5k^2 + 8k - 2) - (4k^3 + 11k + 7) + (2k^2 - 5k)$

18. $(8x - 7)(x + 3)$

19. $\dfrac{8z^3 - 16z^2 + 24z}{8z^2}$

20. $\dfrac{6y^4 - 3y^3 + 5y^2 + 6y - 9}{2y + 1}$

Factor each polynomial completely.

21. $2p^2 - 5pq + 3q^2$ 　**22.** $3k^4 + k^2 - 4$ 　**23.** $x^3 + 512$

Solve by factoring.

24. $2x^2 + 11x + 15 = 0$ 　**25.** $5t(t - 1) = 2(1 - t)$

Perform each operation and express the answer in lowest terms.

26. $\dfrac{y^2 + y - 12}{y^3 + 9y^2 + 20y} \div \dfrac{y^2 - 9}{y^3 + 3y^2}$ 　**27.** $\dfrac{1}{x + y} + \dfrac{3}{x - y}$

Simplify each complex fraction.

28. $\dfrac{\dfrac{-6}{x - 2}}{\dfrac{8}{3x - 6}}$ 　**29.** $\dfrac{\dfrac{1}{a} - \dfrac{1}{b}}{\dfrac{a}{b} - \dfrac{b}{a}}$

30. Natalie can ride her bike 4 mph faster than her husband, Chuck. If Natalie can ride 48 mi in the same time that Chuck can ride 24 mi, what are their speeds?

31. Solve the equation $\dfrac{p + 1}{p - 3} = \dfrac{4}{p - 3} + 6$.

32. If $f(x) = 3x - 7$, find $f(-10)$.

33. Solve the system by elimination or substitution.

$$\begin{aligned} 3x - y &= 23 \\ 2x + 3y &= 8 \end{aligned}$$

34. Solve the system by matrix methods.

$$\begin{aligned} x + y + z &= 1 \\ x - y - z &= -3 \\ x + y - z &= -1 \end{aligned}$$

Solve the problem by using a system of equations.

35. In 1997, if you had sent five 2-oz letters and three 3-oz letters using first-class mail, it would have cost you $5.09. Sending three 2-oz letters and five 3-oz letters would have cost $5.55. What was the 1997 postage rate for one 2-oz letter and for one 3-oz letter? (*Source:* U.S. Postal Service.)

Write each expression in simplest form, using only positive exponents. Assume all variables represent positive real numbers.

36. $27^{-2/3}$

37. $\sqrt{200x^4}$

38. $\sqrt[3]{16x^2y} \cdot \sqrt[3]{3x^3y}$

39. $\sqrt{50} + \sqrt{8}$

40. $\dfrac{1}{\sqrt{10} - \sqrt{8}}$

41. $\left(2\sqrt{x} + \sqrt{y}\right)\left(-3\sqrt{x} - 4\sqrt{y}\right)$

42. Find the distance between the points $(-4, 4)$ and $(-2, 9)$.

43. Solve the equation $\sqrt{3r - 8} = r - 2$.

44. The *fall speed,* in miles per hour, of a vehicle running off the road into a ditch is given by

$$S = \frac{2.74D}{\sqrt{h}},$$

where D is the horizontal distance traveled from the level surface to the bottom of the ditch and h is the height (or depth) of the ditch. What is the fall speed of a vehicle that traveled 32 ft horizontally into a 5-ft-deep ditch?

Write in standard form $a + bi$.

45. $(5 + 7i) - (3 - 2i)$

46. $\dfrac{6 - 2i}{1 - i}$

Quadratic Equations, Inequalities, and Functions

11

In recent years, the number of publicly traded U.S. companies filing for bankruptcy has been at its highest level since the recession of the early 1990s. One casualty of this trend was retailer Montgomery Ward. Started in 1872 as a mail-order catalog business, the company grew to include 250 stores in 30 states. After filing for Chapter 11 bankruptcy protection, the retailer closed for good in 2001. (*Source: USA Today,* December 29, 2000.)

Since then, other high-profile U.S. companies including Enron Corp., Adelphia Communications Corp., Kmart Corp., WorldCom Inc., US Airways, and United Airlines have become victims of this trend. In Sections 11.5 and 11.6, we use *quadratic functions* to model the number of company bankruptcy filings.

11.1 Solving Quadratic Equations by the Square Root Property

We introduced quadratic equations in Section 5.5. Recall that a *quadratic equation* is defined as follows.

> **Quadratic Equation**
>
> An equation that can be written in the form
> $$ax^2 + bx + c = 0,$$
> where a, b, and c are real numbers, with $a \neq 0$, is a **quadratic equation.** The given form is called **standard form.**

A quadratic equation is a *second-degree equation,* that is, an equation with a squared term and no terms of higher degree. For example,
$$4m^2 + 4m - 5 = 0 \quad \text{and} \quad 3x^2 = 4x - 8$$
are quadratic equations, with the first equation in standard form.

OBJECTIVE 1 Review the zero-factor property.   In Section 5.5 we used factoring and the zero-factor property to solve quadratic equations.

> **Zero-Factor Property**
>
> If two numbers have a product of 0, then at least one of the numbers must be 0. That is, if $ab = 0$, then $a = 0$ or $b = 0$.

We solved a quadratic equation such as $3x^2 - 5x - 28 = 0$ using the zero-factor property as follows.

$$3x^2 - 5x - 28 = 0$$
$$(3x + 7)(x - 4) = 0 \qquad \text{Factor.}$$
$$3x + 7 = 0 \quad \text{or} \quad x - 4 = 0 \qquad \text{Zero-factor property}$$
$$3x = -7 \quad \text{or} \qquad x = 4 \qquad \text{Solve each equation.}$$
$$x = -\frac{7}{3}$$

The solution set is $\left\{-\frac{7}{3}, 4\right\}$.

Although factoring is the simplest way to solve quadratic equations, not every quadratic equation can be solved easily by factoring. In this section and Sections 11.2 and 11.3, we develop three other methods of solving quadratic equations.

OBJECTIVE 2 Solve equations of the form $x^2 = k$, where $k > 0$. We can solve a quadratic equation such as $x^2 = 9$ by factoring as follows.

$$x^2 = 9$$

$$x^2 - 9 = 0 \qquad \text{Subtract 9.}$$

$$(x + 3)(x - 3) = 0 \qquad \text{Factor.}$$

$$x + 3 = 0 \quad \text{or} \quad x - 3 = 0 \qquad \text{Zero-factor property}$$

$$x = -3 \quad \text{or} \qquad x = 3$$

The solution set is $\{-3, 3\}$.

We might also have solved $x^2 = 9$ by noticing that x must be a number whose square is 9. Thus, $x = \sqrt{9} = 3$ or $x = -\sqrt{9} = -3$. This is generalized as the **square root property of equations.**

Square Root Property of Equations

If k is a positive number and if $x^2 = k$, then

$$x = \sqrt{k} \qquad \text{or} \qquad x = -\sqrt{k},$$

and the solution set is $\{-\sqrt{k}, \sqrt{k}\}$.

NOTE When we solve an equation, we want to find *all* values of the variable that satisfy the equation. Therefore, we want both the positive and negative square roots of k.

███ **EXAMPLE 1   Solving Quadratic Equations of the Form $x^2 = k$**

Solve each equation. Write radicals in simplified form.

(a) $x^2 = 16$

By the square root property, since $x^2 = 16$,

$$x = \sqrt{16} = 4 \qquad \text{or} \qquad x = -\sqrt{16} = -4.$$

An abbreviation for "$x = 4$ or $x = -4$" is written $x = \pm 4$ (read "positive or negative 4"). Check each solution by substituting it for x in the original equation. The solution set is $\{-4, 4\}$.

(b) $z^2 = 5$

By the square root property, the solutions are $z = \sqrt{5}$ or $z = -\sqrt{5}$. The solution set is $\{\sqrt{5}, -\sqrt{5}\}$, which may be written $\{\pm\sqrt{5}\}$.

(c) $\qquad\qquad m^2 - 8 = 0$

$$m^2 = 8 \qquad\qquad \text{Add 8.}$$

$$m = \sqrt{8} \quad \text{or} \quad m = -\sqrt{8} \qquad \text{Square root property}$$

$$m = 2\sqrt{2} \quad \text{or} \quad m = -2\sqrt{2} \qquad \text{Simplify } \sqrt{8}.$$

The solution set is $\{2\sqrt{2}, -2\sqrt{2}\}$.

(d) $4x^2 - 48 = 0$

Solve for x^2.

$$4x^2 - 48 = 0$$
$$4x^2 = 48 \qquad \text{Add 48.}$$
$$x^2 = 12 \qquad \text{Divide by 4.}$$
$$x = \sqrt{12} \quad \text{or} \quad x = -\sqrt{12} \qquad \text{Square root property}$$
$$x = 2\sqrt{3} \quad \text{or} \quad x = -2\sqrt{3} \qquad \sqrt{12} = \sqrt{4} \cdot \sqrt{3} = 2\sqrt{3}$$

The solutions are $2\sqrt{3}$ and $-2\sqrt{3}$. Check each in the original equation.

Check: $\qquad\qquad 4x^2 - 48 = 0 \qquad$ Original equation

$$4\left(2\sqrt{3}\right)^2 - 48 = 0 \quad ? \qquad\qquad 4\left(-2\sqrt{3}\right)^2 - 48 = 0 \quad ?$$
$$4(12) - 48 = 0 \quad ? \qquad\qquad\qquad 4(12) - 48 = 0 \quad ?$$
$$48 - 48 = 0 \quad ? \qquad\qquad\qquad 48 - 48 = 0 \quad ?$$
$$0 = 0 \quad \text{True} \qquad\qquad\qquad\qquad 0 = 0 \quad \text{True}$$

The solution set is $\left\{2\sqrt{3}, -2\sqrt{3}\right\}$.

(e) $3x^2 + 5 = 11$

Solve for x^2.

$$3x^2 + 5 = 11$$
$$3x^2 = 6 \qquad \text{Subtract 5.}$$
$$x^2 = 2 \qquad \text{Divide by 3.}$$
$$x = \sqrt{2} \quad \text{or} \quad x = -\sqrt{2} \qquad \text{Square root property}$$

Check each solution as shown in part (d). The solution set is $\left\{\sqrt{2}, -\sqrt{2}\right\}$.

Now Try Exercises 5, 7, 11, 15, and 19.

EXAMPLE 2 Using the Square Root Property in an Application

Galileo Galilei (1564–1642) developed a formula for freely falling objects described by

$$d = 16t^2,$$

where d is the distance in feet that an object falls (disregarding air resistance) in t sec, regardless of weight. Galileo dropped objects from the Leaning Tower of Pisa to develop this formula. If the Leaning Tower is about 180 ft tall, use Galileo's formula to determine how long it would take an object dropped from the tower to fall to the ground. (*Source: Microsoft Encarta Encyclopedia 2002.*)

We substitute 180 for d in Galileo's formula.

$$d = 16t^2$$
$$180 = 16t^2 \qquad \text{Let } d = 180.$$
$$11.25 = t^2 \qquad \text{Divide by 16.}$$
$$t = \sqrt{11.25} \quad \text{or} \quad t = -\sqrt{11.25} \qquad \text{Square root property}$$

Since time cannot be negative, we discard the negative solution. In applied problems, we usually prefer approximations to exact values. Using a calculator, $\sqrt{11.25} \approx 3.4$ so $t \approx 3.4$. The object would fall to the ground in about 3.4 sec.

Now Try Exercise 59.

OBJECTIVE 3 Solve equations of the form $(ax + b)^2 = k$, where $k > 0$. In each equation in Example 1, the exponent 2 appeared with a single variable as its base. We can extend the square root property to solve equations where the base is a binomial, as shown in the next examples.

EXAMPLE 3 Solving Quadratic Equations of the Form $(x + b)^2 = k$

Solve each equation.

(a) $(x - 3)^2 = 16$

Apply the square root property, using $x - 3$ as the base.

$$(x - 3)^2 = 16$$

$$x - 3 = \sqrt{16} \quad \text{or} \quad x - 3 = -\sqrt{16}$$

$$x - 3 = 4 \quad \text{or} \quad x - 3 = -4 \qquad \sqrt{16} = 4$$

$$x = 7 \quad \text{or} \quad x = -1 \qquad \text{Add 3.}$$

Check each solution in the original equation.

$$(x - 3)^2 = 16 \qquad\qquad (x - 3)^2 = 16$$

$$(7 - 3)^2 = 16 \ ? \quad \text{Let } x = 7. \qquad (-1 - 3)^2 = 16 \ ? \quad \text{Let } x = -1.$$

$$4^2 = 16 \ ? \qquad\qquad (-4)^2 = 16 \ ?$$

$$16 = 16 \quad \text{True} \qquad\qquad 16 = 16 \quad \text{True}$$

The solution set is $\{7, -1\}$.

(b) $(x + 1)^2 = 6$

By the square root property,

$$x + 1 = \sqrt{6} \quad \text{or} \quad x + 1 = -\sqrt{6}$$

$$x = -1 + \sqrt{6} \quad \text{or} \quad x = -1 - \sqrt{6}.$$

Check:

$$\left(-1 + \sqrt{6} + 1\right)^2 = \left(\sqrt{6}\right)^2 = 6;$$

$$\left(-1 - \sqrt{6} + 1\right)^2 = \left(-\sqrt{6}\right)^2 = 6.$$

The solution set is $\left\{-1 + \sqrt{6}, -1 - \sqrt{6}\right\}$.

Now Try Exercises 21 and 27.

NOTE The solutions in Example 3(b) may be written in abbreviated form as $-1 \pm \sqrt{6}$. If they are written this way, keep in mind that *two* solutions are indicated, one with the $+$ sign and the other with the $-$ sign.

EXAMPLE 4 Solving a Quadratic Equation of the Form $(ax + b)^2 = k$

Solve $(3r - 2)^2 = 27$.

$$3r - 2 = \sqrt{27} \quad \text{or} \quad 3r - 2 = -\sqrt{27} \qquad \text{Square root property}$$

$$3r - 2 = 3\sqrt{3} \quad \text{or} \quad 3r - 2 = -3\sqrt{3} \qquad \sqrt{27} = \sqrt{9 \cdot 3} = 3\sqrt{3}$$

$$3r = 2 + 3\sqrt{3} \quad \text{or} \quad 3r = 2 - 3\sqrt{3} \qquad \text{Add 2.}$$

$$r = \frac{2 + 3\sqrt{3}}{3} \quad \text{or} \quad r = \frac{2 - 3\sqrt{3}}{3} \qquad \text{Divide by 3.}$$

We show the check for the first solution. The check for the second solution is similar.

Check:
$$(3r - 2)^2 = 27 \qquad \text{Original equation}$$

$$\left[3\left(\frac{2 + 3\sqrt{3}}{3} \right) - 2 \right]^2 = 27 \qquad ?$$

$$\left(2 + 3\sqrt{3} - 2 \right)^2 = 27 \qquad ?$$

$$\left(3\sqrt{3} \right)^2 = 27 \qquad ?$$

$$27 = 27 \qquad \text{True}$$

The solution set is $\left\{ \dfrac{2 + 3\sqrt{3}}{3}, \dfrac{2 - 3\sqrt{3}}{3} \right\}$.

Now Try Exercise 31.

CAUTION   The solutions in Example 4 are fractions that cannot be simplified, since 3 is *not* a common factor in the numerator.

OBJECTIVE 4   Solve quadratic equations with solutions that are not real numbers.   So far, all the equations we have solved using the square root property have had two real solutions. In the equation $x^2 = k$, if $k < 0$, there will be two imaginary solutions.

EXAMPLE 5   Solving Quadratic Equations with Imaginary Solutions

Solve each equation.

(a) $x^2 = -15$

$$x = \sqrt{-15} \quad \text{or} \quad x = -\sqrt{-15} \qquad \text{Square root property}$$

$$x = i\sqrt{15} \quad \text{or} \quad x = -i\sqrt{15} \qquad \sqrt{-1} = i$$

The solution set is $\left\{ i\sqrt{15}, -i\sqrt{15} \right\}$. Check by substitution.

(b) $(t + 2)^2 = -16$

$$t + 2 = \sqrt{-16} \quad \text{or} \quad t + 2 = -\sqrt{-16} \qquad \text{Square root property}$$

$$t + 2 = 4i \quad \text{or} \quad t + 2 = -4i \qquad \sqrt{-16} = 4i$$

$$t = -2 + 4i \quad \text{or} \quad t = -2 - 4i$$

Check:
$$(t + 2)^2 = -16 \qquad \text{Original equation}$$

$$(-2 + 4i + 2)^2 = -16 \qquad ?$$

$$(4i)^2 = -16 \qquad ?$$

$$-16 = -16 \qquad i^2 = -1$$

The check for the other solution is similar. The solution set is
$$\{-2 + 4i, -2 - 4i\}.$$

Now Try Exercises 49 and 51.

11.1 EXERCISES

1. A student was asked to solve the quadratic equation $x^2 = 16$ and did not get full credit for the solution set {4}. Why?

2. Why can't the zero-factor property be used to solve every quadratic equation?

3. Give a one-sentence description or explanation of each phrase.

(a) Quadratic equation in standard form (b) Zero-factor property
(c) Square root property

4. What is wrong with the following "solution"?

$$x^2 - x - 2 = 5$$
$$(x - 2)(x + 1) = 5$$
$$x - 2 = 5 \quad \text{or} \quad x + 1 = 5 \qquad \text{Zero-factor property}$$
$$x = 7 \quad \text{or} \qquad x = 4$$

Use the square root property to solve each equation. See Example 1.

5. $x^2 = 81$ **6.** $z^2 = 225$ **7.** $t^2 = 17$ **8.** $x^2 = 19$

9. $m^2 = 32$ **10.** $x^2 = 54$ **11.** $r^2 - 21 = 0$ **12.** $x^2 - 14 = 0$

13. $t^2 - 20 = 0$ **14.** $p^2 - 50 = 0$ **15.** $3n^2 - 72 = 0$ **16.** $5z^2 - 200 = 0$

17. $5a^2 + 4 = 8$ **18.** $4p^2 - 3 = 7$ **19.** $2t^2 + 7 = 61$ **20.** $3x^2 - 8 = 64$

Solve each equation by using the square root property. See Examples 3 and 4.

21. $(x - 3)^2 = 25$ **22.** $(k - 7)^2 = 16$ **23.** $(3k + 2)^2 = 49$

24. $(5t + 3)^2 = 36$ **25.** $(4x - 3)^2 = 9$ **26.** $(7z - 5)^2 = 25$

27. $(x - 4)^2 = 3$ **28.** $(x + 3)^2 = 11$ **29.** $(t + 5)^2 = 48$

30. $(m - 6)^2 = 27$ **31.** $(3x - 1)^2 = 7$ **32.** $(2x + 4)^2 = 10$

33. $(4p + 1)^2 = 24$ **34.** $(5t - 2)^2 = 12$ **35.** $(5 - 2x)^2 = 30$

36. $(3 - 2a)^2 = 70$ **37.** $(3k + 1)^2 = 18$ **38.** $(5z + 6)^2 = 75$

39. $\left(\dfrac{1}{2}x + 5\right)^2 = 12$ **40.** $\left(\dfrac{1}{3}m + 4\right)^2 = 27$ **41.** $(4k - 1)^2 - 48 = 0$

42. $(2s - 5)^2 - 180 = 0$

43. Johnny solved the equation in Exercise 35 and wrote his answer as $\left\{\frac{5 + \sqrt{30}}{2}, \frac{5 - \sqrt{30}}{2}\right\}$.

Linda solved the same equation and wrote her answer as $\left\{\frac{-5 + \sqrt{30}}{-2}, \frac{-5 - \sqrt{30}}{-2}\right\}$. The teacher gave them both full credit. Explain why both students were correct, although their answers look different.

44. In the solutions found in Example 4 of this section, why is it not valid to simplify the answers by dividing out the 3s in the numerator and denominator?

Use a calculator with a square root key to solve each equation. Round your answers to the nearest hundredth.

45. $(k + 2.14)^2 = 5.46$

46. $(r - 3.91)^2 = 9.28$

47. $(2.11p + 3.42)^2 = 9.58$

48. $(1.71m - 6.20)^2 = 5.41$

Find the imaginary solutions of each equation. See Example 5.

49. $x^2 = -12$

50. $x^2 = -18$

51. $(r - 5)^2 = -3$

52. $(t + 6)^2 = -5$

53. $(6k - 1)^2 = -8$

54. $(4m - 7)^2 = -27$

RELATING CONCEPTS (EXERCISES 55–58)

For Individual or Group Work

*In Section 5.4 we saw how certain trinomials can be factored as squares of binomials. Use this idea and **work Exercises 55–58 in order,** considering the equation*

$$x^2 + 6x + 9 = 100.$$

55. Factor the left side of the equation as the square of a binomial, and write the resulting equation.

56. Use the square root property to solve the equation from Exercise 55.

57. What is the solution set of the original equation?

58. Solve the equation $4k^2 - 12k + 9 = 81$ using the method described in Exercises 55–57.

Solve Exercises 59 and 60 using Galileo's formula, $d = 16t^2$. Round answers to the nearest tenth. See Example 2.

59. The Gateway Arch in St. Louis, Missouri, is 630 ft tall. How long would it take an object dropped from the top of it to fall to the ground? (*Source: Home & Away,* November/December 2000.)

60. Mount Rushmore National Memorial in South Dakota features a sculpture of four of America's favorite presidents carved into the rim of the mountain, 500 ft above the valley floor. How long would it take a rock dropped from the top of the sculpture to fall to the ground? (*Source: Microsoft Encarta Encyclopedia 2002.*)

Solve each problem. See Example 2.

61. The area A of a circle with radius r is given by the formula

$$A = \pi r^2.$$

If a circle has area 81π in.^{2}, what is its radius?

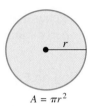

$A = \pi r^2$

62. The surface area S of a sphere with radius r is given by the formula

$$S = 4\pi r^2.$$

If a sphere has surface area 36π ft^2, what is its radius?

$S = 4\pi r^2$

The amount A that P dollars invested at an annual rate of interest r will grow to in 2 yr is

$$A = P(1 + r)^2.$$

63. At what interest rate will $100 grow to $110.25 in 2 yr?

64. At what interest rate will $500 grow to $572.45 in 2 yr?

11.2 Solving Quadratic Equations by Completing the Square

OBJECTIVES

1 Solve quadratic equations by completing the square when the coefficient of the squared term is 1.

2 Solve quadratic equations by completing the square when the coefficient of the squared term is not 1.

3 Simplify an equation before solving.

OBJECTIVE 1   Solve quadratic equations by completing the square when the coefficient of the squared term is 1.   The methods we have studied so far are not enough to solve the equation

$$x^2 + 6x + 7 = 0.$$

If we could write the equation in the form $(x + b)^2 = k$, we could solve it with the square root property discussed in the previous section. To do that, we need to have a perfect square trinomial on the left side of the equation.

Recall from Section 5.4 that the perfect square trinomial

$$x^2 + 6x + 9$$

can be factored as $(x + 3)^2$. In the trinomial, the coefficient of x (the first-degree term) is 6 and the constant term is 9. Notice that if we take half of 6 and square it, we get the constant term, 9.

Coefficient of x Constant

$$\left[\frac{1}{2}(6)\right]^2 = 3^2 = 9$$

Similarly, in

$$x^2 + 12x + 36, \qquad \left[\frac{1}{2}(12)\right]^2 = 6^2 = 36,$$

and in

$$m^2 - 10m + 25, \qquad \left[\frac{1}{2}(-10)\right]^2 = (-5)^2 = 25.$$

This relationship is true in general and is the idea behind rewriting a quadratic equation so the square root property can be applied.

EXAMPLE 1   Rewriting an Equation to Use the Square Root Property

Solve $x^2 + 6x + 7 = 0$.

   This quadratic equation cannot be solved by factoring, and it is not in the correct form to solve using the square root property. To get this form, we need a perfect square trinomial on the left side of the equation. We first subtract 7 from each side.

$$x^2 + 6x + 7 = 0$$
$$x^2 + 6x = -7 \qquad \text{Subtract 7.}$$

We must add a constant to get a perfect square trinomial on the left.

$$x^2 + 6x + \underbrace{\quad ? \quad}$$

Needs to be a perfect
square trinomial

To find this constant, we apply the ideas preceding this example—we take half the coefficient of the first-degree term and square the result.

$$\left[\frac{1}{2}(6)\right]^2 = 3^2 = 9 \quad \leftarrow \text{Desired constant}$$

We add this constant, 9, to *each* side of the equation. (Why?)

$$x^2 + 6x + 9 = -7 + 9$$

Now we factor the perfect square trinomial on the left and add on the right.

$$(x + 3)^2 = 2$$

We can solve this equation using the square root property.

$$x + 3 = \sqrt{2} \qquad \text{or} \quad x + 3 = -\sqrt{2}$$
$$x = -3 + \sqrt{2} \quad \text{or} \qquad x = -3 - \sqrt{2}$$

Check:
$$x^2 + 6x + 7 = 0 \qquad\qquad \text{Original equation}$$
$$\left(-3 + \sqrt{2}\right)^2 + 6\left(-3 + \sqrt{2}\right) + 7 = 0 \qquad ? \qquad \text{Let } x = -3 + \sqrt{2}.$$
$$9 - 6\sqrt{2} + 2 - 18 + 6\sqrt{2} + 7 = 0 \qquad ?$$
$$0 = 0 \qquad\qquad \text{True}$$

The check for the second solution is similar. The solution set is

$$\left\{-3 + \sqrt{2}, -3 - \sqrt{2}\right\}.$$

Now Try Exercise 11.

The process of changing the form of the equation in Example 1 from

$$x^2 + 6x + 7 = 0 \qquad \text{to} \qquad (x + 3)^2 = 2$$

is called **completing the square.** Completing the square changes only the form of the equation. To see this, multiply out the left side of $(x + 3)^2 = 2$ and combine terms. Then subtract 2 from both sides to see that the result is $x^2 + 6x + 7 = 0$.

▪ EXAMPLE 2 Completing the Square to Solve a Quadratic Equation

Solve $x^2 - 8x = 5$.

To complete the square on $x^2 - 8x$, take half the coefficient of x and square it.

$$\frac{1}{2}(-8) = -4 \qquad \text{and} \qquad (-4)^2 = 16$$

↑
Coefficient of x

Add the result, 16, to both sides of the equation.

$x^2 - 8x = 5$	Given equation
$x^2 - 8x + 16 = 5 + 16$	Add 16.
$(x - 4)^2 = 21$	Factor the left side as the square of a binomial; add on the right.

Now apply the square root property.

$x - 4 = \sqrt{21} \qquad \text{or} \quad x - 4 = -\sqrt{21}$	Square root property
$x = 4 + \sqrt{21} \quad \text{or} \qquad x = 4 - \sqrt{21}$	Add 4.

A check indicates that the solution set is

$$\left\{ 4 + \sqrt{21}, 4 - \sqrt{21} \right\}.$$

Now Try Exercise 13.

The steps to solve a quadratic equation $ax^2 + bx + c = 0$ $(a \neq 0)$ by completing the square are summarized here.

Solving a Quadratic Equation by Completing the Square

Step 1 **Be sure the squared term has coefficient 1.** If the coefficient of the squared term is 1, go to Step 2. If it is not 1 but some other nonzero number a, divide both sides of the equation by a.

Step 2 **Write in correct form.** Make sure that all variable terms are on one side of the equation and that all constant terms are on the other.

Step 3 **Complete the square.** Take half the coefficient of the first-degree term, and square it. Add the square to both sides of the equation. Factor the variable side and combine terms on the other side.

Step 4 **Solve.** Use the square root property to solve the equation.

▨ **EXAMPLE 3** Solving a Quadratic Equation by Completing the Square $(a = 1)$

Solve $k^2 + 5k - 1 = 0$.

Follow the steps in the box. Since the coefficient of the squared term is 1, begin with Step 2. Write the equation in correct form.

Step 2 $\qquad\qquad k^2 + 5k = 1 \qquad$ Add 1 to each side.

Step 3 Complete the square. Take half the coefficient of the first-degree term and square the result.

$$\left[\frac{1}{2}(5)\right]^2 = \left(\frac{5}{2}\right)^2 = \frac{25}{4}$$

Add the square to each side of the equation to get

$$k^2 + 5k + \frac{25}{4} = 1 + \frac{25}{4}.$$

$$\left(k + \frac{5}{2}\right)^2 = \frac{29}{4} \qquad \text{Factor on the left; add on the right.}$$

Step 4 Use the square root property and solve for k.

$$k + \frac{5}{2} = \sqrt{\frac{29}{4}} \qquad \text{or} \qquad k + \frac{5}{2} = -\sqrt{\frac{29}{4}}$$

$$k + \frac{5}{2} = \frac{\sqrt{29}}{2} \qquad \text{or} \qquad k + \frac{5}{2} = -\frac{\sqrt{29}}{2}$$

$$k = -\frac{5}{2} + \frac{\sqrt{29}}{2} \qquad \text{or} \qquad k = -\frac{5}{2} - \frac{\sqrt{29}}{2}$$

$$k = \frac{-5 + \sqrt{29}}{2} \qquad \text{or} \qquad k = \frac{-5 - \sqrt{29}}{2}$$

Check that the solution set is $\left\{ \dfrac{-5 + \sqrt{29}}{2}, \dfrac{-5 - \sqrt{29}}{2} \right\}$.

Now Try Exercise 15.

OBJECTIVE 2 Solve quadratic equations by completing the square when the coefficient of the squared term is not 1. To complete the square, the coefficient of the squared term must be 1 (Step 1). The next examples show what to do when this coefficient is not 1.

▨ **EXAMPLE 4** Solving a Quadratic Equation by Completing the Square $(a \neq 1)$

Solve $4z^2 + 16z - 9 = 0$.

Step 1   Divide each side by 4 so that the coefficient of z^2 is 1.

$$4z^2 + 16z - 9 = 0$$

$$z^2 + 4z - \frac{9}{4} = 0 \qquad \text{Divide by 4.}$$

Step 2   Add $\frac{9}{4}$ to each side to get the variable terms on the left and the constant on the right.

$$z^2 + 4z = \frac{9}{4}$$

Step 3   To complete the square, take half the coefficient of z and square the result: $\left[\frac{1}{2}(4)\right]^2 = 2^2 = 4$. Add 4 to both sides of the equation.

$$z^2 + 4z + 4 = \frac{9}{4} + 4 \qquad \text{Add 4.}$$

$$(z + 2)^2 = \frac{25}{4} \qquad \text{Factor on the left; add on the right.}$$

Step 4   Use the square root property and solve for z.

$$z + 2 = \sqrt{\frac{25}{4}} \qquad \text{or} \qquad z + 2 = -\sqrt{\frac{25}{4}}$$

$$z + 2 = \frac{5}{2} \qquad \text{or} \qquad z + 2 = -\frac{5}{2} \qquad \text{Square root property}$$

$$z = -2 + \frac{5}{2} \qquad \text{or} \qquad z = -2 - \frac{5}{2} \qquad \text{Subtract 2.}$$

$$z = \frac{1}{2} \qquad \text{or} \qquad z = -\frac{9}{2}$$

Check by substituting each solution into the original equation. The solution set is $\left\{\frac{1}{2}, -\frac{9}{2}\right\}$.

Now Try Exercise 17.

EXAMPLE 5   Solving a Quadratic Equation by Completing the Square ($a \neq 1$)

Solve $2x^2 - 4x - 5 = 0$.

First divide each side of the equation by 2 to get 1 as the coefficient of the squared term.

$$x^2 - 2x - \frac{5}{2} = 0 \qquad \text{Step 1}$$

$$x^2 - 2x = \frac{5}{2} \qquad \text{Step 2}$$

$$\left[\frac{1}{2}(-2)\right]^2 = (-1)^2 = 1 \qquad \text{Step 3}$$

$$x^2 - 2x + 1 = \frac{5}{2} + 1 \qquad \text{Add 1.}$$

$$(x - 1)^2 = \frac{7}{2} \qquad \text{Factor; add.}$$

$$x - 1 = \sqrt{\frac{7}{2}} \qquad \text{or} \quad x - 1 = -\sqrt{\frac{7}{2}} \qquad \text{Step 4}$$

$$x = 1 + \sqrt{\frac{7}{2}} \qquad \text{or} \qquad x = 1 - \sqrt{\frac{7}{2}}$$

$$x = 1 + \frac{\sqrt{14}}{2} \qquad \text{or} \qquad x = 1 - \frac{\sqrt{14}}{2} \qquad \begin{array}{l}\text{Rationalize} \\ \text{denominators.}\end{array}$$

Add the two terms in each solution as follows.

$$1 + \frac{\sqrt{14}}{2} = \frac{2}{2} + \frac{\sqrt{14}}{2} = \frac{2 + \sqrt{14}}{2}$$

$$1 - \frac{\sqrt{14}}{2} = \frac{2}{2} - \frac{\sqrt{14}}{2} = \frac{2 - \sqrt{14}}{2}.$$

Check that the solution set is $\left\{ \dfrac{2 + \sqrt{14}}{2}, \dfrac{2 - \sqrt{14}}{2} \right\}$.

Now Try Exercise 21.

■ EXAMPLE 6 Solving a Quadratic Equation by Completing the Square (Imaginary Solutions)

Solve $4p^2 + 8p + 5 = 0$.

$$p^2 + 2p + \frac{5}{4} = 0 \qquad \text{Divide by 4.}$$

$$p^2 + 2p = -\frac{5}{4} \qquad \text{Subtract } \tfrac{5}{4}.$$

The coefficient of p is 2. Take half of 2, square the result, and add it to both sides: $\left[\frac{1}{2}(2) \right]^2 = 1^2 = 1$. The left side can then be factored as a perfect square.

$$p^2 + 2p + 1 = -\frac{5}{4} + 1 \qquad \text{Add 1.}$$

$$(p + 1)^2 = -\frac{1}{4} \qquad \text{Factor; add.}$$

Because the constant on the right side of the equation is negative, this equation will have imaginary solutions.

$$p + 1 = \sqrt{-\frac{1}{4}} \qquad \text{or} \quad p + 1 = -\sqrt{-\frac{1}{4}} \qquad \text{Square root property}$$

$$p + 1 = \frac{1}{2}i \qquad \text{or} \quad p + 1 = -\frac{1}{2}i \qquad \qquad \sqrt{-\frac{1}{4}} = \frac{1}{2}i$$

$$p = -1 + \frac{1}{2}i \quad \text{or} \qquad p = -1 - \frac{1}{2}i \qquad \text{Subtract 1.}$$

Check: $\qquad\qquad\qquad\qquad 4p^2 + 8p + 5 = 0 \qquad\qquad$ Original equation

$$4\left(-1 + \frac{1}{2}i\right)^2 + 8\left(-1 + \frac{1}{2}i\right) + 5 = 0 \qquad \text{?}$$

$$4\left[1 + 2(-1)\left(\frac{1}{2}i\right) - \frac{1}{4}\right] - 8 + 4i + 5 = 0 \qquad \text{?}$$

$$4\left[\frac{3}{4} - i\right] - 3 + 4i = 0 \qquad \text{?}$$

$$3 - 4i - 3 + 4i = 0 \qquad \text{?}$$

$$0 = 0 \qquad \text{True}$$

The check of the other solution is similar. The solution set is

$$\left\{-1 + \frac{1}{2}i, -1 - \frac{1}{2}i\right\}.$$

Now Try Exercise 39.

OBJECTIVE 3 Simplify an equation before solving.   Sometimes an equation must be simplified before completing the square. The next example illustrates this.

EXAMPLE 7   Simplifying an Equation Before Completing the Square

Solve $(x + 3)(x - 1) = 2$.

$$(x + 3)(x - 1) = 2$$

$$x^2 + 2x - 3 = 2 \qquad \text{Use FOIL.}$$

$$x^2 + 2x = 5 \qquad \text{Add 3.}$$

$$x^2 + 2x + 1 = 5 + 1 \qquad \text{Add } \left[\frac{1}{2}(2)\right]^2 = 1.$$

$$(x + 1)^2 = 6 \qquad \text{Factor on the left; add on the right.}$$

$$x + 1 = \sqrt{6} \qquad \text{or} \qquad x + 1 = -\sqrt{6} \qquad \text{Square root property}$$

$$x = -1 + \sqrt{6} \qquad \text{or} \qquad x = -1 - \sqrt{6} \qquad \text{Subtract 1.}$$

The solution set is $\left\{-1 + \sqrt{6}, -1 - \sqrt{6}\right\}$.

Now Try Exercise 25.

NOTE   The procedure for completing the square is also used in other areas of mathematics. For example, we use it in Section 11.7 when we graph quadratic equations and again in Chapter 13 when we work with circles.

11.2 EXERCISES

1. Decide what number must be added to make each expression a perfect square trinomial.

(a) $x^2 + 6x + \underline{\quad}$ (b) $x^2 + 14x + \underline{\quad}$ (c) $p^2 - 12p + \underline{\quad}$

(d) $x^2 + 3x + \underline{\quad}$ (e) $q^2 - 9q + \underline{\quad}$ (f) $t^2 - \frac{1}{2}t + \underline{\quad}$

2. Which one of the following steps is an appropriate way to begin solving the quadratic equation

$$2x^2 - 4x = 9$$

by completing the square?

A. Add 4 to both sides of the equation.   **B.** Factor the left side as $2x(x - 2)$.
C. Factor the left side as $x(2x - 4)$.   **D.** Divide both sides by 2.

Determine the number that will complete the square to solve each equation after the constant term has been written on the right side. Do not actually solve. See Examples 1–7.

3. $x^2 + 4x - 2 = 0$ **4.** $t^2 + 2t - 1 = 0$ **5.** $x^2 + 10x + 18 = 0$

6. $x^2 + 8x + 11 = 0$ **7.** $3w^2 - w - 24 = 0$ **8.** $4z^2 - z - 39 = 0$

Solve each equation by completing the square. Use the results of Exercises 3–6 to solve Exercises 11–14. See Examples 1–3.

9. $x^2 - 2x - 24 = 0$ **10.** $m^2 - 4m - 32 = 0$ **11.** $x^2 + 4x - 2 = 0$

12. $t^2 + 2t - 1 = 0$ **13.** $x^2 + 10x + 18 = 0$ **14.** $x^2 + 8x + 11 = 0$

15. $x^2 + 3x - 2 = 0$ **16.** $p^2 + 5p + 5 = 0$

Solve each equation by completing the square. Use the results of Exercises 7 and 8 to solve Exercises 17 and 18. See Examples 4, 5, and 7.

17. $3w^2 - w = 24$ **18.** $4z^2 - z = 39$ **19.** $2k^2 + 5k - 2 = 0$

20. $3r^2 + 2r - 2 = 0$ **21.** $5x^2 - 10x + 2 = 0$ **22.** $2x^2 - 16x + 25 = 0$

23. $9x^2 - 24x = -13$ **24.** $25n^2 - 20n = 1$ **25.** $(x + 2)(x + 1) = 10$

26. $z(z + 6) + 4 = 0$ **27.** $z^2 - \frac{4}{3}z = -\frac{1}{9}$ **28.** $p^2 - \frac{8}{3}p = -1$

29. $.1x^2 - .2x - .1 = 0$ **30.** $.1p^2 - .4p + .1 = 0$

*Solve each equation by completing the square. Give **(a)** exact solutions and **(b)** solutions rounded to the nearest thousandth.*

31. $3r^2 - 2 = 6r + 3$ **32.** $4p + 3 = 2p^2 + 2p$

33. $(x + 1)(x + 3) = 2$ **34.** $(x - 3)(x + 1) = 1$

35. In using the method of completing the square to solve $2x^2 - 10x = -8$, a student began by adding the square of half the coefficient of x $\left(\text{that is, } \left[\frac{1}{2}(-10)\right]^2 = 25\right)$ to both sides of the equation. He then encountered difficulty in his later steps. What was his error? Explain the steps needed to solve the problem, and give the solution set.

36. The equation $x^4 - 2x^2 = 8$ can be solved by completing the square, even though it is not a quadratic equation. What number should be added to both sides of the equation so that the left side can be factored as the square of a binomial? Solve the equation for its real solutions.

Find the imaginary solutions of each equation. See Example 6.

37. $m^2 + 4m + 13 = 0$ **38.** $t^2 + 6t + 10 = 0$ **39.** $3r^2 + 4r + 4 = 0$

40. $4x^2 + 5x + 5 = 0$ **41.** $-m^2 - 6m - 12 = 0$ **42.** $-k^2 - 5k - 10 = 0$

RELATING CONCEPTS (EXERCISES 43–48)

For Individual or Group Work

The Greeks had a method of completing the square geometrically in which they literally changed a figure into a square. For example, to complete the square for $x^2 + 6x$, we begin with a square of side x, as in the figure on the left. We add three rectangles of width 1 to the right side and the bottom to get a region with area $x^2 + 6x$. To fill in the corner (complete the square), we must add 9 1-by-1 squares as shown in the figure on the right.

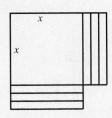 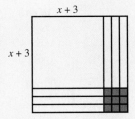

Work Exercises 43–48 in order.

43. What is the area of the original square?

44. What is the area of each strip?

45. What is the total area of the six strips?

46. What is the area of each small square in the corner of the second figure?

47. What is the total area of the small squares?

48. What is the area of the new, larger square?

11.3   Solving Quadratic Equations by the Quadratic Formula

OBJECTIVES

1 Derive the quadratic formula.

2 Solve quadratic equations using the quadratic formula.

3 Use the discriminant to determine the number and type of solutions.

The examples in the previous section showed that any quadratic equation can be solved by completing the square; however, completing the square can be tedious and time consuming. In this section, we complete the square to solve the general quadratic equation

$$ax^2 + bx + c = 0,$$

where a, b, and c are complex numbers and $a \neq 0$. The solution of this general equation gives a formula for finding the solution of any specific quadratic equation.

OBJECTIVE 1 Derive the quadratic formula. To solve $ax^2 + bx + c = 0$ by completing the square (assuming $a > 0$), we follow the steps given in Section 11.2.

$$ax^2 + bx + c = 0$$

$$x^2 + \frac{b}{a}x + \frac{c}{a} = 0 \qquad \text{Divide by } a. \text{ (Step 1)}$$

$$x^2 + \frac{b}{a}x = -\frac{c}{a} \qquad \text{Subtract } \tfrac{c}{a}. \text{ (Step 2)}$$

$$\left[\frac{1}{2}\left(\frac{b}{a}\right)\right]^2 = \left(\frac{b}{2a}\right)^2 = \frac{b^2}{4a^2} \qquad \text{(Step 3)}$$

$$x^2 + \frac{b}{a}x + \frac{b^2}{4a^2} = -\frac{c}{a} + \frac{b^2}{4a^2} \qquad \text{Add } \tfrac{b^2}{4a^2} \text{ to each side.}$$

Write the left side as a perfect square and rearrange the right side.

$$\left(x + \frac{b}{2a}\right)^2 = \frac{b^2}{4a^2} + \frac{-c}{a}$$

$$\left(x + \frac{b}{2a}\right)^2 = \frac{b^2}{4a^2} + \frac{-4ac}{4a^2} \qquad \text{Write with a common denominator.}$$

$$\left(x + \frac{b}{2a}\right)^2 = \frac{b^2 - 4ac}{4a^2} \qquad \text{Add fractions.}$$

$$x + \frac{b}{2a} = \sqrt{\frac{b^2 - 4ac}{4a^2}} \quad \text{or} \quad x + \frac{b}{2a} = -\sqrt{\frac{b^2 - 4ac}{4a^2}} \qquad \begin{array}{l}\text{Square root} \\ \text{property} \\ \text{(Step 4)}\end{array}$$

Since

$$\sqrt{\frac{b^2 - 4ac}{4a^2}} = \frac{\sqrt{b^2 - 4ac}}{\sqrt{4a^2}} = \frac{\sqrt{b^2 - 4ac}}{2a},$$

the right side of each equation can be expressed as

$$x + \frac{b}{2a} = \frac{\sqrt{b^2 - 4ac}}{2a} \qquad \text{or} \quad x + \frac{b}{2a} = \frac{-\sqrt{b^2 - 4ac}}{2a}$$

$$x = \frac{-b}{2a} + \frac{\sqrt{b^2 - 4ac}}{2a} \quad \text{or} \qquad x = \frac{-b}{2a} - \frac{\sqrt{b^2 - 4ac}}{2a}$$

$$x = \frac{-b + \sqrt{b^2 - 4ac}}{2a} \quad \text{or} \qquad x = \frac{-b - \sqrt{b^2 - 4ac}}{2a}.$$

If $a < 0$, the same two solutions are obtained. The result is the **quadratic formula,** which is abbreviated as follows.

Quadratic Formula

The solutions of $ax^2 + bx + c = 0$ $(a \neq 0)$ are given by

$$x = \frac{-b \pm \sqrt{b^2 - 4ac}}{2a}.$$

CAUTION In the quadratic formula, $x = \dfrac{-b \pm \sqrt{b^2 - 4ac}}{2a}$, the square root is added to or subtracted from the value of $-b$ *before* dividing by $2a$.

OBJECTIVE 2 Solve quadratic equations using the quadratic formula. To use the quadratic formula, first write the given equation in standard form $ax^2 + bx + c = 0$; then identify the values of a, b, and c and substitute them into the formula.

EXAMPLE 1   Using the Quadratic Formula (Rational Solutions)

Solve $6x^2 - 5x - 4 = 0$.

First, identify the values of a, b, and c of the general quadratic equation, $ax^2 + bx + c = 0$. Here a, the coefficient of the second-degree term, is 6, while b, the coefficient of the first-degree term, is -5, and the constant c is -4. Substitute these values into the quadratic formula.

$$x = \frac{-b \pm \sqrt{b^2 - 4ac}}{2a}$$

$$x = \frac{-(-5) \pm \sqrt{(-5)^2 - 4(6)(-4)}}{2(6)} \qquad a = 6, \, b = -5, \, c = -4$$

$$x = \frac{5 \pm \sqrt{25 + 96}}{12}$$

$$x = \frac{5 \pm \sqrt{121}}{12}$$

$$x = \frac{5 \pm 11}{12}$$

This last statement leads to two solutions, one from $+$ and one from $-$.

$$x = \frac{5 + 11}{12} = \frac{16}{12} = \frac{4}{3} \qquad \text{or} \qquad x = \frac{5 - 11}{12} = \frac{-6}{12} = -\frac{1}{2}$$

Check each solution by substituting it in the original equation. The solution set is $\left\{ -\frac{1}{2}, \frac{4}{3} \right\}$.

Now Try Exercise 5.

We could have used factoring to solve the equation in Example 1.

$$6x^2 - 5x - 4 = 0$$

$(3x - 4)(2x + 1) = 0$ Factor.

$3x - 4 = 0$ or $2x + 1 = 0$ Zero-factor property

$3x = 4$ or $2x = -1$ Solve each equation.

$x = \dfrac{4}{3}$ or $x = -\dfrac{1}{2}$ Same solutions as in Example 1

When solving quadratic equations, it is a good idea to try factoring first. If the equation cannot be factored or if factoring is difficult, then use the quadratic formula.

Later in this section, we will show a way to determine whether factoring can be used to solve a quadratic equation.

▓ **EXAMPLE 2** Using the Quadratic Formula (Irrational Solutions)

Solve $4r^2 = 8r - 1$.

Write the equation in standard form as

$$4r^2 - 8r + 1 = 0,$$

and identify $a = 4$, $b = -8$, and $c = 1$. Now use the quadratic formula.

$$r = \frac{-b \pm \sqrt{b^2 - 4ac}}{2a}$$

$$r = \frac{-(-8) \pm \sqrt{(-8)^2 - 4(4)(1)}}{2(4)} \qquad a = 4, b = -8, c = 1$$

$$= \frac{8 \pm \sqrt{64 - 16}}{8}$$

$$= \frac{8 \pm \sqrt{48}}{8}$$

$$= \frac{8 \pm 4\sqrt{3}}{8} \qquad \sqrt{48} = \sqrt{16} \cdot \sqrt{3} = 4\sqrt{3}$$

$$= \frac{4\left(2 \pm \sqrt{3}\right)}{4(2)} \qquad \text{Factor.}$$

$$= \frac{2 \pm \sqrt{3}}{2} \qquad \text{Lowest terms}$$

The solution set is $\left\{ \dfrac{2 + \sqrt{3}}{2}, \dfrac{2 - \sqrt{3}}{2} \right\}$.

Now Try Exercise 9.

> **CAUTION** *Every* quadratic equation must be expressed in standard form $ax^2 + bx + c = 0$ before we begin to solve it, whether we use factoring or the quadratic formula. Also, when writing solutions in lowest terms, be sure to *factor first;* then divide out the common factor, as shown in the last two steps in Example 2.

▓ **EXAMPLE 3** Using the Quadratic Formula (Imaginary Solutions)

Solve $(9q + 3)(q - 1) = -8$.

To write this equation in standard form, we first multiply and collect all nonzero terms on the left.

$$(9q + 3)(q - 1) = -8$$
$$9q^2 - 6q - 3 = -8$$
$$9q^2 - 6q + 5 = 0 \qquad \text{Standard form}$$

From the equation $9q^2 - 6q + 5 = 0$, we identify $a = 9$, $b = -6$, and $c = 5$, and use the quadratic formula.

$$q = \frac{-(-6) \pm \sqrt{(-6)^2 - 4(9)(5)}}{2(9)}$$

$$= \frac{6 \pm \sqrt{-144}}{18}$$

$$= \frac{6 \pm 12i}{18} \qquad\qquad \sqrt{-144} = 12i$$

$$= \frac{6(1 \pm 2i)}{6(3)} \qquad\qquad \text{Factor.}$$

$$= \frac{1 \pm 2i}{3} \qquad\qquad \text{Lowest terms}$$

The solution set, written in standard form $a + bi$ for complex numbers, is $\left\{ \frac{1}{3} + \frac{2}{3}i, \frac{1}{3} - \frac{2}{3}i \right\}$.

Now Try Exercise 33.

OBJECTIVE 3 Use the discriminant to determine the number and type of solutions. The solutions of the quadratic equation $ax^2 + bx + c = 0$ are given by

$$x = \frac{-b \pm \sqrt{b^2 - 4ac}}{2a}. \quad \longleftarrow \text{Discriminant}$$

If a, b, and c are integers, the type of solutions of a quadratic equation—that is, rational, irrational, or imaginary—is determined by the expression under the radical sign, $b^2 - 4ac$. Because it distinguishes among the three types of solutions, $b^2 - 4ac$ is called the *discriminant*. By calculating the discriminant before solving a quadratic equation, we can predict whether the solutions will be rational numbers, irrational numbers, or imaginary numbers. (This can be useful in an applied problem, for example, where irrational or imaginary solutions are not acceptable.)

Discriminant

The **discriminant** of $ax^2 + bx + c = 0$ is $b^2 - 4ac$. If a, b, and c are integers, then the number and type of solutions are determined as follows.

Discriminant	Number and Type of Solutions
Positive, and the square of an integer	Two rational solutions
Positive, but not the square of an integer	Two irrational solutions
Zero	One rational solution
Negative	Two imaginary solutions

Calculating the discriminant can also help you decide whether to solve a quadratic equation by factoring or by using the quadratic formula. If the discriminant is a perfect square (including 0), then the equation can be solved by factoring. Otherwise, the quadratic formula should be used.

EXAMPLE 4 Using the Discriminant

Find the discriminant. Use it to predict the number and type of solutions for each equation. Tell whether the equation can be solved by factoring or whether the quadratic formula should be used.

(a) $6x^2 - x - 15 = 0$

We find the discriminant by evaluating $b^2 - 4ac$.

$$b^2 - 4ac = (-1)^2 - 4(6)(-15) \qquad a = 6, b = -1, c = -15$$
$$= 1 + 360$$
$$= 361$$

A calculator shows that $361 = 19^2$, a perfect square. Since a, b, and c are integers and the discriminant is a perfect square, there will be two rational solutions and the equation can be solved by factoring.

(b) $3m^2 - 4m = 5$

Write the equation in standard form as $3m^2 - 4m - 5 = 0$ to find that $a = 3$, $b = -4$, and $c = -5$.

$$b^2 - 4ac = (-4)^2 - 4(3)(-5)$$
$$= 16 + 60$$
$$= 76$$

Because 76 is positive but not the square of an integer and a, b, and c are integers, the equation will have two irrational solutions and is best solved using the quadratic formula.

(c) $4x^2 + x + 1 = 0$

Since $a = 4$, $b = 1$, and $c = 1$, the discriminant is

$$1^2 - 4(4)(1) = -15.$$

Since the discriminant is negative and a, b, and c are integers, this quadratic equation will have two imaginary solutions. The quadratic formula should be used to solve it.

(d) $4t^2 + 9 = 12t$

Write the equation as $4t^2 - 12t + 9 = 0$ to find $a = 4$, $b = -12$, and $c = 9$. The discriminant is

$$b^2 - 4ac = (-12)^2 - 4(4)(9)$$
$$= 144 - 144$$
$$= 0.$$

Because the discriminant is 0, the quantity under the radical in the quadratic formula is 0, and there is only one rational solution. Again, the equation can be solved by factoring.

Now Try Exercises 37 and 39.

EXAMPLE 5 Using the Discriminant

Find k so that $9x^2 + kx + 4 = 0$ will have only one rational solution.

The equation will have only one rational solution if the discriminant is 0. Since $a = 9$, $b = k$, and $c = 4$, the discriminant is

$$b^2 - 4ac = k^2 - 4(9)(4) = k^2 - 144.$$

Set the discriminant equal to 0 and solve for k.

$$k^2 - 144 = 0$$

$$k^2 = 144 \qquad \text{Subtract 144.}$$

$$k = 12 \quad \text{or} \quad k = -12 \qquad \text{Square root property}$$

The equation will have only one rational solution if $k = 12$ or $k = -12$.

Now Try Exercise 53.

11.3 EXERCISES

Answer each question in Exercises 1–4.

1. An early version of Microsoft *Word* for Windows included the 1.0 edition of *Equation Editor.* The documentation used the following for the quadratic formula.

$$x = -b \pm \frac{\sqrt{b^2 - 4ac}}{2a}$$

Was this correct? Explain.

2. The Cadillac Bar in Houston, Texas, encourages patrons to write (tasteful) messages on the walls. One person attempted to write the quadratic formula, as shown here.

$$x = \frac{-b\sqrt{b^2 - 4ac}}{2a}$$

Was this correct? Explain.

3. What is wrong with the following "solution" of $5x^2 - 5x + 1 = 0$?

$$x = \frac{5 \pm \sqrt{25 - 4(5)(1)}}{2(5)} \qquad a = 5, b = -5, c = 1$$

$$x = \frac{5 \pm \sqrt{5}}{10}$$

$$x = \frac{1}{2} \pm \sqrt{5}$$

4. A student claimed that the equation $2x^2 - 5 = 0$ cannot be solved using the quadratic formula because there is no first-degree x-term. Was the student correct? Explain.

Use the quadratic formula to solve each equation. (All solutions for these equations are real numbers.) See Examples 1 and 2.

5. $m^2 - 8m + 15 = 0$ **6.** $x^2 + 3x - 28 = 0$ **7.** $2k^2 + 4k + 1 = 0$

8. $2w^2 + 3w - 1 = 0$ **9.** $2x^2 - 2x = 1$ **10.** $9t^2 + 6t = 1$

11. $x^2 + 18 = 10x$ **12.** $x^2 - 4 = 2x$ **13.** $4k^2 + 4k - 1 = 0$

14. $4r^2 - 4r - 19 = 0$ **15.** $2 - 2x = 3x^2$ **16.** $26r - 2 = 3r^2$

17. $\dfrac{x^2}{4} - \dfrac{x}{2} = 1$ **18.** $p^2 + \dfrac{p}{3} = \dfrac{1}{6}$ **19.** $-2t(t + 2) = -3$

20. $-3x(x + 2) = -4$ **21.** $(r - 3)(r + 5) = 2$ **22.** $(k + 1)(k - 7) = 1$

23. $(g + 2)(g - 3) = 1$ **24.** $(x - 5)(x + 2) = 6$ **25.** $p = \dfrac{5(5 - p)}{3(p + 1)}$

26. $k = \dfrac{k + 15}{3(k - 1)}$

Use the quadratic formula to solve each equation. (All solutions for these equations are imaginary numbers.) See Example 3.

27. $x^2 - 3x + 6 = 0$ **28.** $x^2 - 5x + 20 = 0$ **29.** $r^2 - 6r + 14 = 0$

30. $t^2 + 4t + 11 = 0$ **31.** $4x^2 - 4x = -7$ **32.** $9x^2 - 6x = -7$

33. $x(3x + 4) = -2$ **34.** $z(2z + 3) = -2$

35. $(x + 5)(x - 6) = (2x - 1)(x - 4)$ **36.** $(3x - 4)(x + 2) = (2x - 5)(x + 5)$

Use the discriminant to determine whether the solutions for each equation are

    A. *two rational numbers;*   **B.** *one rational number;*
    C. *two irrational numbers;*   **D.** *two imaginary numbers.*

Do not actually solve. See Example 4.

37. $25x^2 + 70x + 49 = 0$ **38.** $4k^2 - 28k + 49 = 0$ **39.** $x^2 + 4x + 2 = 0$

40. $9x^2 - 12x - 1 = 0$ **41.** $3x^2 = 5x + 2$ **42.** $4x^2 = 4x + 3$

43. $3m^2 - 10m + 15 = 0$ **44.** $18x^2 + 60x + 82 = 0$

45. Using the discriminant, which equations in Exercises 37–44 can be solved by factoring?

Based on your answer in Exercise 45, solve the equation given in each exercise.

46. Exercise 37   **47.** Exercise 38   **48.** Exercise 41   **49.** Exercise 42

50. Find the discriminant for each quadratic equation. Use it to tell whether the equation can be solved by factoring or whether the quadratic formula should be used. Then solve each equation.

    (a) $3k^2 + 13k = -12$ **(b)** $2x^2 + 19 = 14x$

51. Is it possible for the solution of a quadratic equation with integer coefficients to include just one irrational number? Why or why not?

52. Can the solution of a quadratic equation with integer coefficients include one real and one imaginary number? Why or why not?

Find the value of a, b, or c so that each equation will have exactly one rational solution. See Example 5.

53. $p^2 + bp + 25 = 0$ **54.** $r^2 - br + 49 = 0$ **55.** $am^2 + 8m + 1 = 0$

56. $at^2 + 24t + 16 = 0$ **57.** $9x^2 - 30x + c = 0$ **58.** $4m^2 + 12m + c = 0$

59. One solution of $4x^2 + bx - 3 = 0$ is $-\frac{5}{2}$. Find b and the other solution.

60. One solution of $3x^2 - 7x + c = 0$ is $\frac{1}{3}$. Find c and the other solution.

11.4   Equations Quadratic in Form

OBJECTIVES

1 Solve an equation with fractions by writing it in quadratic form.

2 Use quadratic equations to solve applied problems.

3 Solve an equation with radicals by writing it in quadratic form.

4 Solve an equation that is quadratic in form by substitution.

We have introduced four methods for solving quadratic equations written in standard form $ax^2 + bx + c = 0$. The following table lists some advantages and disadvantages of each method.

Methods for Solving Quadratic Equations

Method	Advantages	Disadvantages
Factoring	This is usually the fastest method.	Not all polynomials are factorable; some factorable polynomials are hard to factor.
Square root property	This is the simplest method for solving equations of the form $(ax + b)^2 = c$.	Few equations are given in this form.
Completing the square	This method can always be used, although most people prefer the quadratic formula.	It requires more steps than other methods.
Quadratic formula	This method can always be used.	It is more difficult than factoring because of the square root, although calculators can simplify its use.

OBJECTIVE 1 Solve an equation with fractions by writing it in quadratic form.   A variety of nonquadratic equations can be written in the form of a quadratic equation and solved by using one of the methods in the table. As you solve the equations in this section, try to decide which method is best for each equation.

EXAMPLE 1   Solving an Equation with Fractions that Leads to a Quadratic Equation

Solve $\dfrac{1}{x} + \dfrac{1}{x - 1} = \dfrac{7}{12}$.

Clear fractions by multiplying each term by the least common denominator, $12x(x - 1)$. (Note that the domain must be restricted to $x \neq 0$ and $x \neq 1$.)

$$12x(x - 1)\frac{1}{x} + 12x(x - 1)\frac{1}{x - 1} = 12x(x - 1)\frac{7}{12}$$

$$12(x - 1) + 12x = 7x(x - 1)$$

$$12x - 12 + 12x = 7x^2 - 7x \qquad \text{Distributive property}$$

$$24x - 12 = 7x^2 - 7x \qquad \text{Combine terms.}$$

Recall that a quadratic equation must be in standard form before it can be solved by factoring or the quadratic formula. Combine and rearrange terms so that one side is 0. Then factor to solve the resulting equation.

$$7x^2 - 31x + 12 = 0 \qquad \text{Standard form}$$
$$(7x - 3)(x - 4) = 0 \qquad \text{Factor.}$$

Using the zero-factor property gives the solutions $\frac{3}{7}$ and 4. Check by substituting these solutions in the original equation. The solution set is $\left\{\frac{3}{7}, 4\right\}$.

Now Try Exercise 19.

OBJECTIVE 2 **Use quadratic equations to solve applied problems.**   Earlier we solved distance-rate-time (or motion) problems that led to linear equations or rational equations. Now we solve motion problems that lead to quadratic equations. We continue to use the six-step problem-solving method from Chapter 2.

EXAMPLE 2   Solving a Motion Problem

A riverboat for tourists averages 12 mph in still water. It takes the boat 1 hr 4 min to go 6 mi upstream and return. Find the speed of the current. See Figure 1.

FIGURE 1

Step 1   **Read** the problem carefully.

Step 2   **Assign a variable.** Let $x =$ the speed of the current. The current slows down the boat when it is going upstream, so the rate (or speed) of the boat going upstream is its speed in still water less the speed of the current, or $12 - x$. Similarly, the current speeds up the boat as it travels downstream, so its speed downstream is $12 + x$. Thus,

$$12 - x = \text{the rate upstream;}$$
$$12 + x = \text{the rate downstream.}$$

Use the distance formula, $d = rt$, solved for time t.

$$t = \frac{d}{r}$$

This information can be used to complete a table.

	d	r	t
Upstream	6	$12 - x$	$\dfrac{6}{12 - x}$
Downstream	6	$12 + x$	$\dfrac{6}{12 + x}$

Times in hours

Step 3 **Write an equation.** The total time of 1 hr 4 min can be written as

$$1 + \frac{4}{60} = 1 + \frac{1}{15} = \frac{16}{15} \text{ hr.}$$

Because the time upstream plus the time downstream equals $\frac{16}{15}$ hr,

Time upstream   +   Time downstream   =   Total time

$$\frac{6}{12 - x} \quad + \quad \frac{6}{12 + x} \quad = \quad \frac{16}{15}.$$

Step 4 **Solve** the equation. Multiply each side by $15(12 - x)(12 + x)$, the LCD, and solve the resulting quadratic equation.

$$15(12 + x)6 + 15(12 - x)6 = 16(12 - x)(12 + x)$$
$$90(12 + x) + 90(12 - x) = 16(144 - x^2)$$
$$1080 + 90x + 1080 - 90x = 2304 - 16x^2 \qquad \text{Distributive property}$$
$$2160 = 2304 - 16x^2 \qquad \text{Combine terms.}$$
$$16x^2 = 144$$
$$x^2 = 9 \qquad \text{Divide by 16.}$$
$$x = 3 \quad \text{or} \quad x = -3 \qquad \text{Square root property}$$

Step 5 **State the answer.** The speed of the current cannot be -3, so the answer is 3 mph.

Step 6 **Check** that this value satisfies the original problem.

Now Try Exercise 31.

CAUTION As shown in Example 2, when a quadratic equation is used to solve an applied problem, sometimes only *one* answer satisfies the application. *Always* check each answer in the words of the original problem.

In Chapter 6 we solved problems about work rates. Recall that a person's work rate is $\frac{1}{t}$ part of the job per hour, where t is the time in hours required to do the complete job. Thus, the part of the job the person will do in x hr is $\frac{1}{t}x$.

EXAMPLE 3 Solving a Work Problem

In takes two carpet layers 4 hr to carpet a room. If each worked alone, one of them could do the job in 1 hr less time than the other. How long would it take each carpet layer to complete the job alone?

Step 1 **Read** the problem again. There will be two answers.

Step 2 **Assign a variable.** Let x represent the number of hours for the slower carpet layer to complete the job alone. Then the faster carpet layer could do the entire job in $(x - 1)$ hr. The slower person's rate is $\frac{1}{x}$, and the faster person's rate is $\frac{1}{x - 1}$. Together, they do the job in 4 hr. Complete a table as shown on the next page.

	Rate	Time Working Together	Fractional Part of the Job Done	
Slower Worker	$\dfrac{1}{x}$	4	$\dfrac{1}{x}(4)$	← Sum is 1 whole job.
Faster Worker	$\dfrac{1}{x-1}$	4	$\dfrac{1}{x-1}(4)$	←

Step 3 **Write an equation.** The sum of the fractional parts done by the workers should equal 1 (the whole job).

Part done by slower worker   +   Part done by faster worker   =   1 whole job

$$\dfrac{4}{x} \qquad + \qquad \dfrac{4}{x-1} \qquad = \qquad 1$$

Step 4 **Solve** the equation. Multiply each side by the LCD, $x(x-1)$.

$$4(x-1) + 4x = x(x-1)$$

$$4x - 4 + 4x = x^2 - x \qquad \text{Distributive property}$$

$$x^2 - 9x + 4 = 0 \qquad \text{Standard form}$$

This equation cannot be solved by factoring, so use the quadratic formula.

$$x = \dfrac{9 \pm \sqrt{81-16}}{2} = \dfrac{9 \pm \sqrt{65}}{2} \qquad a-1, b=-9, c=4$$

To the nearest tenth,

$$x = \dfrac{9 + \sqrt{65}}{2} \approx 8.5 \quad \text{or} \quad x = \dfrac{9 - \sqrt{65}}{2} \approx .5. \qquad \text{Use a calculator.}$$

Step 5 **State the answer.** Only the solution 8.5 makes sense in the original problem. (Why?) Thus, the slower worker could do the job in about 8.5 hr and the faster in about $8.5 - 1 = 7.5$ hr.

Step 6 **Check** that these results satisfy the original problem.

Now Try Exercise 37.

OBJECTIVE 3 Solve an equation with radicals by writing it in quadratic form.

EXAMPLE 4 Solving Radical Equations That Lead to Quadratic Equations

Solve each equation.

(a) $k = \sqrt{6k-8}$

This equation is not quadratic. However, squaring both sides of the equation gives a quadratic equation that can be solved by factoring.

$$k^2 = 6k - 8 \qquad \text{Square both sides.}$$

$$k^2 - 6k + 8 = 0 \qquad \text{Standard form}$$

$$(k-4)(k-2) = 0 \qquad \text{Factor.}$$

$$k - 4 = 0 \quad \text{or} \quad k - 2 = 0 \qquad \text{Zero-factor property}$$

$$k = 4 \quad \text{or} \qquad k = 2 \qquad \text{Potential solutions}$$

Recall from our work with radical equations in Section 10.6 that squaring both sides of an equation can introduce extraneous solutions that do not satisfy the original equation. Therefore, *all potential solutions must be checked in the original (not the squared) equation.*

Check: If $k = 4$, then

$$k = \sqrt{6k - 8}$$
$$4 = \sqrt{6(4) - 8} \quad ?$$
$$4 = \sqrt{16} \quad ?$$
$$4 = 4. \qquad \text{True}$$

If $k = 2$, then

$$k = \sqrt{6k - 8}$$
$$2 = \sqrt{6(2) - 8} \quad ?$$
$$2 = \sqrt{4} \quad ?$$
$$2 = 2. \qquad \text{True}$$

Both solutions check, so the solution set is $\{2, 4\}$.

(b) $x + \sqrt{x} = 6$

$$\sqrt{x} = 6 - x \qquad \text{Isolate the radical on one side.}$$
$$x = 36 - 12x + x^2 \qquad \text{Square both sides.}$$
$$0 = x^2 - 13x + 36 \qquad \text{Standard form}$$
$$0 = (x - 4)(x - 9) \qquad \text{Factor.}$$
$$x - 4 = 0 \quad \text{or} \quad x - 9 = 0 \qquad \text{Zero-factor property}$$
$$x = 4 \quad \text{or} \qquad x = 9 \qquad \text{Potential solutions}$$

Check both potential solutions in the *original* equation.

If $x = 4$, then

$$x + \sqrt{x} = 6$$
$$4 + \sqrt{4} = 6 \quad ?$$
$$6 = 6. \qquad \text{True}$$

If $x = 9$, then

$$x + \sqrt{x} = 6$$
$$9 + \sqrt{9} = 6 \quad ?$$
$$12 = 6. \qquad \text{False}$$

Only the solution 4 checks, so the solution set is $\{4\}$.

Now Try Exercises 41 and 47.

OBJECTIVE 4 Solve an equation that is quadratic in form by substitution. A nonquadratic equation that can be written in the form $au^2 + bu + c = 0$, for $a \neq 0$ and an algebraic expression u, is called **quadratic in form.**

EXAMPLE 5 Solving Equations That Are Quadratic in Form

Solve each equation.

(a) $x^4 - 13x^2 + 36 = 0$

Because $x^4 = (x^2)^2$, we can write this equation in quadratic form with $u = x^2$ and $u^2 = x^4$. (Any letter except x could be used instead of u.)

$$x^4 - 13x^2 + 36 = 0$$
$$(x^2)^2 - 13x^2 + 36 = 0 \qquad x^4 = (x^2)^2$$
$$u^2 - 13u + 36 = 0 \qquad \text{Let } u = x^2.$$
$$(u - 4)(u - 9) = 0 \qquad \text{Factor.}$$

$$u - 4 = 0 \quad \text{or} \quad u - 9 = 0 \qquad \text{Zero-factor property}$$
$$u = 4 \quad \text{or} \quad u = 9 \qquad \text{Solve.}$$

To find x, we substitute x^2 for u.

$$x^2 = 4 \quad \text{or} \quad x^2 = 9$$
$$x = \pm 2 \quad \text{or} \quad x = \pm 3 \qquad \text{Square root property}$$

The equation $x^4 - 13x^2 + 36 = 0$, a fourth-degree equation, has four solutions.* The solution set is $\{-3, -2, 2, 3\}$, which can be verified by substituting into the original equation for x.

(b) $4x^6 + 1 = 5x^3$

This equation is quadratic in form with $u = x^3$ and $u^2 = x^6$.

$$4x^6 + 1 = 5x^3$$
$$4(x^3)^2 + 1 = 5x^3$$
$$4u^2 + 1 = 5u \qquad \text{Let } u = x^3.$$
$$4u^2 - 5u + 1 = 0 \qquad \text{Standard form}$$
$$(4u - 1)(u - 1) = 0 \qquad \text{Factor.}$$
$$4u - 1 = 0 \quad \text{or} \quad u - 1 = 0 \qquad \text{Zero-factor property}$$
$$u = \frac{1}{4} \quad \text{or} \quad u = 1 \qquad \text{Solve.}$$
$$x^3 = \frac{1}{4} \quad \text{or} \quad x^3 = 1 \qquad u = x^3$$

From these equations,

$$x = \sqrt[3]{\frac{1}{4}} = \frac{\sqrt[3]{1}}{\sqrt[3]{4}} = \frac{1}{\sqrt[3]{4}} \cdot \frac{\sqrt[3]{2}}{\sqrt[3]{2}} = \frac{\sqrt[3]{2}}{2} \qquad \text{or} \qquad x = \sqrt[3]{1} = 1.$$

There are other complex solutions for this equation, but finding them involves trigonometry. The real number solution set of $4x^6 + 1 = 5x^3$ is $\left\{ \dfrac{\sqrt[3]{2}}{2}, 1 \right\}$.

(c) $x^4 = 6x^2 - 3$

First write the equation as

$$x^4 - 6x^2 + 3 = 0 \qquad \text{or} \qquad (x^2)^2 - 6(x^2) + 3 = 0,$$

which is quadratic in form with $u = x^2$. Substitute u for x^2 and u^2 for x^4 to get

$$u^2 - 6u + 3 = 0.$$

---

*In general, an equation in which an nth-degree polynomial equals 0 has n complex solutions, although some of them may be repeated.

Since this equation cannot be solved by factoring, use the quadratic formula.

$$u = \frac{6 \pm \sqrt{36 - 12}}{2} \qquad a = 1, b = -6, c = 3$$

$$u = \frac{6 \pm \sqrt{24}}{2}$$

$$u = \frac{6 \pm 2\sqrt{6}}{2} \qquad \sqrt{24} = \sqrt{4} \cdot \sqrt{6} = 2\sqrt{6}$$

$$u = \frac{2(3 \pm \sqrt{6})}{2} \qquad \text{Factor.}$$

$$u = 3 \pm \sqrt{6} \qquad \text{Lowest terms}$$

Since $u = x^2$, find x by using the square root property.

$$x^2 = 3 + \sqrt{6} \qquad \text{or} \quad x^2 = 3 - \sqrt{6}$$

$$x = \pm\sqrt{3 + \sqrt{6}} \quad \text{or} \quad x = \pm\sqrt{3 - \sqrt{6}}$$

The solution set contains four numbers:

$$\left\{ \sqrt{3 + \sqrt{6}}, -\sqrt{3 + \sqrt{6}}, \sqrt{3 - \sqrt{6}}, -\sqrt{3 - \sqrt{6}} \right\}.$$

Now Try Exercises 55, 79, and 83.

NOTE Some students prefer to solve equations like those in Examples 5(a) and (b) by factoring directly. For example,

$$x^4 - 13x^2 + 36 = 0 \qquad \text{Example 5(a) equation}$$

$$(x^2 - 9)(x^2 - 4) = 0 \qquad \text{Factor.}$$

$$(x + 3)(x - 3)(x + 2)(x - 2) = 0. \qquad \text{Factor again.}$$

Using the zero-factor property gives the same solutions obtained in Example 5(a). Equations that cannot be solved by factoring, like that in Example 5(c), must be solved using the method of substitution and the quadratic formula.

EXAMPLE 6 Solving Equations That Are Quadratic in Form

Solve each equation.

(a) $2(4m - 3)^2 + 7(4m - 3) + 5 = 0$

Because of the repeated quantity $4m - 3$, this equation is quadratic in form with $u = 4m - 3$.

$$2(4m - 3)^2 + 7(4m - 3) + 5 = 0$$

$$2u^2 + 7u + 5 = 0 \qquad \text{Let } 4m - 3 = u.$$

$$(2u + 5)(u + 1) = 0 \qquad \text{Factor.}$$

$$2u + 5 = 0 \qquad \text{or} \qquad u + 1 = 0 \qquad \text{Zero-factor property}$$

$$u = -\frac{5}{2} \qquad \text{or} \qquad u = -1$$

$$4m - 3 = -\frac{5}{2} \qquad \text{or} \quad 4m - 3 = -1 \qquad \text{Substitute } 4m - 3 \text{ for } u.$$

$$4m = \frac{1}{2} \quad \text{or} \quad 4m = 2 \qquad \text{Solve for } m.$$

$$m = \frac{1}{8} \quad \text{or} \quad m = \frac{1}{2}$$

Check that the solution set of the original equation is $\left\{\frac{1}{8}, \frac{1}{2}\right\}$.

(b) $2a^{2/3} - 11a^{1/3} + 12 = 0$

Let $a^{1/3} = u$; then $a^{2/3} = (a^{1/3})^2 = u^2$. Substitute into the given equation.

$$2u^2 - 11u + 12 = 0 \qquad \text{Let } a^{1/3} = u;\ a^{2/3} = u^2.$$

$$(2u - 3)(u - 4) = 0 \qquad \text{Factor.}$$

$$2u - 3 = 0 \quad \text{or} \quad u - 4 = 0 \qquad \text{Zero-factor property}$$

$$u = \frac{3}{2} \quad \text{or} \quad u = 4$$

$$a^{1/3} = \frac{3}{2} \quad \text{or} \quad a^{1/3} = 4 \qquad u = a^{1/3}$$

$$(a^{1/3})^3 = \left(\frac{3}{2}\right)^3 \quad \text{or} \quad (a^{1/3})^3 = 4^3 \qquad \text{Cube each side.}$$

$$a = \frac{27}{8} \quad \text{or} \quad a = 64$$

Check that the solution set is $\left\{\frac{27}{8}, 64\right\}$.

Now Try Exercises 59 and 65.

CAUTION   A common error when solving problems like those in Examples 5 and 6 is to stop too soon. Once you have solved for u, remember to substitute and solve for the values of the *original* variable.

11.4   EXERCISES

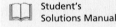

Refer to the box at the beginning of this section. Decide whether factoring, the square root property, *or* the quadratic formula *is most appropriate for solving each quadratic equation. Do not actually solve the equations.*

1. $(2x + 3)^2 = 4$ **2.** $4x^2 - 3x = 1$ **3.** $z^2 + 5z - 8 = 0$

4. $2k^2 + 3k = 1$ **5.** $3m^2 = 2 - 5m$ **6.** $p^2 = 5$

Write a sentence describing the first step you would take to solve each equation. Do not actually solve.

7. $\dfrac{14}{x} = x - 5$ **8.** $\sqrt{1 + x} + x = 5$

9. $(r^2 + r)^2 - 8(r^2 + r) + 12 = 0$ **10.** $3t = \sqrt{16 - 10t}$

11. What is wrong with the following "solution"?

$$x = \sqrt{3x + 4}$$
$$x^2 = 3x + 4 \quad \text{Square both sides.}$$
$$x^2 - 3x - 4 = 0$$
$$(x - 4)(x + 1) = 0$$
$$x - 4 = 0 \quad \text{or} \quad x + 1 = 0$$
$$x = 4 \quad \text{or} \quad x = -1$$

Solution set: $\{4, -1\}$

12. What is wrong with the following "solution"?

$$2(m - 1)^2 - 3(m - 1) + 1 = 0$$
$$2u^2 - 3u + 1 = 0$$
$$\text{Let } u = m - 1.$$
$$(2u - 1)(u - 1) = 0$$
$$2u - 1 = 0 \quad \text{or} \quad u - 1 = 0$$
$$u = \frac{1}{2} \quad \text{or} \quad u = 1$$

Solution set: $\left\{\frac{1}{2}, 1\right\}$

Solve each equation. Check your solutions. See Example 1.

13. $\dfrac{14}{x} = x - 5$

14. $\dfrac{-12}{x} = x + 8$

15. $1 - \dfrac{3}{x} - \dfrac{28}{x^2} = 0$

16. $4 - \dfrac{7}{r} - \dfrac{2}{r^2} = 0$

17. $3 - \dfrac{1}{t} = \dfrac{2}{t^2}$

18. $1 + \dfrac{2}{k} = \dfrac{3}{k^2}$

19. $\dfrac{1}{x} + \dfrac{2}{x + 2} = \dfrac{17}{35}$

20. $\dfrac{2}{m} + \dfrac{3}{m + 9} = \dfrac{11}{4}$

21. $\dfrac{2}{x + 1} + \dfrac{3}{x + 2} = \dfrac{7}{2}$

22. $\dfrac{4}{3 - p} + \dfrac{2}{5 - p} = \dfrac{26}{15}$

23. $\dfrac{3}{2x} - \dfrac{1}{2(x + 2)} = 1$

24. $\dfrac{4}{3x} - \dfrac{1}{2(x + 1)} = 1$

25. $3 = \dfrac{1}{t + 2} + \dfrac{2}{(t + 2)^2}$

26. $1 + \dfrac{2}{3z + 2} = \dfrac{15}{(3z + 2)^2}$

27. $\dfrac{6}{p} = 2 + \dfrac{p}{p + 1}$

28. $\dfrac{k}{2 - k} + \dfrac{2}{k} = 5$

Use the concepts of this section to answer each question.

29. A boat goes 20 mph in still water, and the rate of the current is t mph.

   (a) What is the rate of the boat when it travels upstream?

   (b) What is the rate of the boat when it travels downstream?

30. If it takes m hr to grade a set of papers, what is the grader's rate (in job per hour)?

Solve each problem. See Examples 2 and 3.

31. On a windy day Yoshiaki found that he could go 16 mi downstream and then 4 mi back upstream at top speed in a total of 48 min. What was the top speed of Yoshiaki's boat if the current was 15 mph?

	d	r	t
Upstream	4	x − 15	
Downstream	16		

32. Lekesha flew her plane for 6 hr at a constant speed. She traveled 810 mi with the wind, then turned around and traveled 720 mi against the wind. The wind speed was a constant 15 mph. Find the speed of the plane.

	d	r	t
With Wind	810		
Against Wind	720		

33. In Canada, Medicine Hat and Cranbrook are 300 km apart. Harry rides his Honda 20 km per hr faster than Yoshi rides his Yamaha. Find Harry's average speed if he travels from Cranbrook to Medicine Hat in $1\frac{1}{4}$ hr less time than Yoshi. (*Source: State Farm Road Atlas.*)

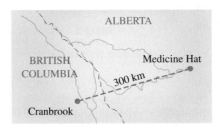

34. In California, the distance from Jackson to Lodi is about 40 mi, as is the distance from Lodi to Manteca. Rico drove from Jackson to Lodi during the rush hour, stopped in Lodi for a root beer, and then drove on to Manteca at 10 mph faster. Driving time for the entire trip was 88 min. Find his speed from Jackson to Lodi. (*Source: State Farm Road Atlas.*)

35. Working together, two people can cut a large lawn in 2 hr. One person can do the job alone in 1 hr less time than the other. How long (to the nearest tenth) would it take the faster person to do the job? (*Hint: x* is the time of the faster person.)

	Rate	Time Working Together	Fractional Part of the Job Done
Faster Worker	$\frac{1}{x}$	2	
Slower Worker		2	

36. A janitorial service provides two people to clean an office building. Working together, the two can clean the building in 5 hr. One person is new to the job and would take 2 hr longer than the other person to clean the building alone. How long (to the nearest tenth) would it take the new worker to clean the building alone?

	Rate	Time Working Together	Fractional Part of the Job Done
Faster Worker			
Slower Worker			

37. Rusty and Nancy Brauner are planting flats of spring flowers. Working alone, Rusty would take 2 hr longer than Nancy to plant the flowers. Working together, they do the job in 12 hr. How long would it have taken each person working alone?

38. Jay Beckenstein can work through a stack of invoices in 1 hr less time than Colleen Manley Jones can. Working together they take $1\frac{1}{2}$ hr. How long would it take each person working alone?

39. A washing machine can be filled in 6 min if both the hot and cold water taps are fully opened. Filling the washer with hot water alone takes 9 min longer than filling it with cold water alone. How long does it take to fill the washer with cold water?

40. Two pipes together can fill a large tank in 2 hr. One of the pipes, used alone, takes 3 hr longer than the other to fill the tank. How long would each pipe take to fill the tank alone?

Solve each equation. Check your solutions. See Example 4.

41. $x = \sqrt{7x - 10}$ **42.** $z = \sqrt{5z - 4}$ **43.** $2x = \sqrt{11x + 3}$ **44.** $4x = \sqrt{6x + 1}$

45. $3x = \sqrt{16 - 10x}$ **46.** $4t = \sqrt{8t + 3}$ **47.** $p - 2\sqrt{p} = 8$ **48.** $k + \sqrt{k} = 12$

49. $m = \sqrt{\dfrac{6 - 13m}{5}}$ **50.** $r = \sqrt{\dfrac{20 - 19r}{6}}$

Solve each equation. Check your solutions. See Examples 5 and 6.

51. $t^4 - 18t^2 + 81 = 0$ **52.** $x^4 - 8x^2 + 16 = 0$ **53.** $4k^4 - 13k^2 + 9 = 0$

54. $9x^4 - 25x^2 + 16 = 0$ **55.** $x^4 + 48 = 16x^2$ **56.** $z^4 = 17z^2 - 72$

57. $(x + 3)^2 + 5(x + 3) + 6 = 0$ **58.** $(k - 4)^2 + (k - 4) - 20 = 0$

59. $3(m + 4)^2 - 8 = 2(m + 4)$ **60.** $(t + 5)^2 + 6 = 7(t + 5)$

61. $2 + \dfrac{5}{3k - 1} = \dfrac{-2}{(3k - 1)^2}$ **62.** $3 - \dfrac{7}{2p + 2} = \dfrac{6}{(2p + 2)^2}$

63. $2 - 6(m - 1)^{-2} = (m - 1)^{-1}$ **64.** $3 - 2(x - 1)^{-1} = (x - 1)^{-2}$

65. $x^{2/3} + x^{1/3} - 2 = 0$ **66.** $x^{2/3} - 2x^{1/3} - 3 = 0$

67. $r^{2/3} + r^{1/3} - 12 = 0$ **68.** $3x^{2/3} - x^{1/3} - 24 = 0$

69. $4k^{4/3} - 13k^{2/3} + 9 = 0$ **70.** $9m^{2/5} = 16 - 10m^{1/5}$

71. $2\left(1 + \sqrt{r}\right)^2 = 13\left(1 + \sqrt{r}\right) - 6$ **72.** $(k^2 + k)^2 + 12 = 8(k^2 + k)$

73. $2x^4 + x^2 - 3 = 0$ **74.** $4k^4 + 5k^2 + 1 = 0$

The equations in Exercises 75–84 are not grouped by type. Decide which method of solution applies, and then solve each equation. Give only real solutions. See Examples 1 and 4–6.

75. $12x^4 - 11x^2 + 2 = 0$ **76.** $\left(x - \dfrac{1}{2}\right)^2 + 5\left(x - \dfrac{1}{2}\right) - 4 = 0$

77. $\sqrt{2x + 3} = 2 + \sqrt{x - 2}$ **78.** $\sqrt{m + 1} = -1 + \sqrt{2m}$

79. $2m^6 + 11m^3 + 5 = 0$ **80.** $8x^6 + 513x^3 + 64 = 0$

81. $6 = 7(2w - 3)^{-1} + 3(2w - 3)^{-2}$ **82.** $m^6 - 10m^3 = -9$

83. $2x^4 - 9x^2 = -2$ **84.** $8x^4 + 1 = 11x^2$

RELATING CONCEPTS (EXERCISES 85–90)

For Individual or Group Work

Consider the following equation, and **work Exercises 85–90 in order.**

$$\frac{x^2}{(x - 3)^2} + \frac{3x}{x - 3} - 4 = 0.$$

85. Why must 3 be excluded from the domain of this equation?

86. Multiply both sides of the equation by the LCD, $(x - 3)^2$, and solve. There is only one solution—what is it?

87. Write the equation so that it is quadratic in form using the rational expression $\frac{x}{x-3}$.

88. Explain why the expression $\frac{x}{x-3}$ cannot equal 1.

89. Solve the equation from Exercise 87 by making the substitution $t = \frac{x}{x-3}$. You should get two values for t. Why is one of them impossible for this equation?

90. Solve the equation $x^2(x-3)^{-2} + 3x(x-3)^{-1} - 4 = 0$ by letting $s = (x-3)^{-1}$. You should get two values for s. Why is this impossible for this equation?

SUMMARY EXERCISES ON SOLVING QUADRATIC EQUATIONS

*Exercises marked * require knowledge of imaginary numbers.*

*We have introduced four algebraic methods for solving quadratic equations written in the form $ax^2 + bx + c = 0$: **factoring, the square root property, completing the square, and the quadratic formula.** Refer to the summary box at the beginning of Section 11.4 to review some of the advantages and disadvantages of each method. Then solve each quadratic equation by the method of your choice.*

1. $p^2 = 7$

2. $6x^2 - x - 15 = 0$

3. $n^2 + 6n + 4 = 0$

4. $(x-4)^2 = 25$

5. $\dfrac{5}{m} + \dfrac{12}{m^2} = 2$

6. $3m^2 = 3 - 8m$

7. $2r^2 - 4r + 1 = 0$

*8. $x^2 = -12$

9. $x\sqrt{2} = \sqrt{5x - 2}$

10. $m^4 - 10m^2 + 9 = 0$

11. $(2k + 3)^2 = 8$

12. $\dfrac{2}{x} + \dfrac{1}{x-2} = \dfrac{5}{3}$

13. $t^4 + 14 = 9t^2$

14. $8x^2 - 4x = 2$

*15. $z^2 + z + 1 = 0$

16. $5x^6 + 2x^3 - 7 = 0$

17. $4t^2 - 12t + 9 = 0$

18. $x\sqrt{3} = \sqrt{2 - x}$

19. $r^2 - 72 = 0$

20. $-3x^2 + 4x = -4$

21. $x^2 - 5x - 36 = 0$

22. $w^2 = 169$

*23. $3p^2 = 6p - 4$

24. $z = \sqrt{\dfrac{5z + 3}{2}}$

25. $2(3k - 1)^2 + 5(3k - 1) = -2$

*26. $\dfrac{4}{r^2} + 3 = \dfrac{1}{r}$

11.5 | Formulas and Further Applications

OBJECTIVE 1 Solve formulas for variables involving squares and square roots. The methods presented earlier can be used to solve such formulas.

■ EXAMPLE 1 Solving for Variables Involving Squares or Square Roots

Solve each formula for the given variable.

(a) $w = \dfrac{kFr}{v^2}$ for v

$w = \dfrac{kFr}{v^2}$ ⟵ Get v alone on one side.

$v^2 w = kFr$ Multiply by v^2.

$v^2 = \dfrac{kFr}{w}$ Divide by w.

$v = \pm\sqrt{\dfrac{kFr}{w}}$ Square root property

$v = \dfrac{\pm\sqrt{kFr}}{\sqrt{w}} \cdot \dfrac{\sqrt{w}}{\sqrt{w}} = \dfrac{\pm\sqrt{kFrw}}{w}$ Rationalize the denominator.

(b) $d = \sqrt{\dfrac{4A}{\pi}}$ for A

$d = \sqrt{\dfrac{4A}{\pi}}$ ⟵ Get A alone on one side.

$d^2 = \dfrac{4A}{\pi}$ Square both sides.

$\pi d^2 = 4A$ Multiply by π.

$\dfrac{\pi d^2}{4} = A$ Divide by 4.

Now Try Exercises 9 and 19.

NOTE In many formulas like $v = \dfrac{\pm\sqrt{kFrw}}{w}$ in Example 1(a), we choose the positive value. In our work here, we will include both positive and negative values.

■ EXAMPLE 2 Solving for a Variable that Appears in First- and Second-Degree Terms

Solve $s = 2t^2 + kt$ for t.

Since the given equation has terms with t^2 and t, write it in standard form $ax^2 + bx + c = 0$, with t as the variable instead of x.

$$2t^2 + kt - s = 0$$

Now use the quadratic formula with $a = 2$, $b = k$, and $c = -s$.

$$t = \frac{-k \pm \sqrt{k^2 - 4(2)(-s)}}{2(2)} \qquad \text{Solve for } t.$$

$$t = \frac{-k \pm \sqrt{k^2 + 8s}}{4}$$

The solutions are $t = \dfrac{-k + \sqrt{k^2 + 8s}}{4}$ and $t = \dfrac{-k - \sqrt{k^2 + 8s}}{4}$.

Now Try Exercise 15.

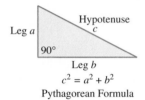

Leg a

Hypotenuse c

$90°$

Leg b

$c^2 = a^2 + b^2$

Pythagorean Formula

OBJECTIVE 2 **Solve applied problems using the Pythagorean formula.** The Pythagorean formula $a^2 + b^2 = c^2$, illustrated by the figure in the margin, was introduced in earlier chapters and is used to solve applications involving right triangles. Such problems often require solving quadratic equations.

EXAMPLE 3 Using the Pythagorean Formula

Two cars left an intersection at the same time, one heading due north, the other due west. Some time later, they were exactly 100 mi apart. The car headed north had gone 20 mi farther than the car headed west. How far had each car traveled?

Step 1 **Read** the problem carefully.

Step 2 **Assign a variable.** Let x be the distance traveled by the car headed west. Then $(x + 20)$ is the distance traveled by the car headed north. See Figure 2. The cars are 100 mi apart, so the hypotenuse of the right triangle equals 100.

Step 3 **Write an equation.** Use the Pythagorean formula.

$$c^2 = a^2 + b^2$$
$$100^2 = x^2 + (x + 20)^2$$

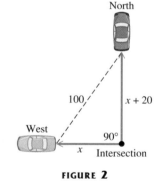

North

100

$x + 20$

West

$90°$

x Intersection

FIGURE 2

Step 4 **Solve.**

$$10,000 = x^2 + x^2 + 40x + 400 \qquad \text{Square the binomial.}$$
$$0 = 2x^2 + 40x - 9600 \qquad \text{Standard form}$$
$$0 = x^2 + 20x - 4800 \qquad \text{Divide by 2.}$$
$$0 = (x + 80)(x - 60) \qquad \text{Factor.}$$
$$x + 80 = 0 \quad \text{or} \quad x - 60 = 0 \qquad \text{Zero-factor property}$$
$$x = -80 \quad \text{or} \qquad x = 60$$

Step 5 **State the answer.** Since distance cannot be negative, discard the negative solution. The required distances are 60 mi and $60 + 20 = 80$ mi.

Step 6 **Check.** Since $60^2 + 80^2 = 100^2$, the answer is correct.

Now Try Exercise 31.

OBJECTIVE 3 Solve applied problems using area formulas.

EXAMPLE 4 Solving an Area Problem

A rectangular reflecting pool in a park is 20 ft wide and 30 ft long. The park gardener wants to plant a strip of grass of uniform width around the edge of the pool. She has enough seed to cover 336 ft². How wide will the strip be?

Step 1 **Read** the problem carefully.

Step 2 **Assign a variable.** The pool is shown in Figure 3. If x represents the unknown width of the grass strip, the width of the large rectangle is given by $20 + 2x$ (the width of the pool plus two grass strips), and the length is given by $30 + 2x$.

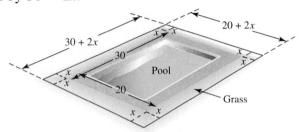

FIGURE 3

Step 3 **Write an equation.** The area of the large rectangle is given by the product of its length and width, $(30 + 2x)(20 + 2x)$. The area of the pool is $30 \cdot 20 = 600$ ft². The area of the large rectangle minus the area of the pool should equal the area of the grass strip. Since the area of the grass strip is to be 336 ft², the equation is

$$\underset{\substack{\text{Area} \\ \text{of} \\ \text{rectangle}}}{} - \underset{\substack{\text{Area} \\ \text{of} \\ \text{pool}}}{} = \underset{\substack{\text{Area} \\ \text{of} \\ \text{grass}}}{}$$

$$(30 + 2x)(20 + 2x) - 600 = 336.$$

Step 4 **Solve.**

$600 + 100x + 4x^2 - 600 = 336$	Multiply.
$4x^2 + 100x - 336 = 0$	Standard form
$x^2 + 25x - 84 = 0$	Divide by 4.
$(x + 28)(x - 3) = 0$	Factor.
$x = -28 \quad \text{or} \quad x = 3$	Zero-factor property

Step 5 **State the answer.** The width cannot be -28 ft, so the grass strip should be 3 ft wide.

Step 6 **Check.** If $x = 3$, then the area of the large rectangle (which includes the grass strip) is

$$(30 + 2 \cdot 3)(20 + 2 \cdot 3) = 36 \cdot 26 = 936 \text{ ft}^2. \qquad \text{Area of pool and strip}$$

The area of the pool is $30 \cdot 20 = 600$ ft². So, the area of the grass strip is $936 - 600 = 336$ ft², which is the area the gardener had enough seed to cover. The answer is correct.

Now Try Exercise 37.

OBJECTIVE 4 Solve applied problems using quadratic functions as models. Some applied problems can be modeled by *quadratic functions,* which can be written in the form

$$f(x) = ax^2 + bx + c,$$

for real numbers a, b, and c, $a \neq 0$.

■ **EXAMPLE 5 Solving an Applied Problem Using a Quadratic Function**

If an object is propelled upward from the top of a 144-ft building at 112 ft per sec, its position (in feet above the ground) is given by

$$s(t) = -16t^2 + 112t + 144,$$

where t is time in seconds after it was propelled. When does it hit the ground?

When the object hits the ground, its distance above the ground is 0. We must find the value of t that makes $s(t) = 0$.

$$0 = -16t^2 + 112t + 144 \qquad \text{Let } s(t) = 0.$$

$$0 = t^2 - 7t - 9 \qquad \text{Divide by } -16.$$

$$t = \frac{7 \pm \sqrt{49 + 36}}{2} \qquad \text{Quadratic formula}$$

$$t = \frac{7 \pm \sqrt{85}}{2} \approx \frac{7 \pm 9.2}{2} \qquad \text{Use a calculator.}$$

The solutions are $t \approx 8.1$ or $t \approx -1.1$. Time cannot be negative, so we discard the negative solution. The object hits the ground about 8.1 sec after it is propelled. ■

Now Try Exercise 43.

■ **EXAMPLE 6 Using a Quadratic Function to Model Company Bankruptcy Filings**

The number of companies filing for bankruptcy was high in the early 1990s due to an economic recession. The number then declined during the middle 1990s, and in recent years has increased again. The quadratic function defined by

$$f(x) = 3.37x^2 - 28.6x + 133$$

approximates the number of company bankruptcy filings during the years 1990–2001, where x is the number of years since 1990. (*Source:* www.BankruptcyData.com)

(a) Use the model to approximate the number of company bankruptcy filings in 1995.

For 1995, $x = 5$, so find $f(5)$.

$$f(5) = 3.37(5)^2 - 28.6(5) + 133 \qquad \text{Let } x = 5.$$

$$= 74.25$$

There were about 74 company bankruptcy filings in 1995.

(b) In what year did company bankruptcy filings reach 150?

Find the value of x that makes $f(x) = 150$.

$$f(x) = 3.37x^2 - 28.6x + 133$$

$$150 = 3.37x^2 - 28.6x + 133 \qquad \text{Let } f(x) = 150.$$

$$0 = 3.37x^2 - 28.6x - 17 \qquad \text{Standard form}$$

Now use $a = 3.37$, $b = -28.6$, and $c = -17$ in the quadratic formula.

$$x = \frac{28.6 \pm \sqrt{(-28.6)^2 - 4(3.37)(-17)}}{2(3.37)}$$

$$x \approx 9.0 \quad \text{or} \quad x \approx -.56 \qquad \text{Use a calculator.}$$

The positive solution is $x \approx 9$, so company bankruptcy filings reached 150 in the year $1990 + 9 = 1999$. (Reject the negative solution since the model is not valid for negative values of x.) Note that company bankruptcy filings doubled from about 74 in 1995 to 150 in 1999.

Now Try Exercises 55 and 57.

11.5 EXERCISES

Answer each question in Exercises 1–4.

1. What is the first step in solving a formula that has the specified variable in the denominator?

2. What is the first step in solving a formula like $gw^2 = 2r$ for w?

3. What is the first step in solving a formula like $gw^2 = kw + 24$ for w?

4. Why is it particularly important to check all proposed solutions to an applied problem against the information in the original problem?

In Exercises 5 and 6, solve for m in terms of the other variables ($m > 0$).

5.

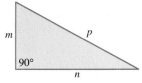

6.

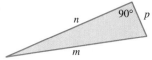

Solve each equation for the indicated variable. (Leave $\pm$ in your answers.) See Examples 1 and 2.

7. $d = kt^2$ for t

8. $s = kwd^2$ for d

9. $I = \dfrac{ks}{d^2}$ for d

10. $R = \dfrac{k}{d^2}$ for d

11. $F = \dfrac{kA}{v^2}$ for v

12. $L = \dfrac{kd^4}{h^2}$ for h

13. $V = \dfrac{1}{3}\pi r^2 h$ for r

14. $V = \pi(r^2 + R^2)h$ for r

15. $At^2 + Bt = -C$ for t

16. $S = 2\pi rh + \pi r^2$ for r

17. $D = \sqrt{kh}$ for h

18. $F = \dfrac{k}{\sqrt{d}}$ for d

19. $p = \sqrt{\dfrac{k\ell}{g}}$ for ℓ

20. $p = \sqrt{\dfrac{k\ell}{g}}$ for g

21. If g is a positive number in the formula of Exercise 19, explain why k and ℓ must have the same sign in order for p to be a real number.

22. Refer to Example 2 of this section. Suppose that k and s both represent positive numbers.

    (a) Which one of the two solutions given is positive?

    (b) Which one is negative?   **(c)** How can you tell?

Solve each equation for the indicated variable.

23. $p = \dfrac{E^2 R}{(r + R)^2}$ for R $(E > 0)$

24. $S(6S - t) = t^2$ for S

25. $10p^2 c^2 + 7pcr = 12r^2$ for r

26. $S = vt + \dfrac{1}{2} gt^2$ for t

27. $LI^2 + RI + \dfrac{1}{c} = 0$ for I

28. $P = EI - RI^2$ for I

Solve each problem. When appropriate, round answers to the nearest tenth. See Example 3.

29. Find the lengths of the sides of the triangle.

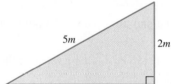

30. Find the lengths of the sides of the triangle.

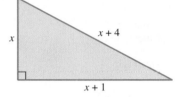

31. Two ships leave port at the same time, one heading due south and the other heading due east. Several hours later, they are 170 mi apart. If the ship traveling south traveled 70 mi farther than the other ship, how many miles did they each travel?

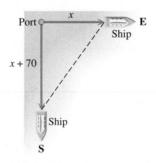

32. Allyson Pellissier is flying a kite that is 30 ft farther above her hand than its horizontal distance from her. The string from her hand to the kite is 150 ft long. How high is the kite?

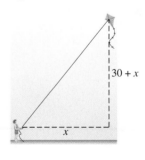

33. A toy manufacturer needs a piece of plastic in the shape of a right triangle with the longer leg 2 cm more than twice as long as the shorter leg, and the hypotenuse 1 cm more than the longer leg. How long should the three sides of the triangular piece be?

34. Michael Fuentes, a developer, owns a piece of land enclosed on three sides by streets, giving it the shape of a right triangle. The hypotenuse is 8 m longer than the longer leg, and the shorter leg is 9 m shorter than the hypotenuse. Find the lengths of the three sides of the property.

35. Two pieces of a large wooden puzzle fit together to form a rectangle with length 1 cm less than twice the width. The diagonal, where the two pieces meet, is 2.5 cm in length. Find the length and width of the rectangle.

36. A 13-ft ladder is leaning against a house. The distance from the bottom of the ladder to the house is 7 ft less than the distance from the top of the ladder to the ground. How far is the bottom of the ladder from the house?

Solve each problem. See Example 4.

37. A couple wants to buy a rug for a room that is 20 ft long and 15 ft wide. They want to leave an even strip of flooring uncovered around the edges of the room. How wide a strip will they have if they buy a rug with an area of 234 ft²?

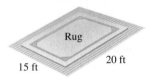

38. A club swimming pool is 30 ft wide and 40 ft long. The club members want an exposed aggregate border in a strip of uniform width around the pool. They have enough material for 296 ft². How wide can the strip be?

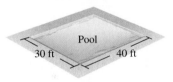

39. Arif's backyard is 20 m by 30 m. He wants to put a flower garden in the middle of the backyard, leaving a strip of grass of uniform width around the flower garden. Arif must have 184 m² of grass. Under these conditions, what will the length and width of the garden be?

40. A rectangle has a length 2 m less than twice its width. When 5 m are added to the width, the resulting figure is a square with an area of 144 m². Find the dimensions of the original rectangle.

41. A rectangular piece of sheet metal has a length that is 4 in. less than twice the width. A square piece 2 in. on a side is cut from each corner. The sides are then turned up to form an uncovered box of volume 256 in.³. Find the length and width of the original piece of metal.

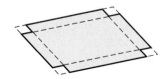

42. Another rectangular piece of sheet metal is 2 in. longer than it is wide. A square piece 3 in. on a side is cut from each corner. The sides are then turned up to form an uncovered box of volume 765 in.³. Find the dimensions of the original piece of metal.

Solve each problem. When appropriate, round answers to the nearest tenth. See Example 5.

43. An object is projected directly upward from the ground. After t sec its distance in feet above the ground is

$$s(t) = 144t - 16t^2.$$

After how many seconds will the object be 128 ft above the ground? (*Hint:* Look for a common factor before solving the equation.)

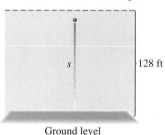

128 ft

Ground level

44. When does the object in Exercise 43 strike the ground?

45. A ball is projected upward from the ground. Its distance in feet from the ground in t sec is given by

$$s(t) = -16t^2 + 128t.$$

At what times will the ball be 213 ft from the ground?

213 ft

46. A toy rocket is launched from ground level. Its distance in feet from the ground in t sec is given by

$$s(t) = -16t^2 + 208t.$$

At what times will the rocket be 550 ft from the ground?

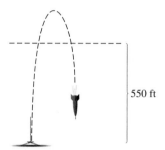

550 ft

47. The function defined by

$$D(t) = 13t^2 - 100t$$

gives the distance in feet a car going approximately 68 mph will skid in t sec. Find the time it would take for the car to skid 180 ft.

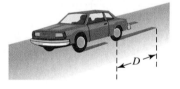

48. The function given in Exercise 47 becomes

$$D(t) = 13t^2 - 73t$$

for a car going 50 mph. Find the time for this car to skid 218 ft.

✍ *A rock is projected upward from ground level, and its distance in feet from the ground in t sec is given by $s(t) = -16t^2 + 160t$. Use algebra and a short explanation to answer Exercises 49 and 50.*

49. After how many seconds does it reach a height of 400 ft? How would you describe in words its position at this height?

50. After how many seconds does it reach a height of 425 ft? How would you interpret the mathematical result here?

Solve each problem using a quadratic equation.

51. A certain bakery has found that the daily demand for bran muffins is $\frac{3200}{p}$, where p is the price of a muffin in cents. The daily supply is $3p - 200$. Find the price at which supply and demand are equal.

52. In one area the demand for compact discs is $\frac{700}{P}$ per day, where P is the price in dollars per disc. The supply is $5P - 1$ per day. At what price does supply equal demand?

53. The formula $A = P(1 + r)^2$ gives the amount A in dollars that P dollars will grow to in 2 yr at interest rate r (where r is given as a decimal), using compound interest. What interest rate will cause \$2000 to grow to \$2142.25 in 2 yr?

54. If a square piece of cardboard has 3-in. squares cut from its corners and then has the flaps folded up to form an open-top box, the volume of the box is given by the formula $V = 3(x - 6)^2$, where x is the length of each side of the original piece of cardboard in inches. What original length would yield a box with volume 432 in.^{3}?

Sales of SUVs (sport utility vehicles) in the United States (in millions) for the years 1990 through 1999 are shown in the bar graph and can be modeled by the quadratic function defined by

$$f(x) = .016x^2 + .124x + .787.$$

Here, $x = 0$ represents 1990, $x = 1$ represents 1991, and so on. Use the graph and the model to work Exercises 55–58. See Example 6.

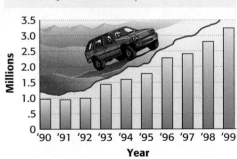

SALES OF SUVs IN THE UNITED STATES (IN MILLIONS)

Source: CNW Marketing Research of Bandon, OR, based on automakers' reported sales.

55. (a) Use the graph to estimate sales in 1997 to the nearest tenth.

(b) Use the model to approximate sales in 1997 to the nearest tenth. How does this result compare to your estimate from part (a)?

56. (a) Use the model to estimate sales in 2000 to the nearest tenth.

(b) Sales through October 2000 were about 2.9 million. Based on this, is the sales estimate for 2000 from part (a) reasonable? Explain.

57. Based on the model, in what year did sales reach 2 million? (Round down to the nearest year.) How does this result compare to the sales shown in the graph?

58. Based on the model, in what year did sales reach 3 million? (Round down to the nearest year.) How does this result compare to the sales shown in the graph?

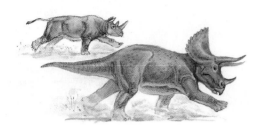

William Froude was a 19th century naval architect who used the expression

$$\frac{v^2}{g\ell}$$

in shipbuilding. This expression, known as the Froude number, was also used by R. McNeill Alexander in his research on dinosaurs. (Source: "How Dinosaurs Ran," Scientific American, *April 1991.) In Exercises 59 and 60, find the value of v (in meters per second), given g* = 9.8 m per sec^2.

59. Rhinoceros: $\ell = 1.2$;
    Froude number = 2.57

60. Triceratops: $\ell = 2.8$;
    Froude number = .16

Recall that corresponding sides of similar triangles are proportional. Use this fact to find the lengths of the indicated sides of each pair of similar triangles. Check all possible solutions in both triangles. Sides of a triangle cannot be negative (and are not drawn to scale here).

61. Side AC

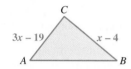

62. Side RQ

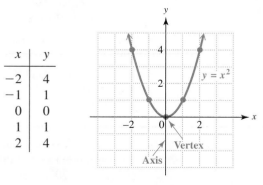

11.6 Graphs of Quadratic Functions

OBJECTIVES

1. Graph a quadratic function.

2. Graph parabolas with horizontal and vertical shifts.

3. Predict the shape and direction of a parabola from the coefficient of x^2.

4. Find a quadratic function to model data.

OBJECTIVE 1 Graph a quadratic function. In Chapter 4, we graphed a few simple second-degree polynomial equations by point-plotting. In Figure 4, we repeat a table of ordered pairs for the simplest quadratic equation, $y = x^2$, and the resulting graph.

x	y
-2	4
-1	1
0	0
1	1
2	4

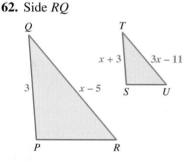

FIGURE 4

As mentioned in Chapter 4, this graph is called a **parabola.** The point $(0, 0)$, the lowest point on the curve, is the **vertex** of this parabola. The vertical line through the vertex is the **axis** of the parabola, here $x = 0$. A parabola is **symmetric about its axis;**

that is, if the graph were folded along the axis, the two portions of the curve would coincide. As Figure 4 suggests, the graph is that of a function. Because x can be any real number, the domain of the function defined by $y = x^2$ is $(-\infty, \infty)$. Since y is always nonnegative, the range is $[0, \infty)$.

In Section 11.5, we solved applications modeled by *quadratic functions*. In this section and the next, we consider graphs of more general quadratic functions as defined here.

Quadratic Function

A function that can be written in the form
$$f(x) = ax^2 + bx + c$$
for real numbers a, b, and c, with $a \neq 0$, is a **quadratic function.**

The graph of any quadratic function is a parabola with a vertical axis.

NOTE We use the variable y and function notation $f(x)$ interchangeably when discussing parabolas. Although we use the letter f most often to name quadratic functions, other letters can be used. We use the capital letter F to distinguish between different parabolas graphed on the same coordinate axes.

Parabolas, which are a type of *conic section* (Chapter 13), have many applications. The large dishes seen on the sidelines of televised football games, which are used by television crews to pick up the shouted signals of players on the field, have cross sections that are parabolas. Cross sections of satellite dishes and automobile headlights also form parabolas. The cables that are used to support suspension bridges are shaped like parabolas.

OBJECTIVE 2 Graph parabolas with horizontal and vertical shifts. Parabolas need not have their vertices at the origin, as does the graph of $f(x) = x^2$. For example, to graph a parabola of the form $F(x) = x^2 + k$, start by selecting sample values of x like those that were used to graph $f(x) = x^2$. The corresponding values of $F(x)$ in $F(x) = x^2 + k$ differ by k from those of $f(x) = x^2$. For this reason, the graph of $F(x) = x^2 + k$ is *shifted*, or *translated*, k units vertically compared with that of $f(x) = x^2$.

EXAMPLE 1 Graphing a Parabola with a Vertical Shift

Graph $F(x) = x^2 - 2$.

This graph has the same shape as that of $f(x) = x^2$, but since k here is -2, the graph is shifted 2 units down, with vertex $(0, -2)$. Every function value is 2 less than the corresponding function value of $f(x) = x^2$. Plotting points on both sides of the vertex gives the graph in Figure 5 on the next page. Notice that since the parabola is symmetric about its axis $x = 0$, the plotted points are "mirror images" of each other. Since x can be any real number, the domain is still $(-\infty, \infty)$; the value of y (or $F(x)$) is always greater than or equal to -2, so the range is $[-2, \infty)$. The graph of $f(x) = x^2$ is shown in Figure 5 for comparison.

x	$f(x) = x^2$	$F(x) = x^2 - 2$
-2	4	2
-1	1	-1
0	0	-2
1	1	-1
2	4	2

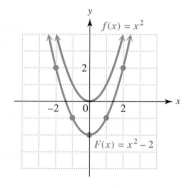

FIGURE 5

Now Try Exercise 23.

Vertical Shift

The graph of $F(x) = x^2 + k$ is a parabola with the same shape as the graph of $f(x) = x^2$. The parabola is shifted k units up if $k > 0$, and $|k|$ units down if $k < 0$. The vertex is $(0, k)$.

The graph of $F(x) = (x - h)^2$ is also a parabola with the same shape as that of $f(x) = x^2$. Because $(x - h)^2 \geq 0$ for all x, the vertex of $F(x) = (x - h)^2$ is the lowest point on the parabola. The lowest point occurs here when $F(x)$ is 0. To get $F(x)$ equal to 0, let $x = h$ so the vertex of $F(x) = (x - h)^2$ is $(h, 0)$. Based on this, the graph of $F(x) = (x - h)^2$ is shifted h units horizontally compared with that of $f(x) = x^2$.

EXAMPLE 2 Graphing a Parabola with a Horizontal Shift

Graph $F(x) = (x - 2)^2$.

If $x = 2$, then $F(x) = 0$, giving the vertex $(2, 0)$. The graph of $F(x) = (x - 2)^2$ has the same shape as that of $f(x) = x^2$ but is shifted 2 units to the right. Plotting several points on one side of the vertex and using symmetry about the axis $x = 2$ to find corresponding points on the other side of the vertex gives the graph in Figure 6. Again, the domain is $(-\infty, \infty)$; the range is $[0, \infty)$.

x	$F(x) = (x - 2)^2$
0	4
1	1
2	0
3	1
4	4

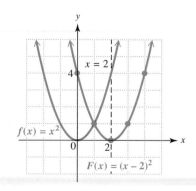

FIGURE 6

Now Try Exercise 27.

Horizontal Shift

The graph of $F(x) = (x - h)^2$ is a parabola with the same shape as the graph of $f(x) = x^2$. The parabola is shifted h units horizontally: h units to the right if $h > 0$, and $|h|$ units to the left if $h < 0$. The vertex is $(h, 0)$.

CAUTION Errors frequently occur when horizontal shifts are involved. To determine the direction and magnitude of a horizontal shift, find the value that would cause the expression $x - h$ to equal 0. For example, the graph of $F(x) = (x - 5)^2$ would be shifted 5 units to the *right*, because $+5$ would cause $x - 5$ to equal 0. On the other hand, the graph of $F(x) = (x + 5)^2$ would be shifted 5 units to the *left*, because -5 would cause $x + 5$ to equal 0.

A parabola can have both horizontal and vertical shifts.

EXAMPLE 3 Graphing a Parabola with Horizontal and Vertical Shifts

Graph $F(x) = (x + 3)^2 - 2$.

This graph has the same shape as that of $f(x) = x^2$, but is shifted 3 units to the left (since $x + 3 = 0$ if $x = -3$) and 2 units down (because of the -2). As shown in Figure 7, the vertex is $(-3, -2)$, with axis $x = -3$. This function has domain $(-\infty, \infty)$ and range $[-2, \infty)$.

x	$F(x)$
-5	2
-4	-1
-3	-2
-2	-1
-1	2

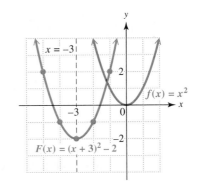

FIGURE 7

Now Try Exercise 29.

The characteristics of the graph of a parabola of the form $F(x) = (x - h)^2 + k$ are summarized as follows.

Vertex and Axis of a Parabola

The graph of $F(x) = (x - h)^2 + k$ is a parabola with the same shape as the graph of $f(x) = x^2$, but with vertex (h, k). The axis is the vertical line $x = h$.

OBJECTIVE 3 Predict the shape and direction of a parabola from the coefficient of x^2. Not all parabolas open up, and not all parabolas have the same shape as the graph of $f(x) = x^2$.

EXAMPLE 4 Graphing a Parabola That Opens Down

Graph $f(x) = -\dfrac{1}{2}x^2$.

This parabola is shown in Figure 8. The coefficient $-\frac{1}{2}$ affects the shape of the graph; the $\frac{1}{2}$ makes the parabola wider $\left(\text{since the values of } \frac{1}{2}x^2 \text{ increase more slowly}\right.$ than those of $x^2\Big)$, and the negative sign makes the parabola open down. The graph is not shifted in any direction; the vertex is still $(0, 0)$ and the axis is $x = 0$. Unlike the parabolas graphed in Examples 1–3, the vertex here has the *largest* function value of any point on the graph. The domain is $(-\infty, \infty)$; the range is $(-\infty, 0]$.

x	$f(x)$
-2	-2
-1	$-\frac{1}{2}$
0	0
1	$-\frac{1}{2}$
2	-2

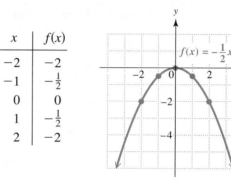

FIGURE 8

Now Try Exercise 21.

Some general principles concerning the graph of $F(x) = a(x - h)^2 + k$ are summarized as follows.

General Principles

1. The graph of the quadratic function defined by

$$F(x) = a(x - h)^2 + k, \quad a \neq 0$$

is a parabola with vertex (h, k) and the vertical line $x = h$ as axis.

2. The graph opens up if a is positive and down if a is negative.

3. The graph is wider than that of $f(x) = x^2$ if $0 < |a| < 1$. The graph is narrower than that of $f(x) = x^2$ if $|a| > 1$.

EXAMPLE 5 Using the General Principles to Graph a Parabola

Graph $F(x) = -2(x + 3)^2 + 4$.

The parabola opens down (because $a < 0$) and is narrower than the graph of $f(x) = x^2$, since $|-2| = 2 > 1$, causing values of $F(x)$ to decrease more quickly than those of $f(x) = -x^2$. This parabola has vertex $(-3, 4)$, as shown in Figure 9. To complete the graph, we plotted the ordered pairs $(-4, 2)$ and, by symmetry, $(-2, 2)$. Symmetry can be used to find additional ordered pairs that satisfy the equation, if desired.

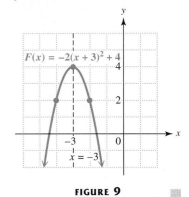

FIGURE 9

Now Try Exercise 33.

OBJECTIVE 4 Find a quadratic function to model data.

EXAMPLE 6 Finding a Quadratic Function to Model the Rise in Multiple Births

The number of higher-order multiple births in the United States is rising. Let x represent the number of years since 1970 and y represent the rate of higher-order multiples born per 100,000 births since 1971. The data are shown in the following table.

Year	x	y
1971	1	29.1
1976	6	35.0
1981	11	40.0
1986	16	47.0
1991	21	100.0
1996	26	152.6

Source: National Center for Health Statistics.

U.S. HIGHER-ORDER MULTIPLE BIRTHS

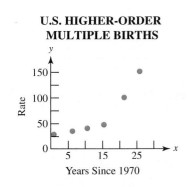

FIGURE 10

U.S. HIGHER-ORDER MULTIPLE BIRTHS

FIGURE 11

Find a quadratic function that models the data.

A scatter diagram of the ordered pairs (x, y) is shown in Figure 10. Notice that the graphed points do not follow a linear pattern, so a linear function would not model the data very well. Instead, the general shape suggested by the scatter diagram indicates that a parabola should approximate these points, as shown by the dashed curve in Figure 11. The equation for such a parabola would have a positive coefficient for x^2 since the graph opens up. To find a quadratic function of the form

$$y = ax^2 + bx + c$$

that models, or *fits,* these data, we choose three representative ordered pairs and use them to write a system of three equations. Using $(1, 29.1)$, $(11, 40)$, and $(21, 100)$, we substitute the x- and y-values from the ordered pairs into the quadratic form $y = ax^2 + bx + c$ to get the following three equations.

$$a(1)^2 + b(1) + c = 29.1 \quad \text{or} \quad a + b + c = 29.1 \quad (1)$$
$$a(11)^2 + b(11) + c = 40 \quad \text{or} \quad 121a + 11b + c = 40 \quad (2)$$
$$a(21)^2 + b(21) + c = 100 \quad \text{or} \quad 441a + 21b + c = 100 \quad (3)$$

We can find the values of a, b, and c by solving this system of three equations in three variables using the methods of Section 8.4. Multiplying equation (1) by -1 and adding the result to equation (2) gives

$$120a + 10b = 10.9. \qquad (4)$$

Multiplying equation (2) by -1 and adding the result to equation (3) gives

$$320a + 10b = 60. \qquad (5)$$

We eliminate b from this system of two equations in two variables by multiplying equation (4) by -1 and adding the result to equation (5) to obtain

$$200a = 49.1$$
$$a = .2455. \qquad \text{Use a calculator.}$$

We substitute .2455 for a in equation (4) or (5) to find that $b = -1.856$. Substituting the values of a and b into equation (1) gives $c = 30.7105$. Using these values of a, b, and c, our model is defined by

$$y = .2455x^2 - 1.856x + 30.7105.$$

Now Try Exercise 49.

NOTE In Example 6, if we had chosen three different ordered pairs of data, a slightly different model would result. The *quadratic regression* feature on a graphing calculator can also be used to generate the quadratic model that best fits given data. See your owner's manual for details.

11.6 EXERCISES

For Extra Help

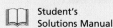

Student's Solutions Manual

MyMathLab

InterAct Math Tutorial Software

Tutor Center AW Math Tutor Center

MathXL MathXL

Digital Video Tutor CD 17/Videotape 17

1. Match each quadratic function with its graph from choices A–D.

(a) $f(x) = (x + 2)^2 - 1$

A.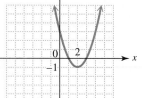

B.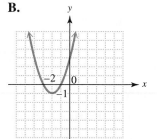

(b) $f(x) = (x + 2)^2 + 1$

(c) $f(x) = (x - 2)^2 - 1$

C.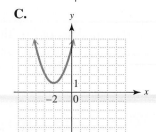

D.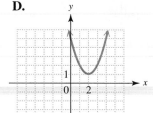

(d) $f(x) = (x - 2)^2 + 1$

2. Match each quadratic function with its graph from choices A–D.

(a) $f(x) = -x^2 + 2$ **A.**

(b) $f(x) = -x^2 - 2$

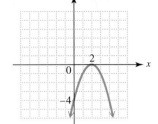

(c) $f(x) = -(x + 2)^2$ **C.**

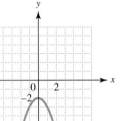

B.

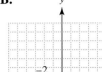

D.

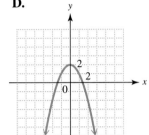

(d) $f(x) = -(x - 2)^2$

Identify the vertex of each parabola. See Examples 1–4.

3. $f(x) = -3x^2$

4. $f(x) = \dfrac{1}{2}x^2$

5. $f(x) = x^2 + 4$

6. $f(x) = x^2 - 4$

7. $f(x) = (x - 1)^2$

8. $f(x) = (x + 3)^2$

9. $f(x) = (x + 3)^2 - 4$

10. $f(x) = (x - 5)^2 - 8$

11. Describe how each of the parabolas in Exercises 9 and 10 is shifted compared to the graph of $f(x) = x^2$.

12. What does the value of a in $F(x) = a(x - h)^2 + k$ tell you about the graph of the function compared to the graph of $f(x) = x^2$?

For each quadratic function, tell whether the graph opens up or down and whether the graph is wider, narrower, or the same shape as the graph of $f(x) = x^2$. See Examples 4 and 5.

13. $f(x) = -\dfrac{2}{5}x^2$

14. $f(x) = -2x^2$

15. $f(x) = 3x^2 + 1$

16. $f(x) = \dfrac{2}{3}x^2 - 4$

17. For $f(x) = a(x - h)^2 + k$, in what quadrant is the vertex if

(a) $h > 0, k > 0$;   **(b)** $h > 0, k < 0$;   **(c)** $h < 0, k > 0$;   **(d)** $h < 0, k < 0$?

18. (a) What is the value of h if the graph of $f(x) = a(x - h)^2 + k$ has vertex on the y-axis?

(b) What is the value of k if the graph of $f(x) = a(x - h)^2 + k$ has vertex on the x-axis?

19. Match each quadratic function with the description of the parabola that is its graph.

(a) $f(x) = (x - 4)^2 - 2$ **A.** Vertex $(2, -4)$, opens down
(b) $f(x) = (x - 2)^2 - 4$ **B.** Vertex $(2, -4)$, opens up
(c) $f(x) = -(x - 4)^2 - 2$ **C.** Vertex $(4, -2)$, opens down
(d) $f(x) = -(x - 2)^2 - 4$ **D.** Vertex $(4, -2)$, opens up

20. Explain in your own words the meaning of each term.

(a) Vertex of a parabola   (b) Axis of a parabola

Graph each parabola. Plot at least two points in addition to the vertex. Give the vertex, axis, domain, and range in Exercises 27–36. See Examples 1–5.

21. $f(x) = -2x^2$ **22.** $f(x) = \dfrac{1}{3}x^2$ **23.** $f(x) = x^2 - 1$

24. $f(x) = x^2 + 3$ **25.** $f(x) = -x^2 + 2$ **26.** $f(x) = 2x^2 - 2$

27. $f(x) = (x - 4)^2$ **28.** $f(x) = -2(x + 1)^2$ **29.** $f(x) = (x + 2)^2 - 1$

30. $f(x) = (x - 1)^2 + 2$ **31.** $f(x) = 2(x - 2)^2 - 4$ **32.** $f(x) = -3(x - 2)^2 + 1$

33. $f(x) = -\dfrac{1}{2}(x + 1)^2 + 2$ **34.** $f(x) = -\dfrac{2}{3}(x + 2)^2 + 1$

35. $f(x) = 2(x - 2)^2 - 3$ **36.** $f(x) = \dfrac{4}{3}(x - 3)^2 - 2$

RELATING CONCEPTS (EXERCISES 37–42)

For Individual or Group Work

The procedures that allow the graph of $y = x^2$ to be shifted vertically and horizontally apply to other types of functions. In Section 7.3 we introduced linear functions of the form $g(x) = ax + b$. Consider the graph of the simplest linear function defined by $g(x) = x$, shown here. **Work Exercises 37–42 in order.**

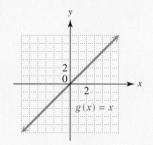

37. Based on the concepts of this section, how does the graph of $F(x) = x^2 + 6$ compare to the graph of $f(x) = x^2$ if a *vertical* shift is considered?

38. Graph the linear function defined by $G(x) = x + 6$.

39. Based on the concepts of Chapter 7, how does the graph of $G(x) = x + 6$ compare to the graph of $g(x) = x$ if a vertical shift is considered? (*Hint:* Look at the y-intercept.)

40. Based on the concepts of this section, how does the graph of $F(x) = (x - 6)^2$ compare to the graph of $f(x) = x^2$ if a *horizontal* shift is considered?

41. Graph the linear function $G(x) = x - 6$.

42. Based on the concepts of Chapter 7, how does the graph of $G(x) = x - 6$ compare to the graph of $g(x) = x$ if a horizontal shift is considered? (*Hint:* Look at the x-intercept.)

In Exercises 43–48, tell whether a linear or quadratic function would be a more appropriate model for each set of graphed data. If linear, tell whether the slope should be positive or negative. If quadratic, tell whether the coefficient a of x^2 should be positive or negative. See Example 6.

43.

U.S. TRADE DEFICIT

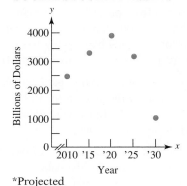

Source: U.S. Department of Commerce.

44. **AVERAGE DAILY VOLUME OF FIRST-CLASS MAIL***

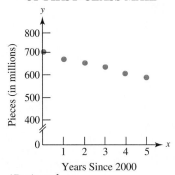

*Projected
Source: General Accounting Office.

45. **SOCIAL SECURITY ASSETS***

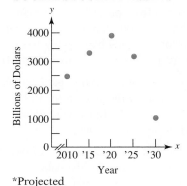

*Projected
Source: Social Security Administration.

46. **CEDAR RAPIDS SCHOOLS— GENERAL RESERVE FUND**

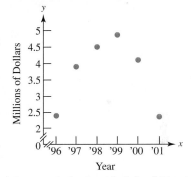

Source: Cedar Rapids School District.

47. **CONSUMER DEMAND FOR ELECTRICITY**

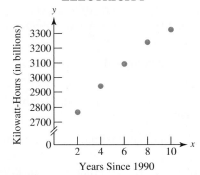

Source: U.S. Department of Energy.

48. **U.S. COMMERCIAL BANK FAILURES**

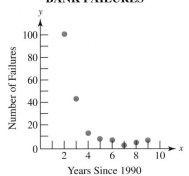

Source: www.ABA.com

Solve each problem. See Example 6.

49. The number of publicly traded companies filing for bankruptcy for selected years between 1990 and 2000 are shown in the table. In the year column, 0 represents 1990, 2 represents 1992, and so on.

Year	Number of Bankruptcies
0	115
2	91
4	70
6	84
8	120
10	176

Source: www.BankruptcyData.com

(a) Use the ordered pairs (year, number of bankruptcies) to make a scatter diagram of the data.

(b) Use the scatter diagram to decide whether a linear or quadratic function would better model the data. If quadratic, should the coefficient a of x^2 be positive or negative?

(c) Use the ordered pairs (0, 115), (4, 70), and (8, 120) to find a quadratic function that models the data. Round the values of a, b, and c in your model to three decimal places, as necessary.

(d) Use your model from part (c) to approximate the number of company bankruptcy filings in 2002. Round your answer to the nearest whole number.

(e) The number of company bankruptcy filings through August 16, 2002 was 129. Based on this, is your estimate from part (d) reasonable? Explain.

50. In a study, the number of new AIDS patients who survived the first year for the years from 1991 through 1997 are shown in the table. In the year column, 1 represents 1991, 2 represents 1992, and so on.

Year	Number of Patients
1	55
2	130
3	155
4	160
5	155
6	150
7	115

Source: HIV Health Services Planning Council.

(a) Use the ordered pairs (year, number of patients) to make a scatter diagram of the data.

(b) Would a linear or quadratic function better model the data?

(c) Should the coefficient a of x^2 in a quadratic model be positive or negative?

(d) Use the ordered pairs (2, 130), (3, 155), and (7, 115) to find a quadratic function that models the data.

(e) Use your model from part (d) to approximate the number of AIDS patients who survived the first year in 1994 and 1996. How well does the model approximate the actual data from the table?

51. In Example 6, we determined that the quadratic function defined by

$$y = .2455x^2 - 1.856x + 30.7105$$

modeled the rate of higher-order multiple births, where x represents the number of years since 1970.

(a) Use this model to approximate the rate of higher-order births in 1999 to the nearest tenth.

(b) The actual rate of higher-order births in 1999 was 184.9. (*Source:* National Center for Health Statistics.) How does the approximation using the model compare to the actual rate for 1999?

TECHNOLOGY INSIGHTS (EXERCISES 52–56)

Recall from Chapters 3 and 7 that the x-value of the x-intercept of the graph of the line $y = mx + b$ is the solution of the linear equation $mx + b = 0$. In the same way, the x-values of the x-intercepts of the graph of the parabola $y = ax^2 + bx + c$ are the real solutions of the quadratic equation $ax^2 + bx + c = 0$.

In Exercises 52–55, the calculator graphs show the x-values of the x-intercepts of the graph of the polynomial in the equation. Use the graphs to solve each equation.

52. $x^2 - x - 20 = 0$

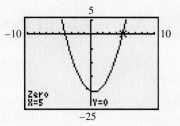

53. $x^2 + 9x + 14 = 0$

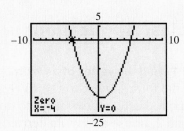
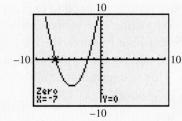

54. $-2x^2 + 5x + 3 = 0$

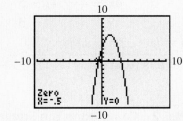
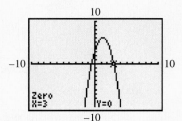

55. $-8x^2 + 6x + 5 = 0$

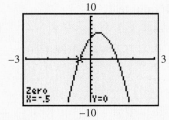
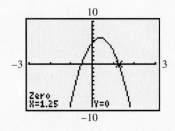

56. The graph of a quadratic function defined by $y = f(x)$ is shown in the standard viewing window, without x-axis tick marks. Which one of the following choices would be the only possible solution set for the equation $f(x) = 0$?

A. $\{-4, 1\}$　　**B.** $\{1, 4\}$　　**C.** $\{-1, -4\}$　　**D.** $\{4, -1\}$

Explain your answer.

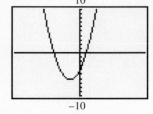

11.7 More About Parabolas; Applications

OBJECTIVES

1 Find the vertex of a vertical parabola.

2 Graph a quadratic function.

3 Use the discriminant to find the number of x-intercepts of a vertical parabola.

4 Use quadratic functions to solve problems involving maximum or minimum value.

5 Graph horizontal parabolas.

OBJECTIVE 1　Find the vertex of a vertical parabola. When the equation of a parabola is given in the form $f(x) = ax^2 + bx + c$, we need to locate the vertex to sketch an accurate graph. There are two ways to do this: complete the square as shown in Examples 1 and 2, or use a formula derived by completing the square.

EXAMPLE 1　Completing the Square to Find the Vertex

Find the vertex of the graph of $f(x) = x^2 - 4x + 5$.

　　To find the vertex, we need to express $x^2 - 4x + 5$ in the form $(x - h)^2 + k$. We do this by completing the square on $x^2 - 4x$, as in Section 11.2. The process is slightly different here because we want to keep $f(x)$ alone on one side of the equation. Instead of adding the appropriate number to each side, we *add and subtract* it on the right. This is equivalent to adding 0.

$$f(x) = x^2 - 4x + 5$$
$$= (x^2 - 4x \quad) + 5 \qquad \text{Group the variable terms.}$$
$$\left[\frac{1}{2}(-4)\right]^2 = (-2)^2 = 4$$
$$= (x^2 - 4x + 4 - 4) + 5 \qquad \text{Add and subtract 4.}$$
$$= (x^2 - 4x + 4) - 4 + 5 \qquad \text{Bring } -4 \text{ outside the parentheses.}$$
$$f(x) = (x - 2)^2 + 1 \qquad \text{Factor; combine terms.}$$

The vertex of this parabola is $(2, 1)$.

Now Try Exercise 5.

EXAMPLE 2　Completing the Square to Find the Vertex When $a \neq 1$

Find the vertex of the graph of $f(x) = -3x^2 + 6x - 1$.

　　We must complete the square on $-3x^2 + 6x$. Because the x^2-term has a coefficient other than 1, we factor that coefficient out of the first two terms and then proceed as in Example 1.

$$f(x) = -3x^2 + 6x - 1$$
$$= -3(x^2 - 2x) - 1 \qquad \text{Factor out } -3.$$
$$\left[\frac{1}{2}(-2)\right]^2 = (-1)^2 = 1$$
$$= -3(x^2 - 2x + 1 - 1) - 1 \qquad \text{Add and subtract 1 within the parentheses.}$$

Bring -1 outside the parentheses; be sure to multiply it by -3.

$$= -3(x^2 - 2x + 1) + (-3)(-1) - 1 \qquad \text{Distributive property}$$

$$= -3(x^2 - 2x + 1) + 3 - 1$$

$$f(x) = -3(x - 1)^2 + 2 \qquad \text{Factor; combine terms.}$$

The vertex is $(1, 2)$.

Now Try Exercise 7.

   To derive a formula for the vertex of the graph of the quadratic function defined by $f(x) = ax^2 + bx + c$ $(a \neq 0)$, complete the square.

$$f(x) = ax^2 + bx + c \qquad \text{Standard form}$$

$$= a\left(x^2 + \frac{b}{a}x\right) + c \qquad \begin{array}{l}\text{Factor } a \text{ from the}\\\text{first two terms.}\end{array}$$

$$\left[\frac{1}{2}\left(\frac{b}{a}\right)\right]^2 = \left(\frac{b}{2a}\right)^2 = \frac{b^2}{4a^2}$$

$$= a\left(x^2 + \frac{b}{a}x + \frac{b^2}{4a^2} - \frac{b^2}{4a^2}\right) + c \qquad \text{Add and subtract } \tfrac{b^2}{4a^2}.$$

$$= a\left(x^2 + \frac{b}{a}x + \frac{b^2}{4a^2}\right) + a\left(-\frac{b^2}{4a^2}\right) + c \qquad \text{Distributive property}$$

$$= a\left(x^2 + \frac{b}{a}x + \frac{b^2}{4a^2}\right) - \frac{b^2}{4a} + c$$

$$= a\left(x + \frac{b}{2a}\right)^2 + \frac{4ac - b^2}{4a} \qquad \text{Factor; combine terms.}$$

$$f(x) = a\left[x - \left(\frac{-b}{2a}\right)\right]^2 + \frac{4ac - b^2}{4a} \qquad f(x) = (x - h)^2 + k$$

$$\underbrace{\phantom{x - \left(\frac{-b}{2a}\right)}}_{h} \qquad \underbrace{\phantom{\frac{4ac - b^2}{4a}}}_{k}$$

Thus, the vertex (h, k) can be expressed in terms of a, b, and c. However, it is not necessary to remember this expression for k, since it can be found by replacing x with $\frac{-b}{2a}$. Using function notation, if $y = f(x)$, then the y-value of the vertex is $f\left(\frac{-b}{2a}\right)$.

Vertex Formula

The graph of the quadratic function defined by $f(x) = ax^2 + bx + c$ $(a \neq 0)$ has vertex

$$\left(\frac{-b}{2a}, f\left(\frac{-b}{2a}\right)\right),$$

and the axis of the parabola is the line

$$x = \frac{-b}{2a}.$$

EXAMPLE 3 Using the Formula to Find the Vertex

Use the vertex formula to find the vertex of the graph of

$$f(x) = x^2 - x - 6.$$

For this function, $a = 1$, $b = -1$, and $c = -6$. The x-coordinate of the vertex of the parabola is given by

$$\frac{-b}{2a} = \frac{-(-1)}{2(1)} = \frac{1}{2}.$$

The y-coordinate is $f\left(\frac{-b}{2a}\right) = f\left(\frac{1}{2}\right)$.

$$f\left(\frac{1}{2}\right) = \left(\frac{1}{2}\right)^2 - \frac{1}{2} - 6 = \frac{1}{4} - \frac{1}{2} - 6 = -\frac{25}{4}$$

The vertex is $\left(\frac{1}{2}, -\frac{25}{4}\right)$.

Now Try Exercise 9.

OBJECTIVE 2 Graph a quadratic function. We give a general approach for graphing any quadratic function here.

Graphing a Quadratic Function f

Step 1 **Determine whether the graph opens up or down.** If $a > 0$, the parabola opens up; if $a < 0$, it opens down.

Step 2 **Find the vertex.** Use either the vertex formula or completing the square.

Step 3 **Find any intercepts.** To find the x-intercepts (if any), solve $f(x) = 0$. To find the y-intercept, evaluate $f(0)$.

Step 4 **Complete the graph.** Plot the points found so far. Find and plot additional points as needed, using symmetry about the axis.

EXAMPLE 4 Using the Steps to Graph a Quadratic Function

Graph the quadratic function defined by

$$f(x) = x^2 - x - 6.$$

Step 1 From the equation, $a = 1$, so the graph of the function opens up.

Step 2 The vertex, $\left(\frac{1}{2}, -\frac{25}{4}\right)$, was found in Example 3 using the vertex formula.

Step 3 Find any intercepts. Since the vertex, $\left(\frac{1}{2}, -\frac{25}{4}\right)$, is in quadrant IV and the graph opens up, there will be two x-intercepts. To find them, let $f(x) = 0$ and solve.

$$f(x) = x^2 - x - 6$$
$$0 = x^2 - x - 6 \qquad \text{Let } f(x) = 0.$$
$$0 = (x - 3)(x + 2) \qquad \text{Factor.}$$
$$x - 3 = 0 \quad \text{or} \quad x + 2 = 0 \qquad \text{Zero-factor property}$$
$$x = 3 \quad \text{or} \qquad x = -2$$

The x-intercepts are $(3, 0)$ and $(-2, 0)$.

To find the y-intercept, evaluate $f(0)$.

$$f(x) = x^2 - x - 6$$
$$f(0) = 0^2 - 0 - 6 \qquad \text{Let } x = 0.$$
$$f(0) = -6$$

The y-intercept is $(0, -6)$.

Step 4 Plot the points found so far and additional points as needed using symmetry about the axis, $x = \frac{1}{2}$. The graph is shown in Figure 12. The domain is $(-\infty, \infty)$, and the range is $\left[-\frac{25}{4}, \infty\right)$.

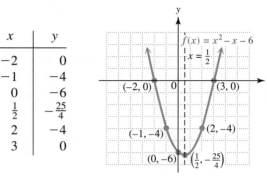

x	y
-2	0
-1	-4
0	-6
$\frac{1}{2}$	$-\frac{25}{4}$
2	-4
3	0

FIGURE 12

Now Try Exercise 17.

OBJECTIVE 3 Use the discriminant to find the number of x-intercepts of a vertical parabola. The graph of a quadratic function may have two x-intercepts, one x-intercept, or no x-intercepts, as shown in Figure 13. Recall from Section 11.3 that $b^2 - 4ac$ is called the *discriminant* of the quadratic equation $ax^2 + bx + c = 0$ and that we can use it to determine the number of real solutions of a quadratic equation.

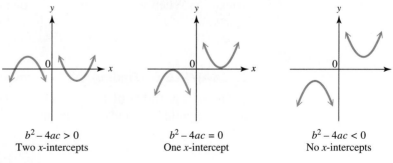

$b^2 - 4ac > 0$ $b^2 - 4ac = 0$ $b^2 - 4ac < 0$
Two x-intercepts   One x-intercept   No x-intercepts

FIGURE 13

In a similar way, we can use the discriminant of a quadratic *function* to determine the number of x-intercepts of its graph. If the discriminant is positive, the parabola will have two x-intercepts. If the discriminant is 0, there will be only one x-intercept, and it will be the vertex of the parabola. If the discriminant is negative, the graph will have no x-intercepts.

▧ **EXAMPLE 5** **Using the Discriminant to Determine the Number of**
x-Intercepts

Find the discriminant and use it to determine the number of x-intercepts of the graph
of each quadratic function.

(a) $f(x) = 2x^2 + 3x - 5$

The discriminant is $b^2 - 4ac$. Here $a = 2$, $b = 3$, and $c = -5$, so

$$b^2 - 4ac = 9 - 4(2)(-5) = 49.$$

Since the discriminant is positive, the parabola has two x-intercepts.

(b) $f(x) = -3x^2 - 1$

Here, $a = -3$, $b = 0$, and $c = -1$. The discriminant is

$$b^2 - 4ac = 0 - 4(-3)(-1) = -12.$$

The discriminant is negative, so the graph has no x-intercepts.

(c) $f(x) = 9x^2 + 6x + 1$

Here, $a = 9$, $b = 6$, and $c = 1$. The discriminant is

$$b^2 - 4ac = 36 - 4(9)(1) = 0.$$

The parabola has only one x-intercept (its vertex) because the value of the discriminant is 0.

Now Try Exercises 11 and 13.

OBJECTIVE 4 **Use quadratic functions to solve problems involving maximum or minimum value.** The vertex of the graph of a quadratic function is either the highest or the lowest point on the parabola. The y-value of the vertex gives the maximum or minimum value of y, while the x-value tells where that maximum or minimum occurs.

▮ **PROBLEM SOLVING** ▮

In many applied problems we must find the largest or smallest value of some quantity. When we can express that quantity in terms of a quadratic function, the value of k in the vertex (h, k) gives that optimum value.

▧ **EXAMPLE 6** **Finding the Maximum Area of a Rectangular Region**

A farmer has 120 ft of fencing. He wants to put a fence around a rectangular field next to a building. Find the maximum area he can enclose.

FIGURE 14

Figure 14 on the preceding page shows the field. Let x represent the width of the field. Since he has 120 ft of fencing,

$$x + x + \text{length} = 120 \qquad \text{Sum of the sides is 120 ft.}$$
$$2x + \text{length} = 120 \qquad \text{Combine terms.}$$
$$\text{length} = 120 - 2x. \qquad \text{Subtract } 2x.$$

The area is given by the product of the width and length, so

$$A(x) = x(120 - 2x)$$
$$= 120x - 2x^2.$$

To determine the maximum area, find the vertex of the parabola given by $A(x) = 120x - 2x^2$ using the vertex formula. Writing the equation in standard form as $A(x) = -2x^2 + 120x$ gives $a = -2$, $b = 120$, and $c = 0$, so

$$h = \frac{-b}{2a} = \frac{-120}{2(-2)} = \frac{-120}{-4} = 30;$$

$$A(30) = -2(30)^2 + 120(30) = -2(900) + 3600 = 1800.$$

The graph is a parabola that opens down, and its vertex is $(30, 1800)$. Thus, the maximum area will be 1800 ft^2. This area will occur if x, the width of the field, is 30 ft.

Now Try Exercise 35.

CAUTION Be careful when interpreting the meanings of the coordinates of the vertex. The first coordinate, x, gives the value for which the *function value* is a maximum or a minimum. Be sure to read the problem carefully to determine whether you are asked to find the value of the independent variable, the function value, or both.

EXAMPLE 7 Finding the Maximum Height Attained by a Projectile

If air resistance is neglected, a projectile on Earth shot straight upward with an initial velocity of 40 m per sec will be at a height s in meters given by

$$s(t) = -4.9t^2 + 40t,$$

where t is the number of seconds elapsed after projection. After how many seconds will it reach its maximum height, and what is this maximum height?

For this function, $a = -4.9$, $b = 40$, and $c = 0$. Use the vertex formula.

$$h = \frac{-b}{2a} = \frac{-40}{2(-4.9)} \approx 4.1 \qquad \text{Use a calculator.}$$

This indicates that the maximum height is attained at 4.1 sec. To find this maximum height, calculate $s(4.1)$.

$$s(4.1) = -4.9(4.1)^2 + 40(4.1)$$
$$\approx 81.6 \qquad \text{Use a calculator.}$$

The projectile will attain a maximum height of approximately 81.6 m.

Now Try Exercise 37.

OBJECTIVE 5 Graph horizontal parabolas. If x and y are interchanged in the equation $y = ax^2 + bx + c$, the equation becomes $x = ay^2 + by + c$. Because of the interchange of the roles of x and y, these parabolas are horizontal (with horizontal lines as axes), compared with the vertical ones graphed previously.

Graph of a Horizontal Parabola

The graph of

$$x = ay^2 + by + c \qquad \text{or} \qquad x = a(y - k)^2 + h$$

is a parabola with vertex (h, k) and the horizontal line $y = k$ as axis. The graph opens to the right if $a > 0$ and to the left if $a < 0$.

EXAMPLE 8 Graphing a Horizontal Parabola

Graph $x = (y - 2)^2 - 3$.

This graph has its vertex at $(-3, 2)$, since the roles of x and y are reversed. It opens to the right because $a = 1 > 0$, and has the same shape as $y = x^2$. Plotting a few additional points gives the graph shown in Figure 15. Note that the graph is symmetric about its axis, $y = 2$. The domain is $[-3, \infty)$, and the range is $(-\infty, \infty)$.

x	y
-3	2
-2	3
-2	1
1	4
1	0

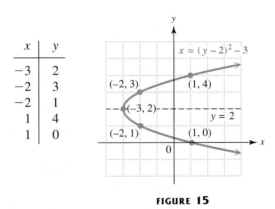

FIGURE 15

Now Try Exercise 21.

When a quadratic equation is given in the form $x = ay^2 + by + c$, completing the square on y allows us to find the vertex.

EXAMPLE 9 Completing the Square to Graph a Horizontal Parabola

Graph $x = -2y^2 + 4y - 3$. Give the domain and range of the relation.

$$x = -2y^2 + 4y - 3$$
$$= -2(y^2 - 2y) - 3 \qquad \text{Factor out } -2.$$
$$= -2(y^2 - 2y + 1 - 1) - 3 \qquad \text{Complete the square within the parentheses; add and subtract 1.}$$
$$= -2(y^2 - 2y + 1) + (-2)(-1) - 3 \qquad \text{Distributive property}$$
$$x = -2(y - 1)^2 - 1 \qquad \text{Factor; simplify.}$$

Because of the negative coefficient (-2) in $x = -2(y - 1)^2 - 1$, the graph opens to the left (the negative x-direction) and is narrower than the graph of $y = x^2$ because $|-2| > 1$. As shown in Figure 16, the vertex is $(-1, 1)$ and the axis is $y = 1$. The domain is $(-\infty, -1]$, and the range is $(-\infty, \infty)$.

x	y
-3	2
-3	0
-1	1

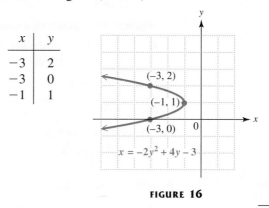

FIGURE 16

Now Try Exercise 25.

CAUTION Only quadratic equations solved for y (whose graphs are vertical parabolas) are examples of functions. *The horizontal parabolas in Examples 8 and 9 are not graphs of functions,* because they do not satisfy the vertical line test. Furthermore, the vertex formula given earlier does not apply to parabolas with horizontal axes.

In summary, the graphs of parabolas studied in this section and the previous one fall into the following categories.

Graphs of Parabolas

Equation	Graph
$y = ax^2 + bx + c$ $y = a(x - h)^2 + k$	*(two graphs shown: upward parabola with vertex (h, k), $a > 0$; downward parabola with vertex (h, k), $a < 0$)* These graphs represent functions.
$x = ay^2 + by + c$ $x = a(y - k)^2 + h$	*(two graphs shown: rightward parabola with vertex (h, k), $a > 0$; leftward parabola with vertex (h, k), $a < 0$)* These graphs are not graphs of functions.

11.7 EXERCISES

1. How can you determine just by looking at the equation of a parabola whether it has a vertical or a horizontal axis?

2. Why can't the graph of a quadratic function be a parabola with a horizontal axis?

3. How can you determine the number of x-intercepts of the graph of a quadratic function without graphing the function?

4. If the vertex of the graph of a quadratic function is $(1, -3)$, and the graph opens down, how many x-intercepts does the graph have?

Find the vertex of each parabola. See Examples 1–3.

5. $f(x) = x^2 + 8x + 10$ **6.** $f(x) = x^2 + 10x + 23$ **7.** $f(x) = -2x^2 + 4x - 5$

8. $f(x) = -3x^2 + 12x - 8$ **9.** $f(x) = -\dfrac{1}{2}x^2 + 2x - 3$ **10.** $f(x) = 4x^2 - x + 5$

Find the vertex of each parabola. For each equation, decide whether the graph opens up, down, to the left, or to the right, and whether it is wider, narrower, or the same shape as the graph of $y = x^2$. If it is a vertical parabola, find the discriminant and use it to determine the number of x-intercepts. See Examples 1–3, 5, 8, and 9.

11. $f(x) = 2x^2 + 4x + 5$ **12.** $f(x) = 3x^2 - 6x + 4$ **13.** $f(x) = -x^2 + 5x + 3$

14. $x = -y^2 + 7y + 2$ **15.** $x = \dfrac{1}{3}y^2 + 6y + 24$ **16.** $x = \dfrac{1}{2}y^2 + 10y - 5$

Use the concepts of this section to match each equation in Exercises 17–22 with its graph in A–F.

17. $y = 2x^2 + 4x - 3$ **18.** $y = -x^2 + 3x + 5$ **19.** $y = -\dfrac{1}{2}x^2 - x + 1$

20. $x = y^2 + 6y + 3$ **21.** $x = -y^2 - 2y + 4$ **22.** $x = 3y^2 + 6y + 5$

A.
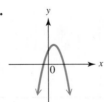

B.

C.

D.
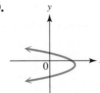

E.
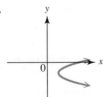

F.
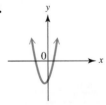

Graph each parabola. (Use the results of Exercises 5–8 to help graph the parabolas in Exercises 23–26.) Give the vertex, axis, domain, and range. See Examples 4, 8, and 9.

23. $f(x) = x^2 + 8x + 10$ **24.** $f(x) = x^2 + 10x + 23$ **25.** $f(x) = -2x^2 + 4x - 5$

26. $f(x) = -3x^2 + 12x - 8$ **27.** $x = (y + 2)^2 + 1$ **28.** $x = (y + 3)^2 - 2$

29. $x = -\dfrac{1}{5}y^2 + 2y - 4$ **30.** $x = -\dfrac{1}{2}y^2 - 4y - 6$ **31.** $x = 3y^2 + 12y + 5$

32. $x = 4y^2 + 16y + 11$

Solve each problem. See Examples 6 and 7.

33. Find the pair of numbers whose sum is 60 and whose product is a maximum. (*Hint:* Let x and $60 - x$ represent the two numbers.)

34. Find the pair of numbers whose sum is 40 and whose product is a maximum.

35. Morgan's Department Store wants to construct a rectangular parking lot on land bordered on one side by a highway. It has 280 ft of fencing that is to be used to fence off the other three sides. What should be the dimensions of the lot if the enclosed area is to be a maximum? What is the maximum area?

36. Keisha Hughes has 100 m of fencing material to enclose a rectangular exercise run for her dog. What width will give the enclosure the maximum area?

37. If an object on Earth is propelled upward with an initial velocity of 32 ft per sec, then its height after t sec is given by

$$h(t) = 32t - 16t^2.$$

Find the maximum height attained by the object and the number of seconds it takes to hit the ground.

38. A projectile on Earth is fired straight upward so that its distance (in feet) above the ground t sec after firing is given by

$$s(t) = -16t^2 + 400t.$$

Find the maximum height it reaches and the number of seconds it takes to reach that height.

39. After experimentation, two Pacific Institute physics students find that when a bottle of California wine is shaken several times, held upright, and uncorked, its cork travels according to the function defined by

$$s(t) = -16t^2 + 64t + 3,$$

where s is its height in feet above the ground t sec after being released. After how many seconds will it reach its maximum height? What is the maximum height?

40. Professor Levy has found that the number of students attending his intermediate algebra class is approximated by

$$S(x) = -x^2 + 20x + 80,$$

where x is the number of hours that the Campus Center is open daily. Find the number of hours that the center should be open so that the number of students attending class is a maximum. What is this maximum number of students?

41. The annual percent increase in the amount pharmacies paid wholesalers for drugs in the years 1990 through 1999 can be modeled by the quadratic function defined by

$$f(x) = .228x^2 - 2.57x + 8.97,$$

where $x = 0$ represents 1990, $x = 1$ represents 1991, and so on. (*Source: IMS Health, Retail and Provider Perspective.*)

(a) Since the coefficient of x^2 in the model is positive, the graph of this quadratic function is a parabola that opens up. Will the y-value of the vertex of this graph be a maximum or minimum?

(b) In what year was the minimum percent increase? (Round down to the nearest year.) Use the actual x-value of the vertex, to the nearest tenth, to find this increase.

42. The U.S. domestic oyster catch (in millions) for the years 1990 through 1998 can be approximated by the quadratic function defined by

$$f(x) = -.566x^2 + 5.08x + 29.2,$$

where $x = 0$ represents 1990, $x = 1$ represents 1991, and so on. (*Source:* National Marine Fisheries Service.)

(a) Since the coefficient of x^2 in the model is negative, the graph of this quadratic function is a parabola that opens down. Will the y-value of the vertex of this graph be a maximum or minimum?

(b) In what year was the maximum domestic oyster catch? (Round down to the nearest year.) Use the actual x-value of the vertex, to the nearest tenth, to find this catch.

43. The graph shows how Social Security assets are expected to change as the number of retirees receiving benefits increases.

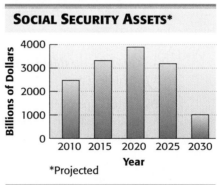

Source: Social Security Administration.

The graph suggests that a quadratic function would be a good fit to the data. The data are approximated by the function defined by

$$f(x) = -20.57x^2 + 758.9x - 3140.$$

In the model, $x = 10$ represents 2010, $x = 15$ represents 2015, and so on, and $f(x)$ is in billions of dollars.

(a) Explain why the coefficient of x^2 in the model is negative, based on the graph.

(b) Algebraically determine the vertex of the graph, with coordinates to four significant digits.

(c) Interpret the answer to part (b) as it applies to the application.

44. The graph shows the performance of investment portfolios with different mixtures of U.S. and foreign investments for the period January 1, 1971, to December 31, 1996.

   ✍ **(a)** Is this the graph of a function? Explain.

   (b) What investment mixture shown on the graph appears to represent the vertex? What relative amount of risk does this point represent? What return on investment does it provide?

   (c) Which point on the graph represents the riskiest investment mixture? What return on investment does it provide?

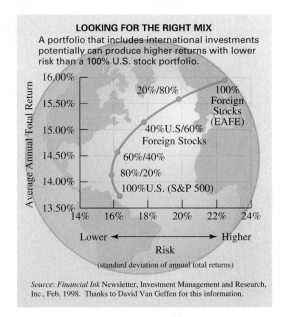

LOOKING FOR THE RIGHT MIX
A portfolio that includes international investments potentially can produce higher returns with lower risk than a 100% U.S. stock portfolio.

Source: *Financial Ink* Newsletter, Investment Management and Research, Inc., Feb. 1998. Thanks to David Van Geffen for this information.

45. A charter flight charges a fare of $200 per person, plus $4 per person for each unsold seat on the plane. If the plane holds 100 passengers and if x represents the number of unsold seats, find the following.

   (a) A function defined by $R(x)$ that describes the total revenue received for the flight (*Hint:* Multiply the number of people flying, $100 - x$, by the price per ticket, $200 + 4x$.)

   (b) The graph of the function from part (a)

   (c) The number of unsold seats that will produce the maximum revenue

   (d) The maximum revenue

46. For a trip to a resort, a charter bus company charges a fare of $48 per person, plus $2 per person for each unsold seat on the bus. If the bus has 42 seats and x represents the number of unsold seats, find the following.

   (a) A function defined by $R(x)$ that describes the total revenue from the trip (*Hint:* Multiply the total number riding, $42 - x$, by the price per ticket, $48 + 2x$)

   (b) The graph of the function from part (a)

   (c) The number of unsold seats that produces the maximum revenue

   (d) The maximum revenue

TECHNOLOGY INSIGHTS (EXERCISES 47–50)

Graphing calculators are capable of determining the coordinates of "peaks" and "valleys" of graphs. In the case of quadratic functions, these peaks and valleys are the vertices and are called maximum and minimum points. For example, the vertex of the graph of $f(x) = -x^2 - 6x - 13$ is $(-3, -4)$, as indicated in the display at the bottom of the screen. In this case, the vertex is a maximum point.

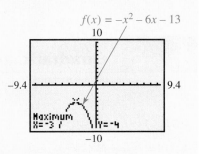

$f(x) = -x^2 - 6x - 13$

(continued)

In Exercises 47–50, match the function with its calculator graph in A–D by determining the vertex and using the display at the bottom of the screen.

47. $f(x) = x^2 - 8x + 18$

48. $f(x) = x^2 + 8x + 18$

49. $f(x) = x^2 - 8x + 14$

50. $f(x) = x^2 + 8x + 14$

A.

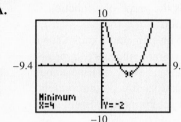

B.

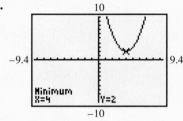

C.

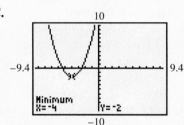

D.

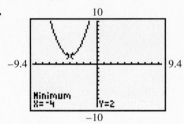

In the following exercise, the distance formula is used to develop the equation of a parabola.

51. A parabola can be defined as the set of all points in a plane equally distant from a given point and a given line not containing the point. (The point is called the *focus* and the line is called the *directrix*.) See the figure.

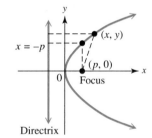

(a) Suppose (x, y) is to be on the parabola. Suppose the directrix has equation $x = -p$. Find the distance between (x, y) and the directrix. (The distance from a point to a line is the length of the perpendicular from the point to the line.)

(b) If $x = -p$ is the equation of the directrix, why should the focus have coordinates $(p, 0)$? (*Hint:* See the figure.)

(c) Find an expression for the distance from (x, y) to $(p, 0)$.

(d) Find an equation for the parabola in the figure. (*Hint:* Use the results of parts (a) and (c) and the fact that (x, y) is equally distant from the focus and the directrix.)

52. Use the equation derived in Exercise 51 to find an equation for a parabola with focus $(3, 0)$ and directrix with equation $x = -3$.

11.8 Quadratic and Rational Inequalities

OBJECTIVES

1 Solve quadratic inequalities.

We discussed methods of solving linear inequalities in Chapter 2 and methods of solving quadratic equations in this chapter. Now we combine these ideas to solve *quadratic inequalities.*

2 Solve polynomial inequalities of degree 3 or more.

3 Solve rational inequalities.

> **Quadratic Inequality**
>
> A **quadratic inequality** can be written in the form
>
> $$ax^2 + bx + c < 0 \qquad \text{or} \qquad ax^2 + bx + c > 0,$$
>
> where a, b, and c are real numbers, with $a \neq 0$.

As before, the symbols $<$ and $>$ may be replaced with $\leq$ and $\geq$.

OBJECTIVE 1 Solve quadratic inequalities. One method for solving a quadratic inequality is by graphing the related quadratic function.

■ EXAMPLE 1 Solving Quadratic Inequalities by Graphing

Solve each inequality.

(a) $x^2 - x - 12 > 0$

To solve the inequality, we graph the related quadratic function defined by $f(x) = x^2 - x - 12$. We are particularly interested in the x-intercepts, which are found as in Section 11.7 by letting $f(x) = 0$ and solving the quadratic equation

$$x^2 - x - 12 = 0.$$
$$(x - 4)(x + 3) = 0 \qquad \text{Factor.}$$
$$x - 4 = 0 \quad \text{or} \quad x + 3 = 0 \qquad \text{Zero-factor property}$$
$$x = 4 \quad \text{or} \qquad x = -3$$

Thus, the x-intercepts are $(4, 0)$, and $(-3, 0)$. The graph, which opens up since the coefficient of x^2 is positive, is shown in Figure 17(a). Notice from this graph that x-values less than -3 or greater than 4 result in y-values *greater than* 0. Therefore, the solution set of $x^2 - x - 12 > 0$, written in interval notation, is

$$(-\infty, -3) \cup (4, \infty).$$

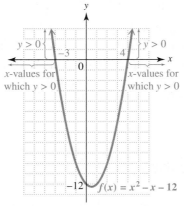

The graph is *above* the x-axis for $(-\infty, -3) \cup (4, \infty)$.

(a)

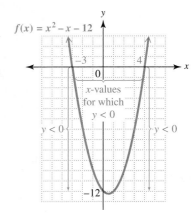

The graph is *below* the x-axis for $(-3, 4)$.

(b)

FIGURE 17

(b) $x^2 - x - 12 < 0$

Here we want values of y that are *less than* 0. Referring to Figure 17(b), we notice from the graph that x-values between -3 and 4 result in y-values less than 0. Therefore, the solution set of the inequality $x^2 - x - 12 < 0$, written in interval notation, is $(-3, 4)$.

Now Try Exercise 1.

> **NOTE** If the inequalities in Example 1 had used $\geq$ and $\leq$, the solution sets would have included the x-values of the intercepts and been written in interval notation as $(-\infty, -3] \cup [4, \infty)$ for Example 1(a) and $[-3, 4]$ for Example 1(b).

In Example 1, we used graphing to divide the x-axis into intervals. Then using the graphs in Figure 17, we determined which x-values resulted in y-values that were either greater than or less than 0. Another method for solving a quadratic inequality uses these basic ideas without actually graphing the related quadratic function.

EXAMPLE 2 Solving a Quadratic Inequality Using Test Numbers

Solve $x^2 - x - 12 > 0$.

Solve the quadratic equation $x^2 - x - 12 = 0$ by factoring, as in Example 1(a).

$$(x - 4)(x + 3) = 0$$
$$x - 4 = 0 \quad \text{or} \quad x + 3 = 0$$
$$x = 4 \quad \text{or} \quad x = -3$$

The numbers 4 and -3 divide a number line into the three intervals shown in Figure 18. Be careful to put the smaller number on the left. (Notice the similarity between Figure 18 and the x-axis with intercepts $(-3, 0)$ and $(4, 0)$ in Figure 17(a).)

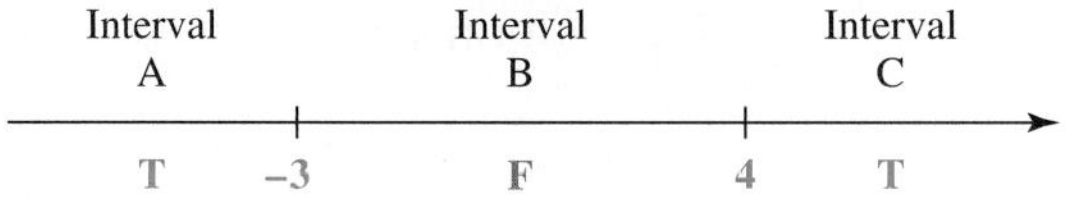

FIGURE 18

The numbers 4 and -3 are the only numbers that make the expression $x^2 - x - 12$ equal to 0. All other numbers make the expression either positive or negative. The sign of the expression can change from positive to negative or from negative to positive only at a number that makes it 0. Therefore, if one number in an interval satisfies the inequality, then all the numbers in that interval will satisfy the inequality.

To see if the numbers in Interval A satisfy the inequality, choose any number from Interval A in Figure 18 (that is, any number less than -3). Substitute this test number for x in the original inequality $x^2 - x - 12 > 0$. If the result is *true,* then all numbers in Interval A satisfy the inequality.

Try -5 from Interval A. Substitute -5 for x.

$$x^2 - x - 12 > 0 \qquad \text{Original inequality}$$
$$(-5)^2 - (-5) - 12 > 0 \qquad ?$$
$$25 + 5 - 12 > 0 \qquad ?$$
$$18 > 0 \qquad \text{True}$$

Because -5 from Interval A satisfies the inequality, all numbers from Interval A are solutions.

Now try 0 from Interval B. If $x = 0$, then

$$0^2 - 0 - 12 > 0 \qquad ?$$

$$-12 > 0. \qquad \text{False}$$

The numbers in Interval B are *not* solutions. Verify that the test number 5 satisfies the inequality, so the numbers in Interval C are also solutions.

Based on these results (shown by the colored letters in Figure 18), the solution set includes the numbers in Intervals A and C, as shown on the graph in Figure 19. The solution set is written in interval notation as

$$(-\infty, -3) \cup (4, \infty).$$

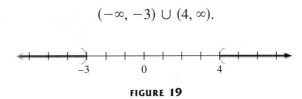

FIGURE 19

This agrees with the solution set we found by graphing the related quadratic function in Example 1(a).

Now Try Exercise 11.

In summary, follow these steps to solve a quadratic inequality.

Solving a Quadratic Inequality

Step 1   **Write the inequality as an equation and solve it.**

Step 2   **Use the solutions from Step 1 to determine intervals.** Graph the numbers found in Step 1 on a number line. These numbers divide the number line into intervals.

Step 3   **Find the intervals that satisfy the inequality.** Substitute a test number from each interval into the original inequality to determine the intervals that satisfy the inequality. All numbers in those intervals are in the solution set. A graph of the solution set will usually look like one of these. (Square brackets might be used instead of parentheses.)

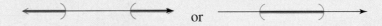

Step 4   **Consider the endpoints separately.** The numbers from Step 1 are included in the solution set if the inequality symbol is $\leq$ or $\geq$; they are not included if it is $<$ or $>$.

Special cases of quadratic inequalities may occur, as in the next example.

▇ **EXAMPLE 3** Solving Special Cases

Solve $(2t - 3)^2 > -1$. Then solve $(2t - 3)^2 < -1$.

Because $(2t - 3)^2$ is never negative, it is always greater than -1. Thus, the solution for $(2t - 3)^2 > -1$ is the set of all real numbers, $(-\infty, \infty)$. In the same way, there is no solution for $(2t - 3)^2 < -1$ and the solution set is $\emptyset$.

▇

Now Try Exercises 25 and 27.

OBJECTIVE 2 **Solve polynomial inequalities of degree 3 or more.** Higher-degree polynomial inequalities that can be factored are solved in the same way as quadratic inequalities.

▇ **EXAMPLE 4** Solving a Third-Degree Polynomial Inequality

Solve $(x - 1)(x + 2)(x - 4) \leq 0$.

This is a *cubic* (third-degree) inequality rather than a quadratic inequality, but it can be solved using the method shown in the box by extending the zero-factor property to more than two factors. Begin by setting the factored polynomial *equal* to 0 and solving the equation. (Step 1)

$$(x - 1)(x + 2)(x - 4) = 0$$

$$x - 1 = 0 \quad \text{or} \quad x + 2 = 0 \quad \text{or} \quad x - 4 = 0$$

$$x = 1 \quad \text{or} \quad x = -2 \quad \text{or} \quad x = 4$$

Locate the numbers -2, 1, and 4 on a number line, as in Figure 20, to determine the Intervals A, B, C, and D. (Step 2)

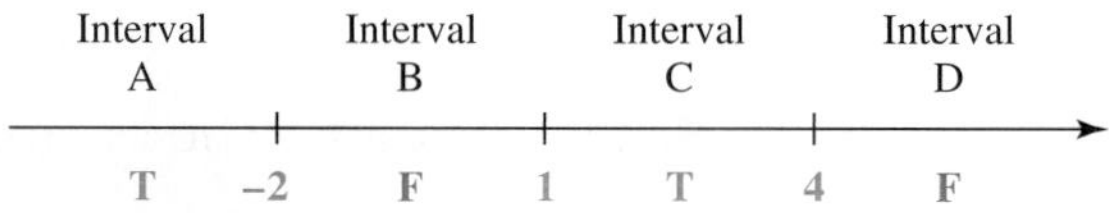

FIGURE 20

Substitute a test number from each interval in the *original* inequality to determine which intervals satisfy the inequality. (Step 3) Use a table to organize this information.

Interval	Test Number	Test of Inequality	True or False?
A	−3	$-28 \leq 0$	T
B	0	$8 \leq 0$	F
C	2	$-8 \leq 0$	T
D	5	$28 \leq 0$	F

Verify the information given in the table and graphed in Figure 21 on the next page. The numbers in Intervals A and C are in the solution set, which is written in interval notation as

$$(-\infty, -2] \cup [1, 4].$$

The three endpoints are included since the inequality symbol is $\leq$. (Step 4)

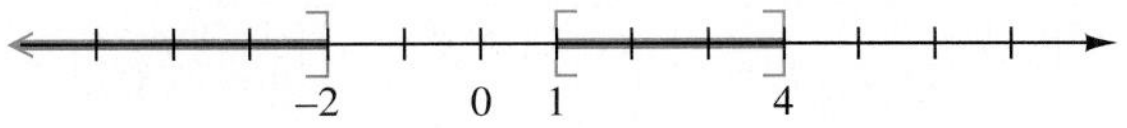

FIGURE 21

Now Try Exercise 29.

OBJECTIVE 3 Solve rational inequalities. Inequalities that involve rational expressions, called **rational inequalities,** are solved similarly using the following steps.

Solving a Rational Inequality

Step 1 **Write the inequality** so that 0 is on one side and there is a single fraction on the other side.

Step 2 **Determine the numbers that make the numerator or denominator equal to 0.**

Step 3 **Divide a number line into intervals.** Use the numbers from Step 2.

Step 4 **Find the intervals that satisfy the inequality.** Test a number from each interval by substituting it into the *original* inequality.

Step 5 **Consider the endpoints separately.** Exclude any values that make the denominator 0.

CAUTION As indicated in Step 5, any number that makes the denominator 0 *must* be excluded from the solution set.

EXAMPLE 5   Solving a Rational Inequality

Solve $\dfrac{-1}{p - 3} > 1$.

Write the inequality so that 0 is on one side. (Step 1)

$$\frac{-1}{p - 3} - 1 > 0 \qquad \text{Subtract 1.}$$

$$\frac{-1}{p - 3} - \frac{p - 3}{p - 3} > 0 \qquad \text{Use } p - 3 \text{ as the common denominator.}$$

$$\frac{-1 - p + 3}{p - 3} > 0 \qquad \begin{array}{l}\text{Write the left side as a single fraction;}\\ \text{be careful with signs in the numerator.}\end{array}$$

$$\frac{-p + 2}{p - 3} > 0 \qquad \text{Combine terms.}$$

The sign of the rational expression $\dfrac{-p + 2}{p - 3}$ will change from positive to negative or negative to positive only at those numbers that make the numerator or denominator 0. The number 2 makes the numerator 0, and 3 makes the denominator 0. (Step 2) These two numbers, 2 and 3, divide a number line into three intervals. See Figure 22. (Step 3)

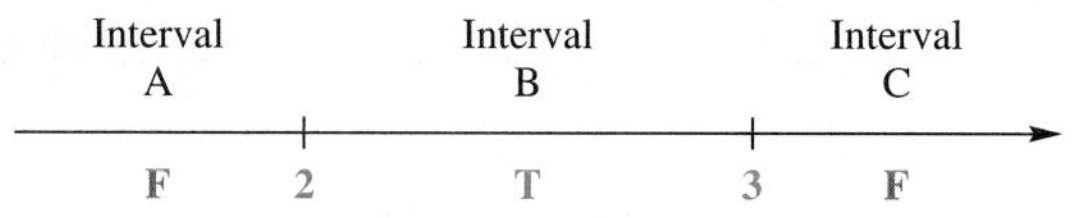

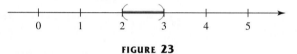

FIGURE 22

Testing a number from each interval in the *original* inequality, $\frac{-1}{p-3} > 1$, gives the results shown in the table. (Step 4)

Interval	Test Number	Test of Inequality	True or False?
A	0	$\frac{1}{3} > 1$	**F**
B	2.5	$2 > 1$	**T**
C	4	$-1 > 1$	**F**

The solution set is the interval (2, 3). This interval does not include 3 since it would make the denominator of the original equality 0; 2 is not included either since the inequality symbol is >. (Step 5) A graph of the solution set is given in Figure 23.

FIGURE 23

Now Try Exercise 37.

EXAMPLE 6 Solving a Rational Inequality

Solve $\frac{m-2}{m+2} \le 2$.

Write the inequality so that 0 is on one side. (Step 1)

$$\frac{m-2}{m+2} - 2 \le 0 \qquad \text{Subtract 2.}$$

$$\frac{m-2}{m+2} - \frac{2(m+2)}{m+2} \le 0 \qquad \text{Use } m+2 \text{ as the common denominator.}$$

$$\frac{m-2-2m-4}{m+2} \le 0 \qquad \text{Write as a single fraction.}$$

$$\frac{-m-6}{m+2} \le 0 \qquad \text{Combine terms.}$$

The number −6 makes the numerator 0, and −2 makes the denominator 0. (Step 2) These two numbers determine three intervals. (Step 3) Test one number from each interval (Step 4) to see that the solution set is

$$(-\infty, -6] \cup (-2, \infty).$$

The number −6 satisfies the original inequality, but −2 cannot be used as a solution since it makes the denominator 0. (Step 5) A graph of the solution set is shown in Figure 24.

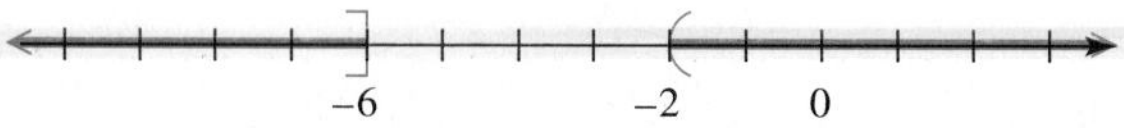

FIGURE 24

Now Try Exercise 41.

11.8 EXERCISES

For Extra Help

 Student's
Solutions Manual

 MyMathLab

 InterAct Math
Tutorial Software

 AW Math
Tutor Center

*Math*XP MathXL

 Digital Video Tutor
CD 17/Videotape 17

In Example 1, we determined the solution sets of the quadratic inequalities $x^2 - x - 12 > 0$ and $x^2 - x - 12 < 0$ by graphing $f(x) = x^2 - x - 12$. The x-intercepts of this graph indicated the solutions of the equation $x^2 - x - 12 = 0$. The x-values of the points on the graph that were above the x-axis formed the solution set of $x^2 - x - 12 > 0$, and the x-values of the points on the graph that were below the x-axis formed the solution set of $x^2 - x - 12 < 0$.

In Exercises 1–4, the graph of a quadratic function f is given. Use the graph to find the solution set of each equation or inequality. See Example 1.

1. (a) $x^2 - 4x + 3 = 0$
   (b) $x^2 - 4x + 3 > 0$
   (c) $x^2 - 4x + 3 < 0$

2. (a) $3x^2 + 10x - 8 = 0$
   (b) $3x^2 + 10x - 8 \geq 0$
   (c) $3x^2 + 10x - 8 < 0$

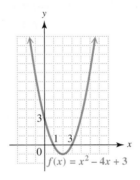

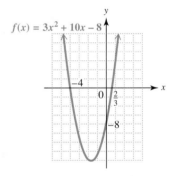

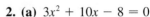

3. (a) $-2x^2 - x + 15 = 0$
   (b) $-2x^2 - x + 15 \geq 0$
   (c) $-2x^2 - x + 15 \leq 0$

4. (a) $-x^2 + 3x + 10 = 0$
   (b) $-x^2 + 3x + 10 \geq 0$
   (c) $-x^2 + 3x + 10 \leq 0$

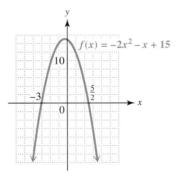

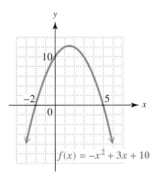

5. Explain how to determine whether to include or exclude endpoints when solving a quadratic or higher-degree inequality.

6. The solution set of the inequality $x^2 + x - 12 < 0$ is the interval $(-4, 3)$. Without actually performing any work, give the solution set of the inequality $x^2 + x - 12 \geq 0$.

Solve each inequality and graph the solution set. See Example 2. (Hint: In Exercises 23 and 24, use the quadratic formula.)

7. $(x + 1)(x - 5) > 0$ **8.** $(m + 6)(m - 2) > 0$ **9.** $(r + 4)(r - 6) < 0$

10. $(x + 4)(x - 8) < 0$ **11.** $x^2 - 4x + 3 \geq 0$ **12.** $m^2 - 3m - 10 \geq 0$

13. $10t^2 + 9t \geq 9$ **14.** $3r^2 + 10r \geq 8$ **15.** $4x^2 - 9 \leq 0$

16. $9x^2 - 25 \leq 0$ **17.** $6x^2 + x \geq 1$ **18.** $4p^2 + 7p \geq -3$

19. $z^2 - 4z \geq 0$ **20.** $x^2 + 2x < 0$ **21.** $3k^2 - 5k \leq 0$

22. $2z^2 + 3z > 0$ **23.** $x^2 - 6x + 6 \geq 0$ **24.** $3k^2 - 6k + 2 \leq 0$

Solve each inequality. See Example 3.

25. $(4 - 3x)^2 \geq -2$ **26.** $(6z + 7)^2 \geq -1$

27. $(3x + 5)^2 \leq -4$ **28.** $(8t + 5)^2 \leq -5$

Solve each inequality and graph the solution set. See Example 4.

29. $(p - 1)(p - 2)(p - 4) < 0$ **30.** $(2r + 1)(3r - 2)(4r + 7) < 0$

31. $(x - 4)(2x + 3)(3x - 1) \geq 0$ **32.** $(z + 2)(4z - 3)(2z + 7) \geq 0$

Solve each inequality and graph the solution set. See Examples 5 and 6.

33. $\dfrac{x - 1}{x - 4} > 0$ **34.** $\dfrac{x + 1}{x - 5} > 0$ **35.** $\dfrac{2n + 3}{n - 5} \leq 0$

36. $\dfrac{3t + 7}{t - 3} \leq 0$ **37.** $\dfrac{8}{x - 2} \geq 2$ **38.** $\dfrac{20}{x - 1} \geq 1$

39. $\dfrac{3}{2t - 1} < 2$ **40.** $\dfrac{6}{m - 1} < 1$ **41.** $\dfrac{g - 3}{g + 2} \geq 2$

42. $\dfrac{m + 4}{m + 5} \geq 2$ **43.** $\dfrac{x - 8}{x - 4} < 3$ **44.** $\dfrac{2t - 3}{t + 1} > 4$

45. $\dfrac{4k}{2k - 1} < k$ **46.** $\dfrac{r}{r + 2} < 2r$ **47.** $\dfrac{2x - 3}{x^2 + 1} \geq 0$

48. $\dfrac{9x - 8}{4x^2 + 25} < 0$ **49.** $\dfrac{(3x - 5)^2}{x + 2} > 0$ **50.** $\dfrac{(5x - 3)^2}{2x + 1} \leq 0$

RELATING CONCEPTS (EXERCISES 51–54)

For Individual or Group Work

A rock is projected vertically upward from the ground. Its distance s in feet above the ground after t sec is given by the quadratic function defined by

$$s(t) = -16t^2 + 256t.$$

Work Exercises 51–54 in order, *to see how quadratic equations and inequalities are related.*

51. At what times will the rock be 624 ft above the ground? (*Hint:* Let $s(t) = 624$ and solve the quadratic *equation.*)

52. At what times will the rock be more than 624 ft above the ground? (*Hint:* Set $s(t) > 624$ and solve the quadratic *inequality.*)

53. At what times will the rock be at ground level? (*Hint:* Let $s(t) = 0$ and solve the quadratic *equation.*)

54. At what times will the rock be less than 624 ft above the ground? (*Hint:* Set $s(t) < 624$, solve the quadratic *inequality,* and observe the solutions in Exercises 52 and 53 to determine the smallest and largest possible values of *t.*)

Chapter **Group Activity**

Finding the Path of a Comet

OBJECTIVE Find and graph an equation of a parabola with a given focus.

The orbit that a comet takes as it approaches the sun depends on its velocity, as well as other factors. If the velocity of a comet equals escape velocity—that is, it is going just fast enough to get away from the sun—then its orbit will be parabolic. (Other possible orbits are *hyperbolic* or *elliptical*. These figures are discussed in Chapter 13.) The sun is at the focus of the parabola. The vertex is the point where the comet is closest to the sun.

A. Refer to Section 11.7, Exercise 51, and make a sketch of the comet and the sun.

   1. Place the vertex of the parabola at the origin, with the focus on the *y*-axis and a horizontal directrix.

   2. Assume that the comet is .75 astronomical unit* from the sun at its closest point.

B. What are the coordinates of the focus?

C. What is the equation of the directrix?

D. Using the information from Section 11.7, Exercise 51, find an equation of the parabola.

E. Graph the parabola (as a vertical parabola). Include the focus and directrix on your graph.

---

*One astronomical unit (AU) is the average distance between Earth and the sun.

CHAPTER **11** SUMMARY

▰ KEY TERMS

11.1 quadratic equation	**11.4** quadratic in form	axis	**11.8** quadratic inequality
11.3 quadratic formula	**11.6** parabola	quadratic function	rational inequality
discriminant	vertex		

▰ TEST YOUR WORD POWER

See how well you have learned the vocabulary in this chapter. Answers, with examples, follow the Quick Review.

1. The **quadratic formula** is
   A. a formula to find the number of solutions of a quadratic equation
   B. a formula to find the type of solutions of a quadratic equation
   C. the standard form of a quadratic equation
   D. a general formula for solving any quadratic equation.

2. A **quadratic function** is a function that can be written in the form
   A. $f(x) = mx + b$ for real numbers m and b
   B. $f(x) = \frac{P(x)}{Q(x)}$, where $Q(x) \neq 0$
   C. $f(x) = ax^2 + bx + c$ for real numbers a, b, and c ($a \neq 0$)
   D. $f(x) = \sqrt{x}$ for $x \geq 0$.

3. A **parabola** is the graph of
   A. any equation in two variables
   B. a linear equation
   C. an equation of degree 3
   D. a quadratic equation in two variables.

4. The **vertex** of a parabola is
   A. the point where the graph intersects the y-axis
   B. the point where the graph intersects the x-axis
   C. the lowest point on a parabola that opens up or the highest point on a parabola that opens down
   D. the origin.

5. The **axis** of a parabola is
   A. either the x-axis or the y-axis
   B. the vertical line (of a vertical parabola) or the horizontal line (of a horizontal parabola) through the vertex
   C. the lowest or highest point on the graph of a parabola
   D. a line through the origin.

6. A parabola is **symmetric about its axis** since
   A. its graph is near the axis
   B. its graph is identical on each side of the axis
   C. its graph looks different on each side of the axis
   D. its graph intersects the axis.

▰ QUICK REVIEW

CONCEPTS	EXAMPLES

11.1 SOLVING QUADRATIC EQUATIONS BY THE SQUARE ROOT PROPERTY

Square Root Property
If x and k are complex numbers and $x^2 = k$, then
$$x = \sqrt{k} \quad \text{or} \quad x = -\sqrt{k}.$$

Solve $(x - 1)^2 = 8$.
$$x - 1 = \sqrt{8} \qquad \text{or} \quad x - 1 = -\sqrt{8}$$
$$x = 1 + 2\sqrt{2} \quad \text{or} \qquad x = 1 - 2\sqrt{2}$$
Solution set: $\left\{1 + 2\sqrt{2}, 1 - 2\sqrt{2}\right\}$

11.2 SOLVING QUADRATIC EQUATIONS BY COMPLETING THE SQUARE

Completing the Square
To solve $ax^2 + bx + c = 0$ ($a \neq 0$):

Step 1 If $a \neq 1$, divide each side by a.

Step 2 Write the equation with the variable terms on one side and the constant on the other.

Solve $2x^2 - 4x - 18 = 0$.
$$x^2 - 2x - 9 = 0 \qquad \text{Divide by 2.}$$
$$x^2 - 2x = 9 \qquad \text{Add 9.}$$

CONCEPTS	EXAMPLES

Step 3 Take half the coefficient of x and square it. Add the square to each side. Factor the perfect square trinomial, and write it as the square of a binomial. Simplify the other side.

Step 4 Use the square root property to complete the solution.

$$\left[\frac{1}{2}(-2)\right]^2 = (-1)^2 = 1$$

$$x^2 - 2x + 1 = 9 + 1$$

$$(x - 1)^2 = 10$$

$$x - 1 = \sqrt{10} \qquad \text{or} \quad x - 1 = -\sqrt{10}$$

$$x = 1 + \sqrt{10} \quad \text{or} \qquad x = 1 - \sqrt{10}$$

Solution set: $\left\{1 + \sqrt{10},\ 1 - \sqrt{10}\right\}$

11.3 SOLVING QUADRATIC EQUATIONS BY THE QUADRATIC FORMULA

Quadratic Formula

The solutions of $ax^2 + bx + c = 0$ $(a \neq 0)$ are given by

$$x = \frac{-b \pm \sqrt{b^2 - 4ac}}{2a}.$$

Solve $3x^2 + 5x + 2 = 0$.

$$x = \frac{-5 \pm \sqrt{5^2 - 4(3)(2)}}{2(3)} = \frac{-5 \pm 1}{6}$$

$$x = -1 \quad \text{or} \quad x = -\frac{2}{3}$$

Solution set: $\left\{-1, -\frac{2}{3}\right\}$

The Discriminant

If a, b, and c are integers, then the discriminant, $b^2 - 4ac$, of $ax^2 + bx + c = 0$ determines the number and type of solutions as follows.

Discriminant	Number and Type of Solutions
Positive, the square of an integer	Two rational solutions
Positive, not the square of an integer	Two irrational solutions
Zero	One rational solution
Negative	Two imaginary solutions

For $x^2 + 3x - 10 = 0$, the discriminant is

$$3^2 - 4(1)(-10) = 49. \qquad \text{Two rational solutions}$$

For $4x^2 + x + 1 = 0$, the discriminant is

$$1^2 - 4(4)(1) = -15. \qquad \text{Two imaginary solutions}$$

11.4 EQUATIONS QUADRATIC IN FORM

A nonquadratic equation that can be written in the form

$$au^2 + bu + c = 0,$$

for $a \neq 0$ and an algebraic expression u, is called quadratic in form. Substitute u for the expression, solve for u, and then solve for the variable in the expression.

Solve $3(x + 5)^2 + 7(x + 5) + 2 = 0$.

$$3u^2 + 7u + 2 = 0 \qquad \text{Let } u = x + 5.$$

$$(3u + 1)(u + 2) = 0$$

$$u = -\frac{1}{3} \quad \text{or} \qquad u = -2$$

$$x + 5 = -\frac{1}{3} \quad \text{or} \quad x + 5 = -2 \qquad x + 5 = u$$

$$x = -\frac{16}{3} \quad \text{or} \qquad x = -7$$

Solution set: $\left\{-7, -\frac{16}{3}\right\}$

(continued)

CONCEPTS	EXAMPLES

11.5 FORMULAS AND FURTHER APPLICATIONS

To solve a formula for a squared variable, proceed as follows.

(a) If the variable appears only to the second power: Isolate the squared variable on one side of the equation, and then use the square root property.

Solve $A = \dfrac{2mp}{r^2}$ for r.

$$r^2 A = 2mp \qquad \text{Multiply by } r^2.$$

$$r^2 = \frac{2mp}{A} \qquad \text{Divide by } A.$$

$$r = \pm\sqrt{\frac{2mp}{A}} \qquad \text{Square root property}$$

$$r = \frac{\pm\sqrt{2mpA}}{A} \qquad \text{Rationalize the denominator.}$$

(b) If the variable appears to the first and second powers: Write the equation in standard form, and then use the quadratic formula.

Solve $m^2 + rm = t$ for m.

$$m^2 + rm - t = 0 \qquad \text{Standard form}$$

$$m = \frac{-r \pm \sqrt{r^2 - 4(1)(-t)}}{2(1)} \qquad a = 1, b = r, c = -t$$

$$m = \frac{-r \pm \sqrt{r^2 + 4t}}{2}$$

11.6 GRAPHS OF QUADRATIC FUNCTIONS

1. The graph of the quadratic function defined by $F(x) = a(x - h)^2 + k$, $a \neq 0$, is a parabola with vertex at (h, k) and the vertical line $x = h$ as axis.

2. The graph opens up if a is positive and down if a is negative.

3. The graph is wider than the graph of $f(x) = x^2$ if $0 < |a| < 1$ and narrower if $|a| > 1$.

Graph $f(x) = -(x + 3)^2 + 1$.

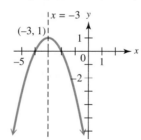

The graph opens down since $a < 0$. It is shifted 3 units left and 1 unit up, so the vertex is $(-3, 1)$, with axis $x = -3$. The domain is $(-\infty, \infty)$; the range is $(-\infty, 1]$.

11.7 MORE ABOUT PARABOLAS; APPLICATIONS

The vertex of the graph of $f(x) = ax^2 + bx + c$, $a \neq 0$, may be found by completing the square. The vertex has coordinates

$$\left(\frac{-b}{2a}, f\left(\frac{-b}{2a}\right) \right).$$

Graphing a Quadratic Function

Step 1 Determine whether the graph opens up or down.

Step 2 Find the vertex.

Step 3 Find the x-intercepts (if any). Find the y-intercept.

Step 4 Find and plot additional points as needed.

Graph $f(x) = x^2 + 4x + 3$.

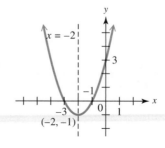

The graph opens up since $a > 0$. The vertex is $(-2, -1)$. The solutions of $x^2 + 4x + 3 = 0$ are -1 and -3, so the x-intercepts are $(-1, 0)$ and $(-3, 0)$. Since $f(0) = 3$, the y-intercept is $(0, 3)$. The domain is $(-\infty, \infty)$; the range is $[-1, \infty)$.

CONCEPTS	EXAMPLES

Horizontal Parabolas

The graph of $x = ay^2 + by + c$ is a horizontal parabola, opening to the right if $a > 0$ or to the left if $a < 0$. Horizontal parabolas do not represent functions.

Graph $x = 2y^2 + 6y + 5$.

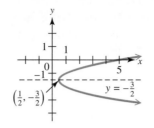

The graph opens to the right since $a > 0$. The vertex is $\left(\frac{1}{2}, -\frac{3}{2}\right)$. The axis is $y = -\frac{3}{2}$. The domain is $\left[\frac{1}{2}, \infty\right)$; the range is $(-\infty, \infty)$.

11.8   QUADRATIC AND RATIONAL INEQUALITIES

Solving a Quadratic (or Higher-Degree Polynomial) Inequality

Step 1   Write the inequality as an equation and solve.

Solve $2x^2 + 5x + 2 < 0$.

$$2x^2 + 5x + 2 = 0$$

$$x = -\frac{1}{2} \quad \text{or} \quad x = -2$$

Step 2   Use the numbers found in Step 1 to divide a number line into intervals.

```
       A          B          C
  ─────┼──────────┼──────────────►
     F  -2     T    -½   F
```

Step 3   Substitute a test number from each interval into the original inequality to determine the intervals that belong to the solution set.

$x = -3$ makes the original inequality false; $x = -1$ makes it true; $x = 0$ makes it false.

Step 4   Consider the endpoints separately.

The solution set is the interval $\left(-2, -\frac{1}{2}\right)$.

Solving a Rational Inequality

Step 1   Write the inequality so that 0 is on one side and there is a single fraction on the other side.

Solve $\dfrac{x}{x + 2} \geq 4$.

$$\frac{x}{x + 2} - 4 \geq 0$$

$$\frac{x}{x + 2} - \frac{4(x + 2)}{x + 2} \geq 0$$

$$\frac{-3x - 8}{x + 2} \geq 0$$

Step 2   Determine the numbers that make the numerator or denominator 0.

$-\frac{8}{3}$ makes the numerator 0; -2 makes the denominator 0.

Step 3   Use the numbers from Step 2 to divide a number line into intervals.

```
       A          B          C
  ─────┼──────────┼──────────────►
     F  -8/3    T   -2    F
```

Step 4   Substitute a test number from each interval into the original inequality to determine the intervals that belong to the solution set.

-4 makes the original inequality false; $-\frac{7}{3}$ makes it true; 0 makes it false.

Step 5   Consider the endpoints separately.

The solution set is the interval $\left[-\frac{8}{3}, -2\right)$. The endpoint -2 is not included since it makes the denominator 0.

(continued)

Answers to Test Your Word Power

1. D; *Example:* The solutions of $ax^2 + bx + c = 0$ $(a \neq 0)$ are given by $x = \dfrac{-b \pm \sqrt{b^2 - 4ac}}{2a}$. **2.** C; *Examples:* $f(x) = x^2 - 2$,

$f(x) = (x + 4)^2 + 1$, $f(x) = x^2 - 4x + 5$ **3.** D; *Examples:* See the figures in the Quick Review for Sections 11.6 and 11.7.
4. C; *Example:* The graph of $y = (x + 3)^2$ has vertex $(-3, 0)$, which is the lowest point on the graph. **5.** B; *Example:* The axis of
$y = (x + 3)^2$ is the vertical line $x = -3$. **6.** B; *Example:* Since the graph of $y = (x + 3)^2$ is symmetric about its axis $x = -3$, the points
$(-2, 1)$ and $(-4, 1)$ are on the graph.

CHAPTER **11** REVIEW EXERCISES

*Exercises marked * require knowledge of imaginary numbers.*

[11.1] *Solve each equation by using the square root property.*

1. $t^2 = 121$ **2.** $p^2 = 3$ **3.** $m^2 - 128 = 0$

4. $(r - 3)^2 = 10$ **5.** $(2x + 5)^2 = 100$ ***6.** $(3k - 2)^2 = -25$

7. A student gave the following "solution" to the equation $x^2 = 12$.

$$x^2 = 12$$
$$x = \sqrt{12} \qquad \text{Square root property}$$
$$x = 2\sqrt{3}$$

What is wrong with this solution?

8. Navy Pier Center in Chicago, Illinois, features a
150-ft tall Ferris wheel. Use Galileo's formula
$d = 16t^2$ to find how long it would take a wallet
dropped from the top of the Ferris wheel to fall
to the ground. Round your answer to the nearest
tenth of a second. (*Source: Microsoft Encarta
Encyclopedia 2002.*)

[11.2] *Solve each equation by completing the square.*

9. $m^2 + 6m + 5 = 0$ **10.** $p^2 + 4p = 7$ **11.** $-x^2 + 5 = 2x$

12. $2z^2 - 3 = -8z$ **13.** $5k^2 - 3k - 2 = 0$ ***14.** $(4a + 1)(a - 1) = -7$

[11.3] *Solve each equation by using the quadratic formula.*

15. $2x^2 + x - 21 = 0$ **16.** $k^2 + 5k = 7$ **17.** $(t + 3)(t - 4) = -2$

***18.** $2x^2 + 3x + 4 = 0$ ***19.** $3p^2 = 2(2p - 1)$ **20.** $m(2m - 7) = 3m^2 + 3$

Use the discriminant to predict whether the solutions to the equations in Exercises 21–24 are

A. *two rational numbers;* **B.** *one rational number;*
C. *two irrational numbers;* **D.** *two imaginary numbers.*

21. $x^2 + 5x + 2 = 0$

22. $4t^2 = 3 - 4t$

23. $4x^2 = 6x - 8$

24. $9z^2 + 30z + 25 = 0$

[11.4] *Solve each equation.*

25. $\dfrac{15}{x} = 2x - 1$

26. $\dfrac{1}{n} + \dfrac{2}{n+1} = 2$

27. $-2r = \sqrt{\dfrac{48 - 20r}{2}}$

28. $8(3x + 5)^2 + 2(3x + 5) - 1 = 0$

29. $2x^{2/3} - x^{1/3} - 28 = 0$

30. $p^4 - 10p^2 + 9 = 0$

Solve each problem. Round answers to the nearest tenth, as necessary.

31. Phong paddled his canoe 20 mi upstream, then paddled back. If the speed of the current was 3 mph and the total trip took 7 hr, what was Phong's speed?

32. Maureen O'Connor drove 8 mi to pick up her friend Laurie, and then drove 11 mi to a mall at a speed 15 mph faster. If Maureen's total travel time was 24 min, what was her speed on the trip to pick up Laurie?

33. An old machine processes a batch of checks in 1 hr more time than a new one. How long would it take the old machine to process a batch of checks that the two machines together process in 2 hr?

34. Greg Tobin can process a stack of invoices 1 hr faster than Carter Fenton can. Working together, they take 1.5 hr. How long would it take each person working alone?

[11.5] *Solve each formula for the indicated variable. (Give answers with ±.)*

35. $k = \dfrac{rF}{wv^2}$ for v

36. $p = \sqrt{\dfrac{yz}{6}}$ for y

37. $mt^2 = 3mt + 6$ for t

Solve each problem. Round answers to the nearest tenth, as necessary.

38. A large machine requires a part in the shape of a right triangle with a hypotenuse 9 ft less than twice the length of the longer leg. The shorter leg must be $\frac{3}{4}$ the length of the longer leg. Find the lengths of the three sides of the part.

39. A square has an area of 256 cm². If the same amount is removed from one dimension and added to the other, the resulting rectangle has an area 16 cm² less. Find the dimensions of the rectangle.

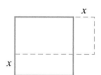

40. Nancy wants to buy a mat for a photograph that measures 14 in. by 20 in. She wants to have an even border around the picture when it is mounted on the mat. If the area of the mat she chooses is 352 in.², how wide will the border be?

41. A searchlight moves horizontally back and forth along a wall with the distance of the light from a starting point at t min given by the quadratic function defined by

$$f(t) = 100t^2 - 300t.$$

How long will it take before the light returns to the starting point?

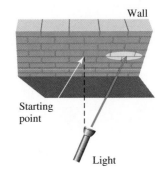

42. The Mart Hotel in Dallas, Texas, is 400 ft high. Suppose that a ball is projected upward from the top of the Mart, and its position in feet above the ground is given by the quadratic function defined by

$$f(t) = -16t^2 + 45t + 400,$$

where t is the number of seconds elapsed. How long will it take for the ball to reach a height of 200 ft above the ground? (*Source: World Almanac and Book of Facts, 2002.*)

43. The Toronto Dominion Center in Winnipeg, Manitoba, is 407 ft high. Suppose that a ball is projected upward from the top of the center, and its position in feet above the ground is given by the quadratic function defined by

$$s(t) = -16t^2 + 75t + 407,$$

where t is the number of seconds elapsed. How long will it take for the ball to reach a height of 450 ft above the ground? (*Source: World Almanac and Book of Facts, 2002.*)

44. The manager of a restaurant has determined that the demand for frozen yogurt is $\frac{25}{p}$ units per day, where p is the price (in dollars) per unit. The supply is $70p + 15$ units per day. Find the price at which supply and demand are equal.

45. Use the formula $A = P(1 + r)^2$ to find the interest rate r at which a principal P of $10,000 will increase to $10,920.25 in 2 yr.

46. The number of e-mail boxes in North America (in millions) for the years 1995 through 2001 are shown in the graph and can be modeled by the quadratic function defined by

$$f(x) = 3.29x^2 - 10.4x + 21.6.$$

In the model, $x = 5$ represents 1995, $x = 10$ represents 2000, and so on.

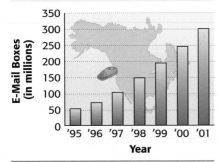

Source: IDC research.

(a) Use the model to approximate the number of e-mail boxes in 2001 to the nearest whole number. How does this result compare to the number shown in the graph?

(b) Based on the model, in what year did the number of e-mail boxes reach 200 million? (Round down to the nearest year.) How does this result compare to the number shown in the graph?

[11.6–11.7] *Identify the vertex of each parabola.*

47. $y = 6 - 2x^2$ **48.** $f(x) = -(x - 1)^2$ **49.** $f(x) = (x - 3)^2 + 7$

50. $y = -3x^2 + 4x - 2$ **51.** $x = (y - 3)^2 - 4$

52. If the discriminant of a quadratic function is negative, what do you know about the graph of the function?

Graph each parabola. Give the vertex, axis, domain, and range.

53. $y = 2(x - 2)^2 - 3$ **54.** $f(x) = -2x^2 + 8x - 5$

55. $x = 2(y + 3)^2 - 4$ **56.** $x = -\dfrac{1}{2}y^2 + 6y - 14$

Solve each problem.

57. Consumer spending for home video games in dollars per person per year is given in the table. Let $x = 0$ represent 1990, $x = 2$ represent 1992, and so on.

CONSUMER SPENDING FOR HOME VIDEO GAMES

Year	Dollars
1990	12.39
1992	13.08
1994	15.78
1996	19.43
1997	22.71
1998	24.14
1999	25.08

Source: Statistical Abstract of the United States.

  (a) Use the data for 1990, 1994, and 1997 in the quadratic form $ax^2 + bx + c = y$ to write a system of three equations.

  (b) Solve the system from part (a) to find a quadratic function f that models the data.

  (c) Use the model found in part (b) to approximate consumer spending for home video games in 1998 to the nearest cent. How does your answer compare to the actual data from the table?

58. The height (in feet) of a projectile t sec after being fired from Earth into the air is given by

$$f(t) = -16t^2 + 160t.$$

Find the number of seconds required for the projectile to reach maximum height. What is the maximum height?

59. Find the length and width of a rectangle having a perimeter of 200 m if the area is to be a maximum.

[11.8] *Solve each inequality and graph the solution set.*

60. $(x - 4)(2x + 3) > 0$ **61.** $x^2 + x \le 12$

62. $(x + 2)(x - 3)(x + 5) \le 0$ **63.** $(4m + 3)^2 \le -4$

64. $\dfrac{6}{2z - 1} < 2$ **65.** $\dfrac{3t + 4}{t - 2} \le 1$

MIXED REVIEW EXERCISES

Solve.

66. $V = r^2 + R^2h$ for R

*67. $3t^2 - 6t = -4$

*68. $x^4 - 1 = 0$

69. $(x^2 - 2x)^2 = 11(x^2 - 2x) - 24$

70. $(r - 1)(2r + 3)(r + 6) < 0$

71. $2x - \sqrt{x} = 6$

72. $(3k + 11)^2 = 7$

73. $S = \dfrac{Id^2}{k}$ for d

74. $(8k - 7)^2 \geq -1$

75. $6 + \dfrac{15}{s^2} = -\dfrac{19}{s}$

76. $x^4 - 8x^2 = -1$

77. $\dfrac{-2}{x + 5} \leq -5$

Match each equation with the figure that most closely resembles its graph.

78. $g(x) = x^2 - 5$

79. $h(x) = -x^2 + 4$

80. $F(x) = (x - 1)^2$

81. $G(x) = (x + 1)^2$

82. $H(x) = (x - 1)^2 + 1$

83. $K(x) = (x + 1)^2 + 1$

A.
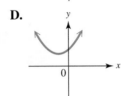

B.
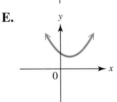

C.
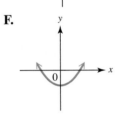

D.

E.

F.

84. Graph $f(x) = 4x^2 + 4x - 2$. Give the vertex, axis, domain, and range.

Solve each problem.

85. Natural gas use in the United States in trillions of cubic feet (ft^3) from 1970 through 1999 can be modeled by the quadratic function defined by

$$f(x) = .014x^2 - .396x + 21.2,$$

where $x = 0$ represents 1970, $x = 5$ represents 1975, and so on. (*Source:* Energy Information Administration.)

(a) Use the model to approximate natural gas use in 2000.

(b) Based on the model, in what year will natural gas use reach 25 trillion ft^3? (Round down to the nearest year.)

86. In 4 hr, Kerrie can go 15 mi upriver and come back. The speed of the current is 5 mph. Find the speed of the boat in still water.

87. Refer to Exercise 42. Suppose that a wire is attached to the top of the Mart and pulled tight. It is attached to the ground 100 ft from the base of the building. How long is the wire?

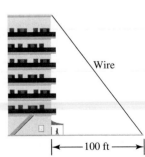

┌───┐
RELATING CONCEPTS (EXERCISES 88–92)
└───┘

For Individual or Group Work

Work Exercises 88–92 in order, to see the connections between equations and inequalities.

88. Use the methods of Chapter 2 to solve the equation or inequality, and graph the solution set.

(a) $3x - (4x + 2) = 0$ (b) $3x - (4x + 2) > 0$ (c) $3x - (4x + 2) < 0$

89. Use the methods of this chapter to solve the equation or inequality, and graph the solution set.

(a) $x^2 - 6x + 5 = 0$ (b) $x^2 - 6x + 5 > 0$ (c) $x^2 - 6x + 5 < 0$

90. Use the methods of Sections 6.6 and 11.8 to solve the equation or inequality, and graph the solution set.

(a) $\dfrac{-5x + 20}{x - 2} = 0$ (b) $\dfrac{-5x + 20}{x - 2} > 0$ (c) $\dfrac{-5x + 20}{x - 2} < 0$

91. Fill in the blanks in the following statement: If we solve a linear, quadratic, or rational equation and the two inequalities associated with it, the union of the three solution sets will be _____; the only exception will be in the case of the rational equation and inequalities, where the number or numbers that cause the _____ to be 0 will be excluded.

92. Suppose that the solution set of a quadratic equation is $\{-5, 3\}$ and the solution set of one of the associated inequalities is $(-\infty, -5) \cup (3, \infty)$. What is the solution set of the other associated inequality?

CHAPTER **11** TEST

*Problems marked * require knowledge of imaginary numbers.*

Solve by using the square root property.

1. $t^2 = 54$

2. $(7x + 3)^2 = 25$

3. Solve $2x^2 + 4x = 8$ by completing the square.

Solve by using the quadratic formula.

4. $2x^2 - 3x - 1 = 0$

***5.** $3t^2 - 4t = -5$

***6.** If k is a negative number, then which one of the following equations will have two imaginary solutions?

A. $x^2 = 4k$ **B.** $x^2 = -4k$ **C.** $(x + 2)^2 = -k$ **D.** $x^2 + k = 0$

7. What is the discriminant for $2x^2 - 8x - 3 = 0$? How many and what type of solutions does this equation have? (Do not actually solve.)

Solve by any method.

8. $3x = \sqrt{\dfrac{9x + 2}{2}}$

9. $3 - \dfrac{16}{x} - \dfrac{12}{x^2} = 0$

10. $4x^2 + 7x - 3 = 0$

11. $9x^4 + 4 = 37x^2$

12. $12 = (2n + 1)^2 + (2n + 1)$

13. Solve $S = 4\pi r^2$ for r. (Leave $\pm$ in your answer.)

Solve each problem.

14. Andrew and Kent do word processing. For a certain prospectus, Kent can prepare it 2 hr faster than Andrew can. If they work together, they can do the entire prospectus in 5 hr. How long will it take each of them working alone to prepare the prospectus? Round your answers to the nearest tenth of an hour.

15. Abby Tanenbaum paddled her canoe 10 mi upstream and then paddled back to her starting point. If the rate of the current was 3 mph and the entire trip took $3\frac{1}{2}$ hr, what was Abby's rate?

16. Tyler McGinnis has a pool 24 ft long and 10 ft wide. He wants to construct a concrete walk around the pool. If he plans for the walk to be of uniform width and cover 152 ft², what will the width of the walk be?

17. At a point 30 m from the base of a tower, the distance to the top of the tower is 2 m more than twice the height of the tower. Find the height of the tower.

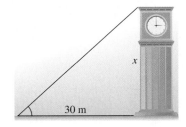

18. Which one of the following most closely resembles the graph of $f(x) = a(x - h)^2 + k$ if $a < 0$, $h > 0$, and $k < 0$?

A.

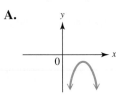

B.

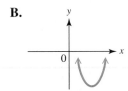

C.

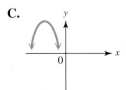

D.

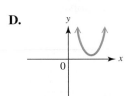

Graph each parabola. Identify the vertex, axis, domain, and range.

19. $f(x) = \dfrac{1}{2}x^2 - 2$

20. $f(x) = -x^2 + 4x - 1$

21. $x = -(y - 2)^2 + 2$

22. The percent increase for in-state tuition at Iowa public universities during the years 1992 through 2002 can be modeled by the quadratic function defined by

$$f(x) = .156x^2 - 2.05x + 10.2,$$

where $x = 2$ represents 1992, $x = 3$ represents 1993, and so on. (*Source:* Iowa Board of Regents.)

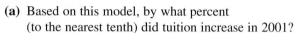

(a) Based on this model, by what percent (to the nearest tenth) did tuition increase in 2001?

(b) In what year was the minimum tuition increase? (Round down to the nearest year.) To the nearest tenth, by what percent did tuition increase that year?

23. Palo Alto College is planning to construct a rectangular parking lot on land bordered on one side by a highway. The plan is to use 640 ft of fencing to fence off the other three sides. What should the dimensions of the lot be if the enclosed area is to be a maximum?

Solve, and graph each solution set.

24. $2x^2 + 7x > 15$

25. $\dfrac{5}{t - 4} \leq 1$

CUMULATIVE REVIEW EXERCISES CHAPTERS 1–11

1. Let $S = \left\{ -\frac{7}{3}, -2, -\sqrt{3}, 0, .7, \sqrt{12}, \sqrt{-8}, 7, \frac{32}{3} \right\}$. List the elements of S that are elements of each set.

(a) Integers   (b) Rational numbers   (c) Real numbers   (d) Complex numbers

Simplify each expression.

2. $|-3| + 8 - |-9| - (-7 + 3)$

3. $2(-3)^2 + (-8)(-5) + (-17)$

Solve each equation.

4. $7 - (4 + 3t) + 2t = -6(t - 2) - 5$

5. $|6x - 9| = |-4x + 2|$

6. $2x = \sqrt{\dfrac{5x + 2}{3}}$

7. $\dfrac{3}{x - 3} - \dfrac{2}{x - 2} = \dfrac{3}{x^2 - 5x + 6}$

8. $(r - 5)(2r + 3) = 1$

9. $x^4 - 5x^2 + 4 = 0$

Solve each inequality.

10. $-2x + 4 \leq -x + 3$

11. $|3x - 7| \leq 1$

12. $x^2 - 4x + 3 < 0$

13. $\dfrac{3}{p + 2} > 1$

Graph each relation. Tell whether or not each is a function, and if it is, give its domain and range.

14. $4x - 5y = 15$

15. $4x - 5y < 15$

16. $f(x) = -2(x - 1)^2 + 3$

17. Find the slope and intercepts of the line with equation $-2x + 7y = 16$.

18. Write an equation for the specified line. Express each equation in slope-intercept form.

    (a) Through $(2, -3)$ and parallel to the line with equation $5x + 2y = 6$

    (b) Through $(-4, 1)$ and perpendicular to the line with equation $5x + 2y = 6$

Write with positive exponents only. Assume variables represent positive real numbers.

19. $\left(\dfrac{x^{-3}y^2}{x^5y^{-2}}\right)^{-1}$

20. $\dfrac{(4x^{-2})^2(2y^3)}{8x^{-3}y^5}$

Perform the indicated operations.

21. $(7x + 4)(2x - 3)$

22. $\left(\dfrac{2}{3}t + 9\right)^2$

23. $(3t^3 + 5t^2 - 8t + 7) - (6t^3 + 4t - 8)$

24. Divide $4x^3 + 2x^2 - x + 26$ by $x + 2$.

Factor completely.

25. $16x - x^3$

26. $24m^2 + 2m - 15$

27. $8x^3 + 27y^3$

28. $9x^2 - 30xy + 25y^2$

Perform the indicated operations and express each answer in lowest terms. Assume denominators are nonzero.

29. $\dfrac{x^2 - 3x - 10}{x^2 + 3x + 2} \cdot \dfrac{x^2 - 2x - 3}{x^2 + 2x - 15}$

30. $\dfrac{3}{2 - k} - \dfrac{5}{k} + \dfrac{6}{k^2 - 2k}$

31. $\dfrac{\dfrac{r}{s} - \dfrac{s}{r}}{\dfrac{r}{s} + 1}$

32. The record track-qualifying speeds at North Carolina Motor Speedway since Richard Petty captured the first pole in 1965 are given in the table. Let $x = 0$ represent 1965, $x = 10$ represent 1975, and so on.

    (a) Use the ordered pairs (year, speed) to make a scatter diagram of the data.

    (b) A linear equation can be used to model the data. Will its slope be positive or negative?

    (c) Use the ordered pairs $(0, 116.26)$ and $(20, 141.85)$ to write a linear equation that models the data.

    (d) Use your model to approximate the record speed for 1998 to the nearest hundredth. How does it compare to the actual value from the table?

QUALIFYING RECORDS

Year	Speed (in mph)
1965	116.26
1975	132.02
1985	141.85
1995	155.38
1998	156.36

Source: NASCAR.

33. Does the relation $x = 5$ define a function? Explain.

34. For the function defined by $f(x) = 2(x - 1)^2 - 5$, find

    (a) $f(-2)$;   **(b)** the domain and range.

Solve each system of equations.

35. $\begin{aligned} 2x - 4y &= 10 \\ 9x + 3y &= 3 \end{aligned}$

36. $\begin{aligned} x + y + 2z &= 3 \\ -x + y + z &= -5 \\ 2x + 3y - z &= -8 \end{aligned}$

37. The merger in 2000 of America Online and Time Warner was the largest in U.S. history. The two companies had combined sales of \$34.2 billion. Sales for AOL were \$.3 billion less than 4 times the sales for Time Warner. What were sales for each company? (*Source:* Company reports.)

(a) Write a system of equations to solve the problem.
(b) Solve the problem.

Simplify each radical expression.

38. $\sqrt[3]{\dfrac{27}{16}}$

39. $\dfrac{2}{\sqrt{7} - \sqrt{5}}$

Solve each problem.

40. Tri rode his bicycle for 12 mi and then walked an additional 8 mi. The total time for the trip was 5 hr. If his rate while walking was 10 mph less than his rate while riding, what was each rate?

41. Two cars left an intersection at the same time, one heading due south and the other due east. Later they were exactly 95 mi apart. The car heading east had gone 38 mi less than twice as far as the car heading south. How far had each car traveled?

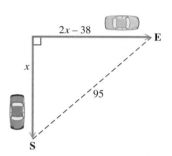

In this day of Automated Teller Machines (ATMs), people often find themselves doing what they have done for years when faced with a soft drink machine that won't respond: They talk to it. According to one report, the following are percentages of people in the United States, the United Kingdom (UK), and Germany who talk to ATMs and what they say.

	United States	UK	Germany
Thanking the ATM	22%	24%	14%
Cursing the ATM	31%	41%	53%
Telling the ATM to Hurry Up	47%	36%	33%

Source: BMRB International for NCR.

In a random sample of 3000 people, how many would there be in each category?

42. People in the United States who curse the ATM

43. People in the UK who thank the ATM

44. People in Germany who tell the ATM to hurry up

45. How many more German cursers would there be than United States thankers?

Inverse, Exponential, and Logarithmic Functions

12

The exponential and logarithmic functions introduced in this chapter are used to model a wide variety of situations, including environmental issues, compound interest, earthquake intensity, fossil dating, and sound levels.

Recently, there has been concern about the level of sound Americans are subjected to daily. For example, action sequences in *Pearl Harbor, The Movie* reached 107 decibels, while the sound levels in *Lethal Weapon 4* often reached 100 decibels or more, compared to an average of 95 decibels for a motorcycle. In Section 12.5, Exercise 41, we give a logarithmic function to measure sound levels and to find the decibel levels of other recent movies. (*Source: World Almanac and Book of Facts, 2001;* www.lhh.org/noise/)

12.1 Inverse Functions

In this chapter we study two important types of functions, *exponential* and *logarithmic*. These functions are related in a special way: They are *inverses* of one another. We begin by discussing inverse functions in general.

OBJECTIVE 1 Decide whether a function is one-to-one and, if it is, find its inverse. Suppose we define the function

$$G = \{(-2, 2), (-1, 1), (0, 0), (1, 3), (2, 5)\}.$$

We can form another set of ordered pairs from G by interchanging the x- and y-values of each pair in G. We can call this set F, so

$$F = \{(2, -2), (1, -1), (0, 0), (3, 1), (5, 2)\}.$$

To show that these two sets are related, F is called the *inverse* of G. For a function f to have an inverse, f must be a *one-to-one function*.

One-to-One Function

In a **one-to-one function,** each x-value corresponds to only one y-value, and each y-value corresponds to only one x-value.

The function shown in Figure 1(a) is not one-to-one because the y-value 7 corresponds to *two* x-values, 2 and 3. That is, the ordered pairs (2, 7) and (3, 7) both belong to the function. The function in Figure 1(b) is one-to-one.

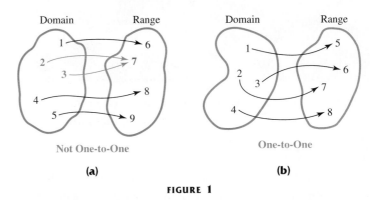

FIGURE 1

The *inverse* of any one-to-one function f is found by interchanging the components of the ordered pairs of f. The inverse of f is written f^{-1}. Read f^{-1} as "the inverse of f" or "f-inverse."

CAUTION The symbol $f^{-1}(x)$ does not represent $\dfrac{1}{f(x)}$.

The definition of the inverse of a function follows.

Inverse of a Function

The **inverse** of a one-to-one function f, written f^{-1}, is the set of all ordered pairs of the form (y, x), where (x, y) belongs to f. Since the inverse is formed by interchanging x and y, the domain of f becomes the range of f^{-1} and the range of f becomes the domain of f^{-1}.

For inverses f and f^{-1}, it follows that

$$f(f^{-1}(x)) = x \quad \text{and} \quad f^{-1}(f(x)) = x.$$

▪ EXAMPLE 1 Finding the Inverses of One-to-One Functions

Find the inverse of each one-to-one function.

(a) $F = \{(-2, 1), (-1, 0), (0, 1), (1, 2), (2, 2)\}$

Each x-value in F corresponds to just one y-value. However, the y-value 2 corresponds to two x-values, 1 and 2. Also, the y-value 1 corresponds to both -2 and 0. Because some y-values correspond to more than one x-value, F is not one-to-one and does not have an inverse.

(b) $G = \{(3, 1), (0, 2), (2, 3), (4, 0)\}$

Every x-value in G corresponds to only one y-value, and every y-value corresponds to only one x-value, so G is a one-to-one function. The inverse function is found by interchanging the x- and y-values in each ordered pair.

$$G^{-1} = \{(1, 3), (2, 0), (3, 2), (0, 4)\}$$

Notice how the domain and range of G become the range and domain, respectively, of G^{-1}.

(c) The U.S. Environmental Protection Agency has developed an indicator of air quality called the Pollutant Standard Index (PSI). If the PSI exceeds 100 on a particular day, then that day is classified as unhealthy. The table shows the number of unhealthy days in Chicago for the years 1991 through 1997.

Year	Number of Unhealthy Days
1991	21
1992	4
1993	3
1994	8
1995	21
1996	6
1997	9

Source: U.S. Environmental Protection Agency.

Let f be the function defined in the table, with the years forming the domain and the numbers of unhealthy days forming the range. Then f is not one-to-one, because in two different years (1991 and 1995), the number of unhealthy days was the same, 21.

Now Try Exercises 1, 9, and 11.

OBJECTIVE 2 **Use the horizontal line test to determine whether a function is one-to-one.** It may be difficult to decide whether a function is one-to-one just by looking at the equation that defines the function. However, by graphing the function and observing the graph, we can use the *horizontal line test* to tell whether the function is one-to-one.

Horizontal Line Test

If any horizontal line intersects the graph of a function in no more than one point, then the function is one-to-one.

The horizontal line test follows from the definition of a one-to-one function. Any two points that lie on the same horizontal line have the same y-coordinate. No two ordered pairs that belong to a one-to-one function may have the same y-coordinate, and therefore no horizontal line will intersect the graph of a one-to-one function more than once.

EXAMPLE 2 **Using the Horizontal Line Test**

Use the horizontal line test to determine whether the graphs in Figures 2 and 3 are graphs of one-to-one functions.

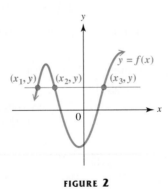

FIGURE 2

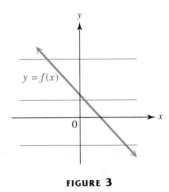

FIGURE 3

Because the horizontal line shown in Figure 2 intersects the graph in more than one point (actually three points), the function is not one-to-one.

Every horizontal line will intersect the graph in Figure 3 in exactly one point. This function is one-to-one.

Now Try Exercise 7.

OBJECTIVE 3 **Find the equation of the inverse of a function.** By definition, the inverse of a function is found by interchanging the x- and y-values of each of its ordered pairs. The equation of the inverse of a function defined by $y = f(x)$ is found in the same way.

Finding the Equation of the Inverse of $y = f(x)$

For a one-to-one function f defined by an equation $y = f(x)$, find the defining equation of the inverse as follows.

Step 1 **Interchange** x and y.

Step 2 **Solve** for y.

Step 3 **Replace** y with $f^{-1}(x)$.

EXAMPLE 3 **Finding Equations of Inverses**

Decide whether each equation defines a one-to-one function. If so, find the equation of the inverse.

(a) $f(x) = 2x + 5$

The graph of $y = 2x + 5$ is a nonvertical line, so by the horizontal line test, f is a one-to-one function. To find the inverse, let $y = f(x)$ so that

$$y = 2x + 5$$
$$x = 2y + 5 \qquad \text{Interchange } x \text{ and } y. \text{ (Step 1)}$$
$$2y = x - 5 \qquad \text{Solve for } y. \text{ (Step 2)}$$
$$y = \frac{x - 5}{2}$$
$$f^{-1}(x) = \frac{x - 5}{2}. \qquad \text{(Step 3)}$$

Thus, f^{-1} is a linear function. In the function defined by $y = 2x + 5$, the value of y is found by starting with a value of x, multiplying by 2, and adding 5. The equation for the inverse has us *subtract* 5, and then *divide* by 2. This shows how an inverse is used to "undo" what a function does to the variable x.

(b) $y = x^2 + 2$

This equation has a vertical parabola as its graph, so some horizontal lines will intersect the graph at two points. For example, both $x = 3$ and $x = -3$ correspond to $y = 11$. Because of the x^2-term, there are many pairs of x-values that correspond to the same y-value. This means that the function defined by $y = x^2 + 2$ is not one-to-one and does not have an inverse.

If this is not noticed, then following the steps for finding the equation of an inverse leads to

$$y = x^2 + 2$$
$$x = y^2 + 2 \qquad \text{Interchange } x \text{ and } y.$$
$$x - 2 = y^2 \qquad \text{Solve for } y.$$
$$\pm\sqrt{x - 2} = y. \qquad \text{Square root property}$$

The last step shows that there are two y-values for each choice of $x > 2$, so the given function is not one-to-one and cannot have an inverse.

(c) $f(x) = (x - 2)^3$

Because of the cube, each value of x produces a different value of y, so this is a one-to-one function.

$$y = (x - 2)^3 \qquad \text{Replace } f(x) \text{ with } y.$$

$$x = (y - 2)^3 \qquad \text{Interchange } x \text{ and } y.$$

$$\sqrt[3]{x} = \sqrt[3]{(y - 2)^3} \qquad \text{Take the cube root on each side.}$$

$$\sqrt[3]{x} = y - 2$$

$$\sqrt[3]{x} + 2 = y \qquad \text{Solve for } y.$$

$$f^{-1}(x) = \sqrt[3]{x} + 2 \qquad \text{Replace } y \text{ with } f^{-1}(x).$$

Now Try Exercises 13, 17, and 19.

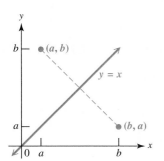

FIGURE 4

OBJECTIVE 4 Graph f^{-1} from the graph of f. One way to graph the inverse of a function f whose equation is known is to find some ordered pairs that belong to f, interchange x and y to get ordered pairs that belong to f^{-1}, plot those points, and sketch the graph of f^{-1} through the points. A simpler way is to select points on the graph of f and use symmetry to find corresponding points on the graph of f^{-1}.

For example, suppose the point (a, b) shown in Figure 4 belongs to a one-to-one function f. Then the point (b, a) belongs to f^{-1}. The line segment connecting (a, b) and (b, a) is perpendicular to, and cut in half by, the line $y = x$. The points (a, b) and (b, a) are "mirror images" of each other with respect to $y = x$. For this reason we can find the graph of f^{-1} from the graph of f by locating the mirror image of each point in f with respect to the line $y = x$.

EXAMPLE 4 Graphing the Inverse

Graph the inverses of the functions f (shown in blue) in Figure 5.

In Figure 5 the graphs of two functions f are shown in blue. Their inverses are shown in red. In each case, the graph of f^{-1} is a reflection of the graph of f with respect to the line $y = x$.

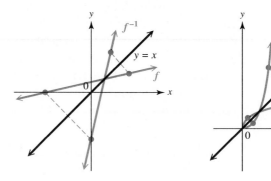

FIGURE 5

Now Try Exercises 25 and 29.

OBJECTIVE 5 Use a graphing calculator to graph inverse functions. We have described how inverses of one-to-one functions may be determined algebraically. We also explained how the graph of a one-to-one function f compares to the graph of its inverse f^{-1}: It is a reflection of the graph of f^{-1} across the line $y = x$. In Example 3 we showed that the inverse of the one-to-one function defined by $f(x) = 2x + 5$ is given by $f^{-1}(x) = \frac{x - 5}{2}$. If we use a square viewing window of a graphing calculator and graph $y_1 = f(x) = 2x + 5$, $y_2 = f^{-1}(x) = \frac{x - 5}{2}$, and $y_3 = x$, we can see how this reflection appears on the screen. See Figure 6.

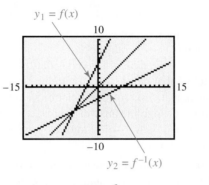

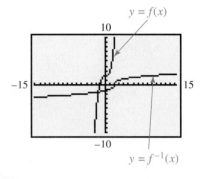

FIGURE 6   **FIGURE 7**

Some graphing calculators have the capability to "draw" the inverse of a function. Figure 7 shows the graphs of $f(x) = x^3 + 2$ and its inverse in a square viewing window.

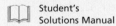 **12.1 EXERCISES**

✐ *In Exercises 1–4, write a few sentences of explanation. See Example 1.*

1. The table shows the number of uncontrolled hazardous waste sites that require further investigation to determine whether remedies are needed under the Superfund program. The seven states listed are ranked in the top ten in the United States.

   If this correspondence is considered to be a function that pairs each state with its number of uncontrolled waste sites, is it one-to-one? If not, explain why. (See Example 1(c).)

State	Number of Sites
New Jersey	108
Pennsylvania	101
California	94
New York	79
Florida	53
Illinois	40
Wisconsin	40

Source: U.S. Environmental Protection Agency.

2. The table shows emissions of a major air pollutant, carbon monoxide, in the United States for the years 1992 through 1998.

   If this correspondence is considered to be a function that pairs each year with its emissions amount, is it one-to-one? If not, explain why.

Year	Amount of Emissions (in thousands of tons)
1992	97,630
1993	98,160
1994	102,643
1995	93,353
1996	95,479
1997	94,410
1998	89,454

Source: U.S. Environmental Protection Agency.

3. The road mileage between Denver, Colorado, and several selected U.S. cities is shown in the table. If we consider this as a function that pairs each city with a distance, is it a one-to-one function? How could we change the answer to this question by adding 1 mile to one of the distances shown?

City	Distance to Denver (in miles)
Atlanta	1398
Dallas	781
Indianapolis	1058
Kansas City, MO	600
Los Angeles	1059
San Francisco	1235

4. Suppose you consider the set of ordered pairs (x, y) such that x represents a person in your mathematics class and y represents that person's mother. Explain how this function might not be a one-to-one function.

In Exercises 5–8, choose the correct response from the given list.

5. If a function is made up of ordered pairs in such a way that the same y-value appears in a correspondence with two different x-values, then

   A. the function is one-to-one
   B. the function is not one-to-one
   C. its graph does not pass the vertical line test
   D. it has an inverse function associated with it.

6. Which equation defines a one-to-one function? Explain why the others are not, using specific examples.

   A. $f(x) = x$ **B.** $f(x) = x^2$ **C.** $f(x) = |x|$ **D.** $f(x) = -x^2 + 2x - 1$

7. Only one of the graphs illustrates a one-to-one function. Which one is it? (See Example 2.)

   A. 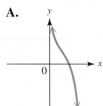   **B.** 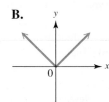   **C.** 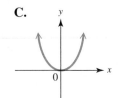   **D.**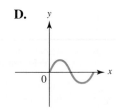

8. If a function f is one-to-one and the point (p, q) lies on the graph of f, then which point *must* lie on the graph of f^{-1}?

   A. $(-p, q)$ **B.** $(-q, -p)$ **C.** $(p, -q)$ **D.** (q, p)

If the function is one-to-one, find its inverse. See Examples 1–3.

9. $\{(3, 6), (2, 10), (5, 12)\}$ **10.** $\left\{ (-1, 3), (0, 5), (5, 0), \left(7, -\frac{1}{2} \right) \right\}$

11. $\{(-1, 3), (2, 7), (4, 3), (5, 8)\}$

12. $\{(-8, 6), (-4, 3), (0, 6), (5, 10)\}$

13. $f(x) = 2x + 4$

14. $f(x) = 3x + 1$

15. $g(x) = \sqrt{x - 3}, \quad x \geq 3$

16. $g(x) = \sqrt{x + 2}, \quad x \geq -2$

17. $f(x) = 3x^2 + 2$

18. $f(x) = -4x^2 - 1$

19. $f(x) = x^3 - 4$

20. $f(x) = x^3 - 3$

Let $f(x) = 2^x$. We will see in the next section that this function is one-to-one. Find each value, always working part (a) before part (b).

21. (a) $f(3)$ **(b)** $f^{-1}(8)$

22. (a) $f(4)$ **(b)** $f^{-1}(16)$

23. (a) $f(0)$ **(b)** $f^{-1}(1)$

24. (a) $f(-2)$ **(b)** $f^{-1}\left(\dfrac{1}{4}\right)$

*The graphs of some functions are given in Exercises 25–30. (**a**) Use the horizontal line test to determine whether the function is one-to-one. (**b**) If the function is one-to-one, then graph the inverse of the function. (Remember that if f is one-to-one and (a, b) is on the graph of f, then (b, a) is on the graph of f^{-1}.) See Example 4.*

25.

26.

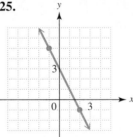

27.

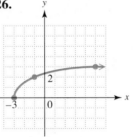

28.

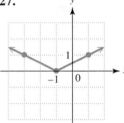

29.

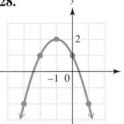

30.

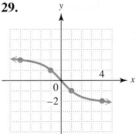

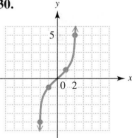

Each function defined in Exercises 31–38 is a one-to-one function. Graph the function as a solid line (or curve) and then graph its inverse on the same set of axes as a dashed line (or curve). In Exercises 35–38 you are given a table to complete so that graphing the function will be a bit easier. See Example 4.

31. $f(x) = 2x - 1$ **32.** $f(x) = 2x + 3$ **33.** $g(x) = -4x$ **34.** $g(x) = -2x$

35. $f(x) = \sqrt{x}, \quad x \geq 0$

36. $f(x) = -\sqrt{x}, \quad x \geq 0$

x	$f(x)$
0	
1	
4	

x	$f(x)$
0	
1	
4	

37. $f(x) = x^3 - 2$

x	$f(x)$
-1	
0	
1	
2	

38. $f(x) = x^3 + 3$

x	$f(x)$
-2	
-1	
0	
1	

RELATING CONCEPTS (EXERCISES 39–42)

For Individual or Group Work

Inverse functions are used by government agencies and other businesses to send and receive coded information. The functions they use are usually very complicated. A simple example might use the function defined by $f(x) = 2x + 5$. *(Note that it is one-to-one.) Suppose that each letter of the alphabet is assigned a numerical value according to its position, as follows:*

A	1	G	7	L	12	Q	17	V	22
B	2	H	8	M	13	R	18	W	23
C	3	I	9	N	14	S	19	X	24
D	4	J	10	O	15	T	20	Y	25
E	5	K	11	P	16	U	21	Z	26
F	6								

This is an Enigma machine, used by the Germans in World War II to send coded messages.

Using the function, the word ALGEBRA *would be encoded as*

$$7 \quad 29 \quad 19 \quad 15 \quad 9 \quad 41 \quad 7,$$

because

$$f(A) = f(1) = 2(1) + 5 = 7, \quad f(L) = f(12) = 2(12) + 5 = 29,$$

and so on. The message would then be decoded by using the inverse of f, defined by $f^{-1}(x) = \frac{x-5}{2}$. *For example,*

$$f^{-1}(7) = \frac{7-5}{2} = 1 = A, \quad f^{-1}(29) = \frac{29-5}{2} = 12 = L,$$

and so on. **Work Exercises 39–42 in order.**

39. Suppose that you are an agent for a detective agency and you know that today's function for your code is defined by $f(x) = 4x - 5$. Find the rule for f^{-1} algebraically.

40. You receive the following coded message today. (Read across from left to right.)

47 95 23 67 -1 59 27 31 51 23 7 -1 43 7 79 43 -1 75 55 67

31 71 75 27 15 23 67 15 -1 75 15 71 75 75 27 31 51

23 71 31 51 7 15 71 43 31 7 15 11 3 67 15 -1 11

Use the letter/number assignment described earlier to decode the message.

41. Why is a one-to-one function essential in this encoding/decoding process?

42. Use $f(x) = x^3 + 4$ to encode your name, using the letter/number assignment described earlier.

Each function defined is one-to-one. Find the inverse algebraically, and then graph both the function and its inverse on the same graphing calculator screen. Use a square viewing window. See Objective 5.

43. $f(x) = 2x - 7$

44. $f(x) = -3x + 2$

45. $f(x) = x^3 + 5$

46. $f(x) = \sqrt[3]{x + 2}$

 Some graphing calculators have the capability to draw the "inverse" of a function even if the function is not one-to-one; therefore, the inverse is not technically a function, but it is a relation. For example, the graphs of $y = x^2$ and $x = y^2$ are shown in the screen using a square viewing window.

    Read your instruction manual to see if your model has this capability. If so, draw both Y_1 and its inverse in the same square window.

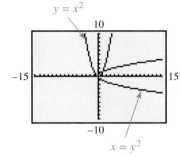

47. $Y_1 = X^2 + 3X + 4$

48. $Y_1 = X^3 - 9X$

49. Explain why the "inverse" of the function in Exercise 47 does not actually satisfy the definition of inverse as given in this section.

50. At what points do the graphs of $y = x^2$ and $x = y^2$ intersect? (See the graph above.) Verify this using algebraic methods.

12.2   Exponential Functions

OBJECTIVES

1 Define an exponential function.

2 Graph an exponential function.

3 Solve exponential equations of the form $a^x = a^k$ for *x*.

4 Use exponential functions in applications involving growth or decay.

OBJECTIVE 1 Define an exponential function. In Section 10.2 we showed how to evaluate 2^x for rational values of *x*. For example,

$$2^3 = 8, \qquad 2^{-1} = \frac{1}{2}, \qquad 2^{1/2} = \sqrt{2}, \qquad \text{and} \qquad 2^{3/4} = \sqrt[4]{2^3} = \sqrt[4]{8}.$$

In more advanced courses it is shown that 2^x exists for all real number values of *x*, both rational and irrational. (Later in this chapter, we will see how to approximate the value of 2^x for irrational *x*.) The following definition of an exponential function assumes that a^x exists for all real numbers *x*.

> **Exponential Function**
>
> For $a > 0$, $a \neq 1$, and all real numbers *x*,
>
> $$f(x) = a^x$$
>
> defines the **exponential function with base *a*.**

NOTE The two restrictions on a in the definition of an exponential function are important. The restriction that a must be positive is necessary so that the function can be defined for all real numbers x. For example, letting a be negative ($a = -2$, for instance) and letting $x = \frac{1}{2}$ would give the expression $(-2)^{1/2}$, which is not real. The other restriction, $a \neq 1$, is necessary because 1 raised to any power is equal to 1, and the function would then be the linear function defined by $f(x) = 1$.

OBJECTIVE 2 Graph an exponential function. We graph an exponential function by finding several ordered pairs that belong to the function, plotting these points, and connecting them with a smooth curve.

CAUTION Be sure to plot enough points to see how rapidly the graph rises.

EXAMPLE 1 Graphing an Exponential Function with $a > 1$

Graph $f(x) = 2^x$.

Choose some values of x, and find the corresponding values of $f(x)$.

x	-3	-2	-1	0	1	2	3	4
$f(x) = 2^x$	$\frac{1}{8}$	$\frac{1}{4}$	$\frac{1}{2}$	1	2	4	8	16

Plotting these points and drawing a smooth curve through them gives the blue graph shown in Figure 8. This graph is typical of the graphs of exponential functions of the form $F(x) = a^x$, where $a > 1$. The larger the value of a, the faster the graph rises. To see this, compare the red graph of $F(x) = 5^x$ with the graph of $f(x) = 2^x$ in Figure 8.

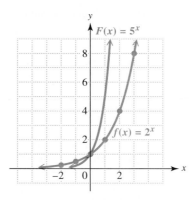

FIGURE 8

By the vertical line test, the graphs in Figure 8 represent functions. As these graphs suggest, the domain of an exponential function includes all real numbers. Because y is always positive, the range is $(0, \infty)$. Figure 8 also shows an important characteristic of exponential functions where $a > 1$: As x gets larger, y increases at a faster and faster rate.

Now Try Exercise 5.

EXAMPLE 2 Graphing an Exponential Function with $a < 1$

Graph $g(x) = \left(\dfrac{1}{2}\right)^x$.

Again, find some points on the graph.

x	-3	-2	-1	0	1	2	3
$g(x) = \left(\frac{1}{2}\right)^x$	8	4	2	1	$\frac{1}{2}$	$\frac{1}{4}$	$\frac{1}{8}$

The graph, shown in Figure 9, is very similar to that of $f(x) = 2^x$ (Figure 8) with the same domain and range, except that here as x gets larger, y *decreases*. This graph is typical of the graph of a function of the form $F(x) = a^x$, where $0 < a < 1$.

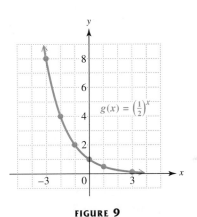

FIGURE 9

Now Try Exercise 7.

Based on Examples 1 and 2, we make the following generalizations about the graphs of exponential functions of the form $F(x) = a^x$.

Graph of $F(x) = a^x$

1. The graph contains the point $(0, 1)$.

2. When $a > 1$, the graph will *rise* from left to right. When $0 < a < 1$, the graph will *fall* from left to right. In both cases, the graph goes from the second quadrant to the first.

3. The graph will approach the x-axis, but never touch it. (Such a line is called an **asymptote**.)

4. The domain is $(-\infty, \infty)$, and the range is $(0, \infty)$.

EXAMPLE 3 Graphing a More Complicated Exponential Function

Graph $f(x) = 3^{2x-4}$.

Find some ordered pairs.

$$\text{If } x = 0, \text{ then } y = 3^{2(0)-4} = 3^{-4} = \frac{1}{81}.$$

$$\text{If } x = 2, \text{ then } y = 3^{2(2)-4} = 3^{0} = 1.$$

These ordered pairs, $\left(0, \frac{1}{81}\right)$ and $(2, 1)$, along with the other ordered pairs shown in the table, lead to the graph in Figure 10. The graph is similar to the graph of $f(x) = 3^x$ except that it is shifted to the right and rises more rapidly.

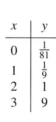

x	y
0	$\frac{1}{81}$
1	$\frac{1}{9}$
2	1
3	9

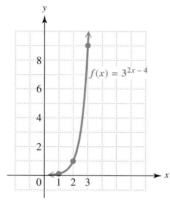

FIGURE 10

Now Try Exercise 11.

OBJECTIVE 3 Solve exponential equations of the form $a^x = a^k$ for x. Until this chapter, we have solved only equations that had the variable as a base, like $x^2 = 8$; all exponents have been constants. An **exponential equation** is an equation that has a variable in an exponent, such as

$$9^x = 27.$$

By the horizontal line test, the exponential function defined by $F(x) = a^x$ is a one-to-one function, so we can use the following property to solve many exponential equations.

Property for Solving an Exponential Equation

For $a > 0$ and $a \neq 1$,   if $a^x = a^y$ then $x = y$.

This property would not necessarily be true if $a = 1$.

To solve an exponential equation using this property, follow these steps.

Solving an Exponential Equation

Step 1 **Each side must have the same base.** If the two sides of the equation do not have the same base, express each as a power of the same base if possible.

Step 2 **Simplify exponents** if necessary, using the rules of exponents.

Step 3 **Set exponents equal** using the property given in this section.

Step 4 **Solve** the equation obtained in Step 3.

N O T E These steps cannot be applied to an exponential equation like

$$3^x = 12$$

because Step 1 cannot easily be done. A method for solving such equations is given in Section 12.6.

■ **EXAMPLE 4** Solving an Exponential Equation

Solve the equation $9^x = 27$.

We can use the property given at the bottom of the previous page if both sides are written with the same base.

$$9^x = 27$$
$$(3^2)^x = 3^3 \qquad \text{Write with the same base;}$$
$$\qquad\qquad\qquad 9 = 3^2 \text{ and } 27 = 3^3. \text{ (Step 1)}$$
$$3^{2x} = 3^3 \qquad \text{Power rule for exponents (Step 2)}$$
$$2x = 3 \qquad \text{If } a^x = a^y, \text{ then } x = y. \text{ (Step 3)}$$
$$x = \frac{3}{2} \qquad \text{(Step 4)}$$

Check that the solution set is $\left\{\frac{3}{2}\right\}$ by substituting $\frac{3}{2}$ for x in the original equation. ■

Now Try Exercise 17.

■ **EXAMPLE 5** Solving Exponential Equations

Solve each equation.

(a)
$$4^{3x-1} = 16^{x+2}$$
$$(2^2)^{3x-1} = (2^4)^{x+2} \qquad \text{Write with the same base;}$$
$$\qquad\qquad\qquad\qquad 4 = 2^2 \text{ and } 16 = 2^4.$$
$$2^{6x-2} = 2^{4x+8} \qquad \text{Power rule for exponents}$$
$$6x - 2 = 4x + 8 \qquad \text{Set exponents equal.}$$
$$2x = 10 \qquad \text{Subtract } 4x; \text{ add 2.}$$
$$x = 5 \qquad \text{Divide by 2.}$$

Verify that the solution set is $\{5\}$.

(b)
$$6^x = \frac{1}{216}$$

$$6^x = \frac{1}{6^3} \qquad 216 = 6^3$$

$$6^x = 6^{-3} \qquad \text{Write with the same base; } \frac{1}{6^3} = 6^{-3}.$$

$$x = -3 \qquad \text{Set exponents equal.}$$

Verify that the solution set is $\{-3\}$.

(c)
$$\left(\frac{2}{3}\right)^x = \frac{9}{4}$$

$$\left(\frac{2}{3}\right)^x = \left(\frac{4}{9}\right)^{-1} \qquad \frac{9}{4} = \left(\frac{4}{9}\right)^{-1}$$

$$\left(\frac{2}{3}\right)^x = \left[\left(\frac{2}{3}\right)^2\right]^{-1} \qquad \text{Write with the same base.}$$

$$\left(\frac{2}{3}\right)^x = \left(\frac{2}{3}\right)^{-2} \qquad \text{Power rule for exponents}$$

$$x = -2 \qquad \text{Set exponents equal.}$$

Check that the solution set is $\{-2\}$.

Now Try Exercises 19, 21, and 25.

OBJECTIVE 4 Use exponential functions in applications involving growth or decay.

EXAMPLE 6 Solving an Application Involving Exponential Growth

One result of the rapidly increasing world population is an increase of carbon dioxide in the air, which scientists believe may be contributing to global warming. Both population and amounts of carbon dioxide in the air are increasing exponentially. This means that the growth rate is continually increasing. The graph in Figure 11 shows the concentration of carbon dioxide (in parts per million) in the air.

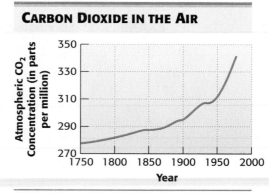

CARBON DIOXIDE IN THE AIR

Source: Sacramento Bee, Monday, September 13, 1993.

FIGURE 11

The data are approximated by the function defined by

$$f(x) = 278(1.00084)^x,$$

where x is the number of years since 1750. Use this function and a calculator to approximate the concentration of carbon dioxide in parts per million for each year.

(a) 1900

Because x represents the number of years since 1750, in this case

$$x = 1900 - 1750 = 150.$$

Thus, evaluate $f(150)$.

$$f(150) = 278(1.00084)^{150} \qquad \text{Let } x = 150.$$
$$\approx 315 \text{ parts per million} \qquad \text{Use a calculator.}$$

(b) 1950

Use $x = 1950 - 1750 = 200$.

$$f(200) = 278(1.00084)^{200}$$
$$\approx 329 \text{ parts per million}$$

Now Try Exercise 39.

EXAMPLE 7   Applying an Exponential Decay Function

The atmospheric pressure (in millibars) at a given altitude x, in meters, can be approximated by the function defined by

$$f(x) = 1038(1.000134)^{-x},$$

for values of x between 0 and 10,000. Because the base is greater than 1 and the coefficient of x in the exponent is negative, the function values decrease as x increases. This means that as the altitude increases, the atmospheric pressure decreases. (*Source:* Miller, A. and J. Thompson, *Elements of Meteorology,* Fourth Edition, Charles E. Merrill Publishing Company, 1993.)

(a) According to this function, what is the pressure at ground level?

At ground level, $x = 0$, so

$$f(0) = 1038(1.000134)^{-0} = 1038(1) = 1038.$$

The pressure is 1038 millibars.

(b) What is the pressure at 5000 m?

Use a calculator to find $f(5000)$.

$$f(5000) = 1038(1.000134)^{-5000}$$
$$\approx 531$$

The pressure is approximately 531 millibars.

Now Try Exercise 41.

12.2 EXERCISES

Choose the correct response in Exercises 1–4.

1. Which point lies on the graph of $f(x) = 2^x$?

    A. $(1, 0)$ **B.** $(2, 1)$ **C.** $(0, 1)$ **D.** $\left(\sqrt{2}, \dfrac{1}{2}\right)$

2. Which statement is true?

    A. The y-intercept of the graph of $f(x) = 10^x$ is $(0, 10)$.
    B. For any $a > 1$, the graph of $f(x) = a^x$ falls from left to right.
    C. The point $\left(\frac{1}{2}, \sqrt{5}\right)$ lies on the graph of $f(x) = 5^x$.
    D. The graph of $y = 4^x$ rises at a faster rate than the graph of $y = 10^x$.

3. The asymptote of the graph of $F(x) = a^x$

    A. is the x-axis   **B.** is the y-axis
    C. has equation $x = 1$ **D.** has equation $y = 1$.

4. Which equation is graphed here?

    A. $y = 1000\left(\dfrac{1}{2}\right)^{.3x}$ **B.** $y = 1000\left(\dfrac{1}{2}\right)^{x}$

    C. $y = 1000(2)^{.3x}$ **D.** $y = 1000^x$

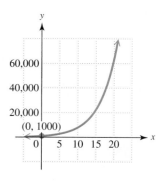

Graph each exponential function. See Examples 1–3.

5. $f(x) = 3^x$ **6.** $f(x) = 5^x$ **7.** $g(x) = \left(\dfrac{1}{3}\right)^x$ **8.** $g(x) = \left(\dfrac{1}{5}\right)^x$

9. $y = 4^{-x}$ **10.** $y = 6^{-x}$ **11.** $y = 2^{2x-2}$ **12.** $y = 2^{2x+1}$

13. (a) For an exponential function defined by $f(x) = a^x$, if $a > 1$, the graph _____
                                                               (rises/falls)
      from left to right. If $0 < a < 1$, the graph _____ from left to right.
                                                (rises/falls)

    (b) Based on your answers in part (a), make a conjecture (an educated guess)
        concerning whether an exponential function defined by $f(x) = a^x$ is one-to-one.
        Then decide whether it has an inverse based on the concepts of Section 12.1.

14. In your own words, describe the characteristics of the graph of an exponential function.
    Use the exponential function defined by $f(x) = 3^x$ (Exercise 5) and the words
    asymptote, domain, and *range* in your explanation.

Solve each equation. See Examples 4 and 5.

15. $6^x = 36$ **16.** $8^x = 64$ **17.** $100^x = 1000$ **18.** $8^x = 4$

19. $16^{2x+1} = 64^{x+3}$ **20.** $9^{2x-8} = 27^{x-4}$ **21.** $5^x = \dfrac{1}{125}$ **22.** $3^x = \dfrac{1}{81}$

23. $5^x = .2$ **24.** $10^x = .1$ **25.** $\left(\dfrac{3}{2}\right)^x = \dfrac{8}{27}$ **26.** $\left(\dfrac{4}{3}\right)^x = \dfrac{27}{64}$

Use the exponential key of a calculator to find an approximation to the nearest thousandth.

27. $12^{2.6}$ **28.** $13^{1.8}$ **29.** $.5^{3.921}$ **30.** $.6^{4.917}$ **31.** $2.718^{2.5}$ **32.** $2.718^{-3.1}$

33. Try to evaluate $(-2)^4$ on a scientific calculator. You may get an error message, since the exponential function key on many calculators does not allow negative bases. Discuss the concept introduced in this section that is closely related to this "peculiarity" of many scientific calculators.

34. Explain why the exponential equation $4^x = 6$ cannot be solved using the method explained in this section. Change 6 to another number that *will* allow the method of this section to be used, and then solve the equation.

The graph shown here accompanied the article "Is Our World Warming?" which appeared in the October 1990 issue of National Geographic. *It shows projected temperature increases using two graphs: one an exponential-type curve, and the other linear. From the graph, approximate the increase (a) for the exponential curve and (b) for the linear graph for each year.*

35. 2000

36. 2010

37. 2020

38. 2040

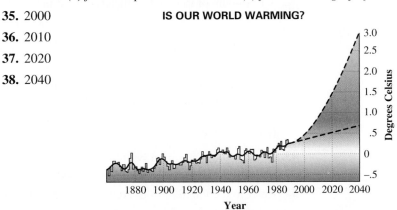

IS OUR WORLD WARMING?

Graph, "Zero Equals Average Global Temperature for the Period 1950–1979." Dale D. Glasgow, © National Geographic Society. Reprinted by permission.

Solve each problem. See Examples 6 and 7.

39. Based on figures from 1970 through 1998, the worldwide carbon monoxide emissions in thousands of tons are approximated by the exponential function defined by

$$f(x) = 132{,}359(1.0124)^{-x},$$

where $x = 0$ corresponds to 1970, $x = 5$ corresponds to 1975, and so on. (*Source:* U.S. Environmental Protection Agency.)

(a) Use this model to approximate the emissions in 1970.

(b) Use this model to approximate the emissions in 1995.

(c) In 1998, the actual amount of emissions was 89,454 million tons. How does this compare to the number that the model provides?

40. Based on figures from 1980 through 1999, the municipal solid waste generated in millions of tons can be approximated by the exponential function defined by

$$f(x) = 157.28(1.0204)^x,$$

where $x = 0$ corresponds to 1980, $x = 5$ corresponds to 1985, and so on. (*Source:* U.S. Environmental Protection Agency.)

    (a) Use the model to approximate the number of tons of this waste in 1980.

    (b) Use the model to approximate the number of tons of this waste in 1995.

    (c) In 1999, the actual number of millions of tons of this waste was 229.9. How does this compare to the number that the model provides?

41. A small business estimates that the value $V(t)$ of a copy machine is decreasing according to the function defined by

$$V(t) = 5000(2)^{-.15t},$$

where t is the number of years that have elapsed since the machine was purchased, and $V(t)$ is in dollars.

    (a) What was the original value of the machine?

    (b) What is the value of the machine 5 yr after purchase? Give your answer to the nearest dollar.

    (c) What is the value of the machine 10 yr after purchase? Give your answer to the nearest dollar.

    (d) Graph the function.

42. The amount of radioactive material in an ore sample is given by the function defined by

$$A(t) = 100(3.2)^{-.5t},$$

where $A(t)$ is the amount present, in grams, of the sample t months after the initial measurement.

    (a) How much was present at the initial measurement? (*Hint:* $t = 0$.)

    (b) How much was present 2 months later?

    (c) How much was present 10 months later?

    (d) Graph the function.

43. Refer to the function in Exercise 41. When will the value of the machine be $2500? (*Hint:* Let $V(t) = 2500$, divide both sides by 5000, and use the method of Example 4.)

44. Refer to the function in Exercise 41. When will the value of the machine be $1250?

RELATING CONCEPTS (EXERCISES 45–50)

For Individual or Group Work

In these exercises we examine several methods of simplifying the expression $16^{3/4}$. Work Exercises 45–50 in order.

45. Write $16^{3/4}$ as a radical expression with the exponent outside the radical. Then simplify the expression.

46. Write $16^{3/4}$ as a radical expression with the exponent under the radical. Then simplify the expression.

47. Use a calculator to find the square root of 16^3. Now find the square root of that result.

48. Explain why the result in Exercise 47 is equal to $16^{3/4}$.

49. Predict the result a calculator will give when 16 is raised to the .75 power. Then check your answer by actually performing the operation on your calculator.

50. Write $\sqrt[100]{16^{75}}$ as an exponential expression. Then write the exponent in lowest terms, rewrite as a radical, and evaluate this radical expression.

12.3 Logarithmic Functions

OBJECTIVES

1 Define a logarithm.

2 Convert between exponential and logarithmic forms.

3 Solve logarithmic equations of the form $\log_a b = k$ for a, b, or k.

4 Define and graph logarithmic functions.

5 Use logarithmic functions in applications involving growth or decay.

The graph of $y = 2^x$ is the curve shown in blue in Figure 12. Because $y = 2^x$ defines a one-to-one function, it has an inverse. Interchanging x and y gives $x = 2^y$, the inverse of $y = 2^x$. As we saw in Section 12.1, the graph of the inverse is found by reflecting the graph of $y = 2^x$ about the line $y = x$. The graph of $x = 2^y$ is shown as a red curve in Figure 12.

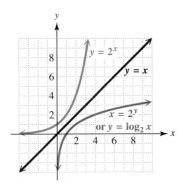

FIGURE 12

OBJECTIVE 1 Define a logarithm. We cannot solve the equation $x = 2^y$ for the dependent variable y with the methods presented up to now. The following definition is used to solve $x = 2^y$ for y.

Logarithm

For all positive numbers a, $a \neq 1$, and all positive numbers x,

$$y = \log_a x \quad \text{means the same as} \quad x = a^y.$$

This key statement should be memorized. The abbreviation **log** is used for the word **logarithm.** Read $\log_a x$ as "the logarithm of x to the base a" or "the base a logarithm of x." To remember the location of the base and the exponent in each form, refer to the following diagrams.

$$
\begin{array}{cc}
\text{Exponent} & \text{Exponent} \\
\downarrow & \downarrow \\
\text{Logarithmic form: } y = \log_a x & \text{Exponential form: } x = a^y \\
\uparrow & \uparrow \\
\text{Base} & \text{Base}
\end{array}
$$

In working with logarithmic form and exponential form, remember the following.

Meaning of $\log_a x$

A logarithm is an exponent; $\log_a x$ is the exponent to which the base a must be raised to obtain x.

OBJECTIVE 2 Convert between exponential and logarithmic forms. We can use the definition of logarithm to write exponential statements in logarithmic form and logarithmic statements in exponential form. The following table shows several pairs of equivalent statements.

Exponential Form	Logarithmic Form
$3^2 = 9$	$\log_3 9 = 2$
$\left(\frac{1}{5}\right)^{-2} = 25$	$\log_{1/5} 25 = -2$
$10^5 = 100{,}000$	$\log_{10} 100{,}000 = 5$
$4^{-3} = \frac{1}{64}$	$\log_4 \frac{1}{64} = -3$

OBJECTIVE 3 Solve logarithmic equations of the form $\log_a b = k$ for a, b, or k. A **logarithmic equation** is an equation with a logarithm in at least one term. We solve logarithmic equations of the form $\log_a b = k$ for any of the three variables by first writing the equation in exponential form.

▬ **EXAMPLE 1 Solving Logarithmic Equations**

Solve each equation.

(a) $\log_4 x = -2$

By the definition of logarithm, $\log_4 x = -2$ is equivalent to $x = 4^{-2}$. Solve this exponential equation.

$$x = 4^{-2} = \frac{1}{16}$$

The solution set is $\left\{\frac{1}{16}\right\}$.

(b)

$$\log_{1/2}(3x + 1) = 2$$

$$3x + 1 = \left(\frac{1}{2}\right)^2 \qquad \text{Write in exponential form.}$$

$$3x + 1 = \frac{1}{4}$$

$$12x + 4 = 1 \qquad \text{Multiply by 4.}$$

$$12x = -3 \qquad \text{Subtract 4.}$$

$$x = -\frac{1}{4} \qquad \text{Divide by 12.}$$

The solution set is $\left\{-\frac{1}{4}\right\}$.

(c)

$$\log_x 3 = 2$$

$$x^2 = 3 \qquad \text{Write in exponential form.}$$

$$x = \pm\sqrt{3} \qquad \text{Take square roots.}$$

Notice that only the principal square root satisfies the equation since the base must be a positive number. The solution set is $\left\{\sqrt{3}\right\}$.

(d)

$$\log_{49} \sqrt[3]{7} = x$$

$$49^x = \sqrt[3]{7} \qquad \text{Write in exponential form.}$$

$$(7^2)^x = 7^{1/3} \qquad \text{Write with the same base.}$$

$$7^{2x} = 7^{1/3} \qquad \text{Power rule for exponents}$$

$$2x = \frac{1}{3} \qquad \text{Set exponents equal.}$$

$$x = \frac{1}{6} \qquad \text{Divide by 2.}$$

The solution set is $\left\{\frac{1}{6}\right\}$.

Now Try Exercises 21, 25, 37, and 39.

For any real number b, we know that $b^1 = b$ and for $b \neq 0$, $b^0 = 1$. Writing these two statements in logarithmic form gives the following two properties of logarithms.

Properties of Logarithms

For any positive real number b, $b \neq 1$,

$$\log_b b = 1 \qquad \text{and} \qquad \log_b 1 = 0.$$

EXAMPLE 2 Using Properties of Logarithms

Use the preceding two properties of logarithms to evaluate each logarithm.

(a) $\log_7 7 = 1$ **(b)** $\log_{\sqrt{2}} \sqrt{2} = 1$

(c) $\log_9 1 = 0$ **(d)** $\log_{.2} 1 = 0$

Now Try Exercise 1.

OBJECTIVE 4 Define and graph logarithmic functions. Now we define the logarithmic function with base a.

Logarithmic Function

If a and x are positive numbers, with $a \neq 1$, then

$$G(x) = \log_a x$$

defines the **logarithmic function with base a.**

The graph of $x = 2^y$ in Figure 12, which is equivalent to $y = g(x) = \log_2 x$, is typical of graphs of logarithmic functions with base $a > 1$. To graph a logarithmic function, it is helpful to write it in exponential form first. Then plot selected ordered pairs to determine the graph.

EXAMPLE 3   Graphing a Logarithmic Function

Graph $f(x) = \log_{1/2} x$.

By writing $y = f(x) = \log_{1/2} x$ in exponential form as $x = \left(\frac{1}{2}\right)^y$, we can identify ordered pairs that satisfy the equation. Here it is easier to choose values for y and find the corresponding values of x. See the table of ordered pairs.

x	y
$\frac{1}{4}$	2
$\frac{1}{2}$	1
1	0
2	−1
4	−2

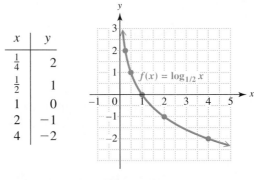

FIGURE 13

Plotting these points (be careful to get the x- and y-values in the right order) and connecting them with a smooth curve gives the graph in Figure 13. This graph is typical of logarithmic functions with $0 < a < 1$.

Now Try Exercise 43.

Based on the graphs of the functions defined by $y = \log_2 x$ in Figure 12 and $y = \log_{1/2} x$ in Figure 13, we make the following generalizations about the graphs of logarithmic functions of the form $G(x) = \log_a x$.

Graph of $G(x) = \log_a x$

1. The graph contains the point $(1, 0)$.

2. When $a > 1$, the graph will *rise* from left to right, from the fourth quadrant to the first. When $0 < a < 1$, the graph will *fall* from left to right, from the first quadrant to the fourth.

3. The graph will approach the y-axis, but never touch it. (The y-axis is an asymptote.)

4. The domain is $(0, \infty)$, and the range is $(-\infty, \infty)$.

Compare these generalizations to the similar ones for exponential functions found in Section 12.2.

OBJECTIVE 5 Use logarithmic functions in applications involving growth or decay.
Logarithmic functions, like exponential functions, can be applied to growth or decay of real-world phenomena.

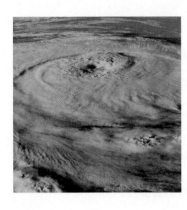

EXAMPLE 4 Solving an Application of a Logarithmic Function

The function defined by

$$f(x) = 27 + 1.105 \log_{10}(x + 1)$$

approximates the barometric pressure in inches of mercury at a distance of x mi from the eye of a typical hurricane. (*Source:* Miller, A. and R. Anthes, *Meteorology,* Fifth Edition, Charles E. Merrill Publishing Company, 1985.)

(a) Approximate the pressure 9 mi from the eye of the hurricane.
   Let $x = 9$, and find $f(9)$.

$$
\begin{aligned}
f(9) &= 27 + 1.105 \log_{10}(9 + 1) &&\text{Let } x = 9. \\
     &= 27 + 1.105 \log_{10} 10 &&\text{Add inside parentheses.} \\
     &= 27 + 1.105(1) &&\log_{10} 10 = 1 \\
     &= 28.105
\end{aligned}
$$

The pressure 9 mi from the eye of the hurricane is 28.105 in.

(b) Approximate the pressure 99 mi from the eye of the hurricane.

$$
\begin{aligned}
f(99) &= 27 + 1.105 \log_{10}(99 + 1) &&\text{Let } x = 99. \\
      &= 27 + 1.105 \log_{10} 100 &&\text{Add inside parentheses.} \\
      &= 27 + 1.105(2) &&\log_{10} 100 = 2 \\
      &= 29.21
\end{aligned}
$$

The pressure 99 mi from the eye of the hurricane is 29.21 in.

Now Try Exercise 55.

| CONNECTIONS |

In the United States, the intensity of an earthquake is rated using the *Richter scale.* The Richter scale rating of an earthquake of intensity x is given by

$$R = \log_{10} \frac{x}{x_0},$$

where x_0 is the intensity of an earthquake of a certain (small) size. The graph here shows Richter scale ratings for major Southern California earthquakes since 1920. As the graph indicates, earthquakes "come in bunches," and the 1990s were an especially busy time.

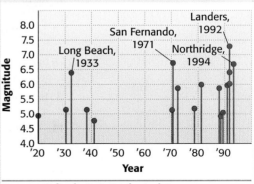

MAJOR SOUTHERN CALIFORNIA EARTHQUAKES

Earthquakes with magnitudes greater than 4.8

Source: Caltech; U.S. Geological Survey.

(continued)

For Discussion or Writing

Writing the given logarithmic equation in exponential form, we obtain

$$10^R = \frac{x}{x_0} \quad \text{or} \quad x = 10^R x_0.$$

1. The 1994 Northridge earthquake had a Richter scale rating of 6.7; the Landers earthquake had a rating of 7.3. How much more powerful was the Landers earthquake than the Northridge earthquake?

2. Compare the smallest rated earthquake in the figure (at 4.8) with the Landers quake. How much more powerful was the Landers quake?

 EXERCISES

For Extra Help

 Student's Solutions Manual

MyMathLab

 InterAct Math Tutorial Software

 AW Math Tutor Center

MathXL

 Digital Video Tutor CD 18/Videotape 18

1. By definition, $\log_a x$ is the exponent to which the base a must be raised in order to obtain x. Use this definition to match the logarithm in Column I with its value in Column II. (*Example:* $\log_3 9$ is equal to 2 because 2 is the exponent to which 3 must be raised in order to obtain 9.)

I	II
(a) $\log_4 16$	**A.** -2
(b) $\log_3 81$	**B.** -1
(c) $\log_3\left(\dfrac{1}{3}\right)$	**C.** 2
(d) $\log_{10} .01$	**D.** 0
(e) $\log_5 \sqrt{5}$	**E.** $\dfrac{1}{2}$
(f) $\log_{13} 1$	**F.** 4

2. Match the logarithmic equation in Column I with the corresponding exponential equation in Column II.

I	II
(a) $\log_{1/3} 3 = -1$	**A.** $8^{1/3} = \sqrt[3]{8}$
(b) $\log_5 1 = 0$	**B.** $\left(\dfrac{1}{3}\right)^{-1} = 3$
(c) $\log_2 \sqrt{2} = \dfrac{1}{2}$	**C.** $4^1 = 4$
(d) $\log_{10} 1000 = 3$	**D.** $2^{1/2} = \sqrt{2}$
(e) $\log_8 \sqrt[3]{8} = \dfrac{1}{3}$	**E.** $5^0 = 1$
(f) $\log_4 4 = 1$	**F.** $10^3 = 1000$

Write in logarithmic form. See the table in Objective 2.

3. $4^5 = 1024$ **4.** $3^6 = 729$ **5.** $\left(\dfrac{1}{2}\right)^{-3} = 8$ **6.** $\left(\dfrac{1}{6}\right)^{-3} = 216$

7. $10^{-3} = .001$ **8.** $36^{1/2} = 6$ **9.** $\sqrt[4]{625} = 5$ **10.** $\sqrt[3]{343} = 7$

Write in exponential form. See the table in Objective 2.

11. $\log_4 64 = 3$ **12.** $\log_2 512 = 9$ **13.** $\log_{10} \dfrac{1}{10,000} = -4$ **14.** $\log_{100} 100 = 1$

15. $\log_6 1 = 0$ **16.** $\log_\pi 1 = 0$ **17.** $\log_9 3 = \dfrac{1}{2}$ **18.** $\log_{64} 2 = \dfrac{1}{6}$

19. When a student asked his teacher to explain how to evaluate $\log_9 3$ without showing any work, his teacher told him, "Think radically." Explain what the teacher meant by this hint.

20. A student told her teacher, "I know that $\log_2 1$ is the exponent to which 2 must be raised in order to obtain 1, but I can't think of any such number." How would you explain to the student that the value of $\log_2 1$ is 0?

Solve each equation for x. See Examples 1 and 2.

21. $x = \log_{27} 3$ **22.** $x = \log_{125} 5$ **23.** $\log_x 9 = \dfrac{1}{2}$ **24.** $\log_x 5 = \dfrac{1}{2}$

25. $\log_x 125 = -3$ **26.** $\log_x 64 = -6$ **27.** $\log_{12} x = 0$ **28.** $\log_4 x = 0$

29. $\log_x x = 1$ **30.** $\log_x 1 = 0$ **31.** $\log_x \dfrac{1}{25} = -2$ **32.** $\log_x \dfrac{1}{10} = -1$

33. $\log_8 32 = x$ **34.** $\log_{81} 27 = x$ **35.** $\log_\pi \pi^4 = x$ **36.** $\log_{\sqrt{2}} \sqrt{2^9} = x$

37. $\log_6 \sqrt{216} = x$ **38.** $\log_4 \sqrt{64} = x$

39. $\log_4(2x + 4) = 3$ **40.** $\log_3(2x + 7) = 4$

If the point (p, q) is on the graph of $f(x) = a^x$ (for $a > 0$ and $a \neq 1$), then the point (q, p) is on the graph of $f^{-1}(x) = \log_a x$. Use this fact, and refer to the graphs required in Exercises 5–8 in Section 12.2 to graph each logarithmic function. See Example 3.

41. $y = \log_3 x$ **42.** $y = \log_5 x$ **43.** $y = \log_{1/3} x$ **44.** $y = \log_{1/5} x$

45. Explain why 1 is not allowed as a base for a logarithmic function.

46. Compare the summary of facts about the graph of $F(x) = a^x$ in Section 12.2 with the similar summary of facts about the graph of $G(x) = \log_a x$ in this section. Make a list of the facts that reinforce the concept that F and G are inverse functions.

47. The domain of $F(x) = a^x$ is $(-\infty, \infty)$, while the range is $(0, \infty)$. Therefore, since $G(x) = \log_a x$ defines the inverse of F, the domain of G is _____, while the range of G is _____.

48. The graphs of both $F(x) = 3^x$ and $G(x) = \log_3 x$ rise from left to right. Which one rises at a faster rate?

Use the graph at the right to predict the value of $f(t)$ for the given value of t.

49. $t = 0$

50. $t = 10$

51. $t = 60$

52. Show that the points determined in Exercises 49–51 lie on the graph of $f(t) = 8 \log_5(2t + 5)$.

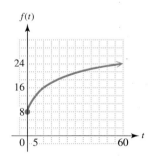

Solve each application of a logarithmic function. See Example 4.

53. According to selected figures from 1981 through 1995, the number of Superfund hazardous waste sites in the United States can be approximated by the function defined by

$$f(x) = 11.34 + 317.01 \log_2 x,$$

where $x = 1$ corresponds to 1981, $x = 2$ to 1982, and so on. (*Source:* U.S. Environmental Protection Agency.) Use the function to approximate the number of sites in each year.

(a) 1984 **(b)** 1988 **(c)** 1996

54. According to selected figures from 1980 through 1993, the number of trillion cubic feet of dry natural gas consumed worldwide can be approximated by the function defined by

$$f(x) = 51.47 + 6.044 \log_2 x,$$

where $x = 1$ corresponds to 1980, $x = 2$ to 1981, and so on. (*Source:* Energy Information Administration.) Use the function to approximate consumption in each year.

(a) 1980   **(b)** 1987   **(c)** 1995

55. Sales (in thousands of units) of a new product are approximated by the function defined by

$$S(t) = 100 + 30 \log_3(2t + 1),$$

where t is the number of years after the product is introduced.

(a) What were the sales after 1 yr?
(b) What were the sales after 13 yr?
(c) Graph $y = S(t)$.

56. A study showed that the number of mice in an old abandoned house was approximated by the function defined by

$$M(t) = 6 \log_4(2t + 4),$$

where t is measured in months and $t = 0$ corresponds to January 1998. Find the number of mice in the house in

(a) January 1998   **(b)** July 1998   **(c)** July 2000.
(d) Graph the function.

57. A supply of hybrid striped bass were introduced into a lake in January 1990. Biologists researching the bass population found that the number of bass in the lake was approximated by the function defined by

$$B(t) = 500 \log_3(2t + 3),$$

where $t = 0$ corresponds to January 1990, $t = 1$ to January 1991, $t = 2$ to January 1992, and so on. Use this function to find the bass population in

(a) January 1990   **(b)** January 1993   **(c)** January 2002.
(d) Graph the function for $0 \le t \le 12$.

58. Use the exponential key of your calculator to find approximations for the expression $\left(1 + \frac{1}{x}\right)^x$, using x values of 1, 10, 100, 1000, and 10,000. Explain what seems to be happening as x gets larger and larger.

As mentioned in Section 12.1, some graphing calculators have the capability of drawing the inverse of a function. For example, the two screens that follow show the graphs of $f(x) = 2^x$ and $g(x) = \log_2 x$. The graph of g was obtained by drawing the graph of f^{-1}, since $g(x) = f^{-1}(x)$. (Compare to Figure 12 in this section.)

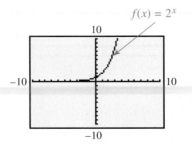
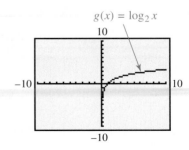

Use a graphing calculator with the capability of drawing the inverse of a function to draw the graph of each logarithmic function. Use the standard viewing window.

59. $g(x) = \log_3 x$ (Compare to Exercise 41.)

60. $g(x) = \log_5 x$ (Compare to Exercise 42.)

61. $g(x) = \log_{1/3} x$ (Compare to Exercise 43.)

62. $g(x) = \log_{1/5} x$ (Compare to Exercise 44.)

12.4   Properties of Logarithms

OBJECTIVES

1 Use the product rule for logarithms.

2 Use the quotient rule for logarithms.

3 Use the power rule for logarithms.

4 Use properties to write alternative forms of logarithmic expressions.

Logarithms have been used as an aid to numerical calculation for several hundred years. Today the widespread use of calculators has made the use of logarithms for calculation obsolete. However, logarithms are still very important in applications and in further work in mathematics.

OBJECTIVE 1 Use the product rule for logarithms.   One way in which logarithms simplify problems is by changing a problem of multiplication into one of addition. We know that $\log_2 4 = 2$, $\log_2 8 = 3$, and $\log_2 32 = 5$. Since $2 + 3 = 5$,

$$\log_2 32 = \log_2 4 + \log_2 8$$
$$\log_2 (4 \cdot 8) = \log_2 4 + \log_2 8.$$

This is true in general.

Product Rule for Logarithms

If x, y, and b are positive real numbers, where $b \neq 1$, then

$$\log_b xy = \log_b x + \log_b y.$$

That is, the logarithm of a product is the sum of the logarithms of the factors.

NOTE   The word statement of the product rule can be restated by replacing "logarithm" with "exponent." The rule then becomes the familiar rule for multiplying exponential expressions: The *exponent* of a product is the sum of the *exponents* of the factors.

To prove this rule, let $m = \log_b x$ and $n = \log_b y$, and recall that

$$\log_b x = m \quad \text{means} \quad b^m = x.$$
$$\log_b y = n \quad \text{means} \quad b^n = y.$$

Now consider the product xy.

$xy = b^m \cdot b^n$	Substitute.
$xy = b^{m+n}$	Product rule for exponents
$\log_b xy = m + n$	Convert to logarithmic form.
$\log_b xy = \log_b x + \log_b y$	Substitute.

The last statement is the result we wished to prove.

EXAMPLE 1 Using the Product Rule

Use the product rule to rewrite each expression. Assume $x > 0$.

(a) $\log_5(6 \cdot 9)$

By the product rule,

$$\log_5(6 \cdot 9) = \log_5 6 + \log_5 9.$$

(b) $\log_7 8 + \log_7 12 = \log_7(8 \cdot 12) = \log_7 96$

(c) $\log_3(3x) = \log_3 3 + \log_3 x$

$\qquad\qquad = 1 + \log_3 x \qquad\qquad \log_3 3 = 1$

(d) $\log_4 x^3 = \log_4(x \cdot x \cdot x) \qquad\qquad x^3 = x \cdot x \cdot x$

$\qquad\quad = \log_4 x + \log_4 x + \log_4 x \qquad$ Product rule

$\qquad\quad = 3 \log_4 x$

Now Try Exercises 7 and 21.

OBJECTIVE 2 Use the quotient rule for logarithms. The rule for division is similar to the rule for multiplication.

Quotient Rule for Logarithms

If x, y, and b are positive real numbers, where $b \neq 1$, then

$$\log_b \frac{x}{y} = \log_b x - \log_b y.$$

That is, the logarithm of a quotient is the difference between the logarithm of the numerator and the logarithm of the denominator.

The proof of this rule is very similar to the proof of the product rule.

EXAMPLE 2 Using the Quotient Rule

Use the quotient rule to rewrite each logarithm.

(a) $\log_4 \dfrac{7}{9} = \log_4 7 - \log_4 9$

(b) $\log_5 6 - \log_5 x = \log_5 \dfrac{6}{x}, \quad x > 0$

(c) $\log_3 \dfrac{27}{5} = \log_3 27 - \log_3 5$

$\qquad\qquad = 3 - \log_3 5 \qquad\qquad \log_3 27 = 3$

Now Try Exercises 9 and 23.

CAUTION Remember that there is no property of logarithms to rewrite the logarithm of a *sum* or *difference*. For example, we *cannot* write $\log_b(x + y)$ in terms of $\log_b x$ and $\log_b y$. Also,

$$\log_b \frac{x}{y} \neq \frac{\log_b x}{\log_b y}.$$

OBJECTIVE 3 Use the power rule for logarithms. An exponential expression such as 2^3 means $2 \cdot 2 \cdot 2$; the base is used as a factor 3 times. Thus, it seems reasonable that the product rule can be extended to rewrite the logarithm of a power as the product of the exponent and the logarithm of the base. For example, by the product rule for logarithms,

$$\log_5 2^3 = \log_5(2 \cdot 2 \cdot 2)$$
$$= \log_5 2 + \log_5 2 + \log_5 2$$
$$= 3 \log_5 2.$$

Also,

$$\log_2 7^4 = \log_2(7 \cdot 7 \cdot 7 \cdot 7)$$
$$= \log_2 7 + \log_2 7 + \log_2 7 + \log_2 7$$
$$= 4 \log_2 7.$$

Furthermore, we saw in Example 1(d) that $\log_4 x^3 = 3 \log_4 x$. These examples suggest the following rule.

Power Rule for Logarithms

If x and b are positive real numbers, where $b \neq 1$, and if r is any real number, then

$$\log_b x^r = r \log_b x.$$

That is, the logarithm of a number to a power equals the exponent times the logarithm of the number.

As further examples of this result,

$$\log_b m^5 = 5 \log_b m \qquad \text{and} \qquad \log_3 5^4 = 4 \log_3 5.$$

To prove the power rule, let

$$\log_b x = m.$$

$b^m = x$	Convert to exponential form.
$(b^m)^r = x^r$	Raise to the power r.
$b^{mr} = x^r$	Power rule for exponents
$\log_b x^r = mr$	Convert to logarithmic form.
$\log_b x^r = rm$	
$\log_b x^r = r \log_b x$	$m = \log_b x$

This is the statement to be proved.

As a special case of the power rule, let $r = \frac{1}{p}$, so

$$\log_b \sqrt[p]{x} = \log_b x^{1/p} = \frac{1}{p} \log_b x.$$

For example, using this result, with $x > 0$,

$$\log_b \sqrt[5]{x} = \log_b x^{1/5} = \frac{1}{5} \log_b x \qquad \text{and} \qquad \log_b \sqrt[3]{x^4} = \log_b x^{4/3} = \frac{4}{3} \log_b x.$$

Another special case is

$$\log_b \frac{1}{x} = \log_b x^{-1} = -\log_b x.$$

NOTE For a review of rational exponents, refer to Section 10.2.

EXAMPLE 3 Using the Power Rule

Use the power rule to rewrite each logarithm. Assume $b > 0$, $x > 0$, and $b \neq 1$.

(a) $\log_5 4^2 = 2 \log_5 4$

(b) $\log_b x^5 = 5 \log_b x$

(c) $\log_b \sqrt{7}$

Begin by rewriting the radical expression with a rational exponent.

$$\log_b \sqrt{7} = \log_b 7^{1/2} \qquad \sqrt{x} = x^{1/2}$$

$$= \frac{1}{2} \log_b 7 \qquad \text{Power rule}$$

(d) $\log_2 \sqrt[5]{x^2} = \log_2 x^{2/5} \qquad \sqrt[5]{x^2} = x^{2/5}$

$$= \frac{2}{5} \log_2 x \qquad \text{Power rule}$$

Now Try Exercise 11.

Two special properties involving both exponential and logarithmic expressions come directly from the fact that logarithmic and exponential functions are inverses of each other.

Special Properties

If $b > 0$ and $b \neq 1$, then

$$b^{\log_b x} = x, \quad x > 0 \qquad \text{and} \qquad \log_b b^x = x.$$

To prove the first statement, let

$$y = \log_b x.$$

$$b^y = x \qquad \text{Convert to exponential form.}$$

$$b^{\log_b x} = x \qquad \text{Replace } y \text{ with } \log_b x.$$

The proof of the second statement is similar.

EXAMPLE 4 Using the Special Properties

Find the value of each logarithmic expression.

(a) $\log_5 5^4 = 4$, since $\log_b b^x = x$.

(b) $\log_3 9 = \log_3 3^2 = 2$

(c) $4^{\log_4 10} = 10$

Now Try Exercises 3 and 5.

Here is a summary of the properties of logarithms.

Properties of Logarithms

If x, y, and b are positive real numbers, where $b \neq 1$, and r is any real number, then

Product Rule	$\log_b xy = \log_b x + \log_b y$
Quotient Rule	$\log_b \dfrac{x}{y} = \log_b x - \log_b y$
Power Rule	$\log_b x^r = r \log_b x$
Special Properties	$b^{\log_b x} = x$ and $\log_b b^x = x$.

OBJECTIVE 4 Use properties to write alternative forms of logarithmic expressions. Applying the properties of logarithms is important for solving equations with logarithms and in calculus.

EXAMPLE 5 Writing Logarithms in Alternative Forms

Use the properties of logarithms to rewrite each expression if possible. Assume all variables represent positive real numbers.

(a) $\log_4 4x^3 = \log_4 4 + \log_4 x^3$ Product rule

$\quad\quad\quad\quad\; = 1 + 3 \log_4 x$ $\log_4 4 = 1$; power rule

(b) $\log_7 \sqrt{\dfrac{m}{n}} = \log_7 \left(\dfrac{m}{n}\right)^{1/2}$

$\quad\quad\quad\quad\;\; = \dfrac{1}{2} \log_7 \dfrac{m}{n}$ Power rule

$\quad\quad\quad\quad\;\; = \dfrac{1}{2} (\log_7 m - \log_7 n)$ Quotient rule

(c) $\log_5 \dfrac{a^2}{bc} = \log_5 a^2 - \log_5 bc$ Quotient rule

$\quad\quad\quad\quad = 2 \log_5 a - \log_5 bc$ Power rule

$\quad\quad\quad\quad = 2 \log_5 a - (\log_5 b + \log_5 c)$ Product rule

$\quad\quad\quad\quad = 2 \log_5 a - \log_5 b - \log_5 c$

Notice the careful use of parentheses in the third step. Since we are subtracting the logarithm of a product and rewriting it as a sum of two terms, we must place parentheses around the sum.

(d) $4 \log_b m - \log_b n = \log_b m^4 - \log_b n$ Power rule

$$= \log_b \frac{m^4}{n} \qquad \text{Quotient rule}$$

(e) $\log_b(x + 1) + \log_b(2x - 1) - \dfrac{2}{3} \log_b x$

$$= \log_b(x + 1) + \log_b(2x - 1) - \log_b x^{2/3} \qquad \text{Power rule}$$

$$= \log_b \frac{(x + 1)(2x - 1)}{x^{2/3}} \qquad \text{Product and quotient rules}$$

$$= \log_b \frac{2x^2 + x - 1}{x^{2/3}}$$

(f) $\log_8(2p + 3r)$ cannot be rewritten using the properties of logarithms.

> **Now Try Exercises 13, 15, 27, and 31.**

In the next example, we use numerical values for $\log_2 5$ and $\log_2 3$. While we use the equals sign to give these values, they are actually just approximations since most logarithms of this type are irrational numbers. We use $=$ with the understanding that the values are correct to four decimal places.

EXAMPLE 6 **Using the Properties of Logarithms with Numerical Values**

Given that $\log_2 5 = 2.3219$ and $\log_2 3 = 1.5850$, evaluate the following.

(a) $\log_2 15 = \log_2(3 \cdot 5)$

$$= \log_2 3 + \log_2 5 \qquad \text{Product rule}$$

$$= 1.5850 + 2.3219$$

$$= 3.9069$$

(b) $\log_2 .6 = \log_2 \dfrac{3}{5}$ $.6 = \frac{6}{10} = \frac{3}{5}$

$$= \log_2 3 - \log_2 5 \qquad \text{Quotient rule}$$

$$= 1.5850 - 2.3219$$

$$= -.7369$$

(c) $\log_2 27 = \log_2 3^3$

$$= 3 \log_2 3 \qquad \text{Power rule}$$

$$= 3(1.5850)$$

$$= 4.7550$$

> **Now Try Exercises 33, 35, and 43.**

▦ **EXAMPLE 7** Deciding Whether Statements about Logarithms Are True

Decide whether each statement is *true* or *false*.

(a) $\log_2 8 - \log_2 4 = \log_2 4$

Evaluate both sides.

Left side: $\log_2 8 - \log_2 4 = \log_2 2^3 - \log_2 2^2 = 3 - 2 = 1$

Right side: $\log_2 4 = \log_2 2^2 = 2$

The statement is false because $1 \neq 2$.

(b) $\log_3(\log_2 8) = \dfrac{\log_7 49}{\log_8 64}$

Evaluate both sides.

Left side: $\log_3(\log_2 8) = \log_3 3 = 1$

Right side: $\dfrac{\log_7 49}{\log_8 64} = \dfrac{\log_7 7^2}{\log_8 8^2} = \dfrac{2}{2} = 1$

The statement is true because $1 = 1$.

Now Try Exercises 45 and 51.

Napier's Rods

| **CONNECTIONS** |

Long before the days of calculators and computers, the search for making calculations easier was an ongoing process. Machines built by Charles Babbage and Blaise Pascal, a system of "rods" used by John Napier, and slide rules were the forerunners of today's electronic marvels. The invention of logarithms by John Napier in the sixteenth century was a great breakthrough in the search for easier methods of calculation.

Since logarithms are exponents, their properties allowed users of tables of common logarithms to multiply by adding, divide by subtracting, raise to powers by multiplying, and take roots by dividing. Although logarithms are no longer used for computations, they play an important part in higher mathematics.

For Discussion or Writing

1. To multiply 458.3 by 294.6 using logarithms, we add $\log_{10} 458.3$ and $\log_{10} 294.6$, then find 10 to the sum. Perform this multiplication using the ⬭log x⬭ key* and the ⬭10ˣ⬭ key on your calculator. Check your answer by multiplying directly with your calculator.

2. Try division, raising to a power, and taking a root by this method.

---

*In this text, the notation log x is used to mean $\log_{10} x$. This is also the meaning of the log key on calculators.

12.4 EXERCISES

Use the indicated rule of logarithms to complete each equation in Exercises 1–5.

1. $\log_{10}(3 \cdot 4) =$ _____ (product rule)

2. $\log_{10} \dfrac{3}{4} =$ _____ (quotient rule)

3. $3^{\log_3 4} =$ _____ (special property)

4. $\log_{10} 3^4 =$ _____ (power rule)

5. $\log_3 3^4 =$ _____ (special property)

6. Evaluate $\log_2(8 + 8)$. Then evaluate $\log_2 8 + \log_2 8$. Are the results the same? How could you change the operation in the first expression to make the two expressions equal?

Use the properties of logarithms to express each logarithm as a sum or difference of logarithms, or as a single number if possible. Assume all variables represent positive real numbers. See Examples 1–5.

7. $\log_7(4 \cdot 5)$

8. $\log_8(9 \cdot 11)$

9. $\log_5 \dfrac{8}{3}$

10. $\log_3 \dfrac{7}{5}$

11. $\log_4 6^2$

12. $\log_5 7^4$

13. $\log_3 \dfrac{\sqrt[3]{4}}{x^2 y}$

14. $\log_7 \dfrac{\sqrt[3]{13}}{pq^2}$

15. $\log_3 \sqrt{\dfrac{xy}{5}}$

16. $\log_6 \sqrt{\dfrac{pq}{7}}$

17. $\log_2 \dfrac{\sqrt[3]{x} \cdot \sqrt[5]{y}}{r^2}$

18. $\log_4 \dfrac{\sqrt[4]{z} \cdot \sqrt[5]{w}}{s^2}$

19. A student erroneously wrote $\log_a(x + y) = \log_a x + \log_a y$. When his teacher explained that this was indeed wrong, the student claimed that he had used the distributive property. Write a few sentences explaining why the distributive property does not apply in this case.

20. Write a few sentences explaining how the rules for multiplying and dividing powers of the same base are similar to the rules for finding logarithms of products and quotients.

Use the properties of logarithms to write each expression as a single logarithm. Assume all variables are defined in such a way that the variable expressions are positive, and bases are positive numbers not equal to 1. See Examples 1–5.

21. $\log_b x + \log_b y$

22. $\log_b 2 + \log_b z$

23. $\log_a m - \log_a n$

24. $\log_b x - \log_b y$

25. $(\log_a r - \log_a s) + 3 \log_a t$

26. $(\log_a p - \log_a q) + 2 \log_a r$

27. $3 \log_a 5 - 4 \log_a 3$

28. $3 \log_a 5 + \dfrac{1}{2} \log_a 9$

29. $\log_{10}(x + 3) + \log_{10}(x - 3)$

30. $\log_{10}(y + 4) + \log_{10}(y - 4)$

31. $3 \log_p x + \dfrac{1}{2} \log_p y - \dfrac{3}{2} \log_p z - 3 \log_p a$

32. $\dfrac{1}{3} \log_b x + \dfrac{2}{3} \log_b y - \dfrac{3}{4} \log_b s - \dfrac{2}{3} \log_b t$

To four decimal places, the values of $\log_{10} 2$ *and* $\log_{10} 9$ *are*

$$\log_{10} 2 = .3010 \qquad \log_{10} 9 = .9542.$$

Evaluate each logarithm by applying the appropriate rule or rules from this section. DO NOT USE A CALCULATOR. See Example 6.

33. $\log_{10} 18$ **34.** $\log_{10} \dfrac{9}{2}$ **35.** $\log_{10} \dfrac{2}{9}$ **36.** $\log_{10} 4$

37. $\log_{10} 36$ **38.** $\log_{10} 162$ **39.** $\log_{10} 3$ **40.** $\log_{10} \sqrt[5]{2}$

41. $\log_{10} \sqrt[4]{9}$ **42.** $\log_{10} \dfrac{1}{9}$ **43.** $\log_{10} 9^5$ **44.** $\log_{10} 2^{19}$

Decide whether each statement is true or false. See Example 7.

45. $\log_2(8 + 32) = \log_2 8 + \log_2 32$ **46.** $\log_2(64 - 16) = \log_2 64 - \log_2 16$

47. $\log_3 7 + \log_3 7^{-1} = 0$ **48.** $\log_9 14 - \log_{14} 9 = 0$

49. $\log_6 60 - \log_6 10 = 1$ **50.** $\log_3 8 + \log_3 \dfrac{1}{8} = 0$

51. $\dfrac{\log_{10} 7}{\log_{10} 14} = \dfrac{1}{2}$ **52.** $\dfrac{\log_{10} 10}{\log_{10} 100} = \dfrac{1}{10}$

53. Refer to the Note following the word statement of the product rule for logarithms in this section. Now, state the quotient rule in words, replacing "logarithm" with "exponent."

54. Explain why the statement for the power rule for logarithms requires that x be a positive real number.

55. Refer to Example 7(a). Change the left side of the equation using the quotient rule so that the statement becomes true, and simplify.

56. What is wrong with the following "proof" that $\log_2 16$ does not exist? Explain.

$$\log_2 16 = \log_2(-4)(-4)$$
$$= \log_2(-4) + \log_2(-4)$$

Since the logarithm of a negative number is not defined, the final step cannot be evaluated, and so $\log_2 16$ does not exist.

RELATING CONCEPTS (EXERCISES 57–62)

For Individual or Group Work

Work Exercises 57–62 in order.

57. Evaluate $\log_3 81$.

58. Write the *meaning* of the expression $\log_3 81$.

59. Evaluate $3^{\log_3 81}$.

60. Write the *meaning* of the expression $\log_2 19$.

61. Evaluate $2^{\log_2 19}$.

62. Keeping in mind that a logarithm is an exponent and using the results from Exercises 57–61, what is the simplest form of the expression $k^{\log_k m}$?

12.5 Common and Natural Logarithms

As mentioned earlier, logarithms are important in many applications of mathematics to everyday problems, particularly in biology, engineering, economics, and social science. In this section we find numerical approximations for logarithms. Traditionally, base 10 logarithms were used most often because our number system is base 10. Logarithms to base 10 are called **common logarithms,** and $\log_{10} x$ is abbreviated as simply $\log x$, where the base is understood to be 10.

OBJECTIVE 1 Evaluate common logarithms using a calculator. We use calculators to evaluate common logarithms. In the next example we give the results of evaluating some common logarithms using a calculator with a (LOG) key. (This may be a second function key on some calculators.) For simple scientific calculators, just enter the number, then press the (LOG) key. For graphing calculators, these steps are reversed. We give all logarithms to four decimal places.

▌ EXAMPLE 1 Evaluating Common Logarithms

Evaluate each logarithm using a calculator.

(a) $\log 327.1 \approx 2.5147$ **(b)** $\log 437{,}000 \approx 5.6405$

(c) $\log .0615 \approx -1.2111$

Now Try Exercises 7, 9, and 11.

Figure 14 shows how a graphing calculator displays the common logarithms in Example 1. The calculator is set to give four decimal places.

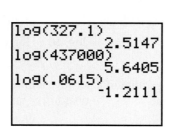

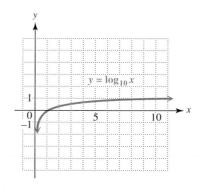

| FIGURE 14 | FIGURE 15 |

Notice that $\log .0615 \approx -1.2111$, a negative result. The common logarithm of a number between 0 and 1 is always negative because the logarithm is the exponent on 10 that produces the number. For example,

$$10^{-1.2111} \approx .0615.$$

If the exponent (the logarithm) were positive, the result would be greater than 1 because $10^0 = 1$. See Figure 15.

OBJECTIVE 2 Use common logarithms in applications. In chemistry, pH is a measure of the acidity or alkalinity of a solution; pure water, for example, has pH 7. In general, acids have pH numbers less than 7, and alkaline solutions have pH values greater than 7. The **pH** of a solution is defined as

$$pH = -\log[H_3O^+],$$

where $[H_3O^+]$ is the hydronium ion concentration in moles per liter. It is customary to round pH values to the nearest tenth.

■ EXAMPLE 2 Using pH in an Application

Wetlands are classified as *bogs, fens, marshes,* and *swamps.* These classifications are based on pH values. A pH value between 6.0 and 7.5, such as that of Summerby Swamp in Michigan's Hiawatha National Forest, indicates that the wetland is a "rich fen." When the pH is between 4.0 and 6.0, the wetland is a "poor fen," and if the pH falls to 3.0 or less, it is a "bog." (*Source:* Mohlenbrock, R., "Summerby Swamp, Michigan," *Natural History,*
March 1994.) Suppose that the hydronium ion concentration of a sample of water from a wetland is 6.3×10^{-3}. How would this wetland be classified?

Use the definition of pH.

$$
\begin{aligned}
pH &= -\log(6.3 \times 10^{-3}) \\
&= -(\log 6.3 + \log 10^{-3}) \qquad \text{Product rule} \\
&= -[.7993 - 3(1)] \qquad \text{Use a calculator to find log 6.3.} \\
&= -.7993 + 3 \\
&\approx 2.2
\end{aligned}
$$

Since the pH is less than 3.0, the wetland is a bog.

Now Try Exercise 29.

■ EXAMPLE 3 Finding Hydronium Ion Concentration

Find the hydronium ion concentration of drinking water with pH 6.5.

$$
\begin{aligned}
pH &= -\log[H_3O^+] \\
6.5 &= -\log[H_3O^+] \qquad \text{Let pH = 6.5.} \\
\log[H_3O^+] &= -6.5 \qquad \text{Multiply by −1.}
\end{aligned}
$$

Solve for $[H_3O^+]$ by writing the equation in exponential form, remembering that the base is 10.

$$
\begin{aligned}
[H_3O^+] &= 10^{-6.5} \\
[H_3O^+] &\approx 3.2 \times 10^{-7} \qquad \text{Use a calculator.}
\end{aligned}
$$

Now Try Exercise 35.

OBJECTIVE 3 Evaluate natural logarithms using a calculator. The most important logarithms used in applications are **natural logarithms,** which have as base the number e. The number e is a fundamental number in our universe. For this reason e, like π, is called a *universal constant.* The letter e is used to honor Leonhard Euler,

Leonhard Euler (1707–1783)

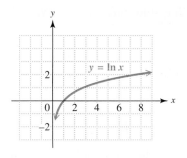

FIGURE 16

who published extensive results on the number in 1748. Since it is an irrational number, its decimal expansion never terminates and never repeats. The first few digits of the decimal value of e are 2.7182818285. A calculator key $\boxed{e^x}$ or the two keys $\boxed{\text{INV}}$ and $\boxed{\ln x}$ are used to approximate powers of e. For example, a calculator gives

$$e^2 \approx 7.389056099, \quad e^3 \approx 20.08553692, \quad \text{and} \quad e^{.6} \approx 1.8221188.$$

Logarithms to base e are called natural logarithms because they occur in biology and the social sciences in natural situations that involve growth or decay. The base e logarithm of x is written $\ln x$ (read "el en x"). A graph of $y = \ln x$, the equation that defines the natural logarithmic function, is given in Figure 16.

A calculator key labeled $\boxed{\ln x}$ is used to evaluate natural logarithms. If your calculator has an $\boxed{e^x}$ key, but not a key labeled $\boxed{\ln x}$, find natural logarithms by entering the number, pressing the $\boxed{\text{INV}}$ key, and then pressing the $\boxed{e^x}$ key. This works because $y = e^x$ defines the inverse function of $y = \ln x$ (or $y = \log_e x$).

```
ln(.5841)
             -.5377
ln(192.7)
            5.2611
ln(10.84)
            2.3832
```

FIGURE 17

EXAMPLE 4 Finding Natural Logarithms

Evaluate each logarithm with a calculator.

(a) $\ln .5841 \approx -.5377$

As with common logarithms, a number between 0 and 1 has a negative natural logarithm.

(b) $\ln 192.7 \approx 5.2611$ **(c)** $\ln 10.84 \approx 2.3832$

Figure 17 shows how a graphing calculator displays these natural logarithms to four decimal places.

Now Try Exercises 15, 17, and 19.

OBJECTIVE 4 Use natural logarithms in applications. Some applications involve functions that use natural logarithms, as seen in the next example.

EXAMPLE 5 Applying a Natural Logarithmic Function

The altitude in meters that corresponds to an atmospheric pressure of x millibars is given by the logarithmic function defined by

$$f(x) = 51,600 - 7457 \ln x.$$

(*Source:* Miller, A. and J. Thompson, *Elements of Meteorology,* Fourth Edition, Charles E. Merrill Publishing Company, 1993.) Use this function to find the altitude when atmospheric pressure is 400 millibars.

Let $x = 400$ and substitute in the expression for $f(x)$.

$$f(400) = 51,600 - 7457 \ln 400$$
$$\approx 6900$$

Atmospheric pressure is 400 millibars at approximately 6900 m.

Now Try Exercise 39.

NOTE In Example 5, the final answer was obtained using a calculator *without* rounding the intermediate values. In general, it is best to wait until the final step to round the answer; otherwise, a buildup of round-off error may cause the final answer to have an incorrect final decimal place digit.

OBJECTIVE 5 Use the change-of-base rule. We have used a calculator to approximate the values of common logarithms (base 10) and natural logarithms (base e). However, some applications involve logarithms to other bases. For example, for the years 1980–1996, the percentage of women who had a baby in the last year and returned to work is given by

$$f(x) = 38.83 + 4.208 \log_2 x,$$

for year x. (*Source:* U.S. Bureau of the Census.) To use this function, we need to find a base 2 logarithm. The following rule is used to convert logarithms from one base to another.

Change-of-Base Rule

If $a > 0$, $a \neq 1$, $b > 0$, $b \neq 1$, and $x > 0$, then

$$\log_a x = \frac{\log_b x}{\log_b a}.$$

NOTE Any positive number other than 1 can be used for base b in the change-of-base rule, but usually the only practical bases are e and 10 because calculators give logarithms only for these two bases.

To derive the change-of-base rule, let $\log_a x = m$.

$$\log_a x = m$$
$$a^m = x \qquad \text{Change to exponential form.}$$

Since logarithmic functions are one-to-one, if all variables are positive and if $x = y$, then $\log_b x = \log_b y$.

$$\log_b(a^m) = \log_b x$$
$$m \log_b a = \log_b x \qquad \text{Power rule}$$
$$(\log_a x)(\log_b a) = \log_b x \qquad \text{Substitute for } m.$$
$$\log_a x = \frac{\log_b x}{\log_b a} \qquad \text{Divide by } \log_b a.$$

The last step gives the change-of-base rule.

EXAMPLE 6 Using the Change-of-Base Rule

Find $\log_5 12$.

Use common logarithms and the change-of-base rule.

$$\log_5 12 = \frac{\log 12}{\log 5}$$

$$\approx 1.5440 \qquad \text{Use a calculator.}$$

Verify that the same value is found when using natural logarithms.

Now Try Exercise 47.

EXAMPLE 7 Using the Change-of-Base Rule in an Application

Use natural logarithms in the change-of-base rule and the function

$$f(x) = 38.83 + 4.208 \log_2 x$$

(given earlier) to find the percent of women who returned to work after having a baby in 1995. In the equation, $x = 0$ represents 1980.

Substitute $1995 - 1980 = 15$ for x in the equation.

$$f(15) = 38.83 + 4.208 \log_2 15$$

$$= 38.83 + 4.208 \left(\frac{\ln 15}{\ln 2} \right) \qquad \text{Change-of-base rule}$$

$$\approx 55.3\% \qquad \text{Use a calculator.}$$

This is very close to the actual value of 55%.

Now Try Exercise 59.

CONNECTIONS

As previously mentioned, the number $e \approx 2.718281828$ is a fundamental number in our universe. If there are intelligent beings elsewhere, they too will have to use e to do higher mathematics.

The properties of e are used extensively in calculus and in higher mathematics. In Section 12.6 we see how it applies to growth and decay in the physical world.

For Discussion or Writing

The value of e can be expressed as

$$e = 1 + \frac{1}{1} + \frac{1}{1 \cdot 2} + \frac{1}{1 \cdot 2 \cdot 3} + \frac{1}{1 \cdot 2 \cdot 3 \cdot 4} + \cdots.$$

Approximate e using two terms of this expression, then three terms, four terms, five terms, and six terms. How close is the approximation to the value of e given above with six terms? Does this infinite sum approach the value of e very quickly?

12.5 EXERCISES

Choose the correct response in Exercises 1–4.

1. What is the base in the expression $\log x$?

   A. e **B.** 1   **C.** 10   **D.** x

2. What is the base in the expression $\ln x$?

   A. e **B.** 1   **C.** 10   **D.** x

3. Since $10^0 = 1$ and $10^1 = 10$, between what two consecutive integers is the value of $\log 5.6$?

   A. 5 and 6   **B.** 10 and 11   **C.** 0 and 1   **D.** -1 and 0

4. Since $e^1 \approx 2.718$ and $e^2 \approx 7.389$, between what two consecutive integers is the value of $\ln 5.6$?

   A. 5 and 6   **B.** 2 and 3   **C.** 1 and 2   **D.** 0 and 1

5. Without using a calculator, give the value of $\log 10^{19.2}$.

6. Without using a calculator, give the value of $\ln e^{\sqrt{2}}$.

You will need a calculator for the remaining exercises in this set.

Find each logarithm. Give an approximation to four decimal places. See Examples 1 and 4.

7. $\log 43$

8. $\log 98$

9. $\log 328.4$

10. $\log 457.2$

11. $\log .0326$

12. $\log .1741$

13. $\log(4.76 \times 10^9)$

14. $\log(2.13 \times 10^4)$

15. $\ln 7.84$

16. $\ln 8.32$

17. $\ln .0556$

18. $\ln .0217$

19. $\ln 388.1$

20. $\ln 942.6$

21. $\ln(8.59 \times e^2)$

22. $\ln(7.46 \times e^3)$

23. $\ln 10$

24. $\log e$

25. Use your calculator to find approximations of the following logarithms:

   (a) $\log 356.8$ **(b)** $\log 35.68$ **(c)** $\log 3.568$.

   (d) Observe your answers and make a conjecture concerning the decimal values of the common logarithms of numbers greater than 1 that have the same digits.

26. Let k represent the number of letters in your last name.

   (a) Use your calculator to find $\log k$.

   (b) Raise 10 to the power indicated by the number you found in part (a). What is your result?

   (c) Use the concepts of Section 12.1 to explain why you obtained the answer you found in part (b). Would it matter what number you used for k to observe the same result?

27. Try to find $\log(-1)$ using a calculator. (If you have a graphing calculator, it should be in real number mode.) What happens? Explain.

Refer to Example 2. In Exercises 28–30, suppose that water from a wetland area is sampled and found to have the given hydronium ion concentration. Determine whether the wetland is a rich fen, a poor fen, or a bog.

28. 2.5×10^{-5} **29.** 2.5×10^{-2} **30.** 2.5×10^{-7}

Find the pH of the substance with the given hydronium ion concentration. See Example 2.

31. Ammonia, 2.5×10^{-12} **32.** Sodium bicarbonate, 4.0×10^{-9}

33. Grapes, 5.0×10^{-5} **34.** Tuna, 1.3×10^{-6}

Use the formula for pH to find the hydronium ion concentration of the substance with the given pH. See Example 3.

35. Human blood plasma, 7.4 **36.** Human gastric contents, 2.0

37. Spinach, 5.4 **38.** Bananas, 4.6

Solve each problem. See Example 5.

39. The number of years, $N(r)$, since two independently evolving languages split off from a common ancestral language is approximated by

$$N(r) = -5000 \ln r,$$

where r is the percent of words (in decimal form) from the ancestral language common to both languages now. Find the number of years since the split for each percent of common words.

(a) 85% (or .85)   **(b)** 35% (or .35)   **(c)** 10% (or .10)

40. The time t in years for an amount increasing at a rate of r (in decimal form) to double is given by

$$t(r) = \frac{\ln 2}{\ln(1 + r)}.$$

This is called *doubling time.* Find the doubling time to the nearest tenth for an investment at each interest rate.

(a) 2% (or .02)   **(b)** 5% (or .05)   **(c)** 8% (or .08)

41. The loudness of sounds is measured in a unit called a *decibel,* abbreviated dB. A very faint sound, called the *threshold sound,* is assigned an intensity I_0. If a particular sound has intensity I, then the decibel level of this louder sound is

$$D = 10 \log\left(\frac{I}{I_0}\right).$$

Find the average decibel level for each popular movie with the given intensity I. For comparison, a motorcycle or power saw has a decibel level of about 95 dB, and the sound of a jackhammer or helicopter is about 105 dB. (*Source: World Almanac and Book of Facts,* 2001; www.lhh.org/noise/)

(a) *Armageddon;* $5.012 \times 10^{10} I_0$

(b) *Godzilla;* $10^{10} I_0$

(c) *Saving Private Ryan;* $6{,}310{,}000{,}000\ I_0$

42. The concentration of a drug injected into the bloodstream decreases with time. The intervals of time T when the drug should be administered are given by

$$T = \frac{1}{k} \ln \frac{C_2}{C_1},$$

where k is a constant determined by the drug in use, C_2 is the concentration at which the drug is harmful, and C_1 is the concentration below which the drug is ineffective. (*Source:* Horelick, Brindell and Sinan Koont, "Applications of Calculus to Medicine: Prescribing Safe and Effective Dosage," *UMAP Module 202*, 1977.) Thus, if $T = 4$, the drug should be administered every 4 hr. For a certain drug, $k = \frac{1}{3}$, $C_2 = 5$, and $C_1 = 2$. How often should the drug be administered? (*Hint:* Round down.)

43. The growth of outpatient surgeries as a percent of total surgeries at hospitals is approximated by

$$f(x) = -1317 + 304 \ln x,$$

where x represents the number of years since 1900. (*Source:* American Hospital Association.)

(a) What does this function predict for the percent of outpatient surgeries in 1998?
(b) When did outpatient surgeries reach 50%? (*Hint:* Substitute for y, then write the equation in exponential form to solve it.)

44. In the central Sierra Nevada of California, the percent of moisture p that falls as snow rather than rain is approximated reasonably well by

$$f(x) = 86.3 \ln x - 680,$$

where x is the altitude in feet.

(a) What percent of the moisture at 5000 ft falls as snow?
(b) What percent at 7500 ft falls as snow?

45. The *cost-benefit equation*

$$T = -.642 - 189 \ln(1 - p)$$

describes the approximate tax T, in dollars per ton, that would result in a $p\%$ (in decimal form) reduction in carbon dioxide emissions.

(a) What tax will reduce emissions 25%?
(b) Explain why the equation is not valid for $p = 0$ or $p = 1$.

46. The age in years of a female blue whale is approximated by

$$t = -2.57 \ln\left(\frac{87 - L}{63}\right),$$

where L is its length in feet.

(a) How old is a female blue whale that measures 80 ft?
(b) The equation that defines t has domain $24 < L < 87$. Explain why.

Use the change-of-base rule (with either common or natural logarithms) to find each logarithm to four decimal places. See Example 6.

47. $\log_3 12$ **48.** $\log_4 18$ **49.** $\log_5 3$

50. $\log_7 4$ **51.** $\log_3 \sqrt{2}$ **52.** $\log_6 \sqrt[3]{5}$

53. $\log_\pi e$ **54.** $\log_\pi 10$ **55.** $\log_e 12$

56. Explain why the answer to Exercise 55 is the same one that you get when you use a calculator to approximate ln 12.

57. Let m be the number of letters in your first name, and let n be the number of letters in your last name.

   (a) In your own words, explain what $\log_m n$ means.
   (b) Use your calculator to find $\log_m n$.
   (c) Raise m to the power indicated by the number you found in part (b). What is your result?

58. The equation $5^x = 7$ cannot be solved using the methods described in Section 12.2. However, in solving this equation, we must find the exponent to which 5 must be raised in order to obtain 7: this is $\log_5 7$.

   (a) Use the change-of-base rule and your calculator to find $\log_5 7$.
   (b) Raise 5 to the number you found in part (a). What is your result?
   (c) Using as many decimal places as your calculator gives, write the solution set of $5^x = 7$. (Equations of this type will be studied in more detail in Section 12.6.)

Solve each application of a logarithmic function. See Example 7.

59. Refer to Exercise 53 in Section 12.3. Determine the number of waste sites in 1998.

60. Refer to Exercise 54 in Section 12.3. Determine the approximate consumption in 1998.

TECHNOLOGY INSIGHTS (EXERCISES 61–64)

Because graphing calculators are equipped with log *x and* ln *x keys, it is possible to graph the functions defined by* $f(x) = \log x$ *and* $g(x) = \ln x$ *directly, as shown in the figures that follow.*

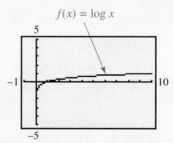

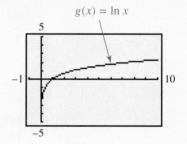

To graph functions defined by logarithms to bases other than 10 or e, however, we must use the change-of-base rule. For example, to graph $y = \log_2 x$, *we may enter* Y_1 *as* $\dfrac{\log X}{\log 2}$ *or* $\dfrac{\ln X}{\ln 2}$. *This is shown in the figure at the right. (Compare it to the figure in Exercises 59–62 of Section 12.3, where it was drawn using the fact that* $y = \log_2 x$ *is the inverse of* $y = 2^x$.)

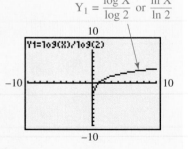

Use the change-of-base rule to graph each logarithmic function with a graphing calculator. Use a viewing window with Xmin $= -1$, Xmax $= 10$, Ymin $= -5$, *and* Ymax $= 5$.

   61. $g(x) = \log_3 x$ **62.** $g(x) = \log_5 x$ **63.** $g(x) = \log_{1/3} x$ **64.** $g(x) = \log_{1/5} x$

12.6 Exponential and Logarithmic Equations; Further Applications

As mentioned earlier, exponential and logarithmic functions are important in many applications of mathematics. Using these functions in applications requires solving exponential and logarithmic equations. Some simple equations were solved in Sections 12.2 and 12.3. More general methods for solving these equations depend on the following properties.

Properties for Solving Exponential and Logarithmic Equations

For all real numbers $b > 0$, $b \neq 1$, and any real numbers x and y:

1. If $x = y$, then $b^x = b^y$.
2. If $b^x = b^y$, then $x = y$.
3. If $x = y$, and $x > 0$, $y > 0$, then $\log_b x = \log_b y$.
4. If $x > 0$, $y > 0$, and $\log_b x = \log_b y$, then $x = y$.

We used Property 2 to solve exponential equations in Section 12.2.

OBJECTIVE 1 Solve equations involving variables in the exponents. The first two examples illustrate the method for solving exponential equations using Property 3.

EXAMPLE 1 Solving an Exponential Equation

Solve $3^x = 12$.

$$3^x = 12$$
$$\log 3^x = \log 12 \qquad \text{Property 3}$$
$$x \log 3 = \log 12 \qquad \text{Power rule}$$
$$x = \frac{\log 12}{\log 3} \qquad \text{Divide by log 3.}$$

This quotient is the exact solution. To get a decimal approximation for the solution, use a calculator.

$$x \approx 2.262$$

The solution set is $\{2.262\}$. Check that $3^{2.262} \approx 12$.

Now Try Exercise 5.

CAUTION Be careful: $\frac{\log 12}{\log 3}$ is *not* equal to log 4 because log 4 $\approx$ .6021, but $\frac{\log 12}{\log 3} \approx 2.262$.

When an exponential equation has e as the base, it is easiest to use base e logarithms.

EXAMPLE 2   Solving an Exponential Equation with Base e

Solve $e^{.003x} = 40$.

Take base e logarithms on both sides.

$$\ln e^{.003x} = \ln 40$$

$$.003x \ln e = \ln 40 \qquad \text{Power rule}$$

$$.003x = \ln 40 \qquad \ln e = \ln e^1 = 1$$

$$x = \frac{\ln 40}{.003} \qquad \text{Divide by } .003.$$

$$x \approx 1230 \qquad \text{Use a calculator.}$$

The solution set is $\{1230\}$. Check that $e^{.003(1230)} \approx 40$.

Now Try Exercise 15.

General Method for Solving an Exponential Equation

Take logarithms to the same base on both sides and then use the power rule of logarithms or the special property $\log_b b^x = x$. (See Examples 1 and 2.)

As a special case, if both sides can be written as exponentials with the same base, do so, and set the exponents equal. (See Section 12.2.)

OBJECTIVE 2   Solve equations involving logarithms.   The properties of logarithms from Section 12.4 are useful here, as is using the definition of a logarithm to change the equation to exponential form.

EXAMPLE 3   Solving a Logarithmic Equation

Solve $\log_2(x + 5)^3 = 4$. Give the exact solution.

$$(x + 5)^3 = 2^4 \qquad \text{Convert to exponential form.}$$

$$(x + 5)^3 = 16$$

$$x + 5 = \sqrt[3]{16} \qquad \text{Take the cube root on each side.}$$

$$x = -5 + \sqrt[3]{16} \qquad \text{Subtract 5.}$$

$$x = -5 + 2\sqrt[3]{2} \qquad \text{Simplify the radical.}$$

Verify that the solution satisfies the equation, so the solution set is $\left\{-5 + 2\sqrt[3]{2}\right\}$.

Now Try Exercise 29.

CAUTION   Recall that the domain of $y = \log_b x$ is $(0, \infty)$. For this reason, it is always necessary to check that the solution of an equation with logarithms yields only logarithms of positive numbers in the original equation.

▨ EXAMPLE 4 Solving a Logarithmic Equation

Solve $\log_2(x + 1) - \log_2 x = \log_2 7$.

$$\log_2(x + 1) - \log_2 x = \log_2 7$$

$$\log_2 \frac{x + 1}{x} = \log_2 7 \qquad \text{Quotient rule}$$

$$\frac{x + 1}{x} = 7 \qquad \text{Property 4}$$

$$x + 1 = 7x \qquad \text{Multiply by } x.$$

$$1 = 6x$$

$$\frac{1}{6} = x$$

Check this solution by substituting in the original equation. Here, both $x + 1$ and x must be positive. If $x = \frac{1}{6}$, this condition is satisfied, so the solution set is $\left\{\frac{1}{6}\right\}$. ▨

Now Try Exercise 35.

▨ EXAMPLE 5 Solving a Logarithmic Equation

Solve $\log x + \log(x - 21) = 2$.

Write the left side as a single logarithm, write in exponential form, and solve the equation.

$$\log x + \log(x - 21) = 2$$

$$\log x(x - 21) = 2 \qquad \text{Product rule}$$

$$x(x - 21) = 10^2 \qquad \begin{array}{l}\log x = \log_{10} x; \text{ write in} \\ \text{exponential form.}\end{array}$$

$$x^2 - 21x = 100$$

$$x^2 - 21x - 100 = 0 \qquad \text{Standard form}$$

$$(x - 25)(x + 4) = 0 \qquad \text{Factor.}$$

$$x - 25 = 0 \quad \text{or} \quad x + 4 = 0 \qquad \text{Zero-factor property}$$

$$x = 25 \quad \text{or} \qquad x = -4$$

The value -4 must be rejected as a solution since it leads to the logarithm of a negative number in the original equation:

$$\log(-4) + \log(-4 - 21) = 2. \qquad \text{The left side is undefined.}$$

The only solution, therefore, is 25, and the solution set is $\{25\}$. ▨

Now Try Exercise 39.

CAUTION Do not reject a potential solution just because it is nonpositive. Reject any value that *leads to* the logarithm of a nonpositive number.

In summary, we use the following steps to solve a logarithmic equation.

Solving a Logarithmic Equation

Step 1   **Transform the equation so that a single logarithm appears on one side.** Use the product rule or quotient rule of logarithms to do this.

Step 2   **(a)** **Use Property 4.** If $\log_b x = \log_b y$, then $x = y$. (See Example 4.)

(b) **Write the equation in exponential form.** If $\log_b x = k$, then $x = b^k$. (See Examples 3 and 5.)

OBJECTIVE 3 **Solve applications of compound interest.**   So far in this book, problems involving applications of interest have been limited to simple interest using the formula $I = prt$. In most cases, interest paid or charged is *compound interest* (interest paid on both principal and interest). The formula for compound interest is an important application of exponential functions.

Compound Interest Formula (for a Finite Number of Periods)

If a principal of P dollars is deposited at an annual rate of interest r compounded (paid) n times per year, the account will contain

$$A = P\left(1 + \frac{r}{n}\right)^{nt}$$

dollars after t years. (In this formula, r is expressed as a decimal.)

EXAMPLE 6   Solving a Compound Interest Problem for A

How much money will there be in an account at the end of 5 yr if $1000 is deposited at 6% compounded quarterly? (Assume no withdrawals are made.)

Because interest is compounded quarterly, $n = 4$. The other values given in the problem are $P = 1000$, $r = .06$ (because $6\% = .06$), and $t = 5$. Substitute into the compound interest formula to find the value of A.

$$A = 1000\left(1 + \frac{.06}{4}\right)^{4 \cdot 5}$$

$$A = 1000(1.015)^{20}$$

Now use the $\boxed{y^x}$ key on a calculator and round the answer to the nearest cent.

$$A = 1346.86$$

The account will contain $1346.86. (The actual amount of interest earned is $1346.86 - \$1000 = \$346.86$. Why?)

Now Try Exercise 45(a).

Interest can be compounded annually, semiannually, quarterly, daily, and so on. The number of compounding periods can get larger and larger. If the value of n is allowed to approach infinity, we have an example of *continuous compounding*. However, the compound interest formula above cannot be used for continuous compounding since there is no finite value for n. The formula for continuous compounding is an example of exponential growth involving the number e.

Continuous Compound Interest Formula

If a principal of P dollars is deposited at an annual rate of interest r compounded continuously for t years, the final amount on deposit is

$$A = Pe^{rt}.$$

▨ **EXAMPLE 7** Solving a Continuous Compound Interest Problem

In Example 6 we found that $1000 invested for 5 yr at 6% interest compounded quarterly would grow to $1346.86.

(a) How much would this same investment grow to if interest were compounded continuously?

Use the formula for continuous compounding with $P = 1000$, $r = .06$, and $t = 5$.

$$\begin{aligned}
A &= Pe^{rt} && \text{Formula} \\
&= 1000e^{.06(5)} && \text{Substitute.} \\
&= 1000e^{.30} \\
&= 1349.86 && \text{Use a calculator and round to the nearest cent.}
\end{aligned}$$

Continuous compounding would cause the investment to grow to $1349.86. Notice that this is $3.00 more than the amount in Example 6, when interest was compounded quarterly.

(b) How long would it take for the initial investment to double its original amount? (This is called the *doubling time*.)

We must find the value of t that will cause A to be $2(\$1000) = \$2000$.

$$\begin{aligned}
A &= Pe^{rt} \\
2000 &= 1000e^{.06t} && \text{Let } A = 2P = 2000. \\
2 &= e^{.06t} && \text{Divide by 1000.} \\
\ln 2 &= .06t && \text{Take natural logarithms; } \ln e^k = k. \\
t &= \frac{\ln 2}{.06} && \text{Divide by .06.} \\
t &\approx 11.55 && \text{Use a calculator.}
\end{aligned}$$

It would take about 11.55 yr for the original investment to double.

Now Try Exercise 47.

OBJECTIVE 4 Solve applications involving exponential growth and decay. One of the most common applications of exponential functions depends on the fact that in many situations involving growth or decay of a population, the amount or number of some quantity present at time t can be closely approximated by

$$y = y_0 e^{kt},$$

where y_0 is the amount or number present at time $t = 0$, k is a constant, and e is the base of natural logarithms.

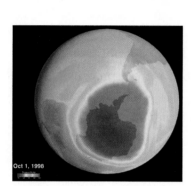

Oct 1, 1998

EXAMPLE 8   Applying an Exponential Function

The *greenhouse effect* refers to the phenomenon whereby emissions of gases such as carbon dioxide, methane, and chlorofluorocarbons (CFCs) have the potential to alter the climate of the earth and destroy the ozone layer. Concentrations of CFC-12, used in refrigeration technology, in parts per billion (ppb) can be modeled by the exponential function defined by

$$f(x) = .48e^{.04x},$$

where $x = 0$ represents 1990. Use this function to approximate the concentration in 1998.

Since $x = 0$ represents 1990, $x = 8$ represents 1998. Evaluate $f(8)$ using a calculator.

$$f(8) = .48e^{.04(8)} = .48e^{.32} \approx .66$$

In 1998, the concentration of CFC-12 was about .66 ppb.

Now Try Exercise 53.

You have probably heard of the carbon 14 dating process used to determine the age of fossils. The method used involves a base e exponential decay function.

EXAMPLE 9   Solving an Exponential Decay Problem

Carbon 14 is a radioactive form of carbon that is found in all living plants and animals. After a plant or animal dies, the radioactive carbon 14 disintegrates according to the function defined by

$$y = y_0 e^{-.000121t},$$

where t is time in years, y is the amount of the sample at time t, and y_0 is the initial amount present at $t = 0$.

(a) If an initial sample contains $y_0 = 10$ g of carbon 14, how many grams will be present after 3000 yr?

Let $y_0 = 10$ and $t = 3000$ in the formula, and use a calculator.

$$y = 10e^{-.000121(3000)} \approx 6.96 \text{ g}$$

(b) How long would it take for the initial sample to decay to half of its original amount? (This is called the *half-life.*)

Let $y = \frac{1}{2}(10) = 5$, and solve for t.

$5 = 10e^{-.000121t}$	Substitute.
$\dfrac{1}{2} = e^{-.000121t}$	Divide by 10.
$\ln \dfrac{1}{2} = -.000121t$	Take natural logarithms; $\ln e^k = k$.
$t = \dfrac{\ln \frac{1}{2}}{-.000121}$	Divide by $-.000121$.
$t \approx 5728$	Use a calculator.

The half-life is just over 5700 yr.

Now Try Exercise 59.

OBJECTIVE 5 Use a graphing calculator to solve exponential and logarithmic equations. Earlier we saw that the x-intercepts of the graph of a function f correspond to the real solutions of the equation $f(x) = 0$. This idea was applied to linear and quadratic equations and can be extended to exponential and logarithmic equations as well. In Example 1, we solved the equation $3^x = 12$ algebraically using rules for logarithms and found the solution set to be $\{2.262\}$. This can be supported graphically by showing that the x-intercept of the graph of the function defined by $y = 3^x - 12$ corresponds to this solution. See Figure 18.

In Example 5, we solved $\log x + \log(x - 21) = 2$ and found the solution set to be $\{25\}$. (We rejected the apparent solution -4 since it led to the logarithm of a negative number.) Figure 19 shows that the x-intercept of the graph of the function defined by $y = \log x + \log(x - 21) - 2$ supports this result.

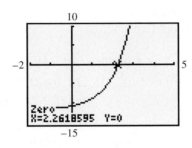

FIGURE 18   **FIGURE 19**

12.6 EXERCISES

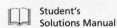

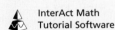

RELATING CONCEPTS (EXERCISES 1–4)

For Individual or Group Work

In Section 12.2 we solved an equation such as $5^x = 125$ by writing each side as a power of the same base, setting exponents equal, and then solving the resulting equation as follows.

$5^x = 125$	Original equation
$5^x = 5^3$	$125 = 5^3$
$x = 3$	Set exponents equal.

Solution set: $\{3\}$

The method described in this section can also be used to solve this equation. **Work Exercises 1–4 in order,** *to see how this is done.*

1. Take common logarithms on both sides, and write this equation.

2. Apply the power rule for logarithms on the left.

3. Write the equation so that x is alone on the left.

4. Use a calculator to find the decimal form of the solution. What is the solution set?

Many of the problems in the remaining exercises require a scientific calculator.

Solve each equation. Give solutions to three decimal places. See Example 1.

5. $7^x = 5$ **6.** $4^x = 3$ **7.** $9^{-x+2} = 13$

8. $6^{-t+1} = 22$ **9.** $3^{2x} = 14$ **10.** $5^{.3x} = 11$

11. $2^{x+3} = 5^x$ **12.** $6^{m+3} = 4^m$ **13.** $2^{x+3} = 3^{x-4}$

Solve each equation. Use natural logarithms. When appropriate, give solutions to three decimal places. See Example 2.

14. $e^{.006x} = 30$ **15.** $e^{.012x} = 23$ **16.** $e^{-.103x} = 7$

17. $e^{-.205x} = 9$ **18.** $\ln e^x = 4$ **19.** $\ln e^{3x} = 9$

20. $\ln e^{.04x} = \sqrt{3}$ **21.** $\ln e^{.45x} = \sqrt{7}$ **22.** $\ln e^{2x} = \pi$

23. Try solving one of the equations in Exercises 14–17 using common logarithms rather than natural logarithms. (You should get the same solution.) Explain why using natural logarithms is a better choice.

24. If you were asked to solve $10^{.0025x} = 75$, would natural or common logarithms be a better choice? Explain.

Solve each equation. Give the exact solution. See Example 3.

25. $\log_3(6x + 5) = 2$ **26.** $\log_5(12x - 8) = 3$ **27.** $\log_2(2x - 1) = 5$

28. $\log_6(4x + 2) = 2$ **29.** $\log_7(x + 1)^3 = 2$ **30.** $\log_4(x - 3)^3 = 4$

31. Suppose that in solving a logarithmic equation having the term $\log(x - 3)$ you obtain an apparent solution of 2. All algebraic work is correct. Explain why you must reject 2 as a solution of the equation.

32. Suppose that in solving a logarithmic equation having the term $\log(3 - x)$ you obtain an apparent solution of -4. All algebraic work is correct. Should you reject -4 as a solution of the equation? Explain why or why not.

Solve each equation. Give exact solutions. See Examples 4 and 5.

33. $\log(6x + 1) = \log 3$ **34.** $\log(7 - x) = \log 12$

35. $\log_5(3t + 2) - \log_5 t = \log_5 4$ **36.** $\log_2(x + 5) - \log_2(x - 1) = \log_2 3$

37. $\log 4x - \log(x - 3) = \log 2$ **38.** $\log(-x) + \log 3 = \log(2x - 15)$

39. $\log_2 x + \log_2(x - 7) = 3$ **40.** $\log(2x - 1) + \log 10x = \log 10$

41. $\log 5x - \log(2x - 1) = \log 4$ **42.** $\log_3 x + \log_3(2x + 5) = 1$

43. $\log_2 x + \log_2(x - 6) = 4$ **44.** $\log_2 x + \log_2(x + 4) = 5$

Solve each problem. See Examples 6 and 7.

45. **(a)** How much money will there be in an account at the end of 6 yr if $2000 is deposited at 4% compounded quarterly? (Assume no withdrawals are made.)
(b) To one decimal place, how long will it take for the account to grow to $3000?

46. **(a)** How much money will there be in an account at the end of 7 yr if $3000 is deposited at 3.5% compounded quarterly? (Assume no withdrawals are made.)
(b) To one decimal place, when will the account grow to $5000?

47. **(a)** What will be the amount A in an account with initial principal $4000 if interest is compounded continuously at an annual rate of 3.5% for 6 yr?
(b) How long will it take for the initial amount to double?

48. Refer to Exercise 46. Does the money grow to a larger value under those conditions, or when invested for 7 yr at 3% compounded continuously?

49. Find the amount of money in an account after 12 yr if $5000 is deposited at 7% annual interest compounded as follows.

(a) Annually   **(b)** Semiannually   **(c)** Quarterly
(d) Daily (Use $n = 365$.)   **(e)** Continuously

50. How much money will be in an account at the end of 8 yr if $4500 is deposited at 6% annual interest compounded as follows?

(a) Annually   **(b)** Semiannually   **(c)** Quarterly
(d) Daily (Use $n = 365$.)   **(e)** Continuously

51. How much money must be deposited today to amount to $1850 in 40 yr at 6.5% compounded continuously?

52. How much money must be deposited today to amount to $1000 in 10 yr at 5% compounded continuously?

Solve each problem. See Examples 8 and 9.

53. The total expenditures in millions of current dollars for pollution abatement and control during the period from 1985 through 1993 can be approximated by the function defined by

$$P(x) = 70{,}967e^{.0526x},$$

where $x = 0$ corresponds to 1985, $x = 1$ to 1986, and so on. Approximate the expenditures for each year. (*Source:* U.S. Bureau of Economic Analysis, *Survey of Current Business, May 1995.*)

(a) 1987   **(b)** 1990   **(c)** 1993
(d) What were the approximate expenditures for 1985?

54. The emission of the greenhouse gas nitrous oxide increased yearly during the first half of the 1990s. Based on figures during the period from 1990 through 1994, the emissions in thousands of metric tons can be modeled by the function defined by

$$N(x) = 446.5e^{.0118x},$$

where $x = 0$ corresponds to 1990, $x = 1$ to 1991, and so on. Approximate the emissions for each year. (*Source:* U.S. Energy Information Administration, *Emission of Greenhouse Gases in the United States, annual.*)

(a) 1991   **(b)** 1992   **(c)** 1994
(d) What were the approximate emissions in 1990?

55. Based on selected figures obtained during the 1980s and 1990s, consumer expenditures on all types of books in the United States can be modeled by the function defined by

$$B(x) = 8768e^{.072x},$$

where $x = 0$ represents 1980, $x = 1$ represents 1981, and so on, and $B(x)$ is in millions of dollars. Approximate consumer expenditures for 1998. (*Source:* Book Industry Study Group.)

56. Based on selected figures obtained during the 1970s, 1980s, and 1990s, the total number of bachelor's degrees earned in the United States can be modeled by the function defined by

$$D(x) = 815{,}427e^{.0137x},$$

where $x = 1$ corresponds to 1971, $x = 10$ corresponds to 1980, and so on. Approximate the number of bachelor's degrees earned in 1994. (*Source:* U.S. National Center for Education Statistics.)

57. Suppose that the amount, in grams, of plutonium 241 present in a given sample is determined by the function defined by

$$A(t) = 2.00e^{-.053t},$$

where t is measured in years. Find the amount present in the sample after the given number of years.

(a) 4   **(b)** 10   **(c)** 20
(d) What was the initial amount present?

58. Suppose that the amount, in grams, of radium 226 present in a given sample is determined by the function defined by

$$A(t) = 3.25e^{-.00043t},$$

where t is measured in years. Find the amount present in the sample after the given number of years.

(a) 20   **(b)** 100   **(c)** 500
(d) What was the initial amount present?

59. A sample of 400 g of lead 210 decays to polonium 210 according to the function defined by

$$A(t) = 400e^{-.032t},$$

where t is time in years.

(a) How much lead will be left in the sample after 25 yr?
(b) How long will it take the initial sample to decay to half of its original amount?

60. The concentration of a drug in a person's system decreases according to the function defined by

$$C(t) = 2e^{-.125t},$$

where $C(t)$ is in appropriate units, and t is in hours.

(a) How much of the drug will be in the system after 1 hr?
(b) Find the time that it will take for the concentration to be half of its original amount.

61. Refer to Exercise 53. Assuming that the function continued to apply past 1993, in what year could we have expected total expenditures to have been 133,500 million dollars? (*Source:* U.S. Bureau of Economic Analysis, *Survey of Current Business, May 1995.*)

62. Refer to Exercise 54. Assuming that the function continued to apply past 1994, in what year could we have expected nitrous oxide emissions to have been 485 thousand metric tons? (*Source:* U.S. Energy Information Administration, *Emission of Greenhouse Gases in the United States, annual.*)

63. The number of ants in an anthill grows according to the function defined by

$$f(t) = 300e^{.4t},$$

where t is time measured in days. Find the time it will take for the number of ants to double.

■■ TECHNOLOGY INSIGHTS (EXERCISES 64–67) ■■

64. The function defined by $P(x) = 70,967e^{.0526x}$, described in Exercise 53, is graphed in the screen at the right. Interpret the meanings of X and Y in the display at the bottom of the screen in the context of Exercise 53.

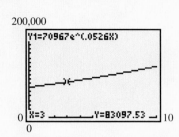

65. The function defined by $A(x) = 3.25e^{-.00043x}$, with $x = t$, described in Exercise 58, is graphed in the following figure. Interpret the meanings of X and Y in the display at the bottom of the screen in the context of Exercise 58.

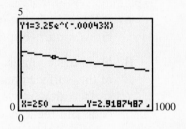

66. The screen shows a table of selected values for the function defined by $Y_1 = \left(1 + \frac{1}{X}\right)^X$.

   (a) Why is there an error message for X = 0?
   (b) What number does the function value seem to approach as X takes on larger and larger values?
   (c) Use a calculator to evaluate this function for X = 1,000,000. What value do you get? Now evaluate $e = e^1$. How close are these two values?
   (d) Make a conjecture: As the values of x approach infinity, the value of $\left(1 + \frac{1}{x}\right)^x$ approaches _____.

67. Here is another property of logarithms: For $b > 0$, $x > 0$, $b \neq 1$, $x \neq 1$,

$$\log_b x = \frac{1}{\log_x b}.$$

Now observe the following calculator screen.

```
log(e^(1))
           .4342944819
ln(10^(1))
           2.302585093

```

   (a) Without using a calculator, give a decimal representation for $\frac{1}{.4342944819}$. Then support your answer using the reciprocal key of your calculator.
   (b) Without using a calculator, give a decimal representation for $\frac{1}{2.302585093}$. Then support your answer using the reciprocal key of your calculator.

Chapter **12**   **Group Activity**

How Much Space Do We Need?

OBJECTIVE   Use natural logarithms and exponential equations to calculate how long it will take to fully populate Earth with people.

Applications of exponential growth and decay were introduced in this chapter. Using function notation, the formula for exponential population growth is

$$P(t) = P_0 e^{kt},$$

where $P(t)$ is population after t yr, P_0 is initial population, k is annual growth rate, and t is number of years elapsed.

A. If Earth's population will double in 30 yr at the current growth rate, what is this growth rate? (Express your answer as a percent.)

B. Earth has a total surface area of approximately 5.1×10^{14} m^2. Seventy percent of this surface area is rock, ice, sand, and open ocean. Another 8% of the total surface area, made up of tundra, lakes and streams, continental shelves, algae beds and reefs, and estuaries, is unfit for living space. The remaining area is suitable for growing food and for living space.

   1. Determine the surface area available for growing food.

   2. Determine the surface area available for living space. Notice that the surface area available for living space is also considered available for growing food.

C. Suppose that each person needs 100 m^2 of Earth's surface for living space. Use 5.5×10^9 for Earth's population.

   1. If none of the surface area available for living space is used for food, how long will it take for the livable surface of Earth to be covered with people? (Use the growth rate you found in part A and the surface area you found in part B2.)

   2. How much surface area would be left to grow food?

D. Measure a space that is 1 m^2 in area. Discuss with your partner whether or not you would want to be packed this closely together on Earth. Take into account that many people live in high-rise apartment buildings and how that translates into surface area used per person.

E. Now suppose that for each person 100 m^2 of Earth's surface is needed for living space and growing food.

1. Using the same population and growth rate as in part C, determine how long it will take to fill Earth with people. (Use the surface area from part B2.)

2. Does 100 m^2 per person for living space and growing food seem reasonable? Consider the following questions in your discussion.

- How much space do you think it takes to raise animals for food? To grow grains, nuts, fruits, and vegetables?
- Would food grow as well in desert areas, mountainous areas, or jungle areas?
- Would there be any space left for wild animals or natural plant life?
- Would there be any space left for shopping malls, movie theaters, concert halls, factories, office buildings, or parking lots?

3. Write a paragraph summarizing your results and your discussion.

CHAPTER 12 SUMMARY

KEY TERMS

12.1 one-to-one function inverse of a function	**12.2** exponential function asymptote exponential equation	**12.3** logarithm logarithmic equation logarithmic function with base a	**12.5** common logarithm natural logarithm

NEW SYMBOLS

$f^{-1}(x)$ the inverse of $f(x)$
$\log_a x$ the logarithm of x to the base a

$\log x$ common (base 10) logarithm of x

$\ln x$ natural (base e) logarithm of x

e a constant, approximately 2.7182818

TEST YOUR WORD POWER

See how well you have learned the vocabulary in this chapter. Answers, with examples, follow the Quick Review.

1. In a **one-to-one function**
   A. each x-value corresponds to only one y-value
   B. each x-value corresponds to one or more y-values
   C. each x-value is the same as each y-value
   D. each x-value corresponds to only one y-value and each y-value corresponds to only one x-value.

2. If f is a one-to-one function, then the **inverse** of f is
   A. the set of all solutions of f
   B. the set of all ordered pairs formed by interchanging the coordinates of the ordered pairs of f
   C. the set of all ordered pairs that are the opposite (negative) of the coordinates of the ordered pairs of f

   D. an equation involving an exponential expression.

3. An **exponential function** is a function defined by an expression of the form
   A. $f(x) = ax^2 + bx + c$ for real numbers a, b, c ($a \neq 0$)
   B. $f(x) = \log_a x$ for positive numbers a and x ($a \neq 1$)
   C. $f(x) = a^x$ for all real numbers x ($a > 0, a \neq 1$)
   D. $f(x) = \sqrt{x}$ for $x \geq 0$.

4. An **asymptote** is
   A. a line that a graph intersects just once
   B. a line that the graph of a function more and more closely approaches as the x-values increase or decrease

   C. the x-axis or y-axis
   D. a line about which a graph is symmetric.

5. A **logarithm** is
   A. an exponent
   B. a base
   C. an equation
   D. a polynomial.

6. A **logarithmic function** is a function that is defined by an expression of the form
   A. $f(x) = ax^2 + bx + c$ for real numbers a, b, c ($a \neq 0$)
   B. $f(x) = \log_a x$ for positive numbers a and x ($a \neq 1$)
   C. $f(x) = a^x$ for all real numbers x ($a > 0, a \neq 1$)
   D. $f(x) = \sqrt{x}$ for $x \geq 0$.

QUICK REVIEW

CONCEPTS	EXAMPLES

12.1 INVERSE FUNCTIONS

Horizontal Line Test

If any horizontal line intersects the graph of a function in no more than one point, then the function is one-to-one.

Find f^{-1} if $f(x) = 2x - 3$.
The graph of f is a straight line, so f is one-to-one by the horizontal line test.

CONCEPTS	EXAMPLES

Inverse Functions

For a one-to-one function f defined by an equation $y = f(x)$, the equation that defines the inverse function f^{-1} is found by interchanging x and y, solving for y, and replacing y with $f^{-1}(x)$.

To find $f^{-1}(x)$, interchange x and y in the equation $y = 2x - 3$.

$$x = 2y - 3$$

Solve for y to get $\qquad y = \dfrac{x + 3}{2}.$

Therefore, $\qquad f^{-1}(x) = \dfrac{x + 3}{2}.$

In general, the graph of f^{-1} is the mirror image of the graph of f with respect to the line $y = x$.

The graphs of a function f and its inverse f^{-1} are shown here.

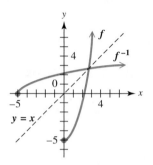

12.2   EXPONENTIAL FUNCTIONS

For $a > 0$, $a \neq 1$, $f(x) = a^x$ defines the exponential function with base a.

$F(x) = 3^x$ defines the exponential function with base 3.

Graph of $F(x) = a^x$
1. The graph contains the point $(0, 1)$.
2. When $a > 1$, the graph rises from left to right. When $0 < a < 1$, the graph falls from left to right.
3. The x-axis is an asymptote.
4. The domain is $(-\infty, \infty)$; the range is $(0, \infty)$.

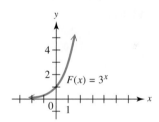

12.3   LOGARITHMIC FUNCTIONS

$y = \log_a x$ means $x = a^y.$

For $b > 0$, $b \neq 1$, $\log_b b = 1$ and $\log_b 1 = 0$.

For $a > 0$, $a \neq 1$, $x > 0$, $G(x) = \log_a x$ defines the logarithmic function with base a.

$y = \log_2 x$ means $x = 2^y$.

$$\log_3 3 = 1 \qquad \log_5 1 = 0$$

$G(x) = \log_3 x$ defines the logarithmic function with base 3.

Graph of $G(x) = \log_a x$
1. The graph contains the point $(1, 0)$.
2. When $a > 1$, the graph rises from left to right. When $0 < a < 1$, the graph falls from left to right.
3. The y-axis is an asymptote.
4. The domain is $(0, \infty)$; the range is $(-\infty, \infty)$.

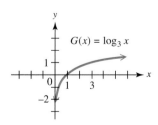

(continued)

CONCEPTS	EXAMPLES

12.4 PROPERTIES OF LOGARITHMS

Product Rule

$$\log_a xy = \log_a x + \log_a y$$

$$\log_2 3m = \log_2 3 + \log_2 m$$

Quotient Rule

$$\log_a \frac{x}{y} = \log_a x - \log_a y$$

$$\log_5 \frac{9}{4} = \log_5 9 - \log_5 4$$

Power Rule

$$\log_a x^r = r \log_a x$$

$$\log_{10} 2^3 = 3 \log_{10} 2$$

Special Properties

$$b^{\log_b x} = x \quad \text{and} \quad \log_b b^x = x$$

$$6^{\log_6 10} = 10 \qquad \log_3 3^4 = 4$$

12.5 COMMON AND NATURAL LOGARITHMS

Common logarithms (base 10) are used in applications such as pH, sound level, and intensity of an earthquake. Use the (LOG) key of a calculator to evaluate common logarithms.

Use the formula $pH = -\log[H_3O^+]$ to find the pH (to one decimal place) of grapes with hydronium ion concentration 5.0×10^{-5}.

$$
\begin{aligned}
pH &= -\log(5.0 \times 10^{-5}) & \text{Substitute.} \\
&= -(\log 5.0 + \log 10^{-5}) & \text{Property of logarithms} \\
&\approx 4.3 & \text{Evaluate.}
\end{aligned}
$$

Natural logarithms (base e) are often used in applications of growth and decay, such as time for money invested to double, decay of chemical compounds, and biological growth. Use the (ln x) key or both the (INV) and (e^x) keys to evaluate natural logarithms.

Use the formula for doubling time (in years) $t = \frac{\ln 2}{\ln(1 + r)}$ to find the doubling time to the nearest tenth at an interest rate of 4%.

$$
\begin{aligned}
t &= \frac{\ln 2}{\ln(1 + .04)} & \text{Substitute.} \\
&\approx 17.7 & \text{Evaluate.}
\end{aligned}
$$

The doubling time is about 17.7 yr.

Change-of-Base Rule
If $a > 0$, $a \neq 1$, $b > 0$, $b \neq 1$, $x > 0$, then

$$\log_a x = \frac{\log_b x}{\log_b a}.$$

$$\log_3 17 = \frac{\ln 17}{\ln 3} = \frac{\log 17}{\log 3} \approx 2.5789$$

12.6 EXPONENTIAL AND LOGARITHMIC EQUATIONS; FURTHER APPLICATIONS

To solve exponential equations, use these properties ($b > 0$, $b \neq 1$).

1. If $b^x = b^y$, then $x = y$.

Solve

$$
\begin{aligned}
2^{3x} &= 2^5. \\
3x &= 5 \\
x &= \frac{5}{3}
\end{aligned}
$$

The solution set is $\left\{\frac{5}{3}\right\}$.

CONCEPTS	EXAMPLES
2. If $x = y$, $x > 0$, $y > 0$, then $\log_b x = \log_b y$.	Solve $$5^m = 8.$$ $$\log 5^m = \log 8$$ $$m \log 5 = \log 8$$ $$m = \frac{\log 8}{\log 5} \approx 1.2920$$ The solution set is $\{1.2920\}$.
To solve logarithmic equations, use these properties, where $b > 0$, $b \neq 1$, $x > 0$, $y > 0$. First use the properties of Section 12.4, if necessary, to write the equation in the proper form.	
1. If $\log_b x = \log_b y$, then $x = y$.	Solve $$\log_3 2x = \log_3(x + 1).$$ $$2x = x + 1$$ $$x = 1$$ The solution set is $\{1\}$.
2. If $\log_b x = y$, then $b^y = x$.	Solve $$\log_2(3a - 1) = 4.$$ $$3a - 1 = 2^4$$ $$3a - 1 = 16$$ $$3a = 17$$ $$a = \frac{17}{3}$$ The solution set is $\left\{\frac{17}{3}\right\}$.

Answers to Test Your Word Power

1. D; *Example:* The function $f = \{(0, 2), (1, -1), (3, 5), (-2, 3)\}$ is one-to-one.　　**2.** B; *Example:* The inverse of the one-to-one function f defined in Answer 1 is $f^{-1} = \{(2, 0), (-1, 1), (5, 3), (3, -2)\}$.　　**3.** C; *Examples:* $f(x) = 4^x$, $g(x) = \left(\frac{1}{2}\right)^x$, $h(x) = 2^{-x+3}$

4. B; *Example:* The graph of $F(x) = 2^x$ has the x-axis ($y = 0$) as an asymptote.　　**5.** A; *Example:* $\log_a x$ is the exponent to which a must be raised to obtain x; $\log_3 9 = 2$ since $3^2 = 9$.　　**6.** B; *Examples:* $y = \log_3 x$, $y = \log_{1/3} x$

CHAPTER **12** REVIEW EXERCISES

[12.1] *Determine whether each graph is the graph of a one-to-one function.*

1.

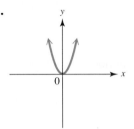

2.

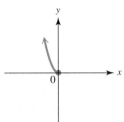

3. The table lists caffeine amounts in several popular 12-oz sodas. If the set of sodas is the domain and the set of caffeine amounts is the range of the function consisting of the six pairs listed, is it a one-to-one function? Why or why not?

Soda	Caffeine (mg)
Mountain Dew	55
Diet Coke	45
Dr. Pepper	41
Sunkist Orange Soda	41
Diet Pepsi-Cola	36
Coca-Cola Classic	34

Source: National Soft Drink Association.

Determine whether each function is one-to-one. If it is, find its inverse.

4. $f(x) = -3x + 7$ **5.** $f(x) = \sqrt[3]{6x - 4}$ **6.** $f(x) = -x^2 + 3$

Each function graphed is one-to-one. Graph its inverse.

7.

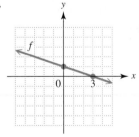

8.

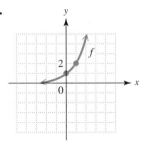

[12.2] *Graph each function.*

9. $f(x) = 3^x$ **10.** $f(x) = \left(\dfrac{1}{3}\right)^x$ **11.** $y = 3^{x+1}$ **12.** $y = 2^{2x+3}$

Solve each equation.

13. $4^{3x} = 8^{x+4}$ **14.** $\left(\dfrac{1}{27}\right)^{x-1} = 9^{2x}$

15. The gross wastes generated in plastics, in millions of tons, from 1960 through 1990 can be approximated by the exponential function defined by

$$W(x) = .67(1.123)^x,$$

where $x = 0$ corresponds to 1960, $x = 5$ to 1965, and so on. Use this function to approximate the plastic waste amounts for each year. (*Source:* U.S. Environmental Protection Agency, *Characterization of Municipal Solid Waste in the United States: 1994 Update,* 1995.)

   (a) 1965   **(b)** 1975   **(c)** 1990

[12.3] *Graph each function.*

16. $g(x) = \log_3 x$ (*Hint:* See Exercise 9.)   **17.** $g(x) = \log_{1/3} x$ (*Hint:* See Exercise 10.)

Solve each equation.

18. $\log_8 64 = x$ **19.** $\log_2 \sqrt{8} = x$ **20.** $\log_x\left(\dfrac{1}{49}\right) = -2$

21. $\log_4 x = \dfrac{3}{2}$ **22.** $\log_k 4 = 1$ **23.** $\log_b b^2 = 2$

24. In your own words, explain the meaning of $\log_b a$.

25. Based on the meaning of $\log_b a$, what is the simplest form of $b^{\log_b a}$?

26. A company has found that total sales, in thousands of dollars, are given by the function defined by

$$S(x) = 100 \log_2(x + 2),$$

where x is the number of weeks after a major advertising campaign was introduced.

   (a) What were the total sales 6 weeks after the campaign was introduced?

   (b) Graph the function.

[12.4] *Apply the properties of logarithms to express each logarithm as a sum or difference of logarithms. Assume all variables represent positive real numbers.*

27. $\log_2 3xy^2$ **28.** $\log_4 \dfrac{\sqrt{x} \cdot w^2}{z}$

Apply the properties of logarithms to write each expression as a single logarithm. Assume all variables represent positive real numbers, $b \neq 1$.

29. $\log_b 3 + \log_b x - 2 \log_b y$ **30.** $\log_3(x + 7) - \log_3(4x + 6)$

[12.5] *Evaluate each logarithm. Give approximations to four decimal places.*

31. $\log 28.9$ **32.** $\log .257$ **33.** $\ln 28.9$ **34.** $\ln .257$

Use the change-of-base rule (with either common or natural logarithms) to find each logarithm. Give approximations to four decimal places.

35. $\log_{16} 13$ **36.** $\log_4 12$

Use the formula $\text{pH} = -\log[H_3O^+]$ *to find the* pH *of each substance with the given hydronium ion concentration.*

37. Milk, 4.0×10^{-7}

38. Crackers, 3.8×10^{-9}

39. If orange juice has pH 4.6, what is its hydronium ion concentration?

40. Suppose the quantity, measured in grams, of a radioactive substance present at time t is given by

$$Q(t) = 500e^{-.05t},$$

where t is measured in days. Find the quantity present at the following times.

  (a) $t = 0$ **(b)** $t = 4$

41. Section 12.5, Exercise 40 introduced the *doubling function* defined by

$$t(r) = \frac{\ln 2}{\ln(1 + r)},$$

that gives the number of years required to double your money when it is invested at interest rate r (in decimal form) compounded annually. How long does it take to double your money at each rate? Round answers to the nearest year.

  (a) 4%   **(b)** 6%   **(c)** 10%   **(d)** 12%

  (e) Compare each answer in parts (a)–(d) with these numbers:

$$\frac{72}{4}, \frac{72}{6}, \frac{72}{10}, \frac{72}{12}.$$

What do you find?

42. The graph shows the percent change in commercial rents in California from 1992 through 1999. The percent change in rents is approximated by the logarithmic function defined by

$$g(x) = -650 + 143 \ln x,$$

where x represents the number of years since 1900.

  (a) Find $g(92)$ and $g(99)$.

  (b) Compare your results with the corresponding values from the graph.

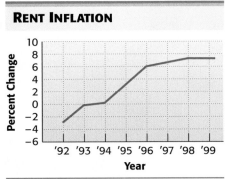

RENT INFLATION

Source: CB Commercial/Torto Wheaton Research.

[12.6] *Solve each equation. Give solutions to three decimal places.*

43. $3^x = 9.42$

44. $2^{x-1} = 15$

45. $e^{.06x} = 3$

Solve each equation. Give exact solutions.

46. $\log_3(9x + 8) = 2$

47. $\log_5(y + 6)^3 = 2$

48. $\log_3(p + 2) - \log_3 p = \log_3 2$

49. $\log(2x + 3) = 1 + \log x$

50. $\log_4 x + \log_4(8 - x) = 2$

51. $\log_2 x + \log_2(x + 15) = \log_2 16$

52. Explain the error in the following "solution" of the equation $\log x^2 = 2$.

$\log x^2 = 2$	Original equation
$2 \log x = 2$	Power rule for logarithms
$\log x = 1$	Divide both sides by 2.
$x = 10^1$	Write in exponential form.
$x = 10$	$10^1 = 10$

Solution set: $\{10\}$

Solve each problem. Use a calculator as necessary.

53. If \$20,000 is deposited at 7% annual interest compounded quarterly, how much will be in the account after 5 yr, assuming no withdrawals are made?

54. How much will \$10,000 compounded continuously at 6% annual interest amount to in 3 yr?

55. Which is a better plan?

> Plan A: Invest \$1000 at 4% compounded quarterly for 3 yr
>
> Plan B: Invest \$1000 at 3.9% compounded monthly for 3 yr

56. What is the half-life of the radioactive substance described in Exercise 40?

57. A machine purchased for business use *depreciates,* or loses value, over a period of years. The value of the machine at the end of its useful life is called its *scrap value.* By one method of depreciation (where it is assumed a constant percentage of the value depreciates annually), the scrap value, S, is given by

$$S = C(1 - r)^n,$$

where C is the original cost, n is the useful life in years, and r is the constant percent of depreciation.

(a) Find the scrap value of a machine costing \$30,000, having a useful life of 12 yr and a constant annual rate of depreciation of 15%.

(b) A machine has a "half-life" of 6 yr. Find the constant annual rate of depreciation.

58. Recall from Exercise 39 in Section 12.5 that the number of years, $N(r)$, since two independently evolving languages split off from a common ancestral language is approximated by

$$N(r) = -5000 \ln r,$$

where r is the percent of words from the ancestral language common to both languages now. Find r if the split occurred 2000 yr ago.

59. Which one of the following is *not* equal to the solution of $7^x = 23$?

A. $\dfrac{\log 23}{\log 7}$ **B.** $\dfrac{\ln 23}{\ln 7}$ **C.** $\log_7 23$ **D.** $\log_{23} 7$

60. Consider the logarithmic equation

$$\log(2x + 3) = \log x + 1.$$

(a) Solve the equation using properties of logarithms.

(b) If $Y_1 = \log(2X + 3)$ and $Y_2 = \log X + 1$, then the graph of $Y_1 - Y_2$ looks like that shown. Explain how the display at the bottom of the screen confirms the solution set found in part (a).

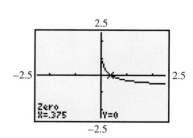

RELATING CONCEPTS (EXERCISES 61–72)

For Individual or Group Work

Work Exercises 61–72 in order, to see some of the relationships between exponential and logarithmic properties and functions.

61. Complete the table, and graph the function defined by $f(x) = 2^x$.

x	$f(x)$
-2	
-1	
0	
1	
2	
3	

62. Complete the table, and graph the function defined by $g(x) = \log_2 x$.

x	$g(x)$
$\frac{1}{4}$	
$\frac{1}{2}$	
1	
2	
4	
8	

63. What do you notice about the ordered pairs found in Exercises 61 and 62? What do we call the functions f and g in relationship to each other?

64. Fill in the blanks with the word *vertical* or *horizontal:* The graph of f in Exercise 61 has a _____ asymptote, while the graph of g in Exercise 62 has a _____ asymptote.

65. Using properties of exponents, $2^2 \cdot 2^3 = 2^{\underset{?}{}}$, because __?__ + __?__ = __?__ .

66. It is a fact that $32 = 4 \cdot 8$. Therefore, using properties of logarithms,

$$\log_2 32 = \log_2 \underline{\quad\quad} + \log_2 \underline{\quad\quad}.$$

67. Use the change-of-base rule to find an approximation for $\log_2 13$. Give as many digits as your calculator displays, and store this approximation in memory.

68. In your own words, explain what $\log_2 13$ means.

69. Simplify without using a calculator: $2^{\log_2 13}$.

70. Use the exponential key of your calculator to raise 2 to the power obtained in Exercise 67. What is the result? Why is this so?

71. Based on your result in Exercise 67, the point $(13, \underline{\quad\quad})$ lies on the graph of $g(x) = \log_2 x$.

72. Use the method of Section 12.2 to solve the equation $2^{x+1} = 8^{2x+3}$.

MIXED REVIEW EXERCISES

Evaluate.

73. $\log_2 128$

74. $5^{\log_5 36}$

75. $e^{\ln 4}$

76. $10^{\log e}$

77. $\log_3 3^{-5}$

78. $\ln e^{5.4}$

Solve.

79. $\log_3(x + 9) = 4$

80. $\ln e^x = 3$

81. $\log_x \dfrac{1}{81} = 2$

82. $27^x = 81$

83. $2^{2x-3} = 8$

84. $5^{x+2} = 25^{2x+1}$

85. $\log_3(x + 1) - \log_3 x = 2$ **86.** $\log(3x - 1) = \log 10$ **87.** $\ln(x^2 + 3x + 4) = \ln 2$

88. A small business estimates that the value of a copy machine is decreasing according to the function defined by

$$f(t) = 5000(2)^{-.15t},$$

where t is the number of years that have elapsed since the machine was purchased and $f(t)$ is in dollars.

(a) What was the original value of the machine? (*Hint:* Find $f(0)$.)

(b) What is the value of the machine 5 yr after purchase? Give your answer to the nearest dollar.

(c) What is the value of the machine 10 yr after purchase? Give your answer to the nearest dollar.

89. Find the useful life of the machine in Exercise 57 if the scrap value is $10,000, the cost is $30,000, and the depreciation rate is 15%.

90. Based on selected figures from 1970 through 1995, the fractional part of the generation of municipal solid waste recovered can be approximated by the function defined by

$$R(x) = .0597e^{.0553x},$$

where $x = 0$ corresponds to 1970, $x = 10$ to 1980, and so on. Based on this model, what *percent* of municipal solid waste was recovered in 1990? (*Source:* Franklin Associates, Ltd., Prairie Village, KS, *Characterization of Municipal Solid Waste in the United States: 1995.*)

One measure of the diversity of the species in an ecological community is the index of diversity, *the logarithmic expression*

$$-(p_1 \ln p_1 + p_2 \ln p_2 + \cdots + p_n \ln p_n),$$

where $p_1, p_2, \ldots, p_n$ are the proportions of a sample belonging to each of n species in the sample. (Source: Ludwig, John and James Reynolds, Statistical Ecology: A Primer on Methods and Computing, *New York, John Wiley and Sons, 1988.) Find the index of diversity to three decimal places if a sample of 100 from a community produces the following numbers.*

91. 90 of one species, 10 of another **92.** 60 of one species, 40 of another

CHAPTER **12** TEST

1. Decide whether each function is one-to-one.

(a) $f(x) = x^2 + 9$ (b)

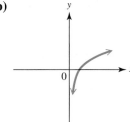

2. Find $f^{-1}(x)$ for the one-to-one function defined by $f(x) = \sqrt[3]{x+7}$.

3. Graph the inverse of f, given the graph of f at the right.

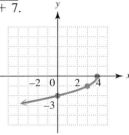

Graph each function.

4. $f(x) = 6^x$

5. $g(x) = \log_6 x$

 6. Explain how the graph of the function in Exercise 5 can be obtained from the graph of the function in Exercise 4.

Solve each equation. Give the exact solution.

7. $5^x = \dfrac{1}{625}$

8. $2^{3x-7} = 8^{2x+2}$

9. A recent report predicts that the U.S. Hispanic population will increase from 26.7 million in 1995 to 96.5 million in 2050. (*Source:* U.S. Bureau of the Census.) Assuming an exponential growth pattern, the population is approximated by

$$f(x) = 26.7e^{.023x},$$

where x represents the number of years since 1995. Use this function to estimate the Hispanic population in each year.

(a) 2000   **(b)** 2010

10. Write in logarithmic form: $4^{-2} = .0625$.

11. Write in exponential form: $\log_7 49 = 2$.

Solve each equation.

12. $\log_{1/2} x = -5$

13. $x = \log_9 3$

14. $\log_x 16 = 4$

15. Fill in the blanks with the correct responses: The value of $\log_2 32$ is _____. This means that if we raise _____ to the _____ power, the result is _____.

Use properties of logarithms to write each expression as a sum or difference of logarithms. Assume variables represent positive real numbers.

16. $\log_3 x^2 y$

17. $\log_5\left(\dfrac{\sqrt{x}}{yz}\right)$

Use properties of logarithms to write each expression as a single logarithm. Assume variables represent positive real numbers, $b \neq 1$.

18. $3\log_b s - \log_b t$

19. $\dfrac{1}{4}\log_b r + 2\log_b s - \dfrac{2}{3}\log_b t$

20. Use a calculator to approximate each logarithm to four decimal places.

(a) $\log 23.1$ **(b)** $\ln .82$

21. Use the change-of-base rule to express $\log_3 19$

   (a) in terms of common logarithms;   **(b)** in terms of natural logarithms;
   (c) correct to four decimal places.

22. Solve, giving the correct solution to four decimal places.
$$3^x = 78$$

23. Solve $\log_8(x + 5) + \log_8(x - 2) = 1$.

24. Suppose that \$10,000 is invested at 4.5% annual interest, compounded quarterly. How much will be in the account in 5 yr if no money is withdrawn?

25. Suppose that \$15,000 is invested at 5% annual interest, compounded continuously.

   (a) How much will be in the account in 5 yr if no money is withdrawn?
   (b) How long will it take for the initial principal to double?

CUMULATIVE REVIEW EXERCISES CHAPTERS **1–12**

Let $S = \left\{-\frac{9}{4}, -2, -\sqrt{2}, 0, .6, \sqrt{11}, \sqrt{-8}, 6, \frac{30}{3}\right\}$. *List the elements of S that are members of each set.*

1. Integers   **2.** Rational numbers

3. Irrational numbers   **4.** Real numbers

Simplify each expression.

5. $|-8| + 6 - |-2| - (-6 + 2)$ **6.** $-12 - |-3| - 7 - |-5|$

7. $2(-5) + (-8)(4) - (-3)$

Solve each equation or inequality.

8. $7 - (3 + 4a) + 2a = -5(a - 1) - 3$ **9.** $2m + 2 \le 5m - 1$

Perform the indicated operations.

10. $(2p + 3)(3p - 1)$ **11.** $(4k - 3)^2$

12. $(3m^3 + 2m^2 - 5m) - (8m^3 + 2m - 4)$

13. Divide $6t^4 + 17t^3 - 4t^2 + 9t + 4$ by $3t + 1$.

Factor.

14. $8x + x^3$ **15.** $24y^2 - 7y - 6$ **16.** $5z^3 - 19z^2 - 4z$

17. $16a^2 - 25b^4$ **18.** $8c^3 + d^3$ **19.** $16r^2 + 56rq + 49q^2$

Perform the indicated operations.

20. $\dfrac{(5p^3)^4(-3p^7)}{2p^2(4p^4)}$ **21.** $\dfrac{x^2 - 9}{x^2 + 7x + 12} \div \dfrac{x - 3}{x + 5}$

22. $\dfrac{2}{k + 3} - \dfrac{5}{k - 2}$ **23.** $\dfrac{3}{p^2 - 4p} - \dfrac{4}{p^2 + 2p}$

Graph.

24. $5x + 2y = 10$

25. $-4x + y \leq 5$

26. The graph indicates that timber harvests by Sierra Pacific Industries dropped from 17,716 acres in 1997 to 9733 acres in 1999.

    (a) Is this the graph of a function?

    (b) What is the slope of the line in the graph? Interpret the slope in the context of the timber harvests.

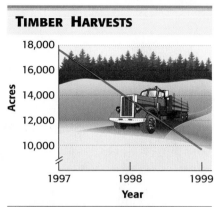

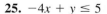

TIMBER HARVESTS

Source: Department of Forestry and Fire Protection.

27. Find an equation of the line through $(5, -1)$ and parallel to the line with equation $3x - 4y = 12$. Write the equation in slope-intercept form.

Solve each system.

28. $5x - 3y = 14$
$2x + 5y = 18$

29. $2x - 7y = 8$
$4x - 14y = 3$

30. $x + 2y + 3z = 11$
$3x - y + z = 8$
$2x + 2y - 3x = -12$

31. Candy worth \$1.00 per lb is to be mixed with 10 lb of candy worth \$1.96 per lb to get a mixture that will be sold for \$1.60 per lb. How many pounds of the \$1.00 candy should be used?

Number of Pounds	Price per Pound	Value
x	\$1.00	$1x$
	\$1.60	

Solve each equation or inequality.

32. $|2x - 5| = 9$

33. $|3p| - 4 = 12$

34. $|3k - 8| \leq 1$

35. $|4m + 2| > 10$

Simplify.

36. $\sqrt{288}$

37. $2\sqrt{32} - 5\sqrt{98}$

38. Solve $\sqrt{2x + 1} - \sqrt{x} = 1$.

39. Multiply $(5 + 4i)(5 - 4i)$.

Solve each equation or inequality.

40. $3x^2 - x - 1 = 0$

41. $k^2 + 2k - 8 > 0$

42. $x^4 - 5x^2 + 4 = 0$

43. Find two numbers whose sum is 300 and whose product is a maximum.

44. Graph $f(x) = \dfrac{1}{3}(x - 1)^2 + 2$.

45. Graph $f(x) = 2^x$.

46. Solve $5^{x+3} = \left(\dfrac{1}{25}\right)^{3x+2}$.

47. Graph $f(x) = \log_3 x$.

48. Given that $\log_2 9 = 3.1699$, what is the value of $\log_2 81$?

49. Rewrite the following using the product, quotient, and power properties of logarithms.

$$\log \frac{x^3\sqrt{y}}{z}$$

50. Let the number of bacteria present in a certain culture be given by

$$B(t) = 25{,}000e^{2t},$$

where t is time measured in hours, and $t = 0$ corresponds to noon. Find, to the nearest hundred, the number of bacteria present at

(a) noon;   **(b)** 1 P.M.;   **(c)** 2 P.M.;
(d) When will the population double?

Nonlinear Functions, Conic Sections, and Nonlinear Systems

13

13.1 Additional Graphs of Functions; Operations and Composition

13.2 The Circle and the Ellipse

13.3 The Hyperbola and Functions Defined by Radicals

13.4 Nonlinear Systems of Equations

13.5 Second-Degree Inequalities and Systems of Inequalities

When a plane intersects an infinite cone at different angles, it produces curves called *conic sections*. In Chapter 9 we studied one conic section, the *parabola*. In 1609, Johann Kepler (1571–1630) established the importance of another conic section, the *ellipse*, when he discovered that the orbits of the planets around the sun are elliptical, not circular. Exercises 51 and 52 of Section 13.2 involve the equations of the elliptical orbits formed by the planets Mars and Venus.

13.1 Additional Graphs of Functions; Operations and Composition

In earlier chapters we introduced the function defined by $f(x) = x^2$, sometimes called the **squaring function.** This is one of the most important elementary functions in algebra.

OBJECTIVE 1 Recognize the graphs of the elementary functions defined by $|x|$, $\frac{1}{x}$, and $\sqrt{x}$, and graph their translations. Another one of the elementary functions, defined by $f(x) = |x|$, is called the **absolute value function.** Its graph, along with a table of selected ordered pairs, is shown in Figure 1. Its domain is $(-\infty, \infty)$, and its range is $[0, \infty)$.

x	y
0	0
± 1	1
± 2	2
± 3	3

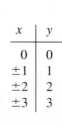

FIGURE 1

The **reciprocal function,** defined by $f(x) = \frac{1}{x}$, is a *rational function.* The graph of this function is shown in Figure 2, along with a table of selected ordered pairs. Notice that x can never equal 0 for this function, and as a result, as x gets closer and closer to 0, $\frac{1}{x}$ approaches either ∞ or $-\infty$. Also, $\frac{1}{x}$ can never equal 0, and as x approaches ∞ or $-\infty$, $\frac{1}{x}$ approaches 0. The axes are called **asymptotes** for the function. (Asymptotes are studied in more detail in college algebra courses.) For the reciprocal function, the domain and the range are both $(-\infty, 0) \cup (0, \infty)$.

x	y
$\frac{1}{3}$	3
$\frac{1}{2}$	2
1	1
2	$\frac{1}{2}$
3	$\frac{1}{3}$

x	y
$-\frac{1}{3}$	-3
$-\frac{1}{2}$	-2
-1	-1
-2	$-\frac{1}{2}$
-3	$-\frac{1}{3}$

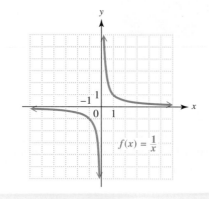

FIGURE 2

x	y
0	0
1	1
4	2

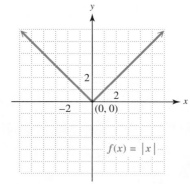

FIGURE 3

The **square root function,** defined by $f(x) = \sqrt{x}$, was introduced in Chapter 10. Its graph is shown in Figure 3. Notice that since we restrict function values to be real numbers, x cannot take on negative values. Thus, the domain of the square root function is $[0, \infty)$. Because the principal square root is always nonnegative, the range is also $[0, \infty)$. A table of values is shown along with the graph.

Just as the graph of $f(x) = x^2$ can be shifted, or translated, as we saw in Section 11.6, so can the graphs of these other elementary functions.

EXAMPLE 1 Applying a Horizontal Shift

Graph $f(x) = |x - 2|$.

The graph of $y = (x - 2)^2$ is obtained by shifting the graph of $y = x^2$ two units to the right. In a similar manner, the graph of $f(x) = |x - 2|$ is found by shifting the graph of $y = |x|$ two units to the right, as shown in Figure 4. The table of ordered pairs accompanying the graph supports this, as you can see by comparing it to the table with Figure 1. The domain of this function is $(-\infty, \infty)$, and its range is $[0, \infty)$.

x	y
0	2
1	1
2	0
3	1
4	2

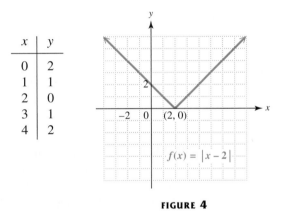

FIGURE 4

Now Try Exercise 9.

EXAMPLE 2 Applying a Vertical Shift

Graph $f(x) = \dfrac{1}{x} + 3$.

The graph of this function is found by shifting the graph of $y = \frac{1}{x}$ three units up. See Figure 5. The domain is $(-\infty, 0) \cup (0, \infty)$, and the range is $(-\infty, 3) \cup (3, \infty)$.

x	y
$\frac{1}{3}$	6
$\frac{1}{2}$	5
1	4
2	3.5

x	y
$-\frac{1}{3}$	0
$-\frac{1}{2}$	1
-1	2
-2	2.5

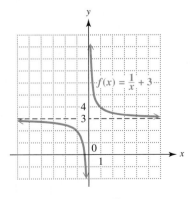

FIGURE 5

Now Try Exercise 11.

EXAMPLE 3 Applying Both Horizontal and Vertical Shifts

Graph $f(x) = \sqrt{x + 1} - 4$.

The graph of $y = (x + 1)^2 - 4$ is obtained by shifting the graph of $y = x^2$ one unit to the left and four units down. Following this pattern here, we shift the graph of $y = \sqrt{x}$ one unit to the left and four units down to get the graph of $f(x) = \sqrt{x + 1} - 4$. See Figure 6. The domain is $[-1, \infty)$, and the range is $[-4, \infty)$.

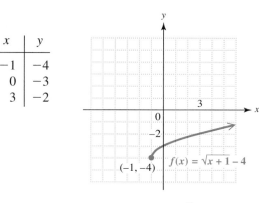

x	y
-1	-4
0	-3
3	-2

FIGURE 6

Now Try Exercise 17.

OBJECTIVE 2 Recognize and graph step functions. The **greatest integer function,** usually written $f(x) = [\![x]\!]$, is defined by saying that $[\![x]\!]$ denotes the largest integer that is less than or equal to x. For example, $[\![8]\!] = 8$, $[\![7.45]\!] = 7$, $[\![\pi]\!] = 3$, $[\![-1]\!] = -1$, $[\![-2.6]\!] = -3$, and so on.

EXAMPLE 4 Graphing the Greatest Integer Function

Graph $f(x) = [\![x]\!]$.

For $[\![x]\!]$, if $-1 \leq x < 0$, then $[\![x]\!] = -1$. If $0 \leq x < 1$, then $[\![x]\!] = 0$. If $1 \leq x < 2$, then $[\![x]\!] = 1$, and so on. Thus, the graph, as shown in Figure 7, consists of a series of horizontal line segments. In each one, the left endpoint is included and the right endpoint is excluded. These segments continue infinitely following this pattern to the left and right. Since x can take any real number value, the domain is $(-\infty, \infty)$. The range is the set of integers $\{\ldots, -4, -3, -2, -1, 0, 1, 2, 3, 4, \ldots\}$. The shape of the graph is the reason that this function is called a **step function.**

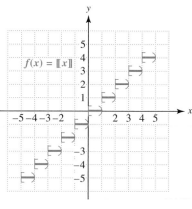

FIGURE 7

The graph of a step function also may be shifted. For example, the graph of $h(x) = [\![x - 2]\!]$ is the same as the graph of $f(x) = [\![x]\!]$ shifted two units to the right. Similarly, the graph of $g(x) = [\![x]\!] + 2$ is the graph of $f(x)$ shifted two units up.

Now Try Exercise 19.

■ **EXAMPLE 5   Applying a Greatest Integer Function**

An overnight delivery service charges $25 for a package weighing up to 2 lb. For each additional pound or fraction of a pound there is an additional charge of $3. Let $D(x)$ represent the cost to send a package weighing x lb. Graph $D(x)$ for x in the interval $(0, 6]$.

For x in the interval $(0, 2]$, $y = 25$. For x in $(2, 3]$, $y = 25 + 3 = 28$. For x in $(3, 4]$, $y = 28 + 3 = 31$, and so on. The graph, which is that of a step function, is shown in Figure 8.

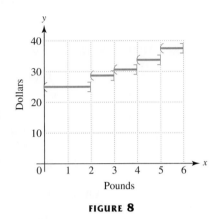

FIGURE 8

Now Try Exercise 21.

OBJECTIVE 3   Perform operations on functions.   The operations of addition, subtraction, multiplication, and division are also defined for functions. For example, businesses use the equation "profit equals revenue minus cost," written using function notation as

$$P(x) = R(x) - C(x),$$

where x is the number of items produced and sold. Thus, the profit function is found by subtracting the cost function from the revenue function.

We define the following **operations on functions.**

Operations on Functions

If $f(x)$ and $g(x)$ define functions, then

$$(f + g)(x) = f(x) + g(x),$$ Sum

$$(f - g)(x) = f(x) - g(x),$$ Difference

$$(fg)(x) = f(x) \cdot g(x),$$ Product

and $$\left(\frac{f}{g}\right)(x) = \frac{f(x)}{g(x)}, \quad g(x) \neq 0.$$ Quotient

In each case, the domain of the new function is the intersection of the domains of $f(x)$ and $g(x)$. Additionally, the domain of the quotient function must exclude any values of x for which $g(x) = 0$. (Why?)

EXAMPLE 6 Adding and Subtracting Functions

For the functions defined by

$$f(x) = 10x^2 - 2x \quad \text{and} \quad g(x) = 2x,$$

find each of the following.

(a) $(f + g)(2)$

$$
\begin{aligned}
(f + g)(2) &= f(2) + g(2) && \text{Use the definition.} \\
&= [10(2)^2 - 2(2)] + 2(2) && \text{Substitute.} \\
&= 40
\end{aligned}
$$

Alternatively, we could first find $(f + g)(x)$.

$$
\begin{aligned}
(f + g)(x) &= f(x) + g(x) && \text{Use the definition.} \\
&= (10x^2 - 2x) + 2x && \text{Substitute.} \\
&= 10x^2 && \text{Combine like terms.}
\end{aligned}
$$

Then,

$$(f + g)(2) = 10(2)^2 = 40. \qquad \text{The result is the same.}$$

(b) $(f - g)(x)$ and $(f - g)(1)$

$$
\begin{aligned}
(f - g)(x) &= f(x) - g(x) && \text{Use the definition.} \\
&= (10x^2 - 2x) - 2x && \text{Substitute.} \\
&= 10x^2 - 4x && \text{Combine like terms.}
\end{aligned}
$$

Then,

$$(f - g)(1) = 10(1)^2 - 4(1) = 6. \qquad \text{Substitute.}$$

Confirm that $f(1) - g(1)$ gives the same result.

Now Try Exercises 35, 37, and 39.

EXAMPLE 7 Multiplying Functions

For $f(x) = 3x + 4$ and $g(x) = 2x^2 + x$, find $(fg)(x)$ and $(fg)(-1)$.

$$
\begin{aligned}
(fg)(x) &= f(x) \cdot g(x) && \text{Use the definition.} \\
&= (3x + 4)(2x^2 + x) \\
&= 6x^3 + 3x^2 + 8x^2 + 4x && \text{FOIL} \\
&= 6x^3 + 11x^2 + 4x && \text{Combine like terms.}
\end{aligned}
$$

Then

$$
\begin{aligned}
(fg)(-1) &= 6(-1)^3 + 11(-1)^2 + 4(-1) && \text{Let } x = -1. \\
&= -6 + 11 - 4 \\
&= 1.
\end{aligned}
$$

(What does $f(-1) \cdot g(-1)$ equal?)

Now Try Exercises 41 and 43.

EXAMPLE 8   Dividing Functions

For $f(x) = 2x^2 + x - 10$ and $g(x) = x - 2$, find $\left(\frac{f}{g}\right)(x)$ and $\left(\frac{f}{g}\right)(-3)$.

$$
\begin{aligned}
\left(\frac{f}{g}\right)(x) &= \frac{f(x)}{g(x)} \qquad &\text{Use the definition.}\\[2mm]
&= \frac{2x^2 + x - 10}{x - 2}\\[2mm]
&= \frac{(2x + 5)(x - 2)}{x - 2} \qquad &\text{Factor.}\\[2mm]
&= 2x + 5, \quad x \neq 2
\end{aligned}
$$

Since $g(x)$, in this case $x - 2$, cannot equal 0, $x \neq 2$. Then

$$
\left(\frac{f}{g}\right)(-3) = 2(-3) + 5 = -1. \qquad \text{Let } x = -3.
$$

(Which is easier to find here—$\left(\frac{f}{g}\right)(-3)$ or $\frac{f(-3)}{g(-3)}$?)

Now Try Exercises 47 and 49.

OBJECTIVE 4   Find the composition of functions.   The diagram in Figure 9 shows a function f that assigns to each element x of set X some element y of set Y. Suppose that a function g takes each element of set Y and assigns a value z of set Z. Using both f and g, then, an element x in X is assigned to an element z in Z. The result of this process is a new function h, which takes an element x in X and assigns it an element z in Z.

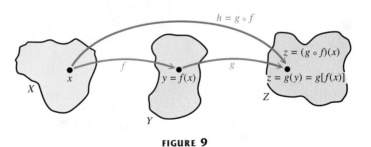

FIGURE 9

This function h is called the *composition* of functions g and f, written $g \circ f$, and is defined as follows.

Composition of Functions

If f and g are functions, then the **composite function,** or **composition,** of g and f is defined by

$$(g \circ f)(x) = g[f(x)]$$

for all x in the domain of f such that $f(x)$ is in the domain of g.

Read $g \circ f$ as "g of f."

FIGURE 10

As a real-life example of how composite functions occur, suppose an oil well off the California coast is leaking, with the leak spreading oil in a circular layer over the surface. See Figure 10. At any time t, in minutes, after the beginning of the leak, the radius of the circular oil slick is given by $r(t) = 5t$ ft. Since $A(r) = \pi r^2$ gives the area of a circle of radius r, the area can be expressed as a function of time by substituting $5t$ for r in $A(r) = \pi r^2$ to get

$$A(r) = \pi r^2$$
$$A[r(t)] = \pi(5t)^2 = 25\pi t^2.$$

The function $A[r(t)]$ is a composite function of the functions A and r.

EXAMPLE 9   Finding a Composite Function

Let $f(x) = x^2$ and $g(x) = x + 3$. Find $(f \circ g)(4)$.

$$
\begin{aligned}
(f \circ g)(4) &= f[g(4)] & &\text{Definition} \\
&= f(4 + 3) & &\text{Use the rule for } g(x); g(4) = 4 + 3. \\
&= f(7) & &\text{Add.} \\
&= 7^2 & &\text{Use the rule for } f(x); f(7) = 7^2. \\
&= 49
\end{aligned}
$$

Now Try Exercise 55.

Notice in Example 6 that if we interchange the order of the functions, the composition of g and f is defined by $g[f(x)]$. Once again, letting $x = 4$, we have

$$
\begin{aligned}
(g \circ f)(4) &= g[f(4)] & &\text{Definition} \\
&= g(4^2) & &\text{Use the rule for } f(x); f(4) = 4^2. \\
&= g(16) & &\text{Square 4.} \\
&= 16 + 3 & &\text{Use the rule for } g(x); g(16) = 16 + 3. \\
&= 19.
\end{aligned}
$$

Here we see that $(f \circ g)(4) \neq (g \circ f)(4)$ because $49 \neq 19$. In general,

$$(f \circ g)(x) \neq (g \circ f)(x).$$

EXAMPLE 10   Finding Composite Functions

Let $f(x) = 4x - 1$ and $g(x) = x^2 + 5$. Find the following.

(a) $(f \circ g)(2)$

$$
\begin{aligned}
(f \circ g)(2) &= f[g(2)] \\
&= f(2^2 + 5) \\
&= f(9) \\
&= 4(9) - 1 \\
&= 35
\end{aligned}
$$

(b) $(f \circ g)(x)$

Here, use $g(x)$ as the input for the function f.

$$(f \circ g)(x) = f[g(x)]$$
$$= 4(g(x)) - 1 \qquad \text{Use the rule for } f(x); f(x) = 4x - 1.$$
$$= 4(x^2 + 5) - 1 \qquad g(x) = x^2 + 5$$
$$= 4x^2 + 20 - 1 \qquad \text{Distributive property}$$
$$= 4x^2 + 19 \qquad \text{Combine terms.}$$

(c) Find $(f \circ g)(2)$ again, this time using the rule obtained in part (b).

$$(f \circ g)(x) = 4x^2 + 19 \qquad \text{From part (b)}$$
$$(f \circ g)(2) = 4(2)^2 + 19$$
$$= 4(4) + 19$$
$$= 16 + 19$$
$$= 35$$

The result, 35, is the same as the result in part (a).

Now Try Exercises 57 and 61.

13.1 EXERCISES

Fill in each blank with the correct response.

1. For the reciprocal function defined by $f(x) = \frac{1}{x}$, _____ is the only real number not in the domain.

2. The range of the square root function, given by $f(x) = \sqrt{x}$, is _____.

3. The lowest point on the graph of $f(x) = |x|$ has coordinates (_____, _____).

4. The range of $f(x) = [\![x]\!]$, the greatest integer function, is _____.

Without actually plotting points, match each function defined by the absolute value expression with its graph. See Example 1.

5. $f(x) = |x - 2| + 2$ **A.**
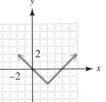

6. $f(x) = |x + 2| + 2$ **B.**
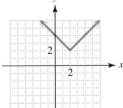

7. $f(x) = |x - 2| - 2$ **C.**
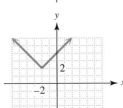

8. $f(x) = |x + 2| - 2$ **D.**
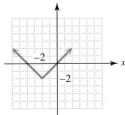

Graph each function. Give the domain and range. See Examples 1–3.

9. $f(x) = |x + 1|$

10. $f(x) = |x - 1|$

11. $f(x) = \dfrac{1}{x} + 1$

12. $f(x) = \dfrac{1}{x} - 1$

13. $f(x) = \sqrt{x - 2}$

14. $f(x) = \sqrt{x + 5}$

15. $f(x) = \dfrac{1}{x - 2}$

16. $f(x) = \dfrac{1}{x + 2}$

17. $f(x) = \sqrt{x + 3} - 3$

18. Explain how the graph of $f(x) = \dfrac{1}{x - 3} + 2$ is obtained from the graph of $g(x) = \dfrac{1}{x}$.

Graph each step function. See Examples 4 and 5.

19. $f(x) = [\![x - 3]\!]$

20. $g(x) = [\![x + 2]\!]$

21. Assume that postage rates are 37¢ for the first ounce, plus 23¢ for each additional ounce, and that each letter carries one 37¢ stamp and as many 23¢ stamps as necessary. Graph the function defined by $p(x) =$ the number of stamps on a letter weighing x oz. Use the interval $(0, 5]$.

22. The cost of parking a car at an airport hourly parking lot is $3 for the first half-hour and $2 for each additional half-hour or fraction thereof. Graph the function defined by $f(x) =$ the cost of parking a car for x hr. Use the interval $(0, 2]$.

*For each pair of functions, find **(a)** $(f + g)(x)$ and **(b)** $(f - g)(x)$. See Example 6.*

23. $f(x) = 5x - 10, \ g(x) = 3x + 7$

24. $f(x) = -4x + 1, \ g(x) = 6x + 2$

25. $f(x) = 4x^2 + 8x - 3, \ g(x) = -5x^2 + 4x - 9$

26. $f(x) = 3x^2 - 9x + 10, \ g(x) = -4x^2 + 2x + 12$

For each pair of functions, find the product $(fg)(x)$. See Example 7.

27. $f(x) = 2x, \ g(x) = 5x - 1$

28. $f(x) = 3x, \ g(x) = 6x - 8$

29. $f(x) = x + 1, \ g(x) = 2x - 3$

30. $f(x) = x - 7, \ g(x) = 4x + 5$

For each pair of functions, find the quotient $\left(\dfrac{f}{g}\right)(x)$ and give any x-values that are not in the domain of the quotient function. See Example 8.

31. $f(x) = 10x^2 - 2x, \ g(x) = 2x$

32. $f(x) = 18x^2 - 24x, \ g(x) = 3x$

33. $f(x) = 2x^2 - x - 3, \ g(x) = x + 1$

34. $f(x) = 4x^2 - 23x - 35, \ g(x) = x - 7$

Let $f(x) = x^2 - 9$, $g(x) = 2x$, and $h(x) = x - 3$. Find each of the following. See Examples 6–8.

35. $(f + g)(x)$ **36.** $(f - g)(x)$ **37.** $(f + g)(3)$

38. $(f - g)(-3)$ **39.** $(f - h)(x)$ **40.** $(f - h)(-3)$

41. $(fg)(x)$ **42.** $(fh)(x)$ **43.** $(fg)(2)$

44. $(fh)(1)$ **45.** $(gh)(x)$ **46.** $(gh)(-3)$

47. $\left(\dfrac{f}{g}\right)(x)$ **48.** $\left(\dfrac{f}{h}\right)(x)$ **49.** $\left(\dfrac{f}{g}\right)(2)$

50. $\left(\dfrac{f}{h}\right)(1)$ **51.** $\left(\dfrac{h}{g}\right)(3)$ **52.** $\left(\dfrac{h}{g}\right)(x)$

53. Construct two polynomial functions defined by $f(x)$, a polynomial of degree 3, and $g(x)$, a polynomial of degree 4. Find $(f - g)(x)$ and $(g - f)(x)$. Use your answers to decide whether subtraction of polynomial functions is a commutative operation. Explain.

54. Find two polynomial functions defined by $f(x)$ and $g(x)$ such that
$$(f + g)(x) = 3x^3 - x + 3.$$

Let $f(x) = x^2 + 4$, $g(x) = 2x + 3$, and $h(x) = x + 5$. Find each value or expression. See Examples 9 and 10.

55. $(h \circ g)(4)$ **56.** $(f \circ g)(4)$ **57.** $(g \circ f)(6)$ **58.** $(h \circ f)(6)$

59. $(f \circ h)(-2)$ **60.** $(h \circ g)(-2)$ **61.** $(f \circ g)(x)$ **62.** $(g \circ h)(x)$

63. $(f \circ h)(x)$ **64.** $(g \circ f)(x)$ **65.** $(h \circ g)(x)$ **66.** $(h \circ f)(x)$

Solve each problem.

67. The function defined by $f(x) = 12x$ computes the number of inches in x ft and the function defined by $g(x) = 5280x$ computes the number of feet in x mi. What is $(f \circ g)(x)$ and what does it compute?

68. The perimeter x of a square with sides of length s is given by the formula $x = 4s$.

(a) Solve for s in terms of x.

(b) If y represents the area of this square, write y as a function of the perimeter x.

(c) Use the composite function of part (b) to find the area of a square with perimeter 6.

69. When a thermal inversion layer is over a city (as happens often in Los Angeles), pollutants cannot rise vertically but are trapped below the layer and must disperse horizontally. Assume that a factory smokestack begins emitting a pollutant at 8 A.M. Assume that the pollutant disperses horizontally over a circular area. Suppose that t represents the time, in hours, since the factory began emitting pollutants ($t = 0$ represents 8 A.M.), and assume that the radius of the circle of pollution is $r(t) = 2t$ mi. Let $A(r) = \pi r^2$ represent the area of a circle of radius r. Find and interpret $(A \circ r)(t)$.

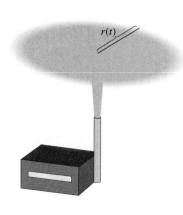

70. An oil well off the Gulf Coast is leaking, with the leak spreading oil over the surface as a circle. At any time t, in minutes, after the beginning of the leak, the radius of the circular oil slick on the surface is $r(t) = 4t$ ft. Let $A(r) = \pi r^2$ represent the area of a circle of radius r. Find and interpret $(A \circ r)(t)$.

13.2 The Circle and the Ellipse

When an infinite cone is intersected by a plane, the resulting figure is called a **conic section.** The parabola is one example of a conic section; circles, ellipses, and hyperbolas may also result. See Figure 11.

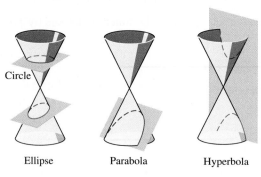

FIGURE 11

OBJECTIVE 1 Find the equation of a circle given the center and radius. A **circle** is the set of all points in a plane that lie a fixed distance from a fixed point. The fixed point is called the **center,** and the fixed distance is called the **radius.** We use the distance formula to find an equation of a circle.

EXAMPLE 1 Finding the Equation of a Circle and Graphing It

Find an equation of the circle with radius 3 and center at $(0, 0)$, and graph it.

If the point (x, y) is on the circle, then the distance from (x, y) to the center $(0, 0)$ is 3. By the distance formula,

$$\sqrt{(x_2 - x_1)^2 + (y_2 - y_1)^2} = d$$
$$\sqrt{(x - 0)^2 + (y - 0)^2} = 3$$
$$x^2 + y^2 = 9. \qquad \text{Square both sides.}$$

An equation of this circle is $x^2 + y^2 = 9$. The graph is shown in Figure 12.

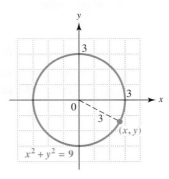

FIGURE 12

Now Try Exercise 1.

A circle may not be centered at the origin, as seen in the next example.

EXAMPLE 2 Finding an Equation of a Circle and Graphing It

Find an equation of the circle with center at $(4, -3)$ and radius 5, and graph it.

Use the distance formula again.

$$\sqrt{(x - 4)^2 + [y - (-3)]^2} = 5$$
$$(x - 4)^2 + (y + 3)^2 = 25 \qquad \text{Square both sides.}$$

To graph the circle, plot the center $(4, -3)$, then move 5 units right, left, up, and down from the center. Draw a smooth curve through these four points, sketching one quarter of the circle at a time. The graph of this circle is shown in Figure 13.

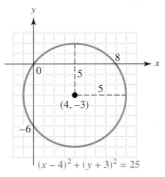

FIGURE 13

Now Try Exercises 7 and 23.

Examples 1 and 2 suggest the form of an equation of a circle with radius r and center at (h, k). If (x, y) is a point on the circle, then the distance from the center (h, k) to the point (x, y) is r. By the distance formula,

$$\sqrt{(x - h)^2 + (y - k)^2} = r.$$

Squaring both sides gives the **center-radius form** of the equation of a circle.

Equation of a Circle (Center-Radius Form)

$$(x - h)^2 + (y - k)^2 = r^2$$

is an equation of the circle with radius r and center at (h, k).

EXAMPLE 3 Using the Center-Radius Form of the Equation of a Circle

Find an equation of the circle with center at $(-1, 2)$ and radius 4.
    Use the center-radius form, with $h = -1$, $k = 2$, and $r = 4$.

$$(x - h)^2 + (y - k)^2 = r^2$$
$$[x - (-1)]^2 + (y - 2)^2 = 4^2$$
$$(x + 1)^2 + (y - 2)^2 = 16$$

Now Try Exercise 9.

OBJECTIVE 2 Determine the center and radius of a circle given its equation. In the equation found in Example 2, multiplying out $(x - 4)^2$ and $(y + 3)^2$ gives

$$(x - 4)^2 + (y + 3)^2 = 25$$
$$x^2 - 8x + 16 + y^2 + 6y + 9 = 25$$
$$x^2 + y^2 - 8x + 6y = 0.$$

This general form suggests that an equation with both x^2- and y^2-terms with equal coefficients may represent a circle. The next example shows how to tell, by completing the square. This procedure was introduced in Chapter 11.

EXAMPLE 4 Completing the Square to Find the Center and Radius

Find the center and radius of the circle $x^2 + y^2 + 2x + 6y - 15 = 0$, and graph it.
    Since the equation has x^2- and y^2-terms with equal coefficients, its graph might be that of a circle. To find the center and radius, complete the squares on x and y.

$$x^2 + y^2 + 2x + 6y = 15$$	Get the constant on the right.
$$(x^2 + 2x \quad) + (y^2 + 6y \quad) = 15$$	Rewrite in anticipation of completing the square.
$$\left[\frac{1}{2}(2)\right]^2 = 1 \qquad \left[\frac{1}{2}(6)\right]^2 = 9$$	Square half the coefficient of each middle term.
$$(x^2 + 2x + 1) + (y^2 + 6y + 9) = 15 + 1 + 9$$	Complete the squares on both x and y.
$$(x + 1)^2 + (y + 3)^2 = 25$$	Factor on the left; add on the right.
$$[x - (-1)]^2 + [y - (-3)]^2 = 5^2$$	Center-radius form

The final equation

$$[x - (-1)]^2 + [y - (-3)]^2 = 5^2$$

or

$$(x + 1)^2 + (y + 3)^2 = 5^2$$

indicates that the graph is a circle with center at $(-1, -3)$ and radius 5. The graph is shown in Figure 14.

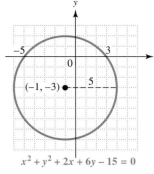

$$x^2 + y^2 + 2x + 6y - 15 = 0$$

FIGURE 14

Now Try Exercise 11.

> **NOTE** If the procedure of Example 4 leads to an equation of the form $(x - h)^2 + (y - k)^2 = 0$, then the graph is the single point (h, k). If the constant on the right side is negative, then the equation has no graph.

OBJECTIVE 3 Recognize the equation of an ellipse. An **ellipse** is the set of all points in a plane the *sum* of whose distances from two fixed points is constant. These fixed points are called **foci** (singular: *focus*). Figure 15 shows an ellipse whose foci are $(c, 0)$ and $(-c, 0)$, with x-intercepts $(a, 0)$ and $(-a, 0)$ and y-intercepts $(0, b)$ and $(0, -b)$. It is shown in more advanced courses that $c^2 = a^2 - b^2$ for an ellipse of this type. The origin is the **center** of the ellipse.

An ellipse has the following equation.

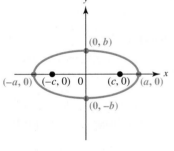

FIGURE 15

Equation of an Ellipse

The ellipse whose x-intercepts are $(a, 0)$ and $(-a, 0)$ and whose y-intercepts are $(0, b)$ and $(0, -b)$ has an equation of the form

$$\frac{x^2}{a^2} + \frac{y^2}{b^2} = 1.$$

> **NOTE** A circle is a special case of an ellipse, where $a^2 = b^2$.

The paths of Earth and other planets around the sun are approximately ellipses; the sun is at one focus and a point in space is at the other. The orbits of communication satellites and other space vehicles are elliptical. Elliptical bicycle gears are designed to respond to the legs' natural strengths and weaknesses. At the top and bottom of the powerstroke, where the legs have the least leverage, the gear offers little resistance, but as the gear rotates, the resistance increases. This allows the legs to apply more power where it is most naturally available. See Figure 16.

FIGURE 16

OBJECTIVE 4 Graph ellipses. To graph an ellipse centered at the origin, we plot the four intercepts and then sketch the ellipse through those points.

EXAMPLE 5 Graphing Ellipses

Graph each ellipse.

(a) $\dfrac{x^2}{49} + \dfrac{y^2}{36} = 1$

Here, $a^2 = 49$, so $a = 7$, and the x-intercepts for this ellipse are $(7, 0)$ and $(-7, 0)$. Similarly, $b^2 = 36$, so $b = 6$, and the y-intercepts are $(0, 6)$ and $(0, -6)$. Plotting the intercepts and sketching the ellipse through them gives the graph in Figure 17.

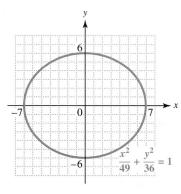

FIGURE 17   **FIGURE 18**

(b) $\dfrac{x^2}{36} + \dfrac{y^2}{121} = 1$

The x-intercepts for this ellipse are $(6, 0)$ and $(-6, 0)$, and the y-intercepts are $(0, 11)$ and $(0, -11)$. Join these with the smooth curve of an ellipse. The graph has been sketched in Figure 18.

Now Try Exercises 27 and 31.

As with the graphs of functions and circles, the graph of an ellipse may be shifted horizontally and vertically, as in the next example.

EXAMPLE 6 Graphing an Ellipse Shifted Horizontally and Vertically

Graph $\dfrac{(x - 2)^2}{25} + \dfrac{(y + 3)^2}{49} = 1$.

Just as $(x - 2)^2$ and $(y + 3)^2$ would indicate that the center of a circle would be $(2, -3)$, so it is with this ellipse. Figure 19 shows that the graph goes through the four points $(2, 4)$, $(7, -3)$, $(2, -10)$, and $(-3, -3)$. The x-values of these points are found by adding $\pm a = \pm 5$ to 2, and the y-values come from adding $\pm b = \pm 7$ to -3.

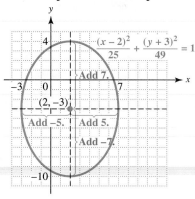

FIGURE 19

Now Try Exercise 35.

OBJECTIVE 5 Graph circles and ellipses using a graphing calculator. The only conic section whose graph is a function is the vertical parabola with equation $f(x) = ax^2 + bx + c$. Therefore, a graphing calculator in function mode cannot directly graph a circle or an ellipse. We must first solve the equation for y, getting two functions y_1 and y_2. The union of these two graphs is the graph of the entire figure. For example, to graph $(x + 3)^2 + (y + 2)^2 = 25$, begin by solving for y.

$$(x + 3)^2 + (y + 2)^2 = 25$$

$$(y + 2)^2 = 25 - (x + 3)^2 \qquad \text{Subtract } (x + 3)^2.$$

$$y + 2 = \pm\sqrt{25 - (x + 3)^2} \qquad \text{Take square roots.}$$

$$y = -2 \pm \sqrt{25 - (x + 3)^2} \qquad \text{Subtract 2.}$$

The two functions to be graphed are

$$y_1 = -2 + \sqrt{25 - (x + 3)^2} \qquad \text{and} \qquad y_2 = -2 - \sqrt{25 - (x + 3)^2}.$$

To get an undistorted screen, a *square viewing window* must be used. (Refer to your instruction manual for details.) See Figure 20. The two semicircles seem to be disconnected. This is because the graphs are nearly vertical at those points, and the calculator cannot show a true picture of the behavior there.

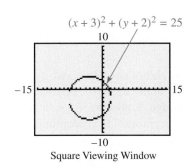

FIGURE 20
Square Viewing Window

13.2 EXERCISES

1. See Example 1. Consider the circle whose equation is $x^2 + y^2 = 25$.

   (a) What are the coordinates of its center?   **(b)** What is its radius?
   (c) Sketch its graph.

2. Explain why a set of points defined by a circle does not satisfy the definition of a function.

Match each equation with the correct graph. See Examples 1–3.

3. $(x - 3)^2 + (y - 2)^2 = 25$ **A.**

4. $(x - 3)^2 + (y + 2)^2 = 25$

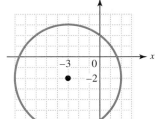
B.
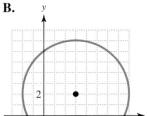

5. $(x + 3)^2 + (y - 2)^2 = 25$ **C.**

6. $(x + 3)^2 + (y + 2)^2 = 25$

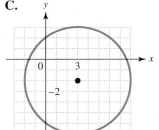
D.
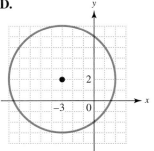

Find the equation of a circle satisfying the given conditions. See Examples 2 and 3.

7. Center: $(-4, 3)$; radius: 2   **8.** Center: $(5, -2)$; radius: 4

9. Center: $(-8, -5)$; radius: $\sqrt{5}$ **10.** Center: $(-12, 13)$; radius: $\sqrt{7}$

Find the center and radius of each circle. (Hint: In Exercises 15 and 16, divide each side by a common factor.) See Example 4.

11. $x^2 + y^2 + 4x + 6y + 9 = 0$ **12.** $x^2 + y^2 - 8x - 12y + 3 = 0$

13. $x^2 + y^2 + 10x - 14y - 7 = 0$ **14.** $x^2 + y^2 - 2x + 4y - 4 = 0$

15. $3x^2 + 3y^2 - 12x - 24y + 12 = 0$ **16.** $2x^2 + 2y^2 + 20x + 16y + 10 = 0$

17. A circle can be drawn on a piece of posterboard by fastening one end of a string with a thumbtack, pulling the string taut with a pencil, and tracing a curve, as shown in the figure. Explain why this method works.

18. This figure shows how the crawfish race is held at the Crawfish Festival in Breaux Bridge, Louisiana. Explain why a circular "racetrack" is appropriate for such a race.

Graph each circle. Identify the center if it is not at the origin. See Examples 1, 2, and 4.

19. $x^2 + y^2 = 9$ **20.** $x^2 + y^2 = 4$

21. $2y^2 = 10 - 2x^2$ **22.** $3x^2 = 48 - 3y^2$

23. $(x + 3)^2 + (y - 2)^2 = 9$ **24.** $(x - 1)^2 + (y + 3)^2 = 16$

25. $x^2 + y^2 - 4x - 6y + 9 = 0$ **26.** $x^2 + y^2 + 8x + 2y - 8 = 0$

Graph each ellipse. See Examples 5 and 6.

27. $\dfrac{x^2}{9} + \dfrac{y^2}{25} = 1$ **28.** $\dfrac{x^2}{9} + \dfrac{y^2}{16} = 1$ **29.** $\dfrac{x^2}{36} = 1 - \dfrac{y^2}{16}$

30. $\dfrac{x^2}{9} = 1 - \dfrac{y^2}{4}$ **31.** $\dfrac{y^2}{25} = 1 - \dfrac{x^2}{49}$ **32.** $\dfrac{y^2}{9} = 1 - \dfrac{x^2}{16}$

33. $\dfrac{x^2}{16} + \dfrac{y^2}{4} = 1$ **34.** $\dfrac{x^2}{49} + \dfrac{y^2}{81} = 1$

35. $\dfrac{(x + 1)^2}{64} + \dfrac{(y - 2)^2}{49} = 1$ **36.** $\dfrac{(x - 4)^2}{9} + \dfrac{(y + 2)^2}{4} = 1$

37. $\dfrac{(x - 2)^2}{16} + \dfrac{(y - 1)^2}{9} = 1$ **38.** $\dfrac{(x + 3)^2}{25} + \dfrac{(y + 2)^2}{36} = 1$

39. It is possible to sketch an ellipse on a piece of posterboard by fastening two ends of a length of string, pulling the string taut with a pencil, and tracing a curve, as shown in the figure. Explain why this method works.

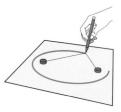

40. Discuss the similarities and differences between the equations of a circle and an ellipse.

41. Explain why a set of ordered pairs whose graph forms an ellipse does not satisfy the definition of a function.

42. (a) How many points are there on the graph of $(x - 4)^2 + (y - 1)^2 = 0$? Explain.
(b) How many points are there on the graph of $(x - 4)^2 + (y - 1)^2 = -1$? Explain.

TECHNOLOGY INSIGHTS (EXERCISES 43 AND 44)

---

43. The circle shown in the calculator graph was created using function mode, with a square viewing window. It is the graph of $(x + 2)^2 + (y - 4)^2 = 16$. What are the two functions y_1 and y_2 that were used to obtain this graph?

44. The ellipse shown in the calculator graph was graphed using function mode, with a square viewing window. It is the graph of $\dfrac{x^2}{4} + \dfrac{y^2}{9} = 1$. What are the two functions y_1 and y_2 that were used to obtain this graph?

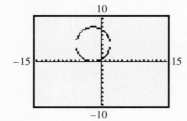

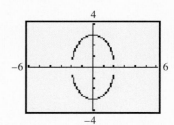

Use a graphing calculator in function mode to graph each circle or ellipse. Use a square viewing window. See Objective 5.

45. $x^2 + y^2 = 36$

46. $(x - 2)^2 + y^2 = 49$

47. $\dfrac{x^2}{16} + \dfrac{y^2}{4} = 1$

48. $\dfrac{(x - 3)^2}{25} + \dfrac{y^2}{9} = 1$

Solve each problem.

49. An arch has the shape of half an ellipse. The equation of the ellipse is $100x^2 + 324y^2 = 32{,}400$, where x and y are in meters.

(a) How high is the center of the arch?
(b) How wide is the arch across the bottom?

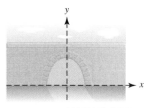

NOT TO SCALE

50. A one-way street passes under an overpass, which is in the form of the top half of an ellipse, as shown in the figure. Suppose that a truck 12 ft wide passes directly under the overpass. What is the maximum possible height of this truck?

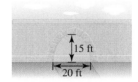

In Exercises 51 and 52, see Figure 15 and use the fact that $c^2 = a^2 - b^2$ where $a^2 > b^2$.

51. The orbit of Mars is an ellipse with the sun at one focus. For x and y in millions of miles, the equation of the orbit is

$$\frac{x^2}{141.7^2} + \frac{y^2}{141.1^2} = 1.$$

(*Source:* Kaler, James B., *Astronomy!*, Addison-Wesley, 1997.)

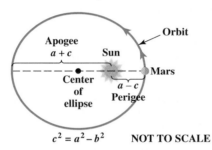

$c^2 = a^2 - b^2$ **NOT TO SCALE**

(a) Find the greatest distance (the *apogee*) from Mars to the sun.

(b) Find the smallest distance (the *perigee*) from Mars to the sun.

52. The orbit of Venus around the sun (one of the foci) is an ellipse with equation

$$\frac{x^2}{5013} + \frac{y^2}{4970} = 1,$$

where x and y are measured in millions of miles. (*Source:* Kaler, James B., *Astronomy!*, Addison-Wesley, 1997.)

(a) Find the greatest distance between Venus and the sun.

(b) Find the smallest distance between Venus and the sun.

A lithotripter is a machine used to crush kidney stones using shock waves. The patient is placed in an elliptical tub with the kidney stone at one focus of the ellipse. A beam is projected from the other focus to the tub, so that it reflects to hit the kidney stone.

53. Suppose a lithotripter is based on the ellipse with equation

$$\frac{x^2}{36} + \frac{y^2}{9} = 1.$$

How far from the center of the ellipse must the kidney stone and the source of the beam be placed?

54. Rework Exercise 53 if the equation of the ellipse is $9x^2 + 4y^2 = 36$. (*Hint:* Write the equation in fraction form by dividing each term by 36, and use $c^2 = b^2 - a^2$, since $b > a$ here.)

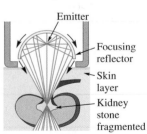

The top of an ellipse is illustrated in this depiction of how a lithotripter crushes kidney stones.

Source: Adapted drawing of an ellipse in illustration of a lithotripter. The American Medical Association, *Encyclopedia of Medicine*, 1989.

55. (a) Suppose that $(c, 0)$ and $(-c, 0)$ are the foci of an ellipse and that the sum of the distances from any point (x, y) on the ellipse to the two foci is $2a$. See the figure. Show that the equation of the resulting ellipse is

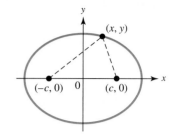

$$\frac{x^2}{a^2} + \frac{y^2}{a^2 - c^2} = 1.$$

(b) Show that in the equation in part (a), the x-intercepts are $(a, 0)$ and $(-a, 0)$.

(c) Let $b^2 = a^2 - c^2$, and show that $(0, b)$ and $(0, -b)$ are the y-intercepts in the equation in part (a).

56. Use the result of Exercise 55(a) to find an equation of an ellipse with foci $(3, 0)$ and $(-3, 0)$, where the sum of the distances from any point of the ellipse to the two foci is 10.

13.3 The Hyperbola and Functions Defined by Radicals

OBJECTIVES

1 Recognize the equation of a hyperbola.

2 Graph hyperbolas by using asymptotes.

3 Identify conic sections by their equations.

4 Graph certain square root functions.

OBJECTIVE 1 Recognize the equation of a hyperbola. A **hyperbola** is the set of all points in a plane such that the absolute value of the *difference* of the distances from two fixed points (called *foci*) is constant. Figure 21 shows a hyperbola; using the distance formula and the definition above, we can show that this hyperbola has equation

$$\frac{x^2}{16} - \frac{y^2}{12} = 1.$$

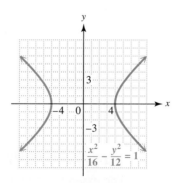

FIGURE 21

To graph hyperbolas centered at the origin, we need to find their intercepts. For the hyperbola in Figure 21, we proceed as follows.

x-Intercepts	**y-Intercepts**
Let $y = 0$.	Let $x = 0$.
$\dfrac{x^2}{16} - \dfrac{0^2}{12} = 1$ Let $y = 0$.	$\dfrac{0^2}{16} - \dfrac{y^2}{12} = 1$ Let $x = 0$.
$\dfrac{x^2}{16} = 1$	$-\dfrac{y^2}{12} = 1$
$x^2 = 16$ Multiply by 16.	$y^2 = -12$ Multiply by -12.
$x = \pm 4$	

The x-intercepts are $(4, 0)$ and $(-4, 0)$.

Because there are no *real* solutions to $y^2 = -12$, the graph has no y-intercepts.

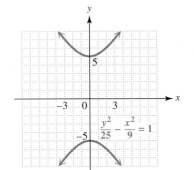

FIGURE 22

The graph of $\dfrac{x^2}{16} - \dfrac{y^2}{12} = 1$ has no y-intercepts. On the other hand, the hyperbola in Figure 22 has no x-intercepts. Its equation is

$$\frac{y^2}{25} - \frac{x^2}{9} = 1,$$

with y-intercepts $(0, 5)$ and $(0, -5)$.

Equations of Hyperbolas

A hyperbola with x-intercepts $(a, 0)$ and $(-a, 0)$ has an equation of the form

$$\frac{x^2}{a^2} - \frac{y^2}{b^2} = 1,$$

and a hyperbola with y-intercepts $(0, b)$ and $(0, -b)$ has an equation of the form

$$\frac{y^2}{b^2} - \frac{x^2}{a^2} = 1.$$

OBJECTIVE 2 Graph hyperbolas by using asymptotes.   The two branches of the graph of a hyperbola approach a pair of intersecting straight lines, which are its asymptotes. See Figure 23 on the next page. The asymptotes are useful for sketching the graph of the hyperbola.

Asymptotes of Hyperbolas

The extended diagonals of the rectangle with vertices (corners) at the points (a, b), $(-a, b)$, $(-a, -b)$, and $(a, -b)$ are the **asymptotes** of the hyperbolas

$$\frac{x^2}{a^2} - \frac{y^2}{b^2} = 1 \qquad \text{and} \qquad \frac{y^2}{b^2} - \frac{x^2}{a^2} = 1.$$

This rectangle is called the **fundamental rectangle.** Using the methods of Chapter 3, we could show that the equations of these asymptotes are

$$y = \frac{b}{a}x \quad \text{and} \quad y = -\frac{b}{a}x.$$

To graph hyperbolas, follow these steps.

Graphing a Hyperbola

Step 1 **Find the intercepts.** Locate the intercepts at $(a, 0)$ and $(-a, 0)$ if the x^2-term has a positive coefficient, or at $(0, b)$ and $(0, -b)$ if the y^2- term has a positive coefficient.

Step 2 **Find the fundamental rectangle.** Locate the vertices of the fundamental rectangle at (a, b), $(-a, b)$, $(-a, -b)$, and $(a, -b)$.

Step 3 **Sketch the asymptotes.** The extended diagonals of the rectangle are the asymptotes of the hyperbola, and they have equations $y = \pm\frac{b}{a}x$.

Step 4 **Draw the graph.** Sketch each branch of the hyperbola through an intercept and approaching (but not touching) the asymptotes.

■ EXAMPLE 1 Graphing a Horizontal Hyperbola

Graph $\dfrac{x^2}{16} - \dfrac{y^2}{25} = 1$.

Step 1 Here $a = 4$ and $b = 5$. The x-intercepts are $(4, 0)$ and $(-4, 0)$.

Step 2 The four points $(4, 5)$, $(-4, 5)$, $(-4, -5)$, and $(4, -5)$ are the vertices of the fundamental rectangle, as shown in Figure 23.

Steps 3 and 4 The equations of the asymptotes are $y = \pm\frac{5}{4}x$, and the hyperbola approaches these lines as x and y get larger and larger in absolute value.

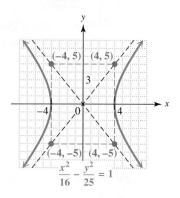

FIGURE 23

Now Try Exercise 7.

CAUTION When sketching the graph of a hyperbola, be sure that the branches do not touch the asymptotes.

EXAMPLE 2 Graphing a Vertical Hyperbola

Graph $\dfrac{y^2}{49} - \dfrac{x^2}{16} = 1$.

This hyperbola has y-intercepts $(0, 7)$ and $(0, -7)$. The asymptotes are the extended diagonals of the rectangle with vertices at $(4, 7)$, $(-4, 7)$, $(-4, -7)$, and $(4, -7)$. Their equations are $y = \pm \frac{7}{4}x$. See Figure 24.

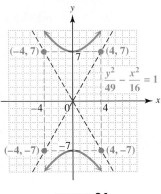

FIGURE 24

Now Try Exercise 9.

Hyperbolas are graphed with a graphing calculator in much the same way as circles and ellipses, by first writing the equations of two root functions whose union is equivalent to the equation of the hyperbola. A square window gives a truer shape for hyperbolas, too.

CONNECTIONS

A hyperbola and a parabola are used together in one kind of microwave antenna system. The cross sections of the system consist of a parabola and a hyperbola with the focus of the parabola coinciding with one focus of the hyperbola. See the figure.

The incoming microwaves that are parallel to the axis of the parabola are reflected from the parabola up toward the hyperbola and back to the other focus of the hyperbola, where the cone of the antenna is located to capture the signal.

For Discussion or Writing

The property of the parabola and the hyperbola that is used here is a "reflection property" of the foci. Explain why this name is appropriate.

OBJECTIVE 3 Identify conic sections by their equations. Rewriting a second-degree equation in one of the forms given for ellipses, hyperbolas, circles, or parabolas makes it possible to determine when the graph is one of these.

Summary of Conic Sections

Equation	Graph	Description	Identification
$y = ax^2 + bx + c$ or $y = a(x - h)^2 + k$	Parabola	It opens up if $a > 0$, down if $a < 0$. The vertex is (h, k).	It has an x^2-term. y is not squared.
$x = ay^2 + by + c$ or $x = a(y - k)^2 + h$	Parabola	It opens to the right if $a > 0$, to the left if $a < 0$. The vertex is (h, k).	It has a y^2-term. x is not squared.
$(x - h)^2 + (y - k)^2 = r^2$	Circle	The center is (h, k), and the radius is r.	x^2- and y^2-terms have the same positive coefficient.
$\dfrac{x^2}{a^2} + \dfrac{y^2}{b^2} = 1$	Ellipse	The x-intercepts are $(a, 0)$ and $(-a, 0)$. The y-intercepts are $(0, b)$ and $(0, -b)$.	x^2- and y^2-terms have different positive coefficients.
$\dfrac{x^2}{a^2} - \dfrac{y^2}{b^2} = 1$	Hyperbola	The x-intercepts are $(a, 0)$ and $(-a, 0)$. The asymptotes are found from (a, b), $(a, -b)$, $(-a, -b)$, and $(-a, b)$.	x^2 has a positive coefficient. y^2 has a negative coefficient.
$\dfrac{y^2}{b^2} - \dfrac{x^2}{a^2} = 1$	Hyperbola	The y-intercepts are $(0, b)$ and $(0, -b)$. The asymptotes are found from (a, b), $(a, -b)$, $(-a, -b)$, and $(-a, b)$.	y^2 has a positive coefficient. x^2 has a negative coefficient.

EXAMPLE 3 Identifying the Graphs of Equations

Identify the graph of each equation.

(a) $9x^2 = 108 + 12y^2$

Both variables are squared, so the graph is either an ellipse or a hyperbola. (This situation also occurs for a circle, which is a special case of an ellipse.) To see whether the graph is an ellipse or a hyperbola, rewrite the equation so that the x^2- and y^2-terms are on one side of the equation and 1 is on the other.

$$9x^2 - 12y^2 = 108 \qquad \text{Subtract } 12y^2.$$

$$\frac{x^2}{12} - \frac{y^2}{9} = 1 \qquad \text{Divide by 108.}$$

Because of the minus sign, the graph of this equation is a hyperbola.

(b) $x^2 = y - 3$

Only one of the two variables, x, is squared, so this is the vertical parabola $y = x^2 + 3$.

(c) $x^2 = 9 - y^2$

Get the variable terms on the same side of the equation.

$$x^2 + y^2 = 9 \qquad \text{Add } y^2.$$

The graph of this equation is a circle with center at the origin and radius 3.

Now Try Exercises 17 and 21.

OBJECTIVE 4 Graph certain square root functions. Recall that no vertical line will intersect the graph of a function in more than one point. Thus, horizontal parabolas, all circles and ellipses, and most hyperbolas discussed in this chapter are examples of graphs that do not satisfy the conditions of a function. However, by considering only a part of the graph of each of these we have the graph of a function, as seen in Figure 25.

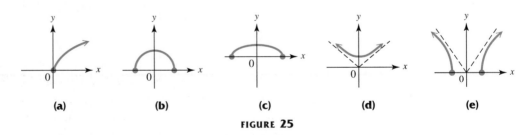

(a)   (b)   (c)   (d)   (e)

FIGURE 25

In parts (a), (b), (c), and (d) of Figure 25, the top portion of a conic section is shown (parabola, circle, ellipse, and hyperbola, respectively). In part (e), the top two portions of a hyperbola are shown. In each case, the graph is that of a function since the graph satisfies the conditions of the vertical line test.

In Sections 10.1 and 13.1 we observed the square root function defined by $f(x) = \sqrt{x}$. To find equations for the types of graphs shown in Figure 25, we extend its definition.

Square Root Function

A function of the form

$$f(x) = \sqrt{u}$$

for an algebraic expression u, with $u \geq 0$, is called a **square root function.**

▨ EXAMPLE 4   Graphing a Semicircle

Graph $f(x) = \sqrt{25 - x^2}$. Give the domain and range.

Replace $f(x)$ with y and square both sides to get the equation

$$y^2 = 25 - x^2 \quad \text{or} \quad x^2 + y^2 = 25.$$

This is the graph of a circle with center at $(0, 0)$ and radius 5. Since $f(x)$, or y, represents a principal square root in the original equation, $f(x)$ must be nonnegative. This restricts the graph to the upper half of the circle, as shown in Figure 26. Use the graph and the vertical line test to verify that it is indeed a function. The domain is $[-5, 5]$, and the range is $[0, 5]$.

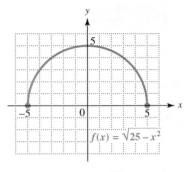

FIGURE 26

Now Try Exercise 25.

▨ EXAMPLE 5   Graphing a Portion of an Ellipse

Graph $\dfrac{y}{6} = -\sqrt{1 - \dfrac{x^2}{16}}$. Give the domain and range.

Square both sides to get an equation whose form is known.

$$\frac{y^2}{36} = 1 - \frac{x^2}{16}$$

$$\frac{x^2}{16} + \frac{y^2}{36} = 1 \qquad \text{Add } \tfrac{x^2}{16}.$$

This is the equation of an ellipse with x-intercepts $(4, 0)$ and $(-4, 0)$ and y-intercepts $(0, 6)$ and $(0, -6)$. Since $\tfrac{y}{6}$ equals a negative square root in the original equation, y must be nonpositive, restricting the graph to the lower half of the ellipse, as shown in Figure 27. Verify that this is the graph of a function, using the vertical line test. The domain is $[-4, 4]$, and the range is $[-6, 0]$.

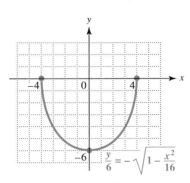

FIGURE 27

Now Try Exercise 27.

Root functions, since they are functions, can be entered and graphed directly with a graphing calculator.

13.3 EXERCISES

Based on the discussions of ellipses in the previous section and of hyperbolas in this section, match each equation with its graph.

1. $\dfrac{x^2}{25} + \dfrac{y^2}{9} = 1$

2. $\dfrac{x^2}{9} + \dfrac{y^2}{25} = 1$

3. $\dfrac{x^2}{9} - \dfrac{y^2}{25} = 1$

4. $\dfrac{x^2}{25} - \dfrac{y^2}{9} = 1$

A.
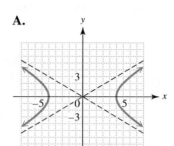

B.
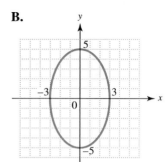

C.
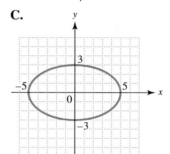

D.
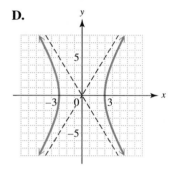

 5. Write an explanation of how you can tell from the equation whether the branches of a hyperbola open up and down or left and right.

6. Describe how the fundamental rectangle is used to sketch a hyperbola.

Graph each hyperbola. See Examples 1 and 2.

7. $\dfrac{x^2}{16} - \dfrac{y^2}{9} = 1$

8. $\dfrac{y^2}{4} - \dfrac{x^2}{25} = 1$

9. $\dfrac{y^2}{9} - \dfrac{x^2}{9} = 1$

10. $\dfrac{x^2}{49} - \dfrac{y^2}{16} = 1$

11. $\dfrac{x^2}{25} - \dfrac{y^2}{36} = 1$

12. $\dfrac{y^2}{9} - \dfrac{x^2}{4} = 1$

13. $\dfrac{y^2}{16} - \dfrac{x^2}{16} = 1$

14. $\dfrac{x^2}{25} - \dfrac{y^2}{9} = 1$

Identify the graph of each equation as a parabola, circle, ellipse, *or* hyperbola, *and then sketch. See Example 3.*

15. $x^2 - y^2 = 16$

16. $x^2 + y^2 = 16$

17. $4x^2 + y^2 = 16$

18. $x^2 - 2y = 0$

19. $y^2 = 36 - x^2$

20. $9x^2 + 25y^2 = 225$

21. $9x^2 = 144 + 16y^2$

22. $x^2 + 9y^2 = 9$

23. $y^2 = 4 + x^2$

24. State in your own words the major difference between the definitions of *ellipse* and *hyperbola*.

Graph each function defined by a radical expression. Give the domain and range. See Examples 4 and 5.

25. $f(x) = \sqrt{16 - x^2}$

26. $f(x) = \sqrt{9 - x^2}$

27. $f(x) = -\sqrt{36 - x^2}$

28. $f(x) = -\sqrt{25 - x^2}$

29. $\dfrac{y}{3} = \sqrt{1 + \dfrac{x^2}{9}}$

30. $y = \sqrt{\dfrac{x + 4}{2}}$

In Section 13.2, Example 6, we saw that the center of an ellipse may be shifted away from the origin. The same process applies to hyperbolas. For example, the hyperbola

$$\frac{(x + 5)^2}{4} - \frac{(y - 2)^2}{9} = 1,$$

shown at the right, has the same graph as $\dfrac{x^2}{4} - \dfrac{y^2}{9} = 1$, *but it is centered at* $(-5, 2)$. *Graph each hyperbola with center shifted away from the origin.*

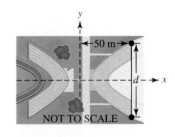

31. $\dfrac{(x - 2)^2}{4} - \dfrac{(y + 1)^2}{9} = 1$

32. $\dfrac{(x + 3)^2}{16} - \dfrac{(y - 2)^2}{25} = 1$

33. $\dfrac{y^2}{36} - \dfrac{(x - 2)^2}{49} = 1$

34. $\dfrac{(y - 5)^2}{9} - \dfrac{x^2}{25} = 1$

Solve each problem.

35. Two buildings in a sports complex are shaped and positioned like a portion of the branches of the hyperbola with equation

$$400x^2 - 625y^2 = 250{,}000,$$

where x and y are in meters.

(a) How far apart are the buildings at their closest point?

(b) Find the distance d in the figure.

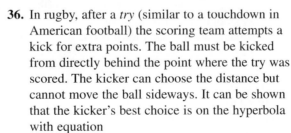

36. In rugby, after a *try* (similar to a touchdown in American football) the scoring team attempts a kick for extra points. The ball must be kicked from directly behind the point where the try was scored. The kicker can choose the distance but cannot move the ball sideways. It can be shown that the kicker's best choice is on the hyperbola with equation

$$\frac{x^2}{g^2} - \frac{y^2}{g^2} = 1,$$

where $2g$ is the distance between the goal posts. Since the hyperbola approaches its asymptotes, it is easier for the kicker to estimate points on the asymptotes instead of on the hyperbola. What are the asymptotes of this hyperbola? Why is it relatively easy to estimate them? (*Source:* Isaksen, Daniel C., "How to Kick a Field Goal," *The College Mathematics Journal*, September 1996.)

37. When a satellite is launched into orbit, the shape of its trajectory is determined by its velocity. The trajectory will be hyperbolic if the velocity V, in meters per second, satisfies the inequality

$$V > \frac{2.82 \times 10^7}{\sqrt{D}},$$

where D is the distance, in meters, from the center of Earth. For what values of V will the trajectory be hyperbolic if $D = 4.25 \times 10^7$ m? (*Source:* Kaler, James B., *Astronomy!*, Addison-Wesley, 1997.)

38. The percent of women in the work force has increased steadily for many years. The line graph shows the change for the period from 1975 through 1999, where $x = 75$ represents 1975, $x = 80$ represents 1980, and so on. The graph resembles the upper branch of a horizontal hyperbola. Using statistical methods, we found the corresponding square root equation

$$y = .607\sqrt{383.9 + x^2},$$

which closely approximates the line graph.

(a) According to the graph, what percent of women were in the work force in 1985?

(b) According to the equation, what percent of women worked in 1985? (Round to the nearest percent.)

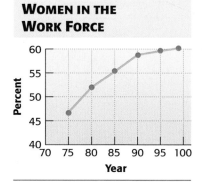

WOMEN IN THE WORK FORCE

Source: U.S. Bureau of Labor Statistics.

TECHNOLOGY INSIGHTS (EXERCISES 39 AND 40)

39. The hyperbola shown in the figure was graphed in function mode, with a square viewing window. It is the graph of $\frac{x^2}{9} - y^2 = 1$. What are the two functions y_1 and y_2 that were used to obtain this graph?

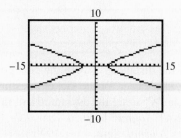

40. Repeat Exercise 39 for the graph of $\frac{y^2}{9} - x^2 = 1$, shown in the figure.

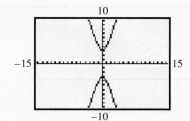

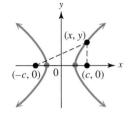 *Use a graphing calculator in function mode to graph each hyperbola. Use a square viewing window.*

41. $\dfrac{x^2}{25} - \dfrac{y^2}{49} = 1$ **42.** $\dfrac{x^2}{4} - \dfrac{y^2}{16} = 1$ **43.** $\dfrac{y^2}{9} - x^2 = 1$ **44.** $\dfrac{y^2}{36} - \dfrac{x^2}{4} = 1$

45. Suppose that a hyperbola has center at the origin, foci at $(-c, 0)$ and $(c, 0)$, and the absolute value of the difference between the distances from any point (x, y) of the hyperbola to the two foci is $2a$. See the figure. Let $b^2 = c^2 - a^2$, and show that an equation of the hyperbola is

$$\frac{x^2}{a^2} - \frac{y^2}{b^2} = 1.$$

46. Use the result of Exercise 45 to find an equation of a hyperbola with center at the origin, foci at $(-2, 0)$ and $(2, 0)$, and the absolute value of the difference between the distances from any point of the hyperbola to the two foci equal to 2.

13.4 Nonlinear Systems of Equations

OBJECTIVES

1 Solve a nonlinear system by substitution.

2 Use the elimination method to solve a system with two second-degree equations.

3 Solve a system that requires a combination of methods.

4 Use a graphing calculator to solve a nonlinear system.

An equation in which some terms have more than one variable or a variable of degree 2 or greater is called a **nonlinear equation**. A **nonlinear system of equations** includes at least one nonlinear equation.

When solving a nonlinear system, it helps to visualize the types of graphs of the equations of the system to determine the possible number of points of intersection. For example, if a system includes two equations where the graph of one is a parabola and the graph of the other is a line, then there may be zero, one, or two points of intersection, as illustrated in Figure 28.

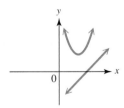
No points of intersection

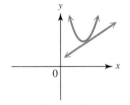
One point of intersection

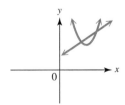
Two points of intersection

FIGURE 28

OBJECTIVE 1 Solve a nonlinear system by substitution. We solve nonlinear systems by the elimination method, the substitution method, or a combination of the two. The substitution method is usually best when one of the equations is linear.

EXAMPLE 1   Solving a Nonlinear System by Substitution

Solve the system

$$x^2 + y^2 = 9 \qquad (1)$$
$$2x - y = 3. \qquad (2)$$

The graph of (1) is a circle and the graph of (2) is a line. Visualizing the possible ways the graphs could intersect indicates that there may be zero, one, or two points of intersection. It is best to solve the linear equation first for one of the two variables; then substitute the resulting expression into the nonlinear equation to obtain an equation in one variable.

$$2x - y = 3 \qquad (2)$$
$$y = 2x - 3 \qquad (3)$$

Substitute $2x - 3$ for y in equation (1).

$$x^2 + (2x - 3)^2 = 9$$
$$x^2 + 4x^2 - 12x + 9 = 9$$
$$5x^2 - 12x = 0$$
$$x(5x - 12) = 0 \qquad \text{GCF is } x.$$
$$x = 0 \quad \text{or} \quad x = \frac{12}{5} \qquad \text{Zero-factor property}$$

Let $x = 0$ in equation (3) to get $y = -3$. If $x = \frac{12}{5}$, then $y = \frac{9}{5}$. The solution set of the system is $\left\{(0, -3), \left(\frac{12}{5}, \frac{9}{5}\right)\right\}$. The graph in Figure 29 confirms the two points of intersection.

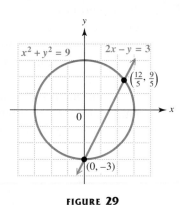

FIGURE 29

Now Try Exercise 19.

EXAMPLE 2   Solving a Nonlinear System by Substitution

Solve the system

$$6x - y = 5 \qquad (1)$$
$$xy = 4. \qquad (2)$$

The graph of (1) is a line. We have not specifically mentioned equations like (2); however, it can be shown by plotting points that its graph is a hyperbola. Visualizing a line and a hyperbola indicates that there may be zero, one, or two points of intersection. Since neither equation has a squared term, we can solve either equation for one of the variables and then substitute the result into the other equation. Solving $xy = 4$ for x gives $x = \frac{4}{y}$. We substitute $\frac{4}{y}$ for x in equation (1).

$$6\left(\frac{4}{y}\right) - y = 5 \qquad \text{Let } x = \frac{4}{y} \text{ in Equation (1).}$$

$$\frac{24}{y} - y = 5$$

$$24 - y^2 = 5y \qquad \text{Multiply by } y, \, y \neq 0.$$

$$0 = y^2 + 5y - 24 \qquad \text{Standard form}$$

$$0 = (y - 3)(y + 8) \qquad \text{Factor.}$$

$$y = 3 \quad \text{or} \quad y = -8 \qquad \text{Zero-factor property}$$

We substitute these results into $x = \frac{4}{y}$ to obtain the corresponding values of x.

If $y = 3$, then $x = \dfrac{4}{3}$. If $y = -8$, then $x = -\dfrac{1}{2}$.

The solution set of the system is $\left\{\left(\frac{4}{3}, 3\right), \left(-\frac{1}{2}, -8\right)\right\}$. The graph in Figure 30 shows that there are two points of intersection.

Now Try Exercise 21.

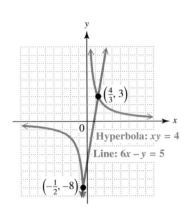

FIGURE 30

OBJECTIVE 2 Use the elimination method to solve a system with two second-degree equations. The elimination method is often used when both equations are second degree.

EXAMPLE 3 Solving a Nonlinear System by Elimination

Solve the system

$$x^2 + y^2 = 9 \qquad (1)$$
$$2x^2 - y^2 = -6. \qquad (2)$$

The graph of (1) is a circle, while the graph of (2) is a hyperbola. By analyzing the possibilities we conclude that there may be zero, one, two, three, or four points of intersection. Adding the two equations will eliminate y, leaving an equation that can be solved for x.

$$
\begin{array}{rcl}
x^2 + y^2 &=& 9 \\
2x^2 - y^2 &=& -6 \\
\hline
3x^2 &=& 3 \\
x^2 &=& 1 \\
x = 1 \quad \text{or} \quad x &=& -1
\end{array}
$$

Each value of x gives corresponding values for y when substituted into one of the original equations. Using equation (1) gives the following.

If $x = 1$, then
$$1^2 + y^2 = 9$$
$$y^2 = 8$$
$$y = \sqrt{8} \quad \text{or} \quad y = -\sqrt{8}$$
$$y = 2\sqrt{2} \quad \text{or} \quad y = -2\sqrt{2}.$$

If $x = -1$, then
$$(-1)^2 + y^2 = 9$$
$$y^2 = 8$$
$$y = 2\sqrt{2} \quad \text{or} \quad y = -2\sqrt{2}.$$

The solution set is $\left\{\left(1, 2\sqrt{2}\right),\left(1, -2\sqrt{2}\right),\left(-1, 2\sqrt{2}\right),\left(-1, -2\sqrt{2}\right)\right\}$. Figure 31 shows the four points of intersection.

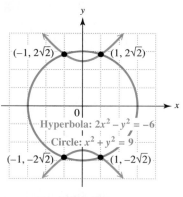

FIGURE 31

Now Try Exercise 35.

OBJECTIVE 3 Solve a system that requires a combination of methods. Solving a system of second-degree equations may require a combination of methods.

EXAMPLE 4 Solving a Nonlinear System by a Combination of Methods

Solve the system

$$x^2 + 2xy - y^2 = 7 \qquad (1)$$
$$x^2 - y^2 = 3. \qquad (2)$$

While we have not graphed equations like (1), its graph is a hyperbola. The graph of (2) is also a hyperbola. Two hyperbolas may have zero, one, two, three, or four points of intersection. We use the elimination method here in combination with the substitution method. We begin by eliminating the squared terms by multiplying each side of equation (2) by -1 and then adding the result to equation (1).

$$
\begin{array}{rcl}
x^2 + 2xy - y^2 &=& 7 \\
-x^2 \qquad\;\; + y^2 &=& -3 \\
\hline
2xy &=& 4
\end{array}
$$

Next, we solve $2xy = 4$ for y. (Either variable would do.)

$$2xy = 4$$
$$y = \frac{2}{x} \qquad (3)$$

Now, we substitute $y = \frac{2}{x}$ into one of the original equations. It is easier to do this with equation (2).

$$x^2 - y^2 = 3 \qquad (2)$$
$$x^2 - \left(\frac{2}{x}\right)^2 = 3$$
$$x^2 - \frac{4}{x^2} = 3$$

$$x^4 - 4 = 3x^2 \qquad \text{Multiply by } x^2, \, x \neq 0.$$
$$x^4 - 3x^2 - 4 = 0 \qquad \text{Subtract } 3x^2.$$
$$(x^2 - 4)(x^2 + 1) = 0 \qquad \text{Factor.}$$
$$x^2 - 4 = 0 \quad \text{or} \quad x^2 + 1 = 0$$
$$x^2 = 4 \quad \text{or} \qquad x^2 = -1$$
$$x = 2 \quad \text{or} \quad x = -2 \quad \text{or} \quad x = i \quad \text{or} \quad x = -i$$

Substituting these four values of x into equation (3) gives the corresponding values for y.

If $x = 2$, then $y = 1$.　　　　If $x = i$, then $y = -2i$.

If $x = -2$, then $y = -1$.　　　If $x = -i$, then $y = 2i$.

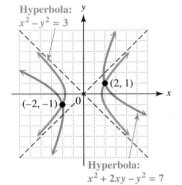

Hyperbola:
$x^2 - y^2 = 3$

(2, 1)

(−2, −1)

Hyperbola:
$x^2 + 2xy - y^2 = 7$

FIGURE 32

Note that if we substitute the x-values we found into equation (1) or (2) instead of into equation (3), we get extraneous solutions. It is always wise to check all solutions in both of the given equations. There are four ordered pairs in the solution set, two with real values and two with imaginary values. The solution set is

$$\{(2, 1), (-2, -1), (i, -2i), (-i, 2i)\}.$$

The graph of the system, shown in Figure 32, shows only the two real intersection points because the graph is in the real number plane. The two ordered pairs with imaginary components are solutions of the system, but do not appear on the graph.

Now Try Exercise 39.

NOTE　In the examples of this section, we analyzed the possible number of points of intersection of the graphs in each system. However, in Examples 2 and 4, we worked with equations whose graphs had not been studied. Keep in mind that it is not absolutely essential to visualize the number of points of intersection in order to solve the system. Furthermore, as in Example 4, there are sometimes imaginary solutions to nonlinear systems that do not appear as points of intersection in the real plane. Visualizing the geometry of the graphs is only an aid to solving these systems.

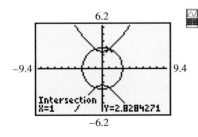

6.2

−9.4 ┼ 9.4

Intersection
X=1 　　Y=2.8284271

−6.2

FIGURE 33

OBJECTIVE 4 Use a graphing calculator to solve a nonlinear system.   If the equations in a nonlinear system can be solved for y, then we can graph the equations of the system with a graphing calculator and use the capabilities of the calculator to identify all intersection points. For instance, the two equations in Example 3 would require graphing the four separate functions

$$Y_1 = \sqrt{9 - X^2}, \quad Y_2 = -\sqrt{9 - X^2}, \quad Y_3 = \sqrt{2X^2 + 6}, \quad \text{and} \quad Y_4 = -\sqrt{2X^2 + 6}.$$

Figure 33 indicates the coordinates of one of the points of intersection.

13.4 EXERCISES

 1. Write an explanation of the steps you would use to solve the system

$$x^2 + y^2 = 25$$
$$y = x - 1$$

by the substitution method. Why would the elimination method not be appropriate for this system?

2. Write an explanation of the steps you would use to solve the system

$$x^2 + y^2 = 12$$
$$x^2 - y^2 = 13$$

by the elimination method.

Each sketch represents the graphs of a pair of equations in a system. How many points are in each solution set?

3.
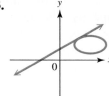

4.
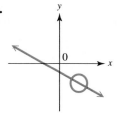

5.
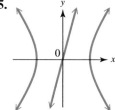

6.
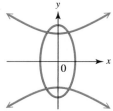

Suppose that a nonlinear system is composed of equations whose graphs are those described, and the number of points of intersection of the two graphs is as given. Make a sketch satisfying these conditions. (There may be more than one way to do this.)

7. A line and a circle; no points

8. A line and a circle; one point

9. A line and a hyperbola; one point

10. A line and an ellipse; no points

11. A circle and an ellipse; four points

12. A parabola and an ellipse; one point

13. A parabola and an ellipse; four points

14. A parabola and a hyperbola; two points

Solve each system by the substitution method. See Examples 1 and 2.

15. $y = 4x^2 - x$
$y = x$

16. $y = x^2 + 6x$
$3y = 12x$

17. $y = x^2 + 6x + 9$
$x + y = 3$

18. $y = x^2 + 8x + 16$
$x - y = -4$

19. $x^2 + y^2 = 2$
$2x + y = 1$

20. $2x^2 + 4y^2 = 4$
$x = 4y$

21. $xy = 4$
$3x + 2y = -10$

22. $xy = -5$
$2x + y = 3$

23. $xy = -3$
$x + y = -2$

24. $xy = 12$
$x + y = 8$

25. $y = 3x^2 + 6x$
$y = x^2 - x - 6$

26. $y = 2x^2 + 1$
$y = 5x^2 + 2x - 7$

27. $2x^2 - y^2 = 6$
$y = x^2 - 3$

28. $x^2 + y^2 = 4$
$y = x^2 - 2$

29. $x^2 - xy + y^2 = 0$
$x - 2y = 1$

30. $x^2 - 3x + y^2 = 4$
$2x - y = 3$

Solve each system by the elimination method or a combination of the elimination and substitution methods. See Examples 3 and 4.

31. $3x^2 + 2y^2 = 12$
$x^2 + 2y^2 = 4$

32. $2x^2 + y^2 = 28$
$4x^2 - 5y^2 = 28$

33. $2x^2 + 3y^2 = 6$
$x^2 + 3y^2 = 3$

34. $6x^2 + y^2 = 9$
$3x^2 + 4y^2 = 36$

35. $5x^2 - 2y^2 = -13$
$3x^2 + 4y^2 = 39$

36. $x^2 + 6y^2 = 9$
$4x^2 + 3y^2 = 36$

37. $2x^2 = 8 - 2y^2$
$3x^2 = 24 - 4y^2$

38. $5x^2 = 20 - 5y^2$
$2y^2 = 2 - x^2$

39. $x^2 + xy + y^2 = 15$
$x^2 + y^2 = 10$

40. $2x^2 + 3xy + 2y^2 = 21$
$x^2 + y^2 = 6$

41. $3x^2 + 2xy - 3y^2 = 5$
$-x^2 - 3xy + y^2 = 3$

42. $-2x^2 + 7xy - 3y^2 = 4$
$2x^2 - 3xy + 3y^2 = 4$

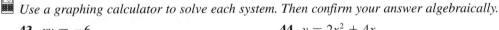 *Use a graphing calculator to solve each system. Then confirm your answer algebraically.*

43. $xy = -6$
$x + y = -1$

44. $y = 2x^2 + 4x$
$y = -x^2 - 1$

Solve each problem by using a nonlinear system.

45. The area of a rectangular rug is 84 ft^2 and its perimeter is 38 ft. Find the length and width of the rug.

46. Find the length and width of a rectangular room whose perimeter is 50 m and whose area is 100 m^2.

47. A company has found that the price p (in dollars) of its scientific calculator is related to the supply x (in thousands) by the equation

$$px = 16.$$

The price is related to the demand x (in thousands) for the calculator by the equation

$$p = 10x + 12.$$

The *equilibrium price* is the value of p where demand equals supply. Find the equilibrium price and the supply/demand at that price by solving a system of equations. (*Hint:* Demand, price, and supply must all be positive.)

48. The calculator company in Exercise 47 has also determined that the cost y to make x (thousand) calculators is

$$y = 4x^2 + 36x + 20,$$

while the revenue y from the sale of x (thousand) calculators is

$$36x^2 - 3y = 0.$$

Find the *break-even point,* where cost equals revenue, by solving a system of equations.

49. In the 1970s, the number of bachelor's degrees earned by men began to decrease. It stayed fairly constant in the 1980s, and then in the 1990s slowly began to increase again. Meanwhile, the number of bachelor's degrees earned by women continued to rise steadily throughout this period. Functions that model the situation are defined by the following equations, where y is the number of degrees (in thousands) granted in year x, with $x = 0$ corresponding to 1970.

$$\text{Men:} \qquad y = .138x^2 + .064x + 451$$
$$\text{Women:} \quad y = 12.1x + 334$$

Solve this system of equations to find the year when the same number of bachelor's degrees were awarded to men and women. How many bachelor's degrees were awarded to each sex in that year? Give the answer to the nearest ten thousand. (*Source:* U.S. National Center for Education Statistics, *Digest of Education Statistics,* annual.)

50. Andy Grove, chairman of chip maker Intel Corp., once noted that decreasing prices for computers and stable prices for Internet access implied that the trend lines for these costs either have crossed or soon will. He predicted that the time is not far away when computers, like cell phones, may be given away to sell on-line time. To see this, assume a price of $1000 for a computer, and let x represent the number of months it will be used. (*Source:* Corcoran, Elizabeth, "Can Free Computers Be That Far Away?", *Washington Post,* from *Sacramento Bee,* February 3, 1999.)

(a) Write an equation for the monthly cost y of the computer over this period.
(b) The average monthly on-line cost is about $20. Assume this will remain constant and write an equation to express this cost.
(c) Solve the system of equations from parts (a) and (b). Interpret your answer in relation to the situation.

13.5 Second-Degree Inequalities and Systems of Inequalities

OBJECTIVES

1 Graph second-degree inequalities.

2 Graph the solution set of a system of inequalities.

OBJECTIVE 1 Graph second-degree inequalities. The linear inequality $3x + 2y \leq 5$ is graphed by first graphing the boundary line $3x + 2y = 5$. *Second-degree inequalities* are graphed in the same way. A **second-degree inequality** is an inequality with at least one variable of degree 2 and no variable with degree greater than 2. An example is $x^2 + y^2 \leq 36$. The boundary of the inequality $x^2 + y^2 \leq 36$ is the graph of the equation $x^2 + y^2 = 36$, a circle with radius 6 and center at the origin, as shown in Figure 34.

The inequality $x^2 + y^2 \leq 36$ will include either the points outside the circle or the points inside the circle, as well as the boundary. We decide which region to shade by substituting any test point not on the circle, such as $(0, 0)$, into the original inequality. Since $0^2 + 0^2 \leq 36$ is a true statement, the original inequality includes the points inside the circle, the shaded region in Figure 34, and the boundary.

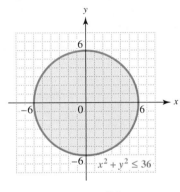

$x^2 + y^2 \leq 36$

FIGURE 34

EXAMPLE 1 Graphing a Second-Degree Inequality

Graph $y < -2(x - 4)^2 - 3$.

The boundary, $y = -2(x - 4)^2 - 3$, is a parabola that opens down with vertex at $(4, -3)$. Using $(0, 0)$ as a test point gives

$$0 < -2(0 - 4)^2 - 3 \qquad ?$$
$$0 < -32 - 3 \qquad ?$$
$$0 < -35. \qquad \text{False}$$

Because the final inequality is a false statement, the points in the region containing $(0, 0)$ do not satisfy the inequality. Figure 35 shows the final graph; the parabola is drawn as a dashed curve since the points of the parabola itself do not satisfy the inequality, and the region inside (or below) the parabola is shaded.

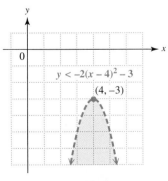

$y < -2(x - 4)^2 - 3$

$(4, -3)$

FIGURE 35

Now Try Exercise 11.

NOTE Since the substitution is easy, the origin is the test point of choice unless the graph actually passes through $(0, 0)$.

EXAMPLE 2 Graphing a Second-Degree Inequality

Graph $16y^2 \leq 144 + 9x^2$.

First rewrite the inequality as follows.

$$16y^2 - 9x^2 \leq 144 \qquad \text{Subtract } 9x^2.$$

$$\frac{y^2}{9} - \frac{x^2}{16} \leq 1 \qquad \text{Divide by 144.}$$

This form shows that the boundary is the hyperbola given by

$$\frac{y^2}{9} - \frac{x^2}{16} = 1.$$

Since the graph is a vertical hyperbola, the desired region will be either the region between the branches or the regions above the top branch and below the bottom branch. Choose $(0, 0)$ as a test point. Substituting into the original inequality leads to $0 \leq 144$, a true statement, so the region between the branches containing $(0, 0)$ is shaded, as shown in Figure 36.

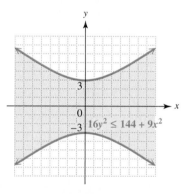

FIGURE 36

Now Try Exercise 17.

OBJECTIVE 2 Graph the solution set of a system of inequalities. If two or more inequalities are considered at the same time, we have a **system of inequalities.** To find the solution set of the system, we find the intersection of the graphs (solution sets) of the inequalities in the system.

EXAMPLE 3 Graphing a System of Two Inequalities

Graph the solution set of the system

$$2x + 3y > 6$$
$$x^2 + y^2 < 16.$$

Begin by graphing the solution set of $2x + 3y > 6$. The boundary line is the graph of $2x + 3y = 6$ and is a dashed line because of the symbol $>$. The test point $(0, 0)$ leads to a false statement in the inequality $2x + 3y > 6$, so shade the region above the line, as shown in Figure 37. The graph of $x^2 + y^2 < 16$ is the interior of a dashed circle centered at the origin with radius 4. This is shown in Figure 38.

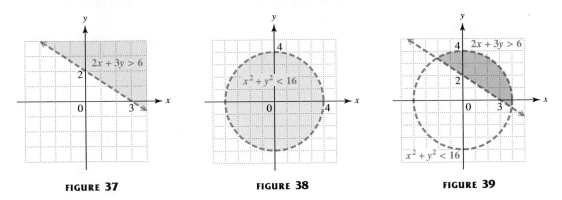

FIGURE 37   **FIGURE 38**   **FIGURE 39**

Finally, to get the graph of the solution set of the system, determine the intersection of the graphs of the two inequalities. The overlapping region in Figure 39 is the solution set.

Now Try Exercise 29.

EXAMPLE 4   Graphing a Linear System with Three Inequalities

Graph the solution set of the system

$$x + y < 1$$
$$y \le 2x + 3$$
$$y \ge -2.$$

Graph each inequality separately, on the same axes. The graph of $x + y < 1$ consists of all points below the dashed line $x + y = 1$. The graph of $y \le 2x + 3$ is the region that lies below the solid line $y = 2x + 3$. Finally, the graph of $y \ge -2$ is the region above the solid horizontal line $y = -2$.

The graph of the system, the intersection of these three graphs, is the triangular region enclosed by the three boundary lines in Figure 40, including two of its boundaries.

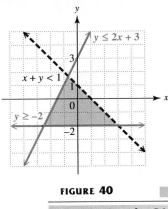

FIGURE 40

Now Try Exercise 31.

EXAMPLE 5   Graphing a System with Three Inequalities

Graph the solution set of the system

$$y \ge x^2 - 2x + 1$$
$$2x^2 + y^2 > 4$$
$$y < 4.$$

The graph of $y = x^2 - 2x + 1$ is a parabola with vertex at $(1, 0)$. Those points above (or in the interior of) the parabola satisfy the condition $y > x^2 - 2x + 1$.

Thus, points on the parabola or in the interior are the solution set of $y \geq x^2 - 2x + 1$. The graph of the equation $2x^2 + y^2 = 4$ is an ellipse. We draw it as a dashed curve. To satisfy the inequality $2x^2 + y^2 > 4$, a point must lie outside the ellipse. The graph of $y < 4$ includes all points below the dashed line $y = 4$. Finally, the graph of the system is the shaded region in Figure 41, which lies outside the ellipse, inside or on the boundary of the parabola, and below the line $y = 4$.

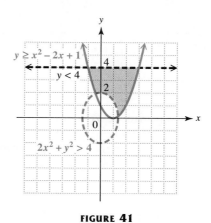

FIGURE 41

Now Try Exercise 33.

13.5 EXERCISES

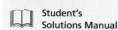

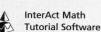

1. Which one of the following is a description of the graph of the solution set of the following system?

$$x^2 + y^2 < 25$$
$$y > -2$$

A. All points outside the circle $x^2 + y^2 = 25$ and above the line $y = -2$
B. All points outside the circle $x^2 + y^2 = 25$ and below the line $y = -2$
C. All points inside the circle $x^2 + y^2 = 25$ and above the line $y = -2$
D. All points inside the circle $x^2 + y^2 = 25$ and below the line $y = -2$

2. Fill in each blank with the appropriate response. The graph of the system

$$y > x^2 + 1$$
$$\frac{x^2}{9} + \frac{y^2}{4} > 1$$
$$y < 5$$

consists of all points ＿＿＿＿＿＿＿ the parabola $y = x^2 + 1$, ＿＿＿＿＿＿＿ the
                  (above/below)   (inside/outside)

ellipse $\frac{x^2}{9} + \frac{y^2}{4} = 1$, and ＿＿＿＿＿＿＿ the line $y = 5$.
                          (above/below)

3. Explain how to graph the solution set of a nonlinear inequality.

4. Explain how to graph the solution set of a system of nonlinear inequalities.

Match each nonlinear inequality with its graph.

5. $y \geq x^2 + 4$ **6.** $y \leq x^2 + 4$ **7.** $y < x^2 + 4$ **8.** $y > x^2 + 4$

A. 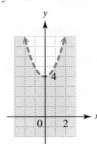   **B.** 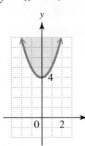   **C.** 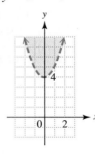   **D.**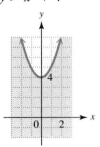

Graph each nonlinear inequality. See Examples 1 and 2.

9. $y^2 > 4 + x^2$ **10.** $y^2 \leq 4 - 2x^2$ **11.** $y + 2 \geq x^2$

12. $x^2 \leq 16 - y^2$ **13.** $2y^2 \geq 8 - x^2$ **14.** $x^2 \leq 16 + 4y^2$

15. $y \leq x^2 + 4x + 2$ **16.** $9x^2 < 16y^2 - 144$ **17.** $9x^2 > 16y^2 + 144$

18. $4y^2 \leq 36 - 9x^2$ **19.** $x^2 - 4 \geq -4y^2$ **20.** $x \geq y^2 - 8y + 14$

21. $x \leq -y^2 + 6y - 7$ **22.** $y^2 - 16x^2 \leq 16$

Graph each system of inequalities. See Examples 3–5.

23. $2x + 5y < 10$ **24.** $3x - y > -6$ **25.** $5x - 3y \leq 15$ **26.** $4x - 3y \leq 0$
       $x - 2y < 4$ $4x + 3y > 12$ $4x + y \geq 4$ $x + y \leq 5$

27. $x \leq 5$ **28.** $x \geq -2$ **29.** $y > x^2 - 4$ **30.** $x^2 - y^2 \geq 9$
      $y \leq 4$ $y \leq 4$ $y < -x^2 + 3$ $\dfrac{x^2}{16} + \dfrac{y^2}{9} \leq 1$

31. $x^2 + y^2 \geq 4$ **32.** $y^2 - x^2 \geq 4$ **33.** $y \leq -x^2$ **34.** $y < x^2$
       $x + y \leq 5$ $-5 \leq y \leq 5$ $y \geq x - 3$ $y > -2$
       $x \geq 0$ $y \leq -1$ $x + y < 3$
       $y \geq 0$ $x < 1$ $3x - 2y > -6$

For each nonlinear inequality in Exercises 35–42, a restriction is placed on one or both variables. For example, the graph of

$$x^2 + y^2 \leq 4, \quad x \geq 0$$

would be as shown in the figure. Only the right half of the interior of the circle and its boundary is shaded, because of the restriction that x must be nonnegative. Graph each nonlinear inequality with the given restrictions.

35. $x^2 + y^2 > 36, \quad x \geq 0$ **36.** $4x^2 + 25y^2 < 100, \quad y < 0$

37. $x < y^2 - 3, \quad x < 0$ **38.** $x^2 - y^2 < 4, \quad x < 0$

39. $4x^2 - y^2 > 16, \quad x < 0$ **40.** $x^2 + y^2 > 4, \quad y < 0$

41. $x^2 + 4y^2 \geq 1, \quad x \geq 0, y \geq 0$ **42.** $2x^2 - 32y^2 \leq 8, \quad x \leq 0, y \geq 0$

 Use the shading feature of a graphing calculator to graph each system.

43. $y \geq x - 3$ **44.** $y \geq -x^2 + 5$
       $y \leq -x + 4$ $y \leq x^2 - 3$

45. $y < x^2 + 4x + 4$ **46.** $y > (x - 4)^2 - 3$
       $y > -3$ $y < 5$

Chapter **13** **Group Activity**

Finding the Paths of Natural Satellites

OBJECTIVE Write and graph equations of ellipses from given data.

The moon, which orbits Earth, and Halley's comet, which orbits the sun, are both natural satellites. In Section 13.2, you solved problems where you were given equations of ellipses for the orbits of planets and were asked to find apogees (greatest distance from the sun) and perigees (smallest distance from the sun). This activity reverses the process; that is, given apogees and perigees you must find equations of ellipses.

A. Have each student choose a natural satellite from the table. Predict the shape of the orbital ellipse for your satellite.

Natural Satellite	Apogee	Perigee
Moon	406.7 thousand km from Earth	356.4 thousand km from Earth
Halley's comet	35 astronomical units* from the sun	.6 astronomical unit* from the sun

Source: World Book Encyclopedia.
*One astronomical unit (AU) is the distance from Earth to the sun.

B. For your satellite, do the following.

1. Find values for a, b, and c. Note that apogee $= a + c$, perigee $= a - c$, and $c^2 = a^2 - b^2$.

2. Write the equation of the ellipse in the form $\dfrac{x^2}{a^2} + \dfrac{y^2}{b^2} = 1$.

3. Rewrite the equation so it can be graphed on a graphing calculator. (See Section 13.2, Objective 5.)

4. Graph your equation on a graphing calculator. Adjust the window setting in order to see the entire graph. Once the window is set correctly, get a square window to see the true shape of the ellipse.

C. Compare your graph with your partner's graph.

1. Do the graphs reflect the shapes you predicted in part A?

2. What window was used to graph each ellipse?

CHAPTER **13** SUMMARY

KEY TERMS

13.1 asymptotes greatest integer function step function composition	**13.2** conic section circle center radius center-radius form ellipse foci	**13.3** hyperbola asymptotes of a hyperbola fundamental rectangle square root function	**13.4** nonlinear equation nonlinear system of equations **13.5** second-degree inequality system of inequalities

TEST YOUR WORD POWER

See how well you have learned the vocabulary in this chapter. Answers, with examples, follow the Quick Review.

1. Conic sections are
   A. graphs of first-degree equations
   B. the result of two or more intersecting planes
   C. graphs of first-degree inequalities
   D. figures that result from the intersection of an infinite cone with a plane.

2. A circle is the set of all points in a plane
   A. such that the absolute value of the difference of the distances from two fixed points is constant
   B. that lie a fixed distance from a fixed point
   C. the sum of whose distances from two fixed points is constant
   D. that make up the graph of any second-degree equation.

3. An **ellipse** is the set of all points in a plane
   A. such that the absolute value of the difference of the distances from two fixed points is constant
   B. that lie a fixed distance from a fixed point
   C. the sum of whose distances from two fixed points is constant
   D. that make up the graph of any second-degree equation.

4. A **hyperbola** is the set of all points in a plane
   A. such that the absolute value of the difference of the distances from two fixed points is constant
   B. that lie a fixed distance from a fixed point
   C. the sum of whose distances from two fixed points is constant

   D. that make up the graph of any second-degree equation.

5. A **nonlinear equation** is an equation
   A. in which some terms have more than one variable or a variable of degree 2 or greater
   B. in which the terms have only one variable
   C. of degree 1
   D. of a linear function.

6. A **nonlinear system of equations** is a system
   A. with at least one linear equation
   B. with two or more inequalities
   C. with at least one nonlinear equation
   D. with at least two linear equations.

QUICK REVIEW

CONCEPTS	EXAMPLES

13.1 ADDITIONAL GRAPHS OF FUNCTIONS; COMPOSITION

Other Functions

In addition to the squaring function, some other important elementary functions in algebra are the absolute value function, defined by $f(x) = |x|$; the reciprocal function, defined by $f(x) = \frac{1}{x}$; and the square root function, defined by $f(x) = \sqrt{x}$.

Step functions, such as the greatest integer function, defined by $f(x) = [\![x]\!]$, are useful in applications.

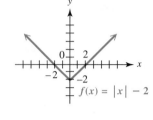

$f(x) = |x| - 2$

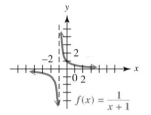

$f(x) = \frac{1}{x+1}$

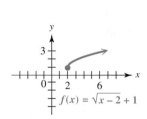

$f(x) = \sqrt{x-2} + 1$

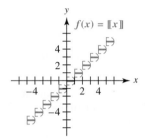

$f(x) = [\![x]\!]$

Operations on Functions

If $f(x)$ and $g(x)$ define functions, then

$$(f + g)(x) = f(x) + g(x),$$
$$(f - g)(x) = f(x) - g(x),$$
$$(fg)(x) = f(x) \cdot g(x),$$

and $\quad \left(\dfrac{f}{g}\right)(x) = \dfrac{f(x)}{g(x)}, \quad g(x) \neq 0.$

If $f(x) = x^2$ and $g(x) = 2x + 1$, then

$$(f + g)(x) = f(x) + g(x) = x^2 + 2x + 1,$$
$$(f - g)(x) = f(x) - g(x) = x^2 - 2x - 1,$$
$$(fg)(x) = f(x) \cdot g(x) = 2x^3 + x^2,$$

and $\quad \left(\dfrac{f}{g}\right)(x) = \dfrac{f(x)}{g(x)} = \dfrac{x^2}{2x + 1}, \quad x \neq -\dfrac{1}{2}.$

Composition of f and g

$$(f \circ g)(x) = f[g(x)]$$

If $f(x) = x^2$ and $g(x) = 2x + 1$, then

$$(f \circ g)(x) = f[g(x)]$$
$$= (2x + 1)^2 = 4x^2 + 4x + 1$$

and $\quad (g \circ f)(x) = g[f(x)]$
$$= 2x^2 + 1.$$

13.2 THE CIRCLE AND THE ELLIPSE

Circle

The circle with radius r and center at (h, k) has an equation of the form

$$(x - h)^2 + (y - k)^2 = r^2.$$

The circle with equation $(x + 2)^2 + (y - 3)^2 = 25$, which can be written $[x - (-2)]^2 + (y - 3)^2 = 5^2$, has center $(-2, 3)$ and radius 5.

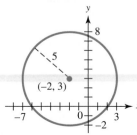

CONCEPTS	EXAMPLES

Ellipse

The ellipse whose x-intercepts are $(a, 0)$ and $(-a, 0)$ and whose y-intercepts are $(0, b)$ and $(0, -b)$ has an equation of the form

$$\frac{x^2}{a^2} + \frac{y^2}{b^2} = 1.$$

Graph $\dfrac{x^2}{9} + \dfrac{y^2}{4} = 1.$

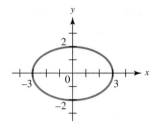

13.3 THE HYPERBOLA AND FUNCTIONS DEFINED BY RADICALS

Hyperbola

A hyperbola with x-intercepts $(a, 0)$ and $(-a, 0)$ has an equation of the form

$$\frac{x^2}{a^2} - \frac{y^2}{b^2} = 1,$$

and a hyperbola with y-intercepts $(0, b)$ and $(0, -b)$ has an equation of the form

$$\frac{y^2}{b^2} - \frac{x^2}{a^2} = 1.$$

Graph $\dfrac{x^2}{4} - \dfrac{y^2}{4} = 1.$

The graph has x-intercepts $(2, 0)$ and $(-2, 0)$.

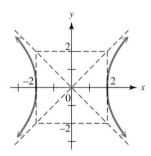

The extended diagonals of the fundamental rectangle with vertices at the points (a, b), $(-a, b)$, $(-a, -b)$, and $(a, -b)$ are the asymptotes of these hyperbolas.

The fundamental rectangle has vertices at $(2, 2)$, $(-2, 2)$, $(-2, -2)$, and $(2, -2)$.

Graphing a Square Root Function

To graph a square root function defined by

$$f(x) = u$$

for an algebraic expression u, with $u \geq 0$, square both sides so that the equation can be easily recognized. Then graph only the part indicated by the original equation.

Graph $y = -\sqrt{4 - x^2}$.

Square both sides and rearrange terms to get

$$x^2 + y^2 = 4.$$

This equation has a circle as its graph. However, graph only the lower half of the circle, since the original equation indicates that y cannot be positive.

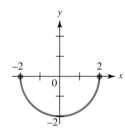

(continued)

CONCEPTS	EXAMPLES

13.4 NONLINEAR SYSTEMS OF EQUATIONS

Solving a Nonlinear System

A nonlinear system can be solved by the substitution method, the elimination method, or a combination of the two.

Solve the system

$$x^2 + 2xy - y^2 = 14 \quad (1)$$
$$x^2 - y^2 = -16. \quad (2)$$

Multiply equation (2) by -1 and use elimination.

$$
\begin{array}{rl}
x^2 + 2xy - y^2 &= 14 \\
-x^2 \qquad\; + y^2 &= 16 \\
\hline
2xy \qquad &= 30 \\
xy &= 15
\end{array}
$$

Solve for y to obtain $y = \frac{15}{x}$, and substitute into equation (2).

$$x^2 - \left(\frac{15}{x}\right)^2 = -16$$

$$x^2 - \frac{225}{x^2} = -16$$

$x^4 + 16x^2 - 225 = 0$ Multiply by x^2; add $16x^2$.

$(x^2 - 9)(x^2 + 25) = 0$ Factor.

$x = \pm 3 \quad \text{or} \quad x = \pm 5i$ Zero-factor property

Find corresponding y-values to get the solution set

$$\{(3, 5), (-3, -5), (5i, -3i), (-5i, 3i)\}.$$

13.5 SECOND-DEGREE INEQUALITIES AND SYSTEMS OF INEQUALITIES

Graphing a Second-Degree Inequality

To graph a second-degree inequality, graph the corresponding equation as a boundary and use test points to determine which region(s) form the solution set. Shade the appropriate region(s).

Graphing a System of Inequalities

The solution set of a system of inequalities is the intersection of the solution sets of the individual inequalities.

Graph $y \geq x^2 - 2x + 3$.

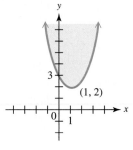

Graph the solution set of the system

$$3x - 5y > -15$$
$$x^2 + y^2 \leq 25.$$

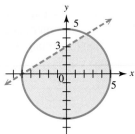

Answers to Test Your Word Power

1. D; *Example:* Parabolas, circles, ellipses, and hyperbolas are conic sections.   **2.** B; *Example:* See the graph of $x^2 + y^2 = 9$ in Figure 12 of Section 13.2.   **3.** C; *Example:* See the graph of $\frac{x^2}{49} + \frac{y^2}{36} = 1$ in Figure 17 of Section 13.2.   **4.** A; *Example:* See the graph of $\frac{x^2}{16} - \frac{y^2}{12} = 1$ in Figure 21 of Section 13.3.   **5.** A; *Examples:* $y = x^2 + 8x + 16$, $xy = 5$, $2x^2 - y^2 = 6$ **6.** C; *Example:* $x^2 + y^2 = 2$
$2x + y = 1$

CHAPTER **13** REVIEW EXERCISES

[13.1]   *Graph each function.*

1. $f(x) = |x + 4|$ **2.** $f(x) = \dfrac{1}{x - 4}$ **3.** $f(x) = \sqrt{x} + 3$ **4.** $f(x) = [\![x]\!] - 2$

5. For $f(x) = 2x + 3$ and $g(x) = 5x^2 - 3x + 2$, find each of the following.

   (a) $(f + g)(x)$ **(b)** $(f - g)(x)$ **(c)** $(f + g)(-1)$ **(d)** $(f - g)(-1)$

6. For $f(x) = 12x^2 - 3x$ and $g(x) = 3x$, find each of the following.

   (a) $(fg)(x)$ **(b)** $\left(\dfrac{f}{g}\right)(x)$ **(c)** $(fg)(-1)$ **(d)** $\left(\dfrac{f}{g}\right)(2)$

Let $f(x) = 3x^2 + 2x - 1$ and $g(x) = 5x + 7$. Find the following.

7. $(g \circ f)(3)$ **8.** $(f \circ g)(3)$ **9.** $(f \circ g)(-2)$

10. $(g \circ f)(-2)$ **11.** $(f \circ g)(x)$ **12.** $(g \circ f)(x)$

[13.2]   *Write an equation for each circle.*

13. Center $(-2, 4)$, $r = 3$ **14.** Center $(-1, -3)$, $r = 5$ **15.** Center $(4, 2)$, $r = 6$

Find the center and radius of each circle.

16. $x^2 + y^2 + 6x - 4y - 3 = 0$ **17.** $x^2 + y^2 - 8x - 2y + 13 = 0$

18. $2x^2 + 2y^2 + 4x + 20y = -34$ **19.** $4x^2 + 4y^2 - 24x + 16y = 48$

Graph each equation.

20. $x^2 + y^2 = 16$ **21.** $\dfrac{x^2}{16} + \dfrac{y^2}{9} = 1$ **22.** $\dfrac{x^2}{49} + \dfrac{y^2}{25} = 1$

23. A satellite is in an elliptical orbit around Earth with perigee altitude of 160 km and apogee altitude of 16,000 km. See the figure. (*Source:* Kastner, Bernice, *Space Mathematics,* NASA.) Find the equation of the ellipse.

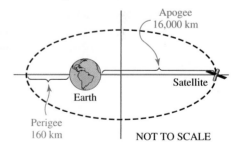

24. (a) The Roman Colosseum is an ellipse with $a = 310$ ft and $b = \dfrac{513}{2}$ ft. Find the distance between the foci of this ellipse.

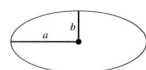

   (b) A formula for the approximate circumference of an ellipse is

$$C \approx 2\pi \sqrt{\dfrac{a^2 + b^2}{2}},$$

where a and b are the lengths given in part (a). Use this formula to find the approximate circumference of the Roman Colosseum.

[13.3] *Graph each equation.*

25. $\dfrac{x^2}{16} - \dfrac{y^2}{25} = 1$ 　　　　 **26.** $\dfrac{y^2}{25} - \dfrac{x^2}{4} = 1$ 　　　　 **27.** $f(x) = -\sqrt{16 - x^2}$

Identify the graph of each equation as a parabola, circle, ellipse, *or* hyperbola.

28. $x^2 + y^2 = 64$ 　　　　 **29.** $y = 2x^2 - 3$ 　　　　 **30.** $y^2 = 2x^2 - 8$

31. $y^2 = 8 - 2x^2$ 　　　　 **32.** $x = y^2 + 4$ 　　　　 **33.** $x^2 - y^2 = 64$

34. Ships and planes often use a location-finding system called LORAN. With this system, a radio transmitter at *M* sends out a series of pulses. (See the figure.) When each pulse is received at transmitter *S*, it then sends out a pulse. A ship at *P* receives pulses from both *M* and *S*. A receiver on the ship measures the difference in the arrival times of the pulses. A special map gives hyperbolas that correspond to the differences in arrival times (which give the distances d_1 and d_2 in the figure.) The ship can then be located as lying on a branch of a particular hyperbola.

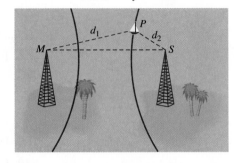

 Suppose $d_1 = 80$ mi and $d_2 = 30$ mi, and the distance between transmitters *M* and *S* is 100 mi. Use the definition to find an equation of the hyperbola the ship is located on.

[13.4] *Solve each system.*

35. $2y = 3x - x^2$ 　　　　 **36.** $y + 1 = x^2 + 2x$ 　　　　 **37.** $x^2 + 3y^2 = 28$
$\quad\; x + 2y = -12$ 　　　　　　　 $y + 2x = 4$ 　　　　　　　　 $\quad\; y - x = -2$

38. $xy = 8$ 　　　　　　 **39.** $x^2 + y^2 = 6$ 　　　　 **40.** $3x^2 - 2y^2 = 12$
$\quad\; x - 2y = 6$ 　　　　　　 $x^2 - 2y^2 = -6$ 　　　　　 $\quad\; x^2 + 4y^2 = 18$

41. How many solutions are possible for a system of two equations whose graphs are a circle and a line?

42. How many solutions are possible for a system of two equations whose graphs are a parabola and a hyperbola?

[13.5] *Graph each inequality.*

43. $9x^2 \geq 16y^2 + 144$ 　　　 **44.** $4x^2 + y^2 \geq 16$ 　　　　 **45.** $y < -(x + 2)^2 + 1$

Graph each system of inequalities.

46. $2x + 5y \leq 10$ 　　　　 **47.** $|x| \leq 2$ 　　　　　 **48.** $9x^2 \leq 4y^2 + 36$
$\quad\; 3x - \;\; y \leq 6$ 　　　　　　 $|y| > 1$ 　　　　　　　 $\quad\; x^2 + y^2 \leq 16$
　　　　　　　　　　　　　　　 $4x^2 + 9y^2 \leq 36$

RELATING CONCEPTS (EXERCISES 49–53)

For Individual or Group Work

In Chapter 8 we discussed several methods of solving systems of linear equations in three variables. These methods can be used to find an equation of a circle through three points in a plane that are not on the same line. The equation of a circle can be written in the form $x^2 + y^2 + ax + by + c = 0$ for some values of a, b, and c. **Work Exercises 49–53 in order,** *to find the equation of the circle through the points* (2, 4)*,* (5, 1)*, and* (−1, 1)*.*

49. Determine one equation in *a*, *b*, and *c* by letting $x = 2$ and $y = 4$ in the general form given above. Write it with *a*, *b*, and *c* on the left and the constant on the right.

50. Repeat Exercise 49 for the point (5, 1).

51. Repeat Exercise 49 for the point (−1, 1).

52. Solve the system formed by the equations found in Exercises 49–51, and give the equation of the circle that satisfies these conditions.

53. Use the methods of this chapter to find the center and the radius of the circle in Exercise 52.

MIXED REVIEW EXERCISES

Graph.

54. $\dfrac{x^2}{64} + \dfrac{y^2}{25} = 1$

55. $\dfrac{y^2}{4} - 1 = \dfrac{x^2}{9}$

56. $x^2 + y^2 = 25$

57. $x^2 + 9y^2 = 9$

58. $x^2 - 9y^2 = 9$

59. $f(x) = \sqrt{4 - x}$

60. $3x + 2y \ge 0$
$y \le 4$
$x \le 4$

61. $4y > 3x - 12$
$x^2 < 16 - y^2$

62. Explain why a set of points that form an ellipse does not satisfy the definition of a function.

The orbit of Mercury around the sun (a focus) is an ellipse with equation

$$\frac{x^2}{3352} + \frac{y^2}{3211} = 1,$$

where x and y are measured in million kilometers.

63. Find its apogee, its greatest distance from the sun. (*Hint:* Refer to Section 13.2, Exercise 51.)

64. Find its perigee, its smallest distance from the sun.

CHAPTER **13** TEST

Match each function with its graph from choices A–D.

1. $f(x) = \sqrt{x - 2}$

2. $f(x) = \sqrt{x + 2}$

3. $f(x) = \sqrt{x} + 2$

4. $f(x) = \sqrt{x} - 2$

A.
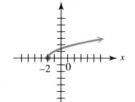

B.
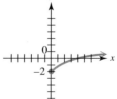

C.
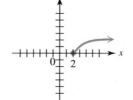

D.
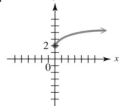

5. Sketch the graph of $f(x) = |x - 3| + 4$.

6. For $f(x) = 3x + 5$ and $g(x) = x^2 + 2$, find the following.

   (a) $(f + g)(x)$ **(b)** $(f - g)(2)$ **(c)** $(fg)(x)$
   (d) $(f \circ g)(-2)$ **(e)** $(f \circ g)(x)$ **(f)** $(g \circ f)(x)$

7. Find the center and radius of the circle whose equation is $(x - 2)^2 + (y + 3)^2 = 16$. Sketch the graph.

8. Find the center and radius of the circle whose equation is $x^2 + y^2 + 8x - 2y = 8$.

Graph.

9. $f(x) = \sqrt{9 - x^2}$

10. $4x^2 + 9y^2 = 36$

11. $16y^2 - 4x^2 = 64$

12. $\dfrac{y}{2} = -\sqrt{1 - \dfrac{x^2}{9}}$

Identify the graph of each equation as a parabola, hyperbola, ellipse, *or* circle.

13. $6x^2 + 4y^2 = 12$

14. $16x^2 = 144 + 9y^2$

15. $4y^2 + 4x = 9$

Solve each nonlinear system.

16. $2x - y = 9$
    $xy = 5$

17. $x - 4 = 3y$
    $x^2 + y^2 = 8$

18. $x^2 + y^2 = 25$
    $x^2 - 2y^2 = 16$

19. Graph the inequality $y < x^2 - 2$.

20. Graph the system $\begin{aligned} x^2 + 25y^2 &\le 25 \\ x^2 + y^2 &\le 9. \end{aligned}$

CUMULATIVE REVIEW EXERCISES CHAPTERS 1–13

1. Simplify $-10 + |-5| - |3| + 4$.

Solve.

2. $4 - (2x + 3) + x = 5x - 3$

3. $-4k + 7 \ge 6k + 1$

4. Find the slope of the line through $(2, 5)$ and $(-4, 1)$.

5. Find an equation of the line through $(-3, -2)$ and perpendicular to the graph of $2x - 3y = 7$. Write the equation in standard form.

Perform the indicated operations.

6. $(5y - 3)^2$

7. $(2r + 7)(6r - 1)$

8. $\dfrac{8x^4 - 4x^3 + 2x^2 + 13x + 8}{2x + 1}$

Factor.

9. $12x^2 - 7x - 10$

10. $2y^4 + 5y^2 - 3$

11. $z^4 - 1$

12. $a^3 - 27b^3$

Perform the indicated operations.

13. $\dfrac{5x - 15}{24} \cdot \dfrac{64}{3x - 9}$

14. $\dfrac{y^2 - 4}{y^2 - y - 6} \div \dfrac{y^2 - 2y}{y - 1}$

15. $\dfrac{5}{c + 5} - \dfrac{2}{c + 3}$

16. $\dfrac{p}{p^2 + p} + \dfrac{1}{p^2 + p}$

Solve.

17. Kareem and Jamal want to clean their office. Kareem can do the job alone in 3 hr, while Jamal can do it alone in 2 hr. How long will it take them if they work together?

18. The president of InstaTune, a chain of franchised automobile tune-up shops, reports that people who buy a franchise and open a shop pay a weekly fee (in dollars) to company headquarters, according to the linear function defined by

$$f(x) = .07x + 135,$$

where $f(x)$ is the fee and x is the total amount of money taken in during the week by the shop. Find the weekly fee if $2000 is taken in for the week. (*Source: Business Week.*)

Solve each system.

19. $3x - y = 12$
$2x + 3y = -3$

20. $x + y - 2z = 9$
$2x + y + z = 7$
$3x - y - z = 13$

21. $xy = -5$
$2x + y = 3$

22. Al and Bev traveled from their apartment to a picnic 20 mi away. Al traveled on his bike while Bev, who left later, took her car. Al's average speed was half of Bev's average speed. The trip took Al $\frac{1}{2}$ hr longer than Bev. What was Bev's average speed?

Simplify. Assume all variables represent positive real numbers.

23. $\left(\dfrac{4}{3}\right)^{-1}$

24. $\dfrac{(2a)^{-2}a^4}{a^{-3}}$

25. $4\sqrt[3]{16} - 2\sqrt[3]{54}$

26. $\dfrac{3\sqrt{5x}}{\sqrt{2x}}$

27. $\dfrac{5 + 3i}{2 - i}$

Solve.

28. $|5m| - 6 = 14$

29. $|2p - 5| > 15$

30. $2\sqrt{k} = \sqrt{5k + 3}$

31. $10q^2 + 13q = 3$

32. $(4x - 1)^2 = 8$

33. $3k^2 - 3k - 2 = 0$

34. $2(x^2 - 3)^2 - 5(x^2 - 3) = 12$

35. $F = \dfrac{kwv^2}{r}$ for v

36. If $f(x) = x^3 + 4$, find $f^{-1}(x)$.

37. Evaluate $3^{\log_3 4}$.

38. Evaluate $e^{\ln 7}$.

39. Use properties of logarithms to write $2 \log(3x + 7) - \log 4$ as a single logarithm.

40. Solve $\log(x + 2) + \log(x - 1) = 1$.

41. If $10,000 is invested at 5% for 4 yr, how much will there be in the account if interest is compounded

(a) quarterly;   (b) continuously?

The bar graph shows on-line retail sales (in billions of dollars) over the Internet. A reasonable model for sales y in billions of dollars is the exponential function defined by

$$y = 1.38(1.65)^x.$$

The years are coded such that x is the number of years since 1995.

42. Use the model to estimate sales in 2000. (*Hint:* Let $x = 5$.)

43. Use the model to estimate sales in 2003.

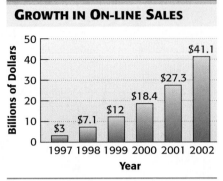

GROWTH IN ON-LINE SALES

Source: Jupiter Communications.

44. If $f(x) = x^2 + 2x - 4$ and $g(x) = 3x + 2$, find

(a) $(g \circ f)(1);$ (b) $(f \circ g)(x).$

Graph.

45. $f(x) = -3x + 5$

46. $f(x) = -2(x - 1)^2 + 3$

47. $\dfrac{x^2}{25} + \dfrac{y^2}{16} \leq 1$

48. $f(x) = \sqrt{x - 2}$

49. $\dfrac{x^2}{4} - \dfrac{y^2}{16} = 1$

50. $f(x) = 3^x$

Sequences and Series

14

Amazing as it may seem, the male honeybee hatches from an unfertilized egg, while the female hatches from a fertilized one. The "family tree" of a male honeybee is shown here, where M represents male and F represents female. If we start with the male honeybee at the top, and count the number of bees in each generation, we obtain the following numbers in the order shown.

$$1, 1, 2, 3, 5, 8$$

Do you see the pattern here? After the first two terms (1 and 1), each successive term is obtained by adding the two previous terms. Thus, the term following 8 is $5 + 8 = 13$. The sequence of numbers described here is called the *Fibonacci sequence,* named after the 13th century Italian mathematician Leonardo of Pisa, who was also known as Fibonacci. This fascinating sequence has countless interesting properties and appears in many places in nature.

In this chapter we study *sequences* and sums of terms of sequences, known as *series.*

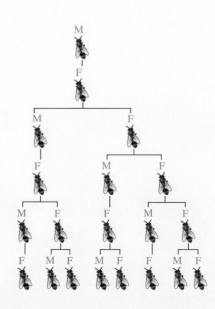

929

14.1 Sequences and Series

OBJECTIVES

1 Find the terms of a sequence given the general term.
2 Find the general term of a sequence.
3 Use sequences to solve applied problems.
4 Use summation notation to evaluate a series.
5 Write a series using summation notation.
6 Find the arithmetic mean (average) of a group of numbers.

A **sequence** is a function whose domain is the set of natural numbers. Intuitively, a sequence is a list of numbers in which the order of their appearance is important. Sequences appear in many places in daily life. For instance, the interest portions of monthly loan payments made to pay off an automobile or home loan form a sequence.

In the Palace of the Alhambra, residence of the Moorish rulers of Granada, Spain, the Sultana's quarters feature an interesting architectural pattern. There are 2 matched marble slabs inlaid in the floor, 4 walls, an octagon (8-sided) ceiling, 16 windows, 32 arches, and so on. If this pattern is continued indefinitely, the set of numbers forms an *infinite sequence.*

Infinite Sequence

An **infinite sequence** is a function with the set of positive integers as the domain.

OBJECTIVE 1 Find the terms of a sequence given the general term. For any positive integer n, the function value (y-value) of a sequence is written as a_n (read "a sub-n") instead of $a(n)$ or $f(n)$. The function values $a_1, a_2, a_3, \ldots,$ written in order, are the **terms** of the sequence, with a_1 the first term, a_2 the second term, and so on. The expression a_n, which defines the sequence, is called the **general term** of the sequence.

In the Palace of the Alhambra example, the first five terms of the sequence are

$$a_1 = 2, \quad a_2 = 4, \quad a_3 = 8, \quad a_4 = 16, \quad \text{and} \quad a_5 = 32.$$

The general term for this sequence is $a_n = 2^n$.

EXAMPLE 1 Writing the Terms of Sequences from the General Term

Given an infinite sequence with $a_n = n + \frac{1}{n}$, find the following.

(a) The second term of the sequence
To get a_2, the second term, replace n with 2.

$$a_2 = 2 + \frac{1}{2} = \frac{5}{2}$$

(b) $a_{10} = 10 + \frac{1}{10} = \frac{101}{10}$

(c) $a_{12} = 12 + \frac{1}{12} = \frac{145}{12}$

Now Try Exercises 1 and 11.

As mentioned earlier, a sequence is a special kind of function. Graphing calculators can be used to generate and graph sequences, as shown in Figure 1. The calculator must be in graphing dot mode, so the discrete points on the graph are not connected. Remember that the domain of a sequence consists only of natural numbers.

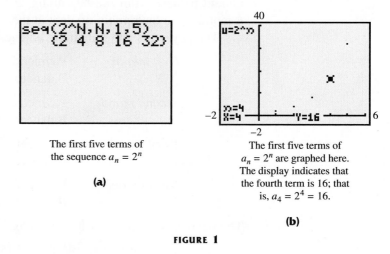

The first five terms of the sequence $a_n = 2^n$

(a)

The first five terms of $a_n = 2^n$ are graphed here. The display indicates that the fourth term is 16; that is, $a_4 = 2^4 = 16$.

(b)

FIGURE 1

OBJECTIVE 2 Find the general term of a sequence. Sometimes we need to find a general term to fit the first few terms of a given sequence. There are no rules for finding the general term of a sequence from the first few terms. In fact, it is possible to give more than one general term that produce the same first three or four terms. However, in many examples, the terms may suggest a general term.

EXAMPLE 2 Finding the General Term of a Sequence

Find an expression for the general term a_n of the sequence.

$$5, 10, 15, 20, 25, \ldots$$

The first term is 5(1), the second is 5(2), and so on. By inspection, $a_n = 5n$ will produce the given first five terms.

▨

Now Try Exercise 17.

CAUTION One problem with using just a few terms to suggest a general term, as in Example 2, is that there may be more than one general term that gives the same first few terms.

OBJECTIVE 3 Use sequences to solve applied problems. Practical problems often involve *finite sequences*.

Finite Sequence

A **finite sequence** has a domain that includes only the first n positive integers.

For example, if n is 5, the domain is $\{1, 2, 3, 4, 5\}$, and the sequence has five terms.

▨ EXAMPLE 3   Using a Sequence in an Application

Keshon borrows $5000 and agrees to pay $500 monthly, plus interest of 1% on the unpaid balance from the beginning of that month. Find the payments for the first 4 months and the remaining debt at the end of this period.

The payments and remaining balances are calculated as follows.

First month	Payment:	$500 + .01(5000) = 550$ dollars
	Balance:	$5000 - 500 = 4500$ dollars
Second month	Payment:	$500 + .01(4500) = 545$ dollars
	Balance:	$5000 - 2 \cdot 500 = 4000$ dollars
Third month	Payment:	$500 + .01(4000) = 540$ dollars
	Balance:	$5000 - 3 \cdot 500 = 3500$ dollars
Fourth month	Payment:	$500 + .01(3500) = 535$ dollars
	Balance:	$5000 - 4 \cdot 500 = 3000$ dollars

The payments for the first four months, in dollars, are

$$550, \; 545, \; 540, \; 535$$

and the remaining debt at the end of this period is 3000 dollars.

Now Try Exercise 21.

OBJECTIVE 4 Use summation notation to evaluate a series. By adding the terms of a sequence, we obtain a *series*.

Series

The indicated sum of the terms of a sequence is called a **series.**

For example, if we consider the sum of the payments listed in Example 3,

$$550 + 545 + 540 + 535,$$

we obtain a series that represents the total amount of payments for the first four months.

Since a sequence can be finite or infinite, there are finite or infinite series. One type of infinite series is discussed in Section 14.3, and the binomial theorem discussed in Section 14.4 defines an important finite series. In this section we discuss only finite series.

We use a compact notation, called **summation notation,** to write a series from the general term of the corresponding sequence. For example, the sum of the first six terms of the sequence with general term $a_n = 3n + 2$ is written with the Greek letter Σ (sigma) as

$$\sum_{i=1}^{6} (3i + 2).$$

We read this as "the sum from $i = 1$ to 6 of $3i + 2$." To find this sum, we replace the letter i in $3i + 2$ with 1, 2, 3, 4, 5, and 6, as follows.

$$\sum_{i=1}^{6} (3i + 2) = (3 \cdot 1 + 2) + (3 \cdot 2 + 2) + (3 \cdot 3 + 2)$$
$$+ (3 \cdot 4 + 2) + (3 \cdot 5 + 2) + (3 \cdot 6 + 2)$$
$$= 5 + 8 + 11 + 14 + 17 + 20$$
$$= 75$$

The letter i is called the **index of summation.**

CAUTION   This use of i has no connection with the complex number i.

■ **EXAMPLE 4**   **Evaluating Series Written in Summation Notation**

Write out the terms and evaluate each series.

(a) $\displaystyle\sum_{i=1}^{5} (i - 4) = (1 - 4) + (2 - 4) + (3 - 4) + (4 - 4) + (5 - 4)$

$$= -3 - 2 - 1 + 0 + 1$$
$$= -5$$

(b) $\displaystyle\sum_{i=3}^{7} 3i^2 = 3(3)^2 + 3(4)^2 + 3(5)^2 + 3(6)^2 + 3(7)^2$

$$= 27 + 48 + 75 + 108 + 147$$
$$= 405$$

Now Try Exercises 25 and 27.

```
sum(seq(I-4,I,1,
5)
               -5
sum(seq(3I²,I,3,
7)
              405
```

FIGURE 2

Figure 2 shows how a graphing calculator can be used to obtain the results found in Example 4.

OBJECTIVE 5 Write a series using summation notation. In Example 4, we started with summation notation and wrote each series using + signs. It is possible to go in the other direction; that is, given a series, we can write it using summation notation. To do this, we observe a pattern in the terms and write the general term accordingly.

■ **EXAMPLE 5**   **Writing Series with Summation Notation**

Write each sum with summation notation.

(a) $2 + 5 + 8 + 11$

First, find a general term a_n that will give these four terms for a_1, a_2, a_3, and a_4. Inspection (and trial and error) shows that $3i - 1$ will work for these four terms, since

$$3(1) - 1 = 2$$
$$3(2) - 1 = 5$$
$$3(3) - 1 = 8$$
$$3(4) - 1 = 11.$$

(Remember, there may be other expressions that also work. These four terms may be the first terms of more than one sequence.) Since i ranges from 1 to 4, write the sum as

$$2 + 5 + 8 + 11 = \sum_{i=1}^{4} (3i - 1).$$

(b) $8 + 27 + 64 + 125 + 216$

Since these numbers are the cubes of 2, 3, 4, 5, and 6,

$$8 + 27 + 64 + 125 + 216 = \sum_{i=2}^{6} i^3.$$

Now Try Exercises 37 and 41.

OBJECTIVE 6 Find the arithmetic mean (average) of a group of numbers.

Arithmetic Mean or Average

The **arithmetic mean,** or **average,** of a group of numbers is symbolized $\bar{x}$ and is found by dividing the sum of the numbers by the number of numbers. That is,

$$\bar{x} = \frac{\sum_{i=1}^{n} x_i}{n}.$$

Here the values of x_i represent the individual numbers in the group, and n represents the number of numbers.

EXAMPLE 6 Finding the Arithmetic Mean or Average

The following table shows the number of companies listed on the New York Stock Exchange for each year during the period 1994 through 2000. What was the average number of listings per year for this 7-yr period?

Year	Number of Listings
1994	2570
1995	2675
1996	2907
1997	3047
1998	3114
1999	3025
2000	2862

Source: New York Stock Exchange.

Let $x_1 = 2570$, $x_2 = 2675$, and so on. Since there are 7 numbers in the group, $n = 7$. Therefore,

$$\bar{x} = \frac{\sum_{i=1}^{7} x_i}{7}$$

$$= \frac{2570 + 2675 + 2907 + 3047 + 3114 + 3025 + 2862}{7}$$

$$= 2886 \quad \text{(rounded to the nearest unit).}$$

The average number of listings per year for this 7-yr period was 2886.

Now Try Exercise 51.

14.1 EXERCISES

Write out the first five terms of each sequence. See Example 1.

1. $a_n = n + 1$

2. $a_n = n - 4$

3. $a_n = \dfrac{n + 3}{n}$

4. $a_n = \dfrac{n + 2}{n + 1}$

5. $a_n = 3^n$

6. $a_n = 1^{n-1}$

7. $a_n = \dfrac{1}{n^2}$

8. $a_n = \dfrac{n^2}{n + 1}$

9. $a_n = (-1)^n$

10. $a_n = (-1)^{2n-1}$

Find the indicated term for each sequence. See Example 1.

11. $a_n = -9n + 2; \ a_8$

12. $a_n = 3n - 7; \ a_{12}$

13. $a_n = \dfrac{3n + 7}{2n - 5}; \ a_{14}$

14. $a_n = \dfrac{5n - 9}{3n + 8}; \ a_{16}$

15. $a_n = (n + 1)(2n + 3); \ a_8$

16. $a_n = (5n - 2)(3n + 1); \ a_{10}$

Find a general term, a_n, for the given terms of each sequence. See Example 2.

17. $4, 8, 12, 16, \ldots$

18. $-10, -20, -30, -40, \ldots$

19. $\dfrac{1}{3}, \dfrac{1}{9}, \dfrac{1}{27}, \dfrac{1}{81}, \ldots$

20. $\dfrac{1}{2}, \dfrac{2}{3}, \dfrac{3}{4}, \dfrac{4}{5}, \ldots$

Solve each applied problem by writing the first few terms of a sequence. See Example 3.

21. Anne borrows $1000 and agrees to pay $100 plus interest of 1% on the unpaid balance each month. Find the payments for the first 6 months and the remaining debt at the end of this period.

22. Larissa Perez is offered a new modeling job with a salary of $20,000 + 2500n$ dollars per year at the end of the nth year. Write a sequence showing her salary at the end of each of the first 5 yr. If she continues in this way, what will her salary be at the end of the tenth year?

23. Suppose that an automobile loses $\frac{1}{5}$ of its value each year; that is, at the end of any given year, the value is $\frac{4}{5}$ of the value at the beginning of that year. If a car costs $20,000 new, what is its value at the end of 5 yr?

24. A certain car loses $\frac{1}{2}$ of its value each year. If this car cost $40,000 new, what is its value at the end of 6 yr?

Write out each series and evaluate it. See Example 4.

25. $\displaystyle\sum_{i=1}^{5} (i + 3)$

26. $\displaystyle\sum_{i=1}^{6} (i + 9)$

27. $\displaystyle\sum_{i=1}^{3} (i^2 + 2)$

28. $\displaystyle\sum_{i=1}^{4} i(i + 3)$

29. $\displaystyle\sum_{i=1}^{6} (-1)^i$

30. $\displaystyle\sum_{i=1}^{5} (-1)^i \cdot i$

31. $\displaystyle\sum_{i=3}^{7} (i - 3)(i + 2)$

32. $\displaystyle\sum_{i=2}^{6} \dfrac{i^2 + 1}{2}$

Write out the terms of each series.

33. $\displaystyle\sum_{i=1}^{5} 2x \cdot i$

34. $\displaystyle\sum_{i=1}^{6} x^i$

35. $\displaystyle\sum_{i=1}^{5} i \cdot x^i$

36. $\displaystyle\sum_{i=2}^{6} \dfrac{x + i}{x - i}$

Write each series using summation notation. See Example 5.

37. $3 + 4 + 5 + 6 + 7$

38. $1 + 4 + 9 + 16$

39. $\dfrac{1}{2} + \dfrac{1}{3} + \dfrac{1}{4} + \dfrac{1}{5} + \dfrac{1}{6}$

40. $-1 + 2 - 3 + 4 - 5 + 6$

41. $1 + 4 + 9 + 16 + 25$

42. $1 + 16 + 81 + 256$

43. Suppose that f is a function with domain all real numbers, where $f(x) = 2x + 4$. Suppose that an infinite sequence is defined by $a_n = 2n + 4$. Discuss the similarities and differences between the function and the sequence. Give examples using each.

44. What is wrong with the following?
For the sequence defined by $a_n = 2n + 4$, find $a_{1/2}$.

45. Explain the basic difference between a sequence and a series.

46. Evaluate $\displaystyle\sum_{i=1}^{3} 5i$ and $5 \displaystyle\sum_{i=1}^{3} i$. Notice that the sums are the same. Explain how the distributive property plays a role in assuring us that the two sums are equal.

Find the arithmetic mean for each collection of numbers. See Example 6.

47. 8, 11, 14, 9, 3, 6, 8

48. 10, 12, 8, 19, 23

49. 5, 9, 8, 2, 4, 7, 3, 2

50. 2, 1, 4, 8, 3, 7

Solve each problem. See Example 6.

51. The number of mutual funds available to investors for each year during the period 1996 through 2000 is given in the table.

Year	Number of Funds Available
1996	6254
1997	6684
1998	7314
1999	7791
2000	8171

Source: Investment Company Institute.

To the nearest whole number, what was the average number of funds available per year during this period?

52. The total assets of mutual funds, in billions of dollars, for each year during the period 1992 through 1996 are shown in the table. To the nearest tenth (in billions of dollars), what were the average assets per year during this period?

Year	Assets (in billions of dollars)
1992	1646.3
1993	2075.4
1994	2161.5
1995	2820.4
1996	3539.2

Source: Investment Company Institute.

> ◼ **RELATING CONCEPTS** (EXERCISES 53–60) ◼
>
> **For Individual or Group Work**
>
> *The following properties of series provide useful shortcuts for evaluating series.*
>
> *If $a_1, a_2, a_3, \ldots, a_n$ and $b_1, b_2, b_3, \ldots, b_n$ are two sequences, and c is a constant, then for every positive integer n,*
>
> **(a)** $\displaystyle\sum_{i=1}^{n} c = nc$ ⠀⠀⠀⠀⠀ **(b)** $\displaystyle\sum_{i=1}^{n} ca_i = c\sum_{i=1}^{n} a_i$
>
> **(c)** $\displaystyle\sum_{i=1}^{n} (a_i + b_i) = \sum_{i=1}^{n} a_i + \sum_{i=1}^{n} b_i$ ⠀⠀ **(d)** $\displaystyle\sum_{i=1}^{n} (a_i - b_i) = \sum_{i=1}^{n} a_i - \sum_{i=1}^{n} b_i.$
>
> **Work Exercises 53–60 in order,** *to see how these shortcuts can make work easier.*
>
> **53.** Use property (c) to write $\displaystyle\sum_{i=1}^{6} (i^2 + 3i + 5)$ as the sum of three summations.
>
> **54.** Use property (b) to rewrite the second summation from Exercise 53.
>
> **55.** Use property (a) to rewrite the third summation from Exercise 53.
>
> **56.** Rewrite $1 + 2 + 3 + 4 + \cdots + n = \dfrac{n(n+1)}{2}$ using summation notation.
>
> **57.** Rewrite $1^2 + 2^2 + 3^2 + 4^2 + \cdots + n^2 = \dfrac{n(n+1)(2n+1)}{6}$ using summation notation.
>
> **58.** Use the summations you wrote in Exercises 56 and 57 and the given properties to evaluate the three summations from Exercises 53–55. This gives the value of $\displaystyle\sum_{i=1}^{6} (i^2 + 3i + 5)$ without writing out all six terms.
>
> **59.** Use the given properties and summations to evaluate $\displaystyle\sum_{i=1}^{12} (i^2 - i).$
>
> **60.** Use the given properties and summations to evaluate $\displaystyle\sum_{i=1}^{20} (2 + i - i^2).$

14.2 ⠀ Arithmetic Sequences

OBJECTIVES

1 Find the common difference of an arithmetic sequence.

2 Find the general term of an arithmetic sequence.

3 Use an arithmetic sequence in an application.

4 Find any specified term or the number of terms of an arithmetic sequence. ⠀(continued)

OBJECTIVE 1 Find the common difference of an arithmetic sequence. In this section we introduce a special type of sequence that has many applications.

Arithmetic Sequence

A sequence in which each term after the first differs from the preceding term by a constant amount is called an **arithmetic sequence** or **arithmetic progression.**

For example, the sequence

$$6, 11, 16, 21, 26, \ldots$$

is an arithmetic sequence, since the difference between any two adjacent terms is always 5. The number 5 is called the **common difference** of the arithmetic sequence.

OBJECTIVES (continued)

5 Find the sum of a specified number of terms of an arithmetic sequence.

The common difference, d, is found by subtracting any pair of terms a_n and a_{n+1}. That is,

$$d = a_{n+1} - a_n.$$

EXAMPLE 1 Finding the Common Difference

Find d for the arithmetic sequence.

$$-11, -4, 3, 10, 17, 24, \ldots$$

Since the sequence is arithmetic, d is the difference between any two adjacent terms. Choosing the terms 10 and 17 gives

$$d = 17 - 10$$
$$= 7.$$

The terms -11 and -4 would give $d = -4 - (-11) = 7$, the same result.

Now Try Exercise 7.

EXAMPLE 2 Writing the Terms of a Sequence from the First Term and Common Difference

Write the first five terms of the arithmetic sequence with first term 3 and common difference -2.

The second term is found by adding -2 to the first term 3, getting 1. For the next term, add -2 to 1, and so on. The first five terms are

$$3, 1, -1, -3, -5.$$

Now Try Exercise 9.

OBJECTIVE 2 Find the general term of an arithmetic sequence. Generalizing from Example 2, if we know the first term a_1 and the common difference d of an arithmetic sequence, then the sequence is completely defined as

$$a_1, \quad a_2 = a_1 + d, \quad a_3 = a_1 + 2d, \quad a_4 = a_1 + 3d, \ldots.$$

Writing the terms of the sequence in this way suggests the following rule.

General Term of an Arithmetic Sequence

The general term of an arithmetic sequence with first term a_1 and common difference d is

$$a_n = a_1 + (n - 1)d.$$

Since $a_n = a_1 + (n - 1)d = dn + (a_1 - d)$ is a linear function in n, any linear expression of the form $kn + c$, where k and c are real numbers, defines an arithmetic sequence.

▦ **EXAMPLE 3** Finding the General Term of an Arithmetic Sequence

Find the general term for the arithmetic sequence.

$$-9, -6, -3, 0, 3, 6, \ldots$$

Then use the general term to find a_{20}.

Here the first term is $a_1 = -9$. To find d, subtract any two adjacent terms. For example,

$$d = -3 - (-6) = 3.$$

Now find a_n.

$$
\begin{aligned}
a_n &= a_1 + (n-1)d && \text{Formula for } a_n \\
&= -9 + (n-1)(3) && \text{Let } a_1 = -9, d = 3. \\
&= -9 + 3n - 3 && \text{Distributive property} \\
a_n &= 3n - 12 && \text{Combine terms.}
\end{aligned}
$$

Thus, the general term is $a_n = 3n - 12$. To find a_{20}, let $n = 20$.

$$a_{20} = 3(20) - 12 = 60 - 12 = 48$$

▦

Now Try Exercise 13.

OBJECTIVE 3 Use an arithmetic sequence in an application.

▦ **EXAMPLE 4** Applying an Arithmetic Sequence

Howie Sorkin's uncle decides to start a fund for Howie's education. He makes an initial contribution of $3000 and each month deposits an additional $500. Thus, after one month there will be $3000 + $500 = $3500. How much will there be after 24 months? (Disregard any interest.)

The contributions can be described using an arithmetic sequence. After n months, the fund will contain

$$a_n = 3000 + 500n \text{ dollars.}$$

To find the amount in the fund after 24 months, find a_{24}.

$$
\begin{aligned}
a_{24} &= 3000 + 500(24) && \text{Let } n = 24. \\
&= 3000 + 12{,}000 && \text{Multiply.} \\
&= 15{,}000 && \text{Add.}
\end{aligned}
$$

The account will contain $15,000 (disregarding interest) after 24 months.

▦

Now Try Exercise 47.

OBJECTIVE 4 Find any specified term or the number of terms of an arithmetic sequence.
The formula for the general term has four variables: a_n, a_1, n, and d. If we know any three of these, the formula can be used to find the value of the fourth variable. The next example shows how to find a particular term.

▧ EXAMPLE 5 Finding Specified Terms

Find the indicated term for each arithmetic sequence.

(a) $a_1 = -6$, $d = 12$; a_{15}

We use the formula $a_n = a_1 + (n - 1)d$. Since we want $a_n = a_{15}$, $n = 15$.

$$\begin{aligned} a_{15} &= a_1 + (15 - 1)d & \text{Let } n = 15. \\ &= -6 + 14(12) & \text{Let } a_1 = -6, d = 12. \\ &= 162 \end{aligned}$$

(b) $a_5 = 2$ and $a_{11} = -10$; a_{17}

Any term can be found if a_1 and d are known. Use the formula for a_n with the two given terms.

$a_5 = a_1 + (5 - 1)d$	$a_{11} = a_1 + (11 - 1)d$
$a_5 = a_1 + 4d$	$a_{11} = a_1 + 10d$
$2 = a_1 + 4d \qquad \text{Let } a_5 = 2.$	$-10 = a_1 + 10d \qquad \text{Let } a_{11} = -10.$

This gives a system of two equations with two variables, a_1 and d. Find d by adding -1 times one equation to the other to eliminate a_1.

$$\begin{aligned} -10 &= \quad a_1 + 10d \\ \underline{-2} &= \underline{-a_1 - \quad 4d} \qquad \text{Multiply } 2 = a_1 + 4d \text{ by } -1. \\ -12 &= \qquad\quad 6d \qquad \text{Add.} \\ -2 &= d \qquad\qquad\quad \text{Divide by 6.} \end{aligned}$$

Now find a_1 by substituting -2 for d into either equation.

$$\begin{aligned} -10 &= a_1 + 10(-2) \qquad \text{Let } d = -2. \\ -10 &= a_1 - 20 \\ 10 &= a_1 \end{aligned}$$

Use the formula for a_n to find a_{17}.

$$\begin{aligned} a_{17} &= a_1 + (17 - 1)d \qquad \text{Let } n = 17. \\ &= a_1 + 16d \\ &= 10 + 16(-2) \qquad\quad \text{Let } a_1 = 10, d = -2. \\ &= -22 \end{aligned}$$

Now Try Exercises 19 and 23.

Sometimes we need to find out how many terms are in a sequence, as shown in the following example.

▧ EXAMPLE 6 Finding the Number of Terms in a Sequence

Find the number of terms in the arithmetic sequence.

$$-8, -2, 4, 10, \ldots, 52$$

Let n represent the number of terms in the sequence. Since $a_n = 52$, $a_1 = -8$, and $d = -2 - (-8) = 6$, use the formula $a_n = a_1 + (n - 1)d$ to find n. Substituting the known values into the formula gives

$$a_n = a_1 + (n - 1)d$$
$$52 = -8 + (n - 1)6 \qquad \text{Let } a_n = 52, a_1 = -8, d = 6.$$
$$52 = -8 + 6n - 6 \qquad \text{Distributive property}$$
$$66 = 6n \qquad \text{Combine terms.}$$
$$n = 11. \qquad \text{Divide by 6.}$$

The sequence has 11 terms.

Now Try Exercise 25.

OBJECTIVE 5 **Find the sum of a specified number of terms of an arithmetic sequence.** To find a formula for the sum, S_n, of the first n terms of an arithmetic sequence, we can write out the terms as

$$S_n = a_1 + (a_1 + d) + (a_1 + 2d) + \cdots + [a_1 + (n - 1)d].$$

This same sum can be written in reverse as

$$S_n = a_n + (a_n - d) + (a_n - 2d) + \cdots + [a_n - (n - 1)d].$$

Now add the corresponding terms of these two expressions for S_n to get

$$2S_n = (a_1 + a_n) + (a_1 + a_n) + (a_1 + a_n) + \cdots + (a_1 + a_n).$$

The right-hand side of this expression contains n terms, each equal to $a_1 + a_n$, so

$$2S_n = n(a_1 + a_n)$$

$$S_n = \frac{n}{2}(a_1 + a_n).$$

EXAMPLE 7 **Finding the Sum of the First n Terms**

Find the sum of the first five terms of the arithmetic sequence in which $a_n = 2n - 5$.

We can use the formula $S_n = \frac{n}{2}(a_1 + a_n)$ to find the sum of the first five terms. Here $n = 5$, $a_1 = 2(1) - 5 = -3$, and $a_5 = 2(5) - 5 = 5$. From the formula,

$$S_5 = \frac{5}{2}(-3 + 5) = \frac{5}{2}(2) = 5.$$

Now Try Exercise 39.

It is sometimes useful to express the sum of an arithmetic sequence, S_n, in terms of a_1 and d, the quantities that define the sequence. We can do this as follows. Since

$$S_n = \frac{n}{2}(a_1 + a_n) \qquad \text{and} \qquad a_n = a_1 + (n - 1)d,$$

by substituting the expression for a_n into the expression for S_n we obtain

$$S_n = \frac{n}{2}(a_1 + [a_1 + (n - 1)d])$$

$$S_n = \frac{n}{2}[2a_1 + (n - 1)d].$$

The summary on the next page gives both of the alternative forms that may be used to find the sum of the first n terms of an arithmetic sequence.

Sum of the First n Terms of an Arithmetic Sequence

The sum of the first n terms of the arithmetic sequence with first term a_1, nth term a_n, and common difference d is

$$S_n = \frac{n}{2}(a_1 + a_n) \qquad \text{or} \qquad S_n = \frac{n}{2}[2a_1 + (n - 1)d].$$

EXAMPLE 8 Finding the Sum of the First n Terms

Find the sum of the first eight terms of the arithmetic sequence having first term 3 and common difference -2.

Since the known values, $a_1 = 3$, $d = -2$, and $n = 8$, appear in the second formula for S_n, we use it.

$$S_n = \frac{n}{2}[2a_1 + (n - 1)d]$$

$$S_8 = \frac{8}{2}[2(3) + (8 - 1)(-2)] \qquad \text{Let } a_1 = 3, d = -2, n = 8.$$

$$= 4[6 - 14]$$

$$= -32$$

Now Try Exercise 35.

As mentioned earlier, linear expressions of the form $kn + c$, where k and c are real numbers, define an arithmetic sequence. For example, the sequences defined by $a_n = 2n + 5$ and $a_n = n - 3$ are arithmetic sequences. For this reason,

$$\sum_{i=1}^{n} (ki + c)$$

represents the sum of the first n terms of an arithmetic sequence having first term $a_1 = k(1) + c = k + c$ and general term $a_n = k(n) + c = kn + c$. We can find this sum with the first formula for S_n, as shown in the next example.

EXAMPLE 9 Using S_n to Evaluate a Summation

Find $\sum_{i=1}^{12} (2i - 1)$.

This is the sum of the first 12 terms of the arithmetic sequence having $a_n = 2n - 1$. This sum, S_{12}, is found with the first formula for S_n,

$$S_n = \frac{n}{2}(a_1 + a_n).$$

Here $n = 12$, $a_1 = 2(1) - 1 = 1$, and $a_{12} = 2(12) - 1 = 23$. Substitute these values into the formula to find

$$S_{12} = \frac{12}{2}(1 + 23) = 6(24) = 144.$$

Now Try Exercise 41.

```
sum(seq(2I-1,I,1
,12)
              144
```

FIGURE 3

Figure 3 shows how a graphing calculator supports the result of Example 9.

14.2   EXERCISES

1. Using several examples, explain the meaning of *arithmetic sequence*.

2. Can any two terms of an arithmetic sequence be used to find the common difference? Explain.

If the given sequence is arithmetic, find the common difference, d. If the sequence is not arithmetic, say so. See Example 1.

3. $1, 2, 3, 4, 5, \ldots$

4. $2, 5, 8, 11, \ldots$

5. $2, -4, 6, -8, 10, -12, \ldots$

6. $-6, -10, -14, -18, \ldots$

7. $-10, -5, 0, 5, 10, \ldots$

8. $1, 2, 4, 7, 11, 16, \ldots$

Write the first five terms of each arithmetic sequence. See Example 2.

9. $a_1 = 5, d = 4$

10. $a_1 = 6, d = 7$

11. $a_1 = -2, d = -4$

12. $a_1 = -3, d = -5$

Use the formula for a_n to find the general term for each arithmetic sequence. See Example 3.

13. $a_1 = 2, d = 5$

14. $a_1 = 5, d = -3$

15. $3, \dfrac{15}{4}, \dfrac{9}{2}, \dfrac{21}{4}, \ldots$

16. $4, 14, 24, \ldots$

17. $-3, 0, 3, \ldots$

18. $-10, -5, 0, 5, 10, \ldots$

Find the indicated term for each arithmetic sequence. See Examples 2 and 5.

19. $a_1 = 4, d = 3; a_{25}$

20. $a_1 = 1, d = -\dfrac{1}{2}; a_{12}$

21. $2, 4, 6, \ldots; a_{24}$

22. $1, 5, 9, \ldots; a_{50}$

23. $a_{12} = -45, a_{10} = -37; a_1$

24. $a_{10} = -2, a_{15} = -8; a_3$

Find the number of terms in each arithmetic sequence. See Example 6.

25. $3, 5, 7, \ldots, 33$

26. $2, \dfrac{3}{2}, 1, \dfrac{1}{2}, \ldots, -5$

27. $\dfrac{3}{4}, 3, \dfrac{21}{4}, \ldots, 12$

28. $4, 1, -2, \ldots, -32$

29. In the formula for S_n, what does n represent?

30. Explain when you would use each of the two formulas for S_n.

RELATING CONCEPTS (EXERCISES 31–34)

For Individual or Group Work

Exercises 31–34 show how to find the sum $1 + 2 + 3 + \cdots + 99 + 100$ in an ingenious way. **Work them in order.**

31. Consider the following:

$$S = 1 + 2 + 3 + \cdots + 99 + 100$$
$$S = 100 + 99 + 98 + \cdots + 2 + 1.$$

Add the left sides of this equation. The result is _____. Add the columns on the right side. The sum _____ appears _____ times, so by multiplication, the sum of the right sides of the equations is _____.

(continued)

32. Form an equation by setting the sum of the left sides equal to the sum of the right sides.

33. Solve the equation from Exercise 32 to find that the desired sum, S, is _____.

34. Find the sum $S = 1 + 2 + 3 + \cdots + 199 + 200$ using the procedure described in Exercises 31–33.

Find S_6 for each arithmetic sequence. See Examples 7 and 8.

35. $a_1 = 6, d = 3$
36. $a_1 = 5, d = 4$
37. $a_1 = 7, d = -3$

38. $a_1 = -5, d = -4$
39. $a_n = 4 + 3n$
40. $a_n = 9 + 5n$

Use a formula for S_n to evaluate each series. See Example 9.

41. $\displaystyle\sum_{i=1}^{10} (8i - 5)$
42. $\displaystyle\sum_{i=1}^{17} (i - 1)$
43. $\displaystyle\sum_{i=1}^{20} (2i - 5)$

44. $\displaystyle\sum_{i=1}^{10} \left(\frac{1}{2}i - 1\right)$
45. $\displaystyle\sum_{i=1}^{250} i$
46. $\displaystyle\sum_{i=1}^{2000} i$

*Solve each applied problem. (*Hint: *Determine whether you need to find a specific term of a sequence or the sum of the terms of a sequence immediately after reading the problem.) See Example 4.*

47. Nancy Bondy's aunt has promised to deposit $1 in her account on the first day of her birthday month, $2 on the second day, $3 on the third day, and so on for 30 days. How much will this amount to over the entire month?

48. Repeat Exercise 47, but assume that the deposits are $2, $4, $6, and so on, and that the month is February of a leap year.

49. Suppose that Randy Morgan is offered a job at $1600 per month with a guaranteed increase of $50 every 6 months for 5 yr. What will his salary be at the end of this period of time?

50. Repeat Exercise 49, but assume that the starting salary is $2000 per month, and the guaranteed increase is $100 every 4 months for 3 yr.

51. A seating section in a theater-in-the-round has 20 seats in the first row, 22 in the second row, 24 in the third row, and so on for 25 rows. How many seats are there in the last row? How many seats are there in the section?

52. José Valdevielso has started on a fitness program. He plans to jog 10 min per day for the first week, and then add 10 min per day each week until he is jogging an hour each day. In which week will this occur? What is the total number of minutes he will run during the first 4 weeks?

53. A child builds with blocks, placing 35 blocks in the first row, 31 in the second row, 27 in the third row, and so on. Continuing this pattern, can she end with a row containing exactly 1 block? If not, how many blocks will the last row contain? How many rows can she build this way?

54. A stack of firewood has 28 pieces on the bottom, 24 on top of those, then 20, and so on. If there are 108 pieces of wood, how many rows are there? (*Hint: $n \leq 7$.*)

14.3 Geometric Sequences

In an arithmetic sequence, each term after the first is found by *adding* a fixed number to the previous term. A *geometric sequence* is defined as follows.

> **Geometric Sequence**
>
> A **geometric sequence** or **geometric progression** is a sequence in which each term after the first is a constant multiple of the preceding term.

OBJECTIVE 1 Find the common ratio of a geometric sequence. We find the constant multiplier, called the **common ratio,** by dividing any term after the first by the preceding term. That is, the common ratio is

$$r = \frac{a_{n+1}}{a_n}.$$

For example,

$$2, 6, 18, 54, 162, \ldots$$

is a geometric sequence in which the first term, a_1, is 2 and the common ratio is

$$r = \frac{6}{2} = \frac{18}{6} = \frac{54}{18} = \frac{162}{54} = 3.$$

EXAMPLE 1 Finding the Common Ratio

Find r for the geometric sequence.

$$15, \frac{15}{2}, \frac{15}{4}, \frac{15}{8}, \ldots$$

To find r, choose any two successive terms and divide the second one by the first. Choosing the second and third terms of the sequence gives

$$r = \frac{a_3}{a_2} = \frac{15}{4} \div \frac{15}{2} = \frac{1}{2}.$$

Any other two successive terms could have been used to find r. Additional terms of the sequence can be found by multiplying each successive term by $\frac{1}{2}$.

Now Try Exercise 3.

OBJECTIVE 2 Find the general term of a geometric sequence. The general term a_n of a geometric sequence $a_1, a_2, a_3, \ldots$ is expressed in terms of a_1 and r by writing the first few terms as

$$a_1, \quad a_2 = a_1 r, \quad a_3 = a_1 r^2, \quad a_4 = a_1 r^3, \ldots,$$

which suggests the rule on the next page.

General Term of a Geometric Sequence

The general term of the geometric sequence with first term a_1 and common ratio r is

$$a_n = a_1 r^{n-1}.$$

CAUTION Be careful to use the correct order of operations when finding $a_1 r^{n-1}$. The value of r^{n-1} must be found first. Then multiply the result by a_1.

EXAMPLE 2 Finding the General Term

Find the general term of the sequence in Example 1.

The first term is $a_1 = 15$ and the common ratio is $r = \frac{1}{2}$. Substituting into the formula for the general term gives

$$a_n = a_1 r^{n-1} = 15\left(\frac{1}{2}\right)^{n-1},$$

the required general term. Notice that it is not possible to simplify further, because the exponent must be applied before the multiplication can be done.

Now Try Exercise 11.

OBJECTIVE 3 Find any specified term of a geometric sequence. We can use the formula for the general term to find any particular term.

EXAMPLE 3 Finding Specified Terms

Find the indicated term for each geometric sequence.

(a) $a_1 = 4$, $r = -3$; a_6

Let $n = 6$. From the general term $a_n = a_1 r^{n-1}$,

$$
\begin{aligned}
a_6 &= a_1 \cdot r^{6-1} && \text{Let } n = 6. \\
&= 4 \cdot (-3)^5 && \text{Let } a_1 = 4, r = -3. \\
&= -972. && \text{Evaluate } (-3)^5 \text{ first.}
\end{aligned}
$$

(b) $\dfrac{3}{4}, \dfrac{3}{8}, \dfrac{3}{16}, \ldots; a_7$

Here, $r = \frac{1}{2}$, $a_1 = \frac{3}{4}$, and $n = 7$.

$$a_7 = \frac{3}{4} \cdot \left(\frac{1}{2}\right)^6 = \frac{3}{4} \cdot \frac{1}{64} = \frac{3}{256}$$

Now Try Exercises 17 and 19.

EXAMPLE 4 Writing the Terms of a Sequence

Write the first five terms of the geometric sequence whose first term is 5 and whose common ratio is $\frac{1}{2}$.

Using the formula $a_n = a_1 r^{n-1}$,

$$a_1 = 5, \quad a_2 = 5\left(\frac{1}{2}\right) = \frac{5}{2}, \quad a_3 = 5\left(\frac{1}{2}\right)^2 = \frac{5}{4},$$

$$a_4 = 5\left(\frac{1}{2}\right)^3 = \frac{5}{8}, \quad a_5 = 5\left(\frac{1}{2}\right)^4 = \frac{5}{16}.$$

Now Try Exercise 23.

OBJECTIVE 4 Find the sum of a specified number of terms of a geometric sequence. It is convenient to have a formula for the sum of the first n terms of a geometric sequence, S_n. We can develop a formula by first writing out S_n.

$$S_n = a_1 + a_1 r + a_1 r^2 + a_1 r^3 + \cdots + a_1 r^{n-1}$$

Next, we multiply both sides by r.

$$rS_n = a_1 r + a_1 r^2 + a_1 r^3 + a_1 r^4 + \cdots + a_1 r^n$$

We subtract the first result from the second.

$$rS_n - S_n = (a_1 r - a_1) + (a_1 r^2 - a_1 r) + (a_1 r^3 - a_1 r^2)$$
$$+ (a_1 r^4 - a_1 r^3) + \cdots + (a_1 r^n - a_1 r^{n-1})$$

Using the commutative, associative, and distributive properties, we obtain

$$rS_n - S_n = (a_1 r - a_1 r) + (a_1 r^2 - a_1 r^2)$$
$$+ (a_1 r^3 - a_1 r^3) + \cdots + (a_1 r^n - a_1)$$
$$S_n(r - 1) = a_1 r^n - a_1.$$

If $r \neq 1$, then

$$S_n = \frac{a_1 r^n - a_1}{r - 1} = \frac{a_1(r^n - 1)}{r - 1}. \qquad \text{Divide by } r - 1.$$

A summary of this discussion follows.

Sum of the First n Terms of a Geometric Sequence

The sum of the first n terms of the geometric sequence with first term a_1 and common ratio r is

$$S_n = \frac{a_1(r^n - 1)}{r - 1} \quad (r \neq 1).$$

If $r = 1$, then $S_n = a_1 + a_1 + a_1 + \cdots + a_1 = na_1$.

Multiplying the formula for S_n by $\frac{-1}{-1}$ gives an alternative form that is sometimes preferable.

$$S_n = \frac{a_1(r^n - 1)}{r - 1} \cdot \frac{-1}{-1} = \frac{a_1(1 - r^n)}{1 - r}$$

■ **EXAMPLE 5** Finding the Sum of the First *n* Terms

Find the sum of the first six terms of the geometric sequence with first term -2 and common ratio 3.

Substitute $n = 6$, $a_1 = -2$, and $r = 3$ into the formula for S_n.

$$S_n = \frac{a_1(r^n - 1)}{r - 1}$$

$$S_6 = \frac{-2(3^6 - 1)}{3 - 1} \qquad \text{Let } n = 6, a_1 = -2, r = 3.$$

$$= \frac{-2(729 - 1)}{2} \qquad \text{Evaluate } 3^6.$$

$$= -728$$

Now Try Exercise 27.

A series of the form

$$\sum_{i=1}^{n} a \cdot b^i$$

represents the sum of the first *n* terms of a geometric sequence having first term $a_1 = a \cdot b^1 = ab$ and common ratio *b*. The next example illustrates this form.

■ **EXAMPLE 6** Using the Formula for S_n to Find a Summation

Find $\displaystyle\sum_{i=1}^{4} 3 \cdot 2^i$.

Since the series is in the form

$$\sum_{i=1}^{n} a \cdot b^i,$$

it represents the sum of the first *n* terms of the geometric sequence with $a_1 = a \cdot b^1$ and $r = b$. The sum is found by using the formula

$$S_n = \frac{a_1(r^n - 1)}{r - 1}.$$

Here $n = 4$. Also, $a_1 = 6$ and $r = 2$. Now substitute into the formula for S_n.

$$S_4 = \frac{6(2^4 - 1)}{2 - 1} \qquad \text{Let } n = 4, a_1 = 6, r = 2.$$

$$= \frac{6(16 - 1)}{1} \qquad \text{Evaluate } 2^4.$$

$$= 90$$

Now Try Exercise 31.

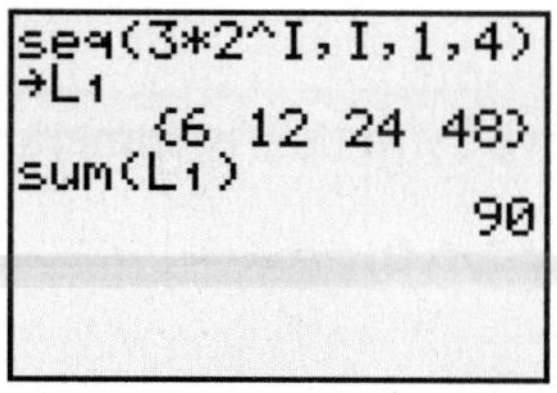

```
seq(3*2^I,I,1,4)
→L₁
     {6 12 24 48}
sum(L₁)
                90
```

FIGURE 4

Figure 4 shows how a graphing calculator can store the terms in a list, and then find the sum of these terms. This supports the result of Example 6.

OBJECTIVE 5 Apply the formula for the future value of an ordinary annuity. A sequence of equal payments made at equal periods of time is called an **annuity.** If the payments are made at the end of the time period, and if the frequency of payments is the same as the frequency of compounding, the annuity is called an **ordinary annuity.** The time between payments is the **payment period,** and the time from the beginning of the first payment period to the end of the last period is called the **term of the annuity.** The **future value of the annuity,** the final sum on deposit, is defined as the sum of the compound amounts of all the payments, compounded to the end of the term.

For example, suppose $1500 is deposited at the end of the year for the next 6 yr in an account paying 8% per yr compounded annually. To find the future value of this annuity, look separately at each of the $1500 payments. The first of these payments will produce a compound amount of

$$1500(1 + .08)^5 = 1500(1.08)^5.$$

Use 5 as the exponent instead of 6 since the money is deposited at the *end* of the first year and earns interest for only 5 yr. The second payment of $1500 will produce a compound amount of $1500(1.08)^4$. Continuing in this way and finding the sum of all the terms gives

$$1500(1.08)^5 + 1500(1.08)^4 + 1500(1.08)^3 + 1500(1.08)^2 + 1500(1.08)^1 + 1500.$$

(The last payment earns no interest at all.) Reading in reverse order, we see that this expression is the sum of the first six terms of a geometric sequence with $a_1 = 1500$, $r = 1.08$, and $n = 6$. Therefore, the sum is

$$\frac{a_1(r^n - 1)}{r - 1} = \frac{1500[(1.08)^6 - 1]}{1.08 - 1} = 11{,}003.89$$

or $11,003.89.

We state the following formula without proof.

Future Value of an Ordinary Annuity

$$S = R\left[\frac{(1 + i)^n - 1}{i}\right]$$

where

S is future value,

R is the payment at the end of each period,

i is the interest rate per period, and

n is the number of periods.

EXAMPLE 7 Applying the Formula for the Future Value of an Annuity

(a) Rocky Rhodes is an athlete who feels that his playing career will last 7 yr. To prepare for his future, he deposits $22,000 at the end of each year for 7 yr in an account paying 6% compounded annually. How much will he have on deposit after 7 yr?

His payments form an ordinary annuity with $R = 22{,}000$, $n = 7$, and $i = .06$. The future value of this annuity (using the formula) is

$$S = 22{,}000 \left[\frac{(1.06)^7 - 1}{.06} \right] = 184{,}664.43, \qquad \text{Use a calculator.}$$

or $184,664.43.

(b) Experts say that the baby boom generation (born between 1946 and 1960) cannot count on a company pension or Social Security to provide a comfortable retirement, as their parents did. It is recommended that they start to save early and regularly. Judy Zahrndt, a baby boomer, has decided to deposit $200 at the end of each month in an account that pays interest of 7.2% compounded monthly for retirement in 20 yr. How much will be in the account at that time?

Because the interest is compounded monthly, $i = \frac{.072}{12}$. Also, $R = 200$ and $n = 12(20)$. The future value is

$$S = 200 \left[\frac{\left(1 + \dfrac{.072}{12}\right)^{12(20)} - 1}{\dfrac{.072}{12}} \right] = 106{,}752.47,$$

or $106,752.47.

Now Try Exercise 35.

OBJECTIVE 6 **Find the sum of an infinite number of terms of certain geometric sequences.** Consider an infinite geometric sequence such as

$$\frac{1}{3}, \frac{1}{6}, \frac{1}{12}, \frac{1}{24}, \frac{1}{48}, \ldots .$$

Can the sum of the terms of such a sequence be found somehow? The sum of the first two terms is

$$S_2 = \frac{1}{3} + \frac{1}{6} = \frac{1}{2} = .5.$$

In a similar manner,

$$S_3 = S_2 + \frac{1}{12} = \frac{1}{2} + \frac{1}{12} = \frac{7}{12} \approx .583, \quad S_4 = S_3 + \frac{1}{24} = \frac{7}{12} + \frac{1}{24} = \frac{15}{24} = .625,$$

$$S_5 = \frac{31}{48} \approx .64583, \quad S_6 = \frac{21}{32} = .65625, \quad S_7 = \frac{127}{192} \approx .6614583.$$

Each term of the geometric sequence is smaller than the preceding one, so each additional term is contributing less and less to the sum. In decimal form (to the nearest thousandth) the first seven terms and the tenth term are given in the table.

Term	a_1	a_2	a_3	a_4	a_5	a_6	a_7	a_{10}
Value	.333	.167	.083	.042	.021	.010	.005	.001

As the table suggests, the value of a term gets closer and closer to 0 as the number of the term increases. To express this idea, we say that as n increases without bound (written $n \to \infty$), the limit of the term a_n is 0, written

$$\lim_{n \to \infty} a_n = 0.$$

A number that can be defined as the sum of an infinite number of terms of a geometric sequence can be found by starting with the expression for the sum of a finite number of terms:

$$S_n = \frac{a_1(r^n - 1)}{r - 1}.$$

If $|r| < 1$, then as n increases without bound the value of r^n gets closer and closer to 0. For example, in the infinite sequence just discussed, $r = \frac{1}{2} = .5$. The following table shows how $r^n = (.5)^n$, given to the nearest thousandth, gets smaller as n increases.

n	1	2	3	4	5	6	7	10
r^n	.5	.25	.125	.063	.031	.016	.008	.001

As r^n approaches 0, $r^n - 1$ approaches $0 - 1 = -1$, and S_n approaches the quotient $\frac{-a_1}{r - 1}$. Thus,

$$\lim_{r^n \to 0} S_n = \lim_{r^n \to 0} \frac{a_1(r^n - 1)}{r - 1} = \frac{a_1(0 - 1)}{r - 1} = \frac{-a_1}{r - 1} = \frac{a_1}{1 - r}.$$

This limit is defined to be the sum of the infinite geometric sequence:

$$a_1 + a_1r + a_1r^2 + a_1r^3 + \cdots = \frac{a_1}{1 - r}, \quad \text{if } |r| < 1.$$

What happens if $|r| > 1$? For example, suppose the sequence is

$$6, 12, 24, \ldots, 3(2)^n, \ldots.$$

In this kind of sequence, as n increases, the value of r^n also increases and so does the sum S_n. Since each new term adds a larger and larger amount to the sum, there is no limit to the value of S_n, and the sum S_n does not exist. A similar situation exists if $r = 1$.

In summary, the sum of the terms of an infinite geometric sequence is defined as follows.

Sum of the Terms of an Infinite Geometric Sequence

The sum S of the terms of an infinite geometric sequence with first term a_1 and common ratio r, where $|r| < 1$, is

$$S = \frac{a_1}{1 - r}.$$

If $|r| \geq 1$, then the sum does not exist.

■ **EXAMPLE 8** **Finding the Sum of the Terms of an Infinite Geometric Sequence**

Find the sum of the terms of the infinite geometric sequence with $a_1 = 3$ and $r = -\frac{1}{3}$.

From the preceding rule, the sum is

$$S = \frac{a_1}{1-r} = \frac{3}{1-\left(-\frac{1}{3}\right)} = \frac{3}{\frac{4}{3}} = \frac{9}{4}.$$

Now Try Exercise 39.

Using summation notation, the sum of an infinite geometric sequence is written as

$$\sum_{i=1}^{\infty} a_i.$$

For instance, the sum in Example 8 would be written

$$\sum_{i=1}^{\infty} 3\left(-\frac{1}{3}\right)^{i-1}.$$

■ **EXAMPLE 9** **Finding the Sum of the Terms of an Infinite Geometric Series**

Find $\displaystyle\sum_{i=1}^{\infty} \left(\frac{1}{2}\right)^i.$

This is the infinite geometric series

$$\frac{1}{2} + \frac{1}{4} + \frac{1}{8} + \cdots,$$

with $a_1 = \frac{1}{2}$ and $r = \frac{1}{2}$. Since $|r| < 1$, we find the sum as follows.

$$S = \frac{a_1}{1-r} = \frac{\frac{1}{2}}{1-\frac{1}{2}} = \frac{\frac{1}{2}}{\frac{1}{2}} = 1$$

Now Try Exercise 43.

14.3 EXERCISES

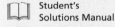

✐ **1.** Using several examples, explain the meaning of *geometric sequence*.

✐ **2.** Explain why the sequence 5, 5, 5, 5, . . . can be considered either arithmetic or geometric.

If the given sequence is geometric, find the common ratio, r. If the sequence is not geometric, say so. See Example 1.

3. 4, 8, 16, 32, . . . **4.** 5, 15, 45, 135, . . . **5.** $\frac{1}{3}, \frac{2}{3}, \frac{3}{3}, \frac{4}{3}, \frac{5}{3}, \ldots$

6. $\frac{1}{3}, \frac{2}{3}, \frac{4}{3}, \frac{8}{3}, \ldots$ **7.** 1, −3, 9, −27, 81, . . . **8.** 1, −3, 7, −11, . . .

9. 1, $-\frac{1}{2}, \frac{1}{4}, -\frac{1}{8}, \frac{1}{16}, \ldots$ **10.** $\frac{2}{3}, \frac{2}{15}, \frac{2}{75}, \frac{2}{375}, \ldots$

Find a general term for each geometric sequence. See Example 2.

11. $5, 10, \ldots$

12. $-2, -6, \ldots$

13. $\dfrac{1}{9}, \dfrac{1}{3}, \ldots$

14. $-3, \dfrac{3}{2}, \ldots$

15. $10, -2, \ldots$

16. $-4, 8, \ldots$

Find the indicated term for each geometric sequence. See Example 3.

17. $a_1 = 2, r = 5; a_{10}$

18. $a_1 = -1, r = 3; a_{15}$

19. $\dfrac{1}{2}, \dfrac{1}{6}, \dfrac{1}{18}, \ldots; a_{12}$

20. $\dfrac{2}{3}, -\dfrac{1}{3}, \dfrac{1}{6}, \ldots; a_{18}$

21. $a_3 = \dfrac{1}{2}, a_7 = \dfrac{1}{32}; a_{25}$

22. $a_5 = 48, a_8 = -384; a_{10}$

Write the first five terms of each geometric sequence. See Example 4.

23. $a_1 = 2, r = 3$

24. $a_1 = 4, r = 2$

25. $a_1 = 5, r = -\dfrac{1}{5}$

26. $a_1 = 6, r = -\dfrac{1}{3}$

Use the formula for S_n to find the sum for each geometric sequence. See Examples 5 and 6. In Exercises 29–34, give the answer to the nearest thousandth.

27. $\dfrac{1}{3}, \dfrac{1}{9}, \dfrac{1}{27}, \dfrac{1}{81}, \dfrac{1}{243}$

28. $\dfrac{4}{3}, \dfrac{8}{3}, \dfrac{16}{3}, \dfrac{32}{3}, \dfrac{64}{3}, \dfrac{128}{3}$

29. $-\dfrac{4}{3}, -\dfrac{4}{9}, -\dfrac{4}{27}, -\dfrac{4}{81}, -\dfrac{4}{243}, -\dfrac{4}{729}$

30. $\dfrac{5}{16}, -\dfrac{5}{32}, \dfrac{5}{64}, -\dfrac{5}{128}, \dfrac{5}{256}$

31. $\displaystyle\sum_{i=1}^{7} 4\left(\dfrac{2}{5}\right)^i$

32. $\displaystyle\sum_{i=1}^{8} 5\left(\dfrac{2}{3}\right)^i$

33. $\displaystyle\sum_{i=1}^{10} (-2)\left(\dfrac{3}{5}\right)^i$

34. $\displaystyle\sum_{i=1}^{6} (-2)\left(-\dfrac{1}{2}\right)^i$

Solve each problem involving an ordinary annuity. See Example 7.

35. A father opened a savings account for his daughter on the day she was born, depositing $1000. Each year on her birthday he deposits another $1000, making the last deposit on her twenty-first birthday. If the account pays 9.5% interest compounded annually, how much is in the account at the end of the day on the daughter's twenty-first birthday?

36. A 45-year-old man puts $1000 in a retirement account at the end of each quarter $\left(\frac{1}{4}\text{ of a year}\right)$ until he reaches age 60. If the account pays 11% annual interest compounded quarterly, how much will be in the account at that time?

37. At the end of each quarter a 50-year-old woman puts $1200 in a retirement account that pays 7% interest compounded quarterly. When she reaches age 60, she withdraws the entire amount and places it in a mutual fund that pays 9% interest compounded monthly. From then on she deposits $300 in the mutual fund at the end of each month. How much is in the account when she reaches age 65?

38. John Bray deposits $10,000 at the beginning of each year for 12 yr in an account paying 5% compounded annually. He then puts the total amount on deposit in another account paying 6% compounded semiannually for another 9 yr. Find the final amount on deposit after the entire 21-yr period.

Find the sum, if it exists, of the terms of each infinite geometric sequence. See Examples 8 and 9.

39. $a_1 = 6, r = \dfrac{1}{3}$

40. $a_1 = 10, r = \dfrac{1}{5}$

41. $a_1 = 1000, r = -\dfrac{1}{10}$

42. $a_1 = 8500, r = \dfrac{3}{5}$

43. $\displaystyle\sum_{i=1}^{\infty} \dfrac{9}{8}\left(-\dfrac{2}{3}\right)^i$

44. $\displaystyle\sum_{i=1}^{\infty} \dfrac{3}{5}\left(\dfrac{5}{6}\right)^i$

45. $\displaystyle\sum_{i=1}^{\infty} \dfrac{12}{5}\left(\dfrac{5}{4}\right)^i$

46. $\displaystyle\sum_{i=1}^{\infty} \left(-\dfrac{16}{3}\right)\left(-\dfrac{9}{8}\right)^i$

Solve each application. (Hint: Determine whether you need to find a specific term of a sequence or the sum of the terms of a sequence immediately after reading the problem.)

47. A certain ball when dropped from a height rebounds $\dfrac{3}{5}$ of the original height. How high will the ball rebound after the fourth bounce if it was dropped from a height of 10 ft?

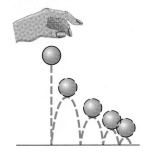

48. A fully wound yo-yo has a string 40 in. long. It is allowed to drop and on its first rebound, it returns to a height 15 in. lower than its original height. Assuming this "rebound ratio" remains constant until the yo-yo comes to rest, how far does it travel on its third trip up the string?

49. A particular substance decays in such a way that it loses half its weight each day. In how many days will 256 g of the substance be reduced to 32 g? How much of the substance is left after 10 days?

50. A tracer dye is injected into a system with an input and an excretion. After one hour $\dfrac{2}{3}$ of the dye is left. At the end of the second hour $\dfrac{2}{3}$ of the remaining dye is left, and so on. If one unit of the dye is injected, how much is left after 6 hr?

51. In a certain community the consumption of electricity has increased about 6% per yr.

   (a) If a community uses 1.1 billion units of electricity now, how much will it use 5 yr from now?

   (b) Find the number of years it will take for the consumption to double.

52. Suppose the community in Exercise 51 reduces its increase in consumption to 2% per yr.

   (a) How much will it use 5 yr from now?

   (b) Find the number of years it will take for the consumption to double.

53. A machine depreciates by $\frac{1}{4}$ of its value each year. If it cost $50,000 new, what is its value after 8 yr?

54. Refer to Exercise 48. Theoretically, how far does the yo-yo travel before coming to rest?

RELATING CONCEPTS (EXERCISES 55–60)

For Individual or Group Work

In Chapter 1 we learned that any repeating decimal is a rational number; that is, it can be expressed as a quotient of integers. Thus, the repeating decimal

$$.99999\ldots,$$

an endless string of 9*s, must be a rational number.* **Work Exercises 55–60 in order,** *to discover the surprising simplest form of this rational number.*

55. Use long division or your previous experience to write a repeating decimal representation for $\frac{1}{3}$.

56. Use long division or your previous experience to write a repeating decimal representation for $\frac{2}{3}$.

57. Because $\frac{1}{3} + \frac{2}{3} = 1$, the sum of the decimal representations in Exercises 55 and 56 must also equal 1. Line up the decimals in the usual vertical method for addition, and obtain the repeating decimal result. The value of this decimal is exactly 1.

58. The repeating decimal $.99999\ldots$ can be written as the sum of the terms of a geometric sequence with $a_1 = .9$ and $r = .1$:

$$.99999\ldots = .9 + .9(.1) + .9(.1)^2 + .9(.1)^3 + .9(.1)^4 + .9(.1)^5 + \cdots.$$

Since $|.1| < 1$, this sum can be found using the formula $S = \dfrac{a_1}{1 - r}$. Use this formula to support the result you found another way in Exercises 55–57.

59. Which one of the following is true, based on your results in Exercises 57 and 58?

   A. $.99999\ldots < 1$ **B.** $.99999\ldots = 1$ **C.** $.99999\ldots \approx 1$

60. Show that $.49999\ldots = \frac{1}{2}$.

14.4 The Binomial Theorem

OBJECTIVE 1 Expand a binomial raised to a power. Writing out the binomial expression $(x + y)^n$ for nonnegative integer values of n gives a family of expressions that is important in the study of mathematics and its applications. For example,

$$(x + y)^0 = 1,$$
$$(x + y)^1 = x + y,$$
$$(x + y)^2 = x^2 + 2xy + y^2,$$
$$(x + y)^3 = x^3 + 3x^2y + 3xy^2 + y^3,$$
$$(x + y)^4 = x^4 + 4x^3y + 6x^2y^2 + 4xy^3 + y^4,$$
$$(x + y)^5 = x^5 + 5x^4y + 10x^3y^2 + 10x^2y^3 + 5xy^4 + y^5.$$

Inspection shows that these expansions follow a pattern. By identifying the pattern, we can write a general expression for $(x + y)^n$.

First, if n is a positive integer, each expansion after $(x + y)^0$ begins with x raised to the same power to which the binomial is raised. That is, the expansion of $(x + y)^1$ has a first term of x^1, the expansion of $(x + y)^2$ has a first term of x^2, the expansion of $(x + y)^3$ has a first term of x^3, and so on. Also, the last term in each expansion is y to this same power, so the expansion of $(x + y)^n$ should begin with the term x^n and end with the term y^n.

The exponents on x decrease by 1 in each term after the first, while the exponents on y, beginning with y in the second term, increase by 1 in each succeeding term. Thus, the *variables* in the expansion of $(x + y)^n$ have the following pattern.

$$x^n, \quad x^{n-1}y, \quad x^{n-2}y^2, \quad x^{n-3}y^3, \dots, xy^{n-1}, \quad y^n$$

This pattern suggests that the sum of the exponents on x and y in each term is n. For example, in the third term above, the variable part is $x^{n-2}y^2$ and the sum of the exponents, $n - 2$ and 2, is n.

Now examine the pattern for the *coefficients* in the terms of the preceding expansions. Writing the coefficients alone in a triangular pattern gives **Pascal's triangle,** named in honor of the 17th century mathematician Blaise Pascal, one of the first to use it extensively.

Blaise Pascal (1623–1662)

Pascal's Triangle

```
                    1
                 1     1
              1     2     1
           1     3     3     1
        1     4     6     4     1
     1     5    10    10     5     1      and so on
```

Arranging the coefficients in this way shows that each number in the triangle is the sum of the two numbers just above it (one to the right and one to the left). For example, in the fifth row from the top, 1 is the sum of 1 (the only number above it), 4 is the sum of 1 and 3, 6 is the sum of 3 and 3, and so on.

To obtain the coefficients for $(x + y)^6$, we attach the seventh row to the table by adding pairs of numbers from the sixth row.

$$1 \quad 6 \quad 15 \quad 20 \quad 15 \quad 6 \quad 1$$

We then use these coefficients to expand $(x + y)^6$ as

$$(x + y)^6 = x^6 + 6x^5y + 15x^4y^2 + 20x^3y^3 + 15x^2y^4 + 6xy^5 + y^6.$$

CONNECTIONS

Over the years, many interesting patterns have been discovered in Pascal's triangle. In the following figure, the triangular array is written in a different form. The indicated sums along the diagonals shown are the terms of the *Fibonacci sequence,* mentioned in the chapter introduction. The presence of this sequence in the triangle apparently was not recognized by Pascal.

```
1   1
1   1   2   3
1   2   1   5   8
1   3   3   1   13
1   4   6   4   1
1   5   10  10  5   1
1   6   15  20  15  6   1
```

Triangular numbers are found by counting the number of points in triangular arrangements of points. The first few triangular numbers are shown in the figure below.

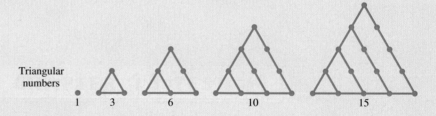

Triangular numbers   1   3   6   10   15

The number of points in these figures form the sequence 1, 3, 6, 10, . . . , a sequence that is found in Pascal's triangle, as shown in the next figure.

```
        1
      1   1
    1   2   1
  1   3   3   1
1   4   6   4   1
1   5   10  10  5   1
```

For Discussion or Writing

1. Predict the next two numbers in the sequence of sums of the diagonals of Pascal's triangle.

2. Predict the next five numbers in the list of triangular numbers.

3. Describe other sequences that can be found in Pascal's triangle.

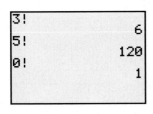

(a)

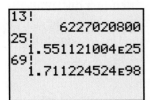

A graphing calculator with a 10-digit display will give the exact value of $n!$ for $n \leq 13$ and approximate values of $n!$ for $14 \leq n \leq 69$.

(b)

FIGURE 5

Although it is possible to use Pascal's triangle to find the coefficients of $(x + y)^n$ for any positive integer value of n, it is impractical for large values of n. A more efficient way to determine these coefficients uses a notational shorthand with the symbol $n!$ (read "n factorial") defined as follows.

---

n Factorial ($n!$)

For any positive integer n,

$$n(n - 1)(n - 2)(n - 3) \cdots (2)(1) = n!.$$

For example,

$$3! = 3 \cdot 2 \cdot 1 = 6 \quad \text{and} \quad 5! = 5 \cdot 4 \cdot 3 \cdot 2 \cdot 1 = 120.$$

From the definition of n factorial, $n[(n - 1)!] = n!$. If $n = 1$, then $1(0!) = 1! = 1$. Because of this, $0!$ is defined as

$$0! = 1.$$

Now Try Exercise 1.

Scientific and graphing calculators can compute factorials. The three example factorial expressions above are shown in Figure 5(a). Figure 5(b) shows some larger factorials.

■ EXAMPLE 1 Evaluating Expressions with $n!$

Find the value of each expression.

(a) $\dfrac{5!}{4!1!} = \dfrac{5 \cdot 4 \cdot 3 \cdot 2 \cdot 1}{(4 \cdot 3 \cdot 2 \cdot 1)(1)} = 5$

(b) $\dfrac{5!}{3!2!} = \dfrac{5 \cdot 4 \cdot 3 \cdot 2 \cdot 1}{(3 \cdot 2 \cdot 1)(2 \cdot 1)} = \dfrac{5 \cdot 4}{2 \cdot 1} = 10$

(c) $\dfrac{6!}{3!3!} = \dfrac{6 \cdot 5 \cdot 4 \cdot 3 \cdot 2 \cdot 1}{(3 \cdot 2 \cdot 1)(3 \cdot 2 \cdot 1)} = \dfrac{6 \cdot 5 \cdot 4}{3 \cdot 2 \cdot 1} = 20$

(d) $\dfrac{4!}{4!0!} = \dfrac{4 \cdot 3 \cdot 2 \cdot 1}{(4 \cdot 3 \cdot 2 \cdot 1)(1)} = 1$

Now Try Exercises 3 and 7.

Now look again at the coefficients of the expansion

$$(x + y)^5 = x^5 + 5x^4y + 10x^3y^2 + 10x^2y^3 + 5xy^4 + y^5.$$

The coefficient of the second term is 5 and the exponents on the variables in that term are 4 and 1. From Example 1(a), $\frac{5!}{4!1!} = 5$. The coefficient of the third term is 10, and the exponents are 3 and 2. From Example 1(b), $\frac{5!}{3!2!} = 10$. Similar results hold true for the remaining terms. The first term can be written as $1x^5y^0$, and the last term can be written as $1x^0y^5$. Then the coefficient of the first term should be $\frac{5!}{5!0!} = 1$, and the coefficient of the last term would be $\frac{5!}{0!5!} = 1$. Generalizing, the coefficient for a term of $(x + y)^n$ in which the variable part is $x^r y^{n-r}$ will be

$$\frac{n!}{r!(n - r)!}.$$

NOTE The denominator factorials in the coefficient of a term are the same as the exponents on the variables in that term.

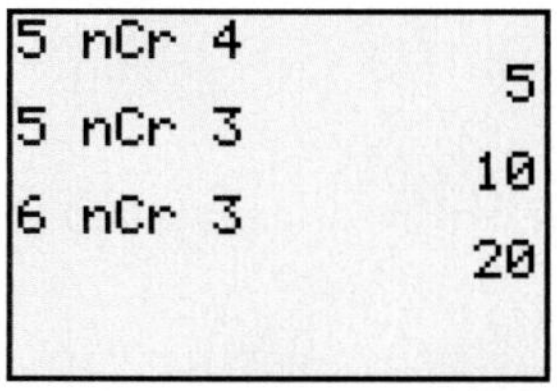

```
5 nCr 4
            5
5 nCr 3
           10
6 nCr 3
           20
```

FIGURE 6

The expression $\frac{n!}{r!(n - r)!}$ is often represented by the symbol $_nC_r$. This comes from the fact that if we choose *combinations* of n things taken r at a time, the result is given by that expression. A graphing calculator can evaluate this expression for particular values of n and r. Figure 6 shows how a calculator evaluates $_5C_4$, $_5C_3$, and $_6C_3$. Compare these results to parts (a), (b), and (c) of Example 1.

Now Try Exercise 5.

Summarizing this work gives the **binomial theorem,** or the **general binomial expansion.**

Binomial Theorem

For any positive integer n,

$$(x + y)^n = x^n + \frac{n!}{(n - 1)!1!}x^{n-1}y + \frac{n!}{(n - 2)!2!}x^{n-2}y^2$$

$$+ \frac{n!}{(n - 3)!3!}x^{n-3}y^3 + \cdots + \frac{n!}{1!(n - 1)!}xy^{n-1} + y^n.$$

The binomial theorem can be written in summation notation as

$$(x + y)^n = \sum_{i=0}^{n} \frac{n!}{(n - i)!i!}x^{n-i}y^i.$$

NOTE The letter i is used here instead of r because we are using summation notation. It is not the imaginary number i.

EXAMPLE 2 Using the Binomial Theorem

Expand $(2m + 3)^4$.

$$(2m + 3)^4 = (2m)^4 + \frac{4!}{3!1!}(2m)^3(3) + \frac{4!}{2!2!}(2m)^2(3)^2 + \frac{4!}{1!3!}(2m)(3)^3 + 3^4$$

$$= 16m^4 + 4(8m^3)(3) + 6(4m^2)(9) + 4(2m)(27) + 81$$

$$= 16m^4 + 96m^3 + 216m^2 + 216m + 81$$

Now Try Exercise 17.

▨ **EXAMPLE 3** Using the Binomial Theorem

Expand $\left(a - \dfrac{b}{2}\right)^5$.

$$\left(a - \frac{b}{2}\right)^5 = a^5 + \frac{5!}{4!1!}a^4\left(-\frac{b}{2}\right) + \frac{5!}{3!2!}a^3\left(-\frac{b}{2}\right)^2 + \frac{5!}{2!3!}a^2\left(-\frac{b}{2}\right)^3$$

$$+ \frac{5!}{1!4!}a\left(-\frac{b}{2}\right)^4 + \left(-\frac{b}{2}\right)^5$$

$$= a^5 + 5a^4\left(-\frac{b}{2}\right) + 10a^3\left(\frac{b^2}{4}\right) + 10a^2\left(-\frac{b^3}{8}\right)$$

$$+ 5a\left(\frac{b^4}{16}\right) + \left(-\frac{b^5}{32}\right)$$

$$= a^5 - \frac{5}{2}a^4b + \frac{5}{2}a^3b^2 - \frac{5}{4}a^2b^3 + \frac{5}{16}ab^4 - \frac{1}{32}b^5$$

Now Try Exercise 19.

CAUTION When the binomial is the *difference* of two terms as in Example 3, the signs of the terms in the expansion will alternate. Those terms with odd exponents on the second variable expression $\left(-\frac{b}{2}\right.$ in Example 3$\left.\right)$ will be negative, while those with even exponents on the second variable expression will be positive.

OBJECTIVE 2 **Find any specified term of the expansion of a binomial.** Any single term of a binomial expansion can be determined without writing out the whole expansion. For example, if $n \geq 10$, then the tenth term of $(x + y)^n$ has y raised to the ninth power (since y has the power of 1 in the second term, the power of 2 in the third term, and so on). Since the exponents on x and y in any term must have a sum of n, the exponent on x in the tenth term is $n - 9$. These quantities, 9 and $n - 9$, determine the factorials in the denominator of the coefficient. Thus,

$$\frac{n!}{(n - 9)!9!}x^{n-9}y^9$$

is the tenth term of $(x + y)^n$. A generalization of this idea follows.

---

rth Term of the Binomial Expansion

If $n \geq r - 1$, then the rth term of the expansion of $(x + y)^n$ is

$$\frac{n!}{[n - (r - 1)]!(r - 1)!}x^{n-(r-1)}y^{r-1}.$$

In this general expression, remember to start with the exponent on y, which is 1 less than the term number r. Then subtract that exponent from n to get the exponent on x: $n - (r - 1)$. The two exponents are then used as the factorials in the denominator of the coefficient.

■ **EXAMPLE 4**   Finding a Single Term of a Binomial Expansion

Find the fourth term of $(a + 2b)^{10}$.

In the fourth term, $2b$ has an exponent of $4 - 1 = 3$ and a has an exponent of $10 - 3 = 7$. The fourth term is

$$\frac{10!}{7!3!}(a^7)(2b)^3 = \frac{10 \cdot 9 \cdot 8}{3 \cdot 2 \cdot 1}(a^7)(8b^3)$$

$$= 120a^7(8b^3)$$

$$= 960a^7b^3.$$

Now Try Exercise 29.

14.4   EXERCISES

For Extra Help

 Student's
Solutions Manual

 MyMathLab

 InterAct Math
Tutorial Software

Tutor Center AW Math
Tutor Center

MathXL MathXL

 Digital Video Tutor
CD 20/Videotape 20

Evaluate each expression. See Example 1.

1. $6!$ **2.** $4!$ **3.** $\dfrac{6!}{4!2!}$ **4.** $\dfrac{7!}{3!4!}$

5. $_6C_2$ **6.** $_7C_4$ **7.** $\dfrac{4!}{0!4!}$ **8.** $\dfrac{5!}{5!0!}$

9. $4! \cdot 5$ **10.** $6! \cdot 7$ **11.** $_{13}C_{11}$ **12.** $_{13}C_2$

Use the binomial theorem to expand each expression. See Examples 2 and 3.

13. $(m + n)^4$ **14.** $(x + r)^5$ **15.** $(a - b)^5$ **16.** $(p - q)^4$

17. $(2x + 3)^3$ **18.** $\left(\dfrac{x}{3} + 2y\right)^5$ **19.** $\left(\dfrac{x}{2} - y\right)^4$ **20.** $(x^2 + 1)^4$

21. $(mx - n^2)^3$ **22.** $(2p^2 - q^2)^3$

Write the first four terms of each binomial expansion.

23. $(r + 2s)^{12}$ **24.** $(m - n)^{20}$ **25.** $(3x - y)^{14}$

26. $(2p + 3q)^{11}$ **27.** $(t^2 + u^2)^{10}$ **28.** $(x^2 - y^2)^{15}$

Find the indicated term of each binomial expansion. See Example 4.

29. $(2m + n)^{10}$; fourth term   **30.** $(a - 3b)^{12}$; fifth term

31. $\left(x + \dfrac{y}{2}\right)^8$; seventh term   **32.** $(3p - 2q)^{15}$; eighth term

33. $(k - 1)^9$; third term   **34.** $(-4 - s)^{11}$; fourth term

35. The middle term of $(x^2 - 2y)^6$ **36.** The middle term of $(m^3 + 3)^8$

37. The term with x^9y^4 in $(3x^3 - 4y^2)^5$ **38.** The term with x^{10} in $\left(x^3 - \dfrac{2}{x}\right)^6$

Investing for the Future

OBJECTIVE   Calculate compound interest; understand the effects of monthly, quarterly, and annual compounding.

In this chapter you have seen many different types of financing and investing options including loans, mutual funds, savings accounts, and annuities. In this activity you will analyze annuities with different periods of compound interest.

   Consider a family that has a 10-year-old child, for whom they want to save for college. The family is considering three different savings options.

Option 1:   An annuity that compounds quarterly

Option 2:   An annuity that compounds monthly

Option 3:   Put off saving until high school (that is, the last 4 yr); then put money into an annuity that compounds annually

A. Have each member of your group calculate total savings for one of the three options. Use the following formula, where S is future value, R is payment amount made each *period, i* is annual interest rate divided by the number of periods per year, and n is total number of compounding periods, along with the specific information given below for each option.

$$S = R\left[\frac{(1 + i)^n - 1}{i}\right]$$

Option 1:   The family plans to save \$300 per quarter (3 months), the interest rate is 5%, and the annuity compounds quarterly. Savings will be for 8 yr.

Option 2:   The family plans to save \$100 per month, the interest rate is 5%, and the annuity compounds monthly. Again, savings will be for 8 yr.

Option 3:   The family plans to wait and save only the last 4 yr. They will save \$2400 a yr, the interest rate is 5%, and the annuity compounds annually.

B. Compare your answers.

   1. How much is being invested using each option?

   2. Which option resulted in the largest amount of savings?

   3. Explain why this option produced more savings.

   4. What other considerations might be involved in deciding how to save?

CHAPTER **14** SUMMARY

KEY TERMS

14.1	sequence infinite sequence terms of a sequence general term finite sequence series summation notation		index of summation arithmetic mean (average) **14.2** arithmetic sequence (arithmetic progression) common difference	**14.3** geometric sequence (geometric progression) common ratio annuity ordinary annuity payment period	future value of an annuity term of an annuity **14.4** Pascal's triangle binomial theorem (general binomial expansion)

NEW SYMBOLS

a_n	nth term of a sequence	S_n	sum of first n terms of a sequence	$\displaystyle\sum_{i=1}^{\infty} a_i$ sum of an infinite number of terms	$_nC_r$ binomial coefficient (combinations of n things taken r at a time)
$\displaystyle\sum_{i=1}^{n} a_i$	summation notation	$\displaystyle\lim_{n\to\infty} a_n$	limit of a_n as n gets larger and larger	$n!$ n factorial	

TEST YOUR WORD POWER

See how well you have learned the vocabulary in this chapter. Answers, with examples, follow the Quick Review.

1. An **infinite sequence** is
   A. the values of a function
   B. a function whose domain is the set of natural numbers
   C. the sum of the terms of a function
   D. the average of a group of numbers.

2. A **series** is
   A. the sum of the terms of a sequence
   B. the product of the terms of a sequence
   C. the average of the terms of a sequence
   D. the function values of a sequence.

3. An **arithmetic sequence** is a sequence in which
   A. each term after the first is a constant multiple of the preceding term

   B. the numbers are written in a triangular array
   C. the terms are added
   D. each term after the first differs from the preceding term by a common amount.

4. A **geometric sequence** is a sequence in which
   A. each term after the first is a constant multiple of the preceding term
   B. the numbers are written in a triangular array
   C. the terms are multiplied
   D. each term after the first differs from the preceding term by a common amount.

5. The **common difference** is
   A. the average of the terms in a sequence

   B. the constant multiplier in a geometric sequence
   C. the difference between any two adjacent terms in an arithmetic sequence
   D. the sum of the terms of an arithmetic sequence.

6. The **common ratio** is
   A. the average of the terms in a sequence
   B. the constant multiplier in a geometric sequence
   C. the difference between any two adjacent terms in an arithmetic sequence
   D. the product of the terms of a geometric sequence.

QUICK REVIEW

CONCEPTS	EXAMPLES

14.1 SEQUENCES AND SERIES

Sequence

General Term a_n

Series

$1, \dfrac{1}{2}, \dfrac{1}{3}, \dfrac{1}{4}, \ldots, \dfrac{1}{n}$ has general term $\dfrac{1}{n}$.

The corresponding series is the *sum*

$$1 + \frac{1}{2} + \frac{1}{3} + \frac{1}{4} + \cdots + \frac{1}{n}.$$

14.2 ARITHMETIC SEQUENCES

Assume a_1 is the first term, a_n is the nth term, and d is the common difference.

The arithmetic sequence 2, 5, 8, 11, . . . has $a_1 = 2$.

Common Difference

$$d = a_{n+1} - a_n$$

$$d = 5 - 2 = 3$$

(Any two successive terms could have been used.)

nth Term

$$a_n = a_1 + (n - 1)d$$

Suppose that $n = 10$. Then the 10th term is

$$a_{10} = 2 + (10 - 1)3$$
$$= 2 + 9 \cdot 3 = 29.$$

Sum of the First n Terms

$$S_n = \frac{n}{2}(a_1 + a_n)$$

The sum of the first 10 terms is

$$S_{10} = \frac{10}{2}(2 + a_{10})$$
$$= 5(2 + 29) = 5(31) = 155$$

or $\qquad S_n = \dfrac{n}{2}[2a_1 + (n - 1)d]$

or $\qquad S_{10} = \dfrac{10}{2}[2(2) + (10 - 1)3]$
$$= 5(4 + 9 \cdot 3)$$
$$= 5(4 + 27) = 5(31) = 155.$$

14.3 GEOMETRIC SEQUENCES

Assume a_1 is the first term, a_n is the nth term, and r is the common ratio.

The geometric sequence 1, 2, 4, 8, . . . has $a_1 = 1$.

Common Ratio

$$r = \frac{a_{n+1}}{a_n}$$

$$r = \frac{8}{4} = 2$$

(Any two successive terms could have been used.)

nth Term

$$a_n = a_1 r^{n-1}$$

Suppose that $n = 6$. Then the sixth term is

$$a_6 = (1)(2)^{6-1} = 1(2)^5 = 32.$$

CONCEPTS	EXAMPLES

Sum of the First n Terms

$$S_n = \frac{a_1(1 - r^n)}{1 - r} \quad \text{or} \quad S_n = \frac{a_1(r^n - 1)}{r - 1} \quad (r \neq 1)$$

The sum of the first six terms is

$$S_6 = \frac{1(2^6 - 1)}{2 - 1} = \frac{64 - 1}{1} = 63.$$

Future Value of an Ordinary Annuity

$$S = R\left[\frac{(1 + i)^n - 1}{i}\right],$$

where S is future value, R is the payment at the end of each period, i is the interest rate per period, and n is the number of periods.

If \$5800 is deposited into an ordinary annuity at the end of each quarter for 4 yr and interest is earned at 6.4% compounded quarterly, then

$$R = \$5800, \quad i = \frac{.064}{4} = .016, \quad n = 4(4) = 16,$$

and

$$S = 5800\left[\frac{(1 + .016)^{16} - 1}{.016}\right]$$

$$= \$104,812.44.$$

Sum of the Terms of an Infinite Geometric Sequence with $|r| < 1$

$$S = \frac{a_1}{1 - r}$$

The sum S of the terms of an infinite geometric sequence with $a_1 = 1$ and $r = \frac{1}{2}$ is

$$S = \frac{1}{1 - \frac{1}{2}} = \frac{1}{\frac{1}{2}} = 2.$$

14.4   THE BINOMIAL THEOREM

For any positive integer n,

$$n(n - 1)(n - 2) \cdots (2)(1) = n!.$$
$$0! = 1$$
$${}_nC_r = \frac{n!}{r!(n - r)!}$$

$$4! = 4 \cdot 3 \cdot 2 \cdot 1 = 24$$

$$\begin{aligned} {}_5C_3 &= \frac{5!}{3!(5 - 3)!} \\ &= \frac{5!}{3!2!} \\ &= \frac{5 \cdot 4 \cdot 3 \cdot 2 \cdot 1}{3 \cdot 2 \cdot 1 \cdot 2 \cdot 1} \\ &= 10 \end{aligned}$$

General Binomial Expansion

For any positive integer n,

$$\begin{aligned} (x + y)^n = {}&x^n + \frac{n!}{(n - 1)!1!}x^{n-1}y \\ &+ \frac{n!}{(n - 2)!2!}x^{n-2}y^2 \\ &+ \frac{n!}{(n - 3)!3!}x^{n-3}y^3 + \cdots \\ &+ \frac{n!}{1!(n - 1)!}xy^{n-1} + y^n. \end{aligned}$$

$$\begin{aligned} (2m + 3)^4 = {}&(2m)^4 + \frac{4!}{3!1!}(2m)^3(3) + \frac{4!}{2!2!}(2m)^2(3)^2 \\ &+ \frac{4!}{1!3!}(2m)(3)^3 + 3^4 \\ = {}&2^4m^4 + 4(2)^3m^3(3) + 6(2)^2m^2(9) \\ &+ 4(2m)(27) + 81 \\ = {}&16m^4 + 12(8)m^3 + 54(4)m^2 + 216m + 81 \\ = {}&16m^4 + 96m^3 + 216m^2 + 216m + 81 \end{aligned}$$

(continued)

CONCEPTS	EXAMPLES
rth Term of the Binomial Expansion of $(x + y)^n$ $$\frac{n!}{[n - (r - 1)]!(r - 1)!}x^{n-(r-1)}y^{r-1}$$	The eighth term of $(a - 2b)^{10}$ is $$\frac{10!}{3!7!}a^3(-2b)^7 = \frac{10 \cdot 9 \cdot 8}{3 \cdot 2 \cdot 1}a^3(-2)^7b^7$$ $$= 120(-128)a^3b^7$$ $$= -15{,}360a^3b^7.$$

Answers to Test Your Word Power

1. B; *Example:* The ordered list of numbers 3, 6, 9, 12, 15, . . . is an infinite sequence.   **2.** A; *Example:* $3 + 6 + 9 + 12 + 15$, written in summation notation as $\sum\limits_{i=1}^{5} 3i$, is a series.   **3.** D; *Example:* The sequence $-3, 2, 7, 12, 17, \ldots$ is arithmetic.   **4.** A; *Example:* The sequence 1, 4, 16, 64, 256, . . . is geometric.   **5.** C; *Example:* The common difference of the arithmetic sequence in Answer 3 is 5 since $2 - (-3) = 5, 7 - 2 = 5, 12 - 7 = 5$, and so on.   **6.** B; *Example:* The common ratio of the geometric sequence in Answer 4 is 4 since $\frac{4}{1} = \frac{16}{4} = \frac{64}{16} = \frac{256}{64} = 4$.

CHAPTER 14 REVIEW EXERCISES

[14.1]   *Write out the first four terms of each sequence.*

1. $a_n = 2n - 3$

2. $a_n = \dfrac{n - 1}{n}$

3. $a_n = n^2$

4. $a_n = \left(\dfrac{1}{2}\right)^n$

5. $a_n = (n + 1)(n - 1)$

Write each series as a sum of terms.

6. $\sum\limits_{i=1}^{5} i^2x$

7. $\sum\limits_{i=1}^{6} (i + 1)x^i$

Evaluate each series.

8. $\sum\limits_{i=1}^{4} (i + 2)$

9. $\sum\limits_{i=1}^{6} 2^i$

10. $\sum\limits_{i=4}^{7} \dfrac{i}{i + 1}$

11. Find the arithmetic mean, or average, of the mutual fund retirement assets for the years 1996 through 2000 shown in the table. Give your answer to the nearest tenth (in billions of dollars).

Year	Assets (in billions of dollars)
1996	1166
1997	1509
1998	1899
1999	2462
2000	2408

Source: Investment Company Institute.

[14.2–14.3] *Decide whether each sequence is* arithmetic, geometric, *or* neither. *If the sequence is arithmetic, find the common difference, d. If it is geometric, find the common ratio, r.*

12. $2, 5, 8, 11, \ldots$

13. $-6, -2, 2, 6, 10, \ldots$

14. $\dfrac{2}{3}, -\dfrac{1}{3}, \dfrac{1}{6}, -\dfrac{1}{12}, \ldots$

15. $-1, 1, -1, 1, -1, \ldots$

16. $64, 32, 8, \dfrac{1}{2}, \ldots$

17. $64, 32, 16, 8, \ldots$

18. $10, 8, 6, 4, \ldots$

[14.2] *Find the indicated term for each arithmetic sequence.*

19. $a_1 = -2, d = 5; a_{16}$

20. $a_6 = 12, a_8 = 18; a_{25}$

Find the general term for each arithmetic sequence.

21. $a_1 = -4, d = -5$

22. $6, 3, 0, -3, \ldots$

Find the number of terms in each arithmetic sequence.

23. $7, 10, 13, \ldots, 49$

24. $5, 1, -3, \ldots, -79$

Find S_8 for each arithmetic sequence.

25. $a_1 = -2, d = 6$

26. $a_n = -2 + 5n$

[14.3] *Find the general term for each geometric sequence.*

27. $-1, -4, \ldots$

28. $\dfrac{2}{3}, \dfrac{2}{15}, \ldots$

Find the indicated term for each geometric sequence.

29. $2, -6, 18, \ldots; a_{11}$

30. $a_3 = 20, a_5 = 80; a_{10}$

Find each sum, if it exists.

31. $\displaystyle\sum_{i=1}^{5} \left(\dfrac{1}{4}\right)^i$

32. $\displaystyle\sum_{i=1}^{8} \dfrac{3}{4}(-1)^i$

33. $\displaystyle\sum_{i=1}^{\infty} 4\left(\dfrac{1}{5}\right)^i$

34. $\displaystyle\sum_{i=1}^{\infty} 2(3)^i$

[14.4] *Use the binomial theorem to expand each binomial.*

35. $(2p - q)^5$

36. $(x^2 + 3y)^4$

37. $\left(\sqrt{m} + \sqrt{n}\right)^4$

38. Write the fourth term of the expansion of $(3a + 2b)^{19}$.

39. Write the twenty-third term of the expansion of $(-2k + 3)^{25}$.

MIXED REVIEW EXERCISES

Find the indicated term and S_{10} for each sequence.

40. a_{40}: arithmetic; $1, 7, 13, \ldots$

41. a_{10}: geometric; $-3, 6, -12, \ldots$

42. a_9: geometric; $a_1 = 1, r = -3$

43. a_{15}: arithmetic; $a_1 = -4, d = 3$

Find the general term for each arithmetic or geometric sequence.

44. 2, 7, 12, . . .

45. 2, 8, 32, . . .

46. 27, 9, 3, . . .

47. 12, 9, 6, . . .

Solve each problem.

48. When Mary's sled goes down the hill near her home, she covers 3 ft in the first second; then for each second after that she goes 4 ft more than in the preceding second. If the distance she covers going down is 210 ft, how long does it take her to reach the bottom?

49. An ordinary annuity is set up so that $672 is deposited at the end of each quarter for 7 yr. The money earns 8% annual interest compounded quarterly. What is the future value of the annuity?

50. The school population in Pfleugerville has been dropping 3% per yr. The current population is 50,000. If this trend continues, what will the population be in 6 yr?

51. A pump removes $\frac{1}{2}$ of the liquid in a container with each stroke. What fraction of the liquid is left in the container after 7 strokes?

52. Consider the repeating decimal number .55555 . . . .

(a) Write it as the sum of the terms of an infinite geometric sequence.
(b) What is r for this sequence?
(c) Find this infinite sum, if it exists, and write it as a common fraction in lowest terms.

53. Can the sum of the terms of the infinite geometric sequence with $a_n = 5(2)^n$ be found? Explain.

54. Can any two terms of a geometric sequence be used to find the common ratio? Explain.

CHAPTER **14** TEST

Write the first five terms of each sequence described.

1. $a_n = (-1)^n + 1$

2. Arithmetic, with $a_1 = 4$ and $d = 2$

3. Geometric, with $a_4 = 6$ and $r = \frac{1}{2}$

Find a_4 for each sequence described.

4. Arithmetic, with $a_1 = 6$ and $d = -2$

5. Geometric, with $a_5 = 16$ and $a_7 = 9$

Find S_5 for each sequence described.

6. Arithmetic, with $a_2 = 12$ and $a_3 = 15$

7. Geometric, with $a_5 = 4$ and $a_7 = 1$

8. The number of commercial banks in the United States for the years 1996 through 2000 are given in the table. What was the average number of banks per year for this period?

Year	Number
1996	66,733
1997	69,468
1998	70,731
1999	72,265
2000	72,998

Source: U.S. Federal Deposit Insurance Corporation, Statistics on Banking, annual.

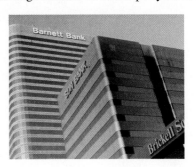

9. If $4000 is deposited in an ordinary annuity at the end of each quarter for 7 yr and earns 6% interest compounded quarterly, how much will be in the account at the end of this term?

10. Under what conditions does an infinite geometric series have a sum?

Find each sum that exists.

11. $\displaystyle\sum_{i=1}^{5} (2i + 8)$

12. $\displaystyle\sum_{i=1}^{6} (3i - 5)$

13. $\displaystyle\sum_{i=1}^{500} i$

14. $\displaystyle\sum_{i=1}^{3} \frac{1}{2}(4^i)$

15. $\displaystyle\sum_{i=1}^{\infty} \left(\frac{1}{4}\right)^i$

16. $\displaystyle\sum_{i=1}^{\infty} 6\left(\frac{3}{2}\right)^i$

Evaluate.

17. $8!$

18. $0!$

19. $\dfrac{6!}{4!2!}$

20. $_{12}C_{10}$

21. Expand $(3k - 5)^4$.

22. Write the fifth term of $\left(2x - \dfrac{y}{3}\right)^{12}$.

Solve each problem.

23. Cheryl bought a new sewing machine for $300. She agreed to pay $20 per month for 15 months plus interest of 1% each month on the unpaid balance. Find the total cost of the machine.

24. During the summer months, the population of a certain insect colony triples each week. If there are 20 insects in the colony at the end of the first week in July, how many are present by the end of September? (Assume exactly 4 weeks in a month.)

CUMULATIVE REVIEW EXERCISES CHAPTERS **1–14**

This set of exercises may be considered a final examination for the course.

Let $P = \left\{-\frac{8}{3}, 10, 0, \sqrt{13}, -\sqrt{3}, \frac{45}{15}, \sqrt{-7}, .82, -3\right\}$. *List the elements of P that are members of each set.*

1. Integers

2. Rational numbers

3. Irrational numbers

4. Real numbers

Simplify each expression.

5. $|-7| + 6 - |-10| - (-8 + 3)$

6. $-15 - |-4| - 10 - |-6|$

7. $4(-6) + (-8)(5) - (-9)$

Solve each equation or inequality.

8. $9 - (5 + 3a) + 5a = -4(a - 3) - 7$

9. $7m + 18 \le 9m - 2$

10. $|4x - 3| = 21$

11. $\dfrac{x + 3}{12} - \dfrac{x - 3}{6} = 0$

12. $2x > 8$ or $-3x > 9$

13. $|2m - 5| \ge 11$

Perform the indicated operations.

14. $(4p + 2)(5p - 3)$

15. $(3k - 7)^2$

16. $(2m^3 - 3m^2 + 8m) - (7m^3 + 5m - 8)$

17. Divide $6t^4 + 5t^3 - 18t^2 + 14t - 1$ by $3t - 2$.

Factor.

18. $7x + x^3$

19. $14y^2 + 13y - 12$

20. $6z^3 + 5z^2 - 4z$

21. $49a^4 - 9b^2$

22. $c^3 + 27d^3$

23. $64r^2 + 48rq + 9q^2$

Simplify.

24. $\left(\dfrac{2}{3}\right)^{-2}$

25. $\dfrac{(3p^2)^3(-2p^6)}{4p^3(5p^7)}$

26. Find any values for which the rational expression $\dfrac{2}{x^2 - 81}$ is undefined.

Simplify.

27. $\dfrac{x^2 - 16}{x^2 + 2x - 8} \div \dfrac{x - 4}{x + 7}$

28. $\dfrac{5}{p^2 + 3p} - \dfrac{2}{p^2 - 4p}$

29. Find the slope of the line through $(4, -5)$ and $(-12, -17)$.

30. Find the standard form of the equation of the line through $(-2, 10)$ and parallel to the line with equation $3x + y = 7$.

Graph.

31. $x - 3y = 6$

32. $4x - y < 4$

33. Consider the set of ordered pairs

$$\{(-3, 2), (-2, 6), (0, 4), (1, 2), (2, 6)\}.$$

(a) Is this a function?

(b) What is its domain?

(c) What is its range?

Solve each system of equations using the method indicated.

34. $\begin{aligned} 2x + 5y &= -19 \\ -3x + 2y &= -19 \end{aligned}$ (Elimination)

35. $\begin{aligned} y &= 5x + 3 \\ 2x + 3y &= -8 \end{aligned}$ (Substitution)

36. $\begin{aligned} x + 2y + z &= 8 \\ 2x - y + 3z &= 15 \\ -x + 3y - 3z &= -11 \end{aligned}$ (Row operations)

37. Nuts worth $3 per lb are to be mixed with 8 lb of nuts worth $4.25 per lb to obtain a mixture that will be sold for $4 per lb. How many pounds of the $3 nuts should be used?

Solve each equation or inequality.

38. $2x^2 + x = 10$

39. $k^2 - k - 6 \le 0$

40. $\dfrac{4}{x-3} - \dfrac{6}{x+3} = \dfrac{24}{x^2-9}$

41. $6x^2 + 5x = 8$

42. $\sqrt{3x-2} = x$

43. $3^{2x-1} = 81$

44. $\log_8 x + \log_8(x+2) = 1$

45. Simplify $5\sqrt{72} - 4\sqrt{50}$.

46. Multiply $(8 + 3i)(8 - 3i)$.

47. Find $f^{-1}(x)$, if $f(x) = 9x + 5$.

Graph.

48. $f(x) = 2(x-2)^2 - 3$

49. $g(x) = \left(\dfrac{1}{3}\right)^x$

50. $y = \log_{1/3} x$

51. $\dfrac{x^2}{9} + \dfrac{y^2}{25} = 1$

52. $x^2 - y^2 = 9$

53. Solve the system $\begin{aligned} xy &= -5 \\ 2x + y &= 3. \end{aligned}$

54. Find the equation of a circle with center at $(-5, 12)$ and radius 9.

55. Write the first five terms of the sequence defined by $a_n = 5n - 12$.

56. Find each sum.

  (a) The sum of the first six terms of the arithmetic sequence with $a_1 = 8$ and $d = 2$

  (b) The sum of the geometric series $15 - 6 + \frac{12}{5} - \frac{24}{25} + \cdots$

57. Find the sum: $\displaystyle\sum_{i=1}^{4} 3i$.

58. Evaluate $9!$.

59. Use the binomial theorem to expand $(2a - 1)^5$.

60. What is the fourth term in the expansion of $\left(3x^4 - \frac{1}{2}y^2\right)^5$?

Appendix A
An Introduction to Calculators

There is little doubt that the appearance of handheld calculators three decades ago and the later development of scientific and graphing calculators have changed the methods of learning and studying mathematics forever. For example, computations with tables of logarithms and slide rules made up an important part of mathematics courses prior to 1970. Today, with the widespread availability of calculators, these topics are studied only for their historical significance.

Calculators come in a large array of different types, sizes, and prices. *For the course for which this textbook is intended, the most appropriate type is the scientific calculator,* which costs $10–$20.

In this introduction, we explain some of the features of scientific and graphing calculators. However, remember that calculators vary among manufacturers and models, and that while the methods explained here apply to many of them, they may not apply to your specific calculator. *This introduction is only a guide and is not intended to take the place of your owner's manual.* Always refer to the manual in the event you need an explanation of how to perform a particular operation.

Scientific Calculators

Scientific calculators are capable of much more than the typical four-function calculator that you might use for balancing your checkbook. Most scientific calculators use *algebraic logic.* (Models sold by Texas Instruments, Sharp, Casio, and Radio Shack, for example, use algebraic logic.) A notable exception is Hewlett-Packard, a company whose calculators use *Reverse Polish Notation* (RPN). In this introduction, we explain the use of calculators with algebraic logic.

Arithmetic Operations   To perform an operation of arithmetic, simply enter the first number, press the operation key, $(+)$, $(-)$, $(\times)$, or $(\div)$, enter the second number, and then press the $(=)$ key. For example, to add 4 and 3, use the following keystrokes.

$$(4) \ (+) \ (3) \ (=) \qquad (\quad 7\quad)$$

Change Sign Key   The key marked $(+/-)$ allows you to change the sign of a display. This is particularly useful when you wish to enter a negative number. For example, to enter -3, use the following keystrokes.

$$\boxed{3} \quad \boxed{+/-} \quad \boxed{\qquad -3}$$

Memory Key   Scientific calculators can hold a number in memory for later use. The label of the memory key varies among models; two of these are $\boxed{M}$ and $\boxed{STO}$. The $\boxed{M+}$ and $\boxed{M-}$ keys allow you to add to or subtract from the value currently in memory. The memory recall key, labeled $\boxed{MR}$, $\boxed{RM}$, or $\boxed{RCL}$, allows you to retrieve the value stored in memory.

Suppose that you wish to store the number 5 in memory. Enter 5, then press the key for memory. You can then perform other calculations. When you need to retrieve the 5, press the key for memory recall.

If a calculator has a constant memory feature, the value in memory will be retained even after the power is turned off. Some advanced calculators have more than one memory. Read the owner's manual for your model to see exactly how memory is activated.

Clearing/Clear Entry Keys   The keys $\boxed{C}$ or $\boxed{CE}$ allow you to clear the display or clear the last entry entered into the display. In some models, pressing the $\boxed{C}$ key once will clear the last entry, while pressing it twice will clear the entire operation in progress.

Second Function Key   This key, usually marked $\boxed{2nd}$, is used in conjunction with another key to activate a function that is printed *above* an operation key (and not on the key itself). For example, suppose you wish to find the square of a number, and the squaring function (explained in more detail later) is printed above another key. You would need to press $\boxed{2nd}$ before the desired squaring function can be activated.

Square Root Key   Pressing $\boxed{\sqrt{}}$ or $\boxed{\sqrt{x}}$ will give the square root (or an approximation of the square root) of the number in the display. On many newer scientific calculators, the square root key is pressed *before* entering the number, while other calculators use the opposite order. Experiment with your calculator to see which method it uses. For example, to find the square root of 36, use the following keystrokes.

$$\boxed{\sqrt{}} \quad \boxed{3} \quad \boxed{6} \quad \boxed{\qquad 6} \qquad \text{or} \qquad \boxed{3} \quad \boxed{6} \quad \boxed{\sqrt{}} \quad \boxed{\qquad 6}$$

The square root of 2 is an example of an irrational number (Chapter 10). The calculator will give an approximation of its value, since the decimal for $\sqrt{2}$ never terminates and never repeats. The number of digits shown will vary among models. To find an approximation for $\sqrt{2}$, use the following keystrokes.

$$\boxed{\sqrt{}} \quad \boxed{2} \quad \boxed{1.4142136} \qquad \text{or} \qquad \boxed{2} \quad \boxed{\sqrt{}} \quad \boxed{1.4142136}$$

An approximation for $\sqrt{2}$

Squaring Key   The $\boxed{x^2}$ key allows you to square the entry in the display. For example, to square 35.7, use the following keystrokes.

$$\boxed{3} \quad \boxed{5} \quad \boxed{\cdot} \quad \boxed{7} \quad \boxed{x^2} \quad \boxed{1274.49}$$

The squaring key and the square root key are often found on the same key, with one of them being a second function (that is, activated by the second function key previously described).

Reciprocal Key   The key marked $\boxed{1/x}$ is the reciprocal key. (When two numbers have a product of 1, they are called *reciprocals*. See Chapter 1.) Suppose that you wish to find the reciprocal of 5. Use the following keystrokes.

$$\boxed{5} \; \boxed{1/x} \qquad \boxed{\qquad 0.2}$$

Inverse Key   Some calculators have an inverse key, marked $\boxed{\text{INV}}$. Inverse operations are operations that "undo" each other. For example, the operations of squaring and taking the square root are inverse operations. The use of the $\boxed{\text{INV}}$ key varies among different models of calculators, so read your owner's manual carefully.

Exponential Key   The key marked $\boxed{x^y}$ or $\boxed{y^x}$ allows you to raise a number to a power. For example, if you wish to raise 4 to the fifth power (that is, find 4^5, as explained in Chapter 1), use the following keystrokes.

$$\boxed{4} \; \boxed{x^y} \; \boxed{5} \; \boxed{=} \qquad \boxed{\qquad 1024}$$

Root Key   Some calculators have a key specifically marked $\boxed{\sqrt[x]{x}}$ or $\boxed{\sqrt[x]{y}}$; with others, the operation of taking roots is accomplished by using the inverse key in conjunction with the exponential key. Suppose, for example, your calculator is of the latter type and you wish to find the fifth root of 1024. Use the following keystrokes.

$$\boxed{1} \; \boxed{0} \; \boxed{2} \; \boxed{4} \; \boxed{\text{INV}} \; \boxed{x^y} \; \boxed{5} \; \boxed{=} \qquad \boxed{\qquad 4}$$

Notice how this "undoes" the operation explained in the exponential key discussion.

Pi Key   The number π is an important number in mathematics. It occurs, for example, in the area and circumference formulas for a circle. By pressing the $\boxed{\pi}$ key, you can display the first few digits of π. (Because π is irrational, the display shows only an approximation.) One popular model gives the following display when the $\boxed{\pi}$ key is pressed.

$$\boxed{3.1415927} \qquad \text{An approximation for } \pi$$

Methods of Display   When decimal approximations are shown on scientific calculators, they are either *truncated* or *rounded*. To see how a particular model is programmed, evaluate $1/18$ as an example. If the display shows .0555555 (last digit 5), it truncates the display. If it shows .0555556 (last digit 6), it rounds the display.

When very large or very small numbers are obtained as answers, scientific calculators often express these numbers in scientific notation (Chapter 4). For example, if you multiply 6,265,804 by 8,980,591, the display might look like this:

$$\boxed{5.6270623 \; 13}$$

The 13 at the far right means that the number on the left is multiplied by 10^{13}. This means that the decimal point must be moved 13 places to the right if the answer is to be expressed in its usual form. Even then, the value obtained will only be an approximation: 56,270,623,000,000.

Graphing Calculators

While you are not expected to have a graphing calculator to study from this book, we include the following as background information and reference should your course or future courses require the use of graphing calculators.

Basic Features In addition to the typical keys found on scientific calculators, graphing calculators have keys that can be used to create graphs, make tables, analyze data, and change settings. One of the major differences between graphing and scientific calculators is that a graphing calculator has a larger viewing screen with graphing capabilities. The screens below illustrate the graphs of $Y = X$ and $Y = X^2$.

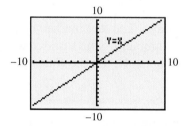 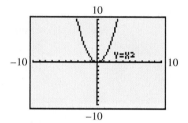

If you look closely at the screens, you will see that the graphs appear to be jagged rather than smooth, as they should be. The reason for this is that graphing calculators have much lower resolution than computer screens. Because of this, graphs generated by graphing calculators must be interpreted carefully.

Editing Input The screen of a graphing calculator can display several lines of text at a time. This feature allows you to view both previous and current expressions. If an incorrect expression is entered, an error message is displayed. The erroneous expression can be viewed and corrected by using various editing keys, much like a word-processing program. You do not need to enter the entire expression again. Many graphing calculators can also recall past expressions for editing or updating. The screen on the left shows how two expressions are evaluated. The final line is entered incorrectly, and the resulting error message is shown in the screen on the right.

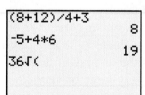 

Order of Operations Arithmetic operations on graphing calculators are usually entered as they are written in mathematical expressions. For example, to evaluate $\sqrt{36}$ you would first press the square root key, and then enter 36. See the left screen at the top of the next page. The order of operations on a graphing calculator is also important, and current models insert parentheses when typical errors might occur. The open parenthesis that follows the square root symbol is automatically entered by the calculator so that an expression such as $\sqrt{2 \times 8}$ will not be calculated incorrectly as $\sqrt{2} \times 8$. Compare the two entries and their results in the screen on the right.

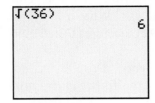
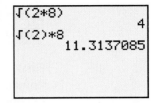

Viewing Windows   The viewing window for a graphing calculator is similar to the viewfinder in a camera. A camera usually cannot take a photograph of an entire view of a scene. The camera must be centered on some object and can capture only a portion of the available scenery. A camera with a zoom lens can photograph different views of the same scene by zooming in and out. Graphing calculators have similar capabilities. The *xy*-coordinate plane is infinite. The calculator screen can only show a finite, rectangular region in the plane, and it must be specified before the graph can be drawn. This is done by setting both minimum and maximum values for the *x*- and *y*-axes. The scale (distance between tick marks) is usually specified as well. Determining an appropriate viewing window for a graph is often a challenge, and many times it will take a few attempts before a satisfactory window is found.

The screen on the left shows a standard viewing window, and the graph of Y = 2X + 1 is shown on the right. Using a different window would give a different view of the line.

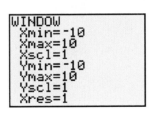
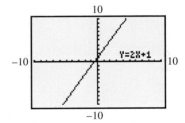

Locating Points on a Graph: Tracing and Tables   Graphing calculators allow you to trace along the graph of an equation and display the coordinates of points on the graph. See the screen on the left below, which indicates that the point (2, 5) lies on the graph of Y = 2X + 1. Tables for equations can also be displayed. The screen on the right shows a partial table for this same equation. Note the middle of the screen, which indicates that when X = 2, Y = 5.

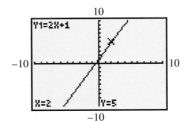
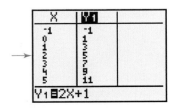

Additional Features   There are many features of graphing calculators that go far beyond the scope of this book. These calculators can be programmed, much like computers. Many of them can solve equations at the stroke of a key, analyze statistical data, and perform symbolic algebraic manipulations. Calculators also provide the opportunity to ask "What if . . . ?" more easily. Values in algebraic expressions can be altered and conjectures tested quickly.

Final Comments   Despite the power of today's calculators, they cannot replace human thought. *In the entire problem-solving process, your brain is the most important component.* Calculators are only tools and, like any tool, they must be used appropriately in order to enhance our ability to understand mathematics. Mathematical insight may often be the quickest and easiest way to solve a problem; a calculator may neither be needed nor appropriate. By applying mathematical concepts, you can make the decision whether or not to use a calculator.

Appendix B
Review of Decimals and Percents

OBJECTIVES

1 Add and subtract decimals.

2 Multiply and divide decimals.

3 Convert percents to decimals and decimals to percents.

4 Find percentages by multiplying.

OBJECTIVE 1 Add and subtract decimals.   A **decimal** is a number written with a decimal point, such as 4.2. The operations on decimals—addition, subtraction, multiplication, and division—are reviewed in the next examples.

▦ EXAMPLE 1   Adding and Subtracting Decimals

Add or subtract as indicated.

(a) $6.92 + 14.8 + 3.217$

Place the numbers in a column, with decimal points lined up, then add. If you like, attach 0s to make all the numbers the same length; this is a good way to avoid errors.

$$
\begin{array}{r}
6.92 \\
14.8 \\
+\ 3.217 \\
\hline
24.937
\end{array}
\quad \text{or} \quad
\begin{array}{r}
6.920 \\
14.800 \\
+\ 3.217 \\
\hline
24.937
\end{array}
$$

Decimal points lined up

(b) $47.6 - 32.509$

Write the numbers in a column, attaching 0s to 47.6.

$$
\begin{array}{r}
47.6 \\
-32.509
\end{array}
\quad \text{becomes} \quad
\begin{array}{r}
47.600 \\
-32.509 \\
\hline
15.091
\end{array}
$$

(c) $3 - .253$

$$
\begin{array}{r}
3.000 \\
-\ .253 \\
\hline
2.747
\end{array}
$$

Now Try Exercises 1 and 5.

OBJECTIVE 2 Multiply and divide decimals.   Multiplication and division of decimals are similar to the same operations with whole numbers.

▦ EXAMPLE 2   Multiplying Decimals

Multiply.

(a) 29.3×4.52

Multiply as if the numbers were whole numbers. To find the number of decimal places in the answer, add the numbers of decimal places in the factors.

$$
\begin{array}{r}
29.3 \\
\times\ 4.52 \\
\hline
5\ 86 \\
14\ 6\ 5 \\
117\ 2 \\
\hline
132.4\ 36
\end{array}
$$

1 decimal place in first factor
2 decimal places in second factor

$1 + 2 = 3$

3 decimal places in answer

(b) 7.003×55.8

$$
\begin{array}{r}
7.003 \\
\times\ 55.8 \\
\hline
5\ 602\ 4 \\
35\ 015 \\
350\ 15 \\
\hline
390.767\ 4
\end{array}
$$

3 decimal places
1 decimal place

$3 + 1 = 4$

4 decimal places in answer

Now Try Exercises 11 and 15.

EXAMPLE 3 Dividing Decimals

Divide: $279.45 \div 24.3$.

Move the decimal point in 24.3 one place to the right, to get 243. Move the decimal point the same number of places in 279.45. By doing this, 24.3 is converted into the whole number 243.

$$243.\overline{)2794.5}$$

Bring the decimal point straight up and divide as with whole numbers.

$$
\begin{array}{r}
11.5 \\
243.\overline{)2794.5} \\
\underline{243} \\
364 \\
\underline{243} \\
121\ 5 \\
\underline{121\ 5} \\
0
\end{array}
$$

Now Try Exercise 19.

OBJECTIVE 3 Convert percents to decimals and decimals to percents. One of the main uses of decimals is in percent problems. The word **percent** means "per one hundred." Percent is written with the sign %. One percent means "one per one hundred."

Percent

$$1\% = .01 \quad \text{or} \quad 1\% = \frac{1}{100}$$

■ EXAMPLE 4 Converting between Decimals and Percents

Convert.

(a) 75% to a decimal

Since $1\% = .01$,

$$75\% = 75 \cdot 1\% = 75 \cdot .01 = .75.$$

The fraction form $1\% = \frac{1}{100}$ can also be used to convert 75% to a decimal.

$$75\% = 75 \cdot 1\% = 75 \cdot \frac{1}{100} = .75$$

(b) 2.63 to a percent

$$2.63 = 263 \cdot .01 = 263 \cdot 1\% = 263\%$$

Now Try Exercises 23 and 33.

OBJECTIVE 4 Find percentages by multiplying. A part of a whole is called a **percentage.** For example, since 50% represents $\frac{50}{100} = \frac{1}{2}$ of a whole, 50% of 800 is half of 800, or 400. Multiply to find percentages, as in the next example.

■ EXAMPLE 5 Finding Percentages

Find each percentage.

(a) 15% of 600

The word *of* indicates multiplication here. For this reason, 15% of 600 is found by multiplying.

$$15\% \cdot 600 = .15 \cdot 600 = 90$$

(b) 125% of 80

$$125\% \cdot 80 = 1.25 \cdot 80 = 100$$

(c) What percent of 52 is 7.8?

We can translate this sentence to symbols word by word.

What percent	of	52	is	7.8?
↓	↓	↓	↓	↓
p	$\cdot$	52	=	7.8

$$52p = 7.8$$
$$p = .15 \qquad \text{Divide both sides by 52.}$$
$$p = 15\% \qquad \text{Change to percent.}$$

Now Try Exercises 39 and 43.

■ EXAMPLE 6 Using Percent in a Consumer Problem

A DVD movie with a regular price of $18 is on sale at 22% off. Find the amount of the discount.

The discount is 22% of $18. Change 22% to .22 and use the fact that *of* indicates multiplication.

$$22\% \cdot 18 = .22 \cdot 18 = 3.96$$

The discount is $3.96.

Now Try Exercise 55.

APPENDIX B EXERCISES

Perform each indicated operation. See Examples 1–3.

1. $14.23 + 9.81 + 74.63 + 18.715$

2. $89.416 + 21.32 + 478.91 + 298.213$

3. $19.74 - 6.53$

4. $27.96 - 8.39$

5. $219 - 68.51$

6. $283 - 12.42$

7.
$$
\begin{array}{r}
48.96 \\
37.421 \\
+\ 9.72 \\
\hline
\end{array}
$$

8.
$$
\begin{array}{r}
9.71 \\
4.8 \\
3.6 \\
5.2 \\
+8.17 \\
\hline
\end{array}
$$

9.
$$
\begin{array}{r}
8.6 \\
-3.751 \\
\hline
\end{array}
$$

10.
$$
\begin{array}{r}
27.8 \\
-13.582 \\
\hline
\end{array}
$$

11. 39.6×4.2

12. 18.7×2.3

13. 42.1×3.9

14. 19.63×4.08

15. $.042 \times 32$

16. 571×2.9

17. $24.84 \div 6$

18. $32.84 \div 4$

19. $7.6266 \div 3.42$

20. $14.9202 \div 2.43$

21. $2496 \div .52$

22. $.56984 \div .034$

Convert each percent to a decimal. See Example 4(a).

23. 53%

24. 38%

25. 129%

26. 174%

27. 96%

28. 11%

29. $.9\%$

30. $.1\%$

Convert each decimal to a percent. See Example 4(b).

31. $.80$

32. $.75$

33. $.007$

34. 1.4

35. $.67$

36. $.003$

37. $.125$

38. $.983$

Respond to each statement or question. Round your answer to the nearest hundredth if appropriate. See Example 5.

39. What is 14% of 780?

40. Find 12% of 350.

41. Find 22% of 1086.

42. What is 20% of 1500?

43. 4 is what percent of 80?

44. 1300 is what percent of 2000?

45. What percent of 5820 is 6402?

46. What percent of 75 is 90?

47. 121 is what percent of 484?

48. What percent of 3200 is 64?

49. Find 118% of 125.8.

50. Find 3% of 128.

51. What is 91.72% of 8546.95?

52. Find 12.741% of 58.902.

53. What percent of 198.72 is 14.68?

54. 586.3 is what percent of 765.4?

Solve each problem. See Example 6.

55. A retailer has $23,000 invested in her business. She finds that she is earning 12% per year on this investment. How much money is she earning per year?

56. Harley Dabler recently bought a duplex for $144,000. He expects to earn 16% per year on the purchase price. How many dollars per year will he earn?

57. For a recent tour of the eastern United States, a travel agent figured that the trip totaled 2300 mi, with 35% of the trip by air. How many miles of the trip were by air?

58. Capitol Savings Bank pays 3.2% interest per year. What is the annual interest on an account of $3000?

59. An ad for steel-belted radial tires promises 15% better mileage when the tires are used. Alexandria's Escort now goes 420 mi on a tank of gas. If she switched to the new tires, how many extra miles could she drive on a tank of gas?

60. A home worth $77,000 is located in an area where home prices are increasing at a rate of 12% per year. By how much would the value of this home increase in 1 year?

61. A family of four with a monthly income of $2000 spends 90% of its earnings and saves the rest. Find the *annual* savings of this family.

Appendix C
Sets

OBJECTIVES

1 List the elements of a set.

2 Learn the vocabulary and symbols used to discuss sets.

3 Decide whether a set is finite or infinite.

4 Decide whether a given set is a subset of another set.

5 Find the complement of a set.

6 Find the union and the intersection of two sets.

OBJECTIVE 1 List the elements of a set. A **set** is a collection of things. The objects in a set are called the **elements** of the set. A set is represented by listing its elements between **set braces,** { }. The order in which the elements of a set are listed is unimportant.

EXAMPLE 1 Listing the Elements of Sets

Represent each set by listing the elements.

(a) The set of states in the United States that border on the Pacific Ocean is

{California, Oregon, Washington, Hawaii, Alaska}.

(b) The set of all counting numbers less than 6 = {1, 2, 3, 4, 5}.

Now Try Exercises 1 and 3.

OBJECTIVE 2 Learn the vocabulary and symbols used to discuss sets. Capital letters are used to name sets. To state that 5 is an element of

$$S = \{1, 2, 3, 4, 5\},$$

write $5 \in S$. The statement $6 \notin S$ means that 6 is not an element of S.

A set with no elements is called the **empty set,** or the **null set.** The symbols $\emptyset$ or { } are used for the empty set. If we let A be the set of all cats that fly, then A is the empty set.

$$A = \emptyset \qquad \text{or} \qquad A = \{ \ \}$$

CAUTION Do not make the common error of writing the empty set as $\{\emptyset\}$.

In any discussion of sets, there is some set that includes all the elements under consideration. This set is called the **universal set** for that situation. For example, if the discussion is about presidents of the United States, then the set of all presidents of the United States is the universal set. The universal set is denoted U.

OBJECTIVE 3 Decide whether a set is finite or infinite. In Example 1, there are five elements in the set in part (a), and five in part (b). If the number of elements in a set is either 0 or a counting number, then the set is **finite.** On the other hand, the set of natural numbers, for example, is an **infinite** set, because there is no final natural number. We can list the elements of the set of natural numbers as

$$N = \{1, 2, 3, 4, \ldots\},$$

985

where the three dots indicate that the set continues indefinitely. Not all infinite sets can be listed in this way. For example, there is no way to list the elements in the set of all real numbers between 1 and 2.

▮ EXAMPLE 2 Distinguishing between Finite and Infinite Sets

List the elements of each set, if possible. Decide whether each set is finite or infinite.

(a) The set of all integers
One way to list the elements is $\{\ldots, -2, -1, 0, 1, 2, \ldots\}$. The set is infinite.

(b) The set of all natural numbers between 0 and 5
$\{1, 2, 3, 4\}$ The set is finite.

(c) The set of all irrational numbers
This is an infinite set whose elements cannot be listed.

Now Try Exercise 11.

Two sets are equal if they have exactly the same elements. Thus, the set of natural numbers and the set of positive integers are equal sets. Also, the sets

$$\{1, 2, 4, 7\} \qquad \text{and} \qquad \{4, 2, 7, 1\}$$

are equal. The order of the elements does not make a difference.

OBJECTIVE 4 Decide whether a given set is a subset of another set.   If all elements of a set A are also elements of another set B, then we say A is a **subset** of B, written $A \subseteq B$. We use the symbol $A \nsubseteq B$ to mean that A is not a subset of B.

▮ EXAMPLE 3 Using Subset Notation

Let $A = \{1, 2, 3, 4\}$, $B = \{1, 4\}$, and $C = \{1\}$. Then

$$B \subseteq A, \qquad C \subseteq A, \qquad \text{and} \qquad C \subseteq B,$$

but $\qquad\qquad A \nsubseteq B, \qquad A \nsubseteq C, \qquad \text{and} \qquad B \nsubseteq C.$

Now Try Exercises 21 and 25.

The set $M = \{a, b\}$ has four subsets: $\{a, b\}$, $\{a\}$, $\{b\}$, and $\emptyset$. The empty set is defined to be a subset of any set. How many subsets does $N = \{a, b, c\}$ have? There is one subset with 3 elements: $\{a, b, c\}$. There are three subsets with 2 elements:

$$\{a, b\}, \qquad \{a, c\}, \qquad \text{and} \qquad \{b, c\}.$$

There are three subsets with 1 element:

$$\{a\}, \qquad \{b\}, \qquad \text{and} \qquad \{c\}.$$

There is one subset with 0 elements: $\emptyset$. Thus, set N has eight subsets.

The following generalization can be made.

Number of Subsets of a Set

A set with n elements has 2^n subsets.

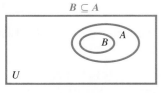

$B \subseteq A$

U

FIGURE 1

To illustrate the relationships between sets, **Venn diagrams** are often used. A rectangle represents the universal set, U. The sets under discussion are represented by regions within the rectangle. The Venn diagram in Figure 1 shows that $B \subseteq A$.

OBJECTIVE 5 **Find the complement of a set.** For every set A, there is a set A', the **complement** of A, that contains all the elements of U that are not in A. The shaded region in the Venn diagram in Figure 2 represents A'.

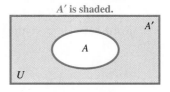

A' is shaded.

A'

A

U

FIGURE 2

■ EXAMPLE 4   Determining Complements of a Set

Given $U = \{a, b, c, d, e, f, g\}$, $A = \{a, b, c\}$, $B = \{a, d, f, g\}$, and $C = \{d, e\}$, then

$$A' = \{d, e, f, g\}, \quad B' = \{b, c, e\}, \quad \text{and} \quad C' = \{a, b, c, f, g\}.$$

Now Try Exercises 45 and 47.

OBJECTIVE 6 **Find the union and the intersection of two sets.** The **union** of two sets A and B, written $A \cup B$, is the set of all elements of A together with all elements of B. Thus, for the sets in Example 4,

$$A \cup B = \{a, b, c, d, f, g\}$$

and

$$A \cup C = \{a, b, c, d, e\}.$$

In Figure 3 the shaded region is the union of sets A and B.

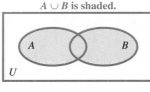

$A \cup B$ is shaded.

A B

U

FIGURE 3

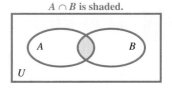

$A \cap B$ is shaded.

A B

U

FIGURE 4

■ EXAMPLE 5   Finding the Union of Two Sets

If $M = \{2, 5, 7\}$ and $N = \{1, 2, 3, 4, 5\}$, then

$$M \cup N = \{1, 2, 3, 4, 5, 7\}.$$

Now Try Exercise 55.

The **intersection** of two sets A and B, written $A \cap B$, is the set of all elements that belong to both A and B. For example if,

$$A = \{\text{Jose, Ellen, Marge, Kevin}\}$$

and

$$B = \{\text{Jose, Patrick, Ellen, Sue}\},$$

then

$$A \cap B = \{\text{Jose, Ellen}\}.$$

The shaded region in Figure 4 represents the intersection of the two sets A and B.

▇ EXAMPLE 6 Finding the Intersection of Two Sets

Suppose that $P = \{3, 9, 27\}$, $Q = \{2, 3, 10, 18, 27, 28\}$, and $R = \{2, 10, 28\}$. List the elements in each set.

(a) $P \cap Q = \{3, 27\}$ **(b)** $Q \cap R = \{2, 10, 28\} = R$ **(c)** $P \cap R = \emptyset$ ▇

Now Try Exercises 49 and 51.

Sets like P and R in Example 6 that have no elements in common are called **disjoint sets.** The Venn diagram in Figure 5 shows a pair of disjoint sets.

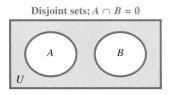

Disjoint sets; $A \cap B = \emptyset$

FIGURE 5

▇ EXAMPLE 7 Using Set Operations

Let $U = \{2, 5, 7, 10, 14, 20\}$, $A = \{2, 10, 14, 20\}$, $B = \{5, 7\}$, and $C = \{2, 5, 7\}$. Find each set.

(a) $A \cup B = \{2, 5, 7, 10, 14, 20\} = U$ **(b)** $A \cap B = \emptyset$

(c) $B \cup C = \{2, 5, 7\} = C$ **(d)** $B \cap C = \{5, 7\} = B$

(e) $A' = \{5, 7\} = B$ ▇

Now Try Exercises 53 and 57.

APPENDIX C EXERCISES

List the elements of each set. See Examples 1 and 2.

1. The set of all natural numbers less than 8

2. The set of all integers between 4 and 10

3. The set of seasons

4. The set of months of the year

5. The set of women presidents of the United States

6. The set of all living humans who are more than 200 years old

7. The set of letters of the alphabet between K and M

8. The set of letters of the alphabet between D and H

9. The set of positive even integers

10. The set of all multiples of 5

11. Which of the sets described in Exercises 1–10 are infinite sets?

12. Which of the sets described in Exercises 1–10 are finite sets?

Tell whether each statement is true *or* false.

13. $5 \in \{1, 2, 5, 8\}$

14. $6 \in \{1, 2, 3, 4, 5\}$

15. $2 \in \{1, 3, 5, 7, 9\}$

16. $1 \in \{6, 2, 5, 1\}$

17. $7 \notin \{2, 4, 6, 8\}$

18. $7 \notin \{1, 3, 5, 7\}$

19. $\{2, 4, 9, 12, 13\} = \{13, 12, 9, 4, 2\}$

20. $\{7, 11, 4\} = \{7, 11, 4, 0\}$

Let

$$A = \{1, 3, 4, 5, 7, 8\}, \quad B = \{2, 4, 6, 8\},$$
$$C = \{1, 3, 5, 7\}, \quad D = \{1, 2, 3\},$$
$$E = \{3, 7\}, \quad \text{and} \quad U = \{1, 2, 3, 4, 5, 6, 7, 8, 9, 10\}.$$

Tell whether each statement is true *or* false. *See Examples 3, 5, 6, and 7.*

21. $A \subseteq U$　　　**22.** $D \subseteq A$　　　**23.** $\emptyset \subseteq A$　　　**24.** $\{1, 2\} \subseteq D$　　　**25.** $C \subseteq A$

26. $A \subseteq C$　　　**27.** $D \subseteq B$　　　**28.** $E \subseteq C$　　　**29.** $D \nsubseteq E$　　　**30.** $E \nsubseteq A$

31. There are exactly 4 subsets of E.

32. There are exactly 8 subsets of D.

33. There are exactly 12 subsets of C.

34. There are exactly 16 subsets of B.

35. $\{4, 6, 8, 12\} \cap \{6, 8, 14, 17\} = \{6, 8\}$

36. $\{2, 5, 9\} \cap \{1, 2, 3, 4, 5\} = \{2, 5\}$

37. $\{3, 1, 0\} \cap \{0, 2, 4\} = \{0\}$

38. $\{4, 2, 1\} \cap \{1, 2, 3, 4\} = \{1, 2, 3\}$

39. $\{3, 9, 12\} \cap \emptyset = \{3, 9, 12\}$

40. $\{3, 9, 12\} \cup \emptyset = \emptyset$

41. $\{4, 9, 11, 7, 3\} \cup \{1, 2, 3, 4, 5\} = \{1, 2, 3, 4, 5, 7, 9, 11\}$

42. $\{1, 2, 3\} \cup \{1, 2, 3\} = \{1, 2, 3\}$

43. $\{3, 5, 7, 9\} \cup \{4, 6, 8\} = \emptyset$

44. $\{5, 10, 15, 20\} \cup \{5, 15, 30\} = \{5, 15\}$

Let

$$U = \{a, b, c, d, e, f, g, h\}, \quad A = \{a, b, c, d, e, f\},$$
$$B = \{a, c, e\}, \quad C = \{a, f\},$$

and　　　$$D = \{d\}.$$

List the elements in each set. See Examples 4–7.

45. A'　　　　　**46.** B'　　　　　**47.** C'　　　　　**48.** D'

49. $A \cap B$　　　**50.** $B \cap A$　　　**51.** $A \cap D$　　　**52.** $B \cap D$

53. $B \cap C$　　　**54.** $A \cup B$　　　**55.** $B \cup D$　　　**56.** $B \cup C$

57. $C \cup B$　　　**58.** $C \cup D$　　　**59.** $A \cap \emptyset$　　　**60.** $B \cup \emptyset$

61. Name every pair of disjoint sets among sets A–D above.

Appendix D
Mean, Median, and Mode

OBJECTIVES

1 Find the mean of a list of numbers.

2 Find a weighted mean.

3 Find the median.

4 Find the mode.

OBJECTIVE 1   Find the mean of a list of numbers.   Making sense of a long list of numbers can be difficult. So when we analyze data, one of the first things to look for is a *measure of central tendency*—a single number that we can use to represent the entire list of numbers. One such measure is the *average* or **mean.** The mean can be found with the following formula.

Finding the Mean (Average)

$$\text{mean} = \frac{\textbf{sum of all values}}{\textbf{number of values}}$$

EXAMPLE 1   Finding the Mean (Average)

David had test scores of 84, 90, 95, 98, and 88. Find his mean (average) score.

Use the formula for finding the mean. Add up all the test scores and then divide the sum by the number of tests.

$$\text{mean} = \frac{84 + 90 + 95 + 98 + 88}{5} \quad \begin{matrix} \leftarrow \text{Sum of test scores} \\ \leftarrow \text{Number of tests} \end{matrix}$$

$$\text{mean} = \frac{455}{5}$$

$$\text{mean} = 91 \qquad \text{Divide.}$$

David has a mean (average) score of 91.

Now Try Exercise 1.

EXAMPLE 2   Applying the Mean (Average)

The sales of photo albums at Sarah's Card Shop for each day last week were $86, $103, $118, $117, $126, $158, and $149. Find the mean daily sales of photo albums.

To find the mean, add all the daily sales amounts and then divide the sum by the number of days (7).

$$\text{mean} = \frac{\$86 + \$103 + \$118 + \$117 + \$126 + \$158 + \$149}{7} \quad \begin{matrix} \leftarrow \text{Sum of sales} \\ \leftarrow \text{Number of days} \end{matrix}$$

$$\text{mean} = \frac{\$857}{7}$$

$$\text{mean} \approx \$122.43 \qquad \text{Nearest cent}$$

Now Try Exercise 7.

OBJECTIVE 2 Find a weighted mean. Some items in a list of data might appear more than once. In this case, we find a **weighted mean,** in which each value is "weighted" by multiplying it by the number of times it occurs.

EXAMPLE 3 Finding a Weighted Mean

The table shows the amount of contribution and the number of times the amount was given (frequency) to a food pantry. Find the weighted mean.

Contribution Value	Frequency	
$ 3	4	← 4 people each contributed $3.
$ 5	2	
$ 7	1	
$ 8	5	
$ 9	3	
$10	2	
$12	1	
$13	2	

In most cases, the same amount was given by more than one person: for example, $3 was given by four people, and $8 was given by five people. Other amounts, such as $12, were given by only one person.

To find the mean, multiply each contribution value by its frequency. Then add the products. Next, add the numbers in the *frequency* column to find the total number of values, that is, the total number of people who contributed money.

Value	Frequency	Product
$ 3	4	($3 · 4) = $12
$ 5	2	($5 · 2) = $10
$ 7	1	($7 · 1) = $ 7
$ 8	5	($8 · 5) = $40
$ 9	3	($9 · 3) = $27
$10	2	($10 · 2) = $20
$12	1	($12 · 1) = $12
$13	2	($13 · 2) = $26
Totals	20	$154

Finally, divide the totals.

$$\text{mean} = \frac{\$154}{20} = \$7.70$$

The mean contribution to the food pantry was $7.70.

Now Try Exercise 11.

A common use of the weighted mean is to find a student's *grade point average (GPA),* as shown in the next example.

EXAMPLE 4 Applying the Weighted Mean

Find the GPA (grade point average) for a student who earned the following grades last semester. Assume A = 4, B = 3, C = 2, D = 1, and F = 0. The number of credits determines how many times the grade is counted (the frequency).

Course	Credits	Grade	Credits · Grade
Mathematics	4	A (= 4)	4 · 4 = 16
Speech	3	C (= 2)	3 · 2 = 6
English	3	B (= 3)	3 · 3 = 9
Computer Science	2	A (= 4)	2 · 4 = 8
Theater	2	D (= 1)	2 · 1 = 2
Totals	14		41

It is common to round grade point averages to the nearest hundredth, so

$$\text{GPA} = \frac{41}{14} \approx 2.93.$$

Now Try Exercise 15.

OBJECTIVE 3 Find the median. Because it can be affected by extremely high or low numbers, the mean is often a poor indicator of central tendency for a list of numbers. In cases like this, another measure of central tendency, called the *median,* can be used. The **median** divides a group of numbers in half; half the numbers lie above the median, and half lie below the median.

Find the median by listing the numbers *in order* from *smallest* to *largest.* If the list contains an *odd* number of items, the median is the *middle number.*

EXAMPLE 5 Finding the Median

Find the median for this list of prices.

$$\$7, \$23, \$15, \$6, \$18, \$12, \$24$$

First arrange the numbers in numerical order from smallest to largest.

Smallest → 6, 7, 12, 15, 18, 23, 24 ← Largest

Next, find the middle number in the list.

6, 7, 12, 15, 18, 23, 24

Three are below.　　Three are above.
Middle number

The median price is $15.

Now Try Exercise 19.

If a list contains an *even* number of items, there is no single middle number. In this case, the median is defined as the mean (average) of the *middle two* numbers.

EXAMPLE 6 Finding the Median

Find the median for this list of ages, in years.

$$74, 7, 15, 13, 25, 28, 47, 59, 32, 68$$

First arrange the numbers in numerical order from smallest to largest. Then, because the list has an even number of ages, find the middle *two* numbers.

Smallest $\rightarrow$ 7, 13, 15, 25, 28, 32, 47, 59, 68, 74 $\leftarrow$ Largest

Middle two numbers

The median age is the mean of the two middle numbers.

$$\text{median} = \frac{28 + 32}{2} = \frac{60}{2} = 30 \text{ yr}$$

Now Try Exercise 21.

OBJECTIVE 4 Find the mode. Another statistical measure is the **mode,** which is the number that occurs *most often* in a list of numbers. For example, if the test scores for 10 students were

74, 81, 39, 74, 82, 80, 100, 92, 74, and 85,

then the mode is 74. Three students earned a score of 74, so 74 appears more times on the list than any other score. It is *not* necessary to place the numbers in numerical order when looking for the mode, although that may help you find it more easily.

A list can have two modes; such a list is sometimes called **bimodal.** If no number occurs more frequently than any other number in a list, the list has *no mode.*

EXAMPLE 7 Finding the Mode

Find the mode for each list of numbers.

(a) 51, 32, 49, 51, 49, 90, 49, 60, 17, 60

The number 49 occurs three times, which is more often than any other number. Therefore, 49 is the mode.

(b) 482, 485, 483, 485, 487, 487, 489, 486

Because both 485 and 487 occur twice, each is a mode. This list is binomial.

(c) 10,708; 11,519; 10,972; 12,546; 13,905; 12,182

No number occurs more than once. This list has no mode.

Now Try Exercises 29, 31, and 33.

Measures of Central Tendency

The **mean** is the sum of all the values divided by the number of values. It is the mathematical *average.*

The **median** is the middle number in a group of values that are listed from smallest to largest. It divides a group of numbers in half.

The **mode** is the value that occurs most often in a group of values.

APPENDIX D EXERCISES

Find the mean for each list of numbers. Round answers to the nearest tenth when necessary. See Example 1.

1. Ages of infants at the child care center (in months): 4, 9, 6, 4, 7, 10, 9

2. Monthly electric bills: $53, $77, $38, $29, $49, $48

3. Final exam scores: 92, 51, 59, 86, 68, 73, 49, 80

4. Quiz scores: 18, 25, 21, 8, 16, 13, 23, 19

5. Annual salaries: $31,900; $32,850; $34,930; $39,712; $38,340; $60,000

6. Numbers of people attending baseball games: 27,500; 18,250; 17,357; 14,298; 33,110

Solve each problem. See Examples 2 and 3.

7. The Athletic Shoe Store sold shoes at the following prices: $75.52, $36.15, $58.24, $21.86, $47.68, $106.57, $82.72, $52.14, $28.60, $72.92. Find the mean shoe sales amount.

8. In one evening, a waitress collected the following checks from her dinner customers: $30.10, $42.80, $91.60, $51.20, $88.30, $21.90, $43.70, $51.20. Find the mean dinner check amount.

9. The table shows the face value (policy amount) of life insurance policies sold and the number of policies sold for each amount by the New World Life Company during one week. Find the weighted mean amount for the policies sold.

Policy Amount	Number of Policies Sold
$ 10,000	6
$ 20,000	24
$ 25,000	12
$ 30,000	8
$ 50,000	5
$100,000	3
$250,000	2

10. Detroit Metro-Sales Company prepared the following table showing the gasoline mileage obtained by each of the cars in their automobile fleet. Find the weighted mean to determine the miles per gallon for the fleet of cars.

Miles per Gallon	Number of Autos
15	5
20	6
24	10
30	14
32	5
35	6
40	4

Find each weighted mean. Round answers to the nearest tenth when necessary. See Example 3.

11.

Quiz Scores	Frequency
3	4
5	2
6	5
8	5
9	2

12.

Credits per Student	Frequency
9	3
12	5
13	2
15	6
18	1

13.

Hours Worked	Frequency
12	4
13	2
15	5
19	3
22	1
23	5

14.

Students per Class	Frequency
25	1
26	2
29	5
30	4
32	3
33	5

Find the GPA (grade point average) for students earning the following grades. Assume A = 4, B = 3, C = 2, D = 1, and F = 0. Round answers to the nearest hundredth. See Example 4.

15.

Course	Credits	Grade
Biology	4	B
Biology Lab	2	A
Mathematics	5	C
Health	1	F
Psychology	3	B

16.

Course	Credits	Grade
Chemistry	3	A
English	3	B
Mathematics	4	B
Theater	2	C
Astronomy	3	C

17. Look again at the grades in Exercise 15. Find the student's GPA in each of these situations.

(a) The student earned a B instead of an F in the 1-credit class.

(b) The student earned a B instead of a C in the 5-credit class.

(c) Both (a) and (b) happened.

18. List the credits for the courses you're taking at this time. List the lowest grades you think you will earn in each class and find your GPA. Then list the highest grades you think you will earn and find your GPA.

Find the median for each list of numbers. See Examples 5 and 6.

19. Number of e-mail messages received: 9, 12, 14, 15, 23, 24, 28

20. Deliveries by a newspaper distributor: 99, 108, 109, 123, 126, 129, 146, 168, 170

21. Students enrolled in algebra each semester: 328, 549, 420, 592, 715, 483

22. Number of cars in the parking lot each day: 520, 523, 513, 1283, 338, 509, 290, 420

23. Number of computer service calls taken each day: 51, 48, 96, 40, 47, 23, 95, 56, 34, 48

24. Number of gallons of paint sold per week: 1072, 1068, 1093, 1042, 1056, 205, 1009, 1081

The table lists the cruising speed and distance flown without refueling for several types of larger airplanes used to carry passengers. Use the table to answer Exercises 25–28.

Type of Airplane	Cruising Speed (miles per hour)	Distance without Refueling (miles)
747-400	565	7650
747-200	558	6450
DC-9	505	1100
DC-10	550	5225
727	530	1550
757	530	2875

Source: Northwest Airlines *WorldTraveler.*

25. What is the mean distance flown without refueling, to the nearest mile?

26. Find the mean cruising speed.

27. **(a)** Find the median distance.

    (b) Is the median distance similar to the mean distance from Exercise 25? Explain why or why not.

28. **(a)** Find the median cruising speed.

    (b) Is the median speed similar to the mean speed from Exercise 26? Explain why or why not.

Find the mode(s) for each list of numbers. Indicate whether a list is bimodal or has no mode. See Example 7.

29. Number of samples taken each hour: 3, 8, 5, 1, 7, 6, 8, 4, 5, 8

30. Monthly water bills: $21, $32, $46, $32, $49, $32, $49, $25, $32

31. Ages of retirees (in years): 74, 68, 68, 68, 75, 75, 74, 74, 70, 77

32. Patients admitted to the hospital each week: 30, 19, 25, 78, 36, 20, 45, 85, 38

33. The number of boxes of candy sold by each child: 5, 9, 17, 3, 2, 8, 19, 1, 4, 20, 10, 6

34. The weights of soccer players (in pounds): 158, 161, 165, 162, 165, 157, 163, 162

Appendix E
The Metric System and Conversions

OBJECTIVES

1 Learn the basic metric units of length.

2 Use unit fractions to convert among units.

3 Move the decimal point to convert among units.

4 Learn the basic metric units of capacity.

5 Convert among metric capacity units.

6 Learn the basic metric units of weight (mass).

7 Convert among metric weight (mass) units.

OBJECTIVE 1 Learn the basic metric units of length.   Around 1790, a group of French scientists developed the metric system of measurement. It is an organized system based on multiples of 10, like our number system and our money. The basic unit of length in the metric system is the **meter** (also spelled *metre*). Use the symbol **m** for meter; do not put a period after it. Look at a yardstick—a meter is just a little longer. A yard is 36 inches long; a meter is about 39 inches long.

To make longer or shorter length units in the metric system, **prefixes** are written in front of the word *meter.* For example, the prefix *kilo* means 1000, so a *kilo*meter is 1000 meters. The table below shows how to use the prefixes for length measurements. It is helpful to memorize the prefixes because they are also used with weight and capacity measurements. The blue boxes indicate the units we use most often in daily life.

Prefix	kilo-meter	hecto-meter	deka-meter	meter	deci-meter	centi-meter	milli-meter
Meaning	1000 meters	100 meters	10 meters	1 meter	$\frac{1}{10}$ of a meter	$\frac{1}{100}$ of a meter	$\frac{1}{1000}$ of a meter
Symbol	km	hm	dam	m	dm	cm	mm

Units that are used most often

Here are some comparisons that involve the most commonly used length units: km, m, cm, mm.

*Kilo*meters are used instead of miles. A kilometer is 1000 meters. It is about .6 mile (a little more than half a mile) or about 5 to 6 city blocks. If you participate in a 10 km run, you'll run about 6 miles.

A meter is divided into 100 smaller pieces called *centi*meters. Each centimeter is $\frac{1}{100}$ of a meter. A centimeter is a little shorter than $\frac{1}{2}$ inch. A nickel is about 2 cm across. The width of your little finger is probably about 1 cm.

A meter is divided into 1000 smaller pieces called *milli*meters. Each millimeter is $\frac{1}{1000}$ of a meter. It takes 10 mm to equal 1 cm, so it is a very small length. The thickness of a dime is about 1 mm.

999

▓ **EXAMPLE 1** Using Metric Length Units

Write the most reasonable metric unit in each blank. Choose from km, m, cm, and mm.

(a) The distance from home to work is 20 _____ .

Choose 20 km because kilometers are used instead of miles; 20 km is about 12 miles.

(b) The wedding ring is 4 _____ wide.

Choose 4 mm because the width of a ring is very small.

(c) The newborn baby is 50 _____ long.

Choose 50 cm, which is half of a meter; a meter is about 39 inches, so half a meter is around 20 inches.

Now Try Exercises 5, 7, 9, and 11.

OBJECTIVE 2 Use unit fractions to convert among units. We can convert among metric length units using unit fractions. Keep these relationships in mind when setting up the unit fractions.

Metric Length Relationships

1 km = 1000 m, so the unit fractions are:

$$\frac{1\text{ km}}{1000\text{ m}} \quad \text{or} \quad \frac{1000\text{ m}}{1\text{ km}}$$

1 m = 1000 mm, so the unit fractions are:

$$\frac{1\text{ m}}{1000\text{ mm}} \quad \text{or} \quad \frac{1000\text{ mm}}{1\text{ m}}$$

1 m = 100 cm, so the unit fractions are:

$$\frac{1\text{ m}}{100\text{ cm}} \quad \text{or} \quad \frac{100\text{ cm}}{1\text{ m}}$$

1 cm = 10 mm, so the unit fractions are:

$$\frac{1\text{ cm}}{10\text{ mm}} \quad \text{or} \quad \frac{10\text{ mm}}{1\text{ cm}}$$

▓ **EXAMPLE 2** Using Unit Fractions to Convert Length Measurements

Convert the following.

(a) 5 km to m

Put the unit for the answer (meters) in the numerator of the unit fraction; put the unit you want to change (km) in the denominator.

Unit fraction equivalent to 1 $\left\{ \dfrac{1000\text{ m}}{1\text{ km}} \right.$ ← Unit for answer
← Unit being changed

Multiply. Divide out common units where possible.

$$5\text{ km} \cdot \frac{1000\text{ m}}{1\text{ km}} = \frac{5\text{ km}}{1} \cdot \frac{1000\text{ m}}{1\text{ km}} = \frac{5 \cdot 1000\text{ m}}{1} = 5000\text{ m}$$

These units should match.

The answer makes sense because a kilometer is much longer than a meter, so 5 km will contain many meters.

(b) 18.6 cm to m

Multiply by a unit fraction that allows you to divide out centimeters.

$$\frac{18.6 \ \cancel{cm}}{1} \cdot \overbrace{\frac{1 \ m}{100 \ \cancel{cm}}}^{\text{Unit fraction}} = \frac{18.6}{100} \ m = .186 \ m$$

There are 100 cm in a meter, so 18.6 cm will be a small part of a meter. The answer makes sense.

Now Try Exercises 23 and 25.

OBJECTIVE 3 Move the decimal point to convert among units. By now you have probably noticed that conversions among metric units are made by multiplying or dividing by 10, 100, or 1000. A quick way to *multiply* by 10 is to move the decimal point one place to the *right*. Move it two places to the right to multiply by 100, three places to multiply by 1000. *Dividing* is done by moving the decimal point to the *left* in the same manner.

An alternate conversion method to unit fractions is moving the decimal point using this **metric conversion line.**

Here are the steps for using the conversion line.

Using the Metric Conversion Line

Step 1   Find the given unit on the metric conversion line.

Step 2   Count the number of places to get from the given unit to the unit needed in the answer.

Step 3   Move the decimal point the *same number of places* and in the *same direction* as on the conversion line.

EXAMPLE 3   Using the Metric Conversion Line

Use the metric conversion line to make the following conversions.

(a) 5.702 km to m

Find **km** on the metric conversion line. To get to **m** , move *three places* to the *right*. So move the decimal point in 5.702 *three places* to the *right*.

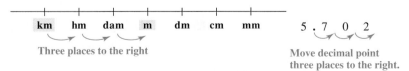

5.702 km = 5702 m

(b) 69.5 cm to m

Find **cm** on the conversion line. To get to **m** , move *two places* to the *left*.

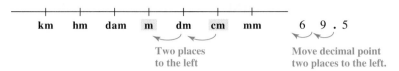

Two places
to the left

Move decimal point
two places to the left.

69.5 cm = .695 m

(c) 8.1 cm to mm

From **cm** to **mm** is *one place* to the *right*.

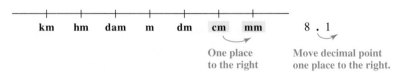

One place
to the right

Move decimal point
one place to the right.

8.1 cm = 81 mm

Now Try Exercises 19, 21, and 27.

OBJECTIVE 4 Learn the basic metric units of capacity.   The basic metric unit for capacity is the **liter** (also spelled *litre*). The capital letter **L** is the symbol for liter, to avoid confusion with the numeral 1. The liter is related to metric length in this way: A box that measures 10 cm on every side holds exactly one liter. (The volume of the box is 1000 cubic centimeters.) A liter is just a little more than 1 quart.

To make larger or smaller capacity units, we use the same prefixes as we did with length units. For example, *kilo* means 1000 so a *kilo*meter is 1000 meters. In the same way, a *kilo*liter is 1000 liters.

Prefix	kilo-liter	hecto-liter	deka-liter	liter	deci-liter	centi-liter	milli-liter
Meaning	1000 liters	100 liters	10 liters	1 liter	$\frac{1}{10}$ of a liter	$\frac{1}{100}$ of a liter	$\frac{1}{1000}$ of a liter
Symbol	kL	hL	daL	L	dL	cL	mL

Used most often

The capacity units we use most often in daily life are liters (L) and *milli*liters (mL). A tiny box that measures 1 cm on every side holds exactly one milliliter. (In medicine, this small amount is also called 1 cubic centimeter, or 1 cc for short.) It takes 1000 mL to make 1 L.

EXAMPLE 4 Using Metric Capacity Units

Write the most reasonable metric unit in each blank. Choose from L and mL.

(a) The bottle of shampoo held 500 _____.

Choose 500 mL because 500 L would be about 500 quarts, which is too much.

(b) I bought a 2 _____ carton of orange juice.

Choose 2 L because 2 mL would be less than a teaspoon.

Now Try Exercises 37 and 39.

OBJECTIVE 5 **Convert among metric capacity units.** Just as with length units, we can convert between milliliters and liters using unit fractions.

Metric Capacity Relationships

$$1 \text{ L} = 1000 \text{ mL, so the unit fractions are:}$$

$$\frac{1 \text{ L}}{1000 \text{ mL}} \quad \text{or} \quad \frac{1000 \text{ mL}}{1 \text{ L}}$$

We can also use a metric conversion line to decide how to move the decimal point.

EXAMPLE 5 **Converting among Metric Capacity Units**

Convert using the metric conversion line or unit fractions.

(a) 2.5 L to mL

Using the metric conversion line:

From **L** to **mL** is *three places* to the *right*.

2.5̲00̲   Write two zeros as placeholders.

2.5 L = 2500 mL

Using unit fractions:

Multiply by a unit fraction that allows you to divide out liters.

$$\frac{2.5 \cancel{L}}{1} \cdot \frac{1000 \text{ mL}}{1 \cancel{L}} = 2500 \text{ mL}$$

(b) 80 mL to L

Using the metric conversion line:

From **mL** to **L** is *three places* to the *left*.

80.   080.

Decimal point starts here.   Move three places left.

Using unit fractions:

Multiply by a unit fraction that allows you to divide out mL.

$$\frac{80 \cancel{\text{mL}}}{1} \cdot \frac{1 \text{ L}}{1000 \cancel{\text{mL}}} = \frac{80}{1000} \text{ L} = .08 \text{ L}$$

80 mL = .080 L or .08 L

Now Try Exercises 59 and 61.

OBJECTIVE 6 **Learn the basic metric units of weight (mass).** The **gram** is the basic metric unit for *mass*. Although we often call it "weight," there is a difference. Weight is a measure of the pull of gravity; the farther you are from the center of the earth, the less you weigh. In outer space you become weightless, but your mass, the amount of matter in your body, stays the same regardless of where you are. For everyday purposes, we will use the word *weight*.

The gram is related to metric length in this way: The weight of the water in a box measuring 1 cm on every side is 1 gram. This is a very tiny amount of water (1 mL) and a very small weight. One gram is also the weight of a dollar bill or a single raisin. A nickel weighs 5 grams. A regular hamburger weighs from 175 to 200 grams.

To make larger or smaller weight units, we use the same prefixes as we did with length and capacity units. For example, *kilo* means 1000 so a *kilo*meter is 1000 meters, a *kilo*liter is 1000 liters, and a *kilo*gram is 1000 grams.

Prefix	kilo- gram	hecto- gram	deka- gram	gram	deci- gram	centi- gram	milli- gram
Meaning	1000 grams	100 grams	10 grams	1 gram	$\frac{1}{10}$ of a gram	$\frac{1}{100}$ of a gram	$\frac{1}{1000}$ of a gram
Symbol	kg	hg	dag	g	dg	cg	mg

Units that are used most often

The units we use most often in daily life are kilograms (kg), grams (g), and milligrams (mg). *Kilo*grams are used instead of pounds. A kilogram is 1000 grams. It is about 2.2 pounds. Two one-pound packages of butter plus one stick of butter weigh about 1 kg. An average newborn baby weighs 3 to 4 kg; a college football player might weigh 100 to 130 kg.

Extremely small weights are measured in *milli*grams. It takes 1000 mg to make 1 g. Recall that a dollar bill weighs about 1 g. Imagine cutting it into 1000 pieces; the weight of one tiny piece would be 1 mg. Dosages of medicine and vitamins are given in milligrams.

▨ EXAMPLE 6 Using Metric Weight Units

Write the most reasonable metric unit in each blank. Choose from kg, g, and mg.

(a) Ramon's suitcase weighed 20 _____.

Choose 20 kg because kilograms are used instead of pounds; 20 kg is about 44 pounds.

(b) LeTia took a 350 _____ aspirin tablet.

Choose 350 mg because 350 g would be more than the weight of a hamburger, which is too much.

(c) Jenny mailed a letter that weighed 30 _____.

Choose 30 g because 30 kg would be much too heavy and 30 mg is less than the weight of a dollar bill.

Now Try Exercises 41, 43, and 47.

OBJECTIVE 7 Convert among metric weight (mass) units. As with length and capacity, we can convert among metric weight units by using unit fractions. The unit fractions we need are as follows.

Metric Weight (Mass) Relationships

1 kg = 1000 g, so the unit fractions are:	**1 g = 1000 mg,** so the unit fractions are:
$\dfrac{1\text{ kg}}{1000\text{ g}}$ or $\dfrac{1000\text{ g}}{1\text{ kg}}$	$\dfrac{1\text{ g}}{1000\text{ mg}}$ or $\dfrac{1000\text{ mg}}{1\text{ g}}$

Or we can use a metric conversion line to decide how to move the decimal point.

1000	100	10	1	$\frac{1}{10}$	$\frac{1}{100}$	$\frac{1}{1000}$
kg	hg	dag	g	dg	cg	mg

▓ EXAMPLE 7 Converting among Metric Weight Units

Convert using the metric conversion line or unit fractions.

(a) 7 mg to g

Using the metric conversion line:

From **mg** to **g** is *three places* to the *left*.

$$7.\qquad\qquad 007.$$

Decimal point starts here.   Move three places left.

7 mg = .007 mg

Using unit fractions:

Multiply by a unit fraction that allows you to divide out mg.

$$\frac{7\text{ mg}}{1}\cdot\frac{1\text{ g}}{1000\text{ mg}}=\frac{7}{1000}\text{ g}=.007\text{ g}$$

(b) 13.72 kg to g

Using the metric conversion line:

From **kg** to **g** is *three* places to the *right*.

13.720 ← Decimal point moves three places to the right.

13.72 kg = 13,720 g

A comma (not a decimal point)

Using unit fractions:

Multiply by a unit fraction that allows you to divide out kg.

$$\frac{13.72\text{ kg}}{1}\cdot\frac{1000\text{ g}}{1\text{ kg}}=13,720\text{ g}$$

A comma (not a decimal point)

13.72 kg = 13,720 g

Now Try Exercises 71 and 75.

APPENDIX E EXERCISES

Use this ruler to measure your thumb and hand. Determine each measurement.

cm 1 2 3 4 5 6 7 8 9 10 11 12 13 14 15

1. The width of your hand in centimeters
2. The width of your hand in millimeters
3. The width of your thumb in millimeters
4. The width of your thumb in centimeters

Determine the most reasonable metric length unit for each blank. Choose from km, m, cm, and mm. See Example 1.

5. The child was 91 _____ tall.
6. The cardboard was 3 _____ thick.
7. Ming-Na swam in the 200 _____ backstroke race.
8. The bookcase is 75 _____ wide.
9. Adriana drove 400 _____ on her vacation.
10. The door is 2 _____ high.
11. An aspirin tablet is 10 _____ across.
12. Lamard jogs 4 _____ every morning.
13. A paper clip is about 3 _____ long.
14. My pen is 145 _____ long.
15. Dave's truck is 5 _____ long.
16. Wheelchairs need doorways that are at least 80 _____ wide.

17. Describe at least three examples of metric length units that you have come across in your daily life.

18. Explain one reason why the metric system would be easier for a child to learn than the English system.

Convert each measurement. Use unit fractions or the metric conversion line. See Examples 2 and 3.

19. 7 m to cm
20. 18 m to cm
21. 40 mm to m
22. 6 mm to m
23. 9.4 km to m
24. .7 km to m
25. 509 cm to m
26. 30 cm to m
27. 400 mm to cm
28. 25 mm to cm
29. .91 m to mm
30. 4 m to mm

Solve each problem.

31. Many cameras use film that is 35 mm wide. Film for movie theaters may be 70 mm wide. Using the ruler given at the top of this page, draw a line segment that is 35 mm long and a line segment 70 mm long. Then convert each measurement to centimeters.

32. Gold wedding bands may be very narrow or quite wide. Common widths are 3 mm, 5 mm, and 10 mm. Using the ruler given at the top of this page, draw line segments that are 3 mm, 5 mm, and 10 mm long. Then convert each measurement to centimeters.

5 mm

33. The Roe River near Great Falls, Montana, is the shortest river in the world, with a north fork that is just under 18 m long. (*Source: Guinness Book of Amazing Nature.*) How many kilometers long is the north fork of the river?

34. There are 60,000 km of blood vessels in the human body. (*Source: Big Book of Knowledge.*) How many meters of blood vessels are in the body?

35. Use two unit fractions to convert 5.6 mm to km.

36. Use two unit fractions to convert 16.5 km to mm.

Determine the most reasonable metric unit for each blank. Choose from L, mL, kg, g, and mg. See Examples 4 and 6.

37. The glass held 250 _____ of water.

38. Hiromi used 20 _____ of water to wash the kitchen floor.

39. Dolores can make 10 _____ of soup in that pot.

40. Jay gave 2 _____ of vitamin drops to the baby.

41. Our yellow Labrador dog grew up to weigh 40 _____ .

42. A small safety pin weighs 750 _____ .

43. Lori caught a small sunfish weighing 150 _____ .

44. One dime weighs 2 _____ .

45. Andre donated 500 _____ of blood today.

46. Barbara bought the 2 _____ bottle of cola.

47. The patient received a 250 _____ tablet of medication each hour.

48. The 8 people on the elevator weighed a total of 500 _____ .

49. The gas can for the lawn mower holds 4 _____ .

50. Kevin poured 10 _____ of vanilla into the mixing bowl.

Today, medical measurements are usually given in the metric system. Since we convert among metric units of measure by moving the decimal point, it is possible that mistakes can be made. Examine the following dosages and indicate whether they are reasonable or unreasonable. If a dose is unreasonable, indicate whether it is too much or too little.

51. Drink 4.1 L of Kaopectate after each meal.

52. Drop 1 mL of solution into the eye twice a day.

53. Soak your feet in 5 kg of Epsom salts per liter of water.

54. Inject .5 L of insulin each morning.

55. Take 15 mL of cough syrup every four hours.

56. Take 200 mg of vitamin C each day.

57. Describe at least two examples of metric capacity units and two examples of metric weight units that you have come across in your daily life.

58. Explain in your own words how the meter, liter, and gram are related.

Convert each measurement. Use unit fractions or the metric conversion line. See Examples 5 and 7.

59. 15 L to mL

60. 6 L to mL

61. 3000 mL to L

62. 18,000 mL to L

63. 925 mL to L

64. 200 mL to L

65. 8 mL to L

66. 25 mL to L

67. 4.15 L to mL

68. 11.7 L to mL

69. 8000 g to kg

70. 25,000 g to kg

71. 5.2 kg to g

72. 12.42 kg to g

73. .85 g to mg

74. .2 g to mg

75. 30,000 mg to g

76. 7500 mg to g

Solve each application. (Source for Exercises 77–82: Top 10 of Everything, 2000.)

77. Human skin has about 3 million sweat glands, which release an average of 300 mL of sweat per day. How many liters of sweat are released each day?

78. In hot climates, the sweat glands in a person's skin may release up to 3.5 L of sweat in one day. How many milliliters is that?

79. The average weight of a human brain is 1.34 kg. How many grams is that?

80. In the Victorian era, people believed that heavier brains meant greater intelligence. They were impressed that Otto von Bismarck's brain weighted 1907 g, which is how many kilograms?

81. A healthy human heart pumps about 70 mL of blood per beat. How many liters of blood does it pump per beat?

82. On average, we breath in and out roughly 900 mL of air every 10 seconds. How many liters of air is that?

83. A small adult cat weighs from 3000 g to 4000 g. How many kilograms is that? (*Source: Lyndale Animal Hospital.*)

84. If the letter you are mailing weighs 29 g, you must put additional postage on it. (*Source:* U.S. Postal Service.) How many kilograms does the letter weigh?

85. One nickel weighs 5 g. How many nickels are in 1 kg of nickels?

86. Seawater contains about 3.5 g of salt per 1000 mL of water. How many grams of salt would be in 1 L of seawater?

Appendix F
Review of Exponents, Polynomials, and Factoring

(Transition from Beginning to Intermediate Algebra)

OBJECTIVES

1 Review the basic rules for exponents.

2 Review addition, subtraction, and multiplication of polynomials.

3 Review factoring techniques.

OBJECTIVE 1 Review the basic rules for exponents. In Sections 4.1 and 4.2 we introduced the following definitions and rules for working with exponents.

Definitions and Rules for Exponents

If no denominators are 0, for any integers m and n:

		Examples
Product rule	$a^m \cdot a^n = a^{m+n}$	$7^4 \cdot 7^5 = 7^9$
Zero exponent	$a^0 = 1$	$(-3)^0 = 1$
Negative exponent	$a^{-n} = \dfrac{1}{a^n}$	$5^{-3} = \dfrac{1}{5^3}$
Quotient rule	$\dfrac{a^m}{a^n} = a^{m-n}$	$\dfrac{2^2}{2^5} = 2^{-3} = \dfrac{1}{2^3}$
Power rules (a)	$(a^m)^n = a^{mn}$	$(4^2)^3 = 4^6$
(b)	$(ab)^m = a^m b^m$	$(3k)^4 = 3^4 k^4$
(c)	$\left(\dfrac{a}{b}\right)^m = \dfrac{a^m}{b^m}$	$\left(\dfrac{2}{3}\right)^2 = \dfrac{2^2}{3^2}$
Negative to positive rules	$\dfrac{a^{-m}}{b^{-n}} = \dfrac{b^n}{a^m}$	$\dfrac{2^{-4}}{5^{-3}} = \dfrac{5^3}{2^4}$
	$\left(\dfrac{a}{b}\right)^{-m} = \left(\dfrac{b}{a}\right)^m$	$\left(\dfrac{4}{7}\right)^{-2} = \left(\dfrac{7}{4}\right)^2$

▨ EXAMPLE 1 Applying Definitions and Rules for Exponents

Simplify each expression. Write answers using only positive exponents. Assume all variables represent nonzero real numbers.

(a) $(x^2 y^{-3})(x^{-5} y^7) = (x^{2+(-5)})(y^{-3+7})$

$\qquad\qquad\qquad\quad = x^{-3} y^4$

$\qquad\qquad\qquad\quad = \dfrac{1}{x^3} y^4$

$\qquad\qquad\qquad\quad = \dfrac{y^4}{x^3}$

(b) $(-5)^0 + (-5^0) = 1 + (-1)$ $-5^0 = -1 \cdot 5^0 = -1 \cdot 1 = -1$
$$= 0$$

(c) $\dfrac{(t^5 s^{-4})^2}{(t^{-3} s^5)^3} = \dfrac{t^{10} s^{-8}}{t^{-9} s^{15}} = \dfrac{t^{10} t^9}{s^{15} s^8} = \dfrac{t^{19}}{s^{23}}$

(d) $\left(\dfrac{-3x^{-4}y}{x^5 y^{-4}}\right)^{-2} = \left(\dfrac{x^5 y^{-4}}{-3x^{-4}y}\right)^2 = \dfrac{x^{10} y^{-8}}{9x^{-8} y^2} = \dfrac{x^{18}}{9y^{10}}$

(e) $(2x^2 y^3 z)^2 (x^4 y^2)^3 = (4x^4 y^6 z^2)(x^{12} y^6) = 4x^{16} y^{12} z^2$

Now Try Exercises 1, 5, 7, 9, and 11.

OBJECTIVE 2 **Review addition, subtraction, and multiplication of polynomials.** These arithmetic operations with polynomials were covered in Sections 4.4–4.6.

Adding and Subtracting Polynomials

To add polynomials, add like terms. To subtract polynomials, change all signs on the second polynomial and add the result to the first polynomial.

EXAMPLE 2 **Adding and Subtracting Polynomials**

Add or subtract as indicated.

(a) $(-4x^3 + 3x^2 - 8x + 2) + (5x^3 - 8x^2 + 12x - 3)$
$$= (-4 + 5)x^3 + (3 - 8)x^2 + (-8 + 12)x + (2 - 3)$$
$$= x^3 - 5x^2 + 4x - 1$$

(b) $-4(x^2 + 3x - 6) - (2x^2 - 3x + 7)$
$$= -4x^2 - 12x + 24 - 2x^2 + 3x - 7$$
$$= -6x^2 - 9x + 17$$

(c) Subtract.
$$\begin{array}{r} 2t^2 - 3t - 4 \\ -8t^2 + 4t - 1 \end{array}$$

Change the sign of each term in $-8t^2 + 4t - 1$, and add.
$$\begin{array}{r} 2t^2 - 3t - 4 \\ 8t^2 - 4t + 1 \qquad \text{Change signs.} \\ \hline 10t^2 - 7t - 3 \qquad \text{Add.} \end{array}$$

Now Try Exercises 13, 19, and 21.

Multiplying Polynomials

To multiply two polynomials, multiply each term of the second polynomial by each term of the first polynomial and add the products. In particular, when multiplying two binomials, use the FOIL method. (See Section 4.5.)

There are also several special product rules that are useful when multiplying binomials.

Special Product Rules

$$(x + y)^2 = x^2 + 2xy + y^2$$
$$(x - y)^2 = x^2 - 2xy + y^2$$
$$(x + y)(x - y) = x^2 - y^2$$

■ EXAMPLE 3   Multiplying Polynomials

Find each product.

(a) $(4y - 1)(3y + 2) = 4y(3y) + 4y(2) - 1(3y) - 1(2)$ FOIL
$$= 12y^2 + 8y - 3y - 2$$
$$= 12y^2 + 5y - 2$$

(b) $(3x + 5y)(3x - 5y) = (3x)^2 - (5y)^2$ $(x + y)(x - y) = x^2 - y^2$
$$= 9x^2 - 25y^2$$

(c) $(2t + 3)^2 = (2t)^2 + 2(2t)(3) + 3^2$ $(x + y)^2 = x^2 + 2xy + y^2$
$$= 4t^2 + 12t + 9$$

(d) $(5x - 1)^2 = (5x)^2 - 2(5x)(1) + 1^2$ $(x - y)^2 = x^2 - 2xy + y^2$
$$= 25x^2 - 10x + 1$$

(e) $(3x + 2)(9x^2 - 6x + 4)$
Multiply vertically.

$$
\begin{array}{r}
9x^2 - 6x + 4 \\
3x + 2 \\
\hline
18x^2 - 12x + 8 \\
27x^3 - 18x^2 + 12x \\
\hline
27x^3 \qquad\qquad + 8
\end{array}
$$

⟵ $2(9x^2 - 6x + 4)$
⟵ $3x(9x^2 - 6x + 4)$
Add like terms.

The product is the sum of cubes, $27x^3 + 8$.

Now Try Exercises 23, 27, 31, and 35.

OBJECTIVE 3 Review factoring techniques. Factoring, which involves writing a polynomial as a product, was covered in Chapter 5. Here are some general guidelines to use when factoring.

Factoring a Polynomial

1. Is there a common factor? If so, factor it out.

2. How many terms are in the polynomial?

Two terms: Check to see whether it is a difference of squares or the sum or difference of cubes. If so, factor as in Section 5.4.

(continued)

Three terms: Is it a perfect square trinomial? If the trinomial is not a perfect square, check to see whether the coefficient of the squared term is 1. If so, use the method of Section 5.2. If the coefficient of the squared term of the trinomial is not 1, use the general factoring methods of Section 5.3.

Four terms: Try to factor the polynomial by grouping using the method of Section 5.1.

3. Can any factors be factored further? If so, factor them.

▌ EXAMPLE 4   Factoring Polynomials

Factor each polynomial completely.

(a) $6x^2y^3 - 12x^3y^2 = 6x^2y^2(y - 2x)$ $6x^2y^2$ is the greatest common factor.

(b) $3x^2 - x - 2$

To find the factors, find two terms that multiply to give $3x^2$ ($3x$ and x) and two terms that multiply to give -2 ($+2$ and -1). Make sure that the sum of the outer and inner products in the factored form is $-x$.

$$3x^2 - x - 2 = (3x + 2)(x - 1)$$

To check, multiply the factors using the FOIL method.

(c) $100t^2 - 81 = (10t)^2 - 9^2$ Difference of squares
$$= (10t + 9)(10t - 9)   x^2 - y^2 = (x + y)(x - y)$$

(d) $4x^2 + 20xy + 25y^2$

The terms $4x^2$ and $25y^2$ are both perfect squares, so factor as a perfect square trinomial.

$$4x^2 + 20xy + 25y^2 = (2x + 5y)^2$$

To check, take twice the product of the two terms in the squared binomial.

$$2(2x)(5y) = 20xy$$

Twice ——— ↑ ↑ ↑ ——— Last term
First term

Since $20xy$ is the middle term of the trinomial, the trinomial is a perfect square and can be factored as $(2x + 5y)^2$.

(e) $1000x^3 - 27 = (10x)^3 - 3^3$ Difference of cubes
$$= (10x - 3)[(10x)^2 + 10x(3) + 3^2]   x^3 - y^3 =$$
$$= (10x - 3)(100x^2 + 30x + 9)   (x - y)(x^2 + xy + y^2)$$

(f) $6xy - 3x + 4y - 2$

Since there are four terms, try factoring by grouping.

$$6xy - 3x + 4y - 2 = (6xy - 3x) + (4y - 2)   \text{Group terms.}$$
$$= 3x(2y - 1) + 2(2y - 1)   \text{Factor each group.}$$
$$= (2y - 1)(3x + 2)   \text{Factor out } 2y - 1.$$

In the final step, factor out the greatest common factor, the binomial $2y - 1$.

Now Try Exercises 41, 45, 47, 49, and 59.

APPENDIX F EXERCISES

Simplify each expression. Write the answers using only positive exponents. Assume all variables represent positive real numbers. See Example 1.

1. $(a^4 b^{-3})(a^{-6} b^2)$

2. $(t^{-3} s^{-5})(t^8 s^{-2})$

3. $(5x^{-2}y)^2(2xy^4)^2$

4. $(7x^{-3}y^4)^3(2x^{-1}y^{-4})^2$

5. $-6^0 + (-6)^0$

6. $(-12)^0 - 12^0$

7. $\dfrac{(2w^{-1}x^2y^{-1})^3}{(4w^5x^{-2}y)^2}$

8. $\dfrac{(5p^{-3}q^2r^{-4})^2}{(10p^4q^{-1}r^5)^{-1}}$

9. $\left(\dfrac{-4a^{-2}b^4}{a^3b^{-1}}\right)^{-3}$

10. $\left(\dfrac{r^{-3}s^{-8}}{-6r^2s^{-4}}\right)^{-2}$

11. $(7x^{-4}y^2z^{-2})^{-2}(7x^4y^{-1}z^3)^2$

12. $(3m^{-5}n^2p^{-4})^3(3m^4n^{-3}p^5)^{-2}$

Add or subtract as indicated. See Example 2.

13. $(2a^4 + 3a^3 - 6a^2 + 5a - 12) + (-8a^4 + 8a^3 - 14a^2 + 21a - 3)$

14. $(-6r^4 - 3r^3 + 12r^2 - 9r + 9) + (8r^4 - 13r^3 - 14r^2 - 10r - 3)$

15. $(6x^3 - 12x^2 + 3x - 4) - (-2x^3 + 6x^2 - 3x + 12)$

16. $(10y^3 - 4y^2 + 8y + 7) - (7y^3 + 5y^2 - 2y - 13)$

17. Add.

$\begin{array}{r} 5x^2y + 2xy^2 + y^3 \\ -4x^2y - 3xy^2 + 5y^3 \\ \hline \end{array}$

18. Add.

$\begin{array}{r} 6ab^3 - 2a^2b^2 + 3b^5 \\ 8ab^3 + 12a^2b^2 - 8b^5 \\ \hline \end{array}$

19. $3(5x^2 - 12x + 4) - 2(9x^2 + 13x - 10)$

20. $-4(2t^3 - 3t^2 + 4t - 1) - 3(-8t^3 + 3t^2 - 2t + 9)$

21. Subtract.

$\begin{array}{r} 6x^3 - 2x^2 + 3x - 1 \\ -4x^3 + 2x^2 - 6x + 3 \\ \hline \end{array}$

22. Subtract.

$\begin{array}{r} -9y^3 - 2y^2 + 3y - 8 \\ -8y^3 + 4y^2 + 3y + 1 \\ \hline \end{array}$

Find each product. See Example 3.

23. $(3x + 1)(2x - 7)$

24. $(5z + 3)(2z - 3)$

25. $(4x - 1)(x - 2)$

26. $(7t - 3)(t - 4)$

27. $(4t + 3)(4t - 3)$

28. $(6x + 1)(6x - 1)$

29. $(2y^2 + 4)(2y^2 - 4)$

30. $(3b^3 + 2t)(3b^3 - 2t)$

31. $(4x - 3)^2$

32. $(9t + 2)^2$

33. $(6r + 5y)^2$

34. $(8m - 3n)^2$

35. $(c + 2d)(c^2 - 2cd + 4d^2)$

36. $(f + 3g)(f^2 - 3fg + 9g^2)$

37. $(4x - 1)(16x^2 + 4x + 1)$

38. $(5r - 2)(25r^2 + 10r + 4)$

39. $(7t + 5s)(2t^2 + 5st - s^2)$

40. $(8p + 3q)(2p^2 - 4pq + q^2)$

Factor each polynomial completely. See Example 4.

41. $8x^3y^4 + 12x^2y^3 + 36xy^4$

42. $10m^5n + 4m^2n^3 + 18m^3n^2$

43. $x^2 - 2x - 15$

44. $x^2 + x - 12$

45. $2x^2 - 9x - 18$

46. $3x^2 + 2x - 8$

47. $36t^2 - 25$

48. $49r^2 - 9$

49. $16t^2 + 24t + 9$

50. $25t^2 + 90t + 81$

51. $4m^2p - 12mnp + 9n^2p$

52. $16p^2r - 40pqr + 25q^2r$

53. $x^3 + 1$

54. $x^3 + 27$

55. $8t^3 + 125$

56. $27s^3 + 64$

57. $t^6 - 125$

58. $w^6 - 27$

59. $5xt + 15xr + 2yt + 6yr$

60. $3am + 18mb + 2an + 12nb$

61. $6ar + 12br - 5as - 10bs$

62. $7mt + 35ms - 2nt - 10ns$

63. $t^4 - 1$

64. $r^4 - 81$

65. $4x^2 + 12xy + 9y^2 - 1$

66. $81t^2 + 36ty + 4y^2 - 9$

Appendix G
Synthetic Division

OBJECTIVES

1 Use synthetic division to divide by a polynomial of the form $x - k$.

2 Use the remainder theorem to evaluate a polynomial.

3 Decide whether a given number is a solution of an equation.

OBJECTIVE 1 Use synthetic division to divide by a polynomial of the form $x - k$.
Often, when one polynomial is divided by a second, the second polynomial is of the form $x - k$, where the coefficient of the x term is 1. To see how a shortcut for these divisions works, look first below left, where the division of $3x^3 - 2x + 5$ by $x - 3$ is shown. Notice that 0 was inserted for the missing x^2-term.

$$
\begin{array}{r}
3x^2 + 9x + 25 \\
x - 3\overline{)3x^3 + 0x^2 - 2x + 5} \\
\underline{3x^3 - 9x^2} \\
9x^2 - 2x \\
\underline{9x^2 - 27x} \\
25x + 5 \\
\underline{25x - 75} \\
80
\end{array}
$$

$$
\begin{array}{r}
 3 \quad\; 9 \quad\; 25 \\
1 - 3\overline{)3 \quad\; 0 \;\; -2 \quad\; 5} \\
\underline{3 \;\; -9} \\
9 \quad\; -2 \\
\underline{9 \;\; -27} \\
25 \quad\; 5 \\
\underline{25 \;\; -75} \\
80
\end{array}
$$

On the right, exactly the same division is shown written without the variables. This is why it is *essential* to use 0 as a placeholder in synthetic division. All the numbers in color on the right are repetitions of the numbers directly above them, so we omit them, as shown on the left below.

$$
\begin{array}{r}
3 \quad\; 9 \quad\; 25 \\
1 - 3\overline{)3 \quad\; 0 \;\; -2 \quad\; 5} \\
\underline{-9} \\
9 \quad\; -2 \\
\underline{-27} \\
25 \quad\; 5 \\
\underline{-75} \\
80
\end{array}
\qquad
\begin{array}{r}
3 \quad\; 9 \quad\; 25 \\
1 - 3\overline{)3 \quad\; 0 \;\; -2 \quad\; 5} \\
\underline{-9} \\
9 \\
\underline{-27} \\
25 \\
\underline{-75} \\
80
\end{array}
$$

The numbers in color on the left are again repetitions of the numbers directly above them; they too are omitted, as shown on the right above.

Now we can condense the problem. If we bring the 3 in the dividend down to the beginning of the bottom row, the top row can be omitted, since it duplicates the bottom row.

$$
\begin{array}{r}
1 - 3\overline{)3 \quad\;\; 0 \quad -2 \quad\;\; 5} \\
\underline{-9 \;\; -27 \;\; -75} \\
3 \quad\;\; 9 \quad\; 25 \quad\; 80
\end{array}
$$

Finally, we omit the 1 at the upper left. Also, to simplify the arithmetic, we replace subtraction in the second row by addition. To compensate for this, we change the -3 at the upper left to its additive inverse, 3, as shown on the next page.

Additive inverse → 3)3 0 −2 5

        9 27 75 ← Signs changed

     3 9 25 80 ← Remainder

The quotient is read $3x^2 + 9x + 25 + \dfrac{80}{x-3}$
from the bottom row.

The first three numbers in the bottom row are the coefficients of the quotient polynomial with degree 1 less than the degree of the dividend. The last number gives the remainder.

Synthetic Division

This shortcut procedure is called **synthetic division.** It is used only when dividing a polynomial by a binomial of the form $x - k$.

▮ EXAMPLE 1 Using Synthetic Division

Use synthetic division to divide $5x^2 + 16x + 15$ by $x + 2$.

 As mentioned, we use synthetic division only when dividing by a polynomial of the form $x - k$. We change $x + 2$ into this form by writing it as

$$x + 2 = x - (-2),$$

where $k = -2$. Now write the coefficients of $5x^2 + 16x + 15$, placing -2 to the left.

$x + 2$ leads to -2. → $-2)5$ 16 15 ← Coefficients

Bring down the 5, and multiply: $-2 \cdot 5 = -10$.

$$-2)\overline{5 \quad 16 \quad 15}$$
$$\quad\quad -10$$
$$\quad 5$$

Add 16 and -10, getting 6, and multiply 6 and -2 to get -12.

$$-2)\overline{5 \quad 16 \quad 15}$$
$$\quad\quad -10 \quad -12$$
$$\quad 5 \quad\quad 6$$

Add 15 and -12, getting 3.

$$-2)\overline{5 \quad\quad 16 \quad\quad 15}$$
$$\quad\quad\quad -10 \quad -12$$
$$\quad 5 \quad\quad 6 \quad\quad 3 \quad ← \text{Remainder}$$

The result is read from the bottom row.

$$\frac{5x^2 + 16x + 15}{x + 2} = 5x + 6 + \frac{3}{x + 2}$$

Now Try Exercise 7.

EXAMPLE 2 Using Synthetic Division with a Missing Term

Use synthetic division to find $(-4x^5 + x^4 + 6x^3 + 2x^2 + 50) \div (x - 2)$.

Use the steps given above, inserting a 0 for the missing x-term.

$$
\begin{array}{r|rrrrrr}
2) & -4 & 1 & 6 & 2 & 0 & 50 \\
 & & -8 & -14 & -16 & -28 & -56 \\
\hline
 & -4 & -7 & -8 & -14 & -28 & -6
\end{array}
$$

Read the result from the bottom row.

$$
\frac{-4x^5 + x^4 + 6x^3 + 2x^2 + 50}{x - 2} = -4x^4 - 7x^3 - 8x^2 - 14x - 28 + \frac{-6}{x - 2}
$$

Now Try Exercise 13.

OBJECTIVE 2 Use the remainder theorem to evaluate a polynomial. We can use synthetic division to evaluate polynomials. For example, in the synthetic division of Example 2, where the polynomial was divided by $x - 2$, the remainder was -6.

Replacing x in the polynomial with 2 gives

$$
\begin{aligned}
-4x^5 + x^4 + 6x^3 + 2x^2 + 50 &= -4 \cdot 2^5 + 2^4 + 6 \cdot 2^3 + 2 \cdot 2^2 + 50 \\
&= -4 \cdot 32 + 16 + 6 \cdot 8 + 2 \cdot 4 + 50 \\
&= -128 + 16 + 48 + 8 + 50 \\
&= -6,
\end{aligned}
$$

the same number as the remainder; that is, dividing by $x - 2$ produced a remainder equal to the result when x is replaced with 2. This always happens, as the following **remainder theorem** states.

Remainder Theorem

If the polynomial $P(x)$ is divided by $x - k$, then the remainder is equal to $P(k)$.

This result is proved in more advanced courses.

EXAMPLE 3 Using the Remainder Theorem

Let $P(x) = 2x^3 - 5x^2 - 3x + 11$. Find $P(-2)$.

Use the remainder theorem; divide $P(x)$ by $x - (-2)$.

$$
\text{Value of } x \rightarrow \begin{array}{r|rrrr}
-2) & 2 & -5 & -3 & 11 \\
 & & -4 & 18 & -30 \\
\hline
 & 2 & -9 & 15 & -19
\end{array} \leftarrow \text{Remainder}
$$

By this result, $P(-2) = -19$.

Now Try Exercise 21.

OBJECTIVE 3 Decide whether a given number is a solution of an equation. We can also use the remainder theorem to show that a given number is a solution of an equation.

EXAMPLE 4 Using the Remainder Theorem

Show that -5 is a solution of the equation

$$2x^4 + 12x^3 + 6x^2 - 5x + 75 = 0.$$

One way to show that -5 is a solution is by substituting -5 for x in the equation. However, an easier way is to use synthetic division and the remainder theorem.

$$
\begin{array}{r}
\text{Proposed solution} \rightarrow \quad -5)\overline{)2 \quad\ 12 \quad\ 6 \quad -5 \quad\ 75} \\
-10 \ -10 \quad 20 \ -75 \\
\hline
2 \quad\ 2 \ -4 \quad 15 \quad\ 0 \leftarrow \text{Remainder}
\end{array}
$$

Since the remainder is 0, the polynomial has value 0 when $k = -5$, and so -5 is a solution of the given equation.

Now Try Exercises 29 and 31.

The synthetic division in Example 4 shows that $x - (-5)$ divides the polynomial with 0 remainder. Thus $x - (-5) = x + 5$ is a *factor* of the polynomial and

$$2x^4 + 12x^3 + 6x^2 - 5x + 75 = (x + 5)(2x^3 + 2x^2 - 4x + 15).$$

The second factor is the quotient polynomial found in the last row of the synthetic division.

CONNECTIONS

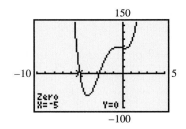

The procedure in Example 4 is exactly how we use a graphing calculator to find real solutions of an equation by determining the x-intercepts of the graph. The screen shows the graph of $P(x) = 2x^4 + 12x^3 + 6x^2 - 5x + 75$ and shows that one value of x that makes $P(x) = 0$ is -5. This agrees with our result in Example 4.

For Discussion or Writing

Estimate the other x-intercept to the nearest tenth. Verify your answer using synthetic division.

APPENDIX G EXERCISES

 1. What is the purpose of synthetic division?

 2. What type of polynomial divisors may be used with synthetic division?

Use synthetic division to find each quotient. See Examples 1 and 2.

3. $\dfrac{x^2 - 6x + 5}{x - 1}$ **4.** $\dfrac{x^2 - 4x - 21}{x + 3}$ **5.** $\dfrac{4m^2 + 19m - 5}{m + 5}$

6. $\dfrac{3k^2 - 5k - 12}{k - 3}$ **7.** $\dfrac{2a^2 + 8a + 13}{a + 2}$ **8.** $\dfrac{4y^2 - 5y - 20}{y - 4}$

9. $(p^2 - 3p + 5) \div (p + 1)$ **10.** $(z^2 + 4z - 6) \div (z - 5)$

11. $\dfrac{4a^3 - 3a^2 + 2a - 3}{a - 1}$ **12.** $\dfrac{5p^3 - 6p^2 + 3p + 14}{p + 1}$

13. $(x^5 - 2x^3 + 3x^2 - 4x - 2) \div (x - 2)$

14. $(2y^5 - 5y^4 - 3y^2 - 6y - 23) \div (y - 3)$

15. $(-4r^6 - 3r^5 - 3r^4 + 5r^3 - 6r^2 + 3r + 3) \div (r - 1)$

16. $(2t^6 - 3t^5 + 2t^4 - 5t^3 + 6t^2 - 3t - 2) \div (t - 2)$

17. $(-3y^5 + 2y^4 - 5y^3 - 6y^2 - 1) \div (y + 2)$

18. $(m^6 + 2m^4 - 5m + 11) \div (m - 2)$

19. $\dfrac{y^3 + 1}{y - 1}$
20. $\dfrac{z^4 + 81}{z - 3}$

Use the remainder theorem to find $P(k)$. See Example 3.

21. $P(x) = 2x^3 - 4x^2 + 5x - 3; k = 2$
22. $P(y) = y^3 + 3y^2 - y + 5; k = -1$

23. $P(r) = -r^3 - 5r^2 - 4r - 2; k = -4$
24. $P(z) = -z^3 + 5z^2 - 3z + 4; k = 3$

25. $P(y) = 2y^3 - 4y^2 + 5y - 33; k = 3$
26. $P(x) = x^3 - 3x^2 + 4x - 4; k = 2$

27. Explain why a 0 remainder in synthetic division of $P(x)$ by $x - k$ indicates that k is a solution of the equation $P(x) = 0$.

28. Explain why it is important to insert 0s as placeholders for missing terms before performing synthetic division.

Use synthetic division to decide whether the given number is a solution of the equation. See Example 4.

29. $x^3 - 2x^2 - 3x + 10 = 0; x = -2$

30. $x^3 - 3x^2 - x + 10 = 0; x = -2$

31. $m^4 + 2m^3 - 3m^2 + 8m - 8 = 0; m = -2$

32. $r^4 - r^3 - 6r^2 + 5r + 10 = 0; r = -2$

33. $3a^3 + 2a^2 - 2a + 11 = 0; a = -2$

34. $3z^3 + 10z^2 + 3z - 9 = 0; z = -2$

35. $2x^3 - x^2 - 13x + 24 = 0; x = -3$

36. $5p^3 + 22p^2 + p - 28 = 0; p = -4$

RELATING CONCEPTS (EXERCISES 37–41)

For Individual or Group Work

We can show a connection between dividing one polynomial by another and factoring the first polynomial. Let $P(x) = 2x^2 + 5x - 12$. **Work Exercises 37–41 in order.**

37. Factor $P(x)$.

38. Solve $P(x) = 0$.

39. Find $P(-4)$ and $P\left(\frac{3}{2}\right)$.

40. Complete the following sentence. If $P(a) = 0$, then $x -$ _____ is a factor of $P(x)$.

41. Use the conclusion in Exercise 40 to decide whether $x - 3$ is a factor of $Q(x) = 3x^3 - 4x^2 - 17x + 6$. Factor $Q(x)$ completely.

TECHNOLOGY INSIGHTS (EXERCISES 42–45)

Use the graph to determine a solution of each equation.

42. $2x^3 + 12x^2 + 24x + 16 = 0$

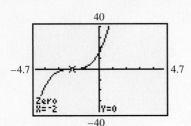

43. $x^3 - x^2 - 21x + 45 = 0$

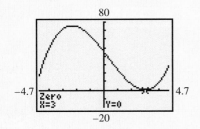

44. $x^3 + 3x^2 - 10x - 24 = 0$

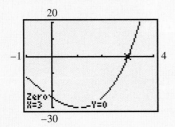

45. $x^3 + 3x^2 - 13x - 15 = 0$

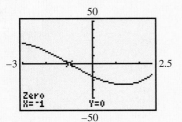

Appendix H
Determinants and Cramer's Rule

OBJECTIVES

1 Evaluate 2 × 2 determinants.

2 Use expansion by minors to evaluate 3 × 3 determinants.

3 Use a graphing calculator to evaluate determinants.

4 Understand the derivation of Cramer's rule.

5 Apply Cramer's rule to solve linear systems.

Recall from Section 8.6 that an ordered array of numbers within square brackets is called a *matrix* (plural *matrices*). Matrices are named according to the number of rows and columns they contain. A *square matrix* has the same number of rows and columns.

$$\text{Rows} \rightarrow \begin{bmatrix} 2 & 3 & 5 \\ 7 & 1 & 2 \end{bmatrix} \begin{array}{l} 2 \times 3 \\ \text{matrix} \end{array} \qquad \begin{bmatrix} -1 & 0 \\ 1 & -2 \end{bmatrix} \begin{array}{l} 2 \times 2 \\ \text{square matrix} \end{array}$$

Associated with every *square matrix* is a real number called the **determinant** of the matrix. A determinant is symbolized by the entries of the matrix placed between two vertical lines, such as

$$\begin{vmatrix} 2 & 3 \\ 7 & 1 \end{vmatrix} \begin{array}{l} 2 \times 2 \\ \text{determinant} \end{array} \qquad \begin{vmatrix} 7 & 4 & 3 \\ 0 & 1 & 5 \\ 6 & 0 & 1 \end{vmatrix} . \begin{array}{l} 3 \times 3 \\ \text{determinant} \end{array}$$

Like matrices, determinants are named according to the number of rows and columns they contain.

OBJECTIVE 1 Evaluate 2 × 2 determinants.   As mentioned above, the value of a determinant is a *real number.* The value of the 2 × 2 determinant

$$\begin{vmatrix} a & b \\ c & d \end{vmatrix}$$

is defined as follows.

Value of a 2 × 2 Determinant

$$\begin{vmatrix} a & b \\ c & d \end{vmatrix} = ad - bc$$

EXAMPLE 1   Evaluating a 2 × 2 Determinant

Evaluate the determinant.

$$\begin{vmatrix} -1 & -3 \\ 4 & -2 \end{vmatrix}$$

Here $a = -1$, $b = -3$, $c = 4$, and $d = -2$, so

$$\begin{vmatrix} -1 & -3 \\ 4 & -2 \end{vmatrix} = -1(-2) - (-3)4 = 2 + 12 = 14.$$

Now Try Exercise 3.

A 3×3 determinant can be evaluated in a similar way.

Value of a 3 × 3 Determinant

$$\begin{vmatrix} a_1 & b_1 & c_1 \\ a_2 & b_2 & c_2 \\ a_3 & b_3 & c_3 \end{vmatrix} = \begin{array}{l} (a_1b_2c_3 + b_1c_2a_3 + c_1a_2b_3) \\ - (a_3b_2c_1 + b_3c_2a_1 + c_3a_2b_1) \end{array}$$

This rule for evaluating a 3×3 determinant is hard to remember. A method for calculating a 3×3 determinant that is easier to use is based on the rule. Rearranging terms and using the distributive property gives

$$\begin{vmatrix} a_1 & b_1 & c_1 \\ a_2 & b_2 & c_2 \\ a_3 & b_3 & c_3 \end{vmatrix} = a_1(b_2c_3 - b_3c_2) - a_2(b_1c_3 - b_3c_1) + a_3(b_1c_2 - b_2c_1). \qquad (1)$$

Each of the quantities in parentheses represents a 2×2 determinant that is the part of the 3×3 determinant remaining when the row and column of the multiplier are eliminated, as shown below.

$$a_1(b_2c_3 - b_3c_2) \qquad \begin{vmatrix} a_1 & b_1 & c_1 \\ a_2 & b_2 & c_2 \\ a_3 & b_3 & c_3 \end{vmatrix}$$

$$a_2(b_1c_3 - b_3c_1) \qquad \begin{vmatrix} a_1 & b_1 & c_1 \\ a_2 & b_2 & c_2 \\ a_3 & b_3 & c_3 \end{vmatrix}$$

$$a_3(b_1c_2 - b_2c_1) \qquad \begin{vmatrix} a_1 & b_1 & c_1 \\ a_2 & b_2 & c_2 \\ a_3 & b_3 & c_3 \end{vmatrix}$$

These 2×2 determinants are called **minors** of the elements in the 3×3 determinant. In the determinant above, the minors of a_1, a_2, and a_3 are, respectively,

$$\begin{vmatrix} b_2 & c_2 \\ b_3 & c_3 \end{vmatrix}, \qquad \begin{vmatrix} b_1 & c_1 \\ b_3 & c_3 \end{vmatrix}, \qquad \text{and} \qquad \begin{vmatrix} b_1 & c_1 \\ b_2 & c_2 \end{vmatrix}.$$

OBJECTIVE 2   Use expansion by minors to evaluate 3 × 3 determinants.   A 3×3 determinant can be evaluated by multiplying each element in the first column by its minor and combining the products as indicated in equation (1). This is called **expansion of the determinant by minors** about the first column.

▌ **EXAMPLE 2**   Evaluating a 3 × 3 Determinant

Evaluate the determinant using expansion by minors about the first column.

$$\begin{vmatrix} 1 & 3 & -2 \\ -1 & -2 & -3 \\ 1 & 1 & 2 \end{vmatrix}$$

In this determinant, $a_1 = 1$, $a_2 = -1$, and $a_3 = 1$. Multiply each of these numbers by its minor, and combine the three terms using the definition. Notice that the second term in the definition is *subtracted*.

$$\begin{vmatrix} 1 & 3 & -2 \\ -1 & -2 & -3 \\ 1 & 1 & 2 \end{vmatrix} = 1 \begin{vmatrix} -2 & -3 \\ 1 & 2 \end{vmatrix} - (-1) \begin{vmatrix} 3 & -2 \\ 1 & 2 \end{vmatrix} + 1 \begin{vmatrix} 3 & -2 \\ -2 & -3 \end{vmatrix}$$

$$= 1[-2(2) - (-3)1] + 1[3(2) - (-2)1]$$
$$+ 1[3(-3) - (-2)(-2)]$$
$$= 1(-1) + 1(8) + 1(-13)$$
$$= -1 + 8 - 13$$
$$= -6$$

Now Try Exercise 9.

To obtain equation (1), we could have rearranged terms in the definition of the determinant and used the distributive property to factor out the three elements of the second or third column or of any of the three rows. Therefore, expanding by minors about any row or any column results in the same value for a 3 × 3 determinant. To determine the correct signs for the terms of other expansions, the following **array of signs** is helpful.

Array of Signs for a 3 × 3 Determinant

$$\begin{matrix} + & - & + \\ - & + & - \\ + & - & + \end{matrix}$$

The signs alternate for each row and column beginning with a + in the first row, first column position. For example, if the expansion is to be about the second column, the first term would have a minus sign associated with it, the second term a plus sign, and the third term a minus sign.

▨ **EXAMPLE 3** Evaluating a 3 × 3 Determinant

Evaluate the determinant of Example 2 using expansion by minors about the second column.

$$\begin{vmatrix} 1 & 3 & -2 \\ -1 & -2 & -3 \\ 1 & 1 & 2 \end{vmatrix} = -3 \begin{vmatrix} -1 & -3 \\ 1 & 2 \end{vmatrix} + (-2) \begin{vmatrix} 1 & -2 \\ 1 & 2 \end{vmatrix} - 1 \begin{vmatrix} 1 & -2 \\ -1 & -3 \end{vmatrix}$$

$$= -3(1) - 2(4) - 1(-5)$$

$$= -3 - 8 + 5$$

$$= -6$$

As expected, the result is the same as in Example 2.

Now Try Exercise 15.

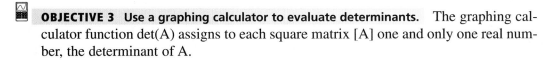 **OBJECTIVE 3** Use a graphing calculator to evaluate determinants. The graphing calculator function det(A) assigns to each square matrix [A] one and only one real number, the determinant of A.

▨ **EXAMPLE 4** Evaluating Determinants Using a Graphing Calculator

Evaluate the determinants in Examples 1 and 2 using a graphing calculator.

Figure 1 shows how a graphing calculator displays the correct value for the determinant in Example 1. Similarly, Figure 2 supports the result of Example 2.

```
[A]
        [[-1  -3]
         [4  -2]]
det([A])
              14
```

```
[B]
        [[1   3  -2]
         [-1  -2  -3]
         [1   1   2 ]]
det([B])
              -6
```

FIGURE 1   FIGURE 2

Now Try Exercise 23.

| **CONNECTIONS** |

⌁ Determinants of larger dimensions (such as 4 × 4) can be evaluated by extending the concepts presented thus far. However, because of the tedious calculations and chance for error, they are usually evaluated by computer or graphing calculator. For example, the determinant

$$\begin{vmatrix} -1 & -2 & 3 & 2 \\ 0 & 1 & 4 & -2 \\ 3 & -1 & 4 & 0 \\ 2 & 1 & 0 & 3 \end{vmatrix}$$

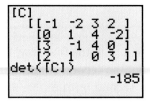

```
[C]
   [[-1  -2  3  2 ]
    [0   1   4  -2]
    [3   -1  4  0 ]
    [2   1   0  3 ]]
det([C])
              -185
```

FIGURE 3

is equal to −185, as shown in the graphing calculator screen in Figure 3.

For Discussion or Writing

1. Use the array of signs

$$
\begin{array}{cccc}
+ & - & + & - \\
- & + & - & + \\
+ & - & + & - \\
- & + & - & +
\end{array}
$$

to evaluate the preceding determinant by hand, expanding about the fourth row.

2. Explain how finding a determinant illustrates the function concept.

OBJECTIVE 4 Understand the derivation of Cramer's rule.   Determinants can be used to solve a system of the form

$$
\begin{aligned}
a_1 x + b_1 y &= c_1 \quad (1) \\
a_2 x + b_2 y &= c_2. \quad (2)
\end{aligned}
$$

The result will be a formula that can be used to solve any system of two equations with two variables. To get this general solution, we eliminate y and solve for x by first multiplying each side of equation (1) by b_2 and each side of equation (2) by $-b_1$. Then we add these results and solve for x.

$$
\begin{array}{ll}
a_1 b_2 x + b_1 b_2 y = c_1 b_2 & \text{Multiply equation (1) by } b_2. \\
\underline{-a_2 b_1 x - b_1 b_2 y = -c_2 b_1} & \text{Multiply equation (2) by } -b_1. \\
(a_1 b_2 - a_2 b_1) x = c_1 b_2 - c_2 b_1 & \\
x = \dfrac{c_1 b_2 - c_2 b_1}{a_1 b_2 - a_2 b_1} & \text{(if } a_1 b_2 - a_2 b_1 \neq 0 \text{)}
\end{array}
$$

To solve for y, we multiply each side of equation (1) by $-a_2$ and each side of equation (2) by a_1 and add.

$$
\begin{array}{ll}
-a_1 a_2 x - a_2 b_1 y = -a_2 c_1 & \text{Multiply equation (1) by } -a_2. \\
\underline{a_1 a_2 x + a_1 b_2 y = a_1 c_2} & \text{Multiply equation (2) by } a_1. \\
(a_1 b_2 - a_2 b_1) y = a_1 c_2 - a_2 c_1 & \\
y = \dfrac{a_1 c_2 - a_2 c_1}{a_1 b_2 - a_2 b_1} & \text{(if } a_1 b_2 - a_2 b_1 \neq 0 \text{)}
\end{array}
$$

Both numerators and the common denominator of these values for x and y can be written as determinants because

$$
a_1 c_2 - a_2 c_1 = \begin{vmatrix} a_1 & c_1 \\ a_2 & c_2 \end{vmatrix},
$$

$$
c_1 b_2 - c_2 b_1 = \begin{vmatrix} c_1 & b_1 \\ c_2 & b_2 \end{vmatrix},
$$

and

$$
a_1 b_2 - a_2 b_1 = \begin{vmatrix} a_1 & b_1 \\ a_2 & b_2 \end{vmatrix}.
$$

Using these results, the solutions for x and y become

$$x = \frac{\begin{vmatrix} c_1 & b_1 \\ c_2 & b_2 \end{vmatrix}}{\begin{vmatrix} a_1 & b_1 \\ a_2 & b_2 \end{vmatrix}} \quad \text{and} \quad y = \frac{\begin{vmatrix} a_1 & c_1 \\ a_2 & c_2 \end{vmatrix}}{\begin{vmatrix} a_1 & b_1 \\ a_2 & b_2 \end{vmatrix}}, \quad \begin{vmatrix} a_1 & b_1 \\ a_2 & b_2 \end{vmatrix} \neq 0.$$

For convenience, we denote the three determinants in the solution as

$$\begin{vmatrix} a_1 & b_1 \\ a_2 & b_2 \end{vmatrix} = D, \quad \begin{vmatrix} c_1 & b_1 \\ c_2 & b_2 \end{vmatrix} = D_x, \quad \text{and} \quad \begin{vmatrix} a_1 & c_1 \\ a_2 & c_2 \end{vmatrix} = D_y.$$

Notice that the elements of D are the four coefficients of the variables in the given system; the elements of D_x are obtained by replacing the coefficients of x by the respective constants; the elements of D_y are obtained by replacing the coefficients of y by the respective constants.

These results are summarized as **Cramer's rule.**

Cramer's Rule for 2 × 2 Systems

Given the system

$$a_1x + b_1y = c_1$$
$$a_2x + b_2y = c_2 \quad \text{with } a_1b_2 - a_2b_1 = D \neq 0,$$

then

$$x = \frac{\begin{vmatrix} c_1 & b_1 \\ c_2 & b_2 \end{vmatrix}}{\begin{vmatrix} a_1 & b_1 \\ a_2 & b_2 \end{vmatrix}} = \frac{D_x}{D} \quad \text{and} \quad y = \frac{\begin{vmatrix} a_1 & c_1 \\ a_2 & c_2 \end{vmatrix}}{\begin{vmatrix} a_1 & b_1 \\ a_2 & b_2 \end{vmatrix}} = \frac{D_y}{D}.$$

OBJECTIVE 5 Apply Cramer's rule to solve linear systems. To use Cramer's rule to solve a system of equations, find the three determinants, D, D_x, and D_y, and then write the necessary quotients for x and y.

CAUTION As indicated in the box, Cramer's rule does not apply if $D = a_1b_2 - a_2b_1 = 0$. When $D = 0$, the system is inconsistent or has dependent equations. For this reason, it is a good idea to evaluate D first.

EXAMPLE 5 Using Cramer's Rule to Solve a 2 × 2 System

Use Cramer's rule to solve the system

$$5x + 7y = -1$$
$$6x + 8y = 1.$$

By Cramer's rule, $x = \dfrac{D_x}{D}$ and $y = \dfrac{D_y}{D}$. As previously mentioned, it is a good idea to find D first since if $D = 0$, Cramer's rule does not apply. If $D \neq 0$, then find D_x and D_y.

$$D = \begin{vmatrix} 5 & 7 \\ 6 & 8 \end{vmatrix} = 5(8) - 7(6) = -2$$

$$D_x = \begin{vmatrix} -1 & 7 \\ 1 & 8 \end{vmatrix} = -1(8) - 7(1) = -15$$

$$D_y = \begin{vmatrix} 5 & -1 \\ 6 & 1 \end{vmatrix} = 5(1) - (-1)6 = 11$$

From Cramer's rule,

$$x = \frac{D_x}{D} = \frac{-15}{-2} = \frac{15}{2} \qquad \text{and} \qquad y = \frac{D_y}{D} = \frac{11}{-2} = -\frac{11}{2}.$$

The solution set is $\left\{\left(\frac{15}{2}, -\frac{11}{2}\right)\right\}$, as can be verified by checking in the given system.

Now Try Exercise 27.

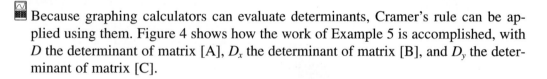 Because graphing calculators can evaluate determinants, Cramer's rule can be applied using them. Figure 4 shows how the work of Example 5 is accomplished, with D the determinant of matrix [A], D_x the determinant of matrix [B], and D_y the determinant of matrix [C].

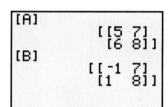

FIGURE 4

In a similar manner, Cramer's rule can be applied to systems of three equations with three variables.

Cramer's Rule for 3 × 3 Systems

Given the system

$$a_1 x + b_1 y + c_1 z = d_1$$
$$a_2 x + b_2 y + c_2 z = d_2$$
$$a_3 x + b_3 y + c_3 z = d_3 \qquad \text{(continued)}$$

with

$$D_x = \begin{vmatrix} d_1 & b_1 & c_1 \\ d_2 & b_2 & c_2 \\ d_3 & b_3 & c_3 \end{vmatrix}, \qquad D_y = \begin{vmatrix} a_1 & d_1 & c_1 \\ a_2 & d_2 & c_2 \\ a_3 & d_3 & c_3 \end{vmatrix},$$

$$D_z = \begin{vmatrix} a_1 & b_1 & d_1 \\ a_2 & b_2 & d_2 \\ a_3 & b_3 & d_3 \end{vmatrix}, \qquad D = \begin{vmatrix} a_1 & b_1 & c_1 \\ a_2 & b_2 & c_2 \\ a_3 & b_3 & c_3 \end{vmatrix} \neq 0,$$

then

$$x = \frac{D_x}{D}, \qquad y = \frac{D_y}{D}, \qquad \text{and} \qquad z = \frac{D_z}{D}.$$

EXAMPLE 6   Using Cramer's Rule to Solve a 3 × 3 System

Use Cramer's rule to solve the system

$$\begin{aligned} x + y - z + 2 &= 0 \\ 2x - y + z + 5 &= 0 \\ x - 2y + 3z - 4 &= 0. \end{aligned}$$

To use Cramer's rule, first rewrite the system in the form

$$\begin{aligned} x + y - z &= -2 \\ 2x - y + z &= -5 \\ x - 2y + 3z &= 4. \end{aligned}$$

Expand by minors about row 1 to find D.

$$\begin{aligned} D &= \begin{vmatrix} 1 & 1 & -1 \\ 2 & -1 & 1 \\ 1 & -2 & 3 \end{vmatrix} \\ &= 1\begin{vmatrix} -1 & 1 \\ -2 & 3 \end{vmatrix} - 1\begin{vmatrix} 2 & 1 \\ 1 & 3 \end{vmatrix} + (-1)\begin{vmatrix} 2 & -1 \\ 1 & -2 \end{vmatrix} \\ &= 1(-1) - 1(5) - 1(-3) \\ &= -3 \end{aligned}$$

Expanding D_x by minors about row 1 gives

$$\begin{aligned} D_x &= \begin{vmatrix} -2 & 1 & -1 \\ -5 & -1 & 1 \\ 4 & -2 & 3 \end{vmatrix} \\ &= -2\begin{vmatrix} -1 & 1 \\ -2 & 3 \end{vmatrix} - 1\begin{vmatrix} -5 & 1 \\ 4 & 3 \end{vmatrix} + (-1)\begin{vmatrix} -5 & -1 \\ 4 & -2 \end{vmatrix} \\ &= -2(-1) - 1(-19) - 1(14) \\ &= 7. \end{aligned}$$

In the same way, $D_y = -22$ and $D_z = -21$, so that

$$x = \frac{D_x}{D} = \frac{7}{-3} = -\frac{7}{3}, \qquad y = \frac{D_y}{D} = \frac{-22}{-3} = \frac{22}{3}, \qquad z = \frac{D_z}{D} = \frac{-21}{-3} = 7.$$

Check that the solution set is $\left\{ \left(-\frac{7}{3}, \frac{22}{3}, 7 \right) \right\}$.

Now Try Exercise 33.

As mentioned earlier, Cramer's rule does not apply when $D = 0$. The next example illustrates this case.

▨ EXAMPLE 7   Determining When Cramer's Rule Does Not Apply

Use Cramer's rule to solve the system

$$\begin{aligned} 2x - 3y + 4z &= 8 \\ 6x - 9y + 12z &= 24 \\ x + 2y - 3z &= 5. \end{aligned}$$

First, find D.

$$D = \begin{vmatrix} 2 & -3 & 4 \\ 6 & -9 & 12 \\ 1 & 2 & -3 \end{vmatrix}$$

$$= 2 \begin{vmatrix} -9 & 12 \\ 2 & -3 \end{vmatrix} - 6 \begin{vmatrix} -3 & 4 \\ 2 & -3 \end{vmatrix} + 1 \begin{vmatrix} -3 & 4 \\ -9 & 12 \end{vmatrix}$$

$$= 2(3) - 6(1) + 1(0)$$

$$= 0$$

Since $D = 0$ here, Cramer's rule does not apply and we must use another method to solve the system. Multiplying each side of the first equation by 3 shows that the first two equations have the same solution set, so this system has dependent equations and an infinite solution set.

Now Try Exercise 35.

Cramer's rule can be extended to 4×4 or larger systems. See a standard college algebra text for details.

APPENDIX H EXERCISES

1. Which one of the following is the expression for the determinant $\begin{vmatrix} -2 & -3 \\ 4 & -6 \end{vmatrix}$?

    A. $-2(-6) + (-3)4$ **B.** $-2(-6) - 3(4)$

    C. $-3(4) - (-2)(-6)$ **D.** $-2(-6) - (-3)4$

▨ **2.** Evaluate $\begin{vmatrix} 0 & 0 \\ 3 & -4 \end{vmatrix}$ and $\begin{vmatrix} 0 & 1 & 2 \\ 0 & -3 & 4 \\ 0 & 2 & 6 \end{vmatrix}$ and make a conjecture (educated guess) about the value of a determinant that has all 0s in a row or a column.

Evaluate each determinant. See Example 1.

3. $\begin{vmatrix} -2 & 5 \\ -1 & 4 \end{vmatrix}$

4. $\begin{vmatrix} 3 & -6 \\ 2 & -2 \end{vmatrix}$

5. $\begin{vmatrix} 1 & -2 \\ 7 & 0 \end{vmatrix}$

6. $\begin{vmatrix} -5 & -1 \\ 1 & 0 \end{vmatrix}$

7. $\begin{vmatrix} 0 & 4 \\ 0 & 4 \end{vmatrix}$

8. $\begin{vmatrix} 8 & -3 \\ 0 & 0 \end{vmatrix}$

Evaluate each determinant by expansion by minors about the first column. See Example 2.

9. $\begin{vmatrix} -1 & 2 & 4 \\ -3 & -2 & -3 \\ 2 & -1 & 5 \end{vmatrix}$

10. $\begin{vmatrix} 2 & -3 & -5 \\ 1 & 2 & 2 \\ 5 & 3 & -1 \end{vmatrix}$

11. $\begin{vmatrix} 1 & 0 & -2 \\ 0 & 2 & 3 \\ 1 & 0 & 5 \end{vmatrix}$

12. $\begin{vmatrix} 2 & -1 & 0 \\ 0 & -1 & 1 \\ 1 & 2 & 0 \end{vmatrix}$

13. Explain in your own words how to evaluate a 2 × 2 determinant. Illustrate with an example.

14. Explain in your own words how to evaluate a 3 × 3 determinant. Illustrate with an example.

Evaluate each determinant by expansion by minors about any row or column. (Hint: The work is easier if you choose a row or a column with 0s.) See Example 3.

15. $\begin{vmatrix} 4 & 4 & 2 \\ 1 & -1 & -2 \\ 1 & 0 & 2 \end{vmatrix}$

16. $\begin{vmatrix} 3 & -1 & 2 \\ 1 & 5 & -2 \\ 0 & 2 & 0 \end{vmatrix}$

17. $\begin{vmatrix} 3 & 5 & -2 \\ 1 & -4 & 1 \\ 3 & 1 & -2 \end{vmatrix}$

18. $\begin{vmatrix} 0 & 0 & 3 \\ 4 & 0 & -2 \\ 2 & -1 & 3 \end{vmatrix}$

19. $\begin{vmatrix} 3 & 0 & -2 \\ 1 & -4 & 1 \\ 3 & 1 & -2 \end{vmatrix}$

20. $\begin{vmatrix} 1 & 1 & 2 \\ 5 & 5 & 7 \\ 3 & 3 & 1 \end{vmatrix}$

21. Explain why a determinant with a row or column of 0s has a value of 0.

Use a graphing calculator with matrix capabilities to find each determinant. See Example 4.

22. $\begin{vmatrix} .68 & .94 \\ .31 & -.56 \end{vmatrix}$

23. $\begin{vmatrix} 1.5 & 2.6 & 9.3 \\ 5.2 & -1.4 & 8.6 \\ 0 & .7 & 1.2 \end{vmatrix}$

24. $\begin{vmatrix} \sqrt{5} & \sqrt{2} & -\sqrt{3} \\ \sqrt{7} & -\sqrt{6} & \sqrt{10} \\ -\sqrt{5} & -\sqrt{2} & \sqrt{17} \end{vmatrix}$ (To as many places as the calculator shows)

25. Consider the system

$$4x + 3y - 2z = 1$$
$$7x - 4y + 3z = 2$$
$$-2x + y - 8z = 0.$$

Match each determinant in parts (a)–(d) with its correct representation from choices A–D.

(a) D

A. $\begin{vmatrix} 1 & 3 & -2 \\ 2 & -4 & 3 \\ 0 & 1 & -8 \end{vmatrix}$ **B.** $\begin{vmatrix} 4 & 3 & 1 \\ 7 & -4 & 2 \\ -2 & 1 & 0 \end{vmatrix}$

(b) D_x

(c) D_y

(d) D_z

C. $\begin{vmatrix} 4 & 1 & -2 \\ 7 & 2 & 3 \\ -2 & 0 & -8 \end{vmatrix}$ **D.** $\begin{vmatrix} 4 & 3 & -2 \\ 7 & -4 & 3 \\ -2 & 1 & -8 \end{vmatrix}$

26. For the system

$$x + 3y - 6z = 7$$
$$2x - y + z = 1$$
$$x + 2y + 2z = -1,$$

$D = -43$, $D_x = -43$, $D_y = 0$, and $D_z = 43$. What is the solution set of the system?

Use Cramer's rule to solve each linear system in two variables. See Example 5.

27. $3x + 5y = -5$
$-2x + 3y = 16$

28. $5x + 2y = -3$
$4x - 3y = -30$

29. $8x + 3y = 1$
$6x - 5y = 2$

30. $3x - y = 9$
$2x + 5y = 8$

31. $2x + 3y = 4$
$5x + 6y = 7$

32. $4x + 5y = 6$
$7x + 8y = 9$

Use Cramer's rule where applicable to solve each linear system in three variables. See Examples 6 and 7.

33. $2x + 3y + 2z = 15$
$x - y + 2z = 5$
$x + 2y - 6z = -26$

34. $x - y + 6z = 19$
$3x + 3y - z = 1$
$x + 9y + 2z = -19$

35. $2x - 3y + 4z = 8$
$6x - 9y + 12z = 24$
$-4x + 6y - 8z = -16$

36. $7x + y - z = 4$
$2x - 3y + z = 2$
$-6x + 9y - 3z = -6$

37. $3x + 5z = 0$
$2x + 3y = 1$
$-y + 2z = -11$

38. $-x + 2y = 4$
$3x + y = -5$
$2x + z = -1$

39. $x - 3y = 13$
$2y + z = 5$
$-x + z = -7$

40. $-5x - y = -10$
$3x + 2y + z = -3$
$-y - 2z = -13$

Use a graphing calculator and the approach described with Figure 4 to solve each system using Cramer's rule.

41. $x + 2y + z = 10$
$2x - y - 3z = -20$
$-x + 4y + z = 18$

42. $2x + y + 3z = 1$
$x - 2y + z = -3$
$-3x + y - 2z = -4$

43.
$$-8w + 4x - 2y + z = -28$$
$$-w + x - y + z = -10$$
$$w + x + y + z = -4$$
$$27w + 9x + 3y + z = 2$$

44.
$$5w + 2x - 3y + z = 4.7$$
$$-2w + x + 2y - z = -3.2$$
$$w + 3x - y + 2z = 2.1$$
$$2w + x - 5y + 3z = 3.4$$

There is another method for evaluating a 3×3 determinant. Refer to Example 2, and copy the first two columns to the right of the original determinant to obtain

$$\begin{vmatrix} 1 & 3 & -2 & 1 & 3 \\ -1 & -2 & -3 & -1 & -2 \\ 1 & 1 & 2 & 1 & 1 \end{vmatrix}.$$

Multiply along the diagonals as shown, placing the product at the end of the arrow.

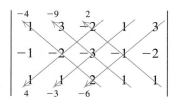

Add the top numbers: $\qquad -4 - 9 + 2 = -11.$

Add the bottom numbers: $\qquad 4 - 3 - 6 = -5.$

Find the *difference* between these sums to obtain the final answer:

$$-11 - (-5) = -6.$$

Use this method to find each determinant in the indicated exercise.

45. Exercise 15 **46.** Exercise 16 **47.** Exercise 17

48. Exercise 18 **49.** Exercise 19 **50.** Exercise 20

Solve each equation by finding an expression for the determinant on the left, and then solving using the methods of Chapter 2.

51. $\begin{vmatrix} 4 & x \\ 2 & 3 \end{vmatrix} = 8$ **52.** $\begin{vmatrix} 5 & 3 \\ x & x \end{vmatrix} = 20$ **53.** $\begin{vmatrix} x & 4 \\ x & -3 \end{vmatrix} = 0$

54. Look at the coefficients and constants in the systems in Exercises 31 and 32. Notice that in both cases, the six numbers are consecutive integers. Make up a system having this same pattern for its coefficients and constants, and solve it using Cramer's rule. Compare the solutions in Exercises 31, 32, and here. What do you notice?

55. Use Cramer's rule to prove that the following system has solution set $\{(-1, 2)\}$.

$$ax + (a + 1)y = a + 2$$
$$(a + 3)x + (a + 4)y = a + 5, \quad \text{where } D \neq 0.$$

56. Under what conditions can a system *not* be solved using Cramer's rule?

RELATING CONCEPTS (EXERCISES 57–64)

For Individual or Group Work

In this section we have seen how determinants can be used to solve systems of equations. There are other applications of determinants. Here, we show how a determinant can be used to find the area of a triangle if we know the coordinates of its vertices.

Suppose that $A(x_1, y_1)$, $B(x_2, y_2)$, and $C(x_3, y_3)$ are the coordinates of the vertices of triangle ABC in the coordinate plane. Then it can be shown that the area of the triangle is given by the absolute value of

$$\frac{1}{2} \begin{vmatrix} x_1 & y_1 & 1 \\ x_2 & y_2 & 1 \\ x_3 & y_3 & 1 \end{vmatrix}.$$

Work Exercises 57–60 in order.

57. Sketch triangle *ABC* in the coordinate plane, given that the coordinates of *A* are $(0, 0)$, of *B* are $(-3, -4)$, and of *C* are $(2, -2)$.

58. Write the determinant expression described above that gives the area of triangle *ABC* described in Exercise 57.

59. Evaluate the absolute value of the determinant expression in Exercise 58 to find the area.

60. Use the determinant expression described above to find the area of the triangle with vertices at $(3, 8)$, $(-1, 4)$, and $(0, 1)$.

*Here is yet another application of determinants. Recall the formula for slope and the point-slope form of the equation of a line from Chapters 3 and 7. Use these formulas to **work Exercises 61–64 in order** and see how a determinant can be used in writing the equation of a line.*

61. Write the expression for the slope of a line passing through the points (x_1, y_1) and (x_2, y_2).

62. Using the expression from Exercise 61 as *m*, and the point (x_1, y_1), write the point-slope form of the equation of the line.

63. Using the equation obtained in Exercise 62, multiply both sides by $x_2 - x_1$, and write the equation so that 0 is on the right side.

64. Consider the *determinant equation*

$$\begin{vmatrix} x & y & 1 \\ x_1 & y_1 & 1 \\ x_2 & y_2 & 1 \end{vmatrix} = 0.$$

Expand by minors on the left and show that this determinant equation yields the same result that you obtained in Exercise 63.

Answers to Selected Exercises

In this section we provide the answers that we think most students will obtain when they work the exercises using the methods explained in the text. If your answer does not look exactly like the one given here, it is not necessarily wrong. In many cases there are equivalent forms of the answer that are correct. For example, if the answer section shows $\frac{3}{4}$ and your answer is .75, you have obtained the right answer but written it in a different (yet equivalent) form. Unless the directions specify otherwise, .75 is just as valid an answer as $\frac{3}{4}$.

In general, if your answer does not agree with the one given in the text, see whether it can be transformed into the other form. If it can, then it is the correct answer. If you still have doubts, talk with your instructor. You might also want to obtain a copy of the *Student's Solutions Manual* that goes with this book. Your college bookstore either has this manual or can order it for you.

CHAPTER 1 THE REAL NUMBER SYSTEM

Section 1.1   (page 10)

Exercises   **1.** true   **3.** false; The fraction $\frac{17}{51}$ is written in lowest terms as $\frac{1}{3}$.   **5.** false; *Product* refers to multiplication, so the product of 8 and 2 is 16.   **7.** prime   **9.** composite; $2 \cdot 2 \cdot 2 \cdot 2 \cdot 2 \cdot 2$ **11.** composite; $2 \cdot 7 \cdot 13 \cdot 19$ **13.** neither   **15.** composite; $2 \cdot 3 \cdot 5$ **17.** composite; $2 \cdot 2 \cdot 5 \cdot 5 \cdot 5$ **19.** composite; $2 \cdot 2 \cdot 31$ **21.** prime   **23.** $\frac{1}{2}$ **25.** $\frac{5}{6}$ **27.** $\frac{3}{10}$ **29.** $\frac{6}{5}$ **31.** C   **33.** $\frac{24}{35}$ **35.** $\frac{6}{25}$ **37.** $\frac{6}{5}$ or $1\frac{1}{5}$ **39.** $\frac{232}{15}$ or $15\frac{7}{15}$ **41.** $\frac{10}{3}$ or $3\frac{1}{3}$ **43.** 12   **45.** $\frac{1}{16}$ **47.** $\frac{84}{47}$ or $1\frac{37}{47}$ **49.** To multiply two fractions, multiply their numerators to get the numerator of the product and multiply their denominators to get the denominator of the product. For example, $\frac{2}{3} \cdot \frac{8}{5} = \frac{2 \cdot 8}{3 \cdot 5} = \frac{16}{15}$. To divide two fractions, replace the divisor with its reciprocal and then multiply. For example, $\frac{2}{5} \div \frac{7}{9} = \frac{2}{5} \cdot \frac{9}{7} = \frac{2 \cdot 9}{5 \cdot 7} = \frac{18}{35}$.   **51.** $\frac{2}{3}$ **53.** $\frac{8}{9}$ **55.** $\frac{43}{8}$ or $5\frac{3}{8}$ **57.** $\frac{2}{3}$ **59.** $\frac{17}{36}$ **61.** $\frac{11}{12}$ **63.** 6 cups   **65.** 34 dollars   **67.** $\frac{9}{16}$ in.   **69.** $618\frac{3}{4}$ ft   **71.** $5\frac{5}{24}$ in.   **73.** $\frac{1}{3}$ cup   **75.** $\frac{1}{20}$ **77.** more than $1\frac{1}{25}$ million   **79. (a)** $\frac{1}{2}$ **(b)** $\frac{1}{4}$ **(c)** $\frac{1}{3}$ **(d)** $\frac{1}{6}$

Section 1.2   (page 20)

Exercises   **1.** false; $4 + 3(8 - 2) = 4 + 3 \cdot 6 = 4 + 18 = 22$. The common error leading to 42 is adding 4 to 3 and then multiplying by 6. One must follow the order of operations.   **3.** false; The correct interpretation is $4 = 16 - 12$.   **5.** 49   **7.** 144   **9.** 64   **11.** 1000   **13.** 81   **15.** 1024   **17.** $\frac{16}{81}$ **19.** .000064   **21.** Write the base as a factor the number of times indicated by the exponent. For example, $6^3 = 6 \cdot 6 \cdot 6 = 216$.   **23.** 32   **25.** $\frac{49}{30}$ or $1\frac{19}{30}$ **27.** 12   **29.** 42   **31.** 95   **33.** 90   **35.** 14   **37.** 9

39. Begin by squaring 2. Then subtract 1, to get a result of $4 - 1 = 3$ within the parentheses. Next, raise 3 to the third power to get $3^3 = 27$. Multiply this result by 3 to obtain 81. Finally, add this result to 4 to get the final answer, 85.　　**41.** $16 \le 16$; true　　**43.** $61 \le 60$; false **45.** $0 \ge 0$; true　　**47.** $45 \ge 46$; false　　**49.** $66 > 72$; false　　**51.** $2 \ge 3$; false　　**53.** $3 \ge 3$; true　　**55.** $15 = 5 + 10$　　**57.** $9 > 5 - 4$ **59.** $16 \ne 19$　　**61.** $2 \le 3$　　**63.** Seven is less that nineteen; true　　**65.** Three is not equal to six; true　　**67.** Eight is greater than or equal to eleven; false　　**69.** Answers will vary. One example is $5 + 3 \ge 2 \cdot 2$.　　**71.** $30 > 5$　　**73.** $3 \le 12$　　**75.** is younger than　　**77.** The inequality symbol $\ge$ implies a true statement if 12 equals 12 *or* if 12 is greater than 12.　　**79.** 1998, 1999　　**81. (a)** \$16.96 **(b)** approximately 31.5%　　**83.** $3 \cdot (6 + 4) \cdot 2 = 60$　　**85.** $10 - (7 - 3) = 6$　　**87.** $(8 + 2)^2 = 100$

Section 1.3　(page 27)

Exercises　**1.** 10　　**3.** $12 + x$; 21　　**5.** no　　**7.** $2x^3 = 2 \cdot x \cdot x \cdot x$, while $2x \cdot 2x \cdot 2x = (2x)^3$.　　**9.** The exponent 2 applies only to its base, which is x. (The expression $(4x)^2$ would require multiplying 4 by $x = 3$ first.)　　**11.** Answers will vary. Two such pairs are $x = 0$, $y = 6$ and $x = 1$, $y = 4$. To determine them, choose a value for x, substitute it into the expression $2x + y$, and then subtract the value of $2x$ from 6.

13. (a) 13 **(b)** 15　　**15. (a)** 20 **(b)** 30　　**17. (a)** 64 **(b)** 144　　**19. (a)** $\dfrac{5}{3}$ **(b)** $\dfrac{7}{3}$　　**21. (a)** $\dfrac{7}{8}$ **(b)** $\dfrac{13}{12}$　　**23. (a)** 52 **(b)** 114

25. (a) 25.836 **(b)** 38.754　　**27. (a)** 24 **(b)** 28　　**29. (a)** 12 **(b)** 33　　**31. (a)** 6 **(b)** $\dfrac{9}{5}$　　**33. (a)** $\dfrac{4}{3}$ **(b)** $\dfrac{13}{6}$　　**35. (a)** $\dfrac{2}{7}$ **(b)** $\dfrac{16}{27}$

37. (a) 12 **(b)** 55　　**39. (a)** 1 **(b)** $\dfrac{28}{17}$　　**41. (a)** 3.684 **(b)** 8.841　　**43.** $12x$　　**45.** $x + 7$　　**47.** $x - 2$　　**49.** $7 - x$　　**51.** $x - 6$

53. $\dfrac{12}{x}$　　**55.** $6(x - 4)$　　**57.** No, it is a connective word that joins the two factors: the number and 6.　　**59.** yes　　**61.** no　　**63.** yes

65. yes　　**67.** yes　　**69.** $x + 8 = 18$; 10　　**71.** $16 - \dfrac{3}{4}x = 13$; 4　　**73.** $2x + 1 = 5$; 2　　**75.** $3x = 2x + 8$; 8　　**77.** expression

79. equation　　**81.** equation　　**83.** 128.02 ft　　**85.** 187.08 ft

Section 1.4　(page 36)

Exercises　**1.** 1,198,000　　**3.** 925　　**5.** −5074　　**7.** −11.35　　**9.** 4　　**11.** 0　　**13.** One example is $\sqrt{12}$. There are others.　　**15.** true

17. true　　**19. (a)** 3, 7 **(b)** 0, 3, 7 **(c)** −9, 0, 3, 7 **(d)** $-9, -1\dfrac{1}{4}, -\dfrac{3}{5}, 0, 3, 5.9, 7$ **(e)** $-\sqrt{7}, \sqrt{5}$ **(f)** All are real numbers.　　**21.** The *natural numbers* are the numbers with which we count. An example is 1. The *whole numbers* are the natural numbers with 0 also included. An example is 0. The *integers* are the whole numbers and their negatives. An example is −1. The *rational numbers* are the numbers that can be represented by a quotient of integers, such as $\dfrac{1}{2}$. The *irrational numbers*, such as $\sqrt{2}$, cannot be represented as a quotient of integers. The *real numbers* include all positive numbers, negative numbers, and zero. All the numbers discussed are real.　　**23.**
$$-6 \quad -4 \quad -2 \quad 0 \quad 2$$

25.
$$-6 \quad -4 \quad -2 \quad 0 \quad 2 \quad 4$$
27.
$$-3\tfrac{4}{5} \quad -1\tfrac{5}{8} \quad \tfrac{1}{4} \quad 2\tfrac{1}{2}$$
$$-4 \quad -2 \quad 0 \quad 2 \quad 4$$
29. (a) A **(b)** A **(c)** B **(d)** B　　**31. (a)** 2 **(b)** 2　　**33. (a)** −6 **(b)** 6　　**35.** 6

37. −12　　**39.** 3　　**41.** −12　　**43.** −8　　**45.** 3　　**47.** $|-3|$ or 3　　**49.** $-|-6|$ or −6　　**51.** $|5 - 3|$ or 2　　**53.** true　　**55.** true **57.** true　　**59.** false　　**61.** true　　**63.** false　　**65.** softwood plywood, 1998 to 1999　　**67.** paving mixtures and blocks, 1998 to 1999 In Exercises 69–73, answers will vary.

69. $\dfrac{1}{2}, \dfrac{5}{8}, 1\dfrac{3}{4}$　　**71.** $-3\dfrac{1}{2}, -\dfrac{2}{3}, \dfrac{3}{7}$　　**73.** $\sqrt{5}, \pi, -\sqrt{3}$　　**75.** This is not true. The absolute value of 0 is 0, and 0 is not positive. A more accurate way of describing absolute value is to say that *absolute value is never negative*, or *absolute value is always nonnegative*.

Section 1.5   (page 46)

Exercises   **1.** negative   **3.** negative   **5.** To add two numbers with the same sign, add their

absolute values and keep the same sign for the sum. For example, $3 + 4 = 7$ and $-3 + (-4) = -7$. To add two numbers with different signs, subtract the smaller absolute value from the larger absolute value, and use the sign of the number with the larger absolute value. For example, $6 + (-4) = 2$ and $(-6) + 4 = -2$.   **7.** -8 **9.** -12 **11.** 2   **13.** -2 **15.** 4   **17.** 12   **19.** 5   **21.** 2   **23.** -9 **25.** 0

27. $\dfrac{1}{2}$ **29.** $-\dfrac{19}{24}$ **31.** $-\dfrac{3}{4}$ **33.** -7.7 **35.** -8 **37.** 0   **39.** -20 **41.** -3 **43.** -4 **45.** -8 **47.** -14 **49.** 9

51. -4 **53.** 4   **55.** $\dfrac{3}{4}$ **57.** $-\dfrac{11}{8}$ or $-1\dfrac{3}{8}$ **59.** $\dfrac{15}{8}$ or $1\dfrac{7}{8}$ **61.** 11.6   **63.** -9.9 **65.** 10   **67.** -5 **69.** 11   **71.** -10

73. 22   **75.** -2 **77.** -6 **79.** -12 **81.** -5.90617 **83.** $-5 + 12 + 6; 13$ **85.** $[-19 + (-4)] + 14; -9$
87. $[-4 + (-10)] + 12; -2$ **89.** $[8 + (-18)] + 4; -6$ **91.** $4 - (-8); 12$ **93.** $-2 - 8; -10$ **95.** $[9 + (-4)] - 7; -2$
97. $[8 - (-5)] - 12; 1$ **99.** -3.4 (billion dollars)   **101.** -2.7 (billion dollars)   **103.** 50,395 ft   **105.** 1345 ft   **107.** 136 ft
109. -12 **111.** 45°F   **113.** -58°F   **115.** 27 ft   **117.** $+31,900$ ft   **119.** $-\$107$

Section 1.6   (page 60)

Exercises   **1.** greater than 0   **3.** greater than 0   **5.** less than 0   **7.** greater than 0   **9.** equal to 0   **11.** 12   **13.** -12 **15.** 120

17. -33 **19.** -165 **21.** $\dfrac{5}{12}$ **23.** $-\dfrac{1}{6}$ **25.** 6   **27.** $-32, -16, -8, -4, -2, -1, 1, 2, 4, 8, 16, 32$ **29.** $-40, -20, -10, -8,$

$-5, -4, -2, -1, 1, 2, 4, 5, 8, 10, 20, 40$ **31.** $-31, -1, 1, 31$ **33.** 3   **35.** -5 **37.** 7   **39.** -6 **41.** $\dfrac{32}{3}$ or $10\dfrac{2}{3}$ **43.** -4

45. 0   **47.** undefined   **49.** -11 **51.** -2 **53.** 35   **55.** 6   **57.** -18 **59.** 67   **61.** -8 **63.** 3   **65.** 7   **67.** 4   **69.** -3

71. 10   **73.** negative   **75.** 47   **77.** 72   **79.** $-\dfrac{78}{25}$ **81.** 0   **83.** -23 **85.** 2   **87.** $9 + (-9)(2); -9$ **89.** $-4 - 2(-1)(6); 8$

91. $(1.5)(-3.2) - 9; -13.8$ **93.** $12[9 - (-8)]; 204$ **95.** $\dfrac{-12}{-5 + (-1)}; 2$ **97.** $\dfrac{15 + (-3)}{4(-3)}; -1$ **99.** $2(8 + 9); 34$

101. $.20(-5 \cdot 6); -6$ **103.** $\dfrac{x}{3} = -3; -9$ **105.** $x - 6 = 4; 10$ **107.** $x + 5 = -5; -10$ **109.** $8\dfrac{2}{5}$ **111.** 2   **113.** 0

115. (a) 6 is divisible by 2. **(b)** 9 is not divisible by 2.   **117. (a)** 64 is divisible by 4. **(b)** 35 is not divisible by 4.   **119. (a)** 2 is divisible
by 2 and $1 + 5 + 2 + 4 + 8 + 2 + 2 = 24$ is divisible by 3. **(b)** While 0 is divisible by 2, $2 + 8 + 7 + 3 + 5 + 9 + 0 = 34$ is not
divisible by 3.   **121. (a)** $4 + 1 + 1 + 4 + 1 + 0 + 7 = 18$ is divisible by 9. **(b)** $2 + 2 + 8 + 7 + 3 + 2 + 1 = 25$ is not divisible by 9.

Summary Exercises on Operations with Real Numbers   (page 63)

1. -16 **2.** 4   **3.** 0   **4.** -24 **5.** -17 **6.** 76   **7.** -18 **8.** 90   **9.** 38   **10.** 4   **11.** -5 **12.** 5   **13.** $-\dfrac{7}{2}$ or $-3\dfrac{1}{2}$

14. 4   **15.** 13   **16.** $\dfrac{5}{4}$ or $1\dfrac{1}{4}$ **17.** 9   **18.** $\dfrac{37}{10}$ or $3\dfrac{7}{10}$ **19.** 0   **20.** 25   **21.** 14   **22.** 0   **23.** -4 **24.** $\dfrac{6}{5}$ or $1\dfrac{1}{5}$ **25.** -1

26. $\dfrac{52}{37}$ or $1\dfrac{15}{37}$ **27.** $\dfrac{17}{16}$ or $1\dfrac{1}{16}$ **28.** $-\dfrac{2}{3}$ **29.** 3.33   **30.** 1.02   **31.** -13 **32.** 0   **33.** 24   **34.** -7 **35.** 37   **36.** -3

37. -1 **38.** $\dfrac{1}{2}$ **39.** $-\dfrac{5}{13}$ **40.** 5

Section 1.7 (page 70)

Exercises **1.** -12; commutative property **3.** 3; commutative property **5.** 7; associative property **7.** 8; associative property
9. (a) B (b) F (c) C (d) I (e) B (f) D, F (g) B (h) A (i) G (j) H **11.** commutative property **13.** associative property
15. associative property **17.** inverse property **19.** inverse property **21.** identity property **23.** commutative property
25. distributive property **27.** identity property **29.** distributive property **31.** identity property **33.** The identity properties allow us
to perform an operation so that the result is the number we started with. The inverse properties allow us to perform an operation that gives an
identity element as a result. **35.** 150 **37.** 2010 **39.** 400 **41.** 1400 **43.** 11 **45.** 0 **47.** $-.38$ **49.** 1 **51.** Subtraction is
not associative. **53.** The expression following the first equals sign should be $-3(4) - 3(-6)$. The student forgot that 6 should be preceded
by a $-$ sign. The correct work is $-3(4 - 6) = -3(4) - 3(-6) = -12 + 18 = 6$. **55.** 85 **57.** $4t + 12$ **59.** $-8r - 24$
61. $-5y + 20$ **63.** $-16y - 20z$ **65.** $8(z + w)$ **67.** $7(2v + 5r)$ **69.** $24r + 32s - 40y$ **71.** $-24x - 9y - 12z$
73. $5(x + 3)$ **75.** $-4t - 3m$ **77.** $5c + 4d$ **79.** $3q - 5r + 8s$ **81.** Answers will vary; for example, "putting on your socks" and
"putting on your shoes." **83.** 0 **84.** $-3(5) + (-3)(-5)$ **85.** -15 **86.** We must interpret $(-3)(-5)$ as 15, since it is the additive
inverse of -15. **87.** (a) no (b) distributive property

Section 1.8 (page 76)

Exercises **1.** $4r + 11$ **3.** $5 + 2x - 6y$ **5.** $-7 + 3p$ **7.** $2 - 3x$ **9.** -12 **11.** 5 **13.** 1 **15.** -1 **17.** 74 **19.** like
21. unlike **23.** like **25.** unlike **27.** $17y$ **29.** $-6a$ **31.** $13b$ **33.** $7k + 15$ **35.** $-4y$ **37.** $2x + 6$ **39.** $14 - 7m$
41. $-17 + x$ **43.** $23x$ **45.** $9y^2$ **47.** $-14p^3 + 5p^2$ **49.** $8x + 15$ **51.** $5x + 15$ **53.** $-4y + 22$ **55.** $-16y + 63$
57. $4r + 15$ **59.** $12k - 5$ **61.** $-2k - 3$ **63.** $4k - 7$ **65.** $-23.7y - 12.6$ **67.** $(x + 3) + 5x$; $6x + 3$ **69.** $(13 + 6x) - (-7x)$;
$13 + 13x$ **71.** $2(3x + 4) - (-4 + 6x)$; 12 **73.** 2, 3, 4, 5 **74.** 1 **75.** (a) 1, 2, 3, 4 (b) 3, 4, 5, 6 (c) 4, 5, 6, 7 **76.** The value
of $x + b$ also increases by 1 unit. **77.** (a) 2, 4, 6, 8 (b) 2, 5, 8, 11 (c) 2, 6, 10, 14 **78.** m **79.** (a) 7, 9, 11, 13 (b) 5, 8, 11, 14
(c) 1, 5, 9, 13; In comparison, we see that while the values themselves are different, the number of units of increase is the same as the
corresponding parts of Exercise 77. **80.** m **81.** Apples and oranges are examples of unlike fruits, just like x and y are unlike terms. We
cannot add x and y to get an expression any simpler than $x + y$; we cannot add, for example, 2 apples and 3 oranges to obtain 5 fruits that are
all alike. **83.** Wording will vary. One example is "the difference between 9 times a number and the sum of the number and 2."

Chapter 1 Review Exercises (page 84)

1. $\dfrac{3}{4}$ **3.** $\dfrac{9}{40}$ **5.** 625 **7.** .0000000032 **9.** 27 **11.** 39 **13.** true **15.** false **17.** $5 + 2 \neq 10$ **19.** 30 **21.** 14

23. $x + 6$ **25.** $6x - 9$ **27.** yes **29.** $2x - 6 = 10$; 8 **31.** **33.** rational numbers, real numbers **35.** -10

37. $-\dfrac{3}{4}$ **39.** true **41.** true **43.** (a) 9 (b) 9 **45.** (a) -6 (b) 6 **47.** 12 **49.** -19 **51.** -6 **53.** -17 **55.** -21.8

57. -10 **59.** -11 **61.** 7 **63.** 10.31 **65.** 2 **67.** $(-31 + 12) + 19$; 0 **69.** $-4 - (-6)$; 2 **71.** -2 **73.** \$26.25

75. $-\$29$ **77.** It gained 4 yd. **79.** 36 **81.** $\dfrac{1}{2}$ **83.** -20 **85.** -24 **87.** 4 **89.** $-\dfrac{3}{4}$ **91.** -1 **93.** 1 **95.** -18

97. 125 **99.** $-4(5) - 9$; -29 **101.** $\dfrac{12}{8 + (-4)}$; 3 **103.** $8x = -24$; -3 **105.** 32 **107.** identity property **109.** inverse property

111. associative property **113.** distributive property **115.** $7(y + 2)$ **117.** $3(2s + 5y)$ **119.** $25 - (5 - 2) = 22$ and
$(25 - 5) - 2 = 18$. Because different groupings lead to different results, we conclude that in general subtraction is not associative.

121. $11m$ **123.** $16p^2 + 2p$ **125.** $-2m + 29$ **127.** C **129.** A **131.** 16 **133.** $\dfrac{8}{3}$ or $2\dfrac{2}{3}$ **135.** 2 **137.** $-\dfrac{3}{2}$ or $-1\dfrac{1}{2}$

139. $-\dfrac{28}{15}$ or $-1\dfrac{13}{15}$ **141.** $8x^2 - 21y^2$ **143.** When dividing 0 by a nonzero number, the quotient will be 0. However, dividing a number by
0 is undefined. **145.** $5(x + 7)$; $5x + 35$

Chapter 1 Test   (page 89)

[1.1] **1.** $\dfrac{7}{11}$ **2.** $\dfrac{241}{120}$ or $2\dfrac{1}{120}$ **3.** $\dfrac{19}{18}$ or $1\dfrac{1}{18}$ **4. (a)** 492 million **(b)** 861 million   [1.2] **5.** true   [1.4] **6.**

7. rational numbers, real numbers   **8.** If -8 and -1 are both graphed on a number line, we see that the point for -8 is to the *left* of the point for -1. This indicates $-8 < -1$.   [1.6] **9.** $\dfrac{-6}{2 + (-8)}; 1$ [1.1, 1.4–1.6] **10.** 4   **11.** $-\dfrac{17}{6}$ or $-2\dfrac{5}{6}$ **12.** 2   **13.** 6   **14.** 108   **15.** 3

16. $\dfrac{30}{7}$ or $4\dfrac{2}{7}$ [1.3, 1.5, 1.6] **17.** 6   **18.** 4   [1.4–1.6] **19.** -70 **20.** 3   **21.** 7000 m   **22.** 15   **23. (a)** $25.1 billion

(b) $-$$11.3 billion **(c)** $5.0 billion **(d)** $-$$2.2 billion   [1.7] **24.** B   **25.** D   **26.** E   **27.** A   **28.** C   **29.** distributive property

30. (a) -18 **(b)** -18 **(c)** The distributive property assures us that the answers must be the same, because $a(b + c) = ab + ac$ for all a, b, c. [1.8] **31.** $21x$ **32.** $15x - 3$

CHAPTER 2 LINEAR EQUATIONS AND INEQUALITIES IN ONE VARIABLE; APPLICATIONS

Section 2.1   (page 98)

Exercises **1.** A and C   **3.** The addition property of equality says that the same number (or expression) added to each side of an equation results in an equivalent equation. Example: $-x$ can be added to each side of $2x + 3 = x - 5$ to get the equivalent equation $x + 3 = -5$.
5. $\{12\}$ **7.** $\{31\}$ **9.** $\{-3\}$ **11.** $\{4\}$ **13.** $\{-9\}$ **15.** $\{-10\}$ **17.** $\{-13\}$ **19.** $\{10\}$ **21.** $\{6.3\}$ **23.** $\{-16.9\}$ **25.** $\{-6\}$
27. $\{-2\}$ **29.** $\{4\}$ **31.** $\{0\}$ **33.** $\{-2\}$ **35.** $\{-7\}$ **37.** A and B; A sample answer might be, "A linear equation in one variable is an

equation that can be written using only one variable term with the variable to the first power."   **39.** $\{13\}$ **41.** $\{-4\}$ **43.** $\{0\}$ **45.** $\left\{\dfrac{7}{15}\right\}$
47. $\{7\}$ **49.** $\{-4\}$ **51.** $\{13\}$ **53.** $\{29\}$ **55.** $\{18\}$ **57.** $\{12\}$ **59.** Answers will vary. One example is $x - 6 = -8$.
61. $3x = 2x + 17; \{17\}$ **63.** $7x - 6x = -9; \{-9\}$

Section 2.2   (page 104)

Exercises **1.** The multiplication property of equality says that the same nonzero number (or expression) multiplied on each side of the

equation results in an equivalent equation. Example: Multiplying each side of $7x = 4$ by $\dfrac{1}{7}$ gives the equivalent equation $x = \dfrac{4}{7}$.   **3.** C

5. To get x alone on the left side, divide each side by 4, the coefficient of x.   **7.** $\dfrac{3}{2}$ **9.** 10   **11.** $-\dfrac{2}{9}$ **13.** -1 **15.** 6   **17.** -4

19. .12   **21.** -1 **23.** $\{6\}$ **25.** $\left\{\dfrac{15}{2}\right\}$ **27.** $\{-5\}$ **29.** $\{-4\}$ **31.** $\left\{-\dfrac{18}{5}\right\}$ **33.** $\{12\}$ **35.** $\{0\}$ **37.** $\{40\}$ **39.** $\{-12.2\}$

41. $\{-48\}$ **43.** $\{72\}$ **45.** $\{-35\}$ **47.** $\{14\}$ **49.** $\{18\}$ **51.** $\left\{-\dfrac{27}{35}\right\}$ **53.** $\{-12\}$ **55.** $\left\{\dfrac{3}{4}\right\}$ **57.** $\{3\}$ **59.** $\{-5\}$ **61.** $\{7\}$

63. $\{0\}$ **65.** $\left\{-\dfrac{3}{5}\right\}$ **67.** Answers will vary. One example is $\dfrac{3}{2}x = -6$.   **69.** $4x = 6; \left\{\dfrac{3}{2}\right\}$ **71.** $\dfrac{x}{-5} = 2; \{-10\}$

Section 2.3   (page 112)

Exercises **1.** *Step 1:* Clear parentheses and combine like terms, as needed. *Step 2:* Use the addition property to get all variable terms on one side of the equation and all numbers on the other. Then combine like terms. *Step 3:* Use the multiplication property to get the equation in the

form $x = $ a number. *Step 4:* Check the solution. Examples will vary.   **3.** D   **5.** $\{-1\}$ **7.** $\{5\}$ **9.** $\left\{-\dfrac{5}{3}\right\}$ **11.** $\left\{\dfrac{4}{3}\right\}$ **13.** $\{5\}$

15. ∅   **17.** {all real numbers}   **19.** {1}   **21.** ∅   **23.** {5}   **25.** {0}   **27.** $\left\{-\dfrac{7}{5}\right\}$ **29.** {120}   **31.** {6}   **33.** {15,000}   **35.** {8}

37. {0}   **39.** {4}   **41.** {20}   **43.** {all real numbers}   **45.** ∅   **47.** $11 - q$ **49.** $x + 7$ **51.** $a + 12; a - 5$ **53.** $\dfrac{t}{5}$

Summary Exercises on Solving Linear Equations   (page 114)

1. {−5}   **2.** {4}   **3.** {−5.1}   **4.** {12}   **5.** {−25}   **6.** {−6}   **7.** {−3}   **8.** {−16}   **9.** {7}   **10.** $\left\{-\dfrac{96}{5}\right\}$ **11.** {5}   **12.** {23.7}

13. {all real numbers}   **14.** {1}   **15.** {−6}   **16.** ∅   **17.** {6}   **18.** {3}   **19.** ∅   **20.** $\left\{\dfrac{7}{3}\right\}$ **21.** {25}   **22.** {−10.8}   **23.** {3}

24. {7}   **25.** {2}   **26.** {all real numbers}   **27.** {−2}   **28.** {70}   **29.** $\left\{\dfrac{14}{17}\right\}$ **30.** $\left\{-\dfrac{5}{2}\right\}$

Section 2.4   (page 122)

Connections   (**page 122**)   Polya's Step 1 corresponds to our Steps 1 and 2. Polya's Step 2 corresponds to our Step 3. Polya's Step 3 corresponds to our Steps 4 and 5. Polya's Step 4 corresponds to our Step 6. Trial and error or guessing and checking fit into Polya's Step 2, devising a plan.

Exercises   **1.** The procedure should include the following steps: read the problem carefully; assign a variable to represent the unknown to be found, and write down variable expressions for any other unknown quantities; translate into an equation; solve the equation; state the answer; check your solution.   **3.** D; there cannot be a fractional number of cars.   **5.** 3   **7.** 6   **9.** −3   **11.** California: 59 screens; New York: 48 screens   **13.** Democrats: 45; Republicans: 55   **15.** U2: $109.7 million; 'N Sync: $86.8 million   **17.** wins: 61; losses: 21   **19.** 1950 Denver nickel: $14.00; 1945 Philadelphia nickel: $12.00   **21.** ice cream: 44,687.9 lb; topping: 537.1 lb   **23.** 18 prescriptions   **25.** peanuts: $22\dfrac{1}{2}$ oz; cashews: $4\dfrac{1}{2}$ oz   **27.** Airborne Express: 3; Federal Express: 9; United Parcel Service: 1   **29.** gold: 10; silver: 13; bronze: 11   **31.** 36 million mi   **33.** A and B: 40°; C: 100°   **35.** $k - m$ **37.** no   **39.** $x - 1$ **41.** 18°   **43.** 39°   **45.** 50°   **47.** 68, 69   **49.** 10, 12   **51.** 101, 102   **53.** 10, 11   **55.** 18   **57.** 15, 17, 19   **59.** $2.78 billion; $3.33 billion; $3.53 billion

Section 2.5   (page 133)

Exercises   **1.** **(a)** The perimeter of a plane geometric figure is the distance around the figure.   **(b)** The area of a plane geometric figure is the measure of the surface covered or enclosed by the figure.   **3.** four   **5.** area   **7.** perimeter   **9.** area   **11.** area   **13.** $P = 26$ **15.** $A = 64$ **17.** $b = 4$ **19.** $t = 5.6$ **21.** $I = 1575$ **23.** $B = 14$ **25.** $r = 2.6$ **27.** $A = 50.24$ **29.** $V = 150$ **31.** $V = 52$ **33.** $V = 7234.56$ **35.** about 154,000 ft²   **37.** perimeter: 13 in., area: 10.5 in.²   **39.** 132.665 ft²   **41.** 23,800.10 ft²

43. length: 36 in.; volume: 11,664 in.³   **45.** 48°, 132°   **47.** 51°, 51°   **49.** 105°, 105°   **51.** $t = \dfrac{d}{r}$ **53.** $b = \dfrac{A}{h}$ **55.** $d = \dfrac{C}{\pi}$

57. $H = \dfrac{V}{LW}$ **59.** $r = \dfrac{I}{pt}$ **61.** $h = \dfrac{2A}{b}$ **63.** $h = \dfrac{3V}{\pi r^2}$ **65.** $b = P - a - c$ **67.** $W = \dfrac{P - 2L}{2}$ or $W = \dfrac{P}{2} - L$ **69.** $m = \dfrac{y - b}{x}$

71. $y = \dfrac{C - Ax}{B}$ **73.** $r = \dfrac{M - C}{C}$

Section 2.6   (page 141)

Exercises   **1.** **(a)** C **(b)** D **(c)** B **(d)** A   **3.** $\dfrac{6}{7}$ **5.** $\dfrac{18}{55}$ **7.** $\dfrac{5}{16}$ **9.** $\dfrac{4}{15}$ **11.** $\dfrac{3}{1}$ **13.** 17-oz size   **15.** 64-oz can   **17.** 500-count

19. 28-oz size   **21.** A ratio is a comparison, while a proportion is a statement that two ratios are equal. For example, $\dfrac{2}{3}$ is a ratio and $\dfrac{2}{3} = \dfrac{8}{12}$ is a proportion.   **23.** true   **25.** false   **27.** true   **29.** {35}   **31.** {7}   **33.** $\left\{\dfrac{45}{2}\right\}$ **35.** {2}   **37.** {−1}   **39.** {5}

41. $\left\{-\dfrac{31}{5}\right\}$ **43.** $67.50   **45.** $28.35   **47.** 4 ft   **49.** 6.875 fluid oz   **51.** $670.48   **53.** 50,000 fish   **55. (a)** $\dfrac{26}{100} = \dfrac{x}{350}$;

$91 million **(b)** $112 million; $11.2 million **(c)** $119 million   **57.** 4   **59.** 1   **61. (a)**   **(b)** 54 ft   **63.** $270

65. $287   **67.** 30   **68. (a)** $5x = 12$ **(b)** $\left\{\dfrac{12}{5}\right\}$ **69.** $\left\{\dfrac{12}{5}\right\}$ **70.** Both methods give the same solution set.

Section 2.7   (page 152)

Exercises   **1.** 35 mL   **3.** $350   **5.** $14.15   **7.** C   **9. (a)** 375,000 **(b)** 575,000 **(c)** 275,000   **11.** 4.5%   **13.** D   **15.** 160 gal

17. $53\dfrac{1}{3}$ kg   **19.** 4 L   **21.** $13\dfrac{1}{3}$ L   **23.** 25 mL   **25.** $5000 at 3%; $1000 at 5%   **27.** $40,000 at 3%; $110,000 at 4%

29. 25 fives   **31.** fives: 84; tens: 42   **33.** 20 lb   **35.** A   **37.** 530 mi   **39.** 4.059 hr   **41.** 7.91 m per sec   **43.** 8.42 m per sec

45. 10 hr   **47.** $2\dfrac{1}{2}$ hr   **49.** 5 hr   **51.** northbound: 60 mph; southbound: 80 mph   **53.** 50 km per hr; 65 km per hr

Section 2.8   (page 168)

Connections   **(page 168)**   The revenue is represented by $5x - 100$. The production cost is $125 + 4x$. The profit is represented by $R - C = (5x - 100) - (125 + 4x) = x - 225$. The solution of $x - 225 > 0$ is $x > 225$. To make a profit, more than 225 cassettes must be produced and sold.

Exercises   **1.** Use a parenthesis if the symbol is $<$ or $>$. Use a square bracket if the symbol is $\le$ or $\ge$.   **3.** $x > -4$ **5.** $x \le 4$

7. $(-\infty, 4]$ **9.** $(-\infty, -3)$ **11.** $(4, \infty)$ **13.** $[8, 10]$

15. $(0, 10]$ **17.** It would imply that $3 < -2$, a false statement.   **19.** $[1, \infty)$

21. $[5, \infty)$ **23.** $(-\infty, -11)$ **25.** It must be reversed when multiplying or dividing by a

negative number.   **27.** $(-\infty, 6)$ **29.** $[-10, \infty)$ **31.** $(-\infty, -3)$

33. $(-\infty, 0]$ **35.** $(20, \infty)$ **37.** $[-3, \infty)$

39. $[-5, \infty)$ **41.** $(-\infty, 1)$ **43.** $(-\infty, 0]$

45. $[4, \infty)$ **47.** $(-\infty, 32)$ **49.** $\left(-\infty, \dfrac{76}{11}\right)$

51. $\left[\dfrac{5}{12}, \infty\right)$ **53.** $(-21, \infty)$ **55.** $-1 < x < 2$ **57.** $-1 < x \le 2$

59. $[-1, 6]$ **61.** $\left(-\dfrac{11}{6}, -\dfrac{2}{3}\right)$ **63.** $(1, 3)$

65. $[-26, 6]$ **67.** $[-3, 6]$ **69.** $\left[-\dfrac{24}{5}, 0\right]$

71. {4} **72.** $(4, \infty)$ **73.** $(-\infty, 4)$ **74.** The graph would be all real

numbers. **75.** The graph would be all real numbers. **76.** If a point on the number line satisfies an equation, the points on one side of that point will satisfy the corresponding less-than inequality, and the points on the other side will satisfy the corresponding greater-than inequality. **77.** 83 or greater **79.** all numbers greater than 16 **81.** It is never more than 86° Fahrenheit. **83.** 32 or greater **85.** 15 min **87.** 9.5 gal **89.** $x \geq 500$ **91. (a)** 140 to 184 lb **(b)** Answers will vary. **93.** from about 2:30 P.M. to 6:00 P.M. **95.** about 84°F–91°F

Chapter 2 Review Exercises (page 178)

1. {6} **3.** {7} **5.** {11} **7.** {5} **9.** {5} **11.** $\left\{\dfrac{64}{5}\right\}$ **13.** {all real numbers} **15.** {all real numbers} **17.** $\emptyset$

19. Democrats: 75; Republicans: 45 **21.** Seven Falls: 300 ft; Twin Falls: 120 ft **23.** 11, 13 **25.** $A = 28$ **27.** $V = 904.32$

29. $h = \dfrac{2A}{b + B}$ **31.** 100°; 100° **33.** diameter: approximately 19.9 ft; radius: approximately 9.95 ft; area: approximately 311 ft^2

35. Not enough information is given. We also need the value of B. **37.** $\dfrac{5}{14}$ **39.** $\dfrac{1}{12}$ **41.** $\left\{-\dfrac{8}{3}\right\}$ **43.** $6\dfrac{2}{3}$ lb **45.** 375 km

47. 25.5-oz size **49.** 3.75 L **51.** 8.2 mph **53.** $2\dfrac{1}{2}$ hr **55.** $(-\infty, 7)$ **57.** $[-3, \infty)$

59. $[3, \infty)$ **61.** $(-\infty, -5)$ **63.** $\left[-2, \dfrac{3}{2}\right]$ **65.** 88 or more

67. {7} **69.** $[-3, 3]$ **71.** {70} **73.** $\emptyset$ **75.** Since $-(8 + 4x) = -1(8 + 4x)$, the first step is to distribute -1 over *both* terms in the parentheses, to get $3 - 8 - 4x$ on the left side of the equation. The student got $3 - 8 + 4x$ instead. The correct solution set is $\{-2\}$. **77.** Golden Gate Bridge: 4200 ft; Brooklyn Bridge: 1595 ft **79.** 8 qt **81.** 44 m

Chapter 2 Test (page 182)

[2.1–2.3] **1.** $\{-6\}$ **2.** {21} **3.** $\emptyset$ **4.** {30} **5.** {all real numbers} [2.4] **6.** wins: 116; losses: 46 **7.** Hawaii: 4021 mi^2; Maui: 728 mi^2; Kauai: 551 mi^2 **8.** 50° [2.5] **9. (a)** $W = \dfrac{P - 2L}{2}$ or $W = \dfrac{P}{2} - L$ **(b)** 18 **10.** 75°, 75° [2.6] **11.** {6} **12.** $\{-29\}$ **13.** 8 slices for $2.19 **14.** 2300 mi [2.7] **15.** $8000 at 3%; $14,000 at 4.5% **16.** 4 hr [2.8] **17.** $(-\infty, 4]$

18. $(-2, 6]$ **19.** 83 or more **20.** When an inequality is multiplied or divided by a negative number, the direction of the inequality symbol must be reversed.

Cumulative Review Exercises Chapters 1–2 (page 184)

[1.1] **1.** $\dfrac{3}{4}$ **2.** $\dfrac{37}{60}$ **3.** $\dfrac{48}{5}$ [1.2] **4.** $\dfrac{1}{2}x - 18$ **5.** $\dfrac{6}{x + 12} = 2$ [1.4] **6.** true [1.5–1.6] **7.** 11 **8.** -8 **9.** 28

[1.3] **10.** $-\dfrac{19}{3}$ [1.7] **11.** distributive property **12.** commutative property [1.8] **13.** $2k - 11$ [2.1–2.3] **14.** $\{-1\}$ **15.** $\{-1\}$

16. $\{-12\}$ [2.6] **17.** {26} [2.5] **18.** $y = \dfrac{24 - 3x}{4}$ **19.** $n = \dfrac{A - P}{iP}$ [2.8] **20.** $(-\infty, 1]$

21. $(-1, 2]$ [2.4] **22.** 4 cm; 9 cm; 27 cm [2.5] **23.** 12.42 cm [2.6] **24.** $\dfrac{25}{6}$ or $4\dfrac{1}{6}$ cups

[2.7] **25.** 40 mph; 60 mph

Section 3.1   (page 196)

Exercises   1. Ohio (OH): about 680 million eggs; Iowa (IA): about 550 million eggs   **3.** Indiana (IN) and Pennsylvania (PA); about 490 million eggs each   **5.** from 1975 to 1980; about $.75   **7.** The price of a gallon of gas was decreasing.   **9.** does; do not   **11.** II   **13.** 3   **15.** A linear equation in one variable can be written in the form $Ax + B = C$, where $A \neq 0$. Examples are $2x + 5 = 0$, $3x + 6 = 2$, and $x = -5$. A linear equation in two variables can be written in the form $Ax + By = C$, where A and B cannot both equal 0. Examples are $2x + 3y = 8$, $3x = 5y$, and $x - y = 0$.   **17.** yes   **19.** yes   **21.** no   **23.** yes   **25.** yes   **27.** no   **29.** No, the ordered pair $(3, 4)$ represents the point 3 units to the right of the origin and 4 units up from the x-axis. The ordered pair $(4, 3)$ represents the point 4 units to the right of the origin and 3 units up from the x-axis.   **31.** 17   **33.** -5 **35.** -1 **37.** -7 **39.** b **41.** 8; 6; 3   **43.** -9; 4; 9   **45.** 12; 12; 12   **47.** -10; -10; -10 **49.–55.**   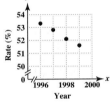   **57.** negative; negative   **59.** positive; negative

61. -3; 6; -2; 4   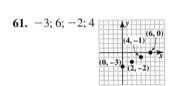   **63.** -3; 4; -6; $-\dfrac{4}{3}$ 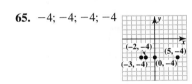   **65.** -4; -4; -4; -4 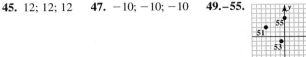   **67.** The points in each graph

appear to lie on a straight line.   **69. (a)** $(5, 45)$ **(b)** $(6, 50)$ **71. (a)** $(1996, 53.3)$, $(1997, 52.8)$, $(1998, 52.1)$, $(1999, 51.6)$ **(b)** $(1995, 54.0)$ means that in 1995, the graduation rate for 4-yr college students within 5 yr was 54.0%. **(c)**   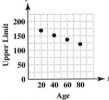   **(d)** The points appear to lie on

a straight line. Graduation rates for 4-yr college students within 5 yr are decreasing.   **73. (a)** 170; 154; 138; 122 **(b)** $(20, 170)$, $(40, 154)$, $(60, 138)$, $(80, 122)$ **(c)**   The points lie in a linear pattern.

Section 3.2   (page 208)

Connections   (page 206)   1. $3x + 4 - 2x - 7 - 4x - 3 = 0$ **2.** $5x - 15 - 3(x - 2) = 0$

Exercises   1. 5; 5; 3   **3.** 1; 3; -1 **5.** -6; -2; -5 **7.** C   **9.** D   **11.** B

13. $(12, 0); (0, -8)$ **15.** $(0, 0); (0, 0)$ **17.** $(4, 0); (0, -10)$ **19.** $(4, 0)$; none **21.** $y = 0; x = 0$ **23.**

25. **27.** **29.** **31.** **33.** **35.**

37. **39.** Find two ordered pairs that satisfy the equation. Plot the corresponding points on a coordinate system. Draw a

straight line through the two points. As a check, find a third ordered pair and verify that it lies on the line you drew. **41.** 8; yes **43.** 7; yes
45. **(a)** 151.5 cm, 174.9 cm, 159.3 cm **(b)** **(c)** 24 cm; 24 cm **47.** **(a)** 130 **(b)** 133 **(c)** They are quite close.

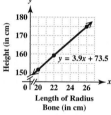

49. between 133 and 162 **51.** **(a)** 1993: 49.8 gal; 1995: 51.4 gal; 1997: 53 gal **(b)** 1993: 50.1 gal; 1995: 51.6 gal; 1997: 53 gal
(c) Corresponding values are quite close. **53.** **(a)** $30,000 **(b)** $15,000 **(c)** $5000 **(d)** After 5 yr, the SUV has a value of $5000.

Section 3.3 (page 219)

Exercises **1.** Rise is the vertical change between two different points on a line. Run is the horizontal change between two different points on
a line. **3.** 4 **5.** $-\dfrac{1}{2}$ **7.** 0 **9.** Yes; it doesn't matter which point you start with. Both differences will be the negatives of the
differences in Exercise 8, and the quotient will be the same. **In Exercises 11 and 13, sketches will vary.** **11.** The line must rise from left
to right. **13.** The line must be horizontal. **15.** Because he found the difference $3 - 5 = -2$ in the numerator, he should have subtracted
in the same order in the denominator to get $-1 - 2 = -3$. The correct slope is $\dfrac{-2}{-3} = \dfrac{2}{3}$. **17.** $\dfrac{5}{4}$ **19.** $\dfrac{3}{2}$ **21.** -3 **23.** 0

25. undefined **27.** $-\dfrac{1}{2}$ **29.** 5 **31.** $\dfrac{1}{4}$ **33.** $\dfrac{3}{2}$ **35.** $\dfrac{3}{2}$ **37.** 0 **39.** undefined **41.** $-3; \dfrac{1}{3}$ **43.** **(a)** negative **(b)** zero

45. **(a)** positive **(b)** negative **47.** **(a)** zero **(b)** negative **49.** A **51.** $-\dfrac{2}{5}; -\dfrac{2}{5}$; parallel **53.** $\dfrac{8}{9}; -\dfrac{4}{3}$; neither **55.** $\dfrac{3}{2}; -\dfrac{2}{3}$;

perpendicular **57.** $5; \dfrac{1}{5}$; neither **59.** $\dfrac{3}{10}$ **61.** 232 thousand or 232,000 **62.** positive; increased **63.** 232,000 students **64.** -1.66

65. negative; decreased **66.** 1.66 students per computer **67.** The change for each year is .1 billion (or 100,000,000) ft², so the graph is a
straight line. **69.** $\dfrac{2}{5}$ **71.** $(0, 4)$

Section 3.4 (page 228)

Exercises **1.** D **3.** B **5.** The slope m of a vertical line is undefined, so it is not possible to write the equation of the line in the form $y = mx + b$. **7.** $y = 3x - 3$ **9.** $y = -x + 3$ **11.** $y = 4x - 3$ **13.** $y = 3$ **15.** $x = 0$ **17.**

19.   **21.**   **23.**   **25.**   **27.** the x-axis **29.** $y = -4x - 1$

31. $y = \dfrac{2}{3}x + \dfrac{19}{3}$ **33.** $y = \dfrac{3}{4}x + 4$ **35.** $y = \dfrac{5}{2}x - 4$ **37.** $y = x$ (There are other forms as well.) **39.** $y = x + 6$

41. $y = -\dfrac{3}{5}x - \dfrac{11}{5}$ **43.** $y = \dfrac{1}{2}x + 2$ **45.** $y = -\dfrac{1}{3}x + \dfrac{22}{9}$ **47.** $(0, 32), (100, 212)$ **48.** $\dfrac{9}{5}$ **49.** $F - 32 = \dfrac{9}{5}(C - 0)$

50. $F = \dfrac{9}{5}C + 32$ **51.** $C = \dfrac{5}{9}(F - 32)$ **52.** $86°$ **53.** $10°$ **54.** $-40°$ **55.** $y = \dfrac{3}{4}x - \dfrac{9}{2}$ **57.** $y = -2x - 3$

59. (a) \$400 (b) \$.25 (c) $y = .25x + 400$ (d) \$425 (e) 1500 **61.** $y = -3x + 6$ **63.** $Y_1 = \dfrac{3}{4}X + 1$

Chapter 3 Review Exercises (page 236)

1. (a) \$1.05 per gal; \$1.75 per gal (b) \$.70 per gal; about 67% (c) between April and June 2000; about \$.40 per gal (d) August–October 1999 and February–April 2000 **3.** $2; \dfrac{3}{2}; \dfrac{14}{3}$ **5.** 7; 7; 7 **7.** no **9.** I Exercises 9 and 11 **11.** none **13.** I or III

15. $\left(\dfrac{8}{3}, 0\right); (0, 4)$ **17.** $-\dfrac{1}{2}$ **19.** 3 **21.** $\dfrac{3}{2}$ **23.** $\dfrac{3}{2}$ **25.** parallel **27.** neither **29.** $y = -\dfrac{1}{2}x + 4$

31. $y = \dfrac{2}{3}x + \dfrac{14}{3}$ **33.** $y = -\dfrac{1}{4}x + \dfrac{3}{2}$ **35.** $x = \dfrac{1}{3}$; It is not possible to express this equation in the form $y = mx + b$.

37. C, D **39.** D **41.** B **43.** $(0, 0); (0, 0); -\dfrac{1}{3}$ **45.** $y = -\dfrac{1}{4}x - \dfrac{5}{4}$ **47.** $y = -\dfrac{4}{7}x - \dfrac{23}{7}$

48. about \$1.5 billion **49.** It will have negative slope since the total spent on video rentals is decreasing over these years.
50. $(1996, 11.1), (2000, 9.6); -.375$ **51.** $-.375$; Yes, the slope is negative as stated in the answer to Exercise 49 and the same as the slope calculated in Exercise 50. **52.** 10.7; 10.4; 10.0 **53.** The actual amounts are fairly close to those given by the equation. **54.** \$8.9 billion

Chapter 3 Test   (page 239)

[3.1] **1.** $-6, -10, -5$ **2.** no   [3.2] **3.** To find the x-intercept, let $y = 0$, and to find the y-intercept, let $x = 0$.   **4.** x-intercept: $(2, 0)$;
y-intercept: $(0, 6)$ **5.** x-intercept: $(0, 0)$; y-intercept: $(0, 0)$ **6.** x-intercept: $(-3, 0)$; y-intercept: none

7. x-intercept: none; y-intercept: $(0, 1)$ **8.** x-intercept: $(4, 0)$; y-intercept: $(0, -4)$

[3.3] **9.** $-\dfrac{8}{3}$ **10.** -2 **11.** undefined   **12.** $\dfrac{5}{2}$ **13.** 0 [3.4] **14.** $y = 2x + 6$ **15.** $y = \dfrac{5}{2}x - 4$ **16.** $y = -9x + 12$

[3.1–3.3] **17.** The slope is positive since food and drink sales are increasing.   **18.** $(0, 43)$, $(30, 376)$; 11.1 **19.** 1990: $265 billion;
1995: $320.5 billion   **20.** In 2000, food and drink sales were $376 billion.

Cumulative Review Exercises Chapters 1–3   (page 240)

[1.1] **1.** $\dfrac{301}{40}$ or $7\dfrac{21}{40}$ **2.** 6 [1.5] **3.** 7 [1.6] **4.** $\dfrac{73}{18}$ or $4\dfrac{1}{18}$ [1.2–1.6] **5.** true   **6.** -43 [1.7] **7.** distributive property

[1.8] **8.** $-p + 2$ [2.5] **9.** $h = \dfrac{3V}{\pi r^2}$ [2.3] **10.** $\{-1\}$ **11.** $\{2\}$ [2.6] **12.** $\{-13\}$ [2.8] **13.** $(-2.6, \infty)$

14. $(0, \infty)$ ———————⟶   **15.** $(-\infty, -4]$ ⟵———————   [2.4, 2.5] **16.** high school diploma: $22,895; bachelor's

degree: $40,478   **17.** 13 mi   [3.1] **18. (a)** 89.45; 81.95; 78.20 **(b)** In 1980, the winning time was 85.7 sec.

19. (a) $7000 **(b)** $10,000 **(c)** about $30,000   [3.2] **20.** $(-4, 0)$; $(0, 3)$ [3.3] **21.** $\dfrac{3}{4}$ [3.2] **22.**

[3.3] **23.** perpendicular   [3.4] **24.** $y = 3x - 11$ **25.** $y = 4$

CHAPTER 4 EXPONENTS AND POLYNOMIALS

Section 4.1   (page 248)

Exercises   1. false   **3.** false   **5.** w^6 **7.** $\dfrac{1}{4^4}$ **9.** $(-7x)^4$ **11.** $\left(\dfrac{1}{2}\right)^6$ **13.** In $(-3)^4$, -3 is the base, while in -3^4, 3 is the base.

$(-3)^4 = 81$, while $-3^4 = -81$.   **15.** base: 3; exponent: 5; 243   **17.** base: -3; exponent: 5; -243 **19.** base: $-6x$; exponent: 4

21. base: x; exponent: 4   **23.** $5^2 + 5^3$ is a sum, not a product. $5^2 + 5^3 = 25 + 125 = 150$.   **25.** 5^8 **27.** 4^{12} **29.** $(-7)^9$ **31.** t^{24}

33. $-56r^7$ **35.** $42p^{10}$ **37.** $-30x^9$ **39.** The product rule does not apply.   **41.** The product rule does not apply.   **43.** 4^6 **45.** t^{20}

47. $7^3 r^3$ **49.** $5^5 x^5 y^5$ **51.** 5^{12} **53.** -8^{15} **55.** $8q^3 r^3$ **57.** $\dfrac{1}{2^3}$ **59.** $\dfrac{a^3}{b^3}$ **61.** $\dfrac{9^8}{5^8}$ **63.** $\dfrac{5^5}{2^5}$ **65.** $\dfrac{9^5}{8^3}$ **67.** $2^{12} x^{12}$ **69.** $-6^5 p^5$

71. $6^5 x^{10} y^{15}$ **73.** x^{21} **75.** $2^2 w^4 x^{26} y^7$ **77.** $-r^{18} s^{17}$ **79.** $\dfrac{5^3 a^6 b^{15}}{c^{18}}$ **81.** This is incorrect. Using the product rule, it is simplified as follows: $(10^2)^3 = 10^{2\cdot3} = 10^6 = 1{,}000{,}000$.   **83.** $12x^5$ **85.** $6p^7$ **87.** $125x^6$ **89.** $-a^4$, $-a^3$, $-(-a)^3$, $(-a)^4$; One way is to choose a positive number greater than 1 and substitute it for a in each expression. Then arrange the terms from smallest to largest.   **91.** \$304.16   **93.** \$1843.88

Section 4.2   (page 257)

Exercises   **1.** 1   **3.** 1   **5.** -1 **7.** 0   **9.** 0   **11.** C   **13.** F   **15.** E   **17.** 2   **19.** $\dfrac{1}{64}$ **21.** 16   **23.** $\dfrac{49}{36}$ **25.** $\dfrac{1}{81}$ **27.** $\dfrac{8}{15}$

29. 5^3 **31.** $\dfrac{5^3}{3^2}$ or $\dfrac{125}{9}$ **33.** 5^2 **35.** x^{15} **37.** 6^3 **39.** $2r^4$ **41.** $\dfrac{5^2}{4^3}$ **43.** $\dfrac{p^5}{q^8}$ **45.** r^9 **47.** $\dfrac{yz^2}{4x^3}$ **49.** $a+b$ **51.** $(x+2y)^2$

53. 1   **54.** $\dfrac{5^2}{5^2}$ **55.** 5^0 **56.** $1 = 5^0$; This supports the definition of 0 as an exponent.   **57.** 7^3 or 343   **59.** $\dfrac{1}{x^2}$ **61.** $\dfrac{64x}{9}$ **63.** $\dfrac{x^2z^4}{y^2}$

65. $6x$ **67.** $\dfrac{1}{m^{10}n^5}$ **69.** $\dfrac{1}{xyz}$ **71.** x^3y^9 **73.** $\dfrac{a^{11}}{2b^5}$ **75.** $\dfrac{108}{y^5z^3}$ **77.** $\dfrac{9z^2}{400x^3}$ **79.** The student attempted to use the quotient rule with

unequal bases. The correct way to simplify this expression is $\dfrac{16^3}{2^2} = \dfrac{(2^4)^3}{2^2} = \dfrac{2^{12}}{2^2} = 2^{10} = 1024$.

Summary Exercises on the Rules for Exponents   (page 259)

1. $\dfrac{6^{12}x^{24}}{5^{12}}$ **2.** $\dfrac{r^6s^{12}}{729t^6}$ **3.** $10^5x^7y^{14}$ **4.** $-128a^{10}b^{15}c^4$ **5.** $\dfrac{729w^3x^9}{y^{12}}$ **6.** $\dfrac{x^4y^6}{16}$ **7.** c^{22} **8.** $\dfrac{1}{k^4t^{12}}$ **9.** $\dfrac{11}{30}$ **10.** $y^{12}z^3$ **11.** $\dfrac{x^6}{y^5}$

12. 0   **13.** $\dfrac{1}{z^2}$ **14.** $\dfrac{9}{r^2s^2t^{10}}$ **15.** $\dfrac{300x^3}{y^3}$ **16.** $\dfrac{3}{5x^6}$ **17.** x^8 **18.** $\dfrac{y^{11}}{x^{11}}$ **19.** $\dfrac{a^6}{b^4}$ **20.** $6ab$ **21.** $\dfrac{61}{900}$ **22.** 1   **23.** $\dfrac{343a^6b^9}{8}$

24. 1   **25.** -1 **26.** 0   **27.** $\dfrac{27y^{18}}{4x^8}$ **28.** $\dfrac{1}{a^8b^{12}c^{16}}$ **29.** $\dfrac{x^{15}}{216z^9}$ **30.** $\dfrac{q}{8p^6r^3}$ **31.** x^6y^6 **32.** 0   **33.** $\dfrac{343}{x^{15}}$ **34.** $\dfrac{9}{x^6}$ **35.** $5p^{10}q^9$

36. $\dfrac{7}{24}$ **37.** $\dfrac{r^{14}t}{2s^2}$ **38.** 1   **39.** $8p^{10}q$ **40.** $\dfrac{1}{mn^3p^3}$ **41.** -1 **42.** $\dfrac{3}{40}$

Section 4.3   (page 264)

Connections   (**page 263**)   **1.** The Erzincan earthquake was ten times as powerful as the Sumatra earthquake.   **2.** The Adana earthquake had one hundredth the power of the San Francisco earthquake.   **3.** The Erzincan earthquake was about 50.12 times as powerful as the Adana earthquake.   **4.** "+3" corresponds to a factor of 1000 times stronger; "−1" corresponds to a factor of one-tenth.

Exercises   **1.** A   **3.** C   **5.** in scientific notation   **7.** not in scientific notation; 5.6×10^6 **9.** not in scientific notation; 8×10^1 **11.** not in scientific notation; 4×10^{-3} **13.** It is written as the product of a power of 10 and a number whose absolute value is between 1 and 10 (inclusive of 1). Some examples are 2.3×10^{-4} and 6.02×10^{23}.   **15.** 5.876×10^9 **17.** 8.235×10^4 **19.** 7×10^{-6} **21.** 2.03×10^{-3} **23.** 750,000   **25.** 5,677,000,000,000   **27.** 6.21   **29.** .00078   **31.** .000000005134   **33.** 600,000,000,000   **35.** 15,000,000   **37.** 60,000   **39.** .0003   **41.** 40   **43.** .000013   **45.** 4.7E−7   **47.** 2E7   **49.** 1E1   **51.** 1×10^{10} **53.** 2,000,000,000   **55.** $\$6.03 \times 10^8$ **57.** $\$5.2466 \times 10^{10}$ **59.** about 15,300 sec   **61.** about 9.2×10^8 acres   **63.** about 3.3   **65.** about .276 lb

Section 4.4   (page 274)

Connections   (**page 274**)   The polynomial gives the following: for 1996, 5730 million (greater than actual); for 1997, 5594 million (less than actual); for 1999, 5260 million (less than actual); for 2000, 5062 million (less than actual); for 2001, 4844 million (greater than actual).

Exercises   **1.** 7; 5   **3.** 8   **5.** 26   **7.** 0   **9.** 1; 6   **11.** 1; 1   **13.** 2; -19, -1 **15.** 3; 1, 8, 5   **17.** $2m^5$ **19.** $-r^5$ **21.** cannot be simplified   **23.** $-5x^5$ **25.** $5p^9 + 4p^7$ **27.** $-2xy^2$ **29.** already simplified; 4; binomial   **31.** $11m^4 - 7m^3$ trinomial   **33.** x^4; 4; monomial   **35.** 7; 0; monomial   **37.** (**a**) 36 (**b**) -12 **39.** (**a**) 14 (**b**) -19 **41.** (**a**) -3 (**b**) 0

45. $5m^2 + 3m + 2$ **47.** $\dfrac{7}{6}x^2 - \dfrac{2}{15}x + \dfrac{5}{6}$ **49.** $6m^3 + m^2 + 12m - 14$ **51.** $3y^3 - 11y^2$ **53.** $4x^4 - 4x^2 + 4x$

55. $15m^3 - 13m^2 + 8m + 11$ **57.** Answers will vary. **59.** $5m^2 - 14m + 6$ **61.** $4x^3 + 2x^2 + 5x$ **63.** $-11y^4 + 8y^2 + y$

65. $a^4 - a^2 + 1$ **67.** $5m^2 + 8m - 10$ **69.** $-6x^2 - 12x + 12$ **71.** -10 **73.** $4b - 5c$ **75.** $6x - xy - 7$

77. $-3x^2y - 15xy - 3xy^2$ **79.** $8x^2 + 8x + 6$ **81.** (a) $23y + 5t$ (b) approximately 37.26°, 79.52°, and 63.22° **83.** $-6x^2 + 6x - 7$

85. $-7x - 1$ **87.** $0, -3, -4, -3, 0$

89. $7, 1, -1, 1, 7$

91. $0, 3, 4, 3, 0$

93. $4, 1, 0, 1, 4$

95. 175 ft **97.** 4; \$5.00 **98.** 6; \$27 **99.** 2.5; 130 **100.** 459.87 billion

Section 4.5 (page 283)

Exercises **1.** (a) B (b) D (c) A (d) C **3.** $15y^{11}$ **5.** $30a^9$ **7.** $15pq^2$ **9.** $-18m^3n^2$ **11.** $6m^2 + 4m$ **13.** $-6p^4 + 12p^3$

15. $-16z^2 - 24z^3 - 24z^4$ **17.** $6y^3 + 4y^4 + 10y^7$ **19.** $28r^5 - 32r^4 + 36r^3$ **21.** $6a^4 - 12a^3b + 15a^2b^2$

23. $21m^5n^2 + 14m^4n^3 - 7m^3n^5$ **25.** $12x^3 + 26x^2 + 10x + 1$ **27.** $81a^3 + 27a^2 + 11a + 2$ **29.** $20m^4 - m^3 - 8m^2 - 17m - 15$

31. $6x^6 - 3x^5 - 4x^4 + 4x^3 - 5x^2 + 8x - 3$ **33.** $5x^4 - 13x^3 + 20x^2 + 7x + 5$ **35.** $3x^5 + 18x^4 - 2x^3 - 8x^2 + 24x$

37. $m^2 + 12m + 35$ **39.** $x^2 - 25$ **41.** $12x^2 + 10x - 12$ **43.** $9x^2 - 12x + 4$ **45.** $10a^2 + 37a + 7$ **47.** $12 + 8m - 15m^2$

49. $20 - 7x - 3x^2$ **51.** $8xy - 4x + 6y - 3$ **53.** $15x^2 + xy - 6y^2$ **55.** $6y^5 - 21y^4 - 45y^3$ **57.** $-200r^7 + 32r^3$

59. $3y^2 + 10y + 7$ **61.** $6p^2 - \dfrac{5}{2}pq - \dfrac{25}{12}q^2$ **63.** $x^2 + 14x + 49$ **65.** $a^2 - 16$ **67.** $4p^2 - 20p + 25$ **69.** $25k^2 + 30kq + 9q^2$

71. $m^3 - 15m^2 + 75m - 125$ **73.** $8a^3 + 12a^2 + 6a + 1$ **75.** $56m^2 - 14m - 21$ **77.** $-9a^3 + 33a^2 + 12a$

79. $81r^4 - 216r^3s + 216r^2s^2 - 96rs^3 + 16s^4$ **81.** $6p^8 + 15p^7 + 12p^6 + 36p^5 + 15p^4$ **83.** $-24x^8 - 28x^7 + 32x^6 + 20x^5$

85. $14x + 49$ **87.** $\pi x^2 - 9$ **89.** $30x + 60$ **90.** $30x + 60 = 600$ **91.** 18 **92.** 10 yd by 60 yd **93.** \$2100 **94.** 140 yd

95. \$1260 **96.** (a) $30kx + 60k$ (dollars) (b) $6rx + 32r$ **97.** Suppose that we want to multiply $(x + 3)(2x + 5)$. The letter F stands for *first*. We multiply the two first terms to get $2x^2$. The letter O represents *outer*. We next multiply the two outer terms to get $5x$. The letter I represents *inner*. The product of the two inner terms is $6x$. L stands for *last*. The product of the two last terms is 15. Very often the outer and inner products are like terms, as they are in this case. So we simplify $2x^2 + 5x + 6x + 15$ to get the final product, $2x^2 + 11x + 15$.

Section 4.6 (page 289)

Exercises **1.** (a) $4x^2$ (b) $12x$ (c) 9 (d) $4x^2 + 12x + 9$ **3.** $m^2 + 4m + 4$ **5.** $r^2 - 6r + 9$ **7.** $x^2 + 4xy + 4y^2$

9. $25p^2 + 20pq + 4q^2$ **11.** $16a^2 + 40ab + 25b^2$ **13.** $49t^2 + 14ts + s^2$ **15.** $36m^2 - \dfrac{48}{5}mn + \dfrac{16}{25}n^2$ **17.** $9t^3 - 6t^2 + t$

19. $-9y^2 + 48y - 64$ **21.** To find the product of the sum and the difference of two terms, we find the difference of the squares of the two For example, $(5x + 7y)(5x - 7y) = (5x)^2 - (7y)^2 = 25x^2 - 49y^2$. **23.** $a^2 - 64$ **25.** $4 - p^2$ **27.** $4m^2 - 25$ **29.** $9x^2 - 16y^2$

x^2 **33.** $169r^2 - 4z^2$ **35.** $81y^4 - 4$ **37.** $81y^2 - \dfrac{4}{9}$ **39.** $25q^3 - q$ **41.** $x^3 + 3x^2 + 3x + 1$

$- 27$ **45.** $r^3 + 15r^2 + 75r + 125$ **47.** $8a^3 + 12a^2 + 6a + 1$ **49.** $81r^4 - 216r^3t + 216r^2t^2 - 96rt^3 + 16t^4$

53. $2ab$ **54.** b^2 **55.** $a^2 + 2ab + b^2$ **56.** They both represent the area of the entire large square. **57.** 1225

1225 **60.** They are equal. **61.** 9999 **63.** 39,999 **65.** $399\dfrac{3}{4}$ **67.** $\dfrac{1}{2}m^2 - 2n^2$ **69.** $9a^2 - 4$

$6x^2 + 12x + 8$

Section 4.7   (page 298)

Connections   (page 298)   1. -104 **2.** -104 **3.** They are both -104.   **4.** The answers should agree.

Exercises   1. $6x^2 + 8; 2; 3x^2 + 4$ **3.** $3x^2 + 4; 2$ (These may be reversed.); $6x^2 + 8$ **5.** The first is a polynomial divided by a monomial, covered in Objective 1. This section does not cover dividing a monomial by a polynomial of several terms.   **7.** $30x^3 - 10x + 5$

9. $4m^3 - 2m^2 + 1$ **11.** $4t^4 - 2t^2 + 2t$ **13.** $a^4 - a + \dfrac{2}{a}$ **15.** $4x^3 - 3x^2 + 2x$ **17.** $1 + 5x - 9x^2$ **19.** $\dfrac{12}{x} + 8 + 2x$

21. $\dfrac{4x^2}{3} + x + \dfrac{2}{3x}$ **23.** $9r^3 - 12r^2 + 2r + \dfrac{26}{3} - \dfrac{2}{3r}$ **25.** $-m^2 + 3m - \dfrac{4}{m}$ **27.** $4 - 3a + \dfrac{5}{a}$ **29.** $\dfrac{12}{x} - \dfrac{6}{x^2} + \dfrac{14}{x^3} - \dfrac{10}{x^4}$

31. $6x^4y^2 - 4xy + 2xy^2 - x^4y$ **33.** $5x^3 + 4x^2 - 3x + 1$ **35.** $-63m^4 - 21m^3 - 35m^2 + 14m$ **37.** 1423

38. $(1 \times 10^3) + (4 \times 10^2) + (2 \times 10^1) + (3 \times 10^0)$ **39.** $x^3 + 4x^2 + 2x + 3$ **40.** They are similar in that the coefficients of powers of ten are equal to the coefficients of the powers of x. They are different in that one is a constant while the other is a polynomial. They are equal if $x = 10$ (the base of our decimal system).   **41.** $x + 2$ **43.** $2y - 5$ **45.** $p - 4 + \dfrac{44}{p + 6}$ **47.** $r - 5$ **49.** $6m - 1$

51. $2a - 14 + \dfrac{74}{2a + 3}$ **53.** $4x^2 - 7x + 3$ **55.** $4k^3 - k + 2$ **57.** $5y^3 + 2y - 3$ **59.** $3k^2 + 2k - 2 + \dfrac{6}{k - 2}$

61. $2p^3 - 6p^2 + 7p - 4 + \dfrac{14}{3p + 1}$ **63.** $r^2 - 1 + \dfrac{4}{r^2 - 1}$ **65.** $y^2 - y + 1$ **67.** $a^2 + 1$ **69.** $x^2 - 4x + 2 + \dfrac{9x - 4}{x^2 + 3}$

71. $x^3 + 3x^2 - x + 5$ **73.** $\dfrac{3}{2}a - 10 + \dfrac{77}{2a + 6}$ **75.** The process stops when the degree of the remainder is less than the degree of the divisor.   **77.** $x^2 + x - 3$ units   **79.** $5x^2 - 11x + 14$ hr   **81.** A is correct; B is incorrect.   **82.** A is incorrect; B is correct.

83. A is correct; B is incorrect.   **84.** Because any power of 1 is 1, to evaluate a polynomial for 1 we simply add the coefficients. If the divisor is $x - 1$, this method does not apply, since $1 - 1 = 0$ and division by 0 is undefined.

Chapter 4 Review Exercises   (page 307)

1. 4^{11} **3.** $-72x^7$ **5.** 19^5x^5 **7.** $5p^4t^4$ **9.** $6^2x^{16}y^4z^{16}$ **11.** The expression is a *sum* of powers of 7, not a *product*.   **13.** 0 **15.** $-\dfrac{1}{49}$

17. 5^8 **19.** $\dfrac{3}{4}$ **21.** x^2 **23.** $\dfrac{r^8}{81}$ **25.** $\dfrac{1}{a^3b^5}$ **27.** 4.8×10^7 **29.** 8.24×10^{-8} **31.** $78,300,000$ **33.** 800 **35.** $.025$

37. $.0000000000016$ **39.** $9.7 \times 10^4; 5 \times 10^3$ **41. (a)** 1×10^3 **(b)** 2×10^3 **(c)** 5×10^4 **(d)** 1×10^5 **43.** $p^3 - p^2 + 4p + 2$; degree 3; none of these   **45.** $-8y^5 - 7y^4 + 9y$; degree 5; trinomial   **47.** $13x^3y^2 - 5xy^5 + 21x^2$ **49.** $y^2 - 10y + 9$

51. $1, 4, 5, 4, 1$

$y = -x^2 + 5$

53. $a^3 - 2a^2 - 7a + 2$ **55.** $5p^5 - 2p^4 - 3p^3 + 25p^2 + 15p$ **57.** $6k^2 - 9k - 6$

59. $12k^2 - 32kq - 35q^2$ **61.** $2x^2 + x - 6$ **63.** $a^2 + 8a + 16$ **65.** $36m^2 - 25$ **67.** $r^3 + 6r^2 + 12r + 8$ **69. (a)** Answers will vary. For example, let $x = 1$ and $y = 2$. $(1 + 2)^2 \neq 1^2 + 2^2$, because $9 \neq 5$.   **(b)** Answers will vary. For example, let $x = 1$ and $y = 2$. $(1 + 2)^3 \neq 1^3 + 2^3$, because $27 \neq 9$.   **71.** In both cases, $x = 0$ and $y = 1$ lead to 1 on each side of the inequality. This would not be sufficient to show that *in general* the inequality is true. It would be necessary to choose other values of x and y.   **73.** $\dfrac{4}{3}\pi(x + 1)^3$ or $\dfrac{4}{3}\pi x^3 + 4\pi x^2 + 4\pi x + \dfrac{4}{3}\pi$ in.^{3} **75.** $y^3 - 2y + 3$ **77.** $2mn + 3m^4n^2 - 4n$ **79.** $2r + 7$ **81.** $x^2 + 3x - 4$ **83.** $4x - 5$

85. $y^2 + 2y + 4$ **87.** $2y^2 - 5y + 4 + \dfrac{-5}{3y^2 + 1}$ **89.** 2 **91.** $144a^2 - 1$ **93.** $\dfrac{1}{8^{12}}$ **95.** $\dfrac{2}{3m^3}$ **97.** r^{13} **99.** $-y^2 - 4y + 4$

101. $y^2 + 5y + 1$ **103.** $10p^2 - 3p - 5$ **105.** $49 - 28k + 4k^2$

Chapter 4 Test (page 311)

[4.1, 4.2] **1.** $\dfrac{1}{625}$ **2.** 2 **3.** $\dfrac{7}{12}$ **4.** $9x^3y^5$ **5.** 8^5 **6.** x^2y^6 **7. (a)** positive **(b)** positive **(c)** negative **(d)** positive

(e) zero **(f)** negative [4.3] **8. (a)** 4.5×10^{10} **(b)** .0000036 **(c)** .00019 **9. (a)** 1×10^3; 5.89×10^{12} **(b)** 5.89×10^{15} mi

[4.4] **10.** $-7x^2 + 8x$; 2; binomial **11.** $4n^4 + 13n^3 - 10n^2$; 4; trinomial **12.** $4, -2, -4, -2, 4$ **13.** $-2y^2 - 9y + 17$

14. $-21a^3b^2 + 7ab^5 - 5a^2b^2$ **15.** $-12t^2 + 5t + 8$ [4.5] **16.** $-27x^5 + 18x^4 - 6x^3 + 3x^2$ **17.** $t^2 - 5t - 24$ **18.** $8x^2 + 2xy - 3y^2$

[4.6] **19.** $25x^2 - 20xy + 4y^2$ **20.** $100v^2 - 9w^2$ [4.5] **21.** $2r^3 + r^2 - 16r + 15$ [4.6] **22.** $9x^2 + 54x + 81$

[4.7] **23.** $4y^2 - 3y + 2 + \dfrac{5}{y}$ **24.** $-3xy^2 + 2x^3y^2 + 4y^2$ **25.** $3x^2 + 6x + 11 + \dfrac{26}{x - 2}$

Cumulative Review Exercises Chapters 1–4 (page 312)

[1.1] **1.** $\dfrac{7}{4}$ **2.** 5 **3.** $\dfrac{19}{24}$ **4.** $-\dfrac{1}{20}$ **5.** $31\dfrac{1}{4}$ yd³ [1.6] **6.** $1836 **7.** 1, 3, 5, 9, 15, 45 **8.** -8 **9.** $\dfrac{1}{2}$ **10.** -4

[1.7] **11.** associative property **12.** distributive property [1.8] **13.** $-10x^2 + 21x - 29$ [2.1–2.3] **14.** $\left\{ \dfrac{13}{4} \right\}$ **15.** $\varnothing$

[2.5] **16.** $r = \dfrac{d}{t}$ [2.6] **17.** $\{-5\}$ [2.1–2.3] **18.** $\{-12\}$ **19.** $\{20\}$ **20.** {all real numbers} [2.4] **21.** mouse: 160; elephant: 10

22. 4 [2.8] **23.** 11 ft and 22 ft [2.8] **24.** $[10, \infty)$ **25.** $\left(-\infty, -\dfrac{14}{5} \right)$ **26.** $[-4, 2)$ [3.2] **27.**

[3.3, 3.4] **28. (a)** 1 **(b)** $y = x + 6$ [4.1, 4.2] **29.** $\dfrac{5}{4}$ **30.** 2 **31.** 1 **32.** $\dfrac{2b}{a^{10}}$ [4.3] **33.** 3.45×10^4 **34.** .000000536

35. about 10,800,000 km [4.4] **36.** **37.** $11x^3 - 14x^2 - x + 14$ [4.5] **38.** $18x^7 - 54x^6 + 60x^5$

39. $63x^2 + 57x + 12$ [4.6] **40.** $25x^2 + 80x + 64$ [4.7] **41.** $2x^2 - 3x + 1$ **42.** $y^2 - 2y + 6$

CHAPTER 5 FACTORING AND APPLICATIONS

Section 5.1 (page 322)

Exercises **1.** 4 **3.** 6 **5.** First, verify that you have factored completely. Then multiply the factors. The product should be the original polynomial. **7.** 8 **9.** $10x^3$ **11.** $6m^3n^2$ **13.** xy^2 **15.** 6 **17.** factored **19.** not factored **21.** yes; x^3y^2 **23.** $3m^2$ **25.** $2z^4$

27. $2mn^4$ **29.** $y + 2$ **31.** $a - 2$ **33.** $2 + 3xy$ **35.** $9m(3m^2 - 1)$ **37.** $8z^2(2z^2 + 3)$ **39.** $\dfrac{1}{4}d(d - 3)$ **41.** $6x^2(2x + 1)$

43. $5y^6(13y^4 + 7)$ **45.** no common factor (except 1)   **47.** $8m^2n^2(n + 3)$ **49.** $13y^2(y^6 + 2y^2 - 3)$ **51.** $9qp^3(5q^3p^2 + 4p^3 + 9q)$
53. $a^3(a^2 + 2b^2 - 3a^2b^2 + 4ab^3)$ **55.** $(x + 2)(c - d)$ **57.** $(m + 2n)(m + n)$ **59.** not in factored form; $(7t + 4)(8 + x)$
61. in factored form   **63.** not in factored form; $(y + 4)(18x^2 + 7)$ **65.** The quantities in parentheses are not the same, so there is no common factor in the two terms $12k^3(s - 3)$ and $7(s + 3)$.   **67.** $(p + 4)(p + q)$ **69.** $(a - 2)(a + b)$ **71.** $(z + 2)(7z - a)$
73. $(3r + 2y)(6r - x)$ **75.** $(a^2 + b^2)(3a + 2b)$ **77.** $(1 - a)(1 - b)$ **79.** $(4m - p^2)(4m^2 - p)$ **81.** $(5 - 2p)(m + 3)$
83. $(3r + 2y)(6r - t)$ **85.** $(a^5 - 3)(1 + 2b)$ **87.** commutative property   **88.** $2x(y - 1) - 3(y - 1)$ **89.** No, because it is not a product. It is the difference between $2x(y - 1)$ and $3(y - 1)$.   **90.** $(2x - 3)(y - 1)$; yes   **91. (a)** yes **(b)** When either one is multiplied out, the product is $1 - a + ab - b$.

Section 5.2   (page 328)

Exercises   1. 1 and 48, -1 and -48, 2 and 24, -2 and -24, 3 and 16, -3 and -16, 4 and 12, -4 and -12, 6 and 8, -6 and -8; the pair with a sum of -19 is -3 and -16.   **3.** 1 and -24, -1 and 24, 2 and -12, -2 and 12, 3 and -8, -3 and 8, 4 and -6, -4 and 6; the pair with a sum of -5 is 3 and -8.   **5.** a and b must have different signs, one positive and one negative.   **7.** A prime polynomial is one that cannot be factored using only integers in the factors.   **9.** C   **11.** $p + 6$ **13.** $x + 11$ **15.** $x - 8$ **17.** $y - 5$ **19.** $x + 11$
21. $y - 9$ **23.** $(y + 8)(y + 1)$ **25.** $(b + 3)(b + 5)$ **27.** $(m + 5)(m - 4)$ **29.** $(y - 5)(y - 3)$ **31.** prime   **33.** $(z - 7)(z - 8)$
35. $(r - 6)(r + 5)$ **37.** $(a + 4)(a - 12)$ **39.** prime   **41.** Factor $8 + 6x + x^2$ directly to get $(2 + x)(4 + x)$. Alternatively, use the commutative property to write the trinomial as $x^2 + 6x + 8$ and factor to get $(x + 2)(x + 4)$, an equivalent answer.   **43.** $(r + 2a)(r + a)$
45. $(t + 2z)(t - 3z)$ **47.** $(x + y)(x + 3y)$ **49.** $(v - 5w)(v - 6w)$ **51.** $4(x + 5)(x - 2)$ **53.** $2t(t + 1)(t + 3)$
55. $2x^4(x - 3)(x + 7)$ **57.** $5m^2(m^3 + 5m^2 - 8)$ **59.** $mn(m - 6n)(m - 4n)$ **61.** $(2x + 4)(x - 3)$ is incorrect because $2x + 4$ has a common factor of 2, which must be factored out.   **63.** $a^3(a + 4b)(a - b)$ **65.** $yz(y + 3z)(y - 2z)$ **67.** $z^8(z - 7y)(z + 3y)$
69. $(a + b)(x + 4)(x - 3)$ **71.** $(2p + q)(r - 9)(r - 3)$ **73.** $a^2 + 13a + 36$

Section 5.3   (page 335)

Exercises   1. $(2t + 1)(5t + 2)$ **3.** $(3z - 2)(5z - 3)$ **5.** $(2s - t)(4s + 3t)$ **7.** B   **9. (a)** 2, 12, 24, 11 **(b)** 3, 8 (Order is irrelevant.)
(c) $3m, 8m$ **(d)** $2m^2 + 3m + 8m + 12$ **(e)** $(2m + 3)(m + 4)$ **(f)** $(2m + 3)(m + 4) = 2m^2 + 11m + 12$ **11.** B   **13.** A   **15.** $2a + 5b$
17. $x^2 + 3x - 4$; $x + 4, x - 1$ or $x - 1, x + 4$ **19.** The binomial $2x - 6$ cannot be a factor because it has a common factor of 2, which the polynomial does not have.   **21.** $(3a + 7)(a + 1)$ **23.** $(2y + 3)(y + 2)$ **25.** $(3m - 1)(5m + 2)$ **27.** $(3s - 1)(4s + 5)$
29. $(5m - 4)(2m - 3)$ **31.** $(4w - 1)(2w - 3)$ **33.** $(4y + 1)(5y - 11)$ **35.** prime   **37.** $2(5x + 3)(2x + 1)$
39. $3(4x - 1)(2x - 3)$ **41.** $q(5m + 2)(8m - 3)$ **43.** $3n^2(5n - 3)(n - 2)$ **45.** $y^2(5x - 4)(3x + 1)$ **47.** $(5a + 3b)(a - 2b)$
49. $(4s + 5t)(3s - t)$ **51.** $m^4n(3m + 2n)(2m + n)$ **53.** $(5 - x)(1 - x)$ **55.** $(4 + 3x)(4 + x)$ **57.** $-5x(2x + 7)(x - 4)$
59. The student stopped too soon. He needs to factor out the common factor $4x - 1$ to get $(4x - 1)(4x - 5)$ as the correct answer.
61. $-1(x + 7)(x - 3)$ **63.** $-1(3x + 4)(x - 1)$ **65.** $-1(a + 2b)(2a + b)$ **67.** Yes, $(x + 7)(3 - x)$ is equivalent to $-1(x + 7)(x - 3)$ because $-1(x - 3) = -x + 3 = 3 - x$.   **69.** $(m + 1)^3(5q - 2)(5q + 1)$ **71.** $(r + 3)^3(5x + 2y)(3x - 8y)$ **73.** $-4, 4$
75. $-11, -7, 7, 11$

Section 5.4   (page 344)

Exercises   1. 1; 4; 9; 16; 25; 36; 49; 64; 81; 100; 121; 144; 169; 196; 225; 256; 289; 324; 361; 400   **3.** 1; 8; 27; 64; 125; 216; 343; 512; 729; 1000   **5. (a)** both of these **(b)** a perfect cube **(c)** a perfect square **(d)** a perfect square   **7.** $(y + 5)(y - 5)$
9. $\left(p + \dfrac{1}{3}\right)\left(p - \dfrac{1}{3}\right)$ **11.** prime   **13.** $(3r + 2)(3r - 2)$ **15.** $\left(6m + \dfrac{4}{5}\right)\left(6m - \dfrac{4}{5}\right)$ **17.** $4(3x + 2)(3x - 2)$
19. $(14p + 15)(14p - 15)$ **21.** $(4r + 5a)(4r - 5a)$ **23.** prime   **25.** $(p^2 + 7)(p^2 - 7)$ **27.** $(x^2 + 1)(x + 1)(x - 1)$
29. $(p^2 + 16)(p + 4)(p - 4)$ **31.** The teacher was justified. $x^2 - 9$ can be factored as $(x + 3)(x - 3)$. The complete factored form is
$(x^2 + 9)(x + 3)(x - 3)$.   **33.** $(w + 1)^2$ **35.** $(x - 4)^2$ **37.** $\left(t + \dfrac{1}{2}\right)^2$ **39.** $(x - .5)^2$ **41.** $2(x + 6)^2$ **43.** $(4x - 5)^2$

45. $(7x - 2y)^2$ **47.** $(8x + 3y)^2$ **49.** $-2(5h - 2y)^2$ **51.** $k(4k^2 - 4k + 9)$ **53.** $z^2(25z^2 + 5z + 1)$ **55.** 10   **57.** 9
59. $(a - 1)(a^2 + a + 1)$ **61.** $(m + 2)(m^2 - 2m + 4)$ **63.** $(3x - 4)(9x^2 + 12x + 16)$ **65.** $6(p + 1)(p^2 - p + 1)$
67. $5(x + 2)(x^2 - 2x + 4)$ **69.** $2(y - 2x)(y^2 + 2xy + 4x^2)$ **71.** $(2p + 9q)(4p^2 - 18pq + 81q^2)$ **73.** $(3a + 4b)(9a^2 - 12ab + 16b^2)$
75. $(5t + 2s)(25t^2 - 10ts + 4s^2)$ **77.** $x^2 - y^2 = (x + y)(x - y)$ Difference of Squares; See Exercises 7–10, 13–22, and 25–30.
$x^2 + 2xy + y^2 = (x + y)^2$ Perfect Square Trinomial; See Exercises 33, 34, 37, 38, 41, 47, 48, 50. $x^2 - 2xy + y^2 = (x - y)^2$ Perfect Square
Trinomial; See Exercises 35, 36, 39, 40, 42–46, 49. $x^3 - y^3 = (x - y)(x^2 + xy + y^2)$ Difference of Cubes; See Exercises 59, 60, 63, 64,
68–70, 74. $x^3 + y^3 = (x + y)(x^2 - xy + y^2)$ Sum of Cubes; See Exercises 61, 62, 65–67, 71–73, 75, 76.   **78.** $(5x - 2)(2x + 3)$
79. $5x - 2$ **80.** Yes. If $10x^2 + 11x - 6$ factors as $(5x - 2)(2x + 3)$, then when $10x^2 + 11x - 6$ is divided by $2x + 3$, the quotient should
be $5x - 2$.   **81.** $x^2 + x + 1$; $(x - 1)(x^2 + x + 1)$ **83.** $-2b(3a^2 + b^2)$ **85.** $3(r - k)(1 + r + k)$

Summary Exercises on Factoring   (page 346)

1. $(a - 6)(a + 2)$ **2.** $(a + 8)(a + 9)$ **3.** $6(y - 2)(y + 1)$ **4.** $7y^4(y + 6)(y - 4)$ **5.** $6(a + 2b + 3c)$ **6.** $(m - 4n)(m + n)$
7. $(p - 11)(p - 6)$ **8.** $(z + 7)(z - 6)$ **9.** $(5z - 6)(2z + 1)$ **10.** $2(m - 8)(m + 3)$ **11.** $(m + n + 5)(m - n)$ **12.** $5(3y + 1)$
13. $8a^3(a - 3)(a + 2)$ **14.** $(4k + 1)(2k - 3)$ **15.** $(z - 5a)(z + 2a)$ **16.** $50(z^2 - 2)$ **17.** $(x - 5)(x - 4)$
18. $10nr(10nr + 3r^2 - 5n)$ **19.** $(3n - 2)(2n - 5)$ **20.** $(3y - 1)(3y + 5)$ **21.** $4(4x + 5)$ **22.** $(m + 5)(m - 3)$
23. $(3y - 4)(2y + 1)$ **24.** $(m + 9)(m - 9)$ **25.** $(6z + 1)(z + 5)$ **26.** $(5z + 2)(z + 5)$ **27.** $(2k - 3)^2$ **28.** $(8p - 1)(p + 3)$
29. $6(3m + 2z)(3m - 2z)$ **30.** $(4m - 3)(2m + 1)$ **31.** $(3k - 2)(k + 2)$ **32.** $15a^3b^2(3b^3 - 4a + 5a^3b^2)$ **33.** $7k(2k + 5)(k - 2)$
34. $(5 + r)(1 - s)$ **35.** $(y^2 + 4)(y + 2)(y - 2)$ **36.** $10y^4(2y - 3)$ **37.** $8m(1 - 2m)$ **38.** $(k + 4)(k - 4)$
39. $(z - 2)(z^2 + 2z + 4)$ **40.** $(y - 8)(y + 7)$ **41.** prime   **42.** $9p^8(3p + 7)(p - 4)$ **43.** $8m^3(4m^6 + 2m^2 + 3)$
44. $(2m + 5)(4m^2 - 10m + 25)$ **45.** $(4r + 3m)^2$ **46.** $(z - 6)^2$ **47.** $(5h + 7g)(3h - 2g)$ **48.** $5z(z - 7)(z - 2)$
49. $(k - 5)(k - 6)$ **50.** $4(4p - 5m)(4p + 5m)$ **51.** $3k(k - 5)(k + 1)$ **52.** $(y - 6k)(y + 2k)$ **53.** $(10p + 3)(100p^2 - 30p + 9)$
54. $(4r - 7)(16r^2 + 28r + 49)$ **55.** $(2 + m)(3 + p)$ **56.** $(2m - 3n)(m + 5n)$ **57.** $(4z - 1)^2$ **58.** $5m^2(5m - 3n)(5m - 13n)$
59. $3(6m - 1)^2$ **60.** $(10a + 9y)(10a - 9y)$ **61.** prime   **62.** $(2y + 5)(2y - 5)$ **63.** $8z(4z - 1)(z + 2)$ **64.** $5(2m - 3)(m + 4)$
65. $(4 + m)(5 + 3n)$ **66.** $(2 - q)(2 - 3p)$ **67.** $2(3a - 1)(a + 2)$ **68.** $6y^4(3y + 4)(2y - 5)$ **69.** $(a - b)(a^2 + ab + b^2 + 2)$
70. $4(2k - 3)^2$ **71.** $(8m - 5n)^2$ **72.** $12y^2(6yz^2 + 1 - 2y^2z^2)$ **73.** $(4k - 3h)(2k + h)$ **74.** $(2a + 5)(a - 6)$
75. $(m + 2)(m^2 + m + 1)$ **76.** $(2a - 3)(4a^2 + 6a + 9)$ **77.** $(5y - 6z)(2y + z)$ **78.** $(m - 2)^2$ **79.** $(8a - b)(a + 3b)$
80. $(a^2 + 25)(a + 5)(a - 5)$ **81.** $(x^3 - 1)(x^3 + 1)$ **82.** $(x - 1)(x^2 + x + 1)(x + 1)(x^2 - x + 1)$ **83.** $(x^2 - 1)(x^4 + x^2 + 1)$
84. $(x - 1)(x + 1)(x^4 + x^2 + 1)$ **85.** The result in Exercise 82 is completely factored.   **86.** Show that $x^4 + x^2 + 1 =$
$(x^2 + x + 1)(x^2 - x + 1)$.   **87.** difference of squares   **88.** $(x - 3)(x^2 + 3x + 9)(x + 3)(x^2 - 3x + 9)$

Section 5.5   (page 353)

Exercises   **1.** $\{-5, 2\}$ **3.** $\left\{\dfrac{7}{2}, 3\right\}$ **5.** $\left\{-\dfrac{5}{6}, 0\right\}$ **7.** $\left\{0, \dfrac{4}{3}\right\}$ **9.** $\left\{-\dfrac{1}{2}, \dfrac{1}{6}\right\}$ **11.** $\{-.8, 2\}$ **13.** $\{9\}$ **15.** Set each *variable*

factor equal to 0, to get $2x = 0$ or $3x - 4 = 0$. The solution set is $\left\{0, \dfrac{4}{3}\right\}$.   **17.** $\{-2, -1\}$ **19.** $\{1, 2\}$ **21.** $\{-8, 3\}$ **23.** $\{-1, 3\}$

25. $\{-2, -1\}$ **27.** $\{-4\}$ **29.** $\left\{-2, \dfrac{1}{3}\right\}$ **31.** $\left\{-\dfrac{4}{3}, \dfrac{1}{2}\right\}$ **33.** $\left\{-\dfrac{2}{3}\right\}$ **35.** $\{-3, 3\}$ **37.** $\left\{-\dfrac{7}{4}, \dfrac{7}{4}\right\}$ **39.** $\{-11, 11\}$

41. $\{0, 7\}$ **43.** $\left\{0, \dfrac{1}{2}\right\}$ **45.** $\{2, 5\}$ **47.** $\left\{-4, \dfrac{1}{2}\right\}$ **49.** $\left\{-12, \dfrac{11}{2}\right\}$ **51.** $\left\{-\dfrac{5}{2}, \dfrac{1}{3}, 5\right\}$ **53.** $\left\{-\dfrac{7}{2}, -3, 1\right\}$

55. $\left\{-\dfrac{7}{3}, 0, \dfrac{7}{3}\right\}$ **57.** $\{-2, 0, 4\}$ **59.** $\{-5, 0, 4\}$ **61.** $\{-3, 0, 5\}$ **63.** $\{-1, 3\}$ **65.** $\{-1, 3\}$ **67.** $\{3\}$ **69.** $\left\{-\dfrac{4}{3}, -1, \dfrac{1}{2}\right\}$

71. $\left\{-\dfrac{2}{3}, 4\right\}$ **73. (a)** 64; 144; 4; 6   **(b)** No time has elapsed, so the object hasn't fallen (been released) yet.   **(c)** Time cannot be negative.
75. $\{1.5, 2.1\}$ **77.** $\{-2.75, 3\}$

Section 5.6   (page 361)

Exercises   **1.** Read; variable; equation; Solve; answer; Check, original   **3.** *Step 3:* $45 = (2x + 1)(x + 1)$; *Step 4:* $x = 4$ or $x = -\dfrac{11}{2}$;

Step 5: base: 9 units; height: 5 units; *Step 6:* $9 \cdot 5 = 45$ **5.** *Step 3:* $80 = (x + 8)(x - 8)$; *Step 4:* $x = 12$ or $x = -12$; *Step 5:*
length: 20 units; width: 4 units; *Step 6:* $20 \cdot 4 = 80$ **7.** length: 7 in.; width: 4 in.   **9.** length: 13 in.; width: 10 in.   **11.** base: 12 in.;
height: 5 in.   **13.** height: 13 in.; width: 10 in.   **15.** mirror: 7 ft; painting: 9 ft   **17.** $-3, -2$ or $4, 5$ **19.** 7, 9, 11   **21.** $-2, 0, 2$ or

6, 8, 10   **23.** 12 cm   **25.** 12 mi   **27.** 8 ft   **29. (a)** 1 sec **(b)** $\dfrac{1}{2}$ sec and $1\dfrac{1}{2}$ sec **(c)** 3 sec **(d)** The negative solution, -1, does not

make sense since t represents time, which cannot be negative.   **31. (a)** 23.4 million; The result using the model is less than 24 million, the
actual number for 1994. **(b)** 9 **(c)** 82.6 million; The result is less than 86 million, the actual number for 1999. **(d)** 137.6 million
32. 107 billion dollars; 65%   **33.** 1995: 66.9 billion dollars; 1997: 148.5 billion dollars; 1999: 230.1 billion dollars   **34.** The answers
using the linear equation are not at all close to the actual data.   **35.** 1995: 104 billion dollars; 1997: 111.2 billion dollars;
1999: 266.4 billion dollars   **36.** The answers in Exercise 35 are fairly close to the actual data. The quadratic equation models the data better.
37. (0, 97.5), (1, 104.3), (2, 104.7), (3, 164.3), (4, 271.3)   **38.** no   **39.** 399.5 billion dollars

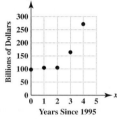

40. (a) The actual deficit is about 30 billion dollars less than the prediction. **(b)** No, the equation is based on data for the years 1995–1999.
Data for later years might not follow the same pattern.

Chapter 5 Review Exercises   (page 370)

1. $7(t + 2)$ **3.** $(x - 4)(2y + 3)$ **5.** $(x + 3)(x + 2)$ **7.** $(q + 9)(q - 3)$ **9.** $(r + 8s)(r - 12s)$ **11.** $8p(p + 2)(p - 5)$
13. $p^5(p - 2q)(p + q)$ **15.** r and $6r$, $2r$ and $3r$ **17.** $(2k - 1)(k - 2)$ **19.** $(3r + 2)(2r - 3)$ **21.** $(v + 3)(8v - 7)$
23. $-3(x + 2)(2x - 5)$ **25.** B   **27.** $(n + 7)(n - 7)$ **29.** $(7y + 5w)(7y - 5w)$ **31.** prime   **33.** $(3t - 7)^2$

35. $(5k + 4x)(25k^2 - 20kx + 16x^2)$ **37.** $\left\{ -\dfrac{3}{4}, 1 \right\}$ **39.** $\left\{ 0, \dfrac{5}{2} \right\}$ **41.** $\{1, 4\}$ **43.** $\left\{ -\dfrac{4}{3}, 5 \right\}$ **45.** $\{0, 8\}$ **47.** $\{7\}$

49. $\left\{ -\dfrac{2}{5}, -2, -1 \right\}$ **51.** length: 10 ft; width: 4 ft   **53.** length: 6 m; width: 2 m   **55.** 6, 7 or $-5, -4$ **57.** 112 ft   **59.** 256 ft

61. (a) \$537 million **(b)** No, the prediction seems high. If eBay revenues in the last half of 2000 are comparable to those for the first half of
the year, annual revenue in 2000 would be about \$366 million.   **63.** The factor $2x + 8$ has a common factor of 2. The complete factored
form is $2(x + 4)(3x - 4)$.   **65.** $(3k + 5)(k + 2)$ **67.** $(y^2 + 25)(y + 5)(y - 5)$ **69.** $8abc(3b^2c - 7ac^2 + 9ab)$
71. $6xyz(2xz^2 + 2y - 5x^2yz^3)$ **73.** $(2r + 3q)(6r - 5)$ **75.** $(7t + 4)^2$ **77.** $\{-5, 2\}$ **79. (a)** 481,000 vehicles **(b)** The estimate may
be unreliable because the conditions that prevailed in the years 1998–2001 may have changed, causing either a greater increase or a greater
decrease in the numbers of alternative-fueled vehicles.   **81.** length: 6 m; width: 4 m   **83. (a)** 256 ft **(b)** 1024 ft   **85.** 6 m

Chapter 5 Test   (page 374)

[5.1–5.4] **1.** D   **2.** $6x(2x - 5)$ **3.** $m^2n(2mn + 3m - 5n)$ **4.** $(2x + y)(a - b)$ **5.** $(x + 3)(x - 8)$ **6.** $(2x + 3)(x - 1)$
7. $(5z - 1)(2z - 3)$ **8.** prime   **9.** prime   **10.** $(2 - a)(6 + b)$ **11.** $(3y + 8)(3y - 8)$ **12.** $(2x - 7y)^2$ **13.** $-2(x + 1)^2$
14. $3t^2(2t + 9)(t - 4)$ **15.** $(r - 5)(r^2 + 5r + 25)$ **16.** $8(k + 2)(k^2 - 2k + 4)$ **17.** $(x^2 + 9)(x + 3)(x - 3)$ **18.** The product

$(p + 3)(p + 3) = p^2 + 6p + 9$, which does not equal $p^2 + 9$. The binomial $p^2 + 9$ is a prime polynomial.   [5.5] **19.** $\left\{ 6, \dfrac{1}{2} \right\}$

20. $\left\{ -\dfrac{2}{5}, \dfrac{2}{5} \right\}$ **21.** $\{10\}$ **22.** $\{-3, 0, 3\}$ [5.6] **23.** 6 ft by 9 ft   **24.** $-2, -1$ **25.** 17 ft   **26.** 181

Cumulative Review Exercises Chapters 1–5   (page 375)

[2.1–2.3] **1.** $\{0\}$ **2.** $\{.05\}$ **3.** $\{6\}$ [2.5] **4.** $P = \dfrac{A}{1 + rt}$ [2.7] **5.** 345; 210; 38%; 15%   [2.4] **6.** gold: 12; silver: 9; bronze: 8

[2.7] **7.** 107 million   [2.5] **8.** 110° and 70°   [3.1] **9. (a)** negative, positive **(b)** negative, negative   [3.2] **10.** $\left(-\dfrac{1}{4}, 0\right)$, $(0, 3)$

[3.3] **11.** 12   [3.2] **12.**

$y = 12x + 3$

[3.3] **13.** 103; A slope of 103 means that the number of radio stations increased by about

103 stations per year.   [3.4] **14.** $y = -\dfrac{1}{2}x + 5$ **15.** $y = -\dfrac{2}{3}$ [4.1, 4.2] **16.** 4   **17.** $\dfrac{16}{9}$ **18.** 256   **19.** $\dfrac{1}{p^2}$ **20.** $\dfrac{1}{m^6}$

[4.4] **21.** $-4k^2 - 4k + 8$ [4.5] **22.** $45x^2 + 3x - 18$ [4.6] **23.** $9p^2 + 12p + 4$ [4.7] **24.** $4x^3 + 6x^2 - 3x + 10$

[4.3] **25.** 5.5×10^4; 2.0×10^6 [5.2, 5.3] **26.** $(2a - 1)(a + 4)$ **27.** $(2m + 3)(5m + 2)$ **28.** $(4t + 3v)(2t + v)$

[5.4] **29.** $(2p - 3)^2$ **30.** $(5r + 9t)(5r - 9t)$ [5.3] **31.** $2pq(3p + 1)(p + 1)$ [5.5] **32.** $\left\{-\dfrac{2}{3}, \dfrac{1}{2}\right\}$ **33.** $\{0, 8\}$

[5.6] **34.** 5 m, 12 m, 13 m

CHAPTER 6 RATIONAL EXPRESSIONS AND APPLICATIONS

Section 6.1   (page 385)

Connections   (page 385)   **1.** $3x^2 + 11x + 8$ cannot be factored, so this quotient cannot be simplified. By long division the quotient is $3x + 5 + \dfrac{-2}{x + 2}$.   **2.** The numerator factors as $(x - 2)(x^2 + 2x + 4)$, so after simplifying the quotient is $x - 2$. Long division gives the same quotient.

Exercises   **1.** A rational expression is a quotient of two polynomials, such as $\dfrac{x^2 + 3x - 6}{x + 4}$. One can think of this as an algebraic fraction.

3. (a) $\dfrac{7}{10}$ **(b)** $\dfrac{8}{15}$ **5. (a)** 0 **(b)** -1 **7. (a)** $-\dfrac{64}{15}$ **(b)** undefined   **9. (a)** undefined **(b)** $\dfrac{8}{25}$ **11.** Division by 0 is undefined, so if the denominator of a rational expression equals 0, the expression is undefined.   **13.** 0   **15.** 6   **17.** $-\dfrac{5}{3}$ **19.** $-3, 2$ **21.** never undefined

23. never undefined   **25. (a)** numerator: x^2, $4x$; denominator: x, 4 **(b)** First factor the numerator, getting $x(x + 4)$, then divide the numerator and denominator by the common factor of $x + 4$ to get $\dfrac{x}{1}$ or x.   **27.** $3r^2$ **29.** $\dfrac{2}{5}$ **31.** $\dfrac{x - 1}{x + 1}$ **33.** $\dfrac{7}{5}$ **35.** $\dfrac{6}{7}$ **37.** $m - n$

39. $\dfrac{2}{t - 3}$ **41.** $\dfrac{3(2m + 1)}{4}$ **43.** $\dfrac{3m}{5}$ **45.** $\dfrac{3r - 2s}{3}$ **47.** $k - 3$ **49.** $\dfrac{x - 3}{x + 1}$ **51.** $\dfrac{x + 1}{x - 1}$ **53.** $\dfrac{z - 3}{z + 5}$ **55.** $\dfrac{m + n}{2}$

57. $-\dfrac{b^2 + ba + a^2}{a + b}$ **59.** $\dfrac{z + 3}{z}$ **61.** B, D   **63.** -1 **65.** $-(m + 1)$ **67.** -1 **69.** already in lowest terms   **71.** B

Answers may vary in Exercises 73, 75, and 77. **73.** $\dfrac{-(x + 4)}{x - 3}$, $\dfrac{-x - 4}{x - 3}$, $\dfrac{x + 4}{-(x - 3)}$, $\dfrac{x + 4}{-x + 3}$ **75.** $\dfrac{-(2x - 3)}{x + 3}$, $\dfrac{-2x + 3}{x + 3}$, $\dfrac{2x - 3}{-(x + 3)}$,

$\dfrac{2x - 3}{-x - 3}$ **77.** $-\dfrac{3x - 1}{5x - 6}$, $\dfrac{-(3x - 1)}{5x - 6}$, $-\dfrac{-3x + 1}{-5x + 6}$, $\dfrac{3x - 1}{-5x + 6}$ **79.** $x^2 + 3$ **81. (a)** 0 **(b)** 1.6 **(c)** 4.1 **(d)** The waiting time also increases.

Section 6.2   (page 392)

Exercises   **1. (a)** B **(b)** D **(c)** C **(d)** A   **3.** $\dfrac{3a}{2}$ **5.** $-\dfrac{4x^4}{3}$ **7.** $\dfrac{2}{c+d}$ **9.** $4(x-y)$ **11.** $\dfrac{t^2}{2}$ **13.** $\dfrac{x+3}{2x}$ **15.** 5   **17.** $-\dfrac{3}{2t^4}$

19. $\dfrac{1}{4}$ **21.** $-\dfrac{35}{8}$ **23.** $\dfrac{2(x+2)}{x(x-1)}$ **25.** $\dfrac{x(x-3)}{6}$ **27.** Suppose I want to multiply $\dfrac{a^2-1}{6}\cdot\dfrac{9}{2a+2}$. I start by factoring where possible:

$\dfrac{(a+1)(a-1)}{6}\cdot\dfrac{9}{2(a+1)}$. Next, I divide out common factors in the numerator and denominator to get $\dfrac{a-1}{2}\cdot\dfrac{3}{2}$. Finally, I multiply

numerator times numerator and denominator times denominator to get the final product, $\dfrac{3(a-1)}{4}$.   **29.** $\dfrac{10}{9}$ **31.** $-\dfrac{3}{4}$ **33.** $-\dfrac{9}{2}$

35. $\dfrac{p+4}{p+2}$ **37.** -1 **39.** $\dfrac{(2x-1)(x+2)}{x-1}$ **41.** $\dfrac{(k-1)^2}{(k+1)(2k-1)}$ **43.** $\dfrac{4k-1}{3k-2}$ **45.** $\dfrac{m+4p}{m+p}$ **47.** $\dfrac{m+6}{m+3}$ **49.** $\dfrac{y+3}{y+4}$

51. $\dfrac{m}{m+5}$ **53.** $\dfrac{r+6s}{r+s}$ **55.** $\dfrac{(q-3)^2(q+2)^2}{q+1}$ **57.** $\dfrac{x+10}{10}$ **59.** $\dfrac{3-a-b}{2a-b}$ **61.** $-\dfrac{(x+y)^2(x^2-xy+y^2)}{3y(y-x)(x-y)}$ or

$\dfrac{(x+y)^2(x^2-xy+y^2)}{3y(x-y)^2}$ **63.** $\dfrac{5xy^2}{4q}$

Section 6.3   (page 397)

Exercises   **1.** C   **3.** C   **5.** 60   **7.** 1800   **9.** x^5 **11.** $30p$ **13.** $180y^4$ **15.** $84r^5$ **17.** $15a^5b^3$ **19.** $12p(p-2)$
21. $28m^2(3m-5)$ **23.** $30(b-2)$ **25.** $2^3\cdot3\cdot5$ **26.** $(t+4)^3(t-3)(t+8)$ **27.** The similarity is that 2 is replaced by $t+4$, 3 is
replaced by $t-3$, and 5 is replaced by $t+8$.   **28.** The procedure used is the same. The only difference is that for algebraic fractions, the
factors may contain variables, while in common fractions, the factors are numbers.   **29.** $18(r-2)$ **31.** $12p(p+5)^2$
33. $8(y+2)(y+1)$ **35.** $c-d$ or $d-c$ **37.** $m-3$ or $3-m$ **39.** $p-q$ or $q-p$ **41.** $k(k+5)(k-2)$ **43.** $a(a+6)(a-3)$
45. $(p+3)(p+5)(p-6)$ **47.** $(k+3)(k-5)(k+7)(k+8)$ **49.** Yes, because $(2x-5)^2=(5-2x)^2$.   **51.** 7   **52.** 1

53. identity property of multiplication   **54.** 7   **55.** 1   **56.** identity property of multiplication   **57.** $\dfrac{20}{55}$ **59.** $\dfrac{-45}{9k}$ **61.** $\dfrac{60m^2k^3}{32k^4}$

63. $\dfrac{57z}{6z-18}$ **65.** $\dfrac{-4a}{18a-36}$ **67.** $\dfrac{6(k+1)}{k(k-4)(k+1)}$ **69.** $\dfrac{36r(r+1)}{(r-3)(r+2)(r+1)}$ **71.** $\dfrac{ab(a+2b)}{2a^3b+a^2b^2-ab^3}$ **73.** $\dfrac{(t-r)(4r-t)}{t^3-r^3}$

75. $\dfrac{2y(z-y)(y-z)}{y^4-z^3y}$ or $\dfrac{-2y(y-z)^2}{y^4-z^3y}$ **77.** *Step 1:* Factor each denominator into prime factors. *Step 2:* List each different denominator

factor the greatest number of times it appears in any of the denominators. *Step 3:* Multiply the factors in the list to get the LCD. For example,

the least common denominator for $\dfrac{1}{(x+y)^3}$ and $\dfrac{-2}{(x+y)^2(p+q)}$ is $(x+y)^3(p+q)$.

Section 6.4   (page 406)

Exercises   **1.** E   **3.** C   **5.** B   **7.** G   **9.** $\dfrac{11}{m}$ **11.** $\dfrac{4}{y+4}$ **13.** 1   **15.** $\dfrac{m-1}{m+1}$ **17.** b **19.** x **21.** $y-6$ **23.** To add or

subtract rational expressions with the same denominators, combine the numerators and keep the same denominator. For example, $\dfrac{3x+2}{x-6}+$

$\dfrac{-2x-8}{x-6}=\dfrac{x-6}{x-6}$. Then write in lowest terms. In this example, the sum simplifies to 1.   **25.** $\dfrac{3z+5}{15}$ **27.** $\dfrac{10-7r}{14}$ **29.** $\dfrac{-3x-2}{4x}$

31. $\dfrac{57}{10x}$ **33.** $\dfrac{x+1}{2}$ **35.** $\dfrac{5x+9}{6x}$ **37.** $\dfrac{7-6p}{3p^2}$ **39.** $\dfrac{-k-8}{k(k+4)}$ **41.** $\dfrac{x+4}{x+2}$ **43.** $\dfrac{6m^2+23m-2}{(m+2)(m+1)(m+5)}$ **45.** $\dfrac{4y^2-y+5}{(y+1)^2(y-1)}$

47. $\dfrac{3}{t}$ **49.** $m-2$ or $2-m$ **51.** $\dfrac{-2}{x-5}$ or $\dfrac{2}{5-x}$ **53.** -4 **55.** $\dfrac{-5}{x-y^2}$ or $\dfrac{5}{y^2-x}$ **57.** $\dfrac{x+y}{5x-3y}$ or $\dfrac{-x-y}{3y-5x}$

59. $\dfrac{-6}{4p - 5}$ or $\dfrac{6}{5 - 4p}$ **61.** $\dfrac{-(m + n)}{2(m - n)}$ **63.** $\dfrac{-x^2 + 6x + 11}{(x + 3)(x - 3)(x + 1)}$ **65.** $\dfrac{-5q^2 - 13q + 7}{(3q - 2)(q + 4)(2q - 3)}$ **67.** $\dfrac{9r + 2}{r(r + 2)(r - 1)}$

69. $\dfrac{2(x^2 + 3xy + 4y^2)}{(x + y)(x + y)(x + 3y)}$ or $\dfrac{2(x^2 + 3xy + 4y^2)}{(x + y)^2(x + 3y)}$ **71.** $\dfrac{15r^2 + 10ry - y^2}{(3r + 2y)(6r - y)(6r + y)}$ **73. (a)** $\dfrac{9k^2 + 6k + 26}{5(3k + 1)}$ **(b)** $\dfrac{1}{4}$ **75.** $\dfrac{10x}{49(101 - x)}$

Section 6.5   (page 414)

Exercises   **1. (a)** $6; \dfrac{1}{6}$ **(b)** $12; \dfrac{3}{4}$ **(c)** $\dfrac{1}{6} \div \dfrac{3}{4}$ **(d)** $\dfrac{2}{9}$ **3.** Choice D is correct, because every sign has been changed in the fraction.

5. Method 1 indicates to write the complex fraction as a division problem, and then perform the division. For example, to simplify $\dfrac{\frac{1}{2}}{\frac{2}{3}}$, write

$\dfrac{1}{2} \div \dfrac{2}{3}$. Then simplify as $\dfrac{1}{2} \cdot \dfrac{3}{2} = \dfrac{3}{4}$.   **7.** -6 **9.** $\dfrac{1}{xy}$ **11.** $\dfrac{2a^2b}{3}$ **13.** $\dfrac{m(m + 2)}{3(m - 4)}$ **15.** $\dfrac{2}{x}$ **17.** $\dfrac{8}{x}$ **19.** $\dfrac{a^2 - 5}{a^2 + 1}$ **21.** $\dfrac{31}{50}$

23. $\dfrac{y^2 + x^2}{xy(y - x)}$ **25.** $\dfrac{40 - 12p}{85p}$ **27.** $\dfrac{5y - 2x}{3 + 4xy}$ **29.** $\dfrac{a - 2}{2a}$ **31.** $\dfrac{z - 5}{4}$ **33.** $\dfrac{-m}{m + 2}$ **35.** $\dfrac{3m(m - 3)}{(m - 1)(m - 8)}$ **37.** division

39. $\dfrac{\frac{3}{8} + \frac{5}{6}}{2}$ **40.** $\dfrac{29}{48}$ **41.** $\dfrac{29}{48}$ **42.** Answers will vary.

Section 6.6   (page 423)

Exercises   **1.** expression; $\dfrac{43}{40}x$ **3.** equation; $\left\{\dfrac{40}{43}\right\}$ **5.** expression; $-\dfrac{1}{10}x$ **7.** equation; $\{-10\}$ **9.** $-2, 0$ **11.** $-3, 4, -\dfrac{1}{2}$

13. $-9, 1, -2, 2$ **15.** When adding and subtracting, the LCD is retained as part of the answer. When solving an equation, the LCD is used

as the multiplier in applying the multiplication property of equality.   **17.** $\left\{\dfrac{1}{4}\right\}$ **19.** $\left\{-\dfrac{3}{4}\right\}$ **21.** $\{-15\}$ **23.** $\{7\}$ **25.** $\{-15\}$

27. $\{-5\}$ **29.** $\{-6\}$ **31.** $\emptyset$ **33.** $\{5\}$ **35.** $\{4\}$ **37.** $\{1\}$ **39.** $\{4\}$ **41.** $\{5\}$ **43.** $\{-4\}$ **45.** $\{-2, 12\}$ **47.** $\emptyset$ **49.** $\{3\}$

51. $\{3\}$ **53.** $\{-3\}$ **55.** $\left\{-\dfrac{1}{5}, 3\right\}$ **57.** $\left\{-\dfrac{1}{2}, 5\right\}$ **59.** $\{3\}$ **61.** $\left\{-\dfrac{1}{3}, 3\right\}$ **63.** $\{-1\}$ **65.** $\{-6\}$ **67.** $\left\{-6, \dfrac{1}{2}\right\}$ **69.** $\{6\}$

71. Transform the equation so that the terms with k are on one side and the remaining term is on the other.   **73.** $F = \dfrac{ma}{k}$ **75.** $a = \dfrac{kF}{m}$

77. $R = \dfrac{E - Ir}{I}$ or $R = \dfrac{E}{I} - r$ **79.** $A = \dfrac{h(B + b)}{2}$ **81.** $a = \dfrac{2S - ndL}{nd}$ or $a = \dfrac{2S}{nd} - L$ **83.** $y = \dfrac{xz}{x + z}$ **85.** $z = \dfrac{3y}{5 - 9xy}$ or

$z = \dfrac{-3y}{9xy - 5}$

Summary Exercises on Operations and Equations with Rational Expressions   (page 427)

1. operation; $\dfrac{10}{p}$ **2.** operation; $\dfrac{y^3}{x^3}$ **3.** operation; $\dfrac{1}{2x^2(x + 2)}$ **4.** equation; $\{9\}$ **5.** operation; $\dfrac{y + 2}{y - 1}$ **6.** operation; $\dfrac{5k + 8}{k(k - 4)(k + 4)}$

7. equation; $\{39\}$ **8.** operation; $\dfrac{t - 5}{3(2t + 1)}$ **9.** operation; $\dfrac{13}{3(p + 2)}$ **10.** equation; $\left\{-1, \dfrac{12}{5}\right\}$ **11.** equation; $\left\{\dfrac{1}{7}, 2\right\}$

12. operation; $\dfrac{16}{3y}$ **13.** operation; $\dfrac{7}{12z}$ **14.** equation; $\{13\}$ **15.** operation; $\dfrac{3m + 5}{(m + 3)(m + 2)(m + 1)}$ **16.** operation; $\dfrac{k + 3}{5(k - 1)}$

17. equation; $\emptyset$ **18.** equation; $\emptyset$ **19.** operation; $\dfrac{t + 2}{2(2t + 1)}$ **20.** equation; $\{-7\}$

Section 6.7   (page 432)

Exercises   **1.** (a) the amount (b) $5 + x$ (c) $\dfrac{5 + x}{6} = \dfrac{13}{3}$ **3.** $\dfrac{12}{18}$ **5.** $\dfrac{12}{3}$ **7.** 12   **9.** $\dfrac{1386}{97}$ **11.** 74.75 sec   **13.** 367.197 m per min

15. 3.090 hr   **17.** $\dfrac{D}{R} = \dfrac{d}{r}$ **19.** $\dfrac{500}{x - 10} = \dfrac{600}{x + 10}$ **21.** 8 mph   **23.** 165 mph   **25.** 3 mph   **27.** 18.5 mph   **29.** $\dfrac{1}{10}$ job per hr

31. $\dfrac{1}{8}x + \dfrac{1}{6}x = 1$ or $\dfrac{1}{8} + \dfrac{1}{6} = \dfrac{1}{x}$ **33.** $4\dfrac{4}{17}$ hr   **35.** $5\dfrac{5}{11}$ hr   **37.** 3 hr   **39.** $2\dfrac{7}{10}$ hr   **41.** $9\dfrac{1}{11}$ min

Chapter 6 Review Exercises   (page 442)

1. (a) $\dfrac{11}{8}$ (b) $\dfrac{13}{22}$ **3.** 3   **5.** $-5, -\dfrac{2}{3}$ **7.** $\dfrac{b}{3a}$ **9.** $\dfrac{-(2x + 3)}{2}$ **11.** (Answers may vary.) $\dfrac{-(4x - 9)}{2x + 3}, \dfrac{-4x + 9}{2x + 3}, \dfrac{4x - 9}{-(2x + 3)},$

$\dfrac{4x - 9}{-2x - 3}$ **13.** $\dfrac{72}{p}$ **15.** $\dfrac{5}{8}$ **17.** $\dfrac{3a - 1}{a + 5}$ **19.** $\dfrac{p + 5}{p + 1}$ **21.** $108y^4$ **23.** $\dfrac{15a}{10a^4}$ **25.** $\dfrac{15y}{50 - 10y}$ **27.** $\dfrac{15}{x}$ **29.** $\dfrac{4k - 45}{k(k - 5)}$

31. $\dfrac{-2 - 3m}{6}$ **33.** $\dfrac{7a + 6b}{(a - 2b)(a + 2b)}$ **35.** $\dfrac{5z - 16}{z(z + 6)(z - 2)}$ **37.** (a) $\dfrac{a}{b}$ (b) $\dfrac{a}{b}$ (c) Answers will vary.   **39.** $\dfrac{4(y - 3)}{y + 3}$ **41.** $\dfrac{xw + 1}{xw - 1}$

43. It would cause the first and third denominators to equal 0.   **45.** $\emptyset$ **47.** $t = \dfrac{Ry}{m}$ **49.** $m = \dfrac{4 + p^2q}{3p^2}$ **51.** $\dfrac{2}{6}$ **53.** $3\dfrac{1}{13}$ hr

55. $\dfrac{m + 7}{(m - 1)(m + 1)}$ **57.** $\dfrac{1}{6}$ **59.** $\dfrac{z + 7}{(z + 1)(z - 1)^2}$ **61.** $\{-2, 3\}$ **63.** $1\dfrac{7}{8}$ hr   **65.** (a) -3 (b) -1 (c) $-3, -1$ **66.** $\dfrac{15}{2x}$

67. If $x = 0$, the divisor R is equal to 0, and division by 0 is undefined.   **68.** $(x + 3)(x + 1)$ **69.** $\dfrac{7}{x + 1}$ **70.** $\dfrac{11x + 21}{4x}$ **71.** $\emptyset$

72. We know that -3 is not allowed because P and R are undefined for $x = -3$.   **73.** Rate is equal to distance divided by time. Here,

distance is 6 mi, and time is $x + 3$ min, so rate $= \dfrac{6}{x + 3}$, which is the expression for P.   **74.** $\dfrac{6}{5}, \dfrac{5}{2}$

Chapter 6 Test   (page 446)

[6.1] **1.** (a) $\dfrac{11}{6}$ (b) undefined   **2.** $-2, 4$ **3.** (Answers may vary.) $\dfrac{-(6x - 5)}{2x + 3}, \dfrac{-6x + 5}{2x + 3}, \dfrac{6x - 5}{-(2x + 3)}, \dfrac{6x - 5}{-2x - 3}$ **4.** $-3x^2y^3$

5. $\dfrac{3a + 2}{a - 1}$ [6.2] **6.** $\dfrac{25}{27}$ **7.** $\dfrac{3k - 2}{3k + 2}$ **8.** $\dfrac{a - 1}{a + 4}$ [6.3] **9.** $150p^5$ **10.** $(2r + 3)(r + 2)(r - 5)$ **11.** $\dfrac{240p^2}{64p^3}$ **12.** $\dfrac{21}{42m - 84}$

[6.4] **13.** 2   **14.** $\dfrac{-14}{5(y + 2)}$ **15.** $\dfrac{-x^2 + x + 1}{3 - x}$ or $\dfrac{x^2 - x - 1}{x - 3}$ **16.** $\dfrac{-m^2 + 7m + 2}{(2m + 1)(m - 5)(m - 1)}$ [6.5] **17.** $\dfrac{2k}{3p}$ **18.** $\dfrac{-2 - x}{4 + x}$

[6.6] **19.** $\left\{-\dfrac{1}{2}\right\}$ **20.** $D = \dfrac{dF - k}{F}$ or $D = \dfrac{k - dF}{-F}$ [6.7] **21.** 3 mph   **22.** $2\dfrac{2}{9}$ hr

Cumulative Review Exercises Chapters 1–6   (page 447)

[1.2, 1.5, 1.6] **1.** 2   [2.3] **2.** $\{17\}$ [2.5] **3.** $b = \dfrac{2A}{h}$ [2.6] **4.** $\left\{-\dfrac{2}{7}\right\}$ [2.8] **5.** $[-8, \infty)$ **6.** $(4, \infty)$ [3.1] **7.** (a) $(-3, 0)$

(b) $(0, -4)$ [3.2] **8.**   [4.4] **9.**   [3.3] **10.** $-\dfrac{3}{2}$ **11.** $-\dfrac{3}{4}$ [4.1, 4.2] **12.** $\dfrac{1}{2^4x^7}$ **13.** $\dfrac{1}{m^6}$

14. $\dfrac{q}{4p^2}$ [4.4] **15.** $k^2 + 2k + 1$ [4.1] **16.** $72x^6y^7$ [4.6] **17.** $4a^2 - 4ab + b^2$ [4.5] **18.** $3y^3 + 8y^2 + 12y - 5$

[4.7] **19.** $6p^2 + 7p + 1 + \dfrac{3}{p-1}$ [4.3] **20.** 1.4×10^5 sec [5.3] **21.** $(4t + 3v)(2t + v)$ **22.** prime

[5.4] **23.** $(4x^2 + 1)(2x + 1)(2x - 1)$ [5.5] **24.** $\{-3, 5\}$ **25.** $\left\{5, -\dfrac{1}{2}, \dfrac{2}{3}\right\}$ [5.6] **26.** -2 or -1 **27.** 6 m [6.1] **28.** A **29.** D

[6.4] **30.** $\dfrac{4}{q}$ **31.** $\dfrac{3r + 28}{7r}$ **32.** $\dfrac{7}{15(q-4)}$ **33.** $\dfrac{-k-5}{k(k+1)(k-1)}$ [6.2] **34.** $\dfrac{7(2z+1)}{24}$ [6.5] **35.** $\dfrac{195}{29}$ [6.6] **36.** 4, 0

37. $\left\{\dfrac{21}{2}\right\}$ **38.** $\{-2, 1\}$ [6.7] **39.** $1\dfrac{1}{5}$ hr **40.** $12\dfrac{2}{3}$ mph

---

CHAPTER 7 EQUATIONS OF LINES; FUNCTIONS

Section 7.1 (page 461)

Connections **(page 455)** We give each equation solved for y, the x-intercept, and the y-intercept. **1.** $y = -2.5x - 5; (-2, 0); (0, -5)$
2. $y = .75x + 1.5; (-2, 0); (0, 1.5)$ **3.** $y = 3.2x + 5.8; (-1.8125, 0); (0, 5.8)$ **4.** $y = -1.5x + 4.2; (2.8, 0); (0, 4.2)$

Exercises **1.** (a) x represents the year; y represents the percent of women in math or computer science professions. (b) 1990–2000

(c) (1990, 36) (d) In 2000, the percent of women in math or computer science professions was 30%. **3.** (a) I (b) III (c) II (d) IV (e) none

(f) none **5.–9.**   **11.** (a) $-3; 3; 2; -1$ (b)   **13.** (a) $-4; 5; -\dfrac{12}{5}; \dfrac{5}{4}$ (b)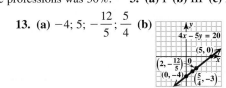

15. In quadrant III, both coordinates of the ordered pairs are negative. If $x + y = k$ and k is positive, then either x or y must be positive
because the sum of two negative numbers is negative. **17.** $(6, 0); (0, 4)$ **19.** $(6, 0); (0, -2)$

21. $\left(\dfrac{21}{2}, 0\right); \left(0, -\dfrac{7}{3}\right)$  **23.** $(0, 0); (0, 0)$ **25.** $(0, 0); (0, 0)$

27. none; $(0, 5)$ **29.** $(-4, 0)$; none **31.** B **33.** The screen on the right is more useful because it shows

the intercepts. **34.** $(6, -2)$ **35.** $(5, -2)$ **36.** $(6, 0)$ **37.** $(5, 0)$ **38.** 5; 0 **39.** The x-coordinate of M is the average of the

x-coordinates of P and Q. The y-coordinate of M is the average of the y-coordinates of P and Q. **41.** $(2, 5)$ **43.** $\left(-\dfrac{3}{2}, \dfrac{5}{2}\right)$

45. $\left(\dfrac{3}{2}, 0\right)$ **47.** $(3.8, 2.6)$ **49.** A, B, and D **51.** (a) B (b) C (c) A **53.** 0 **55.** $-\dfrac{1}{3}$ **57.** B and D are correct. Choice A

is wrong because the order of subtraction must be the same in the numerator and denominator. Choice C is wrong because slope is defined as the change in *y* divided by the change in *x*.   **59.** 1   **61.** $\frac{9}{8}$ **63.** 0   **65.** 2   **67.** B   **69.** A   **71.** $-\frac{1}{2}$

73. $\frac{5}{2}$ **75.** 4 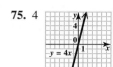   **77.** undefined   **79.** 0 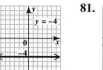   **81.**

83.   **85.**   **87.**   **89.** parallel   **91.** perpendicular   **93.** neither   **95.** neither

97. perpendicular   **99.** $-\$4000$ per yr; The value of the machine is decreasing \$4000 each year during these years.   **101.** 0% per yr (or no change); The percent of pay raise is not changing—it is 3% each year during these years.   **103.** (a) 1000 million per yr; 1000 million per yr; 1000 million per yr (b) The average rate of change is the same. When graphed, the data points lie on a straight line.   **105.** (a) \$200 million per yr (b) The positive slope means expenditures *increased* an average of \$200 million each year.   **107.** $-\$69$ per yr; The price decreased an average of \$69 each year from 1997 to 2002.

Section 7.2   (page 477)

Connections   (page 476)   **1.** {3}   **2.** {1}   **3.** {3}   **4.** {$-.5$}

Exercises   **1.** A   **3.** A   **5.** $3x + y = 10$ **7.** A   **9.** C   **11.** H   **13.** B   **15.** $y = 5x + 15$ **17.** $y = -\frac{2}{3}x + \frac{4}{5}$

19. $y = \frac{2}{5}x + 5$ **21.** $y = \frac{2}{3}x + 1$ **23.** (a) $y = x + 4$ (b) 1 (c) 4 (d)   **25.** (a) $y = -\frac{6}{5}x + 6$ (b) $-\frac{6}{5}$ (c) 6

(d) 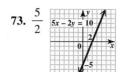   **27.** (a) $y = \frac{4}{5}x - 4$ (b) $\frac{4}{5}$ (c) -4 (d)   **29.** (a) $y = -\frac{1}{2}x - 2$ (b) $-\frac{1}{2}$ (c) -2 **(d)**

31. $3x + 4y = 10$ **33.** $2x + y = 18$ **35.** $x - 2y = -13$ **37.** $4x - y = 12$ **39.** The slope-intercept form, $y = mx + b$, is used when the slope and *y*-intercept are known. The point-slope form, $y - y_1 = m(x - x_1)$, is used when the slope and one point on a line or two points on a line are known. The standard form, $Ax + By = C$, is not useful for writing the equation, but the intercepts can be found quickly and used to graph the equation. The form $y = b$ is used for a horizontal line through the point (a, b). The form $x = a$ is used for a vertical line through the point (a, b).   **41.** $y = 5$ **43.** $x = 9$ **45.** $x = .5$ **47.** $y = 8$ **49.** $2x - y = 2$ **51.** $x + 2y = 8$

53. $2x - 13y = -6$ **55.** $y = 5$ **57.** $x = 7$ **59.** $y = -3$ **61.** $y = 3x - 19$ **63.** $y = \frac{1}{2}x - 1$ **65.** $y = -\frac{1}{2}x + 9$

67. $y = 7$ **69.** $y = 45x$; $(0, 0)$, $(5, 225)$, $(10, 450)$ **71.** $y = 1.50x$; $(0, 0)$, $(5, 7.50)$, $(10, 15.00)$ **73.** (a) $y = 39x + 99$ (b) $(5, 294)$; The cost of a 5-month membership is \$294. (c) \$567   **75.** (a) $y = 50x + 25$ (b) $(5, 275)$; The cost of the plan for 5 months is \$275. (c) \$1225   **77.** (a) $y = .20x + 50$ (b) $(5, 51)$; The charge for driving 5 mi is \$51. (c) 173 mi   **79.** (a) $y = 5.6x + 9$; The percent of households accessing the Internet by broadband is increasing 5.6% per year. (b) 43%   **81.** (a) $y = -103.2x + 28,908$ (b) 28,082; The result using the model is a little high.   **83.** 32; 212   **84.** $(0, 32)$ and $(100, 212)$ **85.** $\frac{9}{5}$ **86.** $F = \frac{9}{5}C + 32$ **87.** $C = \frac{5}{9}(F - 32)$ **88.** When the Celsius temperature is 50°, the Fahrenheit temperature is 122°.

Section 7.3 (page 492)

Exercises **1.** We give one of many possible answers here. A function is a set of ordered pairs in which each first element corresponds to exactly one second element. For example, $\{(0, 1), (1, 2), (2, 3), (3, 4), \ldots\}$ is a function. **3.** independent variable **5.** function **7.** not a function **9.** function **11.** not a function; domain: $\{0, 1, 2\}$; range: $\{-4, -1, 0, 1, 4\}$ **13.** function; domain: $\{2, 3, 5, 11, 17\}$; range: $\{1, 7, 20\}$ **15.** not a function; domain: $\{1\}$; range: $\{5, 2, -1, -4\}$ **17.** function; domain: $(-\infty, \infty)$; range: $(-\infty, \infty)$ **19.** function; domain: $(-\infty, \infty)$; range: $(-\infty, 4]$ **21.** not a function; domain: $[-4, 4]$; range: $[-3, 3]$ **23.** function; domain: $(-\infty, \infty)$ **25.** function; domain: $(-\infty, \infty)$ **27.** function; domain: $(-\infty, \infty)$ **29.** not a function; domain: $[0, \infty)$ **31.** function; domain: $(-\infty, 0) \cup (0, \infty)$

33. function; domain: $(-\infty, 9) \cup (9, \infty)$ **35.** function; domain: $\left(-\infty, -\dfrac{1}{2}\right) \cup \left(-\dfrac{1}{2}, \infty\right)$ **37.** function; domain: $(-\infty, \infty)$

39. B **41.** 4 **43.** -11 **45.** $-3p + 4$ **47.** $3x + 4$ **49.** $-3x - 2$ **51.** $-6m + 13$ **53.** (a) 2 (b) 3 **55.** (a) 15 (b) 10

57. (a) 3 (b) -3 **59.** (a) $f(x) = \dfrac{12 - x}{3}$ (b) 3 **61.** (a) $f(x) = 3 - 2x^2$ (b) -15 **63.** (a) $f(x) = \dfrac{8 - 4x}{-3}$ (b) $\dfrac{4}{3}$

65. line; -2; $-2x + 4$; -2; 3; -2 **67.** domain: $(-\infty, \infty)$; range: $(-\infty, \infty)$

69. domain: $(-\infty, \infty)$; range: $(-\infty, \infty)$ **71.** domain: $(-\infty, \infty)$; range: $(-\infty, \infty)$

73. domain: $(-\infty, \infty)$; range: $\{-4\}$ **75.** (a) 8.25 (dollars) (b) 3 is the value of the independent variable, which represents a

package weight of 3 lb; $f(3)$ is the value of the dependent variable representing the cost to mail a 3-lb package. (c) \$13.75; $f(5) = 13.75$
77. 194.53 cm **79.** 177.41 cm **81.** 1.83 m³ **83.** 4.11 m³ **85.** (a) yes (b) $[0, 24]$ (c) 1200 megawatts
(d) at 17 hr or 5 P.M.; at 4 A.M. (e) $f(12) = 2100$; At 12 noon, electricity use is 2100 megawatts. **87.** $f(3) = 7$

Section 7.4 (page 503)

Exercises **1.** direct **3.** direct **5.** inverse **7.** inverse **9.** inverse **11.** direct **13.** joint **15.** combined **17.** increases; decreases **19.** 36 **21.** $\dfrac{16}{9}$ **23.** .625 **25.** $\dfrac{16}{5}$ **27.** $222\dfrac{2}{9}$ **29.** If y varies inversely as x, then x is in the denominator; however, if y varies directly as x, then x is in the numerator. If $k > 0$, then with inverse variation, as x increases, y decreases. With direct variation, y increases as x increases. **31.** $\$1.69\dfrac{9}{10}$ **33.** about 450 cm³ **35.** 8 lb **37.** 256 ft **39.** $106\dfrac{2}{3}$ mph **41.** 100 cycles per sec

43. $21\dfrac{1}{3}$ foot-candles **45.** \$420 **47.** 448.1 lb **49.** approximately 68,600 calls **51.** 1.105 L **53.** 11.8 lb **55.** $(0, 0), (1, 1.25)$
56. 1.25 **57.** $y = 1.25x + 0$ or $y = 1.25x$ **58.** $a = 1.25, b = 0$ **59.** It is the price per gallon, and it is the slope of the line.
60. It can be written in the form $y = kx$ (where $k = a$). The value of a is called the constant of variation. **61.** It means that 4.6 gal cost \$5.75. **62.** It means that 12 gal cost \$15.00.

Chapter 7 Review Exercises (page 511)

1.

x	y
0	5
$\dfrac{10}{3}$	0
2	2
$\dfrac{14}{3}$	-2

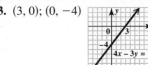

3. $(3, 0); (0, -4)$

5. $(10, 0); (0, 4)$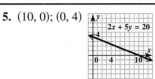

7. If both coordinates are positive, the point lies in quadrant I. If the first coordinate is negative and the second is positive, the point lies in quadrant II. To lie in quadrant III, the point must have both coordinates negative. To lie in quadrant IV, the first coordinate must be positive and the second must be negative. **9.** $-\dfrac{1}{2}$ **11.** $\dfrac{3}{4}$ **13.** $\dfrac{2}{3}$ **15.** undefined **17.** -1 **19.** negative **21.** 0 **23.** 12 ft

25. $y = -\dfrac{1}{3}x - 1$ **27.** $y = -\dfrac{4}{3}x + \dfrac{29}{3}$ **29.** $x = 2$ (Slope-intercept form is not possible.) **31.** $y = \dfrac{7}{5}x + \dfrac{16}{5}$ **33.** $y = 4x - 29$

35. (a) $y = 57x + 159; \$843$ **(b)** $y = 47x + 159; \$723$ **37.** domain: $\{-4, 1\}$; range: $\{2, -2, 5, -5\}$; not a function **39.** domain: $[-4, 4]$; range: $[0, 2]$; function **41.** function; domain: $(-\infty, \infty)$; linear function **43.** function; domain: $(-\infty, 6) \cup (6, \infty)$ **45.** -6 **47.** -8 **49. (a)** yes **(b)** domain: $\{1994, 1995, 1996, 1997, 1998, 1999\}$; range: $\{40, 60, 80, 130, 180, 200\}$ **(c)** Answers will vary. Two possible answers are $(1994, 40)$ and $(1995, 60)$. **(d)** 40; In 1994, CNBC profits were $40 million. **(e)** 1998 **51.** C **53.** Because it falls from left to right, the slope is negative. **54.** $-\dfrac{3}{2}$ **55.** $-\dfrac{3}{2}; \dfrac{2}{3}$ **56.** $\left(\dfrac{7}{3}, 0\right)$ **57.** $\left(0, \dfrac{7}{2}\right)$ **58.** $f(x) = -\dfrac{3}{2}x + \dfrac{7}{2}$ **59.** $f(8) = -\dfrac{17}{2}$ **60.** $x = \dfrac{23}{3}$

61. 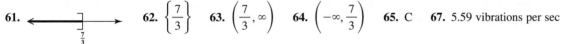 **62.** $\left\{\dfrac{7}{3}\right\}$ **63.** $\left(\dfrac{7}{3}, \infty\right)$ **64.** $\left(-\infty, \dfrac{7}{3}\right)$ **65.** C **67.** 5.59 vibrations per sec

Chapter 7 Test (page 515)

[7.1] **1.** $-\dfrac{10}{3}; -2; 0$ **2.** $\left(\dfrac{20}{3}, 0\right); (0, -10)$ **3.** none; $(0, 5)$ **4.** $(2, 0)$; none

5. $\dfrac{1}{2}$ **6.** It is a vertical line. **7.** perpendicular **8.** neither **9.** -1214 farms per yr; The number of farms decreased by about 1214 each year from 1980 to 2001. [7.2] **10.** $y = -5x + 19$ **11.** $y = 14$ **12.** $y = -\dfrac{3}{5}x - \dfrac{11}{5}$ **13.** $y = -\dfrac{1}{2}x - \dfrac{3}{2}$

14. $y = -\dfrac{1}{2}x + 2$ **15.** B **16. (a)** $y = 1968.75x + 13{,}016.25$ **(b)** $\$26{,}798$; It is close to the actual value. [7.3] **17.** D **18.** D

19. (a) domain: $[0, \infty)$; range: $(-\infty, \infty)$ **(b)** domain: $\{0, -2, 4\}$; range: $\{1, 3, 8\}$ **20.** $0; -a^2 + 2a - 1$
21. domain: $(-\infty, \infty)$; range: $(-\infty, \infty)$ 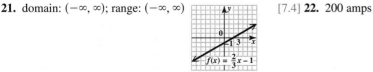 [7.4] **22.** 200 amps **23.** .8 lb

Cumulative Review Exercises Chapters 1–7   (page 517)

[1.4, 1.5] **1.** always true   **2.** always true   **3.** never true   **4.** sometimes true; for example, $3 + (-3) = 0$, but $3 + (-1) = 2 \neq 0$ **5.** 4

[1.8] **6.** $2x^2 + 5x + 4$ [1.5, 1.6] **7.** -24 **8.** undefined   [2.3] **9.** $\left\{\dfrac{7}{6}\right\}$ **10.** $\{-1\}$ [2.5] **11.** $h = \dfrac{3V}{\pi r^2}$ **12.** 6 in.   [2.7] **13.** 2 hr

[2.8] **14.** $\left(-3, \dfrac{7}{2}\right)$ **15.** $(-\infty, 1]$ [3.2, 7.1] **16.** x-intercept: $(4, 0)$;

y-intercept: $\left(0, \dfrac{12}{5}\right)$ [3.3, 7.1] **17. (a)** $-\dfrac{6}{5}$ **(b)** $\dfrac{5}{6}$ [3.4, 7.2] **18.** $y = -\dfrac{3}{4}x - 1$ **19.** $y = -2$

20. $y = -\dfrac{4}{3}x + \dfrac{7}{3}$ [4.1, 4.2] **21.** $\dfrac{y}{18x}$ **22.** $\dfrac{5my^4}{3}$ [4.4] **23.** $x^3 + 12x^2 - 3x - 7$ [4.6] **24.** $49x^2 + 42xy + 9y^2$

[4.5] **25.** $10p^3 + 7p^2 - 28p - 24$ [4.7] **26.** $m^2 - 2m + 3$ [5.1–5.4] **27.** $(2w + 7z)(8w - 3z)$ **28.** $(2x - 1 + y)(2x - 1 - y)$

29. $(2y - 9)^2$ **30.** $(10x^2 + 9)(10x^2 - 9)$ **31.** $(2p + 3)(4p^2 - 6p + 9)$ [5.5] **32.** $\left\{-4, -\dfrac{3}{2}, 1\right\}$ **33.** $\left\{\dfrac{1}{3}\right\}$ [5.6] **34.** 4 ft

35. longer sides: 18 in.; distance between: 16 in.   [6.4] **36.** $\dfrac{6x + 22}{(x + 1)(x + 3)}$ [6.1] **37.** $\dfrac{4(x - 5)}{3(x + 5)}$ **38.** $\dfrac{(x + 3)^2}{3x}$ [6.5] **39.** 6

[6.6] **40.** $\{5\}$ [7.3] **41.** domain: $\{14, 91, 75, 23\}$; range: $\{9, 70, 56, 5\}$; not a function; 75 in the domain is paired with two different values, 70 and 56, in the range.   **42. (a)** domain: $(-\infty, \infty)$; range: $(-\infty, \infty)$ **(b)** 22   [7.1] **43.** 10.5 million per year; The number of U.S. cell phone subscribers increased by 10.5 million per year from 1992 to 2000.   [7.4] **44.** \$9.92

CHAPTER 8 SYSTEMS OF LINEAR EQUATIONS

Section 8.1   (page 527)

Exercises   **1. (a)** B **(b)** C **(c)** D **(d)** A   **3.** no   **5.** yes   **7.** yes   **9.** no   **11.** A; The ordered pair solution must be in quadrant II, and $(-4, -4)$ is in quadrant III.   **13.** $\{(4, 2)\}$ **15.** $\{(0, 4)\}$ **17.** $\{(4, -1)\}$

In Exercises 19–27, we do not show the graphs.   **19.** $\{(1, 3)\}$ **21.** $\{(0, 2)\}$ **23.** $\emptyset$ (inconsistent system)   **25.** $\{(x, y) \mid 3x + y = 5\}$ (dependent equations)   **27.** $\{(4, -3)\}$ **29.** If the coordinates of the point of intersection are not integers, the solution will be difficult to determine from a graph.   **31. (a)** neither **(b)** intersecting lines **(c)** one solution   **33. (a)** dependent **(b)** one line **(c)** infinite number of solutions   **35. (a)** inconsistent **(b)** parallel lines **(c)** no solution   **37. (a)** neither **(b)** intersecting lines **(c)** one solution   **39.** 40   **41.** $(40, 30)$ **43.** 1995–1997   **45. (a)** 1989–1997 **(b)** 1997; NBC; 17% **(c)** 1989: share 20%; 1998: share 16% **(d)** NBC and ABC; $(2000, 16)$ **(e)** Viewership has generally declined during these years.   **47.** B   **49.** A   **51.** $\{(-1, 5)\}$ **53.** $\{(.25, -.5)\}$

Section 8.2   (page 537)

Exercises   **1.** No, it is not correct, because the solution set is $\{(3, 0)\}$. The y-value in the ordered pair must also be determined.   **3.** $\{(3, 9)\}$

5. $\{(7, 3)\}$ **7.** $\{(0, 5)\}$ **9.** $\{(-4, 8)\}$ **11.** $\{(3, -2)\}$ **13.** $\{(x, y) \mid 3x - y = 5\}$ **15.** $\left\{\left(\dfrac{1}{4}, -\dfrac{1}{2}\right)\right\}$ **17.** $\emptyset$

19. $\{(x, y) \mid 2x - y = -12\}$ **21. (a)** A false statement, such as $0 = 3$, occurs. **(b)** A true statement, such as $0 = 0$, occurs.   **23.** $\{(2, 6)\}$

25. $\{(2, -4)\}$ **27.** $\{(-2, 1)\}$ **29.** $\left\{\left(13, -\dfrac{7}{5}\right)\right\}$ **31.** $\{(x, y) \mid x + 2y = 48\}$ **33.** To find the total cost, multiply the number of bicycles (x) by the cost per bicycle ($400), and add the fixed cost ($5000). Thus, $y_1 = 400x + 5000$ gives this total cost (in dollars).

34. $y_2 = 600x$ **35.** $y_1 = 400x + 5000$, $y_2 = 600x$; solution set: $\{(25, 15{,}000)\}$ **36.** 25; 15,000; 15,000

37. $\{(2, 4)\}$ 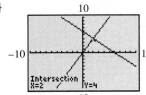 **39.** $\{(1, 5)\}$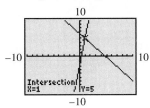

41. $\{(5, -3)\}$; The equations to input are $Y_1 = \dfrac{5 - 4X}{5}$ and $Y_2 = \dfrac{1 - 2X}{3}$.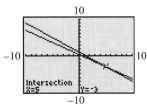

43. Adjust the viewing window so that it does appear.

Section 8.3 (page 543)

Exercises **1.** false; Multiply by -3. **3.** true **5.** $\{(4, 6)\}$ **7.** $\{(-1, -3)\}$ **9.** $\{(-2, 3)\}$ **11.** $\left\{\left(\dfrac{1}{2}, 4\right)\right\}$ **13.** $\{(3, -6)\}$

15. $\{(0, 4)\}$ **17.** $\{(0, 0)\}$ **19.** $\{(7, 4)\}$ **21.** $\{(0, 3)\}$ **23.** $\{(3, 0)\}$ **25.** $\left\{\left(-\dfrac{2}{3}, \dfrac{17}{2}\right)\right\}$ **27.** $\left\{\left(-\dfrac{32}{23}, -\dfrac{17}{23}\right)\right\}$ **29.** $\left\{\left(-\dfrac{6}{5}, \dfrac{4}{5}\right)\right\}$

31. $\left\{\left(\dfrac{1}{8}, -\dfrac{5}{6}\right)\right\}$ **33.** $\{(11, 15)\}$ **35.** $\{(x, y) \mid x - 3y = -4\}$ **37.** $\emptyset$ **39.** $1141 = 1991a + b$ **40.** $1465 = 1999a + b$

41. $1991a + b = 1141$; $1999a + b = 1465$; solution set: $\{(40.5, -79{,}494.5)\}$ **42.** $y = 40.5x - 79{,}494.5$ **43.** 1424.5 (million); This is less than the actual figure. **44.** Since the data do not lie in a perfectly straight line, the quantity obtained from an equation determined in this way will probably be "off" a bit. We cannot put too much faith in models such as this one, because not all sets of data points are linear in nature.

Summary Exercises on Solving Systems of Linear Equations (page 546)

1. (a) Use substitution since the second equation is solved for y. **(b)** Use elimination since the coefficients of the y-terms are opposites. **(c)** Use elimination since the equations are in standard form with no coefficients of 1 or -1. Solving by substitution would involve fractions.
2. The system on the right is easier to solve by substitution because the second equation is already solved for y. **3. (a)** $\{(1, 4)\}$ **(b)** $\{(1, 4)\}$ **(c)** Answers will vary. **4. (a)** $\{(-5, 2)\}$ **(b)** $\{(-5, 2)\}$ **(c)** Answers will vary. **5.** $\{(3, 12)\}$ **6.** $\{(-3, 2)\}$
7. $\left\{\left(\dfrac{1}{3}, \dfrac{1}{2}\right)\right\}$ **8.** $\emptyset$ **9.** $\{(3, -2)\}$ **10.** $\{(-1, -11)\}$ **11.** $\{(x, y) \mid 2x - 3y = 5\}$ **12.** $\{(9, 4)\}$ **13.** $\left\{\left(\dfrac{45}{31}, \dfrac{4}{31}\right)\right\}$
14. $\{(4, -5)\}$ **15.** $\emptyset$ **16.** $\{(-4, 6)\}$ **17.** $\{(-3, 2)\}$ **18.** $\left\{\left(\dfrac{22}{13}, -\dfrac{23}{13}\right)\right\}$ **19.** $\{(5, 3)\}$ **20.** $\{(2, -3)\}$ **21.** $\{(24, -12)\}$
22. $\{(10, -12)\}$ **23.** $\{(3, 2)\}$ **24.** $\{(-4, 2)\}$

Section 8.4 (page 552)

Exercises **1.** The statement means that when -1 is substituted for x, 2 is substituted for y, and 3 is substituted for z in the three equations, the resulting three statements are true. **3.** $\{(3, 2, 1)\}$ **5.** $\{(1, 4, -3)\}$ **7.** $\{(0, 2, -5)\}$ **9.** $\{(1, 0, 3)\}$ **11.** $\{(-12, 18, 0)\}$
13. $\left\{\left(1, \dfrac{3}{10}, \dfrac{2}{5}\right)\right\}$ **15.** $\left\{\left(-\dfrac{7}{3}, \dfrac{22}{3}, 7\right)\right\}$ **17.** $\{(4, 5, 3)\}$ **19.** $\{(2, 2, 2)\}$ **21.** $\left\{\left(\dfrac{8}{3}, \dfrac{2}{3}, 3\right)\right\}$ **23.** $\{(-1, 0, 0)\}$
25. $\{(-3, 5, -6)\}$ **27.** Answers will vary. Some possible answers are **(a)** two perpendicular walls and the ceiling in a normal room, **(b)** the floors of three different levels of an office building, and **(c)** three pages of this book (since they intersect in the spine). **29.** $\emptyset$; inconsistent system **31.** $\{(x, y, z) \mid x - y + 4z = 8\}$; dependent equations **33.** $\{(x, y, z) \mid 2x + y - z = 6\}$; dependent equations **35.** $\{(0, 0, 0)\}$
37. $\{(2, 1, 5, 3)\}$ **39.** $\{(-2, 0, 1, 4)\}$ **41.** $128 = a + b + c$ **42.** $140 = 2.25a + 1.5b + c$ **43.** $80 = 9a + 3b + c$

44.
$$a + b + c = 128$$
$$2.25a + 1.5b + c = 140$$
$$9a + 3b + c = 80; \{(-32, 104, 56)\}$$
45. $f(x) = -32x^2 + 104x + 56$ **46.** height; time **47.** 56 ft **48.** 140.5 ft
49. $a = 3, b = 1, c = -2; f(x) = 3x^2 + x - 2$ **50.** $a = 1, b = 4, c = 3; Y_1 = X^2 + 4X + 3$ **51.** If one were to eliminate *different* variables in the first two steps, the result would be two equations in three variables, and it would not be possible to solve for a single variable in the next step.

Section 8.5 (page 562)

Connections (**page 558**) "Mixed price" refers to the price of a mixture of the two products. The system is $9x + 7y = 107, 7x + 9y = 101$, where x represents the price of a citron and y represents the price of a wood apple.

Exercises **1.** wins: 95; losses: 67 **3.** length: 78 ft; width: 36 ft **5.** ExxonMobil: \$214 billion; General Motors: \$185 billion **7.** $x = 40$ and $y = 50$, so the angles measure $40°$ and $50°$. **9.** NHL: \$219.74; NBA: \$203.38 **11.** single: \$2.09; double: \$3.19 **13.** **(a)** 6 oz **(b)** 15 oz **(c)** 24 oz **(d)** 30 oz **15.** $\$.99x$ **17.** 6 gal of 25%; 14 gal of 35% **19.** pure acid: 6 L; 10% acid: 48 L **21.** nuts: 14 kg; cereal: 16 kg **23.** \$1000 at 2%; \$2000 at 4% **25.** $25y$ **27.** train: 60 km per hr; plane: 160 km per hr **29.** boat: 21 mph; current: 3 mph **31.** Turner: \$80.4 million; 'N Sync: \$76.6 million **33.** general admission: 76; with student ID: 108 **35.** 8 for a citron; 5 for a wood apple **37.** $x + y + z = 180$; angle measures: $70°, 30°, 80°$ **39.** first: $20°$; second: $70°$; third: $90°$ **41.** shortest: 12 cm; middle: 25 cm; longest: 33 cm **43.** Independent: 38; Democrat: 34; Republican: 28 **45.** \$10 tickets: 350; \$18 tickets: 250; \$30 tickets: 50 **47.** type A: 80; type B: 160; type C: 250 **49.** first chemical: 50 kg; second chemical: 400 kg; third chemical: 300 kg **51.** wins: 48; losses: 26; ties: 8

Section 8.6 (page 575)

Connections (**page 574**) **1.** $\begin{bmatrix} 1 & 0 & | & -2 \\ 0 & 1 & | & -1 \end{bmatrix}$ **2.** a system of dependent equations

Exercises **1.** **(a)** $0, 5, -3$ **(b)** $1, -3, 8$ **(c)** yes; The number of rows is the same as the number of columns (three). **(d)** $\begin{bmatrix} 1 & 4 & 8 \\ 0 & 5 & -3 \\ -2 & 3 & 1 \end{bmatrix}$

(e) $\begin{bmatrix} 1 & -\frac{3}{2} & -\frac{1}{2} \\ 0 & 5 & -3 \\ 1 & 4 & 8 \end{bmatrix}$ **(f)** $\begin{bmatrix} 1 & 15 & 25 \\ 0 & 5 & -3 \\ 1 & 4 & 8 \end{bmatrix}$ **3.** $\begin{bmatrix} 1 & 2 & | & 11 \\ 2 & -1 & | & -3 \end{bmatrix}; \begin{bmatrix} 1 & 2 & | & 11 \\ 0 & -5 & | & -25 \end{bmatrix}; \begin{bmatrix} 1 & 2 & | & 11 \\ 0 & 1 & | & 5 \end{bmatrix}; x + 2y = 11, y = 5; \{(1, 5)\}$

5. $\{(4, 1)\}$ **7.** $\{(1, 1)\}$ **9.** $\{(-1, 4)\}$ **11.** $\emptyset$ **13.** $\{(x, y) \,|\, 2x + y = 4\}$ **15.** $\begin{bmatrix} 1 & 1 & -1 & | & -3 \\ 0 & -1 & 3 & | & 10 \\ 0 & -6 & 7 & | & 38 \end{bmatrix}; \begin{bmatrix} 1 & 1 & -1 & | & -3 \\ 0 & 1 & -3 & | & -10 \\ 0 & -6 & 7 & | & 38 \end{bmatrix};$

$\begin{bmatrix} 1 & 1 & -1 & | & -3 \\ 0 & 1 & -3 & | & -10 \\ 0 & 0 & -11 & | & -22 \end{bmatrix}; \begin{bmatrix} 1 & 1 & -1 & | & -3 \\ 0 & 1 & -3 & | & -10 \\ 0 & 0 & 1 & | & 2 \end{bmatrix}; x + y - z = -3, y - 3z = -10, z = 2; \{(3, -4, 2)\}$ **17.** $\{(4, 0, 1)\}$

19. $\{(-1, 23, 16)\}$ **21.** $\{(3, 2, -4)\}$ **23.** $\{(x, y, z) \,|\, x - 2y + z = 4\}$ **25.** $\emptyset$ **27.** Examples will vary. **(a)** A matrix is a rectangular array of numbers. **(b)** A horizontal arrangement of elements in a matrix is a row of a matrix. **(c)** A vertical arrangement of elements in a matrix is a column of a matrix. **(d)** A square matrix contains the same number of rows as columns. **(e)** A matrix formed by the coefficients and constants of a linear system is an augmented matrix for the system. **(f)** Row operations on a matrix allow it to be transformed into another matrix in which the solution of the associated system can be found more easily. **29.** $\{(1, 1)\}$ **31.** $\{(-1, 2, 1)\}$ **33.** $\{(1, 7, -4)\}$

Chapter 8 Review Exercises (page 584)

1. yes **3.** $\{(3, 1)\}$ **5.** $\emptyset$ **7.** No, this is not correct. A false statement indicates that the solution set is $\emptyset$. **9.** $\{(3, 5)\}$ **11.** His answer was incorrect since the system has infinitely many solutions (as indicated by the true statement $0 = 0$). **13.** C **15.** $\{(7, 1)\}$

17. $\{(x, y) \mid 3x - 4y = 9\}$ **19.** $\{(-4, 1)\}$ **21.** $\{(9, 2)\}$ **23.** Answers will vary.   **25.** $\{(1, -5, 3)\}$ **27.** $\emptyset$; inconsistent system
29. (a) *Harry Potter and the Sorcerer's Stone:* \$294 million; *Shrek:* \$268 million **(b)** \$241 million **(c)** \$803 million   **31.** \$2-per-lb nuts:
30 lb; \$1-per-lb candy: 70 lb   **33.** \$40,000 at 10%; \$100,000 at 6%; \$140,000 at 5%   **35.** Mantle: 54; Maris: 61; Blanchard: 21
37. $\{(-1, 5)\}$ **39.** $\{(1, 2, -1)\}$ **41.** $\emptyset$ **43.** $\{(5, 3)\}$ **45.** $\left\{\left(\dfrac{82}{23}, -\dfrac{4}{23}\right)\right\}$ **47.** Subway: 13,247 restaurants; McDonald's:
13,099 restaurants   **49.** U.S.: 97; Russia: 88; China: 59   **50.** $2a + b + c = -5$ **51.** $-a + c = -1$ **52.** $3a + 3b + c = -18$
53. $a = 1, b = -7, c = 0$; $x^2 + y^2 + x - 7y = 0$ **54.** The relation is not a function because a vertical line intersects its graph more than once.

Chapter 8 Test   (page 588)

[8.1] **1.** $x = 8$ or 800 parts   **2.** \$3000   **3. (a)** no **(b)** no **(c)** yes   **4.** $\{(6, 1)\}$ [8.2] **5.** $\{(6, -4)\}$

6. $\{(x, y) \mid 12x - 5y = 8\}$; dependent equations   [8.3] **7.** $\{(3, 3)\}$ **8.** $\{(0, -2)\}$ **9.** $\emptyset$; inconsistent system   **10.** $\{(-15, 6)\}$
[8.4] **11.** $\left\{\left(-\dfrac{2}{3}, \dfrac{4}{5}, 0\right)\right\}$ **12.** $\{(3, -2, 1)\}$ [8.5] **13.** *Pretty Woman:* \$178.4 million; *Runaway Bride:* \$152.3 million   **14.** 45 mph,
75 mph   **15.** 20% solution: 4 L; 50% solution: 8 L   **16.** AC adaptor: \$8; rechargeable flashlight: \$15   **17.** Orange Pekoe: 60 oz; Irish
Breakfast: 30 oz; Earl Grey: 10 oz   [8.6] **18.** $\left\{\left(\dfrac{2}{5}, \dfrac{7}{5}\right)\right\}$ **19.** $\{(-1, 2, 3)\}$ [8.4, 8.6] **20.** $\{(-3, -2, -4)\}$

Cumulative Review Exercises Chapters 1–8   (page 589)

[1.3] **1.** 1   [1.7] **2.** commutative property   **3.** distributive property   [1.6] **4.** 46   [2.3] **5.** $\left\{-\dfrac{13}{11}\right\}$ **6.** $\left\{\dfrac{9}{11}\right\}$ [2.5] **7.** $T = \dfrac{PV}{k}$

[2.8] **8.** $(-18, \infty)$ **9.** $\left(-\dfrac{11}{2}, \infty\right)$ [2.7] **10.** 2010; 1813; 62.8%; 57.2%   [2.4] **11.** not guilty: 105; guilty: 95   [2.5] **12.** 46°, 46°, 88°

13. width: 8.25 in.; length: 10.75 in.   [3.2, 7.1] **14.**   **15.**   [3.3, 7.1] **16.** $-\dfrac{4}{3}$ **17.** $-\dfrac{1}{4}$

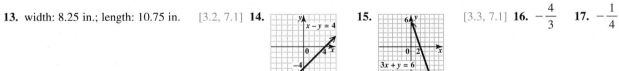

[3.4, 7.2] **18.** $y = \dfrac{1}{2}x + 3$ **19.** $y = 2x + 1$ **20. (a)** $x = 9$ **(b)** $y = -1$ [4.1, 4.2] **21.** $\dfrac{a^{10}}{b^{10}}$ **22.** $\dfrac{m}{n}$ [4.4] **23.** $4y^2 - 7y - 6$

[4.5] **24.** $12f^2 + 5f - 3$ [4.6] **25.** $\dfrac{1}{16}x^2 + \dfrac{5}{2}x + 25$ [4.7] **26.** $x^2 + 4x - 7$ [5.2] **27.** $(2x + 5)(x - 9)$

[5.4] **28.** $25(2t^2 + 1)(2t^2 - 1)$ **29.** $(2p + 5)(4p^2 - 10p + 25)$ [5.5] **30.** $\left\{-\dfrac{7}{3}, 1\right\}$ [6.1] **31.** $\dfrac{y + 4}{y - 4}$ [6.2] **32.** $\dfrac{a(a - b)}{2(a + b)}$

[6.4] **33.** $\dfrac{2(x + 2)}{2x - 1}$ [6.6] **34.** $\{-4\}$ [7.3] **35.** function; domain: $\{1990, 1992, 1994, 1996, 1998, 2000\}$;
range: $\{1.25, 1.61, 1.80, 1.21, 1.94, 2.26\}$ **36.** not a function; domain: $[-2, \infty)$; range: $(-\infty, \infty)$ **37. (a)** $f(x) = \dfrac{5x - 8}{3}$ or
$f(x) = \dfrac{5}{3}x - \dfrac{8}{3}$ **(b)** -1 [7.4] **38.** 17.5   [8.1–8.3, 8.6] **39.** $\{(3, -3)\}$ **40.** $\{(x, y) \mid x - 3y = 7\}$ [8.4, 8.6] **41.** $\{(5, 3, 2)\}$
[8.5] **42.** Tickle Me Elmo: \$27.63; Snacktime Kid: \$36.26   **43.** peanuts: \$2 per lb; cashews: \$4 per lb   [8.1] **44. (a)** years 0 to 6
(b) year 6; about \$650

CHAPTER 9 INEQUALITIES AND ABSOLUTE VALUE

Section 9.1   (page 599)

Exercises   **1.** true   **3.** false; The union is $(-\infty, 8) \cup (8, \infty)$.   **5.** false; The intersection is $\emptyset$.   **7.** $\{1, 3, 5\}$ or B **9.** $\{4\}$ or D **11.** $\emptyset$

13. $\{1, 2, 3, 4, 5, 6\}$ or A **15.** Answers will vary. One example is: The intersection of two streets is the region common to *both* streets.

17.   **19.**   **21.** $(-3, 2)$ **23.** $(-\infty, 2]$

25. $\emptyset$ **27.** $[5, 9]$ **29.** $(-3, -1)$ **31.** $(-\infty, 4]$ **33.**

35.   **37.** $(-\infty, 8]$ **39.** $[-2, \infty)$ **41.** $(-\infty, \infty)$

43. $(-\infty, -5) \cup (5, \infty)$ **45.** $(-\infty, -1) \cup (2, \infty)$ **47.** $[-4, -1]$ **49.** $[-9, -6]$ **51.** $(-\infty, 3)$

53. $[3, 9)$ **55.** intersection; $(-5, -1)$ **57.** union; $(-\infty, 4)$

59. union; $(-\infty, 0] \cup [2, \infty)$ **61.** intersection; $[4, 12]$ **63.** Maria, Joe   **64.** none of them

65. none of them   **66.** Luigi, Than   **67.** Maria, Joe   **68.** all of them   **69.** {Tuition and fees}   **71.** {Tuition and fees, Board rates}

Section 9.2   (page 608)

Connections   **(page 608)**   The filled carton may contain between 30.4 and 33.6 oz, inclusive.

Exercises   **1.** E; C; D; B; A   **3.** Use *or* for the equality statement and the $>$ statement. Use *and* for the $<$ statement.   **5.** $\{-12, 12\}$

7. $\{-5, 5\}$ **9.** $\{-6, 12\}$ **11.** $\{-5, 6\}$ **13.** $\left\{-3, \dfrac{11}{2}\right\}$ **15.** $\left\{-\dfrac{19}{2}, \dfrac{9}{2}\right\}$ **17.** $\{-10, -2\}$ **19.** $\left\{-\dfrac{32}{3}, 8\right\}$

21. $(-\infty, -3) \cup (3, \infty)$ **23.** $(-\infty, -4] \cup [4, \infty)$

25. $(-\infty, -25] \cup [15, \infty)$ **27.** $(-\infty, -12) \cup (8, \infty)$

29. $(-\infty, -2) \cup (8, \infty)$ **31.** $\left(-\infty, -\dfrac{9}{5}\right] \cup [3, \infty)$

33. (a)   (b)   **35.** $[-3, 3]$ **37.** $(-4, 4)$

39. $[-25, 15]$ **41.** $[-12, 8]$ **43.** $[-2, 8]$

45. $\left[-\dfrac{9}{5}, 3\right]$ **47.** $(-\infty, -5) \cup (13, \infty)$ **49.** $(-\infty, -25) \cup (15, \infty)$

51. $\{-6, -1\}$ **53.** $\left[-\dfrac{10}{3}, 4\right]$ **55.** $\left[-\dfrac{7}{6}, -\dfrac{5}{6}\right]$

57. $(-\infty, -3] \cup [4, \infty)$ **59.** $\{-5, 5\}$ **61.** $\{1, -5\}$ **63.** $\{-5, -3\}$ **65.** $(-\infty, -3) \cup (2, \infty)$ **67.** $[-10, 0]$

69. $(-\infty, -1) \cup (5, \infty)$ **71.** $\{-1, 3\}$ **73.** $\left\{-3, \dfrac{5}{3}\right\}$ **75.** $\left\{-\dfrac{1}{3}, -\dfrac{1}{15}\right\}$ **77.** $\left\{-\dfrac{5}{4}\right\}$ **79.** $\emptyset$ **81.** $\left\{-\dfrac{1}{4}\right\}$ **83.** $\emptyset$

85. $(-\infty, \infty)$ **87.** $\left\{-\dfrac{3}{7}\right\}$ **89.** $\left\{\dfrac{2}{5}\right\}$ **91.** $\emptyset$ **93.** $|x - 1000| \le 100;\ 900 \le x \le 1100$ **95.** 472.9 ft   **96.** 1201 Walnut,

Fidelity Bank and Trust Building, City Hall, Kansas City Power and Light, Hyatt Regency   **97.** City Center Square, Commerce Tower,
Federal Office Building, 1201 Walnut, Fidelity Bank and Trust Building, City Hall, Kansas City Power and Light, Hyatt Regency
98. (a) $|x - 472.9| \ge 75$ **(b)** $x \ge 547.9$ or $x \le 397.9$ **(c)** AT&T Town Pavilion, One Kansas City Place **(d)** It makes sense because it
includes all buildings *not* listed earlier.

Section 9.3   (page 616)

Connections   (page 614)   1. Answers will vary.   **2. (a)** $(-\infty, 2]$ **(b)** $[2, \infty)$ **(c)** $(5, \infty)$ **(d)** $(-\infty, 5)$

Connections   (page 616)   1. $x \le 200,\ x \ge 100,\ y \ge 3000$ **2.**   **3.** $C = 50x + 100y$ **4.** Some examples are

$(100, 5000), (150, 3000),$ and $(150, 5000).$ The corner points are $(100, 3000)$ and $(200, 3000).$ **5.** The least cost occurs when $x = 100$ and
$y = 3000.$ The company should use 100 workers and manufacture 3000 units to achieve the least possible cost.

Exercises   1. solid; below   **3.** dashed; above   **5.** The graph of $Ax + By = C$ divides the plane into two regions. In one of the regions,
the ordered pairs satisfy $Ax + By < C$; in the other, they satisfy $Ax + By > C.$ **7.**   **9.**

11. 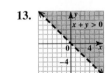   **13.**   **15.**   **17.** 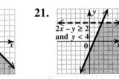   **19.**   **21.**

23. 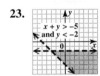   **25.** $-3 < x < 3$ 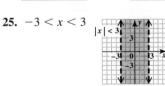   **27.** $-2 < x + 1 < 2$ **29.**

31. 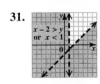   **33.** 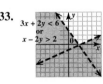   **35.** C   **37.** A   **39. (a)** $\{-4\}$ **(b)** $(-\infty, -4)$ **(c)** $(-4, \infty)$

41. (a) $\{3.5\}$ **(b)** $(3.5, \infty)$ **(c)** $(-\infty, 3.5)$ **We include a calculator graph and supporting explanation only with the answer to Exercise 43.**
43. (a) $\{-.6\}$ **(b)** $(-.6, \infty)$ **(c)** $(-\infty, -.6)$ The graph of $y_1 = 5x + 3$ has x-intercept $(-.6, 0)$, supporting the result of part (a). The graph of y_1
lies *above* the x-axis for values of x *greater than* $-.6$, supporting the result of part (b). The graph of y_1 lies *below* the x-axis for values of x *less
than* $-.6$, supporting the result of part (c). $y_1 = 5x + 3$ **45. (a)** $\{-1.2\}$ **(b)** $(-\infty, -1.2]$ **(c)** $[-1.2, \infty)$

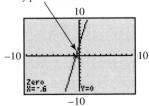

47. (a)   **(b)** (500, 0), (200, 400), (100, 450) Answers may vary.   **49. (a)**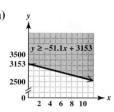

(b) (0, 3153), (2, 3050.8), (10, 2642) Answers may vary.

Chapter 9 Review Exercises (page 623)

1. {a, c}   **3.** {a, c, e, f, g}   **5.** (6, 9)    **7.** $(-\infty, -3] \cup (5, \infty)$    **9.** $\emptyset$ **11.** $(-3, 4)$

13. $(4, \infty)$ **15.** {−7, 7}   **17.** $\left\{-\dfrac{1}{3}, 5\right\}$ **19.** {0, 7}   **21.** $\left\{-\dfrac{3}{4}, \dfrac{1}{2}\right\}$ **23.** $(-14, 14)$ **25.** $[-3, -2]$

27. $\left(-\infty, -\dfrac{8}{5}\right) \cup (2, \infty)$ **29.** $(-\infty, \infty)$ **31.**   **33.**   **35.**   **37.** $[-2, 3)$

39. $\left\{-\dfrac{7}{3}, 1\right\}$ **41.** $\left\{-4, -\dfrac{2}{3}\right\}$ **43.** $(-\infty, \infty)$ **45.** $[-4, -2]$ **47.** 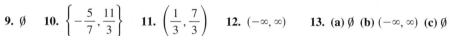   **49.**

51. Managerial/Professional, Technical/Sales/Administrative Support, Service, Operators/Fabricators/Laborers

Chapter 9 Test (page 625)

[9.1] **1.** none   **2.** 1946, 1960, 1968, 1976, 1996, 1998, 1999   **3.** {1, 5}   **4.** {1, 2, 5, 7, 9, 12}

5. $[2, 9)$ 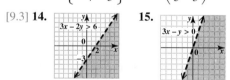   **6.** $(-\infty, 3) \cup [6, \infty)$ [9.2] **7.** $\left\{-1, \dfrac{5}{2}\right\}$ **8.** $\left(-\infty, -\dfrac{7}{6}\right) \cup \left(\dfrac{17}{6}, \infty\right)$

9. $\emptyset$ **10.** $\left\{-\dfrac{5}{7}, \dfrac{11}{3}\right\}$ **11.** $\left(\dfrac{1}{3}, \dfrac{7}{3}\right)$ **12.** $(-\infty, \infty)$ **13.** (a) $\emptyset$ (b) $(-\infty, \infty)$ (c) $\emptyset$

[9.3] **14.**   **15.**   **16.**   **17.**

Cumulative Review Exercises Chapters 1–9 (page 626)

[1.4] **1. (a)** A, B, C, D, F **(b)** B, C, D, F **(c)** D, F **(d)** E, F **(e)** E, F **(f)** D, F   [1.2] **2.** 32   [1.4, 1.5] **3.** 0

[2.1–2.3] **4.** {−65}   **5.** $(-\infty, \infty)$ [2.5] **6.** $t = \dfrac{A - p}{pr}$ [2.8] **7.** $(-\infty, 6)$ [2.7] **8.** 32%; 390; 270; 10%   [2.4] **9.** 15°, 35°, 130°

[3.3, 7.1] **10.** $-\dfrac{4}{3}$ **11.** 0   [3.4, 7.2] **12.** $y = -4x + 15$ **13.** $y = 4x$ [3.2, 7.1] **14.**   [4.1, 4.2] **15.** $\dfrac{8m^9 n^3}{p^6}$ **16.** $\dfrac{y^7}{x^{13} z^2}$

[4.4] **17.** $2x^2 - 4x + 38$ [4.5] **18.** $15x^2 + 7xy - 2y^2$ [4.6] **19.** $64m^2 - 25n^2$ **20.** $x^3 + 8y^3$ [4.7] **21.** $m^2 - 2m + 3$

[5.2, 5.3] **22.** $(m + 8)(m + 4)$ [5.4] **23.** $(5t^2 + 6)(5t^2 - 6)$ **24.** $(9z + 4)^2$ [5.5] **25.** $\{-2, -1\}$ [6.1] **26.** $-7, 2$

[6.2] **27.** $\dfrac{x + 1}{x}$ **28.** $(t + 5)(t + 3)$ or $t^2 + 8t + 15$ [6.4] **29.** $\dfrac{-2x - 14}{(x + 3)(x - 1)}$ [6.5] **30.** -21 [6.6] **31.** $\{19\}$

[7.1, 7.2] **32. (a)** 2505.1 per yr; The number of twin births increased an average of 2505.1 per yr. **(b)** $y = 2505.1x + 93,865$ **(c)** about

123,926   [7.3] **33.** domain: $\{-4, -1, 2, 5\}$; range: $\{-2, 0, 2\}$; function   [7.4] **34.** 800 lb   [8.1–8.3, 8.6] **35.** $\{(3, 2)\}$

36. $\emptyset$ [8.4, 8.6] **37.** $\{(1, 0, -1)\}$ [8.5] **38.** length: 42 ft; width: 30 ft   **39.** 15% solution: 6 L; 30% solution: 3 L

[9.1] **40.** $(-4, 4)$ **41.** $(-\infty, 0] \cup (2, \infty)$ [9.2] **42.** $\left\{ -\dfrac{1}{3}, 1 \right\}$ **43.** $\left(-\infty, -\dfrac{8}{3} \right] \cup [2, \infty)$

[9.3] **44.** 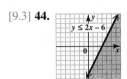   **45.**   **46.**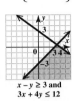

$x - y \geq 3$ and
$3x + 4y \leq 12$

CHAPTER 10 ROOTS, RADICALS, AND ROOT FUNCTIONS

Section 10.1   (page 637)

Exercises   **1.** true   **3.** false; Zero has only one square root.   **5.** true   **7.** $-3, 3$ **9.** $-8, 8$ **11.** $-12, 12$ **13.** $-\dfrac{5}{14}, \dfrac{5}{14}$

15. $-30, 30$ **17.** 1   **19.** 7   **21.** -11 **23.** $-\dfrac{12}{11}$ **25.** not a real number   **27.** 19   **29.** 19   **31.** $\dfrac{2}{3}$ **33.** $3x^2 + 4$

35. a must be positive.   **37.** a must be negative.   **39.** rational; 5   **41.** irrational; 5.385   **43.** rational; -8 **45.** irrational; -17.321

47. not a real number   **49.** irrational; 34.641   **51.** E   **53.** D   **55.** A   **57.** C   **59.** C   **61. (a)** not a real number **(b)** negative

(c) 0   **63.** 1   **65.** -5 **67.** -4 **69.** 8   **71.** 5   **73.** -5 **75.** not a real number   **77.** 2   **79.** not a real number   **81.** $\dfrac{4}{3}$

83. $\dfrac{2}{3}$ **85.** $\dfrac{1}{2}$ **In Exercises 87–93, we give the domain and then the range.**   **87.** $[-3, \infty)$; $[0, \infty)$

$f(x) = \sqrt{x + 3}$

89. $[0, \infty)$; $[-2, \infty)$ 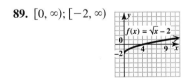   **91.** $(-\infty, \infty)$; $(-\infty, \infty)$ 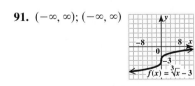   **93.** $(-\infty, \infty)$; $(-\infty, \infty)$ 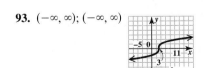   **95.** 12   **97.** 10

$f(x) = \sqrt{x - 2}$ $f(x) = \sqrt[3]{x} - 3$ $f(x) = \sqrt[3]{x} - 3$

99. 2   **101.** -9 **103.** $|x|$ **105.** x **107.** x^5 **109.** -9.055 **111.** 7.507   **113.** 3.162   **115.** 1.885   **117.** 1,183,000 cycles

per sec   **119.** 392,000 mi^2 **121.** 1.732 amps

Section 10.2   (page 646)

Exercises   **1.** C   **3.** A   **5.** H   **7.** B   **9.** D   **11.** 13   **13.** 9   **15.** 2   **17.** $\dfrac{8}{9}$ **19.** -3 **21.** 1000   **23.** -1024 **25.** not a

real number   **27.** $\dfrac{1}{512}$ **29.** $\dfrac{9}{4}$ **31.** $(-64)^{1/2}$ is an even root of a negative number. No real number squared will give -64. On the other hand, $-64^{1/2} = -\sqrt{64} = -8$, which is a real number. ($-64^{1/2}$ is the opposite of $64^{1/2}$.) **33.** $\sqrt{12}$ **35.** $\left(\sqrt[4]{8}\right)^3$ **37.** $\left(\sqrt[8]{9q}\right)^5 - \left(\sqrt[3]{2x}\right)^2$

39. $\dfrac{1}{\left(\sqrt{2m}\right)^3}$ **41.** $\left(\sqrt[3]{2y+x}\right)^2$ **43.** $\dfrac{1}{\left(\sqrt[3]{3m^4+2k^2}\right)^2}$ **45.** $\sqrt{a^2+b^2} = \sqrt{3^2+4^2} = 5$; $a + b = 3 + 4 = 7$; $5 \neq 7$ **47.** 64

49. 64 **51.** x^{10} **53.** $\sqrt[6]{x^5}$ **55.** $\sqrt[15]{t^8}$ **57.** 9 **59.** 4 **61.** y **63.** $k^{2/3}$ **65.** x^3y^8 **67.** $\dfrac{1}{x^{10/3}}$ **69.** $\dfrac{1}{m^{1/4}n^{3/4}}$ **71.** p^2 **73.** $\dfrac{c^{11/3}}{b^{11/4}}$

75. $\dfrac{q^{5/3}}{9p^{7/2}}$ **77.** $p + 2p^2$ **79.** $k^{7/4} - k^{3/4}$ **81.** $6 + 18a$ **83.** $x^{17/20}$ **85.** $\dfrac{1}{x^{3/2}}$ **87.** $y^{5/6}z^{1/3}$ **89.** $m^{1/12}$ **91.** $x^{1/24}$ **93.** 4.5 hr

95. $x^{-1/2}(3 - 4x)$ **96.** $m^{5/2}(m^{1/2} - 3)$ **97.** $t^{-1/2}(4 + 7t^2)$ **98.** $x^{-1/3}(8x + 5)$ **99.** $p^{3/4}(4p^{1/4} - 1)$ **100.** $m^{1/8}(2 - m^{1/2})$
101. $k^{-3/4}(9 - 2k^{1/2})$ **102.** $z^{-3/4}(7z^{1/8} - 1)$

Section 10.3 (page 656)

Connections (**page 654**) no; no; Answers will vary.

Exercises **1.** true; Both are equal to $4\sqrt{3}$ and approximately 6.92820323. **3.** true; Both are equal to $6\sqrt{2}$ and approximately 8.485281374.
5. D **7.** $\sqrt{30}$ **9.** $\sqrt{14x}$ **11.** $\sqrt{42pqr}$ **13.** $\sqrt[3]{14xy}$ **15.** $\sqrt[4]{33}$ **17.** $\sqrt[4]{6xy^2}$ **19.** cannot be simplified using the product rule
21. To multiply two radical expressions with the same index, multiply the radicands and keep the same index. For example,

$\sqrt[3]{3} \cdot \sqrt[3]{5} = \sqrt[3]{15}$. **23.** $\dfrac{8}{11}$ **25.** $\dfrac{\sqrt{3}}{5}$ **27.** $\dfrac{\sqrt{x}}{5}$ **29.** $\dfrac{p^3}{9}$ **31.** $\dfrac{3}{4}$ **33.** $-\dfrac{\sqrt[3]{r^2}}{2}$ **35.** $-\dfrac{3}{x}$ **37.** $\dfrac{1}{x^3}$ **39.** $2\sqrt{3}$ **41.** $12\sqrt{2}$

43. $-4\sqrt{2}$ **45.** $-2\sqrt{7}$ **47.** not a real number **49.** $4\sqrt[3]{2}$ **51.** $-2\sqrt[3]{2}$ **53.** $2\sqrt[3]{5}$ **55.** $-4\sqrt[4]{2}$ **57.** $2\sqrt[5]{2}$ **59.** His reasoning

was incorrect. Here 8 is a term, not a factor. **61.** $6k\sqrt{2}$ **63.** $12xy^4\sqrt{xy}$ **65.** $11x^3$ **67.** $-3t^4$ **69.** $-10m^4z^2$ **71.** $5a^2b^3c^4$

73. $\dfrac{1}{2}r^2t^5$ **75.** $5x\sqrt{2x}$ **77.** $-10r^5\sqrt{5r}$ **79.** $x^3y^4\sqrt{13x}$ **81.** $2z^2w^3$ **83.** $-2zt^2\sqrt[3]{2z^2t}$ **85.** $3x^3y^4$ **87.** $-3r^3s^2\sqrt[3]{2r^3s^2}$

89. $\dfrac{y^5\sqrt{y}}{6}$ **91.** $\dfrac{x^5\sqrt[3]{x}}{3}$ **93.** $4\sqrt{3}$ **95.** $\sqrt{5}$ **97.** $x^2\sqrt{x}$ **99.** $\sqrt[6]{432}$ **101.** $\sqrt[12]{6912}$ **103.** $\sqrt[6]{x^5}$ **105.** 1 **107.** 1 **109.** 5

111. $8\sqrt{2}$ **113.** $\sqrt{29}$ ft; 5.4 ft **115.** .003 **117.** $8\sqrt{5}$ ft; 17.9 ft **119.** 13 **121.** $9\sqrt{2}$ **123.** $\sqrt{17}$ **125.** 5 **127.** $6\sqrt{2}$
129. $\sqrt{5y^2 - 2xy + x^2}$ **131.** $d = [(x_2 - x_1)^2 + (y_2 - y_1)^2]^{1/2}$ **133.** $2\sqrt{106} + 4\sqrt{2}$

Section 10.4 (page 664)

Connections (**page 663**) **1.** 1.618033989 **2.** As one goes farther and farther into the sequence, the successive ratios appear to become
closer and closer to the golden ratio. This is indeed the case.

Exercises **1.** B **3.** 15; Each radical expression simplifies to a whole number. **5.** -4 **7.** $7\sqrt{3}$ **9.** $14\sqrt[3]{2}$ **11.** $5\sqrt[4]{2}$
13. $24\sqrt{2}$ **15.** cannot be simplified further **17.** $20\sqrt{5}$ **19.** $12\sqrt{2x}$ **21.** $-11m\sqrt{2}$ **23.** $\sqrt[3]{2}$ **25.** $2\sqrt[3]{x}$ **27.** $-\sqrt[3]{x^2y}$

29. $-x\sqrt[3]{xy^2}$ **31.** $19\sqrt[4]{2}$ **33.** $x\sqrt[4]{xy}$ **35.** $9\sqrt[4]{2a^3}$ **37.** $(4 + 3xy)\sqrt[3]{xy^2}$ **39.** $2\sqrt{2} - 2$ **41.** $\dfrac{5\sqrt{5}}{6}$ **43.** $\dfrac{7\sqrt{2}}{6}$ **45.** $\dfrac{5\sqrt{2}}{3}$

47. $5\sqrt{2} + 4$ **49.** $\dfrac{5 - 3x}{x^4}$ **51.** $\dfrac{m\sqrt[3]{m^2}}{2}$ **53.** $\dfrac{3x\sqrt[3]{2} - 4\sqrt[3]{5}}{x^3}$ **55.** Both are approximately 11.3137085. **57.** A; 42 m

59. $\left(12\sqrt{5} + 5\sqrt{3}\right)$ in. **61.** $\left(24\sqrt{2} + 12\sqrt{3}\right)$ in.

Section 10.5 (page 672)

Connections (**page 672**) **1.** $\dfrac{319}{6(8\sqrt{5} + 1)}$ **2.** $\dfrac{9a - b}{b(3\sqrt{a} - \sqrt{b})}$ **3.** $\dfrac{9a - b}{\left(\sqrt{b} - \sqrt{a}\right)(3\sqrt{a} - \sqrt{b})}$ **4.** $\dfrac{\left(3\sqrt{a} + \sqrt{b}\right)\left(\sqrt{b} + \sqrt{a}\right)}{b - a}$;

Instead of multiplying by the conjugate of the numerator, we use the conjugate of the denominator.

Exercises **1.** E **3.** A **5.** D **7.** $3\sqrt{6} + 2\sqrt{3}$ **9.** $20\sqrt{2}$ **11.** -2 **13.** -1 **15.** 6 **17.** $\sqrt{6} - \sqrt{2} + \sqrt{3} - 1$
19. $\sqrt{22} + \sqrt{55} - \sqrt{14} - \sqrt{35}$ **21.** $8 - \sqrt{15}$ **23.** $9 + 4\sqrt{5}$ **25.** $26 - 2\sqrt{105}$ **27.** $4 - \sqrt[3]{36}$ **29.** 10
31. $6x + 3\sqrt{x} - 2\sqrt{5x} - \sqrt{5}$ **33.** $9r - s$ **35.** $4\sqrt[3]{4y^2} - 19\sqrt[3]{2y} - 5$ **37.** $3x - 4$ **39.** $4x - y$ **41.** $2\sqrt{6} - 1$ **43.** $\sqrt{7}$

45. $5\sqrt{3}$ **47.** $\dfrac{\sqrt{6}}{2}$ **49.** $\dfrac{9\sqrt{15}}{5}$ **51.** $-\sqrt{2}$ **53.** $\dfrac{\sqrt{14}}{2}$ **55.** $-\dfrac{\sqrt{14}}{10}$ **57.** $\dfrac{2\sqrt{6x}}{x}$ **59.** $\dfrac{-8\sqrt{3k}}{k}$ **61.** $\dfrac{-5m^2\sqrt{6mn}}{n^2}$

63. $\dfrac{12x^3\sqrt{2xy}}{y^5}$ **65.** $\dfrac{5\sqrt{2my}}{y^2}$ **67.** $-\dfrac{4k\sqrt{3z}}{z}$ **69.** $\dfrac{\sqrt[3]{18}}{3}$ **71.** $\dfrac{\sqrt[3]{12}}{3}$ **73.** $\dfrac{\sqrt[3]{18}}{4}$ **75.** $-\dfrac{\sqrt[3]{2pr}}{r}$ **77.** $\dfrac{x^2\sqrt[3]{y^2}}{y}$ **79.** $\dfrac{2\sqrt[4]{x^3}}{x}$

81. $\dfrac{\sqrt[4]{2yz^3}}{z}$ **83.** Multiply both the numerator and the denominator by $4-\sqrt{5}$. No, it would not. The new denominator would be

$\left(4+\sqrt{5}\right)^2 = 21 + 8\sqrt{5}$, which is not rational.   **85.** $\dfrac{3(4-\sqrt{5})}{11}$ **87.** $\dfrac{6\sqrt{2}+4}{7}$ **89.** $\dfrac{2(3\sqrt{5}-2\sqrt{3})}{33}$

91. $2\sqrt{3}+\sqrt{10}-3\sqrt{2}-\sqrt{15}$ **93.** $\sqrt{m}-2$ **95.** $\dfrac{4\sqrt{x}(\sqrt{x}+2\sqrt{y})}{x-4y}$ **97.** $\dfrac{x-2\sqrt{xy}+y}{x-y}$ **99.** $\dfrac{5\sqrt{k}(2\sqrt{k}-\sqrt{q})}{4k-q}$

101. Square both sides to show that each is equal to $\dfrac{2-\sqrt{3}}{4}$.   **103.** $3-2\sqrt{6}$ **105.** $1-\sqrt{5}$ **107.** $\dfrac{4-2\sqrt{2}}{3}$ **109.** $\dfrac{6+2\sqrt{6p}}{3}$

111. $\dfrac{\sqrt{x+y}}{x+y}$ **113.** $\dfrac{p\sqrt{p+2}}{p+2}$ **115.** Each expression is approximately equal to .2588190451.   **117.** $\left(\sqrt{x}+\sqrt{7}\right)\left(\sqrt{x}-\sqrt{7}\right)$

118. $\left(\sqrt[3]{x}-\sqrt[3]{7}\right)\left(\sqrt[3]{x^2}+\sqrt[3]{7x}+\sqrt[3]{49}\right)$ **119.** $\left(\sqrt[3]{x}+\sqrt[3]{7}\right)\left(\sqrt[3]{x^2}-\sqrt[3]{7x}+\sqrt[3]{49}\right)$ **120.** $\dfrac{(x+3)\left(\sqrt{x}+\sqrt{7}\right)}{x-7}$

121. $\dfrac{(x+3)\left(\sqrt[3]{x^2}+\sqrt[3]{7x}+\sqrt[3]{49}\right)}{x-7}$ **122.** $\dfrac{(x+3)\left(\sqrt[3]{x}+\sqrt[3]{7}\right)}{x+7}$ **123.** $\left(\sqrt[3]{5}-\sqrt[3]{3}\right)\left(\sqrt[3]{25}+\sqrt[3]{15}+\sqrt[3]{9}\right)$ **124.** $\sqrt[3]{25}+\sqrt[3]{15}+\sqrt[3]{9}$

125. $\dfrac{17}{2(6+\sqrt{2})}$ **127.** $\dfrac{9a-b}{b(3\sqrt{a}-\sqrt{b})}$

Summary Exercises on Operations with Radicals   (page 676)

1. $-6\sqrt{10}$ **2.** $7-\sqrt{14}$ **3.** $2+\sqrt{6}-2\sqrt{3}-3\sqrt{2}$ **4.** $4\sqrt{2}$ **5.** $73+12\sqrt{35}$ **6.** $\dfrac{-\sqrt{6}}{2}$ **7.** $4\left(\sqrt{7}-\sqrt{5}\right)$ **8.** $3\sqrt[3]{2x^2}$

9. $-3+2\sqrt{2}$ **10.** -2 **11.** -44 **12.** $\dfrac{\sqrt{x}+\sqrt{5}}{x-5}$ **13.** $2abc^3\sqrt[3]{b^2}$ **14.** $5\sqrt[3]{3}$ **15.** $3\left(\sqrt{5}-2\right)$ **16.** $\dfrac{\sqrt{15x}}{5x}$ **17.** $\dfrac{8}{5}$

18. $\dfrac{\sqrt{2}}{8}$ **19.** $-\sqrt[3]{100}$ **20.** $11+2\sqrt{30}$ **21.** $-3\sqrt{3x}$ **22.** $52-30\sqrt{3}$ **23.** 1 **24.** $\dfrac{\sqrt[3]{117}}{9}$ **25.** $t^2\sqrt[4]{t}$

Section 10.6   (page 681)

Exercises   1. (a) yes (b) no   **3.** (a) yes (b) no   **5.** no; There is no solution. The radical expression, which is positive, cannot equal a

negative number.   **7.** $\{11\}$ **9.** $\left\{\dfrac{1}{3}\right\}$ **11.** $\emptyset$ **13.** $\{5\}$ **15.** $\{18\}$ **17.** $\{5\}$ **19.** $\{4\}$ **21.** $\{17\}$ **23.** $\{5\}$ **25.** $\emptyset$ **27.** $\{0\}$

29. $\{0\}$ **31.** $\left\{-\dfrac{1}{3}\right\}$ **33.** $\emptyset$ **35.** You cannot just square each term. The right side should be $(8-x)^2 = 64 - 16x + x^2$. The correct

first step is $3x + 4 = 64 - 16x + x^2$, and the solution set is $\{4\}$.   **37.** $\{1\}$ **39.** $\{-1\}$ **41.** $\{14\}$ **43.** $\{8\}$ **45.** $\{0\}$ **47.** $\emptyset$ **49.** $\{7\}$

51. $\{7\}$ **53.** $\{4, 20\}$ **55.** $\emptyset$ **57.** $\left\{\dfrac{5}{4}\right\}$ **59.** 6 **61.** $\emptyset$; domain: $\left[-\dfrac{2}{3}, 1\right]$ **63.** $\{9, 17\}$ **65.** $\left\{\dfrac{1}{4}, 1\right\}$ **67.** $K = \dfrac{V^2m}{2}$

69. $L = \dfrac{1}{4\pi^2 f^2 C}$ **71.** (a) $r = \dfrac{a}{4\pi^2 N^2}$ (b) 62.5 m (c) 155.1 m   **73.** 7 billion ft³; 14 billion ft³; 22.5 billion ft³; 22.5 billion ft³

75. 12 billion ft³; 12 billion ft³; 14 billion ft³; 17.5 billion ft³   **77.** fairly good; 1920

Section 10.7   (page 690)

Exercises   1. i **3.** -1 **5.** $-i$ **7.** $13i$ **9.** $-12i$ **11.** $i\sqrt{5}$ **13.** $4i\sqrt{3}$ **15.** $-\sqrt{105}$ **17.** -10 **19.** $i\sqrt{33}$ **21.** $\sqrt{3}$

23. $5i$ **25.** (a) Any real number a can be written as $a + 0i$, and this is a complex number with imaginary part 0. (b) A complex number

such as $2 + 3i$, with nonzero imaginary part, is not real.   **27.** $-1 + 7i$ **29.** 0 **31.** $7 + 3i$ **33.** -2 **35.** $1 + 13i$ **37.** $6 + 6i$
39. $4 + 2i$ **41.** -81 **43.** -16 **45.** $-10 - 30i$ **47.** $10 - 5i$ **49.** $-9 + 40i$ **51.** $-16 + 30i$ **53.** 153 **55.** 97
57. $a - bi$ **59.** $1 + i$ **61.** $-1 + 2i$ **63.** $2 + 2i$ **65.** $-\dfrac{5}{13} - \dfrac{12}{13}i$ **67.** $1 - 3i$ **69.** $30 + 5i$ **71.** $\dfrac{5}{41} + \dfrac{4}{41}i$ **73.** -1
75. i **77.** -1 **79.** $-i$ **81.** $-i$ **83.** Since $i^{20} = (i^4)^5 = 1^5 = 1$, the student multiplied by 1, which is justified by the identity property
for multiplication.   **85.** $\dfrac{1}{2} + \dfrac{1}{2}i$ **87.** Substitute both $1 + 5i$ and $1 - 5i$ for x and show that the result is $0 = 0$ in each case.
89. (a) $4x + 1$ (b) $4 + i$ **90.** (a) $-2x + 3$ (b) $-2 + 3i$ **91.** (a) $3x^2 + 5x - 2$ (b) $5 + 5i$ **92.** (a) $-\sqrt{3} + \sqrt{6} + 1 - \sqrt{2}$
(b) $\dfrac{1}{5} - \dfrac{7}{5}i$ **93.** In parts (a) and (b) of Exercises 89 and 90, real and imaginary parts are added, just like coefficients of similar terms in the
binomials, and the answers correspond. In Exercise 91, introducing $i^2 = -1$ when a product is found leads to answers that do not correspond.
94. In parts (a) and (b) of Exercises 89 and 90, real and imaginary parts are added, just like coefficients of similar terms in binomials, and the
answers correspond. In Exercise 92, introducing $i^2 = -1$ when performing the division leads to answers that do not correspond.
95. $\dfrac{37}{10} - \dfrac{19}{10}i$ **97.** $-\dfrac{13}{10} + \dfrac{11}{10}i$

Chapter 10 Review Exercises   (page 698)

1. 42 **3.** 6 **5.** -3 **7.** $\sqrt[n]{a}$ is not a real number if n is even and a is negative.   **9.** -6.856 **11.** 4.960 **13.** -3968.503
15. domain: $[1, \infty)$; range: $[0, \infty)$ **17.** B   **19.** A   **21.** It is not a real number.   **23.** -11 **25.** -4 **27.** -32

29. It is not a real number.   **31.** The radical $\sqrt[n]{a^m}$ is equivalent to $a^{m/n}$. For example, $\sqrt[3]{8^2} = \sqrt[3]{64} = 4$, and $8^{2/3} = (8^{1/3})^2 = 2^2 = 4$.
33. $\dfrac{1}{\left(\sqrt[3]{3a + b}\right)^5}$ or $\dfrac{1}{\sqrt[3]{(3a + b)^5}}$ **35.** $p^{4/5}$ **37.** 96 **39.** $\dfrac{1}{y^{1/2}}$ **41.** $r^{1/2} + r$ **43.** $r^{3/2}$ **45.** $k^{9/4}$ **47.** $z^{1/12}$ **49.** $x^{1/15}$
51. The product rule for exponents applies only if the bases are the same.   **53.** $\sqrt{5r}$ **55.** $\sqrt[4]{21}$ **57.** $5\sqrt{3}$ **59.** $-3\sqrt[3]{4}$
61. $4pq^2\sqrt[3]{p}$ **63.** $2r^2t\sqrt[3]{79r^2t}$ **65.** $\dfrac{m^5}{3}$ **67.** $\dfrac{a^2\sqrt[4]{a}}{3}$ **69.** $p\sqrt{p}$ **71.** $\sqrt[10]{x^7}$ **73.** $\sqrt{197}$ **75.** $23\sqrt{5}$ **77.** $26m\sqrt{6m}$
79. $-8\sqrt[4]{2}$ **81.** $\dfrac{16 + 5\sqrt{5}}{20}$ **83.** $\left(12\sqrt{3} + 5\sqrt{2}\right)$ ft   **85.** 2 **87.** $15 - 2\sqrt{26}$ **89.** $2\sqrt[3]{2y^2} + 2\sqrt[3]{4y} - 3$
91. The denominator would become $\sqrt[3]{6^2} = \sqrt[3]{36}$, which is not rational.   **93.** $-3\sqrt{6}$ **95.** $\dfrac{\sqrt{22}}{4}$ **97.** $\dfrac{3m\sqrt[3]{4n}}{n^2}$ **99.** $\dfrac{5\left(\sqrt{6} + 3\right)}{3}$
101. $\dfrac{1 - 4\sqrt{2}}{3}$ **103.** $\{2\}$ **105.** $\emptyset$ **107.** $\{9\}$ **109.** $\{7\}$ **111.** $\{-13\}$ **113.** $\{14\}$ **115.** $\emptyset$ **117.** $\{7\}$ **119.** $\{3\}$ **120.** $\{-3\}$
121. $\{\pm 3\}$ **122.** $\{3\}$ **123.** $\{\pm 3\}$ **124.** $\{-3\}$ **125.** (a) more than (b) the same as   **127.** $5i$ **129.** no   **131.** $14 + 7i$
133. -45 **135.** $5 + i$ **137.** $1 - i$ **139.** $-i$ **141.** -1 **143.** -4 **145.** $\dfrac{1}{z^{3/5}}$ **147.** $3z^3t^2\sqrt[3]{2t^2}$ **149.** $6x\sqrt[3]{y^2}$ **151.** $-\dfrac{\sqrt{3}}{6}$
153. 1 **155.** $3 - 7i$ **157.** $\dfrac{1 + \sqrt{6}}{2}$ **159.** $\{5\}$ **161.** $\left\{\dfrac{3}{2}\right\}$ **163.** $\{1\}$ **165.** $\{9\}$ **167.** $\{7\}$

Chapter 10 Test   (page 703)

[10.1] **1.** -29 **2.** -8 [10.2] **3.** 5 [10.1] **4.** C   **5.** 21.863 **6.** -9.405 **7.** domain: $[-6, \infty)$; range: $[0, \infty)$

[10.2] **8.** $\dfrac{125}{64}$ **9.** $\dfrac{1}{256}$ **10.** $\dfrac{9y^{3/10}}{x^2}$ **11.** $x^{4/3}y^6$ **12.** $7^{1/2}$ or $\sqrt{7}$ [10.3] **13.** $a^3\sqrt[3]{a^2}$ or $a^{11/3}$ **14.** $\sqrt{145}$ **15.** 10   **16.** $3x^2y^3\sqrt{6x}$

17. $2ab^3\sqrt[4]{2a^3b}$ **18.** $\sqrt[6]{200}$ [10.4] **19.** $26\sqrt{5}$ **20.** $(2ts - 3t^2)\sqrt[3]{2s^2}$ [10.5] **21.** $66 + \sqrt{5}$ **22.** $23 - 4\sqrt{15}$ **23.** $-\dfrac{\sqrt{10}}{4}$

24. $\dfrac{2\sqrt[3]{25}}{5}$ **25.** $-2\left(\sqrt{7} - \sqrt{5}\right)$ **26.** $3 + \sqrt{6}$ [10.6] **27.** **(a)** 59.8 **(b)** $T = \dfrac{V_0^2 - V^2}{-V^2k}$ or $T = \dfrac{V^2 - V_0^2}{V^2k}$ **28.** $\{-1\}$ **29.** $\{3\}$

30. $\{-3\}$ [10.7] **31.** $-5 - 8i$ **32.** $-2 + 16i$ **33.** $3 + 4i$ **34.** i **35.** **(a)** true **(b)** true **(c)** false **(d)** true

Cumulative Review Exercises Chapters 1–10   (page 704)

[1.4–1.6] **1.** 1   **2.** $-\dfrac{14}{9}$ [2.1–2.3] **3.** $\{-4\}$ **4.** $\{-12\}$ **5.** $\{6\}$ [2.8] **6.** $(-6, \infty)$ [9.1] **7.** $(2, 3)$ **8.** $(-\infty, 2) \cup (3, \infty)$

[9.2] **9.** $\left\{-\dfrac{10}{3}, 1\right\}$ **10.** $\left\{\dfrac{1}{4}\right\}$ **11.** $(-\infty, -2] \cup [7, \infty)$ [2.5] **12.** Both angles measure 80°.   [2.7] **13.** 18 nickels; 32 quarters

14. $2\dfrac{2}{39}$ L   [3.2, 7.1] **15.**

$4x - 3y = 12$

[3.3, 3.4, 7.1, 7.2] **16.** $-\dfrac{3}{2}$; $y = -\dfrac{3}{2}x$ [4.4] **17.** $-k^3 - 3k^2 - 8k - 9$

[4.5] **18.** $8x^2 + 17x - 21$ [4.7] **19.** $z - 2 + \dfrac{3}{z}$ **20.** $3y^3 - 3y^2 + 4y + 1 + \dfrac{-10}{2y + 1}$ [5.2, 5.3] **21.** $(2p - 3q)(p - q)$

[5.4] **22.** $(3k^2 + 4)(k - 1)(k + 1)$ **23.** $(x + 8)(x^2 - 8x + 64)$ [5.5] **24.** $\left\{-3, -\dfrac{5}{2}\right\}$ **25.** $\left\{-\dfrac{2}{5}, 1\right\}$ [6.2] **26.** $\dfrac{y}{y + 5}$

[6.4] **27.** $\dfrac{4x + 2y}{(x + y)(x - y)}$ [6.5] **28.** $-\dfrac{9}{4}$ **29.** $\dfrac{-1}{a + b}$ [6.7] **30.** Natalie: 8 mph; Chuck: 4 mph   [6.6] **31.** $\emptyset$ [7.3] **32.** -37

[8.2, 8.3] **33.** $\{(7, -2)\}$ [8.6] **34.** $\{(-1, 1, 1)\}$ [8.5] **35.** 2-oz letter: \$.55; 3-oz letter: \$.78   [10.2] **36.** $\dfrac{1}{9}$ [10.3] **37.** $10x^2\sqrt{2}$

38. $2x\sqrt[3]{6x^2y^2}$ [10.4] **39.** $7\sqrt{2}$ [10.5] **40.** $\dfrac{\sqrt{10} + 2\sqrt{2}}{2}$ **41.** $-6x - 11\sqrt{xy} - 4y$ [10.3] **42.** $\sqrt{29}$ [10.6] **43.** $\{3, 4\}$

[10.1] **44.** 39.2 mph   [10.7] **45.** $2 + 9i$ **46.** $4 + 2i$

CHAPTER **11** QUADRATIC EQUATIONS, INEQUALITIES, AND FUNCTIONS

Section 11.1   (page 713)

Exercises   1. The equation is also true for $x = -4$.   **3. (a)** A quadratic equation in standard form has a second-degree polynomial in decreasing powers equal to 0.   **(b)** The zero-factor property states that if a product equals 0, then at least one of the factors equals 0.   **(c)** The square root property states that if the square of a quantity equals a number, then the quantity equals the positive or negative square root of the number.   **5.** $\{9, -9\}$ **7.** $\{\sqrt{17}, -\sqrt{17}\}$ **9.** $\{4\sqrt{2}, -4\sqrt{2}\}$ **11.** $\{\sqrt{21}, -\sqrt{21}\}$ **13.** $\{2\sqrt{5}, -2\sqrt{5}\}$ **15.** $\{2\sqrt{6}, -2\sqrt{6}\}$

17. $\left\{\dfrac{2\sqrt{5}}{5}, -\dfrac{2\sqrt{5}}{5}\right\}$ **19.** $\{3\sqrt{3}, -3\sqrt{3}\}$ **21.** $\{-2, 8\}$ **23.** $\left\{-3, \dfrac{5}{3}\right\}$ **25.** $\left\{0, \dfrac{3}{2}\right\}$ **27.** $\{4 + \sqrt{3}, 4 - \sqrt{3}\}$

29. $\left\{-5 + 4\sqrt{3}, -5 - 4\sqrt{3}\right\}$ **31.** $\left\{\dfrac{1 + \sqrt{7}}{3}, \dfrac{1 - \sqrt{7}}{3}\right\}$ **33.** $\left\{\dfrac{-1 + 2\sqrt{6}}{4}, \dfrac{-1 - 2\sqrt{6}}{4}\right\}$ **35.** $\left\{\dfrac{5 + \sqrt{30}}{2}, \dfrac{5 - \sqrt{30}}{2}\right\}$

37. $\left\{\dfrac{-1 + 3\sqrt{2}}{3}, \dfrac{-1 - 3\sqrt{2}}{3}\right\}$ **39.** $\left\{-10 + 4\sqrt{3}, -10 - 4\sqrt{3}\right\}$ **41.** $\left\{\dfrac{1 + 4\sqrt{3}}{4}, \dfrac{1 - 4\sqrt{3}}{4}\right\}$ **43.** Johnny's first solution,

$\dfrac{5 + \sqrt{30}}{2}$, is equivalent to Linda's second solution, $\dfrac{-5 - \sqrt{30}}{-2}$. This can be verified by multiplying $\dfrac{5 + \sqrt{30}}{2}$ by 1 in the form $\dfrac{-1}{-1}$. Similarly,

Johnny's second solution is equivalent to Linda's first one.   **45.** $\{-4.48, .20\}$ **47.** $\{-3.09, -.15\}$ **49.** $\left\{2i\sqrt{3}, -2i\sqrt{3}\right\}$

51. $\left\{5 + i\sqrt{3}, 5 - i\sqrt{3}\right\}$ **53.** $\left\{\dfrac{1 + 2i\sqrt{2}}{6}, \dfrac{1 - 2i\sqrt{2}}{6}\right\}$ **55.** $(x + 3)^2 = 100$ **56.** $\{-13, 7\}$ **57.** $\{-13, 7\}$ **58.** $\{-3, 6\}$

59. 6.3 sec   **61.** 9 in.   **63.** 5%

Section 11.2   (page 722)

Exercises   1. (a) 9 (b) 49 (c) 36 (d) $\dfrac{9}{4}$ (e) $\dfrac{81}{4}$ (f) $\dfrac{1}{16}$ **3.** 4   **5.** 25   **7.** $\dfrac{1}{36}$ **9.** $\{-4, 6\}$ **11.** $\left\{-2 + \sqrt{6}, -2 - \sqrt{6}\right\}$

13. $\left\{-5 + \sqrt{7}, -5 - \sqrt{7}\right\}$ **15.** $\left\{\dfrac{-3 + \sqrt{17}}{2}, \dfrac{-3 - \sqrt{17}}{2}\right\}$ **17.** $\left\{-\dfrac{8}{3}, 3\right\}$ **19.** $\left\{\dfrac{-5 + \sqrt{41}}{4}, \dfrac{-5 - \sqrt{41}}{4}\right\}$

21. $\left\{\dfrac{5 + \sqrt{15}}{5}, \dfrac{5 - \sqrt{15}}{5}\right\}$ **23.** $\left\{\dfrac{4 + \sqrt{3}}{3}, \dfrac{4 - \sqrt{3}}{3}\right\}$ **25.** $\left\{\dfrac{-3 + \sqrt{41}}{2}, \dfrac{-3 - \sqrt{41}}{2}\right\}$ **27.** $\left\{\dfrac{2 + \sqrt{3}}{3}, \dfrac{2 - \sqrt{3}}{3}\right\}$

29. $\left\{1 + \sqrt{2}, 1 - \sqrt{2}\right\}$ **31.** (a) $\left\{\dfrac{3 + 2\sqrt{6}}{3}, \dfrac{3 - 2\sqrt{6}}{3}\right\}$ (b) $\{-.633, 2.633\}$ **33.** (a) $\left\{-2 + \sqrt{3}, -2 - \sqrt{3}\right\}$ (b) $\{-3.732, -.268\}$

35. The student should have divided both sides of the equation by 2 as his first step. The correct solution set is $\{1, 4\}$.

37. $\{-2 + 3i, -2 - 3i\}$ **39.** $\left\{\dfrac{-2 + 2i\sqrt{2}}{3}, \dfrac{-2 - 2i\sqrt{2}}{3}\right\}$ **41.** $\left\{-3 + i\sqrt{3}, -3 - i\sqrt{3}\right\}$ **43.** x^2 **44.** x **45.** $6x$ **46.** 1

47. 9   **48.** $(x + 3)^2$ or $x^2 + 6x + 9$

Section 11.3   (page 729)

Exercises   1. The documentation was incorrect, since the fraction bar should extend under the term $-b$.   **3.** The last step is wrong.

Because 5 is not a common factor in the numerator, the fraction cannot be simplified. The solutions are $\dfrac{5 \pm \sqrt{5}}{10}$.   **5.** $\{3, 5\}$

7. $\left\{\dfrac{-2 + \sqrt{2}}{2}, \dfrac{-2 - \sqrt{2}}{2}\right\}$ **9.** $\left\{\dfrac{1 + \sqrt{3}}{2}, \dfrac{1 - \sqrt{3}}{2}\right\}$ **11.** $\left\{5 + \sqrt{7}, 5 - \sqrt{7}\right\}$ **13.** $\left\{\dfrac{-1 + \sqrt{2}}{2}, \dfrac{-1 - \sqrt{2}}{2}\right\}$

15. $\left\{\dfrac{-1 + \sqrt{7}}{3}, \dfrac{-1 - \sqrt{7}}{3}\right\}$ **17.** $\left\{1 + \sqrt{5}, 1 - \sqrt{5}\right\}$ **19.** $\left\{\dfrac{-2 + \sqrt{10}}{2}, \dfrac{-2 - \sqrt{10}}{2}\right\}$ **21.** $\left\{-1 + 3\sqrt{2}, -1 - 3\sqrt{2}\right\}$

23. $\left\{\dfrac{1 + \sqrt{29}}{2}, \dfrac{1 - \sqrt{29}}{2}\right\}$ **25.** $\left\{\dfrac{-4 + \sqrt{91}}{3}, \dfrac{-4 - \sqrt{91}}{3}\right\}$ **27.** $\left\{\dfrac{3}{2} + \dfrac{\sqrt{15}}{2}i, \dfrac{3}{2} - \dfrac{\sqrt{15}}{2}i\right\}$ **29.** $\left\{3 + i\sqrt{5}, 3 - i\sqrt{5}\right\}$

31. $\left\{\dfrac{1}{2} + \dfrac{\sqrt{6}}{2}i, \dfrac{1}{2} - \dfrac{\sqrt{6}}{2}i\right\}$ **33.** $\left\{-\dfrac{2}{3} + \dfrac{\sqrt{2}}{3}i, -\dfrac{2}{3} - \dfrac{\sqrt{2}}{3}i\right\}$ **35.** $\left\{4 + 3i\sqrt{2}, 4 - 3i\sqrt{2}\right\}$ **37.** B   **39.** C   **41.** A   **43.** D

45. The equations in Exercises 37, 38, 41, and 42 can be solved by factoring.   **47.** $\left\{\dfrac{7}{2}\right\}$ **49.** $\left\{-\dfrac{1}{2}, \dfrac{3}{2}\right\}$ **51.** No, because an

irrational solution occurs only if the discriminant is positive, but not the square of an integer. In that case, there will be two irrational solutions.

53. -10 or 10   **55.** 16   **57.** 25   **59.** $b = \dfrac{44}{5}; \dfrac{3}{10}$

Section 11.4   (page 738)

Exercises   **1.** square root property   **3.** quadratic formula   **5.** factoring   **7.** Multiply by the LCD, x.   **9.** Substitute a variable for $r^2 + r$.   **11.** The potential solution -1 does not check. The solution set is $\{4\}$.   **13.** $\{-2, 7\}$ **15.** $\{-4, 7\}$ **17.** $\left\{-\dfrac{2}{3}, 1\right\}$

19. $\left\{-\dfrac{14}{17}, 5\right\}$ **21.** $\left\{-\dfrac{11}{7}, 0\right\}$ **23.** $\left\{\dfrac{-1 + \sqrt{13}}{2}, \dfrac{-1 - \sqrt{13}}{2}\right\}$ **25.** $\left\{-\dfrac{8}{3}, -1\right\}$ **27.** $\left\{\dfrac{2 + \sqrt{22}}{3}, \dfrac{2 - \sqrt{22}}{3}\right\}$

29. (a) $(20 - t)$ mph (b) $(20 + t)$ mph   **31.** 25 mph   **33.** 80 km per hr   **35.** 3.6 hr   **37.** Rusty: 25.0 hr; Nancy: 23.0 hr   **39.** 9 min

41. $\{2, 5\}$ **43.** $\{3\}$ **45.** $\left\{\dfrac{8}{9}\right\}$ **47.** $\{16\}$ **49.** $\left\{\dfrac{2}{5}\right\}$ **51.** $\{-3, 3\}$ **53.** $\left\{-\dfrac{3}{2}, -1, 1, \dfrac{3}{2}\right\}$ **55.** $\left\{-2\sqrt{3}, -2, 2, 2\sqrt{3}\right\}$

57. $\{-6, -5\}$ **59.** $\left\{-\dfrac{16}{3}, -2\right\}$ **61.** $\left\{-\dfrac{1}{3}, \dfrac{1}{6}\right\}$ **63.** $\left\{-\dfrac{1}{2}, 3\right\}$ **65.** $\{-8, 1\}$ **67.** $\{-64, 27\}$ **69.** $\left\{-\dfrac{27}{8}, -1, 1, \dfrac{27}{8}\right\}$

71. $\{25\}$ **73.** $\left\{-1, 1, -\dfrac{\sqrt{6}}{2}i, \dfrac{\sqrt{6}}{2}i\right\}$ **75.** $\left\{-\dfrac{\sqrt{6}}{3}, -\dfrac{1}{2}, \dfrac{1}{2}, \dfrac{\sqrt{6}}{3}\right\}$ **77.** $\{3, 11\}$ **79.** $\left\{-\sqrt[3]{5}, -\dfrac{\sqrt[3]{4}}{2}\right\}$ **81.** $\left\{\dfrac{4}{3}, \dfrac{9}{4}\right\}$

83. $\left\{\dfrac{\sqrt{9 + \sqrt{65}}}{2}, -\dfrac{\sqrt{9 + \sqrt{65}}}{2}, \dfrac{\sqrt{9 - \sqrt{65}}}{2}, -\dfrac{\sqrt{9 - \sqrt{65}}}{2}\right\}$ **85.** It would cause both denominators to be 0, and division by 0 is undefined.   **86.** $\dfrac{12}{5}$ **87.** $\left(\dfrac{x}{x - 3}\right)^2 + 3\left(\dfrac{x}{x - 3}\right) - 4 = 0$ **88.** The numerator can never equal the denominator, since the denominator is 3 less than the numerator.   **89.** $\left\{\dfrac{12}{5}\right\}$; The values for t are -4 and 1. The value 1 is impossible because it leads to a contradiction $\left(\text{since } \dfrac{x}{x - 3} \text{ is never equal to } 1\right)$.   **90.** $\left\{\dfrac{12}{5}\right\}$; The values for s are $\dfrac{1}{x}$ and $\dfrac{-4}{x}$. The value $\dfrac{1}{x}$ is impossible, since $\dfrac{1}{x} \neq \dfrac{1}{x - 3}$ for all x.

Summary Exercises on Solving Quadratic Equations   (page 742)

1. $\left\{\sqrt{7}, -\sqrt{7}\right\}$ **2.** $\left\{-\dfrac{3}{2}, \dfrac{5}{3}\right\}$ **3.** $\left\{-3 + \sqrt{5}, -3 - \sqrt{5}\right\}$ **4.** $\{-1, 9\}$ **5.** $\left\{-\dfrac{3}{2}, 4\right\}$ **6.** $\left\{-3, \dfrac{1}{3}\right\}$ **7.** $\left\{\dfrac{2 + \sqrt{2}}{2}, \dfrac{2 - \sqrt{2}}{2}\right\}$

8. $\left\{2i\sqrt{3}, -2i\sqrt{3}\right\}$ **9.** $\left\{\dfrac{1}{2}, 2\right\}$ **10.** $\{-3, -1, 1, 3\}$ **11.** $\left\{\dfrac{-3 + 2\sqrt{2}}{2}, \dfrac{-3 - 2\sqrt{2}}{2}\right\}$ **12.** $\left\{\dfrac{4}{5}, 3\right\}$ **13.** $\left\{-\sqrt{7}, -\sqrt{2}, \sqrt{2}, \sqrt{7}\right\}$

14. $\left\{\dfrac{1 + \sqrt{5}}{4}, \dfrac{1 - \sqrt{5}}{4}\right\}$ **15.** $\left\{-\dfrac{1}{2} + \dfrac{\sqrt{3}}{2}i, -\dfrac{1}{2} - \dfrac{\sqrt{3}}{2}i\right\}$ **16.** $\left\{-\dfrac{\sqrt[3]{175}}{5}, 1\right\}$ **17.** $\left\{\dfrac{3}{2}\right\}$ **18.** $\left\{\dfrac{2}{3}\right\}$ **19.** $\left\{6\sqrt{2}, -6\sqrt{2}\right\}$

20. $\left\{-\dfrac{2}{3}, 2\right\}$ **21.** $\{-4, 9\}$ **22.** $\{13, -13\}$ **23.** $\left\{1 + \dfrac{\sqrt{3}}{3}i, 1 - \dfrac{\sqrt{3}}{3}i\right\}$ **24.** $\{3\}$ **25.** $\left\{-\dfrac{1}{3}, \dfrac{1}{6}\right\}$

26. $\left\{\dfrac{1}{6} + \dfrac{\sqrt{47}}{6}i, \dfrac{1}{6} - \dfrac{\sqrt{47}}{6}i\right\}$

Section 11.5   (page 747)

Exercises   **1.** Find a common denominator, and then multiply both sides by the common denominator.   **3.** Write it in standard form (with 0 on one side, in decreasing powers of w).   **5.** $m = \sqrt{p^2 - n^2}$ **7.** $t = \dfrac{\pm\sqrt{dk}}{k}$ **9.** $d = \dfrac{\pm\sqrt{skI}}{I}$ **11.** $v = \dfrac{\pm\sqrt{kAF}}{F}$ **13.** $r = \dfrac{\pm\sqrt{3\pi Vh}}{\pi h}$

15. $t = \dfrac{-B \pm \sqrt{B^2 - 4AC}}{2A}$ **17.** $h = \dfrac{D^2}{k}$ **19.** $\ell = \dfrac{p^2 g}{k}$ **21.** If g is positive, the only way to have a real value for p is to have $k\ell$ positive, since the quotient of two positive numbers is positive. If k and ℓ have different signs, their product is negative, leading to a negative radicand.

23. $R = \dfrac{E^2 - 2pr \pm E\sqrt{E^2 - 4pr}}{2p}$ **25.** $r = \dfrac{5pc}{4}$ or $r = -\dfrac{2pc}{3}$ **27.** $I = \dfrac{-cR \pm \sqrt{c^2R^2 - 4cL}}{2cL}$ **29.** 2.3, 5.3, 5.8

31. eastbound ship: 80 mi; southbound ship: 150 mi   **33.** 5 cm, 12 cm, 13 cm   **35.** length: 2 cm; width: 1.5 cm   **37.** 1 ft
39. length: 26 m; width: 16 m   **41.** 20 in. by 12 in.   **43.** 1 sec and 8 sec   **45.** 2.4 sec and 5.6 sec   **47.** 9.2 sec   **49.** It reaches its
maximum height at 5 sec because this is the only time it reaches 400 ft.   **51.** $.80   **53.** .035 or 3.5%   **55.** **(a)** 2.4 million **(b)** 2.4 million;
They are the same.   **57.** 1995; The graph indicates that sales reached 2 million in 1996.   **59.** 5.5 m per sec   **61.** 5 or 14

Section 11.6   (page 758)

Exercises   1. (a) B **(b)** C **(c)** A **(d)** D   **3.** $(0, 0)$ **5.** $(0, 4)$ **7.** $(1, 0)$ **9.** $(-3, -4)$ **11.** In Exercise 9, the parabola is shifted 3 units
to the left and 4 units down. The parabola in Exercise 10 is shifted 5 units to the right and 8 units down.   **13.** down; wider   **15.** up; narrower
17. (a) I **(b)** IV **(c)** II **(d)** III   **19. (a)** D **(b)** B **(c)** C **(d)** A   **21.**   **23.**   **25.**

27. vertex: $(4, 0)$; axis: $x = 4$; domain: $(-\infty, \infty)$; range: $[0, \infty)$ **29.** vertex: $(-2, -1)$; axis: $x = -2$; domain: $(-\infty, \infty)$;

range: $[-1, \infty)$ **31.** vertex: $(2, -4)$; axis: $x = 2$; domain: $(-\infty, \infty)$; range: $[-4, \infty)$ **33.** vertex: $(-1, 2)$;

axis: $x = -1$; domain: $(-\infty, \infty)$; range: $(-\infty, 2]$ **35.** vertex: $(2, -3)$; axis: $x = 2$; domain: $(-\infty, \infty)$; range: $[-3, \infty)$

37. It is shifted 6 units up.   **38.**   **39.** It is shifted 6 units up.   **40.** It is shifted 6 units to the right.

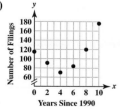

41.   **42.** It is shifted 6 units to the right.   **43.** quadratic; positive   **45.** quadratic; negative   **47.** linear; positive

49. (a)   **(b)** quadratic; positive **(c)** $y = 2.969x^2 - 23.125x + 115$ **(d)** 265 **(e)** No. About 16 companies filed for

bankruptcy each month, so at this rate, filings for 2002 would be about 192. The approximation from the model seems high.   **51. (a)** 183.4
(b) The approximation using the model is quite close.   **53.** $\{-7, -2\}$ **55.** $\{-.5, 1.25\}$

Section 11.7   (page 772)

Exercises   1. If x is squared, it has a vertical axis; if y is squared, it has a horizontal axis.   **3.** Use the discriminant of the function. If it is positive, there are two x-intercepts. If it is 0, there is one x-intercept (at the vertex), and if it is negative, there is no x-intercept.   **5.** $(-4, -6)$

7. $(1, -3)$ **9.** $(2, -1)$ **11.** $(-1, 3)$; up; narrower; no x-intercepts   **13.** $\left(\dfrac{5}{2}, \dfrac{37}{4}\right)$; down; same; two x-intercepts   **15.** $(-3, -9)$; to the right; wider   **17.** F   **19.** C   **21.** D   **23.** vertex: $(-4, -6)$; axis: $x = -4$; domain: $(-\infty, \infty)$; range: $[-6, \infty)$

25. vertex: $(1, -3)$; axis: $x = 1$; domain: $(-\infty, \infty)$; range: $(-\infty, -3]$

27. vertex: $(1, -2)$; axis: $y = -2$; domain: $[1, \infty)$; range: $(-\infty, \infty)$

29. vertex: $(1, 5)$; axis: $y = 5$; domain: $(-\infty, 1]$; range: $(-\infty, \infty)$

31. vertex: $(-7, -2)$; axis: $y = -2$; domain: $[-7, \infty)$; range: $(-\infty, \infty)$

33. 30 and 30   **35.** 140 ft by 70 ft; 9800 ft^2 **37.** 16 ft; 2 sec

39. 2 sec; 67 ft   **41. (a)** minimum **(b)** 1995; 1.7%   **43. (a)** The coefficient of x^2 is negative because the parabola opens down. **(b)** (18.45, 3860) **(c)** In 2018 Social Security assets will reach their maximum value of \$3860 billion.

45. (a) $R(x) = (100 - x)(200 + 4x) = 20,000 + 200x - 4x^2$ **(b)**

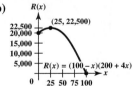

(c) 25 **(d)** \$22,500   **47.** B   **49.** A

51. (a) $|x + p|$ **(b)** The distance from the focus to the origin should equal the distance from the directrix to the origin. **(c)** $\sqrt{(x - p)^2 + y^2}$ **(d)** $y^2 = 4px$

Section 11.8   (page 783)

Exercises   1. (a) $\{1, 3\}$ **(b)** $(-\infty, 1) \cup (3, \infty)$ **(c)** $(1, 3)$ **3. (a)** $\left\{-3, \dfrac{5}{2}\right\}$ **(b)** $\left[-3, \dfrac{5}{2}\right]$ **(c)** $(-\infty, -3] \cup \left[\dfrac{5}{2}, \infty\right)$

5. Include the endpoints if the symbol is $\geq$ or $\leq$. Exclude the endpoints if the symbol is $>$ or $<$.

7. $(-\infty, -1) \cup (5, \infty)$ **9.** $(-4, 6)$ **11.** $(-\infty, 1] \cup [3, \infty)$

13. $\left(-\infty, -\dfrac{3}{2}\right] \cup \left[\dfrac{3}{5}, \infty\right)$ **15.** $\left[-\dfrac{3}{2}, \dfrac{3}{2}\right]$

17. $\left(-\infty, -\dfrac{1}{2}\right] \cup \left[\dfrac{1}{3}, \infty\right)$ **19.** $(-\infty, 0] \cup [4, \infty)$ **21.** $\left[0, \dfrac{5}{3}\right]$

23. $\left(-\infty, 3 - \sqrt{3}\,\right] \cup \left[3 + \sqrt{3}, \infty\right)$ **25.** $(-\infty, \infty)$ **27.** $\emptyset$ **29.** $(-\infty, 1) \cup (2, 4)$

31. $\left[-\dfrac{3}{2}, \dfrac{1}{3}\right] \cup [4, \infty)$ **33.** $(-\infty, 1) \cup (4, \infty)$ **35.** $\left[-\dfrac{3}{2}, 5\right)$

37. $(2, 6]$ **39.** $\left(-\infty, \dfrac{1}{2}\right) \cup \left(\dfrac{5}{4}, \infty\right)$ **41.** $[-7, -2)$

43. $(-\infty, 2) \cup (4, \infty)$ **45.** $\left(0, \dfrac{1}{2}\right) \cup \left(\dfrac{5}{2}, \infty\right)$ **47.** $\left[\dfrac{3}{2}, \infty\right)$

49. $\left(-2, \dfrac{5}{3}\right) \cup \left(\dfrac{5}{3}, \infty\right)$ **51.** 3 sec and 13 sec   **52.** between 3 sec and 13 sec   **53.** at 0 sec (the time when it is

initially projected) and at 16 sec (the time when it hits the ground)   **54.** between 0 and 3 sec and between 13 and 16 sec

Chapter 11 Review Exercises   (page 790)

1. $\{11, -11\}$ **3.** $\{8\sqrt{2}, -8\sqrt{2}\}$ **5.** $\left\{-\dfrac{15}{2}, \dfrac{5}{2}\right\}$ **7.** By the square root property, the first step should be $x = \sqrt{12}$ or

$x = -\sqrt{12}$.   **9.** $\{-5, -1\}$ **11.** $\{-1 + \sqrt{6}, -1 - \sqrt{6}\}$ **13.** $\left\{-\dfrac{2}{5}, 1\right\}$ **15.** $\left\{-\dfrac{7}{2}, 3\right\}$ **17.** $\left\{\dfrac{1 + \sqrt{41}}{2}, \dfrac{1 - \sqrt{41}}{2}\right\}$

19. $\left\{\dfrac{2 + i\sqrt{2}}{3}, \dfrac{2 - i\sqrt{2}}{3}\right\}$ **21.** C   **23.** D   **25.** $\left\{-\dfrac{5}{2}, 3\right\}$ **27.** $\{-4\}$ **29.** $\left\{-\dfrac{343}{8}, 64\right\}$ **31.** 7 mph   **33.** 4.6 hr

35. $v = \dfrac{\pm \sqrt{rFkw}}{kw}$ **37.** $t = \dfrac{3m \pm \sqrt{9m^2 + 24m}}{2m}$ **39.** 12 cm by 20 cm   **41.** 3 min   **43.** .7 sec and 4.0 sec   **45.** 4.5%   **47.** $(0, 6)$

49. $(3, 7)$ **51.** $(-4, 3)$ **53.** vertex: $(2, -3)$; axis: $x = 2$; domain: $(-\infty, \infty)$; range: $[-3, \infty)$ **55.** vertex: $(-4, -3)$;

axis: $y = -3$; domain: $[-4, \infty)$; range: $(-\infty, \infty)$ **57. (a)** $c = 12.39$, $16a + 4b + c = 15.78$, $49a + 7b + c = 22.71$

(b) $f(x) = .2089x^2 + .0118x + 12.39$ **(c)** \$25.85; The result using the model is a little high.   **59.** length: 50 m; width: 50 m

61. $[-4, 3]$ **63.** $\emptyset$ **65.** $[-3, 2)$ **67.** $\left\{\dfrac{3 + i\sqrt{3}}{3}, \dfrac{3 - i\sqrt{3}}{3}\right\}$ **69.** $\{-2, -1, 3, 4\}$ **71.** $\{4\}$

73. $d = \dfrac{\pm \sqrt{SkI}}{I}$ **75.** $\left\{-\dfrac{5}{3}, -\dfrac{3}{2}\right\}$ **77.** $\left(-5, -\dfrac{23}{5}\right]$ **79.** B   **81.** A   **83.** D   **85. (a)** 21.92 trillion ft^3 **(b)** 2005   **87.** 412.3 ft

88. (a) $\{-2\}$ **(b)** $(-\infty, -2)$ **(c)** $(-2, \infty)$ **89. (a)** $\{1, 5\}$

(b) $(-\infty, 1) \cup (5, \infty)$ **(c)** $(1, 5)$ **90. (a)** $\{4\}$ **(b)** $(2, 4)$

(c) $(-\infty, 2) \cup (4, \infty)$ **91.** $(-\infty, \infty)$; denominator   **92.** $(-5, 3)$

Chapter 11 Test   (page 795)

[11.1] **1.** $\left\{3\sqrt{6}, -3\sqrt{6}\right\}$ **2.** $\left\{-\dfrac{8}{7}, \dfrac{2}{7}\right\}$ [11.2] **3.** $\left\{-1 + \sqrt{5}, -1 - \sqrt{5}\right\}$ [11.3] **4.** $\left\{\dfrac{3 + \sqrt{17}}{4}, \dfrac{3 - \sqrt{17}}{4}\right\}$

5. $\left\{\dfrac{2 + i\sqrt{11}}{3}, \dfrac{2 - i\sqrt{11}}{3}\right\}$ [11.1] **6.** A   [11.3] **7.** discriminant: 88; There are two irrational solutions.   [11.1–11.4] **8.** $\left\{\dfrac{2}{3}\right\}$

9. $\left\{-\dfrac{2}{3}, 6\right\}$ **10.** $\left\{\dfrac{-7 + \sqrt{97}}{8}, \dfrac{-7 - \sqrt{97}}{8}\right\}$ **11.** $\left\{-2, -\dfrac{1}{3}, \dfrac{1}{3}, 2\right\}$ **12.** $\left\{-\dfrac{5}{2}, 1\right\}$ [11.5] **13.** $r = \dfrac{\pm\sqrt{\pi S}}{2\pi}$

[11.4] **14.** Andrew: 11.1 hr; Kent: 9.1 hr   **15.** 7 mph   [11.5] **16.** 2 ft   **17.** 16 m   [11.6] **18.** A   **19.** vertex: $(0, -2)$; axis: $x = 0$;

domain: $(-\infty, \infty)$; range: $[-2, \infty)$ [11.7] **20.** vertex: $(2, 3)$; axis: $x = 2$; domain: $(-\infty, \infty)$;

range: $(-\infty, 3]$ **21.** vertex: $(2, 2)$; axis: $y = 2$; domain: $(-\infty, 2]$; range: $(-\infty, \infty)$ **22. (a)** 6.5%

(b) 1996; 3.5%   **23.** 160 ft by 320 ft   [11.8] **24.** $(-\infty, -5) \cup \left(\dfrac{3}{2}, \infty\right)$ **25.** $(-\infty, 4) \cup [9, \infty)$

Cumulative Review Exercises Chapters 1–11   (page 797)

[1.4, 10.7] **1. (a)** $-2, 0, 7$ **(b)** $-\dfrac{7}{3}, -2, 0, .7, 7, \dfrac{32}{3}$ **(c)** all except $\sqrt{-8}$ **(d)** All are complex numbers.   [1.4–1.6] **2.** 6   **3.** 41

[2.3] **4.** $\left\{\dfrac{4}{5}\right\}$ [9.2] **5.** $\left\{\dfrac{11}{10}, \dfrac{7}{2}\right\}$ [10.6] **6.** $\left\{\dfrac{2}{3}\right\}$ [6.6] **7.** $\emptyset$ [11.2, 11.3] **8.** $\left\{\dfrac{7 + \sqrt{177}}{4}, \dfrac{7 - \sqrt{177}}{4}\right\}$ [11.4] **9.** $\{-2, -1, 1, 2\}$

[2.8] **10.** $[1, \infty)$ [9.2] **11.** $\left[2, \dfrac{8}{3}\right]$ [11.8] **12.** $(1, 3)$ **13.** $(-2, 1)$ [3.2, 7.1, 7.3] **14.** function; domain: $(-\infty, \infty)$;

range: $(-\infty, \infty)$ [7.3, 9.3] **15.** not a function   [11.6] **16.** function; domain: $(-\infty, \infty)$;

range: $(-\infty, 3]$ [3.3, 7.1] **17.** $m = \dfrac{2}{7}$; x-intercept: $(-8, 0)$; y-intercept: $\left(0, \dfrac{16}{7}\right)$ [3.4, 7.2] **18. (a)** $y = -\dfrac{5}{2}x + 2$

(b) $y = \dfrac{2}{5}x + \dfrac{13}{5}$ [4.1, 4.2] **19.** $\dfrac{x^8}{y^4}$ **20.** $\dfrac{4}{xy^2}$ [4.5] **21.** $14x^2 - 13x - 12$ [4.6] **22.** $\dfrac{4}{9}t^2 + 12t + 81$

[4.4] **23.** $-3t^3 + 5t^2 - 12t + 15$ [4.7] **24.** $4x^2 - 6x + 11 + \dfrac{4}{x + 2}$ [5.1–5.4] **25.** $x(4 + x)(4 - x)$ **26.** $(4m - 3)(6m + 5)$

27. $(2x + 3y)(4x^2 - 6xy + 9y^2)$ **28.** $(3x - 5y)^2$ [6.2] **29.** $\dfrac{x - 5}{x + 5}$ [6.4] **30.** $-\dfrac{8}{k}$ [6.5] **31.** $\dfrac{r - s}{r}$

[7.2] **32.** (a)

(b) positive (c) $y = 1.279x + 116.26$ (d) 158.47; It is a little too high. [7.3, 11.6] **33.** No, because

the graph is a vertical line, which is not the graph of a function by the vertical line test. **34.** (a) 13 (b) domain: $(-\infty, \infty)$; range: $[-5, \infty)$
[8.1–8.3] **35.** $\{(1, -2)\}$ [8.4] **36.** $\{(3, -4, 2)\}$ [8.5] **37.** (a) $x + y = 34.2$; $x = 4y - .3$ (b) AOL: $27.3 billion; Time Warner:

$6.9 billion [10.1] **38.** $\dfrac{3\sqrt[3]{4}}{4}$ [10.5] **39.** $\sqrt{7} + \sqrt{5}$ [6.7] **40.** biking: 12 mph; walking: 2 mph [11.5] **41.** southbound car: 57 mi;

eastbound car: 76 mi [2.7] **42.** 930 **43.** 720 **44.** 990 **45.** 930

CHAPTER 12 INVERSE, EXPONENTIAL, AND LOGARITHMIC FUNCTIONS

Section 12.1 (page 807)

Exercises **1.** It is not one-to-one because both Illinois and Wisconsin are paired with the same range element, 40. **3.** Yes. By adding 1 to
1058 two distances would be the same, so the function would not be one-to-one. **5.** B **7.** A **9.** $\{(6, 3), (10, 2), (12, 5)\}$ **11.** not

one-to-one **13.** $f^{-1}(x) = \dfrac{x - 4}{2}$ **15.** $g^{-1}(x) = x^2 + 3$, $x \geq 0$ **17.** not one-to-one **19.** $f^{-1}(x) = \sqrt[3]{x + 4}$ **21.** (a) 8 (b) 3

23. (a) 1 (b) 0 **25.** (a) one-to one (b) **27.** (a) not one-to one **29.** (a) one-to one (b)

31. **33.** **35.**

x	$f(x)$
0	0
1	1
4	2

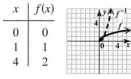 **37.**

x	$f(x)$
-1	-3
0	-2
1	-1
2	6

 39. $f^{-1}(x) = \dfrac{x + 5}{4}$

40. My graphing calculator is the greatest thing since sliced bread. **41.** If the function were not one-to-one, there would be ambiguity
in some of the characters, as they could represent more than one letter. **42.** Answers will vary. For example, Jane Doe is

1004 5 2748 129 68 3379 129. **43.** $f^{-1}(x) = \dfrac{x + 7}{2}$

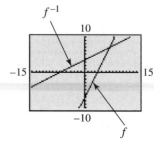

45. $f^{-1}(x) = \sqrt[3]{x} - 5$ **47.**   **49.** It is not a one-to-one function.

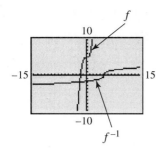

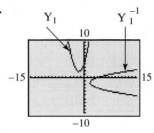

Section 12.2   (page 818)

Exercises   **1.** C   **3.** A   **5.**   **7.**   **9.**   **11.**   **13. (a)** rises; falls

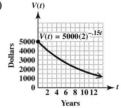

$f(x) = 3^x$

$g(x) = \left(\frac{1}{3}\right)^x$

$y = 4^{-x}$

$y = 2^{2x-2}$

(b) It is one-to-one and thus has an inverse.   **15.** $\{2\}$ **17.** $\left\{\dfrac{3}{2}\right\}$ **19.** $\{7\}$ **21.** $\{-3\}$ **23.** $\{-1\}$ **25.** $\{-3\}$ **27.** 639.545

29. .066   **31.** 12.179   **33.** In the definition of the exponential function defined by $F(x) = a^x$, a must be a positive number. This corresponds to the peculiarity that some scientific calculators do not allow negative bases.   **35. (a)** .5°C **(b)** .35°C   **37. (a)** 1.6°C **(b)** .5°C
39. (a) 132,359 thousand tons **(b)** 97,264 thousand tons **(c)** It is slightly less than what the model provides (93,733 thousand tons).
41. (a) $5000 **(b)** $2973 **(c)** $1768 **(d)**   **43.** 6.67 yr after it was purchased   **45.** $\left(\sqrt[4]{16}\right)^3$; 8   **46.** $\sqrt[4]{16^3}$; 8

$V(t)$

$V(t) = 5000(2)^{-.15t}$

Dollars

Years

47. 64; 8   **48.** Because $\sqrt{\sqrt{x}} = (x^{1/2})^{1/2} = x^{1/4} = \sqrt[4]{x}$, the fourth root of 16^3 can be found by taking the square root twice.   **49.** 8
50. $16^{75/100}$; $16^{3/4} = \left(\sqrt[4]{16}\right)^3 = 2^3 = 8$

Section 12.3   (page 826)

Connections   **(page 826)**   **1.** almost 4 times as powerful   **2.** about 300 times as powerful

Exercises   **1. (a)** C **(b)** F **(c)** B **(d)** A **(e)** E **(f)** D   **3.** $\log_4 1024 = 5$ **5.** $\log_{1/2} 8 = -3$ **7.** $\log_{10} .001 = -3$ **9.** $\log_{625} 5 = \dfrac{1}{4}$

11. $4^3 = 64$ **13.** $10^{-4} = \dfrac{1}{10,000}$ **15.** $6^0 = 1$ **17.** $9^{1/2} = 3$ **19.** By using the word "radically," the teacher meant for him to consider

roots. Because 3 is the square (2nd) root of 9, $\log_9 3 = \dfrac{1}{2}$.   **21.** $\left\{\dfrac{1}{3}\right\}$ **23.** $\{81\}$ **25.** $\left\{\dfrac{1}{5}\right\}$ **27.** $\{1\}$ **29.** $\{x \mid x > 0, x \neq 1\}$

31. $\{5\}$ **33.** $\left\{\dfrac{5}{3}\right\}$ **35.** $\{4\}$ **37.** $\left\{\dfrac{3}{2}\right\}$ **39.** $\{30\}$ **41.**   **43.**   **45.** Every power of 1 is equal to 1,

$y = \log_3 x$

$y = \log_{1/3} x$

and thus it cannot be used as a base. **47.** $(0, \infty)$; $(-\infty, \infty)$ **49.** 8 **51.** 24 **53. (a)** 645 sites **(b)** 962 sites **(c)** 1279 sites

55. (a) 130 thousand units **(b)** 190 thousand units **(c)**

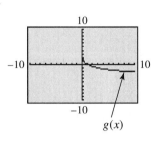

57. (a) 500 **(b)** 1000 **(c)** 1500 **(d)**

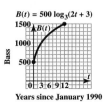

59.

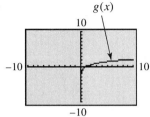

61.

Section 12.4 (page 836)

Connections **(page 835)** **1.**
$$\log_{10} 458.3 \approx 2.661149857$$
$$+ \log_{10} 294.6 \approx 2.469232743$$
$$\approx 5.130382600$$
$$10^{5.130382600} \approx 135,015.18$$
A calculator gives $(458.3)(294.6) = 135,015.18$.

2. Answers will vary.

Exercises **1.** $\log_{10} 3 + \log_{10} 4$ **3.** 4 **5.** 4 **7.** $\log_7 4 + \log_7 5$ **9.** $\log_5 8 - \log_5 3$ **11.** $2 \log_4 6$ **13.** $\frac{1}{3} \log_3 4 - 2 \log_3 x -$

$\log_3 y$ **15.** $\frac{1}{2} \log_3 x + \frac{1}{2} \log_3 y - \frac{1}{2} \log_3 5$ **17.** $\frac{1}{3} \log_2 x + \frac{1}{5} \log_2 y - 2 \log_2 r$ **19.** The distributive property tells us that the *product*

$a(x + y)$ equals the sum $ax + ay$. In the notation $\log_a(x + y)$, the parentheses do not indicate multiplication. They indicate that $x + y$ is the result

of raising a to some power. **21.** $\log_b xy$ **23.** $\log_a \frac{m}{n}$ **25.** $\log_a \frac{rt^3}{s}$ **27.** $\log_a \frac{125}{81}$ **29.** $\log_{10}(x^2 - 9)$ **31.** $\log_p \frac{x^3 y^{1/2}}{z^{3/2} a^3}$ **33.** 1.2552

35. $-.6532$ **37.** 1.5562 **39.** .4771 **41.** .2386 **43.** 4.7710 **45.** false **47.** true **49.** true **51.** false **53.** The exponent of

a quotient is the difference between the exponent of the numerator and the exponent of the denominator. **55.** $\log_2 8 - \log_2 4 = \log_2 \frac{8}{4} =$

$\log_2 2 = 1$ **57.** 4 **58.** It is the exponent to which 3 must be raised to obtain 81. **59.** 81 **60.** It is the exponent to which 2 must be

raised to obtain 19. **61.** 19 **62.** m

Section 12.5 (page 843)

Connections **(page 842)** 2; 2.5; $2.\overline{6}$; $2.70\overline{83}$; $2.71\overline{6}$; The difference is .0016151618. It approaches e fairly quickly.

Exercises **1.** C **3.** C **5.** 19.2 **7.** 1.6335 **9.** 2.5164 **11.** -1.4868 **13.** 9.6776 **15.** 2.0592 **17.** -2.8896 **19.** 5.9613

21. 4.1506 **23.** 2.3026 **25. (a)** 2.552424846 **(b)** 1.552424846 **(c)** .552424846 **(d)** The whole number parts will vary but the decimal

parts are the same. **27.** An error message appears, because we cannot find the common logarithm of a negative number. **29.** bog

31. 11.6 **33.** 4.3 **35.** 4.0×10^{-8} **37.** 4.0×10^{-6} **39. (a)** 800 yr **(b)** 5200 yr **(c)** 11,500 yr **41. (a)** 107 dB **(b)** 100 dB

(c) 98 dB **43. (a)** 77% **(b)** 1989 **45. (a)** \$54 per ton **(b)** If $p = 0$, then $\ln(1 - p) = \ln 1 = 0$, so T would be negative. If $p = 1$, then

$\ln(1 - p) = \ln 0$, but the domain of $\ln x$ is $(0, \infty)$. **47.** 2.2619 **49.** .6826 **51.** .3155 **53.** .8736 **55.** 2.4849 **57.** Answers will

vary. Suppose the name is Jeffery Cole, with $m = 7$ and $n = 4$. **(a)** $\log_7 4$ is the exponent to which 7 must be raised to obtain 4.

(**b**) .7124143742 (**c**) 4   **59.** 1333 sites   **61.**

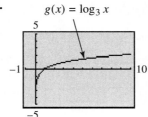

$$g(x) = \log_3 x$$

63.

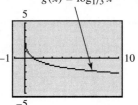

$$g(x) = \log_{1/3} x$$

Section 12.6   (page 853)

Exercises   **1.** $\log 5^x = \log 125$ **2.** $x \log 5 = \log 125$ **3.** $x = \dfrac{\log 125}{\log 5}$ **4.** $\dfrac{\log 125}{\log 5} = 3$; $\{3\}$ **5.** $\{.827\}$ **7.** $\{.833\}$ **9.** $\{1.201\}$

11. $\{2.269\}$ **13.** $\{15.967\}$ **15.** $\{261.291\}$ **17.** $\{-10.718\}$ **19.** $\{3\}$ **21.** $\{5.879\}$ **23.** Natural logarithms are a better choice

because e is the base.   **25.** $\left\{\dfrac{2}{3}\right\}$ **27.** $\left\{\dfrac{33}{2}\right\}$ **29.** $\left\{-1 + \sqrt[3]{49}\right\}$ **31.** 2 cannot be a solution because $\log(2 - 3) = \log(-1)$, and -1

is not in the domain of $\log x$.   **33.** $\left\{\dfrac{1}{3}\right\}$ **35.** $\{2\}$ **37.** $\emptyset$ **39.** $\{8\}$ **41.** $\left\{\dfrac{4}{3}\right\}$ **43.** $\{8\}$ **45.** (**a**) \$2539.47 (**b**) 10.2 yr

47. (**a**) \$4934.71 (**b**) 19.8 yr   **49.** (**a**) \$11,260.96 (**b**) \$11,416.64 (**c**) \$11,497.99 (**d**) \$11,580.90 (**e**) \$11,581.83   **51.** \$137.41
53. (**a**) 78,840 million dollars (**b**) 92,316 million dollars (**c**) 108,095 million dollars (**d**) 70,967 million dollars   **55.** 32,044 million dollars
57. (**a**) 1.62 g (**b**) 1.18 g (**c**) .69 g (**d**) 2.00 g   **59.** (**a**) 179.73 g (**b**) 21.66 yr   **61.** 1997   **63.** 1.733 days   **65.** It means that after
250 yr, approximately 2.9 g of the original sample remain.   **67.** (**a**) 2.302585093 (**b**) .4342944819

Chapter 12 Review Exercises   (page 864)

1. not one-to-one   **3.** This function is not one-to-one because two sodas in the list have 41 mg of caffeine.   **5.** $f^{-1}(x) = \dfrac{x^3 + 4}{6}$

7.

9.

$f(x) = 3^x$

11.

$y = 3^{x+1}$

13. $\{4\}$ **15.** (**a**) 1.2 million tons (**b**) 3.8 million tons (**c**) 21.8 million tons

17.

$g(x) = \log_{1/3} x$

19. $\left\{\dfrac{3}{2}\right\}$ **21.** $\{8\}$ **23.** $\{b \,|\, b > 0, b \neq 1\}$ **25.** a **27.** $\log_2 3 + \log_2 x + 2 \log_2 y$ **29.** $\log_b \dfrac{3x}{y^2}$ **31.** 1.4609

33. 3.3638   **35.** .9251   **37.** 6.4   **39.** 2.5×10^{-5} **41.** (**a**) 18 yr (**b**) 12 yr (**c**) 7 yr (**d**) 6 yr (**e**) Each comparison shows

approximately the same number. For example, in part (a) the doubling time is 18 yr (rounded) and $\dfrac{72}{4} = 18$. Thus, the formula $t = \dfrac{72}{100r}$ (called

the *rule of 72*) is an excellent approximation of the doubling time formula. (It is used by bankers for that purpose.)   **43.** $\{2.042\}$

45. $\{18.310\}$ **47.** $\left\{-6 + \sqrt[3]{25}\right\}$ **49.** $\left\{\dfrac{3}{8}\right\}$ **51.** $\{1\}$ **53.** \$28,295.56   **55.** Plan A is better, since it would pay \$2.92 more.

57. (**a**) about \$4267 (**b**) about 11%   **59.** D   **61.** $\dfrac{1}{4}, \dfrac{1}{2}, 1, 2, 4, 8$

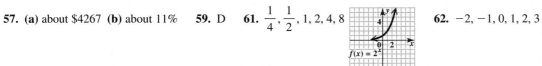

$f(x) = 2^x$

62. $-2, -1, 0, 1, 2, 3$

$g(x) = \log_2 x$

63. The roles of x and y are interchanged. They are inverses.   **64.** horizontal; vertical   **65.** 5; 2; 3; 5   **66.** 4; 8 (or 8; 4)

67. 3.700439718 (The number of displayed digits may vary.)   **68.** $\log_2 13$ is the exponent to which 2 must be raised to obtain 13.   **69.** 13

70. 13; The number in Exercise 67 is the exponent to which 2 must be raised to obtain 13.   **71.** 3.700439718   **72.** $\left\{-\dfrac{8}{5}\right\}$ **73.** 7

75. 4   **77.** -5 **79.** $\{72\}$ **81.** $\left\{\dfrac{1}{9}\right\}$ **83.** $\{3\}$ **85.** $\left\{\dfrac{1}{8}\right\}$ **87.** $\{-2, -1\}$ **89.** 6.8 yr   **91.** .325

Chapter 12 Test   (page 869)

[12.1] **1. (a)** not one-to-one **(b)** one-to-one   **2.** $f^{-1}(x) = x^3 - 7$ **3.**   [12.2] **4.**

[12.3] **5.** 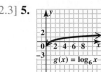   [12.1–12.3] **6.** Once the graph of $f(x) = 6^x$ is sketched, interchange the x- and y-values of its ordered pairs. The

resulting points will be on the graph of $g(x) = \log_6 x$ since f and g are inverses.   [12.2] **7.** $\{-4\}$ **8.** $\left\{-\dfrac{13}{3}\right\}$ [12.5] **9. (a)** 30.0 million

(b) 37.7 million   [12.3] **10.** $\log_4 .0625 = -2$ **11.** $7^2 = 49$ **12.** $\{32\}$ **13.** $\left\{\dfrac{1}{2}\right\}$ **14.** $\{2\}$ **15.** 5; 2; 5th; 32

[12.4] **16.** $2\log_3 x + \log_3 y$ **17.** $\dfrac{1}{2}\log_5 x - \log_5 y - \log_5 z$ **18.** $\log_b \dfrac{s^3}{t}$ **19.** $\log_b \dfrac{r^{1/4}s^2}{t^{2/3}}$ [12.5] **20. (a)** 1.3636 **(b)** $-.1985$

21. (a) $\dfrac{\log 19}{\log 3}$ **(b)** $\dfrac{\ln 19}{\ln 3}$ **(c)** 2.6801   [12.6] **22.** $\{3.9656\}$ **23.** $\{3\}$ **24.** \$12,507.51   **25. (a)** \$19,260.38 **(b)** approximately 13.9 yr

Cumulative Review Exercises Chapters 1–12   (page 871)

[1.4] **1.** $-2, 0, 6, \dfrac{30}{3}$ (or 10)   **2.** $-\dfrac{9}{4}, -2, 0, .6, 6, \dfrac{30}{3}$ (or 10)   **3.** $-\sqrt{2}, \sqrt{11}$ **4.** $-\dfrac{9}{4}, -2, -\sqrt{2}, 0, .6, \sqrt{11}, 6, \dfrac{30}{3}$ (or 10)

[1.4–1.6] **5.** 16   **6.** -27 **7.** -39 [2.3] **8.** $\left\{-\dfrac{2}{3}\right\}$ [2.8] **9.** $[1, \infty)$ [4.5] **10.** $6p^2 + 7p - 3$ [4.6] **11.** $16k^2 - 24k + 9$

[4.4] **12.** $-5m^3 + 2m^2 - 7m + 4$ [4.7] **13.** $2t^3 + 5t^2 - 3t + 4$ [5.1] **14.** $x(8 + x^2)$ [5.2, 5.3] **15.** $(3y - 2)(8y + 3)$

16. $z(5z + 1)(z - 4)$ [5.4] **17.** $(4a + 5b^2)(4a - 5b^2)$ **18.** $(2c + d)(4c^2 - 2cd + d^2)$ **19.** $(4r + 7q)^2$ [4.1, 4.2] **20.** $-\dfrac{1875p^{13}}{8}$

[6.2] **21.** $\dfrac{x + 5}{x + 4}$ [6.3, 6.4] **22.** $\dfrac{-3k - 19}{(k + 3)(k - 2)}$ **23.** $\dfrac{22 - p}{p(p - 4)(p + 2)}$ [3.2, 7.1] **24.**   [9.3] **25.**

[3.3, 7.1, 7.3] **26. (a)** yes **(b)** approximately -4000; The number of acres harvested decreased by approximately 4000 acres per year during

1997–1999.   [3.4, 7.2] **27.** $y = \dfrac{3}{4}x - \dfrac{19}{4}$ [8.1–8.3, 8.6] **28.** $\{(4, 2)\}$ **29.** $\emptyset$ [8.4, 8.6] **30.** $\{(1, -1, 4)\}$ [8.5] **31.** 6 lb

[9.2] **32.** $\{-2, 7\}$ **33.** $\left\{\pm\dfrac{16}{3}\right\}$ **34.** $\left[\dfrac{7}{3}, 3\right]$ **35.** $(-\infty, -3) \cup (2, \infty)$ [10.3] **36.** $12\sqrt{2}$ [10.4] **37.** $-27\sqrt{2}$ [10.6] **38.** $\{0, 4\}$

[10.7] **39.** 41   [11.1–11.3] **40.** $\left\{\dfrac{1 \pm \sqrt{13}}{6}\right\}$ [11.8] **41.** $(-\infty, -4) \cup (2, \infty)$ [11.4] **42.** $\{\pm 1, \pm 2\}$ [11.7] **43.** 150 and 150

[11.6] **44.**   [12.2] **45.**   **46.** $\{-1\}$ [12.3] **47.**   [12.4] **48.** 6.3398

49. $3 \log x + \dfrac{1}{2} \log y - \log z$ [12.6] **50.** **(a)** 25,000 **(b)** 30,500 **(c)** 37,300 **(d)** in about 3.5 hr, or at about 3:30 P.M.

CHAPTER **13** NONLINEAR FUNCTIONS, CONIC SECTIONS, AND NONLINEAR SYSTEMS

Section 13.1   (page 883)

Exercises   **1.** 0   **3.** 0, 0   **5.** B   **7.** A   **9.** domain: $(-\infty, \infty)$; range: $[0, \infty)$ **11.** domain: $(-\infty, 0) \cup (0, \infty)$;

range: $(-\infty, 1) \cup (1, \infty)$ **13.** domain: $[2, \infty)$; range: $[0, \infty)$ **15.** domain: $(-\infty, 2) \cup (2, \infty)$;

range: $(-\infty, 0) \cup (0, \infty)$ **17.** domain: $[-3, \infty)$; range: $[-3, \infty)$ **19.**

21.

23. **(a)** $8x - 3$ **(b)** $2x - 17$ **25.** **(a)** $-x^2 + 12x - 12$ **(b)** $9x^2 + 4x + 6$ **27.** $10x^2 - 2x$

29. $2x^2 - x - 3$ **31.** $5x - 1; 0$ **33.** $2x - 3; -1$ **35.** $x^2 + 2x - 9$ **37.** 6   **39.** $x^2 - x - 6$ **41.** $2x^3 - 18x$ **43.** -20

45. $2x^2 - 6x$ **47.** $\dfrac{x^2 - 9}{2x}, \quad x \neq 0$ **49.** $-\dfrac{5}{4}$ **51.** 0   **53.** For example, let $f(x) = 2x^3 + 3x^2 + x + 4$ and

$g(x) = 2x^4 + 3x^3 - 9x^2 + 2x - 4$; $(f - g)(x) = -2x^4 - x^3 + 12x^2 - x + 8$, and $(g - f)(x) = 2x^4 + x^3 - 12x^2 + x - 8$. Because the two differences are not equal, subtraction of polynomial functions is not commutative.   **55.** 16   **57.** 83   **59.** 13   **61.** $4x^2 + 12x + 13$

63. $x^2 + 10x + 29$ **65.** $2x + 8$ **67.** $(f \circ g)(x) = 63{,}360x$; It computes the number of inches in x mi.   **69.** $(A \circ r)(t) = 4\pi t^2$; This is the area of the circular layer as a function of time.

Section 13.2   (page 891)

Exercises   **1. (a)** $(0, 0)$ **(b)** 5 **(c)**   **3.** B   **5.** D   **7.** $(x + 4)^2 + (y - 3)^2 = 4$ **9.** $(x + 8)^2 + (y + 5)^2 = 5$

11. center: $(-2, -3)$; $r = 2$ **13.** center: $(-5, 7)$; $r = 9$ **15.** center: $(2, 4)$; $r = 4$ **17.** The thumbtack acts as the center and the length of the string acts as the radius.   **19.**   **21.**   **23.** center: $(-3, 2)$

25. center: $(2, 3)$ **27.** 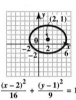   **29.**   **31.**   **33.**

35. 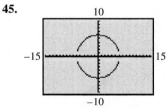   **37.** 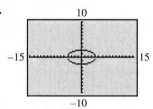   **39.** The fixed ends of the string are at the foci, and the constant length of the string represents

the sum of the distances from any point on the curve to the foci.   **41.** By the vertical line test the set is not a function, because a vertical line may intersect the graph of an ellipse in two points.   **43.** $y_1 = 4 + \sqrt{16 - (x + 2)^2}$, $y_2 = 4 - \sqrt{16 - (x + 2)^2}$

45.

47.

49. (a) 10 m **(b)** 36 m   **51. (a)** 154.7 million mi
(b) 128.7 million mi (Answers are rounded.)   **53.** $3\sqrt{3}$ units
55. Answers will vary.

Section 13.3   (page 902)

Connections   (page 898)   Answers will vary.

Exercises   **1.** C   **3.** D   **5.** When written in one of the forms given in the box titled "Equations of Hyperbolas" in this section, it will open

up and down if the − sign precedes the x^2-term; it will open left and right if the − sign precedes the y^2-term.   **7.**

9.
$$\frac{y^2}{9} - \frac{x^2}{9} = 1$$

11.
$$\frac{x^2}{25} - \frac{y^2}{36} = 1$$

13.
$$\frac{y^2}{16} - \frac{x^2}{16} = 1$$

15. hyperbola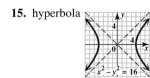
$x^2 - y^2 = 16$

17. ellipse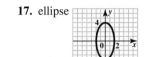
$4x^2 + y^2 = 16$

19. circle
$y^2 = 36 - x^2$

21. hyperbola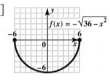
$9x^2 = 144 + 16y^2$

23. hyperbola
$y^2 = 4 + x^2$

25. domain: $[-4, 4]$;

range: $[0, 4]$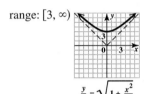
$f(x) = \sqrt{16 - x^2}$

27. domain: $[-6, 6]$; range: $[-6, 0]$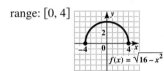
$f(x) = -\sqrt{36 - x^2}$

29. domain: $(-\infty, \infty)$;

range: $[3, \infty)$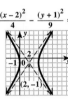
$$\frac{y}{3} = \sqrt{1 + \frac{x^2}{9}}$$

31. $\dfrac{(x-2)^2}{4} - \dfrac{(y+1)^2}{9} = 1$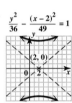

33. $\dfrac{y^2}{36} - \dfrac{(x-2)^2}{49} = 1$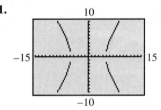

35. (a) 50 m (b) 69.3 m **37.** for V greater than 4325.68 m per sec

39. $y_1 = \sqrt{\dfrac{x^2}{9} - 1},\ y_2 = -\sqrt{\dfrac{x^2}{9} - 1}$ **41.**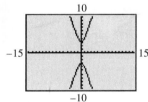

43.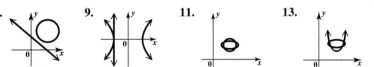

45. Answers will vary.

Section 13.4 (page 910)

Exercises **1.** Substitute $x - 1$ for y in the first equation. Then solve for x. Find the corresponding y-values by substituting back into $y = x - 1$. In the first equation, both variables are squared and in the second, both variables are to the first power, so the elimination method is not appropriate. **3.** one **5.** none **7.** 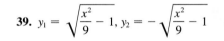 **9.** **11.** **13.** **15.** $\left\{ (0, 0), \left(\dfrac{1}{2}, \dfrac{1}{2} \right) \right\}$

17. $\{(-6, 9), (-1, 4)\}$ **19.** $\left\{ \left(-\dfrac{1}{5}, \dfrac{7}{5} \right), (1, -1) \right\}$ **21.** $\left\{ (-2, -2), \left(-\dfrac{4}{3}, -3 \right) \right\}$ **23.** $\{(-3, 1), (1, -3)\}$

25. $\left\{ \left(-\dfrac{3}{2}, -\dfrac{9}{4} \right), (-2, 0) \right\}$ **27.** $\{(-\sqrt{3}, 0), (\sqrt{3}, 0), (-\sqrt{5}, 2), (\sqrt{5}, 2)\}$ **29.** $\left\{ \left(\dfrac{i\sqrt{3}}{3}, \dfrac{-3 + i\sqrt{3}}{6} \right), \left(\dfrac{-i\sqrt{3}}{3}, \dfrac{-3 - i\sqrt{3}}{6} \right) \right\}$

31. $\{(-2, 0), (2, 0)\}$ **33.** $\left\{ \left(\sqrt{3}, 0 \right), \left(-\sqrt{3}, 0 \right) \right\}$ **35.** $\{(1, 3), (1, -3), (-1, 3), (-1, -3)\}$ **37.** $\{(-2i\sqrt{2}, -2\sqrt{3}), (-2i\sqrt{2}, 2\sqrt{3}),$
$(2i\sqrt{2}, -2\sqrt{3}), (2i\sqrt{2}, 2\sqrt{3})\}$ **39.** $\left\{ \left(-\sqrt{5}, -\sqrt{5} \right), \left(\sqrt{5}, \sqrt{5} \right) \right\}$ **41.** $\{(i, 2i), (-i, -2i), (2, -1), (-2, 1)\}$

43. $\{(2, -3), (-3, 2)\}$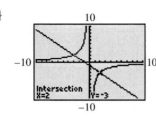

45. length: 12 ft; width: 7 ft **47.** $20; $\frac{4}{5}$ thousand or 800 calculators **49.** 1981; 470 thousand

Section 13.5 (page 916)

Exercises **1.** C **3.** Answers will vary. **5.** B **7.** A **9.** $y > 4 + x^2$ **11.** $y + 2 \geq x^2$ **13.** $2y^2 \geq 8 - x^2$

15. $y \leq x^2 + 4x + 2$ **17.** $9x^2 > 16y^2 + 144$ **19.** $x^2 - 4 \geq -4y^2$ **21.** $x \leq -y^2 + 6y - 7$ **23.** $2x + 5y < 10$; $x - 2y < 4$ **25.** $5x - 3y \leq 15$; $4x + y \geq 4$

27. $x \leq 5$; $y \leq 4$ **29.** $y > x^2 - 4$; $y < -x^2 + 3$ **31.** $x \geq 0$; $y \geq 0$; $x^2 + y^2 \geq 4$; $x + y \leq 5$ **33.** $y \leq -x^2$; $y \geq x - 3$; $y \leq -1$; $x < 1$ **35.** $x^2 + y^2 > 36$; $x \geq 0$ **37.** $x < y^2 - 3$; $x < 0$

39. $4x^2 - y^2 > 16$; $x < 0$ **41.** $x^2 + 4y^2 \geq 1$; $x \geq 0$; $y \geq 0$ **43.** **45.**

Chapter 13 Review Exercises (page 923)

1. $f(x) = |x + 4|$ **3.** $f(x) = \sqrt{x} + 3$ **5. (a)** $5x^2 - x + 5$ **(b)** $-5x^2 + 5x + 1$ **(c)** 11 **(d)** -9 **7.** 167 **9.** 20

11. $75x^2 + 220x + 160$ **13.** $(x + 2)^2 + (y - 4)^2 = 9$ **15.** $(x - 4)^2 + (y - 2)^2 = 36$ **17.** center: $(4, 1)$; $r = 2$

19. center: $(3, -2)$; $r = 5$ **21.** $\frac{x^2}{16} + \frac{y^2}{9} = 1$ **23.** $\frac{x^2}{65{,}286{,}400} + \frac{y^2}{2{,}560{,}000} = 1$ **25.** $\frac{x^2}{16} - \frac{y^2}{25} = 1$ **27.** $f(x) = -\sqrt{16 - x^2}$

29. parabola   **31.** ellipse   **33.** hyperbola   **35.** $\{(6, -9), (-2, -5)\}$ **37.** $\{(4, 2), (-1, -3)\}$

39. $\{(-\sqrt{2}, 2), (-\sqrt{2}, -2), (\sqrt{2}, -2), (\sqrt{2}, 2)\}$ **41.** 0, 1, or 2   **43.**   **45.**   **47.**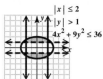

49. $2a + 4b + c = -20$ **50.** $5a + b + c = -26$ **51.** $-a + b + c = -2$ **52.** $\{(-4, -2, -4)\}$; $x^2 + y^2 - 4x - 2y - 4 = 0$

53. center: $(2, 1)$; radius: 3   **55.**   **57.** 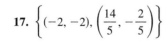   **59.**   **61.**   **63.** 69.8 million km

Chapter 13 Test   (page 925)

[13.1] **1.** C   **2.** A   **3.** D   **4.** B   **5.**   **6.** (a) $x^2 + 3x + 7$ (b) 5 (c) $3x^3 + 5x^2 + 6x + 10$ (d) 23 (e) $3x^2 + 11$

(f) $9x^2 + 30x + 27$ [13.2] **7.** center: $(2, -3)$; radius: 4   **8.** center: $(-4, 1)$; radius: 5   [13.3] **9.**

[13.2] **10.**   [13.3] **11.**   **12.**   **13.** ellipse   **14.** hyperbola   **15.** parabola

[13.4] **16.** $\left\{\left(-\dfrac{1}{2}, -10\right), (5, 1)\right\}$ **17.** $\left\{(-2, -2), \left(\dfrac{14}{5}, -\dfrac{2}{5}\right)\right\}$ **18.** $\{(-\sqrt{22}, -\sqrt{3}), (-\sqrt{22}, \sqrt{3}), (\sqrt{22}, -\sqrt{3}), (\sqrt{22}, \sqrt{3})\}$

[13.5] **19.**   **20.**

Cumulative Review Exercises Chapters 1–13   (page 926)

[1.4, 1.5] **1.** -4 [2.3] **2.** $\left\{\dfrac{2}{3}\right\}$ [2.8] **3.** $\left(-\infty, \dfrac{3}{5}\right]$ [3.3, 7.1] **4.** $\dfrac{2}{3}$ [3.4, 7.2] **5.** $3x + 2y = -13$ [4.6] **6.** $25y^2 - 30y + 9$

[4.5] **7.** $12r^2 + 40r - 7$ [4.7] **8.** $4x^3 - 4x^2 + 3x + 5 + \dfrac{3}{2x + 1}$ [5.2, 5.3] **9.** $(3x + 2)(4x - 5)$ **10.** $(2y^2 - 1)(y^2 + 3)$

[5.4] **11.** $(z^2 + 1)(z + 1)(z - 1)$ **12.** $(a - 3b)(a^2 + 3ab + 9b^2)$ [6.2] **13.** $\dfrac{40}{9}$ **14.** $\dfrac{y - 1}{y(y - 3)}$ [6.4] **15.** $\dfrac{3c + 5}{(c + 5)(c + 3)}$ **16.** $\dfrac{1}{p}$

[6.7] **17.** $1\frac{1}{5}$ hr   [7.3] **18.** \$275   [8.1–8.3, 8.6] **19.** $\{(3, -3)\}$ [8.4, 8.6] **20.** $\{(4, 1, -2)\}$ [13.4] **21.** $\left\{(-1, 5), \left(\frac{5}{2}, -2\right)\right\}$

[8.5] **22.** 40 mph   [4.1, 4.2] **23.** $\frac{3}{4}$ **24.** $\frac{a^5}{4}$ [10.4] **25.** $2\sqrt[3]{2}$ [10.5] **26.** $\frac{3\sqrt{10}}{2}$ [10.7] **27.** $\frac{7}{5} + \frac{11}{5}i$ [9.2] **28.** $\{-4, 4\}$

29. $(-\infty, -5) \cup (10, \infty)$ [10.6] **30.** $\emptyset$ [5.5] **31.** $\left\{\frac{1}{5}, -\frac{3}{2}\right\}$ [11.1–11.3] **32.** $\left\{\frac{1 + 2\sqrt{2}}{4}, \frac{1 - 2\sqrt{2}}{4}\right\}$

33. $\left\{\frac{3 + \sqrt{33}}{6}, \frac{3 - \sqrt{33}}{6}\right\}$ [11.4] **34.** $\left\{-\frac{\sqrt{6}}{2}, \frac{\sqrt{6}}{2}, -\sqrt{7}, \sqrt{7}\right\}$ [11.5] **35.** $v = \frac{\pm\sqrt{rFkw}}{kw}$ [12.1] **36.** $f^{-1}(x) = \sqrt[3]{x - 4}$

[12.4] **37.** 4   [12.5] **38.** 7   [12.4] **39.** $\log\frac{(3x + 7)^2}{4}$ [12.6] **40.** $\{3\}$ **41.** (a) \$12,198.90 (b) \$12,214.03   **42.** \$16.9 billion

43. \$75.8 billion   [13.1] **44.** (a) -1 (b) $9x^2 + 18x + 4$ [7.3] **45.**

$f(x) = -3x + 5$

[11.6] **46.**

$f(x) = -2(x - 1)^2 + 3$

[13.5] **47.**

$\frac{x^2}{25} + \frac{y^2}{16} \le 1$

[13.1] **48.**

$f(x) = \sqrt{x - 2}$

[13.3] **49.**

$\frac{x^2}{4} - \frac{y^2}{16} = 1$

[12.2] **50.**

$f(x) = 3^x$

CHAPTER **14** SEQUENCES AND SERIES

Section 14.1   (page 935)

Exercises   1. 2, 3, 4, 5, 6   **3.** $4, \frac{5}{2}, 2, \frac{7}{4}, \frac{8}{5}$ **5.** 3, 9, 27, 81, 243   **7.** $1, \frac{1}{4}, \frac{1}{9}, \frac{1}{16}, \frac{1}{25}$ **9.** $-1, 1, -1, 1, -1$ **11.** -70 **13.** $\frac{49}{23}$

15. 171   **17.** $4n$ **19.** $\frac{1}{3^n}$ **21.** \$110, \$109, \$108, \$107, \$106, \$105; \$400   **23.** \$6554   **25.** $4 + 5 + 6 + 7 + 8 = 30$

27. $3 + 6 + 11 = 20$ **29.** $-1 + 1 - 1 + 1 - 1 + 1 = 0$ **31.** $0 + 6 + 14 + 24 + 36 = 80$ **33.** $2x + 4x + 6x + 8x + 10x$

35. $x + 2x^2 + 3x^3 + 4x^4 + 5x^5$ **Answers may vary for Exercises 37–41.**   **37.** $\sum_{i=1}^{5} (i + 2)$ **39.** $\sum_{i=1}^{5} \frac{1}{i + 1}$ **41.** $\sum_{i=1}^{5} i^2$

43. The similarities are that both are defined by the same linear expression and that points satisfying both lie in a straight line. The difference is that the domain of f consists of all real numbers, but the domain of the sequence is $\{1, 2, 3, \ldots\}$. An example of a similarity is $f(1) = 6$ and $a_1 = 6$. An example of a difference is $f\left(\frac{3}{2}\right) = 7$ but $a_{3/2}$ is not allowed.   **45.** A sequence is a list of terms in a specific order, while a series is the indicated sum of the terms of a sequence.   **47.** $\frac{59}{7}$ **49.** 5   **51.** 7243   **53.** $\sum_{i=1}^{6} i^2 + \sum_{i=1}^{6} 3i + \sum_{i=1}^{6} 5$ **54.** $3\sum_{i=1}^{6} i$ **55.** $6 \cdot 5 = 30$

56. $\sum_{i=1}^{n} i = \frac{n(n + 1)}{2}$ **57.** $\sum_{i=1}^{n} i^2 = \frac{n(n + 1)(2n + 1)}{6}$ **58.** $91 + 63 + 30 = 184$ **59.** 572   **60.** -2620

Section 14.2   (page 943)

Exercises   1. An arithmetic sequence is a sequence (list) of numbers in a specific order such that there is a common difference between any two successive terms. For example, the sequence $1, 5, 9, 13, \ldots$ is arithmetic with difference $d = 5 - 1 = 9 - 5 = 13 - 9 = 4$. As another example, 2, −1, −4, −7, … is an arithmetic sequence with $d = -3$. **3.** $d = 1$ **5.** not arithmetic **7.** $d = 5$ **9.** 5, 9, 13, 17, 21

11. −2, −6, −10, −14, −18 **13.** $a_n = 5n - 3$ **15.** $a_n = \frac{3}{4}n + \frac{9}{4}$ **17.** $a_n = 3n - 6$ **19.** 76 **21.** 48 **23.** −1 **25.** 16

27. 6 **29.** n represents the number of terms. **31.** $2S$; 101; 100; 10,100 **32.** $2S = 10,100$ **33.** 5050 **34.** 20,100 **35.** 81

37. −3 **39.** 87 **41.** 390 **43.** 320 **45.** 31,375 **47.** \$465 **49.** \$2100 per month **51.** 68; 1100 **53.** no; 3; 9

Section 14.3 (page 952)

Exercises **1.** A geometric sequence is an ordered list of numbers such that each term after the first is obtained by multiplying the previous term by a constant, r, called the common ratio. For example, if the first term is 3 and $r = 4$, then the sequence is 3, 12, 48, 192, …. If the first term is 2 and $r = -1$, the sequence is 2, −2, 2, −2, …. **3.** $r = 2$ **5.** not geometric **7.** $r = -3$ **9.** $r = -\frac{1}{2}$ **11.** $a_n = 5(2)^{n-1}$

13. $a_n = \frac{3^{n-1}}{9}$ **15.** $a_n = 10\left(-\frac{1}{5}\right)^{n-1}$ **17.** $2(5)^9 = 3,906,250$ **19.** $\frac{1}{2}\left(\frac{1}{3}\right)^{11}$ **21.** $2\left(\frac{1}{2}\right)^{24} = \frac{1}{2^{23}}$ **23.** 2, 6, 18, 54, 162

25. $5, -1, \frac{1}{5}, -\frac{1}{25}, \frac{1}{125}$ **27.** $\frac{121}{243}$ **29.** −1.997 **31.** 2.662 **33.** −2.982 **35.** \$66,988.91 **37.** \$130,159.72 **39.** 9

41. $\frac{10,000}{11}$ **43.** $-\frac{9}{20}$ **45.** does not exist **47.** $10\left(\frac{3}{5}\right)^4 \approx 1.3$ ft **49.** 3 days; $\frac{1}{4}$ g **51. (a)** $1.1(1.06)^5 \approx 1.5$ billion units

(b) approximately 12 yr **53.** $\$50,000\left(\frac{3}{4}\right)^8 \approx \$5000$ **55.** .33333… **56.** .66666… **57.** .99999…

58. $\frac{a_1}{1-r} = \frac{.9}{1-.1} = \frac{.9}{.9} = 1$; therefore, .99999… = 1 **59.** B

60. $.49999… = .4 + .09999… = \frac{4}{10} + \frac{1}{10}(.9999…) = \frac{4}{10} + \frac{1}{10}(1) = \frac{5}{10} = \frac{1}{2}$

Section 14.4 (page 961)

Connections **(page 957)** **1.** 21 and 34 **2.** 15, 21, 28, 36, 45 **3.** Answers will vary.

Exercises **1.** 720 **3.** 15 **5.** 15 **7.** 1 **9.** 120 **11.** 78 **13.** $m^4 + 4m^3n + 6m^2n^2 + 4mn^3 + n^4$

15. $a^5 - 5a^4b + 10a^3b^2 - 10a^2b^3 + 5ab^4 - b^5$ **17.** $8x^3 + 36x^2 + 54x + 27$ **19.** $\frac{x^4}{16} - \frac{x^3y}{2} + \frac{3x^2y^2}{2} - 2xy^3 + y^4$

21. $m^3x^3 - 3m^2n^2x^2 + 3mn^4x - n^6$ **23.** $r^{12} + 24r^{11}s + 264r^{10}s^2 + 1760r^9s^3$ **25.** $3^{14}x^{14} - 14(3^{13})x^{13}y + 91(3^{12})x^{12}y^2 - 364(3^{11})x^{11}y^3$

27. $t^{20} + 10t^{18}u^2 + 45t^{16}u^4 + 120t^{14}u^6$ **29.** $120(2^7)m^7n^3$ **31.** $\frac{7x^2y^6}{16}$ **33.** $36k^7$ **35.** $-160x^6y^3$ **37.** $4320x^9y^4$

Chapter 14 Review Exercises (page 966)

1. −1, 1, 3, 5 **3.** 1, 4, 9, 16 **5.** 0, 3, 8, 15 **7.** $2x + 3x^2 + 4x^3 + 5x^4 + 6x^5 + 7x^6$ **9.** 126 **11.** 1888.8 billion dollars

13. arithmetic; $d = 4$ **15.** geometric; $r = -1$ **17.** geometric; $r = \frac{1}{2}$ **19.** 73 **21.** $a_n = -5n + 1$ **23.** 15 **25.** 152

27. $a_n = -1(4)^{n-1}$ **29.** $2(-3)^{10} = 118,098$ **31.** $\frac{341}{1024}$ **33.** 1 **35.** $32p^5 - 80p^4q + 80p^3q^2 - 40p^2q^3 + 10pq^4 - q^5$

37. $m^2 + 4m\sqrt{mn} + 6mn + 4n\sqrt{mn} + n^2$ **39.** $-18,400(3)^{22}k^3$ **41.** $a_{10} = 1536$; $S_{10} = 1023$ **43.** $a_{15} = 38$; $S_{10} = 95$

45. $a_n = 2(4)^{n-1}$ **47.** $a_n = -3n + 15$ **49.** \$24,898.41 **51.** $\frac{1}{128}$ **53.** No, the sum cannot be found because $r = 2$, and this value of r does not satisfy $|r| < 1$.

Chapter 14 Test   (page 968)

[14.1] **1.** 0, 2, 0, 2, 0   [14.2] **2.** 4, 6, 8, 10, 12   [14.3] **3.** 48, 24, 12, 6, 3   [14.2] **4.** 0   [14.3] **5.** $\dfrac{64}{3}$ or $-\dfrac{64}{3}$ [14.2] **6.** 75

[14.3] **7.** 124 or 44   [14.1] **8.** 70,439   [14.3] **9.** \$137,925.91   **10.** It has a sum if $|r| < 1$.   [14.2] **11.** 70   **12.** 33   **13.** 125,250

[14.3] **14.** 42   **15.** $\dfrac{1}{3}$ **16.** The sum does not exist.   [14.4] **17.** 40,320   **18.** 1   **19.** 15   **20.** 66

21. $81k^4 - 540k^3 + 1350k^2 - 1500k + 625$ **22.** $\dfrac{14{,}080x^8y^4}{9}$ [14.1] **23.** \$324   [14.3] **24.** $20(3^{11}) = 3{,}542{,}940$

Cumulative Review Exercises Chapters 1–14   (page 969)

[1.4] **1.** $10, 0, \dfrac{45}{15}$ (or 3), -3 **2.** $-\dfrac{8}{3}, 10, 0, \dfrac{45}{15}$ (or 3), .82, -3 **3.** $\sqrt{13}, -\sqrt{3}$ **4.** all except $\sqrt{-7}$ [1.4–1.6] **5.** 8   **6.** -35

7. -55 [2.3] **8.** $\left\{\dfrac{1}{6}\right\}$ [2.8] **9.** $[10, \infty)$ [9.2] **10.** $\left\{-\dfrac{9}{2}, 6\right\}$ [2.3, 2.6] **11.** $\{9\}$ [9.1] **12.** $(-\infty, -3) \cup (4, \infty)$

[9.2] **13.** $(-\infty, -3] \cup [8, \infty)$ [4.5] **14.** $20p^2 - 2p - 6$ [4.6] **15.** $9k^2 - 42k + 49$ [4.4] **16.** $-5m^3 - 3m^2 + 3m + 8$

[4.7] **17.** $2t^3 + 3t^2 - 4t + 2 + \dfrac{3}{3t - 2}$ [5.1] **18.** $x(7 + x^2)$ [5.2, 5.3] **19.** $(7y - 4)(2y + 3)$ **20.** $z(3z + 4)(2z - 1)$

[5.4] **21.** $(7a^2 + 3b)(7a^2 - 3b)$ **22.** $(c + 3d)(c^2 - 3cd + 9d^2)$ **23.** $(8r + 3q)^2$ [4.1, 4.2] **24.** $\dfrac{9}{4}$ **25.** $-\dfrac{27p^2}{10}$ [6.1] **26.** $-9, 9$

[6.2] **27.** $\dfrac{x + 7}{x - 2}$ [6.4] **28.** $\dfrac{3p - 26}{p(p + 3)(p - 4)}$ [3.3, 7.1] **29.** $\dfrac{3}{4}$ [7.2] **30.** $3x + y = 4$ [3.2, 7.1] **31.**

[9.3] **32.**   [7.3] **33.** (a) yes (b) $\{-3, -2, 0, 1, 2\}$ (c) $\{2, 6, 4\}$ [8.3] **34.** $\{(3, -5)\}$ [8.2] **35.** $\{(-1, -2)\}$

[8.6] **36.** $\{(2, 1, 4)\}$ [8.5] **37.** 2 lb   [5.5] **38.** $\left\{-\dfrac{5}{2}, 2\right\}$ [11.8] **39.** $[-2, 3]$ [6.6] **40.** $\varnothing$ [11.3] **41.** $\left\{\dfrac{-5 + \sqrt{217}}{12}, \dfrac{-5 - \sqrt{217}}{12}\right\}$

[10.6] **42.** $\{1, 2\}$ [12.2] **43.** $\left\{\dfrac{5}{2}\right\}$ [12.6] **44.** $\{2\}$ [10.4] **45.** $10\sqrt{2}$ [10.7] **46.** 73   [12.1] **47.** $f^{-1}(x) = \dfrac{x - 5}{9}$

[11.6] **48.**   [12.2] **49.**   [12.3] **50.**   [13.2] **51.**

[13.3] **52.**   [13.4] **53.** $\left\{(-1, 5), \left(\dfrac{5}{2}, -2\right)\right\}$ [13.2] **54.** $(x + 5)^2 + (y - 12)^2 = 81$ [14.1] **55.** $-7, -2, 3, 8, 13$

[14.2, 14.3] **56.** (a) 78 (b) $\dfrac{75}{7}$ [14.2] **57.** 30   [14.4] **58.** 362,880   **59.** $32a^5 - 80a^4 + 80a^3 - 40a^2 + 10a - 1$ **60.** $-\dfrac{45x^8y^6}{4}$

APPENDIXES

Appendix B   (page 982)

1. 117.385   **3.** 13.21   **5.** 150.49   **7.** 96.101   **9.** 4.849   **11.** 166.32   **13.** 164.19   **15.** 1.344   **17.** 4.14   **19.** 2.23
21. 4800   **23.** .53   **25.** 1.29   **27.** .96   **29.** .009   **31.** 80%   **33.** .7%   **35.** 67%   **37.** 12.5%   **39.** 109.2   **41.** 238.92
43. 5%   **45.** 110%   **47.** 25%   **49.** 148.44   **51.** 7839.26   **53.** 7.39%   **55.** $2760   **57.** 805 mi   **59.** 63 mi   **61.** $2400

Appendix C   (page 988)

1. $\{1, 2, 3, 4, 5, 6, 7\}$ **3.** {winter, spring, summer, fall}   **5.** $\emptyset$ **7.** {L}   **9.** $\{2, 4, 6, 8, 10, \dots\}$ **11.** The sets in Exercises 9 and 10 are
infinite sets.   **13.** true   **15.** false   **17.** true   **19.** true   **21.** true   **23.** true   **25.** true   **27.** false   **29.** true   **31.** true
33. false   **35.** true   **37.** true   **39.** false   **41.** true   **43.** false   **45.** {g, h}   **47.** {b, c, d, e, g, h}   **49.** $\{a, c, e\} = B$ **51.** $\{d\} = D$
53. {a}   **55.** {a, c, d, e}   **57.** {a, c, e, f}   **59.** $\emptyset$ **61.** B and D; C and D

Appendix D   (page 995)

1. 7   **3.** 69.8 (rounded)   **5.** $39,622   **7.** $58.24   **9.** $35,500   **11.** 6.1   **13.** 17.2   **15.** 2.60   **17.** **(a)** 2.80 **(b)** 2.93 (rounded)
(c) 3.13 (rounded)   **19.** 15   **21.** 516   **23.** 48   **25.** 4142 mi   **27.** **(a)** 4050 mi **(b)** The median is somewhat different from the mean;
the mean is more affected by the high and low numbers.   **29.** 8   **31.** 68 and 74; bimodal   **33.** no mode

Appendix E   (page 1006)

1. Answers will vary; about 9 to 11 cm.   **3.** Answers will vary; about 20 to 25 mm.   **5.** cm   **7.** m   **9.** km   **11.** mm   **13.** cm   **15.** m
17. Some possible answers are: 35 mm film for cameras, track and field events, metric auto parts, and lead refills for mechanical pencils.
19. 700 cm   **21.** .040 m or .04 m   **23.** 9400 m   **25.** 5.09 m   **27.** 40 cm   **29.** 910 mm
31. _____ 35 mm = 3.5 cm   **33.** .018 km   **35.** .0000056 km   **37.** mL   **39.** L   **41.** kg   **43.** g   **45.** mL

70 mm = 7 cm

47. mg   **49.** L   **51.** unreasonable; too much   **53.** unreasonable; too much   **55.** reasonable   **57.** Some capacity examples are 2 L bottles of
soda and shampoo bottles marked in mL; weight examples are grams of fat listed on cereal boxes and vitamin doses in milligrams.
59. 15,000 mL   **61.** 3 L   **63.** .925 L   **65.** .008 L   **67.** 4150 mL   **69.** 8 kg   **71.** 5200 g   **73.** 850 mg   **75.** 30 g   **77.** .3 L
79. 1340 g   **81.** .07 L   **83.** 3 kg to 4 kg   **85.** 200 nickels

Appendix F   (page 1013)

1. $\dfrac{1}{a^2 b}$ **3.** $\dfrac{100y^{10}}{x^2}$ **5.** 0   **7.** $\dfrac{x^{10}}{2w^{13}y^5}$ **9.** $\dfrac{a^{15}}{-64b^{15}}$ **11.** $\dfrac{x^{16}z^{10}}{y^6}$ **13.** $-6a^4 + 11a^3 - 20a^2 + 26a - 15$ **15.** $8x^3 - 18x^2 + 6x - 16$
17. $x^2 y - xy^2 + 6y^3$ **19.** $-3x^2 - 62x + 32$ **21.** $10x^3 - 4x^2 + 9x - 4$ **23.** $6x^2 - 19x - 7$ **25.** $4x^2 - 9x + 2$ **27.** $16t^2 - 9$
29. $4y^4 - 16$ **31.** $16x^2 - 24x + 9$ **33.** $36r^2 + 60ry + 25y^2$ **35.** $c^3 + 8d^3$ **37.** $64x^3 - 1$ **39.** $14t^3 + 45st^2 + 18s^2 t - 5s^3$
41. $4xy^3(2x^2 y + 3x + 9y)$ **43.** $(x + 3)(x - 5)$ **45.** $(2x + 3)(x - 6)$ **47.** $(6t + 5)(6t - 5)$ **49.** $(4t + 3)^2$ **51.** $p(2m - 3n)^2$
53. $(x + 1)(x^2 - x + 1)$ **55.** $(2t + 5)(4t^2 - 10t + 25)$ **57.** $(t^2 - 5)(t^4 + 5t^2 + 25)$ **59.** $(5x + 2y)(t + 3r)$ **61.** $(6r - 5s)(a + 2b)$
63. $(t^2 + 1)(t + 1)(t - 1)$ **65.** $(2x + 3y - 1)(2x + 3y + 1)$

Appendix G (page 1018)

Connections (page 1018) -2.7

Exercises **1.** Synthetic division provides a quick, easy way to divide a polynomial by a binomial of the form $x - k$. **3.** $x - 5$

5. $4m - 1$ **7.** $2a + 4 + \dfrac{5}{a + 2}$ **9.** $p - 4 + \dfrac{9}{p + 1}$ **11.** $4a^2 + a + 3$ **13.** $x^4 + 2x^3 + 2x^2 + 7x + 10 + \dfrac{18}{x - 2}$

15. $-4r^5 - 7r^4 - 10r^3 - 5r^2 - 11r - 8 + \dfrac{-5}{r - 1}$ **17.** $-3y^4 + 8y^3 - 21y^2 + 36y - 72 + \dfrac{143}{y + 2}$ **19.** $y^2 + y + 1 + \dfrac{2}{y - 1}$ **21.** 7

23. -2 **25.** 0 **27.** By the remainder theorem, a 0 remainder means that $P(k) = 0$; that is, k is a number that makes $P(x) = 0$. **29.** yes

31. no **33.** no **35.** yes **37.** $(2x - 3)(x + 4)$ **38.** $\left\{\dfrac{3}{2}, -4\right\}$ **39.** $P(-4) = 0, P\left(\dfrac{3}{2}\right) = 0$ **40.** a **41.** Yes, $x - 3$ is a factor.

$Q(x) = (x - 3)(3x - 1)(x + 2)$ **43.** 3 **45.** -1

Appendix H (page 1029)

Connections (page 1025) **1.** -185 **2.** Every square matrix corresponds to *one and only one* real number, called its determinant.

Exercises **1.** D **3.** -3 **5.** 14 **7.** 0 **9.** 59 **11.** 14 **13.** Multiply the upper left and lower right entries. Then multiply the

upper right and lower left entries. Subtract the second product from the first to obtain the determinant. For example, $\begin{vmatrix} 4 & 2 \\ 7 & 1 \end{vmatrix} = 4 \cdot 1 - 2 \cdot 7 =$

$4 - 14 = -10$. **15.** -22 **17.** 20 **19.** -5 **21.** By choosing that row or column to expand about, all terms will have a factor of 0,

and so the sum of all these terms will be 0. **23.** 6.078 **25. (a)** D **(b)** A **(c)** C **(d)** B **27.** $\{(-5, 2)\}$ **29.** $\left\{\left(\dfrac{11}{58}, -\dfrac{5}{29}\right)\right\}$

31. $\{(-1, 2)\}$ **33.** $\{(-2, 3, 5)\}$ **35.** Cramer's rule does not apply. **37.** $\{(20, -13, -12)\}$ **39.** $\left\{\left(\dfrac{62}{5}, -\dfrac{1}{5}, \dfrac{27}{5}\right)\right\}$ **41.** $\{(-1, 3, 5)\}$

43. $\{(1, -3, 2, -4)\}$ **45.** -22 **47.** 20 **49.** -5 **51.** $\{2\}$ **53.** $\{0\}$ **55.** Answers will vary. **57.**

58. $\dfrac{1}{2} \begin{vmatrix} 0 & 0 & 1 \\ -3 & -4 & 1 \\ 2 & -2 & 1 \end{vmatrix}$ **59.** 7 **60.** 8 **61.** $\dfrac{y_2 - y_1}{x_2 - x_1}$ **62.** $y - y_1 = \dfrac{y_2 - y_1}{x_2 - x_1}(x - x_1)$ **63.** $x_2 y - x_1 y - x_2 y_1 - xy_2 + x_1 y_2 + xy_1 = 0$

64. The result is the same as in Exercise 63.

A

absolute value The absolute value of a number is the distance between 0 and the number on a number line. (Section 1.4)

absolute value equation An absolute value equation is an equation that involves the absolute value of a variable expression. (Section 9.2)

absolute value function The function defined by $f(x) = |x|$ with a graph that includes portions of two lines is called the absolute value function. (Section 13.1)

absolute value inequality An absolute value inequality is an inequality that involves the absolute value of a variable expression. (Section 9.2)

addition property of equality The addition property of equality states that the same number can be added to (or subtracted from) both sides of an equation to obtain an equivalent equation. (Section 2.1)

addition property of inequality The addition property of inequality states that the same number can be added to (or subtracted from) both sides of an inequality without changing the solution set. (Section 2.8)

additive inverse (negative, opposite) Two numbers that are the same distance from 0 on a number line but on opposite sides of 0 are called additive inverses. (Section 1.4)

algebraic expression Any collection of numbers or variables joined by the basic operations of addition, subtraction, multiplication, or division (except by 0), or the operations of raising to powers or taking roots is called an algebraic expression. (Section 1.3)

annuity An annuity is a sequence of equal payments made at equal periods of time. (Section 14.3)

area Area is a measure of the surface covered by a two-dimensional (flat) figure. (Section 2.5)

arithmetic mean The arithmetic mean (average) of a group of numbers is the sum of all the numbers divided by the number of numbers. (Section 14.1, Appendix D)

arithmetic sequence (arithmetic progression) An arithmetic sequence is a sequence in which each term after the first differs from the preceding term by a constant amount. (Section 14.2)

array of signs An array of signs is used when evaluating a determinant using expansion by minors. The signs alternate for each row and column, beginning with $+$ in the first row, first column position. (Appendix H)

associative property of addition The associative property of addition states that the way in which numbers being added are grouped does not change the sum. (Section 1.7)

associative property of multiplication The associative property of multiplication states that the way in which numbers being multiplied are grouped does not change the product. (Section 1.7)

asymptote A line that a graph more and more closely approaches as the graph gets farther away from the origin is called an asymptote of the graph. (Sections 12.2, 13.1)

asymptotes of a hyperbola The two intersecting straight lines that the branches of a hyperbola approach are called asymptotes of the hyperbola. (Section 13.3)

augmented matrix An augmented matrix is a matrix that has a vertical bar that separates the columns of the matrix into two groups. (Section 8.6)

axis (axis of symmetry) The axis of a parabola is the vertical or horizontal line through the vertex of the parabola. (Sections 4.4, 11.6)

B

base The base is the number that is a repeated factor when written with an exponent. (Sections 1.2, 4.1)

bimodal A group of numbers with two modes is bimodal. (Appendix D)

binomial A binomial is a polynomial with exactly two terms. (Section 4.4)

binomial theorem (general binomial expansion) The binomial theorem is a formula used to expand a binomial raised to a power. (Section 14.4)

boundary line In the graph of a linear inequality, the boundary line separates the region that satisfies the inequality from the region that does not satisfy the inequality. (Sections 9.3, 13.5)

C

center of a circle The fixed point that is a fixed distance from all the points that form a circle is the center of the circle. (Section 13.2)

center of an ellipse The center of an ellipse is the fixed point located exactly halfway between the two foci. (Section 13.2)

center-radius form of the equation of a circle The center-radius form of the equation of a circle with center (h, k) and radius r is $(x - h)^2 + (y - k)^2 = r^2$. (Section 13.2)

circle A circle is the set of all points in a plane that lie a fixed distance from a fixed point. (Section 13.2)

coefficient (numerical coefficient) A coefficient is the numerical factor of a term. (Sections 1.8, 4.2)

column of a matrix A column of a matrix is a group of elements that are read vertically. (Section 8.6)

combined variation If a problem involves a combination of direct and inverse variation, then it is called a combined variation problem. (Section 7.4)

combining like terms Combining like terms is a method of adding or subtracting like terms by using the properties of real numbers. (Section 1.8)

common difference The common difference d is the difference between any two adjacent terms of an arithmetic sequence. (Section 14.2)

common factor An integer that is a factor of two or more integers is called a common factor of those integers. (Section 5.1)

common logarithm A common logarithm is a logarithm to base 10. (Section 12.5)

common ratio   A common ratio r is the constant multiplier between adjacent terms in a geometric sequence. (Section 14.3)

commutative property of addition   The commutative property of addition states that the order of numbers in an addition problem can be changed without changing the sum. (Section 1.7)

commutative property of multiplication   The commutative property of multiplication states that the product in a multiplication problem remains the same regardless of the order of the factors. (Section 1.7)

complement of a set   The set of elements in the universal set that are not in a set A is the complement of A, written A'. (Appendix C)

complementary angles (complements)   Complementary angles are angles whose measures have a sum of 90°. (Section 2.4)

completing the square   The process of adding to a binomial the number that makes it a perfect square trinomial is called completing the square. (Section 11.2)

complex conjugates   The complex conjugate of $a + bi$ is $a - bi$. (Section 10.7)

complex fraction   A complex fraction is an expression with one or more fractions in the numerator, denominator, or both. (Section 6.5)

complex number   A complex number is any number that can be written in the form $a + bi$, where a and b are real numbers. (Section 10.7)

composite function   A function in which some quantity depends on a variable that, in turn, depends on another variable is called a composite function. (Section 13.1)

composite number   A composite number has at least one factor other than itself and 1. (Section 1.1)

composition of functions   Replacing a variable with an algebraic expression is called composition of functions. (Section 13.1)

compound inequality   A compound inequality consists of two inequalities linked by a connective word such as *and* or *or*. (Section 9.1)

conditional equation   A conditional equation is true for some replacements of the variable and false for others. (Section 2.3)

conic section   When a plane intersects an infinite cone at different angles, the figures formed by the intersections are called conic sections. (Section 13.2)

conjugate   The conjugate of $a + b$ is $a - b$. (Section 10.5)

consecutive integers   Two integers that differ by one are called consecutive integers. (Section 2.4)

consistent system   A system of equations with a solution is called a consistent system. (Section 8.1)

constant function   A linear function of the form $f(x) = b$, where b is a constant, is called a constant function. (Section 7.3)

constant of variation   In the variation equations $y = kx$, or $y = \frac{k}{x}$, or $y = kxz$, the number k is called the constant of variation. (Section 7.4)

contradiction   A contradiction is an equation that is never true. It has no solution. (Section 2.3)

coordinate on a number line   Each number on a number line is called the coordinate of the point that it labels. (Section 1.4)

coordinates of a point   The numbers in an ordered pair are called the coordinates of the corresponding point in the plane. (Sections 3.1, 7.1)

Cramer's rule   Cramer's rule uses determinants to solve systems of linear equations. (Appendix H)

cross products   The cross products in the proportion $\frac{a}{b} = \frac{c}{d}$ are ad and bc. (Section 2.6)

cube root   A number b is the cube root of a if $b^3 = a$. (Section 10.1)

cube root function   The function defined by $f(x) = \sqrt[3]{x}$ is called the cube root function. (Section 10.1)

D

decimal   A decimal is a number written with a decimal point. (Appendix B)

degree   A degree is a basic unit of measure for angles in which one degree (1°) is $\frac{1}{360}$ of a complete revolution. (Section 2.4)

degree of a polynomial   The degree of a polynomial is the greatest degree of any of the terms in the polynomial. (Section 4.4)

degree of a term   The degree of a term is the sum of the exponents on the variables in the term. (Section 4.4)

denominator   The number below the fraction bar in a fraction is called the denominator. It shows the number of equal parts in a whole. (Section 1.1)

dependent equations   Equations of a system that have the same graph (because they are different forms of the same equation) are called dependent equations. (Section 8.1)

dependent variable   In an equation relating x and y, if the value of the variable y depends on the variable x, then y is called the dependent variable. (Section 7.3)

descending powers   A polynomial in one variable is written in descending powers of the variable if the degree of the terms of the polynomial decreases from left to right. (Section 4.4)

determinant   Associated with every square matrix is a real number called the determinant of the matrix, symbolized by the entries of the matrix placed between two vertical lines. (Appendix H)

difference   The answer to a subtraction problem is called the difference. (Section 1.1)

difference of cubes   The difference of cubes, $x^3 - y^3$, can be factored as $x^3 - y^3 = (x - y)(x^2 + xy + y^2)$. (Section 5.4)

difference of squares   The difference of squares, $x^2 - y^2$, can be factored as the product of the sum and difference of two terms, or $x^2 - y^2 = (x + y)(x - y)$. (Section 5.4)

direct variation y varies directly as x if there exists a real number k such that $y = kx$. (Section 7.4)

discriminant   The discriminant is the quantity under the radical, $b^2 - 4ac$, in the quadratic formula. (Section 11.3)

disjoint sets   Sets that have no elements in common are disjoint sets. (Appendix C)

distributive property   For any real numbers a, b, and c, the distributive property states that $a(b + c) = ab + ac$ and $(b + c)a = ba + ca$. (Section 1.7)

domain   The set of all first components (x-values) in the ordered pairs of a relation is the domain. (Section 7.3)

E

element of a matrix   The numbers in a matrix are called the elements of the matrix. (Section 8.6)

elements (members)   Elements are the objects that belong to a set. (Section 1.3, Appendix C)

elimination method The elimination method is an algebraic method used to solve a system of equations in which the equations of the system are combined so that one or more variables is eliminated. (Section 8.3)

ellipse An ellipse is the set of all points in a plane the sum of whose distances from two fixed points is constant. (Section 13.2)

empty set (null set) The empty set, denoted by { } or ∅, is the set containing no elements. (Section 2.3, Appendix C)

equation An equation is a statement that two algebraic expressions are equal. (Section 1.3)

equivalent equations Equivalent equations are equations that have the same solution set. (Section 2.1)

equivalent inequalities Equivalent inequalities are inequalities that have the same solution set. (Section 2.8)

expansion by minors A method of evaluating a 3×3 or larger determinant is called expansion by minors. (Appendix H)

exponent (power) An exponent is a number that indicates how many times a factor is repeated. (Sections 1.2, 4.1)

exponential equation An exponential equation is an equation that has a variable as an exponent. (Section 12.2)

exponential expression A number or letter (variable) written with an exponent is an exponential expression. (Sections 1.2, 4.1)

exponential function An exponential function is a function defined by an expression of the form $f(x) = a^x$, where $a > 0$ and $a \neq 1$ for all real numbers x. (Section 12.2)

extraneous solution A solution to a new equation that does not satisfy the original equation is called an extraneous solution. (Section 10.6)

extremes of a proportion In the proportion $\frac{a}{b} = \frac{c}{d}$, the a- and d-terms are called the extremes. (Section 2.6)

F

factor A factor of a given number is any number that divides evenly (without remainder) into the given number. (Sections 1.1, 5.1)

factored A number is factored by writing it as the product of two or more numbers. (Section 1.1)

factored form An expression is in factored form when it is written as a product. (Section 5.1)

factoring Writing a polynomial as the product of two or more simpler polynomials is called factoring. (Section 5.1)

factoring by grouping Factoring by grouping is a method of grouping the terms of a polynomial in such a way that the polynomial can be factored even though its greatest common factor is 1. (Section 5.1)

factoring out the greatest common factor Factoring out the greatest common factor is the process of using the distributive property to write a polynomial as a product of the greatest common factor and a simpler polynomial. (Section 5.1)

finite sequence A finite sequence has a domain that includes only the first n positive integers. (Section 14.1)

first-degree equation A first-degree (linear) equation has no term with the variable to a power greater than 1. (Section 7.1)

foci (singular, **focus**) Foci are fixed points used to determine the points that form a parabola, an ellipse, or a hyperbola. (Sections 13.2, 13.3)

FOIL FOIL is a method for multiplying two binomials $(A + B)(C + D)$. Multiply **F**irst terms AC, **O**uter terms AD, **I**nner terms BC, and **L**ast terms BD. Then combine like terms. (Section 4.5)

formula A formula is a mathematical expression in which letters are used to describe a relationship. (Section 2.5)

fourth root A number b is a fourth root of a if $b^4 = a$. (Section 10.1)

function A function is a set of ordered pairs (relation) in which each value of the first component x corresponds to exactly one value of the second component y. (Section 7.3)

function notation Function notation $f(x)$ represents the value of the function at x, that is, the y-value that corresponds to x. (Section 7.3)

fundamental rectangle The asymptotes of a hyperbola are the extended diagonals of its fundamental rectangle, with corners at the points (a, b), $(-a, b)$, $(-a, -b)$, and $(a, -b)$. (Section 13.3)

future value of an annuity The future value of an annuity is the sum of the compound amounts of all the payments, compounded to the end of the term. (Section 14.3)

G

general term of a sequence The expression a_n, which defines a sequence, is called the general term of the sequence. (Section 14.1)

geometric sequence (geometric progression) A geometric sequence is a sequence in which each term after the first is a constant multiple of the preceding term. (Section 14.3)

gram The basic metric unit for mass is the gram, written using the symbol g. (Appendix E)

graph of an equation The graph of an equation is the set of all points that correspond to all of the ordered pairs that satisfy the equation. (Sections 3.2, 7.1)

graph of a number The point on a number line that corresponds to a number is its graph. (Section 1.4)

graph of a relation The graph of a relation is the graph of its ordered pairs. (Section 7.3)

greatest common factor (GCF) The greatest common factor of a list of integers is the largest common factor of those integers. The greatest common factor of a polynomial is the largest term that is a factor of all terms in the polynomial. (Sections 1.1, 5.1)

greatest integer function The function defined by $f(x) = [\![x]\!]$, where the symbol $[\![x]\!]$ is used to represent the greatest integer less than or equal to x, is called the greatest integer function. (Section 13.1)

grouping symbols Grouping symbols are parentheses ( ), square brackets [ ], or fraction bars. (Section 1.2)

H

horizontal line test The horizontal line test states that a function is one-to-one if every horizontal line intersects the graph of the function at most once. (Section 12.1)

hyperbola A hyperbola is the set of all points in a plane such that the absolute value of the difference of the distances from two fixed points is constant. (Section 13.3)

hypotenuse The hypotenuse is the longest side in a right triangle. It is the side opposite the right angle. (Sections 5.6, 10.3)

I

identity An identity is an equation that is true for all replacements of the variable. It has an infinite number of solutions. (Section 2.3)

identity element for addition Since adding 0 to a number does not change the number, 0 is called the identity element for addition. (Section 1.7)

identity element for multiplication Since multiplying a number by 1 does not change the number, 1 is called the identity element for multiplication. (Section 1.7)

identity properties The identity properties state that the sum of 0 and any number equals the number, and the product of 1 and any number equals the number. (Section 1.7)

imaginary number A complex number $a + bi$ with $b \neq 0$ is called an imaginary number. (Section 10.7)

imaginary part The imaginary part of the complex number $a + bi$ is b. (Section 10.7)

inconsistent system An inconsistent system of equations is a system with no solution. (Section 8.1)

independent equations Equations of a system that have different graphs are called independent equations. (Section 8.1)

independent variable In an equation relating x and y, if the value of the variable y depends on the variable x, then x is called the independent variable. (Section 7.3)

index (order) In a radical of the form $\sqrt[n]{a}$, n is called the index or order. (Section 10.1)

index of summation When using summation notation, $\sum_{i=1}^{n} f(i)$, the letter i is called the index of summation. (Section 14.1)

inequality An inequality is a statement that two expressions are not equal. (Section 1.2)

infinite sequence An infinite sequence is a function with the set of positive integers as the domain. (Section 14.1)

inner product When using the FOIL method to multiply two binomials $(A + B)(C + D)$, the inner product is BC. (Section 4.5)

integers The set of integers is $\{\ldots, -3, -2, -1, 0, 1, 2, 3, \ldots\}$. (Section 1.4)

intersection The intersection of two sets A and B, written $A \cap B$, is the set of elements that belong to both A and B. (Section 9.1, Appendix C)

interval An interval is a portion of a number line. (Section 2.8)

interval notation Interval notation is a simplified notation that uses parentheses ( ) and/or brackets [ ] to describe an interval on a number line. (Section 2.8)

inverse of a function f If f is a one-to-one function, then the inverse of f is the set of all ordered pairs of the form (y, x) where (x, y) belongs to f. (Section 12.1)

inverse properties The inverse properties state that a number added to its opposite is 0, and a number multiplied by its reciprocal is 1. (Section 1.7)

inverse variation y varies inversely as x if there exists a real number k such that $y = \frac{k}{x}$. (Section 7.4)

irrational numbers Irrational numbers cannot be written as the quotient of two integers but can be represented by points on the number line. (Section 1.4)

J

joint variation y varies jointly as x and z if there exists a real number k such that $y = kxz$. (Section 7.4)

L

least common denominator (LCD) Given several denominators, the smallest expression that is divisible by all the denominators is called the least common denominator. (Sections 1.1, 6.3)

legs of a right triangle The two shorter sides of a right triangle are called the legs. (Sections 5.6, 10.3)

like terms Terms with exactly the same variables raised to exactly the same powers are called like terms. (Sections 1.8, 4.4)

linear equation in one variable A linear equation in one variable can be written in the form $Ax + B = C$, where A, B, and C are real numbers, with $A \neq 0$. (Section 2.1)

linear equation in two variables A linear equation in two variables is an equation that can be written in the form $Ax + By = C$, where A, B, and C are real numbers and A and B are not both 0. (Sections 3.1, 7.1)

linear function A function defined by an equation of the form $f(x) = mx + b$, for real numbers m and b, is a linear function. (Section 7.3)

linear inequality in one variable A linear inequality in one variable can be written in the form $Ax + B < C$ or $Ax + B > C$ (or with $\leq$ or $\geq$), where A, B, and C are real numbers, with $A \neq 0$. (Section 2.8)

linear inequality in two variables A linear inequality in two variables can be written in the form $Ax + By < C$ or $Ax + By > C$ (or with $\leq$ or $\geq$), where A, B, and C are real numbers and A and B are not both 0. (Section 9.3)

linear system (system of linear equations) Two or more linear equations form a linear system. (Section 8.1)

line of symmetry The axis of a parabola is a line of symmetry for the graph. It is a line that can be drawn through the graph in such a way that the part of the graph on one side of the line is an exact reflection of the part on the opposite side. (Sections 4.4, 11.6)

liter The basic metric unit for capacity is the liter, written using the symbol L. (Appendix E)

logarithm A logarithm is an exponent; $\log_a x$ is the exponent on the base a that gives the number x. (Section 12.3)

logarithmic equation A logarithmic equation is an equation with a logarithm in at least one term. (Section 12.3)

logarithmic function with base a If a and x are positive numbers with $a \neq 1$, then $f(x) = \log_a x$ defines the logarithmic function with base a. (Section 12.3)

lowest terms A fraction is in lowest terms when there are no common factors in the numerator and denominator (except 1). (Sections 1.1, 6.1)

M

mathematical model In a real-world problem, a mathematical model is one or more equations (or inequalities) that describe the situation. (Sections 1.3 Exercises, 3.1)

matrix (plural, matrices) A matrix is a rectangular array of numbers, consisting of horizontal rows and vertical columns. (Section 8.6, Appendix H)

mean The mean (average) of a group of numbers is the sum of the numbers divided by the number of numbers. (Section 14.1, Appendix D)

means of a proportion In the proportion $\frac{a}{b} = \frac{c}{d}$, the b- and c-terms are called the means. (Section 2.6)

median The median divides a group of numbers listed in order from smallest to largest in half; that is, half the numbers lie above the median and half lie below the median. (Appendix D)

meter The basic unit of length in the metric system is the meter, written using the symbol m. (Appendix E)

minors The minor of an element in a 3×3 determinant is the 2×2 determinant remaining when a row and a column of the 3×3 determinant are eliminated. (Appendix H)

mixed number A mixed number includes a whole number and a fraction written together and is understood to be the sum of the whole number and the fraction. (Section 1.1)

mode The mode is the number (or numbers) that occurs most often in a group of numbers. (Appendix D)

monomial A monomial is a polynomial with only one term. (Section 4.4)

multiplication property of equality The multiplication property of equality states that the same nonzero number can be multiplied by (or divided into) both sides of an equation to obtain an equivalent equation. (Section 2.2)

multiplication property of inequality The multiplication property of inequality states that both sides of an inequality may be multiplied (or divided) by a positive number without changing the direction of the inequality symbol. Multiplying (or dividing) by a negative number reverses the inequality symbol. (Section 2.8)

multiplicative inverse (reciprocal) The multiplicative inverse of a nonzero real number a is $\frac{1}{a}$. (Section 1.6)

N

n-factorial ($n!$) For any positive integer n, $n(n-1)(n-2)(n-3)\cdots(2)(1) = n!$. (Section 14.4)

natural logarithm A natural logarithm is a logarithm to base e. (Section 12.5)

natural numbers (counting numbers) The set of natural numbers includes the numbers used for counting: $\{1, 2, 3, 4, \ldots\}$. (Sections 1.1, 1.4)

negative number A negative number is located to the left of 0 on a number line. (Section 1.4)

nonlinear equation A nonlinear equation is an equation in which some terms have more than one variable or a variable of degree 2 or higher. (Section 13.4)

nonlinear system of equations A nonlinear system of equations is a system that includes at least one nonlinear equation. (Section 13.4)

nonlinear system of inequalities A nonlinear system of inequalities is two or more inequalities to be considered at the same time, at least one of which is nonlinear. (Section 13.5)

number line A number line is a line with a scale that is used to show how numbers relate to each other. (Section 1.4)

numerator The number above the fraction bar in a fraction is called the numerator. It shows how many of the equivalent parts are being considered. (Section 1.1)

numerical coefficient The numerical factor in a term is its numerical coefficient. (Sections 1.8, 4.4)

O

one-to-one function A one-to-one function is a function in which each x-value corresponds to only one y-value and each y-value corresponds to just one x-value. (Section 12.1)

ordered pair An ordered pair is a pair of numbers written within parentheses in which the order of the numbers is important. (Sections 3.1, 7.1)

ordered triple A solution of an equation in three variables, written (x, y, z), is called an ordered triple. (Section 8.4)

ordinary annuity An ordinary annuity is an annuity in which the payments are made at the end of each time period and the frequency of payments is the same as the frequency of compounding. (Section 14.3)

origin The point at which the x-axis and y-axis of a rectangular coordinate system intersect is called the origin. (Sections 3.1, 7.1)

outer product When using the FOIL method to multiply two binomials $(A + B)(C + D)$, the outer product is AD. (Section 4.5)

P

parabola The graph of a second-degree (quadratic) equation in two variables is called a parabola. (Sections 4.4, 11.6)

parallel lines Parallel lines are two lines in the same plane that never intersect. (Sections 3.3, 7.1)

Pascal's triangle Pascal's triangle is a triangular array of numbers that is helpful in expanding binomials. (Section 14.4)

payment period In an annuity, the time between payments is called the payment period. (Section 14.3)

percent Percent, written with the sign %, means "per one hundred." (Appendix B)

percentage A percentage is a part of a whole. (Appendix B)

perfect cube A perfect cube is a number with a rational cube root. (Section 10.1)

perfect square A perfect square is a number with a rational square root. (Section 10.1)

perfect square trinomial A perfect square trinomial is a trinomial that can be factored as the square of a binomial. (Section 5.4)

perimeter The perimeter of a two-dimensional figure is a measure of the distance around the outside edges of the figure, or the sum of the lengths of its sides. (Section 2.5)

perpendicular lines Perpendicular lines are two lines that intersect to form a right (90°) angle. (Sections 3.3, 7.1)

plot To plot an ordered pair is to locate it on a rectangular coordinate system. (Sections 3.1, 7.1)

point-slope form A linear equation is written in point-slope form if it is in the form $y - y_1 = m(x - x_1)$, where m is the slope of the line and (x_1, y_1) is a point on the line. (Sections 3.4, 7.2)

polynomial A polynomial is a term or a finite sum of terms in which all coefficients are real, all variables have whole number exponents, and no variables appear in denominators. (Section 4.4)

polynomial in x A polynomial containing only the variable x is called a polynomial in x. (Section 4.4)

positive number A positive number is located to the right of 0 on a number line. (Section 1.4)

prime number A natural number (except 1) is prime if it has only 1 and itself as factors. (Section 1.1)

prime polynomial A prime polynomial is a polynomial that cannot be factored using only integer coefficients. (Section 5.2)

principal root (principal nth root) For even indexes, the symbols $\sqrt{\ }$, $\sqrt[4]{\ }$, $\sqrt[6]{\ }$, ..., $\sqrt[n]{\ }$, are used for nonnegative roots, which are called principal roots. (Section 10.1)

product The answer to a multiplication problem is called the product. (Section 1.1)

product of the sum and difference of two terms The product of the sum and difference of two terms is the difference of the squares of the terms: $(x + y)(x - y) = x^2 - y^2$. (Section 4.6)

proportion A proportion is a statement that two ratios are equal. (Section 2.6)

proportional If y varies directly as x and there exists some number (constant) k such that $y = kx$, then y is said to be proportional to x. (Section 7.4)

Pythagorean formula The Pythagorean formula states that the square of the length of the hypotenuse of a right triangle equals the sum of the squares of the lengths of the two legs. (Sections 5.6, 10.3)

Q

quadrant A quadrant is one of the four regions in the plane determined by a rectangular coordinate system. (Sections 3.1, 7.1)

quadratic equation A quadratic equation is an equation that can be written in the form $ax^2 + bx + c = 0$, where a, b, and c are real numbers, with $a \neq 0$. (Sections 5.5, 11.1)

quadratic formula The quadratic formula is a general formula used to solve any quadratic equation. (Section 11.3)

quadratic function A function defined by an equation of the form $f(x) = ax^2 + bx + c$, for real numbers a, b, and c, with $a \neq 0$, is a quadratic function. (Section 11.6)

quadratic inequality A quadratic inequality can be written in the form $ax^2 + bx + c < 0$ or $ax^2 + bx + c > 0$ (or with $\leq$ or $\geq$), where a, b, and c are real numbers, with $a \neq 0$. (Section 11.8)

quadratic in form An equation that is written in the form $a[f(x)]^2 + b[f(x)] + c = 0$, for $a \neq 0$, is called quadratic in form. (Section 11.4)

quotient The answer to a division problem is called the quotient. (Section 1.1)

R

radical A radical sign with a radicand is called a radical. (Section 10.1)

radical equation An equation that includes one or more radical expressions with a variable is called a radical equation. (Section 10.6)

radical expression A radical expression is an algebraic expression that contains radicals. (Section 10.1)

radical sign The symbol $\sqrt{\ }$ is called a radical sign. (Section 10.1)

radicand The number or expression under a radical sign is called the radicand. (Section 10.1)

radius The radius of a circle is the fixed distance between the center and any point on the circle. (Section 13.2)

range The set of all second components (y-values) in the ordered pairs of a relation is the range. (Section 7.3)

ratio A ratio is a comparison of two quantities with the same units. (Section 2.6)

rational expression The quotient of two polynomials with denominator not 0 is called a rational expression, or algebraic fraction. (Section 6.1)

rational function A function that is defined by a rational expression is called a rational function. (Section 13.1)

rational inequality An inequality that involves fractions is called a rational inequality. (Section 11.8)

rational numbers Rational numbers can be written as the quotient of two integers, with denominator not 0. (Section 1.4)

rationalizing the denominator The process of removing radicals from a denominator so that the denominator contains only

rational numbers is called rationalizing the denominator. (Section 10.5)

real numbers Real numbers include all numbers that can be represented by points on the number line, that is, all rational and irrational numbers. (Section 1.4)

real part The real part of a complex number $a + bi$ is a. (Section 10.7)

reciprocal Pairs of numbers whose product is 1 are called reciprocals of each other. (Sections 1.1, 1.6)

reciprocal function The reciprocal function is defined by $f(x) = \frac{1}{x}$. (Section 13.1)

rectangular (Cartesian) coordinate system The x-axis and y-axis placed at a right angle at their zero points form a rectangular coordinate system, also called the Cartesian coordinate system. (Sections 3.1, 7.1)

reduced row echelon form The reduced row echelon form is an extension of row echelon form that has 0s above and below the diagonal of 1s. (Section 8.6)

relation A relation is a set of ordered pairs. (Section 7.3)

right angle A right angle measures 90°. (Section 2.4)

rise Rise is the vertical change between two points on a line, that is, the change in y-values. (Sections 3.3, 7.1)

row echelon form If a matrix is written with 1s on the diagonal from upper left to lower right and 0s below the 1s, it is said to be in row echelon form. (Section 8.6)

row of a matrix A row of a matrix is a group of elements that are read horizontally. (Section 8.6)

row operations Row operations are operations on a matrix that produce equivalent matrices leading to the solution of a system of equations. (Section 8.6)

run Run is the horizontal change between two points on a line, that is, the change in x-values. (Sections 3.3, 7.1)

S

scatter diagram A scatter diagram is a graph of ordered pairs of data. (Section 3.1)

scientific notation A number is written in scientific notation when it is expressed in the form $a \times 10^n$, where $1 \leq |a| < 10$ and n is an integer. (Section 4.3)

second-degree inequality A second-degree inequality is an inequality with at least one variable of degree 2 and no variable with degree greater than 2. (Section 13.5)

sequence A sequence is a function whose domain is the set of natural numbers. (Section 14.1)

series The indicated sum of the terms of a sequence is called a series. (Section 14.1)

set A set is a collection of objects. (Section 1.3, Appendix C)

set-builder notation Set-builder notation is used to describe a set of numbers without actually having to list all of the elements. (Section 1.4)

signed numbers Signed numbers are numbers that can be written with a positive or negative sign. (Section 1.4)

simplified radical A simplified radical meets four conditions:

1. The radicand has no factor raised to a power greater than or equal to the index.

2. The radicand has no fractions.

3. No denominator contains a radical.

4. Exponents in the radicand and the index of the radical have no common factor (except 1).

(Section 10.3)

slope The ratio of the change in y to the change in x along a line is called the slope of the line. (Sections 3.3, 7.1)

slope-intercept form A linear equation is written in slope-intercept form if it is in the form $y = mx + b$, where m is the slope and $(0, b)$ is the y-intercept. (Sections 3.4, 7.2)

solution of an equation A solution of an equation is any replacement for the variable that makes the equation true. (Section 1.3)

solution set The solution set of an equation is the set of all solutions of the equation. (Section 2.1)

solution set of a linear system The solution set of a linear system of equations includes all ordered pairs that satisfy all the equations of the system at the same time. (Section 8.1)

solution set of a system of linear inequalities The solution set of a system of linear inequalities includes all ordered pairs that

make all inequalities of the system true at the same time. (Section 13.5)

square matrix A square matrix is a matrix that has the same number of rows as columns. (Section 8.6, Appendix H)

square of a binomial The square of a binomial is the sum of the square of the first term, twice the product of the two terms, and the square of the last term: $(x + y)^2 = x^2 + 2xy + y^2$ or $(x - y)^2 = x^2 - 2xy + y^2$. (Section 4.6)

square root The opposite of squaring a number is called taking its square root; that is, a number b is a square root of a if $b^2 = a$. (Section 10.1)

square root function The function defined by $f(x) = \sqrt{x}$, with $x \geq 0$, is called the square root function. (Sections 10.1, 13.3)

square root property The square root property states that if $x^2 = k$, then $x = \sqrt{k}$ or $x = -\sqrt{k}$. (Section 11.1)

standard form of a complex number The standard form of a complex number is $a + bi$. (Section 10.7)

standard form of a linear equation A linear equation in two variables written in the form $Ax + By = C$, where A, B, and C are integers with no common factor (except 1) and $A \geq 0$, is in standard form. (Sections 3.4, 7.2)

standard form of a quadratic equation A quadratic equation written in the form $ax^2 + bx + c = 0$, where a, b, and c are real numbers with $a \neq 0$, is in standard form. (Sections 5.5, 11.1)

step function A function with a graph that looks like a series of steps is called a step function. (Section 13.1)

straight angle A straight angle measures 180°. (Section 2.4)

subscript notation Subscript notation is a way of indicating nonspecific values, such as x_1 and x_2. (Sections 3.3, 7.1)

subset If all elements of set A are in set B, then A is a subset of B, written $A \subseteq B$. (Appendix C)

substitution method The substitution method is an algebraic method for solving a system of equations in which one equation is solved for one of the variables and

the result is substituted in the other equation. (Section 8.2)

sum The answer to an addition problem is called the sum. (Section 1.1)

sum of cubes The sum of cubes, $x^3 + y^3$, can be factored as $x^3 + y^3 = (x + y) \cdot (x^2 - xy + y^2)$. (Section 5.4)

summation (sigma) notation Summation notation is a compact way of writing a series using the general term of the corresponding sequence. (Section 14.1)

supplementary angles (supplements) Supplementary angles are angles whose measures have a sum of 180°. (Section 2.4)

synthetic division Synthetic division is a shortcut procedure for dividing a polynomial by a binomial of the form $x - k$. (Appendix G)

system of inequalities A system of inequalities consists of two or more inequalities to be solved at the same time. (Section 13.5)

system of linear equations (linear system) A system of linear equations consists of two or more linear equations to be solved at the same time. (Section 8.1)

T

table of values A table of values is an organized way of displaying ordered pairs. (Section 3.1)

term A term is a number, a variable, or the product or quotient of a number and one or more variables raised to powers. (Section 1.8)

term of an annuity The time from the beginning of the first payment period to the end of the last period is called the term of an annuity. (Section 14.3)

terms of a proportion The terms of the proportion $\frac{a}{b} = \frac{c}{d}$ are a, b, c, and d. (Section 2.6)

terms of a sequence The function values written in order are called the terms of the sequence. (Section 14.1)

three-part inequality An inequality that says that one number is between two other numbers is called a three-part inequality. (Section 2.8)

trinomial A trinomial is a polynomial with exactly three terms. (Section 4.4)

U

union The union of two sets A and B, written $A \cup B$, is the set of elements that belong to either A or B (or both). (Section 9.1, Appendix C)

universal set The set that includes all elements under consideration is the universal set, written U. (Appendix C)

unlike terms Unlike terms are terms that do not have the same variable or the variables are not raised to the same powers. (Section 1.8)

V

variable A variable is a symbol, usually a letter, used to represent an unknown number. (Section 1.3)

vary directly (is proportional to) y varies directly as x if there exists a real number (constant) k such that $y = kx$. (Section 7.4)

vary inversely y varies inversely as x if there exists a real number (constant) k such that $y = \frac{k}{x}$. (Section 7.4)

vary jointly If one variable varies as the product of several other variables (sometimes raised to powers), then the first variable is said to vary jointly as the others. (Section 7.4)

Venn diagram A Venn diagram represents the relationships between sets. (Appendix C)

vertex The point on a parabola that has the smallest y-value (if the parabola opens up) or the largest y-value (if the parabola opens down) is called the vertex of the parabola. (Sections 4.4, 11.6)

vertical angles When two intersecting lines are drawn, the angles that lie opposite each other have the same measure and are called vertical angles. (Section 2.5)

vertical line test The vertical line test states that any vertical line drawn through the graph of a function must intersect the graph in at most one point. (Section 7.3)

volume The volume of a three-dimensional figure is a measure of the space occupied by the figure. (Section 2.5)

W

weighted mean A weighted mean is one in which each number is weighted by multiplying it by the number of times it occurs. (Appendix D)

whole numbers The set of whole numbers is $\{0, 1, 2, 3, 4, \ldots\}$. (Section 1.4)

X

x-axis The horizontal number line in a rectangular coordinate system is called the x-axis. (Sections 3.1, 7.1)

x-intercept A point where a graph intersects the x-axis is called an x-intercept. (Sections 3.1, 7.1)

Y

y-axis The vertical number line in a rectangular coordinate system is called the y-axis. (Sections 3.1, 7.1)

y-intercept A point where a graph intersects the y-axis is called the y-intercept. (Sections 3.1, 7.1)

Z

zero-factor property The zero-factor property states that if two numbers have a product of 0, then at least one of the numbers must be 0. (Sections 5.5, 11.1)

Index

Videotape and CD Index

Text/Video/CD Section	Exercise Numbers	Text/Video/CD Section	Exercise Numbers
Section 1.1	65	Section 6.1	33, 49
Section 1.2	5, 9	Section 6.2	49
Section 1.3	3	Section 6.3	none
Section 1.4	17	Section 6.4	none
Section 1.5	49	Section 6.5	none
Section 1.6	25, 95	Section 6.6	none
Section 1.7	none	Section 6.7	23, 35
Section 1.8	none		
		Section 7.1	33, 47
Section 2.1	7, 13, 19, 25, 37, 49, 55, 63	Section 7.2	19, 23, 35, 53, 65, 75, 77
Section 2.2	27, 31, 45, 47, 49, 61, 69	Section 7.3	7, 11, 19, 33, 49, 75(c)
		Section 7.4	21, 39
Section 2.3	23, 47, 53		
Section 2.4	7, 19, 37	Section 8.1	7, 11
Section 2.5	17, 23, 37, 45, 49, 71, 73	Section 8.2	19
		Section 8.3	13
Section 2.6	25, 29, 61	Section 8.4	13
Section 2.7	11, 21, 27, 31, 33, 39	Section 8.5	7
Section 2.8	67, 77	Section 8.6	7
Section 3.1	35	Section 9.1	37
Section 3.2	5, 31	Section 9.2	37, 57
Section 3.3	31	Section 9.3	21
Section 3.4	45		
		Section 10.1	37
Section 4.1	79	Section 10.2	none
Section 4.2	none	Section 10.3	57, 87
Section 4.3	none	Section 10.4	37, 45
Section 4.4	31, 67	Section 10.5	45, 51
Section 4.5	21, 49	Section 10.6	7, 11, 45, 49, 55, 63
Section 4.6	47, 51	Section 10.7	63
Section 4.7	23, 29, 45, 65		
		Section 11.1	29, 55
Section 5.1	none	Section 11.2	3, 13, 23, 27, 39
Section 5.2	33	Section 11.3	9, 11, 37, 43
Section 5.3	none	Section 11.4	none
Section 5.4	13, 17, 25, 45, 59	Section 11.5	none
Section 5.5	none	Section 11.6	none
Section 5.6	7, 29	Section 11.7	none
		Section 11.8	none